SECOND EDITION

Biomedical Photonics Handbook

Volume III

Therapeutics and Advanced Biophotonics

Biomedical Photonics Handbook, Second Edition

Volume I: Fundamentals, Devices, and Techniques

Volume II: Biomedical Diagnostics

Volume III: Therapeutics and Advanced Biophotonics

SECOND EDITION

Biomedical Photonics Handbook

Volume III

Therapeutics and
Advanced Biophotonics

Edited by

Tuan Vo-Dinh

Duke University
Durham, North Carolina, USA

CRC Press
Taylor & Francis Group
Boca Raton London New York

CRC Press is an imprint of the
Taylor & Francis Group, an informa business

CRC Press
Taylor & Francis Group
6000 Broken Sound Parkway NW, Suite 300
Boca Raton, FL 33487-2742

First issued in paperback 2019

ISBN-13: 978-1-4200-8516-7 (hbk)
ISBN-13: 978-0-367-37847-9 (pbk)

Library of Congress Cataloging-in-Publication Data

Biomedical photonics handbook / edited by Tuan Vo-Dinh. -- Second edition.
    p. ; cm.
    Includes bibliographical references and indexes.
    Summary: "Biomedical photonics is defined as the science of harnessing light and other forms of radiant energy to address problems in medicine and biology. The field has experienced explosive growth due to the noninvasive or minimally invasive nature and cost-effectiveness of photonic modalities in medical diagnostics and therapy. The first volume of the Biomedical Photonics Handbook, Second Edition focuses on the fundamentals and advanced optical techniques and devices. It is an authoritative reference source for those involved in the research, teaching, learning, and practice of medical technologies"--Provided by publisher.
    ISBN 978-1-4398-0444-5 (set : alk. paper) -- ISBN 978-1-4200-8512-9 (v. 1 : hardcover : alk. paper) -- ISBN 978-1-4200-8514-3 (v. 2 : hardcover : alk. paper) -- ISBN 978-1-4200-8516-7 (v. 3 : hardcover : alk. paper)
    I. Vo-Dinh, Tuan, editor. II. Title: Fundamentals, devices, and techniques. III. Title: Biomedical diagnostics. IV. Title: Therapeutics and advanced biophotonics.
    [DNLM: 1. Diagnostic Imaging--instrumentation. 2. Diagnostic Imaging--methods. 3. Biosensing Techniques--instrumentation. 4. Biosensing Techniques--methods. 5. Photons--diagnostic use. WN 150]

R857.O6
610'.28--dc23
                                                                                    2014008504

Visit the Taylor & Francis Web site at
http://www.taylorandfrancis.com

and the CRC Press Web site at
http://www.crcpress.com

*Inspired by the love and
infinite patience of
my wife, Kim-Chi, and
my daughter, Jade*

*This book is dedicated to the
memory of my parents,
Vo Dinh Kinh and Dang Thi Dinh*

Contents

SECTION I   Therapeutic and Interventional Techniques

SECTION II   Advanced Biophotonics and Nanophotonics

Preface

In the tradition of the *Biomedical Photonics Handbook*, the second edition is intended to serve as an authoritative reference source for a broad audience involved in the research, teaching, learning, and practice of medical technologies. Biomedical photonics is defined as the science that harnesses light and other forms of radiant energy to provide the solution of problems arising in medicine and biology. This research field has recently experienced an explosive growth due to its noninvasive or minimally invasive nature and the cost-effectiveness of photonic modalities in medical diagnostics and therapy.

The field of biomedical photonics did not emerge as a well-defined, single research discipline like chemistry, physics, or biology. Its development and growth have been shaped by the convergence of three scientific and technological revolutions of the twentieth century: the *quantum theory revolution*, the *technology revolution*, and the *genomics revolution*.

The quantum theory of atomic phenomena provides a fundamental framework for molecular biology and genetics because of its unique understanding of electrons, atoms, molecules, and light itself. Out of this new scientific framework emerged the discovery of the structure of DNA, the molecular nature of cell machinery, and the genetic cause of diseases, all of which form the basis of molecular medicine. The formulation of quantum theory not only gave birth to the field of molecular spectroscopy but also led to the development of a powerful set of photonics tools—lasers, scanning tunneling microscopes, and near-field nanoprobes—for exploring nature and understanding the cause of disease at the fundamental level.

Advances in technology also played, and continue to play, an essential role in the development of biomedical photonics. The invention of the laser was an important milestone. Laser is now the light source most widely used to excite tissues for disease diagnosis as well as to irradiate tumors for tissue removal in interventional surgery (*optical scalpels*). The microchip is another important technological development that has significantly accelerated the evolution of biomedical photonics. While the laser has provided a new technology for excitation, the miniaturization and mass production of integrated circuits, sensor devices, and their associated electronic circuitry made possible the development of the microchip, which has radically transformed the ways detection and imaging of molecules, tissues, and organs can be performed in vivo and ex vivo. Recently, nanotechnology, which involves research on materials and species at length scales between 1 and 100 nm, has been revolutionizing important areas in biomedical photonics, especially diagnostics and therapy at the molecular and cellular level. The combination of photonics and nanotechnology has already led to a new generation of devices for probing the cell machinery and elucidating intimate life processes occurring at the molecular level that were heretofore invisible to human inquiry. This will open the possibility of detecting and manipulating atoms and molecules using nanodevices, which have the potential for a wide variety of medical uses at the cellular level. The marriage of electronics, biomaterials, and photonics is expected to revolutionize many areas of medicine in the twenty-first century.

A wide variety of biomedical photonic technologies have already been developed for clinical monitoring of early disease states or physiological parameters such as blood pressure, blood chemistry, pH, temperature, and the presence of pathological organisms or biochemical species of clinical importance. Advanced optical concepts using various spectroscopic modalities (e.g., fluorescence, scattering, reflection, and optical coherence tomography) are emerging in the important area of functional imaging. Many photonic technologies originally developed for other applications (e.g., lasers and sensor systems in defense, energy, and aerospace) have now found important uses in medical applications. From the brain to the sinuses to the abdomen, precision navigation and tracking techniques are critical to position medical instruments precisely within the three-dimensional surgical space. For instance, optical stereotactic systems are being developed for brain surgery, and flexible micronavigation devices are being engineered for medical laser ablation treatments.

With the completion of the sequencing of the human genome, one of the greatest impacts of genomics and proteomics is the establishment of an entirely new approach to biomedical research. With whole-genome sequences and new automated, high-throughput systems, photonic technologies such as biochips and microarrays can address biological and medical problems systematically and on a large scale in a massively parallel manner. They provide the tools to study how tens of thousands of genes and proteins work together in interconnected networks to orchestrate the chemistry of life. Specific genes have been deciphered and linked to numerous diseases and disorders, including breast cancer, muscle disease, deafness, and blindness. Furthermore, advanced biophotonics has contributed dramatically to the field of diagnostics, therapy, and drug discovery in the postgenomic area. Genomics and proteomics present the drug discovery community with a wealth of new potential targets. Biomedical photonics can provide tools capable of identifying specific subsets of genes encoded within the human genome that can cause the development of diseases. Photonic techniques based on molecular probes are being developed to identify the molecular alterations that distinguish a diseased cell from a normal cell. Such technologies will ultimately aid in characterizing and predicting the pathologic behavior of that diseased cell, as well as the cell's responsiveness to drug treatment. Information from the human genome project will one day make personal, molecular medicine an exciting reality.

The second edition of this handbook is intended to present the most recent scientific and technological advances in biomedical photonics, as well as their practical applications, in a single source. The three-book handbook represents the collective work of over 150 scientists, engineers, and clinicians. It includes many new topics and chapters such as fiber-optics probes design, laser and optical radiation safety, photothermal detection, multidimensional fluorescence imaging, surface plasmon resonance imaging, molecular contrast optical coherence tomography, multiscale photoacoustics, polarized light for medical diagnostics, quantitative diffuse reflectance imaging, interferometric light scattering, nonlinear interferometric vibrational imaging, nanoscintillator-based therapy, SERS molecular sentinel nanoprobes, and plasmonic coupling interference nanoprobes.

The handbook includes 71 chapters grouped in 8 sections:

1. Volume I: *Biomedical Photonics Handbook*, Second Edition: *Fundamentals, Devices, and Techniques*
2. Volume II: *Biomedical Photonics Handbook*, Second Edition: *Biomedical Diagnostics*
3. Volume III: *Biomedical Photonics Handbook*, Second Edition: *Therapeutics and Advanced Biophotonics*

In Volume I, Section I (Photonics and Tissue Optics) contains introductory chapters on the fundamental optical properties of tissue, light–tissue interactions, and theoretical models for optical imaging. Section II (Basic Instrumentation) deals with basic instrumentation and hardware systems and contains chapters on lasers and excitation sources, basic optical instrumentation, optical fibers, probe designs, laser use, and optical radiation safety. Section III (Photonic Detection and Imaging Techniques) deals with methodologies and contains chapters on various detection techniques and systems (such as lifetime imaging, microscopy, two-photon detection, photothermal detection, interferometry, Doppler imaging, light scattering, and thermal imaging). Finally, Section IV (Spectroscopic Data) provides a comprehensive compilation of useful information on spectroscopic data of biologically and medically relevant species for over 1000 compounds and systems.

In Volume II, Section I (Biomedical Analysis, Sensing, and Imaging) contains chapters describing in vitro diagnostics (e.g., glucose diagnostics, in vitro instrumentation, biosensors, surface plasmon resonance, and flow cytometry) and in vivo diagnostics (optical coherence tomography, polarized light diagnostics, functional imaging and photon migration spectroscopy, and multiscale photoacoustics). Section II (Biomedical Diagnostics and Optical Biopsy) is mainly devoted to novel optical techniques for cancer diagnostics, often referred to as *optical biopsy* (such as fluorescence, scattering, reflectance, interferometric light scattering, optoacoustics, and ultrasonically modulated optical imaging).

In Volume III, Section I (Therapeutic and Interventional Techniques) covers photodynamic therapy as well as various laser-based treatment techniques that are applied to different organs and disease endpoints (dermatology, pulmonology, neurosurgery, ophthalmology, otolaryngology, gastroenterology, and dentistry). There are several chapters dealing with nanotechnology for theranostics, that is, the modality combining diagnostics and therapy. Section II (Advanced Biophotonics and Nanophotonics) is devoted to the most recent advances in methods and instrumentation for biomedical and biotechnology applications. This section contains chapters on emerging photonic technologies (e.g., biochips, nanosensors, quantum dots, molecular probes, molecular beacons, molecular sentinels, plasmonic coupling nanoprobes, bioluminescent reporters, optical tweezers) that are being developed for gene expression research, gene diagnostics, protein profiling, and molecular biology investigations as well as for early diagnostics of disease biomarkers for the *new medicine*.

The goal of the second edition of this handbook is to provide a comprehensive forum that integrates interdisciplinary research and development of interest to scientists, engineers, manufacturers, teachers, students, and clinical providers. Each chapter provides introductory material with an overview of the topic of interest as well as a collection of published data with an extensive list of references for further details. The handbook is designed to present the most recent advances in instrumentation and methods as well as clinical applications in important areas of biomedical photonics. Because light is rapidly becoming an important diagnostic tool and a powerful weapon in the armory of the modern physician, it is our hope that this handbook will stimulate a greater appreciation of the usefulness, efficiency, and potential of photonics in medicine.

Tuan Vo-Dinh
Duke University
Durham, North Carolina

Acknowledgments

The completion of this work has been made possible with the assistance of many friends and colleagues. I wish to express my gratitude to members of the Scientific Advisory Board of the first edition. Their thoughtful suggestions and useful advice in the planning phase of the first edition have been important in achieving the breadth and depth of this handbook. It is a great pleasure for me to acknowledge, with deep gratitude, the contribution of over 150 contributors for the 71 chapters in this handbook. I wish to thank my coworkers at Duke University and the Oak Ridge National Laboratory, and many colleagues in academia, federal laboratories, and industry, for their kind help in reading and commenting on various chapters of the manuscript. My gratitude is extended to all my present and past students, postdoctoral associates, colleagues, and collaborators, who have been traveling with me on this exciting journey of discovery with the ultimate vision of bringing research at the intersection of photonics and medicine to the service of society.

I gratefully acknowledge the support of the US Department of Energy Office of Biological and Environmental Research, the National Institutes of Health, the Defense Advanced Research Projects Agency, the Department of the Army, the Army Medical Research and Materiel Command, the Department of Justice, the Federal Bureau of Investigation, the Office of Naval Research, the Environmental Protection Agency, the Fitzpatrick Foundation, the R. Eugene and Susie E. Goodson Endowment Fund, and the Wallace Coulter Foundation.

The completion of this work has been made possible with the love, encouragement, and inspiration of my wife, Kim-Chi, and my daughter, Jade.

Editor

Tuan Vo-Dinh is R. Eugene and Susie E. Goodson Distinguished Professor of Biomedical Engineering, professor of chemistry, and director of the Fitzpatrick Institute for Photonics at Duke University. A native of Vietnam and a naturalized US citizen, he completed high school education in Saigon (now Ho Chi Minh City). He continued his studies in Europe, where he received his BS in physics in 1970 from EPFL (Ecole Polytechnique Federal de Lausanne) in Lausanne and his PhD in physical chemistry in 1975 from ETH (Swiss Federal Institute of Technology) in Zurich, Switzerland. Before joining Duke University in 2006, Dr. Vo-Dinh was director of the Center for Advanced Biomedical Photonics, group leader of Advanced Biomedical Science and Technology Group, and a corporate fellow, one of the highest honors for distinguished scientists at Oak Ridge National Laboratory (ORNL). His research has focused on the development of advanced technologies for the protection of the environment and the improvement of human health. His research activities involve biophotonics, plasmonics, nanobiotechnology, laser spectroscopy, molecular imaging, medical theranostics, cancer detection, nanosensors, chemical sensors, biosensors, and biochips.

Dr. Vo-Dinh has authored over 350 publications in peer-reviewed scientific journals. He is the author of a textbook on spectroscopy and the editor of six books. He holds over 37 US and international patents, 5 of which have been licensed to private companies for commercial development. Dr. Vo-Dinh has presented over 200 invited lectures at international meetings in universities and research institutions. He has chaired over 30 international conferences in his field of research and served on various national and international scientific committees. He also serves the scientific community through his participation in a wide range of governmental and industrial boards and advisory committees.

Dr. Vo-Dinh has received seven R&D 100 Awards for Most Technologically Significant Advance in Research and Development for his pioneering research and inventions of innovative technologies. He has received the Gold Medal Award, Society for Applied Spectroscopy (1988); the Languedoc–Roussillon Award, France (1989); the Scientist of the Year Award, ORNL (1992); the Thomas Jefferson Award, Martin Marietta Corporation (1992); two Awards for Excellence in Technology Transfer, Federal Laboratory Consortium (1995, 1986); the Inventor of the Year Award, Tennessee Inventors Association (1996); the Lockheed Martin Technology Commercialization Award (1998); the Distinguished Inventors Award, UT-Battelle (2003); and the Distinguished Scientist of the Year Award, ORNL (2003). In 1997, he was presented the Exceptional Services Award for distinguished contribution to a healthy citizenry from the US Department of Energy. In 2011, he received the Award for Spectrochemical Analysis from the American Chemical Society (ACS) Division of Analytical Chemistry.

Contributors

R. Rox Anderson
Harvard Medical School
and
Department of Dermatology
Massachusetts General Hospital
Boston, Massachusetts

Terry Beck
TriLink BioTechnologies, Inc.
San Diego, California

Wladimir Benalcazar
Beckman Institute for Advanced
    Science and Technology
University of Illinois at
    Urbana–Champaign
Urbana, Illinois

Stephen A. Boppart
Beckman Institute for Advanced
    Science and Technology
University of Illinois at
    Urbana–Champaign
Urbana, Illinois

Darryl J. Bornhop
Department of Chemistry
Vanderbilt University
Nashville, Tennessee

David Boyer
The Retina Vitreous Associates
    Medical Group
Doheny Eye Institute
and
Keck School of Medicine
University of Southern California
Los Angeles, California

Murphy Brasuel
Chemistry & Biochemistry
    Department
Colorado College
Colorado Springs, Colorado

Richard D. Bucholz
Department of Neurosurgery
Saint Louis University
Saint Louis, Missouri

Eric J. Chaney
Beckman Institute for Advanced
    Science and Technology
University of Illinois at
    Urbana–Champaign
Urbana, Illinois

Bernard Choi
Department of Surgery
and
Department of Biomedical
    Engineering
University of California, Irvine
Irvine, California

Praveen D. Chowdary
Beckman Institute for Advanced
    Science and Technology
University of Illinois at
    Urbana–Champaign
Urbana, Illinois

Thomas G. Chu
The Retina Vitreous Associates
    Medical Group
Doheny Eye Institute
and
Keck School of Medicine
University of Southern California
Los Angeles, California

Christopher H. Contag
Department of Pediatrics
and
Department of Radiology
and
Department of Microbiology
    and Immunology
Stanford University Medical
    Center
Stanford, California

Pamela R. Contag
ConcentRx Inc.
San Jose, California

Georgeta Crivat
Department of Chemistry
University of New Orleans
New Orleans, Louisiana

Brian M. Cullum
Department of Chemistry and
    Biochemistry
University of Maryland,
    Baltimore County
Baltimore, Maryland

Sandra M. Da Silva
Biochemical Science Division
National Institute of Standards
    and Technology
Gaithersburg, Maryland

Andrew M. Fales
Department of Biomedical
    Engineering
Duke University
Durham, North Carolina

Xiaohong Fang
Institute of Chemistry
Chinese Academy of Science
Beijing, People's Republic
    of China

Daniel Fried
University of California,
    San Francisco
San Francisco, California

Lilit Garibyan
Harvard Medical School
and
Department of Dermatology
Massachusetts General Hospital
Boston, Massachusetts

Sandra O. Gollnick
Photodynamic Therapy Center
Roswell Park Cancer Institute
Buffalo, New York

Molly K. Gregas
Department of Biomedical
    Engineering
Duke University
Durham, North Carolina

Guy D. Griffin
Department of Biomedical
    Engineering
Duke University
Durham, North Carolina

Martin Gruebele
Beckman Institute for Advanced
    Science and Technology
University of Illinois at
    Urbana–Champaign
Urbana, Illinois

Barbara W. Henderson
Photodynamic Therapy Center
Roswell Park Cancer Institute
Buffalo, New York

Henry Hirschberg
Beckman Laser Institute and
    Medical Clinic
University of California, Irvine
Irvine, California

Richard Hogrefe
TriLink BioTechnologies, Inc.
San Diego, California

H. Ray Jalian
Department of Dermatology
Massachusetts General Hospital
Boston, Massachusetts

Zhi Jiang
Beckman Institute for Advanced
    Science and Technology
University of Illinois at
    Urbana–Champaign
Urbana, Illinois

Penny Joshi
Photodynamic Therapy Center
Roswell Park Cancer Institute
Buffalo, New York

Brad A. Kairdolf
Department of Biomedical
    Engineering
Georgia Institute of Technology
and
Emory University
Atlanta, Georgia

Tiina I. Karu
Institute of Laser and
    Information Technologies
Russian Academy of Sciences
Moscow, Russian Federation

Venkata R. Kethineedi
Department of Chemistry
University of New Orleans
New Orleans, Louisiana

Charles K. Klutse
Ghana Atomic Energy
    Commission
Accra, Ghana

Yong-Eun Koo Lee
Department of Chemistry
Hanyang University
Seoul, Korea

Raoul Kopelman
Department of Chemistry
University of Michigan
Ann Arbor, Michigan

Suzanne J. Lassiter
Louisiana State University
Baton Rouge, Louisiana

B. Lauly
Department of Chemistry and
    Biochemistry
Miami University
Oxford, Ohio

Keith A. Laycock
Department of Neurosurgery
Saint Louis University
Saint Louis, Missouri

Kai Licha
Mivenion GmbH
and
Free University
Berlin, Germany

Steen J. Madsen
Department of Health Physics
  and Diagnostic Sciences
University of Nevada, Las Vegas
Las Vegas, Nevada

Ezra Maguen
Ophthalmology Research
  Laboratories
Cedars-Sinai Medical
  Center
and
Geffen School of Medicine
University of California,
  Los Angeles
Los Angeles, California

Daniel L. Marks
Beckman Institute for Advanced
  Science and Technology
University of Illinois at
  Urbana–Champaign
Urbana, Illinois

Anubhav N. Mathur
Department of Dermatology
Indiana University Medical
  School
Indianapolis, Indiana

Praveen N. Mathur
Department of Medicine
Indiana University Medical
  School
Indianapolis, Indiana

Timothy E. McKnight
Oak Ridge National Laboratory
Oak Ridge, Tennessee

Eric Monson
Art, Art History and Visual
  Studies
Duke University
Raleigh, North Carolina

Hoan Thanh Ngo
Department of Biomedical
  Engineering
and
Department of Chemistry
Duke University
Durham, North Carolina

Shuming Nie
Department of Biomedical
  Engineering
Georgia Institute of Technology
and
Emory University
Atlanta, Georgia

Stephen J. Norton
Department of Biomedical
  Engineering
Duke University
Durham, North Carolina

Bergein F. Overholt
Fort Sanders Regional Medical
  Center
Thompson Cancer Survival
  Center
Knoxville, Tennessee

Clyde V. Owens
Louisiana State University
Baton Rouge, Louisiana

Ravindra K. Pandey
Photodynamic Therapy Center
Roswell Park Cancer Institute
Buffalo, New York

Masoud Panjehpour
Fort Sanders Regional Medical
  Center
Thompson Cancer Survival
  Center
Knoxville, Tennessee

Martin A. Philbert
Department of Environmental
  Health Sciences
University of Michigan
Ann Arbor, Michigan

Ashley D. Quach
Department of Chemistry
University of New Orleans
New Orleans, Louisiana

Lou Reinisch
School of Arts and Sciences
Farmingdale State College
Farmingdale, New York

Zeev Rosenzweig
Department of Chemistry and
  Biochemistry
University of Maryland,
  Baltimore County
Baltimore, Maryland

Lynn E. Samuelson
Department of Cancer Biology
Vanderbilt University
Nashville, Tennessee

John P. Scaffidi
Department of Biomedical
  Engineering
Duke University
Durham, North Carolina

Steven A. Soper
Louisiana State University
Baton Rouge, Louisiana

Dimitra N. Stratis-Cullum
Electro-Optics and Photonics
  Division
US Army Research Laboratory
Adelphi, Maryland

Kittisak Suthamjariya
Faculty of Medicine
Mahidol University
Bangkok, Thailand

Weihong Tan
Department of Chemistry
and
Department of Physiology and
Functional Genomics
University of Florida
Gainesville, Florida

and

Hunan University
Changsha, People's Republic of
China

Matthew A. Tarr
Department of Chemistry
University of New Orleans
New Orleans, Louisiana

Haohua Tu
Beckman Institute for Advanced
Science and Technology
University of Illinois at
Urbana–Champaign
Urbana, Illinois

Pierre M. Viallet
Laboratory of Physicochemical
Biology of Integrated Systems
University of Perpignan
Perpignan, France

Tuan Vo-Dinh
Department of Biomedical
Engineering
and
Department of Chemistry
Duke University
Durham, North Carolina

Emanuel Waddell
Louisiana State University
Baton Rouge, Louisiana

Hsin-Neng Wang
Department of Biomedical
Engineering
and
Department of Chemistry
Duke University
Durham, North Carolina

Kemin Wang
College of Biology
and
College of Chemistry and
Chemical Engineering
Hunan University
Changsha, People's Republic of
China

M. Wendy Williams (Retired)
Advanced Biomedical Science
and Technology Group
Oak Ridge National Laboratory
Oak Ridge, Tennessee

Cuichen Wu
Department of Chemistry
and
Department of Physiology and
Functional Genomics
University of Florida
Gainesville, Florida

Yichuan Xu
Louisiana State University
Baton Rouge, Louisiana

Chaoyong James Yang
Department of Chemical
Biology
Xiamen University
Xiamen, People's Republic of
China

Kenji Yasuda
Department of Biomedical
Information
Tokyo Medical and Dental
University
Tokyo, Japan

Hsiangkuo Yuan
Department of Biomedical
Engineering
Duke University
Durham, North Carolina

Y. Zhang
Nanometrics Incorporated
Hillsboro, Oregon

I

Therapeutic and Interventional Techniques

Mechanistic Principles of Photodynamic Therapy

Barbara W.
Henderson
*Roswell Park Cancer
Institute*

Sandra O. Gollnick
*Roswell Park Cancer
Institute*

1.1 Introduction

Photodynamic therapy (PDT) exploits the biological consequences of localized oxidative damage inflicted by photodynamic processes. A schematic outline of the major steps that lead to tumor destruction by PDT is given in Figure 1.1. Three critical elements are required for the initial photodynamic processes to occur: a drug that can be activated by a photosensitizer, light, and oxygen. Interaction of light at the appropriate wavelength with a photosensitizer produces an excited triplet state photosensitizer that can interact with ground state oxygen via two different pathways, designated as type I and type II. The individual steps of these pathways are shown in Figure 1.2. The type II reaction that gives rise to singlet oxygen ($^1O_2$) is believed to be the dominant pathway since the elimination of oxygen or scavenging of $^1O_2$ from the system essentially eliminates the cytocidal effects of PDT.[1-3] Type I reactions, however, may become important under hypoxic conditions or where photosensitizers are highly concentrated.[1] The highly reactive $^1O_2$ has a short lifetime (<0.04 μs) in the biological milieu and therefore a short radius of action (<0.02 μm).[4] Consequently, $^1O_2$-mediated oxidative damage will occur in the immediate vicinity of the subcellular site of photosensitizer localization. Depending on photosensitizer pharmacokinetics, these sites can be varied and numerous, resulting in a large and complex array of cellular effects. Similarly, on a tissue level, tumor cells as well as various normal cells can take up photosensitizer, which, upon activation by light, can lead to effects upon such targets as the tumor cells, the tumor and normal microvasculature, and the inflammatory and immune host system. PDT effects

Photodynamic effects

$$^1P + \text{light} \rightarrow {}^3P + {}^3O_2 \rightarrow {}^1O_2 + S \rightarrow S(O)$$

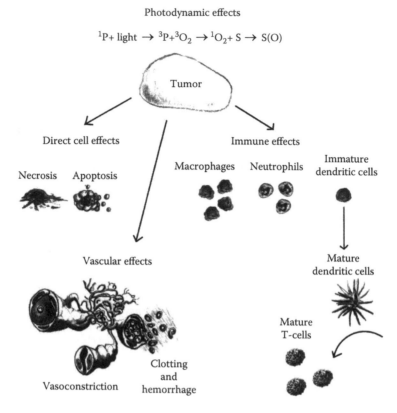

FIGURE 1.1 Illustration of the three major tissue targets affected by the photodynamic effect. Tumor cells can be damaged and/or killed directly by the effects of singlet oxygen generated within them, they can succumb to oxygen and nutrient deprivation due to vascular damage inflicted by PDT, or they can be attacked by the inflammatory/immune system activated by PDT.

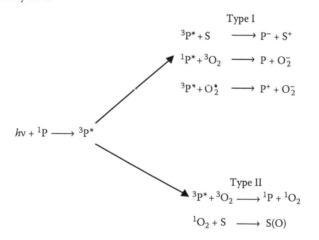

FIGURE 1.2 Photoreaction pathways emanating from the interaction of a photosensitizer with light where ^1P is a photosensitizer in a singlet ground state, $^3P^*$ is a photosensitizer in a triplet excited state, S is a substrate molecule, P^- is a reduced photosensitizer molecule, S^+ is an oxidized substrate molecule, $^3O_2$ is molecular oxygen (triplet ground state), O_2^- is the superoxide anion, $_1O_2^{\bullet}$ is the superoxide radical, P^+ is the oxidized photosensitizer, $^1O_2$ is oxygen in a singlet excited state, and S(O) is an oxygen adduct of a substrate. (Adapted from MacDonald, I.J. and Dougherty, T.J., *J. Porphyrins Phthalocyanines*, 5, 105, 2001. With permission.)

on all these targets may influence each other, producing a plethora of responses; the relative importance of each has yet to be fully defined. It seems clear, however, that the combination of all these components is required for long-term tumor control.

1.2 Subcellular Targets for Photosensitization

The potential cellular targets of PDT are shown schematically in Figure 1.3. They depend on the specific photosensitizer structure and pharmacokinetic characteristics, such as lipophilicity, amphiphilicity, aggregation, and serum protein interactions, and therefore the localization of the photosensitizer, but appear to be largely independent of cell type (Table 1.1). Localization studies have generally been carried out in in vitro cell systems where exposure conditions to the photosensitizer can be easily controlled or varied. Alternative drug uptake mechanisms determine cellular photosensitizer accumulation: Diffusion results in predominantly mitochondrial accumulation, while endocytosis of surface-bound agents leads to deposition in the endosomal/lysosomal compartments.[5] Studies have also revealed that cellular photosensitizer distribution can be a dynamic process, influenced by such parameters as length of exposure and drug concentration.[6–8] Photosensitizers may even relocalize after photodamage to an initial site of accumulation, such as from lysosomes to other, possibly more sensitive, cellular locations where they will then be available for activation.[9] That subcellular localization of a photosensitizer, and consequently the target of PDT, may influence in vivo treatment outcome was demonstrated in a series of studies that could relate structure and activity of a congeneric series of photosensitizing compounds to their subcellular localization (Figure 1.4).[10] Pyropheophorbide-a ether derivatives, designed to possess progressively increasing degrees of lipophilicity, exhibited drug uptake in tumor cells that increased linearly with lipophilicity, while PDT activity tested in an vivo murine tumor system showed a parabolic quantitative structure–activity relationship (QSAR). Comparison of the subcellular localization of the most and least active compounds in this series revealed mitochondrial localization for the former and lysosomal localization for the latter.

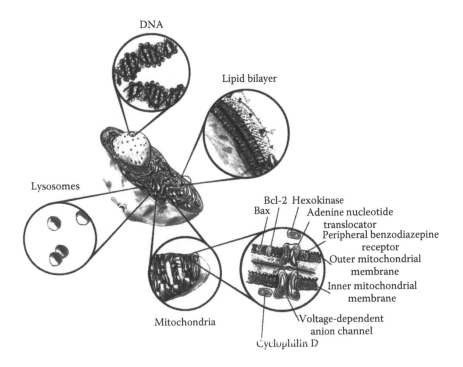

FIGURE 1.3 Illustration of the major cellular targets affected by PDT.

TABLE 1.1   Intracellular Localization Sites of Major Photosensitizers in Cells In Vitro

Photosensitizer	Cell Line	Localization Site	References
BPD-MA (Verteporfin®a)	NHIK3025	ER/Golgi	[36]
Phthalocyanines	NHIK3025, RIF, LOX, V79	Lysosomes Cytoplasm	[23,226–228]
Hematoporphyrin/ Photofrin®	V79, L1210	Plasma membrane Mitochondria	[229,230]
Lutex	EMT6	Lysosomes	[231,232]
mTHPC	V79	Cytoplasm, diffuse	[233]
Npe$_6$ and analogs	CHO, L1210	Lysosomes Mitochondria Plasma membrane	[16,234,235]
PpIX from ALA	WiDr, NHIK3025, V79 FaDu, RIF	Mitochondria	[15,236]
Pyropheophorbide derivatives	FaDu, RIF	Mitochondria Lysosomes	[10]
TPPS analogs	NHIK3025	Lysosomes Mitochondria Perinuclear	[37,237–239]
Purpurin analogs SnEt$_2$	P388, OVCAR5	Lysosomes	[49]

ᵃ Registered trademark of QLT Inc., Vancouver, British Columbia, Canada.

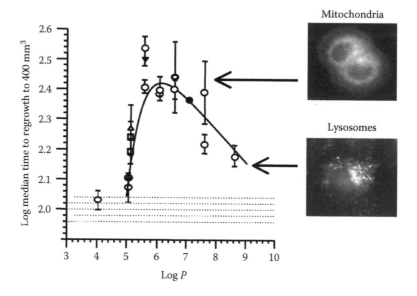

FIGURE 1.4   Relationship of the log median time of tumor regrowth to 400 mm³ tumor volume to log *P* for a series of congeneric pyropheophorbide photosensitizers and to their subcellular distribution. Optimal compounds accumulate in mitochondria and suboptimal compounds in lysosomes. (Adapted from Henderson, B.W. et al., *Cancer Res.*, 57, 4000, 1997. With permission.)

An important discovery was the realization that the adenosine triphosphate (ATPase)-dependent transporter ABCG2, a multidrug-resistant pump expressed to varying degrees in cancer cells and cancer stem cells, can bind and efflux certain photosensitizers, greatly affecting photosensitizer retention.[11,12] Substrates for ABCG2 include certain porphyrins, chlorins, and hypericin. Tyrosine kinase inhibitors, such as imatinib mesylate (Gleevec), can block ABCG2-mediated photosensitizer efflux and thus affect PDT efficacy for agents that are ABCG2 substrates.[5,13]

All cellular sites that accumulate photosensitizer can be effective targets for cell destruction. Mitochondria have long been known to be sensitive sites for PDT (also see Table 1.1).[14-17] Especially the mitochondrial permeability transition pore, a protein complex consisting of hexokinase, peripheral benzodiazepine receptor, voltage-dependent anion channel, creatine kinase, adenine nucleotide translocator (ANT), and cyclophilin D, has been implicated as PDT target.[18-21] PDT-induced rupture of lysosomes can lead to cell death through the release of damaging hydrolytic enzymes into the cytoplasm.[22] Selective lysosomal photodamage that was associated with the release of cathepsin B was found to be followed by a gradual loss of mitochondrial membrane potential, release of cytochrome c into the cytosol, caspase 3 activation, and a limited apoptotic response. Thus, lysosomal damage may secondarily mediate mitochondrial damage. It has also been suggested that lysosomes might serve as reservoirs, from which photosensitizers might be released after vesicle rupture and migrate to more sensitive sites.[23] All these studies assume that initial sensitizer accumulation is restricted to the lysosomes, but the possibility cannot be entirely excluded that high lysosomal concentration may mask small but highly effective sensitizer accumulations at other sites.[24]

When photosensitizers are confined to or specifically targeted to the plasma membrane, it can become a highly lethal PDT target.[25] Membrane damage manifests itself rapidly through blebbing,[26,27] as well as leakage of cytosolic enzymes[28,29] and chromium.[30] Leakage of lactate dehydrogenase showed the same kinetics as generation of prostaglandin E_2 from tumor cells in vitro, directly relating membrane disruption to production of eicosanoids.[29] The biosynthesis of eicosanoids, which play an important role in mediating the PDT tissue response (discussed in the succeeding texts), is set in motion when phospholipases catalyze the liberation of free fatty acids from membrane phospholipids. Unsaturated phospholipids and cholesterol are important membrane targets of photodamage, and lipid peroxidation by PDT has been extensively studied and linked with cell lethality.[31,32] Oxidation of intrinsic membrane proteins has also been observed.[33] PDT can inhibit the plasma membrane enzymes Na^+K^+-ATPase and Mg^{2+}-ATPase.[34] Ca^{2+} flux may be affected and the plasma membrane may become depolarized.[35] Certain membrane-localized photosensitizers, such as sulfonatedtetraphenylporphines (TPPS), can destroy microtubules in interphase cells and lead to arrest of cells in mitosis.[36,37]

While protein cross-linking has been reported by Moan and Vistnes in 1986,[38] it has taken on new emphasis with the discovery that cross-linking of the signal transducer and activator of transcription 3 (STAT3) correlates strongly with the strength of the local PDT reaction and thus can be used as a molecular reporter for that reaction in tissue, including in human biopsy samples collected immediately following PDT.[39]

Photosensitizers used in PDT generally do not accumulate in the cell nucleus, and therefore frank DNA damage such as strand breaks, sister chromatid exchanges, and chromatid aberrations are much less frequently observed in PDT than in ionizing or UV irradiation.[40-42] Consistent with these findings, the mutation potential of PDT was found to be significantly less than that of ionizing radiation or UV.[43]

1.3 Pathways to Cell Death and Survival

Three cell death pathways can be induced by PDT—apoptotic, autophagic, and necrotic (reviewed in Ref. 44). The apoptotic pathway is an intrinsic physiological process that is dependent upon active cellular metabolism and is characterized by chromatin condensation, DNA fragmentation, and the formation of apoptotic bodies.[45] Autophagy is a lysosomal pathway of self-degradation of damaged cellular proteins and organelles, including those damaged by oxidative stress insults.[46-48] (Macro) autophagy, found to be relevant in PDT, is marked by the formation of double-membrane vesicles that sequester and transport materials from the cytoplasm to the lysosomal system.[44,47,48] Necrosis is mainly metabolism independent and a result of a massive insult, especially to the plasma membrane, characterized by vacuolization of the cytoplasm, swelling, and breakdown of the plasma membrane.[44] Recent evidence suggests that under certain circumstances necrosis may be actively induced through signaling pathways.[44] The dominant mode of cell death is dependent upon the photosensitizer used, the localization of the photosensitizer, and the treatment protocol. Studies by

Kessel and Luo[16] demonstrated that photosensitizers that localize to the mitochondria resulted in apoptotic cell death while those that localize to the lysosome or cell membrane did not. Kessel et al. also demonstrated that photodamage of the cell membrane can inhibit the induction of apoptosis by photodamage to the mitochondria and that the mode of cell death shifts in response to the photodynamic dose, such that high doses result in a necrotic cell death, while lower doses result in an apoptotic mode of death.[49,50]

The rapid induction of apoptosis suggests that PDT triggers late-stage apoptotic processes directly. Release of cytochrome c from the mitochondria triggers caspase 3 activation and results in initiation of apoptosis at a late stage in the pathway.[51] Cytochrome c release is associated with a loss of mitochondrial potential. PDT results in the loss of mitochondrial potential and the rapid release of cytochrome.[52–55] Several studies have demonstrated that PDT also induces rapid activation of caspases 3, 6, 7, and 8 and cleavage of poly(ADP-ribose) polymerase (PARP).[52,56] Kessel and Luo have shown that PDT-induced release of cytochrome c is sufficient to directly initiate a caspase-dependent apoptotic cell death.[57] However, while caspase 3 is required for the late stages of apoptosis, it is not the critical lethal event in PDT-induced apoptosis.[58]

Bcl-2, the antiapoptotic member of the Bcl-2 family, has been shown to block the release of cytochrome c from the mitochondria and thus prevent apoptosis.[51] Several groups have altered the expression of bcl-2, through overexpression or antisense technology, and demonstrated that increasing the levels of bcl-2 enhances cellular resistance to PDT-induced apoptotic cell death.[59–61] Xue et al. have demonstrated that bcl-2 is a target for photochemical destruction.[62,63] Interestingly, a study by Kim et al.[63] reported that overexpression of bcl-2 enhanced the apoptotic response. In these studies, overexpression of bcl-2 was accompanied by an increase in bax, a proapoptotic member of the bcl-2 family. PDT resulted in the selective destruction of bcl-2 but had no effect on bax. The greater apoptotic response in the cells overexpressing bcl-2 was attributed to a higher bax–bcl-2 ratio after PDT. A PDT-induced shift in the bax–bcl-2 ratio toward apoptosis has also been reported by others.[59,64]

PDT also affects other proteins implicated in apoptosis. Phospholipases C and A_2, which are involved in the transient increases in intracellular calcium levels and DNA fragmentation, are activated by Photofrin® PDT.[65] Ceramide, which has been linked to apoptosis in several malignant cell lines, accumulates following PDT and has been associated with PDT-induced apoptosis and cytotoxicity.[66] Gupta et al.[67] have demonstrated that Pc 4-PDT induces expression of nitric oxide and have suggested that it may be involved in PDT-induced apoptosis.

Autophagy was first recognized by Kessel and Arroyo[68] as a prominent PDT-induced cell death pathway that occurs in many normal and transformed cell types and that can precede, concur with, or follow the development of apoptosis.[47] Autophagy can contribute to cell survival through sequestration and removal of damaged cellular material but may also have a role in stress-induced cell death.[48] Moreover, extensive cross talk between apoptosis and autophagy exists as they share the same stimuli and similar effectors and regulators.[48] Thus, in apoptosis-deficient cells, such as cells without Bcl-2, autophagy was increased,[68] while overexpression of Bcl-2 only stimulated apoptosis without affecting autophagy.[69] Conversely, silencing of essential autophagy genes, such as *Atg5* and *Atg7*, promotes apoptotic photokilling.[48]

Numerous other signaling pathways are engaged in the cellular response to PDT that can either promote cell death or survival and proliferation. Foremost among these is the stress MAPK pathway, with the activation of JNK and p38 the most consistently observed immediate reaction to PDT.[70–74] PDT has been shown to stimulate stress kinase signaling pathways (SAPK/JNK, p38/HOG1)[70,75] and HS1 phosphorylation[76] as well as stress response transcriptional activators. Hypoxia-inducible factor-1 (HIF-1) is a key regulator of the cellular response to hypoxia and is induced in response to PDT.[77,78] PDT also enhanced the expression of the HIF-1 target gene, vascular endothelial growth factor (VEGF),[78] and it has been proposed that HIF-1 expression may act as a predictor of PDT responsiveness.[77] Finally, PDT also activates the heat shock protein (HSP) family promoters[79] and induces the expression of HSPs and the related glucose-regulated proteins,[80–84] as well as heme oxygenase (HSP 32).[84]

PDT has been shown to alter the expression of the redox-regulated transcription factor, AP-1.[85,86] AP-1 is induced by changes in the redox potential[87] and hypoxia.[88] It is composed of homodimers of the products of the c-jun gene family or heterodimeric combinations of c-jun and c-fos family members. PDT induces prolonged expression of both c-fos and c-jun as a result of oxidative stress.[89,90] Therefore, modulation of AP-1 activity by PDT might be mediated by changes in oxidative potential as a result of the generation of singlet oxygen or PDT-induced hypoxia.

NF-κB plays a critical role in the expression of immunomodulatory and proinflammatory genes.[91] Activated NF-κB is a heterodimeric protein most commonly comprised of the p50 and p65 (Rel A) species and is activated in response to cellular oxidative stress.[92] NF-κB is sequestered in the cytoplasm by IκB proteins; phosphorylation of IκB results in its proteasomal degradation and release of NF-κB.[93,94] NF-κB binding activity was found to be induced by PDT[95–97] and has been proposed to play a role in determining the cellular response to PDT.[98] Granville et al.[95] demonstrated that cellular levels of IκB were transiently depressed following Verteporfin® PDT. Pyropheophorbide-a methyl ester PDT also leads to IκB degradation.[97] In contrast to these studies, other groups have failed to demonstrate NF-κB activation following Photofrin PDT.[85,86] Thus, NF-κB activation may be photosensitizer specific. It has also been suggested that activation is related to PDT dose in that higher doses do not result in decreases in IκB.[95]

1.4 Tissue Targets of Photosensitization

1.4.1 Tumor Cells

The capacity of PDT to eliminate tumor cells through direct photodamage has been most effectively studied by in vivo–in vitro tumor explant methodology.[99] Such in-depth analysis has revealed that the full potential of direct photodynamic tumor cell kill, provided by the gross tumor photosensitizer concentration and absorbed light dose, is generally not realized by in vivo PDT treatment.[100] Clonogenic assays carried out with a number of different photosensitizers and tumor systems immediately after potentially curative PDT exposures in vivo have revealed that at most 1–2 logs of direct tumor cell kill have been achieved,[101–106] far less than the 7–8 logs required for tumor cure. Clearly, limitations to direct tumor cell kill exist in vivo, the most important of which may be (1) inhomogeneous photosensitizer distribution within the tumor, including a gradual decrease of photosensitizer concentration with distance from blood vessels,[107] (2) insufficient light penetration through the tissue (light is attenuated exponentially with depth of tissue penetration),[108,109] and (3) insufficient oxygen availability (discussed in the following under separate heading). That intrinsic tumor cell sensitivity contributes to the overall PDT response was suggested by studies using tumor cells selected to be resistant to PDT.[110] Tumors of resistant phenotype were less responsive to PDT than sensitive tumors while exhibiting equal vascular responses. The studies did not exclude the possibility, however, that tumor immunogenicity might have differed in the two tumor lines, thus affecting the response. On the other hand, tumors expressing a multidrug resistance phenotype that prevented the uptake of a cationic photosensitizer were nevertheless found responsive to PDT carried out with that agent, the antitumor effect being attributed to vascular disruption.[111] Studies like these illustrate the capacity of PDT to exert its antitumor action through several different tissue targets.

Initial studies trying to establish a pattern of selectivity of photosensitizer uptake in malignant versus normal cells in vitro were largely unsuccessful.[112] Recent studies employing a coculture system of primary human lung epithelial tumor cells and corresponding stromal fibroblasts revealed cell-type-specific retention of the pyropheophorbide-a derivative hexyl pyropheophorbide ether (HPPH), which was both independent and dependent on the ABCG2 transporter and could be exploited for highly selective killing of either tumor cells or fibroblasts.[5] In vivo, moderately favorable tumor to normal tissue ratios can be found for almost all photosensitizers, with establishment of these ratios depending on the specific pharmacokinetics of the compound as well as the pathophysiology of the tumor (for detailed reviews, see Refs. 113,114). Mechanisms invoked for this *selectivity* range from leaky vasculature and impaired lymphatic drainage in tumors to low tumor pH and an increase in low-density lipoprotein and/or other membrane receptors on tumor cells.[115,116] Carrier systems, such as antibody conjugates, have been designed to direct the photosensitizer directly to the tumor cells. Such immunoconjugates have been directed against epitopes on ovarian[117] and colon cancer cells[118] as well as against the epidermal growth factor receptor (EGFR) that is overexpressed in many cancers.[119]

1.4.2 Microvasculature

The microvasculature was one of the earliest tissue targets identified because vascular PDT effects are rapid and dramatic, especially with the use of the sensitizer Photofrin and its forerunners. Reduction and/or cessation of tumor microcirculation following in vivo PDT exposure employing a variety of photosensitizers has been demonstrated in preclinical models through numerous different techniques, summarized in Table 1.2. Since the kinetics of vascular shutdown and tumor cell death have been found to coincide[102,120] and inhibition of shutdown retards tumor response,[121] it has been argued that disruption of the tumor microcirculation is a major factor contributing to tumor control by PDT. Differences in response between tumor and normal microvasculature are subtle and difficult to discern.[122] Careful dose-ranging studies have revealed that the tumor vasculature is slightly more susceptible to shutdown than its normal counterpart.[123,124] It has also been demonstrated that Photofrin PDT at high fluence rate can protect the normal skin microvasculature, but the same treatment fails to protect tumor vessels.[106,125] Occlusion of the tumor-surrounding vasculature can contribute to tumor control, at least in preclinical models, presumably by adding to the nutrient deprivation and retardation of vascular resupply of the tumor.[123,126] Another approach toward retardation of vessel regrowth after PDT is the use of anti-angiogenic agents in combination with PDT, which has been shown to enhance long-term tumor control.[78,127]

The acute manifestations of PDT-induced vascular damage greatly depend on the type and dose of photosensitizer used. On the microscopic level, vascular changes in a rat cremaster muscle preparation after Photofrin PDT included vessel constriction; occlusive platelet thrombi in arteries, arterioles, vein, and venules; edema; and neutrophil margination and migration.[128,129] Ultrastructural features included damage to numerous endothelial cell organelles, as well as perivascular changes such as degranulation of perivascular mast cells and damage to myocytes.[128] Npe6 (mono-L-aspartyl chlorin e6) PDT in a rat chondrosarcoma model resulted in obstructive platelet thrombi but no vasoconstriction.[130] PDT with variously substituted zinc Pcs, tested in the same tumor model, showed a spectrum of effects, including vessel constriction and leakage, with one compound (disulfonated zinc Pc) exhibiting no apparent effects.[131] In preclinical studies, PDT-induced vascular leakage has been exploited to enhance the uptake in tumors of the liposomally formulated chemotherapeutic agent doxorubicin.[132]

Numerous vascular response mediators seem to be involved in these processes, summarized in Table 1.3. Most studied and prominent among them are the eicosanoids. They are products of the release from the cell membrane of arachidonic acid that is subsequently metabolized by cyclooxygenase to generate prostaglandins and thromboxanes and by lipoxygenase to form leukotrienes and hydroxy acids. The generation of a wide spectrum of arachidonic acid metabolites by PDT has been described in mast cells, macrophages, endothelial cells, platelets, and tumor cells, as well as in animals. Interestingly, in vitro PDT exposure of platelets blocks their capacity to aggregate[133,134] and in vitro exposure of endothelial cells is dominated by the release of prostacyclin (PGI_2) that inhibits platelet aggregation.[134] These in vitro findings contradict the well-established, proaggregating mechanisms observed in vivo and demonstrate the difficulty of translating in vitro data to the

TABLE 1.2 Tumor Perfusion and Oxygenation Following PDT

Photosensitizer	Tumor Model	Technique	References
Hematoporphyrin derivative	RMA rat mammary tumor	Window chamber observation	[240]
Photofrin®	Cremaster muscle, rat	Window chamber observation	[129]
Phthalocyanine analogs	Chondrosarcoma, rat	Window chamber observation	[131]
Npe6	Chondrosarcoma, rat	Window chamber observation	[130]
Photofrin	RIF-1 mouse fibrosarcoma	Radiobiological assay	[241]
Hematoporphyrin derivative	AY-27 rat urothelial tumor	^{103}Ru, ^{141}Ca	[242]
Photofrin	RIF-1 mouse fibrosarcoma	^{86}Rb extraction Hoechst 33342	[243]
Polyhematoporphyrin ALA	LSBD$_1$ rat fibrosarcoma	Microspheres ^{86}Rb extraction	[244]
HpD	T50 80 mouse mammary tumor	NMR imaging	[245]
Photofrin	R3230A rat mammary tumor	NMR imaging	[246]
Bacteriochlorophyll–serine	M2R melanotic melanoma xenograft	Contrast-enhanced MRI	[247]
Photofrin	Chondrosarcoma, rat	Oxygen microelectrode	[248]
Photofrin	VX-2 rabbit skin carcinoma	Transcutaneous oxygen electrodes	[149]
Photofrin	Mammary carcinoma, mouse	Oxygen microelectrode	[150]
Photofrin	RIF-1 mouse fibrosarcoma	Oxygen microelectrode, ^{86}Rb extraction, fluorescein	[125]
Photofrin	A673 sarcoma, human xenograft	Laser Doppler	[136]
Photofrin	RIF-1 mouse fibrosarcoma	Hypoxia marker	[157]
HPPH	Colo-26 mouse colon carcinoma	Fluorescent microspheres	[132]
HPPH	Ward rat colon carcinoma	MRI	[169]

TABLE 1.3 Vascular Response Mediators Generated/Affected by PDT

Photosensitizer	Cell/Tumor Type	Mediator	References
Hematoporphyrin derivative	Mast cells, rat	Histamine	[249]
Protoporphyrin	Mast cells, rat	PGD$_2$, PGE$_2$, PGI$_2$	[250]
Hematoporphyrin	Platelets, human	Serotonin	[251]
Phthalocyanine (ClAl-S)	Endothelial cells, bovine	Clotting factors	[252]
Photofrin®	Endothelial cells, human	Von Willebrand factor	[253]
Photofrin	EMT6 mammary tumor, mouse; RIF fibrosarcoma, mouse	PGE$_2$	[29]
Photofrin	Macrophages, mouse	PGE$_2$	[29]
Photofrin Phthalocyanine (ZnS)	Endothelial cells, bovine; human	PGI$_2$, AA, F$_2$α, HETES	[134]
Photofrin	Mouse serum	Thromboxane B$_2$	[133]

in vivo situation. The latter represents a much more complex interplay of mechanistic components that likely involves platelets, endothelial cells, leukocytes, macrophages, and other stromal cells. Thromboxane appears to play a major role in mediating the observed vascular effects, at least in rat models.[135] Anti-inflammatory drugs, such as indomethacin or aspirin, can block the release of eicosanoids in vitro[29,134] and in vivo.[131,135-137] Rats made thrombocytopenic and thus deprived of a source for thromboxane generation also show a diminished vascular response.[138] Eicosanoids, as well as serotonin, appear to also be involved in changes of tumor interstitial pressure observed after Photofrin PDT, a consequence of fluid leakage from the vasculature and possibly contributing to occlusion of the tumor microvasculature.[139]

1.5 Tumor Oxygenation and PDT

Any restriction of tissue oxygen supply *during* PDT light delivery will reduce $^1O_2$ production and therefore have negative consequences for treatment outcome. Such restrictions can arise from numerous sources, including preexisting tumor hypoxia, acute vascular damage, and photochemical oxygen depletion. All of these can interact, and the dynamics of these interactions may determine treatment success.

1.5.1 Preexisting Tumor Hypoxia

Preexisting tumor hypoxia, a therapeutic problem still grappled with by radiation oncologists because of the oxygen dependence of sparsely ionizing radiation treatment, may limit the oxygen supply for PDT as well. Solid tumors are prone to develop hypoxic tumor regions due to deteriorating diffusion geometry, structural abnormalities of tumor microvessels, and disturbed microcirculation.[140] Hypoxic, but viable, tumor cells located in these areas may be protected from PDT-induced photodamage. In a preclinical study, Fingar et al.[141] manipulated tumors pharmacologically and physically to induce a wide range of hypoxic tumor fractions (~2% to ~40%) and followed this by aggressive Photofrin PDT. It was found that hypoxic fractions below 5% did not adversely affect tumor control, hypoxic fractions of ~10% slightly diminished tumor control, and hypoxic fractions of ~40% totally blocked tumor control. Vascular shutdown and nutrient deprivation following PDT were believed to be responsible for the elimination of small numbers of initially surviving hypoxic tumor cells. Early preclinical attempts to raise tumor oxygenation prior to PDT through the administration of a perfluorochemical emulsion and carbogen breathing were highly successful in increasing tumor oxygen levels (~10-fold) up to 1 h after PDT. The intervention did not alter long-term tumor control,[142] probably because the vascular damage induced by the aggressive PDT regime overwhelmed any subtle improvement in treatment outcome that might have been attributed to increased tumor oxygenation. Other studies, both preclinical and clinical, have demonstrated significantly improved PDT efficiency with adjuvant administration of hyperbaric oxygen or carbogen.[143-145]

The factor, however, that probably influences preexisting tumor hypoxia most profoundly is one that accompanies most PDT treatments, namely, changes in tumor temperature. Temperature increases due to PDT light delivery have been analyzed by Svaasand et al.[146] and have been recorded in preclinical models[147] and in patient's tumors.[148] Measurements of baseline intratumoral temperature and pO_2 in nodular basal cell carcinomas have demonstrated a linear relationship between increasing tumor pO_2 and lesion temperature in the ranges between 0–20 mm Hg and 30°C–35°C.[156] Upon laser illumination during Photofrin PDT, the temperature in these lesions increased further in a fluence rate–dependent manner. Surprisingly, even low fluence rate light produced significant temperature rises (150 mW/cm², median temperature change +1.9°C [range 1.0–6.2]; 30 mW/cm², median temperature change +1.5°C [range −1.3–3.6]), and these correlated with increased tumor pO_2. It remains to be seen whether these relationships hold true for tumors other than skin tumors.

1.5.2 Oxygen Limitation through Vascular Damage

With photosensitizers that can acutely constrict and/or occlude vessels, blood flow obstruction can be marked, very rapidly limiting the oxygen supply to the tumor. Photofrin PDT in mouse models, for example, rendered up to 10% of tumor cells hypoxic within a very moderate light exposure (45 J/cm^2, 10 min).[10] This hypoxia was persistent and progressive with time, 50% of tumor cells being hypoxic within 1 h of such light exposure. Similar observations were reported for a rabbit skin tumor model, where a series of brief light exposures resulted in induction of irreversible tumor hypoxia that was cumulative with the number of exposures, that is, fluence.[149] Hypoxia induction by PDT depends greatly on the vascular supply of a given tumor and even on the site of tumor implantation in rodent models.[150] Transient reoxygenation may occur, depending on PDT dose.[150] With certain second-generation sensitizers, many of which exert less severe acute effects on the vasculature than Photofrin,[131] and a tendency toward the use of lower drug doses, such acute vascular effects are less likely to occur. As discussed earlier, the extent and timing of vascular damage, and therefore induction of tumor hypoxia, is of significant importance for treatment outcome. Vascular occlusion can be detrimental when occurring *during* treatment but beneficial when occurring *after* completion of the PDT tumor treatment.

1.5.3 Oxygen Limitation through Photochemical Oxygen Depletion

Photochemical oxygen depletion is roughly characterized by instantaneous or near instantaneous development of tumor hypoxia upon light exposure of a photosensitized tumor/tissue and equally rapid reoxygenation upon cessation of light. The theoretical basis for this phenomenon has been provided through mathematical modeling of the dynamic changes to be expected in tissue when oxygen is consumed in the process of $^1O_2$ generation.[151] Photochemical oxygen depletion will occur in tissue if the rate of photodynamic oxygen consumption is faster than the rate of oxygen resupply from the vasculature. The major parameters that determine whether or not photochemical oxygen depletion will occur are (1) the absorption coefficient of the photosensitizer, (2) the tissue concentration of the photosensitizer, (3) the fluence rate of light, and (4) the vascular supply of the tissue.[151–153] If the first three parameters are high, $^1O_2$ production will be rapid and oxygen depletion will be favored; if the vascular supply of the tissue is poor, oxygen depletion will also be favored. The mathematical predictions have been validated in tightly defined in vitro systems[154,155] in tumor models[125,149,156,157] and in humans.[148] Light fluence rate is the most easily controlled parameter, one that can be readily modulated during light delivery. Therefore, much attention has been paid to the effects of fluence rate on PDT oxygen consumption. It is clear that lowering of the fluence rate can diminish or eliminate photochemical oxygen consumption through lowering the $^1O_2$ generation rate. However, the optimal fluence rate will depend on the other parameters listed earlier and therefore vary from situation to situation. In human tumors, the variability is great, both among patients and among lesions in the same patient.[148]

1.5.4 Role of Photobleaching in Photochemical Oxygen Depletion

Photobleaching is the destruction of the photosensitizer by light-mediated processes.[158–161] Since photosensitizer concentration is one of the major determinants for photochemical oxygen depletion, it stands to reason that the destruction of photosensitizer through photobleaching during PDT will reduce the likelihood that oxygen depletion will occur. The most detailed studies of photobleaching have been carried out in an in vitro multicell tumor spheroid model.[162] It was shown that sustained illumination of Photofrin-photosensitized spheroids led to a progressive decrease of photochemical oxygen depletion, implying reduction of photosensitizer levels, and consistent with a theoretical model in which bleaching occurs via a self-sensitized singlet oxygen reaction with the photosensitizer ground state. Similarly, protoporphyrin IX (PpIX) was degraded by $^1O_2$-mediated mechanisms, while another photosensitizer (Nile blue selenium) was degraded by $^1O_2$-independent mechanisms.[163] Oxygen measurements in rodent tumor models and human basal cell carcinomas also implicated photobleaching in influencing photo-chemical oxygen depletion.[125,148] In these studies, significant oxygen depletion was observed during the early time periods of illumination at high fluence rates of Photofrin-photosensitized lesions, but less or no oxygen depletion was detected toward the end of illumination, implying that the Photofrin concentration had been reduced through photobleaching below the threshold needed for oxygen depletion. Noninvasive devices have been developed that can monitor photobleaching in patients during PDT light delivery, allowing correlation of photobleaching kinetics and hemodynamic responses.[164]

1.5.5 Enhancement of PDT Efficiency through Modified Light Delivery Schemes

It is evident that any means by which the well-oxygenated tumor volume can be increased during light exposure should have beneficial effects on treatment outcome. Downward adjustment of treatment fluence rate is one such means, and fractionation of light delivery is another.[125,165–167] While the former allows for continuous maintenance of oxygen levels sufficient for $^1O_2$ production, the latter facilitates reoxygenation of the tissue between light exposures. Significant enhancements of tumor response have been observed with either of these alternatives for PDT with Photofrin, 5-aminolevulinic acid (ALA)/PpIX, and meso-tetra-hydroxyphenyl-chlorin (mTHPC). In part, this may be due to a significant but moderate increase in direct photodamage to tumor cells. Direct tumor cell death increased with low fluence rate PDT by ~1/2–1 log in the RIF mouse model.[168] However, the microvasculature, especially the normal, tumor-surrounding microvasculature, can also be affected by modification of fluence rate.[106] One practical drawback of low fluence rate treatment, as compared to high fluence rate, is the increase in exposure time required to deliver a given fluence. Due to the higher treatment efficiency, this can be somewhat, but not entirely, compensated for by a reduction of the total fluence delivered. In fact, a reduction of the total fluence may be necessary with low fluence rate treatment since the PDT efficiency for causing vascular and normal tissue effects will also increase, thus decreasing treatment selectivity.[106,167,169]

Enhanced responses to light dose fractionation in ALA PDT of murine tumors and normal rat colon may involve relocalization and/or resupply of PpIX during dark periods, in addition to reoxygenation.[170,171]

A two-step irradiance protocol has recently been developed for ALA PDT of nonmelanoma skin cancers that minimizes the severe pain experienced by patients during PDT light delivery.[172,173] It involves an initial short period of illumination at low fluence rate to photobleach 80%–90% of PpIX, followed by high fluence rate for the remainder of the total prescribed fluence. This approach achieves alleviation of pain and excellent results while only minimally prolonging the treatment time.

1.6 Immune Effects of PDT

1.6.1 Immune Suppression

Immune suppressive effects are largely confined to cutaneous and transdermal PDT. Cutaneous PDT can suppress allograft rejection[174–177] and contact hypersensitivity (CHS) reactions.[178–182] The mechanism of PDT-induced immune suppression appears to be associated with the induction of immunosuppressive cytokines. PDT induces tumor necrosis factor (TNF)-α,[183,184] which is involved in some aspects of UV-mediated immunosuppression.[185] However, it is not responsible for PDT-induced CHS suppression.[183] PDT also induces interleukin (IL)-10,[178,186,187] which has been shown to inhibit cell-mediated immune responses, including CHS.[186,188] Gollnick et al. have shown that in vitro PDT induces IL-10 expression from keratinocytes as a result of activation of the IL-10 gene promoter by enhanced expression of AP-1 and prolonged IL-10 mRNA half-life.[86] Direct mechanistic studies of the role of IL-10 in PDT-induced suppression of CHS have yielded contradictory results that can be explained, at least in part, by the treatment regime and dose of PDT. Treatment with cutaneous PDT, using Photofrin and blue light centered at 430 nm, results in irradiation that is mostly limited to the skin. In contrast, transdermal PDT, with benzoporphyrin derivative monoacid ring A (BPD-MA) and illumination with 690 nm light, involves whole body illumination. Also, due to the greater depth of light penetration of this wavelength, some light exposure of internal sites may occur. Simkin et al.[186] implicated IL-10 as the active mediator in transdermal PDT-induced suppression of CHS by demonstrating that IL-10 knockout (KO) mice did not undergo transdermal PDT-induced CHS suppression. In contrast, cutaneous PDT treatment of IL-10 KO mice did induce CHS suppression.[178] The lack of involvement of IL-10 in suppression of CHS by cutaneous PDT was further confirmed when studies using neutralizing anti-IL-10 antibodies failed to inhibit cutaneous PDT-induced CHS suppression.[178,182] Thus, it appears as though the mechanism of PDT-induced suppression of CHS is dependent upon the treatment regime. Regimes that result in large treatment fields and internal exposure, that is, transdermal and potentially peritoneal PDT, mediate CHS suppression via IL-10. PDT regimes that employ lower doses and superficial cutaneous exposure suppress CHS reactions via an IL-10-independent mechanism.

Interestingly, CHS suppression by both transdermal and cutaneous PDT was reversed by administration of exogenous IL-12,[178,186] suggesting that these processes share a common regulatory point, perhaps in the development of Th1 and/or Tc1 cells.

In addition to the effects of cytokines on PDT-induced suppression of immune responses, it is important to consider the effect of PDT on the expression of immune molecules critical to immune system activation. Transdermal PDT has been shown to inhibit the ability of Langerhans cells (LC) to stimulate alloreactive T cells, and LC treated ex vivo expressed lower levels of major histocompatibility complex (MHC) antigens and CD80 and CD86 costimulatory molecules, which are needed for T cell activation.[189] Additionally, in vitro PDT-treated murine dendritic cells (DC) had a reduced ability to stimulate alloreactive T cells and exhibited lower levels of MHC molecules, costimulatory molecules, and adhesion molecules.[190] Thus, in addition to altering the function of antigen presenting cells (APCs), PDT has the ability to disrupt the APC–T cell cognate, which is needed for T cell activation.

Immune suppression has also been reported following topical PDT in humans.[191] In this study, both MAL PDT and ALA PDT significantly suppressed delayed-type hypersensitivity responses. In a follow-up study, Frost et al.[192] showed that the degree of immune suppression was decreased when low flounce rates were used to deliver equivalent light doses. These findings support the hypothesis that the immune effects of PDT are highly dependent upon the treatment regimen employed.

1.6.2 Immune Potentiation

Tumor-directed PDT represents a confined sterile injury that leads to a local inflammatory reaction, the latter usually being the first sign of the PDT tissue response.

The initiation of the inflammatory process has been ascribed to alarmins, molecules that preexist in cells and are released upon cell dissociation.[193] They include damage-associated molecular pattern molecules (DAMPs), cytokines, and metabolites. Several abundant cellular proteins serve as DAMPs, such as HSPs, peroxiredoxin-1, S100 proteins, and high-mobility group protein B1 (HMGB1).[193-196] In addition, intracellular cytokines, released from dying cells, can commence signaling functions via their receptors on neighboring cells. A recent study in a coculture system of primary human epithelial tumor cells and their stromal fibroblast counterparts has identified IL-1α, released from PDT exposed tumor cells, as the major alarmin activating fibroblasts to generate inflammatory mediators such as IL-6.[25]

The first manifestation of the inflammatory response is a rapid influx of neutrophils, which appear to be critical to long-term tumor control.[197-200] The influx of neutrophils into the treatment site is preceded by an induction of chemokines and adhesion molecules critical to neutrophil migrations, and blocking their function retards the PDT response.[200] Neutrophilic infiltration is followed by mast cells and macrophages.[187,201] The PDT regimen can strongly influence the inflammatory response—high PDT doses, while locally curative through mainly vascular mechanisms, suppress inflammation, while oxygen-conserving, low PDT doses enhance it.[202,203] A treatment protocol is currently being developed that combines the two PDT approaches, first inducing the inflammatory reaction to maximize stimulation of the adaptive antitumor immune response, followed by the locally curative treatment (Gollnick, unpublished data).

PDT enhances macrophage tumoricidal activity[204,205] and stimulates macrophage release of TNF-α.[89] It has been suggested that nonspecific killing of tumor cells by inflammatory cells, potentially through the release of reactive oxygen species, contributes to the overall tumor kill by PDT.[201] This hypothesis is supported by studies showing that if the PDT-induced inflammatory response is further stimulated by addition of adjuvants or macrophage activating factors, the overall tumor response to PDT is greater.[206–208] The role of the innate immune response in overall tumor kill by PDT was also shown to involve natural killer (NK) cells. Depletion of NK cells reduced the long-term tumor control by PDT[209] and augmentation of NK activity enhanced PDT tumor control.[210] The ability of cytolytic T lymphocytes (CTLs) and NK cells to destroy tumor cells depends on their recognition of MHC class 1 and MHC class 1–related molecules. PDT induced the MHC class 1–related molecules MICA and increased expression of NKG2DL in human tumor cells and increased their lysis by NK cells.[211]

A plethora of important immune modulators is generated via activation of stress response factors such as AP-1 and NF-κB,[85,89,187] including IL-6, TNF-α, IL-1β, IL-2, and granulocyte macrophage colony-stimulating factor (GM-CSF).[183,184,187,212,213] The role played by IL-6 has been unclear, but a recent study showed that the elimination of IL-6 had no effect on innate cell mobilization into the tumor bed and did not affect primary antitumor T cell activation but did negatively regulate antitumor immune memory.[214]

Taken together, PDT creates a tumor milieu that is conducive to the antitumor immune modulation[215] that has been documented in both preclinical and clinical studies. Tumor draining lymph node cells isolated from PDT-treated mice are able to suppress subsequent tumor challenges when transferred to a naïve host.[216,217] Canti et al.[217] have shown that PDT-treated mice that remain tumor free for 100 days post-PDT are resistant to subsequent tumor challenges, suggesting the presence of immune memory cells. The importance of the immune response in PDT was definitively shown by a series of experiments in severe combined immunodeficient (SCID) and nude mice.[197,209] PDT treatment, at a dose that was curative in immunocompetent BALB/c mice, provided only short-term cures of EMT6 tumors in SCID and nude mice. The ability to provide long-term cures was restored when immunodeficient animals were reconstituted with bone marrow cells from BALB/c mice. Depletion studies showed that the critical cells involved in the PDT-induced immune response were CD8+ cells, with NK cells playing a supportive role.[205,218] Clinical studies have demonstrated an infiltration of CD8+ cells into PDT-treated tumor tissue.[219,220] Kabingu et al.[221] demonstrated in basal cell carcinoma patients that ALA PDT enhanced recognition of MHC-1–antigen complexes by immune cells and activation of tumor-specific CD8+ cells. As in the inflammatory response,[202,203] PDT treatment parameters (fluence rate, fluence) influence the adaptive antitumor immune response.[218,220]

The knowledge gained by these preclinical and clinical studies has raised the hope and possibility that the benefit of PDT, so far restricted to local tumor treatment, might be expanded to attack distant disease. To support this goal, effective antitumor vaccines have been generated by in vitro PDT treatment of tumor cells, which are able to stimulate the maturation and activation of DC as well as activation of tumor-specific CD8+ T cells.[222–225]

1.7 Conclusion

The scientific effort that supports this new cancer therapy has led to numerous significant advances. The development of new photosensitizers is essentially eliminating the problem of prolonged cutaneous photosensitivity and is extending treatment depth, an issue dealt with in detail elsewhere in this volume. The complex dynamics of tumor oxygenation in response to PDT are now largely understood. The oxidative stress effects of PDT on redox-sensitive transcription factors and the genes they control are being uncovered. The complex interplay of biological mechanisms governing the PDT tumor response has been realized. Given these accomplishments, it remains for them to be translated into actual patient benefit. Noninvasive probes need to be perfected that will allow the monitoring of photosensitizer levels, of oxygen status, or, ideally directly, of singlet oxygen, the cytotoxic agent. New light delivery regimes need to be devised for clinical use that will minimize oxygen limitations. The ways in which such regimes might influence redox-sensitive gene regulation and how these genes might affect treatment outcome need to be explored. Finally, our expanding understanding of the complex effects of PDT on host immunity needs to be exploited to modulate the PDT response.

Abbreviations

ALA   5-aminolevulinic acid
APC   antigen presenting cells
BPD-MA   benzoporphyrin derivative, monoacid ring A
CHS   contact hypersensitivity
DC   dendritic cells
HpD   hematoporphyrin derivative
HPPH   hexyl pyropheophorbide ether
LC   Langerhans cells
Lutex   lutetium texaphyrin
mTHPC   meso-tetra-hydroxyphenyl-chlorin
MHC   major histocompatibility complex
NPe6   mono-L-aspartyl chlorin e6
Pc   phthalocyanine
PDT   photodynamic therapy
PpIX   protoporphyrin IX
$SnEt_2$ tin etiopurpurin
Tc   T cells
Th cells   T helper cells
TPPS   sulfonated tetraphenylporphines

Acknowledgment

The authors thank Ian MacDonald for the creation and preparation of figures presented in this chapter.

References

1. Foote CS. Mechanisms of photooxygenation. *Prog Clin Biol Res* 1984;170:3–18.
2. Henderson BW, Miller AC. Effects of scavengers of reactive oxygen and radical species on cell survival following photodynamic treatment in vitro: Comparison to ionizing radiation. *Radiat Res* 1986;108:196–205.
3. Moan J, Sommer S. Oxygen dependence of the photosensitizing effect of hematoporphyrin derivative in NHIK 3025 cells. *Cancer Res* 1985;45:1608–1610.
4. Moan J, Berg K. The photodegradation of porphyrins in cells can be used to estimate the lifetime of singlet oxygen. *Photochem Photobiol* 1991;53:549–553.
5. Tracy EC, Bowman MJ, Pandey RK, Henderson BW, Baumann H. Cell-type selective phototoxicity achieved with chlorophyll-a derived photosensitizers in a co-culture system of primary human tumor and normal lung cells. *Photochem Photobiol* 2011;87:1405–1418.
6. MacDonald IJ, Dougherty TJ. Basic principles of photodynamic therapy. *J Porphyrins Phthalocyanines* 2001;5:105–129.

7. Kessel D, Chang CK, Musselman B. Chemical, biologic and biophysical studies on 'hematoporphyrin derivative'. *Adv Exp Med Biol* 1985;193:213–227.

8. Zheng X, Morgan J, Pandey SK, Chen Y, Tracy E, Baumann H et al. Conjugation of 2-(1′-hexyloxyethyl)-2-devinylpyropheophorbide-a (HPPH) to carbohydrates changes its subcellular distribution and enhances photodynamic activity in vivo. *J Med Chem* 2009;52:4306–4318.

9. Berg K, Madslien K, Bommer JC, Oftebro R, Winkelman JW, Moan J. Light induced relocalization of sulfonated meso-tetraphenylporphines in NHIK 3025 cells and effects of dose fractionation. *Photochem Photobiol* 1991;53:203–210.

10. Henderson BW, Bellnier DA, Greco WR, Sharma A, Pandey RK, Vaughan LA et al. An in vivo quantitative structure-activity relationship for a congeneric series of pyropheophorbide derivatives as photosensitizers for photodynamic therapy. *Cancer Res* 1997;57:4000–4007.

11. Robey RW, Steadman K, Polgar O, Bates SE. ABCG2-mediated transport of photosensitizers: Potential impact on photodynamic therapy. *Cancer Biol Ther* 2005;4:187–194.

12. Morgan J, Jackson JD, Zheng X, Pandey SK, Pandey RK. Substrate affinity of photosensitizers derived from chlorophyll-a: The ABCG2 transporter affects the phototoxic response of side population stem cell-like cancer cells to photodynamic therapy. *Mol Pharm* 2010;7:1789–1804.

13. Liu W, Baer MR, Bowman MJ, Pera P, Zheng X, Morgan J et al. The tyrosine kinase inhibitor imatinib mesylate enhances the efficacy of photodynamic therapy by inhibiting ABCG2. *Clin Cancer Res* 2007;13:2463–2470.

14. Kennedy JC, Pottier RH, Pross DC. Photodynamic therapy with endogenous protoporphyrin IX: Basic principles and present clinical experience. *J Photochem Photobiol B* 1990;6:143–148.

15. Peng Q, Berg K, Moan J, Kongshaug M, Nesland JM. 5-Aminolevulinic acid-based photodynamic therapy: Principles and experimental research. *Photochem Photobiol* 1997;65:235–251.

16. Kessel D, Luo Y. Mitochondrial photodamage and PDT-induced apoptosis. *J Photochem Photobiol B* 1998;42:89–95.

17. Morgan J, Potter WR, Oseroff AR. Comparison of photodynamic targets in a carcinoma cell line and its mitochondrial DNA-deficient derivative. *Photochem Photobiol* 2000;71:747–757.

18. Verma A, Focchina SL, Hirsch S, Dillahey L, Williams J, Snyder SH. Photodynamic tumor therapy: Mitochondrial benzodiazepine receptors as a therapeutic target. *Mol Med* 1998;4:40–45.

19. Miccoli L, Beurdeley-Thomas A, DePinieux G, Sureau F, Oudard S, Dutrillaux B et al. Light induced photoactivation of hypericin affects the energy metabolism of human glioma cells by inhibiting hexokinase bound to mitochondria. *Cancer Res* 1998;58:5777–5786.

20. Perlin DS, Murant RS, Gibson SL, Hilf R. Effects of photosensitization by hematoporphyrin derivative on mitochondrial adenosine triphosphatase-mediated proton transport and membrane integrity of R3230AC mammary adenocarcinoma. *Cancer Res* 1985;45:653–658.

21. Morgan J, Oseroff AR. Mitochondria-based photodynamic anti-cancer therapy. *Adv Drug Deliv Rev* 2001;49:71–86.

22. Geze M, Morliere P, Maziere JC, Smith KM, Santus R. Lysosomes, a key target of hydrophobic photosensitizers proposed for chemotherapeutic applications. *J Photochem Photobiol B* 1993;20:23–25.

23. Moan J, Berg K, Anholt A, Madslien K. Sulfonated aluminum phthalocyanines as sensitizers for photochemotherapy. Effects of small doses on localization, dye fluorescence and photosensitivity in V79 cells. *Int J Cancer* 1994;58:865–870.

24. MacDonald IJ, Morgan J, Bellnier DA, Paszkiewicz G, Whitaker JE, Litchfield DJ et al. Subcellular localization patterns and their relationship to photodynamic activity of pyropheophorbide-a derivatives. *Photochem Photobiol* 1999;70:789–797.

25. Tracy EC, Bowman MJ, Henderson BW, Baumann H. Interleukin-1alpha is the major alarmin of lung epithelial cells released during photodynamic therapy to induce inflammatory mediators in fibroblasts. *Br J Cancer* 2012;107:1534–1546.

26. Kessel D, Woodburn K, Henderson BW, Chang CK. Sites of photodamage in vivo and in vitro by a cationic porphyrin. *Photochem Photobiol* 1995;62(5):875–881.

27. Moan J, Pettersen EO, Christensen T. The mechanism of photodynamic inactivation of human cells in vitro in the presence of haematoporphyrin. *Br J Cancer* 1979;39:398–407.

28. Christensen T, Volden G, Moan J, Sandquist T. Release of lysosomal enzymes and lactate dehydrogenase due to hematoporphyrin derivative and light irradiation of NHIK 3025 cells in vitro. *Ann Clin Res* 1982;14:46–52.

29. Henderson BW, Donovan JM. Release of prostaglandin E2 from cells by photodynamic treatment in vitro. *Cancer Res* 1989;49:6896–6900.

30. Bellnier DA, Dougherty TJ. Membrane lysis in Chinese hamster ovary cells treated with hematoporphyrin derivative plus light. *Photochem Photobiol* 1982;36:43–47.

31. Thomas JP, Girotti AW. Role of lipid peroxidation in hematoporphyrin derivative-sensitized photokilling of tumor cells: Protective effects of glutathione peroxidase. *Cancer Res* 1989;49:1682–1686.

32. Girotti AW. Photodynamic lipid peroxidation in biological systems. *Photochem Photobiol* 1990;51:497–509.

33. Deuticke B, Henseleit U, Haest CW, Heller KB, Dubbelman TM. Enhancement of transbilayer mobility of a membrane lipid probe accompanies formation of membrane leaks during photodynamic treatment of erythrocytes. *Biochim Biophys Acta* 1989;982:53–61.

34. Gibson SL, Murant RS, Hilf R. Photosensitizing effects of hematoporphyrin derivative and photofrin II on the plasma membrane enzymes 5′-nucleotidase, Na⁺K⁺-ATPase, and Mg²⁺-ATPase in R3230AC mammary adenocarcinomas. *Cancer Res* 1988;48:3360–3366.

35. Specht KG, Rodgers MA. Plasma membrane depolarization and calcium influx during cell injury by photodynamic action. *Biochim Biophys Acta* 1991;1070:60–68.

36. Berg K, Moan J. Lysosomes and microtubules as targets for photochemotherapy of cancer. *Photochem Photobiol* 1997;65:403–409.

37. Berg K, Steen HB, Winkelman JW, Moan J. Synergistic effects of photoactivated tetra(4-sulfonatophenyl)porphine and nocodazole on microtubule assembly, accumulation of cells in mitosis and cell survival. *J Photochem Photobiol B* 1992;13:59–70.

38. Moan J, Vistnes AI. Porphyrin photosensitization of proteins in cell membranes as studied by spin-labelling and by quantification of DTNB-reactive SH-groups. *Photochem Photobiol* 1986;44:15–19.

39. Liu W, Oseroff AR, Baumann H. Photodynamic therapy causes cross-linking of signal transducer and activator of transcription proteins and attenuation of interleukin-6 cytokine responsiveness in epithelial cells. *Cancer Res* 2004;64:6579–6587.

40. Gomer CJ, Rucker N, Banerjee A, Benedict WF. Comparison of mutagenicity and induction of sister chromatid exchange in Chinese hamster cells exposed to hematoporphyrin derivative photoradiation, ionizing radiation, or ultraviolet radiation. *Cancer Res* 1983;43:2622–2627.

41. Evensen JF, Moan J. Photodynamic action and chromosomal damage: A comparison of haematoporphyrin derivative (HpD) and light with x-irradiation. *Br J Cancer* 1982;45:456–465.

42. McNair FL, Marples B, West CML, Moore JV. A comet assay of DNA damage and repair in K562 cells after photodynamic therapy using hematoporphyrin derivative, methylene blue and *meso*-tetrahydroxyphenylchlorin. *Br J Cancer* 1997;75:1721–1729.

43. Evans HH, Horng M-F, Ricanati M, Deahl JT, Oleinick NL. Mutagenicity of photodynamic therapy as compared to UVC and ionizing radiation in human and murine lymphoblast cell lines. *Photochem Photobiol* 1997;66:690–696.

44. Buytaert E, Dewaele M, Agostinis P. Molecular effectors of multiple cell death pathways initiated by photodynamic therapy. *Biochim Biophys Acta* 2007;1776:86–107.

45. Henkel T. Apoptosis: Corralling the corpses. *Cell* 2001;104:325–328.

46. Agostinis P, Berg K, Cengel KA, Foster TH, Girotti AW, Gollnick SO et al. Photodynamic therapy of cancer: An update. *CA Cancer J Clin* 2011;61:250–281.

47. Reiners JJ Jr., Agostinis P, Berg K, Oleinick NL, Kessel D. Assessing autophagy in the context of photodynamic therapy. *Autophagy* 2010;6:7–18.

48. Dewaele M, Maes H, Agostinis P. ROS-mediated mechanisms of autophagy stimulation and their relevance in cancer therapy. *Autophagy* 2010;6:838–854.

49. Kessel D, Luo Y, Deng Y, Chang CK. The role of subcellular localization in the initiation of apoptosis by photodynamic therapy. *Photochem Photobiol* 1997;65:422–426.

50. Luo Y, Kessel D. Initiation of apoptosis versus necrosis by photodynamic therapy with chloroaluminum phthalocyanine. *Photochem Photobiol* 1997;66:479–483.

51. Yang J, Bhalla K, Kim CN, Ibrado AM, Peng TI, Jones DP et al. Prevention of apoptosis by Bcl-2: Release of cytochrome c from mitochondria blocked. *Science* 1997;275:1129–1132.

52. Granville DJ, Carthy CM, Jiang H, Shore GC, McManus BM, Hunt DWC. Rapid cytochrome c release, activation of caspases 3, 6, 7 and 8 followed by Bap31 cleavage in HeLa cells treated with photodynamic therapy. *FEBS Lett* 1998;437:5–10.

53. Chiu SM, Evans HH, Lam M, Nieminen A, Oleinick NL. Phthalocyanine 4 photodynamic therapy-induced apoptosis of mouse L5178Y-R cells results from a delayed but extensive release of cytochrome c from mitochondria. *Cancer Lett* 2001;165:51–58.

54. Chiu SM, Oleinick NL. Dissociation of mitochondrial depolarization from cytochrome c release during apoptosis induced by photodynamic therapy. *Br J Cancer* 2001;84:1099–1106.

55. Vantieghem A, Xu Y, Declercq W, Vandenabeele P, Denecker G, Vandenheede JR et al. Different pathways mediate cytochrome c release after photodynamic therapy with hypericin. *Photochem Photobiol* 2001;74:133–142.

56. Granville DJ, Levy JG, Hunt DWC. Photodynamic therapy induces caspase-3 activation in HL-60 cells. *Cell Death Differ* 1997;4:623–628.

57. Kessel D, Luo Y. Photodynamic therapy: A mitochondrial inducer of apoptosis. *Cell Death Differ* 1999;6:28–35.

58. Xue LY, Chiu SM, Oleinick NL. Photodynamic therapy-induced death of MCF-7 human breast cancer cells: A role for caspase-3 in the late steps of apoptosis but not for the critical lethal event. *Exp Cell Res* 2001;263:145–155.

59. Srivastava M, Ahmad N, Gupta S, Mukhtar H. Involvement of bcl-2 and bax in photodynamic therapy-mediated apoptosis. *J Biol Chem* 2001;276:15481–15488.

60. Granville DJ, Jiang H, An MT, Levy JG, McManus BM, Hunt DWC. Bcl-2 overexpression blocks caspase activation and downstream apoptotic events instigated by photodynamic therapy. *Br J Cancer* 1999;79:95–100.

61. He J, Agarwal ML, Larkin HE, Friedman LR, Xue LY, Oleinick NL. The induction of partial resistance to photodynamic therapy by the protooncogene bcl-2. *Photochem Photobiol* 1996;64:845–852.

62. Xue LY, Chiu SM, Oleinick NL. Photochemical destruction of the Bcl-2 oncoprotein during photodynamic therapy with the phthalocyanine photosensitizer Pc 4. *Oncogene* 2001;20:3420–3427.

63. Kim H, Luo Y, Li G, Kessel D. Enhanced apoptotic response to photodynamic therapy after bcl-2 transfection. *Cancer Res* 1999;59:3429–3432.

64. Usuda J, Okunaka T, Furukawa K, Tsuchida T, Kuroiwa Y, Ohe Y et al. Increased cytotoxic effects of photodynamic therapy in IL-6 gene transfected cells via enhanced apoptosis. *Int J Cancer* 2001;93:475–480.

65. Agarwal ML, Larkin HE, Zaidi SIA, Mukhtar H, Oleinick NL. Phospholipase activation triggers apoptosis in photosensitized mouse lymphoma cells. *Cancer Res* 1993;53:5897–5902.

66. Separovic D, Mann KJ, Oleinick NL. Association of ceramide accumulation with photodynamic treatment-induced cell death. *Photochem Photobiol* 1998;68:101–109.

67. Gupta S, Ahmad N, Mukhtar H. Involvement of nitric oxide during phthalocyanine (Pc4) photodynamic therapy-mediated apoptosis. *Cancer Res* 1998;58:1785–1788.

68. Kessel D, Arroyo AS. Apoptotic and autophagic responses to Bcl-2 inhibition and photodamage. *Photochem Photobiol Sci* 2007;6:1290–1295.

69. Xue LY, Chiu SM, Azizuddin K, Joseph S, Oleinick NL. Protection by Bcl-2 against apoptotic but not autophagic cell death after photodynamic therapy. *Autophagy* 2008;4:125–127.

70. Tao J-S, Sanghera S, Pelech SL, Wong G, Levy JG. Stimulation of stress-activated protein kinase and p38 HOG1 kinase in murine keratinocytes following photodynamic therapy with benzoporphyrin derivative. *J Biol Chem* 1996;271:27107–27115.

71. Klotz L-O, Fritsch C, Briviba K, Tsacmacidis N, Schliess F, Sies H. Activation of JNK and p38 but not ERK MAP kinases in human skin cells by 5-aminolevulinate-photodynamic therapy. *Cancer Res* 1998;58:4297–4300.

72. Assefa Z, Vantieghem A, Declercq W, Vandenabelle P, Vandenheede JR, Merlevede W et al. The activation of the c-jun N-terminal kinase and p38 mitogen-activated protein kinase signaling pathways protects HeLa cells from apoptosis following photodynamic therapy with hypericin. *J Biol Chem* 1999;274:8788–8796.

73. Wong TW, Tracy E, Oseroff A, Baumann H. Photodynamic therapy mediates immediate loss of cellular responsiveness to cytokines and growth factors. *Can Res* 2003;63:3812–3818.

74. Klotz LO, Kroncke KD, Sies H. Singlet oxygen-induced signaling effects in mammalian cells. *Photochem Photobiol Sci* 2003;2:88–94.

75. Oleinick NL, He J, Xue LY, Separovic D. Stress-activated signalling responses leading to apoptosis following photodynamic therapy. *Optical Methods for Tumor Treatment and Detection: Mechanisms and Techniques in Photodynamic Therapy VII*, San Jose, CA, January 24–25, 1998, pp. 82–88.

76. Xue LY, He J, Oleinick NL. Rapid tyrosine phosphorylation of HS1 in the response of mouse lymphoma L5178Y-R cells to photodynamic treatment sensitized by the phthalocyanine Pc4. *Photochem Photobiol* 1997;66:105–113.

77. Koukourakis MI, Giatromanolaki A, Skarlatos J, Corti L, Blandamura S, Piazza M et al. Hypoxia inducible factor (HIF-1a and HIF-2a) expression in early esophageal cancer and response to photodynamic therapy and radiotherapy. *Cancer Res* 2001;61:1830–1832.

78. Ferrario A, von Tiehl KF, Rucker N, Schwarz MA, Gill PS, Gomer CJ. Antiangiogenic treatment enhances photodynamic therapy responsiveness in a mouse mammary carcinoma. *Cancer Res* 2000;60(15):4066–4069.

79. Luna MC, Ferrario A, Wong S, Fisher AMR, Gomer CJ. Photodynamic therapy-mediated oxidative stress as a molecular switch for the temporal expression of genes ligated to the human heat shock promoter. *Cancer Res* 2000;60:1637–1644.

80. Gomer CJ, Ryter SW, Ferrario A, Rucker N, Wong S, Fisher AMR. Photodynamic therapy-mediated oxidative stress can induce expression of heat shock proteins. *Cancer Res* 1996;56:2355–2360.

81. Curry PM, Levy JG. Stress protein expression in murine tumor cells following photodynamic therapy with benzoporphyrin derivative. *Photochem Photobiol* 1993;58:374–379.

82. Gomer CJ, Ferrario A, Rucker N, Wong S, Lee AS. Glucose regulated protein induction and cellular resistance to oxidative stress mediated by porphyrin photosensitization. *Cancer Res* 1991;51:6574–6579.

83. Xue LY, Agarwal ML, Varnes ME. Elevation of GRP-78 and loss of HSP-70 following photodynamic treatment of V79 cells: Sensitization by nigericin. *Photochem Photobiol* 1995;62:135–143.

84. Gomer CJ, Luna M, Ferrario A, Rucker N. Increased transcription and translation of heme oxygenase in Chinese hamster fibroblasts following photodynamic stress or Photofrin II incubation. *Photochem Photobiol* 1991;53:275–279.

85. Kick G, Messer G, Goetz A, Plewig G, Kind P. Photodynamic therapy induces expression of interleukin 6 by activation of AP-1 but not NF-kB DNA binding. *Cancer Res* 1995;55:2373–2379.

86. Gollnick SO, Lee BY, Vaughan L, Owczarczak B, Henderson BW. Activation of the IL-10 gene promoter following photodynamic therapy of murine keratinocytes. *Photochem Photobiol* 2001;73:170–177.

87. Abate C, Patel L, Rauscher FJ, Curran T. Redox regulation of fos and jun DNA-binding activity *in vitro*. *Science* 1990;249:1157–1161.

88. Yao K-S, Zanthoudakis S, Curran T, O'Dwyer PJ. Activation of AP-1 and of nuclear redox factor, Ref1, in the response of HT29 colon cancer cells to hypoxia. *Mol Cell Biol* 1994;14:5997–6003.

89. Kick G, Messer G, Plewig G, Kind P, Goetz AE. Strong and prolonged injunction of c-*jun* and c-*fos* proto-oncogenes by photodynamic therapy. *Br J Cancer* 1996;74:30–36.

90. Luna MC, Wong S, Gomer CJ. Photodynamic therapy mediated induction of early response genes. *Cancer Res* 1994;54:1374–1380.

91. Baeuerle PA, Henkel T. Function and activation of NF-B in the immune system. *Annu Rev Immunol* 1994;12:141–179.

92. Schreck R, Albermann K, Baeuerle PA. Nuclear factor kB: An oxidative stress-responsive transcription factor of eukaryotic cells. *Free Radic Res Commun* 1992;17:221–237.

93. DiDonato JA, Mercurio R, Rosette C, Wu-Li J, Suyang H, Ghosh S et al. Mapping of the inducible IkB phosphorylation sites that signal its ubiquitination and degradation. *Mol Cell Biol* 1996;16:1295–1304.

94. DiDonato JA, Hayakawa M, Rothwarf DM, Zandi E, Karin M. A cytokine-responsive IkB kinase that activates the transcription factor NFkB. *Nature* 1997;388:548–554.

95. Granville DJ, Carthy CM, Jiang H, Levy JG, McManus BM, Matroule JY et al. Nuclear factor-kB activation by the photochemotherapeutic agent verteporfin. *Blood* 2000;95:256–262.

96. Ryter SW, Gomer CJ. Nuclear factor kB binding activity in mouse L1210 cells following Photofrin II-mediated photosensitization. *Photochem Photobiol* 1993;58:753–756.

97. Matroule JY, Bonizzi G, Morliere P, Paillous N, Santus R, Bours V et al. Pyropheophorbide-a methyl ester-mediated photosensitization activates transcription factor NF-kB through the interleukin-1 receptor-dependent signaling pathway. *J Biol Chem* 1999;270:2899–3000.

98. Legrand-Poels S, Schoonbroodts S, Matroule JY, Piette J. NF-kB: An important transcription factor in photobiology. *J Photochem Photobiol B* 1998;45:1–8.

99. Henderson BW. Probing the effects of photodynamic therapy through in vivo-in vitro methods. In: Kessel D, editor. *Photodynamic Therapy of Neoplastic Disease*, Vol. I. Boca Raton, FL: CRC Press, 1990, pp. 169–188.

100. Henderson BW, Fingar VH. Oxygen limitation of direct tumor cell kill during photodynamic treatment of a murine tumor model. *Photochem Photobiol* 1989;49:299–304.

101. Henderson BW, Waldow SM, Mang TS, Potter WR, Malone PB, Dougherty TJ. Tumor destruction and kinetics of tumor cell death in two experimental mouse tumors following photodynamic therapy. *Cancer Res* 1985;45:572–576.

102. Henderson BW, Sumlin AB, Owczarczak BL, Dougherty TJ. Bacteriochlorophyll-a as photosensitizer for photodynamic treatment of transplantable murine tumors. *J Photochem Photobiol B: Biol* 1991;10:303–313.

103. Henderson BW, Vaughan L, Bellnier DA, vanLeengoed H, Johnson PG, Oseroff AR. Photosensitization of murine tumor, vasculature and skin by 5-aminolevulinic acid-induced porphyrin. *Photochem Photobiol* 1995;62:780–789.

104. Cincotta L, Foley JW, Cincotta AH. Novel phenothiazinium photosensitizers for photodynamic therapy. In: Hasan T, editor. *Advances in Photochemotherapy*, Vol. 997. Washington, DC: SPIE, 1988, pp. 145–153.

105. Chan W-S, Brasseur N, La Madeleine C, van Lier JE. Evidence for different mechanisms of EMT-6 tumor necrosis by photodynamic therapy with disulfonated aluminum phthalocyanine or Photofrin: Tumor cell survival and blood flow. *Anticancer Res* 1996;16:1887–1892.

106. Sitnik T, Henderson BW. The effect of fluence rate on tumor and normal tissue responses to photodynamic therapy. *Photochem Photobiol* 1998;67:462–466.

107. Korbelik M, Krosl G. Cellular levels of photosensitisers in tumours: The role of proximity to the blood supply. *Br J Cancer* 1994;70:604–610.

108. Svaasand LO. Optical dosimetry for direct and interstitial photoradiation therapy of malignant tumors. *Prog Clin Biol Res* 1984;170:91–114.

109. Wilson BC, Jeeves WP, Lowe DM, Adam G. Light propagation in animal tissues in the wavelength range 375–825 nanometers. *Prog Clin Biol Res* 1984;170:115–132.

110. Adams K, Rainbow AJ, Wilson BC, Singh G. In vivo resistance to Photofrin-mediated photodynamic therapy in radiation-induced fibrosarcoma cells resistant to in vitro Photofrin-mediated photodynamic therapy. *J Photochem Photobiol B: Biol* 1999;49(2–3):136–141.
111. Kessel D, Hampton J, Fingar V, Morgan A. Tumor versus vascular photodamage in a rat tumor model. *J Photochem Photobiol B: Biol* 1998;45(1):25–27.
112. Pass HI, Evans S, Matthews WA, Perry R, Venzon D, Roth JA et al. Photodynamic therapy of oncogene-transformed cells. *J Thorac Cardiovasc Surg* 1991;101:795–799.
113. Henderson BW, Dougherty TJ. How does photodynamic therapy work? *Photochem Photobiol* 1992;55:145–157.
114. Dougherty TJ, Marcus SL. Photodynamic therapy. *Eur J Cancer* 1992;28A(10):1734–1742.
115. Hamblin MR, Newman EL. Photosensitizer targeting in photodynamic therapy II. Conjugates of hematoporphyrin with serum lipoproteins. *J Photochem Photobiol B: Biol* 1994;26:147–157.
116. Hamblin MR, Newman EL. Photosensitizer targeting in photodynamic therapy I. Conjugates of hematoporphyrin with albumin and transferrin. *J Photochem Photobiol B: Biol* 1994;26:45–56.
117. Molpus KL, Hamblin MR, Rizvi I, Hasan T. Intraperitoneal photoimmunotherapy of ovarian carcinoma xenografts in nude mice using charged photoimmunoconjugates. *Gynecol Oncol* 2000;76(3):397–404.
118. DelGovernatore M, Hamblin MR, Shea CR, Rizvi I, Hasan T. Experimental photoimmunotherapy of hepatic metastases of colorectal cancer with a 17.1A chlorin (e6) immunoconjugate. *Cancer Res* 2000;60(15):4200–4205.
119. Soukos NS, Hamblin MR, Keel S, Fabian RL, Deutsch TF, Hasan T. Epidermal growth factor receptor-targeted immunophotodiagnosis and photoimmunotherapy of oral precancer in vivo. *Cancer Res* 2001;61(11):4490–4496.
120. Henderson BW, Dougherty TJ, Malone PB. Studies on the mechanism of tumor destruction by photoradiation therapy. *Prog Clin Biol Res* 1984;170:601–612.
121. Fingar VH, Siegel KA, Wieman TJ, Doak KW. The effects of thromboxane inhibitors on the microvascular and tumor response to photodynamic therapy. *Photochem Photobiol* 1993;58(3):393–399.
122. Reed MWR, Wieman TJ, Schuschke DA, Tseng MT, Miller FN. A comparison of the effects of photodynamic therapy on normal and tumor blood vessels in the rat microcirculation. *Radiat Res* 1989;119:542–552.
123. Fingar VH, Potter WR, Henderson BW. Drug and light dose dependence of photodynamic therapy: A study of tumor cell clonogenicity and histologic changes. *Photochem Photobiol* 1987;45:643–650.
124. Fingar VH, Kik PK, Haydon PS, Cerrito PB, Tseng M, Abang E et al. Analysis of acute vascular damage after photodynamic therapy using benzoporphyrin derivative (BPD). *Br J Cancer* 1999;79:1702–1708.
125. Sitnik TM, Hampton JA, Henderson BW. Reduction of tumor oxygenation during and after photodynamic therapy *in vivo*: Effects of fluence rate. *Br J Cancer* 1998;77:1386–1394.
126. Henderson BW, Sitnik-Busch TM, Vaughan LA. Potentiation of PDT anti-tumor activity in mice by nitric oxide synthase inhibition is fluence rate dependent. *Photochem Photobiol* 1999;70:64–71.
127. Dimitroff CJ, Klohs W, Sharma A, Pera P, Driscoll D, Smith J et al. Anti-angiogenic activity of selected receptor tyrosine kinase inhibitors, PD166285 and PD173074: Implications for combination treatment with photodynamic therapy. *Invest New Drugs* 1999;17:121–135.
128. Tseng MT, Reed MW, Ackermann DM, Schuschke DA, Wieman TJ, Miller FN. Photodynamic therapy induced ultrastructural alterations in microvasculature of the rat cremaster muscle. *Photochem Photobiol* 1988;48:675–681.
129. Fingar VH, Wieman J, Wiehle SA, Cerrito PB. The role of microvascular damage in photodynamic therapy: The effect of treatment on vessel constriction, permeability, and leukocyte adhesion. *Cancer Res* 1992;53:4914–4921.
130. McMahon KS, Wieman TJ, Moore PH, Fingar VH. Effects of photodynamic therapy using mono-L-aspartyl chlorin e₆ on vessel constriction, vessel leakage, and tumor response. *Cancer Res* 1994;54:5374–5379.

131. Fingar VH, Wieman TJ, Karavolos PS, Doak KW, Ouellet R, van Lier JE. The effects of photodynamic therapy using differently substituted zinc phthalocyanines on vessel constriction, vessel leakage and tumor response. *Photochem Photobiol* 1993;58(2):251–258.

132. Snyder JW, Greco WR, Bellnier DA, Vaughan L, Henderson BW. Photodynamic therapy: A means to enhanced drug delivery to tumors. *Cancer Res* 2003;63:8126–8131.

133. Zieve PD, Solomon HM, Krevans JR. The effect of hematoporphyrin and light on human platelets. I. Morphologic, functional, and biochemical changes. *J Cell Physiol* 1966;67:271–279.

134. Henderson BW, Owczarczak B, Sweeney J, Gessner T. Effects of photodynamic treatment of platelets or endothelial cells in vitro on platelet aggregation. *Photochem Photobiol* 1992;56:513–521.

135. Fingar VH, Weiman TJ, Doak KW. Role of thromboxane and prostacyclin release on photodynamic therapy-induced tumor destruction. *Cancer Res* 1990;50:2599–2603.

136. Reed MW, Schuschke DA, Miller FN. Prostanoid antagonists inhibit the response of the microcirculation to "early" photodynamic therapy. *Radiat Res* 1991;127:292–296.

137. Fingar VH, Taber SW, Haydon PS, Harrison LT, Kempf SJ, Wieman TJ. Vascular damage after photodynamic therapy of solid tumors: A view and comparison of effect in pre-clinical and clinical models at the University of Louisville. *In Vivo* 2000;14:93–100.

138. Fingar VH, Wieman TJ, Haydon PS. The effects of thrombocytopenia on vessel stasis and macromolecular leakage after photodynamic therapy using Photofrin. *Photochem Photobiol* 1997;66:513–517.

139. Fingar VH, Wieman TJ, Doak KW. Changes in tumor interstitial pressure induced by photodynamic therapy. *Photochem Photobiol* 1991;53:763–768.

140. Hockel M, Vaupel P. Tumor hypoxia: Definitions and current clinical, biologic, and molecular aspects. *J Natl Cancer Inst* 2001;93:266–276.

141. Fingar VH, Wieman TJ, Park YJ, Henderson BW. Implications of a pre-existing tumor hypoxic fraction on photodynamic therapy. *J Surg Res* 1992;53:524–528.

142. Fingar VH, Mang TS, Henderson BW. Modification of photodynamic therapy-induced hypoxia by fluosol-DA (20%) and carbogen breathing in mice. *Cancer Res* 1988;48:3350–3354.

143. Schouwink H, Ruevekamp M, Oppelaar H., van Veen R, Bass P, Stewart FA. Photodynamic therapy for malignant mesothelioma: Preclinical studies for optimization of treatment protocols. *Photochem Photobiol* 2001;73:410–417.

144. Jirsa MJ, Pouckova P, Dolezal J, Pospisil J, Jirsa M. Hyperbaric oxygen and photodynamic therapy in tumour-bearing nude mice [letter]. *Eur J Cancer* 1991;27:109.

145. Maier A, Tomaselli F, Anegg U, Rehak P, Fell B, Luznik S et al. Combined photodynamic therapy and hyperbaric oxygenation in carcinoma of the esophagus and the esophago-gastric junction. *Eur J Cardiothorac Surg* 2001;18:649–654.

146. Svaasand LO, Doiron DR, Dougherty TJ. Temperature rise during photoradiation therapy of malignant tumors. *Med Phys* 1983;10:10–17.

147. Mattiello J, Hetzel F, Vandenheede L. Intratumor temperature measurements during photodynamic therapy. *Photochem Photobiol* 1987;46:873–879.

148. Henderson BW, Busch TM, Vaughan LA, Frawley NP, Babich D, Sosa TA et al. Photofrin photodynamic therapy can significantly deplete or preserve oxygenation in human basal cell carcinomas during treatment, depending on fluence rate. *Cancer Res* 2000;60:525–529.

149. Tromberg BJ, Orenstein A, Kimel S, Barker SJ, Hyatt J, Nelson JS et al. In vivo tumor oxygen tension measurements for the evaluation of the efficiency of photodynamic therapy. *Photochem Photobiol* 1990;52:375–385.

150. Chen Q, Chen H, Hetzel FW. Tumor oxygenation changes post-photodynamic therapy. *Photochem Photobiol* 1996;63:128–131.

151. Foster TH, Murant RS, Bryant RG, Knox RS, Gibson SL, Hilf R. Oxygen consumption and diffusion effects in photodynamic therapy. *Radiat Res* 1991;126:296–303.

152. Foster TH, Gao L. Dosimetry in photodynamic therapy: Oxygen and the critical importance of capillary density. *Radiat Res* 1992;130:379–383.

153. Pogue BW, Hasan T. A theoretical study of light fractionation and dose-rate effects in photodynamic therapy. *Radiat Res* 1997;147:551–559.

154 Foster TH, Hartley DF, Nichols MG, Hilf R. Fluence rate effects in photodynamic therapy of multi-cell tumor spheroids. *Cancer Res* 1993;53:1249–1254.

155. Mitra S, Finlay JC, McNeill D, Conover DL, Foster TH. Photochemical oxygen consumption, oxygen evolution and spectral changes during UVA irradiation of EMT6 spheroids. *Photochem Photobiol* 2001;73(6):703–708.

156. Zilberstein J, Bromberg A, Frantz A, Rosenbach-Belkin V, Kritzmann A, Pfefermann R et al. Light-dependent oxygen consumption in bacteriochlorophyll-serine-treated melanoma tumors: On-line determination using a tissue-inserted oxygen microsensor. *Photochem Photobiol* 1997;65:1012–1019.

157. Busch TM, Hahn SM, Evans SM, Koch CJ. Depletion of tumor oxygenation during photodynamic therapy: Detection by the hypoxia marker EF3. *Cancer Res* 2000;60:2636–2642.

158. Mang TS, Dougherty TJ, Potter WR, Boyle DG, Somer S, Moan J. Photobleaching of porphyrins used in photodynamic therapy and implications for therapy. *Photochem Photobiol* 1987;45:501–506.

159. Spikes JD. Quantum yields and kinetics of the photobleaching of hematoporphyrin, Photofrin II, tetra(4-sulfonatophenyl)porphine and uroporphyrin. *Photochem Photobiol* 1992;55(6):797–808.

160. Coutier S, Mitra S, Bezdetnaya LN, Parache RM, Georgakoudi I, Foster TH et al. Effects of fluence rate on cell survival and photobleaching in meta-tetra-(hydroxyphenyl)chlorin-photosensitized Colo 26 multicell tumor spheroids. *Photochem Photobiol* 2001;73:297–303.

161. Finlay JC, Conover DL, Hull EL, Foster TH. Porphyrin bleaching and pdt-induced spectral changes are irradiance dependent in ala-sensitized normal rat skin in vivo. *Photobiochem Photobiophys* 2001;73:54–63.

162. Georgakoudi I, Nichols MG, Foster TH. The mechanism of Photofrin photobleaching and its consequences for photodynamic dosimetry. *Photochem Photobiol* 1997;65:135–144.

163. Georgakoudi I, Foster TH. Singlet oxygen- *versus* nonsinglet oxygen-mediated mechanisms of sensitizer photobleaching and their effects on photodynamic dosimetry. *Photochem Photobiol* 1998;67:612–625.

164. Sunar U, Rohrbach D, Rigual N, Tracy E, Keymel KR, Cooper M et al. Monitoring photobleaching and hemodynamic responses to HPPH-mediated photodynamic therapy of head and neck cancer: A case report *Opt Express.* 2010;18(14):14969–14978.

165. Iinuma S, Schomacker KT, Wagnieres G, Rajadhyaksha M, Bamberg M, Momma T et al. In vivo fluence rate and fractionation effects on tumor response and photobleaching: Photodynamic therapy with two photosensitizers in an orthotopic rat tumor model. *Cancer Res* 1999;59(24):6164–6170.

166. van Geel IPJ, Oppelaar H, Marijnissen JPA, Stewart FA. Influence of fractionation and fluence rate in photodynamic therapy with Photofrin or mTHPC. *Radiat Res* 1996;145:602–609.

167. Blant SA, Woodtli A, Wagnieres G, Fontolliet C, Van den Bergh H, Monnier P. In vivo fluence rate effect in photodynamic therapy of early cancers with tetra(m-hydroxyphenyl)chlorin. *Photochem Photobiol* 1996;64:963–968.

168. Sitnik TM, Henderson BW. Effects of fluence rate on cytotoxicity during photodynamic therapy. *Proc SPIE* 1997;2972:95–102.

169. Seshadri M, Bellnier DA, Vaughan LA, Spernyak JA, Mazurchuk R, Foster TH et al. Light delivery over extended time periods enhances the effectiveness of photodynamic therapy. *Clin Cancer Res* 2008;14:2796–2805.

170. Curnow A, McIlroy BW, Postle-Hacon MJ, MacRobert AJ, Bown SG. Light dose fractionation to enhance photodynamic therapy using 5-aminolevulinic acid in the normal rat colon. *Photochem Photobiol* 1999;69:71 76.

171. DeBruijn HS, van der Veen N, Robinson DJ, Star WM. Improvement of systemic 5-aminolevulinic acid-based photodynamic therapy in vivo using light fractionation with a 75-minute interval. *Cancer Res* 1999;59:901–904.

172. Cottrell WJ, Paquette AD, Keymel KR, Foster TH, Oseroff AR. Irradiance-dependent photobleaching and pain in delta-aminolevulinic acid-photodynamic therapy of superficial basal cell carcinomas. *Clin Cancer Res* 2008;14:4475–4483.

173. Zeitouni NC, Paquette AD, Housel JP, Shi Y, Wilding GE, Foster TH et al. A retrospective review of pain control by a two-step irradiance schedule during topical ALA-photodynamic therapy of nonmelanoma skin cancer. *Lasers Surg Med* 2013;45:89–94.

174. Obochi MO, Ratkay LG, Levy JG. Prolonged skin allograft survival after photodynamic therapy associated with modification of donor skin antigenicity. *Transplantation* 1997;63:810–817.

175. Gruner S, Meffert H, Volk HD, Grunow R, Jahn S. The influence of haematoporphyrin derivative and visible light on murine skin graft survival, epidermal Langerhans cells and stimulation of the allogeneic mixed leucocyte reaction. *Scand J Immunol* 1985;21:267–273.

176. Qin B, Selman SH, Payne KM, Keck RW, Metzger DW. Enhanced allograft survival after photodynamic therapy: Association with lymphocyte inactivation and macrophage stimulation. *Transplantation* 1993;56:1481–1486.

177. Dragieva G, Hafner J, Dummer R, Schmid-Grendelmeier P, Roos M, Prinz BM et al. Topical photodynamic therapy in the treatment of actinic keratoses and Bowen's disease in transplant recipients. *Transplantation* 2004;77:115–121.

178. Gollnick SO, Musser DA, Oseroff AR, Vaughan LA, Owczarczak B, Henderson BW. IL-10 does not play a role in cutaneous Photofrin® photodynamic therapy-induced suppression of the contact hypersensitivity response. *Photochem Photobiol* 2001;74:811–816.

179. Elmets CA, Bowen KD. Immunological suppression in mice treated with hematoporphyrin derivative photoradiation. *Cancer Res* 1986;46:1608–1611.

180. Musser DA, Fiel RJ. Cutaneous photosensitizing and immunosuppressive effects of a series of tumor localizing porphyrins. *Photochem Photobiol* 1991;53:119–123.

181. Simkin G, Obochi M, Hunt DWC, Chan AH, Levy JG. Effect of photodynamic therapy using benzoporphyrin derivative on the cutaneous immune response. *Proc SPIE 2392, Optical Methods for Tumor Treatment and Detection: Mechanisms and Techniques in Photodynamic Therapy IV*, 1995, pp. 23–33.

182. Reddan JC, Anderson C, Xu H, Hrabovsky S, Freye K, Fairchild R et al. Immunosuppressive effects of silicon phthalocyanine photodynamic therapy. *Photochem Photobiol* 1999;70:72–77.

183. Anderson C, Hrabovsky S, McKinley Y, Tubesing K, Tang H-P, Dunbar R et al. Phthalocyanine photodynamic therapy: Disparate effects of pharmacologic inhibitors on cutaneous photosensitivity and on tumor regression. *Photochem Photobiol* 1997;65:895–901.

184. Ziolkowski P, Symonowicz K, Milach J, Szkudlarek T. In vivo tumor necrosis factor-alpha induction following chlorin e_6-photodynamic therapy in Buffalo rats. *Neoplasma* 1996;44:192–196.

185. Rivas JM, Ullrich SE. The role of IL-4, IL-10, and TNF-a in the immune suppression induced by ultraviolet radiation. *J Leukoc Biol* 1994;56:769–775.

186. Simkin G, Tao J-S, Levy JG, Hunt DWC. IL-10 contributes to the inhibition of contact hypersensitivity in mice treated with photodynamic therapy. *J Immunol* 2000;164:2457–2462.

187. Gollnick SO, Liu X, Owczarczak B, Musser DA, Henderson BW. Altered expression of interleukin 6 and interleukin 10 as a result of photodynamic therapy in vivo. *Cancer Res* 1997;57:3904–3909.

188. Moore KW, de Waal Malefyt R, Coffman RL, O'Garra A. Interleukin-10 and the interleukin-10 receptor. *Annu Rev Immunol* 2001;19:683–765.

189. Obochi MOK, Ratkay LG, Levy JG. Prolonged skin allograft survival after photodynamic therapy associated with modification of donor skin antigenicity. *Transplantation* 1997;63:810–817.

190. King DE, Jiang H, Simkin G, Obochi M, Levy JG, Hunt DWC. Photodynamic alteration of the surface receptor expression pattern of murine splenic dendritic cells. *Scand J Immunol* 1999;49:184–192.

191. Matthews YJ, Damian DL. Topical photodynamic therapy is immunosuppressive in humans. *Br J Dermatol* 2010;162:637–641.

192. Frost GA, Halliday GM, Damian DL. Photodynamic therapy-induced immunosuppression in humans is prevented by reducing the rate of light delivery. *J Invest Dermatol* 2011;131:962–968.

193. Gallucci S, Matzinger P. Danger signals: SOS to the immune system. *Curr Opin Immunol* 2001;13:114–119.

194. Moroz OV, Burkitt W, Wittkowski H, He W, Ianoul A, Novitskaya V et al. Both Ca^{2+} and Zn^{2+} are essential for S100A12 protein oligomerization and function. *BMC Biochem* 2009;10:11.

195. Levy RM, Mollen KP, Prince JM, Kaczorowski DJ, Vallabhaneni R, Liu S et al. Systemic inflammation and remote organ injury following trauma require HMGB1. *Am J Physiol Regul Integr Comp Physiol* 2007;293:R1538–R1544.

196. Riddell JR, Wang XY, Minderman H, Gollnick SO. Peroxiredoxin 1 stimulates secretion of proinflammatory cytokines by binding to TLR4. *J Immunol* 2010;184:1022–1030.

197. Korbelik M, Krosl G, Krosl J, Dougherty GJ. The role of host lymphoid populations in the response of mouse EMT6 tumor to photodynamic therapy. *Cancer Res* 1996;56:5647–5652.

198. deVree WJA, Essers MC, DeBruijn HS, Star WM, Koster JF, Sluiter W. Evidence for an important role of neutrophils in the efficacy of photodynamic therapy *in vivo*. *Cancer Res* 1996;56:2908–2911.

199. Korbelik M, Cecic I. Contribution of myeloid and lymphoid host cells to the curative outcome of mouse sarcoma treatment by photodynamic therapy. *Cancer Lett* 1999;137:91–98.

200. Gollnick S, Evans SS, Baumann H, Owczarczak B, Maier P, Vaughan L et al. Role of cytokines in photodynamic therapy-induced local and systemic inflammation. *Br J Cancer* 2003;88:1772–1779.

201. Krosl G, Korbelik M, Dougherty GJ. Induction of immune cell infiltration into murine SCCVII tumour by Photofrin-based photodynamic therapy. *Br J Cancer* 1995;71:549–555.

202. Henderson BW, Gollnick SO, Snyder JW, Busch TM, Kousis PC, Cheney RT et al. Choice of oxygen-conserving treatment regimen determines the inflammatory response and outcome of photodynamic therapy of tumors. *Cancer Res* 2004;64:2120–2126.

203. Kousis PC, Henderson BW, Maier PG, Gollnick SO. Photodynamic therapy enhancement of antitumor immunity is regulated by neutrophils. *Cancer Res* 2007;67:10501–10510.

204. Yamamoto N, Hoober JK, Yamamoto S. Tumoricidal capacities of macrophages photodynamically activated with hematoporphyrin derivative. *Photochem Photobiol* 1992;56:245–250.

205. Korbelik M, Krosl G. Enhanced macrophage cytotoxicity against tumor cells treated with photodynamic therapy. *Photochem Photobiol* 1994;60:497–502.

206. Korbelik M, Sun J, Posakony JJ. Interaction between photodynamic therapy and BCG immunotherapy responsible for the reduced recurrence of treated mouse tumors. *Photochem Photobiol* 2001;73:403–409.

207. Korbelik M, Naraparaju VR, Yamamoto N. Macrophage-directed immunotherapy as adjuvant therapy. *Br J Cancer* 1997;75:202–207.

208. Korbelik M, Cecic I. Enhancement of tumour response to photodynamic therapy by adjuvant mycobacterium cell-wall treatment. *J Photochem Photobiol B* 1998;44:151–158.

209. Korbelik M, Dougherty GJ. Photodynamic therapy-mediated immune response against subcutaneous mouse tumors. *Cancer Res* 1999;59:1941–1946.

210. Korbelik M, Sun J. Cancer treatment by photodynamic therapy combined with adoptive immunotherapy using genetically altered natural killer cell line. *Int J Cancer* 2001;93:269–274.

211. Belicha-Villanueva A, Riddell J, Bangia N, Gollnick SO. The effect of photodynamic therapy on tumor cell expression of major histocompatibility complex (MHC) class I and MHC class I-related molecules. *Lasers Surg Med* 2012;44:60–68.

212. Nseyo UO, Whalen RK, Duncan MR, Berman B, Lundahl SL. Urinary cytokines following photodynamic therapy for bladder cancer. A preliminary report. *Urology* 1990;36:167–171.

213. deVree WJ, Essers MC, Koster JF, Sluiter W. Role of interleukin 1 and granulocyte colony-stimulating factor in Photofrin based photodynamic therapy of rat rhabdomyosarcoma tumors. *Cancer Res* 1997;57:2555–2558.

214. Brackett CM, Owczarczak B, Ramsey K, Maier PG, Gollnick SO. IL-6 potentiates tumor resistance to photodynamic therapy (PDT). *Lasers Surg Med* 2011;43:676–685.

215. Gollnick SO, Brackett CM. Enhancement of anti-tumor immunity by photodynamic therapy. *Immunol Res* 2010;46:216–226.

216. Curry PM, Levy JG. Tumor inhibitory lymphocytes derived from the lymph nodes of mice treated with photodynamic therapy. *Photochem Photobiol* 1995;61S–72S.

217. Canti G, Lattuada D, Nicolin A, Taroni P, Valentini G, Cubeddu R. Immunopharmacology studies on photosensitizers used in photodynamic therapy (PDT). *Proc SPIE 2078, Photodyn Ther Cancer* 1994;268–275.

218. Kabingu E, Vaughan L, Owczarczak B, Ramsey KD, Gollnick SO. CD8⁺ T cell-mediated control of distant tumours following local photodynamic therapy is independent of CD4⁺ T cells and dependent on natural killer cells. *Br J Cancer* 2007;96:1839–1848.

219. Abdel-Hady ES, Martin-Hirsch P, Duggan-Keen M, Stern PL, Moore JV, Corbitt G et al. Immunological and viral factors associated with the response of vulval intraepithelial neoplasia to photodynamic therapy. *Cancer Res* 2001;61:192–196.

220. Thong PS, Olivo M, Kho KW, Bhuvaneswari R, Chin WW, Ong KW et al. Immune response against angiosarcoma following lower fluence rate clinical photodynamic therapy. *J Environ Pathol Toxicol Oncol* 2008;27:35–42.

221. Kabingu E, Oseroff AR, Wilding GE, Gollnick SO. Enhanced systemic immune reactivity to a Basal cell carcinoma associated antigen following photodynamic therapy. *Clin Cancer Res* 2009;15:4460–4466.

222. Gollnick SO, Vaughan L, Henderson BW. Generation of effective anti-tumor vaccines using photodynamic therapy. *Cancer Res* 2002;62:1604–1608.

223. Korbelik M, Stott B, Sun J. Photodynamic therapy-generated vaccines: Relevance of tumour cell death expression. *Br J Cancer* 2007;97:1381–1387.

224. Zhang H, Ma W, Li Y. Generation of effective vaccines against liver cancer by using photodynamic therapy. *Lasers Med Sci* 2009;24:549–552.

225. Jalili A, Makowski M, Switaj T, Nowis D, Wilczynski GM, Wilczek E et al. Effective photoimmunotherapy of murine colon carcinoma induced by the combination of photodynamic therapy and dendritic cells. *Clin Cancer Res* 2004;10:4498–4508.

226. Peng Q, Farrants GW, Madslien K, Bommer JC, Moan J, Danielsen HE et al. Subcellular localization, redistribution and photobleaching of sulfonated aluminum phthalocyanines in a human melanoma cell line. *Int J Cancer* 1991;49:290–295.

227. Peng Q, Moan J, Farrants GW, Danielsen HE, Rimington C. Location of P-II and AlPCS4 in human tumor LOX in vitro and in vivo by means of computer-enhanced video fluorescence microscopy. *Cancer Lett* 1991;58:37–47.

228. Wood SR, Holroyd JA, Brown SB. The subcellular localization of Zn(II) phthalocyanines and their redistribution on exposure to light. *Photochem Photobiol* 1997;65:397–402.

229. Kessel D. Sites of photosensitization by derivatives of hematoporphyrin. *Photochem Photobiol* 1986;44:489–493.

230. Singh G, Jeeves WP, Wilson BC, Jang D. Mitochondrial photosensitization by Photofrin II. *Photochem Photobiol* 1987;46:645–649.

231. Woodburn KW, Fan Q, Miles DR, Kessel D, Luo Y, Young SW. Localization and efficacy analysis of the phototherapeutic lutetium texaphyrin (PCI-0123) in the murine EMT6 sarcoma model. *Photochem Photobiol* 1997;65:410–415.

232. Kessel D, Luo Y, Mathieu P, Reiners JJ Jr. Determinants of the apoptotic response to lysosomal photodamage. *Photochem Photobiol* 2000;71:196–200.

233. Ma L, Moan J, Berg K. Evaluation of a new photosensitizer, meso-tetra-hydroxyphenyl-chlorin, for use in photodynamic therapy: A comparison of its photobiological properties with those of two other photosensitizers. *Int J Cancer* 1994;87:883–888.

234. Kessel D, Woodburn K, Gomer CJ, Jagerovic N, Smith KM. Photosensitization with derivatives of chlorin p6. *J Photochem Photobiol B* 1995;28:13–18.

235. Kessel D. Determinants of photosensitization by mono-L-aspartyl chlorin e6 [published erratum appears in *Photochem Photobiol* Dec 1989;50(6):1]. *Photochem Photobiol* 1989;49:447–452.

236. Kennedy JC, Pottier RH. Endogenous protoporphyrin IX, A clinically useful photosensitizer for photodynamic therapy. *J Photochem Photobiol B: Biol* 1992;14:275–292.

237. Kessel D, Thompson P, Saatio K, Nantwi KD. Tumor localization and photosensitization by sulfonated derivatives of tetraphenylporphine. *Photochem Photobiol* 1987;45:787–790.

238. Berg K, Moan J, Bommer JC, Winkelman JW. Cellular inhibition of microtubule assembly by photoactivated sulphonated meso-tetraphenylporphines. *Int J Radiat Biol* 1990;58:475–487.

239. Berg K, Moan J. Lysosomes as photochemical targets. *Int J Biochem* 1994;59:814–822.

240. Star WM, Marijnissen HP, van den Berg Blok AE, Versteeg JA, Franken KA, Reinhold HS. Destruction of rat mammary tumor and normal tissue microcirculation by hematoporphyrin derivative photoradiation observed in vivo in sandwich observation chambers. *Cancer Res* 1986;46:2532–2540.

241. Henderson BW, Fingar VH. Relationship of tumor hypoxia and response to photodynamic treatment in an experimental mouse tumor. *Cancer Res* 1987;47:3110–3114.

242. Selman SH, Kreimer Birnbaum M, Klaunig JE, Goldblatt PJ, Keck RW, Britton SL. Blood flow in transplantable bladder tumors treated with hematoporphyrin derivative and light. *Cancer Res* 1984;44:1924–1927.

243. vanGeel IPJ, Oppelaar H, Rijken PFJW, Bernsen HJJA, Hagemeier NEM, van der Kogel AJ et al. Vascular perfusion and hypoxic areas in RIF-1 tumours after photodynamic therapy. *Br J Cancer* 1996;73:288–293.

244. Roberts DJH, Cairnduff F, Driver I, Dixon B, Brown SB. Tumour vascular shutdown following photodynamic therapy based on polyhaematoporphyrin or 5-aminolaevulinic acid. *Int J Oncol* 1994;5:763–768.

245. Dodd NFJ, Moore JV, Poppitt DG, Wood B. In vivo magnetic resonance imaging of the effects of photodynamic therapy. *Br J Cancer* 1989;60:164–167.

246. Chapman JD, McPhee MS, Walz N, Chetner MP, Stobbe CC, Soderlind K et al. Nuclear magnetic resonance spectroscopy and sensitizer-adduct measurements of photodynamic therapy-induced ischemia in solid tumors. *J Natl Cancer Inst* 1991;83:1650–1659.

247. Zilberstein J, Schreiber S, Bloemers MC, Bendel P, Neeman M, Schechtman E et al. Antivascular treatment of solid melanoma tumors with bacteriochlorophyll-serine-based photodynamic therapy. *Photochem Photobiol* 2001;73:257–266.

248. Reed MW, Mullins AP, Anderson GL, Miller FN, Wieman TJ. The effect of photodynamic therapy on tumor oxygenation. *Surgery* 1989;106:94–99.

249. Kerdel FA, Soter NA, Lim HW. In vivo mediator release and degranulation of mast cells in hematoporphyrin derivative-induced phototoxicity in mice. *J Invest Dermatol* 1987;88:277–280.

250. Lim HW. Effects of porphyrins on skin. *Photosensitizing Compounds: Their Chemistry, Biology and Clinical Use*. Chichester, England: John Wiley & Sons, 1989, pp. 148–153.

251. Zieve PD, Solomon HM. The effect of hematoporphyrin and light on human platelets. 3. Release of potassium and acid phosphatase. *J Cell Physiol* 1966;68:109–112.

252. Ben-Hur E, Heldman E, Crane SW, Rosenthal I. Release of clotting factors from photosensitized endothelial cells: A positive trigger for blood vessel occlusion by photodynamic therapy. *FEBS Lett* 1988;236:105–108.

253. Foster TH, Primavera MC, Marder VJ, Hilf R, Sporn LA. Photosensitized release of von Willebrand factor from cultured human endothelial cells. *Cancer Res* 1991;51:3261–3266.

2

Synthesis and Biological Significance of Porphyrin-Based Photosensitizers in Photodynamic Therapy

Penny Joshi
Roswell Park Cancer Institute

Ravindra K. Pandey
Roswell Park Cancer Institute

The porphyrins and related tetrapyrrolic systems are among the most widely studied compounds for their use as photosensitizers in photodynamic therapy (PDT).[1] Porphyrins are 18π-electron aromatic macrocycles that exhibit characteristic optical spectra with a strong π–π* transition around 400 nm (Soret band) and usually four Q bands in the visible region. As can be seen in Figure 2.1, two of the peripheral double bonds in opposite pyrrolic rings are cross-conjugated and are not required to maintain aromaticity. Thus, the reduction of one or both of these cross-conjugated double bonds (to give chlorins and bacteriochlorins, respectively) maintains much of the aromaticity, but the change in symmetry results in bathochromically shifted Q bands with high extinction coefficients.[2] Nature uses these optical properties of the reduced porphyrins to harvest solar energy for photosynthesis with chlorophylls and bacteriochlorophylls as both antenna and reaction-center pigments.[3] The long-wavelength absorption of these natural chromophores led to explorations of their use as photosensitizers in PDT.

PDT is a promising cancer treatment that involves the combination of visible light and a photosensitizer.[4] Each factor is harmless by itself, but when combined with oxygen, they can produce lethal cytotoxic agents, initially singlet oxygen, that inactivate the tumor cells.[5] This enables greater selectivity

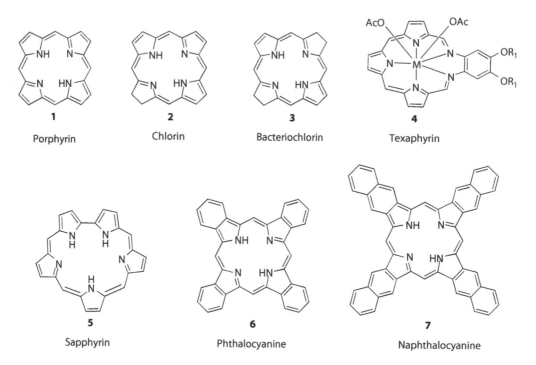

FIGURE 2.1   Basic structures of porphyrins and modified porphyrins.

toward diseased tissue, as only those cells that are simultaneously exposed to the photosensitizer, light, and oxygen are exposed to the cytotoxic effect. The dual selectivity of PDT is produced by both a preferential uptake of the photosensitizer by the diseased tissue and the ability to confine activation of the photosensitizer to this diseased tissue by restricting the illumination to the specific site (Figure 2.1).

As indicated previously, PDT is based on the interaction of a photosensitizer retained in tumors with photons of visible light, resulting in the formation of singlet oxygen ($^1O_2$), the putative lethal agent.[6] To achieve an effective destruction of tumor cells, a high quantum yield of singlet oxygen is required. Even in the absence of heavy atom substitution(s) and coordination of transition-metal ions, porphyrin systems generally satisfy these criteria and that is why most of the sensitizers currently under clinical evaluation for PDT are porphyrins or porphyrin-based molecules.

At present, Photofrin®,* a hematoporphyrin derivative,[7] is the only photosensitizer that has been approved worldwide for the treatment of various types of cancer by PDT. It fits some of the criteria for ideal photosensitizers, but it also suffers from several drawbacks. First, it is a complex mixture of various monomeric, dimeric, and oligomeric forms.[7b-e] Second, its long-wavelength absorption falls at 630 nm, which lies well below the wavelength necessary for the maximum tissue penetration. Finally, it induces prolonged cutaneous phototoxicity, a major adverse effect associated with most of the porphyrin-based photosensitizers.

It is well established that both absorption and scattering of light by tissue increases as the wavelength decreases and that most efficient sensitizers are those that have strong absorption bands from 700 to 800 nm.[8] Light transmission by tissues drops rapidly below 550 nm; however, it doubles from 550 to 630 nm and doubles again from 630 to 700 nm. This is followed by an additional 10% increase in tissue penetration as the wavelength increases toward 800 nm.[3] Another reason to set the ideal wavelength for PDT at 700–800 nm is due to the availability of easy-to-use diode lasers. Although diode lasers are now

---

* Registered trademark of Axcan Scandipharm Inc., Birmingham, AL.

available at 630 nm (where clinically approved Photofrin absorbs), photosensitizers with absorptions between 700 and 800 nm in conjunction with diode lasers are still desirable for treating deeply seated tumors. Therefore, in recent years, a variety of photosensitizers related to chlorins, bacteriochlorins, porphycenes, phthalocyanines, naphthalocyanines, and expanded porphyrins have been synthesized and evaluated for PDT efficacy. However, for designing improved photosensitizers for PDT, it becomes necessary to consider several other factors such as overall lipophilicity (i.e., a proper balance between hydrophilicit and hydrophobicity), pH, lymphatic drainage, and lipoprotein binding, which could influence the biodistribution and localization of sensitizers in tissue and tumors.[9]

The main focus of this review article is to summarize the various synthetic strategies followed by several research groups in designing long-wavelength absorbing photosensitizers related to chlorins, bacteriochlorins, expanded porphyrins, and phthalocyanines. An ongoing interest on developing target-specific photosensitizers has also been briefly reviewed. The majority of chlorins and bacteriochlorins have been generated through three different approaches. One method involves the modification of a preformed porphyrin. The second approach utilizes the use of chlorophyll *a* as the starting material for the synthesis of other chlorins and bacteriochlorins. The third approach utilizes the unstable bacteriochlorophyll *a* as a substrate for the synthesis of stable bacteriochlorins. Each procedure has been used successfully for the preparation of sensitizers that show promise in PDT and is discussed in terms of synthetic methodology and biological significance.

2.1 Chlorins and Bacteriochlorins from Porphyrins

2.1.1 Chlorins and Bacteriochlorins by Diimide Reduction

Almost 30 years ago, Whitlock et al.[10] developed an efficient diimide-reduction method for the synthesis of bacteriochlorins and isobacteriochlorins from porphyrins. Diimide reduction of metal-free tetraphenyl chlorin afforded tetraphenyl bacteriochlorin, while the reduction of the corresponding zinc analog produced the related tetraphenylisobacteriochlorin. It is now accepted that the reduced double bond in chlorins induces a pathway for the delocalized π electrons that *isolates* the diagonal crossing-conjugated pyrrolic double bond, such that the reduction of this double bond is favored due to minimal loss of π energy over the double bond present in the adjacent ring. The presence of a metal changes the delocalization of the π electrons, which makes the adjacent pyrrolic ring more reactive, and diimide reduction produces mainly the corresponding isobacteriochlorin. In order to avoid the formation of an isomeric mixture, this approach is useful only for reduction of symmetrical porphyrins. This diimide-reduction approach was later employed by Bonnett[11] for preparing the *meso*-tetra (*m*-hydroxyphenyl)-chlorin (*m*-THPC) (9) (650 nm) and the bacteriochlorin (*m*-THPBC) (10). The formation of these components was found to depend on the amount of the reductant used.

Although the formation of a bacteriochlorin resulted in further red shift in the electronic absorption spectrum with long-wavelength absorption near 750 nm, these molecules were generally found to be air sensitive. Among various chlorin analogs, *m*-THPC (Foscan®) (9) (Scheme 2.1) appears to be quite effective and is currently under phase III human clinical trials.

2.1.2 Chlorins and Bacteriochlorins by Diels–Alder Reaction

Cycloaddition reactions are among the most powerful reactions available to the organic chemists.[12] The ability to simultaneously form and break several bonds, with a wide variety of atomic substitution patterns and often with high degree of stereocontrol, has made cycloaddition reactions the subject of intense study. In porphyrin chemistry, the [4+2] Diels–Alder reactions have been used by various investigators for converting porphyrins into chlorin systems. Callot et al. were the first to show that protoporphyrin IX dimethyl ester (11) can undergo cycloaddition reactions with various dienophiles[13] (Scheme 2.2). A few years later, Dolphin and coworkers discovered the utility of one of such analogs

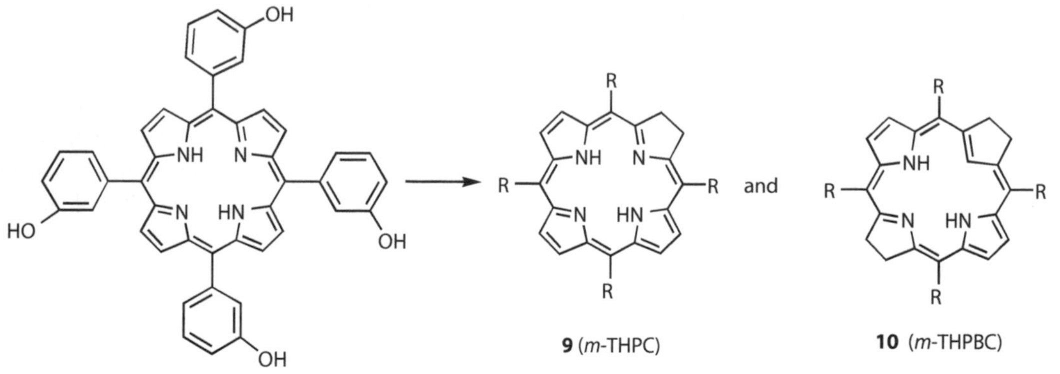

SCHEME 2.1   Conversion of meso-substituted porphyrin to corresponding chlorin and bacteriochlorin.

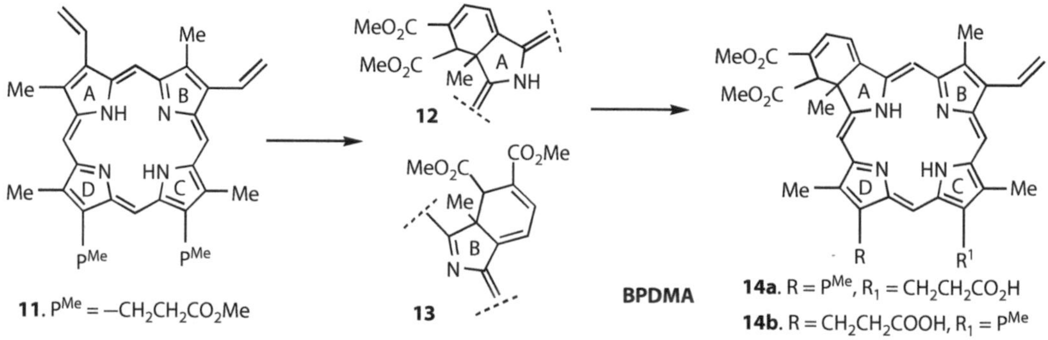

SCHEME 2.2   Conversion of Protoporphyrin IX dimethyl ester to benzoporphyrin derivatives.

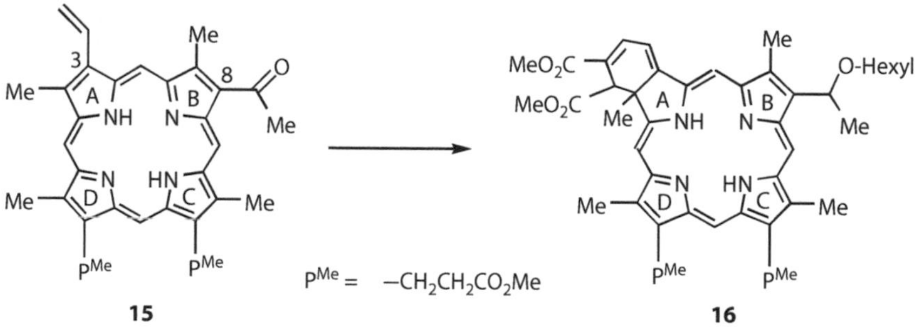

SCHEME 2.3   Synthesis of 8-(1′-hexyloxy)ethylbenzoporphyrin derivative dimethyl ester.

named as benzoporphyrin derivative monocarboxylic acid (BPDMA) **14a** and **14b** for treating age-related macular degeneration (AMD) when activated with light at 690 nm.[14] This treatment has already received approval worldwide. BPDMA has also been used for the treatment of cancer by PDT. However, due to its rapid clearance, it was found to be effective only if the tumors were treated with light at 3 h postinjection of the drug. Pandey et al.[15] developed another approach for preparing these analogs starting from 8-acetyl-3-vinyl deuteroporphyrin IX dimethyl ester (Scheme **2.3**).

The vinyl group was replaced with various alkyl ether functionalities. Among these analogs, the related 8-(1′-hexyloxy)ethyl derivative (**16**) was found to be more effective than BPDMA in eradicating tumors in mice bearing SMT-F tumors.[15a–c]

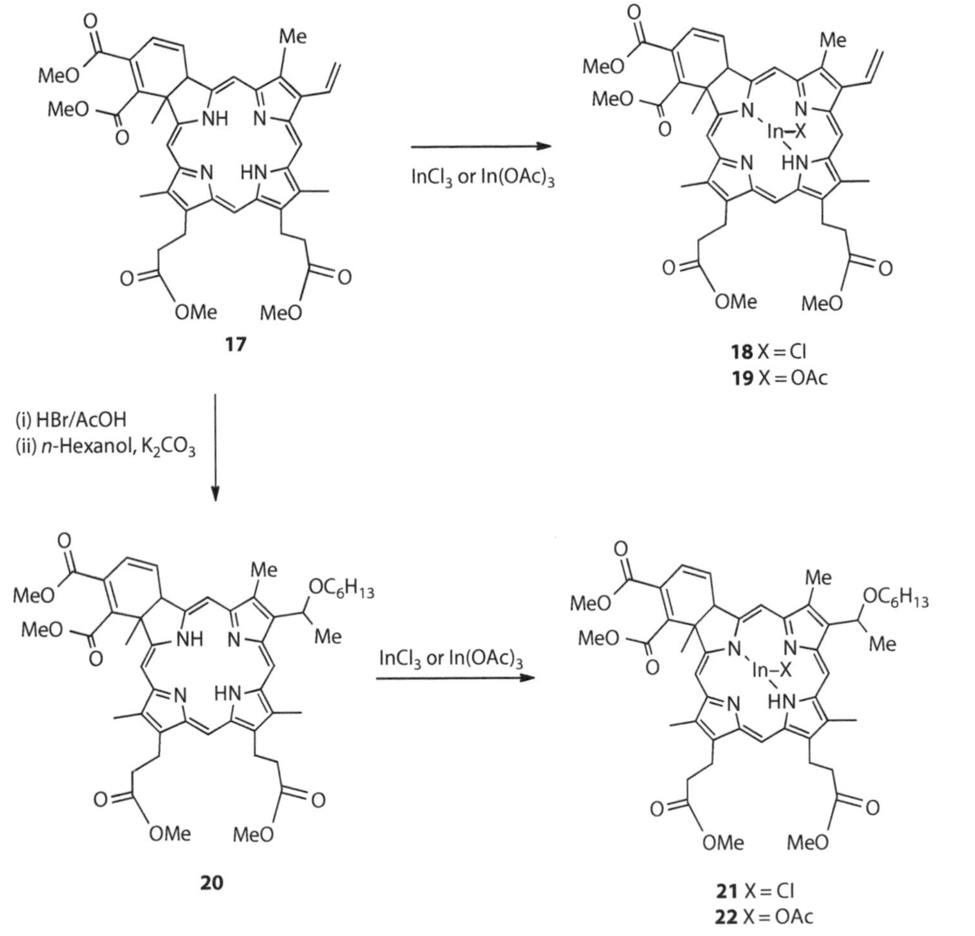

SCHEME 2.4 Synthesis of In(III) complex of 8-(1′-hexyloxy)ethyl benzoporphyrin derivative dimethyl ester.

To investigate the impact of indium as a central metal atom, recently Pandey et al. synthesized In (III) complex of benzoporphyrin dimethyl ester (**18, 19**) and its 8-(1¹-hexyloxy)ethyl analog (**21, 22**), which showed enhanced in vitro photosensitizing ability[15d] (Scheme **2.4**).

The methodology mentioned earlier was further extended independently by Pandey et al.[16] and Yon-Hin et al.[17] for the synthesis of novel bacteriochlorins, which involved a double Diels–Alder reaction on divinylporphyrins **23** and **25** (Scheme **2.5**). These bacteriochlorins **24** and **26** exhibit long-wavelength absorption maxima near 800 nm with PDT efficacy.

Morgan et al.[18] have shown that bacteriochlorin-like macrocycles can also be generated by cyclization of either 5,10- or 5,15-bis[(ethoxycarbonyl)vinyl]-porphyrins. However, the resulting products rapidly decomposed upon exposure to air, thus precluding their use as photosensitizers for PDT. For developing a general synthesis of stable bacteriochlorins, the same authors[19] followed the pinacol–pinacolone approach in preparing ketochlorins **30** and **31**. In brief, dehydration of **27** produced a mixture of **28** and **29**, which on reaction with dimethyl acetylenedicarboxylate (DMAD) produced the corresponding bacteriochlorins **30** and **31** as an isomeric mixture. This isomeric mixture showed some photodynamic activity in a mouse tumor model; 75% of the mice treated at a dose of 1 mg/kg were found to be free from palpable tumor 12 days after the light treatment. However, in this class of compounds, the spectroscopic properties of **30** and **31** resemble those of porphyrinones (long-wavelength absorption near 700 nm) rather than bacteriochlorins (Scheme **2.6**).

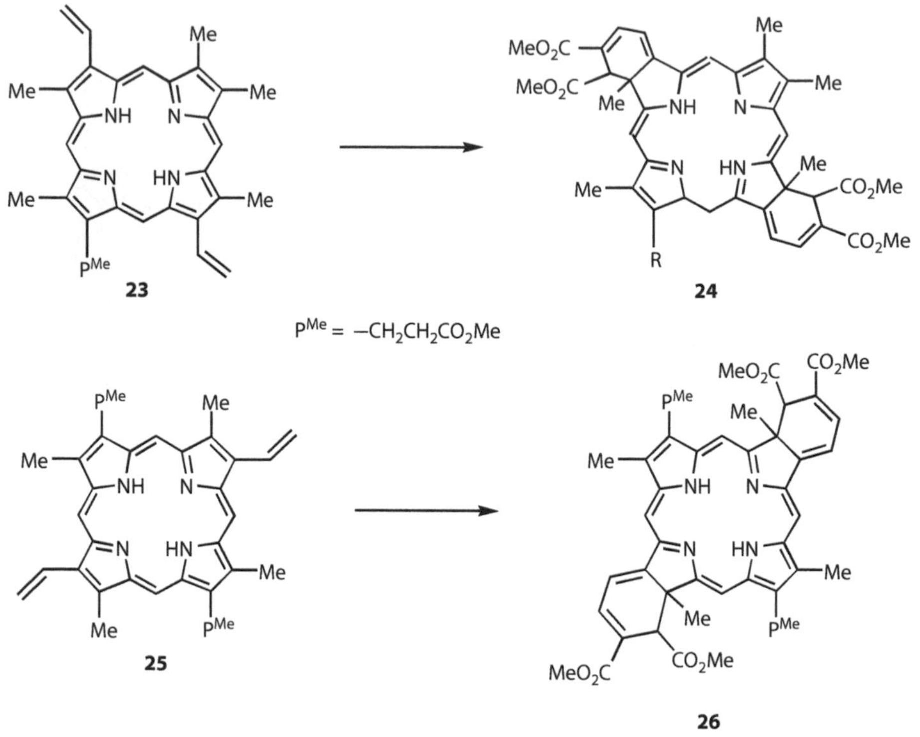

SCHEME 2.5   Synthesis of bacteriochlorins via Diels–Alder reaction.

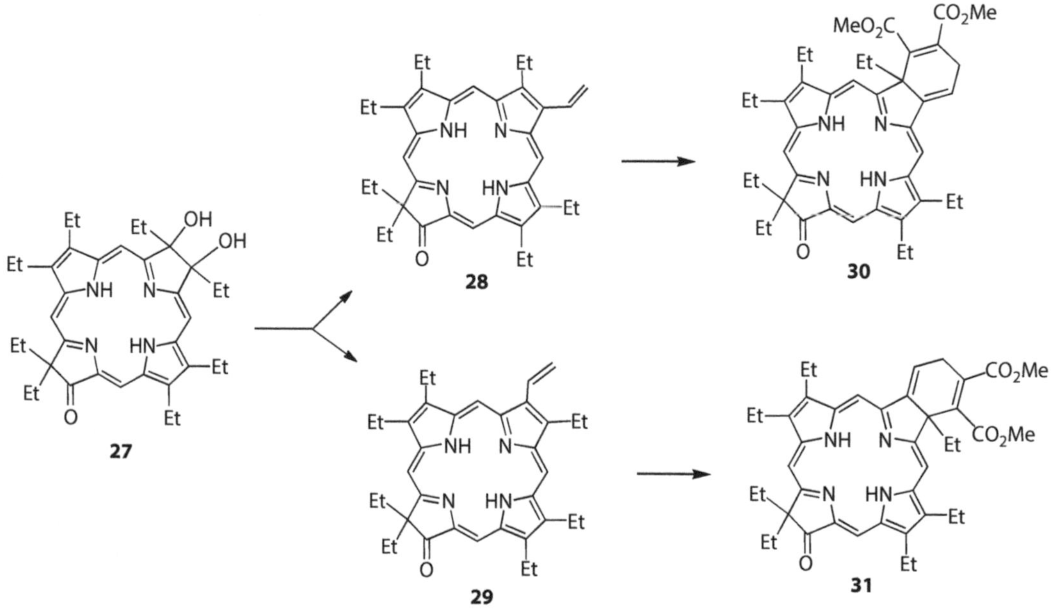

SCHEME 2.6   Synthesis of 17-ketobacteriochlorins from 7,8-di-hydroxy-1-keto-octaethylchlorin.

2.1.3 Benzochlorins and Benzobacteriochlorins

Benzochlorin consists of a benzene ring fused between the *meso-* and the adjacent β-position of the pyrrole ring. In a sequence of reactions, this class of compounds was first reported by Arnold et al. from octaethylporphyrin.[20] Morgan et al.[21] were the first to demonstrate the photosensitizing efficacy of these analogs (e.g., **36**). One of the major problems associated with this preparation is the difficulty in demetalation at the final step of the synthesis, and it is also difficult to chemically modify these benzochlorins. Therefore, this procedure has limited application in preparing a series of analogs with variable lipophilicity. This problem can be avoided by following the method recently reported by Li et al.[22,23] (Scheme **2.7**) In their approach, Ni(II)meso-(2-formylvinyl)octaethylporphyrin (**34**) was reacted with the Grignard's reagent of various fluorinated or nonfluorinated alkyl halides and/or Ruppert's reagent. The corresponding intermediates via intramolecular cyclization under acidic conditions afforded the related free-base benzochlorins (**42**). In this series of compounds, compared to the free-base analogs, the related Zn(II) complexes (671–677 nm) were found to be more effective both in vitro and in vivo. In preliminary screening, the fluorinated analogs showed better efficacy than the corresponding nonfluorinated derivatives.[22b]

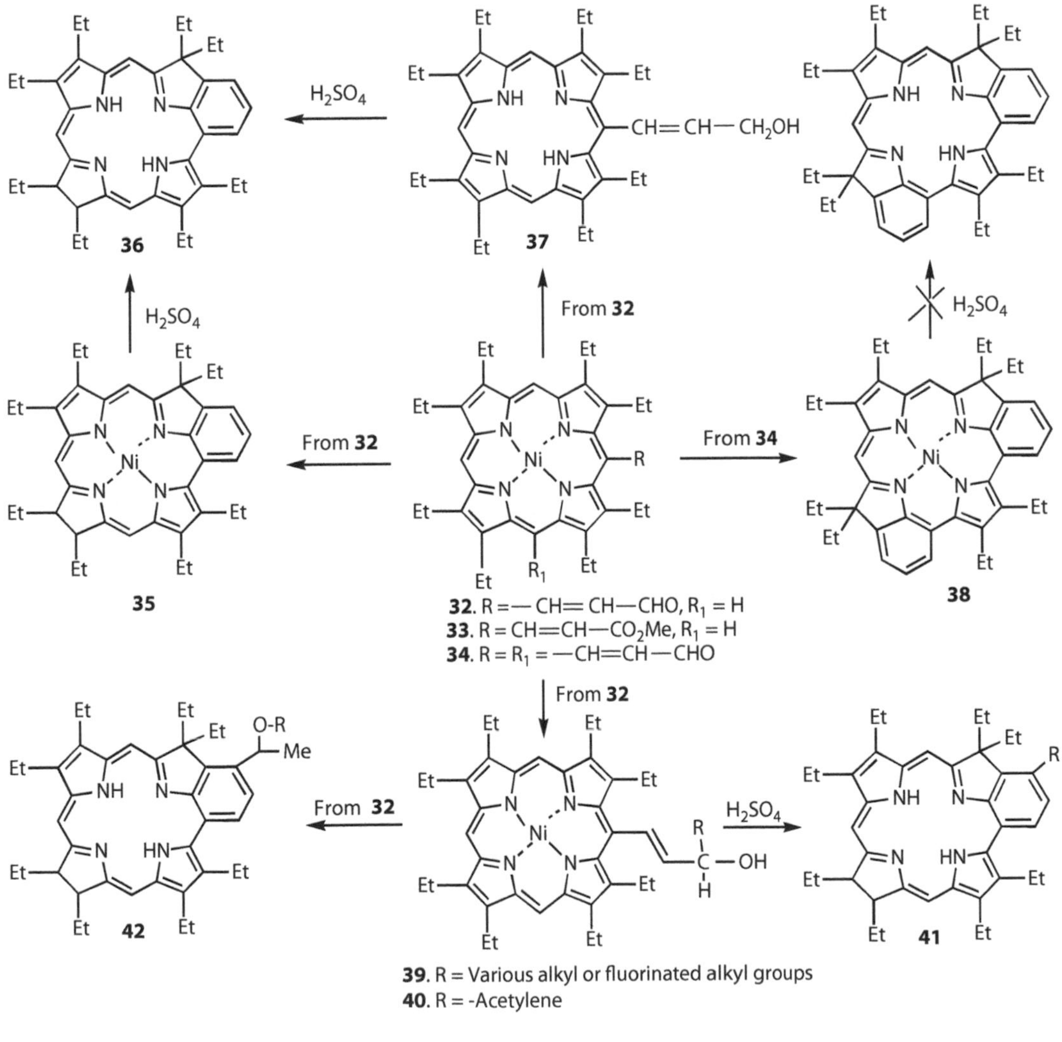

39. R = Various alkyl or fluorinated alkyl groups
40. R = -Acetylene

SCHEME 2.7   Synthesis of octaethyl-based benzochlorins.

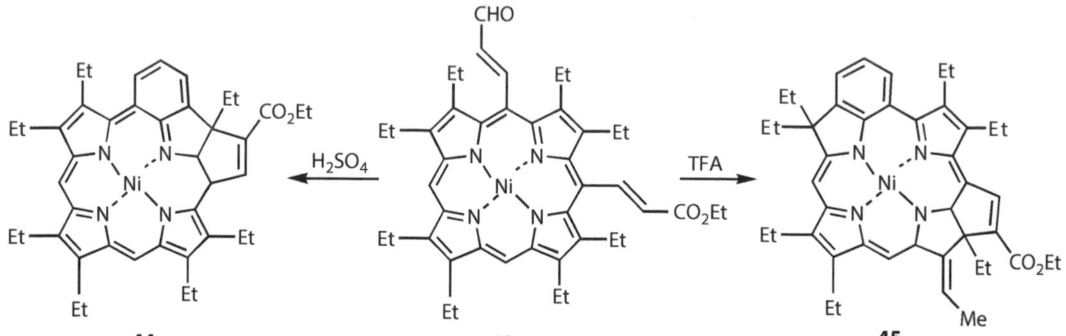

SCHEME 2.8 Synthesis of chlorins and bacteriochlorins with fused ring systems.

This methodology was later extended by Vicente and Smith[24] for the preparation of octaethylporphyrin-based benzochlorin (**38**) by intramolecular cyclization, of Ni(II)5,10 bis-(2-formylvinyl) porphyrin (**34**) (Scheme **2.8**). Unfortunately, attempts to remove the Ni(II) metal were unsuccessful. When Ni(II)5-(2-formyl-vinyl)-10-(2-ethoxycarbonyl-vinyl)octaethyl porphyrin was used as a substrate, the formation of the reaction product was found to depend on the strength of the acid used.[25] For example, reaction of **43** with sulfuric acid produced a chlorin containing both six- and five-member rings fused at the same pyrrole unit (**44**). Replacing sulfuric acid with trifluoroacetic acid (TFA) produced **45**, the Ni (II) complex of bacteriochlorin containing an ethylidene group at the peripheral position (λ_{max} 895 nm). Attempts to prepare the desired free-base analogs for investigating their application as photosensitizers were unsuccessful.

2.1.4 Purpurins (Tin Etiopurpurin Dichloride)

Purpurins have been known as degradation products of chlorophyll for quite some time.

The first synthesis of this class of compounds was reported by Woodward[26] during the synthesis of chlorophyll *a* by intramolecular cyclization of a *meso*acrylate functionality to a β-pyrrolic position. This methodology was later followed by Morgan et al.[27] and others[28] to synthesize a series of octa-ethylporphyrin, etioporphyrin, and 5,10-diphenyl- and 5,10-dipyridylporphyrin-based purpurin analogs. Among all the purpurins evaluated for PDT efficacy, the Sn etiopurpurin (SnEt$_2$) (**48**) (Scheme **2.9**) is considered to be the most effective in vivo (Scheme **2.9**).[27] Its long-wavelength absorption falls at 650 nm and produces a high singlet oxygen quantum yield. This product is currently in phase III clinical trials for the treatment of AMD, a major cause of blindness among people over 50 years of age.

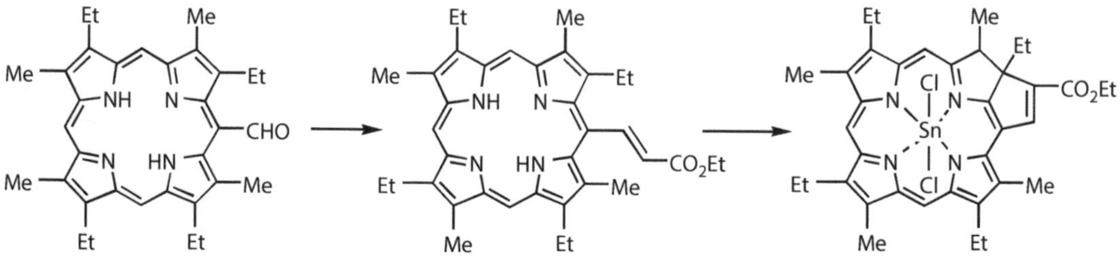

SCHEME 2.9 Synthesis of Sn(II) etiopurpurin.

2.2 Chlorins and Bacteriochlorins from Chlorophyll

Chlorophyll *a*, the green photosynthetic pigment, is one of the prototypes of the chlorin class of natural product. Because of its ready availability, a large amount of work has been done by several investigators to modify and to synthesize other chlorin-like chromophores. The photosensitizers derived from chlorophyll *a* can be divided into three categories in which the five-member isocyclic ring was either cleaved or kept intact or replaced with other ring system(s). Some of the photosensitizers in these series have attracted enormous attention; their description follows.

2.2.1 Aspartic Acid Derivative of Chlorin e_6 (Npe$_6$)

Chlorophyllin, a water-soluble degradation product of chlorophyll *a*, can be obtained by the cleavage of the isocyclic ring of chlorophyll.[29] The removal of magnesium resulted in chlorin e_6 with limited in vivo photosensitizing efficacy. It has been shown that replacing the vinyl group with alkyl ether groups of variable carbon units generally enhances the photosensitizing efficacy.[30] However, in this series, better results were obtained with the monoaspartyl derivative known as Npe$_6$.[31] This photosensitizer appears to clear rapidly from skin, and good tumor response was obtained only after irradiation within 3–4 h of sensitizer administration. Npe$_6$ is in human clinical trials in Japan for treatment of endobronchial lung cancer. The recent extensive NMR studies of Npe$_6$ confirmed that in Npe$_6$, the aspartic acid functionality is linked with an amide bond at position 15 (**49**) of chlorin e_6,[32] instead of at position 17 (**50**) as reported in several publications[33] (Scheme **2.10**).

2.2.2 Alkyl Ether Derivatives of Pyropheophorbide *a*

To understand the effect of various substituents on photosensitizing efficacy, the Roswell Park group synthesized and evaluated a series of pyropheophorbide *a* analogs with variable lipophilicity. In their effort to establish a structure–activity relationship (quantitative structure activity relationship [QSAR]), a congeneric series of the primary and secondary alkyl ether derivatives of pyropheophorbide *a* were synthesized (the isocyclic ring was kept intact). For the preparation of these analogs, methylpheophorbide *a* obtained from *Spirulina Pacifica* was converted into more stable pyropheophorbide *a* (**51**), which on reacting with HBr/AcOH and then the appropriate alcohol(s) produced the corresponding ether analogs in excellent yield[34] (Scheme **2.11**). At the final step, the methyl ester functionality was hydrolyzed into the corresponding carboxylic acid (**52**) (hexyl ether derivative of pyropheophorbide-a

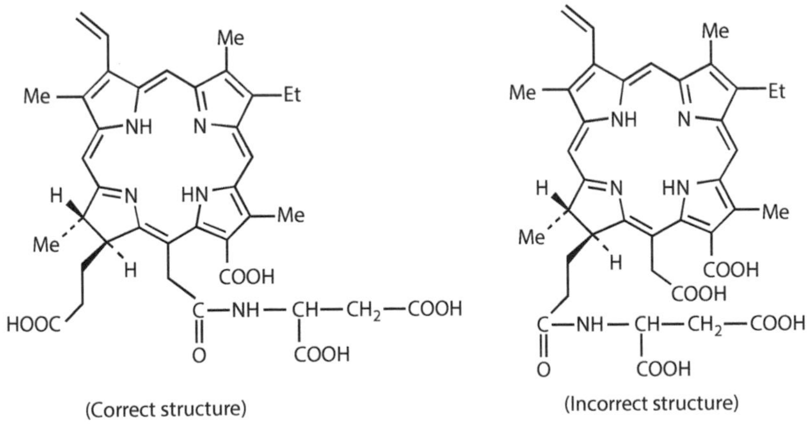

(Correct structure)
49

(Incorrect structure)
50

SCHEME 2.10 Structures of the aspartic acid derivative of chlorin e_6.

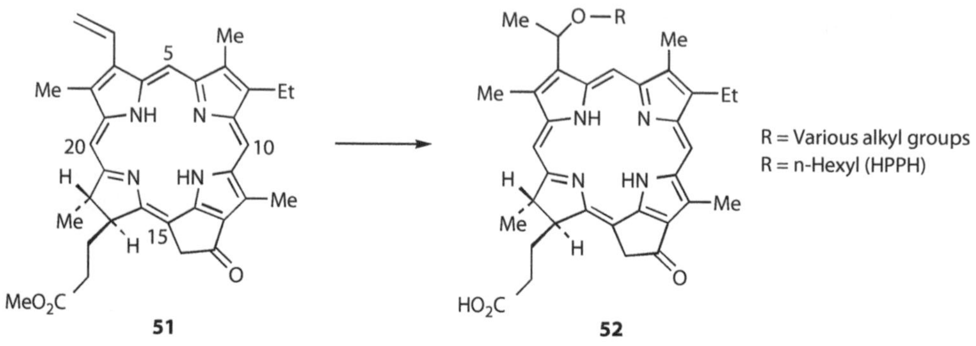

SCHEME 2.11 Synthesis of alkyl ether analogs of pyropheophorbide-a.

[HPPH]: R = *n*-hexyl). These analogs exhibit long-wavelength absorption near 665 nm (in vivo) and showed excellent singlet oxygen-producing efficiency (45%). The results obtained from the in vivo studies in mice demonstrated that the photodynamic efficacy of these photosensitizers increased by increasing the length of the carbon chain, reaching a maximum in compounds with *n*-hexyl and *n*-heptyl chains at position 3. Interestingly, the PDT efficacy decreased by further increasing the length of alkylether carbon units. When compensated for differences in tumor photosensitizer concentration, the *n*-hexyl derivative (HPPH) (optimal lipophilicity) was fivefold more potent than the *n*-dodecyl derivative (more lipophilic) and threefold more potent than the *n*-pentyl analog (less lipophilic). Interestingly, the introduction of the hexyl ether side chain at other positions of the macrocycle (position 8 or position 20) significantly reduced the in vivo efficacy.[35] These data suggest that besides the lipophilicity, the presence and position of the substituent possibly play an important role in drug efficacy. HPPH is currently at phase I/II human clinical trials for the treatment of a variety of cancers. Among the patients treated so far, no long-term skin phototoxicity has been observed.[36a]

To investigate the effect of central metal in PDT, Pandey et al. synthesized and investigated a series of pyropheophorbide and their metal complexes. They converted pyropheophorbide *a* (53) into the corresponding Zn (II), In(III), and Ni (II) complexes.[36b] Among these analogs, In(III) complexes showed the best PDT efficacy. The Ni (II) complexes because of its inability to produce any singlet oxygen didn't show any PDT efficacy.[36c] Also to look into the effect of lipophilicity, a series of In (III) analogs of methyl pyropheophorbide with variable lipophilicity were synthesized in which the vinyl group at position 3 was replaced with hexyl ether (55) and di- and mono-PEG substituents 61 and 63, respectively (Scheme 2.12).[36b]

It was seen that the presence of metals had a significant effect in the peripheral benzodiazepine receptor (PBR) binding and photosensitizing efficacy. On the basis of in vitro screening, indium complexes were found to be the most efficacious, which could be due to its higher singlet oxygen. Out of all the metal analogs, only HPPH (54) and its and indium complex (55) were tested for their in vivo PDT efficacy in C3H mice bearing RIF tumors. The indium complex of HPPH (55) was found to be more potent than the free-base HPPH, and with compound at a dose of 0.2 μmol/kg, 80% of mice were tumor-free after day 90 at a light 135 J/cm², 75 mW/cm².[36b]

2.2.3 Alkyl Ether Analogs of Purpurinimides

Having developed a QSAR for the alkyl ether analogs of pyropheophorbide series, the Roswell Park group extended their approach to photosensitizers with longer-wavelength absorption. For this study, the purpurin-18 methyl ester obtained from methylpheophorbide *a*[37] was converted into purpurin-18-*N*-alkyl imides (65).[38] The vinyl group at position 3 was then replaced with a variety of alkyl ether analogs (66) (Scheme 2.13) with variable carbon units with log *P* values ranging from 5.32 to 16.44 and exhibiting long-wavelength absorption near 700 nm.[39]

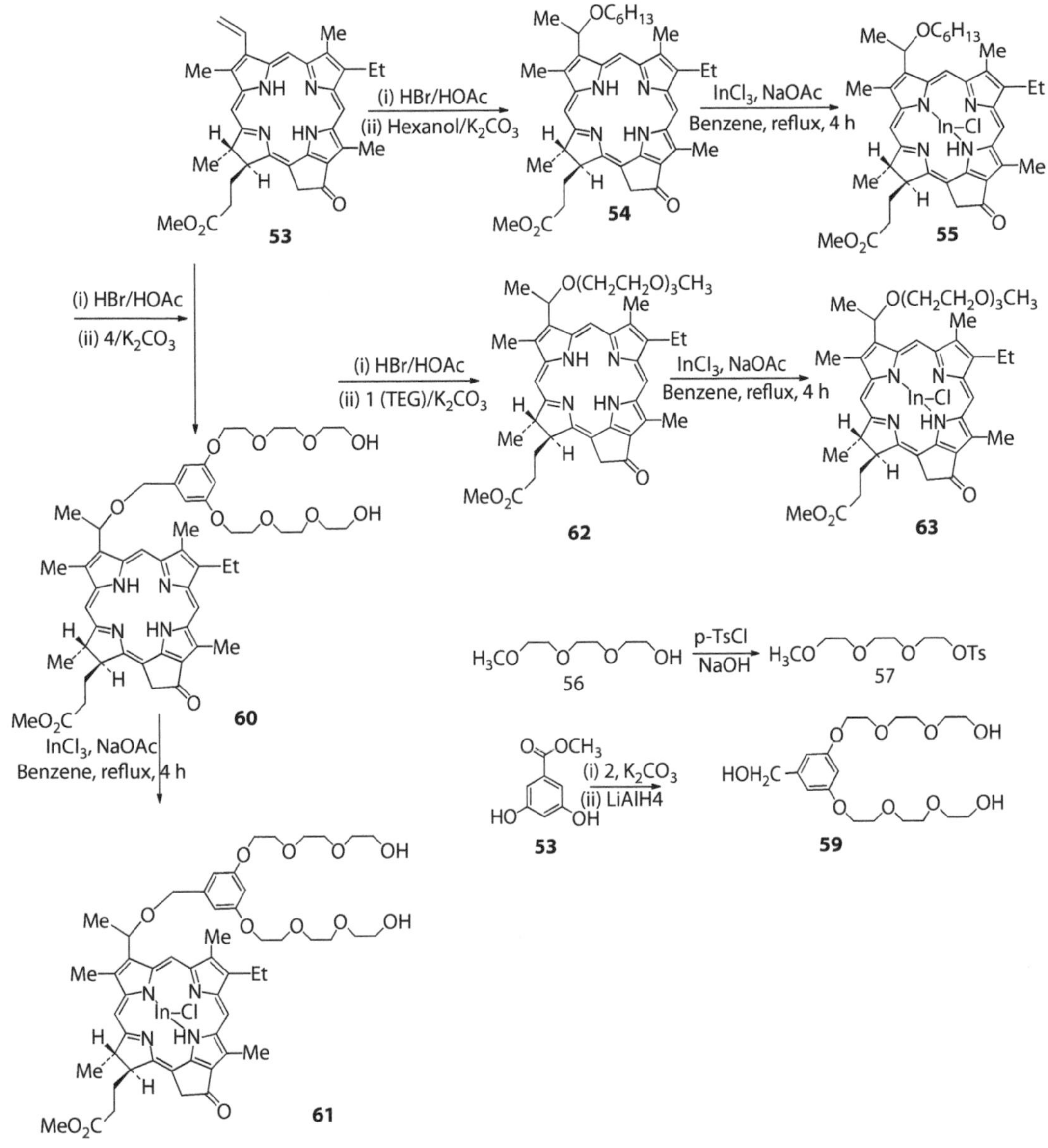

SCHEME 2.12 Synthesis of In(III)3-substituted methyl pyropheophorbide-a.

In animal studies, this class of compounds was found to be quite effective in vivo. The results obtained from a set of photosensitizers with similar lipophilicity (log *P* 10.68–10.88) indicate that similar to the pyropheophorbide series, in addition to the overall lipophilicity, the presence and position of the alkyl groups (*O*-alkyl vs. *N*-alkyl) in a molecule also play an important role in tumor uptake, tumor selectivity, and in vivo PDT efficacy.[39,40a,b]

The importance of fluorine in medicinal chemistry is well known.[40c,d] Fluorine substitutions are known to increase lipid solubility, which could result in increasing the rate of transportation of biologically active compounds across the lipid membrane.[40e] Gryshuk et al. synthesized a series of fluorinated and corresponding nonfluorinated purpurin-based photosensitizer (**70, 71, 73, 74**), and it was observed that fluorinated analogs bearing trifluoromethyl substituents (**71**) showed enhanced photodynamic efficacy[40f] (Scheme **2.14**).

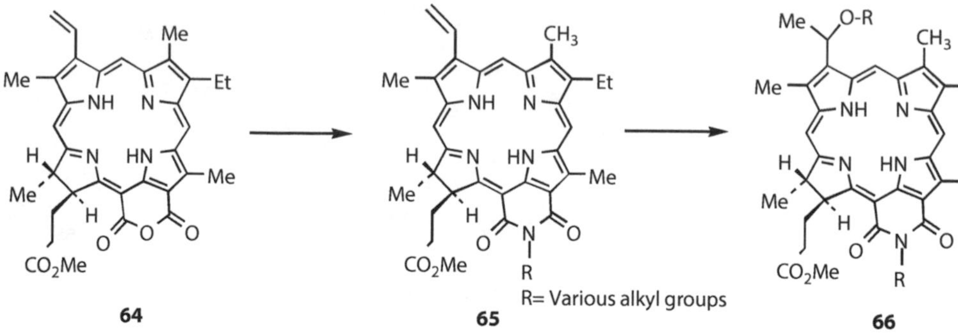

SCHEME 2.13 Synthesis of alkyl ether analogs of N-alkyl substituted purpurinimides.

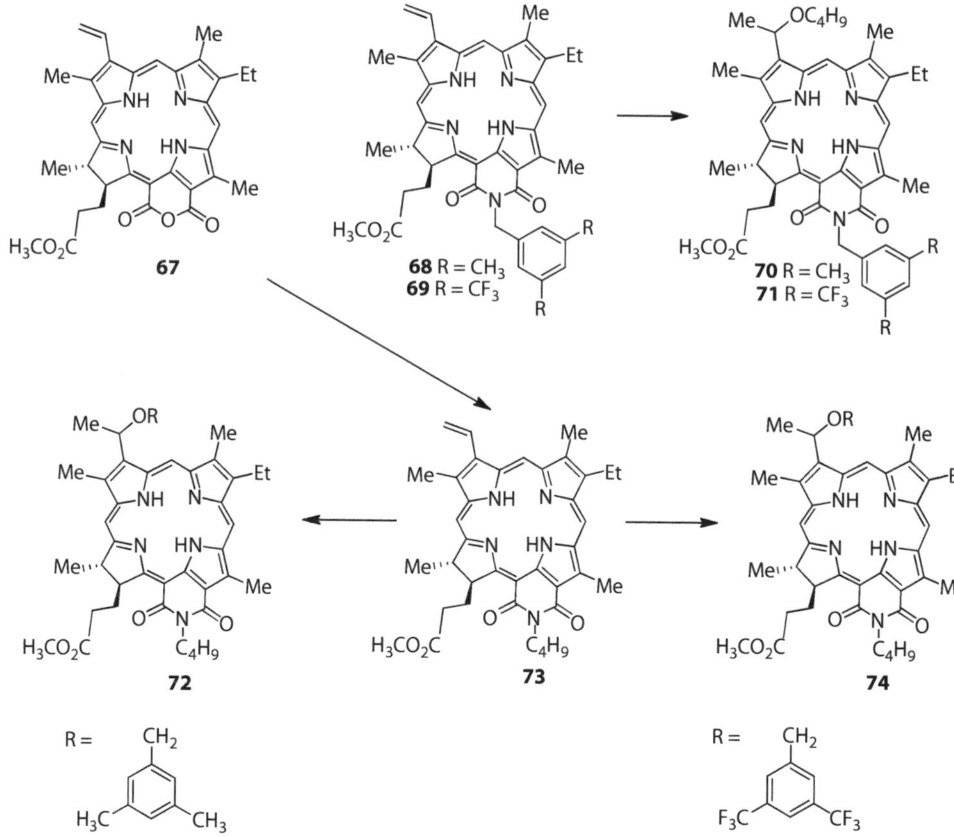

SCHEME 2.14 Synthesis of aryl ether analogs of N-aryl substituted purpurinimides.

2.2.4 Benzoporphyrin Derivatives Derived from Pyropheophorbide *a* and Purpurinimides

One of the main synthetic problems associated with PP-IX-based benzoporphyrin derivatives is to isolate the most effective analog (ring A reduced, monocarboxylic acid) from the complex reaction mixture. In order to solve this problem, the Roswell Park and Vancouver groups[41,42] have reported the preparation of various BPD analogs (e.g., **76**) from phylloerythrin and methyl 9-deoxypyropheophorbide (Scheme **2.15**).

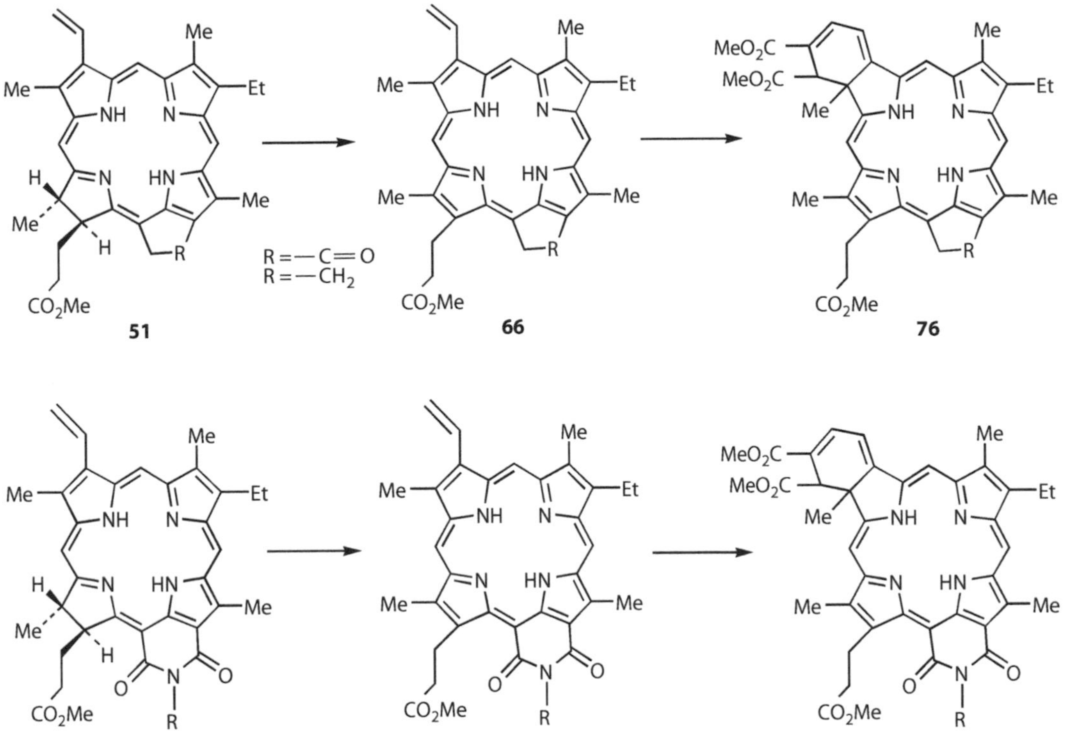

SCHEME 2.15 Benzoporphyrin derivatives derived from pyropheophorbide-a and purpurinimide.

Among these compounds, the benzoporphyrin derivative (*cis*-isomer) obtained from rhodoporphyrin XV di-*tert*-butyl aspartate was found to have PDT efficacy similar to BPDMA. This methodology was also extended in the purpurinimide series, and the lipophilicity was altered by introducing *N*-alkyl groups with variable carbon units at the imide ring system (**77**).[43] In preliminary in vivo testing, the corresponding *N*-hexyl and *N*-dodecyl analogs were found to be quite effective at a dose of 0.5 μM/kg when treated with light (135 J/cm², 75 mW/cm²) at 728 nm and 24 h postinjection. Under similar treatment conditions (treated with light at 690 nm), the BPDMA obtained from protoporphyrin IX dimethyl ester did not produce any photosensitizing efficacy.[44] Therefore, the Diels–Alder approach in purpurinimide system provides a simple approach for generating effective photosensitizers with variable lipophilicity.

2.2.5 *Vic*-Dihydroxy- and Ketobacteriochlorins

Osmium tetroxide has very frequently been used for the conversion of porphyrins to the corresponding *vic*-dihydroxy chlorins and tetrahydroxy bacteriochlorins as a mixture of isomers.[45] The overall lipophilicity of these analogs can be altered by subjecting them to pinacol–pinacolone reaction conditions. The formation of the corresponding ketoanalog is not straightforward and depends not only on the intrinsic nature of the migratory group but also of the electronic and steric factors elsewhere on the porphyrin nucleus.[46] Therefore, the concept of designing chlorin and bacteriochlorin analogs from porphyrins by following this approach was not successful. A few years ago, Chang et al.[47] showed that chlorins under certain conditions can be converted into *vic*-dihydroxybacteriochlorins upon reaction with osmium tetroxide. The Roswell Park group extended this methodology to the pheophorbide and chlorin *e₆*, and a series of *vic*-hydroxy- and ketobacteriochlorins were synthesized. The stable ketobacteriochlorins had strong absorptions in the range of 710–760 nm region, but did not show any significant photosensitizing

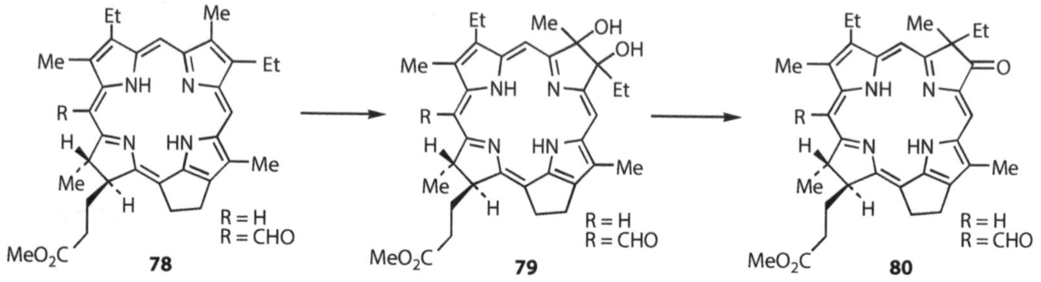

SCHEME 2.16 Synthesis of 8-keto methyl bacterio-pyropheophorbide-a analogs.

activity in mice (DBA/2) transplanted with SMT-F tumors.[48] However, the ketobacteriochlorins obtained from 9-deoxypyropheophorbide *a* (**78**) and the related *meso*formyl derivative with long-wavelength absorption showed long-wavelength absorptions at 734 and 758 nm, respectively (Scheme **2.16**). Among these bacteriochlorins, the triplet states were quenched by ground-state molecular oxygen in a relatively similar manner, yielding comparative singlet oxygen quantum yields. In preliminary in vivo screening, the ketochlorins (**80**) (R = CHO) were found to be more photodynamically active than the related *vic*-dihydroxy analogs. Replacement of the methyl ester functionalities with di-*tert*-butylaspartic acids enhanced the in vivo efficacy.[49a] It appears to be cleared rapidly from skin and good tumor responses can be obtained only after irradiation within 3–4 h of the sensitizer administration.

Joshi et al. synthesized a series of ketobacteriochlorins from ring B and ring D reduced chlorins (Schemes **2.17** and **2.18**). These newly synthesized compounds (**83**, **86**, **91**, **92**) show strong long-wavelength absorption and produce significant in vitro (Colon 26 cells) photosensitizing ability. Among all the compounds, the one containing a ketogroup at position 7 of ring B and bearing a cleaved five-member isocyclic ring (**92**) showed the best efficacy.[49b]

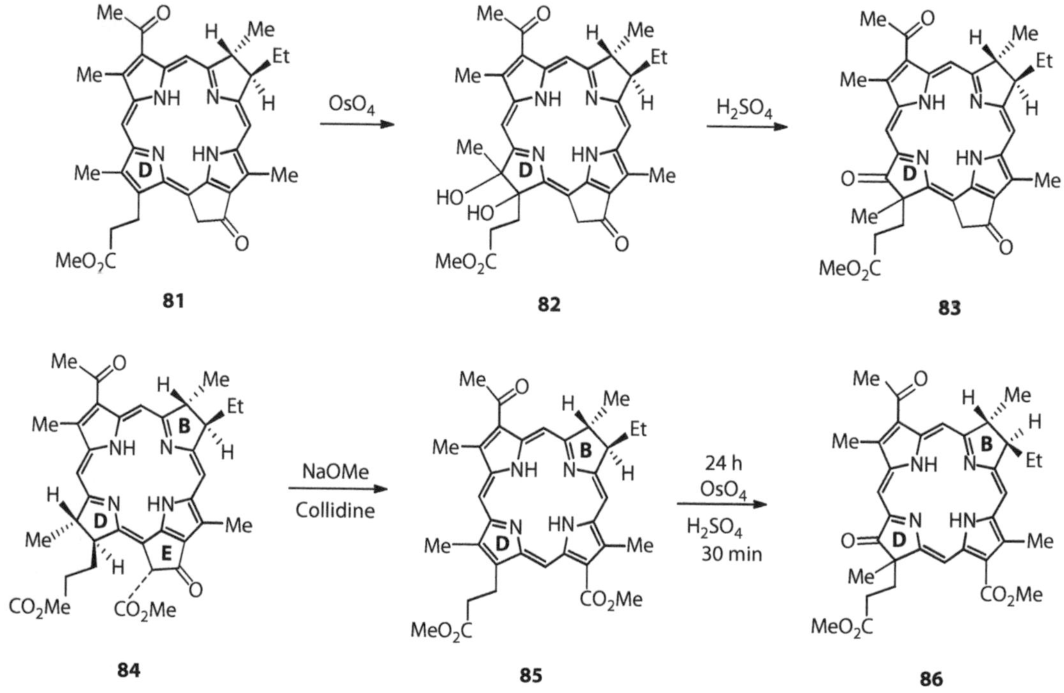

SCHEME 2.17 Synthesis of 18-keto-bacteriopyropheophorbide and bacteriorhodo-bacteriochlorins.

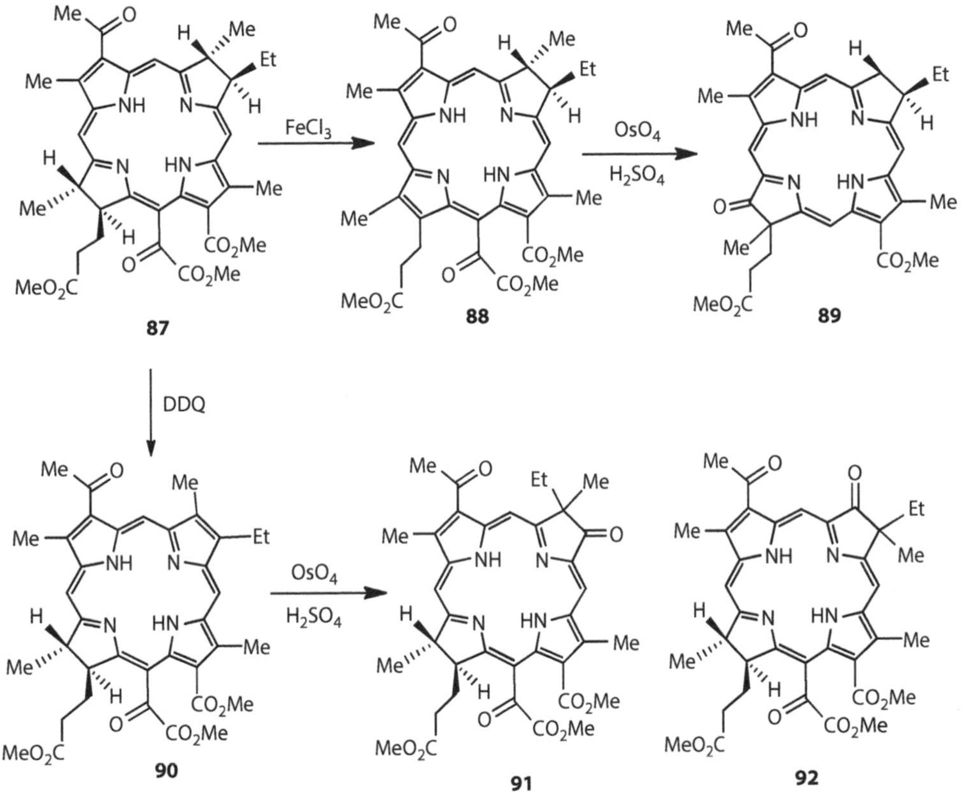

SCHEME 2.18 Synthesis of 3-acetyl-18-keto and 3-acetyl-8-ketobacteriochlorins.

2.2.6 Bacteriochlorins Derived from 8-Vinyl Chlorins

The Roswell Park group combined the use of osmium tetroxide and Diels–Alder approach for the construction of stable bacteriochlorins. In their approach, mesopurpurin-18 methyl ester (**93**) obtained from methylpheophorbide *a* was reacted with osmium tetroxide. The resulting *vic*-dihydroxy bacteriochlorin on reacting with *p*-toluenesulfonic acid in refluxing benzene produced the 8-vinyl derivative (**94**), which on reacting with DMAD under Diels–Alder reaction conditions produced bacteriochlorin (**95**) with long-wavelength absorption near 800 nm.[50] Unfortunately, the utility of this compound for the use in PDT was diminished due to the unstable nature of the six-member anhydride ring system (Scheme **2.19**). In another approach, the anhydride ring is replaced with an *N*-hexyl-imide ring system, and these compounds were found to be quite stable in vivo.[51] This system also possesses a unique opportunity to prepare a series of *N*-alkyl ether derivatives with variable carbon units and to establish the structure/activity relationship in a particular series of compounds.

2.2.7 Bacteriochlorins from Bacteriochlorophyll

Most of the naturally occurring bacteriochlorins have absorptions between 760 and 780 nm and have been studied by various investigators for their use as photosensitizers for PDT.[52] They were found to be extremely sensitive to oxidation, resulting in a rapid transformation into the chlorin state that generally has an absorption maxima at or below 660 nm.[53] Furthermore, if a laser is used to excite the bacteriochlorin in vivo, oxidation may result in the formation of a new chromophore absorbing outside the laser window, reducing the photodynamic efficacy. Due to the desirable photophysical properties and

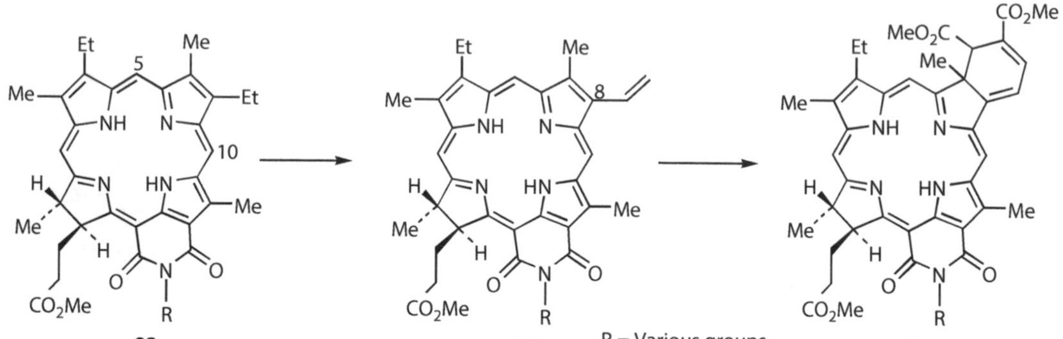

SCHEME 2.19   Synthesis of bacteriopurpurinimides.

promising in vitro/in vivo photosensitizing efficacy of bacteriochlorins, there has been increasing interest in the synthesis of stable bacteriochlorins either from bacteriochlorophyll *a* or from the other related tetrapyrrolic systems.

In general, for designing improved photosensitizers, overall lipophilicity has been proven to be one of the important factors. For example, among porphyrin-based photosensitizers, the hydrophobic por-phyrins are preferentially accumulated and partitioned into corresponding hydrophobic loci in vivo. Moan and coworkers[54] have shown that among diether derivatives of hematoporphyrin, retention in cells increases with decreasing polarity. The Roswell Park group have studied the uptake of a series of alkyl ether derivatives of pyropheophorbide *a* and found that a strong correlation exists between uptake and hydrophobicity, although each correlation cannot be extended to the in vivo PDT efficacy. On the other hand, photosensitizers with high partition coefficient values (increased hydrophobicity) induce sensitizer insolubility, thus preventing drugs from entering the circulation. Therefore, a proper balance between hydrophobicity and hydrophilicity is probably the most important factor that influences tumor localization of sensitizers.

A simple approach used by the Roswell Park group was to vary the overall lipophilicity of various types of photosensitizers such as pyropheophorbide *a*, benzoporphyrin derivatives, benzochlorins, and purpurinimides by altering the length of carbon units in alkyl ether substituents, an approach which has been quite successful. It was demonstrated that replacing an anhydride ring system in purpurin-18 (a chlorophyll *a* analog) with a six-member imide ring substantially enhanced its in vivo stability and retained effective in vivo photodynamic activity.[55] Therefore, in order to investigate the effect of such substitutions in the bacteriochlorin series, bacteriochlorophyll *a*, present in *Rhodobacter sphaeroides*, was first converted (in situ) into bacteriopurpurin-18 (**96**),[56] which in a sequence of reactions was transformed into a series of related *N*-alkyl derivatives (**97**)[57] (Scheme **2.20**). To determine the effect of the presence of these alkyl substituents with variable carbon units, the acetyl-group was first reduced with sodium borohydride, which on reacting with HBr gas and an appropriate alcohol produced the corresponding alkyl ether derivatives of bacteriochlorin (**98**) in high yield. These compounds are stable both in vitro and in vivo, exhibit long-wavelength absorption near 790 nm, and show high tumor uptake. In preliminary in vitro and in vivo studies, some of these compounds have been found to be quite effective at low injected doses.[58a] Earlier, it was shown that fluorinated analogs of purpurin showed enhanced photodynamic efficacy compared to the nonfluorinated analogs.[40f] The same group in order to investigate the effect of fluorine in bacteriopurpurinimide reacted bacteriopurpurinimide methyl ester with 3,5-bis(trifluoromethyl)benzyl amine (Scheme **2.21**).[58b] The compound (**101**) was found to be quite effective in vivo, and a drug dose of 1 µm/kg and light dose of 135/75 produced a 60% long-term tumor cure in C3H mice bearing RIF tumor.[58b]

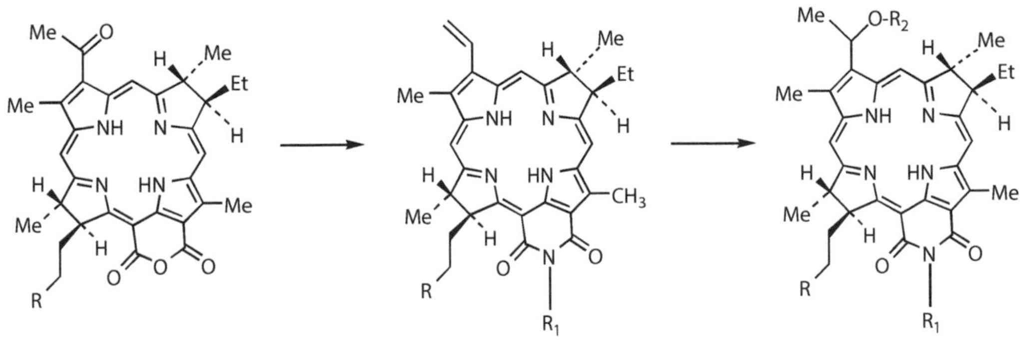

SCHEME 2.20 Synthesis of alkyl ether derivatives of *N*-substituted bacteriopurpurinimides.

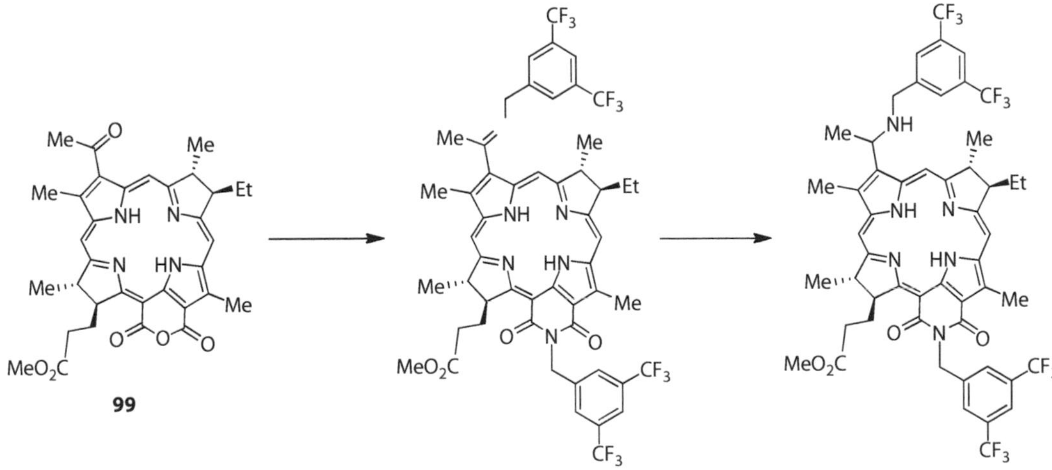

SCHEME 2.21 Synthesis of fluorinated bacteriopurpurimides.

2.3 Expanded Porphyrins

2.3.1 Texaphyrin

The texaphyrins are aromatic tripyrrolic, pentaaza, Schiff-base macrocycles that bear a strong, but *expanded*, resemblance to the porphyrins and other naturally occurring tetrapyrrolic prosthetic groups.[59] Similar to porphyrins, the texaphyrins are fully aromatic and colored compounds (Scheme **2.22**). However, they are 22π-electron electron systems rather than 18π-electron ones. This class of compounds exhibit long-wavelength absorption >700 nm depending on the nature of substituents present at the peripheral position. Also in contrast to porphyrins, the texaphyrins are monoanionic ligands that contain five, rather than four, coordinating nitrogen atoms within the central core that is roughly 20% larger than that of the porphyrins. High-yield production of long-lived triplet states and their remarkable singlet oxygen-producing efficiency are of important features of this class of photosensitizers. Currently, two different water-solubilized lanthanide(III)texaphyrin complexes, namely, the gadolinium(III) (**104**)[60] and lutetium(III) (**105**)[61a] derivatives (Gd-Tex and Lu-Tex, respectively), are being tested clinically. The first of these, XCYTRIN™, is in a pivotal phase III clinical trial as a potential enhancer of radiation therapy for patients with metastatic cancers of the brain receiving whole-brain radiation therapy. The second, in various formulations, is being

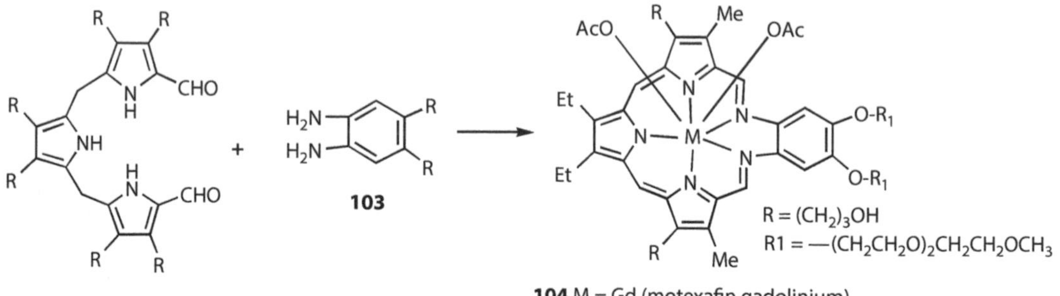

102

103

104 M = Gd (motexafin gadolinium)
105 M = Lu (motexafin lutetium)

SCHEME 2.22 Synthesis of metallated saphyrins.

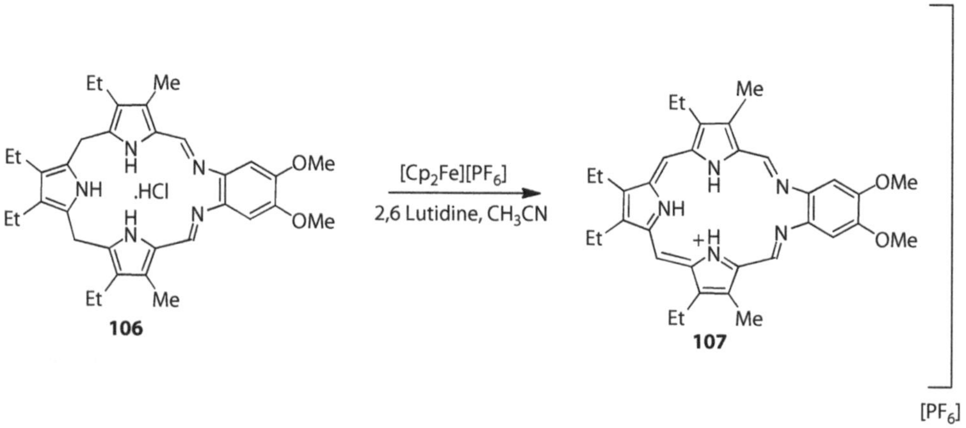

106

107

SCHEME 2.23 Synthesis of substituted nonmetallated saphyrins.

tested as a photosensitizer for use in the treatment of recurrent breast cancer (LUTRIN) and is in phase II clinical trials, photoangioplastic reduction of atherosclerosis involving peripheral arteries (ANTRIN), and light-based treatment of AMD (OPRTIN), currently phase I clinical trials.[59]

Earlier, texaphyrins could only be obtained in the form of metal complexes. In 2001, Sesseler et al. reported the synthesis of a metal-free form of texaphyrin (Scheme **2.23**).[61b] The metal-free oxidized texaphyrin 107 as its HPF6 salt was isolated by using ferrocenium cation as oxidizing agent and using a reduced porphyrinogen-like nonaromatic form of texaphyrin.[61b]

Recently, Lu et al. synthesized a benzotexaphyrin with an extensively delocalized π-electron system.[61c] Benzotexaphyrin absorbs at 810 nm and has high efficiency in generating singlet oxygen in methanol (0.65).

2.3.2 Sapphyrins

Another class of expanded porphyrins are the sapphyrins (**5**), which were discovered accidentally during the synthesis of vitamin B_{12}.[61d] These have a 22π-electron pathway, and as a result, they have more electron affinity than the corresponding porphyrin system.[61e]

They absorb in the near-infrared region and produce high singlet oxygen yields, which makes them potential PDT agent.[61f] Sesseler and coworkers synthesized and characterized a series of novel water-soluble sapphyrins that were found to localize selectively in pancreatic carcinoma tissue in a xenographic murine model (Scheme **2.24**).[61g] Among these, the sapphyrins bearing neutral solubilizing groups (compounds **111–113**) were found to have selectivities for tumor tissue over surrounding tissues. The incorporation of

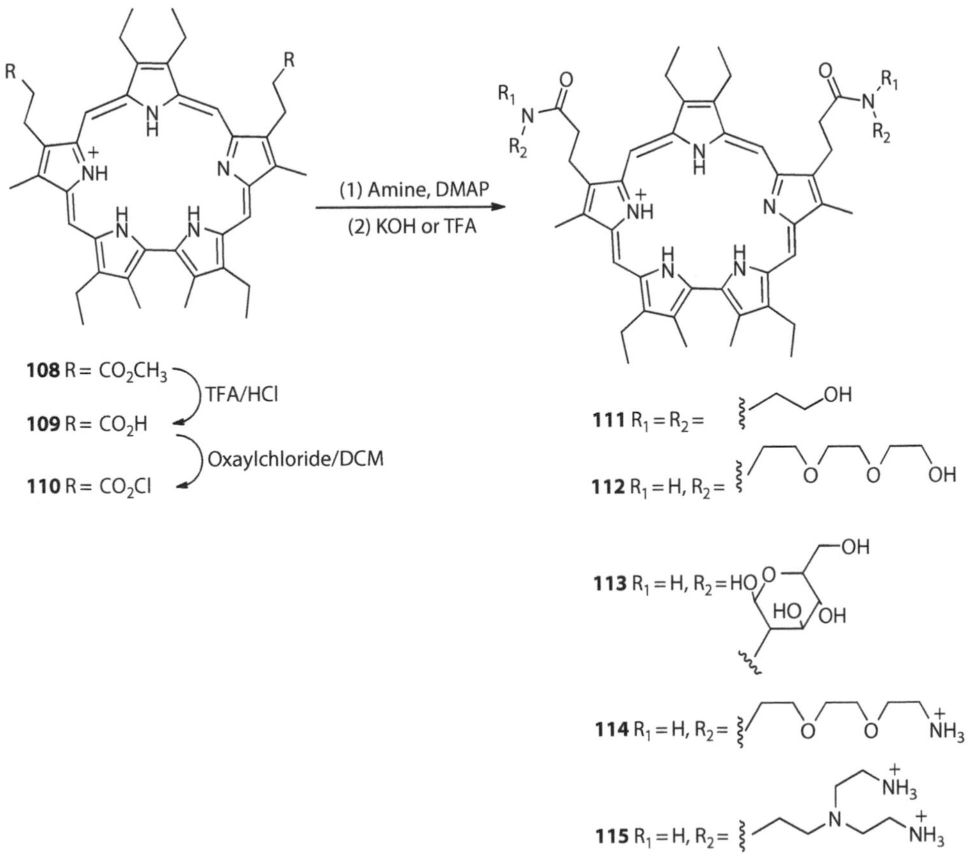

SCHEME 2.24 Synthesis of substituted saphyrins.

charged moieties into the sapphyrin (compounds **114** and **115**) significantly reduced the tumor localization. The tetrahydroxy sapphyrin (**111**) exhibited the best tumor-to-muscle ratio, whereas the incorporation of glucosamine into the sapphyrin (**119**) cores afforded the best tumor-to-liver ratio.

Recently Hooker et al. showed the activity of sapphyrins and heterosapphyrins in the presence of light against *Leishmania* parasites.[61h]

2.4 Phthalocyanines and Naphthalocyanines

Phthalocyanines (**6**, Pc) and naphthalocyanines (**7**, Nc) can be regarded as azaporphyrins containing four isoindoles linked by nitrogen atoms[62] (Scheme **2.25**). Compared to porphyrins, Pc and Nc offer high molar-extinction coefficients and red shift maximums at 680 nm for Pc and 780 nm for Nc resulting from the benzene or naphthalene rings condensed at the periphery of the porphyrin-like macrocycle. They possess high singlet oxygen producing efficiency, and interestingly, chelation of the metal ions such as zinc or aluminum increases the singlet oxygen yield to nearly 100%. Therefore, metal complexes of the Pc and Nc have attracted attention for their use as photosensitizers in PDT. In recent years, a large number of metalated or nonmetalated phthalocyanine-based photosensitizers have been synthesized by introducing a variety of substituents at the peripheral position(s). If the valency of the central metal is higher than 2, it binds various axial ligands. All of these chemical changes of Pc and Nc skeleton alter their PDT efficacy. Aggregated Pc and Nc are inactive photochemically because of a greatly enhanced rate of excited singlet state deactivation by internal conversion of the ground state. In the phthalocyanine series, Olenick and coworkers in collaboration with Kenney[63] synthesized and evaluated four silicon analogs with variable

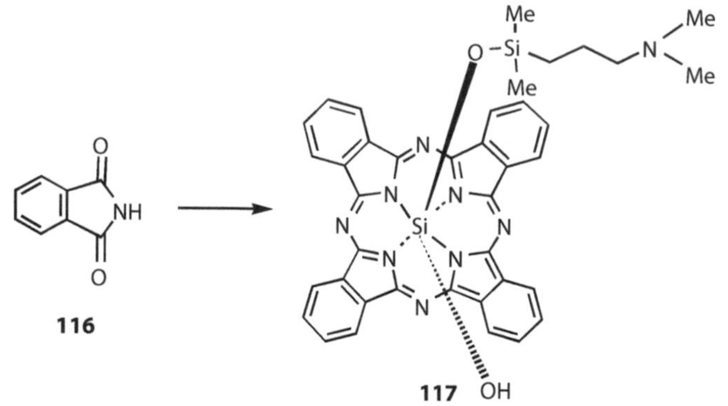

SCHEME 2.25   Substituted silicon phthalocyanines.

lipophilicity to learn more about the structural features that silicon phthalocyanine must have in order to be a good PDT photosensitizer. All these analogs produced similar photophysical properties; however, the photosensitizer denoted as Pc4, bearing a long-chain amino axial ligand (**117**), has shown promising results both in vitro and in vivo and is presently entering clinical trials.[64] Further, it was concluded that the presence of structural features leading to improvement in the association between the photosensitizers and important cellular targets is more useful than those leading to improvements in their already acceptable photophysical and photochemical characteristics. A series of benzyl-substituted phthalonitriles were converted into the corresponding Zn(II) hydroxyphthalocyanines (phthalocyanine phenol analogs). Their efficacy as sensitizers for PDT was evaluated on the EMT-B mammary tumor cell line. In vitro, the 2-hydroxy Zn Pc was the most active, followed by 2,3- and 2,9-dihydroxy ZnPc, with the 2,9,16-trihydroxy ZnPc exhibiting the least activity. In vivo, the monohydroxy derivative and the 2,3-dihydroxy analog were both efficient in inducing tumor necrosis, but complete tumor regression was poor even at high doses. In contrast, the 2,9-dihydroxy isomer at 2 µM/kg induced tumor necrosis in all animals treated, with 75% complete regression. These results underline the importance of the position of the substituents on the Pc macrocycle to optimize tumor response and confirm the PDT potential of the unsymmetrical Pcs bearing functional groups on adjacent benzene rings.[65a]

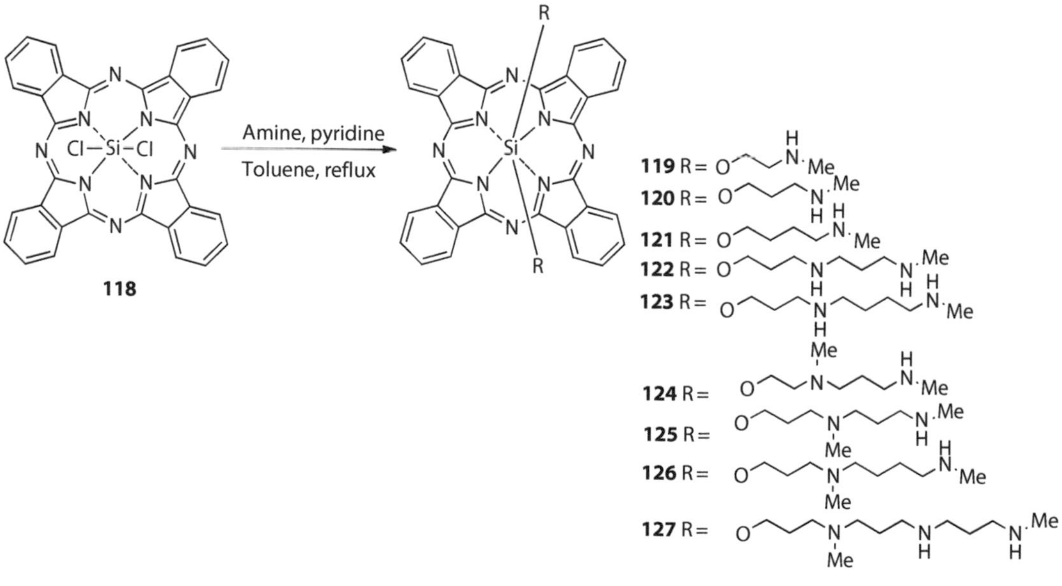

SCHEME 2.26   Substituted silicon phthalocyanines.

Jiang and coworkers synthesized a series of silicon (IV) phthalocyanines with polyamine moieties at axial positions (Scheme **2.26**).[65b] These compounds (**119–127**) were found to be potent PSs toward the HT29 cells with IC_{50} values as low as 1 nm, and also compounds **120** and **123** suppressed the growth of tumors in nude mice bearing HT29 tumor.[65b]

Recently, Dumoulin et al. synthesized a chalcone phthalocyanine conjugate (**130**) to combine the vascular disrupting effect of chalcones with the photodynamic effect of phthalocyanines.[65c] For this, they converted the aminochalcone to the activated isocyanate chalcone (**129**), and then the isocyanate chalcone was coupled to tetrahydroxylated Zn (II) phthalocyanine (**128**) under basic conditions (Scheme **2.27**). The photophysical and biological studies of chalcone phthalocyanine conjugate (**130**) are under progress.[65c]

Recently, certain water-soluble dual-function photosensitizers containing phthalocyanine-ALA (5-aminolevulinic) conjugate (**135, 140**) were synthesized by Oliveira et al. These compounds produce good singlet oxygen yields and could be efficient agents for the use in PDT (Scheme **2.28**).[65d]

The pegylated zinc phthalocyanines were found to be highly cytotoxic toward HT29 human colorectal carcinoma and HepG2 human hepatocarcinoma on illumination with light with IC_{50} as low as 0.02 μM.[65e]

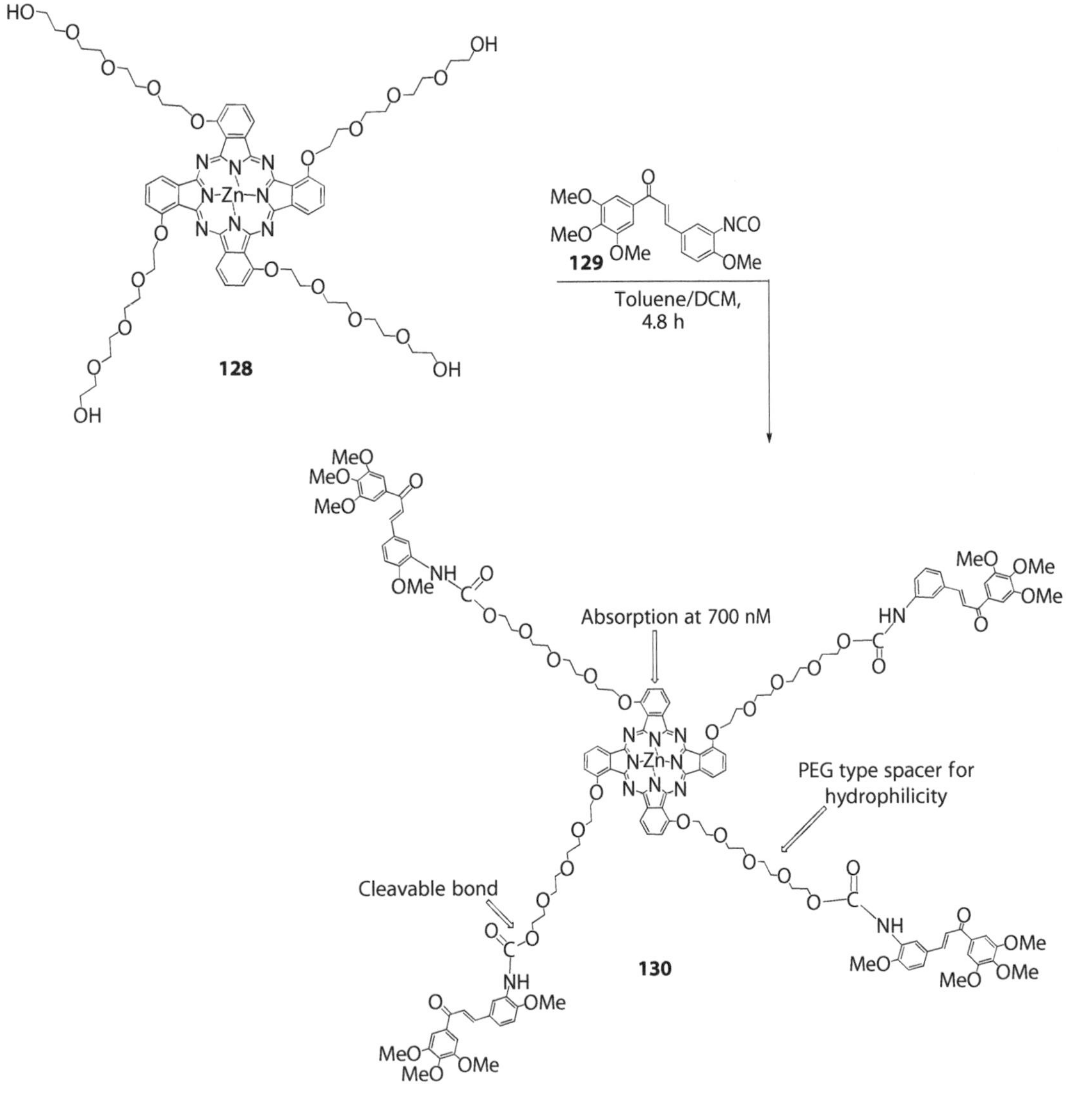

SCHEME 2.27 PEG-substituted tetraphenylporphyrins.

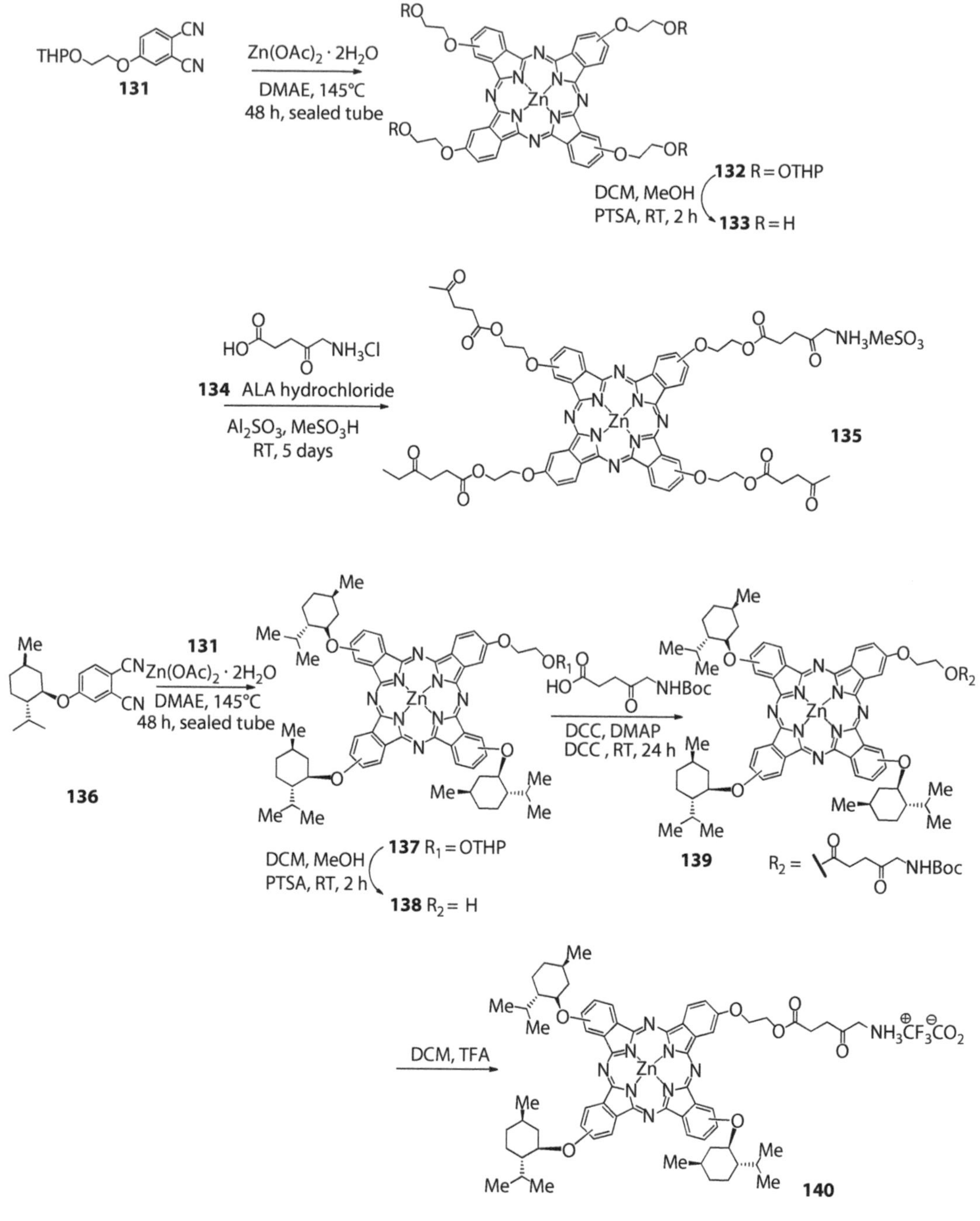

SCHEME 2.28 Phthalocyanines substituted at peripheral positions.

Water-soluble, 3-hydroxypyridin tetrasubstituted indium (III) phthalocyanines (**143, 147**) and their quarternized derivatives (**144, 148**) were synthesized by Durmus and coworkers and showed to have high singlet oxygen yield (>0.55) (Scheme **2.29**).[65f]

Rodgers et al. in collaboration with Kenney and collaborators[66] developed a new route to silicon-substituted phthalocyanines and phthalocyanines-like compounds that is robust and flexible. One

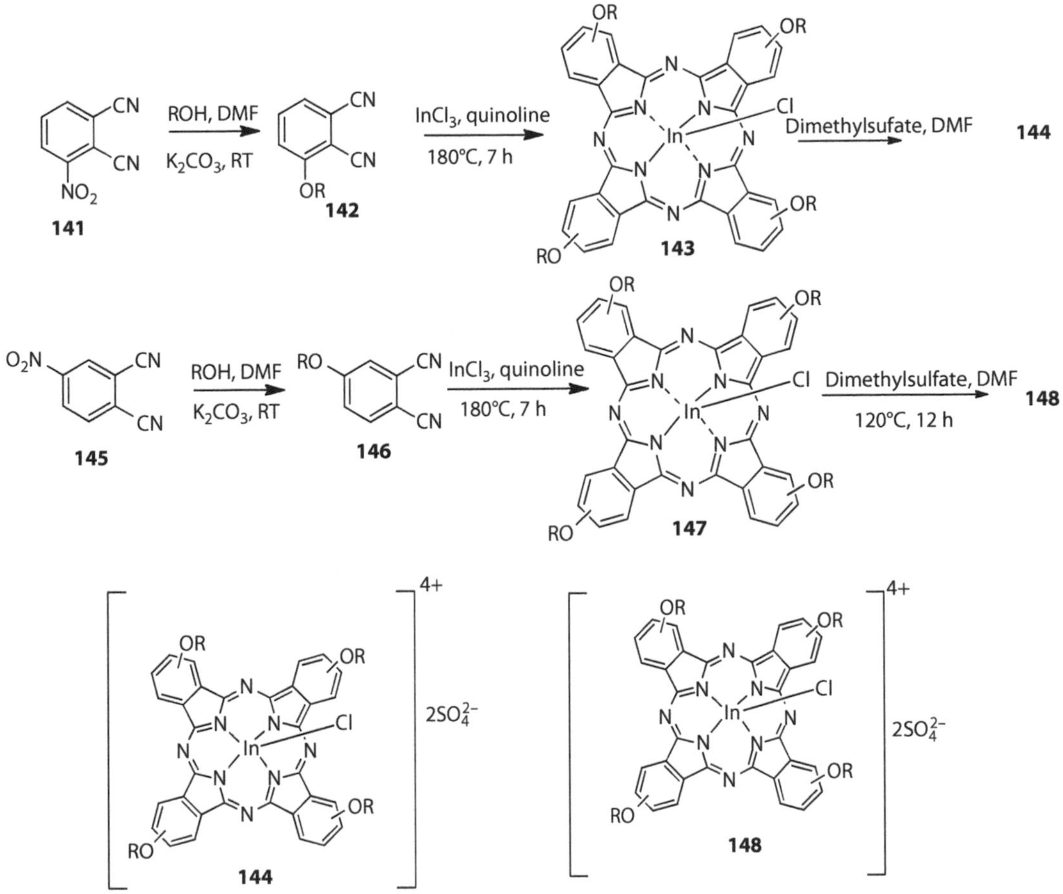

SCHEME 2.29 Synthesis of In(II) phthalocyanines.

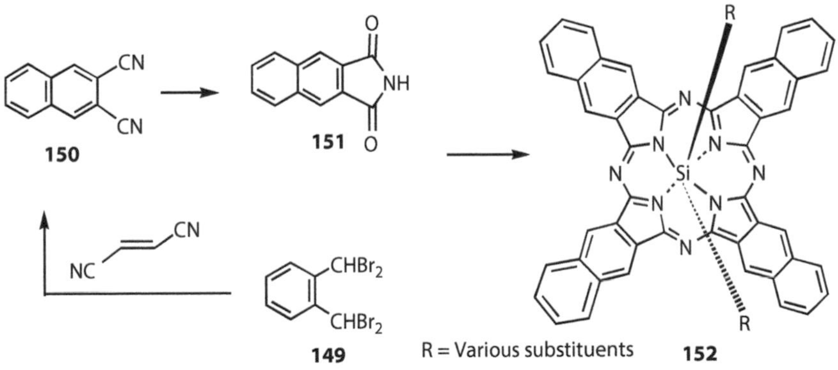

SCHEME 2.30 Synthesis of silicon naphthocyanines.

of the siloxysilicon compounds, that with the ligand 5,9,12,16,23,28,32-octabutoxy-33H, 35H[b,g] dinaphtho[2,3–1:2′,3′-q]porphyrazine (**152**), has a q band at a wavelength of 804 nm and an extinction coefficient of 1.9×10^5 M^{-1} cm^{-1} (Scheme **2.30**). These compounds showed promising in vitro photosensitizing efficacy, however, in animal studies, compared to phthalocyanine (Pc 4) were found to be less effective. A first photophysical study of a member of the family of subnaphthalocyanines is described.[67]

The cone-shaped unsubstituted subnaphthalocyanines, synthesized in 35% yield, showed distinctive photophysical properties that are better than those of the related planar phthalo- and naphthalocyanines. SubNc absorbs in the red part of the spectrum and has substantial fluorescence. Triplet and singlet oxygen quantum yields are substantially higher than those of the related phthalo- and naphthalocyanines. These results, together with their synthetic availability, high solubility, and low tendency to aggregate, make this class of sensitizers to be studied further with a view to investigate their PDT applications.

2.5 Target-Specific Photosensitizers

Since the introduction of the first PDT drug Photofrin, there has not been much success on improving the photosensitizer's tumor selectivity and specificity because tumor cells in general have non-specific affinity to porphyrins.[68] Although the mechanism of porphyrin retention by tumors is not well understood, the balance between lipophilicity and hydrophilicity is recognized as an important factor.[69] Some attempts have been made to direct photosensitizers to known cellular targets by creating a photosensitizer conjugate, where the other molecule is a ligand that is specific for the target. For example, to improve localization to cell membranes, cholesterol[70] and antibody conjugates have also been prepared to direct photosensitizers to specific tumor antigens.[71–73] Certain chemotherapeutic agents have also been attached to porphyrin chromophores to increase the lethality of the PDT treatment.[74] Certain protein and microsphere conjugates were made to improve the pharmacology of the compounds.[75] These strategies seldom work well because the pharmacological properties of both compounds are drastically altered.[76]

Recently, the bovine serum albumin conjugate of a sulfonated phthalocyanine (BSA-AlPcS$_4$) prepared by Brasseur et al. has been shown to target the scavenger receptor of macrophages.[77] Relative photocytotoxicities were reported using a receptor positive and a receptor negative cell line, where their lethal effects correlated with its receptor affinity. Also, certain adenoviral proteins have been employed to target lung cancer cells rich in the appropriate class integrin receptor.[78] Using the EMT-6 murine model, in vivo results were encouraging with Pc-adenoviral protein conjugates. Further, both AlPc and CoPc have been covalently labeled with epidermal growth factor (EGF), and this conjugate produced a fivefold increase in photocytotoxicity as compared to nonconjugated phthalocyanine.[79]

Since oligosaccharides play essential roles in molecular recognition,[80] porphyrins with sugar moieties should not only have good aqueous solubility but also have possible specific membrane interaction. In addition to providing the molecule with polar hydroxy groups, it is possible that the sugar moiety might lead the conjugate to a cell surface target through specific binding to its receptor. Oligosaccharides play essential roles in various cellular activities, as antigens, growth signals, targets of bacterial and viral infection, and glues in cell adhesion and metastasis, where the saccharide–receptor interactions are usually specific and multivalent.[81] This specificity suggests a potential utility of synthetic saccharide derivatives as carriers in directed drug delivery. Therefore, in recent years, a number of carbohydrate derivatives of various photosensitizers have been synthesized. Schell and Hombrecher[82] reported the preparation of a galactopyranosyl–cholesteryloxy substituted porphyrin by following the McDonald approach. Several tetra- and octaglycoconjugated tetraphenylporphyrins were also reported by Yano and coworkers,[83] Cornia et al.,[84] Momenteau and coworkers,[85] and Ono et al.[86] A glycosylated peptide porphyrin synthesized by Krausz and coworkers[87] showed promising in vitro PDT efficacy. Millard et al.[88] reported the synthesis of a glycoconjugated *meso*-monoarylbenzochlorin (related to **154** and **156**). See Scheme **2.30**. This compound displayed good in vitro PDT efficacy in tumor cell lines. A few years ago, Bonnett and coworkers[89] showed that compared to β-hydroxyoctaethylchlorin (**158**), the related glycosylated analog was more effective as a photosensitizing agent in vitro as well as in vivo. Montforts et al.[90]

further explored this approach by attaching hydrophilic carbohydrate structural units to certain chlorins. Such conjugation increased the water solubility of the parent chlorins by introducing an estradiol with a diethyl spacer. The chlorin–estrogen conjugate was then prepared with the hope that it would bind to an estrogen receptor, which could induce destruction of a mammalian carcinoma. In a recent report, Aoyama and coworkers[91] reported the synthesis of certain tetraphenylporphyrin-based saccharide-functionalized porphyrins and demonstrated the importance of hydrophobicity masking for the saccharide-directed cell recognition. Since the saccharide–receptor interactions are ubiquitous, well-defined/well-designed synthetic saccharide clusters may serve as a new tool in glycoscience and glycotechnology (Scheme **2.31**).

It is known that galectins are involved in the modulation of cell adhesion, cell growth, immune response, and angiogenesis; therefore, there is a good possibility that their expression might have a critical role in tumor progression.[92] Gal-1 mRNA levels increase 20-fold in low tumorigenic and up to 100-fold in highly tumorigenic cells.[93] The Roswell Park group has recently reported the synthesis and biological significance of certain β-galactose-conjugated purpurinimides (a class of chlorins containing a six-member fused imide ring system) **161–163** (Scheme **2.32**) as Gal-1 recognized photosensitizer via enyne metathesis.[94] Molecular modeling analysis utilizing model photosensitizers and the available crystal structures of galectin-carbohydrate moiety indicates

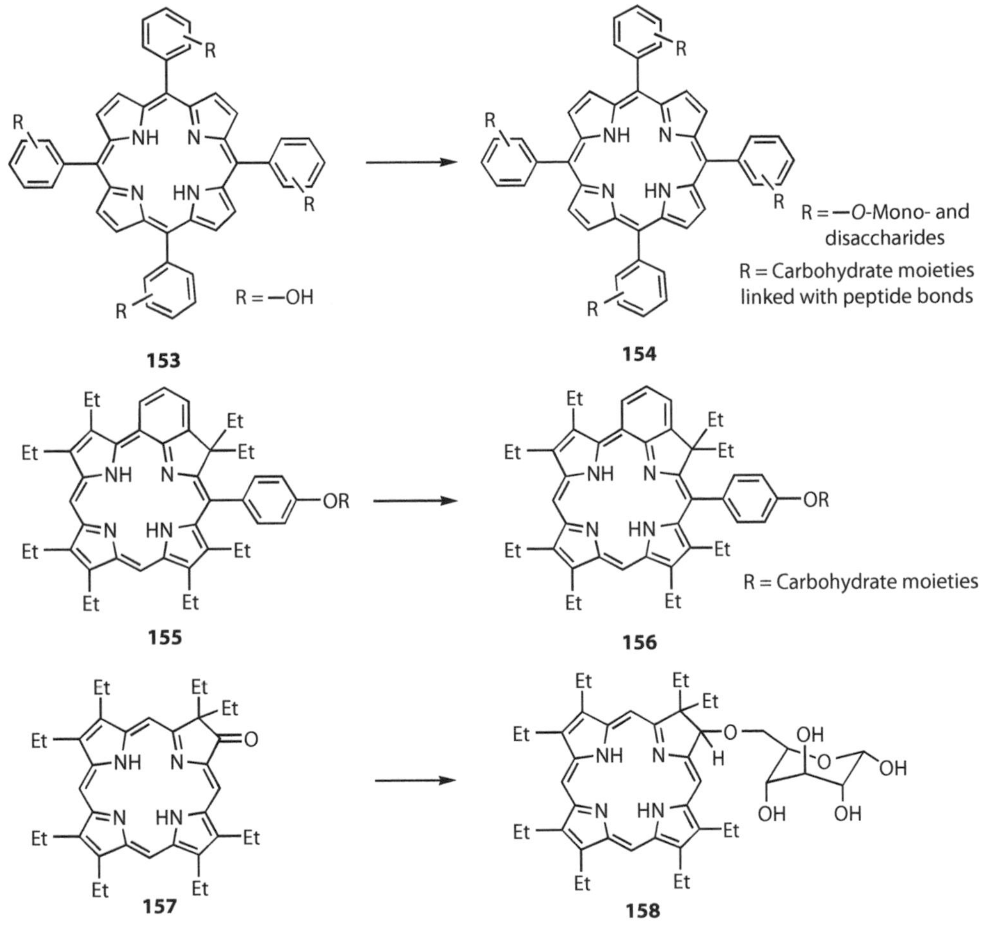

SCHEME 2.31 Carbohydrate analogs of porphyrins and chlorins.

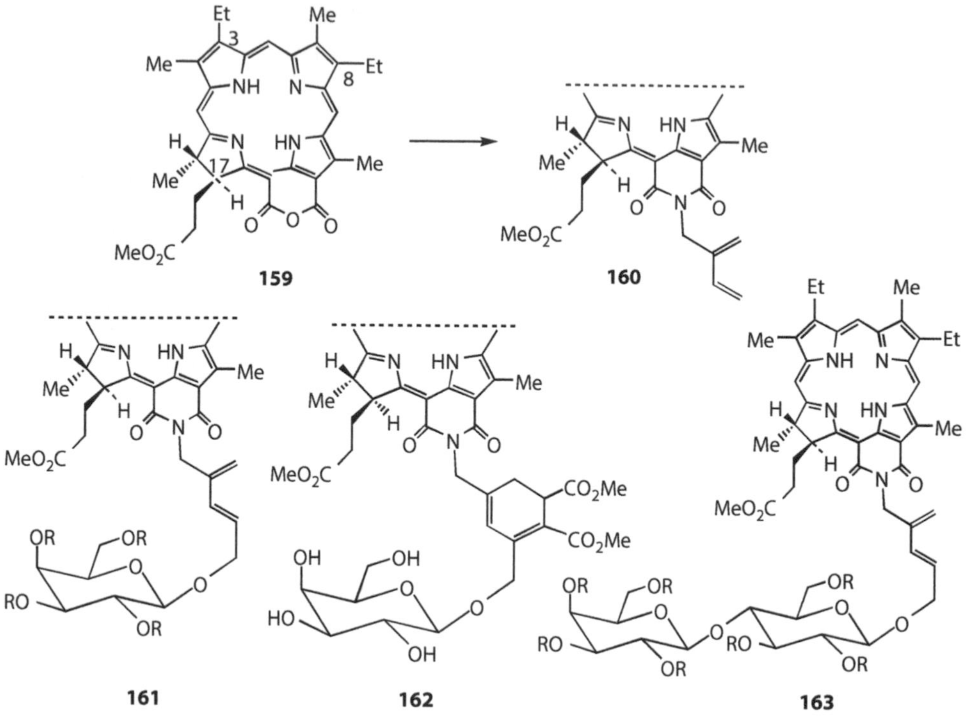

SCHEME 2.32 Galactose and lactose analogs of purpurinimides.

that when placed at appropriate position, the photosensitizer does not interfere with the galectin-carbohydrate recognition.

The intracellular studies with known cell surface counterstains confirmed the cell surface recognition of the conjugates. Under similar drug and light doses, compared to the free purpurinimide analog, the galactose- and lactose-conjugated analog showed a considerable increase both in vitro and in vivo photosensitizing efficacy. These results, therefore, indicate the possibility for development of a new class of tumor-specific photosensitizers for PDT-based on recognition of a cellular receptor.[94a] Zheng et al. examined the in vitro and in vivo photodynamic effects of a series of carbohydrate conjugates of HPPH, **164–171** (Scheme **2.33**). The photosensitizers were conjugated to carbohydrates that were known for their high affinity to galectin-3.[94b]

In more recent work, Pandey et al. illustrated the effectiveness of an HPPH–peptide conjugate. They have showed the in vitro and in vivo behavior of a series of HPPH–peptide conjugates. Among the conjugates, the HPPH–cRGD conjugate, structure **177** shown in Scheme **2.34**, showed improved PDT efficacy, faster clearance, and enhanced tumor imaging when compared to HPPH alone.[95a] Cyclic Arg-Gly-Asp (cRGD) peptide represents a selective $\alpha_v\beta_3$-integrin ligand that has been extensively used for research, therapy, and diagnosis of neoangiogenesis.[95a] However, the presence of a cRGD moiety to HPPH made a significant difference in clearance of the conjugate from the tumor and the other organs, and in contrast to HPPH (the parent analog), it showed improved efficacy on treating the tumor with light at 4 h, instead of 24 h.

Zhang et al. synthesized a target-specific water-soluble PDT agent (**183**) with improved target specificity and delivery efficiency. Their methodology involved the combination of folate-mediated targeting (**182**) and a short peptide (**178**) to achieve better target delivery of PDT agent (Scheme **2.35**).[95b]

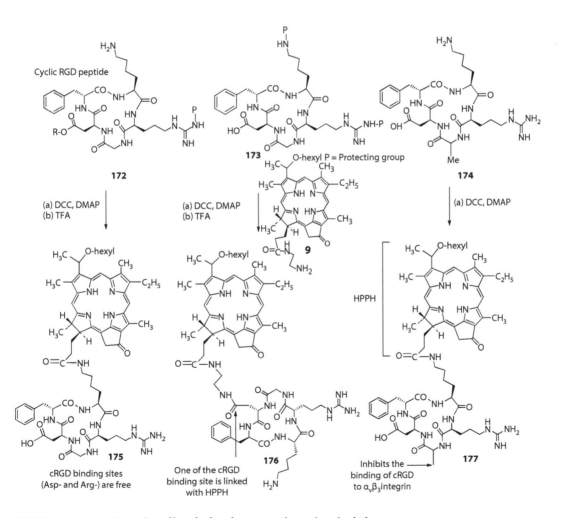

164 R = Peracetylated galactose
165 R = Peracetylated glucose
166 R = Peracetylated lactose
167 R = Peracetylated cellobiose

168 R^1 = Galactose
169 R^1 = Glucose
170 R^1 = Lactose
171 R^1 = Cellobiose

SCHEME 2.33 Carbohydrate analogs of pyropheophorbide-a.

SCHEME 2.34 cRGD analog of hexyl ether derivative of pyropheophorbide-a.

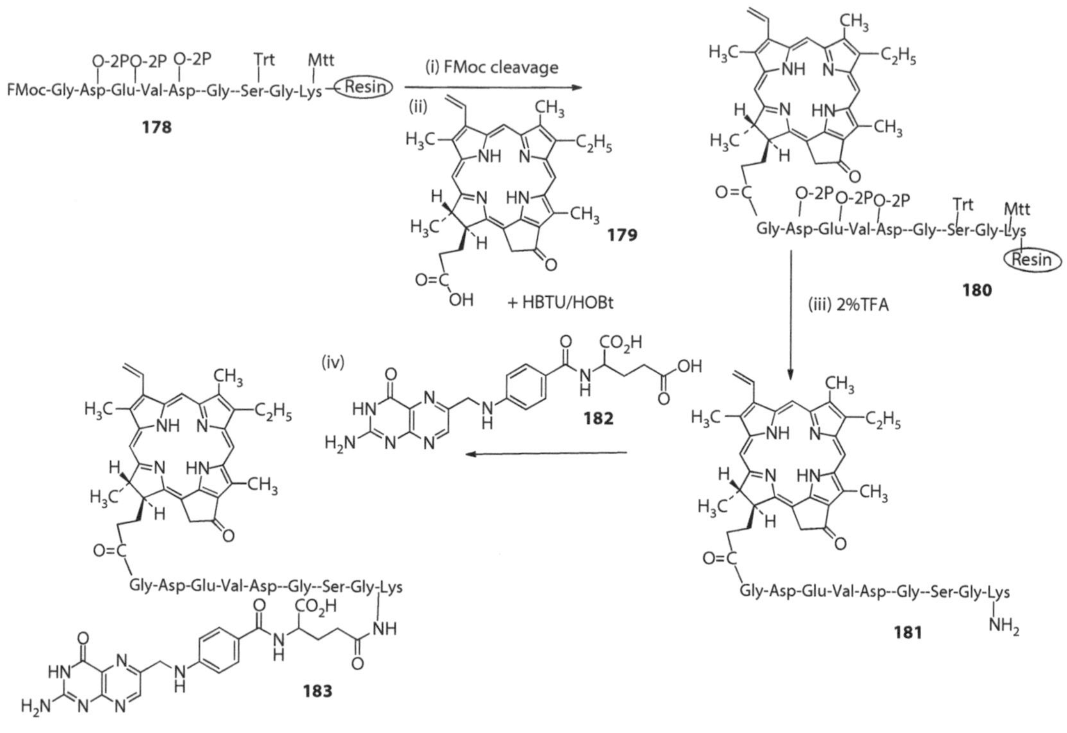

SCHEME 2.35 Folate-mediated targeted photosensitizers.

2.6 Summary

This article focused largely on various synthetic approaches for the preparation of porphyrin-based photosensitizers. However, the biological significance of these compounds is also briefly discussed. It is important to note that porphyrin-based compounds in addition to their use in cancer have also shown potential for application in other areas: treatment of AMD, tumor imaging by magnetic resonance imaging (MRI), psoriasis, bone marrow purging, and purification of blood infected with various viruses, including HIV. Since the discovery of Photofrin, enormous progress has been made in the development of various porphyrin-based compounds with improved photophysical characteristics. In recent years, a number of pharmaceutical companies such as QLT Pharmaceuticals, Vancouver, British Columbia; Miravant, Santa Barbara, California; Scotia Pharmaceuticals, United Kingdom; Ciba Vision, New Jersey; Pharmacia, Sweden; DUSA, New Jersey; Pharmacyclics, California; Light Sciences Corp. Washington; and Photolitec, LLC, Buffalo, New York have shown major interest in porphyrin-based compounds for cancer imaging (positron emission tomography [PET], fluorescence, MRI) and PDT. However, given the possible implications for the use of porphyrin-based compounds (porphyrins, chlorins, bacteriochlorins, expanded porphyrins, phthalocyanines) for the treatment of cancer by PDT, future design strategies for new agents should be directed toward tumor-specific drug molecules. Such compounds might show greater tumor selectivity with reduced skin phototoxicity, a major problem associated with most of the porphyrin-based compounds. In recent years, the use of biodegradable nanoparticles for developing multifunctional platforms has created enormous interest, and the in vivo results using animal models look quite promising. However, due to space limitation, the reported studies of nanoparticles focused on cancer imaging and therapy is not included in this review article.

Acknowledgments

The work summarized in this review article was funded by NIH (CA 55791, CA127369), Photolitec, LLC, and Roswell Park Alliance, which is highly appreciated.

References

1. (a) Ethirajan, M., Chen, Y., Joshi, P., Pandey, R. K. The role of porphyrin chemistry in tumor imaging and photodynamic therapy. *Chem. Soc. Rev.*, 40, 340–362, 2011. (b) Pandey, R. K., Zheng, G. Porphyrins as photosensitizers in photodynamic therapy, in *The Porphyrin Handbook*, vol. 6, Kadish, K. M., Smith, K. M., Guilard, R., Eds. Academic Press, San Diego, CA, 2000, Chap. 43.

2. Chang, C. K. Cation radicals of ferrous and free base isobacteriochlorins: Models for siroheme and sirohydrochlorin. *Proc. Natl. Acad. Sci. U.S.A.*, 78, 2653, 1981.

3. Barkigia, K. M., Fajer, J. *The Photosynthetic Reaction Center*, vol. 2, Deisenhofer, H., Norris, J. R., Eds. Academic Press, San Diego, CA, 1993, p. 514.

4. Dougherty, T. J., Gomer, C., Henderson, B. W., Jori, G., Kessel, D., Korbelik, M., Moan, J., Peng, Q. Photodynamic therapy. *J. Natl. Cancer Inst.*, 90, 889, 1998.

5. Weishaupt, K. R., Gomer, C. J., Dougherty, T. J. Identification of singlet oxygen as the cytotoxic agent in photoinactivation of murine tumor. *Cancer Res.*, 36, 2326, 1976.

6. Sherman, W. M., Allen, C. M., van Lier, J. E. Role of activated oxygen species in photodynamic therapy. *Methods Enzymol.*, 319, 376, 2000.

7. (a) Dougherty, T. J., Kaufman, J. H., Goldfrab, A., Weishaupt, K. R., Boyle, D., Mittleman, A. Photoradiation therapy for the treatment of malignant tumors. *Cancer Res.*, 38, 2628, 1978. (b) Pandey, R. K., Marshall, M. S., Tsao, R., McReynolds, J. H., Dougherty, T. J. Fast atom bombardment mass spectral analyses of Photofrin II and its synthetic analogs. *Biomed. Environ. Mass Spectrom.*, 19, 405, 1990 and references therein. (c) Pandey, R. K., Smith, K. M., Dougherty, T. J. Porphyrin dimers as photosensitizers in photodynamic therapy. *J. Med. Chem.*, 33, 2032, 1990. (d) Pandey, R. K., Dougherty, T. J. Syntheses and photosensitizing activity of porphyrins joined with ester linkages. *Cancer Res.*, 49, 2042, 1989. (e) Pandey, R. K., Shiau, F.-Y., Dougherty, T. J., Smith, K. M. Regioselective syntheses of ether linked porphyrin dimers and trimers related to Photofrin II. *Tetrahedron*, 47, 9571, 1991 and references therein.

8. Dolphin, 1993 Syntex Award Lecture: Photomedicine and photodynamic therapy. D. *Can. J. Chem.*, 72, 1005, 1994 and references therein.

9. MacDonald, I., Dougherty, T. J. Basic principles of photodynamic therapy. *J. Porphyr. Phthalocyanines*, 5, 105, 2001.

10. Whitlock, H. W., Hanauerr, R., Oester, M. Y., Bower, B. K. Diimide reduction of porphyrins. *J. Am. Chem. Soc.*, 91, 7485, 1969.

11. Bonnett, R. Photosensitizers of the porphyrin and phthalocyanine series for photodynamic therapy. *Chem. Soc. Rev.*, 24, 19, 1995.

12. Hamata, M. *Advances in Cycloaddition*, vol. 4, Hamata, M. Ed. Jai Press, London, U.K., 1998.

13. Callot, H. L., Johnson, A. W., Sweeney, A. Addition to porphyrins involving the formation of new carbon-carbon bonds. *J. Chem. Soc. Perkin Trans 1*, 13, 1424–1427, 1973.

14. Morgan, A. R., Pangka, V. S., Dolphin, D. Ready syntheses of benzoporphyrins via Diels-Alder reactions with protoporphyrin IX. *Chem. Commun.*, 1047–1048, 1984.

15. (a) Meunier, I., Pandey, R. K., Walker, M. M., Senge, M. O., Dougherty, T. J., Smith, K. M. New syntheses of benzoporphyrin derivatives and analogs for use in photodynamic therapy. *Bioorg. Med. Chem. Lett.*, 2, 1575, 1992. (b) Meunier, I., Pandey, R. K., Senge, M. O., Dougherty, T. J., Smith, K. M. Benzoporphyrin derivatives: Synthesis, structure and preliminary biological activity. *J. Chem. Soc. Perkin Trans. 1*, 961–969, 1994. (c) Pandey, R. K., Potter, W. R., Meunier, I., Sumlin, A. B., Smith, K. M. Structure activity relationships among benzoporphyrin derivatives. *Photochem. Photobiol.*,

62, 764, 1995. (d) Saenz, C., Ethirajan, M., Iacobucci, G., Pandey, A., Missert, J. R., Dobhal, M. P., Pandey, R. K. Indium as a central metal enhances the photosensitizing efficacy of benzoporphyrin derivatives. *J. Porphyr. Phthalocyanines*, 15, 1, 2011.

16. Pandey, R. K., Shiau, F. Y., Ramachandran, K., Dougherty, T. J., Smith, K. M. Long wavelength photosensitizers related to chlorins and bacteriochlorins for the use in photodynamic therapy. *J. Chem. Soc. Perkin Trans. 1*, 1377, 1992.

17. Yon-Hin, P., Wijesekera, T. P., Dolphin, D. A convenient synthetic route for the bacteriochlorin chromophore. *Tetrahedron Lett.*, 32, 2875, 1991.

18. Morgan, A. R., Skalkos, D., Garbo, G. M., Keck, R. W., Selman, S. H. Synthesis and in vivo photodynamic activity of some bacteriochlorin derivatives against bladder tumors in rodents. *J. Med. Chem.*, 34, 2126, 1991.

19. Morgan, A. R. unpublished results.

20. Arnold, D. P., Johnson, A. W., Williams, G. A. Wittig condensation products from nickel meso-formyl-octaethylporphyrin and aetioporphyrin-I and some cyclization reactions. *J. Chem. Soc. Perkin Trans. 1*, 1660, 1978.

21. (a) Morgan, A. R., Garbo, G. M., Ranapersaud, A., Shalkos, D., Keck, R. W., Selman, S. H. Photodynamic action of benzochlorins. *Proc. SPIE*, 146, 1065, 1989. (b) Morgan, A. R., Skalkos, D., Maguire, G., Ranpersaud, A., Garbo, G., Keck, K., Selman, S. H. Observations of the synthesis and in vivo photodynamic activity of some benzochlorins. *Photochem. Photobiol.*, 55, 133, 1992.

22. (a) Li, G., Graham, A., Potter, W. R., Grossman, Z. D., Oseroff, A., Dougherty, T. J., Pandey, R. K. A simple and efficient approach for the synthesis of fluorinated and non-fluorinated octaethylporphyrin-based benzochlorins with variable lipophilicity. Their in vivo tumor uptake, and the preliminary in vitro photosensitizing efficacy. *J. Org. Chem.*, 66, 1316, 2001. (b) Pandey, R. K. et al., unpublished results.

23. Li, G., Chen, Y., Missert, J. R., Rungta, A., Dougherty, T. J., Pandey, R. K. Application of Ruppert's reagent in preparing novel perfluorinated porphyrins, chlorins and bacteriochlorins. *J. Chem. Soc. Perkin Trans. 1*, 1785, 1999.

24. Vicente, M. G. H., Rezanno, I. N., Smith, K. M. Efficient new syntheses of benzochlorins, isobacteriochlorins and bacteriochlorins. *Tetrahedron Lett.*, 31, 1365, 1990.

25. Morgan, A. R., Gupta, S. Synthesis of benzopurpurins, isobacteriobenzopurpurins and bacteriobenzopurpurins. *Tetrahedron Lett.*, 35, 4291, 1994.

26. Woodward, R. B. et al. The total synthesis of chlorophyll-a. *J. Am. Chem. Soc.*, 82, 3800, 1960; *Tetrahedron*, 46, 7599, 1990.

27. Morgan, A. R., Garbo, G. M., Keck, R. W., Selman, S. H. New photosensitizers for photodynamic therapy: Combined effect of metallopurpurin derivatives and light in transplantable bladder. *Cancer Res.*, 48, 194, 1988.

28. (a) Gunter, M. J., Robinson, B. C. Purpurins bearing functionality at the 6,16 meso-positions: Synthesis from 5,15-disubstituted meso-[β-(methoxycarbonyl)vinyl] porphyrins. *Aust. J. Chem.*, 43, 1839, 1990. (b) Forsyth, T. P., Nurco, D. J., Pandey, R. K., Smith, K. M. Syntheses and structure of 5,15-bis-(4-pyridyl)purpurins. *Tetrahedron Lett.*, 36, 9093, 1995.

29. Smith, K. M., Ed. *Porphyrins and Metalloporphyrins*. Elsevier Science Publishers, Amsterdam, the Netherlands, 1975.

30. Pandey, R. K., Bellnier, D. A., Smith, K. M., Dougherty, T. J. Chlorin and porphyrin derivatives as potential photosensitizers in photodynamic therapy. *Photochem. Photobiol.*, 53, 65, 1991.

31. Bommer, J. C., Burnham, B. F. Tetrapyrrole compounds. Euro Patent 169831, 1986.

32. Gomi, S., Nishizuka, T., Ushiroda, O., Uchida, N., Takahashi, H., Sumi, S. The structures of mono-L-aspartyl chlorin e_6 and its related compounds. *Heterocycles*, 48, 2231, 1998.

33. (a) Nelson, J. S., Roberts, W. G., Berns, M. W. In vivo studies on the utilization of mono-L-aspartyl chlorin (NPe_6) in photodynamic therapy. *Cancer Res.*, 47, 5681, 1987. (b) Roberts, W. G., Shiau, F. Y., Nelson, J. S., Smith, K. M., Berns M. W. In vitro characterization of monoaspartyl chlorin e_6 and diaspartyl chlorin e_6 for photodynamic therapy. *J. Natl. Cancer Inst.*, 80, 330, 1988.

34. Pandey, R. K., Sumlin, A. B., Constantine, S., Potter, W. R., Bellnier, D. A., Henderson, B. W., Rodgers, M. A., Smith, K. M., Dougherty, T. J. Alkyl ether analogs of chlorophyll-a derivatives. Synthesis, photophysical properties and photodynamic efficacy of chlorophyll-a derivatives. *Photochem. Photobiol.*, 64, 194, 1996.

35. Henderson, B. W., Bellnier, D. A., Greco, W. R., Sharma, A., Pandey, R. K., Vaughan, L. A., Weishaupt, K. R., Dougherty, T. J. An in vivo quantitative structure-activity relationship for a congeneric series of pyropheophorbide derivatives as photosensitizers for photodynamic therapy. *Cancer Res.*, 57, 4000, 1997.

36. (a) Dougherty, T. J., Pandey, R. K., Nava, H. R., Smith, J. A., Douglass, H. O., Edge, S. B., Bellnier, D. A., Cooper, M. Preliminary clinical data on a new photodynamic therapy photosensitizer-HPPH. *Proc. SPIE*, 3909, 25, 2000. (b) Chen, Y., Zheng, X., Dobhal, M. P., Gryshuk, A., Morgan, J., Dougherty, T. J., Oseroff, A., Pandey, R. K. Methyl pyropheophorbide-*a* analogues: Potential fluorescent probes for the peripheral-type benzodiazepine receptor. Effect of central metal in photosensitizing efficacy. *J. Med. Chem.*, 48, 3692, 2005. (c) Mosinger, J., Micka, Z. Quantum yields of singlet oxygen of metal complexes of meso-tetrakis(sulphonatophenyl)porphine. *J. Photochem. Photobiol. A: Chem.*, 107, 77, 1997.

37. Lee, S. H., Jagerovic, N., Smith, K. M. Use of chlorophyll derivative purpurin-18 for syntheses of sensitizers for use in photodynamic therapy. *J. Chem. Soc. Perkin Trans. 1*, 2369, 1993.

38. Kozyrev, A. N., Zheng, G., Lazarou, E., Dougherty, T. J., Smith, K. M., Pandey, R. K. Synthesis of emeraldin and purpurin-18 analogs as target-specific photosensitizers for photodynamic therapy. *Tetrahedron Lett.*, 38, 3335, 1997.

39. Zheng, G., Potter, W. R., Camacho, S. H., Missert, J. R., Wang, G., Bellnier, D. A., Henderson, B. W., Rodgers, M. A. J., Dougherty, T. J., Pandey, R. K. Synthesis, photophysical properties, tumor uptake and preliminary in vivo photosensitizing efficacy of a homologous series of 3-(1-alkoloxy)ethyl-3-devinyl-ppurpurin-18-N-alkylimides with variable lipophilicity. *J. Med. Chem.*, 44, 1540, 2001.

40. (a) Zheng, G., Potter, W. R., Sumlin, A. B., Dougherty, T. J., Pandey, R. K. Photosensitizers related to purpurin-18-N-alkylimides. A comparative in vivo tumoricidal ability of ester versus amide functionalities. *Bioorg. Med. Chem. Lett.*, 10, 123, 2000. (b) Rungta, A., Zheng, G., Missert, J. R., Potter, W. R., Dougherty, T. J., Pandey, R. K. Purpurinimides as photosensitizers: Effects of the position and presence of the substituents in the in vivo photodynamic efficacy. *Bioorg. Med. Chem. Lett.*, 10, 1463, 2000. (c) Filler, R., Kobayashi, Y., Yagupolski, L. M., Eds. *Organofluorine Compounds in Medicinal Chemistry and Biomedical Applications*. Elsevier, Amsterdam, the Netherlands, 1993. (d) Surya Prakash, G. K., Yudin, A. K. Perfluoroalkylation with organosilicon reagents. *Chem. Rev.*, 97, 757, 1997. (e) Banks, R. E., Smart, B. E., Tatlow, J. C. *Organofluorine Chemistry: Principles and Commercial Applications*. Plenum Press, New York, 1994. (f) Gryshuk, A., Graham, A., Pandey, S. K., Potter, W. R., Missert, J. R., Oseroff, A., Dougherty, T., Pandey, R. K. A first comparative study of purpurinimide-based fluorinated vs. nonfluorinated photosensitizers for photodynamic therapy. *Photochem. Photobiol.*, 76, 555, 2002.

41. Pandey, R. K., Jagerovic, N., Ryan, J. M., Dougherty, T. J., Smith, K. M. Syntheses and preliminary in vivo photodynamic efficacy of benzoporphyrin derivatives from phylloerythrin and rhodoporphyrin XV methyl ester and aspartyl amides. *Tetrahedron*, 52, 5349, 1996.

42. Ma, L., Dolphin, D. Chemical modification of chlorophyll-a: Synthesis of new regiochemically pure benzoporphyrin and dibenzoporphyrin derivatives. *Can. J. Chem.*, 75, 262, 1997.

43. Mettath, S., Dougherty, T. J., Pandey, R. K. Cycloaddition reaction of 2-vinylemeraldines: Formation of unexpected porphyrins with seven member exocyclic ring systems. *Tetrahedron Lett.*, 40, 6171, 1999.

44. Mettath, S. Synthesis of new photosensitizers and their biological significance. PhD thesis, Roswell Park Graduate Division, SUNY, Buffalo, NY, January 2000.

45. Chang, C. K., Sotiriou, C. Migratory aptitudes in pinacol-pinacolone rearrangement of *vic*-dihydroxychlorins. *J. Heterocycl. Chem.*, 22, 1739, 1985 and references therein.

46. Pandey, R. K., Issac, M., MacDonald, I., Medforth, C. J., Senge, M. O., Dougherty, T. J., Smith, K. M. Pinacol-pinacolone rearrangements in vic-dihydroxychlorins and bacteriochlorins. Effect of substituents at the peripheral positions. *J. Org. Chem.*, 62, 1463, 1997.

47. Chang, C. K., Sotiriou, C., Weishih, W. Differentiation of bacteriochlorin and isobacteriochlorin formation by metallation. High yield synthesis of porphyrindiones via OsO_4 oxidation. *Chem. Commun.*, 1213, 1986.

48. Kessel, D., Smith, K. M., Pandey, R. K., Shiau, F. Y., Henderson, B. W. Photosensitization with bacteriochlorins. *Photochem. Photobiol.*, 58, 200, 1993.

49. (a) Pandey, R. K., Tsuchida, T., Constantine, S., Zheng, G., Medforth, C., Kozyrev, A., Mohammad, A., Rodgers, M. A. J., Smith, K. M., Dougherty, T. J. Synthesis, photophysical properties and in vivo photosensitizing activity of some novel bacteriochlorins. *J. Med. Chem.*, 40, 3770, 1997. (b) Joshi, P., Ethirajan, M., Goswami, L. N., Srivatsan, A., Missert, J. R., Pandey, R. K. Synthesis, spectroscopic, and in vitro photosensitizing efficacy of ketobacteriochlorins derived from ring-B and ring-D reduced chlorins via pinacol-pinacolone rearrangement. *J. Org. Chem.*, 76, 8629, 2011.

50. P. Zheng, G., Kozyrev, A. N., Dougherty, T. J., Smith, K. M., Pandey, R. K. Synthesis of novel bacteriopurpurinimides by Diels-Alder cycloaddition. *Chem. Lett.*, 1119, 1996.

51. Li, G., Graham, A., Potter, W. R., Oseroff, A., Dougherty, T. J., Pandey, R. K. unpublished results.

52. Beems, E. M., Dubbelman, T. M. A. R., Lugtenberg, J., Best, J. A. V., Smeets, M. F. M. A., Boefheim, J. P. J. Photosensitizing properties of bacteriochlorophyll-a and bacteriochlorin-a. *Photochem. Photobiol.*, 46, 639, 1987.

53. Henderson, B. W., Sumlin, A. B., Owczarczak, B. L., Dougherty, T. J. Bacteriochlorophyll-a as photosensitizer for photodynamic treatment of transplantable murine tumors. *J. Photochem. Photobiol.*, 10, 303, 1991.

54. Moan, J., Peng, Q., Evenson, J. F., Berg, K., Western, A., Rimington, C. Photosensitizing efficacy, tumor and cellular uptake of different photosensitizing drugs for photodynamic therapy. *Photochem. Photobiol.*, 46, 713, 1987.

55. Pandey, R. K., Herman, C. Shedding some light on tumors. *Chem. Ind. (Lond.)*, 739, 1998.

56. Kozyrev, A. N., Zheng, G., Dougherty, T. J., Smith, K. M., Pandey, R. K. Syntheses of stable bacteriochlorophyll-a derivatives as potential photosensitizers for PDT. *Tetrahedron Lett.*, 37, 6431, 1996.

57. Chen, Y., Graham, A., Potter, W. R., Morgan, J., Vaughan, L., Bellnier, D. A., Henderson, B. W., Oseroff, A., Dougherty, T. J., Pandey, R. K. Bacteriopurpurinimides: Highly stable and potent photosensitizer for photodynamic therapy. *J. Med. Chem.*, 45, 255, 2002.

58. (a) Chen, Y. et al. unpublished results. (b) Gryshuk, A. L., Chen, Y., Potter, W., Ohulchansky, T., Oseroff, A., Pandey, R. K. In vivo stability and photodynamic efficacy of fluorinated bacteriopurpurinimides derived from bacteriochlorophyll-a. *J. Med. Chem.*, 49, 1874, 2006.

59. Mody, T. R., Sessler, J. L. Texaphyrins: A new approach to drug development. *J. Porphyr. Phthalocyanines*, 5, 892, 1996.

60. Young, S. W., Woodburn, K. W., Wright, M., Modt, T. D., Fan, Q., Sessler, J. L., Dow, W. C., Miller, R. A. Lutetium texaphyrin (PC-0123): A near-infrared water-soluble photosensitizer. *Photochem. Photobiol.*, 63, 692, 1996.

61. (a) Young, S. W., Quing, F., Harriman, A., Sessler, J. L., Dow, W. C., Mody, T. D., Hemmi, G. W., Hao, Y., Miller, A. Gadolinium(III) texaphyrin: A tumor selective radiation sensitizer that is detectable by MRI. *Proc. Natl. Acad. Sci. U.S.A.*, 93, 6610, 1996. (b) Hannah, S., Lynch, V. M., Gerasimchuk, N., Maqda, D., Sessler, J. L. Synthesis of a metal-free texaphyrin. *Org. Lett.*, 3, 3911, 2001. (c) Lu, T., Shao, P., Mathew, I., Sand, A., Sun, W. Synthesis and photophysics of benzotexaphyrin: A near-infrared emitter and photosensitizer. *J. Am. Chem. Soc.*, 130(47), 15782, 2008. (d) Bauer, V. J., Clive, D. L. J., Dolphin, D., Paine, J. B. III, Harris, F. L., King, M. M., Loder, J., Wang, S. W. C., Woodward, R. B. Sapphyrins: Novel aromatic pentapyrrolic macrocycles. *J. Am. Chem. Soc.*, 105, 6429, 1983. (e) Springs, S. L., Gosztola, D., Wasielewski, M., Kral, V., Andreivsky, A., Sessler,

J. L. Picosecond dynamics of energy transfer in porphyrin-sapphyrin noncovalent assemblies. *J. Am. Chem. Soc.*, 1121, 2281, 1999. (f) Maiya, B. G., Cyr, M., Harriman, A., Sessler, J. L. Vitro photodynamic activity of diprotonated sapphyrin: A 22-π-electron pentapyrrolic porphyrin-like macrocycle. *J. Phys. Chem.*, 94, 3597, 1990. (g) Kral, V., Davis, J., Andrievsky, A., Kralova, J., Synytsya, A., Pouckova, P., Sessler, J. L. Synthesis and biolocalization of water-soluble sapphyrins. *J. Med. Chem.*, 45, 1073, 2002. (h) Hooker, J. D., Nguyen, V. H., Taylor, V. M., Cedeno, D. L., Lash, T. D., Jones, M. A., Robledo, S. M., Velez, I. D. New application for expanded porphyrins: Sapphyrin and heterosapphyrins as inhibitors of *Leishmania* parasites. *Photochem. Photobiol.*, 88, 194, 2012.

62. Allen, C. M., Sharman, W. M., van Lier, J. E. Current status of phthalocyanines in the photodynamic therapy of cancer. *J. Porphyr. Phthalocyanines*, 5, 161, 2001.

63. Olenick, N. L., Antunez, A. R., Clay, M. E., Rihter, B. D., Kenney, M. E. New phthalocyanine photosensitizers for photodynamic therapy. *Photochem. Photobiol.*, 57, 242, 1993.

64. Sherman, W. M., Allen, C. M., van Lier, J. E. Photodynamic therapeutics: Basic principles and clinical applications. *Drug Discov. Today*, 4, 507–517, 1999.

65. (a) Hu, Ma, Brasseur, N., Yildiz, S. Z., van Lier, J. E., Leznoff, C. C. Hydroxyphthalocyanines as potential photodynamic agents for cancer therapy. *J. Med. Chem.*, 41, 1789, 1998. (b) Jiang, X. J., Lo, P. C., Tsang, Y. M., Yeung, S. L., Fong, W. P., Ng, D. K. P. Phthalocyanine-polyamine conjugates as pH-controlled photosensitizers for photodynamic therapy. *Chem. Eur. J.*, 16, 4777, 2010. (c) Tuncel, S., Chabert, J. F., Albrieux, F., Ahsen, V., Ducki, S., Dumoulin, F. Towards dual photodynamic and antiangiogenic agents: Design and synthesis of a phthalocyanine-chalcone conjugate. *Org. Biol. Chem.*, 10, 1154, 2012. (d) de Oliveira, K. T., de Assis, F. F., Ribeiro, A. O., Neri, C. R., Fernandes, A. U., Baptista, M. S., Lopes, N. P., Serra, O. A., Iamamoto, Y. Synthesis of phthalocyanines-ALA conjugates: Water-soluble compounds with low aggregation. *J. Org. Chem.*, 74, 7962, 2009. (e) Liu, J. Y., Jiang, X. J., Fong, W. P., Ng, D. K. Highly photocytotoxic 1,4-dipegylated zinc(II) phthalocyanines. Effects of the chain length on the in vitro photodynamic activities. *Org. Biomol. Chem.*, 6, 4560, 2008. (f) Durmus, M., Nyokong, T. Synthesis, photophysical and photochemical studies of new water-soluble indium(III) phthalocyanines. *Photochem. Photobiol. Sci.*, 6, 659, 2007.

66. Aoudia, M., Cheng, G. Z., Kennedy, V. O., Kenney, M. E., Rodgers, M. A. J. Synthesis of a series of octabutoxy- and octabutoxybenzonaphthalocyanines and photophysical properties of two members of the series. *J. Am. Chem. Soc.*, 119, 6029, 1997.

67. He, J., Larkin, J. E., Li, Y. S., Rihter, B. D., Zaidi, S. I. A., Rodgers, M. A. J., Mukhtar, H., Kenney, M. E., Oleinick, N. L. The synthesis, photophysical and photobiological properties and in vitro structure-activity relationships of a set of silicon-phthalocyanine PDT photosensitizers. *Photochem. Photobiol.*, 65, 581, 1997.

68. (a) Schmidt-Erfurth, U., Diddens, H., Birngruber, R., Hasan, T. Photodynamic targeting of human retinoblastoma cells using covalent low density lipoprotein conjugates. *Br. J. Cancer*, 75, 54, 1997. (b) Finger, V. H., Guo, H. H., Lu, Z. H., Peiper, S. C. Expression of chemokine receptors by endothelial cells: Detection by intravital microscopy using chemokine-located fluorescent microspheres. *Methods Enzymol.*, 28, 148, 1997.

69. Pandey, R. K. Photosensitizers related to chlorins and bacteriochlorins. Effect of lipophilicity in PDT efficacy. *First International Conference on Porphyrins and Phthalocyanines*, Dijon, France, June 25–30, 2001, Abstract SYM 149.

70. Hombrecher, H. K., Schell, C., Thiem, J. Synthesis and investigation of galactopyranosyl-cholesteryloxy substituted porphyrins. *Bioorg. Med. Chem. Lett.*, 6, 1199, 1999.

71. Donald, P. J., Cardiff, R. D., He, D., Kendell, K. Monoclonal antibody-porphyrin conjugate for head and neck cancer, the possible magic bullet. *Head Neck Surg.*, 105, 781, 1991.

72. Hemming, A. W. et al. Photodynamic therapy of squamous cell carcinoma. An evaluation of a new photosensitization agent, benzoporphyrin derivatives and new photoimmunoconjugate. *Surg. Oncol.*, 2, 187, 1993.

73. Vrouenraets, M. B. et al. Development of *meta*-tetrahydroxyphenylchlorin-monoclonal antibody conjugate for photoimmunotherapy. *Cancer Res.*, 59, 1505, 1999.

74. Karagianis, G. et al. Biophysical and biological evaluation of porphyrin-bisacridine conjugates. *Anti-Cancer Drug Des.*, 11, 205, 1996.

75. Bachor, B. S., Shea, C. R., Gillies, R., Hasan, T. Photosensitized destruction of human bladder carcinoma cells treated with chlorine e_6-conjugated microspheres. *Proc. Natl. Acad. Sci. U.S.A.*, 88, 1580, 1991.

76. Ali, H., van Lier, J. E. Metal complexes as photo- and radiosensitizers. *Chem. Rev.*, 99, 2379, 1999.

77. Brasseur, N., Langlois, R., La Madeleine, C., Ouellet, R., van Lier, J. E. Receptor-mediated targeting of phthalocyanines to macrophages via covalent coupling to native or maleylated bovine serum albumin. *Photochem. Photobiol.*, 69, 345, 1999.

78. Allen, C. M., Sharman, W. M., La Madeleine, C., Weber, J. M., Langlois, R., Guellet, R., van Lier, J. E. Photodynamic therapy: Tumor targeting with adenoviral proteins. *Photochem. Photobiol.*, 70, 512, 1999.

79. Lutsenko, S. V., Feldman, N. B., Finakova, G. V., Posypanova, G. A., Severin, S. E., Skryabin, K. G., Kirpichnikov, M. P., Lukyanets, E. A., Vorozhtsov, G. N. Targeting phthalocyanines to tumor cells using epidermal growth factor conjugates. *Tumor Biol.*, 20, 218, 1999.

80. Sears, P., Wong, C. H. Carbohydrate mimetics: A new strategy for tackling the problem of carbohydrate-mediated biological recognition. *Angew. Chem. Int. Ed. Eng.*, 38, 2301, 1999.

81. Lee, Y. C., Lee, R. T. Carbohydrate-protein interactions: Basis of glycobiology. *Acc. Chem. Res.*, 28, 321, 1995.

82. Schell, C., Hombrecher, H. K. Synthesis and investigation of glycosylated mono- and diarylporphyrins for photodynamic therapy. *Bioorg. Med. Chem. Lett.*, 6, 1199, 1996.

83. Mikata, Y., Onchi, Y., Tabata, K., Ogure, S., Okura, I., Ono, H., Yano, S. Sugar-dependent photocytotoxic property of tetra- and octa-glycoconjugated tetraphenyl porphyrins. *Tetrahedron Lett.*, 19, 4505, 1998.

84. Cornia, M., Valenti, C., Capacchi, S., Cozzini, P. Synthesis, characterization and conformational studies of lipophilic, amphiphilic and water-soluble c-glyco-conjugated porphyrins. *Tetrahedron*, 54, 8091, 1998.

85. Millard, P., Guerquin-Kern, J.-L., Huel, C., Momemteau, M. Glycoconjugated porphyrins: Synthesis of sterically constructed polyglycosylated compounds derived from tetraphenylporphyrins. *J. Org. Chem.*, 58, 2774, 1993.

86. Ono, N., Bougauchi, M., Marutama, K. Water-soluble porphyrins with four sugar molecules. *Tetrahedron Lett.*, 33, 1629, 1992.

87. Sol, V., Blais, J. C., Bolbach, G., Carre, V., Granet, R., Guillonton, M., Spiro, M., Krausz, P. Toward glycosylated peptide porphyrins: A new strategy for PDT. *Tetrahedron Lett.*, 38, 6391, 1997; Synthesis, spectroscopy, and phototoxicity of glycosylated amino acid porphyrin derivatives as promising molecules for cancer therapy. *J. Org. Chem.*, 64, 4431, 1999.

88. Millard, P., Hery, C., Momemteau, M. Synthesis, characterization and phototoxicity of a glycoconjugated *meso*-monoarylbenzochlorin. *Tetrahedron Lett.*, 38, 3731, 1997.

89. Adams, K. R., Berenbaum, M. C., Bonnett, R., Nizhnik, A. N., Salgado, A., Valles, M. A. Second generation tumour photosensitizers: The syntheses and biological activity of octaethyl chlorines and bacteriochlorins with graded amphiphilic character. *J. Chem. Soc. Perkin Trans. 1*, 1465, 1992.

90. Montforts, F. P., Gerlach, B., Haake, G., Hoper, F., Kusch, D., Meier, A., Scheurich, G., Brauer, H. D., Schiwon, K., Schermann, G. Selective synthesis and photophysical properties of tailor-made chlorins for photodynamic therapy. *Proc. SPIE*, 29, 2325, 1994.

91. Fujimoto, K., Miyata, T., Aoyama, Y. Saccharide-directed cell recognition using monocyclic saccharide clusters: Masking of hydrophobicity to enhance specificity. *J. Am. Chem. Soc.*, 122, 3558, 2000.

92. Chiariotti, L., Berlingieri, M. T., De Rosa, P., Battaglia, C., Berger, N., Bruni, C. B., Fusco, A. Increased expression of the negative growth factor, galactoside-binding gene in transformed thyroid cells and in human thyroid carcinoma. *Oncogene*, 7, 2507, 1992.

93. Chiariotti, L., Salvatore, P., Benvenuto, G., Bruni, C. B. Control of galectin gene expression. *Biochimie*, 81, 381, 1999.

94. (a) Zheng, G., Graham, A., Shibata, M., Missert, J. R., Oseroff, A. R., Dougherty, T. J., Pandey, R. K. Synthesis of b-galactose conjugated chlorines by enyne metathesis as galectin-specific photosensitizers for photodynamic therapy. *J. Org. Chem.*, 66, 8709, 2001. (b) Zheng, X., Morgan, J., Pandey, S. K., Chen, Y., Tracy, E., Baumann, H., Missert, J. R., Pandey, R. K. Conjugation of 2-(1′-hexyloxyethyl)-2-devinylpyropheophorbide-a (HPPH) to carbohydrates changes its subcellular distribution and enhances photodynamic activity in vivo. *J. Med. Chem.*, 52, 4306, 2009.

95. (a) Srivatsan, A., Ethirajan, M., Pandey, S. K., Dubey, S., Zheng, X., Liu, T., Shibata, M., Missert, J., Pandey, R. K. Conjugation of cRGD peptide to chlorophyll-a based photosensitizer (HPPH) alters its pharmacokinetics with enhanced tumor-imaging and photosensitizing (PDT) efficacy. *Mol. Pharm.*, 8, 1186, 2011. (b) Stefflova, K., Li, H., Chen, J., Zheng, G. Peptide based pharmacomodulation of a cancer targeted optical imaging and photodynamic therapy agent. *Bioconjug. Chem.*, 18, 379, 2007.

3

Lasers in Dermatology

Lilit Garibyan
Harvard Medical School
and
Massachusetts General
Hospital

H. Ray Jalian
Massachusetts General
Hospital

Kittisak
Suthamjariya
Mahidol University

R. Rox Anderson
Harvard Medical School
and
Massachusetts General
Hospital

3.1 Historical Introduction

Long after Einstein proposed the theoretical concept of stimulated photon emission, Maiman built the first laser in 1960.[1] Dr. Leon Goldman then pioneered the use of lasers in dermatology.[2,3] Treatment of basal cell carcinoma with ruby and neodymium-doped yttrium aluminum garnet (Nd:YAG) lasers and tattoo removal with Q-switched ruby laser were reported in 1964 and 1965, respectively.[4,5] However, surgical applications were limited because the high-power pulses emitted by these pulsed lasers were unable to perform controlled vaporization or coagulation of tissues. Ruby and other pulsed lasers were essentially abandoned when argon-ion and carbon dioxide (CO_2) lasers were widely applied for coagulation and cutting during the 1970s. A new era began in dermatology during the early 1980s when Anderson and Parrish proposed a theory for target-selective injury with pulsed lasers, called selective photothermolysis.[6] A 450 μs yellow dye laser was initially developed for treating port-wine lesions, the first laser cavity design intrinsically motivated by a medical need.[7] With this, pulsed lasers reentered dermatology. At present, selective photothermolysis is the basis for many low-risk treatments including vascular lesions, pigmented lesions, hypertrichosis, and photoaged skin. During the 1990s, laser *resurfacing* was also developed and became very popular using CO_2 or erbium lasers that vaporize a superficial layer of skin. Despite excellent results in most patients, the occurrence of side effects led to the recent development of *fractional* laser resurfacing. The same lasers, when focused to produce an array of very small zones of injury, stimulate skin remodeling with a lower risk of side effects. At present, lasers are a mainstay in the practice of dermatology. The potentials for lasers that *target* lipid-rich structures, for laser-assisted drug delivery, for normalization of scars, and for diagnostic laser microscopy of skin are current topics for research and development.

3.2 Skin Optics

Optical penetration into the skin is determined by a combination of absorption and scattering. The principal chromophores (absorbing molecules) for optical radiation in the skin are water, melanins, hemoglobins, and lipids. When absorption occurs, photon energy is transferred to the chromophore. The excited chromophore can dissipate this energy in various ways, including a photochemical reaction, heat, or reemission of light. Absorption spectra of the major chromophores in the skin throughout the ultraviolet (UV), visible, and infrared (IR) regions are shown in Figure 3.1. As shown in the figure, each major chromophore in the skin has characteristic bands of absorption at certain wavelengths, and this allows for the delineation of specific targets for laser treatment.

Absorption is needed for tissue effects to occur. The absorption coefficient (μ_a) is defined as the probability per unit path length that a photon at a particular wavelength will be absorbed. Absorption depends on the concentration of chromophores present. Scattering occurs when the photon changes its direction of propagation. About 4%–7% of light is reflected upon striking the skin surface because of the sudden change in refractive index between air and stratum corneum.[8] The remaining light penetrates into the skin and can be either absorbed or scattered by molecules, particles, and structures in the tissue.

Absorption is the primary process limiting penetration of UV and visible light through the epidermis. Proteins, melanin, urocanic acid, and DNA absorb UV wavelengths shorter than about 320 nm. For wavelengths from 320 to 1200 nm, especially at the shorter wavelengths, melanin is the primary absorber

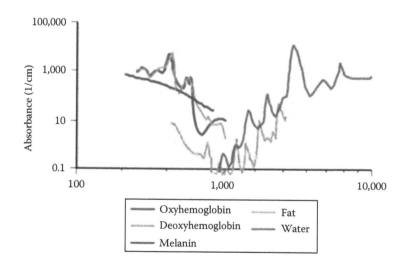

FIGURE 3.1 **(See color insert.)** Absorption spectra of major skin chromophores. (From Hruza, G. and Avram, M., *Lasers and Lights Book*, 3rd edn.; Modified from Sakamoto, F.H. et al., Lasers and flashlamps in dermatology, in Wolff, K. et al., eds., *Fitzpatrick's Dermatology in General Medicine*, vol. II, The McGraw-Hill Companies, Inc., Columbus, OH, 2007, pp. 2263–2279.)

of light in the epidermis. Water is the dominant skin chromophore for mid- and far-IR wavelengths. In the dermis, scattering by collagen fibers is of major importance in determining penetration of light.

In general, scattering is an inverse function of wavelength, which accounts in part for a large increase in optical penetration into the dermis with increasing wavelength from UV through visible and near-IR wavelengths. Absorption in the dermis is low throughout the visible and near-IR spectrum, except in blood vessels and pigmented hair follicles. Blood has strong absorption in the blue, green, and yellow wavelengths and a weak but significant absorption band in the 800–1000 nm region. An optical *window* exists between 600 and 1300 nm, which are the most-penetrating wavelengths. The least-penetrating wavelengths are in the far UV and far IR due to protein and water absorption, respectively. For visible and near-IR wavelengths, the exposure spot size also affects the effective treatment penetration depth of a laser beam into the skin, because of scattering.[9] Over this spectral region, larger spot sizes have a greater effective depth of penetration.

3.3 Laser–Skin Interactions

The Grotthuss–Draper law states that no effect of light occurs without absorption. In other words, laser light can only impose a tissue effect when it is absorbed by chromophores. Three basic effects occur upon absorption of the laser light by the chromophore: photothermal effects, photochemical effects, and photomechanical effects. In practice, these modes of laser interaction frequently coexist, but one or two usually dominate. Photothermal effects are the mainstay of current clinical laser use. The majority of current laser applications in dermatology use photothermal effects through absorption of optical energy with resultant production of heat.

3.3.1 Photothermal Interactions

In contrast to photochemistry, which requires a photon quantum energy larger than the activation energy for a chemical change, heating can occur after absorption at any wavelength. Thermal effects on tissue are both time and temperature dependent. The Arrhenius model states that the rate of thermal denaturation of a given molecule is exponentially related to temperature.[10] Therefore, the accumulation of denatured material rises exponentially with temperature and proportionally to time.[11] Temperature is an expression of average molecular kinetic motion. At high temperatures, the specially configured macromolecules necessary for life are shaken open, resulting in loss of function. Most proteins are denatured above 60°C and DNA above 70°C. Type I collagen, the major protein in the dermis, has a sharp melting transition for the fibrillar form near 70°C, causing irreversible coagulation and shrinkage of the connective tissue matrix. Above 100°C, tissue water is vaporized. If desiccated tissue is heated further, carbonization occurs.

3.3.2 Selective Photothermolysis

The term *selective photothermolysis* was coined to describe site-specific, thermally mediated injury of pigmented tissue targets by pulses of radiation. This technique relies on selective absorption of a brief optical pulse to generate and confine heat at certain pigmented targets. There are three basic requirements to achieve selective photothermolysis. First, the wavelength used must reach and be preferentially absorbed by the targeted structure. This can be achieved by choosing a wavelength within absorption bands for a chromophore associated with the target, for example, melanin in a pigmented hair follicle or hemoglobins in a blood vessel. Second, a sufficient fluence (fluence = energy/area) is required to achieve a damaging temperature in the targets. Third, the pulse duration or exposure time must be about equal to or less than the time needed for the targets to cool. As soon as heat is created in the chromophores, it begins to dissipate by conduction. Thus, a competition between active heating and passive cooling determines how hot the targets become. One useful concept is that of the *thermal relaxation time*, defined as the time needed for substantial cooling of a target structure. When the laser exposure is less than the thermal relaxation time for a given light-absorbing target, maximal thermal confinement occurs. Thermal relaxation time is proportional to the square of the targets' size and also to shape. Thus, small objects cool much faster than large ones. For a given diameter, spheres cool faster than cylinders, which cool faster than planes. For most tissue targets, a simple rule of thumb can be used: the thermal relaxation time in seconds is about equal to the square of the target dimension in millimeters.

Many of the cutaneous target structures treated by selective photothermolysis are heterogeneous, such that the light-absorbing chromophore is not identical to the actual target. For example, blood vessel walls are the actual targets when treating a vascular birthmark, rather than the light-absorbing blood contained within these vessels. Because heat must flow from blood to the surrounding vessel wall, ideally one should use a pulse duration approximately equal to the thermal relaxation time of the entire vessel structure. Laser hair removal is a similar example, in which the heat generated by light absorption in the pigmented hair shaft must flow into surrounding structures of the hair follicle.

3.3.3 Photomechanical Interactions

Photomechanical, sometimes called photoacoustic, interactions are characterized by light-induced stress and strain, leading to mechanical disruption of tissue structures, organelles, membranes, and cells. Rapid thermal expansion, high-pressure stress waves, low-pressure tensile waves, cavitation, and local vaporization can occur when laser pulses are absorbed by tissue. Rapid heating with thermal expansion and vaporization is often involved. For example, hemorrhage occurred when a 1 μs 577 nm pulsed-dye laser (PDL) was tested for the treatment of vascular lesions, prior to the present generation of longer PDLs.[12] In contrast, mechanical effects are a primary therapeutic mechanism for the treatment of pigmented lesions and tattoo removal with Q-switched lasers (QSLs). These lasers emit nanosecond-domain pulses with up to billions of watts of power.

3.3.4 Photochemical Interactions

Classic photobiology of skin is based on photochemical reactions. The major applications for laser-induced photochemistry at present are 308 nm excimer laser UV phototherapy and photodynamic therapy (PDT). PDT involves the therapeutic use of photochemical reactions mediated through the interaction of photosensitizing agents, light, and oxygen for the treatment of malignant or benign diseases. PDT is a two-step procedure. In the first step, the photosensitizer is administered to the patient and is absorbed by the target tissues. The second step involves activation of the photosensitizer, with a specific wavelength region and fluence of light delivered to the target tissue. In general, PDT drugs transfer the excitation energy to oxygen, creating a reactive intermediate, singlet oxygen. Selectivity for tissue damage during PDT therefore depends on the distributions of the three factors necessary for PDT: drug, light, and oxygen.

3.4 Laser Safety

The potential hazards from laser uses are injury to the skin and eye, aerosolization of hazardous biologic materials, ignition of fires, and electrocution. Eye injuries result from absorption of laser light by the structures in the eye, including the retina, lens, sclera, and cornea. Lasers operating between 400 and 1400 nm are particularly dangerous to retina. Wavelengths between 295 and 320 nm and wavelengths between 1 and 2 μm may injure the lens. Injury to the cornea is possible from the wavelengths in the UV and most of the IR at wavelengths beyond 1400 nm.[13] Appropriate safety goggles that filter specific wavelengths of laser light must be worn at all times while the laser is in operation. Skin burns can be caused directly by laser beam exposures or by laser-ignited fires. The smoke plume generated by laser vaporization of tissue may contain bacteria, viral DNA, or viable cells, with a possibility of infective potential.[14-16]

Protective cylinders and shields that attach to the end of the laser handpiece should be used to contain fumes, vaporized particles, and splattered tissue. Specifically designed laser filter facemasks should be used during a tissue-ablating laser procedure, and good local exhaust ventilation has to be utilized. Fires ignited by laser exposure are a major hazard, especially in the presence of oxygen and flammable items. Lasers can also pose electrical and chemical hazards.

3.5 Lasers for Vascular Skin Lesions

3.5.1 Mechanisms of Action

Laser wavelengths absorbed by hemoglobin are used to treat cutaneous vascular lesions such as port-wine stains (PWSs), hemangioma, spider angioma, cherry angioma, venous lake, and telangiectasia of face and legs. The basic objective in treating vascular lesions is to irreversibly damage abnormal blood vessels while sparing normal skin tissue. Argon and other continuous and quasi-continuous visible light lasers were first used to treat PWS. The risk of scarring in the pink PWS of children was high.[17] The PDL was then developed based on the concept of selective photothermolysis.[6] A flashlamp-pumped, tunable dye laser near 595 nm (0.5–3 ms pulse duration) is now a standard method for treating PWS because of a very low risk of scarring when used in single pulses at 6–8 J/cm² fluence. With its short-pulsed duration, this laser produces transient purpura because of hemorrhage and a delayed vasculitis. Purpura occurring immediately after laser treatment is generally due to blood vaporization, vessel rupture, and hemorrhage. A number of other tunable-pulsewidth sources including pulsed dye (585 nm), frequency-doubled Nd:YAG (532 nm), alexandrite (755 nm), Nd:YAG (1064 nm), diode (800 nm), and flashlamps (500–1200 nm) are now available. In general, good results can be achieved with pulse durations, which nearly match the thermal relaxation time of the target vessels, using sufficient fluence in combination with active skin cooling methods, described in the following.

Wavelength affects the depth of vascular lesion treatment and uniformity of heating luminal blood. Generally, the longer the laser wavelength up to 1200 nm, the deeper the penetration depth,[6,18,19] due to lower scattering and absorption. At wavelengths weakly absorbed by hemoglobin, such as 700–1100 nm, high treatment fluences are needed. Spot size also plays a significant role in determining the depth of effective treatment. The intensity of a narrow beam (small spot size) tends to suffer greater loss with depth, as light is scattered outside the beam diameter. Larger exposure spots are therefore desirable when target vessels are deep in the skin. Another significant issue is the number of pulses delivered to a single skin site. The Arrhenius model suggests that thermal injury is cumulative over time; therefore, in theory, multiple lower-fluence pulses that do not cause hemorrhage might be used to accumulate selective, gentler, and more complete damage to microvessels. This has been reported in animal and human studies for treatment of vascular lesions.[20,21]

3.5.2 Port-Wine Stain

The PWS is a congenital malformation of dermal microvasculature present in 0.3%–0.5% of newborns, which persists throughout life. It commonly occurs on the face and is often found in distribution within a single or adjacent sensory nerve territories. Histologically, PWS reveals an increased number of venules in the papillary and upper reticular dermis but may extend throughout the skin and subcutaneous tissues.[22] PWS darken progressively with age, from pink to red to deep violet, and is frequently associated with progressive hypertrophy and raised blebs.[23] The tissue hypertrophy can be quite dramatic and involve underlying structures, leading to marked asymmetry and functional disability. The natural darkening and thickening correlate with vessel ectasia thought to be due to decreased sympathetic innervation.

Currently, treatment of PWS with the PDL is the gold standard. Approximately 50%–75% lightening of PWS is achieved within two to three treatments with PDL therapy. Complete resolution of the lesions can be achieved with repetitive laser treatments in about half of patients.[24] Immediate purpura in the treated areas is the treatment end point and takes 10 or more days to subside. Side effects from PDL treatment of PWS occur infrequently. In a study of 701 PWS patients with 3877 PDL treatments, hyperpigmentation was the most frequent side effect observed (9.1%) followed by blistering, atrophic scar (4.3%), hypopigmentation, crusting, and hypertrophic scar (0.7%).[25]

Although PDLs at present have been established as the treatment of choice for PWS, the results may be variable and unpredictable.[33] Also hypertrophic PWSs, especially in adult patients, are typically less responsive to PDLs. The natural course of PWS is to darken and dilate over years, and residual PWS vessels follow this course.

Even with continuous treatment, some PWSs might become unresponsive to PDL over time, a phenomenon termed *treatment resistance*.[26] One of the factors proposed for treatment resistance or incomplete clearance of PWS with PDL is the inability of this laser to target deeper residing ectatic vessels in the skin. In order to achieve a combination of vascular selectivity and deeper penetration that theoretically may be superior to PDL treatment, near-IR lasers at wavelengths with greater tissue penetration are now being utilized for the treatment of hypertrophic and resistant PWS.[27,28] While more deeply penetrating, these near-IR wavelengths have lower absorption coefficients with hemoglobin, and thus higher fluences are needed for vessel damage.[29] Of the commercially available near-IR lasers with vascular selectivity, reports exist for the use of both the 755 nm alexandrite laser and the 1064 nm Nd:YAG in the treatment of deeply situated PWS. However, use of the Nd:YAG has an increased risk of scarring, explained by its higher absorption coefficient by oxygenated hemoglobin, which is more selective for arterioles.[26] In contrast, the alexandrite laser has relative venous selectivity, and a recent case series demonstrates its utility for the treatment of hypertrophic and treatment-resistant PWS in adult and pediatric patients with a very low rate of complications.[29]

The recurrence of PWS after laser therapy is thought to be due to (1) recanalization of high-flow microvessels and (2) reformation (angiogenesis) of blood vessels. After photocoagulation of blood vessels within the skin, there is relative tissue anoxia. This low oxygen tension in the tissue leads to activation of hypoxia-induced factor-1α, which results in initiation of angiogenesis through various mechanisms, including the upregulation of vascular endothelial growth factor (VEGF). In an effort to attenuate this posttreatment tissue response, medications that block various steps of the angiogenesis pathway have been administered topically or systemically in conjunction with laser irradiation. One promising compound, rapamycin, binds the mammalian target of rapamycin (mTOR), which is integral to hypoxia-induced angiogenesis. One report with promising preliminary evidence compared PDL treatment alone with PDL treatment combined with oral rapamycin. This led to significantly enhanced PWS blanching and slower recurrence over time, even after discontinuation of rapamycin. Rapamycin, which has been shown to inhibit the growth of vascular endothelial cells and smooth muscle cells essential for new blood vessel formation,[30] likely prevents PWS blood vessel reformation and reperfusion after the PDL therapy.[31] Imiquimod, which is a topical immune modulator with antiangiogenic properties, used for treatment of external genital warts, basal cell carcinoma, and vascular proliferative lesions, is another topical agent, which when used in combination with PDL can lead to more enhanced PWS treatment efficacy when compared to PDL alone.[32] Other candidate medications including selective VEGF inhibitors and calcineurin inhibitors are under investigation to inhibit this hypoxia-induced response.

3.5.3 Hemangioma

Infantile hemangiomas are the most common vascular tumors of infancy affecting about 5% of all infants. They typically undergo a rapid growth phase during the first year of life followed by slow complete or partial involution. Histologically, there are dense collections of thin-walled vessels interspersed with sheets of vascular endothelial cells. Because of the natural course with spontaneous involution of hemangioma, conservative management has been advocated for the majority of these lesions. Although most of the hemangiomas will resolve spontaneously, a significant proportion can lead to cosmetic disfigurement and functional impairment, making treatment imperative. Hemangiomas have been treated with variable successes, using compression therapy, ionizing radiation, intralesional or systemic corticosteroids, interferon alfa, cryosurgery, embolization, and various laser therapies. The greatest advance in treating these lesions has not come with optimization of physical treatment modalities, but rather advances in medical therapy. Leaute-Labreze et al. first made the astute but fortuitous observation in 2008 that propranolol, a beta-blocker, is useful in the management of hemangiomas.[33] Since this seminal observation, hundreds of reports and studies have substantiated the efficacy of propranolol.[34,35] Promising results from much smaller studies and individual case reports have also been published showing efficacy of topical timolol, a nonselective topical beta-blocker, for management of hemangiomas.[36–38]

PDL treatment of infantile hemangiomas remains somewhat controversial, although clearly useful for superficial and/or ulcerated lesions and lesions poorly responsive to medical treatment. In one study, total regression was achieved in nearly half of the small superficial hemangiomas; however, a tightly controlled study of these self-regressing lesions is still lacking.[39] Clinically, treatment with PDL appears to be effective and may be the treatment of choice for small superficial cutaneous hemangiomas. Laser treatment is generally accepted for control of ulcerated hemangiomas, leading to a decrease in pain, promotion of healing, and acceleration of involution.[40–42] One to three laser treatments have been reported to heal most ulcerated lesions.[40,41] However, laser treatment can also induce ulceration and scarring, as well as hypopigmentation. Hemangiomas with a deep component do not benefit from PDL treatment due to the limitation of the laser penetration depth. Early therapeutic intervention with PDL may or may not prevent proliferative growth of the deeper or subcutaneous component of the hemangioma.[43,44] More recently, fractional photothermolysis (FP), discussed in the following, has effectively demonstrated clinical improvement in the residual fibrofatty change associated with involuted hemangiomas.[45]

Perhaps the greatest utility of lasers in the treatment of hemangioma will be as an adjunct to systemic treatments, such as propranolol.[46] Several studies are under way to determine if there is synergism between laser and medical therapies (Figure 3.2).

3.5.4 Telangiectasias and Dilated Leg Veins

Telangiectasias are common, dilated superficial blood vessels associated with aging, chronic sun damage, scars, malformations, acne rosacea, and other conditions. Dilated veins are also common in both sexes, due to poor venous return from incompetent vein valves, pregnancy, prolonged standing or sitting, aging, and heredity. Sclerotherapy, which consists of local injection of substances that kill

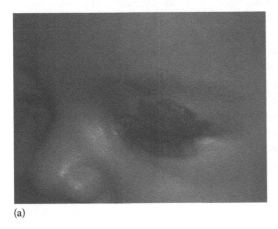
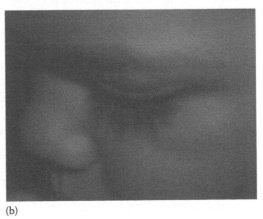

(a) (b)

FIGURE 3.2 (See color insert.) (a) A hemangioma of the left upper eyelid prior to treatment. (b) After a series of eight treatments with the 595 nm pulsed-dye laser at 10–15 J/cm² and 7 mm spot size, the hemangioma cleared significantly.

endothelial cells lining blood vessels, remains the gold standard of treatment for telangiectasias of the lower extremities. Facial telangiectasias are less responsive to sclerotherapy than those located on the leg and are more prone to complications.[47] Therefore, laser treatment in general has a greater role for treating facial telangiectasias.

Red linear and arborizing telangiectasias frequently occur on the face, particularly on the nose, cheeks, and chin. They measure 0.1–1.0 mm in diameter and represent either a dilated venule, a capillary, or an arteriole.[48] Immediately after appropriate laser treatment with green, yellow, or IR laser pulses between about 10 and 50 ms, the abnormal vessels appear subtly darker and coagulated or simply disappear from view. Broadband flashlamp light sources have also emerged as an effective treatment for telangiectasia and facial erythema, with equivalent safety and efficacy outcomes as compared to PDL.[49,50]

Laser treatment of leg telangiectasia is suboptimal. Traditional PDLs are ineffective for vessels greater than 0.2 mm in diameter.[51] Based on theoretical considerations, longer wavelengths and longer pulse durations would be more appropriate for treatment of most leg veins. Longer wavelengths were needed to penetrate deeper into the skin, and longer pulsewidth produces less mechanical injury, with more uniform thermal damage of large vessels and without purpura. Vessel size and depth largely determines the appropriate choice of laser for treatment. Studies have suggested that near-IR, millisecond-domain alexandrite, diode, and Nd:YAG lasers are considerably superior to visible lasers for treating vessels larger than about 1 mm diameter, but less effective for small vessels. The near-IR 1064 nm laser provides good absorption by hemoglobin and deeply penetrates into the skin to the depth of a few millimeters, thus allowing targeting of deeper and larger leg veins. However, this laser tends to target arterial more than venous blood. Selective photothermolysis optimized for venous vessels may be possible by taking advantage of the differences in relative absorption of deoxyhemoglobin compared to oxyhemoglobin. In fact, preferential targeting of venous blood was shown to be possible in treatment with 630–780 nm sources using human blood samples at known oxygen saturation levels in vitro.[52]

3.5.5 Hypertrophic Scars, Keloids, and Striae

There are many reports of hypertrophic scars, keloids, and striae distensae (*stretch marks*) treated with PDL.[53–55] It is generally accepted that the PDL helps treat the erythema associated with these lesions. Striae may respond better to lower fluences (3.0 J/cm^2), in one or two treatments.[56]

3.6 Laser Treatment of Pigmented Lesions (Melanin)

Melanocytes at the base of the epidermis synthesize melanin in the form of 0.5 μm intracellular organelles, called melanosomes. The thermal relaxation time of a melanosome is somewhat less than 1 μs. Selective photothermolysis of melanosomes was first described in 1983,[6] using 351 nm, sub-microsecond excimer laser pulses at less than 1 J/cm^2 fluence. Melanin absorbs across the UV, visible, and near-IR spectrum with decreasing absorption at longer wavelengths.[7] Across this wide spectral range, any short-pulsed laser with sufficient pulse energy can cause thermal injury due to absorption by melanin.

Target specificity of laser pulses depends not only on wavelength but also on pulsewidth. The appropriate pulsewidth depends primarily on the size of the target and its thermal relaxation time. The calculated thermal relaxation time for melanosomes is about 250–1000 ns. Sub-microsecond or even femtosecond laser pulses cause individual pigmented cell death by violent cavitation, a microscopic steam bubble erupting around each melanosome after the laser pulse, but longer pulsewidths, in the hundreds of microsecond domain, do not appear to cause specific melanosome damage.[57] The immediate effect of sub-microsecond near-UV, visible, or near-IR laser pulses in pigmented skin is immediate whitening, which fades away after several minutes. Immediate skin whitening correlates well with melanosome rupture seen by electron microscopy[58] and is presumably due to residual gas bubbles after cavitation of melanosomes. Nearly identical but deeper whitening occurs with QSL exposure of tattoos. This offers a clinically useful immediate end point. Larger pigmented targets, such as hair follicles and *nests* of congenital nevus cells, are more effectively treated with much longer pulses, for example, in the millisecond domain. The clinical importance of matching laser pulsewidth with the intended pigmented targets' thermal relaxation time is well illustrated by comparing skin responses to Q-switched (ns) versus long-pulse (ms) ruby lasers, both at a wavelength of 694 nm. At typical fluences, the Q-switched ruby laser is excellent for treating nevus of Ota, in which individual pigmented cells are the target, but incapable of permanent hair removal. In contrast, the long-pulse ruby laser is capable of permanent hair removal, but has little effect on nevus of Ota.

3.6.1 Epidermal Pigmented Lesions

Epidermal pigmented lesions are typically responsive to many different treatments, which damage the epidermis but spare the dermis, including lasers, cryotherapy, and electrosurgery. These common lesions include freckles, benign lentigo, and café-au-lait macules. Pigment-selective sub-microsecond pulsed lasers include green light lasers (510 nm pulsed dye, 532 nm frequency-double Nd:YAG), red light lasers (694 nm ruby and 755 nm alexandrite), and near-IR lasers (1064 nm Nd:YAG).[59-62] Hyperpigmentation and hypopigmentation are common posttreatment side effects. In general, postinflammatory hyperpigmentation is most frequent[60,63] in darker skin types and sun-tanned individuals. The risk of hypopigmentation may be somewhat higher after Q-switched ruby laser than after Q-switched alexandrite or Nd:YAG lasers (Figure 3.3).

3.6.2 Dermal and Mixed Epidermal/Dermal Pigmented Lesions

Dermal melanocytosis include nevus of Ota, nevus of Ito, and mongolian spot. The Q-switched ruby, Q-switched alexandrite, and Q-switched Nd:YAG lasers are all highly effective and are the treatment of choice for these pigmented lesions. Significant or complete clearing of lesions can be achieved after an average of four to six treatment sessions.[64-67] Hypopigmentation is the most common complication encountered. Once the nevus of Ota has fully cleared after laser treatment, the results are usually permanent. Acquired bilateral nevus of Ota-like macules (ABNOM), also called nevus fuscoceruleus zygomaticus or nevus of Hori, is a dermal melanocytic lesion characterized histopathologically

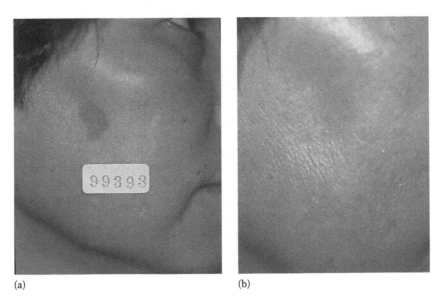

(a) (b)

FIGURE 3.3 (a) Senile lentigo at the right cheek of a 50-year-old woman prior to treatment. (b) Six months after one single treatment with a 532 nm Q-switched Nd:YAG laser, the lesion faded almost completely.

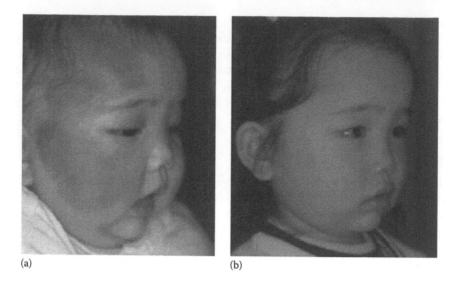

(a)   (b)

FIGURE 3.4 **(See color insert.)** (a) A 4-month-old girl with nevus of Ota on her right face prior to treatment. (b) Almost complete clearing after five treatments with Q-switched alexandrite laser.

by the presence of elongated melanocytes in upper dermis similar to that found in nevus of Ota. They have been reported to be effectively treated with Q-switched ruby and Q-switched Nd:YAG lasers.[68–70] Postinflammatory hyperpigmentation may result from any inflammatory insult causing dermal melanin deposition. Pigment-specific lasers can sometimes improve these lesions, but results are often disappointing because laser treatment itself can trigger postinflammatory hyperpigmentation in susceptible individuals (Figure 3.4).[126]

3.6.3 Nevomelanocytic Nevus

Nevomelanocytic nevi are usually pigmented *moles* consisting of melanocyte-related cells present at the base of the epidermis (junctional nevi), within the dermis (dermal and congenital nevi), or both (compound nevi). The lesions may be acquired or congenital and exhibit cellular atypia ranging from benign to malignant (melanoma arising within a nevus). The use of pigment-specific lasers to remove congenital and acquired melanocytic nevi remains controversial. The standard method of nevus removal is surgical excision with complete histologic evaluation. For the time being, laser treatment should be considered only in selected cases in which surgical excision cannot be performed or it may result in unsightly scars. The largest longitudinal studies for efficacy and safety of treatment of congenital nevi with pigment-specific lasers come from Japan. Numerous studies demonstrate the utility of combined long-pulsed pigment-specific lasers and Q-switched pigment lasers for the treatment of medium to large congenital nevi. In addition to good cosmetic results, no evidence indicates increased incidence of malignant transformation in these lesions.[71,72]

The effects of pigment-specific laser pulses on melanocytes with the potential for malignant transformation are not well understood. Removing the superficial portion of nevi might be beneficial, by decreasing the number of melanocytic cells capable of malignant degeneration.[150,153] On the other hand, healing after laser treatment may stimulate or select for persistent nevus cells with neoplastic change. To date, there have been no reports of melanoma arising in benign pigmented lesions treated with lasers. However, patients who have had a nevomelanocytic lesion removed by laser must continue to be followed, and any signs of recurrence should be approached with biopsy to rule out malignancy. The authors suggest that currently lasers should not be used in the treatment of atypical nevi, which are considered precursor lesions to melanoma.[73]

3.6.4 Tattoos

Tattoos consist of phagocytosed nanoparticles of ink, trapped in the lysosomes of phagocytic dermal cells, mostly fibroblasts, macrophages, and mast cells. Most tattoos can effectively be removed or substantially lightened using QSL, with a low risk of scarring. The mechanisms of tattoo ink removal involve fracture of ink granules, followed by release from cells into the dermis. Some of the ink is removed by lymphatic and transepidermal elimination, but most of it is rephagocytosed by somatic dermal cells within a few days.[74] In general, different laser wavelengths are needed to remove different tattoo ink colors. Black tattoo ink, which is typically carbon or black iron oxide, has a broad absorption spectrum and is well treated by either Q-switched ruby, Q-switched alexandrite, or Q-switched Nd:YAG lasers. Red and orange tattoo inks can be removed with green pulses (510 nm pulsed dye or 532 nm Q-switched frequency-double Nd:YAG lasers). For green colors, Q-switched ruby and to some extent Q-switched alexandrite lasers can usually achieve successful clearing, but the Q-switched Nd:YAG laser is generally ineffective.[75–77] Green and yellow inks are usually the most difficult to remove. Resistant tattoos can be ablated with CO_2 or erbium-doped yttrium aluminum garnet (Er:YAG) lasers, with scarring. In general, the responses of tattoos to laser treatment are fluence dependent, and amateur tattoos require fewer treatments than professional tattoos.[78,79] Traumatic tattoos usually contain less pigment than others and require even fewer treatments. Q-switched Nd:YAG laser is preferred in darkly pigmented patients with black tattoos, because of fewer epidermal side effects due to melanin absorption. Pulse duration also affects the treatment result. For example, picosecond laser pulses were more efficient than nanosecond laser pulses for tattoo removal.[80] The more efficient removal of tattoos by the picosecond laser is explained again by the theory of selective photothermolysis. Since most of the common tattoo ink particles have a particle size of about 0.1 μm, then based on this theory thermal relaxation time of under 10 ns can more selectively target the tattoo ink. Thus, picosecond lasers, which have sub-nanosecond pulsewidth, have been shown to be more effective than the Q-switched nanosecond lasers.[81,82] Despite the known potential of the picosecond laser being more effective in removing tattoos, it has become available for use in clinics only recently because of difficulties associated with production of a commercially viable and stable version of a picosecond laser.[83,84] Another new method recently introduced for more efficient tattoo removal is called the *R20 method*. This method involves performing three to four laser treatments in one session separated by 20 min. The result of the R20 method is 50%–85% clearing in a single session without additional side effects.[85]

Cosmetic tattoos containing red iron oxide (Fe_2O_3) and titanium oxide (TiO_2) can turn black on exposure by QSL.[86] Ink darkening is probably a combination of chemical reduction and changes in particle size. A similar reaction has been reported as localized chrysiasis after Q-switched ruby laser treatment in patients receiving parenteral gold therapy.[87] Ross et al. reported a significant association between the presence of titanium dioxide in tattoos and poor response to laser treatment (Figure 3.5).[88]

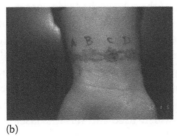
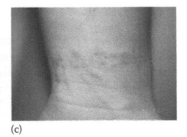

(a)   (b)   (c)

FIGURE 3.5 (a) Multicolored tattoo prior to treatment. (b) Site "C" after three treatments with Q-switched ruby (694 nm) laser shows loss of green and black ink but not red. Other sites treated with Q-switched Nd:YAG (1064, 532 nm) laser show loss of red and black inks, but not green. (c) After six further treatments using both lasers on the entire tattoo, all ink colors have substantially faded.

3.6.5 Laser Hair Removal

There are many medical and surgical settings in which excessive hair growth is a crucial problem. Hirsutism often is associated with ovarian, adrenal, or pituitary disorders. Impressive hypertrichosis may be part of a congenital syndrome or drug-induced.[89] Hair removal methods include shaving, plucking, chemical depilatories, waxing, electrolysis (direct current [dc]), electrothermolysis (RF current), and laser hair removal. Only the last three methods are known to produce permanent hair loss. Women with hirsutism often require hormonal and medical treatment, in addition to laser hair removal.

There are two potentially important targets for permanent hair removal: the *bulb* and *bulge* areas of hair follicles. The bulb is a neurovascular papilla at the base of anagen (active growth phase) hair follicles, responsible for hair shaft growth, and typically about 3–7 mm below the skin surface. The bulge or stem cell area is a region near the insertion of arrector pili muscle, about 1.5 mm below the skin surface, and is the source of new matrix cells with each hair cycle.[90,91] These stem cells appear to be necessary for cycling of the hair follicles into anagen phase. Perifollicular vessel damage may also play a role in the laser hair removal process.[92]

Photothermal destruction of hair follicle relies on melanin in the hair shaft and the matrix to provide local absorption of a pulse of light, which is timed to produce heating and destruction of the follicles. The wavelength region capable of reaching follicular targets, and with sufficiently strong and selective absorption by melanin, lies between about 600 and 1100 nm. Pulsed optical sources that operate in these wavelength region include 694 nm ruby laser, 755 nm alexandrite laser, 800 nm diode laser, 1064 nm Nd:YAG laser, and xenon flashlamp, an incoherent source spectrally filtered to emit in this spectrum. These optical devices for hair removal have pulse durations in the millisecond domain. Laser hair removal requires destructive heating of hair follicles due to melanin absorption, without significant injury to the epidermis, which also contains melanin. For patients with fair skin and darkly pigmented hair, there is a wide range of safe and effective laser fluences using any of these devices. For darkly pigmented or tanned skin however, various combinations of longer wavelength, longer pulse duration, and active skin cooling are used to minimize the risk of unwanted epidermal injury. Dynamic (epidermal) and bulk (epidermal and dermal) cooling devices have been developed, which are most useful with short and long pulses, respectively. Temporary hair removal for a few months is easily achieved even at low fluences and appears to be due to induction of catagen and then telogen (resting) phase. Transition to catagen and telogen is a common follicular response to injury, for example, from stress, drugs, hypoxia, surgery, and after laser treatment, from photothermal injury. Permanent hair reduction requires higher fluences, which are capable of necrosing follicular target regions, and has been defined as "a significant decrease in terminal hairs after a given treatment, which is sustained for a period of time longer than the complete growth cycle of hair follicles at the given body site."

A ruby laser was first reported by Grossman et al.[93] to remove hair from normal human skin. No scarring occurred, and about one-third of the subjects had transient pigmentary changes. Selective thermal injury of superficial and deep follicular epithelium was observed histologically. Ruby lasers were the first devices marketed for permanent laser hair removal and remain useful especially in patients with light-colored hair and relatively fair skin. Alexandrite lasers, with a slightly longer wavelength than ruby lasers, behave very similar to ruby lasers for hair removal. Comparatively, there is an advantage of a slightly greater depth of penetration and a disadvantage of lower absorption by pheomelanin, the dominant form of melanin in *red* hair. A number of reports have shown that alexandrite lasers are efficacious in removing unwanted hair.[94–96] A widely used and versatile laser for hair removal is a pulsed 800 nm diode laser, with variable pulsewidth of 5–100 ms and variable spot sizes up to 22 × 35 mm, delivered through a cold sapphire window or pneumatic suction device. The larger spot size allows for covering given surface areas in a shorter period of time, while the pneumatic device is thought to decrease pain (gated theory of pain) and bring the target chromophore closer to the skin, as well as blanch blood vessels, which may act as a competing chromophore in the skin.[97]

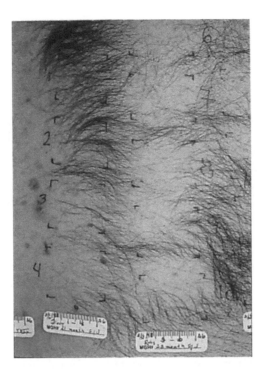

FIGURE 3.6 Hair removal more than 20 months after treatment with 800 nm diode laser (left) and 694 nm ruby laser (right), in test sites on the back.

Millisecond-domain, long-pulsed Nd:YAG (1064 nm) lasers also are capable of inducing long-term hair loss; however, higher fluence is needed to compensate for the lower melanin absorption. These lasers, when delivered in combination with skin cooling, are excellent for controlling hair growth in dark-skinned individuals, for example, for pseudofolliculitis barbae in African Americans.[98-100] Filtered xenon flashlamps, also called *intense pulsed light* (IPL), emit a wide wavelength band of noncoherent light pulses. A variety of these sources have been shown to produce permanent hair removal, which in fair-skinned, dark-haired patients can be of equal efficacy to lasers.[101-104] In general, IPLs are associated with a higher risk of skin burns in darkly pigmented patients.

Before initiating laser hair removal procedure, factors to consider include skin color, hair color, hair diameter, and density of hair. These factors affect laser parameters to be used and in turn affect the efficacy and side effects of treatment. For any given laser type, efficacy is laser fluence, hair color, and hair diameter dependent, but the side effects are laser fluence, skin type, and hair density dependent. Perifollicular erythema and edema are expected as an end point for laser hair removal. In general, laser hair removal is effective and safe in the majority of patients with dark hair and relatively fair skin. With a combination of longer wavelength, longer pulse duration, and parallel cooling, dark skin can be safely and effectively treated (Figure 3.6).[105]

3.7 Lasers Targeting Water as Chromophore

Laser skin resurfacing (LSR) uses water in tissue as a target chromophore, to rapidly deposit energy in a thin layer of tissue, which vaporizes in a controlled way. Due to strong water absorption in the far IR, either CO_2 lasers (10,600 nm) or erbium lasers (2,940 nm) are used. The epidermis and upper dermis are precisely ablated, with a predictable zone of residual thermal damage (RTD) followed by reepithelialization from the residual hair follicles and remodeling of dermal collagen during wound healing.

It is extremely important to limit the depth of residual thermal injury, which is accomplished by vaporizing a thin layer at the surface in a time short enough to limit the amount of heat conducted into the dermis.

3.7.1 CO_2 Lasers

With pulsed CO_2 lasers at fluences of 4–19 J/cm^2 per pulse, precise ablation of tissue (20–40 μm thick) with less than 100 μm RTD can be produced.[106,107] The minimum ablation depth per *pass* of laser exposure is consistent with the optical penetration depth, which is about 20 μm for the CO_2 laser (and is equal to $1/\mu_a$, where μ_a is the tissue absorption coefficient of about 500 cm^{-1}) at this laser wavelength.[107] Selective tissue ablation with minimal RTD occurs when the laser energy is deposited faster than the thermal relaxation time of the heated tissue layer, which for CO_2 lasers is around 1 ms. In order to cleanly vaporize the superficial layer of skin with minimal RTD, a CO_2 laser delivering energy in less than 1 ms must produce a fluence greater than 5 J/cm^2 (the vaporization threshold).[107–109]

It must be emphasized that clinical judgment and careful assessment of ablation depth at the time of treatment is absolutely necessary, as a guide to laser resurfacing. Experienced physicians obtain essentially equivalent clinical results with any of the commercially available lasers for skin resurfacing. Treating patients in a cookbook fashion is, in general, hazardous. One very helpful sign of tissue depth during laser resurfacing is immediate shrinkage. This obvious reaction, due to thermal denaturation of type I collagen at about 70°C, only occurs when the dermis is initially subjected to a layer of RTD. Clinically, it is well established that some dermal injury is necessary for improvement of moderate wrinkles, regardless of the type of laser being used. In contrast, epidermal pigmented lesions such as freckles, lentigines, and café-au-lait macules can be removed without dermal injury. Typically, the first *pass* with pulsed lasers, either CO_2 or Er:YAG, causes little or no immediate shrinkage.

3.7.2 Er:YAG Lasers

The Er:YAG laser, operating at 2940 nm, is even more strongly absorbed by water than CO_2 laser. The μ_a of pure water at this wavelength is approximately 13,000 cm^{-1}. In tissue, which is typically 70% water, the μ_a is approximately 9000 cm^{-1} and the optical penetration depth, $1/\mu_a$, is approximately 1 μm.[110,111] Thermally induced changes in μ_a during Er:YAG laser pulses tend to increase the optical penetration depth, however. The estimated thermal relaxation time of the layer in which the laser energy is absorbed is approximately 1–10 μs. The calculated threshold fluence for tissue ablation is 0.28 J/cm^2, which is close to that measured for sub-microsecond Er:YAG laser pulses. However, longer (e.g., 100–1000 μs) Er:YAG laser pulses are used for LSR, to provide better hemostasis. For the Er:YAG lasers used for skin resurfacing, the tissue ablation threshold is typically 1.5 J/cm^2.[111,112] With sub-microsecond Er:YAG laser pulses, the RTD zone is only 5–10 μm.[110] Short-pulsed Er:YAG lasers are therefore capable of ablating only one or two cell layers at a time with minimal residual injury. This is excellent for extremely fine ablation but a poor choice when hemostasis is needed.[113] When operated to produce longer pulses, hemostasis can be achieved (and thermal injury increased intentionally) by allowing more thermal conduction to occur. Thus, Er:YAG lasers can be made to perform more like a CO_2 laser, and vice versa, by manipulating fluence and pulse duration. When the same absorbed power density (W/cm^3 deposited at the surface of the tissue) is used for both systems, one generally sees similar effects. Some Er:YAG laser manufacturers have made this laser behave like the CO_2 laser (increasing RTD and improved hemostasis) by either adding a simultaneous CO_2 laser or manipulating the pulse duration and fluence of the Er:YAG pulses.

3.7.3 Mechanisms of Action

Tissue removal (ablation) was initially thought to be the most important mechanism in LSR, as if one were literally sculpting the skin. While sculpting may play some role, it is also well established that other, nonselective agents such as dermabrasion, acids, and solvents (chemical *peels*) can achieve equivalent clinical results. Furthermore, when ablation-plus-RTD depth into the dermis is less than the wrinkle depths, marked cosmetic improvement is often observed. It appears that a combination of collagen shrinkage and dermal extracellular matrix remodeling during wound healing contributes to the final cosmetic improvement. When the CO_2 laser interacts with tissue, there are three distinct zones of tissue alteration correlating with the degree of tissue heating. These are as follows: a zone of vaporization; underlying this, a zone of irreversible thermal damage and denaturation resulting in tissue necrosis; and below this, a zone of reversible, sublethal thermal damage.

Immediate collagen shrinkage occurs in the zone of irreversible injury and, to some extent, the zone of reversible injury. This accounts for the visible tissue tightening observable as the CO_2 laser interacts with the dermis, but there is controversy about the role of immediate collagen shrinkage in achieving the final outcome. Using two to five passes with a pulsed CO_2 laser, shrinkage increases somewhat with the number of passes, while the zone of dermal collagen denaturation stays constant (~115 μm).[114] In pigs, pulsed CO_2 laser resurfacing produces immediate tissue contraction (maximum of approximately 38%) and RTD that is saturable for multiple passes and high fluences.[115] A subsequent study in pigs showed that CO_2 laser resurfacing produces short- and long-term wound contraction, greater than that induced by purely ablative methods (mechanical or Er:YAG laser) for the same total depth of injury. Wound contraction over time was dependent on both the depth of RTD and depth of ablation. Wounds with more than 70 μm RTD layer healed with greater fibrosis (microscopic scarring), seen by greater compaction and horizontal orientation of collagen fibers in the superficial dermis after healing. These data suggest that initial collagen contraction and thermal damage modulate wound healing.[116] Also supporting this concept, there is a greater and more sustained elevation in collagen content after CO_2 laser injury versus 50% trichloroacetic acid *peel* in mouse skin. New collagen deposition and remodeling in the dermis appear to be responsible for the long-term clinical improvement (up to 1 year) after LSR, although immediate tissue shrinkage is present for only a short period.[117–120] Another study found increased elements other than collagen, for example, factor 13, vimentin, and actin expression after pulsed carbon dioxide laser injury in a pig.[121] Contraction of these filaments may provide a scaffold for deposition and organization of collagen during wound healing, which may contribute to skin tightening and decreased surface irregularities seen after CO_2 laser resurfacing (Figure 3.7).

It is interesting to contrast mechanisms of LSR with that of a *face lift*, a common surgical procedure for wrinkles and skin laxity in which facial skin is stretched. Two nearby points on facial skin typically become closer together after laser resurfacing, whereas after a face lift, the same two points typically become further apart. The combination of LSR and plastic surgery is often more effective than either modality alone.

3.7.4 Indications for Laser Skin Resurfacing

3.7.4.1 Acne Scars

LSR is moderately beneficial in treating atrophic partial-thickness nondistensible scars,[122] although results are variable.[123] Kye et al. reported 40% improvement of acne scar at 3-month follow-up after Er:YAG laser treatment.[124] With acne scar treatment, a longer postoperative interval (12–18 months) prior to assessment for retreatment was advocated because continued clinical improvement was observed as long as 18 months after CO_2 laser resurfacing, with an 11% increase in improvement observed between 6 and 18 months postoperatively.[125] Kwon et al. treated 12 patients with hypertrophic scars, 20 patients with depressed scars, and 4 patients with burn scars, using Er:YAG laser, 2 mm handpiece, 500–1200 mJ/pulse at 3.5–9 W. In this series, 9 of 12 hypertrophic scars, 17 of 20 depressed scars, and 2 of 4 burn scars were substantially improved.[126]

3.7.4.2 Photoaging

Chronic photoaging, comprised of rhytides, dyschromias, solar lentigines, ephellides, and actinic keratoses, has been shown to respond very favorably to CO_2 LSR. Ross et al. showed that scanning

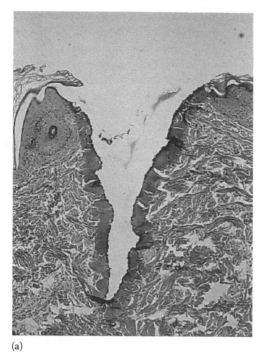

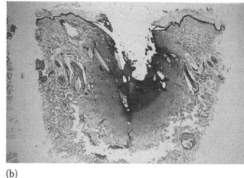

(a)　　　　　　　　　　　　　　　　　　　　(b)

FIGURE 3.7 Photomicrograph of skin biopsy specimens comparing the amount of residual thermal damage (RTD) after CO_2 laser ablation. (a) After submillisecond CO_2 laser. (b) After continuous CO_2 laser exposure. Note minimal amount of RTD in submillisecond exposure.

laser with a short dwell time performed just as well as a pulsed laser, in a side-by-side comparison trial.[127] Er:YAG lasers have also been used successfully in this context.[128–131] It is now apparent that clinical improvement following equivalent depth dermal laser wounding with the Er:YAG laser is less than that observed following CO_2 LSR, although the time of reepithelialization and postoperative erythema are shorter.[132] Side-by-side comparison studies using CO_2 laser alone and combined CO_2/Er:YAG laser resurfacing have demonstrated that the overall healing times and erythema can be reduced without compromising clinical efficacy in the combined CO_2 and Er:YAG laser treatment group.[133,134]

In clinical practice, for deep photodamage, both lasers are useful. The CO_2 laser is used to efficiently remove the epidermis in a single pass, followed by a second pass to produce collagen tightening. A third pass is used in selective areas of deeper photodamage. This may also be followed by one to two passes with Er:YAG laser. For superficial therapy, when the patient desires rapid healing and is willing to accept improvement only in texture and pigmentation, a single pass of the CO_2 laser can be performed without wiping the epidermal debris away. For single pass wounds, not wiping decreased the level of wounding. In contrast, not wiping in multiple pass wounds significantly increased the depth and variability of RTD and necrosis, resulting in prolonged healing.[135] Deeper wrinkles caused by facial expression, for example, the glabella and nasolabial folds, do not respond as well as finer wrinkles, especially, perioral and periorbital rhytides. These are best managed with other techniques, for example, injections of botulinum toxin.

3.7.5 Other Indications

Almost any skin lesion treatable with electrodesiccation, curettage, or exhibiting irregular surface topography can be considered for LSR. Epidermal processes, such as actinic and seborrheic dermatoses, diffuse solar lentigines, actinic cheilitis, superficial basal cell carcinoma, and squamous cell carcinoma in situ, often respond nicely to resurfacing. Dermal processes, including nonmovement-associated rhytides and papular elastosis, xanthelasma, sebaceous hyperplasia, rhinophyma, and angiofibroma, may also be responsive to pulsed CO_2 or Er:YAG laser therapy.[136] Laser-assisted hair transplantation[137] and various unrelated conditions, including xanthelasmas, viral warts, Hailey–Hailey disease, and lymphangioma circumscriptum, have been reported to respond to CO_2 laser treatment.[138,139]

3.7.6 Contraindications

Patients with history of keloid, in general, should not undergo LSR. Any patient with severe systemic disease or diseases complicated by immunosuppression should not be treated, as these may alter wound healing. Isotretinoin has been reported to produce atypical scarring with resurfacing methods. It is recommended that patients discontinue isotretinoin for at least 1 year before laser treatment.[140] Any condition in which the adnexal structures are compromised (e.g., radiation therapy, collagen vascular disease, or the skin on the neck and extremities) may retard reepithelialization.[141]

3.7.7 Side Effects and Complications

The most common short-term side effects include postoperative erythema, edema, pain, pruritus, burning discomfort, a sensation of tightness, and textural changes. Postoperative erythema occurs in all patients and its duration is mainly a reflection of the depth of thermal damage.[136] The risk of complications is largely determined by the depth and extent of the resurfacing. Before reepithelialization occurs, the major complications are bleeding, particularly following Er:YAG laser resurfacing, and infection. *Staphylococcus aureus* and *Pseudomonas aeruginosa* appear to be the most common infective agents, followed by *Staphylococcus epidermidis* and *Candida albicans*.[142] Antiherpetic prophylaxis should be used in patients having full facial or lower facial laser peels, irrespective of a history of herpes simplex virus (HSV). The role of prophylactic antibiotics is controversial. A retrospective study of 130 patients undergoing laser resurfacing revealed a significantly higher rate of infection in patients receiving combination intraoperative and postoperative antibiotic prophylaxis.[143] Late complications include acneiform folliculitis and milia formation, contact dermatitis, pigmentary change, scarring, and ectropion.[144] Generally, pigmentary changes are related to the level of injury. Injury just into the papillary dermis is more likely to cause hyperpigmentation and is dependent on skin type. With deeper passes, hypopigmentation is possible or even common. Normally, hyperpigmentation is seen 2–4 weeks after injury, but hypopigmentation is typically more delayed, sometimes occurring several months after surgery. Two types of hypopigmentation can be recognized. One is a relative hypopigmentation compared to the mottled coloration of background photodamaged untreated skin, which can be prevented by feathering the perimeter of the area or performing a full-face procedure. Another type is the true hypopigmentation, which usually occurs 6–12 months after LSR. This common delayed and apparently permanent side effect may be caused by loss of epidermal melanin or by underlying fibrosis.[145] Permanent hypopigmentation is a significant concern limiting LSR in moderately pigmented skin.

3.7.8 Nonablative Laser Rejuvenation

LSR has excellent efficacy but also requires weeks of recovery and occasionally causes unwanted side effects and complications. A novel nonablative approach for treating wrinkles and scars was recently introduced, variously called *nonablative laser resurfacing* and *nonablative rejuvenation*. As discussed, clinical improvement following LSR is partly due to sculpting by ablation, partly due to remodeling during healing after controlled thermal injury, and perhaps partly due to shrinkage of dermal collagen, immediately tightening the dermis. Nonablative rejuvenation attempts to achieve all of these effects except sculpting, by creating an *upside-down* thermal injury in which the epidermis is spared while the dermis is heated to temperatures that denature collagen and/or stimulate a healing response. This is easily accomplished, by cooling the epidermis while propagating optical energy through it that is absorbed in the dermis. The first devices for nonablative rejuvenation used mid-IR lasers that are weakly or moderately absorbed by water, but it has also been observed that selective photothermolysis using visible light pulses absorbed by blood vessels produces beneficial effects on photoaged skin. The present popularity of this approach is such that many lasers and light sources are being used for this purpose including Q-switched 1064 nm Nd:YAG, 585 nm PDLs (0.3–20 ms, at fluences below the threshold for immediate purpura), 980 nm diode laser, 1320 nm long-pulsed Nd:YAG and 1540 nm Er:glass lasers in combination with skin cooling devices, 1440 nm diode lasers, and IPL sources. All of these seem to offer real but subtle and minimal improvement in wrinkles compared with skin resurfacing. New dermal collagen formation and remodeling has been shown histologically.[146–152]

Currently, several laser systems have clearance by the FDA specifically for this indication including the PDL, 1320 nm Nd:YAG, and the 1450 nm diode laser. The 1320 nm Nd:YAG laser was designed for the purpose of improving rhytides and overall skin texture. Numerous clinical studies have been conducted evaluating the efficacy of the Nd:YAG laser in the treatment of facial rhytides with mild improvement in rhytides.[150,152] Increase in collagen has been histologically confirmed by Goldberg.[153] The 1450 nm diode laser has a similar mechanism. The first study to demonstrate the efficacy of this treatment evaluated split-face treatments of periorbital and perioral rhytides with comparison to cryogen cooling alone.[151] Promising results have also been reported of photorejuvenation using IPL (Figure 3.8).[154,155]

Despite initial excitement, nonablative rejuvenation does not approach the clinical efficacy of classical LSR. It should be noted that there may be nothing unique about optical energy for stimulating an upside-down minor burn of the skin. Radiofrequency current and/or ultrasound are other radiant energy sources capable of dermal heating currently employed to accomplish similar clinical end points.

3.7.9 Fractional Photothermolysis

In 2004, Manstein and colleagues introduced the concept of FP, resulting in significant improvement in the approach for both nonablative and ablative rejuvenation.[156] The common theme in FP involves the delivery of energy to induce microscopic zones of damage. For nonablative fractional resurfacing (NAFR), narrow beams of high-energy light at a variety of wavelengths targeting tissue water result in zones of necrosis known as *microthermal zones* (MTZs). These are distributed, either randomly or in

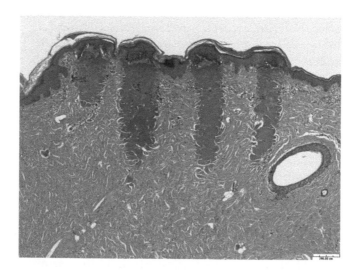

FIGURE 3.8 Histology of fractional laser resurfacing. Photomicrograph of porcine skin specimens treated with NAFR. Note the columns of thermal denaturation and partial-thickness epidermal necrosis with surrounding untreated skin.

a predetermined manner, across the treatment area in a pixilated pattern. This is apparent only when viewed microscopically. The energy delivered produces sequestered thermal damage without spread to adjacent tissue. The normal skin around these *islands* of thermal damage provides a nutritional and structural reservoir enabling more rapid recovery. This concept was extended to ablative fractional resurfacing (AFR), which utilizes narrow columns of thermal energy at wavelengths previously used for classical resurfacing to vaporize columns of tissue. This concept of focal microscopic zones of injury surrounded by a reservoir of spared skin is the underlying premise of all fractional devices and is paramount for the shorter recovery time and improved safety.[156]

3.7.10 Mechanism

For NAFR, microscopic analysis of tissue immediately after treatment reveals clear zones of epidermal and dermal necrosis within the MTZ. Epidermal damage is usually limited to the lower half of epidermal keratinocytes, and the stratum corneum is left intact—hence the designation as nonablative. Lactate dehydrogenase staining demonstrates viability of tissue adjacent to the MTZs. Microscopic epidermal necrotic debris (MEND) overlying the MTZ contains cellular components in addition to melanin that are gradually extruded over the course of 2 weeks. No histologic or clinical evidence of persistence of the MTZs was present at 3 months. Expression of markers for neocollagenesis including heat shock proteins, collagen III, proliferating cell nuclear antigen, and alpha-smooth muscle actin is present after treatment with some persisting for 3 months after treatment. Histology for AFR is similar with biopsies immediately posttreatment demonstrating tapering columns of tissue ablation lined by an eschar and surrounded by a zone of thermal coagulation. This cylindrical column is quickly lined by invagination of the epidermis within 2 days of treatment. Despite rapid reepithelialization, there is histologic evidence of the residual thermal coagulative zone for up to 1 month following treatment indicating ongoing tissue remodeling. Similar to NAFR, immunohistochemical stains demonstrate evidence of prolonged neocollagenesis with upregulation of various heat shock proteins up to 3 months after treatment.[157]

3.7.11 Indications

3.7.11.1 Rhytides

NAFR was initially approved for the treatment of periorbital rhytides.[156] Later, Wanner and colleagues demonstrated improvement in mild-to-moderate photodamage and rhytides on the face, neck, and chest using a 1550 nm erbium-doped fiber fractional laser.[158] NAFR is also effective for improvement of photodamage on nonfacial sites. Reports of successful use in the treatment of photodamage of the hands and poikiloderma of civatte indicate efficacy and safety in nonfacial application.[159,160] AFR has similarly been used for the treatments of facial rhytides; however, care must be taken for the treatment of photoaging on nonfacial skin.

3.7.11.2 Acne Scarring

Studies support the use of both NAFR and AFR for the treatment of acne scars. Initial investigations for the use of fractional lasers for the treatment of acne scars were done with the nonablative devices. Improvement of the appearance of ice pick as well as atrophic scars has been reported following NAFR.[161] AFR can also improve acne scarring. Though initial improvement following treatment is remarkable, only 74% long-term maintenance was noted in long-term follow-up.[162] Darker skin phototypes can be effectively treated with the nonablative device but often require more treatments given the lower treatment densities needed to achieve these treatments.[163] Improvement of postinflammatory scar erythema from acne is also noted and is thought to result from nonselective thermal destruction of blood vessels.[164]

3.7.11.3 Burn Scarring

FP has increasingly been utilized for the treatment of burn scars. Numerous studies demonstrate efficacy for both the nonablative and ablative devices in the improvement of skin texture, dyschromia, atrophy, hypertrophy, and hypopigmentation.[165,166] In addition to cosmetic improvement in burn scars, functional improvement has also been noted after AFR including improvement in scar contracture and range of motion following a series of treatments.[167] Histologic studies on burn scars after treatment with AFR demonstrate a change in the collagen I and collagen III expression posttreatment.[168]

3.8 Pigmentary Disorders

Various forms of pigmentation can be treated with FP. The mechanism for efficacy is related to the transepidermal elimination of melanin with shedding of the MENDs. NAFR for the treatment of melasma, a refractory pigmentary disorder of sun-exposed skin, showed initial promise.[169] Histologically, post-treatment specimens demonstrated a relative decrease in melanin granules within keratinocytes when compared to pretreatment controls. Limitations exist to the use of NAFR for melasma. Not surprisingly, clinical improvements in melasma were less extensive in patients with progressively darker skin types. In fact, following initial improvement, a substantial subset of patients get worse. Moreover, high recurrence rate following termination of treatment calls the longevity of the clinical benefit into question.[170] Reports of clearance of other pigmentary disorders including drug-induced hyperpigmentation[171] underscore the relative utility of NAFR for the nonselective removal of pigment. Exacerbation of melasma can be triggered by any laser, IPL, or other treatment causing inflammation. There is some evidence that low-fluence Q-switched Nd:YAG laser treatment combined with topical medical therapy can improve melasma in the majority of cases.

3.9 Drug Delivery

Because AFR creates microscopic holes in tissue, these openings can be utilized as conduits for delivery of a variety of substances. Animal studies have used fractionated CO_2 to deliver a variety of topical medications including methyl-aminolevulanic acid and imiquimod.[172,173] Not only was drug absorption more rapid through this methodology, higher concentrations and greater distribution were observed. These early animal studies demonstrate significantly enhanced dermal drug delivery following AFR. Preliminary evidence from clinical trials supports the safety and increased efficacy.[174] While far from being optimized, AFR may serve as a channel for the delivery of large molecules that are unable to penetrate intact skin. Perhaps ultimately, AFR will be utilized for drug delivery including targeted biologic peptides and vaccines.

3.10 Tattoos

One of the initial applications for AFR was for the treatment of tattoos. This application is quite intuitive as repeated removal of random array of microscopic columns of tissue will eventually lead to fragmentation and lightening of the tattoo. An initial report demonstrated enhanced tattoo removal when AFR was combined in conjunction with QSL. The likely efficacy of this is multifactorial and is postulated to be related to both vaporization of microscopic zones of the tattoo and providing a conduit for passive ink drainage after fragmentation with treatment with QSL.[175] More recently, Ibrahimi et al. reported the use of AFR for the treatment of allergic tattoos.[176] The lack of color selectivity for AFR may prove to be property worth exploiting to treat tattoos with more vibrant and complex color compositions. Reports for AFR have also indicated its utility in the treatment of cosmetic tattoos that can undergo paradoxical darkening during treatment with QSL.[177]

3.11 Side Effects

HSV is the most common infectious complication following fractional resurfacing and increases the risk of scarring.[178,179] This can be greatly minimized by appropriate antiviral prophylaxis prior to treatment.[178] In contrast, the incidence of bacterial infection is significantly lower. Most common pathogens include *S. aureus* and *P. aeruginosa*; however, a single case of atypical mycobacteria has been reported.[180] Candidal and pityrosporal infections can also occur at a low frequency, especially in the setting of antibacterial prophylaxis.[181] Similar to traditional resurfacing, acneiform eruptions can occur following fractional resurfacing. However, these rates are much lower than that of traditional resurfacing. Milia and frank acne flares can occur. These can be managed by using noncomedogenic moisturizers as well as appropriate pretreatment prophylaxis with tetracyclines in those with a history of acne or flares prior treatment.[161,182] Erythema following fractional resurfacing is expected. Redness that does not resolve 4 days or 1 month after NAFR and AFR, respectively, is termed prolonged erythema. Incidence is low and typically self resolves within 3 months. Many practitioners correlate erythema with prolonged collagen remodeling. Hyperpigmentation following fractional resurfacing has a variable incidence but occurs less frequently than in traditional resurfacing. Patients with darker skin phototypes (Fitzpatrick IV–VI) are at greatest risk. In contrast to full-face carbon dioxide laser resurfacing, delayed onset, permanent hypopigmentation has never been reported. Hypertrophic scarring has rarely been reported with fractional resurfacing. Areas prone to scarring include the neck, mandibular ridge, and periorbital region. One can minimize this by optimizing the parameters and treatment techniques to minimize bulk heating.

3.12 Lipid-Selective Lasers

Selective photothermolysis of lipid-rich tissue has introduced a paradigm shift in targeting lipids. Near-IR light is able to alter the vibrational mode of the C–H bonds within target tissue and therefore can selectively target lipids that are rich in C–H bonds. Anderson et al. measured the absorption spectrum of fat, identifying peaks at 1210 and 1720 nm where the absorption coefficient of lipid is greater than that of water. While clinical studies are under way, ex vivo histologic data demonstrate selective targeting of both subcutaneous fat and sebaceous glands.[183] Exposure of fresh, intact, full-thickness porcine tissue samples at wavelengths near 1210 nm reproducibly caused thermal damage of subcutaneous fat with little or no injury to the overlying skin.[184] This was later confirmed by a pilot study performed using a 1210 nm diode laser on the abdomen of volunteer subjects. Biopsies obtained after delivery of a single pulse demonstrated dose-related damage to the subcutaneous fat and dermis. Lipomembranous changes of the fat were seen in biopsy specimens obtained 4–6 weeks after exposure. Similarly, absorption peaks in sebum identified near 1210, 1728, 1760, 2306, and 2346 nm were observed. Scalp samples exposed to 1700 nm at 100–135 ms pulse durations using a free electron laser confirmed selective photothermolysis of sebaceous glands without epidermal or dermal damage.[185] While there is no clinical data supporting efficacy of lipid-selective wavelengths of light for the treatment of acne, medical therapies that reduce sebaceous gland volume or sebum secretion have a proven track record for efficacy.[186]

3.13 Laser Diagnostics in Dermatology

The skin is one of the most accessible organ systems in the body. The ability to physically see the changes often allows for rapid diagnosis. However, despite this, oftentimes biopsy is needed for histologic diagnosis. The invasive nature of biopsy, the associated morbidity and risk for scar formation, as well as the time and cost associated with histology tissue processing make the procedure far from ideal. Moreover, the *static* nature of histology does not adequately allow for evaluating disease processes in evolution. Recent advancements in laser microscopy imaging techniques allow for noninvasive high-resolution in vivo skin imaging, which is increasingly utilized in clinical dermatology. The most widely used of these techniques in dermatology are optical coherence tomography (OCT), reflectance-mode confocal microscopy, and multiphoton microscopy. These techniques take advantage of the accessibility of the skin and are noninvasive, allowing for rapid, real-time imaging of structural morphology of the skin in vivo. In addition, these modalities allow the same exact target area on the skin to be followed over time.

OCT utilizes IR broadband light for real-time in vivo imaging. OCT is based on the interference of IR laser light. IR light is delivered by fiber optics and split into a probe and a reference beam. The probe beam is backscattered and then remixed with the reference beam. Interference occurs when both beams match within the coherence length of light. Vertical images of the target tissue are generated by the analysis of the optical coherence. The axial resolution of OCT depends on the bandwidth and coherence length of light used, while the lateral resolution is dependent on the focusing objective.[187] The data obtained from OCT resemble an ultrasound image but with better resolution. The axial resolution is about 8 μm and lateral resolution is about 24 μm, but this is still low for visualizing cell morphology.[188] OCT can image tissue down to depth of about 2 mm. OCT is currently being used in dermatology for skin cancer diagnostics.[189,190] OCT can also be used for evaluation of inflammatory skin diseases, quantification of skin changes, and visualization of superficial layers of the skin in vivo in real time. To obtain higher resolution for studying morphologic changes at the cellular level, reflectance-mode confocal laser scanning microscopy (RCLSM) was introduced to dermatology.[191] This is also a noninvasive laser scanning microscopy and real-time in vivo imaging technique used to visualize human skin tissue morphology at a resolution comparable to that of histology. It works by detecting the back-reflected near-IR laser light. The resolution obtained by RCLSM imaging is about 0.5–1.0 μm in the lateral direction and 3–5 μm in the axial direction. RCLSM can produce high-resolution images of the entire epidermis and a superficial layer of dermis allowing physicians to clearly visualize cellular details. The depth of imaging is limited by optical penetration and signal-to-noise ration. Currently, the maximum depth of penetration with confocal scanning laser microscopy is 350 μm into the skin.[192] Numerous papers have been published on diagnostic utility of confocal laser scanning microscopy in diagnosis of melanocytic skin tumors with promising results.[193]

Another emerging diagnostic in vivo imaging technique is multiphoton laser scanning microscopy (MPLSM). This is also a near-IR laser scanning microscopy technique that is based on nonlinear optical process predicted by Maria Goeppert-Meyer in 1931 but not used in microscopy until 1990.[194] Two-photon excitation of an endogenous fluorophores found in the skin, such as NADH, keratin, melanin, collagen, and elastin, is commonly used for imaging. The resolution MPLSM is about 0.5 μm in the lateral direction and 1–2 μm in the axial direction.[195] It is capable of visualizing tissue morphology down to depth of around 200 μm. This technique has promising potential in improving noninvasive in vivo skin cancer diagnosis and for being used as a bedside diagnostic tool for histopathologic analysis.[196]

The different laser-based in vivo imaging modalities are very promising emerging technology, and it is only a matter of time before these innovative techniques overcome their major limitations that include cost and depth of imaging due to optical penetration.

References

1. Maiman T. Stimulated optical radiation in ruby. *Nature* 1960;187:493–494.
2. Goldman L, Blaney DJ, Kindel DJ Jr., Franke EK. Effect of the laser beam on the skin. Preliminary report. *J Invest Dermatol* 1963;40:121–122.
3. Goldman L, Blaney DJ, Kindel DJ Jr., Richfield D, Franke EK. Pathology of the effect of the laser beam on the skin. *Nature* 1963;197:912–914.
4. Goldman L, Wilson RG. Treatment of basal cell epithelioma by laser radiation. *JAMA* 1964;189:773–775.
5. Goldman L, Wilson RG, Hornby P, Meyer RG. Radiation from a Q-switched ruby laser. Effect of repeated impacts of power output of 10 megawatts on a tattoo of man. *J Invest Dermatol* 1965;44:69–71.
6. Anderson RR, Parrish JA. Selective photothermolysis: Precise microsurgery by selective absorption of pulsed radiation. *Science* 1983;220:524–527.

89

7. Anderson RR, Parrish JA. Microvasculature can be selectively damaged using dye lasers: A basic theory and experimental evidence in human skin. *Lasers Surg Med* 1981;1:263–276.

8. Anderson RR, Parrish JA. The optics of human skin. *J Invest Dermatol* 1981;77:13–19.

9. Domankevitz Y, Waldman J, Lin CP, Anderson RR. Selective delivery of laser energy to biological tissue based on refractive-index differences. *Opt Lett* 2000;25:1104–1106.

10. Henriques FC Jr. Studies of thermal injury; the predictability and the significance of thermally induced rate processes leading to irreversible epidermal injury. *Arch Pathol (Chic)* 1947;43:489–502.

11. Welch AJ, Polhamus GD. Measurement and prediction of thermal injury in the retina of the rhesus monkey. *IEEE Trans Biomed Eng* 1984;31:633–643.

12. Garden JM, Tan OT, Kerschmann R et al. Effect of dye laser pulse duration on selective cutaneous vascular injury. *J Invest Dermatol* 1986;87:653–657.

13. Sliney DH. Laser safety. *Lasers Surg Med* 1995;16:215–225.

14. Garden JM, O'Banion MK, Shelnitz LS et al. Papillomavirus in the vapor of carbon dioxide laser-treated verrucae. *JAMA* 1988;259:1199–1202.

15. Baggish MS, Poiesz BJ, Joret D, Williamson P, Refai A. Presence of human immunodeficiency virus DNA in laser smoke. *Lasers Surg Med* 1991;11:197–203.

16. Baggish MS, Elbakry M. The effects of laser smoke on the lungs of rats. *Am J Obstet Gynecol* 1987;156:1260–1265.

17. Silver L. Argon laser photocoagulation of port wine stain hemangiomas. *Lasers Surg Med* 1986;6:24–28, 52–55.

18. van Gemert MJ, Welch AJ, Amin AP. Is there an optimal laser treatment for port wine stains? *Lasers Surg Med* 1986;6:76–83.

19. Tan OT, Sherwood K, Gilchrest BA. Treatment of children with port-wine stains using the flashlamp-pulsed tunable dye laser. *N Engl J Med* 1989;320:416–421.

20. Dierickx C, Farinelli W, Anderson R. Multiple pulse photocoagulation of port wine stain blood vessels with a 585 nm pulsed dye laser. *Lasers Surg Med* 1995;Suppl 7:56.

21. Roider J, Traccoli J, Michaud N, Flotte T, Anderson R, Birngruber R. Selective vascular occlusion by repetitive short laser pulse. *Ophthalmologe* 1994;91:274–279.

22. Barsky SH, Rosen S, Geer DE, Noe JM. The nature and evolution of port wine stains: A computer-assisted study. *J Invest Dermatol* 1980;74:154–157.

23. Noe JM, Barsky SH, Geer DE, Rosen S. Port wine stains and the response to argon laser therapy: Successful treatment and the predictive role of color, age, and biopsy. *Plast Reconstr Surg* 1980;65:130–136.

24. Kauvar AN, Geronemus RG. Repetitive pulsed dye laser treatments improve persistent port-wine stains. *Dermatol Surg* 1995;21:515–521.

25. Seukeran DC, Collins P, Sheehan-Dare RA. Adverse reactions following pulsed tunable dye laser treatment of port wine stains in 701 patients. *Br J Dermatol* 1997;136:725–729.

26. Yang MU, Yaroslavsky AN, Farinelli WA et al. Long-pulsed neodymium:yttrium-aluminum-garnet laser treatment for port-wine stains. *J Am Acad Dermatol* 2005;52:480–490.

27. Jasim ZF, Handley JM. Treatment of pulsed dye laser-resistant port wine stain birthmarks. *J Am Acad Dermatol* 2007;57:677–682.

28. Li L, Kono T, Groff WF, Chan HH, Kitazawa Y, Nozaki M. Comparison study of a long-pulse pulsed dye laser and a long-pulse pulsed alexandrite laser in the treatment of port wine stains. *J Cosmet Laser Ther* 2008;10:12–15.

29. Izikson L, Nelson JS, Anderson RR. Treatment of hypertrophic and resistant port wine stains with a 755 nm laser: A case series of 20 patients. *Lasers Surg Med* 2009;41:427–432.

30. Law BK. Rapamycin: An anti-cancer immunosuppressant? *Crit Rev Oncol Hematol* 2005;56:47–60.

31. Nelson JS, Jia W, Phung TL, Mihm MC Jr. Observations on enhanced port wine stain blanching induced by combined pulsed dye laser and rapamycin administration. *Lasers Surg Med* 2011;43:939–942.

32. Tremaine AM, Armstrong J, Huang YC et al. Enhanced port-wine stain lightening achieved with combined treatment of selective photothermolysis and imiquimod. *J Am Acad Dermatol* 2012;66:634–641.
33. Leaute-Labreze C, Dumas de la Roque E, Hubiche T, Boralevi F, Thambo JB, Taieb A. Propranolol for severe hemangiomas of infancy. *N Engl J Med* 2008;358:2649–2651.
34. Hogeling M, Adams S, Wargon O. A randomized controlled trial of propranolol for infantile hemangiomas. *Pediatrics* 2011;128:e259–e266.
35. Kim LH, Hogeling M, Wargon O, Jiwane A, Adams S. Propranolol: Useful therapeutic agent for the treatment of ulcerated infantile hemangiomas. *J Pediatr Surg* 2011;46:759–763.
36. Ni N, Guo S, Langer P. Current concepts in the management of periocular infantile (capillary) hemangioma. *Curr Opin Ophthalmol* 2011;22:419–425.
37. Pope E, Chakkittakandiyil A. Topical timolol gel for infantile hemangiomas: A pilot study. *Arch Dermatol* 2010;146:564–565.
38. Chakkittakandiyil A, Phillips R, Frieden IJ et al. Timolol maleate 0.5% or 0.1% gel-forming solution for infantile hemangiomas: A retrospective, multicenter, cohort study. *Pediatr Dermatol* 2012;29:28–31.
39. Hohenleutner S, Badur-Ganter E, Landthaler M, Hohenleutner U. Long-term results in the treatment of childhood hemangioma with the flashlamp-pumped pulsed dye laser: An evaluation of 617 cases. *Lasers Surg Med* 2001;28:273–277.
40. Morelli JG, Tan OT, Yohn JJ, Weston WL. Treatment of ulcerated hemangiomas infancy. *Arch Pediatr Adolesc Med* 1994;148:1104–1105.
41. Morelli JG, Tan OT, Weston WL. Treatment of ulcerated hemangiomas with the pulsed tunable dye laser. *Am J Dis Child* 1991;145:1062–1064.
42. Barlow RJ, Walker NP, Markey AC. Treatment of proliferative haemangiomas with the 585 nm pulsed dye laser. *Br J Dermatol* 1996;134:700–704.
43. Poetke M, Philipp C, Berlien HP. Flashlamp-pumped pulsed dye laser for hemangiomas in infancy: Treatment of superficial vs mixed hemangiomas. *Arch Dermatol* 2000;136:628–632.
44. Landthaler M, Hohenleutner U, el-Raheem TA. Laser therapy of childhood haemangiomas. *Br J Dermatol* 1995;133:275–281.
45. Laubach HJ, Anderson RR, Luger T, Manstein D. Fractional photothermolysis for involuted infantile hemangioma. *Arch Dermatol* 2009;145:748–750.
46. Chen TS, Eichenfield LF, Friedlander SF. Infantile hemangiomas: An update on pathogenesis and therapy. *Pediatrics* 2013;131:99–108.
47. Goldman MP, Weiss RA, Brody HJ, Coleman WP 3rd, Fitzpatrick RE. Treatment of facial telangiectasia with sclerotherapy, laser surgery, and/or electrodesiccation: A review. *J Dermatol Surg Oncol* 1993;19:899–906.
48. Goldman MP, Bennett RG. Treatment of telangiectasia: A review. *J Am Acad Dermatol* 1987;17:167–182.
49. Tanghetti EA. Split-face randomized treatment of facial telangiectasia comparing pulsed dye laser and an intense pulsed light handpiece. *Lasers Surg Med* 2012;44:97–102.
50. Neuhaus IM, Zane LT, Tope WD. Comparative efficacy of nonpurpuragenic pulsed dye laser and intense pulsed light for erythematotelangiectatic rosacea. *Dermatol Surg* 2009;35:920–928.
51. Goldman MP, Fitzpatrick RE. Pulsed-dye laser treatment of leg telangiectasia: With and without simultaneous sclerotherapy. *J Dermatol Surg Oncol* 1990;16:338–344.
52. Rubin IK, Farinelli WA, Doukas A, Anderson RR. Optimal wavelengths for vein-selective photothermolysis. *Lasers Surg Med* 2012;44:152–157.
53. Alster TS, Williams CM. Treatment of keloid sternotomy scars with 585 nm flashlamp-pumped pulsed-dye laser. *Lancet* 1995;345:1198–1200.
54. Alster TS, McMeekin TO. Improvement of facial acne scars by the 585 nm flashlamp-pumped pulsed dye laser. *J Am Acad Dermatol* 1996;35:79–81.
55. Manuskiatti W, Fitzpatrick RE, Goldman MP. Energy density and numbers of treatment affect response of keloidal and hypertrophic sternotomy scars to the 585-nm flashlamp-pumped pulsed-dye laser. *J Am Acad Dermatol* 2001;45:557–565.

56. Alster TS. Laser treatment of hypertrophic scars, keloids, and striae. *Dermatol Clin* 1997;15:419–429.

57. Watanabe S, Tomino Y, Inoue W et al. Correlation of renal histopathology duration of diabetes, and control of blood glucose in patients with type II diabetes. *J Diabetes Complications* 1987;1:41–44.

58. Tong AK, Tan OT, Boll J, Parrish JA, Murphy GF. Ultrastructure: Effects of melanin pigment on target specificity using a pulsed dye laser (577 nm). *J Invest Dermatol* 1987;88:747–752.

59. Fitzpatrick RE, Goldman MP, Ruiz-Esparza J. Clinical advantage of the CO_2 laser superpulsed mode. Treatment of verruca vulgaris, seborrheic keratoses, lentigines, and actinic cheilitis. *J Dermatol Surg Oncol* 1994;20:449–456.

60. Goldberg DJ. Benign pigmented lesions of the skin. Treatment with the Q-switched ruby laser. *J Dermatol Surg Oncol* 1993;19:376–379.

61. Fitzpatrick RE, Goldman MP, Ruiz-Esparza J. Laser treatment of benign pigmented epidermal lesions using a 300 nsecond pulse and 510 nm wavelength. *J Dermatol Surg Oncol* 1993;19:341–347.

62. Kilmer SL, Wheeland RG, Goldberg DJ, Anderson RR. Treatment of epidermal pigmented lesions with the frequency-doubled Q-switched Nd:YAG laser. A controlled, single-impact, dose-response, multicenter trial. *Arch Dermatol* 1994;130:1515–1519.

63. Nelson JS, Applebaum J. Treatment of superficial cutaneous pigmented lesions by melanin-specific selective photothermolysis using the Q-switched ruby laser. *Ann Plast Surg* 1992;29:231–237.

64. Alster TS, Williams CM. Treatment of nevus of Ota by the Q-switched alexandrite laser. *Dermatol Surg* 1995;21:592–596.

65. Watanabe S, Takahashi H. Treatment of nevus of Ota with the Q-switched ruby laser. *N Engl J Med* 1994;331:1745–1750.

66. Geronemus RG. Q-switched ruby laser therapy of nevus of Ota. *Arch Dermatol* 1992;128:1618–1622.

67. Apfelberg DB. Argon and Q-switched yttrium-aluminum-garnet laser treatment of nevus of Ota. *Ann Plast Surg* 1995;35:150–153.

68. Kunachak S, Leelaudomlipi P, Sirikulchayanonta V. Q-Switched ruby laser therapy of acquired bilateral nevus of Ota-like macules. *Dermatol Surg* 1999;25:938–941.

69. Polnikorn N, Tanrattanakorn S, Goldberg DJ. Treatment of Hori's nevus with the Q-switched Nd:YAG laser. *Dermatol Surg* 2000;26:477–480.

70. Kunachak S, Leelaudomlipi P. Q-switched Nd:YAG laser treatment for acquired bilateral nevus of Ota-like maculae: A long-term follow-up. *Lasers Surg Med* 2000;26:376–379.

71. Ueda S, Imayama S. Normal-mode ruby laser for treating congenital nevi. *Arch Dermatol* 1997;133:355–359.

72. Kono T, Nozaki M, Chan HH, Sasaki K, Kwon SG. Combined use of normal mode and Q-switched ruby lasers in the treatment of congenital melanocytic naevi. *Br J Plast Surg* 2001;54:640–643.

73. Duke D, Byers HR, Sober AJ, Anderson RR, Grevelink JM. Treatment of benign and atypical nevi with the normal-mode ruby laser and the Q-switched ruby laser: Clinical improvement but failure to completely eliminate nevomelanocytes. *Arch Dermatol* 1999;135:290–296.

74. Taylor CR, Anderson RR, Gange RW, Michaud NA, Flotte TJ. Light and electron microscopic analysis of tattoos treated by Q-switched ruby laser. *J Invest Dermatol* 1991;97:131–136.

75. Kilmer SL, Lee MS, Grevelink JM, Flotte TJ, Anderson RR. The Q-switched Nd:YAG laser effectively treats tattoos. A controlled, dose-response study. *Arch Dermatol* 1993;129:971–978.

76. Stafford TJ, Lizek R, Tan OT. Role of the Alexandrite laser for removal of tattoos. *Lasers Surg Med* 1995;17:32–38.

77. Goyal S, Arndt KA, Stern RS, O'Hare D, Dover JS. Laser treatment of tattoos: A prospective, paired, comparison study of the Q-switched Nd:YAG (1064 nm), frequency-doubled Q-switched Nd:YAG (532 nm), and Q-switched ruby lasers. *J Am Acad Dermatol* 1997;36:122–125.

78. Taylor CR, Gange RW, Dover JS et al. Treatment of tattoos by Q-switched ruby laser. A dose-response study. *Arch Dermatol* 1990;126:893–899.

79. Kilmer SL, Anderson RR. Clinical use of the Q-switched ruby and the Q-switched Nd:YAG (1064 nm and 532 nm) lasers for treatment of tattoos. *J Dermatol Surg Oncol* 1993;19:330–338.

80. Ross V, Naseef G, Lin G et al. Comparison of responses of tattoos to picosecond and nanosecond Q-switched neodymium: YAG lasers. *Arch Dermatol* 1998;134:167–171.

81. Izikson L, Farinelli W, Sakamoto F, Tannous Z, Anderson RR. Safety and effectiveness of black tattoo clearance in a pig model after a single treatment with a novel 758 nm 500 picosecond laser: A pilot study. *Lasers Surg Med* 2010;42:640–646.

82. Herd RM, Alora MB, Smoller B, Arndt KA, Dover JS. A clinical and histologic prospective controlled comparative study of the picosecond titanium:sapphire (795 nm) laser versus the Q-switched alexandrite (752 nm) laser for removing tattoo pigment. *J Am Acad Dermatol* 1999;40:603–606.

83. Brauer JA, Reddy KK, Anolik R et al. Successful and rapid treatment of blue and green tattoo pigment with a novel picosecond laser. *Arch Dermatol* 2012;148:820–823.

84. Saedi N, Metelitsa A, Petrell K, Arndt KA, Dover JS. Treatment of tattoos with a picosecond alexandrite laser: A prospective trial. *Arch Dermatol* 2012;148:1360–1363.

85. Kossida T, Rigopoulos D, Katsambas A, Anderson RR. Optimal tattoo removal in a single laser session based on the method of repeated exposures. *J Am Acad Dermatol* 2012;66:271–277.

86. Anderson RR, Geronemus R, Kilmer SL, Farinelli W, Fitzpatrick RE. Cosmetic tattoo ink darkening. A complication of Q-switched and pulsed-laser treatment. *Arch Dermatol* 1993;129:1010–1014.

87. Trotter MJ, Tron VA, Hollingdale J, Rivers JK. Localized chrysiasis induced by laser therapy. *Arch Dermatol* 1995;131:1411–1414.

88. Ross EV, Yashar S, Michaud N et al. Tattoo darkening and nonresponse after laser treatment: A possible role for titanium dioxide. *Arch Dermatol* 2001;137:33–37.

89. Olsen EA. Methods of hair removal. *J Am Acad Dermatol* 1999;40:143–155.

90. Akiyama M, Smith LT, Holbrook KA. Growth factor and growth factor receptor localization in the hair follicle bulge and associated tissue in human fetus. *J Invest Dermatol* 1996;106:391–396.

91. Akiyama M, Dale BA, Sun TT, Holbrook KA. Characterization of hair follicle bulge in human fetal skin: The human fetal bulge is a pool of undifferentiated keratinocytes. *J Invest Dermatol* 1995;105:844–850.

92. Adrian RM. Vascular mechanisms in laser hair removal. *J Cutan Laser Ther* 2000;2:49–50.

93. Grossman MC, Dierickx C, Farinelli W, Flotte T, Anderson RR. Damage to hair follicles by normal-mode ruby laser pulses. *J Am Acad Dermatol* 1996;35:889–894.

94. Goldberg DJ, Ahkami R. Evaluation comparing multiple treatments with a 2-msec and 10-msec alexandrite laser for hair removal. *Lasers Surg Med* 1999;25:223–228.

95. Finkel B, Eliezri YD, Waldman A, Slatkine M. Pulsed alexandrite laser technology for noninvasive hair removal. *J Clin Laser Med Surg* 1997;15:225–229.

96. Garcia C, Alamoudi H, Nakib M, Zimmo S. Alexandrite laser hair removal is safe for Fitzpatrick skin types IV-VI. *Dermatol Surg* 2000;26:130–134.

97. Ibrahimi OA, Kilmer SL. Long-term clinical evaluation of a 800-nm long-pulsed diode laser with a large spot size and vacuum-assisted suction for hair removal. *Dermatol Surg* 2012;38:912–917.

98. Goldberg DJ, Samady JA. Evaluation of a long-pulse Q-switched Nd:YAG laser for hair removal. *Dermatol Surg* 2000;26:109–113.

99. Bencini PL, Luci A, Galimberti M, Ferranti G. Long-term epilation with long-pulsed neodymium:YAG laser. *Dermatol Surg* 1999;25:175–178.

100. Alster TS, Bryan H, Williams CM. Long-pulsed Nd:YAG laser-assisted hair removal in pigmented skin: A clinical and histological evaluation. *Arch Dermatol* 2001;137:885–889.

101. Gold MH, Bell MW, Foster TD, Street S. Long-term epilation using the EpiLight broad band, intense pulsed light hair removal system. *Dermatol Surg* 1997;23:909–913.

102. Schroeter CA, Raulin C, Thurlimann W, Reineke T, De Potter C, Neumann HA. Hair removal in 40 hirsute women with an intense laser-like light source. *Eur J Dermatol* 1999;9:374–379.

103. Weiss RA, Weiss MA, Marwaha S, Harrington AC. Hair removal with a non-coherent filtered flashlamp intense pulsed light source. *Lasers Surg Med* 1999;24:128–132.

104. Sadick NS, Weiss RA, Shea CR, Nagel H, Nicholson J, Prieto VG. Long-term photoepilation using a broad-spectrum intense pulsed light source. *Arch Dermatol* 2000;136:1336–1340.

105. Battle EF, Suthamjariya K, Alora MB, Palli K, Anderson RR. Very long-pulsed (20–200 ms) diode laser for hair removal on all skin types. *Lasers Surg Med* 2000;Suppl 12:21.

106. Green HA, Domankevitz Y, Nishioka NS. Pulsed carbon dioxide laser ablation of burned skin: In vitro and in vivo analysis. *Lasers Surg Med* 1990;10:476–484.

107. Walsh JT Jr., Flotte TJ, Anderson RR, Deutsch TF. Pulsed CO_2 laser tissue ablation: Effect of tissue type and pulse duration on thermal damage. *Lasers Surg Med* 1988;8:108–118.

108. Green HA, Burd E, Nishioka NS, Bruggemann U, Compton CC. Middermal wound healing. A comparison between dermatomal excision and pulsed carbon dioxide laser ablation. *Arch Dermatol* 1992;128:639–645.

109. Yang CC, Chai CY. Animal study of skin resurfacing using the ultrapulse carbon dioxide laser. *Ann Plast Surg* 1995;35:154–158.

110. Walsh JT Jr., Flotte TJ, Deutsch TF. Er:YAG laser ablation of tissue: Effect of pulse duration and tissue type on thermal damage. *Lasers Surg Med* 1989;9:314–326.

111. Walsh JT Jr., Deutsch TF. Er:YAG laser ablation of tissue: Measurement of ablation rates. *Lasers Surg Med* 1989;9:327–337.

112. Hohenleutner U, Hohenleutner S, Baumler W, Landthaler M. Fast and effective skin ablation with an Er:YAG laser: Determination of ablation rates and thermal damage zones. *Lasers Surg Med* 1997;20:242–247.

113. Kaufmann R, Hibst R. Pulsed Erbium:YAG laser ablation in cutaneous surgery. *Lasers Surg Med* 1996;19:324–330.

114. Gardner ES, Reinisch L, Stricklin GP, Ellis DL. In vitro changes in non-facial human skin following CO_2 laser resurfacing: A comparison study. *Lasers Surg Med* 1996;19:379–387.

115. Ross EV, Yashar SS, Naseef GS et al. A pilot study of in vivo immediate tissue contraction with CO_2 skin laser resurfacing in a live farm pig. *Dermatol Surg* 1999;25:851–856.

116. Ross EV, Naseef GS, McKinlay JR et al. Comparison of carbon dioxide laser, erbium:YAG laser, dermabrasion, and dermatome: A study of thermal damage, wound contraction, and wound healing in a live pig model: Implications for skin resurfacing. *J Am Acad Dermatol* 2000;42:92–105.

117. Khatri KA, Ross V, Grevelink JM, Magro CM, Anderson RR. Comparison of erbium:YAG and carbon dioxide lasers in resurfacing of facial rhytides. *Arch Dermatol* 1999;135:391–397.

118. Lowe NJ, Lask G, Griffin ME, Maxwell A, Lowe P, Quilada F. Skin resurfacing with the ultrapulse carbon dioxide laser. Observations on 100 patients. *Dermatol Surg* 1995;21:1025–1029.

119. Fitzpatrick RE, Goldman MP, Satur NM, Tope WD. Pulsed carbon dioxide laser resurfacing of photo-aged facial skin. *Arch Dermatol* 1996;132:395–402.

120. Shim E, Tse Y, Velazquez E, Kamino H, Levine V, Ashinoff R. Short-pulse carbon dioxide laser resurfacing in the treatment of rhytides and scars. A clinical and histopathological study. *Dermatol Surg* 1998;24:113–117.

121. Smith KJ, Skelton HG, Graham JS, Hamilton TA, Hackley BE Jr., Hurst CG. Depth of morphologic skin damage and viability after one, two, and three passes of a high-energy, short-pulse CO_2 laser (Tru-Pulse) in pig skin. *J Am Acad Dermatol* 1997;37:204–210.

122. Alster TS, West TB. Resurfacing of atrophic facial acne scars with a high-energy, pulsed carbon dioxide laser. *Dermatol Surg* 1996;22:151–154.

123. West TB. Laser resurfacing of atrophic scars. *Dermatol Clin* 1997;15:449–457.

124. Kye YC. Resurfacing of pitted facial scars with a pulsed Er:YAG laser. *Dermatol Surg* 1997;23:880–883.

125. Walia S, Alster TS. Prolonged clinical and histologic effects from CO_2 laser resurfacing of atrophic acne scars. *Dermatol Surg* 1999;25:926–930.

126. Kwon SD, Kye YC. Treatment of scars with a pulsed Er:YAG laser. *J Cutan Laser Ther* 2000;2:27–31.

127. Ross EV, Grossman MC, Duke D, Grevelink JM. Long-term results after CO_2 laser skin resurfacing: A comparison of scanned and pulsed systems. *J Am Acad Dermatol* 1997;37:709–718.

128. Alster TS. Clinical and histologic evaluation of six erbium:YAG lasers for cutaneous resurfacing. *Lasers Surg Med* 1999;24:87–92.

129. Weinstein C. Computerized scanning erbium:YAG laser for skin resurfacing. *Dermatol Surg* 1998;24:83–89.

130. Teikemeier G, Goldberg DJ. Skin resurfacing with the erbium:YAG laser. *Dermatol Surg* 1997;23:685–687.

131. Weiss RA, Harrington AC, Pfau RC, Weiss MA, Marwaha S. Periorbital skin resurfacing using high energy erbium:YAG laser: Results in 50 patients. *Lasers Surg Med* 1999;24:81–86.

132. Ross EV, McKinlay JR, Anderson RR. Why does carbon dioxide resurfacing work? A review. *Arch Dermatol* 1999;135:444–454.

133. McDaniel DH, Lord J, Ash K, Newman J. Combined CO_2/erbium:YAG laser resurfacing of peri-oral rhytides and side-by-side comparison with carbon dioxide laser alone. *Dermatol Surg* 1999;25:285–293.

134. Goldman MP, Manuskiatti W. Combined laser resurfacing with the 950-microsec pulsed CO_2 + Er:YAG lasers. *Dermatol Surg* 1999;25:160–163.

135. Ross EV, Mowlavi A, Barnette D, Glatter RD, Grevelink JM. The effect of wiping on skin resurfacing in a pig model using a high energy pulsed CO_2 laser system. *Dermatol Surg* 1999;25:81–88.

136. Ratner D, Tse Y, Marchell N, Goldman MP, Fitzpatrick RE, Fader DJ. Cutaneous laser resurfacing. *J Am Acad Dermatol* 1999;41:365–389.

137. Grevelink JM. Laser hair transplantation. *Dermatol Clin* 1997;15:479–486.

138. Hruza GJ. Laser treatment of warts and other epidermal and dermal lesions. *Dermatol Clin* 1997;15:487–506.

139. Alster TS, Lewis AB. Dermatologic laser surgery. A review. *Dermatol Surg* 1996;22:797–805.

140. Weinstein C. Carbon dioxide laser resurfacing. Long-term follow-up in 2123 patients. *Clin Plast Surg* 1998;25:109–130.

141. Apfelberg DB, Varga J, Greenbaum SS. Carbon dioxide laser resurfacing of peri-oral rhytids in scleroderma patients. *Dermatol Surg* 1998;24:517–519.

142. Sriprachya-Anunt S, Fitzpatrick RE, Goldman MP, Smith SR. Infections complicating pulsed carbon dioxide laser resurfacing for photoaged facial skin. *Dermatol Surg* 1997;23:527–536.

143. Walia S, Alster TS. Cutaneous CO_2 laser resurfacing infection rate with and without prophylactic antibiotics. *Dermatol Surg* 1999;25:857–861.

144. Bernstein LJ, Kauvar AN, Grossman MC, Geronemus RG. The short- and long-term side effects of carbon dioxide laser resurfacing. *Dermatol Surg* 1997;23:519–525.

145. Laws RA, Finley EM, McCollough ML, Grabski WJ. Alabaster skin after carbon dioxide laser resurfacing with histologic correlation. *Dermatol Surg* 1998;24:633–636.

146. Fournier N, Dahan S, Barneon G et al. Nonablative remodeling: Clinical, histologic, ultrasound imaging, and profilometric evaluation of a 1540 nm Er:glass laser. *Dermatol Surg* 2001;27:799–806.

147. Trelles MA, Allones I, Luna R. Facial rejuvenation with a nonablative 1320 nm Nd:YAG laser: A preliminary clinical and histologic evaluation. *Dermatol Surg* 2001;27:111–116.

148. Muccini JA Jr., O'Donnell FE Jr., Fuller T, Reinisch L. Laser treatment of solar elastosis with epithelial preservation. *Lasers Surg Med* 1998;23:121–127.

149. Menaker GM, Wrone DA, Williams RM, Moy RL. Treatment of facial rhytids with a nonablative laser: A clinical and histologic study. *Dermatol Surg* 1999;25:440–444.

150. Kelly KM, Nelson JS, Lask GP, Geronemus RG, Bernstein LJ. Cryogen spray cooling in combination with nonablative laser treatment of facial rhytides. *Arch Dermatol* 1999;135:691–694.

151. Ross EV, Sajben FP, Hsia J, Barnette D, Miller CH, McKinlay JR. Nonablative skin remodeling: Selective dermal heating with a mid-infrared laser and contact cooling combination. *Lasers Surg Med* 2000;26:186–195.

152. Goldberg DJ. Full-face nonablative dermal remodeling with a 1320 nm Nd:YAG laser. *Dermatol Surg* 2000;26:915–918.

153. Goldberg DJ. Non-ablative subsurface remodeling: Clinical and histologic evaluation of a 1320-nm Nd:YAG laser. *J Cutan Laser Ther* 1999;1:153–157.

154. Bitter PH. Noninvasive rejuvenation of photodamaged skin using serial, full-face intense pulsed light treatments. *Dermatol Surg* 2000;26:835–843.

155. Negishi K, Tezuka Y, Kushikata N, Wakamatsu S. Photorejuvenation for Asian skin by intense pulsed light. *Dermatol Surg* 2001;27:627–632.

156. Manstein D, Herron GS, Sink RK, Tanner H, Anderson RR. Fractional photothermolysis: A new concept for cutaneous remodeling using microscopic patterns of thermal injury. *Lasers Surg Med* 2004;34:426–438.

157. Hantash BM, Bedi VP, Kapadia B et al. In vivo histological evaluation of a novel ablative fractional resurfacing device. *Lasers Surg Med* 2007;39:96–107.

158. Wanner M, Tanzi EL, Alster TS. Fractional photothermolysis: Treatment of facial and nonfacial cutaneous photodamage with a 1,550-nm erbium-doped fiber laser. *Dermatol Surg* 2007;33:23–28.

159. Jih MH, Goldberg LH, Kimyai-Asadi A. Fractional photothermolysis for photoaging of hands. *Dermatol Surg* 2008;34:73–78.

160. Behroozan DS, Goldberg LH, Glaich AS, Dai T, Friedman PM. Fractional photothermolysis for treatment of poikiloderma of civatte. *Dermatol Surg* 2006;32:298–301.

161. Alster TS, Tanzi EL, Lazarus M. The use of fractional laser photothermolysis for the treatment of atrophic scars. *Dermatol Surg* 2007;33:295–299.

162. Ortiz AE, Tremaine AM, Zachary CB. Long-term efficacy of a fractional resurfacing device. *Lasers Surg Med* 2010;42:168–170.

163. Lee HS, Lee JH, Ahn GY et al. Fractional photothermolysis for the treatment of acne scars: A report of 27 Korean patients. *J Dermatol Treat* 2008;19:45–49.

164. Glaich AS, Goldberg LH, Friedman RH, Friedman PM. Fractional photothermolysis for the treatment of postinflammatory erythema resulting from acne vulgaris. *Dermatol Surg* 2007;33:842–846.

165. Waibel J, Wulkan AJ, Lupo M, Beer K, Anderson RR. Treatment of burn scars with the 1,550 nm nonablative fractional Erbium Laser. *Lasers Surg Med* 2012;44:441–446.

166. Glaich AS, Rahman Z, Goldberg LH, Friedman PM. Fractional resurfacing for the treatment of hypopigmented scars: A pilot study. *Dermatol Surg* 2007;33:289–294.

167. Shumaker PR, Kwan JM, Landers JT, Uebelhoer NS. Functional improvements in traumatic scars and scar contractures using an ablative fractional laser protocol. *J Trauma Acute Care Surg* 2012;73:S116–S121.

168. Ozog DM, Liu A, Chaffins ML et al. Evaluation of clinical results, histological architecture, and collagen expression following treatment of mature burn scars with a fractional carbon dioxide laser. *JAMA Dermatol* 2013;149:50–57.

169. Rokhsar CK, Fitzpatrick RE. The treatment of melasma with fractional photothermolysis: A pilot study. *Dermatol Surg* 2005;31:1645–1650.

170. Karsai S, Fischer T, Pohl L et al. Is non-ablative 1550-nm fractional photothermolysis an effective modality to treat melasma? Results from a prospective controlled single-blinded trial in 51 patients. *J Eur Acad Dermatol Venereol* 2012;26:470–476.

171. Izikson L, Anderson RR. Resolution of blue minocycline pigmentation of the face after fractional photothermolysis. *Lasers Surg Med* 2008;40:399–401.

172. Lee WR, Shen SC, Al-Suwayeh SA, Yang HH, Yuan CY, Fang JY. Laser-assisted topical drug delivery by using a low-fluence fractional laser: Imiquimod and macromolecules. *J Control Release* 2011;153:240–248.

173. Haedersdal M, Katsnelson J, Sakamoto FH et al. Enhanced uptake and photoactivation of topical methyl aminolevulinate after fractional CO_2 laser pretreatment. *Lasers Surg Med* 2011;43:804–813.

174. Togsverd-Bo K, Haak CS, Thaysen-Petersen D, Wulf HC, Anderson RR, Haedersdal M. Intensified photodynamic therapy of actinic keratoses with fractional CO_2 laser: A randomized clinical trial. *Br J Dermatol* 2012;166:1262–1269.

175. Weiss ET, Geronemus RG. Combining fractional resurfacing and Q-switched ruby laser for tattoo removal. *Dermatol Surg* 2011;37:97–99.

176. Ibrahimi OA, Syed Z, Sakamoto FH, Avram MM, Anderson RR. Treatment of tattoo allergy with ablative fractional resurfacing: A novel paradigm for tattoo removal. *J Am Acad Dermatol* 2011;64:1111–1114.

177. Wang CC, Huang CL, Sue YM, Lee SC, Leu FJ. Treatment of cosmetic tattoos using carbon dioxide ablative fractional resurfacing in an animal model: A novel method confirmed histopathologically. *Dermatol Surg* 2013;39:571–577.

178. Graber EM, Tanzi EL, Alster TS. Side effects and complications of fractional laser photothermolysis: Experience with 961 treatments. *Dermatol Surg* 2008;34:301–305.

179. Setyadi HG, Jacobs AA, Markus RF. Infectious complications after nonablative fractional resurfacing treatment. *Dermatol Surg* 2008;34:1595–1598.

180. Palm MD, Butterwick KJ, Goldman MP. *Mycobacterium chelonae* infection after fractionated carbon dioxide facial resurfacing (presenting as an atypical acneiform eruption): Case report and literature review. *Dermatol Surg* 2010;36:1473–1481.

181. Shamsaldeen O, Peterson JD, Goldman MP. The adverse events of deep fractional CO(2): A retrospective study of 490 treatments in 374 patients. *Lasers Surg Med* 2011;43:453–456.

182. Tanzi EL, Wanitphakdeedecha R, Alster TS. Fraxel laser indications and long-term follow-up. *Aesthet Surg J* 2008;28:675–678.

183. Anderson RR, Farinelli W, Laubach H et al. Selective photothermolysis of lipid-rich tissues: A free electron laser study. *Lasers Surg Med* 2006;38:913–919.

184. Wanner M, Avram M, Gagnon D et al. Effects of non-invasive, 1,210 nm laser exposure on adipose tissue: Results of a human pilot study. *Lasers Surg Med* 2009;41:401–407.

185. Sakamoto FH, Doukas AG, Farinelli WA et al. Selective photothermolysis to target sebaceous glands: Theoretical estimation of parameters and preliminary results using a free electron laser. *Lasers Surg Med* 2012;44:175–183.

186. Janiczek-Dolphin N, Cook J, Thiboutot D, Harness J, Clucas A. Can sebum reduction predict acne outcome? *Br J Dermatol* 2010;163:683–688.

187. Sattler E, Kastle R, Welzel J. Optical coherence tomography in dermatology. *J Biomed Opt* 2013;18:061224.

188. Mogensen M, Nurnberg BM, Forman JL, Thomsen JB, Thrane L, Jemec GB. In vivo thickness measurement of basal cell carcinoma and actinic keratosis with optical coherence tomography and 20-MHz ultrasound. *Br J Dermatol* 2009;160:1026–1033.

189. Drexler W. Ultrahigh-resolution optical coherence tomography. *J Biomed Opt* 2004;9:47–74.

190. Gambichler T, Orlikov A, Vasa R et al. In vivo optical coherence tomography of basal cell carcinoma. *J Dermatol Sci* 2007;45:167–173.

191. Rajadhyaksha M, Gonzalez S, Zavislan JM, Anderson RR, Webb RH. In vivo confocal scanning laser microscopy of human skin II: Advances in instrumentation and comparison with histology. *J Invest Dermatol* 1999;113:293–303.

192. Wang SQ, Hashemi P. Noninvasive imaging technologies in the diagnosis of melanoma. *Semin Cutan Med Surg* 2010;29:174–184.

193. Gerger A, Hofmann-Wellenhof R, Samonigg H, Smolle J. In vivo confocal laser scanning microscopy in the diagnosis of melanocytic skin tumours. *Br J Dermatol* 2009;160:475–481.

194. Denk W, Strickler JH, Webb WW. Two-photon laser scanning fluorescence microscopy. *Science* 1990;248:73–76.

195. Konig K, Riemann I. High-resolution multiphoton tomography of human skin with subcellular spatial resolution and picosecond time resolution. *J Biomed Opt* 2003;8:432–439.

196. Paoli J, Smedh M, Ericson MB. Multiphoton laser scanning microscopy—A novel diagnostic method for superficial skin cancers. *Semin Cutan Med Surg* 2009;28:190–195.

4

Lasers in Interventional Pulmonology

Anubhav N. Mathur
Indiana University
Medical School

Praveen N. Mathur
Indiana University
Medical School

4.1 Introduction

The epidemic of cancer of the lung has resulted in many patients having malignant obstruction of the central airways. Conventional therapy, including radiation therapy, chemotherapy, or dilation techniques with rigid bronchoscopy, has proved to be inadequate to relieve the symptomatic airway obstruction. Thus, the application of laser technology for the treatment of various tracheobronchial obstructions, using standard rigid and flexible bronchoscopes, has proved to be a major advance in the treatment of these patients and was the original stimulus for the specialty of interventional pulmonology. The successes of laser bronchoscopy have stimulated interest in other endobronchial therapies, such as cryotherapy, electrocautery, and brachytherapy.

Laser bronchoscopists must be experts in diagnostic bronchoscopy techniques, become knowledgeable in laser physics and laser tissue interaction, and be comfortable dealing with patients with respiratory insufficiency, acute hypoxemia, and airway bleeding. This chapter reviews laser physics, tissue interaction, and laser safety as it relates to bronchoscopic applications. It also addresses indications, contraindications, and techniques of laser bronchoscopy.

4.2 Laser Physics

4.2.1 Laser Light

A *light amplification by stimulated emission* of *radiation* (LASER) beam is an intense form of electromagnetic energy channeled into parallel, synchronized rays of light of the same wavelength that are used for many industrial and medical applications. Light represents a portion of the electromagnetic spectrum designated with wavelengths between 100 and 20,000 nm. Included in this range are ultraviolet light, visible light, and infrared light. Visible light is perceived by the human eye as different colors, with the shorter wavelengths of 400 nm seen as violet blue and the longer wavelengths of 700 nm seen as red light. All light is produced by the spontaneous or stimulated emission of energy by atoms, the smallest components of a molecule. The nucleus of the atom contains positively charged particles called protons and neutral particles called neutrons. Orbiting the nucleus are negatively charged electrons, which balance the positively charged protons. Each electron orbit corresponds to a certain energy level and the greater the distance from the nucleus, the higher the energy level of the electron. Light is generated when electrons surrounding an atom spontaneously drop from an orbit with a higher energy level to one with a lower energy level. For this release to occur, the atom must first be *excited* to the higher level by an external photon of energy. Photons are discrete particles of radiant energy that travel in waves and provide the energy necessary to form light. The energy of a photon is related to the frequency (Hz) of the wave multiplied by a constant (Planck's constant = 6.623×10^{-34} J s) and is inversely proportional to the wavelength. This influx of energy propels the electron to a more distant orbit corresponding to a higher energy level. If an atom in an excited state is again stimulated by a photon of the same energy as the one absorbed, two photons of equal energy and frequency will be released when the electron's orbit decays to a lower level [1]. This release is termed stimulated emission of radiation and is the basis for laser light energy. When the process is amplified to an intense level by an external energy source, a specialized form of light energy is created, which we term laser light. Lasers create their special form of light energy by using an optical cylinder containing a medium consisting of solids, gases, or dyes. The cylinder contains a fully reflective mirror at one end and a partially reflective mirror at the other end. Atoms of the active laser medium become stimulated when an external source adds thermal, electrical, or optical energy to the system. Spontaneous emission of energy occurs as electrons change orbit around the atoms within the medium, giving off energy in various directions. As the photons of energy bounce off the walls, mirrors, and other atoms within the cylinder, intensification of the electromagnetic energy occurs. The light energy leaves through the partially transmissive mirror at a rate that is dictated by the incoming external energy source. The energy is emitted as laser light and follows the special properties of stimulated light emission [2,3]. Three unique properties of laser light make lasers particularly useful [1,3]. First, all waves of laser light are in phase with each other and have identical frequencies, wavelengths, and velocities as they travel through space. This property is known as coherence, and it results in a concentration of laser power at the target point. Waves of laser light are also considered monochromatic because they have nearly the same color, wavelength, and energy. This characteristic permits all of the radiant power to be centered in a narrow electromagnetic spectral band. Finally, all waves of laser light are essentially parallel and thus are considered collimated. This property permits lasers to travel great distances with minimal loss of intensity of light energy. Ordinary light energy (i.e., incandescent bulbs) will scatter in multiple directions at varied wavelengths, making it good for illumination but inadequate for more precise medical applications. Laser light, on the other hand, is emitted in a narrow, parallel bundle with minimal scatter into space [3,4], making it particularly useful as a surgical tool. Compared with most other forms of light, lasers also have a high luminosity (brightness), which makes them useful for light therapy.

Two other important characteristics of laser light determine a laser's impact on a target. The transverse electromagnetic mode (TEM) is one measure of a laser's distribution of intensity in a cross-sectional dimension. Although lasers are collimated and coherent, the impact configuration on a target is not perfectly uniform, as illustrated in the ideal mode. Instead, the actual cross-sectional impact configuration of most lasers is symmetric in a bell-shaped or Gaussian distribution, with more beam intensity being concentrated in the center of the target. This impact configuration is termed basic mode.

In addition to the TEM, a laser beam's intensity is also determined by its power density, which is the concentration of power per unit of cross-sectional area of the beam. Optical lenses focus the laser beam into a small focal spot, creating a high power density in that area. For each laser, the beam focus is fixed to prevent divergence and loss of power density. The power density is a function of the size of the focal spot. The smaller the focal spot, the higher the power density. The power density can also be intensified by increasing the incoming power of the beam, which is usually in the range of 20–100 watts (W) for most medical lasers [1]. Therefore, power density is proportional to incoming power and inversely proportional to the area of the focal spot. Although the power consumed by a laser is less than that of an ordinary light bulb, the intensity of a concentrated beam makes it useful for cutting, coagulating, and vaporizing tissue.

A laser's TEM and power density are important determinants of the laser beam's effect on the target tissue. Because the TEM determines the profile of the beam's effective cross-sectional area, it also determines the pattern of tissue interaction. The tissue cavity created by laser destruction has the same configuration as the intensity profile (basic mode) of the laser beam. The extent of tissue destruction is a function of the amount of absorption, the length of exposure, and the power density of the laser beam. When laser light is applied to tissue, some of the energy penetrates into the tissue and is absorbed producing heat, whereas the rest of the energy is reflected back. Power density at the surface of the tissue will be higher than the power density farther inside because of the partial reflection of energy. Absorption attenuates the original power density incrementally as it passes deeper into the tissues. The distance below the surface of the tissue at which the power density has decreased to a percentage of the original level (usually 1%) is termed the attenuation depth [3]. It is differentiated from penetration depth, which is the distance at which the laser energy no longer has significant effects on the tissue. Laser beams can still have a considerable effect on tissues even after 99% attenuation if the exposure is prolonged or if the power density of the original beam is very high. For instance, if the laser beam has an original power density at the surface of 100 W/cm², only 1 W/cm² will be present at the 99% attenuation depth. However, if the original power density at the surface is 10,000 W/cm², 100 W/cm² will be present at the 99% attenuation depth. The attenuation depth is a function of the tissue type and the wavelength of the laser light, whereas the penetration depth is dependent on the tissue type, wavelength, and the power density.

The amount of attenuation that occurs secondary to absorption is termed the attenuation coefficient [3]. This value is the sum of the absorption coefficient (a) and the scattering coefficient (0) [A = a + uI]. Each laser has a different scattering characteristic when the beam penetrates living tissues, but the amount of scattering within tissues is generally dependent on the tissue type and wavelength of the laser beam. Lasers that have high scatter coefficients function well as coagulators, whereas lasers that have high absorption coefficients are more useful for precision cutting. For instance, a carbon dioxide (CO_2) laser has a high absorption coefficient that produces rapid tissue vaporization and makes it an ideal cutting instrument. Its low scattering coefficient, however, limits its ability to photocoagulate bleeding tissues. In contrast, a neodymium, yttrium, aluminum, garnet (Nd:YAG) laser functions well as a coagulator because of its high scattering coefficient, but it is a relatively poor cutting instrument. Nevertheless, the Nd:YAG laser is capable of serving both functions because at higher power settings or within dark tissues, more absorption and vaporization occur. The main determinant of a laser's effect on a particular tissue is the proportion of the scattering and absorption coefficients at the wavelength of the individual laser.

4.2.2 Laser Tissue Interaction

Lasers can interact with living tissues in different ways [2]. Light energy is converted into heat energy, enabling the thermal effects of laser radiation to be used for cutting, vaporization, and coagulation of tissues. Because of the high water content, laser energy is readily absorbed by living tissues. This laser energy is rapidly converted into heat, which results in cell destruction. Lasers can also mechanically disrupt tissues either by pressure waves or by inducing vaporization of intracellular or extracellular water, resulting in separation of tissue layers. This interaction is useful in the removal of cataracts or intravascular atheromatous plaques. Laser light energy can also stimulate various biochemical reactions by interacting with photosensitizing chemicals absorbed by abnormal tissues. This use of laser light is termed photodynamic therapy, which is used in the diagnosis and treatment of patients with tracheobronchial malignancies [5–8].

The performance characteristics of individual lasers are determined by the substance used as the laser medium. Each substance produces varying wavelengths of emitted light. The first laser used a synthetic ruby crystal as a laser medium; now, commonly used materials include argon, krypton, fluorescent organic dyes, gold vapor gas, helium–neon, potassium titanyl phosphate (KTP), CO_2, and a combination of erbium (Er) and Nd:YAG.

The argon gas laser has a wavelength of 514 nm in the visible blue-green spectrum of light and can be transmitted through optical fibers. Because of its blue color, however, much of its energy is absorbed in hemoglobin, which greatly limits its use in photoresection of endobronchial malignancies [9]. The gold vapor and liquid dye lasers are used to excite the photosensitizer dihematoporphyrin ether, which is selectively retained by tumor cells [7,8]. By exposing the tissue to appropriate laser light, clusters of tumor cells can easily be identified and treated. The helium–neon gas laser has no therapeutic role in laser bronchoscopy, but it produces light in the visible spectrum, making it useful as an aiming guide for other lasers, such as the CO_2 and Nd:YAG lasers. The KTP laser uses a solid medium similar to the Nd:YAG laser and has been used in photocoagulation and resection of tracheobronchial stenosis [10].

The CO_2 laser has commonly been used for lesions of the aerodigestive tract [11]. The CO_2 laser has a long wavelength of 10,600 nm, resulting in prompt absorption by most living tissues. This invisible infrared electromagnetic energy is rapidly converted to thermal energy, resulting in vaporization of the water in the tissue. If the tissue has a high water content, absorption is further enhanced and tissue destruction is more precise. The CO_2 laser's high absorption and low scatter coefficients give it a predictable depth of penetration and make it ideal for precise surgical applications [3,12]. Tonsillectomy, adenoidectomy, and uvulopalatopharyngoplasty (UPPP) have been performed with the use of the CO_2 laser [13]. Although well accepted for the treatment of benign and malignant lesions of the head and neck [14], its use in tracheobronchial malignancies has been limited. The CO_2 laser can seal small vessels and lymphatics up to 0.5 mm in diameter [15,16], but it is an ineffective photocoagulator and cannot reliably control bleeding from larger vessels. Another important drawback of the CO_2 laser is its cumbersome articulated mirror delivery system, which requires a micromanipulator and surgical microscope for operation. Use of this system for lower airway tumors requires precise alignment of the laser beam down the entire length of a rigid bronchoscope.

Recently, novel hollow-core optical fibers for CO_2 lasers were developed, which offer high flexibility. It is mechanically robust with good optical performance under tight bends. These fibers have been used with rigid and flexible bronchoscopes as well as various handpieces. This has allowed for delicate and precise laser surgical procedures in a minimally invasive manner. The hollow fiber was invented at the Massachusetts Institute of Technology: the hollow-core fiber had a high reflective surface for optical rays of all angles known as an omnidirectional mirror that is used to confine and guide light through the fiber [17]. Despite very high reflectivity of the mirror structure, some of the guided radiation still penetrates into the fiber walls, especially where the fiber is bent, which causes heating of the fiber. A cooling gas is passed at 8 L/min through the hollow core to cool the fiber. This rapid gas flow is potentially detrimental in the lower airway as it can cause pneumothorax; gas emboli thus have not been used. However, it has been very successfully used in the upper airway [18,19].

4.3 Laser Bronchoscopy

The Nd:YAG laser has been used extensively for the treatment of patients with tracheobronchial malignancies because the laser beam is delivered through a quartz optical fiber, which is easily adaptable for use with either the flexible or rigid bronchoscope [20]. The Nd:YAG laser operates in the invisible near infrared range of the electromagnetic spectrum at a wavelength of 1064 rim. At this wavelength, scattering exceeds absorption for most lighter-colored tissues, which makes the Nd:YAG laser effective at coagulating large amounts of tissue. If the tissue is darker pigmented or if a higher power setting is used, absorption and vaporization can be increased. The wide power range (5–100 W) of the Nd:YAG laser contributes to its versatility by permitting photocoagulation at low settings and vaporization at higher settings. The Nd:YAG laser penetrates to about 6 mm and will coagulate vessels of up to 2 mm in diameter. A new type of Nd:YAG laser has been developed with a wavelength of 1320 nm, which may have advantages over the current system because of slightly less tissue scatter [9]. The Nd:YAG laser fiber can be fitted with a sapphire tip to permit a contact operational mode, but because these tips limit viewing, contact probes are usually not used in tracheobronchial endoscopy [4]. Because the noncontact Nd:YAG laser is the most common laser used for tracheobronchial disease, the rest of this chapter will concentrate on the use of this laser for bronchoscopic applications.

4.3.1 Indications

The Nd:YAG laser has been used to reestablish airway patency in patients with primary bronchogenic and metastatic malignancies, benign tumors, tracheal stenosis, and tumors of uncertain prognosis [21]. Malignant tracheobronchial obstruction from primary bronchogenic carcinoma is the most common indication for Nd:YAG laser bronchoscopy and accounts for 49%–75% of the cases (Table 4.1) [21–23,25–27]. Tracheobronchial obstruction as a result of endobronchial metastasis from renal, thyroid, colon, breast, and esophageal carcinomas accounts for between 6% and 18% of laser bronchoscopies and is the next most common indication for the procedure. Although they are responsible for less than 10% of the cases of tracheobronchial obstruction, benign tumors, such as hamartomas, fibromas, papillomas, lipomas, and amyloidomas, are an excellent indication for use of the Nd:YAG laser [24]. Benign tumors often originate from a pedicle base in the tracheobronchial mucosa and are usually not vascular, making them particularly attractive for flexible endoscopic laser resection [25]. Laser treatment of the base of lesions may help prevent recurrences, and in some cases, treatment of lesions, such as hamartomas and papillomas, can be curative. Endobronchial amyloid and tracheopathia osteoplastica are usually not curable by laser resection, but these lesions are relatively slow growing and often do not require multiple treatments [26].

At many institutions, tracheal stenosis represents a relatively common indication for laser bronchoscopy, accounting for 10%–27% of the cases. Tracheal stenosis usually occurs after translaryngeal intubation or tracheostomy, but it can occur without a known cause or in response to granulomatous or diphtherial infections. Personne et al. [25] have identified three main types of iatrogenic stenoses. The diaphragm type of stenosis (type I) consists of a concentric fibrotic intraluminal stricture without involvement of the tracheal wall and is ideal for laser resection. The second type of stenosis (type II) involves extrinsic, *bottleneck narrowing* of the trachea secondary to collapse of the tracheal wall. Because no intraluminal component is present, use of the Nd:YAG laser for this type of narrowing is contraindicated. The last type of stenosis (type III) is a combination of the first two types and may be acceptable for laser therapy of the intraluminal fibrotic ring in certain situations. If stenosis is suitable for endoscopic therapy, the laser is used to make radial incisions in the tracheal scar before gentle rigid bronchoscopic

TABLE 4.1 Indications for Laser Bronchoscopy

Author	Year	Patients	Bronchogenic Carcinoma	Metastatic Carcinoma	Tracheal Stenosis	Benign Tumor	Uncertain Prognosis	Miscellaneous
Cavaliere et al.[19]	1994	1884	1419 (75%)	105 (5.6%)	Not included	156 (8%)	143 (8%)	61 (3.2%)
Cavaliere et al.[23]	1988	1000	612 (61%)	37 (3.7%)	139 (13.9%)	59 (5.9%)	64 (6.4%)	89 (8.9%)
Brutinel et al.[24]	1987	116	71 (61%)	17 (14.7)	Not included	9 (8%)	13 (11%)	6 (5%)
Beamis et al.[20]	1991	269	200 (74%)	36 (18%)	27 (10%)	6 (2%)	Not included	Not included
Personne et al.[22]	1986	1310	643 (49%)		389 (27%)	75 (5.7%)	57 (4%)	146 (11%)
Dumon[18]	1985	544	278 (51%)		132 (24%)	31 (6%)	20 (4%)	83 (15%)[a]

[a] Suture removal, 29; hemorrhage, 20; foreign body, 4.

TABLE 4.2 Characteristics of Lesions for Laser Bronchoscopy

Favorable Lesions for Laser Resection	Unfavorable Lesions for Laser Resection
1. Tracheal and main stem lesions	1. Extrinsic lesions
2. Polypoid lesions	2. Diffuse lesions with extensive submucosal involvement
3. Short lesion length	3. Upper lobe and segmental lesions
4. Large endobronchial involvement	4. Long tapering lesions
5. Visible distal bronchial lumen	5. Total bronchial obstruction
6. Functional lung distal to the obstruction	

dilation of the tracheal or subglottic stenosis [28,29]. Some patients with severe obstruction of the lower airway may require a permanent tracheostomy or implantation of a silicone stent or a Montgomery tracheal T tube to maintain patency of the airway [30,31]. If implantation of the stent fails to achieve long-term patency, tracheal sleeve resection may be necessary, but symptomatic clinical improvement after relief of tracheal obstruction using laser bronchoscopy has been well documented [28,29,32].

A less common indication for endobronchial laser therapy is treatment of tumors of uncertain prognosis, which include carcinoid, adenoid cystic, spindle cell, mucoepidermoid, and other mixed cell types. These lesions should be resected surgically if possible, but in many instances, surgery is impossible because the lesions are widespread or the patient is a poor operative candidate. The Nd:YAG laser can be used to treat these low-grade malignant tumors effectively, which accounts for approximately 5% of laser bronchoscopies. Other infrequent uses for the Nd:YAG laser include treatment of tracheobronchial granulomas [26], removal of suture threads, control of hemorrhage from biopsy sites or necrotic mucosa [21], and even vaporization of granulation tissue surrounding an aspirated foreign body [33].

Nd:YAG laser bronchoscopy is indicated for almost any centrally obstructing lesion within the airway lumen of the tracheobronchial tree, but some lesions are more favorable for treatment than others (Table 4.2). Tracheal and bronchial main stern lesions are more easily treated than lesions in the distal bronchi, and they represent the primary indication for laser bronchoscopy. Although location of the lesion is a primary concern, the appearance of a lesion is also important. For instance, mechanical resection of short, polypoid lesions is technically easier than resection of lesions that involve a longer portion of the airway. Because much of laser bronchoscopy involves the use of the rigid bronchoscope to physically resect the tumor, long tapering lesions with extensive submucosal infiltration make the procedure technically more difficult and often reduce the chances for a successful outcome. Complete obstruction of the airway lumen also complicates the procedure because it makes it difficult to know the course of the bronchus beyond the area of obstruction. This information is crucial to avoid inadvertent perforation of the airway wall during laser resection. Most patients do not have complete airway obstruction, and a passage beyond the lesion can often be found by gently probing the area with a suction catheter or a flexible bronchoscope. If the course of the bronchus can be determined, mechanical and laser resection can proceed. Otherwise, complete obstruction of the airway lumen is a contraindication to laser bronchoscopy, and it represents the most common reason for not performing laser resection [34].

4.3.2 Contraindications

Because the procedure is often used as palliative therapy for individuals with severe malignant tracheobronchial stenosis, there are actually few contraindications to laser bronchoscopy. The only absolute contraindication to laser bronchoscopy is lack of an intraluminal lesion. Because lesions extrinsic to the airway are not amenable for treatment, it is important to confirm that the airway narrowing is not the result of compression from mediastinal tumors, lymphadenopathy, lobar collapse, or type II tracheal stenosis. Failure to recognize extrinsic airway compression as a cause of airway narrowing can result in inadvertent perforation of the tracheobronchial wall. Except for extraluminal disease, there are few other reasons not to consider laser resection even in patients who are at a higher surgical risk. Personne et al. [25] performed laser resection on 267 patients *in extremis* with severe hypoxia or coma and recorded only a 2.7% mortality rate. A low

TABLE 4.3 Contraindications to Laser Bronchoscopy

Absolute contraindications
  Extraluminal disease
Relative contraindications
  Cardiovascular
    Recent myocardial infarction
    Ventricular arrhythmias
    Conduction abnormalities
    Hypotension
    Decompensated congestive heart failure
  Respiratory
    Severe obstructive lung disease
    Extensive tumor involvement
    Involvement of pulmonary artery
    Unsalvageable lung distal to obstruction
    Chronic collapse
Small cell carcinoma cell type
Posterior tracheal wall location or tracheoesophageal fistula
Extensive preprocedure radiation therapy
Electrolyte abnormalities
Bleeding diathesis
Sepsis
Tracheobronchial malacia

rate of fatal complications of 0.3%–3% has been confirmed by three other large series [21,23,26]. This low perioperative mortality rate indicates that the procedure can be an acceptable palliative treatment.

Although there are few absolute contraindications to laser bronchoscopy, a number of relative contraindications need to be considered (Table 4.3). A preoperative evaluation is often used to identify cardiovascular and respiratory problems that might place the patient at an unacceptably high risk for complications. Although knowledge of underlying cardiopulmonary problems is important and may preclude surgery in some individuals, the palliative benefits of laser bronchoscopy frequently outweigh the small risk of post procedural complications. In many instances, respiratory insufficiency is both the primary indication for the procedure and the reason why the patient is a higher surgical risk. For some patients, bronchoscopic laser resection represents the only chance for rapid relief of dyspnea. Failure to wean patients from mechanical ventilation after bronchoscopy is infrequent and usually occurs as a result of being unable to establish an effective airway and not from underlying obstructive lung disease.

It is important to know whether there is functional lung distal to the obstruction. Lung parenchyma that is severely damaged by radiation treatments or recurrent post obstructive pneumonia is unable to participate in gas exchange even when the obstruction is removed [35,36]. In addition, significant tumor involvement of the pulmonary artery can result in irreversible ventilation–perfusion inequalities, negating any effects of laser resection. Computed tomography of the chest can be helpful in evaluating lung parenchyma in the region of the obstruction. Nonfunctional lung found distal to the region of obstruction represents a strong relative contraindication to laser resection.

The type and location of the lesion may be additional relative contraindications to laser bronchoscopy because they influence the likelihood of procedural complications. Tumors involving the posterior tracheal wall can extend into the esophagus, increasing the risk of perforation or fistula formation during laser applications. The risk of airway erosion and perforation is also increased by extensive preprocedure radiation therapy, which can distort and soften the tracheal wall [21]. Particular caution should be taken with upper lobe lesions because laser treatment can result in excessive bleeding owing to the proximity of major vessels. The cell type of the tumor can influence the success of the procedure and may be an indication to use other methods of treatment. For example, small cell carcinoma of the lung is often diffuse and usually contains a large extra bronchial component, making the lesion unfavorable for laser resection. Coagulopathies, electrolyte abnormalities, hypotension, and sepsis represent other obvious contraindications to laser endoscopy and should be corrected before the procedure.

4.3.3 Anesthesia

The method of anesthesia delivery for Nd:YAG laser bronchoscopy depends on the type of bronchoscopic technique. For flexible bronchoscopies, intravenous conscious sedation is usually employed in a fashion similar to standard diagnostic bronchoscopy. If the patient is intubated, however, intravenous general anesthesia can be given. Anesthesia for the rigid bronchoscopic procedure is usually accomplished by a combination of intravenous agents, including midazolam, propofol, fentanyl, and occasionally a depolarizing neuromuscular blocker. If a closed rigid bronchoscopic method is preferred, ventilation is achieved through the side arm of the bronchoscope. Inhalational anesthetic agents can be used with a closed system, but there may be more potential for cardiopulmonary complications [37]. Some bronchoscopists prefer to operate through the open end of the rigid bronchoscope, and thus a different type of ventilation *is* required. Several authors prefer spontaneous assisted ventilation [38], but other [23,39–42] use a manual Venturi jet ventilation technique. This method of ventilation enables the surgeon to operate through the open end of the rigid bronchoscope while ventilation is maintained by manually injecting an air–oxygen mixture at 20–60 breaths per minute [40,41,43]. A high-frequency jet ventilation technique has also been described wherein a machine delivers approximately 300 breaths per minute [44]. The effectiveness of manual Venturi jet ventilation has been well validated [45,46], and the use of continuous CO_2 monitoring is not required. Pulse oximetry is recommended to avoid inadvertent desaturation events during the procedure. After rigid bronchoscopy, most patients are easily extubated in the operating room after clearance of the intravenous anesthesia. Patients usually stay a few hours in the postanesthesia care unit after which they are either transferred back to their hospital room or discharged. General anesthesia for rigid laser bronchoscopy can be performed safely through various techniques, and most elderly or debilitated patients can tolerate the intervention without difficulty.

4.4 Techniques for Laser Bronchoscopy

4.4.1 Rigid versus Flexible Bronchoscopy

The Nd:YAG laser can be used with either a flexible or rigid bronchoscope, and which instrument is better has been discussed widely [47,48]. Ironically, the original papers in the early 1980s by Toty et al. [49] and Dumon et al. [50] described the use of the Nd:YAG laser with both the flexible and rigid bronchoscopes. For the next few years, the flexible bronchoscope was used alone or in conjunction with an endotracheal tube or rigid scope in the majority of laser resections [27,34,51,52]. Subsequently, reports of major complications, including deaths resulting from massive hemorrhage [27,37,53] and airway fire [54], began to surface, and by the late 1980s, many authors [27,55] advocated using the rigid bronchoscope for most Nd:YAG laser resections. Each method is associated with certain advantages and disadvantages (Table 4.4).

TABLE 4.4   Indications for the Flexible Fiberscope in Laser Bronchoscopy

In combination with a rigid bronchoscope
Benign tumors (granulomas, papillomas)
Small malignant tumors
Distal tumors
Less than 50% obstruction of airway lumen
Recurrent tumors
Narrow based pedunculated tumors
Hemoptysis photocoagulation

Source: Adapted from Ramser, E.R. and Beamis, J.F., Jr., *Clin. Chest Med.*, 16, 415, 1995.

4.4.2 Flexible Bronchoscopy

The major advantage of the flexible bronchoscope is its versatility and familiarity to the pulmonologist. In the Bronchoscopy Survey by the American College of Chest Physicians [56], the flexible scope was used by more than 80% of responding physicians at some point in their practice. The popularity of the flexible scope for laser procedures may partly be related to the limited number of physicians who are trained in rigid bronchoscopy. In this survey, only 8.4% of the responders used the rigid scope in their practice. This trend away from the use of rigid bronchoscopy limits the choice of methods to apply the Nd:YAG laser, but in some instances, flexible bronchoscopy may actually be the preferred technique. The flexible scope is small and maneuverable, which permits easier access to distal airway lesions and improves visualization of the bronchial tree. Aiming the laser beam is also easier with the flexible scope because the operator has more control over the tip of the laser fiber. Because of its size and controllability, it may also be passed through the rigid bronchoscope for dissection and coagulation purposes. Usually, the flexible bronchoscope is reserved for lesions that do not have a propensity for hemorrhage, such as granulomas, papillomas, and small distal malignant tumors. The flexible laser bronchoscope is ideal for recurrent tumors and pedunculated tumors with a narrow base. The versatility and familiarity of the flexible scope make it well adapted for the safe and effective laser resection of specific lesions within the tracheobronchial tree.

Other proposed advantages of flexible bronchoscopy for laser applications include potential cost savings and the avoidance of general anesthesia. Because flexible bronchoscopy is often performed outside of an operating room setting, it can be more convenient and less costly than rigid bronchoscopy. For large obstructing lesions, however, multiple procedures are often required, frequently negating any anticipated cost savings. In one study [57], an average of 1.97 treatments were required to obtain a response compared with only one treatment session using a rigid bronchoscope. Additionally, each procedure may be more time-consuming for the operator because removal of tumor debris is less efficient through the flexible bronchoscope. The frequent need for multiple procedures and the increased investment of time minimize the proposed advantages of flexible bronchoscopy for laser photoresection.

Another theoretical advantage of the flexible bronchoscope is the potential to avoid the inherent risks of general anesthesia, including myocardial infarction, arrhythmia, hemodynamic instability, CO_2 retention, barotrauma, and prolonged neuromuscular weakness [37,39,58]. Unger et al. [48,49] advocate the use of flexible bronchoscopy for most tracheobronchial laser procedures and report no significant anesthesia-related hemodynamic or ventilatory complications in more than 290 procedures performed under topical anesthesia. Cardiovascular complications, such as myocardial infarction, however, can still result from flexible bronchoscopy [57], especially if severe hemorrhage or airway compromise occurs. In fact, the only deaths in some studies using flexible bronchoscopy for laser procedures have not been the result of anesthetic complications but rather due to uncontrollable hemorrhage [27,37,57]. Although complications related to general anesthesia for rigid bronchoscopy have been well documented [23,25,26,38], the overall rate of occurrence is exceedingly low. Cavaliere et al. [22] reported respiratory failure in only 0.5%, cardiac arrest in 0.4%, and myocardial infarction in 0.2% of 2253 cases in which more than 90% were performed under general anesthesia using a rigid bronchoscope. In a separate prospective study [38] of 124 interventional rigid bronchoscopies, no cardiac complications occurred, and transient reversible hypoxemia was seen in only 15% of cases. Avoidance of complications from general anesthesia and potential cost savings represent only theoretical advantages of flexible bronchoscopy, but they may still be important considerations in some patients.

4.4.3 Rigid Bronchoscopy

Although there may be advantages to flexible bronchoscopy for certain situations, the rigid bronchoscope is now generally preferred for most large obstructing tracheobronchial lesions [23,25,26,55]. Rigid laser bronchoscopy has a number of important advantages over flexible bronchoscopy, including efficiency, effectiveness, and, most importantly, safety. It is usually more efficient because larger pieces of tumor material can be removed through the rigid scope, significantly reducing the time of the procedure and the total number of procedures required [47,57]. Often only one procedure is required because the barrel of the bronchoscope can be used to *core* out the obstructing endoluminal lesion and quickly reestablish airway patency. The barrel can also be used to dilate a narrowed trachea and possibly eliminate the need for surgical resection. Other advantages of rigid bronchoscopy include the ability to perform stent implantation or removal at the time of the laser intervention. Both silicone stents and self-expandable metal stents can be placed through a rigid bronchoscope.

Although improved efficiency is desirable, the most important advantage of rigid bronchoscopy is safety. Three safety priorities for laser bronchoscopy [61] are maintenance of ventilation, effective suction, and a clear visual field. They felt the rigid bronchoscope was the most effective instrument to accomplish these goals. With rigid bronchoscopy, the airway is secure, and light general anesthesia can be administered for patient comfort and surgeon convenience. Maintaining a clear field and patent airway is more easily achieved with the rigid bronchoscope because of better suctioning capabilities. With the flexible scope, suction is limited because the working channel is partially occluded by the laser fiber. Control of bleeding is also superior with the rigid bronchoscope because the barrel of the scope can be used to tamponade the hemorrhaging site. An epinephrine soaked gauze pad can even be passed through the rigid scope and placed directly on the bleeding area. Although some physicians [59–62] report that the bleeding rate may actually be higher with the rigid technique because of the tendency to use the scope as a shearing tool, most authors [55,56] believe the ability to maintain a patent airway and control any excessive bleeding is of paramount importance. Thus, the rigid bronchoscope is a safer instrument for laser application in patients at higher risk for procedural complications.

Both rigid and flexible bronchoscopies are associated with advantages and disadvantages that need to be considered when deciding on the best technique to employ the Nd:YAG laser. Each case should be handled individually, but the primary advantages of the rigid bronchoscope are its efficiency, effectiveness, and safety in the treatment of large malignant obstructing lesions of the tracheobronchial tree. For distal small airway lesions or nonvascular benign tumors, the flexible bronchoscope can be used safely and effectively for laser applications. Debate will likely continue as to which is the best technique for laser bronchoscopy, but the answer is more likely a factor of the type and location of the lesion rather than the actual bronchoscopic instrument used. Bronchoscopists performing laser photoresection of tracheobronchial lesions should be proficient in both techniques to assess the risks and benefits of each technique as it applies to a particular patient.

4.5 Safety

Laser safety requires the knowledge of the properties of individual lasers and the potential hazards that can arise with improper use. Each laser has a different capacity for producing injury, and the American National Standards Institute (ANSI) has classified lasers (I–IV) based on the degree of their potential hazard [63]. This classification system uses an accessible emission limit (AEL) and a maximum permissible exposure (MPE) limit for various organs [1]. The AEL is the amount of laser radiation that can be collected by a specific detector per unit of area. The MPE corresponds to the total energy absorbed by the tissue and is a function of the wavelength of the laser and the duration of exposure. A class I rating is given to lasers that do not emit hazardous levels of radiation and represent the lowest risk for injury. Class II lasers are also very safe if exposure duration is limited to less than 0.25 s in the visible band of light because the natural aversion reflex of the eye is rapid enough to avoid injury by this type of laser. Their low power gives an AEL of below 1 mW, which is the arbitrary power level that separates class II lasers from the next laser class. Class III lasers have a wide power range between 1 and 500 mW and can be potentially dangerous even if viewed momentarily. Most surgical lasers are classified as class IV because they have a power output above 500 mW. The main safety risks for surgical lasers are thermal and ocular injuries related to accidental reflection of laser light.

Inadvertent injury to the patient and operating room personnel is an important safety concern with the use of class IV surgical lasers. Misdirected laser light energy can reflect off instruments and specular (shiny) surfaces located near the operating field causing corneal, retinal, and thermal injury to exposed areas. Surgical instruments can be roughened or blackened to help diffuse and absorb these potentially dangerous reflections of laser light [64]. Even if a laser beam is reflected, it retains much of its focus and intensity because of the collimation and coherence properties of laser light [65]. The dangers of reflected laser light are often not readily apparent because Nd:YAG and CO_2 lasers have wavelengths in the invisible range of the electromagnetic spectrum. The bronchoscopist may actually be at the highest risk for injury because Nd:YAG laser light can scatter and reflect back along the endoscope. This is usually not a concern with the CO_2 laser as the optics of the system protect the operator from injury. To avoid inadvertent injury from reflected laser light, safety glasses or eye shields are mandatory for all operating room personnel as well as for the patient. In addition, the windows in the procedure room should be covered to protect anyone outside the room from becoming accidentally exposed. Signs indicating the use of laser light should also be placed on the doors entering the room. Certain built-in precautions, such as foot pedals for laser activation and a standby mode, help to avoid accidental deployment of laser energy.

One of the most well-recognized hazards of class IV surgical lasers is retinal or corneal injury. Although the eye has a natural lid aversion reflex that limits exposure to bright light, the power density of certain lasers is great enough to produce significant damage. Because of their various wavelengths, lasers affect the eye in different ways. Laser beams with short wavelengths in the ultraviolet region (100–400 nm) or long wavelengths in the far infrared region (1,400–10,000 nm) of the electromagnetic spectrum are absorbed by the cornea and lens of the eye. Lasers with wavelengths in the visible and near infrared region (400–1400 nm), such as the Nd:YAG laser, are more dangerous to the retina. This area of the electromagnetic spectrum has been termed the retinal hazard region because laser light of this wavelength can be focused by the lens onto the retina with a magnitude of up to 100,000 times [65]. If the beam becomes focused onto a small point on the retina, significant damage can occur from thermal effects. If the damage is in an area, such as the fovea of the macula, irreversible loss of fine vision can occur. Unfortunately, the loss of vision is usually permanent because neural tissue has limited ability to regenerate itself. For this reason, all lasers in classes II–IV should be considered potentially dangerous to the eye if adequate precautions are not taken. The type of eye protection is important because safety glasses are designed for use with lasers of a specific wavelength. Eye protection manufactured for use with one type of laser may not be protective against lasers with a different wavelength. For instance, clear plastic safety glasses are adequate for brief exposures to CO_2 laser energy because of the limited penetration of CO_2 laser light, but a Nd:YAG laser can easily penetrate plastic or glass. For protection against the Nd:YAG laser, special eye wear with an optical density of 7.0 at 1064 nm is required.

Although ocular damage is the most feared safety risk, the skin and airway mucosa are also susceptible to injury from the laser. The likelihood of injury to the skin may actually be greater than to the eye because of the larger area of exposure. Lasers in the visible and near infrared region (400–1400 nm), such as the Nd:YAG laser, can produce thermal injury severe enough to create third-degree burns [65]. The CO_2 laser, however, has a longer wavelength (10,600 nm) that limits skin penetration and reduces the risk of a severe burn. Nevertheless, thermal injury can still occur if exposure is prolonged or if plastic, paper, or cloth items near the surgical field are accidentally ignited by the laser. The airway mucosa is also vulnerable to thermal injury because airway fires can occur if the flexible bronchoscope, endotracheal tube, or plastic suction catheter is accidentally ignited by the laser [54,66].

A polyvinyl chloride (PVC) endotracheal tube should not be used with a CO_2 laser because it is flammable, and toxic gases can be released from the burned plastic. The use of a red rubber tube with aluminum wrapping can reduce the risk of airway fires caused by the CO_2 laser [67]. This special tube should not be used with a Nd:YAG laser because the darker rubber tube may actually absorb more laser energy. The risk of airway fires can also be reduced by limiting the power of the Nd:YAG laser exposure time and the Fi_{O2} [68]. The use of a metallic rigid bronchoscope will also substantially reduce the risk of airway fires, although with proper precautions, a flexible bronchoscope or endotracheal tube can be used safely during laser applications. Although airway fires and burns are rare events, they are important safety considerations when using most types of surgical lasers.

The composition of the smoke plume generated by the laser has also recently come under investigation [69]. In 1988, a questionnaire study by Lobraico et al. [70] demonstrated an association between CO_2 laser treatment of papillomas and the subsequent development of these lesions in the treating physicians. Furthermore, Garden et al. [71] demonstrated the presence of intact human papillomavirus (HPV) DNA in the vapors of laser-treated verrucae and cautioned practitioners about the potential transmissibility of infection. Abramson et al. [72], however, were unable to demonstrate HPV DNA present in the smoke plume from vaporization of laryngeal papillomas unless direct suction contact was made with the tissue during surgery. The presence of human immunodeficiency virus (HIV) DNA in laser smoke has also been documented, but sustained growth of the cultured cells did not occur, possibly because of damage produced by the laser energy [73]. The risk of clinical disease in surgical personnel from viral transmission has not yet been established, but special safety recommended until further studies can be performed.

Certain guidelines need to be followed for the *safe* operation of the Nd:YAG laser during bronchoscopy. For flexible bronchoscopy, these recommendations concentrate on the prevention of hemorrhage and airway fires. To help avoid these complications and achieve successful results, the location of the tip of the laser fiber should be known at all times. When firing the laser, the tip of the fiber should be at least 4 cm from the distal end of an endotracheal tube and at least 4 mm from the end of the bronchoscope to prevent ignition of the plastic. The concentration of inspired oxygen should also be 40% or less to decrease the risk of airway fires. The tip of the laser fiber is best positioned 0.4–1 cm from the lesion and should be kept clean at all times. When appropriately positioned, the laser should also be fired parallel to the airway wall in pulses of 0.4–1 s in duration with a setting of 20–40 W. The goal of Nd:YAG laser application should not be to vaporize the tumor completely but rather to photocoagulate the lesion to produce hemostasis and necrosis. Residual tumor can be removed mechanically with biopsy forceps, a balloon catheter, or vigorous suctioning, eliminating the need for vaporization. Aggressive use of laser energy should also be avoided because penetration of the Nd:YAG laser is often deeper than is apparent on the surface, and the full extent of tissue necrosis is not complete until 24–48 h after the procedure [48]. Instead, repeat bronchoscopy is recommended in a *few* days to assess the extent of the remaining tumor and need for further treatment. For smaller lesions, subsequent bronchoscopy may not be necessary, but for large obstructing lesions, a multiple stage procedure is often the safest approach. If these guidelines are followed, Nd:YAG laser bronchoscopy can safely and effectively be performed through the flexible bronchoscope with successful palliation in more than 80% of patients [59].

For rigid bronchoscopy, the same Nd:YAG laser fiber is passed down the barrel of the scope along with a suction catheter and telescope. The laser fiber is aimed at the lesion by manipulating the rigid scope itself. The laser power settings remain at 20–40 W, with a pulse duration of 0.4–1 s. The metal rigid bronchoscope is nonflammable, and thus the laser fiber does not have to extend beyond the end of the scope. It is important, however, to keep the flexible suction catheter out of the laser field to avoid any chance of fire. The laser fiber should be in close proximity (<1 cm) to the lesion because laser energy disperses as it exits the tip of the fiber, and the distance to the lesion determines the effective power density. As with flexible laser bronchoscopy, the initial goal should be photocoagulation of the tumor and not vaporization. Debridement of the remaining tissue can be accomplished by alligator forceps or a suction catheter. If needed, the beveled end of the rigid bronchoscope can be also used to shear off parts of the tumor once photocoagulation has devitalized the tissue. Care must be taken not to perforate the airway wall with the rigid scope. It is always important to know the path of the distal airway before aggressively advancing the scope beyond the tumor. The flexible scope can be helpful for blunt dissection of the tumor and in evaluating the airways distal to the main obstruction. After most of the tumor has been removed by these means, hemostasis can be achieved through further laser photocoagulation. Both metal and silicone airway stents, which are used for recanalizing an occluded airway, can be adversely affected by laser. The silicone stent could ignite similar to an endotracheal tube. The metal stent, which is made of nitinol alloy, could be damaged even with low power settings [74,75].

4.6 Outcome

Although laser bronchoscopy has become more commonly used for benign and malignant tracheo-bronchial obstruction [21,22,23,25], no randomized controlled trials exist proving a survival benefit in treated patients (Figure 4.1). Numerous studies, however, have shown a significant palliative benefit with regard to relief of dyspnea, control of hemoptysis, and facilitation of weaning from mechanical ventilation. In 1419 patients with bronchogenic carcinoma, Cavaliere et al. [22] achieved satisfactory immediate results corresponding to a normal airway lumen gauge or significant improvement in ventilation in 93% of patients. Brutinel et al. [27] found that airway caliber was improved in 79.7% of 182 sites and that even completely obstructed airways could be opened 57.7% of the time. In a previous study at the Lahey Clinic [23], subjective symptomatic relief was demonstrated after 313 of 400 procedures (78.3%) in patients with benign and malignant disease.

Although many of the larger studies relied on subjective measurements of improvement, other studies have documented objective benefits in pulmonary function tests, performance scores, and even ventilation–perfusion scans after laser bronchoscopy. Spirometry and flow volume loops have shown immediate functional improvement, with increases in peak expiratory flow rates ranging from 26% to 512% [76,77]. Waller et al. [78] showed a mean overall improvement in FEV_1 of 27% and symptomatic relief in 103 of 116 patients whose main complaint was dyspnea. Gelb and Epstein [52] reported significant improvement in Karnofsky performance scores, dyspnea index, in 23 of 27 patients with incomplete malignant obstruction of the tracheobronchial tree. Jam et al. [79] also used Karnofsky scores in patients treated by Nd:YAG laser photoresection and reported an improved performance status in 13 of 15 patients lasting between 2 and 13 months. Other performance scores, such as the simple 0–4 symptom score used by Clarke et al. [80], confirmed good symptomatic relief in 73% of their first 200 patients consecutively treated with laser bronchoscopy.

Ventilation–perfusion relationships have also been studied in patients undergoing laser bronchoscopy. George et al. [81] found that both ventilation and perfusion improved after laser treatments in 23 of 28 patients (82%) and that spirometric values, 6 mm walking distance, Karnofsky performance index, and breathlessness scores also improved significantly. In addition, it appears that improvements are not simply transient. Although some patients with benign and malignant tracheobronchial lesions may require multiple procedures, Emslander et al. [82] reported complete

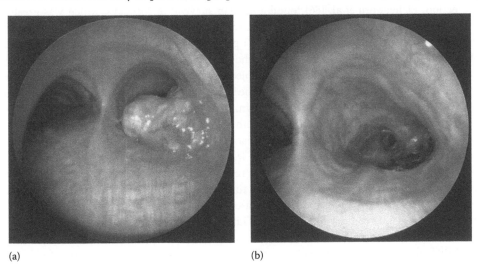

(a)   (b)

FIGURE 4.1 (a) Endobronchial lesion due to non-small-cell lung cancer seen in the right mainstem bronchus arising from the right upper lobe. (b) Endobronchial lesion is removed using the Nd:YAG laser and mechanical resection.

or partial removal of tumor in 74% of 224 laser applications and showed that in 72% of the successful results, the stenoses were still open after 4–6 months.

Nd:YAG laser photocoagulation is also effective at controlling hemoptysis [9,79,80] from malignant endobronchial lesions, with a response rate estimated to be 60% [9]. Although photocoagulation for hemoptysis has shown initial success, long-term benefits were not as easily maintained as with treatment for relief of dyspnea. Clarke et al. [80] reported that patients who were treated principally for hemoptysis tended to have a recurrence of their symptoms within 30 days of treatment, whereas those who were treated for dyspnea or cough tended to experience recurrence much later.

Laser bronchoscopy has also been shown to be beneficial for patients with impending respiratory failure as well as for those who require mechanical ventilation. Radiation therapy can relieve cough, chest pain, and dyspnea in 60%–80% of patients, but it only relieves atelectasis in 25% of patients, and improvement is usually delayed [83]. External beam radiation therapy is a common palliative intervention for patients with bronchogenic carcinoma, but only laser bronchoscopy provides the means for rapid resolution of symptoms. In one series [76], laser bronchoscopy was successful at relieving breathlessness and avoiding mechanical ventilation in 11 of 14 patients who had respiratory distress on presentation. The authors [76] concluded that laser treatment provided an excellent method of resuscitating patients with life-threatening tracheal obstruction, enabling subsequent management, such as radiation therapy, chemotherapy, or tracheobronchial stenting. Laser bronchoscopy may also facilitate weaning from mechanical ventilation in patients with respiratory failure caused by endoluminal obstruction. Stanopoulos et al. [84] reported that laser bronchoscopy improved the clinical status of 9 of 17 mechanically ventilated patients by permitting successful extubation. This subgroup of patients had appreciably shorter requirements for mechanical ventilation and a longer survival. The initial use of Nd:YAG laser bronchoscopy for treatment of an endobronchial lesion can be debated, but the palliative use of laser therapy in an inoperable, symptomatic patient has been shown to be beneficial for relief of dyspnea, control of hemoptysis, and weaning from mechanical ventilation.

Laser bronchoscopy is clearly a valuable tool in providing immediate symptomatic relief, but its survival benefit is more difficult to demonstrate. Randomized studies [85] evaluating this palliative treatment in terminal patients have been considered unethical. Although the issue of survival is a difficult one, it appears that there may be a survival advantage for patients undergoing laser resection of malignant tracheobronchial obstruction. In a preliminary study of 19 patients treated with laser photoresection, Eichenhorn et al. [86] found a median survival of 340 days, which was greater than 198–266 days found in historical control subjects using irradiation alone [87–89]. In another study using historical controls, Brutinel et al. [27] compared 71 patients with bronchogenic carcinoma treated with Nd:YAG laser photoresection with 25 patients who would have received laser treatment had it been available. They found that 76% of the control subjects were dead within 4 months and all were dead within 7 months. By contrast, 60% of patients in the laser-treated group were alive at 7 months and 28% were alive at 1 year. Desai et al. [85] also found a significant increase in survival in a subgroup of patients treated with emergent laser resection compared with historical control subjects who had emergent radiation therapy alone. In mechanically ventilated patients, Stanopoulos et al. [84] showed an increased survival (98 days vs. 8.5 days) in patients who were weaned after emergent laser resection of an obstructing lesion. The addition of brachytherapy to laser resection may also contribute to increased survival. Shea et al. [89] found an improved mean survival time (41 weeks vs. 16 weeks) in patients treated with both endobronchial brachytherapy and Nd:YAG laser photoresection.

Zaric et al. [90] evaluated the effects of Nd:YAG laser resection in combination with high-dose-rate (HDR) brachytherapy and external beam radiotherapy (EBRT) versus combination of HDR brachytherapy and EBRT alone. Eighty one patients in group I were treated with a combination of HDR brachytherapy and EBRT, while 97 patients in group II were treated with Nd:YAG laser in combination with HDR brachytherapy and EBRT. After radiation therapy ± laser treatment, all patients received standard chemotherapy (cisplatin plus etoposide) during the course of disease. Improvement in dyspnea, thoracic pain, body weight loss, and ECOG (Eastern Cooperative Oncology Group) performance status was significantly better in the Nd:YAG laser group ($p < 0.05$). In addition, there was a longer time to disease progression and improved overall survival ($p < 0.05$). Longer time to progression and survival of lung cancer patients could be the result of recanalizing imminently occluded airways accomplished with Nd:YAG laser.

Han et al. [91] confirmed the safety and effectiveness of Nd:YAG laser therapy and investigated the effectiveness of multimodal therapy compared with laser alone. One hundred fifty-three Nd:YAG laser treatments on 110 patients were performed. Symptom scores for dyspnea, hemoptysis, and cough before and after the procedure were compared. Although there were no operative mortalities directly caused by laser intervention, 30-day mortality was 6.5%. After Nd:YAG laser intervention, 76% of patients reported improvement in dyspnea, 94% for hemoptysis, and 75% for cough. Median survival after Nd:YAG laser treatment was 6.64 months; 21% of patients required repeated laser treatment. Compared with Nd:YAG laser treatment alone, multimodality treatments significantly prolonged median time to reintervention by 1.7 months ($p = 0.002$) and prolonged median survival by 4.9 months ($p < 0.001$) in patients with non-small-cell lung cancer (NSCLC). Compared with Nd:YAG laser therapy alone, multimodal treatment prolonged survival.

Nd:YAG has been successfully used in treatment of tracheal obstruction in advanced thyroid cancer [92], adult subglottic stenosis [93], airway venous malformations [94], and subglottic stenosis in Wegener's granulomatosis [95].

Although a survival advantage was suggested by these data, the retrospective design of the studies makes statistical validation difficult. Furthermore, other authors [96] failed to show any overall survival benefit for patients undergoing laser treatment. Although Desai et al. [85] found improved survival in a subgroup of patients treated emergently, they did not find an overall increase in survival in patients who underwent elective laser photoresection. Clarke et al. [80] also gave no evidence that laser treatment prolongs life and suggested that patients die of progression of their primary disease rather than obstruction of the bronchi. Nevertheless, patients on mechanical ventilation or with impending respiratory failure likely represent a subgroup of patients who may have a survival benefit from laser photoresection. A survival advantage may also be seen if endobronchial brachytherapy is combined with laser resection.

Absolute survival, however, may not be the best way to judge the importance of a palliative intervention in symptomatic patients. The rapid relief of dyspnea and improved performance provide justification for laser bronchoscopy in most patients. Laser bronchoscopy has been shown to be an effective method for immediate relief of respiratory symptoms related to both benign and malignant tracheobronchial obstruction. It is complementary to other modes of therapy, such as radiation, chemotherapy, stenting, or surgery, and often provides the time necessary to initiate these treatments. After 11 years and more than 1500 patients, Cavaliere et al. [22] concluded that laser therapy is no longer considered a treatment of last resort but instead has its own role in the therapy of cancer of the lung. Laser bronchoscopy should be an initial consideration for patients with severe tracheobronchial obstruction irrespective of survival analysis.

References

1. Wright CV, Riopelle MA: *Surgical CO₂ Laser Fundamentals*. Houston, TX: Biomedical Communications, 1988.
2. Beamis JF, Shapshay SM: Nd:YAG laser therapy for tracheobronchial disorders. *Postgrad Med* 1984; 75:173–180.
3. Polanyi TG: Laser physics. *Otolaryngol Clin North Am* 1983; 16:753–774.
4. Dumon JF, Corsini A: Cicatricial tracheostenosis. In: *Bronchoscopic Laser Resection Manual*. Houston, TX: Biomedical Communications, 1989.
5. Pass HI, Pogrebniak H: Photodynamic therapy for thoracic malignancies. *Semin Surg Oncol* 1992; 8:217–225.
6. Hayata Y, Kato H, Konaka C, Ono J, Takizawa N: Hematoporphyrin derivative and laser photoradiation in the treatment of lung cancer. *Chest* 1982; 81:269–277.

7. Edell ES, Cortese DA: Bronchoscopic phototherapy with hematoporphyrin derivative for treatment of localized bronchogenic carcinoma: A 5-year experience. *Mayo Clin Proc* 1987; 62:8–14.

8. Kato H, Cortese DA: Early detection of lung cancer by means of hematoporphyrin derivative fluorescence and laser photoradiation. *Clin Chest Med* 1985; 6:237–253.

9. Hetzel MR, Smith SG: Endoscopic palliation of tracheobronchial malignancies. *Thorax* 1991; 46:325–333.

10. Rimell FL, Shapiro AM, Mitskavich MT, Modreck P, Post JC, Maisel RH: Pediatric fiberoptic laser rigid bronchoscopy. *Otolaryngol Head Neck Surg* 1996; 114:413–417.

11. Strong MS, Jako GJ, Polanyi TG, Wallace RA: Laser surgery in the aerodigestive tract. *Am J Surg* 1973; 126:529–533.

12. Polanyi TG: Physics of surgery with lasers. *Clin Chest Med* 1985; 6:179–202.

13. Ossoff RH, Coleman JA, Courey MS, Duncavage JA, Werkhaven JA, Reinisch L: Clinical applications of lasers in otolaryngology—Head and neck surgery. *Lasers Surg Med* 1994; 15:217–248.

14. Coleman JA Jr, Van Duyne MJ, Ossoff RH: Laser treatment of lower airway stenosis. *Otolaryngol Clin North Am* 1995; 28:771–783.

15. Shapshay SM, Simpson GT: 2d: Lasers in bronchology. *Otolaryngol Clin North Am* 1983; 16:879–886.

16. Gillis TM, Strong MS: Surgical lasers and soft tissue interactions. *Otolaryngol Clin North Am* 1983; 16:775–784.

17. Shurgalin M, Anastassiou C: A new modality for minimally invasive CO2 laser surgery: Flexible hollow-core photonic bandgap fibers. *Biomed Instrum Technol* 2008 Jul–Aug; 42(4):318–325.

18. Jacobson AS, Woo P, Shapshay SM: Emerging technology: Flexible CO_2 laser WaveGuide. *Otolaryngol Head Neck Surg* 2006; 135(3):469–470.

19. Halum SL, Moberly AC: Patient tolerance of the flexible CO_2 laser for office-based laryngeal surgery. *Voice* 2010 Nov; 24(6):750–754.

20. Mehta AC: Laser applications in respiratory care. In: Kacmarek RM, Stoller JK, eds., *Current Respiratory Care.* Toronto, Ontario, Canada: BC Decker, 1988, pp. 100–106.

21. Dumon IF: *YAG Laser Bronchoscopy.* New York: Praeger Publishers, 1985.

22. Cavaliere S, Foccoli P, Toninelli C, Feijo S: Nd:YAG laser therapy in lung cancer: An 11-year experience with 2,253 applications in 1,585 patients. *J Bronchol* 1994; 1:105–111.

23. Beamis JF Jr, Vergos K, Rebeiz EE, Shapshay SM: Endoscopic laser therapy for obstructing tracheobronchial lesions. *Ann Otol Rhinol Laryngol* 1991; 100:413–419.

24. Shah H, Garbe L, Nusshauni E, Dumon JF, Chiodera PL, Cayaliere S: Benign tumors of the tracheobronchial tree: Endoscopic characteristics and role of laser resection. *Chest* 1995; 107:1744–1751.

25. Personne C, Colchen A, Leroy M, Vourc'h G, Toty L: Indications and technique for endoscopic laser resections in bronchology: A critical analysis based upon 2,284 resections. *J Thorac Cardiovasc Surg* 1986; 91:710–715.

26. Cavaliere S, Foccoli P, Farina PL: Nd:YAG laser bronchoscopy: A five-year experience with 1,396 applications in 1,000 patients. *Chest* 1988; 94:15–21.

27. Brutinel WM, Cortese DA, McDougall JC, Gillio RG, Bergstralh El: A two-year experience with neodymium-YAG laser in endobronchial obstruction. *Chest* 1987; 91:159–165.

28. Shapshay SM, Beamis JF Jr, Dumon JP: Total cervical tracheal stenosis: Treatment by laser, dilation, and stenting. *Ann Otol Rhinol Laryngol* 1989; 98:890–895.

29. Mehta AC, Lee FY, Cordasco EM, Kirby T, Eliachar I, DeBoer G: Concentric tracheal and subglottic stenosis: Management using the Nd:YAG laser for mucosal sparing followed by gentle dilatation. *Chest* 1993; 104:673–677.

30. Wanamaker JR. Eliachar I: An overview of the treatment options for lower airway obstruction. *Otolaryngol Clin North Am* 1995; 28:751–770.

31. Montgomery WW, Montgomery SK: Manual for use of montgomery laryngeal, tracheal, and esophageal prostheses: Update 1990. *Ann Otol Rhinol Laryngol Suppl* 1990; 150:2–28.

32. Gelb AF, Tashkin DP, Epstein JD, Zamel N: Nd:YAG laser surgery for severe tracheal stenosis physiologically and clinically masked by severe diffuse obstructive pulmonary disease. *Chest* 1987; 91:166–170.

33. Hayashi AH, Gillis DA, Bethune D, Hughes D, O'Neil M: Management of foreign-body bronchial obstruction using endoscopic laser therapy. *J Pediatr Surg* 1990; 25:1174–1176.

34. Kvale PA, Eichenhorn MS, Radke JR, Miks V: YAG laser photoresection of lesions obstructing the central airways. *Chest* 1985; 87:283–288.

35. Dierkesmann R: Indication and results of endobronchial laser therapy. *Lung* 1990; 168(Suppl):1095–1102.

36. Dierkesmann R, Huzly A: The significance of the pulmonary artery in endobronchial laser treatment (abstr). *Lasers Surg Med* 1983; 3:197–198.

37. Warner ME, Warner MA, Leonard PF: Anesthesia for neodymium-YAG (Nd:YAG) laser resection of major airway obstructing tumors. *Anesthesiology* 1984; 60:230–232.

38. Perrin C, Colt HG, Martin C, Mak MA, Dumon JF, Gouin F: Safety of interventional rigid bronchoscopy using intravenous anesthesia and spontaneous assisted ventilation: A prospective study. *Chest* 1992; 102:1526–1530.

39. Duckett JE, McDonnell TJ, Unger M, Parr GV: General anaesthesia for Nd:YAG laser resection of obstructing endobronchial tumours using the rigid bronchoscope. *Can Anaesth Soc J* 1985; 32:67–72.

40. Vourc'h G, Fischler MF, Michon F, Meichior JC, Seigneur F: High frequency jet ventilation manual jet ventilation during bronchoscopy in patients with tracheo-bronchial stenosis. *Br J Anaesth* 1983; 55:969–972.

41. Vourc'h G, Fischler MF, Michon F, Meichior JC, Seigneur F: Manual jet ventilation v. high frequency jet ventilation during laser resection of tracheo-bronchial stenosis. *Br J Anaesth* 1983; 55:973–975.

42. Sanders RD: Two ventilating attachments for bronchoscopes. *Del Med J* 1967; 39:170–175.

43. Ramser ER, Beamis JF Jr: Laser bronchoscopy. *Clin Chest Med* 1995; 6:415–426.

44. Schlenkhoff D, Droste H, Scieszka S. Vogt H: The use of high-frequency jet-ventilation in operative bronchoscopy. *Endoscopy* 1986; 18:192–194.

45. Lennon RL, Hosking MP, Warner MA, Cortese DA, McDougall JC, Brutinel WM, Leonard PF: Monitoring and analysis of oxygenation and ventilation during rigid bronchoscopic neodymium-YAG laser resection of airway tumors. *Mayo Clin Proc* 1987; 62:584–588.

46. Goidhill DR, Hill AJ, Whithurn RH, Feneck RO, George PJ, Keeling P: Carboxyhaemoglobin concentrations, pulse oximetry and arterial blood-gas tensions during jet ventilation for Nd:YAG laser bronchoscopy. *Br J Anaesth* 1990; 65:749–753.

47. Chan AL, Tharratt RS, Siefkin AD, Albertson TE, Volz WG, Allen RP: Nd:YAG laser bronchoscopy: Rigid or fiberoptic mode? *Chest* 1990; 98:271–275.

48. Unger M: Rigid vs flexible bronchoscope in laser bronchoscopy. *J Bronchol* 1994; 3:69–71.

49. Toty L, Personne C, Colchen A, Vourc'h G: Bronchoscopic management of tracheal lesions using the neodymium yttrium aluminum garnet laser. *Thorax* 1981; 36:175–178.

50. Dumon JF, Reboud E, Garbe L, Aucomte F, Meric B: Treatment of tracheobronchial lesions by laser photoresection. *Chest* 1982; 81:278–284.

51. Hetzel MR, Nixon C, Edmondstone WM, Mitchell DM, Millard FJ, Nanson EM, Woodcock AA, Bridges CE, Humberstone AM: Laser therapy in 100 tracheobronchial tumours. *Thorax* 1985; 40:341–345.

52. Gelb AF, Epstein JD: Laser in treatment of lung cancer. *Chest* 1984; 86:662–666.

53. Arabian A, Spagnolo SV: Laser therapy in patients with primary lung cancer. *Chest* 1984; 86:519–523.

54. Casey KR, Fairfax WR, Smith SJ, Nixon JA: Intratracheal fire ignited by the Nd:YAG laser during treatment of tracheal stenosis. *Chest* 1983; 84:295–296.

55. Dumon JF, Shapshay S, Bourcereau J, Cavaliere S, Meric B, Garbi N, Beamis J: Principles for safety in application of neodymium:YAG laser in bronchology. *Chest* 1984; 86:163–168.

56. Prakash UB, Offord KP, Stubbs SE: Bronchoscopy in North America: The ACCP survey. *Chest* 1991; 100:1668–1675.

57. George PJ, Garrett CP, Nixon C, Hetzel MR, Nanson EM, Millard FJ: Laser treatment for tracheobronchial tumours: Local or general anesthesia? *Thorax* 1987; 42:656–660.

58. Hanowell LH, Martin WR, Savelle JE, Foppiano LE: Complications of general anesthesia for Nd:YAG laser resection of endobronchial tumors. *Chest* 1991; 99:72–76.

59. Mehta AC, Golish JA, Ahmad M, Zurick A, Padua NS, O'Donnell J: Palliative treatment of malignant airway obstruction by Nd:YAG laser. *Cleve Clin Q* 1985; 52:513–524.

60. Unger M: Neodymium:YAG laser therapy for malignant and benign endobronchial obstructions. *Clin Chest Med* 1985; 6:227–290.

61. Personne C, Coichen A, Bonnette P, Leroy M, Bisson A: Laser in bronchology: Methods of application. *Lung* 1990; 168(Suppl):1085–1088.

62. Unger M, Parr GV, Lugano E: Various applications of Nd:YAG laser in tracheobronchial pathology. In: Ogura Y, ed., *Nd:YAG Laser in Medicine and Surgery: Fundamental and Clinical Aspects.* Tokyo, Japan: Professional Postgraduate Services, 1986, pp. 261–265.

63. American National Standards Institute, Inc.: *American National Standards for the Safe Use of Lasers in Health Care Facilities.* Orlando, FL: The Laser Institute of America, 1996.

64. Wood RL Jr, Sliney DH, Basye RA: Laser reflections from surgical instruments. *Lasers Surg Med* 1992; 12:675–678.

65. Sliney DH: Laser safety. *Lasers Surg Med* 1995; 16:215–225.

66. Komatsu T, Kaji R, Okazaki S, Miyawaki I, Ishihara K, Takahashi Y. Endotracheal tube ignition during the intratracheal laser treatment *Asian Cardiovasc Thorac Ann* 2008 Dec; 16(6):e49–e51.

67. Schramm VL Jr, Mattox DE, Stool SE: Acute management of laser-ignited intratracheal explosion. *Laryngoscope* 1981; 91:1417–1426.

68. Shapshay SM, Beamis JP Jr: Safety precautions for bronchoscopic Nd:YAG laser surgery *Otolaryngol Head Neck Surg* 1986; 94:175–180.

69. Wenig BL, Stenson KM, Wenig BM, Tracey D: Effects of plume produced by the Nd:YAG laser and electrocautery on the respiratory system. *Lasers Surg Med* 1993; 13:242–245.

70. Lobraico RV, Schifano MJ, Brader KR: A retrospective study on the hazards of the carbon dioxide laser plume. *J Laser Appl* 1988; 1:6–8.

71. Garden JM, O'Banion MK, Shelnitz LS, Pinski KS, Bakus AD, Reichmann ME, Sundberg JP: Papillomavirus in the vapor of carbon dioxide laser-treated verrucae. *JAMA* 1988; 259:1199–1202.

72. Abramson AL, DiLorenzo TP, Steinberg BM: Is papillomavirus detectable in the plume of laser-treated laryngeal papilloma? *Arch Otolaryngol Head Neck Surg* 1990; 116:604–607.

73. Baggish MS, Poiesz BJ, Joret D, Williamson P, Refai A: Presence of human immunodeficiency virus DNA in laser smoke. *Lasers Surg Med* 1991; 11:197–203.

74. Young O, Kirrane F, Hughes JP, Fenton JE. KTP laser and nitinol alloy stents: are they compatible? *Lasers Surg Med.* 2007 Dec; 39(10):803–807.

75. Hautmann H, Huber RM. Laser resistance of expandable metal stents in interventional bronchoscopy: An experimental evaluation. *Lasers Surg Med.* 2001; 29(1):70–72.

76. George PJ, Garrett CP, Hetzel MR: Role of the neodymium YAG laser in the management of tracheal tumours. *Thorax* 1987; 42:440–444.

77. Mohsenifar Z, Jasper AC, Koerner SK: Physiologic assessment of lung function in patients undergoing laser photoresection of tracheobronchial tumors. *Chest* 1988; 93:65–69.

78. Waller DA, Gower A, Kashyap AP, Conacher ID, Morritt GN: Carbon dioxide laser bronchoscopy—A review of its use in the treatment of malignant tracheobronchial tumours in 142 patients. *Respir Med* 1994; 88:737–741.

79. Jam PR, Dedhia HV, Lapp NL, Thompson AB, Frich JC Jr: Nd:YAG laser followed by radiation for treatment of malignant airway lesions. *Lasers Surg Med* 1985; 5:47–53.

80. Clarke CP, Ball DL, Sephton R: Follow-up of patients having Nd:YAG laser resection of bronchostenotic lesions. *J Bronchol* 1994; 1:19–22.

81. George PJ, Clarke G, Tolfree S, Garrett CP, Hetzel MR: Changes in regional ventilation and perfusion of the lung after endoscopic laser treatment. *Thorax* 1990; 45:248–253.

82. Emslander HP, Munteanu J, Prauer HJ, Heinl KW, Hinke KW, Sebenning H, Daum S: Palliative endobronchial tumor reduction by laser therapy: Procedure—Immediate results—Long-term results. *Respiration* 1987; 51:73–79.

83. Slawson RG, Scott RM: Radiation therapy in bronchogenic carcinoma. *Radiology* 1979; 132:175–176.

84. Stanopoulos IT, Beamis JF Jr, Martinez FJ, Vergos K, Shapshay SM: Laser bronchoscopy in respiratory failure from malignant airway obstruction. *Crit Care Med* 1993; 21:286–391.

85. Desai SJ, Mehta AC, VanderBrug Medendorp S, Golish JA, Ahmad M: Survival experience following Nd:YAG laser photo-resection for primary bronchogenic carcinoma. *Chest* 1988; 94:939–944.

86. Eichenhorn MS, Kvale PA, Miks VM, Seydel FIG, Horowitz B, Radke JR: Initial combination therapy with YAG laser photoresection and irradiation for inoperable non-small cell carcinoma of the lung: A preliminary report. *Chest* 1986; 89:782–785.

87. Petrovich Z, Stanley K, Cox JD, Paig C: Radiotherapy in the management of locally advanced lung cancer of all cell types: Final report of randomized trial. *Cancer* 1981; 48:1335–1340.

88. Roswit B, Patno ME, Rapp R, Veinbergs A, Feder B, Stuhlbarg J, Reid CB: The survival of patients with inoperable lung cancer: A large-scale randomized study of radiation therapy versus placebo. *Radiology* 1968; 90:688–697.

89. Shea JM, Allen RP, Tharratt RS, Chan AL, Siefkin AD: Survival of patients undergoing Nd:YAG laser therapy compared with Nd:YAG laser therapy and brachytherapy for malignant airway disease. *Chest* 1993; 103:1028–1031.

90. Zaric B, Canak V, Milovancev A, Jovanovic S, Budisin E, Sarcev T, Nisevic V: The effect of Nd:YAG laser resection on symptom control, time to progression and survival in lung cancer patients. *J BUON* 2007 Jul–Sep; 12(3):361–368.

91. Han CC, Prasetyo D, Wright GM: Endobronchial palliation using Nd:YAG laser is associated with improved survival when combined with multimodal adjuvant treatments. *J Thorac Oncol* 2007 Jan; 2(1):569–64.

92. Ribechini A, Bottici V, Chella A, Elisei R, Vitti P, Pinchera A, Ambrosino N: Interventional bronchoscopy in the treatment of tracheal obstruction secondary to advanced thyroid cancer. *J Endocrinol Invest* 2006 Feb; 29(2):131–135.

93. Leventhal DD, Krebs E, Rosen MR: Flexible laser bronchoscopy for subglottic stenosis in the awake patient. *Arch Otolaryngol Head Neck Surg* 2009 May; 135(5):467–471.

94. Glade R, Vinson K, Richter G, Suen JY, Buckmiller LM: Endoscopic management of airway venous malformations with Nd:YAG laser. *Ann Otol Rhinol Laryngol* 2010 May; 119(5):289–293.

95. Shvero J, Shitrit D, Koren R, Shalomi D, Kramer MR: Endoscopic laser surgery for subglottic stenosis in Wegener's granulomatosis. *Yonsei Med J* 2007 Oct 31; 48(5):748–753.

96. Quinj A, Letsou GV, Tanoue LT, Matthay RA, Higgins RS, Baldwin JC: Use of neodymium yttrium aluminum garnet laser in long-term palliation of airway lesions. *Conn Med* 1995; 59: 407–412.

5

Lasers in Diagnostics and Treatment of Brain Diseases

Steen J. Madsen
University of Nevada,
Las Vegas

Bernard Choi
University of
California, Irvine

Henry Hirschberg
University of
California, Irvine

5.1 Lasers in Diagnostics of Brain Diseases

5.1.1 Introduction

Recent advances in optical diagnostic and imaging technologies have enabled researchers to initiate hypothesis-driven studies on brain function. A primary emphasis has been on cerebral hemodynamics: passive monitoring, response to functional activation, and delineating healthy from diseased cerebral tissue.

The optical technology presented in this section is arranged in order of characteristic-length scales, from diffuse optics (centimeters) to microscopy (micrometers). The section concludes with a discussion of multimodal optical technologies. The overall objective is to describe recent advances in each optical modality and thereby provide the reader with a sense of the current state of each technology for neuroscience-related research.

5.1.2 Near-Infrared Spectroscopy

5.1.2.1 Fundamentals of Near-Infrared Spectroscopy Technology

Near-infrared spectroscopy (NIRS) enables monitoring of cerebral hemodynamics.[1-6] Continuous-wave laser light is used as the source, with two or more wavelengths used. The source light is coupled into individual source *optodes*, which are placed in contact with the subject's head. Light remitted at select points distal to the source optodes is collected with detection optodes. The detected light has propagated through the scalp, skull, and superficial regions of the cerebral cortex. Multiple sources and detectors are used to interrogate different regions of the brain.

Franceschini et al.[7] determined that the detected optical signals using an NIRS instrument correlated well with established methods to assess cardiac cycle, respiration, heart rate, and blood pressure. Changes in detected light intensity are associated with estimates of cerebral blood volume and hemoglobin oxygen saturation.[8,9] The hemodynamics assessed with NIRS and functional magnetic resonance imaging (fMRI) are similar.[10] With model-based analysis of the signals collected at various source–detector optode separation distances, 3D tomographic images of changes in cerebral blood volume and hemoglobin oxygen saturation can be achieved.[11]

5.1.2.2 Specifics of Near-Infrared Spectroscopy Technology for Brain Characterization

For NIRS measurements, the optodes must be placed in direct contact with the subject's head. The presence of hair reduces the signal-to-noise ratio of the measurements. The relative location of source and detector optodes is an important consideration in NIRS studies.[12]

Widespread acceptance of NIRS for functional monitoring of the brain has not occurred, purportedly due to its low spatial resolution and problems with localization of hemodynamic events. To address these limitations, researchers have evaluated the use of a higher density of source and detector optodes and have found that this has resulted in improvements in localization of regions of functional activation.[13] Dehghani et al.[14] used computational modeling to estimate the interrogation depth of NIRS, using a high density of source and detector optodes. Their model results suggest that depths greater than 15 mm into the brain can be characterized with adequate signal-to-noise ratio. Hemodynamics in the scalp are a source of noise in NIRS-based assessment of cerebral hemodynamics.[15] Liao et al.[16] have introduced a method (superficial signal regression) to reduce the contribution of scalp hemodynamics to high-density NIRS-based monitoring of neonates during a functional stimulation exercise.

5.1.2.3 Current Directions in Functional Characterization of the Brain with NIRS

NIRS has been used to study functional activation in normal human brains[7,17–20] as well as directional coupling between brain regions.[21] Recently, Im et al.[21] have applied NIRS to study functional connectivity in the rodent brain.

NIRS has been used to study the brains of neonates[11,15,16,22] with the translational goal of developing a robust NIRS-based bedside monitor. Liao et al.[16] have focused initially on characterization of healthy newborn infants, as a necessary first step toward interpretation of the NIRS signals to enable identification of abnormalities in the signals as a sign of potential problems in at-risk infants. NIRS also has been used to study the differences in cerebral hemodynamics between healthy children and those with cerebral palsy.[23]

NIRS has been used to determine if an asymmetric response of the brain can be assessed in stroke victims.[24] They observed that the resting hemodynamics in each cerebral hemisphere may differ for stroke victims. Lee et al.[25] used NIRS to study the rodent brain during seizures. Electroencephalography was used to determine the time course of the seizure, and NIRS was used to assess the corresponding cerebral hemodynamics.

5.1.3 Laser Speckle Imaging

5.1.3.1 Fundamentals of Laser Speckle Imaging Technology

An excellent review of laser speckle imaging (LSI) (also known as laser speckle contrast analysis [LASCA] or laser speckle contrast imaging [LSCI]) has been published by Boas and Dunn.[26] Briefly, LSI typically involves use of coherent, continuous-wave light to irradiate the surface of the object under evaluation. The irradiated surface is imaged with a charge-coupled device (CCD) camera equipped with a macro lens. Due to the interference of electromagnetic waves superimposed at each camera pixel, the collected image has a grainy appearance (i.e., bright pixels correspond to a high degree of constructive interference, dark pixels a high degree of destructive interference). The resulting pattern is termed laser speckle.

In the absence of motion of or within the object, and with judicious selection of lens aperture and magnification,[27] the collected image can have high local speckle contrast. With motion of the object or with the presence of moving optical scatterers (i.e., intralipid, red blood cells) within the object, local regions of dynamic speckle are present. If the exposure time of the camera is relatively long compared with a characteristic speckle decorrelation time, the local speckle pattern will be blurred, resulting in a decrease in local speckle contrast. The speckle contrast is inversely proportional to the degree of motion. With assumption of either Brownian motion or ordered flow, a Lorentzian or Gaussian line shape, respectively, can be assumed for the autocorrelation function.[28] With use of the assumed model of red blood cell motion, an expression can be derived that relates the local speckle contrast with speckle correlation time.

In vitro phantom experiments suggest that, with judicious selection of camera exposure time, a linear response range exists between a parameter termed the speckle flow index (SFI; equal to the reciprocal of speckle correlation time) and actual flow speed in the phantom.[29]

5.1.3.2 Specifics of Laser Speckle Imaging Technology for Brain Imaging

For neuroscience studies involving LSI of small animals, it is highly desirable to have a model in which the superficial cerebral vasculature is visible. For rats, a thinned-skull preparation enables visualization of cerebral vasculature. A layer of saline or mineral oil is maintained on top of the skull to mitigate dehydration. For mice, retraction of the skin alone is sufficient, as the skull is sufficiently thin to enable LSI-based characterization of cerebral hemodynamics.

The degree to which blood flow modulates speckle contrast depends on the depth at which blood flows and the degree of optical scattering of surrounding, motionless tissue structures (i.e., static scatterers). The detected intensity at any camera pixel can be described as the sum of two independent components: light that has interacted only with static scatterers and light that has interacted with moving scatterers. Parthasarathy et al.[30] employed a clever method to extract a more accurate value for speckle correlation time in the presence of static scattering. Their method employs use of a single camera exposure time and multiple laser irradiation times to achieve multiple camera exposed times with relatively fixed camera noise.

5.1.3.3 Current Directions in Functional Characterization of the Brain with Laser Speckle Imaging

Due to the simplicity and low cost of the associated instrumentation, LSI has been applied in numerous studies of the cerebral vasculature, including functional activation in normal brain,[31] cortical spreading depression,[32-36] ischemic stroke,[32,37-42] and neurosurgery.[43]

5.1.4 Photoacoustics

5.1.4.1 Fundamentals of Photoacoustic Technology

Details of photoacoustic imaging are presented by Xu and Wang.[44] Briefly, short-pulsed (ns regime) light is injected into the medium of interest. Light propagation in the medium is governed by the local tissue optical properties. Optical absorption of light can result in heat generation and thermal expansion. With judicious selection of pulse duration (less than the characteristic stress relaxation time), pressure waves at ultrasound frequencies are generated. The pressure-wave amplitude is proportional to the quantity of absorbed optical energy. The waves propagate from absorption sites with minimal scattering. Ultrasound transducers typically are positioned at multiple sites to detect the generated pressure waves. Algorithms can be used to generate tomographic images of optical absorption.[45,46]

A primary advantage of the photoacoustic method lies in the minimal scattering of the pressure waves. The spatial resolution and imaging depth of all optical imaging methods is limited by the substantial optical scattering present in most biological tissues at visible and near-infrared wavelengths.

With photoacoustics, optical scattering plays a role in modulating the degree of light absorption by specific chromophores. However, once the absorption event occurs and pressure waves are generated, optical scattering no longer affects the performance of photoacoustic imaging. Due to the relatively low degree of acoustic-wave scattering in biological tissue, the pressure waves reach the transducer with minimal alteration, which greatly facilitates precise localization of the pressure wave source.

5.1.4.2 Specifics of Photoacoustic Technology for Brain Imaging

With the use of optical wavelengths (i.e., 532 nm) that are strongly absorbed by hemoglobin molecules, high-resolution images of the cerebral microvasculature can be achieved. Due to the relatively low speed of sound in air, photoacoustic imaging requires contact between the ultrasound transducer and subject. Typically, standard ultrasound gel is used.

Recently, a full-ring photoacoustic tomography instrument was developed.[47] This technological advance addressed the need for shorter data collection times. The authors used a 512-element ring array, although only 64 were active with each excitation laser pulse. An imaging frame rate of 0.625 fps was achieved. By scanning the array vertically, a 3D image was generated.

Li et al.[48] also demonstrated that the intravascular distribution of exogenous contrast agents such as Evans blue can be assessed with photoacoustic tomography. They demonstrated that with a ring-array transducer, the dynamics of the agent distribution could be monitored. Furthermore, with use of two or more excitation wavelengths, photoacoustics can be used to assess accurately hemoglobin oxygenation in the brain.[49]

5.1.5 Optical Coherence Tomography

5.1.5.1 Fundamentals of Optical Coherence Tomography Technology

Optical coherence tomography (OCT) is one of the most successful optical imaging technologies in terms of commercialization and biomedical applications. Specifics of the technology have been described in numerous publications[50] and book chapters,[51,52] and advances in the technology have occurred at a frenetic pace.

5.1.5.2 Recent Advances in Functional Neuroimaging with Optical Coherence Tomography

Optical angiography (OAG)[53–55] enables imaging of cerebral microvasculature with exquisite sensitivity. Ren et al.[56] applied a high-pass filter to the OCT signal to enhance detection of microvasculature. The high-pass filter enables separation of signal frequency components associated with moving red blood cells, from those associated with static scatterers. Wang et al.[55] improved the filter design to further enhance detection of microvasculature. A method to estimate blood-flow velocity has been integrated into OAG instrumentation.[57] The authors report an order-of-magnitude improvement in the precision of the measured blood-flow value, as compared with phase-resolved Doppler OCT. OAG has been used to study hemodynamics in response to a functional activation challenge.[58] Srinivasan et al.[54] integrated a method to perform angle-independent measurements of cerebral blood flow into an OAG instrument.

5.1.6 Intravital Optical Microscopy

An exciting new direction in in vivo microscopy of the brain is the use of fluorescence microscopy in freely moving animals. Flusberg et al.[59] describe a lightweight epifluorescence microscope, with the microscope attached to the cranium. Flexible fiber optics are used to deliver light to and from the microscope.

To improve the imaging depth and resolution of these microscopes, light-sheet microscopy methodology can be employed. Fluorescence light-sheet microscopy involves delivery of excitation light and collection of the fluorescence emission perpendicular to the direction of excitation light. The excitation light is delivered as a *sheet*, either with the use of a cylindrical lens or with a linear scan of a collimated beam.[60] The focus of the collection optics is positioned at the same depth at which the excitation light is delivered. With this method, the collected fluorescence emission is restricted primarily to the focal plane of the collection optics.

Engelbrecht et al.[61] developed a fiber-optic-based instrument designed to perform fluorescence light-sheet microscopy. They performed in vivo imaging in the rodent neocortex and showed images of green fluorescent protein (GFP) expressing neurons. Unfortunately, in their study, the microscopy was performed in an invasive fashion. To achieve side illumination of the neocortex, a section of the brain was removed to enable positioning of the light delivery optics. The authors propose that modifications to the microscope can be made to reduce the degree of invasiveness required for in vivo neuroimaging.

5.1.7 Multimodal Optical Neuroimaging

Luo et al.[62] described the use of a combined LSI/Doppler OCT instrument to characterize cerebral blood-flow dynamics in a rodent model. Two wavelengths (785 and 830 nm) are used with LSI to enable imaging of the changes in blood flow and in oxy- and deoxyhemoglobin content. As a follow-up, Yuan et al.[63] published a paper in which they present a trimodal neuroimaging instrument. This instrument employs dual-wavelength LSI, Doppler OCT, and fluorescence imaging.

Sakadzic et al.[34] developed a multimodal imaging instrument that combines LSI and phosphorescence-based oxygen tension mapping. The latter method enables absolute quantification of oxygen tension, as opposed to multispectral imaging of hemoglobin oxygen saturation. Ponticorvo and Dunn[64] described a similar single-platform instrument that combines LSI and phosphorescence-based oxygen tension mapping. However, they used a digital micromirror device (DMD) to irradiate selectively specific regions of interest with excitation light and quantified the emitted phosphorescence with a photomultiplier tube. Use of the DMD enabled flexible selection of the region of interest and hence allowed for spatial mapping of absolute oxygen tension values.

Durduran et al.[65] have used NIRS and diffuse correlation spectroscopy (DCS) to study cerebral hemodynamics in neonates with congenital heart disease. DCS enables noninvasive characterization of tissue blood flow and has been reviewed previously.[66] The objective of this study was to determine if these optical methods of hemodynamic monitoring correlate with those derived from MRI. Their data suggest that there is good correlation between the optical and MRI-derived metrics of cerebral blood flow. The authors also used NIRS and DCS to study cerebral hemodynamics in stroke victims.[67] With this combined method, they studied the effect of relative head position on oxy- and deoxyhemoglobin content and cerebral blood flow. Their measurements support the hypothesis that autoregulation of the measured parameters is impaired in the cerebral hemisphere in which the ischemic episode occurred, in stroke victims.

5.2 Lasers in Therapy of Brain Diseases

5.2.1 Introduction

For the purposes of therapy, laser–tissue interactions can be divided into three categories (photochemical, photothermal, and photomechanical) depending on the total energy delivered and exposure time. In the vast majority of neurosurgical procedures, lasers have been used to produce heat (photothermal interactions). The following effects are observed with increasing temperature: hyperthermia (37°C–43°C), photothrombosis and tissue welding (45°C–60°C), photocoagulation (60°C–100°C), photocarbonization (100°C–300°C), and photovaporization (>300°C).[68] Neurosurgical lasers have typically been used for hemostasis (50°C–100°C) and vaporization (100°C–300°C).[69] Photomechanical effects have little current application in neurosurgery; however, there have been a number of preclinical and human trials investigating the feasibility of photochemical interactions, such as photodynamic therapy (PDT), for the treatment of malignant gliomas.

5.2.2 Historical Perspective

Therapeutic applications of lasers in the brain were investigated in animal models only 5 years after the invention of the laser in 1960.[70] Results were rather disappointing due to the traumatic effects caused by the high-power pulsed ruby laser used in these experiments. Nevertheless, a low-power ruby laser was used in the first clinical study in 1966 to debulk a malignant glioma.[71] This was a rather cautious attempt and, as such, no improvement in patient survival was observed.

The introduction of continuous-wave lasers provided renewed interest in the use of lasers in neurosurgical applications since accurate cutting and vaporization of brain tissues was now possible.[69] From this perspective, the long-wavelength CO_2 laser (λ = 10.6 μm) was particularly appealing due to its high absorption in tissue and water, which confines damage to the immediate vicinity of the laser beam. Studies in mice demonstrated the potential of the CO_2 laser to vaporize and resect brain tumors,[72] thus providing the impetus for clinical studies.

In 1969, Stellar et al.[72,73] were the first to attempt the resection of a human brain tumor (glioblastoma multiforme [GBM]) with a CO_2 laser. Although the resection was successful, a number of limitations were noted including longer operating times and the cumbersome nature of the articulating arms. From the mid-1970s through the early 1980s, a number of large clinical studies focused on the utility of CO_2 lasers for tumor debulking.[74–77] Although these studies demonstrated the tumor excision capabilities of the CO_2 laser, many investigators remained cautious about its role in neurosurgery and there was a general feeling that this new device would not replace traditional resection techniques using ultrasonic aspirators and bipolar and loop cautery.[70] Nevertheless, the importance of the laser in neurosurgical applications was recognized by the establishment of the *First American Congress on Lasers in Neurosurgery* held in Chicago in 1981.

In addition to tumor debulking, vessel coagulation is a commonly performed neurosurgical procedure. Unfortunately, the poor absorption of hemoglobin at 10.6 μm precludes the effective use of the CO_2 laser for photocoagulation. Such procedures are better suited for the lower-wavelength (λ = 1.064 μm) neodymium/yttrium–aluminum–garnet (Nd:YAG) laser. The 1.064 μm light is absorbed by hemoglobin with a blood–brain absorption ratio of 100:1[78] and therefore selectively heats blood vessels. Thus, the 1.064 μm Nd:YAG ensures excellent hemostasis on veins and arteries of ≥3 mm diameter. The application of the Nd:YAG laser for neurovascular surgery was first suggested by Yahr et al.[79,80] in 1964; however, it wasn't until the late 1970s[81,82] that it was used in neurosurgery. Nd:YAG lasers were found to be particularly useful for the excision of vascular meningiomas and arteriovenous malformations.[83] Since its wavelength is sufficiently short to propagate through standard optical fibers, the Nd:YAG laser is ideally suited for endoscopic applications and, although endoscopy of ventricles has been attempted, the serious risk of damage to adjacent structures has prevented this procedure from gaining widespread acceptance.[70] This illustrates a fundamental problem with the Nd:YAG laser in the treatment of brain diseases, namely, the significant scatter of 1.064 μm light in brain tissue that makes it difficult to confine the beam to the treatment volume thereby jeopardizing adjacent normal structures. The seriousness of this problem can be appreciated by comparing the penetration depth in brain tissue of the Nd:YAG laser (2–3.5 mm) to that of the CO_2 laser (<300 μm.).[78,84,85]

Longer-wavelength Nd:YAG (λ = 1.32 μm) and argon ion (Ar: λ = 0.488–0.514 μm) lasers have had limited application in the treatment of brain diseases. The 1.32 μm Nd:YAG laser has relatively high absorption in water (although less than the CO_2 laser) and can achieve a relatively high quality of hemostasis.[69] Although this laser combines the favorable properties of both 1.064 and 10.6 μm lasers, it has been used in only a few studies: in situations requiring both hemostasis and tissue vaporization, clinicians have favored the use of combined 1.064 and 10.6 μm systems even though this has increased the complexity of the procedures. Due to its strong absorption by hemoglobin and other chromophores, the Ar ion laser has minimal penetration in vascular and pigmented tissues thus making it useful for hemostasis, especially in vascular malformation surgery[86] and for vaporization and cutting when used at higher power densities.[87] Due to the cost, and cumbersome nature of the Ar ion laser, there have been no serious attempts to integrate it into the neurosurgical operating theater.

Traditionally, lasers in neurosurgery have been used primarily for photocoagulation and vaporization, hence the prominence of Nd:YAG and CO_2 lasers. From the mid-1960s through the beginning of this millennium, these lasers have been used primarily in the treatment of brain tumors and vascular malformations—detailed descriptions of these studies can be found in a number of excellent review papers.[69,70,83,88–91] The remainder of the review will concentrate on some of the newer laser procedures for the treatment of brain diseases, including laser-assisted endoscopic third ventriculostomy (LA-ETV), excimer laser-assisted nonocclusive anastomosis (ELANA), laser interstitial thermal therapy (LITT), fluorescence-guided resection (FGR), PDT and photochemical internalization (PCI).

5.2.3 Laser-Assisted Endoscopic Third Ventriculostomy

ETV is used in the treatment of noncommunicating hydrocephalus. The objective of the procedure is to perforate the third ventricular floor to allow circulation of cerebrospinal fluid. A number of different perforation techniques have been tried, including blunt methods using a leucotome, or the endoscope itself, electrocoagulation, balloon catheterization, or laser coagulation.[92–97] There have been a number of small-scale clinical trials investigating the efficacy of laser-assisted third ventriculostomy using either Nd:YAG ($\lambda = 1.064\ \mu m$) or near-infrared ($\lambda = 0.805-0.810\ \mu m$) diode lasers.[69,92,98–100] A significant advantage of the LA-ETV technique is that it may be accomplished with minimal cerebral damage since it can be performed with smaller-diameter (2–3 mm) endoscopes compared with the 5–6 mm neuroendoscopes typically used for ETV with non-laser-based approaches.[101] Damage to vascular structures, in particular the basilar artery, is a serious complication of ETV.[102–104] The risk of damage to adjacent structures is especially high when using Nd:YAG or near-infrared lasers due to the high laser powers required for tissue perforation and the deep penetration of these wavelengths in brain tissues. As a result, many clinicians have been hesitant to adopt LA-ETV.[70] An innovative solution to this problem has been proposed by Vandertop et al.[99] who have developed a special fiber to confine the laser energy thus sparing surrounding normal structures. The 400 μm diameter fiber is terminated with a 800–900 μm diameter carbon-coated sphere that absorbs more than 90% of the incident laser energy. As a result, the tip instantly reaches ablation temperatures, even when relatively low laser powers are used. For example, ablation has been observed at laser powers as low as 1 W delivered over 1 s.[100] In essence, the fiber tip converts optical energy into thermal energy rendering this a contact-based technique with precisely controlled ablation of tissue in layers of 0.3–0.5 mm per incident laser pulse.[99] Since the carbon particles absorb a broad range of wavelengths very efficiently, any continuous-wave laser source that can be transmitted by optical fibers is suitable for this application. Diode lasers are particularly attractive since they are compact, easy to use, and inexpensive.

The safety and efficacy of LA-ETV using a carbon-coated fiber tip/diode laser combination for controlled ablation was recently investigated in a large clinical study consisting of 202 patients.[101] The results demonstrated LA-ETV to be a safe and effective procedure comparable to other ETV techniques. Other indications treated with this approach include cyst fenestration, colloid cyst resection, and fenestration of the septum pellucidum.[99]

5.2.4 Excimer Laser-Assisted Nonocclusive Anastomosis

In conventional cerebrovascular bypass, the recipient vessel is temporarily occluded; however, this is a somewhat risky procedure especially in cases requiring temporary occlusion proximal to a major brain artery, such as the internal cartotid.[91] To overcome the risk of ischemia during the occlusion time, Tulleken and colleagues have developed a laser-based nonocclusive approach.[105–111] The ELANA technique was designed primarily for the treatment of patients with giant or large intracranial aneurysms needing an extracranial-to-intracranial bypass who have either failed balloon occlusion tests or have insufficient collateral circulation to allow a carotid sacrifice.[91]

To construct an ELANA anastomosis, a platinum ring is sutured to a vein graft. This graft is then sutured to the recipient artery with eight microsutures without occluding the recipient artery.[112] The central element of the ELANA technique is the subsequent vessel wall perforation via the vein graft by the specially designed ELANA catheter, which punches out a disc of recipient artery wall from the anastomosis. The catheter tip consists of a 2 mm diameter central metal grid with suction ports for creating vacuum pressure. The metal grid is surrounded by two rings of 170 optical fibers (60 μm diameter) into which light from a XeCl (λ = 308 nm) laser is coupled.[113] After 2 min of vacuum suction, the laser is typically activated for 5 s. The 100 ns laser pulses (frequency of 40 Hz and average energy of 15–18 mJ) produce explosive tissue water vaporization with explosive bubble formation causing tearing of the tissue in contact with the catheter tip.[114] The resultant flap is prevented from entering the recipient lumen by continuous vacuum suction during laser activation.

The primary advantage of the ELANA approach is the avoidance of the risks associated with intraoperative ischemia due to temporary vessel occlusion and the subsequent elimination of the time pressure for the neurosurgeon. In addition, with the lack of flow arrest and cerebral ischemia, ELANA can be performed without using systemic heparin or barbiturate protection, thus eliminating the risks associated with these drugs.[91] ELANA is also less invasive than conventional surgical techniques since only a short segment of vessel exposure is required. Although ELANA is a relatively safe procedure, a number of complications have been reported including intracranial air embolism and aneurysm rupture, both resulting in deaths.[111] Nonfatal complications include postoperative ischemia, aneurysm bleed, and cranial nerve palsy.[111]

A clinical trial involving 32 patients was recently completed to evaluate the efficacy of the ELANA technique for the treatment of patients with noncoilable, nonclippable giant intracranial aneurysms of the middle cerebral artery.[115] The results were promising: patent bypasses were constructed in 94% of the patients while nonfatal complications were observed in 15% of the patients. Overall, the results suggest that ELANA may be a useful tool for vascular neurosurgeons. The technique is gaining acceptance in Europe and an investigational device exemption was recently approved by the U.S. FDA to conduct a clinical study in the United States.[116]

To address shortcomings of the ELANA approach, a number of modifications are under active investigation. ELANA involves eight intracranial microsutures requiring a high level of microsurgical skill. Every suture is accompanied by potential complications such as breaking of the suture, rupturing of the recipient artery wall, or damage to surrounding structures.[112] To address these problems, a sutureless ELANA technique (SELANA) has been developed and tested in human cadavers.[112] Instead of attaching the anastomosis with sutures, it is fixed using a special ring with pins and sealed using bioglue.

Puca and colleagues[117] have recently developed a minimally occlusive technique—minimally occlusive laser vascular anastomosis (MOLVA)—that is currently being tested in animals. This end-to-side anastomosis procedure makes use of an 810 nm diode laser along with a chromophore (indocyanine green) applied to the anastomosis site to be welded. The chromophore, or photosensitizer, induces a spatially confined and controllable rise in the local temperature minimizing the risk of heat damage to the tissue. Although the procedure requires occlusion of the recipient vessel for a short period of time (approximately 6 min), it can be accomplished with fewer suture materials and, unlike ELANA, uses a diode laser that is much cheaper, more compact, and easier to use than an excimer laser.

5.2.5 Laser Interstitial Thermotherapy

Interstitial treatment of malignancies by laser irradiation was first described by Bown in 1983.[118] In 1990, Sugiyama et al.[119] were the first to report on interstitial laser treatment of brain tumors. In LITT, laser light is delivered interstitially by stereotaxy into tumors through one or more optical fibers.[120] The basic principle of LITT is the absorption of light and its conversion into heat and, as such, tumor destruction occurs via coagulation. In order to confine thermal damage to the tumor, laser light has traditionally been delivered at low powers and over long times. Due to the relatively low temperature threshold for damage to normal brain (approx. 43°C), accurate temperature monitoring is a requirement for safe and effective treatment. Two methods for intraoperative control of laser-induced lesions have been attempted: computer-controlled light delivery based on temperature readings from tumor-implanted thermocouples and MRI for real-time visualization of LITT-induced lesions.[120] Of the two approaches, MRI is superior since it allows for real-time damage visualization and the possibility of feedback control. In addition, fiber insertion can also be accomplished under MR guidance. Jolesz[121] was the first to use MRI to visualize laser–tissue interactions in the brain. Shortly thereafter, a number of small clinical trials using the 1.064 μm Nd:YAG laser demonstrated that MRI could be used effectively to monitor LITT for small focal intracranial lesions.[122-129]

In a recent series of preclinical studies, using a canine brain tumor model, near-infrared diode lasers (λ = 0.980 μm) were used to induce coagulative necrosis.[130] Since the absorption coefficient of water at 0.980 μm is four times higher than that at 1.064 μm,[131] energy at the lower wavelength is absorbed much more rapidly in tumor tissues resulting in a faster rate of heating, even at relatively low laser powers, and as such, LITT treatments can be accomplished in a few minutes as opposed to the 30 min required for LITT with the Nd:YAG laser.[132] A potential problem with this approach is that rapid energy absorption increases the risk of uncontrolled tissue heating with the potential of tissue carbonization. Induction of carbonization is problematic since temperature measurements become impossible in the carbonized area. In addition, carbonization produces inhomogeneous temperature distributions that may result in unwanted heating of normal brain and/or lack of heating in the target volume.[133] Rapid heating may also result in the production of gas bubbles, leading to increased intracranial pressure. Water vaporization followed by tissue carbonization and ablation occurs once the temperature exceeds 100°C and therefore the goal of LITT is to maintain a temperature range of between 50°C and 90°C.[134] Rapid tumor heating using high-power (a few watts) near-infrared lasers can be achieved with minimal risk of carbonization through careful design of the optical fiber. To that end, water-cooled fibers with diffusing tips have been developed.[135]

Results of canine studies have demonstrated the feasibility of thermal ablation of small brain tumors (0.5–1.0 cm diameter) using magnetic resonance temperature imaging (MRTI) for feedback-regulated 0.980 μm laser irradiation.[130] MRTI has been shown to have a high degree of accuracy (±0.2°C) in a number of tissue types.[136-138] A prototype computer-controlled laser thermal therapy system (Visualase™, BioTex, Inc., Houston, TX) was used in these studies.[130,139] The Visualase software consists of a number of applications capable of processing MRTI images in near real time and provides a user interface for controlling laser therapy based on MR-derived temperature profiles. The Visualase system has also been used in a small clinical trial evaluating the safety and feasibility of real-time MR-guided laser-induced thermal therapy of metastatic intracranial tumors.[132] Results from four patients were encouraging: the procedure was well tolerated and no tumor recurrences were found within the thermal ablation zones. The study suggests that LITT may be an effective treatment option for the 20% of patients who fail the current standard of care of radiation therapy. The technique is suitable for relatively small focal lesions (<3 cm diameter), but not for invasive tumors such as high-grade gliomas.

An interesting approach involving the use of nanoshells to increase the therapeutic efficacy of LITT has recently been reported by Schwartz and colleagues[140] in an orthotopic canine brain tumor model. The 150 nm diameter nanoshells were composed of a dielectric silica core encased in a thin layer of gold and demonstrated strong resonance absorption in the near infrared. Nanoshells were infused intravenously followed by passive accumulation in the tumor via the enhanced permeability and retention (EPR) effect. Near-infrared light (λ = 808 nm) was provided by a high-power diode laser delivered via a water-cooled diffusing optical fiber applicator. The study showed that passively delivered 150 nm nanoshells preferentially accumulated in canine brain tumors and that laser activation of the nanoshells resulted in selective photothermal ablation of tumor tissue using laser fluences that were tissue sparing in the absence of nanoshells.

5.2.6 Fluorescence-Guided Resection

5.2.6.1 Introduction

According to the Central Brain Tumor Registry of the United States (CBTRUS) 2005–2006 statistical report, there were 43,800 new adult cases of brain tumors diagnosed in 2005. Of these, 25,500 cases were malignant with 12,700 deaths estimated in 2007. Gliomas represent 40% of all primary brain tumors, contributing up to 78% of all malignant brain tumor cases. Although the most common site of involvement is the brain, gliomas can also affect the spinal cord, or any other part of the central nervous system (CNS), such as the optic nerves. Gliomas do not metastasize throughout the body, unlike most forms of cancer, but cause symptoms by invading the brain, and even with modern therapy including surgical tumor resection and radio- and/or chemotherapy, patient prognosis remains poor. Tumor resection is usually the first modality employed in the treatment of gliomas. Past surgical studies have noted that complete resection of contrast-enhancing tumor is achieved in fewer than 20% of patients, highlighting the difficulties in defining tumor/normal brain margins intraoperatively using conventional neurosurgical techniques. Although the impact of the extent of surgical tumor resection on the prognosis of malignant glioma patients remains controversial, some recent studies indicate that gross tumor resection as determined by postoperative MRI is beneficial.[141–145] Additional to the survival benefit associated with gross total resection, maximum cytoreductive surgery is advantageous to improve the efficacy of postoperative adjuvant therapies. Therefore, some form of intraoperative imaging modality like MRI, CT, ultrasound, or visual or fluorescence tagging to detect residual tumor should be of value. Neuronavigation techniques are now standard in neurosurgical procedures.[146] They are, in principle, similar to the global positioning system (GPS) navigation devices found in automobiles (although they do not use orbiting satellites) in that they translate a point in physical space (i.e., surgical field) onto a map, in this case a preoperative MRI image data set. Although extremely useful, they are based on preoperative MRI, and therefore movement of the brain and the resection cavity during surgery greatly limits their ability to define resection margins. This is clearly illustrated in the intraoperative MRI image (Figure 5.1) taken during the closing phase of surgery. The large degree of brain shift from its normal position and the collapse of the resection cavity following tumor removal are apparent.

Intraoperative MRI and ultrasonic scanners have been developed to update the data sets of neuronavigation systems during surgery, in order to achieve as complete a tumor resection as possible, but they are costly and time consuming and have limitations of their own.[147–152] A more direct method of

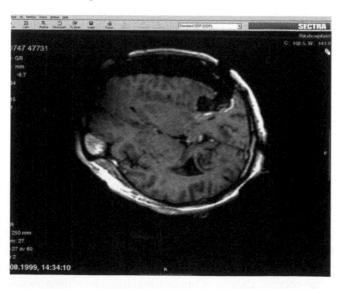

FIGURE 5.1   Intraoperative image showing brain and resection cavity drift.

determining tumor tissue from normal brain is fluorescent labeling of the tumor cells. Due to the extensive work, both experimentally and clinically, by Stummer and coworkers, as well as other groups, FGR employing the prodrug 5-aminolevulinic acid (ALA) has gained considerable interest.[153–164]

5.2.6.2 5-Aminolevulinic Acid

ALA is a naturally occurring precursor in the biosynthetic pathway for heme production.[157,160,165–168] The last step in the biosynthetic route involves conversion of protoporphyrin IX (PpIX) to heme. PpIX is a porphyrin-based photosensitizer that absorbs blue light and has significant emission in the red part of the electromagnetic spectrum (Figure 5.2). Heme biosynthesis starts in the mitochondrion, and under physiologic conditions, cellular heme synthesis is regulated in a negative feedback control of the enzyme ALA synthase by free heme. ALA synthase then becomes the rate-limiting step. When exogenous ALA is added, the control mechanism is bypassed resulting in excess synthesis of downstream metabolites. Under these conditions, ferrochelatase, which catalyzes iron insertion into PpIX, becomes the rate-limiting enzyme. Following the addition of exogenous ALA, the low physiologic rate of iron insertion by ferrochelatase is unable to compensate for the excess PpIX that is formed, leading to an accumulation of PpIX in cells. Shortly after administration of ALA, significant amounts of PpIX accumulate in the mitochondria and cytosol. PpIX selectivity is achieved through an enzyme activity difference between tumor cells and normal cells. Ferrochelatase activity and its capacity for incorporating iron into PpIX are limited in tumor cells and, therefore, result in a greater accumulation of PpIX. Further, the intermediary enzyme deaminase exhibits increased activity in tumor cells leading to faster PpIX production. With more rapid PpIX production in tumor cells and the subsequent inability to efficiently convert PpIX to heme, exogenous ALA administration provides superior tumor selectivity compared to first-generation photosensitizers such as hematoporphyrin derivative (HpD) and its purified version, Photofrin® (Pf).[169] Maximal PpIX production occurs approximately 4 h after administration in most cells. In addition to

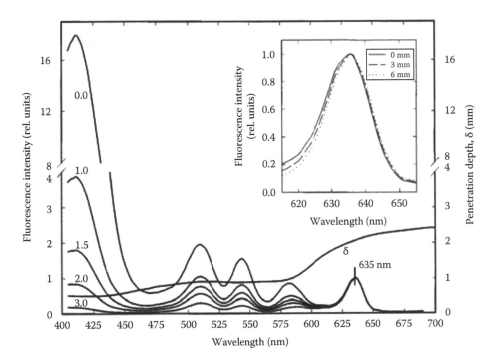

FIGURE 5.2 PpIX fluorescence excitation spectrum. The penetration depth (δ) of light in soft tissue increases with increasing wavelength. The expanded spectrum shows that PpIX fluorescence remains unchanged as a function of depth below the tissue surface (0–6 mm). (From Peng, Q. et al., *Cancer*, 79, 2282, 1997. With permission.)

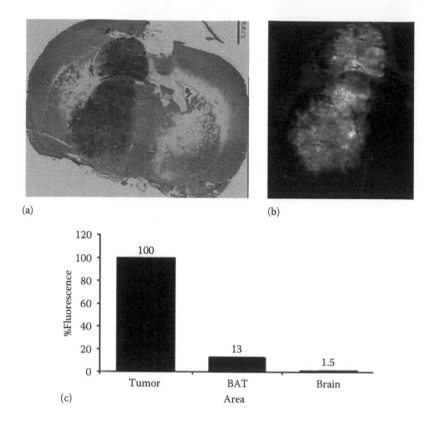

(a)

(b)

(c)

FIGURE 5.3   (See color insert.) PpIX accumulation in an experimental brain tumor: (a) H&E histological section, (b) fluorescence image, and (c) average fluorescence intensity from tumor, BAT, and normal brain. Tumor fluorescence set at 100%.

its superior tumor selectivity, additional advantages of ALA include its relatively short cutaneous photosensitization period of 24–48 h (permitting repeat or fractionated treatments) and easy administration (oral or topical).

Accumulated PpIX levels are sufficiently high that neoplastic tissue can be directly visualized through a surgical microscope adapted with a violet-blue light source to excite PpIX fluorescence and a red observation filter. Figure 5.3 clearly demonstrates the ability of PpIX to accumulate in experimental brain tumors by comparing an H&E histological section (a) with the fluorescence image (b). The ratio of the average intensity of tumor tissue to normal brain can be as high as 200:1. Intracranial tumors are therefore an excellent target for FGR because the tumor-to-brain PpIX concentration ratio is very high due to the blood–brain barrier (BBB).

5.2.6.3 Fluorescence-Guided Resection Surgery

Three to four hours before induction of anesthesia, the patient receives an oral dose of 20 mg/kg 5-ALA dissolved in water or juice. The operative procedure starts with resection of the tumor that is clearly identifiable under white-light illumination. When differentiation between residual tumor and infiltrated or normal brain tissue becomes difficult, illumination is switched to violet-blue light and the low-pass observation filter switched in place. Regions of residual tumor can be identified by their characteristic red PpIX fluorescence with the border between normal brain and tumor recognizable. During the procedure, adequate homeostasis is important since blood absorbs blue light.

In a large, well-designed, and thoroughly analyzed multicenter phase III trial by Stummer and coworkers and the ALA study group in Germany,[159] postoperative MRI contrast-enhancing tumor was completely resected in 65% of patients in the ALA group compared to 36% in the white-light group (p < 0.0001). Progression-free survival was improved in the ALA group compared to white-light patients with cumulative 6 months progression-free survival rates of 41% and 21% (p = 0.0003), respectively (median progression free survival of 5.1 vs. 3.9). Despite the fact that the study lacked sufficient power to detect survival benefit, a trend of increased survival among ALA-resected patients was nevertheless observed. Similar conclusions have also been reached employing intraoperative MRI resection guidance.[170]

5.2.7 Photodynamic Therapy for Malignant Brain Tumors

5.2.7.1 Characteristics of Glioblastoma Multiforme

The resistance of the highest-grade glioma (GBM) to therapeutic intervention is due to a number of factors. Among the most important are significant genetic variations, including numerous deletions, amplifications and point mutations, and variations in BBB patency throughout the tumor volume—it is intact in some regions while leaky in others. This obviously has significant implications for the delivery of therapeutic agents. Additionally, the diffuse and infiltrative nature of GBMs makes complete surgical resection virtually impossible as previously mentioned. The propensity of glioma cells to migrate along white matter tracts makes it clear that a cure is only possible if the migratory cells can be eradicated. Despite the factors already mentioned, 80% of GBMs recur within 2 cm of the resection margin, and therefore, a reasonable starting point for improving the prognosis of GBM patients would be the development of improved local therapies capable of eradicating glioma cells in the margin, or brain adjacent to tumor (BAT). Although a number of therapeutic strategies have been attempted, the most popular include the local delivery of (1) chemotherapeutic agents using polymer wafers and (2) ionizing radiation in the form of brachytherapy.[171,172] Unfortunately, neither of these strategies has resulted in significant prolongation of survival.

5.2.7.2 Photodynamic Therapy Principles

PDT is a treatment modality combining a photosensitizing drug and light to activate the photosensitizer in an oxygen-dependent manner resulting in oxidation of biomolecules in the light-exposed region. PDT has been approved for the treatment of a variety of conditions including cancer, vascular diseases, viral infections, and age-related macular degeneration. Interest in PDT for the treatment of gliomas is due to the fact that PDT does not lead to cumulative toxicity in the patient, and there is no known maximum cumulative dose as exists with both radiation therapy and chemotherapy. Therefore, repeated PDT treatments may be an option over conventional single PDT treatment schemes.[163,173–176]

Most photosensitizers currently used in PDT are porphyrins or porphyrin-related compounds, although there are other sensitizers under development. Although far from ideal, porphyrins such as HpD and Pf have been used almost exclusively in PDT trials of brain tumors. An ideal PDT photosensitizer should exhibit the following properties: (1) strong light absorption in the red or near-infrared part of the spectrum in order to achieve optimal tissue penetration and minimize the photosensitizer dose required to achieve the desired effect; (2) high triplet state quantum yield ($\Phi T > 0.4$) and long triplet state lifetime (>1 ms); (3) high photostability; (4) amphiphilicity; (5) favorable pharmacokinetics, preferably accumulating rapidly and preferentially in diseased tissues; (6) low levels of dark toxicity; and (7) minimal side effects and rapid clearance from the body after PDT.[177–179] Photosensitizers used in PDT accumulate preferentially in neoplastic lesions and the tumor-to-normal tissue accumulation ratio is about 2–3: 1. The mechanisms involved in this selectivity are not fully understood, but several properties of tumors seem to be of importance. The high number of low density lipid (LDL) receptors, low interstitial pH, reduced lymphatic drainage, and large interstitial space with high amounts of collagen and leaky vasculature all tend to favor the accumulation of photosensitizers in tumors. The selectivity may be further enhanced by incorporation of the photosensitizers into delivery vehicles such as liposomes or nanoparticles.[180,181]

A significant drawback of PDT is the limited penetrance of light in brain tissue. For example, the penetration depth of 630 nm light (used in ALA- and HpD/Pf-mediated PDT) is approximately 2.5 mm.[173,174]

Since the effective PDT treatment distance is ca. three times the penetration depth, 630 nm light would appear to be inadequate for eradicating tumor cells in the BAT since the treatment volume only extends to approximately 0.75 cm from the light source. Obvious strategies to increase the treatment volume are (1) employing photosensitizers that absorb at longer wavelengths where light penetration in tissue is higher and (2) increasing the treatment time. The latter is difficult to achieve with traditional *one-shot* intraoperative PDT regimes, but a potential solution to this problem is to use implantable balloon applicators that would facilitate light delivery over extended time periods.[173] Such applicators would also allow for investigations of other light delivery schemes such as fractionation and long-term repeated PDT treatments. Theoretical calculations suggest that it may be possible to attain threshold light fluences in the brain for ALA-PDT at depths approaching 1.4 cm using spherical balloon applicators of 1.5 cm radius or less.[182]

5.2.7.3 In Vitro Studies

Numerous in vitro PDT studies have been carried out employing cell monolayers, multicell tumor spheroids, and the vascularized chick chorioallantoic membrane (CAM) assay. Multicell tumor spheroids, which are 3D aggregates of cells that mimic microtumors, have proved to be valuable in understanding the mechanisms of sensitizer photobleaching and complex PDT treatments such as multiple repetitive schemes and combined treatment protocols (a review is provided in Ref. 183). In particular, the increased efficacy of low fluence rate and repetitive PDT has been demonstrated in human glioma spheroids using ALA.[175,176] The observation that PDT appears to be an effective inhibitor of tumor cell invasion suggests that this type of therapy may be particularly useful, since malignant gliomas are characterized by a large central volume of extensive necrosis surrounded by a dense shell of invasive cells that typically migrate beyond the therapeutic margins resulting in tumor recurrence.[184]

5.2.7.4 Animal Studies

Although a variety of animal brain tumor models have been employed, the vast majority of preclinical brain tumor studies have been conducted in rodents. Several studies suggest that Pf and HpD may not be ideally suited for PDT of gliomas due to the potential of significant damage to normal brain.[169,185,186] Treatment-induced disruption of normal BBB likely results in the spreading of drugs with generated edema, causing sensitization of normal brain. The high sensitivity of normal brain to PDT with first-generation photosensitizers has motivated a search for more clinically appropriate compounds such as ALA. White matter has been reported to be extremely insensitive to ALA-PDT-induced damage, for example, light fluences in white matter must be 5000 times higher than in tumor tissue for equivalent damage, and since nearly all adult brain tumors arise in white matter, ALA is an appealing compound for PDT of gliomas.[160,169]

5.2.7.5 Clinical Studies

The first report on PDT for the treatment of gliomas appeared in 1980, where HpD and intraoperative light irradiation were employed.[187] HpD and its purified versions (Pf and Photosan®) have been the most commonly employed photosensitizers in subsequent trials that have been relatively few in number due, in part, to the rarity of the disease and the ever-growing list of novel therapies that PDT must compete against. A number of reviews have been published that effectively summarizes the status of PDT for the management of high-grade gliomas from the mid-1980s to the present.[157,160,169,188–194] Table 5.1 lists treatment parameters and clinical results for selected PDT clinical trials.

Median survival for the majority of PDT studies ranged between 8 and 19 months for primary GBM and 3–13.5 months for recurrent GBM. In one study, there was a 28% survival of more than 24 months and 22% of patients survived long term beyond 60 months.[193] Many of the studies were done with HpD-mediated PDT that has several drawbacks including systemic and prolonged photosensitivity (6–8 weeks), limited tumor-selective photosensitizer uptake, and the risk of PDT-induced brain edema.[189–191] In the clinical trial using mTHPC, a median survival of 19 months for primary and

TABLE 5.1 Selected PDT Clinical Trials

Study	Photosensitizer	Patients	Results
Perria[187]	5 mg/kg b.w. HP	Recurrent tumor, n = 7	MS: 6.8 m
Origitano[195]	2 mg/kg b.w. HpD	Recurrent GBM, n = 11	MS: 8 m
Stylli[193]	2.5–5 mg/kg b.w. HpD	Primary GBM, n = 31	MS: 14.3 m
		Recurrent GBM, n = 55	MS: 13.5 m
Kostron[189]	0.15 mg/kg b.w. mTHPC	Primary GBM, n = 12	MS: 19 m
		Recurrent GBM, n = 39	MS: 9
Kaye[188]	5 mg/kg b.w. HpD	AA/GBM, n = 130	MS: 27 m
Muller[191]	2 mg/kg b.w. Pf	Primary GBM, n = 12	MS: 8.2 m
		Recurrent GBM, n = 37	MS: 7.2 m
Beck[196]	20 mg/kg b.w. ALA	Recurrent GBM, n = 10	MS: 15 m
Eljamel[163]	2 mg/kg b.w. Pf 20 mg/kg ALA	Primary GBM, n = 13	MS: 12.6 m

Note: MS, median survival.

9 months for recurrent GBM was obtained.[189] PDT of GBM using exogenous porphyrins, or mTHPC, appeared to result in patient survival times comparable to conventional treatment protocols now in use. This, together with safety and complexity issues, has prevented PDT, with exogenous photosensitizers from becoming an accepted treatment alternative. As previously mentioned, ALA-induced PpIX is a photosensitizer that has several advantages including limited toxicity—it has been given to a large number of brain tumor patients for FGR with few side effects. ALA-mediated PDT has been tried as postoperative therapy[93] or employing stereotactic-guided light delivery as a minimally invasive single treatment modality. A median survival of 15 months and a 1-year survival rate of 60% were observed in this group of highly selected patients.[196]

ALA-mediated PDT targets tumor cells, whereas Pf PDT has a significant effect on the tumor vasculature. Preclinical studies by Peng and coworkers have shown that the combination of ALA and Pf yields an increased PDT effect.[197] A small randomized clinical trial has been reported using a standard ALA dosage (20 mg/kg b.w.) combined with i.v. administration of 2 mg/kg b.w. Pf.[163] FGR and irradiation of the surgical cavity at 630 nm and 100 J/cm² was performed at the time of surgery. The light applicator was left in place and an additional four daily doses of light irradiation with 100 J/cm² were administered. The mean survival of the treatment group was 52.8 weeks (n = 13) compared to 24.6 weeks in the control group (n = 14) receiving conventional surgery only. The mean time to tumor progression was 8.6 months in the study group compared to 4.8 months in the control group. Since both FGR and repetitive PDT were used in the study group, it is difficult to separate their individual therapeutic roles in this small series, but the results are promising. Preclinical studies have demonstrated the increased efficacy of low fluence rate and repetitive PDT compared to single treatment protocols both in vitro and in vivo.[174–176,198]

5.2.8 Photochemical Internalization

PCI is a special type of PDT that can be used to enhance the delivery of macromolecules in a site-specific manner.[199] The concept is based on the use of specially designed photosensitizers, which localize preferentially in the membranes of endocytic vesicles. Upon light activation, the photosensitizer interacts with ambient oxygen causing vesicular membrane damage resulting in the release of encapsulated macromolecules into the cell cytosol instead of being transported and degraded in the lysosomes (Figure 5.4). PCI has been shown to potentiate the biological activity of a large variety of macromolecules and other molecules that do not readily penetrate the plasma membrane, including proteins (e.g., protein toxins and immunotoxins), peptides, DNA delivered as a complex with cationic polymers or incorporated in adenovirus or adeno-associated virus, peptide nucleic acids (PNA), and chemotherapeutic agents.[200–203]

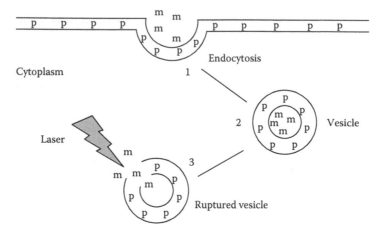

FIGURE 5.4 Illustration of the PCI concept: (a) The photosensitizer, p, and the therapeutic molecule, m, are endocytosed by the cell. (b) Both drugs end up in the same vesicles. (c) Upon exposure to light, a photodynamic reaction results in the rupture of vesicle membranes and the therapeutic molecules are released into the cytosol.

Localized opening of the BBB is a potentially useful application of PCI as it could enhance the delivery of therapeutic agents for the treatment of a wide variety of brain diseases including cancer. The BBB controls the passage of blood-borne agents into the CNS, and as such, it plays a vital role in protecting the brain against pathogens. Although this protective mechanism is essential for normal brain function, it also poses a significant hindrance to the entry of drugs into the brain. In this context, it is hardly surprising that brain diseases account for approximately 30% of the total burden of all diseases.[204] The protective function of the BBB is particularly problematic for the treatment of infiltrating gliomas. Although surgery is used to remove gross tumor, standard adjuvant therapies consisting of radiation and chemotherapy often fail to eliminate infiltrating glioma cells in or beyond the BAT region—a zone that commonly extends several centimeters from the resection margin. This is the reason for the high rate of tumor recurrence (80%) within a 2–3 cm margin of the surgical resection cavity.[205] Infiltrating tumor cells are supplied with nutrients and oxygen by the normal brain vasculature and consequently protected by the BBB: few anticancer drugs are capable of crossing this barrier. Therefore, eradication of gliomas is highly unlikely without addressing the problems posed by the BBB.

Site-specific disruption of the BBB for drug delivery into the brain has been accomplished using a number of approaches including highly focused ultrasound[206] and laser-based techniques such as PDT[207] and PCI.[208] These approaches are appealing for a number of reasons including the highly localized nature of the BBB disruption: unlike the use of hyperosmolar solutions, the BBB is only disrupted at sites subjected to sufficient laser power densities that can be controlled by the user to coincide with the location of the pathology. Through judicious choice of beam parameters, the affected volume can be as small as a few mm³. Equally important are observations showing that these highly focused approaches do not cause permanent damage to the BBB, as long as incident power densities remain below threshold levels. Under these conditions, the BBB may remain open for relatively long periods of time, thus, facilitating multifractionated drug delivery. In contrast, repeated injections of hyperosmotic compounds are required for extended treatment regimens since the BBB remains open for only a few minutes following bradykinin administration.[209]

Localized BBB opening via PCI-mediated delivery of *Clostridium perfringens* epsilon prototoxin (ETXp) was recently investigated in Fischer rats.[208] The rationale for using ETXp is due to the ability of active toxin (ETX) to cause widespread but reversible opening of the BBB.[210–212] Following systemic administration, ETXp is converted to fully active toxin by proteolytic cleavage. Administration of ETXp in rats has also been shown to result in a reduction of the endothelium barrier antigen in rat brain endothelial cells accompanied by a reversible opening of the BBB.[208] The results demonstrated that ETXp-PCI was capable of causing localized BBB disruption at very low light fluences (0.5 and 1 J). Of particular interest was the time duration and evolution of the ETXp-PCI BBB disruption since this represents the therapeutic window for drug delivery. Based on an analysis of MR images, enhancement volumes were observed to peak 3 days following ETXp-PCI suggestive of maximum BBB opening at that time. Thereafter, contrast volumes were observed to decrease, and by day 11, only trace amounts of contrast were observed. In a follow-on study using an orthotopic brain tumor model consisting of F98 glioma cells in Fischer rats,[208] animal survival was significantly extended following ETXp-PCI opening and administration of a chemotherapeutic agent (bleomycin).

5.3 Summary

The development of the diode laser has had a significant impact on laser-based diagnostic applications in the brain. Their low cost, ease of use, and lack of power supplies and external cooling make diode lasers ideally suited to the clinical environment and, as such, they have been used in a number of bedside monitoring applications. The primary focus has been on the monitoring of cerebral hemodynamics that may provide valuable diagnostic information for a number of disease states including ischemia, hemorrhage, seizures, cerebral palsy, and neoplasms. NIRS has been used in the clinic for a number of years, while promising results in preclinical studies suggest that LSI and OCT will soon make the transition from benchtop to bedside. Rapid technological advances in both optical spectroscopy and imaging are expected to lead to an improved understanding of brain function in both healthy and diseased states and therefore the future of laser-based diagnostics in the brain appears promising.

In contrast, the current status of lasers for the treatment of brain diseases is somewhat uncertain: some believe that lasers play a prominent role while others downplay their importance. Although lasers are undoubtedly useful for tumor debulking and coagulation, many surgeons prefer bipolar cautery and microdissection for the removal of tumors adjacent to neural or vascular structures. In the past decade, there have been relatively few publications on lasers in neurosurgery suggesting that they do not play a prominent role in the treatment of brain diseases. Cumbersome ergonomics and cost constitute the two major barriers to the use of lasers, such as the CO_2 and Nd:YAG, in neurosurgery. The large footprint of these lasers is also problematic in operating rooms where space is limited. The development of high-power diode lasers has solved many of these problems. Diode lasers have replaced Nd:YAG lasers in a number of neurosurgical applications including LITT. The wide range of wavelengths possible with these lasers makes them suitable for a number of neurosurgical applications. Notwithstanding the CO_2 laser, it's possible that diode lasers could replace all other neurosurgical lasers and hence make the laser more appealing for the treatment of a wide variety of brain diseases including noncommunicating hydrocephalus (using LA-ETV), neurovascular conditions (using ELANA), and tumors (using LITT and PDT).

The clinical utility of a number of investigational laser-based procedures (PDT, FGR, and PCI) for the treatment of brain diseases is somewhat uncertain. Primarily due to the efforts of Stummer et al., ALA-based FGR has been the most intensively investigated of these procedures. The multicenter phase III trial in Germany has shown that FGR can result in a significant improvement in the degree of resection compared to conventional surgery. In comparison to other techniques for resection guidance, like intraoperative MRI, FGR is easily adapted into the operating theater, is relatively inexpensive, and does not significantly add to the complexity and duration of the surgical procedure. ALA-FGR has recently received clinical approval in the EU for the resection of gliomas and the first single-center ALA-FGR study has recently been completed in the United States. ALA-FGR will no doubt gain increasing acceptance in the neurosurgical community and will certainly undergo further development to increase the sensitivity and selectivity of the technique. Ultimately, a significant improvement in survival for GBM patients will not result from more extensive surgery but from improved adjuvant treatment for infiltrative tumor cells beyond the resection margin. Nevertheless, techniques that provide for the minimum possible tumor burden can enhance the chances of success for new therapeutic modalities.

In the 30 years since the first PDT trial, approximately 500 patients have participated in a number of studies, mostly phase I and II, employing a variety of photosensitizers and different light delivery schemes, making scientific evaluation difficult. Although far from ideal, HpD and Pf have been used almost exclusively in PDT trials of brain tumors. A phase III single light treatment Pf trial failed to demonstrate a survival advantage, while two recent trials, one combining ALA-FGR and multiple Pf–PDT treatments and the other using ALA combined with interstitial light delivery, both showed an increase in the time to tumor progression compared to controls. The clinical implications of these trials are difficult to evaluate due to the limited number of patients. It is clear that multicenter randomized phase III trials, similar to the German FGR trial, are required in order to evaluate the efficacy of PDT for the treatment of malignant gliomas.

Limited light penetration and poor tumor selectivity are obvious shortcomings of PDT that have hampered its use in a number of oncologic applications including malignant gliomas. Optimization of PDT requires the development of photosensitizers with strong absorption at near-infrared wavelengths allowing increased penetration in brain tissues, while improved selectivity may be accomplished through the use of targeted delivery approaches employing liposomes or nanoparticles. Additional optimization may be achieved through the use of different light delivery schemes, such as repetitive low fluence rate protocols, which have been shown to be more effective than standard delivery techniques in preclinical studies.

Relatively little is known about the utility of PCI for the treatment of neurological diseases. Studies in non–brain tumor models suggest that PCI-mediated delivery of chemotherapeutic agents is more effective than PDT and, as such, provides a compelling rationale for its use as an intra- or postoperative technique for eradicating cancer cells in the resection cavity following incomplete resections. Although localized BBB opening via PCI-mediated delivery of a prototoxin was recently demonstrated in rats, neither animal nor clinical studies of the direct effects of PCI and chemotherapeutic drugs on brain tumors have been reported. Clearly, this technique offers many possibilities for the treatment of gliomas and therefore bears further investigation.

References

1. Jobsis, F.F., Noninvasive, infrared monitoring of cerebral and myocardial oxygen sufficiency and circulatory parameters, *Science*, 198, 1264, 1977.
2. Villringer, A. and Chance, B., Non-invasive optical spectroscopy and imaging of human function, *Trends Neurosci.*, 20, 435, 1997.
3. Huppert, T.J., Diamond, S.G., Franceschini, M.A., and Boas, D.A., HomER: A review of time-series analysis methods for near-infrared spectroscopy of the brain, *Appl. Opt.*, 48, D280, 2009.
4. Boas, D.A., Dale, A.M., and Franceschini, M.A., Diffuse optical imaging of brain activation: Approaches to optimizing image sensitivity, resolution, and accuracy, *Neuroimage*, 23, S275, 2004.
5. Gibson, A.P., Hebden, J.C., and Arridge, S.R., Recent advances in diffuse optical imaging, *Phys. Med. Biol.*, 50, R1, 2005.
6. Gratton, E., Toronov, V., Wolf, U., Wolf, M., and Webb, A., Measurement of brain activity by near-infrared light, *J. Biomed. Opt.*, 10, 011008, 2005.
7. Franceschini, M.A., Joseph, D.K., Huppert, T.J., Diamond, S.G., and Boas, D.A., Diffuse optical imaging of the whole head, *J. Biomed. Opt.*, 11, 054007, 2006.
8. Edwards, A.D., Richardson, C., Cope, M., Wyatt, J.S., Delpy, D.T., and Reynolds, E.O.R., Cotside measurement of cerebral blood flow in ill newborn infants by near infrared spectroscopy, *Lancet*, 332, 770, 1988.
9. Yoxall, C.W., Weindling, A.M., Dawani, N.H., and Peart, I., Measurement of cerebral venous oxyhemoglobin saturation in children by near-infrared spectroscopy and partial jugular venous occlusion, *Pediatr. Res.*, 38, 319, 1995.

10. Huppert, T.J., Hoge, R.D., Diamond, S.G., Franceschini, M.A., and Boas, D.A., A temporal comparison of BOLD, ASL, and NIRS hemodynamic responses to motor stimuli in adult humans, *Neuroimage*, 29, 368, 2006.

11. Austin, T., Gibson, A.P., Branco, G., Yusof, R.M., Arridge, S.R., Meek, J.H., Wyatt, J.S., Delpy, D.T., and Hebden, J.C., Three dimensional optical imaging of blood volume and oxygenation in the neonatal brain, *Neuroimage*, 31, 1426, 2006.

12. Tian, F., Alexandrakis, G., and Liu, H., Optimization of probe geometry for diffuse optical brain imaging based on measurement density and distribution, *Appl. Opt.*, 48, 2496, 2009.

13. White, B.R. and Culver, J.P., Quantitative evaluation of high-density diffuse optical tomography: In vivo resolution and mapping performance, *J. Biomed. Opt.*, 15, 026006, 2010.

14. Dehghani, H., White, B.R., Zeff, B.W., Tizzard, A., and Culver, J.P., Depth sensitivity and image reconstruction analysis of dense imaging arrays for mapping brain function with diffuse optical tomography, *Appl. Opt.*, 48, D137, 2009.

15. Zeff, B.W., White, B.R., Dehghani, H., Schlaggar, B.L., and Culver, J.P., Retinotopic mapping of adult human visual cortex with high-density diffuse optical tomography, *Proc. Natl Acad. Sci. U. S. A.*, 104, 12169, 2007.

16. Liao, S.M., Gregg, N.M., White, B.R., Zeff, B.W., Bjerkaas, K.A., Inder, T.E., and Culver, J.P., Neonatal hemodynamic response to visual cortex activity: High-density near-infrared spectroscopy study, *J. Biomed. Opt.*, 15, 026010, 2010.

17. Meek, J.H., Elwell, C.E., Khan, M.J., Romaya, J., Wyatt, J.S., Delpy, D.T., and Zeki, S., Regional changes in cerebral haemodynamics as a result of a visual stimulus measured by near infrared spectroscopy, *Proc. R. Soc. Lond. B: Biol. Sci.*, 261, 351, 1995.

18. Watanabe, E., Maki, A., Kawaguchi, F., Takashiro, K., Yamashita, Y., Koizumi, H., and Mayanagi, Y., Non-invasive assessment of language dominance with near-infrared spectroscopic mapping, *Neurosci. Lett.*, 256, 49, 1998.

19. Franceschini, M.A., Toronov, V., Filiaci, M., Gratton, E., and Fantini, S., On-line optical imaging of the human brain with 160-ms temporal resolution, *Opt. Express*, 6, 49, 2000.

20. Wilcox, T., Bortfeld, H., Woods, R., Wruck, E., and Boas, D.A., Using near-infrared spectroscopy to assess neural activation during object processing in infants, *J. Biomed. Opt.*, 10, 011010, 2005.

21. Im, C.-H., Jung, Y.-J., Lee, S., Koh, D., Kim, D.-W., and Kim, B.-M., Estimation of directional coupling between cortical areas using near-infrared spectroscopy NIRS, *Opt. Express*, 18, 5730, 2010.

22. Obrig, H. and Villringer, A., Beyond the visible—Imaging the human brain with light, *J. Cereb. Blood Flow Metab.*, 23, 1, 2003.

23. Khan, B., Tian, F., Behbehani, K., Romero, M.I., Delgado, M.R., Clegg, N.J., Smith, L., Reid, D., Liu, H., and Alexandrakis, G., Identification of abnormal motor cortex activation patterns in children with cerebral palsy by functional near-infrared spectroscopy, *J. Biomed. Opt.*, 15, 036008, 2010.

24. Muehlschlegel, S., Selb, J., Patel, M., Diamond, S., Franceschini, M., Sorensen, A., Boas, D., and Schwamm, L., Feasibility of NIRS in the neurointensive care unit: A pilot study in stroke using physiological oscillations, *Neurocrit. Care*, 11, 288, 2009.

25. Lee, S., Lee, M., Koh, D., Kim, B.-M., and Choi, J.H., Cerebral hemodynamic responses to seizure in the mouse brain: Simultaneous near-infrared spectroscopy—Electroencephalography study, *J. Biomed. Opt.*, 15, 037010, 2010.

26. Boas, D.A. and Dunn, A.K., Laser speckle contrast imaging in biomedical optics, *J. Biomed. Opt.*, 15, 011109, 2010.

27. Kirkpatrick, S.J., Duncan, D.D., and Wells-Gray, E.M., Detrimental effects of speckle-pixel size matching in laser speckle contrast imaging, *Opt. Lett.*, 33, 2886, 2008.

28. Ramirez-San-Juan, J.C., Ramos-Garcia, R., Guizar-Iturbide, I., Martinez-Niconoff, G., and Choi, B., Impact of velocity distribution assumption on simplified laser speckle imaging equation, *Opt. Express*, 16, 3197, 2008.

29. Choi, B., Ramirez-San-Juan, J.C., Lotfi, J., and Nelson, J.S., Linear response range characterization and in vivo application of laser speckle imaging of blood flow dynamics, *J. Biomed. Opt.*, 11, 041129, 2006.

30. Parthasarathy, A.B., Tom, W.J., Gopal, A., Zhang, X.J., and Dunn, A.K., Robust flow measurement with multi-exposure speckle imaging, *Opt. Express*, 16, 1975, 2008.

31. Dunn, A.K., Devor, A., Dale, A.M., and Boas, D.A., Spatial extent of oxygen metabolism and hemodynamic changes during functional activation of the rat somatosensory cortex, *Neuroimage*, 27, 279, 2005.

32. Dunn, A.K., Bolay, T., Moskowitz, M.A., and Boas, D.A., Dynamic imaging of cerebral blood flow using laser speckle, *J. Cereb. Blood Flow Metab.*, 21, 195, 2001.

33. Dunn, A.K., Devor, A., Bolay, H., Andermann, M.L., Moskowitz, M.A., Dale, A.M., and Boas, D.A., Simultaneous imaging of total cerebral hemoglobin concentration, oxygenation, and blood flow during functional activation, *Opt. Lett.*, 28, 28, 2003.

34. Sakadzic, S., Yuan, S., Dilekoz, E., Ruvinskaya, S., Vinogradov, S.A., Ayata, A., and Boas, D.A., Simultaneous imaging of cerebral partial pressure of oxygen and blood flow during functional activation and cortical spreading depression, *Appl. Opt.*, 48, D169, 2009.

35. Bolay, H., Reuter, U., Dunn, A.K., Huang, Z.H., Boas, D.A., and Moskowitz, M.A., Intrinsic brain activity triggers trigeminal meningeal afferents in a migraine model, *Nat. Med.*, 8, 136, 2002.

36. Ayata, C., Shin, H.K., Salomone, S., Ozdemir-Gursoy, Y., Boas, D.A., Dunn, A.K., and Moskowitz, M.A., Pronounced hypoperfusion during spreading depression in mouse cortex, *J. Cereb. Blood Flow Metab.*, 24, 1172, 2004.

37. Strong, A.J., Bezzina, E.L., Anderson, P.B.J., Boutelle, M.G., Hopwood, S.E., and Dunn, A.K., Evaluation of laser speckle flowmetry for imaging cortical perfusion in experimental stroke studies: Quantitation of perfusion and detection of peri-infarct depolarisations, *J. Cereb. Blood Flow Metab.*, 26, 645, 2006.

38. Strong, A.J., Anderson, P.J., Watts, H.R., Virley, D.J., Lloyd, A., Irving, E.A., Nagafuji, T. et al., Peri-infarct depolarizations lead to loss of perfusion in ischaemic gyrencephalic cerebral cortex, *Brain*, 130, 995, 2007.

39. Shin, H.K., Dunn, A.K., Jones, P.B., Boas, D.A., Moskowitz, M.A., and Ayata, C., Vasoconstrictive neurovascular coupling during focal ischemic depolarizations, *J. Cereb. Blood Flow Metab.*, 26, 1018, 2006.

40. Shin, H.K., Dunn, A.K., Jones, P.B., Boas, D.A., Lo, E.H., Moskowitz, M.A., and Ayata, C., Normobaric hyperoxia improves cerebral blood flow and oxygenation, and inhibits peri-infarct depolarizations in experimental focal ischaemia, *Brain*, 130, 1631, 2007.

41. Zhang, S.X. and Murphy, T.H., Imaging the impact of cortical microcirculation on synaptic structure and sensory-evoked hemodynamic responses in vivo, *PLoS Biol.*, 5, 1152, 2007.

42. Lay, C.C., Davis, M.F., Chen-Bee, C.H., and Frostig, R.D., Mild sensory stimulation completely protects the adult rodent cortex from ischemic stroke, *PLoS One*, 5, e11270, 2010.

43. Hecht, N., Woitzik, J., Dreier, J.P., and Vajkoczy, P., Intraoperative monitoring of cerebral blood flow by laser speckle contrast analysis, *Neurosurg. Focus*, 27, E11, 2009.

44. Xu, M.H. and Wang, L.H.V., Photoacoustic imaging in biomedicine, *Rev. Sci. Instrum.*, 77, 041101, 2006.

45. Xu, M.H. and Wang, L.H.V., Universal back-projection algorithm for photoacoustic computed tomography, *Phys. Rev. E*, 71, 016706, 2005.

46. Provost, J. and Lesage, F., The application of compressed sensing for photo-acoustic tomography, *IEEE Trans. Med. Imaging*, 28, 585, 2009.

47. Gamelin, J., Maurudis, A., Aguirre, A., Huang, F., Guo, P.Y., Wang, L.V., and Zhu, Q., A real-time photoacoustic tomography system for small animals, *Opt. Express*, 17, 10489, 2009.

48. Li, C., Aguirre, A., Gamelin, J., Maurudis, A., Zhu, Q., and Wang, L.V., Real-time photoacoustic tomography of cortical hemodynamics in small animals, *J. Biomed. Opt.*, 15, 010509, 2010.

49. Petrova, Y.I., Petrov, Y.Y., Esenaliev, R.O., Deyo, D.J., Cicenaite, I., and Prough, D.S., Noninvasive monitoring of cerebral blood oxygenation in ovine superior sagittal sinus with novel multi-wavelength optoacoustic system, *Opt. Express*, 17, 7285, 2009.

50. Bouma, B.E., Yun, S.-H., Vakoc, B.J., Suter, M.J., and Tearney, G.J., Fourier-domain optical coherence tomography: Recent advances toward clinical utility, *Curr. Opin. Biotechnol.*, 20, 111, 2009.

51. Fujimoto, J.G. and Brezinksi, M.E., Optical coherence tomography imaging, in *Biomedical Photonics Handbook*, Vo Dinh, T., ed. (CRC Press, Boca Raton, FL, 2003).

52. Bouma, B.E. and Tearney, G.J., *Handbook of Optical Coherence Tomography* (Informa Healthcare, Zug, Switzerland, 2001).

53. Mariampillai, A., Leung, M.K.K., Jarvi, M., Standish, B.A., Lee, K., Wilson, B.C., Vitkin, A., and Yang, V.X.D., Optimized speckle variance OCT imaging of microvasculature, *Opt. Lett.*, 35, 1257, 2010.

54. Srinivasan, V.J., Sakadi, S., Gorczynska, I., Ruvinskaya, S., Wu, W., Fujimoto, J.G., and Boas, D.A., Quantitative cerebral blood flow with optical coherence tomography, *Opt. Express*, 18, 2477, 2010.

55. Wang, R.K., Jacques, S.L., Ma, Z., Hurst, S., Hanson, S.R., and Gruber, A., Three dimensional optical angiography, *Opt. Express*, 15, 4083, 2007.

56. Ren, H., Sun, T., MacDonald, D.J., Cobb, M.J., and Li, X., Real-time in vivo blood-flow imaging by moving-scatterer-sensitive spectral-domain optical Doppler tomography, *Opt. Lett.*, 31, 927, 2006.

57. Wang, R.K. and An, L., Doppler optical micro-angiography for volumetric imaging of vascular perfusion in vivo, *Opt. Express*, 17, 8926, 2009.

58. Srinivasan, V.J., Sakadzic, S., Gorczynska, I., Ruvinskaya, S., Wu, W., Fujimoto, J.G., and Boas, D.A., Depth-resolved microscopy of cortical hemodynamics with optical coherence tomography, *Opt. Lett.*, 34, 3086, 2009.

59. Flusberg, B.A., Nimmerjahn, A., Cocker, E.D., Mukamel, E.A., Barretto, R.P.J., Ko, T.H., Burns, L.D., Jung, J.C., and Schnitzer, M.J., High-speed, miniaturized fluorescence microscopy in freely moving mice, *Nat. Methods*, 5, 935, 2008.

60. Mertz, J. and Kim, J., Scanning light-sheet microscopy in the whole mouse brain with HiLo background rejection, *J. Biomed. Opt.*, 15, 016027, 2010.

61. Engelbrecht, C.J., Voigt, F., and Helmchen, F., Miniaturized selective plane illumination microscopy for high-contrast in vivo fluorescence imaging, *Opt. Lett.*, 35, 1413, 2010.

62. Luo, Z.C., Yuan, Z.J., Pan, Y.T., and Du, C.W., Simultaneous imaging of cortical hemodynamics and blood oxygenation change during cerebral ischemia using dual-wavelength laser speckle contrast imaging, *Opt. Lett.*, 34, 1480, 2009.

63. Yuan, Z., Luo, Z., Volkow, N.D., Pan, Y., and Du, C., Imaging separation of neuronal from vascular effects of cocaine on rat cortical brain in vivo, *Neuroimage*, 54(2), 1130–1139, 2011.

64. Ponticorvo, A. and Dunn, A.K., Simultaneous imaging of oxygen tension and blood flow in animals using a digital micromirror device, *Opt. Express*, 18, 8160, 2010.

65. Durduran, T., Zhou, C., Buckley, E.M., Kim, M.M., Yu, G., Choe, R., Gaynor, J.W. et al., Optical measurement of cerebral hemodynamics and oxygen metabolism in neonates with congenital heart defects, *J. Biomed. Opt.*, 15, 037004, 2010.

66. Yodh, A.G. and Boas, D.A., Functional imaging with diffusing light, in *Biomedical Photonics Handbook*, Vo-Dinh, T., ed. (CRC Press, Boca Raton, FL, 2003).

67. Durduran, T., Zhou, C., Edlow, B.L., Yu, G.Q., Choe, R., Kim, M.N., Cucchiara, B.L. et al., Transcranial optical monitoring of cerebrovascular hemodynamics in acute stroke patients, *Opt. Express*, 17, 3884, 2009.

68. Muller, G., Dorschel, K., and Kar, H., Biophysics of the photoablation process, *Lasers Med. Sci.*, 6, 241, 1991.

69. Devaux, B.C. and Roux, F.X., Experimental and clinical standards, and evolution of lasers in neurosurgery, *Acta Neurochir. (Wien)*, 138, 1135, 1996.

70. Ryan, R.W., Spetzler, R.F., and Preul, M.D., Aura of technology and the cutting edge: A history of lasers in neurosurgery, *Neurosurg. Focus*, 27, E6, 2009.

71. Rosomoff, H.L. and Carroll, F., Reaction of neoplasm and brain to laser, *Arch. Neurol.*, 14, 143, 1966.

72. Stellar, S., Polanyi, T.G., and Diedemeler, H.C., Experimental studies with the carbon dioxide laser as a neurosurgical instrument, *Med. Biol. Eng.*, 8, 549, 1970.

73. Stellar, S. and Polanyi, T.G., Lasers in neurosurgery: A historical overview, *J. Clin. Laser Med. Surg.*, 10, 399, 1992.
74. Heppner, F., The laser scalpel on the nervous system, in *Laser Surgery II*, Kaplan, I., ed. (Jerusalem Academic Press, Jerusalem, 1978, pp. 79–80).
75. Ascher, P.W., Neurosurgery, in *Microscopic and Endoscopic Surgery with the CO₂ Laser*, Kaplan, I., ed. (John Wright-PSG, Boston, MA, 1982, pp. 298–314).
76. Ascher, P.W. and Heppner, F., CO₂ laser in neurosurgery, *Neurosurg. Rev.*, 7, 123, 1984.
77. Takizawa, T., Yamazaki, T., Miura, N., Matsumoto, M., Tanaka, Y., Takeuchi, K., Nakata, Y. et al., Laser surgery of basal, orbital and ventricular meningiomas which are difficult to extirpate by conventional methods, *Neurol. Med. Chir.*, 20, 729, 1980.
78. Wharen, R.E., Anderson, R.E., Scheithauer, B., and Sundt, T.M., The Nd:YAG laser in neurosurgery. Part 1. Laboratory investigations: Dose-related biological response of neural tissue, *J. Neurosurg.*, 60, 531, 1984.
79. Yahr, W.Z., Strully, K.J., and Hurwitt, E.S., Non-occlusive small arterial anastomosis with a neodymium laser, *Surg. Forum*, 15, 224, 1964.
80. Yahr, W.Z. and Strully, K.J., Blood vessel anastomosis by laser and other biomedical applications, *J. Assoc. Adv. Med. Instrum.*, 1, 28, 1966.
81. Beck, O.J., The use of the Nd:YAG and the carbon dioxide laser in neurosurgery, *Neurosurg. Rev.*, 3, 261, 1980.
82. Takeuchi, J., Handa, H., Taki, W., and Yamagami, T., The Nd:YAG laser in neurological surgery, *Surg. Neurol.*, 18, 140, 1982.
83. Krishnamurthy, S. and Powers, S.K., Lasers in neurosurgery, *Lasers Surg. Med.*, 15, 126, 1994.
84. Sterenborg, H.J., van Gemert, M.J., Kamphorst, W., Wolbers, J.G., and Hogervorst, W., The spectral dependence of the optical properties of the human brain, *Lasers Med. Sci.*, 4, 221, 1989.
85. Svaasand, L.O. and Ellingsen, R., Optical properties of human brain, *Photochem. Photobiol.*, 38, 293, 1983.
86. Passano, V.A., The use of laser in neurosurgery, *J. Neurosurg. Sci.*, 26, 245, 1982.
87. Powers, S.K., Edwards, M.S., Boggan, J.E., Pitts, L.H., Gutin, P.H., Hosobuchi, Y., Adams, J.E., and Wilson, C.B., Use of the argon surgical laser in neurosurgery, *J. Neurosurg.*, 60, 523, 1984.
88. Edwards, M.S.B., Boggan, J.E., and Fuller, T.A., The laser in neurological surgery, *J. Neurosurg.*, 59, 555, 1983.
89. Jain, K.K., Lasers in neurosurgery: A review, *Lasers Surg. Med.*, 2, 217, 1983.
90. Powers, S.K., Current status of lasers in neurosurgical oncology, *Semin. Surg. Oncol.*, 8, 226, 1992.
91. Lin, L.-M., Sciubba, D.M., and Jallo, G.I., Neurosurgical applications of laser technology, *Surg. Technol. Int.*, 18, 63, 2009.
92. Devaux, B.C., Joly, L.M., Page, P., Nataf, F., Turak, B., Beuvon, F., Trystram, D., and Roux, F.X., Laser-assisted endoscopic third ventriculostomy for obstructive hydrocephalus: Technique and results in a series of 40 consecutive cases, *Lasers Surg. Med.*, 34, 368, 2004.
93. Farin, A., Aryan, H.E., Ozgur, B.M., Parsa, A.T., and Levy, M.L., Endoscopic third ventriculostomy, *J. Clin. Neurosci.*, 13, 763, 2006.
94. Hopf, N.J., Grunert, P., Fries, G., Resch, K.D., and Perneczky, A., Endoscopic third ventriculostomy: Outcome analysis of 100 consecutive procedures, *Neurosurgery*, 44, 795, 1999.
95. Jones, R.F., Stening, W.A., and Brydon, M., Endoscopic third ventriculostomy, *Neurosurgery*, 26, 86, 1990.
96. Kunz, U., Goldman, A., Bader, C., Waldbaur, H., and Oldenkott, P., Endoscopic fenestration of the 3rd ventricular floor in aqueductal stenosis, *Minim. Invasive Neurosurg.*, 37, 42, 1994.
97. Rhoten, R.L., Luciano, M.G., and Barnett, G.H., Computer-assisted endoscopy for neurosurgical procedures: Technical note, *Neurosurgery*, 40, 632, 1997.
98. Devaux, B.C., Roux, F.X., Natal, F., Turak, B., and Cioloca, C., High-power diode laser in neurosurgery: Clinical experience in 30 cases, *Surg. Neurol.*, 50, 33, 1998.

99. Vandertop, W.P., Verdaasdonk, R.M., and van Swol, C.F., Laser-assisted neuroendoscopy using a neodymium-yttrium aluminum garnet or diode contact laser with pretreated fiber tips, *J. Neurosurg.*, 88, 82, 1998.

100. Willems, P.W., Vandertop, W.P., Verdaasdonk, R.M., van Swol, C.F., and Jansen, G.H., Contact laser-assisted neuroendoscopy can be performed safely by using pre-treated black fibre tip: Experimental data, *Lasers Surg. Med.*, 28, 324, 2001.

101. van Beijnum, J., Hanlo, P.W., Fischer, K., Majidpour, M.M., Kortekaas, M.F., Verdaasdonk, R.M., and Vandertop, W.P., Laser-assisted endoscopic third ventriculostomy: Long-term results in a series of 202 patients, *Neurosurgery*, 62, 437, 2008.

102. Abtin, K., Thompson, B.G., and Walker, M.L., Basilar artery perforation as a complication of endoscopic third ventriculostomy, *Pediatr. Neurosurg.*, 28, 236, 1998.

103. DiRocco, C., Massimi, L., and Tamburrini, G., Shunts vs endoscopic third ventriculostomy in infants: Are there different types and/or rates of complications? A review, *Childs Nerv. Syst.*, 22, 1573, 2006.

104. Schroeder, H.W., Warzok, R.W., Assaf, J.A., and Gaab, M.R., Fatal subarachnoid hemorrhage after endoscopic third ventriculostomy. Case report, *J. Neurosurg.*, 90, 153, 1999.

105. Streefkerk, H.J., Bremmer, J.P., and Tulleken, C.A., The ELANA technique: High flow revascularization of the brain, *Acta Neurochir. Suppl.*, 94, 143, 2005.

106. Tulleken, C.A. and Verdaasdonk, R.M., First clinical experience with excimer assisted high flow bypass surgery of the brain, *Acta Neurochir. (Wien)*, 134, 66, 1995.

107. Tulleken, C.A., van der Zwan, A., and Kappelle, L.J., High-flow transcranial bypass for prevention of brain ischemia, *Ned. Tijdschr. Geneeskd.*, 143, 2281, 1999.

108. Tulleken, C.A., van der Zwan, A., van Rooij, W.J., and Ramos, L.M., High-flow bypass using nonocclusive excimer laser-assisted end-to-side anastomosis of the external carotid artery to the P1 segment of the posterior cerebral artery via the sylvian route. Technical note, *J. Neurosurg.*, 88, 925, 1998.

109. Tulleken, C.A., Verdaasdonk, R.M., Beck, R.J., and Mali, W.P., The modified excimer laser-assisted high-flow bypass operation, *Surg. Neurol.*, 46, 424, 1996.

110. Tulleken, C.A., Verdaasdonk, R.M., Berendsen, W., and Mali, W.P., Use of the excimer laser in high-flow bypass surgery of the brain, *J. Neurosurg.*, 78, 477, 1993.

111. van Doormaal, T.P., van der Zwan, A., VerWeij, B.H., Langer, D.J., and Tulleken, C.A., Treatment of giant and large internal carotid artery aneurysms with a high-flow replacement bypass using the excimer laser-assisted nonocclusive anastomosis technique, *Neurosurgery*, 59, ONS328, 2006.

112. van Doormaal, T.P., van der Zwan, A., Aboud, E., van der Sprenkel, J.W.B., Tulleken, C.A.F., Krisht, A.F., and Regli, L., The sutureless excimer laser assisted non-occlusive anastomosis (SELANA); a feasibility study in a pressurized cadaver model, *Acta Neurochir. (Wien)*, 152, 1603, 2010.

113. Verdaasdonk, R.M. and van Swol, C.F.P., Laser light delivery systems for medical applications, *Phys. Med. Biol.*, 42, 869, 1997.

114. Bremmer, J.P., Verweij, B.H., Klijn, C.J.M., van der Zwan, A., Kappelle, L.J., and Tulleken, C.A.F., Predictors of patency of excimer laser-assisted nonocclusive extracranial-to-intracranial bypasses, *J. Neurosurg.*, 110, 887, 2009.

115. van Doormaal, T.P.C., van der Zwan, A., Verweij, B.H., Regli, L., and Tulleken, C.A.F., Giant aneurysm clipping under protection of an excimer laser-assisted non-occlusive anastomosis bypass, *Neurosurgery*, 66, 439, 2010.

116. van Doormaal, T.P.C., van der Zwan, A., Verweij, B.H., Han, K.S., Langer, D.J., and Tulleken, C.A.F., Treatment of giant middle cerebral artery aneurysms with a flow replacement bypass using the excimer laser-assisted nonocclusive anastomosis technique, *Neurosurgery*, 63, 12, 2008.

117. Puca, A., Esposito, G., Albanese, A., Maira, G., Rossi, F., and Pini, R., Minimally occlusive laser vascular anastomosis (MOLVA): Experimental study, *Acta Neurochir.*, 151, 363, 2009.

118. Bown, S.G., Phototherapy of tumors, *World J. Surg.*, 7, 700, 1903.

119. Sugiyama, K., Sakai, T., Fujishima, I., Ryu, H., Uemura, K., and Yokoyama, T., Stereotactic interstitial laser-hyperthermia using Nd-YAG laser, *Stereotact. Funct. Neurosurg.*, 54, 501, 1990.
120. Menovsky, T., Beek, J.F., Roux, F.X., and Bown, S.G., Interstitial laser thermotherapy: Developments in the treatment of small deep-seated brain tumors, *Surg. Neurol.*, 46, 568, 1996.
121. Jolesz, F.A., Bleier, A.R., Jakab, P., Ruenzel, P.W., Huttl, K., and Jako, G.J., MR imaging of laser-tissue interactions, *Radiology*, 168, 249, 1988.
122. Bettag, M., Ulrich, F., Schober, R., Furst, G., Langen, K.J., Sabel, M., and Kiwit, J.C., Stereotactic laser therapy in cerebral gliomas, *Acta Neurochir. Suppl.*, 52, 81, 1991.
123. Kahn, T., Bettag, M., Ulrich, F., Schwarzmaier, H.J., Schober, R., Furst, G., and Modder, U., MRI-guided laser-induced interstitial thermotherapy of cerebral neoplasms, *J. Comput. Assist. Tomogr.*, 18, 519, 1994.
124. Kahn, T., Schwabe, B., Bettag, M., Harth, T., Ulrich, F., Rassek, M., Schwarzmaier, H.J., and Modder, U., Mapping of the cortical motor hand area with functional MR imaging and MR imaging-guided laser-induced interstitial thermotherapy of brain tumors. Work in progress, *Radiology*, 200, 149, 1996.
125. Kahn, T., Harth T., Bettag, M., Schwabe, B., Ulrich, F., Schwarzmaier, H.J., and Modder, U., Preliminary experience with the application of gadolinium-DTPA before MR imaging-guided laser-induced interstitial thermotherapy of brain tumors, *J. Magn. Reson. Imaging*, 7, 226, 1997.
126. Kahn, T., Harth, T., Kiwit, J.C., Schwarzmaier, H.J., Wald, C., and Modder, U., In vivo MRI thermometry using a phase-sensitive sequence: Preliminary experience during MRI-guided laser-induced interstitial thermotherapy of brain tumors, *J. Magn. Reson. Imaging*, 8, 160, 1998.
127. Reimer, P., Bremer, C., Horch, C., Morgenroth, C., Allkemper, T., and Schuierer, G., MR-monitored LITT as a palliative concept in patients with high grade gliomas: Preliminary clinical experience, *J. Magn. Reson. Imaging*, 8, 240, 1998.
128. Schwabe, B., Kahn, T., Harth, T., Ulrich, F., and Schwarzmaier, H.J., Laser-induced thermal lesions in the human brain: Short- and long-term appearance on MRI, *J. Comput. Assist. Tomogr.*, 21, 818, 1997.
129. Schwarzmaier, H.J., Yaroslavsky, I.V., Fiedler, V, Ulrich, F., and Kahn, T., Treatment planning for MRI-guided laser-induced interstitial thermotherapy of brain tumors—The role of blood perfusion, *J. Magn. Reson. Imaging*, 8, 121, 1998.
130. Kangasniemi, M., McNichols, R.J., Bankson, J.A., Gowda, A., Price, R.E., and Hazle, J.D., Thermal therapy of canine cerebral tumors using a 980 nm diode laser with MR temperature-sensitive imaging feedback, *Lasers Surg. Med.*, 35, 41, 2004.
131. Kou, L., Labrie, D., and Chylek, P., Refractive indices of water and ice in the 0.65–2.5 μm spectral range, *Appl. Opt.*, 32, 3531, 1993.
132. Carpentier, A., McNichols, R.J., Stafford, R.J., Itzcovitz, J., Guichard, J.-P., Reizine, D., Delalogue, S. et al., Real-time magnetic resonance-guided laser thermal therapy for focal metastatic brain tumors, *Neurosurgery*, 63, ONS21, 2008.
133. Atsumi, H., Matsumae, M., Kaneda, M., Muro, I., Mamata, Y., Komiya, T., Tsugu, A., and Tsugane, R., Novel laser system and laser irradiation method reduces the risk of carbonization during laser interstitial thermotherapy: Assessed by MR temperature measurement, *Lasers Surg. Med.*, 29, 108, 2001.
134. Leonardi, M.A., Lumenta, C.B., Gumprecht, H.K., van Einsiedel, G.H., and Wilhelm, T., Stereotactic guided laser-induced interstitial thermotherapy (SLITT) in gliomas with intraoperative morphologic monitoring in an open MR-unit, *Minim. Invasive Neurosurg.*, 44, 37, 2001.
135. McNichols, R.J., Kangasniemi, M., Gowda, A., Bankson, J.A., Price, R.E., and Hazle, J.D., Technical developments for cerebral thermal treatment: Water-cooled diffusing fibre tips and temperature-sensitive MRI using intersecting image planes, *Int. J. Hyperthermia*, 20, 45, 2004.
136. De Poorter, J., Wagter, C.D., Deene, Y.D., Thomson, C., Stahlberg, F., and Achten, E., Noninvasive MRI thermometry with the proton resonance frequency (PRF) method: In vivo results in human muscle, *Magn. Reson. Med.*, 33, 74, 1995.
137. De Poorter, J., Noninvasive MRI thermometry with the proton resonance method: Study of susceptibility effects, *Magn. Reson. Med.*, 34, 359, 1995.

138. Ishihara, Y., Calderon, A., Watanabe, H., Okamoto, K, Suzuki, Y., Kuroda, K., and Suzuki, Y., A precise and fast temperature mapping using water proton chemical shift, *Magn. Reson. Med.*, 34, 814, 1995.

139. McNichols, R.J., Gowda, A., Kangasniemi, M., Bankson, J.A., Price, R.E., and Hazle, J.D., MR thermometry-based feedback control of laser interstitial thermal therapy at 980 nm, *Lasers Surg. Med.*, 34, 48, 2004.

140. Schwartz, J.A., Shetty, A.M., Price, R.E., Stafford, R.J., Wang, J.C., Uthamanthil, R.K., Pham, K., McNichols, R.J., Coleman, C.L., and Payne, J.D., Feasibility study of particle-assisted laser ablation of brain tumors in orthotopic canine model, *Cancer Res.*, 69, 1659, 2009.

141. Vecht, C.J., Avezaat, C.J., van Putten, W.L., Eijkenboom, W.M.., and Stefanko, S.Z., The influence of the extent of surgery on the neurological function and survival in malignant glioma. A retrospective analysis in 243 patients, *J. Neurol. Neurosurg. Psychiatry*, 53, 466, 1990.

142. Albert, F.K., Forsting, M., Sartor, K., Adams, H.P., and Kunze, S., Early postoperative magnetic resonance imaging after resection of malignant glioma: Objective evaluation of residual tumor and its influence on regrowth and prognosis, *Neurosurgery*, 34, 45, 1994.

143. Barker, F.G., Prados, M.D., Chang, S.M., Gutin, P.H., Lamborn, K.R., Larson, D.A., Malec, M.K. et al., Radiation response and survival time in patients with glioblastoma multiforme, *J. Neurosurg.*, 84, 442, 1996.

144. Kowalczuk, A., Macdonald, R.L., Amidei, C., Dohrmann, G. 3rd, Erickson, R.K., Hekmatpanah, J., Krauss, S. et al., Quantitative imaging study of extent of surgical resection and prognosis of malignant astrocytomas, *Neurosurgery*, 41, 1028, 1997.

145. Lacroix, M., Abi-Said, D., Fourney, D.R., Gokaslan, Z.L., Shi, W., DeMonte, F., Lang, F.F. et al., Multivariate analysis of 416 patients with glioblastoma multiforme: Prognosis, extent of resection, and survival, *J. Neurosurg.*, 95, 190, 2001.

146. Willems, P.W.A., Taphoorn, M.J.B., Burger, H., Berkelbach van der Sprenkel, J.W., and Tulleken, C.A., Effectiveness of neuro-navigation in resecting solitary intracerebral contrast enhancing tumors: A randomized controlled trial, *J. Neurosurg.*, 104, 360, 2006.

147. Wirtz, C.R., Bonsanto, M.M., Knauth, M., Tronnier, V.M., Albert, F.K., Staubert, A., and Kunze, S., Intraoperative magnetic resonance imaging to update interactive navigation in neurosurgery. Method and preliminary experience, *Comput. Aided Surg.*, 2, 172, 1997.

148. Samset, E., Høgetveit, J.O., and Hirschberg, H., Integrated neuro-navigation system with intraoperative image updating, *Minim. Invasive Neurosurg.*, 48, 73, 2005.

149. Albayrak, B.A., Samdani, A.F., and Black, P.M., Intra-operative magnetic resonance imaging in neurosurgery, *Acta Neurochir. (Wien)*, 146, 543, 2004.

150. Hammoud, M.A., Ligon, B.L., el Souki, R., Shi, W.M., Schomer, D.F., and Sawaya, R., Use of intraoperative ultrasound for localizing tumors and determining the extent of resection. A comparative study with magnetic resonance imaging, *J. Neurosurg.*, 84, 737, 1996.

151. Hirschberg, H. and Unsgaard, G., Incorporation of ultrasonic imaging in an optically coupled frameless stereotactic system, *Acta Neurochir. Suppl.*, 68, 75, 1997.

152. Gronningsaeter, A., Kleven, A., Ommedal, S., Aarseth, T.E., Lie, T., Lindseth, F., Lango, T., and Unsgaard, G., SonoWand, an ultrasound-based neuronavigation system, *Neurosurgery*, 47, 1373, 2000.

153. Kennedy, J.C., Marcus, S.L., and Pottier, R.H., Photodynamic therapy (PDT) and photodiagnosis (PD) using endogenous photosensitization induced by 5-aminolevulinic acid (ALA): Mechanisms and clinical results, *J. Clin. Laser Med. Surg.*, 14, 289, 1996.

154. Stummer, W., Stocker, S., Wagner, S., Stepp, H., Fritsch, C., Goetz, C. et al., Intraoperative detection of malignant gliomas by 5-aminolevulinic acid-induced porphyrin fluorescence, *Neurosurgery*, 42, 518, 1998.

155. Stummer, W., Stepp, H., Möller, G., Ehrhardt, A., Leonhard, M., and Reulen, H.J., Technical principles for protoporphyrin-IX-fluorescence guided microsurgical resection of malignant gliomas tissue, *Acta Neurochir. (Wien)*, 140, 995, 1998.

156. Stummer, W., Novotny, A., Stepp, H., Goetz, C., Bise, K., and Reulen, H.J., Fluorescence-guided resection of glioblastoma multiforme by using 5-aminolevulinic acid-induced porphyrins: A prospective study in 52 consecutive patients, *J. Neurosurg.*, 93, 1003, 2000.

157. Friesen, S.A., Hjortland, G.O., Madsen, S.J., Hirschberg, H., Engebraten, O., Nesland, M., and Peng, Q., 5-Aminolevulinic acid-based photodynamic detection and therapy of brain tumors (review), *Int. J. Oncol.*, 21, 577, 2002.

158. Stummer, W., Reulen, H.J., Novotny, A., Stepp, H., and Tonn, J.C., Fluorescence-guided resections of malignant gliomas—An overview, *Acta Neurochir. Suppl.*, 88, 9, 2003.

159. Stummer, W., Pichlmeier, U., Meinel, T., Wiestler, O.D., Zanella, F., and Reulen, H.J., Fluorescence-guided surgery with 5-aminolevulinic acid for resection of malignant glioma: A randomized controlled multicentre phase III trial, *Lancet Oncol.*, 7, 392, 2006.

160. Stepp, H., Beck, T., Pongratz, T., Meinel, T., Kreth, F.W., Tonn, J.C., and Stummer, W., ALA and malignant glioma: Fluorescence-guided resection and photodynamic treatment, *J. Environ. Pathol. Toxicol. Oncol.*, 26, 157, 2007.

161. Stummer, W., Reulen, H.J., Meinel, T., Pichlmeier, U., Schumacher, W., Tonn, J.C., Rohde V, Extent of resection and survival in glioblastoma multiforme: Identification of and adjustment for bias, *Neurosurgery*, 62, 564, 2008.

162. Pichlmeier, U., Bink, A., Schackert, G., and Stummer, W., Resection and survival in glioblastoma multiforme: An RTOG recursive partitioning analysis of ALA study patients, *Neurooncol.*, 10, 1025, 2008.

163. Eljamel, M.S., Goodman, C., and Moseley, H., ALA and Photofrin fluorescence-guided resection and repetitive PDT in glioblastoma multiforme: A single centre phase III randomised controlled trial, *Lasers Med. Sci.*, 23, 361, 2008.

164. Roberts, D.W., Valdés, P.A., Harris, B.T., Fontaine, K.M., Hartov, A., Fan, X., Ji, S. et al., Coregistered fluorescence-enhanced tumor resection of malignant glioma: Relationships between δ-aminolevulinic acid-induced protoporphyrin IX fluorescence, magnetic resonance imaging enhancement, and neuropathological parameters, *J. Neurosurg.*, 114(3):595–603, 2010.

165. Peng, Q., Warloe, T., Berg, K., Mohan, J., Kongshaug, M., Giercksky, K.-E., and Nesland, J.M., 5-Aminolevulinic acid-based photodynamic therapy: Clinical research and future challenges, *Cancer*, 79, 2282, 1997.

166. Stummer, W.S., Stocker, A., Novotny, A., Heimann, A., Sauer, O., Kempski, O., Plesnila, N., Wietzorrek, J., and Reulen, H.J., In vitro and in vivo porphyrin accumulation by C6 glioma cells after exposure to 5-aminolevulinic acid, *J. Photochem. Photobiol. B*, 45, 160, 1998.

167. Gibson, S.L., Nguyen, M.L., Havens, J.J., Barbarin, A., and Hilf, R., Relationship of δ-aminolevulinic acid-induced protoporphyrin IX levels to mitochondrial content in neoplastic cells in vitro, *Biochem. Biophys. Res. Commun.*, 265, 315, 1999.

168. Berg, K., Selbo, P., Weyergang, A., Dietze, A., Prasmickaite, L., Bonsted, A., Engesaeter, B. et al., Porphyrin-related photosensitizers for cancer imaging and therapeutic applications, *J. Microsc.*, 218, 133, 2005.

169. Lilge, L. and Wilson, B.C., Photodynamic therapy of intracranial tissues: A preclinical comparative study of four different photosensitizers, *J. Clin. Laser Med. Surg.*, 16, 81, 1998.

170. Hirschberg, H., Samset, E., Hole, P.K., and Lote, K., Impact of intraoperative MRI on the results of surgery for high grade gliomas, *Minim. Invasive Neurosurg.*, 48, 77, 2006.

171. Brem, H., Piantadosi, S., Burger, P.C., Walker, M., Selker, R., Vick, N.A., Black, K. et al., Placebo-controlled trial of safety and efficacy of intraoperative controlled delivery by biodegradable polymers of chemotherapy for recurrent gliomas, *Lancet*, 345, 1008, 1995.

172. Johannesen, T.B., Watne, K., Lote, K., Norum, J., Tvera, K., and Hirschberg, H., Intracavity fractionated balloon brachytherapy in glioblastoma, *Acta Neurochir.*, 141, 127, 1999.

173. Madsen, S.J., Sun, C.H., Tromberg, B.J., and Hirschberg, H., Development of a novel indwelling balloon applicator for optimizing light delivery in photodynamic therapy, *Lasers Surg. Med.*, 29, 406, 2001.

174. Madsen, S.J., Sun, C.H., Tromberg, B.J., and Hirschberg, H., Repetitive ALA mediated photodynamic therapy on human glioma spheroids, *J. Neurooncol.*, 62, 243, 2003.

175. Hirschberg, H., Sørensen, D.R., Angell-Petersen, E., Peng, Q., Tromberg, B.J., Sun, C.H., and Madsen, S.J., Repetitive photodynamic therapy of malignant brain tumors, *J. Environ. Pathol. Toxicol. Oncol.*, 25, 261, 2006.

176. Mathews, M.S., Angell-Petersen, E., Sanchez, R., Sun, C.H., Vo, V., Hirschberg, H., and Madsen, S.J., The effects of ultra low fluence rate photodynamic therapy on glioma spheroids, *Lasers Surg. Med.*, 41, 578, 2009.

177. Castano, A., Demidova, T., and Hamblin, M., Mechanisms in photodynamic therapy: Part one—Photosensitizers, photochemistry and cellular localization, *Photodiagn. Photodyn. Ther.*, 1, 279, 2004.

178. Castano, A., Demidova, T., and Hamblin, M., Mechanisms in photodynamic therapy: Part two—Cellular signaling, cell metabolism and modes of cell death, *Photodiagn. Photodyn. Ther.*, 2, 1, 2005.

179. Plaetzer, K., Krammer, B., Berlanda, J., Berr, F., and Kiesslich, T., Photophysics and photochemistry of photodynamic therapy: Fundamental aspects, *Lasers Med. Sci.*, 24, 259, 2009.

180. Bourré, L., Thibaut, S., Fimiani, M., Ferrand, Y., Simonneaux, G., Patrice, T., In vivo photosensitizing efficiency of a diphenylchlorin sensitizer: Interest of a DMPC liposome formulation, *Pharmacol. Res.*, 47, 253, 2003.

181. Wieder, M.E., Hone, D.C., Cook, M.J., Handsley, M.M., Gavrilovic, J., and Russell, D.A., Intracellular photodynamic therapy with photosensitizer-nanoparticle conjugates: Cancer therapy using a "Trojan horse," *Photochem. Photobiol. Sci.*, 5, 727, 2006.

182. Madsen, S.J., Svaasand, L.O., Tromberg, B.J., and Hirschberg, H., Characterization of optical and thermal distributions from an intracranial balloon applicator for photodynamic therapy, *Proc. SPIE*, 4257, 41, 2001.

183. Madsen, S.J., Sun, C.H., Tromberg, B.J., Cristini, V., DeMagalhaes, N., and Hirschberg, H., Multicell tumor spheroids in photodynamic therapy, *Lasers Surg. Med.*, 38, 555, 2006.

184. Hirschberg, H., Sun, C.H., Krasieva, T., and Madsen, S.J., Effects of ALA-mediated photodynamic therapy on the invasiveness of human glioma cells, *Lasers Surg. Med.*, 38, 939, 2006.

185. Hebeda, K.M., Kamphorst, W., Sterenborg, H.J.C.M., and Wolbers, J.G., Damage to tumour and brain by interstitial photodynamic therapy in the 9L rat tumour model comparing intravenous and intratumoral administration of the photosensitizer, *Acta Neurochir. (Wien)*, 140, 495, 1998.

186. Goetz, C., Hasan, A., Stummer, W., Heimann, A., and Kempski, O., Experimental research photodynamic effects in perifocal, oedematous brain tissue, *Acta Neurochir. (Wien)*, 144, 173, 2002.

187. Perria, C., First attempts at the photodynamic treatment of human gliomas, *J. Neurosurg. Sci.*, 24, 119, 1980.

188. Kaye, A.H. and Hill, J.S., A review of photoradiation therapy in the management of central nervous system tumours, *Aust. N. Z. J. Surg.*, 58, 767, 1998.

189. Kostron, H., Photodynamic therapy in neurosurgery: A review, *J. Photochem. Photobiol. B*, 36, 157, 1996.

190. Stylli, S.S., Photodynamic therapy of cerebral glioma - A review. Part II—Clinical studies, *J. Clin. Neurosci.*, 13, 709, 2006.

191. Muller, P.J., Photodynamic therapy of brain tumors—A work in progress, *Lasers Surg. Med.*, 38, 384, 2006.

192. Madsen, S.J. and Hirschberg, H., Photodynamic therapy and photodynamic detection of high-grade glioma, *J. Environ. Pathol. Toxicol. Oncol.*, 25, 453, 2006.

193. Stylli, S.S., Kaye, A.H., MacGregor, L., Howes, M., and Rajendra, P., Photodynamic therapy of high grade glioma—Long term survival. *J. Clin. Neurosci.*, 12, 385, 2005.

194. Kostron, H., Hochleitner, B.W., Obwegeser, A., and Seiwald, M., Clinical and experimental results of photodynamic therapy in neurosurgery *Proc. SPIE*, 2371, 126, 1995.

195. Origitano, T.C., Caron, M.J., and Reichman, O.H., Photodynamic therapy for intracranial neoplasms, a literature review and institutional experience, *Mol. Chem. Neuropathol.*, 21, 337, 1994.

196. Beck, T.J., Interstitial photodynamic therapy of nonresectable malignant glioma recurrences using 5-aminolevulinic acid induced protoporphyrin IX, *Lasers Surg. Med.*, 39, 386, 2007.

197. Peng, Q., Warloe, T., Moan, J., Godal, A., Apricena, F., Giercksky, K.E., and Nesland, J.M., Antitumor effect of 5-aminolevulinic acid-mediated photodynamic therapy can be enhanced by the use of a low dose of Photofrin in human tumor xenografts, *Cancer Res.*, 61, 5824, 2001.

198. Bisland, S.K., Lilge, L., Lin, A., Rusnov, R., and Wilson, B.C., Metronomic photodynamic therapy as a new paradigm for photodynamic therapy: Rationale and preclinical evaluation of technical feasibility for treating malignant brain tumours, *Photochem. Photobiol.*, 80, 22, 2004.

199. Berg, K., Selbo, P.K., Prasmickaite, L., Tjelle, T.E., Sandvig, K., Moan, J., Gaudernack, G. et al., Photochemical internalization: A novel technology for delivery of macromolecules into cytosol, *Cancer Res.*, 59, 1180, 1999.

200. Dietze, A., Peng, Q., Selbo, P.K., Kaalhus, O., Muller, C., Bown, S., and Berg, K., Enhanced photodynamic destruction of a transplantable fibrosarcoma using photochemical internalization of gelonin, *Br. J. Cancer*, 92, 2004, 2005.

201. Selbo, P.K., Kaalhus, O., Sivam, G., and Berg, K., 5-Aminolevulinic acid-based photochemical internalization of the immunotoxin MOC31-gelonin generates synergistic cytotoxic effects in vitro, *Photochem. Photobiol.*, 74, 303, 2001.

202. Selbo, P.K., Sivam, G., Fodstad, Ø., Sandvig, K., and Berg, K., Photochemical internalization increases the cytotoxic effect of the immunotoxin MOC31 gelonin, *Int. J. Cancer*, 87, 853, 2000.

203. Prasmickaite, L., Høgset, A., Selbo, P., Engesæter, B., Hellum, M., and Berg, K., Photochemical disruption of endocytic vesicles before delivery of drugs: A new strategy for cancer therapy, *Br. J. Cancer*, 86, 652, 2002.

204. Teichberg, V.I., From the liver to the brain across the blood-brain barrier, *Proc. Natl. Acad. Sci. U. S. A.*, 104, 7315, 2007.

205. Wallner, K.E., Galicich, J.H., Krol, G., Arbit, E., and Malkin, M.G., Patterns of failure following treatment for glioblastoma multiforme and anaplastic astrocytoma, *Int. J. Radiat. Oncol. Biol. Phys.*, 16, 1405, 1989.

206. Vykhodtseva, N., McDannold, N., and Hynynen, K., Progress and problems in the application of focused ultrasound for blood-brain barrier disruption, *Ultrasonics*, 48, 279, 2008.

207. Hirschberg, H., Uzal, F.A., Chighvinadze, D., Zhang, M.J., Peng, Q., and Madsen, S.J., Disruption of the blood-brain barrier following ALA-mediated photodynamic therapy, *Lasers Surg. Med.*, 40, 535, 2008.

208. Hirschberg, H., Zhang, M.J., Gach, H.M., Uzal, F.A., Peng, Q., Sun, C.-H., Chighvinadze, D., and Madsen, S.J., Targeted delivery of bleomycin to the brain using photo-chemical internalization of *Clostridium perfringens* epsilon prototoxin, *J. Neurooncol.*, 95, 317, 2009.

209. Murphy, L.J., Hachey, D.L., Oates, J.A., Morrow, J.D., and Brown, N.J., Metabolism of bradykinin in vivo in humans: Identification of BK1-5 as a stable plasma peptide metabolite, *J. Pharmacol. Exp. Ther.*, 294, 263, 2000.

210. Worthington, R. and Mulders, M., The effect of *Clostridium perfringens* epsilon toxin on the blood-brain barrier of mice, *Onderstepoort J. Vet. Res.*, 42, 25, 1975.

211. Nagahama, M. and Sakurai, J., Distribution of labeled *Clostridium perfringens* epsilon toxin in mice, *Toxicon*, 29, 211, 1991.

212. Dorca-Arevalo, J., Soler-Jover, A., Gibert, M., Popoff, M., Martin-Satue, M., and Blasi, J., Binding of epsilon toxin from *Clostridium perfringens* in the nervous system, *Vet. Microbiol.*, 131, 14, 2008.

6
Lasers in Ophthalmology

Ezra Maguen
*Cedars-Sinai Medical Center
and
University of California,
Los Angeles*

Thomas G. Chu
*University of Southern
California*

David Boyer
*University of Southern
California*

6.1 Laser Surgery of the Anterior Segment of the Eye

6.1.1 Introduction

The eye is a transparent organ, and for the past 200 years optical devices such as ophthalmoscopes and slit-lamp biomicroscopes were used to view its structural elements. With the emergence of coherent light sources, early efforts were made to use such sources in order to affect disease processes in the eye. Early on, other properties of laser were found to be helpful in the treatment of eye disease. Small spot sizes made it possible to avoid exposing unwanted structures. Coherence of light was found to be synonymous with a significant decrease of collateral damage to surrounding tissues. The ability to reflect visible laser light off mirrors made it possible to treat ocular structures that would otherwise be unreachable. The ability to drive coherent light through flexible optical fibers also made it possible to use lasers in the operating room, both with *open-sky* and endoscopic techniques.

At the same time that range lasers were developing, intraocular surgery techniques, which had been performed with magnifying eyeglasses, started being performed with operating microscopes. These microscopes, along with slit lamps and ophthalmoscopes, became the obvious delivery systems for coherent light into the eye. They were modified to allow coaxial placement of the light source along with the optical system and an aiming beam, mostly HeNe lasers of power small enough not to damage eye structures.

This chapter focuses on how different structures in the eye can be treated with different lasers; brief descriptions of the most common diseases involved are provided.

6.1.2 Functional Anatomy of the Eye

Figure 6.1 shows a cross section of the eye where the cornea is anterior and the optic nerve is posterior. The eye is arbitrarily divided into an anterior segment and a posterior segment whereby the dividing line between the two is a vertical line tangent to the posterior end of the ciliary body.

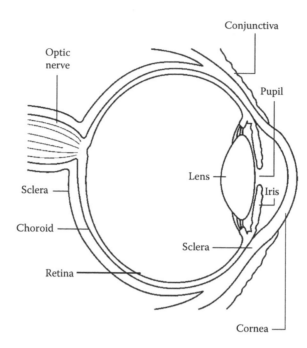

FIGURE 6.1   Cross section of the eye.

The cornea is a transparent organ responsible for most of the refracting power of the eye. There are no blood vessels within the cornea. Transparency is achieved by a structure of collagen-fiber packs that are perfectly parallel to one another. The bundles appear to be at various angles to one another in a 2D view. However, in a 3D view, they are parallel to each other. The cornea is continued by the sclera, which is made of collagen arranged in intersecting fibers. It is therefore nontransparent and serves as a wall protecting the inner structures of the eye. It has several holes within it to allow passage of several nerves and blood vessels, the largest of which is the optic nerve.

Behind the cornea is a space called the anterior chamber (AC). It is filled with fluid leaked from vessels in the ciliary body. These fluids travel in the space between the lens and iris, access the AC through the pupil, and are drained out of the eye via a drainage system located at the anterior chamber angle (ACA). A cross-section view shows that the ACA is located at the junction between the backside of the cornea and the periphery of the iris. A decreased rate of fluid drainage out of the eye does not reduce leakage; this produces a buildup in intraocular pressure called glaucoma. If excessive pressure is allowed to remain, blindness can result. Decrease in drainage may be induced by a disturbance in the structures of the ACA, hence the name open-angle glaucoma. If an eye has an inherent narrow access to the ACA and pressure builds up, it is called narrow-angle glaucoma. With a narrow angle, due to the buildup in pressure, the iris may be pushed forward and obstruct fluid access to the ACA. This condition is called angle-closure glaucoma. This condition is urgent and requires prompt treatment; otherwise, the eye could lose vision. A similar acute condition known as pupillary block can be induced by blockage of the passage between the iris, lens, and pupil.

The iris consists of two muscles designed to vary the diameter of the central aperture, the pupil. The sphincter decreases pupil size, while the dilator increases it. Behind the iris is a system consisting of the ciliary body and the lens. Both are connected by a network of thin fibers called zonules. The base of the ciliary body is anchored solidly to the sclera. The lens is made of a *bag* or capsule containing proteins of different densities. With respect to the cornea, the 3D molecular arrangement of both capsule and contents makes the lens transparent. The contraction and relaxation of the ciliary body makes the lens stretch and relax. This movement induces a change of the curvature of the front surface of the lens,

TABLE 6.1 Optical Properties of the Cornea and Lens Based on Gullstrand's *Schematic Eye*

Diameter	12 mm
Thickness	0.55 mm
Refracting power, anterior surface	48.83 *D*
Refractive power, posterior surface	−5.88 *D*
Total refracting power of the eye	58.64 *D*

making it possible to vary the focal distance of the refracting system of the eye (cornea and lens). This capability decreases with age because of the decreased flexibility of the lens (presbyopia), hence the need for near correction with glasses. The lens can also become more opaque (cataract). When the opacity induces a significant decrease in vision, the lens is removed and replaced with a prosthetic lens. Behind the plane formed by the ciliary body and the back of the lens is a space filled by the vitreous. The posterior segment of the eye is comprised of clear vitreous gel and the back wall of the globe. The anatomic features of the back wall are the retina, choroids, and sclera. The vitreous cavity occupies four–fifths of the volume of the globe. The vitreous gel is transparent, consisting of 99% water, mucopolysaccharide, and hyaluronic acid. It absorbs laser energy negligibly within the visible-light spectrum (Table 6.1).

The retina processes light into electrical impulses, which are transmitted by a network of nerves and fibers to the back of the brain (the occipital lobe), where these data are processed into images. The retina can be divided into the central retina, which includes the optic nerve exit, and the macula. The macula has a higher density of retinal cells than the rest of the retina. As a result, it is responsible for better visual acuity with better resolution and color vision. The peripheral retina provides peripheral vision with less resolution and night vision. The retina is comprised of two layers: (1) an inner transparent neurosensory retina, consisting of neural (photoreceptors), glial, and vascular elements and (2) an outer pigmented layer called retinal pigment epithelium (RPE), consisting of hexagonal-shaped cells important in the metabolic maintenance of the neurosensory retina. Of note, the cytoplasm of RPE cells contains multiple round and ovoid pigment granules or melanosomes. The retina has no physical attachment to the next layer, the choroid. It is maintained in place by capillarity. Therefore, if the retina breaks and fluid is allowed to invade the space between the retina and the choroid, the retina will detach (retinal detachment).

The choroid—the posterior portion of the uveal tract—nourishes the outer portion of the retina. It is a highly vascularized structure consisting of multiple layers of blood vessels and is separated from the RPE by Bruch's membrane, which is not a true membrane but rather a layer of periodic acid–Schiff (PAS)-positive material. Bruch's membrane presents a barrier to large molecules and blood vessels but is permeable to small molecules.

The sclera is an essentially avascular structure consisting of dense connective tissue, fibroblasts, collagen, and ground substance.

6.1.3 Laser Surgery of the Cornea

The goal of laser surgery of the cornea is either to make incisions (linear,[1] curved, or circular) or, most commonly, alter the curvature of the front surface of the cornea. Table 6.1 details the relevant physical properties of the cornea. These numbers make clear that the anterior surface of the cornea has the greatest influence on the refractive power of the eye. This, combined with easy access, makes the anterior surface of the cornea the obvious target organ for surgery designed to alter the refraction of the eye.

6.1.3.1 Ultraviolet Lasers

Srinivasan[2] and Trokel supplied the basic ideas and tools to apply excimer lasers to refractive corneal surgery. Their experiments made it clear that the 193 nm ArF excimer laser provides a smooth ablation pattern, minimal thermal damage, and exquisite control of the depth of ablation. Based on the simplified Munnerlyn[3] formula, an ablation depth (t_0) of 12 μm per diopter ($D = 1$) is carried out for an ablation diameter, or optical zone (Z) of 6 mm that reflects the tight tolerances of such a system in order to be accurate.

$$t_0 = \frac{Z^2 \times D}{3} \tag{6.1}$$

The first clinical models of excimer lasers for refractive surgery were built by Meditec (Oberkochen, Germany),[1] VISX (Santa Clara, CA),[4] and Summit Technologies (Greenwood Village, CO).[5] They included a delivery system similar to an operating microscope and a system of quartz lenses to guide the beam from the laser chamber to the eye. The laser beam was delivered via a scanning mechanism (Meditec) and as a broad beam (VISX and Summit). Beam-shape control was achieved with the use of physical masks (Meditec) and a shutter mechanism designed to vary the diameter of the beam (VISX and Summit). When these were used, a lenslike profile could be obtained, thereby flattening the central cornea and correcting spherical myopia.

An additional challenge in refractive surgery with the excimer laser was the correction of astigmatism. This involved an ablation pattern that would transform a toroidal surface into a spherical surface. This problem was solved by combining a variable slit opening to the shutter mechanism (VISX) and by interposing a proprietary *erodible mask* designed to vary the ablation profile in a customized fashion (Summit).

Additional strategies are needed to be devised to treat high refractive errors. In the case of high myopia, an ablation pattern varying the optical zone has been devised. It allows for the significant reduction of the ablation depth by decreasing the size of the zone as the ablation becomes more central. Indeed, an ablation pattern of 6 mm initially reduced to 5 mm in mid-depth and to 4 mm at the deepest could save significant amounts of corneal tissue. Barraquer[6] recognized the limit of ablation depth beyond which a significant number of complications would occur. The thickness of the residual cornea underlying the laser treatment area cannot be less than 250 μm.

Additional refinements in the delivery of the beam to the target organ included a scanning wide beam (Nidek EC5000) and a combination of the same with an axicon-type prism (VISX)[7] was designed to increase the ablation diameter to 9 mm to treat hyperopia (far sightedness). Indeed, the ablation profile for hyperopia involves steepening the original corneal profile. To accomplish this, most of the ablation must be performed in the periphery of the ablation zone, thereby allowing a minimum of a 4–5 mm optical zone.

Further improvements in the delivery system included the technology of randomly applying a laser spot with a diameter of 1 mm. This method can theoretically create any possible ablation pattern over any size of optical zone. The use of larger optical zones can in many cases reduce night glare if the border of the zone exceeds the diameter of the dilated pupil as it occurs in the dark. This delivery method carries with it the need for precise spot application and centration of the ablation pattern, hence the need for a tracking device of the eye as ablation proceeds.[8] Alcon Autonomous Technologies (Fort Worth, TX) was the first to develop such a system and have it approved by the US Food and Drug Administration (FDA) for all refractive surgical applications. Other manufacturers of lasers with similar technology are Bausch & Lomb (Rochester, NY) (Technolas 217) and LaserSight (Green Bay, WI).

The most recent advance in laser vision correction involves the use of wavefront technology in an attempt to correct higher-order optical aberrations. Most laser manufacturers have included this capability into their systems. In the United States, Alcon Autonomous Technologies and VISX are proceeding with FDA-guided clinical trials. The procedure includes obtaining data pertaining to higher-order aberrations as produced by a Tscherning aberrometer or a Hartmann–Shack device. These data are used to modify the ablation pattern accordingly. Once the prospective trials are complete, an assessment could be made as to whether this technology improves visual acuity or quality.

There are two procedures for performing laser vision correction: photorefractive keratectomy (PRK) and laser-assisted in situ keratomileusis (LASIK). A third procedure, laser-assisted subepithelial keratomileusis (LASEK), has recently become popular.

PRK consists of removal of the superficial layer of the cornea (epithelium), either manually or by laser ablation. The refractive portion of the laser ablation is performed on the bulk of the cornea (stroma). A contact lens is inserted, and under the contact lens, the epithelium regrows over the ablated area.

LASIK consists of using a mechanical device called a microkeratome to perform a partial-thickness round cut of the cornea. The cut is intentionally incomplete so that a hinge of tissue is left securing the flap of tissue thus formed in place. The flap is then lifted and the ablation performed on the exposed corneal stroma. The flap is replaced, and the anterior surface flattens or steepens as the flap conforms to the new profile made by the laser ablation. LASEK requires carefully displacing the corneal epithelium from the stroma after loosening the attachment between the two layers with diluted alcohol. A flap of epithelium secured by a hinge is formed and reflected. Ablation then proceeds on the underlying stroma. The epithelium is then replaced and smoothened.

Both procedures have pros and cons. PRK is simpler, and the surgery is safer because it does not involve the use of a microkeratome. After surgery, significant pain is experienced for 24–48 h. The healing is slow and may take several weeks. Superficial hazing of the cornea is relatively common, especially with higher corrections. LASIK is relatively less safe as it involves the use of a microkeratome with the complications specific to that surgical procedure. This is outweighed by very little pain after surgery and a fast recovery of optimal vision. With LASEK, no microkeratome is used. There is less pain than with PRK, and visual recovery time is better than with PRK but slower than with LASIK.

6.1.3.2 Infrared and Other Lasers

The use of ultraviolet lasers is complicated by the fact that they are gas lasers. Solid-state lasers could be simpler to build and maintain. Such systems could be miniaturized with relative ease. In addition, solid-state lasers within certain ranges of infrared could be used for other ophthalmic procedures on the iris, ACA, and lens.

The Novatec laser[8] is a proprietary solid-state laser with a wavelength of 0.2 μm, fluence of 100 mJ/cm^2, a variable spot size between 10 and 500 μm, and optical zone of up to 10 mm. The laser showed promise in initial clinical trials but ultimately was not produced. Other infrared solid-state lasers were built for PRK and LASIK, but they were not commercially produced.

Other strategies of using infrared lasers for vision correction included shortening the pulse width of the beam. Earlier models showed that in addition to performing corneal surgery on the surface of the end organ, the beam could be focused inside the cornea and tissue altered with relatively little collateral tissue damage. Picosecond[9] lasers were used with some success but never came to market. One of the most recent attempts at using this strategy is the emergence of a femtosecond (Fs) laser (Intralase, Irvine, CA). The system can create a corneal flap without using a mechanical device and remove tissue within the deeper corneal stroma to create a change in curvature. Clinical trials are ongoing to assess the safety and efficacy of the system.

The Hyperion solid-state infrared laser produced by Sunrise Technologies (Winston-Salem, NC)[9] uses a different strategy and is FDA approved to correct hyperopia. It is a solid-state holmium–yttrium aluminum garnet (YAG) (Ho:YAG) laser operating at a wavelength of 2.13 μm and pulse duration of 250 μs. The spot size is 500–600 μm. The treatment strategy (laser thermokeratoplasty) consists of applying 8–16 simultaneous applications on the cornea in a ring pattern at a given distance from the corneal center, thereby inducing central steepening and correcting hyperopia by inducing collagen shrinkage around the laser applications. It is simple to perform and carries relatively low intraoperative risk. At the same time, the amount of correction is not as controlled as with excimer-laser procedures. The range of correction is limited, and in many cases, regression of the correction occurs over time.

6.1.4 Laser Surgery of the Anterior Chamber Angle

Laser procedures here are designed to treat open-angle glaucoma. They target a collagen meshwork called the trabecular meshwork. This is the initial part of the system that drains fluid out of the eye. The surgical technique consists of driving a laser beam through a slit-lamp delivery system into a reflecting mirror. From there, the beam travels at an angle into the target organ. Of the 360° of the ACA, 180° are treated at one time with an average of 50 applications per treatment. The mechanism of action is unclear. The most plausible explanation is that collagen shrinks at the treatment area, stretching wider open the meshwork adjacent to the laser spots. The success rate of the procedure is reportedly up to 90% in the short term and decreases to 50% at 5 years after surgery.

The argon laser has been exclusively used for the past 20 years for this type of treatment, hence the name argon laser trabeculoplasty (ALT).[10] The treatment parameters range as follows: energy = 800–1000 mW, spot size = 50 μm, and exposure time = 0.1 s. Recently, a new treatment modality of the ACA was introduced using a 532 nm, frequency-doubled, Q-switched neodymium-doped YAG (Nd:YAG) laser (Selecta 7000 model, Lumenis, Inc., Santa Clara, CA). The treatment strategy is similar to that of ALT, using 54–55 spots per treatment. Treatment parameters are as follows: spot size = 400 μm, pulse duration = 3 ns, and energy = 0.6–1.2 mJ. The mechanism of action is purported to produce a biological effect by targeting pigmented cells within the trabecular meshwork, hence the name selective laser trabeculoplasty (SLT).[11] During FDA-guided trials, a 70% success rate was reported.

6.1.5 Laser Surgery of the Iris

Several laser surgical procedures are performed on the iris. There are three main goals with this type of surgery: (1) create a bypass between the space behind the iris and the AC to reduce eye pressure, (2) enlarge the approach to the ACA, and (3) enlarge pupil size. The argon laser described earlier and Nd:YAG lasers can be used. The Nd:YAG lasers (wavelength = 1064 nm) are mostly of the fundamental Q-switched type, paired with a HeNe laser for aiming. They induce photodisruption, which in turn releases a shock wave, thereby creating disruption of the iris fibers. The argon lasers induce mostly thermal damage of the tissue. Iris color is directly related to the efficacy of the two lasers. Argon lasers are more efficient in more pigmented (darker color) irides, whereas in lighter irides, the Nd:YAG laser is preferred. A collimating lens placed on the eye can reduce laser energy.

6.1.5.1 Laser Iridotomy

This procedure is designed to create a bypassing hole in the iris to allow fluid trapped behind it to flow to the AC. Fluid blockage occurs due to closure of the ACA or to pupillary block. Both argon[12] and Nd:YAG[13] lasers can be used—separately or in combination. Laser parameters for the argon laser are energy = 800–1000 mW, diameter = 50 μm, and exposure time = 0.1 s. A total of 40–80 applications are used. For the Nd:YAG laser, energy = 3–7 mJ, and four to ten applications are used.

6.1.5.2 Laser Iridoplasty (Gonioplasty)

This procedure[14] is designed to enlarge the approach to the ACA by thinning the peripheral iris using the thermal damage to tissue induced by the argon laser. Thermal damage is followed by atrophy of the tissue and therefore is thinned. Several applications are made in a 360° circular pattern. One or more rows can be applied. Laser parameters are energy = 400–600 mW, diameter = 400 μm, and exposure time = 0.2–0.4 s.

6.1.5.3 Laser Photomydriasis

This procedure is designed to increase the size of the pupil by using a technique similar to laser iridoplasty. Laser applications are guided as close as possible to the margin of the pupil. The same type of laser–tissue interaction eventually induces atrophy of the iris muscle, which closes the pupil (iris sphincter). A small pupil can be induced by drugs or by previous disease, causing adhesion of the iris to the lens. Its enlargement is sought mainly to better view the inside of the eye.

6.1.6 Laser Surgery of the Ciliary Body

The ciliary body is the structure least accessible to viewing and laser treatment because of its location, which is immediately behind the peripheral iris. The goal of laser treatment of the ciliary body is to decrease intraocular pressure. Indeed, vessels in this structure leak fluid into the eye, and a decrease in fluid outflow could regulate better eye pressure. There are three approaches to laser treatment of the ciliary body.

6.1.6.1 Laser Cyclophotocoagulation

Direct treatment of the ciliary body with the argon laser is possible only if a sector of the iris was previously removed by surgery. Typical laser settings are mean energy = 400 mW, diameter = 100 μm, and exposure time = 0.1–0.2 s.

6.1.6.2 Endolaser Cyclophotocoagulation

In this procedure,[15] the ciliary body is visualized and treated via a fiber optic with the argon laser. This involves a major surgical procedure, as the surrounding vitreous must be removed. The treatment cannot be applied if the lens has not been removed.

6.1.6.3 Transscleral Cyclophotocoagulation

The third approach to the ciliary body involves using a laser beam to irradiate the overlying sclera. Laser energy is transmitted via the sclera to the ciliary body. Laser–tissue interaction reflects mostly transformation of laser energy to thermal energy. Two methodologies can be applied: noncontact treatment can be provided with an Nd:YAG laser. A typical treatment consists of 32 applications of 8 J each to the 360° circumference of the ciliary body. The LASAG Microruptor® laser can be used for the treatment.[16] Contact transscleral treatment (the probe is applied to the sclera overlying the ciliary body) can be performed with a gallium–aluminum–arsenide diode laser emitting at 810 nm.[17]

6.1.7 Laser Surgery of the Lens

The lens may become opaque over time, leading to a condition called cataract. Modern cataract surgery involves opening the lens capsule (bag), liquefying and aspirating the protein contents, and inserting a prosthetic lens implant into the remaining bag. After surgery, the back of the bag (posterior capsule) may get opaque, and an opening must be made to recover good vision. Lasers can be used for both liquefying and aspirating (phakoemulsification) and for perforating the posterior capsule (capsulotomy), if needed later.

6.1.7.1 Laser Phakoemulsification

Erbium YAG (Er:YAG)[18] and Nd:YAG[19] laser systems are available for lens emulsification. The Nd:YAG system is manufactured by ARC Laser AG (Jona, Switzerland) and is named the Dodick Photolysis System® after its inventor. The energy is delivered through a handheld probe containing a quartz fiber optic and is focused on a titanium target within the probe tip. Each pulse releases 12 mJ over 14 ns. The pulse impacts the titanium target, leading to plasma formation. The ensuing shock wave fragments the lens proteins, which are aspirated through the same handpiece. The Er:YAG system is made by several manufacturers. It includes a handpiece containing a zirconium fluoride fiber optic to drive laser energy and an irrigation and aspiration system. Because of the water content of the lens, this laser can effectively emulsify this tissue. Data published for the Aesculap-Meditec MCL 29 model show that application frequency was between 20 and 60 Hz. For a typical phakoemulsification, the mean number of pulses was 1740, mean total energy was 38.5 J, and mean treatment time was 3 min.

6.1.7.2 Laser Posterior Capsulotomy

This procedure[20] was the first application of the Nd:YAG laser in ophthalmology and is now a well-established routine procedure. The system consists of an Nd:YAG laser delivered via a slit lamp and guided with a HeNe beam. A collimating lens can be applied to the eye to decrease the energy necessary to induce a central rupture of the posterior capsule. Typical settings are energy = 2–3 mJ and number of applications = 5–20. The laser–tissue interaction includes plasma formation followed by tissue disruption, mostly by the shock wave induced.

6.1.7.3 Femtosecond Laser Applications in Cataract Surgery

Recently, three start-up companies unveiled Fs laser systems capable at this point of performing a peripheral corneal stepped cataract incision, a circular opening of the anterior capsule (capsulorrhexis) and splitting the lens nucleus into a predetermined pattern of fragments. The procedure is then completed with lens phakoemulsification using high-speed ultrasound to emulsify the lens fragments and aspirate them out of the eye. A lens implant is then placed in the lens capsular bag thus formed in the usual manner.

The Opti-Medica system was developed by a team headed by Daniel Planker PhD.[21] It is guided by a proprietary frequency domain optical coherence tomography device operating at 11 μ resolution.

The Fs laser system has been operated within the following range of parameters: pulse duration of approximately 400 fs, wavelength of 1.03 μm, focal spot size of less than 10 μm, pulse energy range of 3–10 μJ, and pulse repetition range of 12–80 kHz. The Fs laser system has been also operated within the following scanning parameters: lateral spot spacing of about 5 μm for capsulotomy and 10 μm for lens segmentation and axial spacing of about 10 μm for capsulotomy and 20 μm for lens segmentation.

Clinical results are forthcoming in the peer review literature with all three systems and generally show similar results obtained with phakoemulsification. It is assumed that because of the precision and reproducibility of the capsulorrhexis, better centration of multifocal lens implants could be achieved. Also, with lens fragmentation by the Fs laser, less energy will be expanded by phakoemulsification hence improving the clinical results.

6.2 Laser Surgery of the Posterior Segment of the Eye

6.2.1 Pathophysiologic Considerations

Diseases within the posterior segment of the eye can be classified as conditions that (1) overlie the neurosensory retina, (2) lie within the neurosensory retina, (3) lie completely through the neurosensory retina, and (4) underlie the neurosensory retina. Various laser therapies have been developed to address all these conditions.

Abnormal tissue growth above the neurosensory retina can be vascular tissue or fibrous tissue. Such disease conditions include epiretinal membranes and abnormal neovascularization (known as neovascular proliferation) associated with retinal ischemia (i.e., diabetic retinopathy, sickle-cell retinopathy, and retinopathy of prematurity). The various laser treatments discussed in the succeeding text have been designed to remove such abnormal tissue.

Disease processes that occur within the substance of the neurosensory retina are typically associated with abnormal leakage of serum or incompetence from retinal vascular. Leakage into the neurosensory retina causes damage to the normal retinal architecture, thereby destroying vision. Many vascular diseases, such as diabetic macular edema, retinal-vein occlusions, retinal-vessel telangiectasia, and retinal-arterial macroaneurysms, can cause this type of direct damage to the neurosensory retina.

Disease processes that involve the entire neurosensory retina are typically retinal holes or retinal tears. These conditions are significant because they can develop into retinal detachments, which are vision-threatening conditions.

Lastly, abnormal conditions can develop below or under the neurosensory retina, causing retinal disease and vision loss. Abnormal vascular tissue (choroidal neovascular membrane [CNVM]) can grow under the retina, causing damage to the normal retinal architecture and resulting in vision loss. Examples of these conditions include exudative macular degeneration, myopic degeneration, and ocular

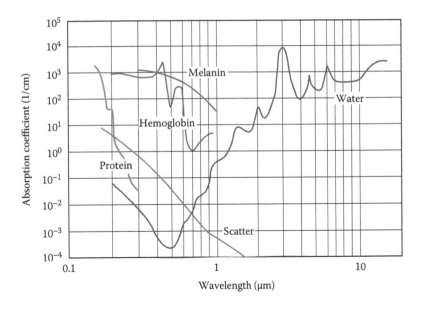

FIGURE 6.2 Laser absorption and tissue penetration.

histoplasmosis. In addition, abnormal tumor cells and other tissues (hamartomas), such as choroidal melanoma, metastatic carcinomas, and choroidal hemangioma, can grow under the neurosensory retina.

6.2.1.1 Light Absorption

Melanin, hemoglobin, and macular xanthophyll represent the three most significant light absorbers in the retina and the choroid. Melanin is the most significant chorioretinal light absorber. Its light absorption gradually decreases with increasing wavelength, thereby allowing longer laser wavelengths, such as krypton, red light, and diode infrared light, to produce deeper chorioretinal lesions than occur with shorter-wavelength argon green laser light (Figure 6.2).

Hemoglobin is the next most effective light absorber. Hemoglobin absorption also decreases with increasing wavelength; however, there are two peaks in the absorption spectrum of oxyhemoglobin (542 nm, green; 577 nm, yellow) and one absorption peak for reduced hemoglobin (555 nm, yellow). Hemoglobin absorbs blue, green, and yellow light well but has poor red-light absorption.

Xanthophyll in the macular region of the retina is the least effective light absorber. Macular xanthophylls absorb blue-wavelength light well, green light minimally, and yellow to red light poorly.

6.2.1.2 Temperature Changes

Various forms of laser photocoagulation cause different temperature rises within the retina. Melanin within the RPE and choroid acts as the primary light absorber in retinal photocoagulation. Light absorption converts laser radiation into heat energy, increasing the temperature of light-absorbing tissues. Temperature rise is proportional to retinal irradiance (laser power/area) for a particular wavelength, spot size, and exposure duration.[22] Heat conduction spreads temperature increases from the light-absorbing RPE and choroid to contiguous tissues. For very long exposures, heat convection, because of choroidal blood flow, moderates chorioretinal temperature rise.

6.2.2 Tissue Photocoagulation

Conventional short-pulse photocoagulation is a highly suprathreshold procedure, associated with temperature increases of 40°C–60°C above normal body temperature of 37°C.[22] Most ophthalmic lasers in use today exemplify this concept of short-pulse tissue photocoagulation. Various laser wavelengths are used to accomplish this: (1) argon blue (488 nm), (2) argon blue/green (514.5 nm), (3) krypton red (647 nm), (4) dye (577–630 nm), (5) diode (810 nm), and (6) frequency-doubled Nd:YAG (532 nm).

This concept of short-pulse tissue photocoagulation has been successfully applied to treat a wide variety of ocular conditions. In essence, they can be divided into five broad categories:

1. Focal treatment of leaking microvascular or macrovascular abnormalities. Focal-tissue obliteration of leaks and grid-laser treatment to damage leaking capillary beds are used. Examples are diabetic macular edema, branch-vein occlusion, radiation retinopathy, and idiopathic perifoveal telangiectasias.[23,24]
2. Panretinal photocoagulation for treatment of proliferative retinopathies or neovascularization of the iris. Examples include proliferative diabetic retinopathy, sickle-cell retinopathy, central retinal-vein occlusion with rubeosis, proliferative disease following radiation retinopathy, and proliferation associated with branch-vein occlusion.[25–27]
3. Treatment or prevention of retinal detachment. Examples include treatment of retinal breaks, demarcation of subclinical retinal detachment, and treatment of lattice degeneration.[28]
4. Treatment of ocular tumors and neoplasms. Examples include choroidal melanoma, retinoblastoma, choroidal cavernous hemangioma, and angioma (Von Hippel disease).[29,30]
5. Treatment of CNVM or focal choroidal lesions. Examples include central serous retinopathy, age-related macular degeneration (AMD), presumed ocular histoplasmosis, myopia degeneration, and idiopathic polypoidal choroidal vasculopathy.[31–34]

Multiple different wavelengths have been used to treat the aforementioned conditions. Several generalities exist regarding the use of various wavelengths. Argon green appears to have advantages over blue light only if there is less damage when treatment is placed near the fovea due to less absorption by the retinal xanthophyll and there is more light scatter. Krypton red may have advantages in patients with hazy ocular media. The deeper the penetration of laser energy into the choroid (krypton red and diode), the more pain is associated with treatment.

Diode lasers emit laser energy at a wavelength of 810 nm. This longer wavelength can be absorbed by all three ocular pigments—hemoglobin, melanin, and xanthophyll—albeit inefficiently. Because of the inefficient absorption of this wavelength, diode-laser energy can penetrate deeper into the choroidal layer of the eye (Figure 6.3). It therefore has theoretical advantages over other wavelengths in treating

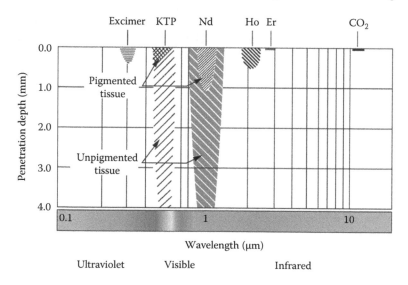

FIGURE 6.3 Tissue penetration characteristics.

retinal pathology located deeper within the eye. This theoretical advantage, however, has not been noted in clinical studies. Diode-laser photocoagulation has been used to treat all the conditions effectively treated by argon green and yellow and krypton red laser photocoagulation, with similar, but not better, results. It has proven particularly effective in the treatment of retinopathy of prematurity, a blinding condition in neonatal infants, and is the laser wavelength of choice.[35] Unfortunately, the deeper penetration of the diode-laser energy in the eye also causes more pain stimulation than with shorter wavelengths.

6.2.3 Thermotherapy

Thermotherapy, the treatment of tissue disorders by causing a rise in tissue temperature, is another novel medical approach. Retinal temperature rise in laser therapy is proportional to retinal irradiance (laser power/area) for a particular spot size, exposure duration, and wavelength.[1] Transpupillary thermotherapy (TTT) is a low-irradiance, large-spot-size, and prolonged-exposure (long-pulse) laser treatment using infrared diode-laser energy. TTT has been successfully used to treat intraocular tumors such as choroidal melanoma and choroidal hemangioma.[36-38] Recently, long-pulse TTT has been advocated in the treatment of choroidal neovascularization in AMD.[39,40]

Maximal chorioretinal temperature rise in short- or long-pulse intraocular or transpupillary photocoagulation occurs at the inner surface of the RPE and in the center of the laser spot. In long-pulse TTT for choroidal neovascularization in macular degeneration, an 810 nm infrared diode laser is used with a 3 mm diameter retinal spot size and a 60 s exposure. For these parameters, maximal temperature rise is roughly 10°C for a typical 800 mW clinical power setting used in treating CNVM in a lightly pigmented fundus.[39] In comparison, even a 50 mW 0.1 s, 200 µm argon laser exposure produces a 42°C RPE temperature rise.[41] In photodynamic therapy (PDT), discussed later in this chapter, a 3 mm diameter, 0.6 W/cm², 83 s, 689 nm exposure causes a maximal RPE temperature rise of less than 2°C.[22] Actual temperature increases are somewhat lower than maximal increases calculated for TTT and PDT because of the cooling effect of choroidal circulation.

6.2.4 Photodynamic Therapy

PDT requires administration of a photosensitizing dye and activation of the dye by light irradiation at the target tissue. This usually causes activation of the dye to a higher state, releasing singlet oxygen and causing tissue–cell injury. This process accomplishes destruction of abnormal blood vessels.

Many photosensitizing dyes are available including rose bengal, hematoporphyrin, Photofrin®, benzoporphyrin, tin etiopurpura, chlorinae 6, bacterochlorinae, and chaloalimium sulfonated phthalocyanine.

The laser light used to activate the photosensitized dye coincides with the absorption maximum of the photosensitizer. A better dye would localize preferentially to the target, minimizing collateral damage. An additional desirable feature would be a relatively short half-life, thereby minimizing the potential general side effects of environmental light exposure.

In ophthalmology, PDT lends itself to treatment of CNVMs and tumors. The photosensitizing dye can be allowed to accumulate in these pathologic tissues without accumulating in normal tissues and structures. In CNVM and tumor treatment, the damage is done by singlet-oxygen damage of the endothelial wall, which leads to vessel closure. PDT is therefore a nonthermal treatment. Today, only verteporfin (Visudyne®, QLT Pharmaceuticals, Vancouver, Canada) has been approved for clinical use for the treatment of classic CNVM in AMD.[42-45]

6.2.5 Selective Retinal Pigment Epithelial Photocoagulation

Micropulse laser treatment, with pulse durations from nanoseconds to microseconds, has been shown to cause selective damage to RPE tissue, sparing the neurosensory retina altogether. Again, this has theoretical advantages over shorter-wavelength treatment, especially in conditions originating in the RPE and choroidal layers. This treatment is limited by the lack of a visible change in the retinal appearance with laser application, making it difficult to assess adequate treatment. Several laser systems have been tested to create this selective type of tissue damage including Q-switched Nd:YAG, micropulsed diode, micropulsed argon, and neodymium-doped yttrium lithium fluoride (Nd:YLF). It has been suggested that treatment above the ablation threshold may reduce untoward thermal and mechanical effects.[46,47]

6.2.6 Photodisruption

Various laser systems have been used to create tissue photodisruption in the posterior segment of the eye. Practical considerations for all these systems include their method of delivery and ease of use. Fiber optics wear out or are not easily bendable without breaking their tips and can become coated, losing their effectiveness. Also, cavitation can occur, making visualization during surgery difficult.

6.2.6.1 Carbon Dioxide Laser

Wavelength of 10.6 μm (infrared spectrum) has been used intraocularly to phototransect and simultaneously coagulate vitreoretinal membranes. Because of the absorption of energy by water, all of the effect is at the point of impact at the target tissue. An advantage of this wavelength is its ability to treat on top of critical structures. The CO_2 laser does not require pigment to accomplish cutting or coagulation. The major disadvantage of the CO_2 laser is that it requires direct contact with the treating tissue. With too much pressure, a tear in the retina may occur. It cannot be delivered easily by fiber optic, an additional disadvantage.[48,49]

6.2.6.2 Ho:YAG Laser

Wavelength of 2.12 μm can be transmitted by flexible optical fibers with little power attenuation. Due to transmission, the beam must be shielded to prevent underlying retinal burns or hemorrhage. It has been used successfully to cut experimental vitreous membranes in rabbits.[50]

6.2.6.3 Sapphire-Tipped Nd:YAG Laser

Sapphire crystal has been used at the tip of various laser probes to deliver energy to the target tissue. The sapphire crystals do not alter the laser energy or transmission. The crystal has a high meeting point and low thermal conductivity and is mechanically strong. This tip can deliver Nd:YAG and argon energy. The Nd:YAG has been used clinically to coagulate retina and internally to transect vitreoretinal membranes.[51]

6.2.6.4 Er:YAG Laser (2.94 μm)

The Er:YAG laser offers almost complete absorption of laser energy in water with little transmission. It has been used through intraocular fiber-optic delivery systems to photovaporize membranes in rabbits. In addition, it has been used in humans in two clinical studies, treating elevated membranes and vascularized diabetic preretinal membranes, creating retinotomies, and ablating epiretinal membranes. Recent trials showed that ablation was time-consuming and coagulation poor. Complications noted included hemorrhage, retinal breaks, and intraocular lens damage.[52,53]

6.2.6.5 Picosecond Nd:YLF Laser

A power level of 1053 nm (operating in picoseconds) may have high power, short pulse, and high repetition rate and may reduce breakdown of ocular tissues adjacent to target tissue. A pulse rate of 1000 per second for up to 10 s allows segmentation of tissue near the retinal surface and has been used clinically in diabetic patients.[54]

6.2.6.6 Nd:YAG Phototransection (1.06 µm)

Vitreoretinal membranes may be cut in vitreous or AC. Because of explosive cutting, this should not be used within 3 mm of the retina. A contact lens is used to focus, and burst mode is used to cut. A 100 per second Nd:YAG laser pulsing at 50–200 Hz achieved optical breakdown and vitreous cutting using only 70 µJ of energy.[55,56]

6.3 Summary

Coherent-light sources with novel laser-energy delivery systems are well suited in the treatment of many ophthalmologic conditions. Various laser-energy wavelengths have been used effectively in ophthalmic medicine and continue to be the mainstay of many treatment therapies. Novel uses of new laser-energy sources and wavelengths will undoubtedly become possible in the future.

Acknowledgment

The research for this chapter was supported by the Discovery Fund for Eye Research, Los Angeles, CA.

References

1. Tenner, A., Neuhann, T., and Schroeder, E., Excimer laser radial keratotomy in the living human eye: A preliminary report, *J. Refract. Surg.*, 4, 5, 1988.
2. Srinivasan, R. and Sutcliffe, E., Dynamics of the ultraviolet laser ablation of corneal tissue, *Am. J. Ophthalmol.*, 103, 470, 1987.
3. Munnerlyn, C.R., Koons, S.J., and Marshall, L., Photorefractive keratectomy: A technique for laser refractive surgery, *J. Cataract Refract. Surg.*, 14, 46, 1988.
4. McDonald, M.B., Frantz, J.M., Klyce, S.D. et al., Central photorefractive keratectomy for myopia: The blind eye study, *Arch. Ophthalmol.*, 108, 799, 1990.
5. Seiler, T., Kahle, G., and Kriegerowski, M., Excimer laser (193 nm) myopic keratomileusis in sighted and blind human eyes, *Refract. Corneal Surg.*, 6, 165, 1990.
6. Barraquer, J.I., *Queratomileusis y Queratofaquia*, Instituto Barraquer de America, Bogota, Colombia, p. 115, 1980.
7. Jackson, W.B., Casson, E., Hodge, W.G., Mintsioulis, G., and Agapitos, P.J., Laser vision correction for low hyperopia. An 18-month assessment of safety and efficacy, *Ophthalmology*, 105, 1727, 1998.
8. Swinger, C.A. and Lai, S.T., Solid state photoablative decomposition, the Novatec laser, in *Corneal Laser Surgery*, Salz, J.J., ed., C.V. Mosby, St. Louis, MO, p. 261, 1995.
9. Rowsey, J.J., Koch, D.D. et al., Alternative lasers and strategies for corneal modification, in *Corneal Laser Surgery*, Salz, J.J., ed., C.V. Mosby, St. Louis, MO, p. 269, 1995.
10. Wise, J.B. and Witter, S.L., Argon laser therapy for open-angle glaucoma, *Arch. Ophthalmol.*, 97, 319, 1979.
11. Latina, M.A., Sibayan, S.A., Shin, D.H., Noecker, R.J., and Marcellino, O., Q-switched 532 nm Nd:YAG laser trabeculoplasty (selective laser trabeculoplasty): A multicenter pilot clinical study, *Ophthalmology*, 105, 2082, 1998.
12. Abraham, R.K. and Miller, G.E., Outpatient argon laser iridectomy for angle closure glaucoma: A 2-year study, *Trans. Am. Acad. Ophthalmol. Otolaryngol.*, 79, 529, 1975.
13. Tomey, K.F., Traverso, C.E., and Shammas, I.V., Neodymium-YAG laser iridotomy in the treatment and prevention of angle closure glaucoma: A review of 373 eyes, *Arch. Ophthalmol.*, 105, 476, 1987.
14. Shin, D., Argon laser iris photocoagulation to relieve acute angle-closure glaucoma, *Am. J. Ophthalmol.*, 93, 348, 1982.

15. Chen, J., Cohn, R.A., and Lin, S.C., Endoscopic photocoagulation of the ciliary body for treatment of refractory glaucomas, *Am. J. Ophthalmol.*, 124, 787, 1997.

16. van der Zypen, E., Kwasniewska, S., Roe, P., and England, C., Transscleral cyclophotocoagulation using a neodymium:YAG laser, *Ophthalmic Surg.*, 17, 94, 1986.

17. Bloom, P.A., Tsai, J.C., Sharma, K. et al., "Cyclodiode": Trans-scleral diode laser cyclophotocoagulation in the treatment of advanced refractory glaucoma, *Ophthalmology*, 104, 1508, 1997.

18. Stevens, G., Jr., Long, B., Hamman, J.M., and Allen, R.C., Erbium:YAG laser-assisted cataract surgery, *Ophthalmic Surg. Lasers*, 29(3), 185, 1998.

19. Kanellopoulos, A.J., Dodick, J.M., Brauweiler, P., and Alzner, E., Dodick photolysis for cataract surgery. Early experience with the Q-switched neodymium:YAG laser in 100 consecutive patients, *Ophthalmology*, 106, 2197, 1999.

20. Francois, J.H. and Aladlouni, T., Treatment of opacification of the posterior crystalline capsule after extracapsular extraction: Surgery or laser, *Bull. Soc. Ophtalmol. Fr.*, 89(11), 1297, 1989.

21. Palanker, D., Blumenkranz, M.S., Andersen, D., Wiltberger, M. et al. Femtosecond laser assisted cataract surgery. To be submitted to Science.

22. Mainster, M.A., White, T.J., Tips, J.H., and Wilson, P.W., Retinal-temperature increases produced by intense light sources, *J. Opt. Soc. Am.*, 60, 264, 1970.

23. Early Treatment Diabetic Retinopathy Study Research Group, Photocoagulation for diabetic macular edema. Early Treatment Diabetic Retinopathy Study report number 1, *Arch. Ophthalmol.*, 103, 1796, 1985.

24. Branch Vein Occlusion Study Group, Argon laser photocoagulation for macular edema in branch vein occlusion, *Am. J. Ophthalmol.*, 98, 271, 1984.

25. Photocoagulation treatment of proliferative diabetic retinopathy: The second report of diabetic retinopathy study findings, *Ophthalmology*, 85, 82, 1978.

26. Kimmel, A.S., Magargal, L.E., and Tasman, W.S., Proliferative sickle retinopathy and neovascularization of the disc: Regression following treatment with peripheral retinal scatter laser photocoagulation, *Ophthalmic Surg.*, 17, 20, 1986.

27. Central Vein Occlusion Study Group, Natural history and clinical management of central retinal vein occlusion, *Arch. Ophthalmol.*, 115, 486, 1997.

28. Pollak, A. and Oliver, M., Argon laser photocoagulation of symptomatic flap tears and retinal breaks of fellow eyes, *Br. J. Ophthalmol.*, 65, 469, 1981.

29. Shields, C.L., Shields, J.A., Kiratli, H., and De Potter, P.V., Treatment of retinoblastoma with indirect ophthalmoscope laser photocoagulation, *J. Pediatr. Ophthalmol. Strabismus*, 32, 317, 1995.

30. Shields, J.A., The expanding role of laser photocoagulation for intraocular tumors: The 1993 H. Christian Zweng Memorial Lecture, *Retina*, 14, 310, 1994.

31. Argon laser photocoagulation for senile macular degeneration: Results of a randomized clinical trial, *Arch. Ophthalmol.*, 100, 912, 1982.

32. Macular Photocoagulation Study Group, Krypton laser photocoagulation for idiopathic neovascular lesions: Results of a randomized clinical trial, *Arch. Ophthalmol.*, 108, 832, 1990.

33. Macular Photocoagulation Study Group, Laser photocoagulation for juxtafoveal choroidal neovascularization: Five-year results from randomized clinical trials, *Arch. Ophthalmol.*, 12, 500, 1994.

34. Macular Photocoagulation Study (MPS) Group, Evaluation of argon green vs. krypton red laser for photocoagulation of subfoveal choroidal neovascularization in the macular photocoagulation study, *Arch. Ophthalmol.*, 112, 1176, 1994.

35. Hunter, D.G. and Repka, M.X., Diode laser photocoagulation for threshold retinopathy of prematurity: A randomized study, *Ophthalmology*, 100, 238, 1993.

36. Journee-de Korver, J.G., Oosterhuis, J.A., Kakebeeke-Kemme, H.M., and de Wolff-Rouendaal, D., Transpupillary thermotherapy (TTT) by infrared irradiation of choroidal melanoma, *Doc. Ophthalmol.*, 82, 185, 1992.

37. Rapizzi, E., Grizzard, W.S., and Capone, A., Jr., Transpupillary thermotherapy in the management of circumscribed choroidal hemangioma, *Am. J. Ophthalmol.*, 127, 481, 1999.

38. Shields, C.L., Shields, J.A., DePotter, P., and Kheterpal, S., Transpupillary thermotherapy in the management of choroidal melanoma, *Ophthalmology*, 103, 1642, 1996.

39. Reichel, E., Berrocal, A.M., Ip, M., Kroll, A.J., Desai, V., Duker, J.S., and Puliafito, C.A., Transpupillary thermotherapy of occult subfoveal choroidal neovascularization in patients with age-related macular degeneration, *Ophthalmology*, 106, 1908, 1999.

40. Newsom, R.S., McAlister, J.C., Saeed, M., and McHugh, J.D., Transpupillary thermotherapy (TTT) for the treatment of choroidal neovascularisation, *Br. J. Ophthalmol.*, 85, 173, 2001.

41. Mainster, M.A., White, T.J., and Allen, R.G., Spectral dependence of retinal damage produced by intense light sources, *J. Opt. Soc. Am.*, 60, 848, 1970.

42. Photodynamic therapy of subfoveal choroidal neovascularization in age-related macular degeneration with verteporfin: One-year results of 2 randomized clinical trials—TAP report. Treatment of age-related macular degeneration with photodynamic therapy (TAP) Study Group, *Arch. Ophthalmol.*, 117, 1329, 1999.

43. Verteporfin therapy of subfoveal choroidal neovascularization in age-related macular degeneration: Two-year results of a randomized clinical trial including lesions with occult with no classic choroidal neovascularization—Verteporfin in photodynamic therapy report 2, *Am. J. Ophthalmol.*, 131, 541, 2001.

44. Photodynamic therapy of subfoveal choroidal neovascularization in pathologic myopia with verteporfin: 1-year results of a randomized clinical trial—VIP report no. 1, *Ophthalmology*, 108, 841, 2001.

45. American Academy of Ophthalmology, Photodynamic therapy with verteporfin for age-related macular degeneration, *Ophthalmology*, 107, 2314, 2000.

46. Roider, J., Hillenkamp, F., Flotte, T., and Birngruber, R., Microphotocoagulation: Selective effects of repetitive short laser pulses, *Proc. Natl. Acad. Sci. U.S.A.*, 90, 8643, 1993.

47. Brinkmann, R., Huttmann, G., Rogener, J., Roider, J., Birngruber, R., and Lin, C.P., Origin of retinal pigment epithelium cell damage by pulsed laser irradiance in the nanosecond to microsecond time regimen, *Lasers Surg. Med.*, 27, 451, 2000.

48. Karlin, D., Jakobiec, F., Harrison, W. et al., Endophotocoagulation in vitrectomy with a carbon dioxide laser, *Am. J. Ophthalmol.*, 101, 445, 1986.

49. Meyers, S.M., Bonner, R.F., Rodrigues, M.M., and Ballintine, E.J., Phototransection of vitreal membranes with the carbon dioxide laser in rabbits, *Ophthalmology*, 90, 563, 1983.

50. Borirakchanyavat, S., Puliafito, C.A., Kliman, G.H., Margolis, T.I., and Galler, E.L., Holmium-YAG laser surgery on experimental vitreous membranes, *Arch. Ophthalmol.*, 109, 1605, 1991.

51. Peyman, G.A., Katoh, N., Tawakol, M., Khoobehi, B., and Desai, A., Contact application of Nd:YAG laser through a fiberoptic and a sapphire tip, *Int. Ophthalmol.*, 11, 3, 1987.

52. Brazitikos, P.D., D'Amico, D.J., Bernal, M.T., and Walsh, A.W., Erbium:YAG laser surgery of the vitreous and retina, *Ophthalmology*, 102, 278, 1995.

53. D'Amico, D.J., Blumenkranz, M.S., Lavin, M.J., Quiroz-Mercado, H., Pallikaris, I.O., Marcellino, G.R., and Brooks, G.E., Multicenter clinical experience using an erbium:YAG laser for vitreoretinal surgery, *Ophthalmology*, 103, 1575, 1996.

54. Cohen, B.Z., Wald, K.J., and Toyama, K., Neodymium:YLF picosecond laser segmentation for retinal traction associated with proliferative diabetic retinopathy, *Am. J. Ophthalmol.*, 123, 515, 1997.

55. Peyman, G.A., Contact lenses for Nd:YAG application in the vitreous, *Retina*, 4, 129, 1984.

56. Lin, C.P., Weaver, Y.K., Birngruber, R., Fujimoto, J.G., and Puliafito, C.A., Intraocular microsurgery with a picosecond Nd:YAG laser, *Lasers Surg. Med.*, 15, 44, 1994.

7

Lasers in Otolaryngology

Lou Reinisch
*Farmingdale State
College*

7.1 Introduction

The laser and its role in otolaryngology actually began just after Theodore Maiman made the first ruby laser on May 16, 1960 [1]. It was not only the ophthalmologists and the dermatologists who first attempted to use the laser in medicine [2–4]. In the early 1960s, otolaryngologists first considered different methods to use pulsed laser systems in the middle ear and labyrinth [5,6]. During this time, Geza Jako began studying the effects of laser energy on human vocal folds [7]. His first attempts at tissue ablation were made with the neodymium-glass laser, with a wavelength of 1.06 μm. The absorption characteristics of the tissue were not suitable for precise excision with this wavelength of light. In 1965, Strully and Yahr tried to enhance the absorption of the tissue by painting the tissue with a copper sulfate solution [8]. The results were still unsatisfactory. They found at least three problems with using the laser: (1) They needed higher intensity levels, (2) they could produce only small lesions, and (3) they were left with significant thermal destruction of the tissue surrounding the laser ablation site.

In 1967, Polanyi experimented with the CO_2 laser in a human cadaver larynx and was encouraged by the ability to produce discrete wounds. The 10.6 μm wavelength of the CO_2 laser is strongly absorbed by water ($\alpha = 250$ cm^{-1}) [9]. Therefore, biological tissue, which is high in water content, absorbs the laser energy well. The energy is concentrated at the point of laser impact and comparatively minimal spread through the surrounding tissue occurs. In addition, the longer wavelength at 10.6 μm shows minimal scattering of the laser light in tissue. Their work spurred the development of an endoscopic delivery system so the laser could be tested in vivo [10–12]. In 1972, Jako reported the initial use of this new equipment in a canine model [7,13].

The most common use of medical laser in otolaryngology is for tissue ablation. It is therefore no surprise that the most commonly used laser in otolaryngology is the CO_2 laser. The 10.6 μm light from the CO_2 laser can create intense localized heating sufficient to vaporize both extra- and intracellular water, producing a coagulative necrosis [14–16].

7.2 Laser Use

The tissue effects produced by the laser vary with the wavelength and pulse structure of the laser (Figure 7.1). The interaction of laser energy with living tissue can produce at least three distinct reactions. First, the laser energy can be absorbed by chromophores within the tissue [17,18]. For example, water is the chromophore for the CO_2 laser and hemoglobin is often the chromophore for the potassium titanyl phosphate (KTP)/532 laser. The absorption of the light energy by the tissue is then converted into heat. This is the thermal effect used today with most conventional surgical laser systems.

The second reaction, the radiant energy of a laser, can stimulate or react with molecules within a cell [19–22]. This molecule, after the absorption of the light energy, causes a biomolecular chemical change to occur within the cell. This effect is termed photochemical. An example of a photochemical process is the reaction that occurs with injection of a photosensitizing drug into the tissue and the subsequent biochemical effect that is produced when the drug is activated by the laser energy.

Third, the use of short pulses of high-intensity laser light can disrupt cellular architecture because of the production of stress transient waves or photoacoustic shock waves. The short burst of light causes rapid heating and thermal expansion of a small volume of the tissue. The expansion causes an acoustic wave to propagate from the source. This mechanical disruption of tissue is an example of a nonthermal tissue effect [23].

Several studies comparing the histological and tensile properties of wounds after laser and scalpel-produced incisions on experimental animals have been performed (Figure 7.2). As early as 1971, Hall demonstrated that the tensile strength in a CO_2 laser-induced incision was less than the scalpel incision up to the 20th day of healing and became the same by the 40th day [14]. Later, in 1981, Norris studied

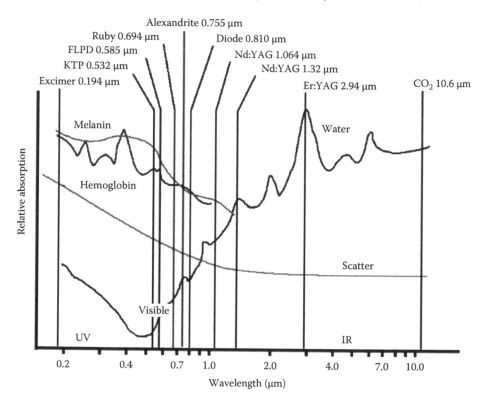

FIGURE 7.1 The electromagnetic spectrum from the ultraviolet, through the visible to the mid-infrared. The absorption spectra of major tissue chromophores, hemoglobin, melanin, and water are shown. Also, the limit of depth of penetration due to scatter is shown. The wavelengths of several important medical lasers have been added.

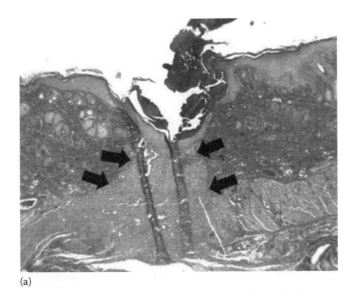

(a)

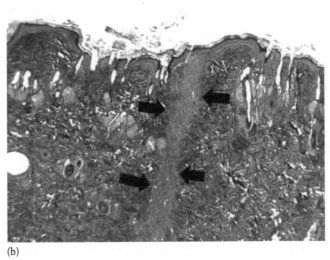

(b)

FIGURE 7.2 Histological sections of incisions made to rat skin after (a) CO_2 laser incision made with 5 W and 0.2 s repeat pulse mode and focused to 125 μm and (b) scalpel incision. The wounds were harvested 7 days after the incisions were made. (a) Shows a wide band of denatured collagen (marked with arrows). The tissue has reepithelialized and a large amount of coagulum is seen on the surface. (b) Shows a narrow band of new collagen (marked with arrows). The laser incision (a) shows the most pronounced delay in wound healing at its early (7 day) time point.

the healing properties of CO_2 laser incisions using a porcine model histologically [24]. He showed that scalpel-induced incisions exhibited better tissue reconstruction than laser-induced incisions up to the 30th day, after which time both incisions exhibited similar results. In a similar study presented in 1983, Buell and Schuller created CO_2 laser incisions in pigs and compared them to scalpel incisions [25]. They found the laser wound to be weaker in tensile strength than the scalpel wounds for the first 3 weeks.

For all the advantages of using the laser, including hands-off operation and hemostasis, the delay in wound healing caused by the lateral thermal damage remains a problem. When the otolaryngologist decides to use the laser, all of the advantages must be weighed against the disadvantages and the potential for increased morbidity. Thus, the surgeon needs to be informed and aware of how the laser interacts with the tissue.

7.2.1 Types of Lasers

Various types of lasers are currently used in otolaryngology—head and neck surgery. Each laser will be explained and the range of applications given. Of course, this list is constantly changing. In general, any laser is first tried in many different applications, and the use becomes better defined and more specific as the surgeon gains experience with the laser.

7.2.1.1 Argon Ion Laser

The argon ion laser (frequently called the argon laser) has wavelengths of 514 nm (green in color) and 488 nm (blue in color). The argon laser is operated in the continuous wave (CW) mode, can be delivered through optical fibers, and is used in cutaneous applications as well as stapedotomies. This laser normally has special electrical power requirements and needs tap water for cooling the laser. The argon laser energy is poorly absorbed by clear liquids. Hemoglobin and melanin strongly absorb the laser light. The argon laser can be used to treat vascular cutaneous lesions because of its absorption by melanin and hemoglobin [26].

Focusing the argon laser beam to a small spot size results in high power densities sufficient to vaporize tissue. Otologists have used the argon laser to perform stapedectomy procedures because of its ability to be focused to the small spot size and the optical fiber delivery system [27,28]. Another otologic application of this laser is the lysis of middle ear adhesions [29].

7.2.1.2 KTP/532 Laser

Similar to the argon laser, the KTP or KTP/532 laser emits light at 532 nm (green color) in a quasi-CW mode and can be delivered through optical fibers. The single wavelength of this KTP laser is centered on a hemoglobin absorption band. The laser normally does not have any special power requirements and does not require external water to cool the laser. The lasing source is a neodymium–yttrium aluminum garnet (Nd:YAG) laser. The Nd:YAG laser rod is continuously pumped with a krypton arc lamp and Q-switched. The Q-switching process changes the CW operation to quasi-CW. The light is emitted in a series of short pulses that repeat so quickly that the light appears to the human eye to be CW. The pulsed 1.06 μm light traverses a frequency-doubling KTP crystal yielding the quasi-CW 532 nm green light.

Like the argon laser, the radiant energy from the KTP laser is readily transmitted through clear aqueous tissues because it has a low water absorption coefficient. Certain tissue pigments, such as melanin and hemoglobin, absorb the KTP laser light effectively. When low levels of green laser light interact with highly pigmented tissues, a localized coagulation takes place within these tissues. The KTP laser can be selected for procedures requiring precise surgical excision with minimal damage to surrounding tissue, vaporization, or photocoagulation. The power density chosen for a given application determines the tissue interaction achieved at the operative site.

The KTP laser is transmitted through a flexible fiber optic delivery system, which can be used in association with a micromanipulator attached to an operating microscope or freehand in association with various handheld delivery probes having several different tip angles. These handheld probes facilitate use of the KTP laser for functional endoscopic sinus surgery and other intranasal applications [30], otologic applications [31], and microlaryngeal applications [32]. In the past decade or so, the KTP laser has experienced a renewed interest in the photocoagulation of microvascular lesions of the vocal folds [33]. There is some controversy concerning the *best* laser for microvascular lesions of the vocal folds. Many of the reported studies reflect a bit of bias. For instance, some investigators claim that the 585 nm wavelength of the flashlamp-excited dye laser (FEDL) is more strongly absorbed by the hemoglobin compared to the 532 nm wavelength of the KTP laser. In Figure 7.3, we show the absorption of oxyhemoglobin. We also added vertical lines at 532 and 585 nm. It is clear from the figure that the absorption coefficients are within 10% of each other (Figure 7.3). Other studies have shown that the KTP laser works as well as the carbon dioxide laser or the flashlamp-pumped dye laser [34]. The optical fiber delivery of the 532 nm laser light can be manipulated through a rigid pediatric bronchoscope as small as 3.0 mm, facilitating lower tracheal and endobronchial lesion treatment in infants and neonates [35].

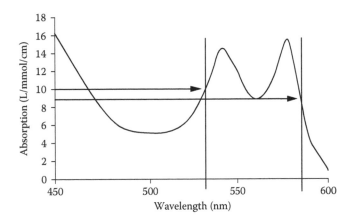

FIGURE 7.3 The absorption profile of oxyhemoglobin from 450 to 600 nm. The vertical lines are placed at 532 and 585 nm (the emission wavelengths of the KTP laser and the FEDL, respectively). The absorption of the oxyhemoglobin is nearly the same for these two wavelengths.

Examples of handheld KTP laser applications include tonsillectomy [36–40], stapedectomy [41–43], excision of acoustic neuroma [42], and excision of benign and malignant laryngeal lesions [44].

7.2.1.3 Nd:YAG Laser

The Nd:YAG laser at 1.06 μm (near-infrared region of the spectrum) has the deepest penetration depth of any of the common surgical lasers. This laser can produce a zone of thermal coagulation and necrosis that extends up to 4 mm from the impact site. The CW light can also be delivered through optical fibers. The laser normally does not have any special power requirements and does not need external cooling water.

The primary applications for the Nd:YAG laser in otolaryngology head and neck surgery include palliation of obstructing tracheobronchial lesions [45–49], palliation of obstructing esophageal lesions [50], photocoagulation of vascular lesions of the head and neck [51,52], and photocoagulation of lymphatic malformations [53]. The Nd:YAG laser has several distinct advantages in the management of obstructing lesions of the tracheobronchial tree. Hemorrhage is a frequent and dangerous complication associated with laser bronchoscopy, and its control is extremely important. Control of hemorrhage is more secure with this laser because of its deep penetration in tissue. Nd:YAG laser application through an open, rigid bronchoscope allows for multiple distal suction capabilities simultaneous with laser application and rapid removal of tumor fragments and debris to prevent hypoxemia.

The major disadvantage of the Nd:YAG laser is its less predictable depth of tissue penetration. This laser is used primarily to photocoagulate tumor masses rapidly at power settings in the upper and lower aerodigestive tract of 40–50 W, 0.5–1.0 s exposures. The laser beam is always applied parallel to the wall of the tracheobronchial tree, whenever possible. The rigid tip of the bronchoscope is used mechanically to separate the devascularized tumor mass from the wall of the tracheobronchial tree.

7.2.1.4 Other Lasers

Other lasers include the FEDL. This laser operates at 585 nm with pulses of light approximately 0.4 ms long. The parameters have been optimized for the selective treatment of vascular lesions with minimal damage to the dermis. The light is delivered through an optical fiber. The laser normally requires 220 V and cooling water. The FEDL was devised to treat dermal vascular lesions. However, this laser has been found to be effective for several laryngeal applications [54,55]. Another laser developed for cutaneous applications is the argon-pumped dye laser [56,57]. Here, the wavelength can be varied from 488 nm through the red region of the spectrum (800 nm). The CW light is delivered through an optical fiber. This laser has special electrical requirements and requires a significant flow of tap water for cooling.

Investigations have considered the diode laser. These small lasers operated in the red to near-infrared region of the spectrum. They are small and require no water for cooling. The lasers operate CW and the light is delivered through an optical fiber. The intensities are relatively low. They have been investigated for photodynamic therapy and tissue welding [58–60]. The diode laser has also been investigated to shrink the tonsillar tissue without damaging the overlying mucosal layer [61]. This is the first step towards tonsillectomies that are relatively painless and with no loss of blood.

7.2.1.5 CO_2 Laser

The workhorse of surgery, the CO_2 laser operates at 10.6 μm. The invisible infrared CO_2 laser beam has a coaxial helium–neon laser or diode laser beam to act as a pointing beam. This laser does not have special electrical requirements and is normally air cooled. The wavelength of the CO_2 laser is strongly absorbed by water [9]. Therefore, the light energy of this laser is well absorbed by all tissues of high water content. The mid-infrared light at 10.6 μm cannot propagate through glass or sapphire optical fibers and is normally delivered through an articulated arm. Additionally, silver halide optical fibers for the CO_2 laser as well as waveguides have been introduced in the market [62]. The laser energy can be delivered to tissue either through a hand-piece for macroscopic surgery or adapted to an operating microscope for microscopic surgery. Additionally, the universal endoscopic coupler allows for delivery of the laser energy through a rigid bronchoscope [63,64].

7.3 Microscopic Applications

Laser technology is being coupled to additional new instrumentation for the laser surgeon; especially newer devices have been introduced to deliver the laser beam. Often, hospitals will update their equipment on a regular basis. It is mandatory for the surgeon to be completely familiar with the laser unit and the delivery system before any patient application is attempted. Power output, spot size, and therefore power density can vary and cannot always be extrapolated from one unit to another.

The lasers and the wavelengths presented earlier may all be adapted for use with the binocular operating microscope. The argon, KTP, diode, and Nd:YAG lasers may be used with optical fiber delivery probes that are threaded through a suction catheter. The optical fibers are used in either a contact or noncontact mode. The CO_2 and KTP laser have the added feature of being delivered by an optical system with a micromanipulator. This allows precise, noncontact, delivery at a predetermined focal length.

Microscopic delivery systems, for the CO_2, have evolved since the initial endoscopic couplers. Several problems, including parallax error and a large spot size have been overcome in the newest generation of micromanipulators. Perhaps the most significant change in the micromanipulator has been the addition of the partially reflective dichroic mirror. This mirror allows reflection of the CO_2 laser while permitting transmission of almost 75% of visible light [65]. The larger beam diameter possible with the dichroic mirror also allows the spot size to be reduced from 800 to 200–250 μm [66]. As the thermal effects of laser applications have become better understood, the desirability of using a smaller spot size for tissue interaction has become evident.

Two devices that work with the micromanipulator to scan the beam in a straight line or over a circular pattern are now available (Figure 7.4). These devices allow the lasers to work the superpulse mode and significantly reduce the lateral thermal damage. They also permit an even fluence of laser energy to impact the tissue. Although it seems that these devices should offer many advantages, there are few clinical reports yet available.

7.3.1 Laryngeal Surgery

Over the past 20 years, the role and limitations of lasers in microlaryngoscopic applications have become better understood. The CO_2 laser remains the laser of first choice for many microlaryngoscopic applications. Certain benign laryngeal diseases, such as the removal of recurrent respiratory papillomatosis, retain the laser as the instrument of choice. More otolaryngologists are, once again, considering the KTP laser

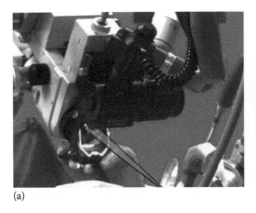
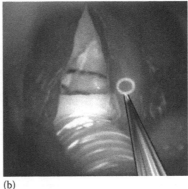

(a)   (b)

FIGURE 7.4   A scanning device is coupled with a microscope micromanipulator for laryngeal surgery. In (a), the device is shown. This scanner scans the beam in a spiral pattern. In (b), the area that will be ablated is shown using the circle drawn by the aiming beam. This area ablation is effective to remove papilloma. The beam is aiming at the vocal fold. A laser safe endotracheal tube is used and the cuff is protected by saline-soaked cottonoids.

and 585 nm FEDL for these applications. Unlike the CO_2 laser, the two visible wavelength lasers (the KTP and FEDL) can be transmitted through optical fibers. This allows patients to be treated in the clinic. This flexibility has caused a flurry of studies that demonstrate these other lasers can be safely used [32,33,54,55]. The surgeon is quick to realize the advantages of microscopic control and decreased postoperative edema. Although there was early disappointment when the laser failed to cure the papillomatosis, the disappointment has been overcome by the laser's ability to precisely remove papilloma and spare normal laryngeal tissue. Other applications appropriate for the CO_2 laser include subglottic stenosis, webs, granuloma, and capillary hemangiomas. Surgery for other benign laryngeal disease processes such as polyps, nodules, leukoplakia, and cysts may also be performed with the CO_2 laser; however, cold knife excision has been shown to produce equal, if not improved, postoperative results. Surgery in the pediatric group for webs, subglottic stenosis, and capillary meningiomas has been significantly improved by the precision, preservation of normal tissue, and decreased postoperative edema associated with the CO_2 laser.

The Nd:YAG laser has much more limited application with regard to intralaryngeal use. However, there are applications where the Nd:YAG is appropriate and often superior to the CO_2 laser. Specifically, the increased tissue absorption with an increased depth of penetration makes the Nd:YAG laser ideal for the treatment of vascular lesions, such as cavernous hemangiomas, where hemostasis and vessel coagulation are the treatment goals.

7.3.1.1 Stenosis

Stenotic lesions appropriate for endoscopic management have certain features in common as determined by retrospective analysis [67]. First, all lesions treated with endoscopic techniques must have intact external cartilaginous support. Attempted endoscopic incision or excision of areas of tracheomalacia can have disastrous results if surrounding structures are perforated. Second, lesions appropriate for endoscopic management are usually less than 1 cm in vertical length. Favorable results, however, have been reported for lesions up to 3 cm in length when endoscopic incision is combined with prolonged stenting [68,69]. Finally, total cervical tracheal or subglottic stenosis does not usually respond well to endoscopic management. Again, successful case reports exist for endoscopic management when it is combined with prolonged stenting of the stenotic area [68].

Still, the management of laryngotracheal stenosis is a difficult problem for the otolaryngologist. The first decision, whether open management is necessary or if endoscopic techniques alone are adequate, is probably the most demanding question. All patients with laryngotracheal stenosis need staging direct laryngoscopy and bronchoscopy to determine the extent and degree of stenosis. The CO_2 laser should

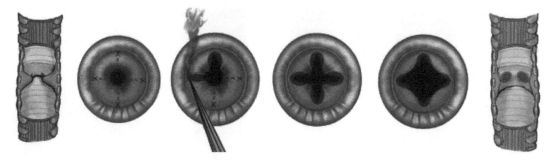

FIGURE 7.5 On the left, a subglottic stenosis is shown in the trachea with an exposed view. The images progressing from left to right represent the stenosis as viewed by the otolaryngologist. The radial incisions are made with the laser and then dilated. The resulting airway is shown on the right with a cutaway view of the trachea.

always be available during the staging laryngoscopy. With the laser on standby, any scarring can be easily removed or incised with the CO_2 laser. Supplemental dilation with the bronchoscope or stent placement may then be beneficial in further management of the stenotic area [68].

Endoscopic management of laryngotracheal stenosis relies on mucosal preservation. The two techniques advocated for this task are radial incision with bronchoscopic dilation [68] and the microtrapdoor flap [70–73]. In bronchoscopic dilation, the laser is used to make radial incision, like the spokes of a wheel, in the stenotic area (Figure 7.5). Bronchoscopes are then sequentially passed through the stenosis to dilate it. The microtrapdoor flap is more complicated and uses a flap of mucosa to drape over the excised tissue.

7.3.1.2 Recurrent Respiratory Papillomatosis

The CO_2 laser has become the standard treatment modality for patients with recurrent respiratory papillomatosis. Some surgeons have reported successful treatment results with the KTP laser for this disorder; however, the less predictable depth of penetration and therefore increased potential for thermal injury in surrounding normal tissue make it less than ideal [74]. As stated earlier, the CO_2 laser cannot cure the disease. The greater absorption and decreased scatter potential in laryngeal tissue of the 10.6 μm wavelength of the CO_2 laser account for its greater effectiveness in preserving normal laryngeal structures and maintaining the airway.

The CO_2 laser treatment should be directed at removing as much papilloma from one vocal cord as possible and then as much as possible from the other. Even though some papilloma may remain, it is important to preserve a 2–3 mm strut of covering over the anterior part of one vocal cord. This lessens the possibility of creating an anterior web. The papilloma overlying the true vocal cord can be vaporized to the vocal ligament. Following the initial CO_2 laser removal of papilloma, a planned repeat operation should be performed in approximately 6 weeks. In a series reported by Ossoff et al., 22 patients underwent 105 CO_2 laser excisions. The intraoperative soft tissue complication rate was zero. The delayed soft tissue complication rate consisting of two patients with slight true vocal fold scarring and one patient with a small posterior web was 13.6% [75]. This compares favorably with other published complication rates of 28.7% and 45% [76,77].

Two further considerations regarding the treatment of recurrent respiratory papillomatosis are worthy of mention. First, considerable attention has been given to the possible detection of papilloma virus in the laser plume. Conflicting reports exist on both sides of the issue [78,79]. Both surgeons and anesthesiologists have been treated for the disease that was manifest only after clinical exposure. Current recommendations to lessen the potential risk of exposure include the use of adequate smoke evacuation and high filtration face masks. The second issue is the management of patients who require multiple repeat laser laryngoscopies for excision on a frequent basis. Treatment intervals then can be based on the rate of reexpression. There are a number of adult and pediatric patients, however, who cannot be placed into clinical remission with this regimen. Therapeutic trials with interferon alpha n-3 have shown no long-term benefit of administration. An investigation with interferon alpha n-1, however, seems to show more promise in producing remission after a 3- to 6-month drug trial [80].

7.4 Laser-Assisted Uvulopalatoplasty

The laser-assisted uvulopalatoplasty (LAUP) is a technique that was developed by Dr. Yves-Victor Kamami in Paris, France, in the late 1980s [81]. The procedure is designed to correct snoring caused by airway obstruction and soft tissue vibration at the level of the soft palate by reducing the amount of tissue in the velum and uvula. The procedure can be performed in an ambulatory setting under local anesthesia and is performed over several stages.

The patient should first be evaluated completely to determine the degree of severity of airway obstruction during sleep. Apnea cannot be ruled out on the basis of history and physical appearance alone [82,83]. If the patient is found to have apnea, then treatment of this condition is mandatory.

One can consider this the equivalent to a dental procedure. If the patient needs subacute bacterial endocarditis prophylaxis, it will need to be administered in conjunction with this procedure. The procedure is performed with the patient sitting in an exam chair in the upright position. A local anesthetic is administered. The procedure is performed using a CO_2 laser. The laser is delivered with a handpiece equipped with a backstop to protect the posterior pharyngeal wall from stray laser energy. The CO_2 laser was selected because of its high efficiency for making incisions. Additionally, the high absorption of the infrared light creates very fine incisions, and the tissue interactions are favorable for this type of procedure. Although the CO_2 laser is not the best coagulating laser available, it has adequate coagulation for the diameter of most vessels encountered in the soft palate. The CO_2 laser is used at a power setting of between 15 and 20 watts in a continuous mode. The beam is focused for cutting and defocused for ablation and vaporization.

Once the adequate level of anesthesia has been achieved in the soft palate, the CO_2 laser is then used to make bilateral vertical incisions through and through the palate at the base of the uvula (Figure 7.6). The uvula can be reduced by approximately 50% of its length and is reshaped in a curved fashion. One can retract the tip of the uvula anteriorly and vaporize the central portion of the uvula muscle while leaving the mucosa intact. Maintaining the mucosa results in less postoperative pain and less delay in wound healing.

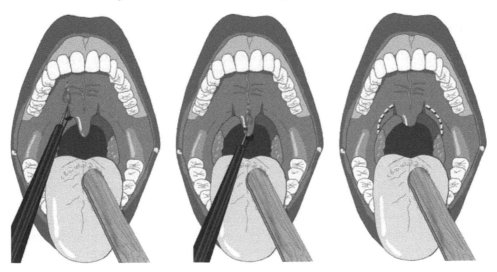

FIGURE 7.6 The steps of the LAUP. From left to right: First, the vertical incisions are made using a laser handpiece with a backstop. Next, the uvula is trimmed. If necessary, incisions can be made along the dashed line if tissue contraction pulls the airway closed.

After the initial visit, the patient is scheduled for a subsequent treatment 4 weeks later. When the patient returns to the clinic, they are requested to report on their amount of pain, any changes in voice or swallowing, as well as bleeding and infection. Subjective improvements in their sleep are also elicited, as well as any objective changes in their snoring. The subsequent procedures are basically the same as the initial procedure. During the treatment course, the otolaryngologist should always examine the palate and determine if attention is needed in other areas. As the retraction occurs, the palate will advance anteriorly and superiorly, and this may tether the posterior tonsillar pillars, causing them to medialize. Should this occur and the pillars themselves begin to obstruct the airway, they can be released by making a horizontal incision at the superior aspect of the pillar, through and through the pillar. If the tonsils are still present, they can be superficially vaporized after first anesthetizing this surface. A crescent-shaped incision can be made vaporizing lateral to the incision in a crescent-shaped manner that may also prevent some of the side-to-side healing that would tend to take place in the palate with simple through and through incisions. By reducing side-to-side healing, more retracting of the palate occurs, and this, over the course of the treatment sessions, will reduce the number of sessions necessary to achieve the end result.

The endpoint for this procedure is determined by any one or combination of the following factors: (1) If the snoring has stopped, (2) if the patient is satisfied (unfortunately, many patients accept a small improvement), and (3) if the patient is unable to make a snoring sound. The patient might continue to have noisy breathing but is unable to make a snoring sound. At this point, the procedure should stop. Further resection of the soft palate after the patient is no longer able to snort runs the risk of excising too much tissue and creating velopharyngeal incompetence. The patient may be having noisy breathing from the vibration of other tissues, such as in the region of the base of the tongue, and not necessarily at the soft palate.

The complication rate for this procedure has been extremely low. There has been some postoperative bleeding in the immediate post-op period and has responded to silver nitrate cauterization. Permanent voice changes, velopharyngeal insufficiency, temporary or permanent nasal reflux, and dehydration from the inability to take food or liquid orally have not been seen with this procedure.

Although there is pain associated with the procedure, it is not of an incapacitating nature, and patients are able to attend classes or work in their normal routine. Most patients report some degree of weight loss, maximum being between 10 and 15 lb over the course of the treatment. In most of our patients, this has been a desirable side effect.

7.5 Limitations on Laser Use

Prior to any surgical procedure, the surgeon must consider whether the laser is the best method to treat a particular laryngeal disorder. As noted earlier, such entities as vocal nodules may require no therapy whatsoever. Although the laser is a highly precise instrument, thermal injury and carbonaceous debris may stimulate submucosal fibrosis that may be unacceptable in certain types of patients, such as in the professional voice user. If the laser is felt to be the optimal method of treatment, limitations may be imposed by the patient's particular anatomic configuration. Mandible size and position as well as spinal column flexibility may limit the type of laryngoscope that can be utilized and thereby affect visualization of the larynx. Patients with cervical arthritis, retrognathia, prominent teeth, or hypertrophy of the base of the tongue may be difficult candidates for endoscopy. Patients with ischemic cardiovascular disease may not withstand the prolonged laryngoscopic suspension that stimulates the vagus nerve and may produce subsequent cardiac arrhythmias as well as silent myocardial infarctions. Even in ideal circumstances, laser laryngoscopy may require more time to complete than other surgical techniques. For example, the patient with unstable cardiac disease suspected of having a laryngeal tumor may best be treated with operative visualization and biopsy by conventional techniques rather than with use of laser [84].

Patients with chronic obstructive or restrictive pulmonary disease may be difficult to ventilate. In such situations, an endotracheal tube may be mandatory and by its presence may limit laryngeal visualization and access.

The precise tissue interaction and the relatively minimal morbidity of endoscopic CO_2 laser surgery result in only three contraindications for use. First, patients with circumferential stenosis of the trachea wider than 1 cm in length will probably have unfavorable results. Therefore, endoscopic management of these patients should be limited to the initial diagnostic procedure. The second contraindication to CO_2 laser bronchoscopy is tracheomalacia. In patients suffering from tracheal stenosis caused by tracheomalacia or loss of tracheal cartilaginous support, use of the laser bronchoscope can be dangerous. It may result in perforation of the trachea wall with subsequent mediastinitis or rupture of a great vessel. This complication obviously can be devastating. The third and final contraindication to CO_2 laser bronchoscopy is in patients with airway obstruction caused by extrinsic compression of the tracheobronchial tree. Extrinsic compression is not amenable to any form of endoscopic treatment.

7.6 Conclusions

The laser continues to be an important tool for the otolaryngologist. In surgery, the importance of the laser results from the ability to direct the beam and make precise incisions. The addition of computer-controlled microscanners will add the usefulness of the laser. The microscanners permit precise control and the ability to ablate lines or patterns with minimal loss of material. They can also reduce lateral thermal damage and thus reduce the delay in wound healing often associated with the laser.

The user needs to be appropriately trained before using any laser and fully understand all the advantages and possible complications. Appropriate safety measures are also necessary to prevent unnecessary morbidity and mortality.

Finally, the laser–tissue interactions need to be understood. The otolaryngologist must always consider if the laser is necessary. The benefits of the laser use must always outweigh any potential problems or risks. Used appropriately, the laser is not only an important tool, but it is also a critical tool for otolaryngology.

References

1. Maiman, T.H., Stimulated optical radiation in Ruby, *Nature*, 187, 493, 1960.
2. Koester, C.J., Snitzer, E., Campbell, C.J., and Rittler, M.C., Experimental laser retina photocoagulation, *J. Opt. Soc. Am.*, 52, 607, 1962.
3. Goldman, L., Blaney, D., Kindel, D., and Franke, E.K., Effect of the laser beam on the skin, *J. Invest. Dermatol.*, 40, 121–122, 1963.
4. Goldman, L., Blaney, D., Kindel, D.J., Jr., Richfield, D., and Franke, E.K., Pathology of the effect of the laser beam on the skin, *Nature*, 197, 912–914, 1963.
5. Sataloff, J., Experimental use of laser in otosclerotic stapes, *Arch. Otolaryngol.*, 85, 614–616, 1967.
6. Hogberg, L., Stahle, J., and Vogel, K., The transmission of high-powered ruby laser beam through bone, *Acta Soc. Medicorum Upsaliensis*, 72, 223–228, 1967.
7. Jako, G.J., Laser surgery of the vocal cords: An experimental study with carbon dioxide laser on dogs, *Laryngoscope*, 82, 2204–2216, 1972.
8. Strully, K.J. and Yahr, W., Biological effects of laser radiation enhancements by selective stains, *Fed. Proc.*, 24, S-81, 1965.
9. Hale, G.M. and Querry, M.R., Optical constants of water in the 200-nm to 200-μm wavelength region, *Appl. Opt.*, 12, 555–563, 1973.
10. Bredemeier, H.C., 1969, Laser accessory for surgical applications, US Patent No. 3,659,613, issued 1972.
11. Polanyi, T.G., Bredemeier, H.C., and Davis, T.W., Jr., Lasers for surgical research, *Med. Biol. Eng.*, 8, 541–548, 1970.
12. Bredemeier, H.C., 1973, Stereo laser endoscope, US Patent No. 3,796,220, issued 1974.
13. Strong, M.S. and Jako, G.J., Laser surgery in the larynx Early clinical experience with continuous CO_2 laser, Annals of otology, *Rhinol. Laryngol.*, 81, 791–798, 1972.

14. Hall, R.R., The healing of tissues incised by carbon-dioxide laser, *Br. J. Surg.*, 58, 222–225, 1971.
15. Fleischer, D., Lasers and gastroenterology, a review, *Am. J. Gastroenterol.*, 79, 406–415, 1984.
16. Cochrane, J.P.S., Beacon, J.P., Creasey, G.H., and Russel, C.G., Wound healing after laser surgery: An experimental study, *Br. J. Surg.*, 67, 740–743, 1980.
17. LeCarpentier, G.L., Motamedi, M., McMath, L.P., Rastegar, S., and Welch, A.J., Continuous wave laser ablation of tissue: Analysis of thermal and mechanical events, *IEEE Trans. Biomed. Eng.*, 40, 188–200, 1993.
18. Verdaasdonk, R.M., Borst, C., and van Gemert, M.J., Explosive onset of continuous wave laser tissue ablation, *Phys. Med. Biol.*, 35, 1129–1144, 1990.
19. Wenig, B.L., Kurtzman, D.M., Grossweiner, L.I., Mafee, M.F., Harris, D.M., Lobraico, R.V., Prycz, R.A. et al., Photodynamic therapy in the treatment of squamous cell carcinoma of the head and neck, *Arch. Otolaryngol. Head Neck Surg.*, 116, 1267–1270, 1990.
20. Rausch, P.C., Rolfs, F., Winkler, M.R., Kottysch, A., Schauer, A., and Steiner, W., Pulsed versus continuous wave excitation mechanisms in photodynamic therapy of differently graded squamous cell carcinomas in tumor-implanted nude mice, *Eur. Arch. Otorhinolaryngol.*, 250, 82–87, 1993.
21. Okunaka, T., Kato, H., Conaka, C., Yamamoto, H., Bonaminio, A., and Eckhauser, M.L., Photodynamic therapy of esophageal carcinoma, *Surg. Endosc.*, 4, 150–153, 1990.
22. Shikowitz, M.J., Comparison of pulsed and continuous wave light in photodynamic therapy of papillomas: An experimental study, *Laryngoscope*, 102, 300–310, 1992.
23. Doukas, A.G., McAuliffe, D.J., and Flotte, T.J., Biological effects of laser-induced shock waves: Structural and functional cell damage in vitro, *Ultrasound Med. Biol.*, 19, 137–146, 1993.
24. Norris, C.W. and Mullarky, M.B., Experimental skin incision made with the carbon dioxide laser, *Laryngoscope*, 92, 416–419, 1982.
25. Buell, B.R. and Schuller, D.E., Comparison of tensile strength in CO_2 laser and scalpel skin incisions, *Arch. Otolaryngol.*, 109, 465–467, 1983.
26. Apfelberg, D.B., Maser, M.R., Lash, H., and Rivers, J. The argon laser for cutaneous lesions, *JAMA*, 245, 2073–2075, 1981.
27. Perkins, R.C., Laser stapedotomy for otosclerosis, *Laryngoscope*, 90, 228–240, 1980.
28. Strunk, C.L., Jr. and Quinn, F.B., Jr., Stapedectomy surgery in residency: KTP-532 laser versus argon laser, *Am. J. Otol.*, 14, 113–117, 1993.
29. DiBartolomeo, J.R. and Ellis, M., The argon laser in otology, *Laryngoscope*, 90, 1786–1796, 1980.
30. Levine, H.L., Endoscopy and the KTP/532 laser for nasal sinus disease, *Ann. Otol. Rhinol. Laryngol.*, 98, 46–51, 1989.
31. Thedinger, B.S., Applications of the KTP laser in chronic ear surgery, *Am. J. Otol.*, 11, 79–84, 1990.
32. Atiyah, R.A., Friedman, C.D., and Sisson, G.A., The KTP/532 laser in glossal surgery. KTP/532 clinical update, Laserscope, San Jose, CA, Report No. 22, 1988.
33. Hirano, S., Yamashita, M., Kitamura, M., and Takagita, S., Photocoagulation of microvascular and hemorrhagic lesions of the vocal fold with the KTP laser, *Ann. Otol. Rhinol. Laryngol.*, 115, 253–259, 2006.
34. Xie, X., Young, J., Kost, K., and McGregor, M., KTP 532 nm laser for laryngeal lesions. A systematic review, *J. Voice*, 27, 245–249, 2013.
35. Ward, R.F., Treatment of tracheal and endobronchial lesions with the potassium titanyl phosphate laser, *Ann. Otol. Rhinol. Laryngol.*, 101, 205–208, 1992.
36. Joseph, M., Reardon, E., and Goodman, M., Lingual tonsillectomy: A treatment for inflammatory lesions of the lingual tonsil, *Laryngoscope*, 94, 179–184, 1984.
37. Krespi, Y.P., Har-El, G., Levine, T.M., Ossoff, R.H., Wurster, C.F., and Paulsen, J.W., Laser laryngeal tonsillectomy, *Laryngoscope*, 99, 131–135, 1989.
38. Kuhn, F., The KTP/532 laser in tonsillectomy. KTP/532 clinical update, Laserscope, San Jose, CA, Report No. 06, 1988.

39. Strunk, C.L. and Nichols, M.L., A comparison of the KTP/532-laser tonsillectomy vs. traditional dissection/snare tonsillectomy, *Otolaryngol. Head Neck Surg.*, 103, 966–971, 1990.

40. Linden, B.E., Gross, C.W., Long, T.E., and Lazar, R.H., Morbidity in pediatric tonsillectomy, *Laryngoscope*, 100, 120–124, 1990.

41. Bartels, L.J., KTP laser stapedotomy: Is it safe? *Otolaryngol. Head Neck Surg.*, 103, 685–692, 1990.

42. Strunk, C.L., Quinn, F.B., Jr., and Bailey, B.J., Stapedectomy techniques in residency training, *Laryngoscope*, 102, 121–124, 1992.

43. McGee, T.M., The KTP/532 laser in otology. KTP/532 clinical update, Laserscope, San Jose, CA, Report No. 08, 1988.

44. Atiyah, R.A., The KTP/532 laser in laryngeal surgery KTP/532 clinical update, Laserscope, San Jose, CA, Report No. 21, 1988.

45. McDougall, J.C. and Cortese, D.A., Neodymium-YAG laser therapy of malignant airway obstruction, *Mayo Clin. Proc.*, 58, 35–39, 1983.

46. Shapshay, S.M. and Simpson, G.T., Lasers in bronchology, *Otolaryngol. Clin. North Am.*, 16, 879–886, 1983.

47. Toty, A., Personne, C., Colchen, A., and Vourc'h, G., Bronchoscopic management of tracheal lesions using the Nd, YAG laser, *Thorax*, 36, 175–178, 1981.

48. Dumon, J.F., Reboud, E., Garbe, L., Aucomte, F., and Meric, B., Treatment of tracheobronchial lesions by laser photoresection, *Chest*, 81, 278–284, 1982.

49. Bemis, J.F., Jr., Vergos, K., Rebeiz, E.E., and Shapshay, S.M., Endoscopic laser therapy for obstructing tracheobronchial lesions, *Ann. Otol. Rhinol. Laryngol.*, 100, 413–419, 1991.

50. Fleischer, D., Endoscopic laser therapy for gastrointestinal neoplasms, *Otolaryngol. Clin. North Am.*, 64, 947–953, 1984.

51. Shapshay, S.M. and Oliver, P., Treatment of hereditary hemorrhagic telangiectasia by Nd-YAG laser photocoagulation, *Laryngoscope*, 94, 1554–1556, 1984.

52. Rebeiz, E., April, M.M., Bohigian, R.K., and Shapshay, S.M., Nd-YAG laser treatment of venous malformations of the head and neck: An update, *Otolaryngol. Head Neck Surg.*, 105, 655–661, 1991.

53. April, M.M., Rebeiz, E.E., Friedman, E.M., Healy, G.B., and Shapshay, S.M., Laser therapy for lymphatic malformations of the upper aerodigestive tract. An evolving experience, *Arch. Otolaryngol. Head Neck Surg.*, 118, 205–208, 1992.

54. Franco, R.A., Jr., Zeitels, S.M., Farinelli, W.A., Faquin, W., and Anderson, R.R., 585-nm pulsed dye laser treatment of glottal dysplasia, *Ann. Otol. Rhinol. Laryngol.*, 112, 751–758, 2003.

55. Franco, R.A., Jr., Zeitels, S.M., Farinelli, W.A., Faquin, W., and Anderson, R.R., 585-nm pulsed dye laser treatment of glottal papillomatosis, *Ann. Otol. Rhinol. Laryngol.*, 111, 486–492, 2002.

56. Cosman, B., Experience in the argon laser therapy for port-wine stains, *Plast. Reconstr. Surg.*, 65, 119–129, 1980.

57. Parkin, J.L. and Dixon, J.A., Argon laser treatment of head and neck vascular lesions, *Otolaryngol. Head Neck Surg.*, 93, 211–216, 1985.

58. Spitzer, M. and Krumholz, B.A., Photodynamic therapy in gynecology, *Obstet. Gynecol. Clin. North Am.*, 18, 649–659, 1991.

59. Wang, Z., Pankratov, M.M., Gleich, L.L., Rebeiz, E.E., and Shapshay, S.M., New technique for laryngotracheal mucosa transplantation. 'Stamp' welding using indocyanine green dye and albumin interaction with diode laser, *Arch. Otolaryngol. Head Neck Surg.*, 121, 773–777, 1995.

60. Wang, Z., Pankratov, M.M., Rebeiz, E.E., Perrault, D.F., Jr., and Shapshay, S.M., Endoscopic diode laser welding of mucosal grafts on the larynx: A new technique, *Laryngoscope*, 105, 49–52, 1995.

61. Volk, M.S., Wang, Z., Pankratov, M.M., Perrault, D.F., Jr., Ingrams, D.R., and Shapshay, S.M., Mucosal intact laser tonsillar ablation, *Arch. Otolaryngol. Head Neck Surg.*, 122, 1355–1359, 1996.

62. Kao, M.C., Video endoscopic sympathectomy using a fiberoptic CO_2 laser to treat palmar hyperhidrosis, *Neurosurgery*, 30, 131–135, 1992.

63. Ossoff, R.H., Duncavage, J.A., Gluckman, J.L., Adkins, J.P., Karlan, M.S., Toohill, R.J., Keane, W.M. et al., The universal endoscopic coupler bronchoscopic carbon dioxide laser surgery: A multi-institutional clinical trial, *Otolaryngol. Head Neck Surg.*, 93, 824–830, 1985.

64. Ossoff, R.H., Sisson, G.A., and Shapshay, S.M., Endoscopic management of selected early vocal cord carcinoma, *Ann. Otol. Rhinol. Laryngol.*, 94, 560–564, 1985.

65. Ossoff, R.H., Werkhaven, J.A., Raif, J., and Abraham, M., Advanced microspot microslad for the CO_2 laser, *Otolaryngol. Head Neck Surg.*, 105, 411–414, 1991.

66. Shapshay, S.M., Wallace, R.A., Kveton, J.F., Hybels, R.L., and Setzer, S.E., New microspot microma-nipulator for CO_2 laser application in otolaryngology—Head and neck surgery, *Otolaryngol. Head Neck Surg.*, 98, 179–181, 1988.

67. Simpson, G.T. and Polanyi, T.G., History of the carbon dioxide laser in otolaryngologic surgery, *Otolaryngol. Clin. North Am.*, 16, 739–752, 1983.

68. Shapshay, S.M., Beamis, J.F., Jr., and Dumon, J.F., Total cervical tracheal stenosis: Treatment by laser, dilation, and stenting, *Ann. Otol. Rhinol. Laryngol.*, 98, 890–895, 1989.

69. Whitehead, E. and Salam, M.A., Use of the carbon dioxide laser with the Montgomery T-tube in the management of extensive subglottic stenosis, *J. Laryngol. Otol.*, 106, 829–831, 1992.

70. Beste, D.J. and Toohill, R.J., Microtrapdoor flap repair of laryngeal and tracheal stenosis, *Ann. Otol. Rhinol. Laryngol.*, 100, 420–423, 1991.

71. Dedo, H.H. and Sooy, C.D., Endoscopic laser repair of posterior glottic, subglottic and tracheal ste-nosis by division or micro-trapdoor flap, *Laryngoscope*, 94, 445–450, 1984.

72. Duncavage, J.A., Ossoff, R.H., and Toohill, R.J., Laryngotracheal reconstruction with composite nasal septal cartilage grafts, *Ann. Otol. Rhinol. Laryngol.*, 98, 565–585, 1989.

73. Duncavage, J.A., Piazza, L.S., Ossoff, R.H., and Toohill, R.J., Microtrapdoor technique for the man-agement of laryngeal stenosis, *Laryngoscope*, 97, 825–828, 1987.

74. Strong, M.S., Vaughan, C.W., Cooperband, S.R., Healy, G.B., and Clemente, M.A., Recurrent respi-ratory papillomatosis management with the CO_2 laser, *Ann. Otol. Rhinol. Laryngol.*, 85, 508–516, 1976.

75. Ossoff, R.H., Werkhaven, J.A., and Dere, H., Soft-tissue complications of laser surgery for recurrent respiratory papillomatosis, *Laryngoscope*, 101, 1162–1166, 1991.

76. Crockett, D.M., McCabe, B.F., and Shive, C.J., Complications of laser surgery for recurrent respira-tory papillomatosis, *Ann. Otol. Rhinol. Laryngol.* 96, 639–644, 1987.

77. Wetmore, S.J., Key, J.M., and Suen, J.Y., Complications of laser surgery for laryngeal papillomatosis, *Laryngoscope*, 95, 798–801, 1985.

78. Abramson, A.L., DiLorenzo, T.P., and Steinberg, B.M., Is papillomavirus detectable in the plume of laser-treated laryngeal papilloma? *Arch. Otol. Head Neck Surg.*, 116, 604–607, 1990.

79. Garden, J.M., O'Banion, M.K., Shelnitz, L.S., Pinski, K.S., Bakus, A.D., Reichmann, M.E., and Sundberg, J.P., Papillomavirus in the vapor of carbon dioxide laser-treated verrucae, *JAMA*, 259, 1199–1202, 1988.

80. Leventhal, B.G., Kashima, H.K., Mounts, P., Thurmond, L., Chapman, S., Buckley, S., and Wold, D., Long-term response of recurrent respiratory papillomatosis to treatment with lymphoblastoid interferon alfa-N1. Papilloma Study Group, *N. Engl. J. Med.*, 325, 613–617, 1991.

81. Kamami, Y-V., Laser CO_2 for snoring: Preliminary results, *Acta Otorhinolaryngol. Belg.*, 44, 451–456, 1990.

82. Croaker, B.D., Allison, G.L., Saunders, N.A., Henley, M.J., McKeon, J.L., Allen, K.M., and Gyulay, S.G., Estimation of the probability of disturbed breathing during sleep before a sleep study, *Am. Rev. Respir. Dis.*, 142, 14–18, 1990.

83. Young, T., Palta, M., Dempsey, J., Skatrud, J., Weber, S., and Badr, S., The occurrence of sleep disor-dered breathing among adults, *N. Engl. J. Med.*, 328, 1230–1235, 1993.

84. Fried, M.P., Kelly, J.H., and Strome, M., *Complications of Laser Surgery of the Head and Neck*, Chicago, IL: Year Book Medical Publishers, 1986.

8

Therapeutic Applications of Lasers in Gastroenterology

Masoud Panjehpour
Thompson Cancer Survival Center

Bergein F. Overholt
Thompson Cancer Survival Center

8.1 Introduction

The applications of lasers in gastroenterology have been evolving mainly due to the availability of flexible endoscopes allowing easy access to the upper and lower gastrointestinal (GI) tract. Laser energy may easily be delivered using an optical fiber passed through the biopsy channel of the flexible endoscope. Lasers have been used for both therapeutic and diagnostic applications. In this chapter, therapeutic applications of lasers in gastroenterology will be discussed.

The major therapeutic applications of lasers in gastroenterology have been for the following: ablation of early and advanced cancers that can be reached via an endoscope, control of hemorrhage from ulcers, treatment of vascular malformations, and lithotripsy.

8.2 Lasers for Destruction of Tumors

Lasers used for destruction of tumors can be categorized as either photodynamic therapy (PDT) lasers or thermal lasers. PDT lasers are used to induce an interaction between a specific photosensitizer and oxygen in the tumor. This interaction results in the formation of singlet oxygen and free radicals that are highly cytotoxic. Thermal lasers deliver a highly focused laser energy to the tissue, resulting in coagulation and/or vaporization of tumors.

8.2.1 Lasers for Photodynamic Therapy

PDT is a class of cancer treatment that uses a combination of photosensitizer and laser light to destroy malignant tissues. First, a photosensitizer is administered, typically intravenously. After accumulation of the photosensitizer in the tumor, light from a laser is delivered to the tumor, resulting in a reaction between the photosensitizer and the oxygen in the tissue. The resulting singlet oxygen and/or free radicals are highly cytotoxic, causing necrosis of the tissue. PDT for palliation of dysphagia in patients with partially or completely obstructing esophageal cancer has been reported [1–4]. Photofrin® (porfimer sodium) from Axcan Pharma, Inc. (Mont-Saint-Hilaire, Quebec, Canada), is currently approved by the U.S. Food and Drug Administration (FDA) for the treatment of esophageal cancer, high-grade dysplasia in Barrett's esophagus [5–8], and endobronchial lung cancer.

A critical issue in PDT is the proper delivery of light to the tumor. Here, we will discuss different lasers and light delivery devices used for PDT of premalignant and malignant conditions in the GI tract. The choice of laser is determined by the photosensitizer in addition to the desired depth of necrosis. For example, porfimer sodium (Photofrin) is activated at the wavelength of 630 nm. Meta-tetrahydroxyphenylchlorin (mTHPC), a second-generation photosensitizer, is typically activated at 652 nm. These wavelengths are in the red region of the spectrum and penetrate the tissue effectively. On the other hand, a particular laser may be employed to limit the depth of necrosis in the treated tissue. An argon laser with a single wavelength of 514 nm may be used to produce a minimal depth of tissue necrosis, thereby minimizing the risk of perforation in the treatment of superficial lesions in the esophagus.

In gastroenterology, laser light must be delivered to the tissue using optical fibers. Different fiber configurations and light delivery devices have been developed for the treatment of different abnormal conditions. Following the discussion of PDT lasers, light delivery devices for the treatment of advanced esophageal cancer, dysplasia and early cancer in Barrett's esophagus, gastric cancer, and colorectal tumors will be reviewed.

8.2.1.1 Argon-Pumped Dye Laser (Argon/Dye Laser) for PDT

The most widely used laser for PDT has been the continuous-wave (CW) argon-pumped dye laser (argon/dye laser). Argon/dye lasers can generate a powerful monochromatic adjustable wavelength of light that can be coupled into an optical fiber for delivery through the scope. The wavelength of the dye laser should be adjusted to match the optimum absorption of the photosensitizer. For example, porfimer sodium and its older versions, hematoporphyrin derivative (HPD) and dihematoporphyrin ether (DHE), are activated at 630 nm generated from an argon/dye laser [3–5,9–15]. Aminolevulinic acid (ALA)-induced protoporphyrin IX (PPIX) is activated at 635 nm [16] that can easily be obtained from the same argon/dye laser with minor adjustment to the birefringent filter in the dye laser. An argon/dye laser tuned at a wavelength of 652 nm can be used to activate mTHPC [17,18].

The amount of power from a dye laser is mostly a function of the power of the pumping argon laser. High-power argon lasers, in the range of 20 W, are typically required to obtain sufficient power conversion from the dye laser [3]. Different laser manufacturers such as Spectra Physics (Mountain View, CA) make scientific argon/dye lasers that have been used extensively [9,17]. Scientific lasers require frequent minor optical adjustments for optimal performance.

Lumenis, Inc. (Santa Clara, CA), formerly known as Coherent, manufactures a clinical argon/dye laser, Lambda Plus PDT laser, specifically designed for PDT with Photofrin [4,19]. This laser can generate 2.7 W of 630 nm light. All optical adjustments are done automatically during the warm-up period, which takes about 2 min.

8.2.1.2 KTP-Pumped Dye Laser (KTP/Dye Laser) for PDT

Potassium–titanyl–phosphate (KTP)/dye laser is a clinical PDT laser system manufactured by Laserscope (San Jose, CA). It is a quasi-CW (25 kHz) pulsed laser system (consisting of a KTP laser and a dye module) that has been shown to be equivalent to a CW light source such as argon/dye laser in several studies [20,21]. The dye module is designed to operate at different wavelengths. This laser has been used at 630 nm for activation of Photofrin [5], at 652 nm for activation of mTHPC [22], and at 635 nm for activation of ALA-induced PPIX [23,24].

The Laserscope system employs a commonly used surgical KTP laser (532 nm). The 532 nm light from the KTP laser is used to optically pump a specially designed dye laser (600 Series dye module) that generates the 630 nm. Two models are available. The standard model has a maximum power of 3.2 W when pumped by a 700 Series KTP laser. The high-power XP 600 Series dye module has a maximum power of 7 W when pumped by 30 W of 532 nm from an 800 Series KTP laser. A high-power system is useful when treating a long segment of Barrett's esophagus using a 7 cm balloon [5]. An advantage of this system is the application of a KTP to pump a dye module. The laser energy from the KTP laser is delivered to the dye module using a fiber that allows independent movement of the dye module and the KTP laser. A PDT dye module can be purchased as an accessory to the KTP laser.

8.2.1.3 Gold Vapor Laser for PDT

Gold vapor lasers generate a monochromatic 627.8 nm pulsed light with a typical repetition rate range of 5–15 kHz [25–28]. Gold vapor lasers have been used for activation of ALA-induced PPIX and Photofrin (and/or HPD and DHE). Nakamura et al. [25] used a gold vapor laser to treat early gastric cancer using HPD PDT. Mlkvy et al. [29,30] used a gold vapor laser to activate Photofrin and ALA-induced PPIX for the treatment of a variety of GI tumors. The use of this laser eliminated the need for a dye laser. However, since the beam diameter of gold vapor laser is large, a 600 μm diameter fiber is typically required for efficient coupling of laser beam to the fiber [31]. Since gold vapor lasers are pulsed, several studies have shown their equivalence to CW argon/dye laser in inducing PDT response [26–28,31].

8.2.1.4 Copper Vapor–Pumped Dye Laser for PDT

While gold vapor lasers can produce high powers at a single wavelength of 627.8 nm for activation of Photofrin, their use for activation of other photosensitizer is limited. However, gold vapor lasers can be modified into copper vapor lasers that in turn can be used to pump a dye laser to generate suitable wavelengths for PDT [31]. Barr et al. [32] showed that a copper vapor–pumped dye laser produced the same PDT effect in the normal rat colon as that produced with a CW argon/dye laser.

8.2.1.5 Diode Lasers for PDT

Diode lasers are semiconductor light sources that are compact, user-friendly, and less expensive than conventional lasers. Diode lasers have been available at longer wavelengths such as 664 nm for the activation of tin ethyl etiopurpurin or SnET2 (Miravant Medical Technologies, Santa Barbara, CA) for the treatment of metastatic breast cancer and Kaposi's sarcoma. Diode lasers with sufficient power at 630 nm (for activation of Photofrin) have been more difficult to manufacture. Diomed (Andover, MA) manufactures a diode laser with a wavelength of 630 nm ± 3 nm with a maximum calibrated power of 2.0 W from the fiber-optic delivery system. An internal power meter allows measurement of power from the fiber delivery system. Diomed also manufactures diode lasers at other wavelengths such as 635, 652, and 730 nm. All diode lasers operate on standard electrical power supply and require no water cooling. They are mobile and rugged, require less maintenance, and are cheaper than other PDT lasers.

8.2.1.6 Argon Laser for PDT

The choice of laser is sometimes made based on the desired depth of tissue necrosis. Most photosensitizers can be effectively activated at both short and long wavelengths. Therefore, a suitable short wavelength that is strongly absorbed by tissue may be used to limit the depth of necrosis in the treated area. Green light (514 nm) from an argon laser has a limited penetration depth in tissue due to its strong absorption by hemoglobin. Using mTHPC, 514 nm green light from an argon laser has been recommended to eliminate the possibility of through-the-wall necrosis in the esophagus [11,17,18]. In contrast, esophageal perforations after mTHPC PDT were reported when using 652 nm red light [11,17,18].

8.2.2 Light Delivery Devices for Photodynamic Therapy

Light delivery devices for GI PDT are fiber optics with special configurations for the treatment of different tissue sites. Historical data on the evolution of devices will be presented with emphasis on the most recent developments.

8.2.2.1 Delivery Devices for PDT of Esophageal Cancer

Before the development of cylindrical diffusers, many investigators used a straight-tip bare fiber that delivered the laser light from the tip of the fiber, illuminating the surface of the lesion [3,4,12,14]. These fibers were passed through the scope and the tip of the fiber was held at a specific distance from the tissue during the treatment. Often, the fiber tip was inserted into the tumor for interstitial treatment of the tumor [4,12,14].

Cylindrical diffusers were developed for endoscopic intraluminal treatment of partially or completely obstructing esophageal cancers [3,12,33]. Cylindrical diffusers circumferentially illuminate a specific length of the inner lumen of a tubular organ such as the esophagus. The distal end of the fiber is modified such that the light is emitted uniformly in a circumferential manner throughout the length of the diffuser section. These fibers are manufactured at different lengths of 1, 1.5, 2.0, 2.5, and 5 cm for the treatment of esophageal cancer and endobronchial cancers (OPTIGUIDE, fibersdirect.com). The choice of diffuser length is made based on the length of the tumor. These fibers are semirigid but can easily be passed through a standard gastroscope and may be implanted into a tumor such as in completely obstructing esophageal cancer. The typical light dose for palliation of dysphagia in advanced esophageal cancer is 300 J/cm. These diffusers are designed for use at the wavelength of 630 nm.

Flexible cylindrical diffusers have been developed for intraluminal illumination of tissue (CardioFocus, Inc., West Yarmouth, MA). Flexible diffusers are available at several lengths (1–5 cm). These fibers are easily passed through the scope regardless of scope angulation. They are specially useful in endobronchial PDT where the bronchoscope channel is much smaller.

Van den Bergh [34] described a reusable light delivery device called esophageal light distributor based on the standard Savary–Gilliard dilator with a diameter of 15 mm. The distal end of the light distributor is shaped like a Savary–Gilliard dilator. The illuminating section is rigid and allows either a 180° or a 240° noncircumferential illumination of the esophageal lumen. This design was recommended over a circumferential illuminator to reduce stenosis after PDT [11,17,34]. The esophageal light distributor is typically used under general anesthesia [34].

8.2.2.2 Delivery Devices for PDT of Barrett's Esophagus and Early Esophageal Cancer

Several techniques have been used to treat superficial cancer of the esophagus and dysplasia in Barrett's esophagus. Before the development of cylindrical diffusers and balloons, a straight-tip fiber was used to treat early esophageal cancer [13,14]. Later, cylindrical diffusers were developed to treat a specific length of the esophagus. Overholt et al. [9] used a 2 cm cylindrical diffuser to treat two patients with early invasive cancer in Barrett's esophagus. Laukka and Wang [10] used a 2 cm cylindrical diffuser for the treatment of patients with dysplastic Barrett's esophagus. Long segments were treated sequentially in 2 cm intervals by repositioning of the diffuser. Barr et al. [23] used a 3 cm cylindrical diffuser in a specially designed 10–14 mm Perspex dilator to provide an even light distribution. Long segments were treated by withdrawing the light source and repeating the treatment with some overlapping of the fields. Gossner et al. [24] used a 2 cm cylindrical diffuser for the treatment of short segments of Barrette's esophagus and early cancer. Long segments were treated sequentially by repositioning of the diffuser.

Grosjean et al. [17] and Savary et al. [11] used a previously described cylindrical light distributor with 180° and 240° windows and a diameter of 15 mm to treat patients with early squamous cell carcinoma of the esophagus. They indicated that by using a noncircumferential light delivery device, the incidence of stenosis was reduced by limiting tissue damage to some part of the lumen. The design of the cylindrical light distributor has been described elsewhere [34].

A balloon light delivery device was specifically developed for the treatment of Barrett's esophagus [35]. The balloon was constructed from a polyurethane membrane and had a diameter of 25 mm to effectively flatten the esophageal folds to allow uniform illumination of the specific length of the esophagus. A balloon with a larger diameter reduced the blood flow in tissue, resulting in no PDT injury even at higher light doses [36].

A balloon with a 2 or 3 cm window was first used clinically to treat four patients [19]. A 2 cm cylindrical diffuser was used in the 2 cm windowed balloon. A 2.5 cm cylindrical diffuser was used in the 3 cm windowed balloon (3 cm diffusers were not available). Long segments were treated sequentially by repositioning of the balloon. This balloon was used to determine the proper light dosimetry and for developing the methodology [37].

PDT of long segments of Barrett's esophagus using a 2 or 3 cm windowed balloon required sequential treatments of the esophagus during each session. Later, 5 and 7 cm long windowed balloons were developed to allow treatment of a longer segment of Barrett's esophagus without repositioning of the balloon [5,38]. Flexible 5 or 7 cm cylindrical diffusers were used in the 5 and 7 cm windowed balloons, respectively. In addition, the balloon design was modified for guidewire positioning of the balloon in the esophagus. The light dose of 175–200 J/cm using these balloons is typically required for destruction of high-grade dysplasia in Barrett's esophagus.

These balloons were modified with a 180° window to illuminate half the circumference of the lumen. While the concept of semicircumferential treatment was demonstrated in the canine esophagus [39], it was never used clinically since suitable cases were not found. We believe that the entire circumference of the esophagus should be treated to eliminate all Barrett's mucosa. While this increases the potential complications from strictures, it improves results in the ultimate goal of eliminating all Barrett's mucosa.

The balloon was modified for the phase III multicenter study for the treatment of high-grade dysplasia in Barrett's esophagus. The interior surface of the distal and proximal capped portions of the balloon was coated with a reflective material (Wilson-Cook Medical Inc., Winston-Salem, NC). A cylindrical diffuser 2 cm longer than the window length was used inside the central channel, extending 1 cm proximally and 1 cm distally beyond the window margins. Using this diffuser/balloon configuration, the uniformity of light emitted from the balloon window was improved. In addition, laboratory testing and canine studies showed that the light intensity emitted from the balloon window was 1.5 times higher using the reflective balloons' design compared to that from the initial balloon design (unpublished data). Using this balloon, a light dose of 130 J/cm was used for the treatment of high-grade dysplasia in Barrett's esophagus. The reflective balloons were available in window lengths of 3, 5, and 7 cm for the study. The balloon material was changed to nonstretchable polyethylene terephthalate. It is critical to inflate the balloon to about 20 mm Hg to reduce the pressure on the esophageal wall, minimizing the reduction of oxygen supply to the tissue [5]. This balloon was eventually approved by the US FDA for PDT of Barrett's esophagus [7,8]. Recently, the application of radiofrequency ablation (RFA) for the treatment of Barrett's esophagus was approved by the FDA. RFA has practically replaced the use of PDT for the management of Barrett's patients [40].

8.2.2.3 Delivery Devices for PDT of Gastric Cancer

Straight-tip bare fibers have been used for the treatment of gastric cancer [4,13,14]. These were simple fiber optics with cleaved and polished ends that allowed surface illumination of the tumor by positioning the fiber tip at a specific distance from the tissue. Often, the fiber tip was inserted into the tumor for intralesional (interstitial) illumination of tumors [5,14]. In most recent works, microlens fibers have been developed to deliver light to the surface of the gastric tumors [16,22]. Microlens fibers generate a circular beam of light with excellent edge-to-edge uniformity and a well-defined illumination field. Microlens fibers are routinely used for cutaneous PDT. It should be noted that the beam diameter (and ultimately the energy density) is strongly dependent on the distance between the tissue and the tip of the fiber when using a microlens fiber (or bare straight-tip fiber). Therefore, endoscopic illumination of a lesion should be performed taking into account the relative movements of fiber and tissue inherent to any endoscopy procedure.

8.2.2.4 Delivery Devices for PDT of Lower GI Tract Cancers

The use of PDT has been reported for the treatment of a variety of lower GI malignancies. Much of the treatments were delivered using a bare straight-tip fiber. Loh et al. [15] used a bare fiber by removing the cladding from the distal end. They inserted the tip of the fiber into villous adenomas and treated the polyps intralesionally. Mlkvy et al. [29] used a bare fiber and inserted the tip 1–2 mm deep into adeno-matous polyps. Patrice et al. [4] used bare fibers to either superficially or intralesionally treat rectosig-moid adenocarcinoma in inoperable patients. In the case of superficial treatment, the tip of the fiber was positioned 2–2.5 cm from the surface of the tumor during the treatment. For intralesional treatment, the tip of the fiber was implanted into the tumor. Mlkvy et al. [30] treated colorectal and duodenal tumors using a bare fiber placed interstitially or intraluminally using a 1 cm cylindrical diffuser.

8.2.3 Lasers for Thermal Ablation of Tumors

8.2.3.1 Nd:YAG Laser and Argon Laser for Thermal Ablation

The initial application of laser in gastroenterology was for thermal ablation. Laser energy is always deliv-ered through the working channel (biopsy channel) of an endoscope using an optical fiber. The tissue effect depends on several parameters including the wavelength, power density, and energy density of the laser delivered to the tissue. Vaporization of tissue is achieved when the power density is sufficiently high. However, tissue ablation is also possible at lower power densities where coagulation is achieved, resulting in the tissue necrosis. The necrotic tissue either is sloughed or can be debrided endoscopically at a later session. While several lasers have been applied, Nd:YAG laser has been the most widely used system. Nd:YAG lasers are solid-state lasers that can generate high powers in the range of 60–120 W at the wavelength of 1064 nm. The laser energy is easily transmitted through a 600 μm quartz fiber that is passed through the biopsy channel of the endoscope.

Fleischer et al. [41] reported the first palliative application of Nd:YAG for esophageal carcinoma. Significant clinical, endoscopic, and radiographical improvements were noted in all patients. Jensen et al. [42] compared the Nd:YAG laser and electrocautery technique for palliation of esophageal cancer. They concluded that electrocautery and Nd:YAG laser were similar for the treatment of circumferential tumors, while the laser was more effective for the treatment of noncircumferential tumors.

Nd:YAG laser has typically been delivered as a noncontact technique using a quartz fiber where the tip of the fiber is held about 1.5 cm from the tissue during the laser application. Later, contact probes were introduced by Joffe [43] to allow more precise application of laser energy to the lesions by touch-ing the probe to the tissue. The contact probe is attached to the distal end of a quartz fiber. The contact probes are constructed from a synthetic sapphire crystal that has proven superior to conventional non-contact quartz fiber for the delivery of Nd:YAG laser. The advantages of using contact technique include greater precision in delivering the laser, protecting the tip of quartz fiber from damage, and lower power requirement from the laser. In addition, there is less smoke generated using the contact technique. Contact probes are geometrically designed for each specific application and desired effect (coagulation, vaporization, cutting). The rounded contact probe is typically used for coagulative endoscopic applica-tions using a laser power of 12–15 W, compared with noncontact technique requiring 80–100 W of laser power [44]. Overholt [45] provided a clinical review of both contact and noncontact laser ablation for esophageal cancer. Regardless of technique, multiple sessions of laser therapy are required to establish an open lumen in esophageal cancer patients. While laser therapy is not intended to increase survival, Karlin et al. [46] reported significantly longer survival in esophageal cancer patients who were treated with Nd:YAG laser.

Another technique was described by Sander and Poesl [47] to deliver Nd:YAG laser energy for coagu-lation of tissue. In this technique, a water jet was used to guide the laser energy to the tissue. The advan-tages of this technique include ease of use, absence of smoke and carbonization, reduction in organ distention, as well as deeper coagulation in tissue. They suggested the use of water jet Nd:YAG laser for the treatment of tumors in the GI tract.

While surgery remains the standard treatment for patients with colorectal cancers, lasers have been used for palliation of symptoms in inoperable or high-risk patients. The majority of treated tumors are in the rectum or rectosigmoid due to better access and less patient preparation. The typical laser power is about 100 W using a noncontact laser [48]. Mathus-Vliegen [49] has thoroughly reviewed the application of lasers for colorectal cancer. In general, lasers were effective for those with bleeding tumors. There was a 75% efficacy in palliation of obstructing colorectal cancers. While the majority of laser applications have been for palliation of symptoms in advanced lower GI cancers, Lambert et al. [50] reported successful treatment of early colorectal cancer using laser.

Thermal lasers have also been used for the treatment of benign lower GI lesions. Brunetaud et al. [51] reported using argon laser and Nd:YAG laser for rectal and rectosigmoid villous adenomas. Total tumor ablation was achieved in 92% of patients.

Yasuda et al. [52] reported on the use of Nd:YAG laser for the treatment of early gastric cancer. They indicated that in 28 cases of early gastric cancer, Nd:YAG laser was curative in 96% of cases. They recommended using laser therapy for those cases that were confirmed as mucosal lesions by endoscopic ultrasound.

8.2.3.2 Mid-Infrared Lasers for Thermal Ablation

Mid-infrared lasers such as erbium:YAG laser (2.94 μm) and thulium–holmium–chromium:YAG laser (THC:YAG laser) (2.15 μm) [53] may also be applied endoscopically since the laser energy can be transmitted through quartz fibers. In addition, the strong absorption peak of water in 2–3 μm range makes these lasers attractive for precision cutting similar to CO_2 lasers. Treat et al. [53] reported preclinical results using these lasers in human colon (in vitro) and in rabbit stomach (in vivo). They indicated that the depth of penetration was controllable with minimal spreading of injury 24 h after the treatment. Bass et al. [54] compared pulsed THC:YAG laser with a clinical Nd:YAG laser in canine colonic mucosa. They concluded that the THC:YAG laser created significantly less collateral thermal damage compared with the Nd:YAG laser. They recommended this laser to reduce the risk of perforation when removing sessile polyps. Similarly, Nishioka et al. [55] tested a flashlamp-excited pulsed holmium–yttrium–scandium–gallium garnet laser (2.1 μm, 250 μs pulses) in rabbit liver, stomach, and colon. They indicated that this laser produced less thermal necrosis than Nd:YAG laser and the ablation rate could be controlled in a more precise manner and recommended it as an alternative method for endoscopic ablation of tissues.

8.3 Lasers for Endoscopic Control of Hemorrhage in Ulcers

Both argon laser and Nd:YAG laser have been used for photocoagulation of hemorrhage from peptic ulcers [56–58]. However, Nd:YAG has proven to be superior to argon laser and has been used most commonly. Typically, Nd:YAG laser is delivered to the hemorrhage via a quartz fiber at a high power of 70–90 W using the noncontact method where several applications are delivered around the ulcer. The efficacy of laser therapy for hemorrhaging ulcers was reported by Swain et al. [58] and Rutgeerts et al. [59] where the need for emergency surgery and mortality was effectively reduced. The exact role of laser for this application is not clear since other nonlaser techniques such as the bipolar electrocoagulation technique have been shown to be as effective at a lower cost [60].

8.4 Lasers for the Treatment of Vascular Malformations

Angiodysplasia and vascular malformations associated with hereditary hemorrhage telangiectasia (HHT) are the major causes of hemorrhage in the GI tract. Surgical treatment of these types of hemorrhage has been associated with considerable morbidity and mortality and is not an option for patients with multiple lesions throughout the GI tract. Argon laser and Nd:YAG laser have been used for endoscopic treatment of such vascular lesions.

Argon laser has an advantage that its blue-green light is strongly absorbed by hemoglobin, resulting in relatively shallow treatment depth. This may reduce the risk of perforation, giving a careful control of delivered energy. However, deeper lesions within the submucosa may not be treated effectively. Waitman et al. [61] used an argon laser for treating hemorrhage secondary to telangiectasia, indicating that two-thirds of patients had no recurrence of bleeding. Jensen et al. [62] used an argon laser for the treatment of GI angioma. They reported a significant reduction in bleeding and transfusions after laser therapy.

Nd:YAG laser has a better penetration in tissue than argon laser. Therefore, Nd:YAG laser can coagulate vessels that are deeper within the submucosa. Rutgeerts et al. [63] used a Nd:YAG laser to treat vascular lesions in the upper and lower GI tract. The laser therapy resulted in significant reduction in bleedings as well as the need for transfusions. The treatment was most effective for patients with angiodysplasia. The treatment was not effective in patients with HHT. Bown et al. [64] used both argon laser and Nd:YAG laser for the treatment of vascular lesions (HHT, single and multiple angiodysplasias). The majority of patients required minimal transfusions after one or more laser therapies. Nd:YAG laser appeared to achieve better long-term results due to its greater depth of penetration to destroy submucosal vessels.

Lasers have also been used for treating hemorrhage from other vascular lesions such as watermelon stomach (also called gastric antral vascular ectasia) and radiation-induced vascular lesions. Gostout et al. [65] reported using a Nd:YAG laser for the treatment of watermelon stomach. Reduction in bleeding was achieved in 92% of patients. Bjorkman and Buchi [66] used an argon laser to effectively treat watermelon stomach within the mucosal layer. Viggiano et al. [67] used a Nd:YAG laser for treating radiation-induced proctopathy. After laser therapy, the number of patients with daily bleeding was reduced from 85% to 5%. It should be noted that other nonlaser techniques such as RFA using the HALO system [68] and argon plasma coagulation [69] have also been used for the treatment of hemorrhage in patients with watermelon stomach.

8.5 Lasers for Lithotripsy

High-energy pulsed lasers may be used to deliver a large amount of optical energy resulting in shattering of the biliary stones. The laser energy is delivered to the stone using a small optical fiber. The laser fiber should be in direct contact with the stone to improve the efficiency of fragmentation. The efficacy of stone fragmentation is also dependent on other laser parameters, such as the pulse duration [70]. Longer pulses tend to melt the renal stones [71] rather than fragment them. However, pulse durations ranging from 20 ns to 2 ms have been used effectively.

Several pulsed lasers have been used for fragmenting stones in the common bile duct or intrahepatic ducts. Ell et al. [72] used a pulsed Nd:YAG laser to treat patients with large common bile duct stones. The laser energy was transmitted through a 200 μm quartz fiber under direct visualization through a choledochoscope. Two-thirds of patients were completely cleared of their stones. In another study by Cotton et al. [73], a pulsed tunable dye laser at a wavelength of 504 nm was used for treating patients who were unresponsive to standard treatments. The pulse duration was 1 μs. Some degree of fragmentation was detected in 92% of patients. Eighty percent of patients had clearance of their stones. The authors concluded that laser lithotripsy was safe but a challenging alternative to the surgery in patients with large bile duct stones. Dawson et al. [74] also used a pulsed tunable dye laser at the wavelength of 504 nm to fragment stones in the hepatic ducts or common bile ducts. The pulse energy was 60 mJ. They concluded that laser lithotripsy was a safe nonsurgical alternative for patients who had failed standard treatments. Ell et al. [75] described a smart system that could detect whether the delivery fiber was in contact with a stone or tissue during treatment. If the tip of the probe was in contact with tissue, the laser treatment was terminated to minimize tissue injury. Using a flashlamp-pumped tunable dye laser to fragment biliary stone in vitro, Nishioka et al. [76] showed that the energy threshold was increased when the wavelength was increased from 450 to 700 nm. However, changing the pulse duration from 0.8 to 360 μs did not affect the efficacy of fragmentation. The efficiency of lithotripsy was enhanced if the procedure was performed in an aqueous medium [76]. Due to complexity of the procedure and the expense of the system, laser lithotripsy is used in a small percentage of patients with common bile duct stone that have not responded to standard treatments.

While not a lithotripsy application, an interesting use of a high-power pulsed laser is worth noting here. Lam et al. [77] reported the use of a holmium:YAG laser for fragmenting a denture that was impacted in the esophagus of a patient. After disimpacting the denture into the stomach, the laser was used to successfully fracture the denture into three pieces. The pieces were then successfully removed from the stomach.

8.6 Closing Remarks

In this chapter, the applications of therapeutic lasers in gastroenterology were reviewed. Emphasis on the use of lasers and techniques for PDT in the esophagus is noted since PDT is the newest development in the use of lasers in gastroenterology. This chapter is not a thorough review of clinical literature and is intended to introduce the reader to different applications of lasers in gastroenterology.

References

1. Marcon, N.E., Photodynamic therapy and cancer of the esophagus, *Seminars in Oncology*, 21, 20, 1994.
2. Lightdale, C.J. et al., Photodynamic therapy with porfimer sodium versus thermal ablation therapy with Nd:YAG laser for palliation of esophageal cancer: A multicenter randomized trial, *Gastrointestinal Endoscopy*, 42, 507, 1995.
3. McCaughan, J.S. et al., Palliation of esophageal malignancy with photoradiation therapy, *Cancer*, 54, 2905, 1984.
4. Patrice, T. et al., Endoscopic photodynamic therapy with hematoporphyrin derivative for primary treatment of gastrointestinal neoplasms in inoperable patients, *Digestive Diseases and Sciences*, 35, 545, 1990.
5. Overholt, B.F., Panjehpour, M., and Haydek, J.M., Photodynamic therapy for Barrett's esophagus: Follow-up in 100 patients, *Gastrointestinal Endoscopy*, 49, 1, 1999.
6. Wang, K.K., Current status of photodynamic therapy of Barrett's esophagus, *Gastrointestinal Endoscopy*, 49, S20, 1999.
7. Overholt, B.F. et al., Five-year efficacy and safety of photodynamic therapy with Photofrin in Barrett's high-grade dysplasia, *Gastrointestinal Endoscopy*, 66, 460, 2007.
8. Overholt, B.F. et al., Photodynamic therapy with porfimer sodium for ablation of high grade dysplasia in Barrett's esophagus: International, partially blinded, randomized phase III trial, *Gastrointestinal Endoscopy*, 62, 488, 2005.
9. Overholt, B.F., Panjehpour, M., Teffteller, E., and Rose, M., Photodynamic therapy for treatment of early adenocarcinoma in Barrett's esophagus, *Gastrointestinal Endoscopy*, 39, 73, 1993.
10. Laukka, M.A. and Wang, K.K., Initial results using low-dose photodynamic therapy in the treatment of Barrett's esophagus, *Gastrointestinal Endoscopy*, 42, 59, 1995.
11. Savary, J.F. et al., Photodynamic therapy of early squamous cell carcinoma of the esophagus: A review of 31 cases, *Endoscopy*, 30, 258, 1998.
12. Okunaka, T. et al., Photodynamic therapy of esophageal carcinoma, *Surgical Endoscopy*, 4, 150, 1990.
13. Tajiri, H. et al., Photoradiation therapy in early gastrointestinal cancer, *Gastrointestinal Endoscopy*, 33, 88, 1987.
14. Hayata, Y. et al., Photodynamic therapy with hematoporphyrin derivative in cancer of the upper gastrointestinal tract, *Seminars in Surgical Oncology*, 1, 1, 1985.
15. Loh, C.S. et al., Photodynamic therapy for villous adenomas of the colon and rectum, *Endoscopy*, 26, 243, 1994.
16. Gossner, L. et al., Photodynamic therapy: Successful destruction of gastrointestinal cancer after oral administration of aminolevulinic acid, *Gastrointestinal Endoscopy*, 41, 55, 1995.
17. Grosjean, P. et al., Photodynamic therapy for cancer of the upper aerodigestive tract using tetra(m-hydroxyphenyl)chlorin, *Journal of Clinical Laser Medicine and Surgery*, 14, 281, 1996.

18. Savary, J.F. et al., Photodynamic therapy for early squamous cell carcinoma of the esophagus, bronchi, and mouth with m-tetra (hydroxyphenyl) chlorin, *Archives of Otolaryngology Head Neck Surgery*, 123, 162, 1997.

19. Overholt, B.F. and Panjehpour, M., Barrett's esophagus: Photodynamic therapy for ablation of dysplasia, reduction of specialized mucosa, and treatment of superficial esophageal cancer, *Gastrointestinal Endoscopy*, 42, 64, 1995.

20. Ferrario, A. et al., Direct comparison of in-vitro and in-vivo Photofrin-II mediated photosensitization using a pulsed KTP pumped dye laser and a continuous wave argon ion pumped dye laser, *Lasers in Surgery and Medicine*, 11, 404, 1991.

21. Panjehpour, M. et al., Comparative study between pulsed and continuous wave lasers for Photofrin photodynamic therapy, *Laser in Surgery and Medicine*, 13, 296, 1993.

22. Ell, C. et al., Photodynamic ablation of early cancers of the stomach by means of mTHPC and laser irradiation: Preliminary clinical experience, *Gut*, 43, 345, 1998.

23. Barr, H. et al., Eradication of high-grade dysplasia in columnar-line (Barrett's) oesophagus by photodynamic therapy with endogenously generated protoporphyrin IX, *Lancet*, 348, 584, 1996.

24. Gossner, L. et al., Photodynamic ablation of high-grade dysplasia and early cancer in Barrett's esophagus by means of 5-aminolevulinic acid, *Gastroenterology*, 114, 448, 1998.

25. Nakamura, T. et al., Photodynamic therapy for early gastric cancer using a pulsed gold vapor laser, *Journal of Clinical Laser Medicine and Surgery*, 8, 63, 1990.

26. Cowled, P.A., Grace, J.R., and Forbes, I.J., Comparison of the efficacy of pulsed and continuous-wave red laser light in induction of photocytotoxicity by haematoporphyrin derivative, *Photochemistry and Photobiology*, 39, 115, 1984.

27. McCaughan, J.S. et al., Gold vapor laser versus tunable argon-dye laser for endobronchial photodynamic therapy, *Laser in Surgery and Medicine*, 19, 347, 1996.

28. LaPlant, M. et al., Comparison of the optical transmission properties of pulsed and continuous wave light in biological tissue, *Lasers in Surgery and Medicine*, 7, 336, 1987.

29. Mlkvy, P. et al., Photodynamic therapy for polyps in familial adenomatous polyposis—A pilot study, *European Journal of Cancer*, 31A, 1160, 1995.

30. Mlkvy, P. et al., Photodynamic therapy for gastrointestinal tumors using three photosensitizers— ALA induced PPIX, Photofrin and MTHPC: A pilot study, *Neoplasma*, 45, 157, 1998.

31. Mckenzie, A.L. and Carruth, J.A.S., A comparison of gold-vapour and dye lasers for photodynamic therapy, *Lasers in Medical Science*, 1, 117, 1986.

32. Barr, H. et al., Comparison of lasers for photodynamic therapy with a phthalocyanine photosensitizer, *Lasers in Medical Science*, 4, 7, 1989.

33. Mimura, S. et al., Cooperative clinical trial of photodynamic therapy with Photofrin II and excimer dye laser for early gastric cancer, *Lasers in Surgery and Medicine*, 19, 168, 1996.

34. van den Bergh, H., On the evolution of some endoscopic light delivery systems for photodynamic therapy, *Endoscopy*, 30, 392, 1998.

35. Panjehpour, M. et al., Centering balloon to improve esophageal photodynamic therapy, *Lasers in Surgery and Medicine*, 12, 631, 1992.

36. Overholt, B.F. et al., Balloon photodynamic therapy of esophageal cancer: Effect of increasing balloon size, *Lasers in Surgery and Medicine*, 18, 248, 1996.

37. Overholt, B.F. and Panjehpour, M., Photodynamic therapy in Barrett's esophagus, *Journal of Clinical Laser Medicine and Surgery*, 14, 245, 1996.

38. Overholt, B.F. and Panjehpour, M., Photodynamic therapy for Barrett's esophagus, *Gastrointestinal Endoscopy Clinics of North America*, 7, 207, 1997.

39. Overholt, B.F. et al., Photodynamic therapy for esophageal cancer using a 180 degree windowed esophageal balloon, *Lasers in Surgery and Medicine*, 14, 27, 1994.

40. Shaheen, N.J., Sharma, P., and Overholt, B.D., Radiofrequency ablation in Barrett's esophagus with dysplasia, *The New England Journal of Medicine*, 360, 2277, 2009.

41. Fleischer, D., Kessler, F., and Haye, O., Endoscopic Nd:YAG laser therapy for carcinoma of the esophagus: A new palliative approach, *American Journal of Surgery*, 143, 280, 1982.

42. Jensen, D.M. et al., Comparison of low power YAG laser and BICAP tumor probe for palliation of esophageal cancer strictures, *Gastroenterology*, 94, 1263, 1988.

43. Joffe, S.N., Contact neodymium:YAG laser surgery in gastroenterology, *Surgical Endoscopy*, 1, 25, 1987.

44. Sander, R.R. and Poesl, H., Cancer of the oesophagus—Palliation—Laser treatment and combined procedures, *Endoscopy*, 25(supplement), 679, 1993.

45. Overholt, B.F., Photodynamic therapy and thermal treatment of esophageal cancer, *Gastrointestinal Endoscopy Clinics of North America*, 2, 433, 1992.

46. Karlin, D.A, Fisher, R.S., and Krevsky, B., Prolonged survival and effective palliation in patients with squamous cell carcinoma of the esophagus following endoscopic laser therapy, *Cancer*, 59, 1969, 1987.

47. Sander, R. and Poesl, H., Water jet guided Nd:YAG laser coagulation–Its application in the field of gastroenterology, *Endoscopic Surgery and Allied Technologies*, 1, 233, 1993.

48. Escourrou, J. et al., Laser for curative treatment of rectal cancer. Indications and follow-up [abstract], *Gastrointestinal Endoscopy*, 34, 195, 1988.

49. Mathus-Vliegen, E.M.H., Treatment modalities in colorectal cancer. In: Krasner N., ed., *Lasers in Gastroenterology*, New York, Wiley-Liss, 1991, p. 151.

50. Lamber, R. et al., cancer of the rectum: Results of laser treatment, *Laser in Surgery and Medicine*, 3, 342, 1984.

51. Brunetaud, J.M., Maunoury, V., and Ducrott, E., Palliative treatment of rectosigmoid carcinoma by laser endoscopic photoablation, *Gastroenterology*, 92, 663, 1987.

52. Yasuda, K., Nakajima M, and Kawai, K., Endoscopic diagnosis and treatment of early gastric cancer, *Gastrointestinal Endoscopy Clinics of North America*, 3, 495, 1992.

53. Treat, M.R. et al., Mid-infrared lasers for endoscopic surgery. A new class of surgical lasers, *The American Surgeon*, 55, 81, 1989.

54. Bass, L.S. et al., Alternative lasers for endoscopic surgery: Comparison of pulsed thulium-holmium-chromium:YAG with continuous-wave neodymium:YAG laser for ablation of colonic mucosa, *Lasers in Surgery and Medicine*, 11, 545, 1991.

55. Nishioka, N.S. et al., Ablation of rabbit liver, stomach, and colon with a pulsed holmium laser, *Gastroenterology*, 96, 831, 1989.

56. Fruhmorgan, P. et al., The first endoscopic laser coagulation in the human GI tract, *Endoscopy*, 7, 156, 1975.

57. Swain, C.P. et al., Controlled trial of argon laser photocoagulation in bleeding peptic ulcers, *Lancet*, 2, 1313, 1981.

58. Swain, C.P. et al., Controlled trial of Nd:YAG laser photocoagulation in bleeding peptic ulcers, *Lancet*, 1, 1113, 1986.

59. Rutgeerts, P. et al., A new and effective technique of YAG laser photocoagulation for severe upper gastrointestinal bleeding, *Endoscopy*, 16, 115, 1984.

60. Rutgeerts, P. et al., Nd:YAG laser photocoagulation versus multipolar electrocoagulation for the treatment of severely bleeding ulcers: A randomized comparison, *Gastrointestinal Endoscopy*, 33, 199, 1987.

61. Waitman, A.M., Graut, D.Z., and Chateau, F., Argon laser photocoagulation treatment of patients with acute and chronic bleeding secondary to telangiectasia, *Gastrointestinal Endoscopy*, 28, 153, 1982.

62. Jensen, D.M., Machicado, G.M., and Silpa, M.L., Treatment of GI angioma with argon laser, heater probe or bipolar electrocoagulation, *Gastrointestinal Endoscopy*, 30, 134, 1984.

63. Rutgeerts, P. et al., Long term result of treatment of vascular malformations of the gastrointestinal tract by neodymium-YAG laser photocoagulation, *Gut*, 26, 586, 1985.

64 Bown, S.G. et al., Endoscopic laser treatment of vascular anomalies of the upper gastrointestinal tract, *Gut*, 26, 1338, 1985.

65. Gostout, C.J. et al., Endoscopic laser therapy for watermelon stomach, *Gastroenterology*, 96, 1462, 1989.
66. Bjorkman, D.J. and Buchi, K.N., Endoscopic laser therapy of watermelon stomach, *Lasers in Surgery and Medicine*, 12, 478, 1992.
67. Viggiano, T.R. et al., Endoscopic Nd:YAG laser coagulation of bleeding from radiation proctopathy, *Gastrointestinal Endoscopy*, 39, 513, 1993.
68. Gross, S.A., Al-Haddad, M., Gill, K.R., Schore, A.N., and Wallace, M.B., Endoscopic mucosal ablation for the treatment of gastric antral vascular ectasia with the HALO90 system: A pilot study, *Gastrointestinal Endoscopy*, 67, 325, 2008.
69. Herrera, S., Bordas, J.M., Llach, J., Gines, M. et al., The beneficial effects of argon plasma coagulation in the management of different types of gastric vascular ectasia lesions in patients admitted for GI hemorrhage, *Gastrointestinal Endoscopy*, 68, 440, 2008.
70. Nishioka, N.S., Laser lithotripsy of biliary calculi, *Seminars in Interventional Radiology*, 5, 202, 1988.
71. Watson, G.M. et al., Laser fragmentation of renal calculi, *British Journal of Urology*, 55, 613, 1983.
72. Ell, C. et al., Laser lithotripsy of common bile duct stones, *Gut*, 29, 746, 1988.
73. Cotton, P.B., Endoscopic laser lithotripsy of large bile duct stones, *Gastroenterology*, 99, 1128, 1990.
74. Dawson, S.L. et al., Treatment of bile duct stones by laser lithotripsy: Results in 12 patients, *AJR American Journal of Roentgenology*, 158, 1007, 1992.
75. Ell, C., Laser lithotripsy of difficult bile duct stones by means of a rhodamine-6G laser and an integrated automatic stone-tissue detection system, *Gastrointestinal Endoscopy*, 39, 755, 1993.
76. Nishioka, N.S. et al., Fragmentation of biliary calculi with tunable dye laser, *Gastroenterology*, 93, 250, 1987.
77. Lam, Y.H. et al., Laser-assisted removal of a foreign body impacted in the esophagus, *Lasers in Surgery and Medicine*, 20, 480, 1997.

9

Low-Power Laser Therapy

Tiina I. Karu
*Russian Academy
of Sciences*

9.1 Introduction

The first publications about low-power laser therapy (then called laser biostimulation) appeared more than 40 years ago. Since then, 4000 studies have been published on this still controversial topic (Tuner and Hode 2010). In the 1960s and 1970s, doctors in Eastern Europe, and especially in the Soviet Union and Hungary, actively developed laser biostimulation. However, scientists around the world harbored an open skepticism about the credibility of studies stating that low-intensity visible laser radiation acts directly on an organism at the molecular level. The coherence of laser radiation for achieving stimulative effects on biological objects was more than suspect. Supporters in Western countries, such as Italy, France, and Spain, as well as in Japan and China also adopted and developed this method, but the method was—and still remains—outside mainstream medicine. During the last years, some excellent experimental work was performed in the United States (Wong-Riley et al. 2001, 2005; Eells et al. 2003, 2004; Pal et al. 2007; Anders 2009; Wu et al. 2009). The controversial points of laser biostimulation (Karu 1987, 1988, 1989a,b, 1998, 1999, 2007), which were topics of great interest at that time, were analyzed in reviews that appeared in the late 1980s. Since then, medical treatment with coherent-light sources (lasers) or noncoherent light (light-emitting diodes [LEDs]) has passed through its childhood and adolescence. Most of the controversial points from *childhood* are no longer topical. Currently, low-power laser therapy—or low-level laser therapy or photobiomodulation—is considered part of light therapy as well as part of physiotherapy. In fact, light therapy is one of the oldest therapeutic methods used by humans (historically as sun therapy, later as color light therapy and UV therapy). A short history of experimental work with colored light on various kinds of biological subjects can be found elsewhere (Karu 1987, 1989a). The use of lasers and LEDs as light sources was the next step in the technological development of light therapy.

It is clear now that laser therapy cannot be considered separately from physiotherapeutic methods that use such physical factors as low-frequency pulsed electromagnetic fields; microwaves; time-varying, static, and combined magnetic fields; focused ultrasound; and direct-current electricity. Some common features of biological responses to physical factors have been briefly analyzed (Karu 1998).

As this handbook makes abundantly clear, by the twenty-first century, a certain level of development of (laser) light use in therapy and diagnostics (e.g., photodynamic therapy and optical tomography) had been achieved. In low-power laser therapy, the question is no longer whether light has biological effects but rather how radiation from therapeutic lasers and LEDs works at the cellular and organism levels and what the optimal light parameters are for different uses of these light sources.

This chapter is organized as follows: First, Section 9.2 briefly reviews clinical applications and considers one of still topical issues in low-power laser medicine today, that is, whether coherent and polarized light has additional benefits in comparison with noncoherent light at the same wavelength and intensity.

Second, direct activation of various types of cells via light absorption in mitochondria is described. Primary photoacceptors and mechanisms of light action on cells as well as mechanisms of cellular signaling are considered (Section 9.3). Section 9.4 describes enhancement of cellular metabolism via activation of nonmitochondrial photoacceptors and possible indirect effects via secondary cellular messengers, which are produced by cells as a result of direct activation. This chapter does not consider systemic effects of low-power laser therapy. These data can be found in Tuner and Hode (2010), Baxter (1994), and Simunovic (2000–2002).

9.2 Clinical Applications and Effects of Light Coherence and Polarization

Low-power laser therapy is used by physiotherapists (to treat a wide variety of acute and chronic musculoskeletal aches and pains), by dentists (to treat inflamed oral tissues and to heal diverse ulcerations), by dermatologists (to treat edema, indolent ulcers, burns, and dermatitis), by rheumatologists (to relieve pain and treat chronic inflammations and autoimmune diseases), and by other specialists, as well as general practitioners. Laser therapy is also widely used in veterinary medicine (especially in racehorse-training centers) and in sports medicine and rehabilitation clinics (to reduce swelling and hematoma, relieve pain, improve mobility, and treat acute soft tissue injuries). Lasers and LEDs are applied directly to the respective areas (e.g., wounds, sites of injuries) or to various points on the body (acupuncture points, muscle trigger points). Several books provide details of clinical applications and techniques used (Baxter 1994; Simunovic 2000–2002; Tuner and Hode 2010).

Clinical applications of low-power laser therapy are diverse. The field is characterized by a variety of methodologies and uses of various light sources (lasers, LEDs) with different parameters (wavelength, output power, continuous-wave [CW] or pulsed operation modes, pulse parameters). Figure 9.1 presents schematically the types of light therapeutic devices, possible wavelengths they can emit, and maximal output power used in therapy.

The GaAlAs diodes are used in both diode lasers and LEDs; the difference is whether the device contains the resonator (as the laser does) or not (LED). In recent years, longer wavelengths (~800 to 900 nm) and higher output powers (to 100 mW) have been preferred in therapeutic devices.

One of the most topical and widely discussed issues in the low-power laser therapy clinical community is whether the coherence and polarization of laser radiation have additional benefits as compared with monochromatic light from a conventional light source or LED with the same wavelength and intensity.

Two aspects of this problem must be distinguished: the *coherence of light* itself and the *coherence of the interaction* of light with matter (biomolecules, tissues).

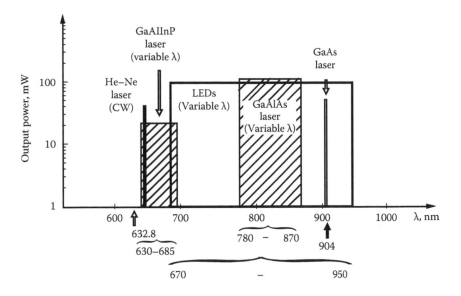

FIGURE 9.1 Wavelength and maximal output power of lasers and LEDs used in low-power laser therapy.

9.2.1 Coherence of Light

The coherent properties of light are described by *temporal* and *spatial* coherence. Temporal coherence of light is determined by the spectral width, $\Delta\nu$, since the coherence time τ_{coh} during which light oscillates at the point of irradiation has a regular and strongly periodical character:

$$\tau_{coh} \cong \frac{1}{\Delta\nu} \tag{9.1}$$

here, $\Delta\nu$ is the spectral width of the beam in Hz. Since light propagates at the rate $c = 3 \times 10^{10}$ cm/s, the light oscillations are matched by the phase (i.e., they are coherent) on the length of light propagation L_{coh}. L_{coh} is called longitudinal coherence:

$$L_{coh} = \frac{c}{\Delta\nu \ [Hz]} \tag{9.2}$$

$$L_{coh} = \frac{1}{\Delta\nu \ [cm^{-1}]} \tag{9.3}$$

The more monochromatic the light, the longer the length where the light field is coherent in volume. For example, for a multimode He–Ne laser with $\Delta\nu = 500$ MHz, $L_{coh} = 60$ cm. But for a LED emitting at $\lambda = 800$ nm (= 12,500 cm^{-1}), $\Delta\nu = 160$ cm^{-1} (or $\Delta\lambda = 10$ nm), and $L_{coh} = 1/160$ cm$^{-1} = 60$ μm, that is, L_{coh} is longer than the thickness of a cell monolayer (≈ 10–30 μm). Spatial coherence describes the correlation between the phases of the light field in a lateral direction. For this reason, spatial coherence is also called lateral coherence. The size of the lateral coherence (ℓ_{coh}) is connected with the divergence (φ) of the light beam at the point of irradiation:

$$\ell_{coh} \cong \frac{\lambda}{\varphi} \tag{9.4}$$

For example, for a He–Ne laser, which operates in the TEM_{00} mode, the divergence of the beam is determined by the diffraction

$$\varphi \cong \frac{\lambda}{D} \tag{9.5}$$

where D is the beam diameter. In this case, ℓ_{coh} coincides with the beam diameter, since for the TEM_{00} laser mode the phase of the field along the wave front is constant.

With conventional light sources, the size of the emitting area is significantly larger than the light wavelength, and various parts of this area emit light independently or noncoherently. In this case, the size of the lateral coherence ℓ_{coh} is significantly less than the diameter of the light beam, and ℓ_{coh} is determined by the light divergence, as shown in Equation 9.4.

An analysis of published clinical results from the point of view of various types of radiation sources does not lead to the conclusion that lasers have a higher therapeutic potential than LEDs. But in certain clinical cases, the therapeutic effect of coherent light is believed to be higher (Tuner and Hode 2010).

However, when human peptic ulcers were irradiated by a He–Ne laser or properly filtered red light was irradiated in a specially designed clinical double-blind study, equally positive results were documented for both types of radiation sources (Sazonov et al. 1985) (for a review, see Karu 1989a).

9.2.2 Coherence of Light Interaction with Biomolecules, Cells, and Tissues

The coherent properties of light are not manifested when the beam interacts with a biotissue on the molecular level. This problem was first considered several years ago (Karu 1987). The question then arose of whether coherent light was needed for *laser biostimulation* or was it simply a photobiological phenomenon. The conclusion was that under physiological conditions, the absorption of low-intensity light by biological systems is of purely noncoherent (i.e., photobiological) nature because the rate of decoherence of excitation is many orders of magnitude higher than the rate of photoexcitation. The time of decoherence of photoexcitation determines the interaction with surrounding molecules (under normal conditions less than 10^{-12} s). The average excitation time depends on the light intensity (at an intensity of 1 mW/cm², this time is around 1 s). At 300 K in condensed matter for compounds absorbing monochromatic visible light, the light intensity at which the interactions between coherent light and matter start to occur was estimated to be above the GW/cm² level (Karu 1987). Note that the light intensities used in clinical practice are not higher than tens or hundreds of mW/cm². Indeed, the stimulative action of various bands of visible light at the level of organisms and cells was known long before the advent of the laser. Also, specially designed experiments at the cellular level have provided evidence that coherent and noncoherent light with the same wavelength, intensity, and irradiation time provide the same biological effect (Karu et al. 1982a,b, 1983a; Bertoloni et al. 1993). Successful use of LEDs in many areas of clinical practice also confirms this conclusion (Simunovic 2000–2002; Tuner and Hode 2010).

Therefore, it is possible that the effects of light coherence are manifested at the macroscopic (e.g., tissue) level at various depths (L) of irradiated matter. Figure 9.2 presents the coherence volumes (V_{coh}) and coherence lengths (L_{coh}) for four different light sources.

Figure 9.2a presents the data for two coherent-light sources (He–Ne and diode lasers as typical examples of therapeutic devices). Figure 9.2b presents the respective data for noncoherent light (LED and spectrally filtered light from a lamp). Figure 9.2 illustrates how large volumes of tissues are irradiated only by laser sources with monochromatic radiation (Figure 9.2a). For noncoherent radiation sources (Figure 9.2b), the length of the coherence, L_{coh}, is small. This means that only surface layers of an irradiated substance can be achieved by coherent light.

The spatial (lateral) coherence of the light source is unimportant due to strong scattering of light in biotissue when propagated to the depth $L \gg \ell_{sc}$, where ℓ_{sc} is the free pathway of light in relation to

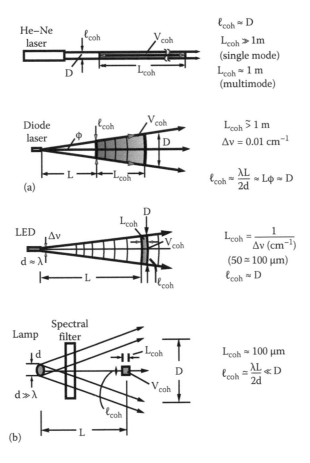

FIGURE 9.2 Coherence volumes and coherence lengths of light from (a) laser and (b) conventional sources when a tissue is irradiated. L_{coh} = length of temporal (longitudinal) coherence, ℓ_{coh} = size of spatial (lateral) coherence, D = diameter of light beam, d = diameter of noncoherent light source, φ = beam divergence, $\Delta \nu$ = beam spectral width.

scattering. This is because every region in a scattering medium is illuminated by radiation with a wide angle ($\varphi \sim 1$ rd). This means that $\ell_{coh} = \lambda$, that is, the size of spatial coherence ℓ_{coh} decreases to the light wavelength (Figure 9.2).

Thus, the length of longitudinal coherence, L_{coh}, is important when bulk tissue is irradiated because this parameter determines the volume of the irradiated tissue, V_{coh}. In this volume, the random interference of scattered light waves and formation of random nonhomogeneities of intensity in space (speckles) occur. For noncoherent light sources, the coherence length is small (tens to hundreds of microns). For laser sources, this parameter is much higher. Thus, the additional therapeutic effect of coherent radiation, if this indeed exists, depends not only on the length of L_{coh} but also, and even mainly, on the penetration depth into the tissue due to absorption and scattering, that is, by the depth of attenuation. Table 9.1 summarizes qualitative characteristics of coherence of various light sources, as discussed earlier.

The difference in the coherence length L_{coh} is unimportant when thin layers are irradiated inasmuch as the longitudinal size of irradiated object $\Delta \ell$ is less than L_{coh} for any source of monochromatic light (filtered lamp light, LED, laser). Examples are the monolayer of cells and optically thin layers of cell suspensions (Figure 9.3a and b). Indeed, experimental results (Karu et al. 1982a,b; Bertoloni et al. 1993) on these models provide clear evidence that the biological responses of coherent and noncoherent light with the same parameters are equal. Also, it was found that the elementary processes in cells after light absorption do not depend on the degree of the beam polarization (Karu et al. 2008a).

TABLE 9.1   Qualitative Characteristics of Coherence

Light Source	Temporal Coherence	Length of Longitudinal (Temporal) Coherence, L_{coh}	Spatial Coherence	Volume of Spatial (Lateral) Coherence, ℓ_{coh}
Laser	Very high	Very long	Very high	Large
LED	Low	Short ($\gg\lambda$)	High	Small (very thin layer)
Lamp with spectral filter	Low	Short ($\gg\lambda$)	Very low	Very small
Lamp	Very low	Very short ($\approx\lambda$)	Very low ($\approx\lambda$)	Extremely small ($\approx\lambda^3$)

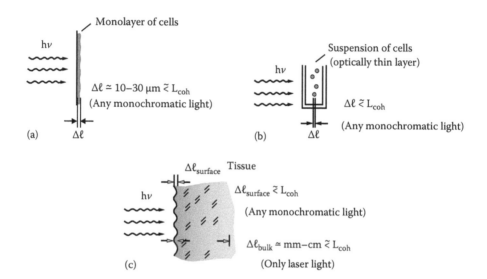

FIGURE 9.3   Depth ($\Delta\ell$), in which the beam coherency is manifested, and coherence length (L_{coh}) in various irradiated systems: (a) monolayer of cells, (b) optically thin suspension of cells, and (c) surface layer of tissue and bulk tissue.

The situation is quite different when a bulk tissue is irradiated (Figure 9.3c). The coherence length, L_{coh}, is very short for noncoherent light sources and can play some role only on surface layers of the tissue with thickness $\Delta\ell_{surface}$. For coherent-light sources, the coherence of the radiation is retained along the entire penetration depth L. The random interference of light waves of various directions occurs over this entire distance in bulk tissue ($\Delta\ell_{bulk}$). As a result, a speckle pattern of intensity appears. Maximum values of the intensity appear at the random constructive interference. The minima (i.e., regions of zero intensity) occur at the random destructive interference. The dimensions of these speckles at every occurrence of directed random interference are approximately within the range of the light wavelength, λ. The coherent effects (speckles) appear only at the depth L_{coh}. These laser-specific speckles cause a spatially nonhomogeneous deposition of light energy and lead to statistically nonhomogeneous photochemical processes, an increase in temperature, changes in local pressure, deformation of cellular membranes, etc.

For nonpolarized coherent light, the random speckles are less pronounced (they have lower contrast) as compared to the speckles caused by coherent polarized light. A special feature of nonpolarized coherent radiation is that the regions with zero intensity appear less often as compared with the action of coherent polarized light. Thus, the polarization of light causes brighter random intensity gradients that can enhance the manifestation of the effects of light coherence when the tissue is irradiated. It was established experimentally that elementary processes in cells after the light absorption do not depend on the degree of beam polarization (Karu et al. 2008a).

Thus, perhaps in scattering biotissue, the main role is played by coherence length (monochromaticity of light) inasmuch as this parameter determines the depth of tissue where the coherent properties of the light beam can potentially be manifested, depending on the attenuation. This is the spatial (lateral) coherence of the beam, that is, its directivity, which plays the main role in the delivery of light into biotissue. In addition, the direction and orientation of laser radiation could be important factors for some types of tissues (e.g., dental tissue) that have fiber-type structures (filaments). In this case, waveguide propagation effects of light can appear that provide an enhancement of penetration depth. Indeed, last experimental data allow to suggest that the coherence length can play a role in laser phototherapy of gingival inflammation (Qadri et al. 2007).

Considered within the framework of this qualitative picture, some additional (i.e., additional to those effects caused by light absorption by photoacceptor molecules) manifestation of light coherence for deeper tissue is quite possible. This qualitative picture also explains why coherent and noncoherent light with the same parameters produce the same biological effects on cell monolayer (Karu et al. 1982a,b), thin layers of cell suspension (Tiphlova and Karu 1991a; Bertoloni et al. 1993; Karu et al. 1996d), and tissue surface (e.g., by healing of peptic ulcers) (Sazonov et al. 1985). Some additional (therapeutic) effects from the coherent and polarized radiation can appear only in deeper layers of the bulk tissue. To date, no experimental work has been performed to qualitatively and quantitatively study these possible additional effects. In any case, the main therapeutic effects occur due to light absorption by cellular photoacceptors.

9.3 Enhancement of Cellular Metabolism via Activation of Respiratory Chain: A Universal Photobiological Action Mechanism

9.3.1 Cytochrome *c* Oxidase as the Photoacceptor in the Visible-to-Near-Infrared Spectral Range

Photobiological reactions involve the absorption of a specific wavelength of light by the functioning photoacceptor molecule. The photobiological nature of low-power laser effects (Karu 1987, 1989a,b, 2007) means that some molecule (photoacceptor) must first absorb the light used for the irradiation. After promotion of electronically excited states, primary molecular processes from these states can lead to a measurable biological effect at the cellular level. The problem is knowing which molecule is the photoacceptor. When considering the cellular effects, this question can be answered by action spectra.

A graph representing photoresponse as a function of wavelength λ, wave number λ^{-1}, frequency v, or photon energy e is called an action spectrum. The action spectrum of a biological response resembles the absorption spectrum of the photoacceptor molecule. The existence of a structured action spectrum is strong evidence that the phenomenon under study is a photobiological one (i.e., primary photoacceptors and cellular signaling pathways exist, Hartman 1983; Lipson 1985).

The first action spectra in the visible light region were recorded in the early 1980s for DNA and RNA synthesis rate (Karu et al. 1981, 1982b, 1983a, 1984b,c), growth stimulation of *Escherichia coli* (Karu et al 1983b, 1984a, 1987; Tiphlova and Karu 1991a), and protein synthesis by yeasts (Karu et al. 1984a; Karu 1989a,b) for the purpose of investigating the photobiological mechanisms of laser biostimulation. In addition, other action spectra were recorded in various ranges of visible wavelengths: photostimulation of formation of E-rosettes by human lymphocytes, mitosis in L cells, exertion of DNA factor from lymphocytes in the violet–green range (Gamaleya et al. 1983), and oxidative phosphorylation by mitochondria in the violet–blue range (Vekshin 1991). All these spectra were recorded for narrow ranges of the optical spectrum and with a limited number of wavelengths, which prevented identification of the photoacceptor molecule.

Full action spectra from 313 to 860 nm for DNA and RNA synthesis rate in both exponentially growing and plateau-phase HeLa cells were also recorded in the early 1980s (Karu et al. 1984b,c) and 1996 (Karu et al. 1996d) (for a review, see Karu 1987, 1989a,b, 1998, 2007; Karu and Kolyakov 2005). The question of the nature of the photoacceptor molecule remained then open. It was suggested in 1988 (Karu 1988) (see also Karu 1987) that the mechanism of low-power laser therapy at the cellular level was based on the absorption of monochromatic visible and near infrared (NIR) radiation by components of the cellular respiratory chain. Absorption and promotion of electronically excited states cause changes in redox properties of these molecules and acceleration of electron transfer (primary reactions). Primary reactions in mitochondria of eukaryotic cells were supposed to be followed by a cascade of secondary reactions (photosignal transduction and amplification chain or cellular signaling) occurring in cell cytoplasm, membrane, and nucleus (Karu 1988) (for a review, see Karu 1999, 2007, 2008).

It is remarkable that the five action spectra that were analyzed in the work of Karu and Kolyakov (2005) had very close (within the confidence limits) peak positions in spite of the fact that these processes occurred in different parts of the cells (nucleus and plasma membrane). However, there were differences in peak intensities. Five of these action spectra only for the red-to-NIR range (wavelengths that are important in low-power laser therapy) are presented in Figure 9.4.

Two conclusions can be drawn from the action spectra. First, the fact that the peak positions were found to be practically the same suggests that the primary photoacceptor is the same. Second, the existence of the action spectra implied the existence of cellular signaling pathways inside the cell between photoacceptor and the nucleus as well as between the photoacceptor and cell membrane.

The bands of the action spectra were identified by analogy with the absorption spectra of the metal–ligand system characteristic of this spectral range (Karu and Afanasyeva 1991) (for a review, see Karu 1998, 1999, 2007). It was concluded that the ranges 400–450 nm and 620–680 nm were characterized by the bands pertaining to a complex associated with charge transfer in a metal–ligand system, and within 760–830 nm, these were d-d transitions in metals, most probably in Cu (II). The range 400–420 nm was found to be typical of a π-π^* transition in a porphyrin ring. A comparative analysis of lines of possible d-d transitions and charge-transfer complexes of Cu with our action spectra suggested that the photoacceptor was the terminal enzyme of the mitochondrial respiratory chain cytochrome *c* oxidase. It was suggested that the main contribution to the 825 nm band was made by the oxidized Cu_A, to the 760 nm band by the reduced Cu_B, to the 680 nm band by the oxidized Cu_B, and to the 620 nm band by the reduced Cu_A. The 400–450 nm band was more likely the envelope of a few absorption bands in the 350–500 nm range (i.e., a superposition of several bands). Analysis of the band shapes in the action spectra and the line-intensity ratios also led to the conclusion that cytochrome *c* oxidase cannot be considered a primary photoacceptor when fully oxidized or fully reduced but only when it is in one of the intermediate forms (partially reduced or mixed-valence enzyme) (Karu and Afanasyeva 1995; for a review, see Karu 1998, 2007) that have not yet been identified. Figure 9.5 illustrates these conclusions for red-to-near IR spectral region.

Taken together, the terminal respiratory-chain oxidases in eukaryotic cells (cytochrome *c* oxidase) and in prokaryotic cells of *E. coli* (cytochrome *bd* complex, Tiphlova and Karu 1991a) were believed to be photoacceptor molecules for red-to-NIR radiation. In the violet-to-blue spectral range, flavoproteins (e.g., NADH-dehydrogenase) (Karu et al. 1984b; Karu 1998) in the beginning of the respiratory chain are also among the possible photoacceptors like the terminal oxidases.

One important step in identifying the photoacceptor molecule is to compare the absorption and action spectra. For recording the absorption of a cell monolayer and investigating the changes in absorption under irradiation at various wavelengths of monochromatic light, a sensitive multichannel registration method was developed (Karu et al. 1998, 2001a, 2005a). Recall that the absorption spectra of individual living cells were recorded at wavelengths of up to 650 nm years ago with the aim to identify the respiratory-chain enzymes. The absorption spectrum of whole cells in the visible region was found to be qualitatively similar to that of isolated mitochondria (Chance 1957). The extension of optical measurements from the visible spectral range to the far-red and NIR regions (650–1000 nm) was undertaken late in the 1970s for the purpose of monitoring the redox behavior of cytochrome *c* oxidase in vivo.

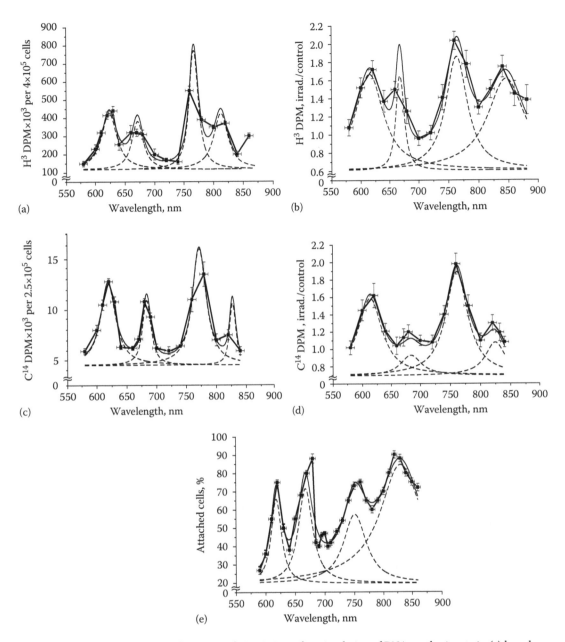

FIGURE 9.4 Action spectra in the region of 580–860 nm for stimulation of DNA synthesis rate in (a) log-phase and (b) plateau-phase HeLa cultures, stimulation of RNA synthesis rate in (c) log-phase and (d) plateau-phase cultures, (e) increase of cell attachment to glass matrix. Experimental curves (-■-■-), curve fittings (—), and Lorentzian fittings (– – –) are shown. (Adapted from Karu, T.I. and Kolyakov, S.F., *Laser Surg.*, 23, 355, 2005.) Dose 100 J/m² (a–d) or 52 J/m² (e).

These studies led to the discovery of a *NIR window* into the body and the development of NIR spectroscopy for monitoring tissue oxygenation (Jöbsis-van der Vliet 1999; Jöbsis-van der Vliet and Jöbsis 1999).

Figure 9.6 presents the intact absorption spectra of HeLa cell monolayer (a–c) and the same spectra after the irradiation at 830 nm (a₁–c₁) as well as two action spectra for the comparison (d and e). Table 9.2 shows the peak positions in all these absorption and action spectra as resolved by Lorentzian fitting. All experimental details can be found in Karu et al. (2005a).

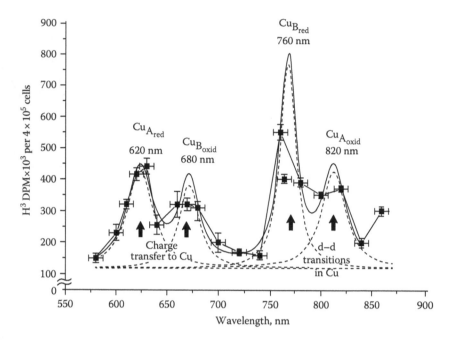

FIGURE 9.5 Action spectrum for stimulation of DNA synthesis rate on cellular level. Suggested absorbing chromophores of the photoacceptor, cytochrome *c* oxidase, are marked. Original curve (-■-), curve fitting (—), and Lorentzian fitting (– – –) are shown. (After Karu, T.I., *Photomed. Laser Surg.*, 28, 159, 2010a; Karu, T.I., *IUBMB Life*, 62, 607, 2010b.)

For quantitative characterization purposes, as well as for comparison between the recorded absorption spectra, we decided to use intensity ratios between certain absorption bands. We used the band present in all absorption spectra near 760 nm (exactly at 754, 756, 767, 765, and 762 nm) (Table 9.2) as a characteristic band for the relatively reduced photoacceptor. The band used by us to characterize the relatively oxidized photoacceptor was the one near 665 nm (exactly at 666, 661, and 665 nm) in spectra B, B_1 and C, C_1 (Table 9.2). This band is so weak it could not be resolved by the Lorentzian fitting method in spectra A and A_1 belonging to the most strongly reduced photoacceptor in our experiments. For this reason, we used in our intensity calculations for spectra A and A_1 absorption on the curve fitting level at 665 nm. The gray vertical lines in Figure 9.6 mark the bands chosen. The intensity ratio I_{760}/I_{665} was calculated to characterize every spectrum. In these simple calculations, we used only the peak intensities (peak heights) and not the integral intensities (peak areas) that are certainly needed for further developments. In the case of equal concentrations of the reduced and oxidized forms of the photoacceptor molecule, the ratio I_{760}/I_{665} should be equal to unity. When the reduced forms prevail, the ratio I_{760}/I_{665} is greater than unity, and it is less than unity in cases where the oxidized forms dominate. Recall that the internal electron transfer within the cytochrome *c* oxidase molecule causes the reduction of the molecular oxygen via several transient intermediates of various redox states (Chance and Hess 1957; Jöbsis-van der Vliet 1999; Jöbsis-van der Vliet and Jöbsis 1999).

The magnitude of the I_{760}/I_{665} criterion is 9.5 for spectrum A, 1.0 for spectrum B, and 0.36 for spectrum C. By this criterion, irradiation of the cells, whose spectrum is marked by A ($I_{760}/I_{665} = 9.5$), causes the reduction of the absorbing molecule (I_{760}/I_{665} for spectrum A_1 is equal to 16). Irradiation of the cells characterized by spectrum B also causes the reduction of the photoacceptor, as evidenced by the increase of the I_{760}/I_{665} ratio from 1.0 to 2.5 in spectrum B_1. In the spectrum of the cells with initially more reduced photoacceptor (spectrum A), irradiation causes reduction to a lesser extent (16/9.5 = 1.7) than in that of the cells with initially less reduced photoacceptor (spectrum B). The intensity ratio in this case is 2.5/1 = 2.5.

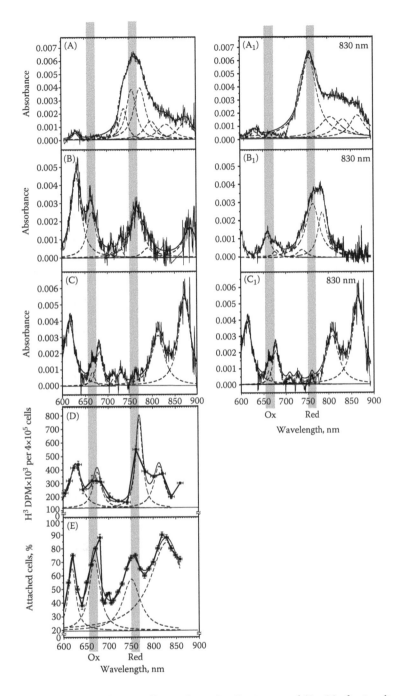

FIGURE 9.6 Absorption spectra of HeLa cell monolayer: (A–C) prior to and (A$_1$–C$_1$) after irradiation at 830 nm. (A, A$_1$) Enclosed cuvette, (B, B$_1$) open cuvette, (C, C$_1$) air-dry monolayer. Original spectrum, curve fitting (—), and Lorentzian fitting (- - -) are shown. (Adapted from Karu, T.I. et al., *J. Photochem. Photobiol. B: Biol.*, 81, 98, 2005a.) Action spectra for (D) stimulation of DNA synthesis and (E) stimulation of cell adhesion to a glass matrix, measured, respectively, 1.5 h after irradiation of HeLa cell monolayer (D = 100 J/m², t = 10 s, I = 10 W/m²) and 30 min after irradiation of HeLa cell suspension (D = 52 J/m², t = 40 s, I = 1.3 W/m²). Experimental curves (-■-■-), curve fitting (—), and Lorentzian fitting (- - -) are shown as described in Karu and Kolyakov (2005). The gray lines mark the bands characteristic for relatively reduced photoacceptor near 770 nm and those for relatively oxidized photoacceptor near 675 nm (for details, see Karu et al. 2005a).

TABLE 9.2   Peak Positions in Absorption and Action Spectra of HeLa Cells in Red-to-NIR Region as Resolved by Lorentzian Fitting

Absorption Spectra						Action Spectra		Characterization
A	A1	B	B1	C	C1	D — DNA Synthesis	E — Adhesion	
$R^2 = 0.99$	$R^2 = 0.98$	$R^2 = 0.95$	$R^2 = 0.98$	$R^2 = 0.95$	$R^2 = 0.95$	$R^2 = 0.97$	$R^2 = 0.91$	
—	—	—	—	616	616	—	618	Oxidized photoacceptor
(630)	(634)	633	—	—	—	624	—	Reduced photoacceptor
—	—	666	661	665	665	672	668	Oxidized photoacceptor (Gray line in Figure 9.2)
—	—	—	681	681	681	—	—	
—	—	(711)	—	(712)	(712)	—	—	Reduced photoacceptor
736	—	(730)	739	(730)	(730)	—	—	
754	756	767	765	(762)	(762)	767	751	Gray line in Figure 9.2
773	—	—	788	—	—	—	—	
797	—	(791)	—	—	—	—	—	
—	807	—	—	813	813	813	—	Oxidized photoacceptor
830	834	—	—	—	—	—	831	
874	867	880	—	872	872	Not measured	Not measured	

Source: Karu, T.I. et al., *J. Photochem. Photobiol. B: Biol.*, 81, 98, 2005a.

Note: A, B, C—absorption spectra before, A1, B1, C1—after irradiation at 830 nm, R2—mean-square deviation of fitting. Weak bands are marked with brackets.

Figure 9.6 also presents two action spectra, one for the stimulation of the DNA synthesis in our HeLa cells (D) and the other for the stimulation of the attachment of the cells to a glass matrix (E). Recall that under ideal conditions, the action spectrum should mimic the absorption spectrum of the light-absorbing molecule whose photochemical alteration causes the effect (Hartman 1983; Lipson 1995).

The two action spectra presented in Figure 9.6d and e are characterized by four bands whose peak positions are situated close to each other, namely, 624 and 618 nm, 672 and 668 nm, and 767 and 751 nm, respectively (Table 9.2). There is a significant difference in the peak positions of spectra D and E at wavelengths over 800 nm (813 and 831 nm, Table 9.2). However, the peak positions at 813 and 831 nm are resolved by deconvolution in absorption spectra A, A_1 and C, C_1 (Table 9.2). Comparison between the absorption and action spectra presented in Figure 9.6 evidences that all the bands present in action spectra D and E are present in the absorption spectra as well (Table 9.2). There are more peaks resolved by the Lorentzian fitting method in the absorption spectra than in the action spectra. This controversy can be explained by the definition of the action spectrum that mirrors the absorption spectrum of the primary photoacceptor. This is an advantage and a specificity of the action spectrum spectroscopy as compared to other types of spectroscopy. It is well known that the transient species of the cytochrome c oxidase turnover are extremely difficult to confidently identify by optical means in physiological conditions. The primary photoacceptor is believed to be one of the turnover intermediates of cytochrome c oxidase that has not as yet been identified (Karu et al. 1998, 2005a).

The I_{760}/I_{665} intensity ratio is 2.4 for spectrum D and 0.74 for spectrum E, in Figure 9.6, which means that the redox state of the photoacceptor molecule differs between these two spectra, it being more reduced in spectrum D. As far as the I_{760}/I_{665} intensity ratio is concerned, spectrum D is close to absorption spectrum B_1. The two photoresponses whose action spectra are presented in Figure 9.6d and e belong to reactions occurring in different parts of the cell, namely, in the nucleus and in the plasma membrane, respectively. It means that the cellular signaling cascades from the photoacceptor (Karu 1988, 2008) can differ as well. It cannot also be ruled out that it is different intermediates of the cytochrome c oxidase turnover that play the role of the photoacceptor for these two cellular responses. Redox-absorbance changes after irradiation at 632.8 nm were also measured in $E.\ coli$ cells (Dube et al. 1997).

Changes in the absorption of HeLa cells were accompanied by conformational changes in the molecule of cytochrome c oxidase (measured by circular dichroism [CD] spectra [Karu et al. 2001b; Kolyakov et al. 2001]). Distinct maxima in CD spectra (the spectra were recorded from 250 to 780 nm) of control cells were found at 566, 634, 680, 712, and 741 nm. After irradiation at 820 nm, the most remarkable changes in peak positions as well as in CD signals were recorded in the range 750–770 nm—an appearance of a new peak at 767 nm and its shift to 757 nm after the second irradiation. Also, the peaks at 712 and 741 nm disappeared, and a new peak at 601 nm appeared. It was suggested that the changes in degree of oxidation of the chromophores of cytochrome c oxidase caused by the irradiation were accompanied by conformational changes in their vicinity. It was further suggested that these changes occurred in the environment of Cu_B (Karu et al. 2001b). It is known that even small structural changes in the binuclear site of cytochrome c oxidase control both rates of the dioxygen reduction and rates of internal electron- and proton-transfer reactions (Chance and Hess 1959). Our suggestion that cytochrome c oxidase is the photoacceptor responsible for various cellular responses connected with light therapy in the red-to-near IR region was later confirmed by the work of Pastore et al. (2000), Wong-Riley et al. (2001, 2005), and Eells et al. (2003, 2004).

Absorption measurements on cell monolayers after irradiation at 632.8 nm evidenced that the shape of dose dependence curve strongly depends on the initial redox state of irradiated cells (Karu et al. 2008b). The irradiation (three times at $\lambda = 632.8$ nm, dose $= 6.3 \times 10^3$ J/m^2, $\tau_{irrad.} = 10$ s, $\tau_{record.} = 600$ ms) of cells initially characterized by relatively oxidized cytochrome c oxidase caused first a reduction of the photoacceptor and then its oxidation (a bell-shaped curve) (Figure 9.7a). The irradiation by the same scheme of the cells with initially relatively reduced cytochrome c oxidase caused first oxidation and then a slight reduction of the enzyme (a curve opposite to the bell-shaped curve, Figure 9.7b). These experimental results demonstrate that the irradiation at 632.8 nm causes either a (transient)

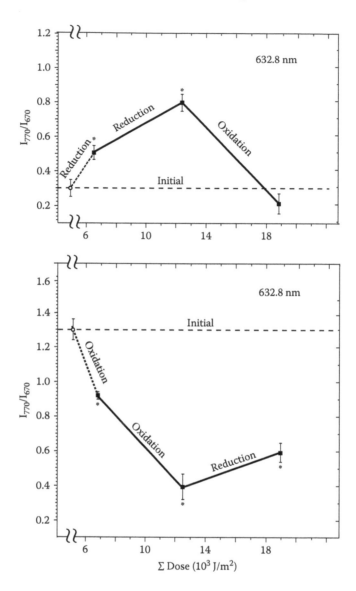

FIGURE 9.7 Dependences of the peak intensity ratio-criterion I_{770}/I_{670} in absorption spectra of HeLa cell monolayer on the dose of radiation ($\lambda = 632.8$ nm). The asterisks indicate statistical significance ($p < 0.05$) from the initial values. The initial redox state of cells differs in two sets of experiments being relatively more oxidized by the same criterion (a) or relatively reduced by the same criterion (b). (Modified from Karu, T.I. et al., *Photomed. Laser Surg.*, 26, 77, 2008a.)

relative reduction of the photoacceptor, putatively cytochrome *c* oxidase, or its (transient) relative oxidation depending on the initial redox status of the photoacceptor. The maximum in the bell-shaped dose dependence curve or the minimum of the reverse curve is the turning point between prevailing of oxidation or reduction processes. Our results evidence that bell-shaped dose dependences usually recorded for various cellular responses (reviews [Karu 1987, 1989a,b, 1998]) are characteristic also for redox changes in the photoacceptor, cytochrome *c* oxidase. Let us emphasize that another type of dose–response curve (Figure 9.7b) can be recorded very rear in studies of cellular responses, for example, in special conditions of cultivation of cells.

The results of various studies (Karu et al. 1998, 2001a, 2005a) support the suggestion made earlier (Karu 1988) that the mechanism of low-power laser therapy at the cellular level is based on the increase of oxidative metabolism in mitochondria, which is caused by electronic excitation of components of the respiratory chain (e.g., cytochrome *c* oxidase). Our results also provide evidence that various wavelengths (670, 632.8, and 820 nm) can be used for increasing respiratory activity. The wavelengths that were used in our experiments (Karu et al. 1998, 2001a,b, 2005a) were chosen in accordance with the maxima in the action spectra (Figure 9.4). Note that 632.8 nm (He–Ne laser) and 820 nm (diode laser or LED) have been until now among the most common wavelengths used in therapeutic light sources (Tuner and Hode 2010).

It must be emphasized that when excitable cells (e.g., neurons, cardiomyocytes) are irradiated with monochromatic visible light, photoacceptors are also believed to be the components of the respiratory chain. Since the publication in 1947 of a study by Arvanitaki and Chalazonitis (1947), it has been known that mitochondria of excitable cells have photosensitivity. Some of the experimental evidence concerning excitable cells is summarized briefly in Figure 9.8. These experiments were not performed in connection with light therapy. Experimental data (Salet 1971, 1972; Berns et al. 1972; Berns and Salet 1972; Salet et al. 1979) made it clear that monochromatic visible radiation could cause (via absorption in mitochondria) physiological and morphological changes in nonpigmented excitable cells, which do not contain specialized photoreceptors. Later, similar irradiation experiments were performed with neurons in connection with low-power laser therapy (Balaban et al. 1992). It was shown experimentally in the 1980s that He–Ne laser radiation altered the firing pattern of nerves. In addition, it was found that transcutaneous irradiation with a He–Ne laser mimicked the effect of peripheral stimulation of a behavioral reflex and that dose-related effects existed (Walker and Akhanjee 1985). And, what is even more important, these findings were found to be connected with pain therapy (Walker 1983, 1987). Later clinical developments of these findings can be found in other publications (Baxter 1994; Simunovic 2000–2002; Tuner and Hode 2010).

(1) Action spectrum = Absorption spectrum of mitochondria
(Arvanitaki, Chalazonitis, *Arch. Sci. Physiol.* 1:385, 1947)

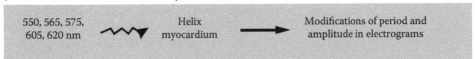

(2) Activation is achieved when the mitochondrial area of a cell is irradiated by microirradiation technique
(Berns et al., *J. Mol. Cell. Cardiol.* 4:71, 1972; 4:427, 1972; Salet, *Exp. Cell Res.* 73:360, 1972; Kitzest et al., *J. Cell physiol.* 93:99, 1977)

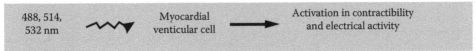

(3) In experiments performed by microirradiation technique, inhibitors of respiratory chain alter the radiation effects
(Salet et al., *Exp. Cell Res.* 120:25, 1979)

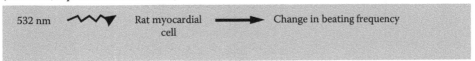

FIGURE 9.8 Experimental data obtained from irradiation of excitable cells indicating photoacceptors are located in the mitochondria.

9.3.2 Primary Reactions after Light Absorption

The primary mechanisms of light action after absorption of light quanta and the promotion of electronically excited states have not been established. The suggestions made to date are summarized in Figure 9.9; for simplicity, only singlet states (S_0 and S_1) are shown. However, triplet states are also involved.

Historically, the first mechanism, proposed in 1981 before recording of the action spectra, was the *singlet oxygen hypothesis* (Karu et al. 1981). It is known that photoabsorbing molecules like porphyrins and flavoproteins (some respiratory-chain components belong to these classes of compounds) can be reversibly converted to photosensitizers (Giese 1980). Based on visible laser-light action on RNA synthesis rates in HeLa cells and spectroscopic data for porphyrins and flavins, the hypothesis was put forward that the absorption of light quanta by these molecules was responsible for the generation of singlet oxygen $^1O_2$ and, therefore, for stimulation of the RNA synthesis rate (Karu et al. 1981) and the DNA synthesis rate (Karu et al. 1982a,b). This possibility has been considered for some time as a predominant suppressive reaction when cells are irradiated at higher doses and intensities (Karu 1989a,b, 1998).

The next mechanism proposed was the *redox properties alteration hypothesis* in 1988 (Karu 1988). Photoexcitation of certain chromophores in the cytochrome c oxidase molecule (like Cu_A and Cu_B or hemes a and a_3) influences the redox state of these centers and, consequently, the rate of electron flow in the molecule (Karu 1988, 1999).

The latest developments indicate that under physiological conditions, the activity of cytochrome c oxidase is also regulated by nitric oxide (NO) (Brown 1999). This regulation occurs via reversible inhibition of mitochondrial respiration. It was hypothesized (Karu et al. 2004a,b) that laser irradiation and activation of electron flow in the molecule of cytochrome c oxidase could reverse the partial inhibition of the catalytic center by NO and in this way increase the O_2 binding and respiration rate (*NO hypothesis*, Figure 9.9). This may be a factor in the increase of the concentration of the oxidized form of Cu_B (as can be integrated by results presented in Figures 9.5 and 9.6). Experimental results on the modification of irradiation effects with donors of NO did not exclude this hypothesis

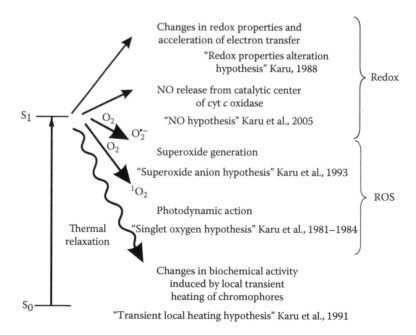

FIGURE 9.9 Possible primary reactions in photoacceptor molecules after promotion of excited electronic states. ROS, reactive oxygen species.

(Karu et al. 2004a, 2005b), as commented in Lane (2006). Note also that under pathological conditions, the concentration of NO is increased (mainly due to the activation of macrophages producing NO) (Hothersall et al. 1997). This circumstance also increases the probability that the respiration activity of various cells will be inhibited by NO. Under these conditions, light activation of cell respiration may have a beneficial effect.

When electronic states are excited with light, a noticeable fraction of the excitation energy is inevitably converted to heat, which causes a local transient increase in the temperature of absorbing chromophores (*transient local heating hypothesis*, Figure 9.9) (Karu et al. 1991a). Any appreciable time- or space-averaged heating of the sample can be prevented by controlling the irradiation intensity and dose appropriately. The local transient rise in temperature of absorbing biomolecules may cause structural (e.g., conformational) changes and trigger biochemical activity (cellular signaling or secondary dark reactions) (Karu et al. 1991a; Karu 1992).

In 1993, it was suggested (Karu et al. 1993a,b) that activation of the respiratory chain by irradiation would also increase production of superoxide anions (*superoxide anion hypothesis*, Figure 9.9). It has been shown that the production of $O_2^{\cdot-}$ depends primarily on the metabolic state of the mitochondria (Forman and Boveris 1982).

The belief that only one of the reactions discussed earlier occurs when a cell is irradiated and excited electronic states are produced is groundless. The question is, which mechanism is decisive? It is entirely possible that all the mechanisms discussed earlier lead to a similar result—a modulation of the redox state of the mitochondria (a shift in the direction of greater oxidation). However, depending on the light dose and intensity used, some of these mechanisms can prevail significantly. Experiments with *E. coli* provided evidence that, at different laser-light doses, different mechanisms were responsible—a photochemical one at low doses and a thermal one at higher doses (Karu et al. 1994).

9.3.3 Cellular Signaling (Secondary Reactions)

If photoacceptors are located in the mitochondria, how then are the primary reactions that occur under irradiation in the respiratory chain connected with DNA and RNA synthesis in the nucleus (the action spectra in Figure 9.4a–d) or with changes in the plasma membrane (Figure. 9.4e)? The principal answer is that between these events are secondary (dark) reactions (cellular signaling cascades or photosignal transduction and amplification chain, Figure 9.10).

Figure 9.10 presents a latest version (2008) of a scheme of cellular signaling cascades (Karu 2008), which was first proposed in Karu (1988) to explain the increase in DNA synthesis rate after the irradiation of HeLa cells with monochromatic visible light and later named as a mitochondrial signaling pathway (Karu et al. 2004a).

Figure 9.10 suggests three regulation pathways. The first one is the control of the photoacceptor over the level of intracellular ATP. It is known that even small changes in ATP level can significantly alter cellular metabolism (Brown 1992; Karu 2010a). However, in many cases, the regulative role of redox homeostasis has proved to be more important than that of ATP. For example, the susceptibility of cells to hypoxic injury depends more on the capacity of cells to maintain the redox homeostasis and less on their capacity to maintain the energy status (Chance 1957; Karu 2010b).

The second and third regulation pathways are mediated through the cellular redox state. This may involve redox-sensitive transcription factors (nuclear factor kappa B [NF-κB] and activator protein [AP]-1 in Figure 9.10) or cellular signaling homeostatic cascades from cytoplasm via cell membrane to nucleus (Figure 9.10). As a whole, the scheme in Figure 9.10 suggests a shift in overall cell redox potential in the direction of greater oxidation. Details can be found in Karu (2008, 2010b).

Experimental results of modification of an irradiation effect (increase of plasma-membrane adhesion when HeLa cells are irradiated at 820 nm) with various chemicals support the suggestions presented in Figure 9.10. Among these chemicals were respiratory-chain inhibitors, donors of NO, oxidants and antioxidants, thiol reactive chemicals, and chemicals that modify the activity of enzymes in the plasma

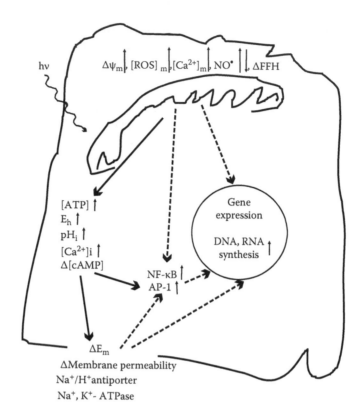

FIGURE 9.10 Schematic explaining putative mitochondrial retrograde signaling pathways after absorption visible and IR-A radiation (marked hv) by the photoacceptor, cytochrome *c* oxidase. Arrows ↑ and ↓ mark increase or decrease of the values; brackets [] mark concentration; ΔFFH, changes in mitochondrial fusion–fission homeostasis; AP-1, activator protein-1; NF-κB, nuclear factor kappa B. Experimentally proved (→) and theoretically suggested (– – –>) pathways are shown. (Adapted from Karu, T.I., *Photochem. Photobiol.*, 84, 1091, 2008.)

membrane (Karu et al. 2001c,d,e,f). Recall that the overall redox state of a cell represents the net balance between stable and unstable reducing and oxidizing equivalents in dynamic equilibrium and is determined by three couples: NAD/NADH, NADP/NADPH, and GSH/GSSG (GSH = glutathione).

It has been established that many cellular signaling pathways are regulated by the intracellular redox state (Sun and Oberley 1996; Nakamura et al. 1997; Gius et al. 1999; Kamata and Hirata 1999). It is believed now that extracellular stimuli elicit cellular responses such as proliferation, differentiation, and even apoptosis through the pathways of cellular signaling. Modulation of the cellular redox state affects gene expression via mechanisms of cellular signaling (via effector molecules like transcription factors and phospholipase A_2) (Sun and Oberley 1996; Nakamura et al. 1997; Gius et al. 1999). There are at least two well-defined transcription factors—NF-κB and AP-1—that have been identified as being regulated by the intracellular redox state (Sun and Oberley 1996; Gius et al. 1999). As a rule, oxidants stimulate cellular signaling systems, and reductants generally suppress the upstream signaling cascades, resulting in suppression of transcription factors (Sun and Oberley 1996). It is believed now that redox-based regulation of gene expression appears to represent a fundamental mechanism in cell biology (Sun and Oberley 1996; Gius et al. 1999). It is important to emphasize that in spite of some similar or even identical steps in cellular signaling, the final cellular responses to irradiation can differ due to the existence of different modes of regulation of transcription factors.

cDNA microarray technique has been used to analyze gene expression profiles in irradiated cells (Zhang et al. 2003; Jaluria et al. 2007; Danel et al. 2010). These experiments provided a more detailed answer how the cellular signaling between mitochondria and the nucleus works. The gene expression profiles of irradiated at 628 nm light evidenced that 111 genes of 10 categories were upregulated (Zhang et al. 2003). The activated genes were grouped into functional categories and were mostly genes that directly or indirectly play roles in the enhancement of cell proliferation and the suppression of apoptosis (Zhang et al. 2003). Gene expression in human skin fibroblasts was found to be altered after irradiation and it depended also at light intensity (Daniel et al. 2010). What is more important is the result that two genes, siat7e and lama4, were found to be involved in the regulation of anchorage-dependent HeLa cell adhesion (Jaluria et al. 2007). Recall that HeLa cell attachment was a model used abundantly in our studies of modification of light effects with chemicals and these results are referred earlier (Karu et al. 1996d; Karu et al. 2001c,d,e,f).

It was suggested in 1980s that activation of cellular metabolism by monochromatic visible light is a redox-regulated phenomenon (Karu 1988, 1989a). Specificity of the light action is as follows: the radiation is absorbed by the components of the respiratory chain, and this is the starting point for redox regulation. The experimental data from following years have supported this suggestion. Irradiated at $\lambda = 647$ nm, human fibroblasts were studied using confocal laser scanning microscopy at the single cell level in real time, changes in the mitochondrial membrane potential ($\Delta\Psi$), intracellular pH (pH_i) and intracellular calcium (Ca_i^{2+}) alterations, and generation of reactive oxygen species (ROS) after the irradiation. After the laser irradiation, a gradual alkalization of pH_i with its later normalization to the basal level occurred during 15 min. The maximal increase in mitochondrial membrane potential, $\Delta\Psi$, was observed 2 min after the 15 s irradiation and reached 30% of its basal value. $\Delta\Psi$ was back on the basal level approximately 4 min after irradiation. Recurrent spikes of Ca_i^{2+} were triggered and ROS were generated as a result of the irradiation (Alexandratou et al. 2002).

Dependencies of various biological responses (i.e., secondary reactions) on the irradiation dose, wavelength, pulsation mode, and intensity are available (for reviews, see Karu 1987, 1989a,b, 1998, 2007). The main features are mentioned in the following text. First, dose–biological response curves are usually bell shaped, characterized by a threshold, a distinct maximum, and a decline phase like the curve in Figure 9.7a. However, at very certain conditions (when the photoacceptors are rather reduced), the dose dependence curve can have the form like in Figure 9.7b. Second, in most cases, the photobiological effects depend only on the radiation dose and not on the radiation intensity and exposure time (the reciprocity rule holds true), but in other cases, the reciprocity rule proves invalid (the irradiation effects depend on light intensity). One example is provided in Figure 9.11.

Third, although the biological responses of various cells may be qualitatively similar, they may have essential quantitative differences. Fourth, the biological effects of irradiation depend on wavelength (action spectra). The biological responses of the same cells to pulsed and CW light of the same wavelength, average intensity, and dose can vary (see Karu [1998] for a detailed review).

Figure 9.12 explains magnitudes of low-power laser effects as being dependent on the initial redox status of a cell. The main idea expressed in Figure 9.12 is that cellular response is weak or absent (the dashed arrows on the right side) when the overall redox potential of a cell is optimal or near optimal for the particular growth conditions. The cellular response is stronger when the redox potential of the target cell is initially shifted (the arrows on left side) to a more reduced state (and pH_i is lowered). This explains why the degrees of cellular responses can differ markedly in different experiments and why they are sometimes nonexistent. A jump in pH_i due to irradiation has been measured experimentally, 0.20 units in mammalian cells (Chopp et al. 1990) and 0.32 units in *E. coli* (Quickenden et al. 1995).

Various magnitudes of low-power laser effects (strong effect, weak effect, or no effect at all) have always been one of the most criticized aspects of low-power laser therapy. An attempt was made to quantify the magnitude of irradiation effects as dependent on the metabolic status of *E. coli* cells (Tiphlova and Karu 1991b) (for a review, see Tiphlova and Karu 1991a). The correlation was found between the amount of ATP in irradiated cells and the initial amount of ATP in control cells (Karu et al. 2001f).

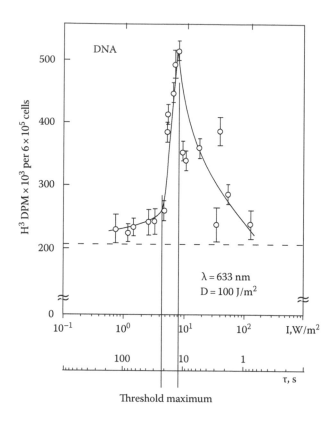

FIGURE 9.11 Dependence of stimulation of DNA synthesis rate on light intensity or irradiation time at a constant dose measured 1.5 h after irradiation of log-phase cells with a CW dye laser pumped by an argon laser ($\lambda = 633$ nm, $I_{max} = 80$ W/m^2). Dashed line shows the control level. (Modified from Karu, T.I. et al., *Laser Chem.*, 5, 19, 1984a; Karu, T.I. et al., *Nuov. Cim. D*, 3, 309, 1984b.)

Thus, variations in the magnitude of low-power laser effects at the cellular level are explained by the overall redox state (and pH$_i$) at the moment of irradiation. Cells with a lowered pH$_i$ (in which redox state is shifted to the reduced side) respond stronger than cells with a normal or close-to-normal pH$_i$ value.

9.3.4 Partial Derepression of Genome of Human Peripheral Lymphocytes: Biological Limitations of Low-Power Laser Effects

Monochromatic visible light cannot always induce full metabolic activation. One such example is considered in this section. Circulating lymphocytes confronted with an immunological stimulus shift from the resting state (G$_0$ phase of cellular cycle) to one of rapid enlargement, culminating in DNA synthesis and mitosis (blast transformation). The characteristics of biochemical and morphological reactions in lymphocytes under the action of mitogens (agents responsible for blast transformation, e.g., phytohemagglutinin [PHA]) have been studied for years (for a review, see Ashman 1984). Cellular responses to a mitogen can be divided into short-term responses without de novo protein synthesis and occurring during the first seconds, minutes, and hours after contact with the mitogen starts and long-term ones connected with protein synthesis hours and days after the beginning of stimulation.

Parallel experiments with PHA treatment and He–Ne laser irradiation were carried out, and the results for these two experimental groups were compared with each other and with those of intact control (Fedoseyeva et al. 1988; Karu et al. 1991b; Smolyaninova et al. 1991; Manteifel et al. 1994, 1997; Shliakhova et al. 1996; Manteifel and Karu 2009). The 10 s irradiation with a He–Ne laser (D = 56 J/m^2)

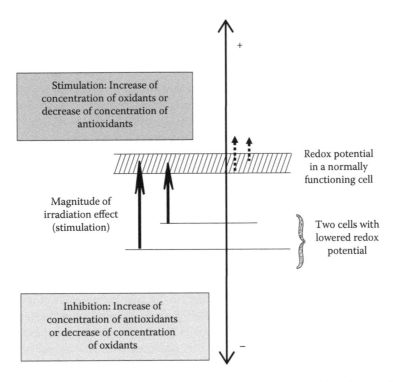

FIGURE 9.12 Schematic illustration of the action principle of monochromatic visible and NIR radiation on a cell. Irradiation shifts the cellular redox potential in a more oxidized direction. The magnitude of cellular response is determined by the cellular redox potential at the moment of irradiation.

induced short-term changes in lymphocytes that were qualitatively similar and quantitatively close to those caused by PHA (which is present in the incubation medium during all experiments). Among the short-term responses compared in this set of experiments were Ca^{2+} influx, RNA synthesis, accessibility of chromatin to acridine orange (a test characterizing the chromatin template activity and its transcription function), and steady-state level of *c-myc* mRNA (Fedoseyeva et al. 1988; Karu et al. 1991b; Smolyaninova et al. 1991; Shliakhova et al. 1996). Also, ultrastructural changes of the nucleus (Manteifel and Karu 2009) and chromatin (Manteifel et al. 1994) were found to be similar in two experimental groups during the first hours after stimulation. These changes were interpreted as an activation of rRNA metabolism, including its synthesis, processing, and transport (Manteifel and Karu 1998).

Two characteristic features of laser-light action were established. First, transcription function was activated in T lymphocytes but not in B lymphocytes. At the same time, PHA was stimulative for both types of lymphocytes (Manteifel and Karu 1998).

Second, despite the similarities in the early responses of lymphocytes to PHA and He–Ne laser radiation, the irradiated lymphocytes did not enter the S phase of the cell cycle (Karu et al. 1991b; Smolyaninova et al. 1991). This means that full mitogenic activation did not occur in the irradiated lymphocytes. It is quite possible that the period of irradiation (10 s in our experiments) was too short to cause the entire cascade of reactions needed for blast transformation. But this time was long enough to cause a boosting effect of blast transformation in PHA-treated lymphocytes (Karu et al. 1991b; Smolyaninova et al. 1991). The number of blast-transformed cells in the sample, which was irradiated before the beginning of PHA treatment, was 120%–170% higher depending on PHA concentration (Karu et al. 1991b; Smolyaninova et al. 1991). It was suggested that the cause of this effect was a partial activation of lymphocytes under irradiation. This may also be a conditioning (priming) effect of certain subpopulations of lymphocytes (Karu et al. 1991). It is possible that this is a redox priming effect. Two lines of evidence allow for this suggestion.

First, it is believed that lymphocyte activation under laser radiation starts with mitochondria as it generally occurs, as described in Section 9.3.1. This suggestion is supported by experiments showing that ATP extrasynthesis and an increase of energy charge occur in irradiated lymphocytes (Herbert et al. 1989). Formation of giant mitochondrial structures in the irradiated cells indicated that a higher level of respiration and energy turnover occurred in these cells (Manteifel et al. 1997). But recording of action spectra is needed for further studies of photoacceptors in lymphocytes.

Second, the early transcriptional activation of lymphocytes both by PHA and He–Ne laser radiation (Figure 9.13a) can be eliminated by a reducing agent, cysteine (Figure 9.13b) (Fedoseyeva et al. 1988). As seen in Figure 9.13b, the effect depends on the concentration of cysteine. Cysteine also cancels blast transformation of lymphocytes (Novogrodski 1975). The activation events in T lymphocytes and monocytes, which are mediated through translocation of the transcription factor NF-κB, depend on the redox state of these cells (Israel et al. 1992). A basal redox equilibrium tending toward oxidation was a prerequisite for the activation of T lymphocytes and U937 monocytes; both constitutive activation and that induced by mitogens were inhibited or even canceled by treatment of cells with reducing agents or antioxidants (Novogrodski 1975; Israel et al. 1992).

Partial mitogenic activation of lymphocytes under He–Ne laser radiation is not the only example of limited activation. For example, silent neurons of *Helix pomatia* did not respond to He–Ne laser irradiation, while the spontaneously active neurons responded strongly under the same experimental conditions (Balaban et al. 1992). Only 25%–27% of 3T3 fibroblasts responded to NIR radiation by extending their pseudopodia toward the monochromatic-light source (Albrecht-Büchler 1991).

The results of experiments with *E. coli* batches showed that they contained a subpopulation that, in response to the irradiation, rapidly began a new cycle of replication and division (Tiphlova and Karu 1991b; Karu 1998). The number of cells in this subpopulation depended on the cultivation conditions,

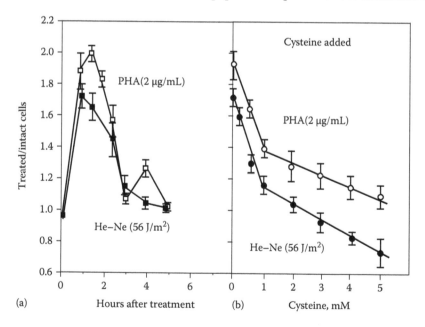

FIGURE 9.13 Transcription activation (measured by binding of acridine orange to chromatin) of human peripheral lymphocytes: (a) after irradiation with He–Ne laser (10 s, 56 J/m²) or treatment with PHA (2 μg/mL); (b) decrease of transcription activation 1 h after irradiation or PHA treatment depending on concentration of cysteine added immediately after the irradiation or adding PHA. (Modified from Fedoseyeva, G.E. et al., *Lasers Life Sci.*, 2, 197, 1988.)

being smaller in faster-growing populations (e.g., the glucose-grown culture) and larger in slower-growing populations (e.g., the arabinose-grown culture). Presumably, in cells of this light-sensitive sub-population, the particular metabolic state necessary for the division could be established. Irradiation is what enables the cells to achieve this active state rapidly. Also, this set of experiments (Tiphlova and Karu 1991b; Karu 1998) clearly proved that there is a limit to the specific growth rate of all populations (0.80 h^{-1}) that does not depend on growth conditions and, in addition, that populations that are already growing at this rate cannot be stimulated. This was also found to be the reason why *E. coli* growth was maximally stimulated not in summer but rather in winter. In autumn and winter, the intact culture featured relatively slow growth. In spring and summer, when the growth of that culture accelerated and the growth rate of the control culture was almost comparable to that of the culture exposed to the optimum dose of red light in the autumn–winter period, irradiation had but little effect (Karu 1989a).

One conclusion from experiments with cultured cells was that only proliferation of slowly growing subpopulations could be stimulated by irradiation. Also, the experiments with HeLa cells demonstrated that one of the effects of He–Ne laser irradiation of these cells was a decrease of the duration of the G period but not other periods of the cell cycle (Karu et al. 1987) (for a review, see Karu 1990). Taken together, there exist certain biological as well as other limitations connected with the physiological status of an irradiated object.

9.4 Enhancement of Cellular Metabolism via Activation of Nonmitochondrial Photoacceptors: Indirect Activation/Suppression

The redox-regulation mechanism cannot occur solely via the respiratory chain (see Section 9.3). Redox chains containing molecules that absorb light in the visible spectral region are usually key structures that can regulate metabolic pathways. One such example is NADPH oxidase of phagocytic cells, which is responsible for nonmitochondrial respiratory burst. This multicomponent enzyme system is a redox chain that generates ROS as a response to the microbicidal or other types of activation. Irradiation with the He–Ne (Karu et al. 1989, 1996a–c, 1997a,b) and semiconductor lasers and LEDs (Karu et al. 1993a,b, 1995, 1997b) can activate this chain. The features of radiation-induced nonmitochondrial respiratory burst, which was in our experiments quantitatively and qualitatively characterized by measurements of luminol-amplified chemiluminescence (CL) (Karu et al. 1989, 1993a,b, 1995, 1996a,b,c,d,e, 1997a,b), must be followed. First, nonmitochondrial respiratory burst can be induced both in homogeneous cell populations and cellular systems (blood, spleen cells, and bone marrow) by both CW and pulsed lasers and LEDs. Figure 9.14 presents some examples. Qualitatively, the kinetics of CL enhancement after irradiation is similar to that after treatment of cells with an object of phagocytosis, *Candida albicans*. Quantitatively, the intensity induced by radiation CL is approximately one order of magnitude lower (Figure 9.14b). This is true for He–Ne laser radiation (Karu et al. 1989) as well as for radiation of various pulsed LEDs (Karu et al. 1993a,b). Second, irradiation effects (stimulation or inhibition of CL) on phagocytic cells strongly depend on the health status of the host organism (Karu et al. 1993a, 1995, 1996a–c, 1997a). This circumstance can be used for diagnostic purposes. Third, there are complex dependencies on irradiation parameters; irradiation can suppress or activate the nonmitochondrial respiratory burst (Karu et al. 1993b, 1995, 1996c, 1997b). These problems have been reviewed in detail elsewhere (Karu 1998).

Finally, ROS, the burst of which is induced by direct irradiation of phagocytes, can activate or deactivate other cells that were not directly irradiated. In this way, indirect activation (or suppression) of metabolic pathways in nonirradiated cells occurs. Cooperative action among various cells via secondary messengers (ROS, lymphokines, cytokines) (Funk et al. 1992) and NO (Naim et al. 1996) requires much more attention when the mechanisms of low-power laser therapy are considered at the organism level.

This chapter did not consider systemic effects of low-power laser therapy at the organism level. The mechanisms of these effects have not yet been established. Perhaps NO plays a role as a secondary

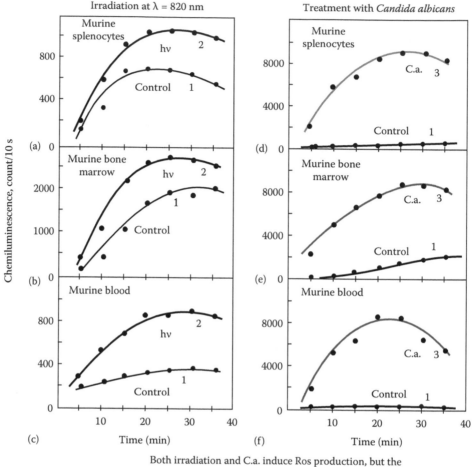

FIGURE 9.14 Kinetic curves of CL of murine splenocytes (a, d), bone marrow (b, e), and blood (c, f) after irradiation (a–c) or treatment (d–f) with *C. albicans*. Curve 1 denotes everywhere the spontaneous CL of control cells; curve 2 is CL after irradiation of samples in dose 800 J/m² (λ = ..., I = 51 W/m², f = 292 Hz); and curve 3 marks the CL induced by the treatment with *C. albicans* (5×10^7 particles/mL). The measurement error is < ±5%. (Adapted from Karu, T.I. et al., *Lasers Surg. Med.*, 21, 485, 1997b.)

messenger for systemic effects of laser irradiation. A possible mechanism connected with the NO–cytochrome *c* oxidase complex was considered earlier (Section 9.3.2). In addition, mechanisms of analgesic effects of laser radiation (Mrowiec et al. 1997) and systemic therapeutic effects that occur via blood irradiation (Vladimirov et al. 2000) could be connected with NO.

Recent studies have demonstrated that a number of nonphagocytic cell types, including fibroblasts, osteoblasts, endothelial cells, chondrocytes, kidney mesangial cells, and others, generate ROS (mainly superoxide anion) in low concentrations in response to stimuli (Sbarra and Strauss 1988). The function of this ROS production is not yet known. It is believed that an NADPH oxidase (probably different from that in phagocytic cells) is present in nonphagocytic cells as well (Sbarra and Strauss 1988). To date, the effects of irradiation on this enzyme have not yet been studied.

Another example of important redox chains is NO synthases, a group of redox-active P450-like flavocytochromes that are responsible for NO generation under physiological conditions (Sharp and Shapman 1999). So far, the irradiation effects on these systems have not been investigated.

9.5 Conclusion

This chapter considered three principal ways of activating individual cells by monochromatic (laser) light. The photobiological action mechanism via activation of the respiratory chain is a universal mechanism. Primary photoacceptors are terminal oxidases (cytochrome *c* oxidase in eukaryotic cells and, e.g., cytochrome *bd* complex in the prokaryotic cell of *E. coli*) as well as NADH dehydrogenase (for the blue-to-red spectral range). Primary reactions in or with a photoacceptor molecule lead to photobiological responses at the cellular level through cascades of biochemical homeostatic reactions (cellular signaling or photosignal transduction and amplification chain). Crucial events of this type of cell-metabolism activation occur due to a shift of cellular redox potential in the direction of greater oxidation. Cell-metabolism activation via the respiratory chain occurs in all cells susceptible to light irradiation. Susceptibility to irradiation and capability for activation depend on the physiological status of irradiated cells; cells whose overall redox potential is shifted to a more reduced state (e.g., certain pathological conditions) are more sensitive to irradiation. The specificity of final photobiological response is determined not at the level of primary reactions in the respiratory chain but at the transcription level during cellular signaling cascades. In some cases, only partial activation of cell metabolism happens (e.g., priming of lymphocytes). All light-induced biological effects depend on the parameters of the irradiation (wavelength, dose, intensity, radiation time, CW or pulsed mode, pulse parameters).

Second principal activation way is the following. Other redox chains in cells can also be activated by irradiation. In phagocytic cells, irradiation initiates a nonmitochondrial respiratory burst (production of ROS, especially superoxide anion) through activation of NADPH oxidase located in the plasma membrane of these cells. The irradiation effects on phagocyting cells depend on the physiological status of the host organism as well as on radiation parameters.

Thirdly, direct activation of cells can lead to the indirect activation of other cells. This occurs via secondary messengers released by directly activated cells: ROS produced by phagocytes, lymphokines and cytokines produced by various subpopulations of lymphocytes, or NO produced by macrophages or as a result of NO–hemoglobin photolysis of red blood cells.

Coherent properties of laser light are not manifested at the molecular level by light interaction with biotissue. The absorption of low-intensity laser light by biological systems is of a purely noncoherent (i.e., photobiological) nature. At the cellular level, biological responses are determined by absorption of light with photoacceptor molecules. Coherent properties of laser light are unimportant when the cellular monolayer, the thin layer of cell suspension, and the thin layer of tissue surface are irradiated. In these cases, the coherent and noncoherent light with the same wavelength, intensity, and dose provide the same biological response. Some additional (therapeutic) effects from coherent and polarized radiation can occur only in deeper layers of bulk tissue.

References

Albrecht-Büchler, G. 1991. Surface extensions of 3T3 cells towards distant infrared light sources, *J. Cell Biol.* 114: 494–501.

Alexandratou, E., Yova, D., Handris, P., Kletsas, D., and Loukas, S. 2002. Human fibroblast alterations induced by low power laser irradiation at the single cell level using confocal microscopy, *Photochem. Photobiol. Sci.* 1: 547–552.

Anders, J.J. 2009. The potential of light therapy for central nervous system injury and disease, *Photomed. Laser Surg.* 27: 379–380.

Arvanitaki, A. and Chalazonitis, N. 1947. Reactiones bioelectriques a la photoactivation des cytochromes, *Arch. Sci. Physiol.* 1: 385–405.

Ashman, R.F. 1984. Lymphocyte activation, in *Fundamental Immunology*, Paul, W.P., ed., Raven Press, New York, pp. 267–302.

Balaban, P., Esenaliev, R., Karu, T., Kutomkina, E., Letokhov, V., Oraevsky, A., and Letokhov, V.S. 1992. He–Ne laser irradiation of single identified neurons, *Lasers Surg. Med.* 12: 329–337.

Baxter, G.D. 1994. *Therapeutic Lasers: Theory and Practice*, Churchill Livingstone, London, U.K.

Berns, M.W., Gross, D.C.L., Cheng, W.K., and Woodring, D. 1972. Argon laser microirradiation of mitochondria in rat myocardial cell in tissue culture. II. Correlation of morphology and function in single irradiated cells, *J. Mol. Cell. Cardiol.* 4: 71–83.

Berns, M.W. and Salet, C. 1972. Laser microbeam for partial cell irradiation, *Int. Rev. Cytol.* 33: 131–155.

Bertoloni, G., Sacchetto, R., Baro, E., Ceccherelli, E., and Jori, G. 1993. Biochemical and morphological changes in *Escherichia coli* irradiated by coherent and non-coherent 632.8 nm light, *J. Photochem. Photobiol. B: Biol.* 18: 191–196.

Brown, G.C. 1992. Control of respiration and ATP synthesis in mammalian mitochondria and cells, *Biochem. J.* 284: 1–213.

Brown, G.C. 1999. Nitric oxide and mitochondrial respiration, *Biochem. Biophys. Acta* 1411: 351–363.

Chance, B. 1957. Cellular oxygen requirements, *Fed. Proc. Fed. Am. Soc. Exp. Biol.* 16: 671–680.

Chance, B. and Hess, B. 1959. Spectroscopic evidence of metabolic control, *Science* 129: 700–708.

Chopp, H., Chen, Q., Dereski, M.O., and Hetzel, F.W. 1990. Chronic metabolic measurement of normal brain tissue response to photodynamic therapy, *Photochem. Photobiol.* 52: 1033–1038.

Daniel, D.H., Weiss, R.A., Geronemus, R.G., Mazur, C., Wilson, S., and Weiss M.A. 2010. Varying rates of wavelengths in dual wavelength LED photomodulation alters gene expression profiles in human skin fibroblasts, *Laser Surg. Med.* 42: 540–546.

Dube, A., Gupta, P.K., and Bharti, S. 1997. Redox absorbance changes of the respiratory chain components of *E. coli* following He–Ne laser irradiation, *Lasers Life Sci.* 7: 173–178.

Eells, J., Wong-Riley, M.T., VerHoeve, J., Henry, M., Buchman, E.V., Kane, M.P., Gould, L.J. et al. 2004. Mitochondrial signal introduction in accelerated wound and retinal healing by near-infrared light therapy, *Mitochondrion* 4: 559–567.

Eells, J.T., Henry, M.M., Summerfelt, P., Wong-Riley, M.T., Buchmann, E.V., Kane, M., Whelan, N.T., and Whelan, H.T. 2003. Therapeutic photobiomodulation for methanol-induced retinal toxicity, *Proc. Natl. Acad. Sci. U.S.A.* 100: 3439–3444.

Fedoseyeva, G.E., Smolyaninova, N.K., Karu, T.I., and Zelenin, A.V. 1988. Human lymphocyte chromatin changes following irradiation with a He–Ne laser, *Lasers Life Sci.* 2: 197–205.

Forman, N.J. and Boveris, A. 1982. Superoxide radical and hydrogen peroxide in mitochondria, in *Free Radicals in Biology*, Vol. 5, Pryor, A., ed., Academic Press, New York, pp. 65–90.

Funk, J.O., Kruse, A., and Kirchner, H. 1992. Cytokine production in cultures of human peripheral blood mononuclear cells, *J. Photochem. Photobiol. B: Biol.* 16: 347–355.

Gamaleya, N.F., Shishko, E.D., and Yanish, G.B. 1983. New data about mammalian cells photosensitivity and laser biostimulation, *Dokl. Akad. Nauk S.S.S.R. (Moscow)* 273: 224–227.

Giese, A.C. 1980. Photosensitization of organisms with special reference to natural photosensitizers, in *Lasers in Biology and Medicine*, Hillenkampf, E., Pratesi, R., and Sacchi, C., eds., Plenum Press, New York, p. 299.

Gius, D., Boreto, A., Shah, S., and Curry, H.A. 1999. Intracellular oxidation reduction status in the regulation of transcription factors NF-κB and AP-1, *Toxicol. Lett.* 106: 93–118.

Hartman, K.M. 1983. Action spectroscopy, in *Biophysics*, Hoppe, W., Lohmann, W., Marke, H., and Ziegler, H., eds., Springer-Verlag, Heidelberg, Germany, pp. 115–134.

Herbert, K.E., Bhusate, L.L., Scott, D.L., Diamantopolos, C., and Perrett, D. 1989. Effect of laser light at 820 nm on adenosine nucleotide levels in human lymphocytes, *Lasers Life Sci.* 3: 37–46.

Hothersall, J.S., Cunha, F.Q., Neild, G.H., and Norohna-Dutra, A. 1997. Induction of nitric oxide synthesis in J774 cell lowers intracellular glutathione: Effect of oxide modulated glutathione redox status on nitric oxide synthase induction, *Biochem. J.* 322: 477–486.

Israel, N., Gougerot-Pocidalo, M.-A., Aillet, F., and Verelizier, J.-L. 1992. Redox status of cells influences constitutive or induced NF-κB translocation and HIV long terminal repeat activity in human T-lymphocytes and monocytic cell lines, *J. Immunol.* 149: 3386–3394.

Jaluria, P., Betenbaugh, M., Kontatopoulos, k., Frank, B., and Shiloah, J. 2007. Application of microarrays to identify and characterize genes involved in attachment dependence in He-La cells, *Metab. Eng.* 9: 241–248.

Jöbsis-vander Vliet, F.F. 1999. Discovery of the near-infrared window in the body and the early development of near-infrared spectroscopy, *J. Biomed. Opt.* 4: 392–396.

Jöbsis-vander Vliet, F.F. and Jöbsis, P.D. 1999. Biochemical and physiological basis of medical near-infrared spectroscopy, *J. Biomed. Opt.* 4: 397–412.

Kamata, H. and Hirata, H. 1999. Redox regulation of cellular signalling, *Cell. Signal.* 11: 1–14.

Karu, T.I. 1987. Photobiological fundamentals of low-power laser therapy, *IEEE J. Quantum Electron.* QE-23: 1703–1717.

Karu, T.I. 1988. Molecular mechanism of the therapeutic effect of low-intensity laser radiation, *Lasers Life Sci.* 2: 53–74.

Karu, T.I. 1989a. *Photobiology of Low-Power Laser Therapy*, Harwood Academic, London, U.K.

Karu, T.I. 1989b. Photobiology of low-power laser effects, *Health Phys.* 56: 691–704.

Karu, T.I. 1990. Effects of visible radiation on cultured cells, *Photochem. Photobiol.* 52: 1089–1098.

Karu, T.I. 1992. Local pulsed heating of absorbing chromophores as a possible primary mechanism of low-power laser effects, in *Laser Applications in Medicine and Surgery*, Galletti, G., Bolognani, L., and Ussia, G., eds., Monduzzi Editore, Bologna, Italy, pp. 253–258.

Karu, T.I. 1998. *The Science of Low Power Laser Therapy*, Gordon & Breach, London, U.K.

Karu, T.I. 1999. Primary and secondary mechanisms of action of visible-to-near IR radiation on cells, *J. Photochem. Photobiol. B: Biol.* 49: 1–17.

Karu, T.I. 2007. *Ten Lectures on Basic Science of Laser Phototherapy*, Prima Books AB, Grängesberg, Sweden.

Karu, T.I. 2008. Mitochondrial signaling in mammalian cells activated by red and near IR radiation, *Photochem. Photobiol.* 84: 1091–1099.

Karu, T.I. 2010a. Mitochondrial mechanisms of photobiomodulation in context of new data about multiple roles of ATP, *Photomed. Laser Surg.* 28: 159–160.

Karu, T.I. 2010b. Multiple roles of cytochrome c oxidase in mammalian cells under action of red and IR-A radiation, *IUBMB Life* 62: 607–610.

Karu, T.I. and Afanasyeva, N.I. 1995. Cytochrome oxidase as primary photoacceptor for cultured cells in visible and near IR regions, *Dokl. Akad. Nauk (Moscow)* 342: 693–695.

Karu, T.I., Afanasyeva, N.I., Kolyakov, S.F., and Pyatibrat, L.V. 1998. Changes in absorption spectra of monolayer of living cells after irradiation with low intensity laser light, *Dokl. Akad. Nauk (Moscow)* 360:267–270.

Karu, T.I., Afanasyeva, N.I., Kolyakov, S.F., Pyatibrat, L.V., and Welser, L. 2001a. Changes in absorbance of monolayer of living cells induced by laser radiation at 633, 670, and 820 nm, *IEEE J. Sel. Top. Quantum Electron.* 7: 982–988.

Karu, T.I., Andreichuk, T., and Ryabykh, T. 1993a. Suppression of human blood chemiluminescence by diode laser radiation at wavelengths 660, 820, 880 or 950 nm, *Laser Ther.* 5: 103–109.

Karu, T.I., Andreichuk, T., and Ryabykh, T. 1993b. Changes in oxidative metabolism of murine spleen following diode laser (660–950nm) irradiation: Effect of cellular composition and radiation parameters, *Lasers Surg. Med.* 13: 453–462.

Karu, T.I., Andreichuk, T.N., and Ryabykh, T.R. 1995. On the action of semiconductor laser radiation (λ = 820 nm) on the chemiluminescence of blood of clinically healthy humans, *Lasers Life Sci.* 6: 277–282.

Karu, T.I., Kalendo, G.S., and Letokhov, V.S. 1981. Control of RNA synthesis rate in tumor cells HeLa by action of low intensity visible light of copper laser. *Lett. Nuov. Cim.* 32. 55–59.

Karu, T.I., Kalendo, G.S., Letokhov, V.S., and Lobko, V.V. 1982a. Biostimulation of HeLa cells by low intensity visible light. *Nuov. Cim. D* 1: 828–840.

Karu, T.I., Kalendo, G.S., Letokhov, V.S., and Lobko, V.V. 1982b. Biological action of low-intensity visible light on HeLa cells as a function of the coherence, dose, wavelength, and irradiation dose. *Sov. J. Quantum Electron.* 12: 1134–1139.

Karu, T.I., Kalendo, G.S., Letokhov, V.S., and Lobko, V.V. 1983a. Biological action of low-intensity visible light on HeLa cells as a function of the coherence, dose, wavelength, and irradiation regime. II., *Sov. J. Quantum Electron.* 13: 1169–1175.

Karu, T.I., Kalendo, G.S., Letokhov, V.S., and Lobko, V.V. 1984b. Biostimulation of HeLa cells by low-intensity visible light. II. Stimulation of DNA and RNA synthesis in a wide spectral range, *Nuov. Cim. D* 3: 309–318.

Karu, T.I., Kalendo, G.S., Letokhov, V.S., and Lobko, V.V. 1984c. Biostimulation of HeLa cells by low intensity visible light. III. Stimulation of nucleic acid synthesis in plateau phase cells, *Nuov. Cim. D* 3: 319–325.

Karu, T.I. and Kolyakov, S.F. 2005. Exact action spectra for cellular responses relevant to phototherapy, *Photomed. Laser Surg.* 23: 355–361.

Karu, T.I., Kolyakov, S.F., Pyatibrat, L.V., Mikhailov, E.L., and Kompanets, O.N. 2001b. Irradiation with a diode at 820 nm induces changes in circular dichroism spectra (250–750 nm) of living cells, *IEEE J. Sel. Top. Quantum Electron.* 7: 976–981.

Karu, T.I., Pyatibrat, L.V., and Afanasyeva, N.I. 2004a. A novel mitochondrial signaling pathway activated by visible-to-near infrared radiation. *Photochem. Photobiol.* 80: 366–372.

Karu, T.I., Pyatibrat, L.V., and Afanasyeva, N.I. 2005b. Cellular effects of low power laser therapy can be mediated by nitric oxide. *Lasers Surg. Med.* 36: 307–314.

Karu, T.I., Pyatibrat, L.V., and Kalendo, G.S. 2001c. Cell attachment modulation by radiation from a pulsed semiconductor light diode (λ = 820 nm) and various chemicals, *Lasers Surg. Med.* 28: 227–236.

Karu, T.I., Pyatibrat, L.V., and Kalendo, G.S. 2001d. Thiol reactive agents and semiconductor light diode radiation (λ = 820 nm) exert effects on cell attachment to extracellular matrix, *Laser Ther.* 11: 177–187.

Karu, T.I., Pyatibrat, L.V., and Kalendo, G.S. 2001e. Cell attachment to extracellular matrices is modulated by pulsed radiation at 820 nm and chemicals that modify the activity of enzymes in the plasma membrane, *Lasers Surg. Med.* 29: 274–281.

Karu, T.I., Pyatibrat, L.V., and Kalendo, G.S. 2001f. Studies into the action specifics of a pulsed GaAlAs laser (λ = 820 nm) on a cell culture. I. Reduction of the intracellular ATP concentration: dependence on initial ATP amount, *Lasers Life Sci.* 9: 203–210.

Karu, T.I., Pyatibrat, L.V., and Kalendo, G.S. 2004b. Photobiological modulation of cell attachment via cytochrome c oxidase. *Photochem. Photobiol Sci.* 3: 211–216.

Karu, T.I., Pyatibrat, L.V., and Kalendo, G.S. 1987. Biostimulation of HeLa cells by low-intensity visible light. V. Stimulation of cell proliferation in vitro by He-Ne laser radiation, *Nuov. Cim. D* 9: 1485–1494.

Karu, T.I., Pyatibrat, L.V., Kalendo, G.S., and Esenaliev, R.O. 1996d. Effects of monochromatic low-intensity light and laser irradiation on adhesion of HeLa cells *in vitro*, *Lasers Surg. Med.* 18: 171–177.

Karu, T.I., Pyatibrat, L.V., Kolyakov, S.F., and Afanasyeva, N.I. 2005a. Absorption measurements of a cell monolayer relevant to phototherapy: Reduction of cytochrome c oxidase under near IR radiation. *J. Photochem. Photobiol. B: Biol.* 81: 98–106.

Karu, T.I., Pyatibrat, L.V., Kolyakov, S., and Afanasyeva, N.I. 2008b. Absorption measurements of cell monolayers relevant to mechanisms of laser phototherapy: Reduction or oxidation of cytochrome c oxidase under laser radiation at 632.8 nm. *Photomed. Laser Surg.* 26: 593–599.

Karu, T.I., Pyatibrat, L.V., Moskvin, S.V., Andreev, S., and Letokhov, V.S. 2008a. Elementary processes in cells after light absorption do not depend on the degree of polarization: Implications for the mechanisms of laser phototherapy. *Photomed. Laser Surg.* 26: 77–82.

Karu, T.I., Pyatibrat, L.V., and Ryabykh, T.R. 1997b. Nonmonotonic behaviour of the dose dependence of the radiation effect on cells in vitro exposed to pulsed laser radiation at $\lambda = 820$ nm, *Lasers Surg. Med.* 21: 485–492.

Karu, T.I., Ryabykh, T.R, and Antonov, S.N. 1996a. Different sensitivity of cells from tumor-bearing organisms to continuous-wave and pulsed laser radiation ($\lambda = 632.8$ nm) evaluated by chemiluminescence test. I. Comparison of responses of murine splenocytes: intact mice and mice with transplanted leukemia EL-4, *Lasers Life Sci.* 7: 91–98.

Karu, T.I., Ryabykh, T.R, and Antonov, S.N. 1996b. Different sensitivity of cells from tumor-bearing organisms to continuous-wave and pulsed laser radiation ($\lambda = 632.8$ nm) evaluated by chemiluminescence test. II. Comparison of responses of human blood: healthy persons and patients with colon cancer, *Lasers Life Sci.* 7: 99–106.

Karu, T.I., Ryabykh, T.R, Fedoseyeva, G.E., and Puchkova, N.I. 1989. Induced by He-Ne laser radiation respiratory burst on phagocytic cells, *Lasers Surg. Med.* 9: 585–588.

Karu, T.I., Ryabykh, T.R, and Letokhov, VS. 1997a. Different sensitivity of cells from tumor-bearing organisms to continuous-wave and pulsed laser radiation ($\lambda = 632.8$ nm) evaluated by chemiluminescence test. III. Effect of dark period between pulses, *Lasers Life Sci.* 7: 141–156.

Karu, T.I., Ryabykh, T.R, Sidorova, T.A., and Dobrynin, Ya.V. 1996c. The use of a chemiluminescence test to evaluate the sensitivity of blast cells in patients with hemoblastoses to antitumor agents and low-intensity laser radiation, *Lasers Life Sci.* 7: 1–11.

Karu, T.I., Smolyaninova, N.K., and Zelenin, A.V. 1991b. Long-term and short-term responses of human lymphocytes to He-Ne laser radiation, *Lasers Life Sci.* 4: 167–178.

Karu, T.I., Tiphlova, O., Esenaliev, R., and Letokhov, V. 1994. Two different mechanisms of low-intensity laser photobiological effects on *Escherichia coli*, *J. Photochem. Photobiol. B: Biol.* 24: 155–161.

Karu, T.I., Tiphlova, O.A., Fedoseyeva, G.E., Kalendo, G.S., Letokhov, V.S., Lobko, V.V., Lyapunova, T.S., Pomoshnikova, N.A., and Meissel, M.N. 1984a. Biostimulating action of low-intensity monochromatic visible light: is it possible? *Laser Chem.* 5: 19–26.

Karu, T.I., Tiphlova, O.A., Letokhov, V.S., and Lobko, V.V. 1983b. Stimulation of *E. coli* growth by laser and incoherent red light, *Nuov. Cim. D* 2: 1138–1144.

Karu, T.I., Tiphlova, O.A., Matveyets, Yu. A., Yartsev, A.P., and Letokhov, V.S. 1991a. Comparison of the effects of visible femtosecond laser pulses and continuous wave laser radiation of low average intensity on the clonogenicity of *Escherichia coli*, *J. Photochem. Photobiol. B: Biol.* 10: 339–345.

Kolyakov, S.F., Pyatibrat, L.V., Mikhailov, E.L., Kompanets, O.N., and Karu, T.I. 2001. Changes in the spectra of circular dichroism of suspension of living cells after low intensity laser radiation at 820 nm, *Dokl. Akad. Nauk (Moscow)* 377: 128–131.

Lane, N. 2006. Power games, *Nature* 443: 901–903.

Lipson, E.D. 1995. Action spectroscopy: Methodology, in *CRC Handbook of Organic Chemistry and Photobiology*, Horspool, W.H. and Song, P.-S., eds., CRC Press, Boca Raton, FL, p. 1257.

Manteifel, V., Bakeeva, L., and Karu, T. 1997. Ultrastructural changes in chondriome of human lymphocytes after irradiation with He–Ne laser: Appearance of giant mitochondria, *J. Photochem. Photobiol. B: Biol.* 38: 25–30.

Manteifel, V.M., Andreichuk, T.N., and Karu, T.I. 1994. Influence of He–Ne laser radiation and phytohemagglutinin on the ultrastructure of chromatin of human lymphocytes, *Lasers Life Sci.* 6: 1–8.

Manteifel, V.M. and Karu, T.I. 1998. Activation of chromatin in T-lymphocytes nuclei under the He–Ne laser radiation, *Lasers Life Sci.* 8: 117–125.

Manteifel, V.M. and Karu, T.I. 2009. Loosening of condensed chromatin in human blood lymphocytes exposed to irradiation with a low-energy He–Ne laser, *Biol. Bull.* 36: 555–561.

Mrowiec, J., Sieron, A., Plech, A., Cieslar, G., Biniszkiewicz, T., and Brus, R. 1997. Analgesic effect of low-power infrared laser radiation in rats, *Proc. SPIE* 3198: 83–87.

Naim, J.O., Yu, W., Ippolito, K.M.L., Gowan, M., and Lanafame, R.J. 1996. The effect of low level laser irradiation on nitric oxide production by mouse macrophages, *Lasers Surg. Med. Suppl.* 8: 7.

Nakamura, H., Nakamura, K., and Yodoi, J. 1997. Redox regulation of cellular activation, *Annu. Rev. Immunol.* 15: 351–368.

Novogrodski, A. 1975. Lymphocyte activation induced by modifications of surface, in *Immune Recognition*, Rosenthal, S., ed., Academic Press, New York, pp. 43–91.

Pal, G., Dutta, A., Mitra, K., Grace, M.S., Romanczyk, T.B., Chakrabarti, K., Anders, J., Gorman, E., Waynant, R.W., and Tata, D.B. 2007. Effect of low intensity laser interaction with human skin fibroblast cells using fiber-optic nano-probes, *J. Photochem. Photobiol. B: Biol.* 86: 252–261.

Pastore, D., Greco, M., and Passarella, S. 2000. Specific helium-neon laser sensitivity of the purified cytochrome c oxidase, *Int. J. Rad. Biol.* 76: 863–870.

Qadri, T., Bohdanecka, P., Tunér, J., Miranda, L., Altamash, M., and Gustafsson, A. 2007. The importance of coherence length in laser phototherapy of gingival inflammation—A pilot study, *Lasers Med. Sci.* 22(4): 245–251. doi: 10.1007/s10103-006-0439-1.

Quickenden, T.R., Daniels, L.L., and Byrne, L.T. 1995. Does low-intensity He–Ne radiation affect the intracellular pH of intact *E. coli*? *Proc. SPIE* 2391: 535–538.

Salet, C. 1971. Acceleration par micro-irradiation laser du rhythme de contraction de cellular cardiaques en culture, *C.R. Acad. Sci. Paris* 272: 2584–2592.

Salet, C. 1972. A study of beating frequency of a single myocardial cell. I. Q-switched laser microirradiation of mitochondria, *Exp. Cell Res.* 73: 360–369.

Salet, C, Moreno, G., and Vinzens, F. 1979. A study of beating frequency of a single myocardial cell. III. Laser microirradiation of mitochondria in the presence of KCN or ATP, *Exp. Cell. Res.* 120: 25–32.

Sazonov, A.M., Romanov, G.A., Portnoy, L.M., Odinokova, V.A., Karu, T.I., Lobko, V.V., and Letokhov, V.S. 1985. Low-intensity noncoherent red light in complex healing of peptic and duodenal ulcers, *Sov. Med.* 12: 42–48 (in Russian).

Sbarra, A.J. and Strauss, R.R., eds. 1988. *The Respiratory Burst and Its Photobiological Significance*, Plenum Press, New York.

Sharp, R.E. and Chapman, S.K. 1999. Mechanisms for regulating electron transfer in multicentre redox proteins, *Biochem. Biophys. Acta* 1432: 143–151.

Shliakhova, L.N., Itkes, A.V., Manteifel, V.M., and Karu, T.I. 1996. Expression of *c-myc* gene in irradiated at 670 nm human lymphocytes: A preliminary report, *Lasers Life Sci.* 7: 107–114.

Simunovic, Z., ed. 2000–2002. *Lasers in Medicine and Dentistry*, Vols. 1–3, Vitgraf, Rijeka, Croatia.

Smolyaninova, N.K., Karu, T.I., Fedoseyeva, G.E., and Zelenin, A.V. 1991. Effect of He–Ne laser irradiation on chromatin properties and nucleic acids synthesis of human blood lymphocytes, *Biomed. Sci.* 2: 121–126.

Sun, Y. and Oberley, L.W. 1996. Redox regulation of transcriptional activators, *Free Radic. Biol. Med.* 21: 335–346.

Tiphlova, O. and Karu, T. 1991a. Action of low-intensity laser radiation on *Escherichia coli*, *CRC Crit. Rev. Biomed. Eng.* 18: 387–412.

Tiphlova, O. and Karu, T. 1991b. Dependence of *Escherichia coli* growth rate on irradiation with He–Ne laser and growth substrates, *Lasers Life Sci.* 4: 161–166.

Tuner, J. and Hode, L. 2010. *New Laser Therapy Handbook*, Prima Books, Grangesberg, Sweden.

Vekshin, N.A. 1991. Light-dependent ATP synthesis in mitochondria, *Mol. Biol. (Moscow)* 25: 54–58.

Vladimirov, Y, Borisenko, G., Boriskina, N., Kazarinov, K., and Osipov, A. 2000. NO-hemoglobin may be a light-sensitive source of nitric oxide both in solution and in red blood cells, *J. Photochem. Photobiol. B: Biol.* 59: 115–121.

Walker, J. 1983. Relief from chronic pain by laser irradiation, *Neurosci. Lett.* 43: 339–341.

Walker, J. and Akhanjee, K. 1985. Laser-induced somatosensory evoked potential: Evidence of photosensitivity of peripheral nerves, *Brain Res.* 344: 281–288.

Wong-Riley, M.T., Bai, X., Buchman, E., and Whelan, H.T. 2001. Light-emitting diode treatment reverses the effect of TTX on cytochrome c oxidase in neurons, *Neuroreport* 12: 3033–3037.

Wong-Riley, M.T., Liang, H.L., Eells, J.T., Chance, B., Henry, M.M., Buchmann, E. et al. 2005. Photobiomodulation directly benefits primary neurons functionally inactivated by toxins: Role of cytochrome c oxidase, *J. Biol. Chem.* 280: 4761–4771.

Wu, X., Dmitriev, A.E., Cardoso, M.J., Viers-Costello, A.G., Borke, R.C., Streeter, J., and Anders, J.J. 2009. 810-nm wavelength light: An effective therapy for transected or contused rat spinal cord, *Lasers Surg. Med.* 41: 36–41.

Zhang, Y., Song, S., Fong, C.-C., Tsang, C.-H., Yang, Z., and Yang, M. 2003. cDNA microarray analysis of gene expression profiles in human fibroblast cells irradiated with red light, *J. Invest. Dermatol.* 120: 849–857.

10

Image-Guided Surgery

Richard D. Bucholz
Saint Louis University

Keith A. Laycock
Saint Louis University

10.1 Introduction

Surgery is based upon a basic inherent conflict. The Hippocratic Oath commands surgeons to abstain from whatever causes harm to the patient; yet the essence of surgery is to invade the body, and thereby inflict injury, to achieve a therapeutic effect. The majority of surgical research has focused upon techniques that alter the balance between improving the outcome for a patient suffering from disease and minimizing the mortality and morbidity attendant to the therapeutic act.

Much of the morbidity associated with surgery is incurred during the development of situational knowledge by the surgeon while performing surgery. For a surgeon to reach a lesion, the opening into the body must afford not only access to the lesion but an appreciation of location. Surgeons use landmarks visualized during dissection to provide course corrections that eventually bring them to the therapeutic target. Once at the target, the opening must be large enough to allow the surgeon to effectively deal with the lesion in an appropriate fashion, often consisting of the removal of pathological tissue. The therapeutic ratio between the efficacy and injury of a procedure is therefore a function of the quality of visualization by the surgeon of anatomy during surgery and the efficacy of the instruments, or effectors, at hand to deal with the pathology once visualized, compared to the damage incurred in developing surgical access to the lesion. Any technology that holds the promise of improving both visualization and the effectiveness of surgery will therefore have a dramatic effect upon the therapeutic ratio and will be eagerly embraced by both surgeons and their patients. A prime example of such a technology is image-guided surgery.

Surgeons have always sought to achieve visualization with the minimum exposure of anatomy. The remarkable developments within medical imaging make it a likely avenue to provide this visualization. Soon after the development of the first imaging modality to be introduced (radiographic x-ray imaging by Roentgen in 1895), surgeons employed the technology within the operating room to reduce the invasiveness of surgery. For the first time, a surgeon was able to visualize skeletal injuries and malformations and plan their repair or correction prior to commencing a surgical procedure. Radiography also enabled localization of radio-opaque masses or foreign objects inside the body cavity, expediting their removal and minimizing iatrogenic trauma arising from exploratory procedures. In addition, the ability to visualize bony anatomy also enabled the surgeon to use bony landmarks as a frame of reference when it was not feasible to expose the target site, such as in neurosurgical procedures in which lesions are made within the brain to restore function.

In such neurosurgical procedures, the highest degree of precision is essential to avoid iatrogenic injury to structures adjacent to the target site, while reaching targets that may be quite small and deeply seated. The need for accuracy in the insertion of instruments inspired neurosurgeons to develop the field of stereotactic surgery (from *stereos*, meaning solid, as in three-dimensional (3D), and either the Greek *taxis*, meaning ordered or arranged, or the Latin *tactus*, meaning touch), in which the surgical instruments are precisely guided to a selected target within the patient by equipment that uses the patient's image data for navigational reference. Zernov first proposed the concept of a mechanical stereotactic guidance device in 1890 (Zernov 1890), and several clinically employed devices were developed following the invention of radiology.

Stereotactic frames are literally bolted to the patient's head by means of screws driven into the skull and thus provide a rigid frame of reference within which the exact 3D coordinates of instruments and target can be defined. The surgeon chooses a target and an entry point, and a small calculator is used to calculate how to set the angles of the moveable arcs attached to the frame. Once these are correctly positioned, an instrument passed through the holder will be restricted to travel only from the entry point to the desired target, with the frame acting to constrain the surgeon to the preset surgical path.

In most of these frame-based coordinate systems, the head ring of the frame defines the plane of origin of the surgical reference system, with the x and y coordinates describing a unique point within that plane. The z-axis is perpendicular to the plane of origin, and any point in the anatomy of the patient can be precisely described using three numbers consisting of the respective x, y, and z coordinates. Examples of frames that are coordinate-based include the Leksell (Leksell 1949), Talairach (Talairach and Bancaud 1973), and Hitchcock (Hitchcock 1988) frames. The Leksell is an arc-centered system, in which navigation during the procedure is confined by a probe carrier held by a series of arcs that are easily adjusted. The target point in such a system is at the center of the arc system. In non-Cartesian frames, such as the popular Brown–Roberts–Wells (BRW) frame (Brown 1979; Apuzzo and Fredericks 1988), targets and trajectories are defined by four angles and a length, and any point within the patient can be reached regardless of its relation to the center of the coordinate system The latter frame was developed after computers became commonly available, allowing Cartesian scanner coordinates to be readily converted into spherical coordinates.

However, plain radiographs of the body, consisting of two-dimensional (2D) images depicting the 3D structure of the body, are inherently limited by the loss of information in reducing three dimensions to two. The use of stereotactic surgery was limited during the first half of the twentieth century as the only image data available for surgical planning purposes were 2D radiographs obtained preoperatively from several perspectives. A further limitation was that such radiographs do not depict the soft tissue within the skull. As radiographs do not show the location of the brain without the introduction of a contrast agent, pneumoencephalography (a technique whereby air is injected into the brain, usually via a lumbar puncture) was commonly performed prior to a stereotactic procedure to show the location of the ventricles. Fortunately, the third ventricle, located precisely in the midline of the brain, has two structures that can be used to define points within space, the anterior (AC) and posterior (PC) commissures, which are easily identified on lateral skull radiographs, allowing the position of the third ventricle to be determined accurately. Stereotactic atlases of normal human cranial anatomy were then developed based upon these points, which related specific brain anatomy to the line joining these structures, the *AC–PC line*. These atlases were adequate for localizing structures close to the ventricles, but their accuracy progressively decreased the further the targeted structure was located from the AC–PC line, given the variability in anatomy of normal brains. Thus, early stereotactic procedures could only be performed with precision on targets located near the center of the brain and in patients with normal anatomy not affected by the presence of space-occupying lesions. The value, and precision, of stereotactic surgery was limited to a few targets close to the midline to restore lost neurological function; it was of no use to the vast majority of neurosurgeons operating on large, superficial lesions. The result was that large openings in the skull were necessary to achieve the degree of visualization necessary to safely remove these cortical, or subcortical, targets.

Stereotactic frames had several other limitations which further restricted their use. Target and trajectory parameters were calculated using preoperative images and did not compensate for intraoperative changes in anatomy. The frames themselves were cumbersome in that they required setting of arcs and calipers and occasionally obstructed surgical visualization. Finally, the use of a frame meant that imaging of the patient had to be time-coupled to the surgical act. These limitations led to the development of devices that could determine position in 3D space without the need for mechanical arcs and calipers. These devices are called 3D digitizers. They assign coordinates to each selected point, enabling coregistration of patients, images, and instruments within the same coordinate system. Finally, the limitation associated with 2D imaging was resolved by the advent of 3D imaging techniques such as computed tomography (CT) and magnetic resonance imaging (MRI). With the advent of 3D imaging and intraoperative localization devices, stereotactic frames have been largely replaced with frameless navigational systems. These frameless systems are discussed later, but it is first necessary to examine the advances in imaging techniques that have allowed image-guided navigation techniques to achieve their current level of ubiquity.

10.2 Advances in Imaging Technology

10.2.1 Computed Tomography and Magnetic Resonance Imaging

While radiographs are clearly still invaluable in many medical applications, they have generally been superseded for image guidance purposes by other imaging modalities, predominantly CT and MRI. CT is optimally employed for imaging vascular and skull base lesions, and is ideal for trauma cases as blood clots (hematomas) are easily seen on this imaging modality. However, MRI offers superior tissue contrast and is therefore extremely useful for imaging brain tumors in soft tissues and organs and structures within the brain itself, such as deep collections of neural tissue called nuclei. The individual 2D images acquired with these modalities can be manipulated by relatively inexpensive computer technology into 3D virtual models of the patient's anatomy. Such virtual models can be rotated and examined from any angle to enable the surgeon to determine the best approach to the target. These visualization techniques have thereby markedly increased the range of applications for stereotactic techniques. Preoperative images can also be supplemented by the use of intraoperative imaging, allowing the surgeon to evaluate the progress of the operation and decide if changes to the surgical plan are necessary due to movement of tissue.

10.2.2 Functional Imaging

Functional imaging consists of the anatomical depiction of specific neurological function. For example, functional images can depict the area of the brain involved in speech or movement. There are three general techniques employed to locate function. One detects changes in metabolism produced by function and overlays the location of such functional activity on structural images. A second technique employs technology that can localize activity resulting from function and then perform a similar overlay on structural images. The third technique involves stimulation of the brain with a concomitant examination of the patient to determine what function results, or is blocked, by the stimulation, again noting the location of this result on structural imaging of the brain. Currently, the main functional imaging modalities using changes in blood flow are positron emission tomography (PET), single-photon emission computed tomography (SPECT), and functional magnetic resonance imaging (fMRI). Techniques that directly detect functional activity are magnetic source imaging (MSI) and quantitative electroencephalography (qEEG). MSI is by far the more accurate of these two techniques and uses magnetoencephalography (MEG) to detect minute magnetic changes that arise from electrical activity in the brain during performance of a specific function. MSI is used for mapping normal and continuous abnormal function. It is already capable of identifying motor and sensory cortex in the brain, and new algorithms may enable detection of speech activity as well. Finally, a form of functional imaging can be created by direct stimulation of the brain through transcranial magnetic stimulation and activation of specific functional areas followed by mapping of these areas of activation on structural images (Krings et al. 2001). As transcranial magnetic stimulation can only stimulate superficial structures, the potential ability of ultrasonic pulses to stimulate deep structures (Tufail et al. 2010) offers an exciting alternative that allows a complete functional map of the brain to be created without resorting to performing awake craniotomies and direct stimulation of the brain to localize, and avoid, functional tissue.

In functional imaging, alterations in function, rather than structure, serve as the basis for the imaging. For example, whereas CT develops an image based upon the absorption of x-rays by the tissue within the x-ray beam, fMRI detects differences in blood flow while the patient is at rest versus performing a specific task. Such changes in blood flow are usually induced by the demands of a specific

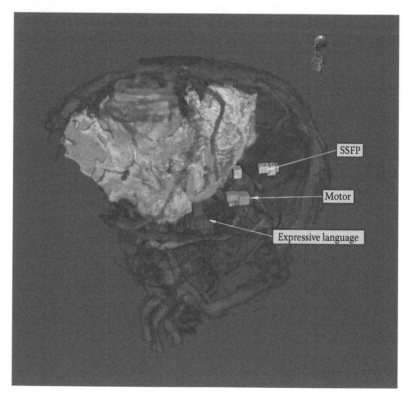

FIGURE 10.1 **(See color insert.)** A 3D reconstruction combining image data of a brain tumor (MRI) and the surrounding vasculature (MRV) with functional data (MEG) to show the relative positions of sensitive areas.

part of the brain performing a specific task. The resultant image is therefore a depiction of the anatomical location of a specific function. Knowledge of the location of function, rather than anatomy, can be particularly useful in procedures either targeting abnormal function for correction (such as surgery for epilepsy intractable to medical intervention) or in the removal of lesions close to critical areas of the brain such as the motor cortex. As neurosurgery evolves from techniques designed to remove lesions to those designed to restore function, it is anticipated that functional imaging will become increasingly employed as the imaging modality of choice for image-guided surgery.

While functional imaging has considerable potential, it does have its limitations. The resolution of functional imaging is far less than that typically seen with structural imaging. The images usually demonstrate only function, and the relationship between function and anatomy is not necessarily obvious. Therefore, to detect an abnormality in a specific patient, it may be necessary to acquire data from normal controls for comparison purposes or refer to an atlas that delineates the relationship of function to anatomy. Thus, the use of functional imaging almost always involves a technique called image fusion, in which two different images of the same 3D object are precisely overlaid onto each other to render an image comprising information from both sources (Figure 10.1). To perform such an overlay, the two images must be related to each other through a process of registration, in which points or geometry depicted on both images are used to align the images with one another. The process of registration is central not only to image-to-image fusion but also to image-to-patient fusion and will be covered later in this chapter.

10.3 Segmentation and Deformation of Images

When imaging a patient for surgical planning purposes, the entire anatomy of part of the patient, such as the head, is normally depicted. Usually, however, only a specific area is of interest for a particular case, and it is desirable to separate this region from the remainder to allow study of the shape and size of the tissue in question. This process is called segmentation and can be accomplished by comparison of the patient's image to a standardized printed atlas or an electronic atlas that has specific structures already segmented. This process can be rendered very inaccurate by the significant differences between individual brains. This discrepancy may be further exacerbated by the presence of tumors, abscesses, or accidental trauma that distort the normal anatomy, making it difficult to identify or orient the patient's image data relative to the atlas example. A solution is to use an elastic deformation technique to resize and warp the atlas to match the patient's anatomy. This approach constrains the points on the patient image and atlas plate that have to be connected but otherwise leaves the structures free to deform to a template. Not only does this enable more precise preoperative planning, but it is also valuable if the surgical plan must be updated intraoperatively to reflect brain shift or if ambiguous structures are encountered that must be identified before proceeding further. Several deformable atlases have now been produced for neurosurgical applications (Nowinski et al. 1997, 1998), although there is persistent concern as to the accuracy of the resultant segmentations.

At first, limited computing capacity meant that the digitized data had to be manually segmented. Manual segmentation of images is time consuming and tedious. Fortunately, studies have indicated that automated segmentation techniques are as reliable and accurate as manual segmentations (Kaus et al. 2001). Automated techniques are much faster and are robust with respect to diverse disease entities (cf. the increasing use of automated hippocampal segmentation in studies of psychiatric disorders [Villarreal et al. 2002; Bergouignan et al. 2009]). A number of automated segmentation methods have now been developed and their respective merits evaluated (Morey et al. 2009). Use of such automated techniques may avoid the need for invasive monitoring prior to functional neurosurgical procedures. For example, when planning a surgical intervention for epilepsy, it has been necessary to surgically insert depth electrodes to record abnormal activity. Automated segmentation coupled with functional-to-anatomical image fusion may eliminate the need for such complex and costly evaluation prior to surgery. The technique could also accommodate structural lesions such as tumors. After segmenting out the lesion, the brain is deformed to what it would look like without the lesion. The atlas is deformed to the brain, then the tumor can be "grown" back into the image dataset. Although this process is still experimental, it might allow the depiction of the location of pathways altered by the presence of mass lesions.

10.4 Registration

The fundamental concept underlying all image-guided surgery is that of *registration*. This is the method for mapping the detailed imaging data of a specific patient to one another and to the physical space/anatomy of the patient. Registration has been traditionally performed by identifying and aligning points seen on the sets of images and their locations on the patient in physical space, a process termed paired-point registration.

In conventional frame-based stereotactic surgery, markers placed on the frame during imaging (called fiducials) allowed the registration of the images to the operative field. Paired-point-based methods have now been supplemented with curve and surface matching methods, moment and principal fixed axes methods, correlation methods, and atlas methods (Maurer and Fitzpatrick 1993).

10.4.1 Paired-Point Methods

The most straightforward method for achieving coregistration of coordinate spaces is to use a set of markers that are visible in the patient's image dataset and to determine the location of these points relative to a reference body in the operating room by the use of a localization device or digitizer (see Section 10.5). While only three points (in a nonlinear arrangement) are required to relate two 3D objects to one another, using more points improves the accuracy of the subsequent registration and lessens the risk of failure should one of the markers move relative to the others or become obscured and undetectable by the digitizer.

Some investigators (notably in the radiosurgical field) have used clearly defined anatomical features as reference markers (Adler 1993). However, as anatomical features do not resemble the points needed for paired-point registration, there is ambiguity associated with feature-based registration, and most neurosurgeons prefer to use artificial markers applied to the patient prior to imaging. Different types and sizes of artificial fiducials are available, and they may be fixed to the patient's skin using adhesive or embedded directly in the bone. Markers applied to the skin cause no discomfort to the patient, but may be dislodged between preoperative imaging and surgery and are more prone to localized movement relative to one another. This can have deleterious effects on the accuracy of both registration and target localization. In contrast, bone-implanted fiducials (Figure 10.2) offer the highest accuracy of any registration technique, including stereotactic frames (Maciunas et al. 1993), since they cannot move relative to the patient's bone and, once implanted, remain in place until the surgery is completed. Implanted markers are most frequently used in procedures involving the skull, spine, or pelvis. However, the trauma and subsequent pain associated with implantation of fiducials in bone militate against their routine use. Until recently, the inability to fixate fiducials to soft tissue essentially prevented the application of image guidance to organs that are not confined to a rigid structure such as the skull or spine. However, procedures have now been developed for implanting fiducials in a variety of organs to enable the motion of tumors to be tracked and compensated for during image-guided radiotherapy (Murphy et al. 2000; Shirato et al. 2003).

Irrespective of the type of marker used for paired-point registration, the resultant accuracy of the registration process depends not only upon the markers but also on how they are distributed over the volume being registered. Regardless of their number, markers arranged in a line will never be useable for a 3D registration. Further, coplanar markers will allow registration but will have little accuracy in the z dimension perpendicular to the plane of the markers. For the best possible registration, the markers should be spread as widely over the entire 3D volume and as far from one another in each coordinate dimension as possible.

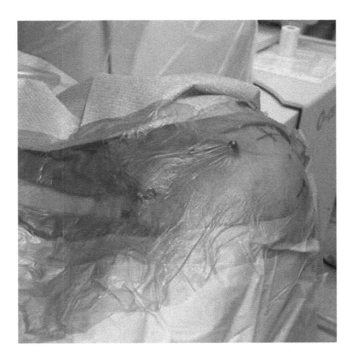

FIGURE 10.2 The head of a patient prepared for image-guided surgery, showing a skull-mounted fiducial marker.

10.4.2 Contour and Surface Mapping

As an alternative to matching isolated detectable points, it is also possible to obtain a registration by matching contours obtained from the surface of the patient. Data from imaging studies depicting a contour can be matched to points on that contour obtained randomly using the digitizer depicting the curvature of a portion of the patient's anatomy. The computer can then align these features, with areas of maximum curvature being the most useful for the process. A further development is to dispense with point matching altogether and simply match large surface areas. In neurosurgical procedures, the surfaces imaged by PET, CT, or MRI can be matched directly with the scalp as imaged in the operating room. This permits rapid registration immediately prior to the surgery, without the need to repeat preoperative diagnostic imaging with fiducials in place. This saves both time and money, though care must be taken to ensure that the diagnostic images are of sufficient quality to justify relying on them for surgical planning and guidance. Diagnostic images frequently lack the fine spacing needed between images to allow for accurate intraoperative guidance. There is some variation in the degree to which surface and contour matching techniques can be automated. The best-known surface matching strategy is that of Pelizzari et al. (1989), but a variety of approaches to surface and contour matching have been developed, including some specifically intended for registration of the cortical surface of the brain (Davatzikos et al. 1993; Miga et al. 2003).

Existing surface-based registration techniques are generally less accurate and less reliable than fiducial marker-based techniques and require considerably more technical support and time. The application accuracy achieved for surface matching is usually in excess of 3.0 mm (West et al. 1997). Further, it is important to note that contour matching is incapable of registering two perfect spheres using their contours; therefore, points for the registration process should be obtained wherever there is a departure of normal anatomy away from a sphere, such as the area around the face and nose.

A combination of rapid contour mapping with fully automated segmentation would result in a system permitting frequent, rapid, and accurate updates of the registration information at each stage of the surgical procedure, though the technology required for this is still under development.

Finally, the use of intraoperative imaging may allow novel registration techniques. For example, attaching a registration device of known geometry to the patient's anatomy and obtaining an intraoperative highly resolved 3D structural image (such as with portable CT) will allow automated registration in a manner similar to framed stereotactic systems of the past. By building this registration device into the reference system for a frameless navigational device, an extremely accurate registration can be obtained with essentially no user intervention whatsoever.

10.5 Frameless Systems

10.5.1 Early Approaches to Frameless Localization

As mentioned earlier, stereotactic frames have now been almost completely replaced by frameless navigation systems for most neurosurgical procedures. Among the earliest attempts at 3D localizers were articulated arms with potentiometers at the joints between the links. These sensors determined the angles of the joints by measuring resistance, enabling the orientation of the distal section and location of the tip to be determined. The first reported use of a passive localization arm of this type for neurosurgical guidance was by Watanabe (1993), who used an arm with six joints. A widely used early commercial model was the ISG Viewing Wand (ISG Technologies, Toronto, Canada). Guthrie and Adler (1992) developed a similar series of arms in which optical encoders replaced the potentiometers to improve accuracy. These arm-based systems were simple and did not require a clear line of sight. However, as they interfered with the surgeon's movements and were difficult to use with surgical instruments, their use has been almost completely abandoned.

Another early approach used ultrasound emitters and microphone arrays in a known configuration to localize the position of the instruments. Position was determined on the basis of the time delay between the emission of an ultrasonic pulse by the emitter and its detection by each microphone. Roberts and colleagues applied a version of this approach to an operating microscope (Roberts et al. 1986), and Barnett et al. (1993a,b) developed a system in which ultrasonic emitters are fitted to a handheld interactive localization device (Picker International, Highland Heights, Ohio) and the detecting microphone arrays are fitted to the side rails of a standard operating table. However, ultrasound-based digitizing systems are limited by their requirement for a clear line of sight between the emitters and detectors, and the systems may be confused by echoes, environmental noise, and the effects of fluctuating air temperature. On the positive side, they are relatively inexpensive, costing no more than half as much as an optical digitizer, can be set up rapidly, and can cover a very large working volume. Also, it is easy to affix the emitters to all kinds of instruments and other equipment.

10.5.2 Optical Digitizers

Modern navigation systems make use of light-emitting diodes (LEDs) to track the location of instruments within the operative space and were first developed by Bucholz. The LEDs used in such systems emit near-infrared light that can be detected by at least two charge-coupled device (CCD) cameras positioned in the operating room. Inside the cameras, light from the LEDs is focused onto a layer of several thousand CCD sensor elements, and the infrared energy is converted into electrical impulses. These digital signals may be further processed or converted to analog form. Through triangulation, the point of light emission can be calculated with 0.3 mm accuracy.

Optical triangulation techniques are highly accurate and robust. A useful comparison of their properties with those of sonic digitizers has been published previously (Bucholz and Smith 1993). Like sonic digitizers, optical systems require a clear line of sight, but they are unaffected by temperature fluctuations and do not appear to be compromised by surgical light sources. The infrared light is not visible to the surgical team and causes no harmful effects.

A variation on the infrared detection approach uses passive reflectors instead of active emitters, with an infrared source illuminating the operative field. The reflected light is detected as discussed previously.

Instruments fitted with passive reflectors do not need to have electrical cables attached and are thus easier to sterilize. However, the passive markers cannot be "turned off" and thus appear simultaneously to the CCD cameras at all times. This can confuse the navigation system if multiple instruments are being used and a large number of reflectors are present in the operative field at once. Another passive approach substitutes an ultraviolet (UV) source as the field illuminator, with fluorescent markers reirradiating the UV light at visible frequency levels. However, there is concern that prolonged exposure to UV in this context may pose a health hazard.

10.5.3 Magnetic Navigation Systems

An alternative to the use of optical tracking is magnetic navigation. The first such system to enter routine clinical use is the AxiEM™ from Medtronic. In this approach, small trackers (only 13 mm across) are embedded percutaneously or subcutaneously in the patient's anatomy, and a magnetic field is then generated around the patient by a transmitter. Sensors mounted on the instruments measure the relative strength of the magnetic field in six directions, thereby providing continuous localization of the instruments in relation to the patient landmarks. The obvious drawback of electromagnetic tracking systems is their sensitivity to ferrous metals within the magnetic field, necessitating exclusion of such materials from the operating room. However, in contrast to optically based tracking systems, there is no requirement for a clear line of sight, making magnetic systems particularly useful in procedures where the markers are embedded within the patient's anatomy, such as in the placement of intraventricular shunts. Following success in cranial applications, the AxiEM system has been successfully applied to orthopedic procedures, notably total knee replacement, and some ENT procedures.

10.5.4 Reference Systems

With stereotactic frames, the mounting of surgical instruments on the base ring ensured that they were accurately positioned relative to the patient's anatomy. With frameless devices, there is no safeguarded connection between the instruments and the patient's anatomy, which can lead to error if the position of the patient is not accurately known.

An important benefit of surgical navigation systems is the ability to change the patient's position following registration. To avoid the need for reregistration, it is necessary to have some means of tracking the head relative to the detection mechanism. This is commonly achieved with a reference arc or similar device fixed directly to the patient's head or to the head holder, assuming that the head does not move in relationship to the holder. The reference device has at least three emitters that are detectable by the digitizer employed in the design of the system; for optically based systems, a multiplicity of LEDs serves this function, allowing the position of the head to be determined prior to localization of the surgical instruments even if one or more LEDs are blocked. As mentioned previously, the use of an electromagnetic tracking system with head-mounted magnetic trackers in place of LEDs avoids the need for the markers to remain visible during the procedure.

By monitoring the relative positions of the instrument and the reference arc, the navigation system correlates the position of the instrument to the patient's anatomy on a continuous basis, regardless of the frequent movements of the patient that occur during a standard operation.

10.5.5 Effectors

The term effector simply refers to those instruments used by a surgeon to perform surgery on the patient. Although surgical instruments comprise the largest portion of such devices, the term *effector* can also be used to refer to anything used to treat the patient surgically, including surgical microscopes, endoscopes, genetic material, drug polymers, and robots. Such effectors all rely upon precise positioning in order to produce the maximal benefit for the patient, so integration of navigational technology is critical to their success.

Essential components of a surgical navigation system are the instruments that, through modifications by the addition of LEDs or reflectors, permit localization. As the LEDs or spheres must be visible to the camera array at all times, they cannot be mounted on the tip of an instrument, as this will not be visible to the cameras during surgery. Instead, the LEDs or spheres are usually located at a given distance from the tip in the handle of the instrument. If the instrument is essentially linear, such as a forceps, then the minimum requirement for localization is two LEDs mounted in alignment with the tip. Reflective spheres generally do not localize well if they are placed in a linear alignment, as reflective systems experience difficulty when the views of the spheres partially overlap, so reflective arrays usually consist of at least three spheres in a nonlinear arrangement. Given the relatively large size of reflective spheres, LEDs are usually the detector of choice for microsurgical instrumentation.

A bayoneted instrument is commonly employed to allow surgery through a small opening. The offset design of a bayonet instrument allows the handle to be placed off the axis of the line of sight of the surgeon, and its geometry is ideal for use in conjunction with a navigational system. Similarly, for instruments with complex 3D shapes, LEDs or spheres must be placed off-axis to allow tracking.

When a specific trajectory into the brain is desired, rigid fixation of instrumentation is preferred to ensure that the instrumentation is aligned along the selected surgical path. Typical situations in which this path is key include tumor biopsy, insertion of depth electrodes for epilepsy, insertion of ventricular catheters, and functional surgery. This function is carried out by a biopsy guide tube adaptor, which is attached to the reference arc using a standard retractor arm with adjustable tension. The tube is equipped with four LEDs to provide redundancy.

The most popular frameless surgical navigation system is the StealthStation® (manufactured by Medtronic, Inc.). It consists of (1) a UNIX-based workstation, which, by communicating with the other components of the system, displays position on a high-resolution monitor or head-mounted display; (2) an infrared optical digitizer with camera array (as described earlier); (3) a reference LED array (e.g., a reference arc) for the patient; and (4) surgical instruments modified by the addition either passive reflectors or active emitters (usually LEDs). Optional components of the system include a robotically controlled locatable surgical microscope and surgical endoscopes modified by the placement of LEDs. As mentioned earlier, an electromagnetic tracking system, AxiEM, has also been developed by Medtronic and is optimized for use in conjunction with the StealthStation as an alternative to optically based tracking.

The StealthStation uses CT or MRI for intraoperative guidance. Images are typically obtained in the axial dimension, and the scanning parameters for each modality are adjusted to achieve roughly cubic voxels: Image files are transmitted from the CT or MRI scanner to the surgical workstation over an Ethernet-based local area network and are then converted to a standard file format. During surgery, three standard views (the original axial projection and reconstructed sagittal and coronal images) are displayed on a monitor at all times. A cross-hair pointer superimposed on these images indicates the position of the surgical instrument, endoscope, or microscope focal point. A fourth window can alternatively display a surface-rendered 3D view of the patient's anatomy, real-time video from the endoscope or microscope, or reference data from an anatomical atlas. An alternative view, the navigational view, produces images orthogonal to the surgical instrument rather than the patient and is particularly useful for aligning the instrument with a surgical path.

An operating microscope can also be tracked relative to the surgical field and the position of the focal point displayed on the preoperative images. Certain microscopes can be robotically controlled by the system: By means of motors that move the microscope head, the workstation can drive the scope to focus upon a specific point in the surgical field chosen by clicking on the spot as viewed on the workstation display.

A rigid straight fiberscope (INCLUSIVE endoscope, Medtronic Sofamor Danek, Inc., Memphis, TN) has also been modified to work with the StealthStation for intraventricular procedures. Four LEDS are attached to the endoscope near the camera mount using a star-shaped adapter, and the geometric configuration of the LEDs is programmed into the surgical navigational system along with the endoscopic dimensions.

10.6 Intraoperative Imaging

Images acquired preoperatively obviously depict the situation prior to opening of the cranium. However, the brain is elastic and usually deforms during a surgical procedure, particularly when a large tumor is being resected or a significant volume of cerebrospinal fluid is drained. Attempts have been made to model how the brain would respond to surgery, but this is mathematically complex. An alternative solution is to use intraoperative imaging to update and correct preoperative images as the surgery progresses. In its simplest form, the same imaging modality used preoperatively could be employed intraoperatively. Alternatively, a different imaging modality could be used during a procedure, and the deformation detected by this imaging modality could be used to elastically deform the preoperative images. Several different modalities are currently available for intraoperative imaging, and their relative advantages and disadvantages are summarized in the following table:

Factor/Type	MRI	CT	Ultrasound	Fluoroscopy
Cost	--	-	+	+
Continuous duty	+	--	++	-
Resolution	+	+	+++	--
Contrast	++	+	+	+
Distortion	-	++	+	+
Signal-to-noise ratio	++	+	--	-

Ultrasound has been proposed as a low-cost alternative to stereotaxis. Ultrasound data may be acquired as a single 2D image, or a 3D dataset may be acquired with newer systems. However, the images are taken at oblique angles to normal anatomical axes and are often difficult to interpret. There is a high noise-to-signal ratio, and some tumors are not visible at all using ultrasound. However, it may be useful if registered to other imaging modalities that enable the recognition of specific structures. Ultrasound-to-MRI registration is normally accomplished by using a modified probe that is equipped with LEDs. In some cases, ultrasound-to-MRI fusion can be accomplished without the need for external fiducials (Porter et al. 2001).

With the StealthStation, ultrasound images are fed in through a video port and preoperative MR or CT images reformatted to match the ultrasound images. Color Doppler units, such as the Aloka Model 5000 (Aloka Co., Ltd., Wallingford, CT), can detect the presence of vessels easily and in color. The system has a modified operative display with five active windows, which enable the axial, coronal, and sagittal views to be displayed as reformatted MR or CT with or without overlaid ultrasound images.

10.6.1 The O-Arm

A major advance in intraoperative imaging technology is the Medtronic O-arm Surgical Imaging System, which received FDA approval in 2005. The O-arm is an integrated mobile imaging platform intended to be used in conjunction with the StealthStation and optimized for neurosurgical and orthopedic applications. The distinctive feature of the O-arm is its "breakable" circular gantry that can be opened to permit lateral access to the patient then closed to form a complete "O" for imaging. Offering both conventional 2D fluoroscopy and high-resolution 3D imaging, the O-arm provides superior imaging within the operating room itself throughout the procedure, including real-time axial views that are not available with standard fluoroscopy. The surgeon is able to perform navigation using images obtained when the surgeon is outside the operative field. The system also has the potential to eliminate the need for postoperative CT, further reducing the radiation exposure for the patient and staff. The system can also be used to monitor in real time the insertion of electrodes into the brain to make certain that the system has accurately placed an electrode in accordance with the surgical plan, providing a feedback loop for a surgical procedure (Figure 10.3).

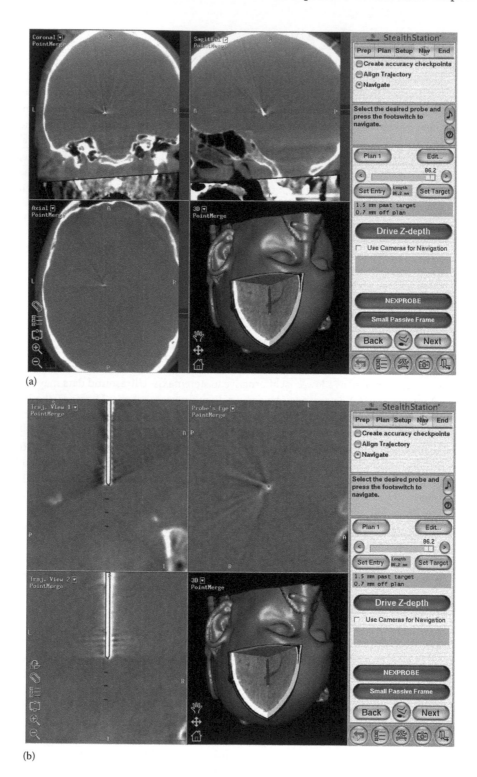

FIGURE 10.3 StealthStation screenshots depicting the implantation of a stimulation electrode in a patient with Parkinson's disease. (a) O-arm images showing the position of the electrode. The desired target position is marked with a red dot. (b) The original surgical plan (in blue) superimposed on the O-arm image of the implanted electrode.

10.7 Intraoperative System Control

As navigational systems become more complex, the need for the surgeon to interact with the system during a procedure increases. This need will intensify as more complex effectors, such as robots, are brought into the operating room. Furthermore, as these units proliferate in community hospitals, it will be imperative to make them capable of being controlled by a small surgical team, without the need to have additional technicians on hand to operate the systems.

Several approaches have been developed that allow the surgeon to control the system while scrubbed. One technique involves the use of touch-sensitive flat panel displays that can be placed in a sterile bag and used for controlling the system as well as indicating position. Another solution is to incorporate voice recognition in the head-mounted display worn by the surgeon during a procedure.

As these systems become more advanced, it will be important to improve the diversity and utility of the information presented. Rather than being limited to simply showing where a surgeon is within the patient's anatomy, these systems can be employed to compare the patient's anatomy to previous patients' functional anatomy. This function is served through the use of an atlas of functional anatomy.

10.8 Current Applications

Image-guided surgery techniques have been found to be valuable in diverse procedures, including tumor removal and ablation, treatment of arteriovenous malformations, functional neurosurgery for the treatment of epilepsy or tremor-inducing conditions, and implantation of devices to provide deep brain stimulation (DBS) or control hydrocephalus, for example. Image guidance also provides significant assistance in various orthopedic procedures, including spine surgery, pelvic fixation, and joint replacement.

10.8.1 Tumors

Image guidance is of great value for planning tumor removal procedures. The actual location of certain types of tumor with distinct margins is usually apparent in preoperative imaging, though some forms of tumor are not so easily delineated. In most cases, image guidance facilitates a more complete resection of the tumor during the initial surgery, and subsequent imaging will reveal whether a follow-up surgery is necessary to remove any residual material or recurrent growth.

A recent development has been the use of agents such as 5-aminolevulinic acid (5-ALA), a compound that fluoresces under UV radiation, to more clearly delineate the margins of contrast-enhancing tumors. An image showing 5-ALA fluorescence can be superimposed on the intraoperative image to enable a more complete resection of a malignant tumor, resulting in extended patient survival (Stummer et al. 2006).

Another application of image guidance for tumor therapy is in brachytherapy, the placement of radioactive seeds directly inside tumors that are otherwise inoperable, or where surgery may result in unwanted side effects. In general, this technique is employed to palliate rather than cure a lesion. One area in which this technique has been employed successfully is in the management of malignant tumors of the prostate that commonly occur in patients incapable of tolerating extensive resective surgery (Rubens et al. 2006).

10.8.2 Vascular Malformations

Intracerebral vascular malformations normally require surgical intervention to prevent a future hemorrhage and the attendant neurological deficits. Excision of such malformations is particularly difficult when they occur in eloquent areas of the brain such as those responsible for speech or motor–sensory functions. Image-guidance technology enables the malformation to be modeled in three dimensions and thus permits the optimal resection to be planned, taking into account the relationships of the feeding and draining vessels.

10.8.3 Ventriculostomy

The most common application of endoscopy within the head is in the performance of third ventriculostomies in patients with obstructive hydrocephalus or aqueductal stenosis (Kelly 1991; Dalrymple and Kelly 1992; Drake 1993). Once the endoscope is successfully inserted, it is frequently possible to proceed on the basis of visual guidance. However, image guidance can ensure a straight trajectory from the burr hole through the foramen of Monro to the floor of the third ventricle and help avoid unseen branches of the basilar artery. Image guidance is also of particular value in cases of abnormal anatomy or where orientation becomes difficult owing to bloody or blurry cerebrospinal fluid (Muacevic and Muller 1999). Developments in the field of virtual endoscopy promise to aid in the planning and conduct of these procedures (Burtscher et al. 2000; Jödicke et al. 2003).

10.8.4 Functional Neurosurgery

Patients with medically intractable epilepsy are now frequently referred for functional neurosurgery to alleviate their condition. In order to treat the condition surgically, it is first necessary to identify and localize the seizure focus within the brain. Once localized, the region can be eradicated. These processes require accurate spatial localization during preoperative imaging and electrophysiological investigations. Accurate coregistration of the patient data with the surgical field is essential to prevent destruction of uninvolved tissue.

While the advent of CT and MRI had a significant impact on epilepsy management, the more recent development of PET, SPECT, fMRI, and MSI has enabled greater understanding of the underlying neurophysiology and has permitted refinement of treatment protocols. The ability to coregister each of the imaging modalities with the others (and with the patient) is fundamental to the process of determining the nature of the pathology in a given patient.

Video-EEG recordings of actual seizure events also provide vital information for determining which cases are suitable for surgery. Invasive recording techniques are used to evaluate more difficult cases, and placement of the required recording hardware can be facilitated by image guidance. A common approach involves the insertion of depth electrodes into critical mesial structures to localize the seizure origins (McCarthy et al. 1991; Murphy et al. 2002). Another evaluation technique requires the placement of subdural strip electrodes on the cortical surface. Accurate placement of subdural strips is crucial, and frameless image guidance is potentially of great assistance in the process (Eröss et al. 2009), though some centers still place the strips without such guidance.

Therapeutic procedures for treating seizure disorders associated with structural lesions have been highly successful in producing seizure-free outcomes (Engel et al. 1993; Piepgras et al. 1993). All such surgical interventions for seizure disorders benefit considerably from image guidance, since it is desirable to fully eliminate the causative lesion, while minimizing damage to surrounding tissue. It is likely that the future will see the development of noninvasive source localization that will obviate the need for surgical implantation of recording electrodes. However, there will still be a requirement for image guidance in therapeutic surgery. Improvements in noncontact registration should enable the registration of patient image data to be accomplished and updated more rapidly.

10.8.5 Deep Brain Stimulation

An increasingly common approach to controlling movement and affective disorders is deep brain stimulation (DBS), in which an implanted pulse generator sends electrical pulses to specific regions of the brain to modulate the neural activity and thus suppress the symptoms. While the generator itself is implanted in the patient's chest or abdomen, the leads and electrodes must be placed precisely in the appropriate brain regions, a process that obviously requires the use of image guidance. For Parkinson's disease, for example, the electrodes are placed in the globus pallidus or subthalamic nucleus. In addition to movement disorders, such as essential tremor and dystonia, DBS has also shown considerable promise in the control of severe obsessive–compulsive disorder (Goodman et al. 2010), treatment-resistant depression (Malone et al. 2009), and Tourette syndrome (Neuner et al. 2009).

In view of the expanding range of possible applications, demand for DBS is expected to increase considerably in the coming years. Until now, DBS electrodes have been placed using frame-based stereotaxy or frameless neuronavigation in conjunction with bone-implanted fiducials. Recently, a less time-consuming method for implantation of DBS leads has been described in which interventional MRI (iMRI) is used to guide placement of the electrodes without the need for fiducials (Starr et al. 2010). The planning, insertion, and MRI confirmation are all performed with the patient in the MRI gantry, with coordinates defined with respect to the MRI isocenter rather than a stereotactic space defined by fiducials. The DBS leads are inserted using a disposable trajectory guide (NexFrame, Medtronic) mounted on the burr hole, rather than a traditional stereotactic frame and arc, in combination with specialized iMRI-compatible alignment stems and probe guides. Accuracy with this approach is reportedly superior to that obtained with frame-based stereotaxy. Streamlined approaches such as this, which eliminate the need for fiducial placement, will improve accessibility of the technique to surgeons and increase the availability to prospective patients.

10.8.6 Stereotactic Radiosurgery

Another area where image guidance has been of considerable assistance is stereotactic radiosurgery, in which one or more highly collimated beams of radiation are directed at a tumor. Until recently, this has usually been an outpatient, nonsurgical procedure using the LINAC Scalpel or Gamma Knife. For certain types of brain tumor, this treatment is an effective alternative to conventional surgery. However, the radiation does not remove the tumor (though it may shrink it), and the resulting scar tissue may make any future surgery more difficult. In this technique, the patient first undergoes the application of a stereotactic frame to establish a reference system upon the patient's anatomy. The target of interest is then defined by imaging the patient with the frame in place, and the coordinates of the target are calculated using conventional stereotactic techniques. A focused beam of radiation is then sent through the target point, the patient or the beam moved, and the beam repeated, as in LINAC-based stereotactic radiosurgery. Alternatively, the patient's head is precisely placed at the center of numerous intersecting beams such that all of the beams intersect within the target, as in Gamma Knife radiosurgery. Both devices embrace the key concept of image guidance in terms of maximal functionality with minimal invasiveness.

The more recent CyberKnife (Chang and Adler 2001) uses CT image data of the patient to direct a small linear accelerator mounted on a precision-controlled robotic arm, enabling the energy beam to be directed at the target area from almost any angle. The use of multiple treatment fractions permits radiosurgery of irregularly shaped tumors while minimizing irradiation of adjacent healthy tissue. In contrast to LINAC and Gamma Knife radiosurgery, CyberKnife radiosurgery is entirely frameless, and the ability of the robot to follow movement of the target means that it is not even necessary to immobilize the patient. X-ray-based skull tracking obviates the need to clamp the patient's head for treatment of intracranial lesions, while the Xsight Spine option tracks the position of spinal tumors using image data of the spinal processes without the need to attach fiducials. The recently introduced Xsight Lung option permits fiducial-free tracking of some lung tumors, but for most tumors that are in motion due to involuntary movement such as respiration, guidance is provided by the Synchrony system, a combination of surgically placed internal fiducials and light-emitting markers attached to the patient's skin that enables the position of the tumor to be predicted at any time.

10.8.7 Functional Endoscopic Sinus Surgery

A fairly recent application of image guidance in the ENT field is functional endoscopic sinus surgery (FESS) (Kennedy et al. 1985), a minimally invasive approach to reopening the paranasal sinuses in patients suffering from chronic or recurring sinusitis. In FESS, the structures and tissues to be operated on are visualized endoscopically, while CT imaging and MRI are used to determine the anatomic relationships of vital structures such as the optic nerve and carotid artery to the diseased tissue and obstructions to be removed so that a safe and effective surgical plan can be devised. FESS represents a considerable improvement over traditional sinus surgery, which is often conducted via external approaches to the sinuses, resulting in scarring, extensive bruising, and significant postoperative discomfort for the patient, and may sometimes damage the nerve supply to the teeth, while overlooking relevant intranasal pathology. In contrast, FESS leaves no scars or nerve damage, enables a more precise and complete excision of pathological tissue, and can often be performed as an outpatient procedure. Originally, surgeons performing FESS had to mentally correlate information from preoperative imaging with the endoscopic view, but the introduction of navigation systems enables these sources of data to be fully integrated, allowing the instruments to be precisely tracked and controlled throughout the procedure and thereby further increasing its safety and efficacy (Koulechov et al. 2006).

10.8.8 Spinal and Orthopedic Procedures

While x-rays have been a valued tool of orthopedic surgeons since their introduction, there has been some resistance to the adoption of more advanced forms of image guidance in orthopedics. This is partly due to concern over the cost-effectiveness of navigation in well-established procedures such as knee replacement (Novak et al. 2007). However, the range of image-guided techniques for orthopedic procedures continues to expand, and there is growing evidence that the use of navigation confers superior accuracy in both hip and knee replacement (Jolles et al. 2004; Stöckl et al. 2004), potentially resulting in reduced need for revision surgery. Similarly, for many years, the primary application of image guidance in spine procedures was the placement of pedicle screws with increased safety and reduced radiation exposure (Merloz et al. 2007). The recent introduction of the O-arm has been particularly helpful in extending the use of image guidance in spinal surgery, resulting in improved accuracy, fewer complications, and better outcomes in the surgical treatment of conditions such as severe scoliosis (Metz and Burch 2008) and spondylolysis (Brennan et al. 2008).

10.9 Conclusions

Image guidance directly supports minimally invasive surgery techniques by enabling the surgeon to view the operative site directly without the need for extensive exposure of normal anatomy. In the hands of a well-trained surgeon, the use of image guidance can markedly reduce the risk associated with critical interventions when compared to the situation before navigation was widely adopted.

Based upon the advances made in imaging, image guidance has gone from a few applications limited to the midline of the brain to being employed in nearly every cranial procedure imaginable, along with an increasing number of orthopedic procedures and even some thoracoabdominal procedures. Fusion of intraoperatively acquired ultrasound images to preoperative MRI data may mean that expensive imaging equipment is unnecessary to bridge the gap between the preoperative and intraoperative situations.

As this technology improves, and as more effectors that rely upon precise placement are built, image guidance will become the standard of care for all cranial interventions and many other surgical procedures.

References

Adler, J. R. Jr. 1993. Image-based frameless stereotactic radiosurgery. In *Interactive Image-Guided Neurosurgery*, ed. R. J. Maciunas, pp. 81–89. Park Ridge, IL: American Association of Neurological Surgeons.

Apuzzo, M. L. J. and C. A. Fredericks. 1988. The Brown-Roberts-Wells system. In *Modern Stereotactic Neurosurgery*, ed. L. D. Lunsford, pp. 63–77. Boston, MA: Martinus Nijhoff.

Barnett, G. H., D. W. Kormos, C. P. Steiner et al. 1993a. Intraoperative localization using an armless, frameless stereotactic wand. *J Neurosurg* 78: 510–514.

Barnett, G. H., D. W. Kormos, C. P. Steiner et al. 1993b. Frameless stereotaxy using a sonic digitizing wand: Development and adaptation to the Picker Vistar medical imaging system. In *Interactive Image-Guided Neurosurgery*, ed. R. J. Maciunas, pp. 113–120. Park Ridge, IL: American Association of Neurological Surgeons.

Bergouignan, L., M. Chupin, Y. Czechowska et al. 2009. Can voxel based morphometry, manual segmentation and automated segmentation equally detect hippocampal volume differences in acute depression? *NeuroImage* 45: 29–37.

Brennan, R. P., P. Y. Smucker, and E. M. Horn. 2008. Minimally invasive image-guided direct repair of bilateral L-5 pars interarticularis defects. *Neurosurg Focus* 25: E13.

Brown, R. A. 1979. A computerized tomography-computer graphics approach to stereotaxic localization. *J Neurosurg* 50: 715–720.

Bucholz, R. D. and K. R. Smith. 1993. A comparison of sonic digitizers versus light-emitting diode-based localization. In *Interactive Image-Guided Neurosurgery*, ed. R. J. Maciunas, pp. 179–200. Park Ridge, IL: American Association of Neurological Surgeons.

Burtscher, J., A. Dessl, R. Bale et al. 2000. Virtual endoscopy for planning endoscopic third ventriculostomy procedures. *Pediatr Neurosurg* 32: 77–82.

Chang, S. D. and J. R. Adler. 2001. Robotics and radiosurgery—The CyberKnife. *Stereotact Funct Neurosurg* 76: 204–208.

Dalrymple, S. J. and P. J. Kelly. 1992. Computer-assisted stereotactic third ventriculostomy in the management of non-communicating hydrocephalus. *Stereotact Funct Neurosurg* 59: 105–110.

Davatzikos, C. A., J. L. Prince, and R. N. Bryan. 1993. Brain image registration based on cortical contour mapping. In *Proceedings of the IEEE 1993 Nuclear Science Symposium and Medical Imaging Conference*, San Francisco, CA, October 31–November 6, 1993, pp. 1823–1826.

Drake, J. M. 1993. Ventriculostomy for treatment of hydrocephalus. *Neurosurg Clin N Am* 4: 657–666.

Engel, J. Jr., P. C. Van Ness, T. B. Rasmussen et al. 1993. Outcome with respect to epileptic seizures. In *Surgical Treatment of the Epilepsies*, ed. J. Engle Jr., pp. 609–621. New York: Raven Press.

Eröss, L., A. G. Bagó, L. E. Entz et al. 2009. Neuronavigation and fluoroscopy-assisted subdural strip electrode positioning: A simple method to increase intraoperative accuracy of strip localization in epilepsy surgery. Technical note. *J Neurosurg* 110: 327–331.

Goodman, W. K., K. D. Foote, B. D. Greenberg et al. 2010. Deep brain stimulation for intractable obsessive compulsive disorder: Pilot study using a blind, staggered-onset design. *Biol Psychiatry* 67: 535–542.

Guthrie, B. L. and J. R. Adler Jr. 1992. Computer-assisted preoperative planning, interactive surgery, and frameless stereotaxy. *Clin Neurosurg* 38: 112–131.

Hitchcock, E. R. 1988. The Hitchcock system. In *Modern Stereotactic Neurosurgery*, ed. L. D. Lunsford, pp. 47–61. Boston, MA: Martinus Nijhoff.

Jödicke, A., V. Accomazzi, I. Reiss, and D. K. Böker. 2003. Virtual endoscopy of the cerebral ventricles based on 3-D ultrasonography. *Ultrasound Med Biol* 29: 339–345.

Jolles, B. M., P. Genoud, and P. Hoffmeyer. 2004. Computer-assisted cup placement techniques in total hip arthroplasty improve accuracy of placement. *Clin Orthop Relat Res* 426: 174–179.

Kaus, M. R., S. K. Warfield, A. Nabavi, P. M. Black, F. A. Jolesz, and R. Kikinis. 2001. Automated segmentation of MR images of brain tumors. *Radiology* 218: 586–591.

Kelly, P. J. 1991. Stereotactic third ventriculostomy in patients with nontumoral adolescent/adult onset aqueductal stenosis and symptomatic hydrocephalus. *J Neurosurg* 75: 865–873.

Kennedy, D. W., S. J. Zinreich, A. E. Rosenbaum, and M. E. Johns. 1985. Functional endoscopic sinus surgery. Theory and diagnostic evaluation. *Arch Otolaryngol* 111: 576–582.

Koulechov, K., G. Strauss, A. Dietz, M. Strauss, M. Hofer, and T. C. Lueth. 2006. FESS control: Realization and evaluation of navigated control for functional endoscopic sinus surgery. *Comput Aided Surg* 11: 147–159.

Krings, T., K. H. Chiappa, H. Foltys, M. H. Reinges, G. R. Cosgrove, and A. Thron. 2001. Introducing navigated transcranial magnetic stimulation as a refined brain mapping methodology. *Neurosurg Rev* 24: 171–179.

Leksell, L. A. 1949. Stereotactic apparatus for intracerebral surgery. *Acta Chir Scand* 99: 229–233.

Maciunas, R. J., J. M. Fitzpatrick, R. L. Galloway et al. 1993. Beyond stereotaxy: Extreme levels of application accuracy are provided by implantable fiducial markers for interactive image-guided neurosurgery. In *Interactive Image-Guided Neurosurgery*, ed. R. J. Maciunas, pp. 259–270. Park Ridge, IL: American Association of Neurological Surgeons.

Malone, D. A. Jr., D. D. Dougherty, A. R. Rezai et al. 2009. Deep brain stimulation of the ventral capsule/ventral striatum for treatment-resistant depression. *Biol Psychiatry* 65: 267–275.

Maurer, C. M. and J. M. Fitzpatrick. 1993. A review of medical image registration. In *Interactive Image-Guided Neurosurgery*, ed. R. J. Maciunas, pp. 17–44. Park Ridge, IL: American Association of Neurological Surgeons.

McCarthy, G., D. D. Spencer, and R. J. Riker. 1991. The stereotaxic placement of depth electrodes in epilepsy. In *Epilepsy Surgery*, ed. H Lüders, pp. 371–384. New York: Raven Press.

Merloz, P., J. Troccaz, H. Vouaillat et al. 2007. Fluoroscopy-based navigation system in spine surgery. *Proc Inst Mech Eng H* 221: 813–820.

Metz, L. N. and S. Burch. 2008. Computer-assisted surgical planning and image-guided surgical navigation in refractory adult scoliosis surgery: Case report and review of the literature. *Spine* 33: E287–E292.

Miga, M. I., T. K. Sinha, D. M. Cash, R. L. Galloway, and R. J. Weil. 2003. Cortical surface registration for image-guided neurosurgery using laser-range scanning. *IEEE Trans Med Imaging* 22: 973–985.

Morey, R. A., C. M. Petty, Y. Xu et al. 2009. A comparison of automated segmentation and manual tracing for quantifying hippocampal and amygdala volumes. *NeuroImage* 45: 855–866.

Muacevic, A. and A. Muller. 1999. Image-guided endoscopic ventriculostomy with a new frameless armless neuronavigation system. *Comput Aid Surg* 4: 87–92.

Murphy, M. A., T. J. O'Brien, and M. J. Cook. 2002. Insertion of depth electrodes with or without subdural grids using frameless stereotactic guidance systems—Technique and outcome. *Br J Neurosurg* 16: 119–125.

Murphy, M. J., J. R. Adler Jr., M. Bodduluri et al. 2000. Image-guided radiosurgery for the spine and pancreas. *Comput Aided Surg* 5: 278–288.

Neuner, I., K. Podoll, H. Janouschek et al. 2009. From psychosurgery to neuromodulation: Deep brain stimulation for intractable Tourette syndrome. *World J Biol Psychiatry* 10: 366–376.

Novak, E. J., M. D. Silverstein, and K. J. Bozic. 2007. The cost-effectiveness of computer-assisted navigation in total knee arthroplasty. *J Bone Joint Surg Am* 89: 2389–2397.

Nowinski, W. L., R. N. Bryan, and R. Raghavan. 1997. *The Electronic Clinical Brain Atlas: Multi-Planar Navigation of the Human Brain*. New York: Thieme.

Nowinski, W. L., T. T. Yeo, and A. Thirunavuukarasuu. 1998. Microelectrode-guided functional neurosurgery assisted by Electronic Clinical Brain Atlas CD-ROM. *Comput Aided Surg* 3: 115–122.

Pelizzari, C. A., G. T. Y. Chen, D. R. Spelbring, R. R. Weichselbaum, and C. Chen. 1989. Accurate three-dimensional registration of CT, PET, and/or MR images of the brain. *J Comput Assist Tomogr* 13: 20–26.

Piepgras, D. G., T. M. Sundt Jr., A. T. Ragoowansi et al. 1993. Seizure outcome in patients with surgically treated cerebral arteriovenous malformations. *J Neurosurg* 78: 5–11.

Porter, B. C., D. J. Rubens, J. G. Strang, J. Smith, S. Totterman, and K. J. Parker. 2001. Three-dimensional registration and fusion of ultrasound and MRI using major vessels as fiducial markers. *IEEE Trans Med Imaging* 20: 354–359.

Roberts, D. W., J. W. Strohbehn, J. F. Hatch et al. 1986. A frameless stereotaxic integration of computerized tomographic imaging and the operating microscope. *J Neurosurg* 64: 545–549.

Rubens, D. J., Y. Yu, A. S. Barnes, J. G. Strang, and R. Brasacchio. 2006. Image-guided brachytherapy for prostate cancer. *Radiol Clin North Am* 44: 735–748.

Shirato, H., T. Harada, T. Harabayashi et al. 2003. Feasibility of insertion/implantation of 2.0-mm-diameter gold internal fiducial markers for precise setup and real-time tumor tracking in radiotherapy. *Int J Radiat Oncol Biol Phys* 56: 240–247.

Starr, P. A., A. J. Martin, J. L. Ostrem, P. Talke, N. Levesque, and P. S. Larson. 2010. Subthalamic nucleus deep brain stimulator placement using high-field interventional magnetic resonance imaging and a skull-mounted aiming device: Technique and application accuracy. *J Neurosurg* 112: 479–490.

Stöckl, B., M. Nogler, R. Rosiek et al. 2004. Navigation improves accuracy of rotational alignment in total knee arthroplasty. *Clin Orthop Relat Res* 426: 180–186.

Stummer, W., U. Pichlmeier, T. Meinel et al. 2006. Fluorescence-guided surgery with 5-aminolevulinic acid for resection of malignant glioma: A randomized controlled multicentre phase III trial. *Lancet Oncol* 7: 392–401.

Talairach, J. and J. Bancaud. 1973. Stereotaxic approach to epilepsy. Methodology of anatomo-functional stereotaxic investigations. *Progr Neurol Surg* 5: 297–354.

Tufail, Y., A. Matyushov, N. Baldwin et al. 2010. Transcranial pulsed ultrasound stimulates intact brain circuits. *Neuron* 66: 681–694.

Villarreal, G., D. A. Hamilton, H. Petropoulos et al. 2002. Reduced hippocampal volume and total white matter volume in posttraumatic stress disorder. *Biol Psychiatry* 52: 119–125.

Watanabe, E. 1993. The neuronavigator: A potentiometer-based localization arm system. In *Interactive Image-Guided Neurosurgery*, ed. R. J. Maciunas, pp. 135–148. Park Ridge, IL: American Association of Neurological Surgeons.

West, J., J. M. Fitzpatrick, M. Y. Wang et al. 1997. Comparison and evaluation of retrospective intermodality image registration techniques. *J Comput Assist Tomogr* 21: 554–566.

Zernov, D. N. 1890. L'encéphalomètre. *Rev Gen Clin Ther* 19: 302.

11

Optical Methods for Caries Detection, Diagnosis, and Therapeutic Intervention

Daniel Fried
*University of California,
San Francisco*

11.1 Optical, Physical, and Thermal Properties of Dental Hard Tissues

11.1.1 Optical Properties of Dental Hard Tissue in the Visible and Near-IR

Dental enamel is an ordered array of rods of inorganic apatite-like crystals surrounded by a protein/lipid/water matrix. The crystals are approximately 30–40 nm in diameter and can be as long as 10 µm. The crystals are clustered together in 4 µm diameter rods (or prisms), which are roughly perpendicular to the tooth surface. Dentin is honeycombed with dentinal tubules of 1–3 µm in diameter. Each of these tubules is surrounded by a matrix of needle-shaped, hydroxyapatite (HAP)-like crystals in a protein matrix largely composed of collagen. The scattering distributions of these complex tissues are generally anisotropic and depend on tissue orientation relative to the irradiating light source[2-5] in addition to the polarization of the incident light.

The optical properties of biological tissue can be completely and quantitatively described by defining the optical constants, the absorption (μ_a), and scattering coefficients (μ_s), which represent the probability

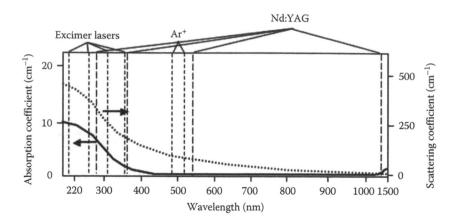

FIGURE 11.1 The scattering coefficient (μ_s), dotted line; and the absorption coefficient (μ_a), solid line; of dental enamel from the UV–near-IR compiled from Refs. [4,10,14,22] are shown. Laser wavelengths of interest are indicated by the vertical dashed lines.

of the incident photons being absorbed or scattered and the scattering phase function $\Phi(\cos(\theta))$, which is a mathematical function that describes the directional nature of scattering.[6–9] With the knowledge of these parameters, light transport in dental hard tissue can be completely characterized and modeled. The optical parameters for normal enamel and dentin have been reported between the wavelength range of 200–700 nm. For enamel, absorption is very weak in the visible range ($\mu_a < 1$ cm^{-1}, $\lambda = 400$–700 nm) and increases in the ultraviolet (UV) ($\mu_a > 10$ cm^{-1}, $\lambda < 240$ nm).[10] For dentin in the 400–700 nm wavelength range, the absorption coefficient is essentially wavelength independent with a value of $\mu_a \sim 4$ cm^{-1}.[13] Scattering in enamel is strong in the near-UV and decreases $\sim \lambda^{-3}$ with increasing wavelength to a value of only 2–3 cm^{-1} at 1550 nm[14,15] (see Figure 11.1). In contrast, scattering in dentin is strong throughout the near-UV, visible, and near-IR.[4,13]

Accurate description of light transport in dental hard tissue relies on knowledge of the exact form of the phase function $\Phi(\cos(\theta))$ for each tissue scatterer at each wavelength.[9] Therefore, direct measurement of the phase function is necessary. It is important to note that the empirically derived Kubelka–Munk (KM) coefficients that are commonly used in dentistry are not fundamental optical constants and are not appropriate for describing light transport in tissue with forward directed scattering such as enamel and dentin, particularly in the near-IR.[6,8] Scattering in most biological tissues can be represented by a Henyey–Greenstein (HG) function with values of (g) greater than 0.8 scattering.[6–9] The scattering anisotropy (g) should be determined within the context of an appropriate phase function based on the nature of the scatterers in the tissue[6,8] and subsequently validated through comparison of simulated scattering distributions with measured distributions of various thickness.[4] Measured angular-resolved scattering distributions could not be represented by a single scattering phase function $\Phi(\cos \theta)$ and required a linear combination of a highly forward peaked phase function, an HG function, and an isotropic phase function represented by the following equation[4]:

$$\Phi(\cos\theta) = f_d + (1 - f_d)\left(\frac{(1 - g^2)}{(1 + g^2 - 2g\cos\theta)^{3/2}}\right). \tag{11.1}$$

The parameter f_d for *fraction diffuse* is defined as the fraction of isotropic scatterers. The average value of the cosine of the scattering angle (θ) is called the scattering anisotropy (g), and $g = 0.96$ and 0.93 for enamel and dentin, respectively, at 1053 nm.[4] Note that Zijp et al.[2,5] reported values of $g = 0.4$ for dentin and $g = 0.68$ for enamel, calculated by taking the ratio of the forward and backward scattered light. This latter approach is prone to error due to the contribution of surface scattering.[4] The fraction of isotropic scatterers, (f_d), was measured to be 36% for enamel and less than 2% for dentin at 1053 nm. The phase function changes with wavelength; therefore, the phase function has to be independently determined at each wavelength.

Increased backscatter from the demineralized region of early caries lesions is the basis for the visual appearance of white spot lesions.[16] Increased porosity of the lesion leads to increased scattering at the lesion surface and higher scattering in the body of the lesion, producing an increase in the magnitude of the diffuse reflectance.[17] Attempts at measuring the optical properties of dental caries have been limited to measurements of backscattered light from optically thick, multilayered sections of simulated caries lesions.[17–19] In those measurements, empirical KM coefficients were calculated representing the diffusion of light through the tissue. Ko et al.[19] recently proposed that the optical scattering power—the product of the KM scattering parameter and the lesion depth—is a good estimate of enamel demineralization.

Enamel exhibits double refraction or birefringence upon illumination in the visible and near-IR.[20] The apatite crystals in dental enamel are highly oriented along the long or c-axis of the enamel prismatic structures. The major source of birefringence in dentin is the highly oriented collagen fibrils in the collagen; the collagen fibers are positively birefringent, while the apatite crystals are negatively birefringent. Demineralized enamel and dentin appears black under observation through crossed polarizers due to depolarization of the incident polarized light. Theuns et al.[21] have related changes in the birefringence of dental enamel during demineralization to the mineral content. Polarized-light microscopy measurements over the past several decades on thin sections have demonstrated that carious lesions rapidly depolarize incident polarized light.

11.1.2 Optical Properties of Dental Hard Tissue in the IR

In the mid-IR, scattering is negligible and with accurate knowledge of the absorption coefficient and the reflectivity, the light deposition can be completely described using Beer's law. The magnitude of the absorption coefficient is high in the mid-IR due to resonant absorption by molecular groups in water, protein, and mineral. The reflectivity can exceed 50% at wavelengths coincident with mineral absorption bands due to the large increase in the imaginary component of the refractive index (see Figures 11.2 and 11.3). Coefficients corresponding to CO_2 laser lines that could be obtained using conventional transmission were determined to be 1168 ± 49 cm^{-1} at 10.3 μm and 819 ± 62 cm^{-1} at 10.6 μm for enamel and 1198 ± 104 cm^{-1} at 10.3 μm and 813 ± 63 cm^{-1} at 10.6 μm for dentin. Enamel absorption coefficients corresponding to Er:YAG (2.94 μm) and Er:YSGG (2.79 μm) were calculated to be 768 ± 27 cm^{-1} and 451 ± 29 cm^{-1}, respectively.

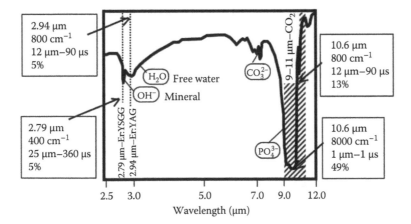

FIGURE 11.2 An IR transmission spectrum of dental enamel is shown with the principal molecular absorption groups indicated along with the relevant laser wavelengths. In each box under the indicated laser wavelength are the measured absorption coefficient, depth of absorption, thermal relaxation time calculated from the absorption depth, and the reflectance taken from Refs. [22,23,244].

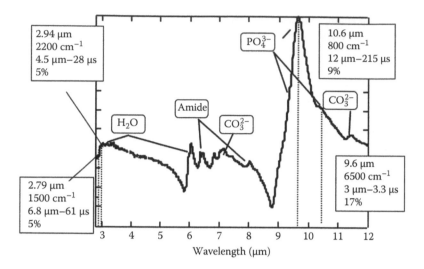

FIGURE 11.3 An IR transmission spectrum of dentin is shown with the principal molecular absorption groups indicated along with the relevant laser wavelengths. In each box under the indicated laser wavelength are the measured absorption coefficient, depth of absorption, thermal relaxation time calculated from the absorption depth, and the reflectance taken from Ref. [244].

The absorption coefficient of dentin at 2.79 μm was calculated to be 988 ± 111 cm^{-1}. These absorption coefficients were based on the Beer–Lambert law accounting for Fresnel reflectance losses. Conventional transmission measurements were not possible for the determination of the optical properties of enamel and dentin at 9.3 and 9.6 μm, and therefore alternative methods such as angular-resolved reflection measurements of polarized-light and time-resolved radiometric measurements were necessary. Duplain et al.[22] measured absorption coefficients of 18,700, 31,300, 6,500, and 5,200 cm^{-1} at 9.3, 9.6, 10.3, and 10.6 μm, respectively, using angular-resolved reflection measurements of polarized light. The magnitude of the coefficients of Duplain et al.[22] in the wavelength range in which direct transmission measurements were possible, 10.6 and 10.3 μm, is not consistent with the dental enamel absorption coefficients determined using direct transmission measurements (1168 and 819 cm^{-1} for 10.3 and 10.6 μm, respectively). Moreover, those values are also not consistent with observed surface modification thresholds based on the known melting range of dental enamel (800°C–1200°C). More recent measurements employing time-resolved radiometry measurements coupled with numerical simulations of thermal relaxation in the tissue[23] are a factor of five to six times lower. Those values are shown in Figures 11.2 and 11.3 for enamel and dentin along with the absorption depth, reflectance, and corresponding thermal relaxation times computed from the absorption depth and the thermal property data shown in Table 11.1. These later values are consistent with time-resolved temperature measurements during laser irradiation and surface melting thresholds.[24]

TABLE 11.1 Thermal and Mechanical Properties of Dental Hard Tissues

Tissue Property	Enamel	Dentin
Density (g/cm³)[1]	2.8	2.0
Specific heat (J/g°C)[1]	0.71	1.59
Thermal conductivity (W/cm°C)[1]	0.0093	0.0057
Coefficient of linear expansion (μm/°C)[11]	27	
Melting point (°C)[12]	1280	
Compressive strength (MPa)[1]	250–550	300–380
Tensile strength (MPa)[1]	10–70	50–60

11.1.3 IR Reflectance of Dental Hard Tissue

At CO_2 wavelengths, the reflection losses during enamel irradiation are substantial and reduce the laser energy absorbed by the target surface; near 3.0 μm losses are minimal. The fraction of incident laser light reflected at the surface of the tooth is described by the Fresnel reflection formula, $R = [(n_r - 1)^2 + k^2]/[(n_r + 1)^2 + k^2]$, where n_r is the real component of the refractive index and k is the attenuation index. The absorption coefficient, μ_a, k, and the wavelength λ are related by the expression $\mu_a = 4\pi k/\lambda$. Thus, the reflectance of materials can increase markedly and approach 100% in regions of strong absorption, for example, some metals have absorption coefficients $>10^6$ cm^{-1} and reflectance >99% in the visible and IR. The reflectance of dentin and enamel was measured at the λ = 2.79, 2.94, 10.6, 10.3, 9.6, and 9.3 μm wavelengths[22,25,26] (see Figures 11.2 and 11.3). The reflectance of enamel is substantially higher at λ = 9.6 and 9.3 μm than at λ = 10.6 μm, near 50%, and must be accounted for when calculating ablation efficiencies, ablation thresholds, and the heat deposition in the tooth. Transient and permanent changes in the reflectance of enamel and dentin were observed during and after laser irradiation.[25] These changes resulted in increased energy coupling during irradiation at λ = 9.3 and 9.6 μm for irradiation intensities that raised the enamel surface temperature several hundred degrees. The reflectance of dentin at λ = 9.6 μm permanently increased by as much as 30% as a result of laser irradiation. This change can be attributed to an increase in the mineral density and loss of organics.[25]

11.2 Optical Caries Diagnostics

During the past century, the nature of dental decay or dental caries in the United States has changed markedly due to the introduction of fluoride to the drinking water, the advent of fluoride dentifrices and rinses, and improved dental hygiene. In spite of these advances, dental decay continues to be the leading cause of tooth loss in the United States.[27–29] By age 17, 80% of children have experienced at least one cavity.[30] In addition, two-thirds of adults age 35–44 have lost at least one permanent tooth to caries. Older adults suffer tooth loss due to the problem of root caries. The nature of the caries problem has changed dramatically with the majority of newly discovered caries lesions being highly localized to the occlusal pits and fissures of the posterior dentition and the interproximal contact sites between teeth (Figure 11.4). These early carious lesions are often obscured or *hidden* in the complex and convoluted topography of the pits and fissures or are concealed by debris that frequently accumulates in those regions of the posterior teeth. Moreover, such lesions are difficult to detect in the early stages of development. By definition, early caries lesions are those lesions confined to the enamel or incident on the dentin–enamel junction (DEJ). In the caries process, demineralization occurs as organic acids generated by bacterial plaque diffuse through the porous enamel of the tooth dissolving the mineral. If the decay process is not arrested, the demineralization spreads through the enamel and reaches the dentin where it rapidly accelerates due to the markedly higher solubility and permeability of dentin. The lesion spreads throughout the underlying dentin to encompass a large area, resulting in loss of integrity of the tissue and cavitation. Caries lesions are usually not detected until after the lesions have progressed to the point at which surgical intervention and restoration are necessary, often resulting in the loss of healthy tissue structure and weakening of the tooth. Carious lesions also occur adjacent to existing restorations, and diagnostic tools are needed to diagnose the severity of those lesions and determine if an existing restoration needs to be replaced. The diagnostic and treatment paradigms that were developed in the past such as radiography are adequate for large, cavitated lesions; however, they do not have sufficient sensitivity or specificity for the diagnosis of early noncavitated lesions, root surface caries, or secondary caries.

New diagnostic tools are needed for the detection and characterization of caries lesions in the early stages of development. In a recent consensus statement released by the NIH entitled "The diagnosis and management of dental caries throughout life," the development of new devices and techniques for caries diagnosis was identified as one of five major areas in which additional research is needed.[30] Caries lesions are routinely detected in the United States using visual/tactile (explorer) methods coupled with radiography.

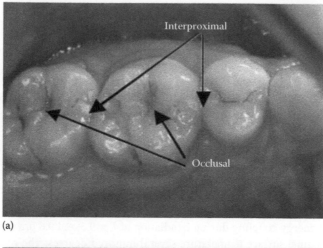

(a)

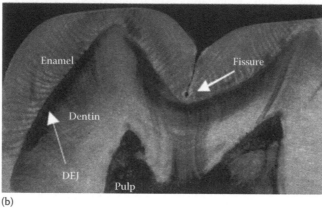

(b)

FIGURE 11.4 (a) The primary (high risk) sites for dental decay are in the pits and fissures of the posterior teeth and the approximal contact sites between teeth. (b) A typical fissure with a diameter of 200 μm is shown in a tooth cross section.

Unfortunately, these methods have numerous shortcomings and are inadequate for the detection of the early stages of the caries process.[31–33] Radiographic methods do not have the sensitivity for early lesions, particularly occlusal lesions, and by the time the lesions are radiolucent, they have often progressed well into the dentin at which point surgical intervention is necessary.[33–35] At that stage in the decay process, it is far too late for preventive and conservative intervention, and it is necessary to remove a large portion of carious and healthy tissue often compromising the mechanical integrity of the tooth. If left untreated, the decay will eventually infect the pulp, leading to the loss of tooth vitality and possible extraction. The caries process is potentially preventable and curable. If carious lesions are detected early enough, it is likely that they can be arrested/reversed by nonsurgical means through fluoride therapy, antibacterial therapy, and dietary changes or by low-intensity laser irradiation.[30,36] Therefore, one cannot overstate the importance of detecting the decay in the early stage of development at which point noninvasive preventive measures can be taken to halt further decay.

In a recent review of conventional methods of caries diagnosis, ten Cate[35] indicated that visual and tactile caries diagnosis was far from ideal and that visual diagnosis of occlusal caries typically has a very low sensitivity. Sensitivities scatter around a value of 0.3, implying that only 20%–48% of the caries present (usually into the dentine) are found.[35,37,38] The specificity typically exceeds 0.95. The poor sensitivity can be attributed to the *hidden* nature of the majority of occlusal lesions. The area of the lesion accessible for visual and tactile inspection is typically confined to the upper region of the fissure. The bulk of the lesion is not accessible and is most often not detected unless it is so extensive that it is resolvable radiographically.

11.2.1 Optical Transillumination (Interproximal Lesions)

Interproximal areas between teeth are generally inaccessible for visual inspection. Bitewing radiographs are the standard method of diagnosis. Unfortunately, as much as 25% of the interproximal areas of bitewing x-rays are unresolved due to the overlap with healthy tooth structure on adjoining teeth.[39,40] Values for sensitivity (0.38 and 0.59) and specificity (0.99 and 0.96) are reported for visual and radiological diagnosis, respectively.[41]

Optical transillumination was used extensively before the discovery of x-rays for detection of dental caries. Recently, there has been renewed interest in this method with the availability of high-intensity fiber-optic-based illumination systems for the detection of interproximal lesions.[39,42–45] During fiber-optic transillumination (FOTI), a carious lesion appears dark upon transillumination because of decreased transmission due to increased scattering and absorption by the lesion. A digital lesion has been developed by Electro-Optical Sciences, Irvington, NY,[46] and recently received FDA approval.

Since enamel is virtually transparent at longer wavelengths in the near-IR with optical attenuation 1–2 orders of magnitude less than in the visible range, this wavelength range is ideally suited for transillumination imaging.[14,47–49] Measurements of light scattering in natural lesions at 1310 nm show that the attenuation increases by 2–3 orders of magnitude due to demineralization, and therefore the optimum contrast between sound and demineralized enamel lies in the near-IR.[50] Figure 11.5 shows a visible light image of two teeth placed in contact along with a near-IR transillumination image of the contact point taken using a 1310 nm light source. The enamel is transparent in the near-IR image and the interproximal lesion and the dentin appear dark due to the higher scattering. The lesion is clearly visible with high contrast. The internal incremental growth lines of the enamel are also visible within the transparent enamel.

Other wavelengths have been investigated including the region accessible to conventional silicon-based CCD cameras, namely, 830 nm; however, the maximum contrast lies in the 1300–1600 nm region. The image contrast of tooth sections up to 7 mm thick with simulated lesions was measured at visible 830 and 1310 nm wavelengths and only at 1310 nm was high contrast acquired through the maximum thickness of enamel.[48] The imaging system operating near 830 nm, utilizing a low-cost

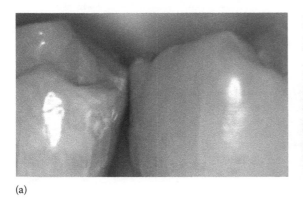

(a)

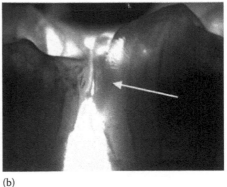

(b)

FIGURE 11.5 (a) Visible light image of the proximal contact point of two extracted teeth placed in contact. (b) Near-IR transillumination image of the same contact point using 1310 nm light source and InGaAs camera showing the interproximal lesion at the position of the arrow with high contrast.

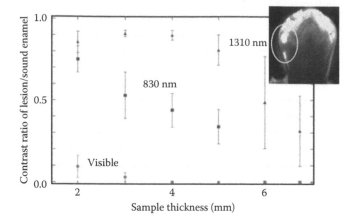

FIGURE 11.6 Plot of the optical contrast between sound enamel and the lesion area measured at three wavelength regions: visible (CCD camera), 830 nm (CCD camera without IR filter), and 1310 nm (InGaAs camera) for enamel sections of 1–7 mm thickness. Inset on upper right shows near-IR image of 3 mm tooth section with lesion in the white circle. (From Jones, G. et al., Transillumination of interproximal caries lesions with 830-nm light, in *Proc SPIE*, Lasers in Dentistry X, SPIE, San Jose, CA, Vol. 5313, pp. 17–22, 2004.)

silicon CCD optimized for the NIR, is capable of significantly higher performance than a visible system, but does not provide as significantly high a contrast as that attainable at 1310 nm as can be seen in Figure 11.6.[48] Imaging in the near-IR has major advantages, due to the markedly higher mean-free path through the enamel; lesions can also be imaged from the occlusal surface in addition to the traditional approach of imaging through the proximal contact points. This is demonstrated in Figure 11.7 where near-IR images are shown taken from the occlusal surface in addition to the more traditional buccal–lingual view. The radiograph is also shown for comparison. The images taken from multiple angles are advantageous since they allow better characterization of the lesion volume than a simple projection image. Many of the chromophores responsible for stains do not absorb in the near-IR and thus do not interfere allowing easier discrimination of caries lesions particularly in the important occlusal surfaces. Near-IR imaging also has great potential for the examination of defects in tooth structure, and cracks in enamel can be clearly resolved with this method due to the high transparency of the enamel in the near-IR.

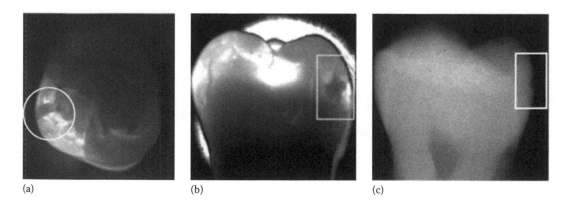

FIGURE 11.7 Multiple views of a tooth with an interproximal lesion acquired using a 1310 nm light source and an InGaAs camera: (a) occlusal near-IR view (lesion in white circle), (b) buccal–lingual near-IR view (lesion in white square), and (c) bitewing radiograph for comparison.

11.2.2 Fluorescence (DIAGNOdent™ and QLF)

Teeth naturally fluoresce upon irradiation with UV and visible light. Alfano[51,52] and Bjelkhagen and Sundstrom[53] demonstrated that laser-induced fluorescence (LIF) of endogenous fluorophores in human teeth could be used as a basis for discrimination between carious and noncarious tissues. Upon illumination with near-UV and visible light and imaging of the emitted fluorescence in the range of 600–700 nm, carious/demineralized areas appear dark. The origin of the endogenous fluorescence in teeth in this particular wavelength range has not been established.

Hafstroem-Bjoerkman et al.[54] established an experimental relationship between the loss of fluorescence intensity and extent of enamel demineralization. The method was subsequently labeled the quantitative light fluorescence (QLF) method, for quantitative laser fluorescence. An empirical relationship between overall lesion demineralization (ΔZ) and loss of fluorescence was established, which can be used to monitor lesion progression on smooth surfaces.[55-57] Recent measurements suggest that QLF may be useful for determination of the degree of hydration of the lesion that in turn may be indicative of the lesion activity.[58] Unfortunately, it is difficult to apply QLF to occlusal and interproximal lesions that constitute the majority of carious lesions. Furthermore, the fluorescence method cannot be used to provide information about the subsurface characteristics of the lesion.

Bacteria produce significant amounts of porphyrins and dental plaque fluoresces upon excitation with red light.[59] A novel caries detection system, the DIAGNOdent (Kavo, Germany), was recently developed and received FDA approval in the United States. This device uses a diode laser and a fiber-optic probe designed to detect the near-IR fluorescence from porphyrins.[60] This low-cost diagnostic tool is a major step toward better caries detection in occlusal surfaces, greatly aiding in the detection of *hidden* occlusal lesions, namely, lesions that have progressed into the dentin and are too small to show up on a radiograph.[61] It is important to note that the DIAGNOdent has poor sensitivity (~0.4) for early lesions confined to enamel and it cannot provide information about the lesion depth and the degree of severity.[62,63]

11.2.3 Optical Coherence Tomography for Caries Imaging

Optical coherence tomography (OCT) is a sensitive method for resolving changes in light scattering in early caries lesions. Excellent texts for explaining the mechanics of OCT and polarization sensitive OCT (PS-OCT) edited by Bouma and Tearney[64] and Brezinski[65] are available, and both these texts include sections on dentistry. The first OCT images of soft and hard tissue structures of the oral cavity were acquired by Colston et al.[66-68] Baumgartner et al.[69,70] presented the first polarization-resolved images of dental caries. Feldchtein et al.[71] presented high-resolution dual-wavelength 830 and 1280 nm images of dental hard tissues, enamel and dentin caries, and restorations in vivo. Wang et al.[72] measured the birefringence in dentin and enamel and suggested that the enamel rods act as waveguides. Everett et al.[73] presented polarization-resolved images using a high-power 1310 nm broadband source and a bulk optic PS-OCT system. In those images, changes in the mineral density of tooth enamel were resolvable to depths of 2–3 mm into the tooth.

There are three important advantages of PS-OCT over conventional OCT: First, the strong surface reflection at the tooth surface can be reduced to allow better visualization of the surface structure of the lesions. Second, depolarization due to strong scattering in demineralized enamel and dentin allows higher contrast of the lesion in the images. Third, subsurface structure (artifacts) caused by the tooth birefringence can be identified. Figure 11.8 shows PS-OCT scans across the occlusal surface of a premolar tooth with a shallow lesion in the area of the fissure. The light was incident on the tooth in one linear polarization (∥), and the reflectivity was measured in both orthogonal polarization states defined as parallel (∥) and perpendicular (⊥) to the incident linear polarization. The reflectivity from the tooth surface is greatly reduced in the orthogonal polarization image (⊥), and the lesion appears with greater contrast. Figure 11.9 contains an image of another tooth with a natural lesion that shows considerable internal structure that can be resolved in the orthogonal polarization image. There are areas of reduced scattering near the surface that may be indicative

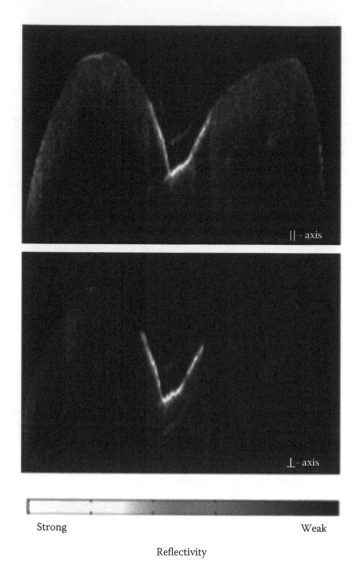

FIGURE 11.8 PS-OCT scans of the occlusal fissure of a tooth with shallow demineralization in the fissure of the occlusal surface. The parallel scan (∥) is shown on top with the reflectivity aligned with the original polarization state incident on the tooth. The scan representing the light reflected in the orthogonal (⊥) polarization is shown in the bottom image. Note higher contrast of the lesion in the (⊥) polarization image and the weak reflectivity from the tooth surface.

of remineralization, and those regions of the lesion may be inactive. In the area between the dotted lines, the scattering near the surface of the lesion is very strong. This may be indicative of an active area of the lesion that is progressing. Being able to access the activity of the lesion, that is, whether it is active and needs intervention or if it has been arrested and can be left alone, is extremely important to the clinician, and no current diagnostic technology is available capable of such a diagnosis. Therefore, one of the most exciting potential applications of OCT is for monitoring the remineralization of existing lesions, and initial studies suggest that it is ideally suited for this task.[74-77] Figure 11.10 shows a PS-OCT image, a polarized-light micrograph (PLM), and a transverse microradiograph (TMR) of a human dentin sample that was exposed to a demineralization solution (left side) and a demineralization solution followed by a remineralization solution (right side). The lower reflectivity surface layer of remineralized dentin is clearly visible in the PS-OCT image.

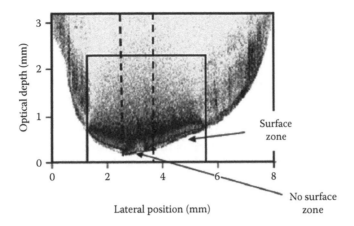

FIGURE 11.9 PS-OCT scan, orthogonal (⊥) polarization, of an obvious smooth surface interproximal lesion (white spot lesion). The scan was taken perpendicular to the long axis of the tooth from the facial to lingual surface. The dark areas on the tooth surface indicate the position of the lesion. The lesion, which is enclosed in the box, appears more active between the dotted lines where it lacks a surface zone of reduced scattering. The surface of the lesion contains a thin surface zone of reduced scattering outside of the dotted lines that may indicate remineralization.

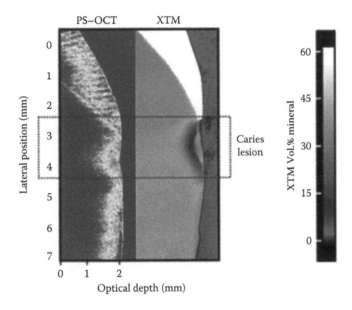

FIGURE 11.10 **(See color insert.)** Comparison of the PS-OCT (⊥) image (orthogonal polarization state) with the corresponding XTM of the mineral density taken from the same region of the tooth (right) (see Ref. [80]). A small root caries lesion is present just below the cementum–enamel junction shown in the box. The intensity of the OCT images ranges from 12 to −45 dB; areas with regions of intensity greater than −5 dB are shown in red and those of intensity less than −35 dB are shown in blue. In the XTM image on the right, normal dentin is yellow, enamel is white, the water outside the tooth is indicated in red, and the demineralized area of the lesion is blue (color bar on right).

According to Kidd,[78] 75% of operative dentistry is the replacement of existing restorations, and secondary caries is the most common reason given for replacing both amalgam and composite fillings. PS-OCT is well suited for the detection of secondary caries since restorative materials have markedly different scattering properties compared to dental hard tissues. Coloton[68] and Feldchtein[71] have shown that OCT can be used to differentiate between restorative materials and enamel and detect decay around the periphery of a restoration.[71] Fried et al.[79,80] have recently demonstrated that composite can be distinguished from dentin and enamel using PS-OCT and that the degree of demineralization underneath composite restorations can be quantified.

Root caries are an increasing problem among our aging population and is the principal reason for tooth loss above the age of 44. The scattering coefficient of dentin is much higher than enamel; therefore, the penetration depth is more limited than for enamel. However, early root caries are localized on the surface of the dentin and cementum. It is difficult to obtain intact thin sections through natural root caries lesions without damage to the fragile, poorly mineralized regions of the lesion. High-resolution x-ray tomography (XTM) can be used to acquire tomograms of the mineral density in the tooth with a resolution of 9 μm.[81,82] XTM and PS-OCT scans taken from the crown to the root across a small root caries lesion located just below the cementum–enamel junction on a human tooth are shown in Figure 11.10. The principal advantage of XTM is that tissue slices representing the mineral density versus position can be extracted from the XTM with any desired orientation to match the OCT scan geometry. The XTM image of the area scanned by the PS-OCT system is shown on the right of Figure 11.10. Taking into account that the high-refractive indices of dentin and enamel result in compression of those areas of the image, the agreement between the two imaging modalities is excellent. The entire lesion morphology is reproduced in the fast-axis OCT image with the characteristic lesion semicircle shape being clearly visible. The fast-axis image provides the best match since the confounding influence of the surface reflection and residual coherence are markedly reduced.

High-speed Fourier domain systems (FD-OCT) have also been investigated for imaging dental caries.[83–86] These systems employing either spectral domain or swept laser approaches can be used to acquire images with high resolution and can be modified to also have large scan ranges even though

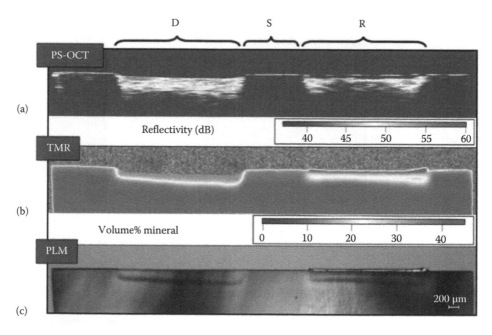

FIGURE 11.11 **(See color insert.)** (a) (⊥-axis) PS-OCT scan of human dentin taken along long axis of a sample before sectioning. Intensity scale is shown in red–white–blue in units of decibels (dB). Remineralized layer is visible as a blue gap above the remineralized lesion. (b) The corresponding TMR of a 260 μm thick section in a similar orientation to the PS-OCT image. The intensity scale is also shown in red–white–blue with units in volume% mineral. (c) The PLM image of the same section is shown at 15× magnification. (From Manesh, S.K. et al., Assessment of dentin remineralization with polarization sensitive optical coherence tomography, in *Proc SPIE*, Lasers in Dentistry XV, SPIE, San Jose, CA, Vol. 7162W, pp. 1–7, 2009.)

most of the turn key systems currently available have a scan range limited to under 4 mm, which is a problem for scanning dental hard tissue due to the occlusal surface topography and the high-refractive index of dental enamel, 1.63 (Figure 11.11).

11.3 Therapeutic Applications on Dental Hard Tissues

Since the initial investigations of Stern[87] over 30 years ago, several unique laser dental applications have evolved for restorative dentistry, namely, laser ablation of dental hard tissue,[88] caries inhibition treatments by localized surface heating,[89] and surface conditioning for bonding.[88] For several years, medical lasers have been approved by the FDA for soft tissue vaporization, and even though hard tissue applications have been investigated for over 30 years, only recently have the first lasers been approved for hard tissue use in the United States. The first lasers to receive approval were the Centuri™ Er:YAG and the Millennium™ Er:YSGG introduced by Premier Laser and Biolase Technology Inc., respectively.

11.3.1 New Preventive and Conservative Approach to Restorative Dentistry

Several developments in dentistry are driving the shift to new more conservative approaches to restorative dentistry than the classical *G. V. Black* approach that has been the standard for cavity preparation since the nineteenth century. These include the development of new more effective adhesive restorative materials, a more thorough understanding of the caries process, the now recognized role of fluoride as a remineralizing agent, new and more sensitive methods of caries detection, and the public's demand for more aesthetic restorations. Over the past 50 years, the nature of dental caries has changed with the majority of new lesions, 90%, occurring in the pits and fissures (occlusal surfaces) of posterior teeth[90,91] (see Figure 11.4).

New caries diagnostic procedures, such as those described earlier, are becoming available that enable the dentist to identify early caries lesions before they have spread extensively into the underlying dentin. For lesions of this magnitude, it is too early to use conventional restorations that require the removal of large amounts of healthy tissue. As part of the more conservative approach to restorative dentistry, some experts have revised the G. V. Black cavity classification scheme to emphasize the importance of early pit and fissure sites.[92] This new conservative approach emphasizes micropreparation with minimal removal of healthy tissue. The laser is ideally suited for this approach for the following reasons:

(1) Laser pulses can preferentially ablate carious tissue due to the higher volatility of water and protein that are present in carious tissue at a higher ratio than in normal tissue. (2) They can be tightly focused to drill holes with very high aspect ratios (depth/height) well beyond those obtainable by the dental drill that is limited by the size of the dental burr (>1 mm). Conservation of enamel structure is paramount for preservation of the natural dentition for retention of sealants and to limit exposure to wear. (3) The laser can be used to open up the neck of the fissures sufficiently to allow the entrance of flowable composite to the base of the fissure. (4) Laser irradiation does not produce a smear layer that needs to be removed; hence, restorative materials can be applied directly to the ablated area without the necessity of further surface preparation and etching.[93] This advantage is of great importance since fissure areas are difficult to etch by conventional means due to the unique enamel morphology. The enamel on the shoulder at the entry to the fissure is often prismless and irregular and may not accept a good etch pattern, so attachment of the resin may be tenuous. Therefore, laser preparation may be superior to the conventional acid etch in pits and fissures. (5) Laser radiation vaporizes water and protein and changes the chemical composition of the remaining mineral of enamel and dentin, thus decreasing the solubility to acids around the periphery of the restoration site to leave a smooth surface with an enhanced resistance to secondary caries.[89,94–96] (6) Lasers are less likely to require anesthesia and induce less noise and pain.[97–100] Therefore, they are advantageous for treating children and patients with dental phobias.

Thus, lasers are ideally suited for minimally invasive surgical intervention and have the potential of markedly reducing or eliminating the loss of adjacent healthy tissue during the removal of carious tissue from early lesions or the removal of existing composite restorations. The laser can be precisely focused into the carious fissure of a molar to remove any debris and bacteria, ablate away the carious tissue, and enlarge the narrow neck of the fissure leaving a smooth crater. If the appropriate irradiation conditions are met, the walls will have enhanced resistance to acid dissolution.[95,96] A flowable composite can subsequently be applied for a highly conservative restoration. By taking this approach, the dentist can intervene in a conservative manner before the lesion has progressed to the point that conventional surgical intervention is necessary.

11.3.2 IR Laser Ablation of Enamel

Enamel is a biological composite containing 12% by volume water, 85% mineral (carbonated HAP [CAP]), and 3% protein and lipid. The mineral component is crystalline in nature comprising hexagonal crystallites approximately 40 nm in diameter, which are aligned in enamel rods roughly 5 μm in diameter that run in an S-shaped pattern from the enamel surface to the dentin/enamel junction. A sheath of protein and water surrounds each rod. Approximately half of the water is bound tightly to the enamel as caged waters of hydration, and the other half is more loosely bound and located in between the individual enamel crystals and the rods.[101] The mechanism of interaction of laser light with dental enamel is inherently complex and varies markedly with the laser irradiation wavelength and the nature of the primary absorber in the tissue, water, protein, or mineral. The principal factor limiting the ablation rate of dental hard tissue is the risk of excessive heat deposition in the tooth that may lead to eventual loss of pulpal vitality. The accumulation of heat in the tooth can be minimized by using a laser wavelength tuned to the maximum absorption coefficient of the tissue irradiated and by judicious selection of the laser pulse duration to be commensurate with the thermal relaxation time of the deposited energy.

Several erbium laser wavelengths have been investigated for the ablation of dental enamel, namely, the Er:YAG, 2.94 μm; Er:YLF, 2.81; Er:YSGG, 2.79 μm; Er:YAP, 2.73 μm; and CTE:YAG, 2.69 μm.[102–106] Those wavelengths give access to the water absorption band from 2.6 to 3.0 μm. These lasers can be operated in either free running mode, in which the pulse duration can be varied from tens of μs to almost a ms, or Q-switched mode in which the pulse duration can vary from tens to hundreds of ns. Altshuler et al.[103] determined that the ablation threshold in free running mode increased in the following order: Er:YSGG, Er:YLF, Er:YAP, and Er:YAG. The CTE:YAG ablation threshold is considerably higher than the other erbium systems since absorption in water is weaker. Albeit, the $\lambda = 2.69$ μm laser wavelength is still of significant interest since it is readily transmitted through robust, inexpensive low-OH fibers.[107,108] Three studies have investigated variation of the pulse duration during the ablation of enamel. Majaran et al.[109] observed a decrease in the ablation efficiency upon stretching the laser pulse from 200 to 1000 μs. Zeck et al.[110] observed a twofold increase in the ablation efficiency of enamel upon reducing 400 μs Er:YAG laser pulses to 30 μs and a fourfold reduction in the ablation threshold. An even greater reduction in the ablation threshold[111] was observed for Q-switched 200 ns laser pulses, particularly for Er:YSGG laser pulses used in conjunction with water.

Carbon dioxide (CO_2) lasers are the most common lasers found in clinics today and have been used for soft tissue surgical procedures for three decades. The CO_2 laser can be designed to operate or *lase* at discrete wavelengths of $\lambda = 9-11$ μm. Those wavelengths correspond to specific rotational–vibrational transitions in the ground state of gas phase CO_2 molecules. Early results with the CO_2 laser were discouraging because these studies used continuous wave lasers operating at $\lambda = 10.6$ μm[112–122] and extensive thermal damage was observed. Recent studies using CO_2 laser pulses of submillisecond duration indicate that enamel can be ablated efficiently without generating peripheral damage.[123–125] Several recent in vitro[123,124,126–131] and in vivo[132–137] studies have demonstrated the great potential for microsecond-pulsed 9.3 and 9.6 μm CO_2 lasers for hard tissue use, namely, caries ablation and caries prevention treatments. However, no company has yet received FDA approval for either procedure. One advantage of the CO_2 laser is that it can be operated at very high pulse repetition rates and scanned over the tooth surface to rapidly and precisely remove dental decay[138,139] or cut bone.[130,140] The low ablation thresholds

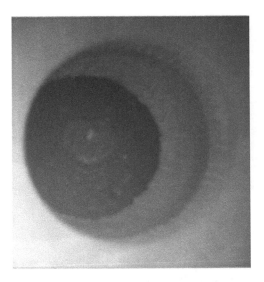

FIGURE 11.12 A cavity preparation (2 mm in diameter ~2 mm deep) produced with a 9.3 μm CO_2 laser at 300 Hz in approximately 10 s with integrated scanner.

for microsecond laser pulses at 9.3 and 9.6 μm facilitate also this approach. Figure 11.12 shows a cylindrical cavity approximately 2 mm in diameter and 2 mm deep that was produced using rapidly scanned 9.3 μm laser pulses of several microsecond duration delivered at 300 Hz. If water cooling is employed, the collagen-rich tissues of dentin and bone can be rapidly ablated without extensive peripheral thermal damage as can be seen in Figure 11.13.

The laser ablation of dental hard tissue is a water-mediated explosive process at 2.94 and 2.79 μm[104] caused by preferential absorption by water that is localized between the enamel rods of enamel or is intrinsic to the protein/mineral matrix of dentin. During rapid heating, the inertially confined water can create enormous subsurface pressures that can lead to the explosive removal of the surrounding mineral matrix.[141] Studies of hard tissue ablation in the $\lambda = 3$ μm region indicate that large intact particles are ejected with high velocity from the irradiated tissue.[104] Moreover, the normal highly ordered structure of dentin and enamel is conserved during the ablation process (see Figure 11.14). In contrast, the mechanism of ablation of dental hard tissue at the highly absorbed CO_2 wavelengths of 9.3 and 9.6 μm is apparently

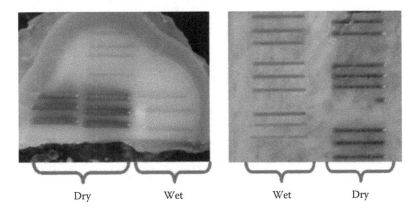

Dry   Wet   Wet   Dry

FIGURE 11.13 (Top) Incisions in dentin (left) and bone (right) produced with a 9.3 μm CO_2 laser operating at a wavelength of 9.3 μm with a pulse duration of 10–15 μs (see Ref. [126]) with and without a water spray at a pulse repetition rate of 400 Hz. No char, cracks, or discoloration is visible if water is used. Cross sections viewed under polarized light indicate the zone of thermal damage is less than 10 μm.

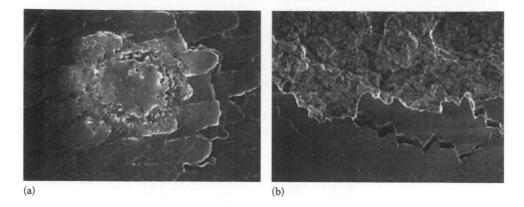

(a)   (b)

FIGURE 11.14   (a) SEM image (5000×) of an enamel surface irradiated by a pulsed 9.3 μm CO_2 laser near the surface modification threshold (see Ref. [26]). The enamel prisms are beginning to fuse together as a result of melting of the mineral phase. (b) SEM image (2000×) of an enamel surface irradiated by a pulsed 2.94 μm Er:YAG laser just above the surface modification threshold. Intact enamel prisms are ejected without melting of the surface.

mineral mediated, and SEM micrographs show that the tissue morphology is markedly changed after irradiation[106] (see Figure 11.14). Surface temperatures at the ablation threshold for enamel indicate that the mechanism of ablation is thermal and occurs at approximately 300°C–400°C for Er:YAG, 800°C for Er:YSGG, and 1200°C for CO_2 lasers.[142] Although it is advantageous to ablate enamel at lower surface temperatures to avoid heat deposition, it may also be useful to heat the tissue peripheral to the ablation site to temperatures exceeding 400°C. This creates a zone of increased acid resistance around the restoration site that may occur to a greater degree after Er:YSGG and CO_2 laser irradiation[143] than with Er:YAG laser irradiation.

11.3.3 Safety Issues

Heat may accumulate in the tooth to dangerous levels during multiple pulse laser irradiation. Hence, it is necessary to measure accumulation of heat in the tooth during ablation. The 1965 Zach and Cohen study of the effect of heat on the pulp of Rhesus monkeys indicated that a temperature rise of 5.5°C in the pulp caused irreversible pulpitis in 15% of the pulps.[144] This 5.5°C temperature rise is typically used as a temperature threshold that should not be exceeded. Higher temperatures may cause thermal damage to the pulp. The accumulation of heat after the use of the dental drill for cavity preparations[145] and for the finishing of restorative materials[146] can raise the pulpal temperatures to dangerous levels if air/water cooling is not used. Similarly, the use of multiple pulse irradiation without air/water cooling may also result in the accumulation of heat to levels dangerous to the pulp. Subsurface thermocouple measurements, simulations of heat conduction, and histological examinations during laser irradiation show that the extent of pulpal heating is determined by the rate of deposition of the laser energy in the tooth, the distance from the laser spot to the pulp, and the rate of energy loss from the tooth.[114,116–118,147–150] Miserendino et al.[151] observed that typically the temperature excursions in the pulpal wall occurred 10–20 s after irradiation of the surface of the tooth. Applying air/water cooling for 5 s after the 5 s continuous wave ($\lambda = 10.6$ μm), CO_2 laser exposures of 10–50 J (2–10 W and a 1 mm spot size) reduced the otherwise excessive temperature rise in the pulp chamber[151] to a safe level.

Strong acoustic waves can be generated during ablation.[152–155] These pressure waves can propagate through the tissue radiating outward from the site of absorption. Transient stress waves may cause mechanical damage to the surface enamel or dentin surrounding the ablation site, generate cracks, and increase porosity. Generally, biological tissue has a relatively high compressive strength; however, the tensile strength is much weaker and the tensile forces can cavitate soft tissue and generate cracks in hard tissue. A unipolar compressive stress incident on the interface with another tissue of lower density becomes a bipolar stress wave. The compressive stress wave propagates through the next medium, and a tensile wave or rarefaction wave is reflected back from the interface. It is the reflected tensile stress wave that has the greatest potential of causing tissue damage. The additional contribution of the explosive release of inertially confined water and carbon dioxide during laser heating and the subsequent ablative recoil may also generate large acoustic transients. The magnitude of the stresses generated in enamel and dentin during CO_2 laser irradiation has not been measured, and the effect of varying the pulse duration, fluence, and absorption depth has not been evaluated.

11.3.4 Water Augmentation of Ablation

One of the most important and somewhat controversial aspects of hard tissue ablation is the critical role of water in the ablation process. Initial studies of the role of water focused on preventing tissue dehydration.[156–159] Absorption and diffusion studies in enamel indicate that approximately half of the water in enamel is actually diffusible.[101] Thermal analysis studies indicate that the tissue has to be heated to greater than 200°C–300°C before most of that mobile water is removed.[12] However, the rate for water diffusion is quite slow, on the order of several hours to days; therefore, it is unlikely that sample rehydration has a significant effect during laser irradiation. Several studies have shown that the ablation rate of erbium lasers can be increased by applying a water spray or a layer of water to the surface.[160–162] Several interesting mechanisms have been proposed to explain this phenomenon. One hypothesis is that cavitation bubbles are formed in the water that cut the sound enamel due to the large tensile stresses generated by the collapse of the bubble.[160] Another proposed mechanism is that water droplets are rapidly accelerated into the enamel by absorption in the laser beam.[161,163] Other studies suggest that solid particles of ablated material are accelerated against the walls of the crater resulting in a polishing effect that removes debris and any protruding sharp edges.[164,165] Real-time near-IR images of ablation crater formation during ablation process[166,167] suggest different mechanisms of crater evolution between the Er:YAG and CO_2 lasers.

During high-intensity laser irradiation, marked chemical and physical changes may be induced in the irradiated dental enamel. These changes can have profound effects on the laser ablation/drilling process and may lead to a reduction in the ablation rate and efficiency, increase peripheral thermal damage, and even lead to stalling without further removal of tissue with subsequent laser pulses. Furthermore, thermal decomposition of the mineral can lead to changes in the susceptibility of the modified mineral to organic acids in the oral environment. Morphological changes may result in the formation of loosely attached layers of modified enamel that can delaminate leading to failure during the bonding to restorative materials.[168,169] Therefore, it is important to thoroughly characterize the laser (thermal)-induced chemical and crystalline changes after laser irradiation. The mineral, CAP, found in bone and teeth contains carbonate inclusions that render it highly susceptible to acid dissolution by organic acids generated from bacteria in dental plaque. Upon heating to temperatures in excess of 400°C, the mineral decomposes to form a new mineral phase that has increased resistance to acid dissolution.[89] Recent studies suggest that as a side effect of laser ablation, the walls around the periphery of a cavity preparation will be transformed through laser heating into a more acid-resistant phase with an enhanced resistance to future decay.[95,96,170] However, poorly crystalline nonapatite phases of calcium phosphate may have an opposite effect on plaque acid resistance[171] and may increase the quantity of poorly attached grains associated with delamination failures.

Zeck et al.[110] and Rechmann et al.[169] observed that a loosely attached layer of fused enamel was formed during Er:YAG laser ablation if a water spray was not used. Zeck labeled this fused enamel *recrystallites*. Some of the likely sources of these phases are recondensation of vaporized enamel, spallated droplets of ejected melted enamel, and repeated melting and recrystallization of the enamel.[172] Poorly crystalline-fused enamel particles and any surface protrusions or asperities in the ablation crater are likely to inhibit efficient ablation for FOTI system, called DiFoti, that utilizes visible light for the detection of caries subsequent laser pulses leading to stalling and excessive heat accumulation. Forces imparted to the enamel surface by recoiling water particles are not of a sufficient magnitude to ablate the normal intact enamel; however, they may have sufficient strength to cleanse the surface of the crater of these poorly crystalline, loosely adherent mineral phases after the preceding laser pulses. The same recoil forces produce cavitation of the water layer[173–175] and energetically propel the remaining water on the surface several cm from the tooth.

11.3.5 Laser Ablation of Dentin and Bone

Dentin is composed of an interwoven network of water, small mineral crystals a few nm across, and collagen fibers. Intertubular dentin is approximately 47% by volume mineral (CAP), 33% protein (mostly collagen), and 20% water. Bone has a lower volume fraction of mineral and is less dense than dentin. Dentin contains long tubules that are 1–3 μm in diameter surrounded by a collar of mineral (peritubular dentin). These tubules extend from the DEJ to the pulp. This uniform orientation of the tubules in dentin facilitates visualization of thermal damage using polarized-light microscopy. In contrast, thermal damage in bone is more difficult to visualize using polarized light due to the lack of uniform orientation of the concentric lamellae. The collagen in dentin and bone is susceptible to thermal denaturation and carbonization (charring). Carbonization and thermal denaturation of dentin may adversely affect the bonding of cements and composite restorative materials and inhibit healing. This is of particular importance for the treatment of root caries that is an increasingly important problem for our aging population. There have been several studies of bonding laser-treated enamel and dentin.[93,176–179]

Although the emphasis of this chapter is on caries ablation, the optical properties of bone are similar to dentin, and the laser conditions suitable for the ablation of dentin apply equally to bone. Several procedures in oral medicine require the removal and contouring of bone. Thermal damage to bone during laser cutting results in delayed healing time and bone defects.[180,181] Conventional technology often results in mechanical trauma, excessive thermal damage, and hemorrhage.[182,183] Specific areas of application include trepanation for implant placement, precise incisions for distraction osteogenesis for facial reconstruction and correction of craniofacial anomalies, and bone removal and recontouring for periodontology and oral surgery.

Free running Er:YAG lasers can be used to ablate tissue fairly efficiently in conjunction with a water spray without significant charring to surrounding tissue.[158,162] Absorption in dentin and bone at erbium laser wavelengths is markedly higher than in enamel due to the higher water content. Absorption is confined to a depth of 4–5 μm with a thermal relaxation time of 20–30 μs. However, it has been established that the absorption coefficient of water at 2.94 μm drops by almost an order of magnitude upon heating from room temperature to the critical point at 374°C.[184–186] Thus, it is reasonable to assume that absorption occurs to depths greater than 5 μm during laser irradiation; however, that needs to be confirmed experimentally.

Early studies using continuous wave CO_2 lasers operated at 10.6 μm reported extensive cracking and charring of surrounding dentin and bone.[180,181,187–189] Based on these disappointing initial observations, many laser researchers overlooked the potential of CO_2 laser-based systems for hard tissue ablation. Pulsed CO_2 lasers, however, are well suited for use on dental hard tissue due to strong absorption of water, protein, and mineral between 9 and 11 μm. Several workers[123,127,190–193] have demonstrated that pulsed CO_2 lasers with a pulse duration of less than 20 μs can be used to avoid carbonization of the biopolymer matrix of dentin with the simultaneous application of water cooling. Figure 11.13 shows incisions in dentin and bone produced with a short-pulsed 9.3 μm CO_2 laser operating at a pulse repetition rate of 400 Hz. A char is only visible for the incisions produced without a water spray, and the peripheral thermal damage was less than 10 μm if a water spray was used. Mineral absorption peaks at 9.6 μm where the absorption coefficient of dentin is approximately 8000 cm^{-1}. This high absorption coefficient indicates that the incident laser light will be absorbed at a depth of under 2–3 μm with a corresponding thermal relaxation time on the order of 5 μs. Ivanenko and Hering[130] demonstrated that a mechanically Q-switched 10.6 μm CO_2 industrial laser could be used to cut bone rapidly at 300 Hz without thermal damage; therefore, rapid processing is feasible.

11.3.6 Alternative Mechanisms and Wavelengths for Hard Tissue Ablation

The optical properties of dental hard tissues from λ = 0.4 to 2.7 μm are characterized by weak absorption and strong scattering and are not well suited for efficient ablation, which requires localization of the energy deposition to the surface of the tissue. Nevertheless, there has been some successful use of lasers emitting within this frequency range for ablation, and they warrant discussion. The absorption of the laser light can be increased by adding exogenous chromophores to the surface.[194–196] However, even with the added chromophore, typical ablation thresholds are still high (200 J/cm²) and ablation efficiencies low (0.03 mm³/J).[195] There have been some successes in the removal of dentin using the fiber-coupled Nd:YAG laser in the contact mode.[197–200] However, this laser wavelength cannot be used to remove sound enamel. Niemz[201] has used picosecond laser pulses to ablate carious and noncarious enamel. The short, highly focused laser pulses generate irradiation in excess of 10^{10} W/cm², and the strong electric fields ionize the tissue forming a plasma.[202,203] Plasma-mediated ablation has several advantages, namely, it produces very clean highly machined craters with minimal peripheral thermal damage, and suitable visible and near-IR lasers are available with high spatial beam quality and short pulse durations. Such laser systems, however, are expensive and sophisticated and yield rather low ablation rates.[204]

UV radiation is highly absorbed by collagen and is efficiently absorbed by dentin.[13,205–209] Hennig and Rechmann[210–212] used the frequency-doubled alexandrite laser (377 nm) to selectively ablate carious dentin. They found that there was at least a fourfold increase in the ablation rate of carious dentin over healthy dentin. At λ = 193 and 248 nm, there is strong absorption and minimal penetration;[206,213] however, the ablation efficiency is very low, 0.001 and 0.002 mm³/J, for enamel and dentin, respectively.[205,214]

11.3.7 Selective Removal of Composite-Secondary Caries

Dental composites are used as adhesives and restorative materials for various dental procedures. They are composed of acrylic resins combined with quartz, zirconia, silica, and fused silica filler materials. The most common application of composites is for restorative procedures, that is, fillings. Composites are color matched to the tooth making it difficult for the dentist to differentiate between the enamel and the restorative material. All composites from an existing restoration must be removed to assure proper bonding of the new composite to the tooth structure, since new composite does not bond well to residual composite. It is difficult for the clinician to differentiate between the composite of the restoration and the surrounding tooth. Most restorations will eventually fail and need to be replaced. Hence, the dentist frequently removes excessive amounts of healthy tooth structure to ensure complete removal of the composite. The applicability of lasers for the removal of dental composite and bone cement has been investigated for several laser systems including the free running Er:YAG,[215,216] Nd:YAG,[217,218] XeCl (308 nm),[219] and millisecond-pulsed CO_2 lasers.[220] All of those laser systems can be used to remove composites, as indicated in the articles listed previously; however, they lack selectivity, have undesirable ablation rates, or produce excessive thermal damage. Recently, greater selectivity has been demonstrated using a transverse excited atmospheric (TEA) CO_2 laser operating at 10.6 μm[221] and the frequency-tripled Nd:YAG laser operating at 355 nm.[222] In those studies, TEA pressure CO_2 laser with laser pulses of approximately 1 μs duration was used to ablate composite at a rate almost an order of magnitude higher than for enamel. Spectral analysis of the emission plume created by ablating composite and enamel identified a number of

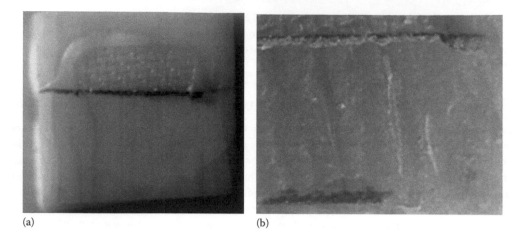

(a)   (b)

FIGURE 11.15 Selective removal of composite from enamel surfaces using nanosecond laser pulses at 355 nm. (a) The original acid etch margin and (b) the fine-structured perikymata are still visible after the removal of composite indicating no damage to the enamel surface (see Ref. [222]).

spectral lines that could be used to distinguish between the ablation of enamel and composite. However, there was measurable damage to the underlying enamel due to the similar ablation thresholds. Laser pulses of the third harmonic of the Q-switched Nd:YAG (355 nm) can be used to completely remove composite from the surface of the tooth with no discernable damage to the underlying enamel, thus achieving the desired selectivity.[222] This can be seen in Figure 11.15 that shows images of bovine and human enamel surfaces after composite was selectively ablated with 355 nm laser pulses. There is no damage to the enamel by the laser, and even the damage caused by the original acid etch to place the composite is still visible after composite removal.

11.3.8 Surface Modification for Caries Prevention: Enamel

The earliest dental laser studies of Stern and Yammamoto showed that high-intensity laser radiation rendered enamel mineral more resistant to acid dissolution.[223–225] However, such studies utilized ruby, Nd:YAG (1064 nm), and continuous CO_2 lasers at energy levels that produced excessive heat accumulation and extensive peripheral damage, that is, stress cracking, that prohibited their clinical implementation. More recently pulsed IR lasers with the wavelength and pulse duration optimally matched to the optical properties of enamel have been used to demonstrate that the transformation of CAP to a more acid-resistant mineral phase is indeed practical with minimal heat deposition in the tooth.[89,225] Several studies over the past 20 years have eluded to the mechanism of inhibition. Thermal analysis studies of Holcomb and Young[12] and subsequent studies by Fowler and Kuroda[171] indicated that there was substantial loss of carbonate and water at temperatures between 100°C and 400°C, which were sufficient to change the crystallinity of the intrinsic mineral. Kuroda and Fowler[226] observed a reduction in the carbonate content of enamel by 66% after irradiation with a continuous wave CO_2 laser. Heat treatment between 350°C and 650°C reduced the solubility of enamel to acid dissolution,[227–230] even though the permeability was increased. Featherstone et al. correlated the carbonate loss from enamel measured with FTIR with the inhibition of the dissolution rate providing strong evidence for the transformation of the mineral of enamel to a purer phase.[231] Therefore, the decreased solubility after IR irradiation can most likely be attributed to the thermal decomposition of the more soluble CAP into the less soluble HAP with corresponding changes in the crystallinity. Other studies have suggested that the production of pyrophosphate, permeability changes, and modification of the protein matrix play a role in inhibition.[227,228,232,233]

Levels of inhibition exceeding 70% have been achieved with CO_2 irradiation,[234] and levels close to 100% have been achieved by combining laser treatments with fluoride application.[89,194,235] Caries inhibition has also been demonstrated for the erbium laser wavelengths when a water spray was not employed.[143] The inhibition after Er:YSGG ($\lambda = 2.79$ μm) laser irradiation (13 J/cm²) was 60%, which approaches the best results obtained for CO_2 laser irradiation or daily application of a fluoride dentifrice (70%–85%).[143] Studies of heat-treated and CO_2 laser-irradiated enamel have shown that there are significant chemical and structural changes induced in enamel at temperatures well below the melting point.[12,171,228,235,236] The principal changes are loss of water and carbonate, crystal sintering and fusion, permeability changes, and changes in the crystallinity. Therefore, it may not be necessary to melt and recrystallize the enamel for effective caries preventive treatments.

Kantola showed 30 years ago that lasers could be used to increase the mineral content and crystallinity of dentin by preferential removal of the inherent water and protein.[237] Since that time, investigators have evaluated the susceptibility of dentin modified by various laser systems to artificial caries-like lesion formation. Nammour et al.[238] and Kinney et al.[239] used continuous wave $\lambda = 10.6$ μm, CO_2, and pulsed Nd:YAG lasers operating at very high intensities to irradiate dentin and create a hypermineralized layer of dentin 50–100 μm thick with greater than 80% mineral composition on the outer surface. This surface layer was resistant to acid dissolution; however, extensive cracking of the surface was observed and there was some subsurface demineralization, albeit at a somewhat reduced rate in the study of Nammour and co-workers.[238] More recent studies using a multitude of CO_2 laser irradiation conditions have not been successful in inhibiting demineralization on dentin surfaces even though there are a couple of published studies claiming there was some inhibition of demineralization on laser-treated dentin.[240,241] Apparently, the laser irradiation does not completely seal the dentinal tubule lumen in those studies, thus allowing acid to diffuse through the large surface cracks created during irradiation and the tubules to a depth below the acid-resistant-modified layer.

11.4 Future

In summary, dentists are already utilizing optical diagnostic technology for caries detection and diagnostics and lasers for therapeutic intervention. By combining these methods, future dentists may usher in a new era of conservative dentistry in which caries lesions are detected and assessed before they have progressed to the point where the removal of significant amounts of tissue is required. Subsequent to diagnosis, lasers will be used to selectively and precisely remove the carious (decayed) tissue, painlessly, while minimizing the loss of healthy tissue. The mineral phase of the surrounding enamel in these high-risk areas will be transformed through laser heating into a more acid-resistant phase to have an enhanced resistance to future decay. Thus, future patients may not have to endure the current standard of treatment that results in the loss of excessive amounts of healthy tissue structure, thus affecting the appearance and compromising the mechanical integrity of the tooth.

Pilot studies on both artificial lesions and natural caries lesions suggest that it is feasible to combine imaging methods with laser ablation for highly selective image-guided ablation, namely, a CAD/CAM system for removing dental caries.[242,243] Near-IR, fluorescent, or OCT images of dental decay will be employed to program a scanned laser beam to remove only decay or composite restorative materials. Figure 11.16 demonstrates this concept. Near-IR images were acquired of demineralized enamel surfaces in which specific lesion patterns were generated and those images were used to program a CO_2 laser to selectively remove them. OCT images were acquired before and after removal and the scans demonstrate the highly selective removal.

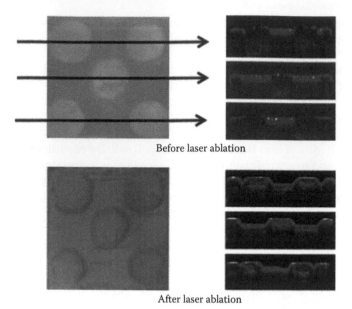

Before laser ablation

After laser ablation

FIGURE 11.16 Demonstration of highly selective ablation of demineralized enamel using near-IR image-guided ablation using a lesion pattern produced on a block of bovine enamel (see Ref. [242]). The lesion areas appear dark in the near-IR image (top left) and the lesion appears with higher intensity in the PS-OCT scans ($\perp$) across each area of the block as shown by the arrows. The near-IR image was used to program the CO_2 laser for selective removal, which removed only the demineralized areas as shown by the near-IR and PS-OCT images on the bottom.

Acknowledgments

The author would like to acknowledge Cynthia L. Darling, Michal Staninec, John D. B. Featherstone, Wolf Seka, Charles Le, John Xie, Chulsung Lee, Ken Fan, Robert Jones, Kathy Tao, Saman Manesh, and Robert Alexander for their many contributions to this work. The author would also like to thank NIDCR for the support of his work.

References

1. Duck, F. A., *Physical Properties of Tissue*, Academic Press, San Diego, CA, 1990.
2. Zijp, J. R. and ten Bosch, J. J., Angular dependence of HeNe laser light scattering by bovine and human dentine, *Archs Oral Biol* 36, 283–289, 1991.
3. Zijp, J. R. and ten Bosch, J. J., Theoretical model for the scattering of light by dentin and comparison with measurements, *Appl Opt* 32, 411–415, 1993.
4. Fried, D., Featherstone, J. D. B., Glena, R. E., and Seka, W., The nature of light scattering in dental enamel and dentin at visible and near-IR wavelengths, *Appl Opt* 34 (7), 1278–1285, 1995.
5. Zijp, J. R., ten Bosch, J. J., and Groenhuis, R. A., HeNe laser light scattering by human dental enamel, *J Dent Res* 74, 1891–1898, 1995.
6. Cheong, W., Prahl, S. A., and Welch, A. J., A review of the optical properties of biological tissues, *IEEE J Quantum Electron* 26, 2166–2185, 1990.
7. Wilson, B. C., Patterson, M. S., and Flock, S. T., Indirect versus direct techniques for the measurement of the optical properties of tissues, *Photochem Photobiol* 46(5), 601–608, 1987.
8. van der Zee, P., Methods for measuring the optical properties of tissue in the visible and near-IR wavelength range, in *Medical Optical Tomography: Functional Imaging and Monitoring*, SPIE, Bellingham, WA, 1993, Vol. IS11, pp. 450–472.

9. Tuchin, V., *Tissue Optics: Light Scattering Methods and Instruments for Medical Diagnostics*, SPIE, Bellingham, WA, 2000.

10. Spitzer, D. and ten Bosch, J. J., The absorption and scattering of light in bovine and human dental enamel, *Calcif Tissue Res* 17, 129–137, 1975.

11. Wang, T., Xu, H. C., and Lin, W. Y., Measurements of thermal expansion coefficient of human teeth, *Aust J Dent* 34(6), 530–535, 1989.

12. Holcomb, D. W. and Young, R. A., Thermal decomposition of human tooth enamel, *Calcif Tissue Int* 31, 189–201, 1980.

13. ten Bosch, J. J. and Zijp, J. R., Optical properties of dentin, in *Dentine and Dentine Research in the Oral Cavity*, A. Thylstrup, ed., IRL Press, Oxford, England, 1987, pp. 59–65.

14. Jones, R. S. and Fried, D., Attenuation of 1310-nm and 1550-nm laser light through sound dental enamel, in *Lasers in Dentistry VIII*, SPIE, San Jose, CA, 2002, Vol. 4610, pp. 187–190.

15. Fried, D., Featherstone, J. D. B., Glena, R. E., Bordyn, B., and Seka, W., The light scattering properties of dentin and enamel at 543, 632, and 1053 nm, in *Lasers in Orthopedic, Dental, and Veterinary Medicine II*, SPIE, Los Angeles, CA, 1993, Vol. 1880, pp. 240–245.

16. ten Bosch, J. J., van der Mei, H. C., and Borsboom, P. C. F., Optical monitor of in vitro caries, *Caries Res* 18, 540–547, 1984.

17. Angmar-Mansson, B. and ten Bosch, J. J., Optical methods for the detection and quantification of caries, *Adv Dent Res* 1(1), 14–20, 1987.

18. ten Bosch, J. J. and Coops, J. C., Tooth color and reflectance as related to light scattering and enamel hardness, *J Dent Res* 74, 374–380, 1995.

19. Ko, C. C., Tantbirojn, D., Wang, T., and Douglas, W. H., Optical scattering power for characterization of mineral loss, *J Dent Res* 79 (8), 1584–1589, 2000.

20. Schmidt, W. J. and Keil, A., *Polarizing Microscopy of Dental Tissues*, Pergamon Press, New York, 1971.

21. Theuns, H. M., Shellis, R. P., Groeneveld, A., van Dijk, J. W. E., and Poole, D. F. G., Relationships between birefringence and mineral content in artificial caries lesions in enamel, *Caries Res* 27, 9–14, 1993.

22. Duplain, G., Boulay, R., and Belanger, P. A., Complex index of refraction of dental enamel at CO_2 wavelengths, *Appl Opt* 26, 4447–4451, 1987.

23. Zuerlein, M., Fried, D., Featherstone, J., and Seka, W., Optical properties of dental enamel at 9–11 μm derived from time-resolved radiometry, special topics, *IEEE J Quantum Electron* 5 (4), 1083–1089, 1999.

24. Zuerlein, M. and Fried, D., Depth profile analysis of the chemical and morphological changes in CO_2-laser irradiated dental enamel, in *Lasers in Dentistry V*, SPIE, San, Jose, CA, 1999, Vol. 3593, pp. 204–210.

25. Fried, D., Glena, R. E., Featherstone, J. D. B., and Seka, W., Permanent and transient changes in the reflectance of CO_2 laser irradiated dental hard tissues at lambda = 9.3, 9.6, 10.3, and 10.6 μm and at fluences of 1–20 J/cm^2, *Lasers Surg Med* 20, 22–31, 1997.

26. Fried, D., Zuerlein, M., Featherstone, J. D. B., Seka, W. D., and McCormack, S. M., IR laser ablation of dental enamel: Mechanistic dependence on the primary absorber, *Appl Surf Sci* 128, 852–856, 1997.

27. Chauncey, H. H., Glass, R. L., and Alman, J. E., Dental caries, principal cause of tooth extraction in a sample of US male adults, *Caries Res* 23, 200–205, 1989.

28. Kaste, L. M., Selwitz, R. H., Oldakowski, R. J., Brunelle, J. A., Winn, D. M., and Brown, L. J., Coronal caries in the primary and permanent dentition of children and adolescents 1–17 years of age: United States, 1988–1991, *J Dent Res* 75, 631–641, 1996.

29. Winn, D. M., Brunelle, J. A., Selwitz, R. H., Kaste, L. M., Oldakowski, R. J., Kingman, A., and Brown, L. J., Coronal and root caries in the dentition of adults in the United States, 1988–1991, *J Dent Res* 75, 642–651, 1996.

30. NIH Consensus Statement 2001, March 26–28, 18(1), 1–24, 2001.

31. Lussi, A., Validity of diagnostic and treatment decisions of fissure caries, *Caries Res* 25, 296–303, 1991.

32. Featherstone, J. D. B. and Young, D., The need for new caries detection methods, in *Lasers in Dentistry V*, SPIE, San Jose, CA, 1999, Vol. 3593, pp. 134–140.

33. Hume, W. R., Need for change in dental caries diagnosis, in *Early Detection of Dental Caries*, Stookey, G. K., ed., Indiana University, Indianapolis, IN, 1996, pp. 1–10.

34. Featherstone, J. D. B., Clinical implications: New strategies for caries prevention, in *Early Detection of Dental Caries*, Stookey, G. K., ed., Indiana University, Indianapolis, IN, 1996, pp. 287–296.

35. ten Cate, J. M. and van Amerongen, J. P., Caries diagnosis: Conventional methods, in *Early Detection of Dental Caries*, Stookey, G. K., ed., Indiana University, Indianapolis, IN, 1996, pp. 27–37.

36. Featherstone, J. D. B., Prevention and reversal of dental caries: Role of low level fluoride, *Commun Dent Oral Epidemiol* 27, 31–40, 1999.

37. Kidd, E. A. M., Ricketts, D. N. J., and Pitts, N. B., Occlusal caries diagnosis: A changing challenge for clinicians and epidemiologists, *J Dent Res* 21, 323–331, 1993.

38. Wenzel, A., New caries diagnostic methods, *J Dent Educ* 57, 428–432, 1993.

39. Pine, C. M., Fiber-optic transillumination (FOTI) in caries diagnosis, in *Early Detection of Dental Caries*, Stookey, G. K., ed., Indiana University, Indianapolis, IN, 1996, pp. 51–66.

40. Pine, C. M. and ten Bosch, J. J., Dynamics of and diagnostic methods for detecting small carious lesions, *Caries Res* 30, 381–388, 1996.

41. Peers, A., Hill, F. J., Mitropoulos, C. M., and Holloway, P. J., Validity and reproducibility of clinical examination, fibre-optic transillumination, and bite-wing radiology for the diagnosis of small approximal carious lesions, *Caries Res* 27, 307–311, 1993.

42. Peltola, J. and Wolf, J., Fiber optics transillumination in caries diagnosis, *Proc Finn Dent Soc* 77, 240–244, 1981.

43. Barenie, J., Leske, G., and Ripa, L. W., The use of fiber optic transillumination for the detection of proximal caries, *Oral Surg* 36, 891–897, 1973.

44. Holt, R. D. and Azeevedo, M. R., Fiber optic transillumination and radiographs in diagnosis of approximal caries in primary teeth, *Commun Dent Health* 6, 239–247, 1989.

45. Mitropoulis, C. M., The use of fiber optic transillumination in the diagnosis of posterior approximal caries in clinical trials, *Caries Res* 19, 379–384, 1985.

46. Schneiderman, A., Elbaum, M., Schultz, T., Keem, S., Greenebaum, M., and Driller, J., Assessment of dental caries with digital imaging fiber-optic transillumination (DIFOTI): In vitro study, *Caries Res* 31, 103–110, 1997.

47. Bühler, C. M., Ngaotheppitak, P., and Fried, D., Imaging of occlusal dental caries (decay) with near-IR light at 1310-nm, *Opt Express* 13(2), 573–582, 2005.

48. Jones, G., Jones, R. S., and Fried, D., Transillumination of interproximal caries lesions with 830-nm light, in *Lasers in Dentistry X*, SPIE, San Jose, CA, 2004, Vol. 5313, pp. 17–22.

49. Jones, R. S., Huynh, G. D., Jones, G. C., and Fried, D., Near-IR transillumination at 1310-nm for the imaging of early dental caries, *Opt Express* 11(18), 2259–2265, 2003.

50. Darling, C. L., Huynh, G. D., and Fried, D., Light scattering properties of natural and artificially demineralized dental enamel at 1310-nm, *J Biomed Opt* 11(3), 034023 (1–11), 2006.

51. Alfano, R. R. and Yao, S. S., Human teeth with and without dental caries studied by visible luminescent spectroscopy, *J Dent Res* 60, 120–122, 1981.

52. Alfano, R. R., Lam, W., Zarrabi, H. J., Alfano, M. A., Cordero, J., and Tata, D. B., Human teeth with and without caries studied by laser scattering, fluorescence and absorption spectroscopy, *IEEE J Quantum Electron* 20, 1512–1515, 1984.

53. Bjelkhagen, H. and Sundstrom, F., A clinically applicable laser luminescence for the early detection of dental caries, *IEEE J Quantum Electron* 17, 266–268, 1981.

54. Hafstroem-Bjoerkman, U., de Josselin de Jong, E., Oliveby, A., and Angmar-Mansson, B., Comparison of laser fluorescence and longitudinal microradiography for quantitative assessment of in vitro enamel caries, *Caries Res* 26, 241–247, 1992.

55. Ando, M., Eckert, A. F., Schemehorn, B. R., Analoui, M., and Stookey, G. K., Relative ability of laser fluorescence techniques to quantitate early mineral loss in vitro, *Caries Res* 31, 125–131, 1997.

56. de Josselin de Jong, E., Sundstrom, F., Westerling, H., Tranaeus, S., ten Bosch, J. J., and Angmar-Mansson, B., A new method for in vivo quantification of changes in initial enamel caries with laser fluorescence, *Caries Res* 29(1), 2–7, 1995.

57. Lagerweij, M. D., van der Veen, M. H., Ando, M., and Lukantsova, L., The validity and repeatability of three light-induced fluorescence systems: An in vitro study, *Caries Res* 33, 220–226, 1999.

58. van der Veen, M. H., de Josselin de Jong, E., and Al-Kateeb, S., Caries activity detection by dehydration with qualitative light fluorescence, in *Early Detection of Dental Caries II*, Indiana University, Indianapolis, IN, 1999, Vol. 4, pp. 251–260.

59. Koenig, K., Schneckenburger, H., Hemmer, J., Tromberg, B. J., Steiner, R. W., and Rudolf, W., In-vivo fluorescence detection and imaging of porphyrin-producing bacteria in the human skin and in the oral cavity for diagnosis of acne vulgaris, caries, and squamous cell carcinoma, in *Advances in Laser and Light Spectroscopy to Diagnose Cancer and Other Diseases*, SPIE, San Jose, CA, 1994, Vol. 2135, pp. 129–138.

60. Hibst, R., Paulus, R., and Lussi, A., Detection of occlusal caries by laser fluorescence: Basic and clinical investigations, *Med Laser Appl* 16, 205–213, 2001.

61. Shi, X. Q., Welander, U., and Angmar-Mansson, B., Occlusal caries detection with Kavo DIAGNOdent and radiography: An in vitro comparison, *Caries Res* 34, 151–158, 2000.

62. Hibst, R. and Paulus, R., New approach on fluorescence spectroscopy for caries detection, in *Lasers in Dentistry V*, SPIE, San Jose, CA, 1999, Vol. 3593, pp. 141–148.

63. Lussi, A., Imwinkelreid, S., Pitts, N. B., Longbottom, C., and Reich, E., Performance and reproducibility of a laser fluorescence system for detection of occlusal caries in vitro, *Caries Res* 33, 261–266, 1999.

64. Bouma, B. E. and Tearney, G. J., *Handbook of Optical Coherence Tomography*, Marcel Dekker, New York, 2002.

65. Brezinski, M., *Optical Coherence Tomography: Principles and Applications*, Elsevier, London, U.K., 2006.

66. Colston, B., Everett, M., Da Silva, L., Otis, L., Stroeve, P., and Nathel, H., Imaging of hard and soft tissue structure in the oral cavity by optical coherence tomography, *Appl Opt* 37(19), 3582–3585, 1998.

67. Colston, B. W., Everett, M. J., Da Silva, L. B., and Otis, L. L., Optical coherence tomography for diagnosis of periodontal diseases, in *Coherence Domain Optical Methods in Biomedical Science and Clinical Applications II*, SPIE, San Jose, CA, 1998, Vol. 3251, pp. 52–58.

68. Colston, B. W., Sathyam, U. S., DaSilva, L. B., Everett, M. J., and Stroeve, P., Dental OCT, *Opt Express* 3(3), 230–238, 1998.

69. Baumgartner, A., Hitzenberger, C. K., Dicht, S., Sattmann, H., Moritz, A., Sperr, W., and Fercher, A. F., Optical coherence tomography for dental structures, in *Lasers in Dentistry IV*, SPIE, San Jose, CA, 1998, Vol. 3248, pp. 130–136.

70. Baumgartner, A., Dicht, S., Hitzenberger, C. K., Sattmann, H., Robi, B., Moritz, A., Sperr, W., and Fercher, A. F., Polarization-sensitive optical optical coherence tomography of dental structures, *Caries Res* 34, 59–69, 2000.

71. Feldchtein, F. I., Gelikonov, G. V., Gelikonov, V. M., Iksanov, R. R., Kuranov, R. V., Sergeev, A. M., Gladkova, N. D., Ourutina, M. N., Warren, J. A., and Reitze, D. H., In vivo OCT imaging of hard and soft tissue of the oral cavity, *Opt Express* 3(3), 239–251, 1998.

72. Wang, X. J., Zhang, J. Y., Milner, T. E., de Boer, J. F., Zhang, Y., Pashley, D. H., and Nelson, J. S., Characterization of dentin and enamel by use of optical coherence tomography, *Appl Opt* 38(10), 586–590, 1999.

73. Everett, M. J., Colston, B. W., Sathyam, U. S., Silva, L. B. D., Fried, D., and Featherstone, J. D. B., Non-invasive diagnosis of early caries with polarization sensitive optical coherence tomography (PS-OCT), in *Lasers in Dentistry V*, SPIE, San Jose, CA, 1999, Vol. 3593, pp. 177–183.

74. Can, A. M., Darling, C. L., and Fried, D., High resolution PS OCT of enamel remineralization, in *Lasers in Dentistry XIV*, SPIE, San Jose, CA, 2008, Vol. 6843, p. 68430T-7.

75. Jones, R. S., Darling, C. L., Featherstone, J. D. B., and Fried, D., Remineralization of in vitro dental caries assessed with polarization sensitive optical coherence tomography, *J Biomed Opt* 11(1), 014016 1–9 2006.

76. Jones, R. S. and Fried, D., Remineralization of enamel caries can decrease optical reflectivity, *J Dent Res* 85(9), 804–808, 2006.

77. Manesh, S. K., Darling, C. L., and Fried, D., Assessment of dentin remineralization with polarization sensitive optical coherence tomography, in *Lasers in Dentistry XV*, SPIE, San Jose, CA, 2009, Vol. 7162W, pp. 1–7.

78. Kidd, E. A. M., Secondary caries, *Int Dent J* 42, 127–138, 1992.

79. Fried, D., Xie, J., Shafi, S., Featherstone, J., Breunig, T. M., and Le, C. Q., Imaging caries lesions and lesion progression with polarization optical coherence tomography, in *Lasers in Dentistry VIII*, SPIE, San Jose, CA, 2002, Vol. 4610, pp. 113–124.

80. Fried, D., Xie, J., Shafi, S., Featherstone, J. D. B., Breunig, T., and Lee, C. Q., Early detection of dental caries and lesion progression with polarization sensitive optical coherence tomography, *J Biomed Opt* 7(4), 618–627, 2002.

81. Kinney, J. H., Balooch, M., Haupt, D. L., Marshall, S. J., and Marshall, G. W., Mineral distribution and dimensional changes in human dentin during demineralization, *J Dent Res* 74(5), 1179–1184, 1995.

82. Kinney, J. H., Haupt, D. L., Nichols, M. C., Breunig, T. M., Marshall, G. W., and Marshall, S. J., The X-ray tomographic microscope: Three dimensional perspectives of evolving structures, *Nucl Instrum Methods* A 347, 480–486, 1994.

83. Madjarova, V. D., Yasuno, Y., Makita, S., Hori, Y., Voeffray, J. B., Itoh, M., Yatagai, T., Tamura, M., and Nanbu, T., Investigations of soft and hard tissues in oral cavity by spectral domain optical coherence tomography, *Coherence Domain Optical Methods and Optical Coherence Tomography in Biomedicine X, Proc SPIE-Int Soc Opt Eng* 6079(1), 60790N-1-7, 2006.

84. Seon, Y. R., Jihoon, N., Hae, Y. C., Woo, J. C., Byeong, H. L., and Gil-Ho, Y., Realization of fiber-based OCT system with broadband photonic crystal fiber coupler, *Coherence Domain Optical Methods and Optical Coherence Tomography in Biomedicine X, Proc SPIE-Int Soc Opt Eng* 6079(1), 60791N-1-7, 2006.

85. Yamanari, M., Makita, S., Violeta, D. M., Yatagai, T., and Yasuno, Y., Fiber-based polarization-sensitive Fourier domain optical coherence tomography using B-scan-oriented polarization modulation method, *Opt Express* 14(14), 6502–6515, 2006.

86. Furukawa, H., Hiro-Oka, H., Amano, T., DongHak, C., Miyazawa, T., Yoshimura, R., Shimizu, K., and Ohbayashi, K., Reconstruction of three-dimensional structure of an extracted tooth by OFDR-OCT, *Coherence Domain Optical Methods and Optical Coherence Tomography in Biomedicine X, Proc SPIE-Int Soc Opt Eng* 6079(1), 60790T-1-7, 2006.

87. Stern, R. H. and Sognnaes, R. F., Laser beam effect on hard dental tissues, *J Dent Res* 43, 873, 1964.

88. Wigdor, H. A., Walsh, J. T., Featherstone, J. D. B., Visuri, S. R., Fried, D., and Waldvogel, J. L., Lasers in dentistry, *Lasers Surg Med* 16, 103–133, 1995.

89. Featherstone, J. D. B. and Nelson, D. G. A., Laser effects on dental hard tissue, *Adv Dent Res* 1(1), 21–26, 1987.

90. Harris, N. and Garcia-Godoy, F., *Primary Preventive Dentistry*, 5th edn., Appleton & Lange, Stamford, CT, 1999.

91. Mertz-Fairhurst, E. J., Pit-and-fissure sealants: A global lack of scientific transfer? *J Dent Res* 115, 1543–1544, 1992.

92. Mount, G. J. and Hume, W. R., *Preservation and Restoration of Tooth Structure*, Mosby, New York, 1998.

93. Cooper, L. F., Myers, M. L., Nelson, D. G. A., and Mowery, A. S., Shear strength of composite resin bonded to laser pretreated dentin, *J Prosthet Dent* 60, 45–49, 1988.

94. Featherstone, J. D. B., Barrett-Vespone, N. A., Fried, D., Kantorowitz, Z., Lofthouse, J., and Seka, W., Rational choice of CO_2 laser conditions for inhibition of caries progression, in *Lasers in Dentistry*, SPIE, San Jose, CA, 1995, Vol. 2394, pp. 57–67.

95. Konishi, N., Fried, D., Featherstone, J. D. B., and Staninec, M., Inhibition of secondary caries by CO_2 laser treatment, *Am J Dent* 12(5), 213–216, 1999.

96. Young, D. A., Fried, D., and Featherstone, J. D. B., Ablation and caries inhibition of pits and fissures by IR laser irradiation, in *Lasers in Dentistry VI*, SPIE, San Jose, CA, 2000, Vol. 3910, pp. 247–253.

97. Dostalova, T., Jelainkova, H., Kuacerova, H., Krejsa, O., Hamal, K., Kubelka, J., and Prochaazka, S., Noncontact Er:YAG laser ablation: Clinical evaluation, *J Clin Laser Med Surg* 16, 273–282, 1998.

98. Keller, U. and Hibst, R., Effects of Er:YAG laser in caries treatment: A clinical pilot study, *Lasers Surg Med* 20, 32–38, 1997.

99. Pelagalli, J., Gimbel, C. B., Hansen, R. T., Swett, A., and Winn, D. W., Investigational study of the use of Er:YAG laser versus the drill for caries removal and cavity preparation: Phase I, *J Clin Laser Med Surg* 15, 109–115, 1997.

100. DenBesten, P. K., White, J. M., Pelino, J., Lee, K., and Parkins, F., A randomized prospective parallel controlled study of the safety and effectiveness of Er:YAG laser use in children for caries removal, in *Lasers in Dentistry VI*, SPIE, San Jose, CA, 2000, Vol. 3910, pp. 171–174.

101. Dibdin, G. H., The water in human dental enamel and its diffusional exchange measured by clearance of tritiated water from enamel slabs of varying thickness, *Caries Res* 27, 81–86, 1993.

102. Altshuler, G. B., Belikov, A. V., Erofeev, A. V., and Sam, R. C., Optimum regimes of laser destruction of human tooth enamel and dentin, in *Lasers in Orthopedic, Dental, and Veterinary Medicine II*, SPIE, Los Angeles, CA, 1993, Vol. 1880, pp. 101–107.

103. Altshuler, G. B., Belikov, A. V., and Erofeev, A. V., Laser treatment of enamel and dentin by different Er-lasers, in *Lasers in Surgery: Advanced Characterization, Therapeutics, and Systems IV*, SPIE, San Jose, CA, 1994, Vol. 2128, pp. 273–281.

104. Hibst, R. and Keller, U., Mechanism of Er:YAG laser induced ablation of dental hard substances, in *Lasers in Orthopedic, Dental, and Veterinary Medicine II*, SPIE, San Jose, CA, 1993, Vol. 1880, pp. 156–162.

105. Hibst, R. and Keller, U., Experimental studies of the application of the Er:YAG laser on dental hard substances: I. Measurement of the ablation rate, *Lasers Surg Med* 9, 338–344, 1989.

106. Keller, U. and Hibst, R., Experimental studies of the application of the Er:YAG laser on dental hard substances: II. Light microscopic and SEM investigations, *Lasers Surg Med* 9, 345–351, 1989.

107. Kermani, O., Lubatschowski, H., Asshauer, T., Ertmer, W., Lukin, A., Ermakov, B., and Krieglstein, G. K., Q-switched CTE:YAG laser ablation: Basic investigation on soft (corneal) and hard (dental) tissues, *Lasers Surg Med* 13, 537–542, 1993.

108. Shori, R., Fried, D., Featherstone, J. D. B., and Duhn, C., CTE:YAG applications in dentistry, in *Lasers in Dentistry IV*, SPIE, San Jose, CA, 1998, Vol. 3248, pp. 86–91.

109. Majaron, M., Sustercic, D., and Lukac, M., Debris screening and heat diffusion in Er:YAG drilling of hard dental tissues, in *Lasers in Dentistry III*, SPIE, San Jose, CA, 1997, Vol. 2973, pp. 11–22.

110. Zeck, M., Benthin, H., Ertl, T., Siebert, G. K., and Muller, G., Scanning ablation of dental hard tissue with erbium laser radiation, in *Medical Applications of Lasers III*, SPIE, Barcelona, Spain, 1996, Vol. 2623, pp. 94–102.

111. Fried, D. and Shori, R., Q-switched Er:YAG ablation of dental hard tissue, in *6th International Congress on Lasers in Dentistry*, University of Utah, Maui, Hawaii, 1998, pp. 77–79.

112. Lenz, P., Glide, H., and Walz, R., Studies on enamel sealing with the CO_2 laser, *Dtsch Zahnarztl Z* 37, 469–478, 1982.

113. Kantola, S., Laine, E., and Tarna, T., Laser-induced effects on tooth structure. VI. X-ray diffraction study of dental enamel exposed to a CO_2 laser, *Acta Odontol Scand* 31, 369–379, 1973.

114. Melcer, J., Chaumette, M. T., and Melcer, F., Dental pulp exposed to the CO_2 laser beam, *Lasers Surg Med* 7, 347–352, 1987.

115. Miserendino, L. J., Neiburger, E. J., Walia, H., Luebke, N., and Brantley, W., Thermal effects of continuous wave CO_2 laser exposure on human teeth: An in vitro study, *J Endod* 15(7), 302, 1989.

116. Leighty, S. M., Pogrel, M. A., Goodis, H. E., and White, J. M., Thermal effects of the carbon dioxide laser on teeth, *Lasers Life Sci* 4(2), 93–102, 1991.

117. Jeffrey, I. W. M., Lawrenson, B., Saunders, E. M., and Longbottom, C., Dentinal temperature transients caused by exposure to CO_2 laser irradiation and possible pulp damage, *J Dent* 18, 31–36, 1990.

118. Neiburger, E. J. and Miserendino, L., Pulp chamber warming due to CO_2 laser exposure, *N Y State Dent J* 54(3), 25–27, 1988.

119. Melcer, J., Latest treatment in dentistry by means of the CO_2 laser beam, *Lasers Surg Med* 6, 396–398, 1986.

120. Palamara, J., Phakey, P. P., Orams, H. J., and Rachinger, W. A., The effect on the ultrastructure of dental enamel of excimer-dye, argon-ion and CO_2 lasers, *Scanning Microsc* 6(4), 1061–1071, 1992.

121. Pogrel, M. A., Muff, D. F., and Marshall, G. W., Structural changes in dental enamel induced by high energy continuous wave carbon dioxide laser, *Lasers Surg Med* 13, 89–96, 1993.

122. Ferreira, J. M., Palamara, J., Phakey, P. P., Rachinger, W. A., and Orams, H. J., Effects of continuous-wave CO_2 laser on the ultrastructure of human dental enamel, *Arch Oral Biol* 34(7), 551–562, 1989.

123. Krapchev, V. B., Rabii, C. D., and Harrington, J. A., Novel CO_2 laser system for hard tissue ablation, in *Lasers in Surgery: Advanced Characterization, Therapeutics, and Systems IV*, SPIE, Los Angeles, CA, 1994, Vol. 2128, pp. 341–348.

124. Ertl, T. and Muller, G., Hard tissue ablation with pulsed CO_2 lasers, in *Lasers in Orthopedic, Dental, and Veterinary Medicine II*, SPIE, Los Angeles, CA, 1993, Vol. 1880, pp. 176–181.

125. Lukac, M., Hocevar, F., Cencic, S., Nemes, K., Keller, U., Hibst, R., Sustercic, D., Gaspirc, B., Skaleric, U., and Funduk, N., Effects of pulsed CO_2 and Er:YAG lasers on enamel and dentin, in *Lasers in Orthopedic, Dental, and Veterinary Medicine II*, SPIE, Los Angeles, CA, 1993, Vol. 1880, pp. 169–175.

126. Fan, K., Bell, P., and Fried, D., The rapid and conservative ablation and modification of enamel, dentin and alveolar bone using a high repetition rate TEA CO_2 laser operating at $\lambda = 9.3$ µm, *J Biomed Opt* 11(6), 064008 1–11, 2006.

127. Forrer, M., Frenz, M., Romano, V., Altermatt, H. J., Weber, H. P., Silenok, A., Istomyn, M., and Konov, V. I., Bone-ablation mechanism using CO_2 lasers of different pulse duration and wavelength., *Appl Phys B* 56, 104–112, 1993.

128. Frentzen, M., Gotz, W., Ivanenko, M., Afilal, S., Werner, M., and Hering, P., Osteotomy with 80-microsecond CO_2 laser pulses—Histological results, *Lasers Med Sci* 18(2), 119–124, 2003.

129. Fried, D., Seka, W., Glena, R. E., and Featherstone, J. D. B., The thermal response of dental hard tissues to (9–11 µm) CO_2 laser irradiation, *Opt Eng* 35(7), 1976–1984, 1996.

130. Ivanenko, M. M. and Hering, P., Wet bone ablation with mechanically Q-switched high-repetition-rate CO_2 laser, *Appl Phys B* 67, 395–397, 1998.

131. Melcer, J., Farcy, J. C., Hellas, G., and Badiane, M., Preparation of cavities using a TEA CO_2 laser, in *3rd International Congress on Lasers in Dentistry*, Salt Lake City, UT, Vol. 1992, Abstract #58.

132. Goodis, H. E., Fried, D., Gansky, S., Rechmann, P., and Featherstone, J. D., Pulpal safety of 9.6 microm TEA CO_2 laser used for caries prevention, *Lasers Surg Med* 35(2), 104–110, 2004.

133. Ivanenko, M. M., Eyrich, G., Bruder, E., and Hering, P., In vitro incision of bone tissue with a Q-switch CO_2 laser. Histological examination, *Lasers Life Sci* 9, 171–179, 2000.

134. Mullejans, R., Eyrich, G., Raab, W. H., and Frentzen, M., Cavity preparation using a superpulsed 9.6-microm CO_2 laser—A histological investigation, *Lasers Surg Med* 30(5), 331–336, 2002.

135. Staninec, M., Darling, C. L., Goodis, H. E., Pierre, D., Cox, D. P., Fan, K., Larson, M. et al., Pulpal effects of enamel ablation with a microsecond pulsed $\lambda = 9.3$-µm CO_2 laser, *Lasers Surg Med* 41, 256–263, 2009.

136. Wigdor, H., Walsh, J. T., and Mostofi, R., The effect of the CO_2 laser (9.6 µm) on the dental pulp in humans, in *Lasers in Dentistry VI*, SPIE, San Jose, CA, 2000, Vol. 3910, pp. 158–163.

137. Wigdor, H. A. and Walsh, J. T. Jr., Histologic analysis of the effect on dental pulp of a 9.6-µm CO_2 laser, *Lasers Surg Med* 30(4), 261–266, 2002.

138. Fan, K. and Fried, D., A high repetition rate TEA CO_2 laser operating at $\lambda = 9.3$-µm for the rapid and conservative ablation and modification of dental hard tissues, in Rechmann, P. and Fried, D., eds., *Lasers in Dentistry XII*, SPIE, San Jose, CA, 2006, Vol. 6137, pp. 61370G (1–9).

139. Assa, S., Meyer, S., and Fried, D., Ablation of dental hard tissues with a microsecond pulsed carbon dioxide laser operating at 9.3-μm with an integrated scanner, in *Lasers in Dentistry XIV*, SPIE, San Jose, CA, 2008, Vol. 6843, pp. 68430E (1–7).

140. Ivanenko, M., Sader, R., Afilal, S., Werner, M., Hartstock, M., von Hanisch, C., Milz, S., Erhardt, W., Zeilhofer, H. F., and Hering, P., In vivo animal trials with a scanning CO_2 laser osteotome, *Lasers Surg Med* 37(2), 144–148, 2005.

141. Albagli, D., Perelman, L. T., Janes, G. S., Rosenburg, C. V., Itzkan, I., and Feld, M. S., Inertially confined ablation of biological tissue, *Lasers Life Sci* 6(1), 55–68, 1994.

142. Fried, D., Visuri, S. R., Featherstone, J. D. B., Seka, W., Glena, R. E., Walsh, J. T., McCormack, S. M., and Wigdor, H. A., Infrared radiometry of dental enamel during Er:YAG and Er:YSGG laser irradiation, *J Biomed Opt* 1(4), 455–465, 1996.

143. Fried, D., Featherstone, J. D. B., Visuri, S. R., Seka, W., and Walsh, J. T., The caries inhibition potential of Er:YAG and Er:YSGG laser radiation, in *Lasers in Dentistry II*, SPIE, San Jose, CA, 1996, Vol. 2672, pp. 73–78.

144. Zach, L. and Cohen, G., Pulp response to externally applied heat, *Oral Surg Oral Med Oral Pathol* 19, 515–530, 1965.

145. Hamilton, A. I. and Kramer, I. R. H., Cavity preparation with and without waterspray: Effects on the human dental pulp and additional effects of further dehydration of the dentine, *Br Dent J* 126, 281–285, 1967.

146. Stewart, G. P., Bachman, T. A., and Hatton, J. F., Temperature rise due to finishing of direct restorative materials, *Am J Dent* 4, 23–28, 1991.

147. Anic, I., Vidovic, D., Luic, M., and Tudja, M., Laser induced molar tooth pulp chamber temperature changes, *Caries Res* 26, 165–169, 1992.

148. Nammour, S. and Pourtois, M., Pulp temperature increases following caries removal by CO_2 laser, *J Clin Laser Med Surg* 13(5), 337–342, 1995.

149. Yu, D., Powell, G. L., Higuchi, W. I., and Fox, J. L., Comparison of three lasers on dental pulp chamber temperature change, *J Clin Laser Med Surg* 11(3), 119–122, 1993.

150. Sandford, M. A. and Walsh, L. J., Differential thermal effects of pulsed vs. continuous CO_2 laser radiation on human molar teeth, *J Clin Laser Med Surg* 12(3), 139–142, 1994.

151. Miserendino, L. J., Abt, E., Wigdor, H., and Miserendino, C. A., Evaluation of thermal cooling mechanisms for laser application to teeth, *Laser Surg Med* 13, 83–88, 1993.

152. Paltauf, G. and Schmidt-Kloiber, H., Modeling and experimental observation of photomechanical effects in tissue-like media, in *Laser-Tissue Interaction VI*, SPIE, San Jose, CA, 1995, Vol. 2391, pp. 403–412.

153. Motamedi, M. and Rastegar, S., Thermal stress distribution in laser irradiated hard dental tissue: Implications for dental applications, in *Laser-Tissue Interaction III*, SPIE, Los Angeles, CA, 1992, Vol. 1646, pp. 316–321.

154. Perelman, L. T., Albagli, D., Dark, M., Schaffer, J., Rosenburg, C. V., Itzkan, I., and Feld, M. S., Physics of laser-induced stress wave propagation, cracking, and cavitation in biological tissue, in *Laser-Tissue Interaction V*, 1994, SPIE, Los Angeles, CA, Vol. 2134, pp. 144–151.

155. Dingus, R. S. and Scammon, R. J., Grüneisen-stress induced ablation of biological tissue, in *Laser-Tissue Interaction II*, SPIE, Los Angeles, CA, 1991, Vol. 1427, pp. 45–54.

156. Vickers, V. A., Jacques, S. L., Schwartz, J., Motamedi, M., Rastegar, S., and Martin, J. W., Ablation of hard dental tissues with the Er:YAG laser, in *Laser-Tissue Interaction III*, SPIE, Los Angeles, CA, 1992, Vol. 1646, pp. 46–55.

157. Walsh, J. T. and Hill, D. A., Erbium laser ablation of bone: Effect of water content, in *Laser-Tissue Interaction II*, SPIE, Los Angeles, CA, 1991, Vol. 1427, pp. 27–33.

158. Wigdor, H. A., Walsh, J. T., and Visuri, S. R., Effect of water on dental material ablation of the Er:YAG laser, in *Lasers in Surgery: Advanced Characterization, Therapeutics, and Systems IV*, SPIE, San Jose, CA, 1994, Vol. 2128, pp. 267–272.

159. Burkes, E. J., Hoke, J., Gomes, E., and Wolbarsht, M., Wet versus dry enamel ablation by Er:YAG laser, *J Prosthet Dent* 67(6), 847–851, 1992.

160. Cozean, C., Arcoria, C. J., Pelagalli, J., and Powell, L., Dentistry for the 21st century: Er:YAG laser for teeth, *J Am Dent Assoc* 128, 1079–1086, 1997.

161. Rizoiu, I. M. and DeShazer, L. G., New laser-matter interaction concept to enhance hard tissue cutting efficiency, in *Laser-Tissue Interaction V*, SPIE, San Jose, CA, 1994, Vol. 2134A, pp. 309–317.

162. Majaron, B., Sustercic, D., and Lukac, M., Influence of water spray on Er:YAG ablation of hard dental tissues, in *Medical Applications of Lasers in Dermatology, Ophthalmology, Dentistry, and Endoscopy*, SPIE, San Remo, Italy, 1997, Vol. 3192, pp. 82–87.

163. Rizoiu, I., Kimmel, A. I., and Eversole, L. R., The effects of an Er, Cr:YSGG laser on canine oral tissues, in *Laser Applications in Medicine and Dentistry*, SPIE, Vienna, Austria, 1996, Vol. 2922, pp. 74–83.

164. Altshuler, G. B., Belikov, A. V., and Erofeev, A. V., Comparative study of contact and noncontact operation mode of hard tooth tissues Er-laser processing, in *5th International Congress of the International Society of Laser Dentistry*, Jerusalem, Israel, 1996, pp. 21–25.

165. Majaron, M., Sustercic, D., and Lukac, M., Fiber-tip drilling of hard dental tissues with Er:YAG laser, in *Lasers in Dentistry IV*, SPIE, San Jose, CA, 1998, Vol. 3248, pp. 69–77.

166. Darling, C. L., Maffei, M. E., Fried, W. A., and Fried, D., Near-IR imaging of erbium laser ablation with a water spray, in *Lasers in Dentistry XIV*, SPIE, San Jose, CA, 2008, Vol. 6843, pp. 684303–684307.

167. Darling, C. L. and Fried, D., Real-time near-IR imaging of laser-ablation crater evolution in dental enamel, in *Lasers in Dentistry XIII*, SPIE, San Jose, CA, 2007, Vol. 6425, pp. 64250I-1–64250I-6.

168. Altshuler, G. B., Belikov, A. V., Erofeev, A. V., and Skrypnik, A. V., Physical aspects of cavity formation of Er-laser radiation, in *Lasers in Dentistry*, SPIE, San Jose, CA, 1995, Vol. 2394, pp. 211–222.

169. Rechmann, P., Glodin, D. S., and Hennig, T., Changes in surface morphology of enamel after Er:YAG irradiation, in *Lasers in Dentistry IV*, SPIE, San Jose, CA, 1998, Vol. 3248, pp. 62–68.

170. Fried, D., IR ablation of dental enamel, in *Lasers in Dentistry VI*, SPIE, San Jose, CA, 2000, Vol. 3910, pp. 136–148.

171. Fowler, B. and Kuroda, S., Changes in heated and in laser-irradiated human tooth enamel and their probable effects on solubility, *Calcif Tissue Int* 38, 197–208, 1986.

172. Fried, D., Ashouri, N., Breunig, T. M., and Shori, R. K., Mechanism of water augmentation during IR laser irradiation of dental enamel, *Lasers Surg Med* 31, 186–193, 2002.

173. Ith, M., Pratisto, H., Altermatt, H. J., and Weber, H. P., Dynamics of laser-induced channel formation in water and influence of pulse duration on the ablation of biotissue under water with pulsed Er:YAG Lasers, *Appl Phys B* 59, 621–629, 1994.

174. Loertscher, H., Shi, W. Q., and Grundfest, W. S., Tissue ablation through water with erbium:YAG lasers, *IEEE Trans Biomed Eng* 39(1), 86–88, 1992.

175. Forrer, M., Ith, M., Frenz, M., Romano, V., Weber, H. P., Silenok, A., and Konov, V. I., Mechanism of channel propagation in water by pulsed erbium laser radiation, in *Laser Interaction with Hard and Soft Tissue*, SPIE, Budapest, Hungary, 1993, Vol. 2077, pp. 104–112.

176. Visuri, S. R., Gilbert, J. L., and Walsh, J. T., Shear test of composite bonded to dentin: Er:YAG laser vs. dental handpiece preparations, in *Lasers in Dentistry*, SPIE, San Jose, CA, 1995, Vol. 2394, pp. 223–227.

177. Attrill, D. C., Farrar, S. R., King, T. A., Dickinson, M. R., Davies, R. M., and Blinkhorn, A. S., Er:YAG (l = 2.94 μm) laser etching of dental enamel as an alternative to acid etching, *Lasers Med Sci* 15, 154–161, 2000.

178. von Fraunhofer, J. A., Allen, D. J., and Orbell, G. M., Laser etching of enamel for direct bonding, *Angle Orthod* 63, 73–76, 1993.

179. Lieberman, R., Segal, T. H., Nordenberg, D., and Serebro, L. I., Adhesion of composite materials to enamel: Comparison between the use of acid and lasing as pretreatment, *Lasers Surg Med* 4, 323–327, 1984.

180. Clayman, L., Fuller, T., and Beckman, H., Healing of continuous wave and rapid superpulsed, carbon dioxide, laser-induced bone defects, *J Oral Surg* 36, 932–937, 1978.

181. Friesen, L. R., Cobb, C. M., Rapley, J. W., Forgas, B. L., and Spencer, P., Laser irradiation of bone. II. Healing response following treatment by CO_2 and Nd:YAG lasers, *J Periodontol* 68(9), 75–83, 1997.

182. Spencer, P., Payne, J. T., Cobb, C. M., Peavy, G. M., and Reinisch, L., Effective laser ablation of bone based on the absorption characteristics of water and protein, *J Periodontol* 70(5), 68–74, 1999.

183. Esposito, M., Hirsh, J. M., Lekholm, U., and Thomsen, P., Review: Biological factors contributing to failures of osseointegrated oral implants. (I) Success criteria and epidemiology, *Eur J Oral Sci* 106, 527–551, 1998.

184. Cummings, J. P. and Walsh, J. T. Jr., Erbium laser ablation: The effect of dynamic optical properties, *Appl Phys Lett* 62(16), 1988–1990, 1993.

185. Vodop'yanov, K. L., Bleaching of water by intense light at the maximum of the 1~3 μm absorption band, *Sov Phys JETP* 70(1), 114–347, 1990.

186. Shori, R. K., Walston, A. A., Stafsudd, O. M., Fried, D., and Walsh, J. T., Quantification and modeling of the non-linear changes in absorption coefficient of water at clinically relevant IR laser wavelengths, Special Topics, *J Quantum Electron* 7(6), 959–970, 2001.

187. Boehm, R. F., Rich, J., Webster, J., and Janke, S., Thermal stress effects and surface cracking associated with laser use on human teeth, *J Biomech Eng* 99(1), 189–194, 1977.

188. Krause, L. S., Cobb, C. M., Rapley, J. W., Killoy, W. J., and Spencer, P., Laser irradiation of bone. I. An in vitro study concerning the effects of the CO_2 laser on oral mucosa and subjacent bone, *J Periodontol* 68(9), 872–880, 1997.

189. Fisher, S. E. and Frame, J. W., The effects of the CO_2 surgical laser on oral tissues, *Br J Oral Maxillofac Surg* 22, 414–425, 1984.

190. Koort, H. J. and Frentzen, M., The effect of TEA-CO_2-Laser on dentine, *3rd International Congress on Lasers in Dentistry*, Salt Lake City, UT, 1992, Abstract #64.

191. Romano, V., Rodriguez, R., Altermatt, H. J., Frenz, M., and Weber, H. P., Bone microsurgery with IR-lasers: A comparative study of the thermal action at different wavelengths, in *Laser Interaction with Hard and Soft Tissue*, SPIE, Budapest, Hungary, 1994, Vol. 2077, pp. 87–96.

192. Fried, N. M. and Fried, D., Comparison of Er:YAG and 9.6-μm TE CO_2 lasers for ablation of skull tissue, *Lasers Surg Med* 28, 335–343, 2001.

193. Lee, C., Ragadio, J., and Fried, D., Influence of wavelength and pulse duration on peripheral thermal and mechanical damage to dentin and alveolar bone, in *Lasers in Dentistry VI*, SPIE, San Jose, CA, 2000, Vol. 3910, pp. 193–203.

194. Tagomori, S. and Morioka, T., Combined effects of laser and fluoride on acid resistance of human dental enamel, *Caries Res* 23, 225–251, 1989.

195. Jennett, E., Motamedi, M., Rastegar, S., Arcoria, C. J., and Fredrickson, C. J., Dye enhanced Alexandrite laser for ablation of dental tissue, *Lasers Surg Med Suppl* 6, 15, 1994.

196. Arcoria, C. J., Fredrickson, C. J., Hayes, D. J., Wallace, D. B., and Judy, M. M., Dye microdrop assisted laser for dentistry, *Lasers Surg Med Suppl* 5, 17, 1993.

197. White, J. M., Goodis, H. E., Setcos, J. C., Eakle, S., Hulscher, B. E., and Rose, C. L., Effects of pulsed Nd:YAG laser on human teeth. A three-year follow-up study, *J Am Dent Assoc* 124, 45–51, 1993.

198. White, J. M., Goodis, H. E., and Daniels, T. E., Effects of Nd:YAG laser on pulps of extracted teeth, *Lasers Life Sci* 4(3), 191–200, 1991.

199. White, J. M., Neev, J., Goodis, H. E., and Berns, M. W., Surface temperature and thermal penetration depth of Nd:YAG laser applied to enamel and dentin, in *Laser Surgery: Advanced Characterization, Therapeutics, and Systems III*, SPIE, Los Angeles, CA, 1992, Vol. 1643, pp. 423–436.

200. White, J. M., Goodis, H. E., Marshall, G. W., and Marshall, S. J., Identification of the physical modification threshold of dentin induced by neodymium and holmium YAG lasers using scanning electron microscopy, *Scanning Microsc* 7(1), 239–246, 1993.

201. Niemz, M. H., Cavity preparation with the Nd:YLF picosecond laser, *J Dent Res* 74(5), 1194–1199, 1995.

202. Boulnois, J. L., Photophysical processes in recent medical laser developments: A review, *Lasers Med Sci* 1, 47–66, 1986.

203. Niemz, M. H., Investigation and spectral analysis of the plasma-induced ablation mechanism of dental hydroxyapatite, *Appl Phys B* 58, 273–281, 1994.

204. Neev, J., Da Silva, L. B., Feit, M. D., Perry, M. D., Rubenchik, A. M., and Stuart, B. C., Ultrashort pulse lasers for hard tissue ablation, *IEEE J Sel Top Quantum Electron* 2(4), 790–808, 1996.

205. Neev, J., Liaw, L. L., Raney, D. V., Fujishige, J. T., Ho, P. D., and Berns, M. W., Selectivity, efficiency, and surface characteristics of hard dental tissues ablated with ArF pulsed excimer lasers, *Lasers Surg Med* 11, 499–510, 1991.

206. Frentzen, M., Koort, H. J., and Thiensiri, I., Excimer lasers in dentistry: Future possibilities with advanced technology, *Quintessence Int* 23, 117–133, 1992.

207. Neev, J., Liaw, L. L., Stabholtz, A., Torabinejad, J. T., Fujishige, J. T., and Berns, M. W., Tissue alteration and thermal characteristics of excimer laser interaction with dentin, in *Laser Surgery: Advanced Characterization, Therapeutics, and Systems III*, SPIE, Los Angeles, CA, 1992, Vol. 1643, pp. 386–397.

208. Neev, J., Stabholtz, A., Liaw, L. L., Torabinejad, M., Fujishige, J. T., Ho, P. D., and Berns, M. W., Scanning electron microscopy and thermal characteristics of dentin ablated by a short-pulse XeCl excimer laser, *Lasers Surg Med* 12, 353–361, 1993.

209. Moss, J. P., Patel, B. C. M., Pearson, G. J., Arthur, G., and Lawes, R. A., Krypton fluoride excimer ablation of tooth tissues: Precision tissue machining, *Biomaterials* 15(12), 1013–1018, 1994.

210. Hennig, T., Rechmann, P., Pilgrim, C., and Kaufmann, R., Basic principles of caries selective ablation by pulsed lasers, in *Proceedings of the Third International Congress on Lasers in Dentistry*, Salt Lake City, UT, 1992, pp. 119–120.

211. Hennig, T., Rechmann, P., Pilgrim, C. G., Schwarzmaier, H.-J., and Kaufmann, R., Caries selective ablation by pulsed lasers, in *Lasers in Orthopedic, Dental, and Veterinary Medicine*, SPIE, Los Angeles, CA, 1991, Vol. 1424, p. 99.

212. Rechmann, P. and Hennig, T., Caries selective ablation: First histological examinations, in *Laser Surgery: Advanced Characterization, Therapeutics, and Systems IV*, SPIE, Los Angeles, CA, 1994, Vol. 2128, p. 389.

213. Feuerstein, O., Palanker, D., Fuxbrunner, A., Lewis, A., and Deutsch, D., Effect of the ArF excimer laser on human enamel, *Lasers Surg Med* 12, 471–477, 1992.

214. Melis, M., Berna, G., Berna, N., and Benvenuti, A., Ablation of hard dental tissues by ArF and XeCl excimer lasers, in *Lasers in Surgery: Advanced Characterization, Therapeutics, and Systems IV*, SPIE, Los Angeles, CA, 1994, Vol. 2128, pp. 349–358.

215. Hibst, R. and Keller, U., Removal of dental filling materials by Er:YAG laser radiation, in *Lasers in Orthopedic, Dental, and Veterinary Medicine*, SPIE, Los Angeles, CA, 1991, Vol. 1424, pp. 120–126.

216. Nelson, J. S., Yow, L., Liaw, L. H., Macleay, L., and Zavar, R. B., Ablation of bone and methacrylate by a prototype Er:YAG laser, *Lasers Surg Med* 8(5), 494–500, 1988.

217. Marshall, S., Marshall, G., Watanabe, L., and White, J., Effects of the Nd:YAG laser on amalgams and composites, *Trans Acad Dent Mater* 2, 297–298, 1989.

218. Thomas, B. W., Hook, C. R., and Draughn, R. A., Laser-aided degradation of composite resin, *Angle Orthod* 66(4), 281–286, 1996.

219. Yow, L., Nelson, J. S., and Berns, M. W., Ablation of bone and PMMA by an XeCl 308 nm excimer laser, *Lasers Surg Med* 9(2), 141–147, 1989.

220. Sherk, H. H., Lane, G., Rhodes, A., and Black, J., Carbon dioxide laser removal of polymethylmethacrylate, *Clin Orthopaed Relat Res* 310, 67–71, 1995.

221. Dumore, T. and Fried, D., Selective ablation of orthodontic composite using sub-microsecond IR laser pulses with optical feedback, *Lasers Surg Med* 27(2), 103–110, 2000.

222. Alexander, R. and Fried, D., Selective removal of orthodontic composite using 355-nm Q-switched laser pulses, *Lasers Surg Med* 30, 240–245, 2002.

223. Yamamoto, H. and Ooya, K., Potential of yttrium-aluminum-garnet laser in caries prevention, *J Oral Pathol* 38, 7–15, 1974.

224. Stern, R. H., Sognnaes, R. F., and Goodman, F., Laser effect on in vitro enamel permeability and solubility, *J Am Dent Assoc* 78, 838–843, 1966.

225. Featherstone, J. D. B. and Fried, D., Fundamental interactions of lasers with dental hard tissue, *Med Laser Appl* 16, 181–195, 2001.

226. Kuroda, S. and Fowler, B. O., Compositional, structural and phase changes in in vitro laser-irradiated human tooth enamel, *Calcif Tissue Int* 36, 361–369, 1984.

227. Yamamoto, H. and Sato, K., Prevention of dental caries by Nd: YAG laser irradiation, *J Dent Res* 59, 1271–1277, 1980.

228. Fox, J. L., Yu, D., Otsuka, M., Higuchi, W. I., Wong, J., and Powell, G. L., Initial dissolution rate studies on dental enamel after CO_2 laser irradiation, *J Dent Res* 71, 1389–1397, 1992.

229. Otsuka, M., Wong, J., Higuchi, W. I., and Fox, J. L., Effects of laser irradiation on the dissolution kinetics of hydroxyapatite preparations, *J Pharm Sci* 79, 510–515, 1990.

230. Otsuka, M., Matsuda, Y., Suwa, Y., Wong, J., Fox, J. L., Powell, G. L., and Higuchi, W. I., Effect of carbon dioxide laser irradiation on the dissolution kinetics of self-setting hydroxyapatite cement, *Lasers Life Sci* 5(3), 199–208, 1993.

231. Featherstone, J. D. B., Fried, D., and Bitten, E. R., Mechanism of laser induced solubility reduction of dental enamel, in *Lasers in Dentistry III*, SPIE, San Jose, CA, 1997, Vol. 2973, pp. 112–116.

232. Hsu, J., Fox, J. L., Wang, Z., Powell, G. L., Otzuka, M., and Higuch, W. I., Combined effects of laser irradiation/solution fluoride on enamel demineralization, *J Clin Laser Med Surg* 16, 93–105, 1998.

233. Borggreven, J. M. P. M., Van Dijk, J. W. E., and Driessens, F. C. M., Effect of laser irradiation on the permeability of bovine dental enamel, *Arch Oral Biol* 25, 831–832, 1980.

234. Featherstone, J. D. B., Barrett-Vespone, N. A., Fried, D., Kantorowitz, Z., and Lofthouse, J., CO_2 laser inhibition of artificial caries-like lesion progression in dental enamel, *J Dent Res* 77(6), 1397–1403, 1998.

235. Fox, J. L., Yu, D., Otsuka, M., Higuchi, W. I., Wong, J., and Powell, G. L., The combined effects of laser irradiation and chemical inhibitors on the dissolution of dental enamel, *Caries Res* 26, 333–339, 1992.

236. Fox, J. L., Wong, J., Yu, D., Otsuka, M., Higuchi, W. I., Hsu, J., and Powell, G. L., Carbonate apatite as a model for the effect of laser irradiation on human dental enamel, *J Dent Res* 73(12), 1848–1853, 1994.

237. Kantola, S., Laine, E., and Tarna, T., Laser-induced effects on tooth structure. VII. X-ray diffraction study of dentine exposed to a CO_2 laser, *Acta Odontol Scand* 31, 381–386, 1973.

238. Nammour, S., Renneboog-Squilbin, C., and Nyssen-Behets, C., Increased resistance to artificial caries-like lesions in dentin treated with CO_2 laser, *Caries Res* 26, 170–175, 1992.

239. Kinney, J. H., Haupt, D. L., Balooch, M., White, J. M., Bell, W. L., Marshall, S. J., and Marshall, G. W., The threshold effects of Nd and Ho:YAG laser-induced surface modification on demineralization of dentin surfaces, *J Dent Res* 75(6), 1388–1395, 1996.

240. Le, C. Q., Fried, D., and Featherstone, J. D. B., Lack of dentin acid resistance following 9.3 um CO_2 laser irradiation, in *Lasers in Dentistry XIV*, SPIE, San Jose, CA, Vol. 6843, 2008, p. 68430J-5.

241. Featherstone, J. D. B., Le, C. Q., Hsu, D., Manesh, S., and Fried, D., Changes in acid resistance of dentin irradiated by a CW 10.6 μm CO_2 laser, in *Lasers in Dentistry XIV*, SPIE, San Jose, CA, 2008, Vol. 6843, pp. 684305 (1–5).

242. Tao, Y.-C., Fan, K., and Fried, D., Near-infrared image-guided laser ablation of artificial caries lesions, in *Lasers in Dentistry XIII*, SPIE, San Jose, CA, 2007, Vol. 6425, pp. 64250T1–8.

243. Tao, Y.-C. and Fried, D., Selective removal of natural occlusal caries by coupling near-infrared imaging with a CO_2 laser, in *Lasers in Dentistry XIV*, SPIE, San Jose, CA, 2008, Vol. 6843, pp. 68430I1–8.

244. Fried, D., Zuerlein, M. J., Le, C. Q., and Featherstone, J., Thermal and chemical modification of dentin by 9–11 μm CO_2 laser pulses of 5–100-μs duration, *Lasers Surg Med* 3, 275–282, 2002.

12

Nonlinear Interferometric Vibrational Imaging and Spectroscopy

Haohua Tu
*University of Illinois
at Urbana–Champaign*

Zhi Jiang
*University of Illinois
at Urbana–Champaign*

Praveen
D. Chowdary
*University of Illinois
at Urbana–Champaign*

Wladimir
Benalcazar
*University of Illinois
at Urbana–Champaign*

Eric J. Chaney
*University of Illinois
at Urbana–Champaign*

Daniel L. Marks
*University of Illinois
at Urbana–Champaign*

Martin Gruebele
*University of Illinois
at Urbana–Champaign*

Stephen A. Boppart
*University of Illinois
at Urbana–Champaign*

12.1 Introduction

Optical imaging, both traditional and emerging, plays a significant and widespread role in medicine and surgery, as noted by the ubiquitous use of microscopes, endoscopes, and video-based imaging systems. The ultimate goal is to rapidly quantify the local concentration of the biomolecule of interest on the microscopic scale, preferably without external labeling. This is intrinsically difficult for common techniques such as confocal microscopy, infrared microscopy, (spontaneous) Raman microspectroscopy, optical coherence tomography (OCT), x-ray fluorescence microscopy, and conventional incoherent or coherent multiphoton microscopy including two-photon fluorescence, second-order harmonic generation, and third-order harmonic generation. In this respect, coherent anti-Stokes Raman scattering (CARS) has attracted a growing interest since 1999 [1] due to three prominent advantages. First, the multiphoton process of CARS permits high-resolution subcellular imaging in three dimensions, termed as nonlinear sectioning capability. Second, fast imaging can be conducted on unstained (unperturbed) samples with chemically specific vibration (Raman) contrast (Figure 12.1). This label-free molecular imaging is enabled by the coherent process of CARS that constructively amplifies the otherwise weak spontaneous signal of Raman microspectroscopy. Third, the optical frequency upconverted CARS signal can be easily separated from the excitation, spontaneous Raman photons, and fluorescence background. Note that the coherence process of CARS requires a phase-matching condition to maintain momentum conservation (Figure 12.1), which typically leads to geometrically complicated noncollinear imaging [2]. Fortunately, in the scenario of CARS microscopy with tight focusing conditions, collinear excitation beams at the focus come from a solid angle, which is sufficiently large to cover the phase-matching condition [1].

From the perspective of spectroscopy vs. imaging, there is a trade-off, and CARS can be broadly classified into multiplex CARS and single-frequency CARS, which emphasize the chemical specificity and fast imaging, respectively. Rather than collect a Raman spectrum at each point in the image (spectral tomogram), as would be done by multiplex CARS, single-frequency CARS [3] and the closely related stimulated Raman scattering (SRS) imaging [4] collect an image for one isolated vibrational frequency in this spectrum. Single-frequency CARS has achieved shot noise-limited performance [5] free of experimental artifacts [6], so that desirable video-rate epi-imaging can be realized [3,5]. However, the spectroscopic Raman information to discriminate different biomolecules is lost. It is possible to multiplex SRS but at the expense of reduced sensitivity [7] or with an unrealistic assumption that the biomolecule is known a priori [8]. Thus, multiplex CARS that collects the structure-rich Raman spectrum is more useful for quantitative molecular imaging, even though the imaging speed must be reduced accordingly.

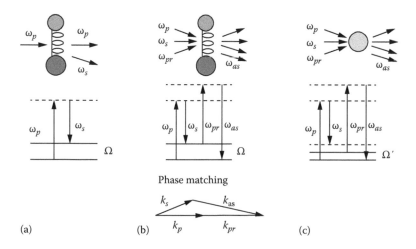

FIGURE 12.1 Comparison of (a) spontaneous Raman scattering, (b) R-CARS, and (c) NR-CARS. Optical frequencies of pump, probe, Stokes, and anti-Stokes photons are related by $\omega_{as} = \omega_p - \omega_s + \omega_{pr}$.

Regular multiplex CARS [9] necessitates the use of broadband ultrafast laser sources, which usually result in producing a strong background signal from the electronic response of molecules [10], manifested as the nonresonant CARS (NR-CARS) (Figure 12.1). The Raman signal useful for quantitative molecular imaging, termed as the resonant CARS (R-CARS), is often dominated (distorted) by this background. A wide variety of techniques have been developed to suppress the nonresonant background, each with noticeable strengths and weaknesses [11]. In this chapter, we focus on an interferometric technique because it may simultaneously remove the nonresonant background and enhance the imaging sensitivity. Although interferometric multiplex CARS can have rather different schemes [12–17] (Table 12.1), they always interfere a specific raw CARS signal from the sample with a preferably stronger reference beam, termed as the local oscillator. The electromagnetic field (amplitude and phase) of the raw signal is then interferometrically retrieved and finally used to derive the Raman spectrum and the local concentration of the Raman oscillators. This is advantageous over the noninterferometric multiplex CARS, which only measures the intensity (amplitude) of the CARS signal.

We compare these schemes in the context of biological imaging (Table 12.1). Each can be either *homodyne* if the local oscillator is from the sample itself [14] or *heterodyne* if the local oscillator is from an external source. In general, the homodyne approach automatically compensates the sample spot-dependent phase variation between the raw signal and the local oscillator [13], so that Raman spectrum reconstruction is straightforward. This is less so in the heterodyne approach in which the compensation must be achieved by sending the local oscillator into the sample [15], by cancelling the instrument phase along the sample arm and the local oscillator arm [16], or by careful calibration (see Section 12.4 for details) [17]. On the other hand, the sensitivity enhancement of the homodyne approach is limited by the magnitude of the nonresonant background (the major component of the local oscillator), which is in turn limited by the photodamage of the biological sample. The heterodyne approach advantageously removes this intrinsic limitation so that the sensitivity enhancement factor is generally higher. More than 1000-fold enhancement can be attained by the heterodyne amplification [15].

The autocorrelation scheme is a Fourier-transform-based implementation of multiplex CARS, wherein the raw signal consisting of both the resonant signal and the nonresonant background is interfered with a replica of itself [12,18]. In contrast, the continuum-enhanced dual-pump scheme first separates the two components and then lets them interfere with each other [13]. In these two schemes, the relation between the interferometric signal and the concentration of the Raman oscillators ranges from linear to quadratic, depending on the ratio of signal/background [19]. A linear relation is preferred for quantitative molecular imaging. In addition, the acquisition of a Raman spectrum requires mechanical time-delay scanning of the excitation beam, which limits the imaging speed. Employing the phase-contrast CARS [20] and a $4f$ pulse shaper [21], the phase-contrast single-beam scheme has overcome these two limitations [14]. More importantly, Raman spectra acquisition rates over 3000 Hz have been achieved using a single-beam alignment-free setup [22]. In this scheme, however, the Raman spectrum reconstruction relies on the assumption of constant nonresonant background, which is questionable for biological samples [23]. Also, a fundamental trade-off exists between interferometric signal magnitude and Raman spectral resolution. The phase-cycling single-beam scheme [15,24,25] largely overcomes these shortcomings, but at the expense of slow speed, which is due to the phase cycling of the excitation beam to acquire the Raman spectrum [26]. In contrast to the two single-beam schemes that have sensitivity vs. speed trade-off, the spectral-domain OCT-like scheme [16] combines the sensitivity advantage of its time-domain heterodyne counterpart [27] and the speed advantage of the spectral-domain OCT [28]. This combination represents the most desirable property for fast imaging, even with the cost of a more complex and expensive setup. However, the raw signal undergoing heterodyne amplification consists of not only the resonant signal but also the nonresonant background. In the important Raman fingerprint region (600–1800 cm^{-1}) of biological samples, the signal is dominated by the background so that accurate Raman spectrum reconstruction becomes problematic.

TABLE 12.1   Plausible Interferometric Multiplex CARS for Biological Samples

Scheme	Autocorrelation [12]	Continuum-Enhanced dual pump [13]	Phase-contrast single beam [14]	Phase-cycling single beam [15]	Spectral-domain OCT-like [16]	NIVI [17]
Raw signal	R-CARS + NR-CARS	R-CARS + partial NR-CARS	R-CARS	R-CARS + suppressed NR-CARS	R-CARS + NR-CARS	R-CARS + suppressed NR-CARS
Nature (local oscillator)	Homodyne (same as raw signal)	Homodyne (NR-CARS + partial R-CARS)	Homodyne (NR-CARS)	Heterodyne (external strong source)	Heterodyne (external NR-CARS)	Heterodyne (external strong NR-CARS)
Sensitivity enhancement	Low to moderate	Low to moderate	Moderate	High	High	High
Signal vs. concentration	Linear to quadratic	Linear to quadratic	Linear	Linear	Linear if R-CARS $\approx$ NR-CARS	Linear
Raman spectrum acquisition rate	Slow, time-delay scan required	Slow, time-delay scan required	Fast, no scanning required	Slow, excitation phase scan	Fast, no scanning required	Fast, no scanning required
Excitation beam(s)	Two	Three	One	One	Two	Two
Interferometer	Required	Required	Not required	Not required	Required	Required
Detection system	Single detector	Array detector	Array detector	Single detector	Array detector	Array detector
Factor dictates Raman spectral resolution	Resolution of time delay	Bandwidth of narrowband probe	Spectral resolution of pulse shaper	Spectral resolution of pulse shaper	Bandwidth of narrowband probe	Linear chirp rate of broadband probe
Strengths	Simple setup, cost-effective, reliable	High spectral resolution, broad spectrum	Easy alignment (i.e., stable phase), fast speed	Easy alignment, simple setup, high sensitivity	High sensitivity, fast speed, epi-imaging	High sensitivity, fast speed, proven bioimaging
Weaknesses	Low sensitivity, not quantitative imaging, slow speed	Low sensitivity, not quantitative imaging, slow speed, complex setup	Possible analysis artifact, trade-off of signal intensity vs. spectral resolution	Slow speed, difficult to measure pulse at microscope objective focus	Complex expensive setup, possible analysis artifact in fingerprint region	Complex expensive setup, narrow spectrum, careful calibration

Sources: Ogilvie, J.P. et al., *Opt. Lett.*, 31, 480, 2006; Kee, T.W. et al., *Opt. Express*, 14, 3631, 2006; Lim, S.-H. et al., *Phys. Rev. A*, 72, 041803, 2005; von Vacano, B. et al., *Opt. Lett.*, 31, 2495, 2006; Evans, C.L. et al., *Opt. Lett.*, 29, 2923, 2004; Jones, G. W. et al., *Opt. Lett.*, 31, 1543, 2006.

Nonlinear interferometric vibrational imaging (NIVI) [17] overcomes this limitation of the spectral-domain OCT-like scheme [16] without compromising either the speed or the sensitivity. NIVI first separates the resonant signal and the nonresonant background by a time-resolved technique [29,30] and then applies the heterodyne amplification to the resonant signal alone to improve Raman spectrum reconstruction and quantitative molecular imaging. Thus, the interferometry in NIVI serves mainly the dual purposes of heterodyne amplification and resonant field reconstruction. In contrast, the interferometry in other schemes (Table 12.1) serves the triple purposes of heterodyne (or homodyne) amplification, resonant signal separation (from nonresonant background), and resonant field reconstruction. NIVI capitalizes on the well-documented strength of the time-resolved technique for clean separation of the resonant signal and the nonresonant background [31,32]. Due to the limitation of the corresponding laser source, NIVI employs a time-delayed broadband pulse to probe the excited broadband Raman vibrations, so that the pairwise correspondence between the anti-Stokes frequency and the Raman resonance is lost. Fortunately, the correspondence can be computationally restored with high spectral resolution by linearly chirping the probe pulse [17,33].

The state-of-the-art NIVI combines three key features of our earlier works, including heterodyne amplification and interferometric field reconstruction [27,34], time-resolved resonant signal isolation [30], and probe-dechirped spectral interferometry [17]. We have experimentally demonstrated that the NIVI spectra of biomolecules are equivalent to the corresponding Raman spectra but can be acquired at least 1000 times faster than Raman microspectroscopy for a comparable signal-to-noise ratio [35,36]. The molecular imaging capability of NIVI has mapped the most prevalent molecular constituents of skin [37]. In a study using a preclinical breast cancer model [38], we have used NIVI to differentiate rat mammary tumors from that of normal rat mammary tissue with a >99% confidence interval and resolve the *molecular* tumor margin within 100 μm. We note that other schemes of interferometric multiplex CARS have also begun to image biological samples [39,40]. Interferometric multiplex CARS will continue to evolve in the future. The ideal scheme is expected to combine many desirable strengths of the current schemes (Table 12.1). It most likely would integrate a fiber continuum source for broadband Raman spectrum acquisition [13], a pulse shaper for single-beam alignment-free setup [14,15], a detection system based on spectral-domain OCT [16,17], and a time-resolved mechanism for resonant signal isolation [17]. We have recently demonstrated a fiber continuum source that can be coherently controlled by a 4f pulse shaper [41,42], providing a starting point to integrate all the four components into a continuum-enhanced single-beam NIVI-like system [43,44].

In the following sections, we describe our current solid-state-laser-based NIVI system for imaging and spectroscopy, including its theory, experimental setup, and measurement results on both material and biological samples.

12.2 Theory

Degenerate multiplex CARS utilizes a narrowband pump/probe pulse and a transform-limited broadband Stokes pulse, as illustrated in Figure 12.2a. Assuming that only one vibrational resonance is excited, the R-CARS signal resembles a decaying sinusoidal oscillation. The power spectrum of this signal is the Lorentzian line shape. In contrast, NIVI uses a positively chirped broadband pump/probe pulse elongated or dispersed in time, with its frequency increasing with time and its leading edge coincident with the Stokes pulse, as shown in Figure 12.2b. The R-CARS signal still resembles a decaying sinusoid but is now chirped identically to the pump/probe pulse. The power spectrum of the R-CARS signal is no longer Lorentzian. Because a single vibrational frequency produces multiple frequencies in the R-CARS signal, the correspondence between vibrational and anti-Stokes frequencies is lost. When multiple vibrations are excited, they produce overlapping and interfering spectra. Therefore, in general, the power spectrum alone is insufficient to deduce the Raman spectrum, as it is in standard multiplex CARS.

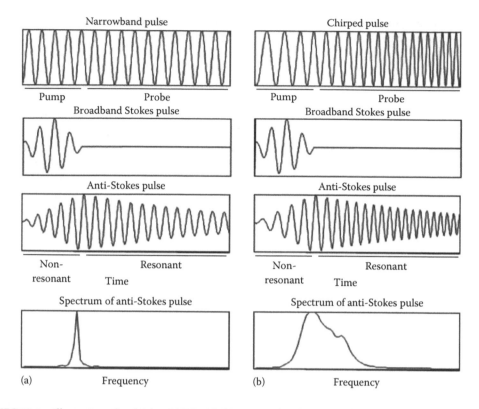

FIGURE 12.2 Illustration of multiplex CARS with (a) a narrowband pump/probe pulse and (b) a chirped broadband pump/probe pulse. (Adapted from Jones, G.W. et al., *Opt. Lett.*, 31, 1543, 2006. With permission.)

If instead of measuring just the power spectrum of the R-CARS signal we measure the cross correlation of the R-CARS signal with a known reference signal. Using an interferometric method such as spectral interferometry, we can time-resolve the R-CARS signal. Because all the excited vibrations are stimulated by the same probe pulse, they all produce R-CARS signal that are chirped identically. We can therefore computationally remove the chirp of the time-resolved R-CARS signal by multiplying the measured signal by the opposite chirp in the time domain. In doing so, we calculate the R-CARS signal that one would have measured if a narrowband pump/probe pulse was used. Because the correspondence between vibrational frequencies and anti-Stokes frequencies is restored, one can determine the complex Raman spectrum from the corrected anti-Stokes spectrum.

We consider a pump/probe pulse and a Stokes pulse, with electric fields given in the frequency domain by $\tilde{E}_P(\omega)$ and $\tilde{E}_S(\omega)$, respectively. The medium has a complex Raman spectrum of $\chi^{(3)}(\Omega)$, where Ω is the Raman frequency. The excited vibration population $N(\Omega)$ and the anti-Stokes electric field $\tilde{E}_{AS}(\omega)$ produced by the CARS process are given by

$$N(\Omega) = \chi^{(3)}(\Omega) \int_0^\infty \tilde{E}_P(\omega + \Omega)\tilde{E}_S(\omega)^* d\omega \qquad (12.1)$$

$$\tilde{E}_{AS}(\omega) = \int_0^\omega N(\Omega)\tilde{E}_P(\omega - \Omega)d\Omega \qquad (12.2)$$

These equations assume the following: (1) The optical field is strong and can be treated classically, (2) there is a perturbative interaction with the sample that begins and ends with the vibrational ground state, and (3) there are no levels directly resonant with any of the individual frequencies in the pulse (resonant one-photon interactions do not occur). Near-infrared light (800–1400 nm), which is only weakly absorbed by biological tissues, contains frequencies typically well above the vibrational frequencies of molecular bonds, but below the electronic excitation frequencies, and is thus well-suited for NIVI. The two equations give the two-step time evolution of a CARS process involving many possible simultaneous Raman-active vibrations. In the first step (Equation 12.1), the molecule is excited by the pump frequency ω_1 and probe frequency ω_2 that are separated by the Raman vibrational frequency $\Omega = \omega_1 - \omega_2$. The use of a short broadband Stokes pulse is to stimulate a broad bandwidth of Raman transitions. A vibration at frequency Ω is excited by every pair of optical frequencies with a frequency difference of Ω. In the second step (Equation 12.2), $N(\Omega)$ is converted to anti-Stokes radiation $\tilde{E}_{AS}(\omega)$ by a second SRS process whereby the probe at frequency ω_1 is upconverted to the anti-Stokes signal at frequency $\omega_{AS} = \omega_1 + \Omega = 2\omega_1 - \omega_2$. For broadband excitation, the anti-Stokes signal must be summed over all probe frequencies, yielding Equation 12.2. The information about the sample is contained in its Raman spectrum $\chi^{(3)}(\Omega)$. By sending in pulses with known electric fields $\tilde{E}_P(\omega)$ and $\tilde{E}_S(\omega)$ and measuring the returned anti-Stokes signal $\tilde{E}_{AS}(\omega)$, we would like to infer $\chi^{(3)}(\Omega)$ in a particular frequency range.

To explain the mathematics of chirped pump pulse CARS, we consider a chirped pump/probe pulse with spectrum

$$\tilde{E}_P(\omega) = \tilde{E}_0(\omega)\exp\left[\frac{-i(\omega - \omega_0)^2}{2\alpha}\right] \tag{12.3}$$

where $\tilde{E}_P(\omega)$ is the spectral amplitude and α is the chirp rate of the pulse. From the stationary phase approximation for a small α, the time-domain signal can be written as

$$E_P(t) = \alpha^{-1/2}\tilde{E}_0(\alpha t + \omega_0)\exp\left[i\left(\frac{\alpha t}{2 + \omega_0}\right)t\right] \tag{12.4}$$

The Stokes pulse spectrum $\tilde{E}_S(\omega)$ is transform limited and centered about $t = 0$ such that $\tilde{E}_S(\omega)$ is real. Because of this, the instantaneous frequency of the pump/probe pulse at the time the Stokes pulse arrives $(t = 0)$ is ω_0, which can be approximated as a $\delta(\bullet)$ function. With these representations of the pump/probe and Stokes pulses, the excited vibration population in Equation 12.1 can be approximated as

$$N(\Omega) = \chi^{(3)}(\Omega)\int_0^\infty \alpha^{-1/2}\tilde{E}_0(\omega_0)\delta(\omega + \Omega - \omega_0)\tilde{E}_S(\omega)^* d\omega$$

$$= \chi^{(3)}(\Omega)\alpha^{-1/2}\tilde{E}_0(\omega_0)\tilde{E}_S(\omega_0 - \Omega) \tag{12.5}$$

Then, the anti-Stokes pulse in the time domain $E_{AS}(t)$ can be calculated by Equation 12.2, yielding

$$E_{AS}(t) = \alpha^{-1/2}\tilde{E}_0(\alpha t + \omega_0)\exp\left[i\left(\frac{\alpha t}{2 + \omega_0}\right)t\right]\frac{1}{2\pi}\int_0^{\omega_0} d\Omega \chi^{(3)}(\Omega)\tilde{E}_S(\omega_0 - \Omega)\tilde{E}_0(\omega_0)\exp(i\Omega t) \tag{12.6}$$

As can be seen, one can multiply $E_{AS}(t)$ by the conjugate phase $\exp\left[-i\left(\alpha t/2 + \omega_0\right)t\right]$ to remove the chirp of the probe that is mixed into the anti-Stokes signal. Because the pump/probe pulse spectrum $\tilde{E}_0(\omega)$ tends to be smooth, one can approximate it as a constant. Therefore, one applies the inverse Fourier transform to $E_{AS}(t)\exp\left[-i\left(\alpha t/2 + \omega_0\right)t\right]$ to yield an estimate of the Raman spectrum $\chi^{(3)}(\Omega)$, weighted by the Stokes spectrum. Because the Stokes pulse spectrum tends to be slowly varying, this $\chi^{(3)}(\Omega)$ strongly resembles the true Raman spectrum.

Noninterferometric experiments usually detect the intensity of $|E_{AS}(t)|^2$, but this produces a distorted Raman line shape from nonresonant contributions to the signal, as well as quadratic concentration dependence. The amplitude and phase information of the anti-Stokes pulse $E_{AS}(t)$ has to be measured on the ultrafast time scale to properly retrieve the Raman spectrum. Unfortunately, at optical frequencies, the electric field $E_{AS}(t)$ is not directly measurable because it oscillates on a time scale too fast to be electronically demodulated. Therefore, interferometric demodulation is employed. To demodulate the anti-Stokes field, we generate a reference field from the oscillator field that contains the same frequencies as in the anti-Stokes radiation. The interferometric demodulator measures the intensity of the cross correlation of the anti-Stokes and reference fields. Spectral interferometry is especially attractive because it allows the signal to be sampled in one shot to minimize transient effects and can achieve a higher signal-to-noise ratio than conventional temporal interferometry.

NIVI works by reconstructing the full complex CARS field from a real-valued spectral interferogram. To create the interferogram, the CARS field is mixed with a transform-limited reference pulse $\tilde{E}_{ref}(\omega)$ in an interferometer. The resulting spectral interferogram $I(\omega)$ contains dc components from the CARS and reference fields and the phase-sensitive cross term that will yield the NIVI signal:

$$I(\omega) = \left|\tilde{E}_{ref}(\omega)\right|^2 + \left|\tilde{E}_{AS}(\omega)\right|^2 + 2\,\mathrm{Re}\left[\tilde{E}_{AS}(\omega)\tilde{E}_{ref}(\omega)^* \exp(-i\omega\tau)\right] \tag{12.7}$$

where τ is the time delay between the CARS and reference pulses, essential for the reconstruction of the complex CARS field. The first term in the right-hand side of Equation 12.7 can be rejected by background subtraction of the known reference pulse (Figure 12.3a). The second term can be also rejected temporally by separating the dc and phase-sensitive components as follows.

Letting $\chi(\omega) = \tilde{E}_{AS}(\omega)\tilde{E}_{ref}(\omega)^*$, we can express the inverse Fourier transform of Equation 12.7 as (Figure 12.3b).

$$FT^{-1}\left(I(\omega)\right) = FT^{-1}(dc) + \chi(t - \tau) + \chi(-t - \tau) \tag{12.8}$$

The first term is a slowly varying dc component symmetric about $t = 0$, while the last two terms are time reversed from each other (Equation 12.7 is real). Because of causality, there is no signal prior to the Stokes pulse initiating vibrational coherence. After shifting the time origin to be the center of the Stokes pulse (Figure 12.3c), one can impose $\chi(t) = 0$ for $t < -T$, where T is greater than the Stokes pulse width. If the delay τ of the interferometer is chosen larger than T, the two time-reversed terms in Equation 12.8 do not overlap with each other or with the dc component (Figure 12.3b). The function $\chi(t - \tau)$ thus obtained by shifting and truncation is Fourier-transformed to recover $\chi(\omega) \propto \chi^{(3)}(\Omega)^* \tilde{E}_S(\omega_0 - \Omega)\tilde{E}_{ref}(\omega)^*$. Both $\chi(\omega)$ and $\chi^{(3)}(\Omega)$ are linear in analyte concentration, but $\chi(\omega)$ is weighted by the known field envelopes $\tilde{E}_S$ and $\tilde{E}_{ref}$. In the absence of electronic resonances at ω, the NR-CARS background is real and can be rejected by discarding the real part of $\chi^{(3)}$. The NIVI spectrum is given by the imaginary part (Figure 12.3d) and matches the Raman spectrum.

Perhaps a more intuitive way to understand why interferometry can isolate the R-CARS signal from the NR-CARS background is to examine Figure 12.2. When the pump and Stokes pulses overlap, SRS

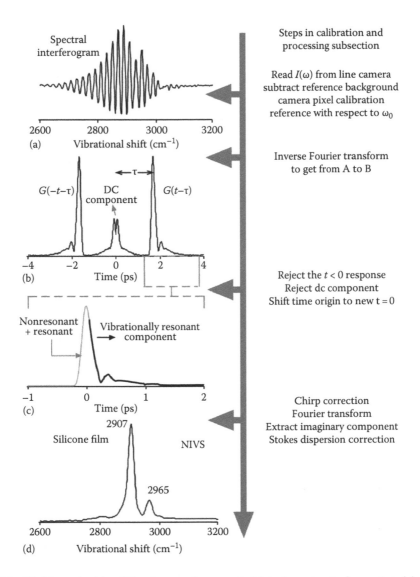

FIGURE 12.3 Working principle of Fourier transform spectral interferometry is demonstrated by extracting the NIVI spectrum of a silicone film starting from the interferogram acquired on the line camera. (a) Wavelength-calibrated spectral interferogram referenced to the pump wavelength. The reference power spectrum is subtracted. (b) Modulus of the time-domain polarization obtained by inverse Fourier transforming the interferogram in A. (c) The $t > 0$ response, identifying the short-time nonresonant (follows the Stokes pulse shape) and long-time resonant (vibrational dephasing) components. The temporal chirp of the pump can be corrected here. (d) NIV spectrum of a silicone film obtained as the imaginary part by Fourier transforming the time response in (c), and Stokes dispersion is corrected. (Adapted from Chowdary, P.D. et al., *Anal. Chem.*, 82, 3812, 2010. With permission.)

populates molecular vibration(s) to generate the R-CARS signal. After the Stokes pulse passes, this R-CARS generation will remain. At the moment of the overlap, four-wave-mixing of the molecule can also be excited. However, because there is no persistent state associated with the nonresonant process, the non-resonant emission will end quickly after the Stokes pulse passes. Therefore, the nonresonant component can be discarded by rejecting any anti-Stokes radiation that occurs coincident with the Stokes pulse, since interferometry provides precise time-resolved information of the CARS signal (Figure 12.3b).

12.3 Instrumentation

A CARS microscope operates by tightly focusing the pump and Stokes beams to a point in the tissue and collecting the scattered anti-Stokes photons. By scanning the focus through the tissue, a 3D image is acquired. Because CARS is a multiphoton process, the anti-Stokes signal is produced mainly at the focus, so that high spatial resolution is achieved. Figure 12.4 shows the general experimental setup. Pulses from a titanium:sapphire mode-locked seed laser (810 nm center wavelength, ~20 nm bandwidth) are amplified by a regenerative amplifier. The pulse repetition rate is dumped down to 250 kHz with approximately microjoule pulse energies to pump an optical parametric amplifier (OPA). Both Stokes and anti-Stokes pulses are generated after the OPA and separated by a dichroic mirror. A portion of the laser beam from the regenerative amplifier is dispersed by an 85 cm BK7 glass bar to generate ~6 ps temporally chirped pulses as both the pump and probe beams. The pump and Stokes beams are combined by another dichroic mirror and focused onto the sample by an objective. This same objective can be used for CARS signal collection in the epidirection, or a second one can be used for collection in the forward direction (shown). Samples are placed on a translation stage for raster scanning (shown), or the incident beams can be scanned across a fixed sample using mirrors mounted on computer-controlled galvanometers. Delay in the pump beam path is tuned to temporally overlap the pump and Stokes beams. Forward CARS signal is collected by the second objective and filtered before being combined with the anti-Stokes reference beam for spectral interferometry. Spectral interferometry is implemented by a grating and line camera, which allows for a significant increase in image acquisition speed, in signal-to-noise, and in interferometric signal stability.

At first glance, the use of the broadband laser beams might be expected to generate a significant amount of nonresonant background and also lose the pairwise correspondence between vibrational frequencies and anti-Stokes frequencies. However, as discussed in Section 12.2, these limitations can be removed by the following considerations in our setup. First, the CARS signal is measured by spectral interferometry; thus, full spectral/temporal information of the CARS signal rather than its total energy (e.g., measured by a photomultiplier tube) can be retrieved. Second, the R-CARS and NR-CARS signals are time resolved and can be easily separated. Third, the broadband pump beam has been heavily chirped by glass and all wavelength components are mapped to different temporal locations. Roughly speaking, the long-wavelength portion of the chirped pump pulses plays the role of the pump beam and overlaps with the Stokes beam from the OPA to excite the molecular vibration in the sample, while the short-wavelength portion of the chirped pulses functions as a probe beam to produce the final CARS signal. Compared with standard multiplex CARS using a narrowband pump and broadband Stokes, our

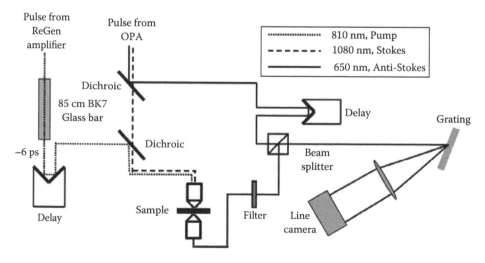

FIGURE 12.4 **(See color insert.)** Experimental setup for the NIVI system.

configuration takes full advantage of all frequency components of the broadband pump beam to maximize the CARS signal. When overlapping with the Stokes beam, only a narrow band of pump beam, as monochromatic light, is used to excite the molecular Raman vibration since the broadband Stokes remains as short pulses. This not only reduces the nonresonant background but also makes it possible to recover the correspondence between vibrational frequencies and anti-Stokes frequencies. As a result, one single measurement can obtain a broad Raman spectrum covering several hundred wavenumbers without tuning laser wavelengths. These improvements are achieved at the expense of experimental setup complexity and computational complexity. The wavenumber range of achievable Raman spectra is determined by the Stokes beam bandwidth (~200 cm⁻¹ in our current setup), and the resolution of the Raman spectra is determined by the amount of chirp applied to the pump beam.

12.4 System Calibration

To calibrate the system, a nonresonant sample (sapphire or glass slide) was used. Both the OPA signal and the anti-Stokes pulse are nearly transform limited and nearly identical. Therefore, the angular dispersion of the wavelengths on the line camera could be characterized by minimizing the measured pulse width. A sample of acetone was placed in the interferometer in order to find the chirp dispersion parameter. Acetone has a strong, isolated resonance at 2925 cm⁻¹. The dispersion introduced into the pump pulse by the BK7 glass bar was calibrated by adjusting chirp dispersion parameter to minimize the bandwidth of the retrieved acetone Raman spectrum. We have calibrated our NIVI system by retrieving the Raman spectrum of other organic solvents (isopropanol, methanol, and ethanol) or liquid lipids [17,35].

Here, we use solid materials to calibrate the NIVI system for both spectroscopy and imaging, which is not possible with the use of the liquid samples. Figures 12.5a and b show the measured spectral fringes

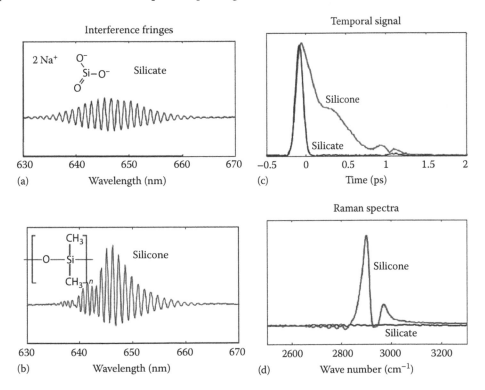

FIGURE 12.5 Spectral fringes for (a) sodium silicate and (b) silicone. (c) Temporal signal after Fourier transform. (d) Retrieved Raman spectra (weighted by the Stokes spectrum).

(continued)

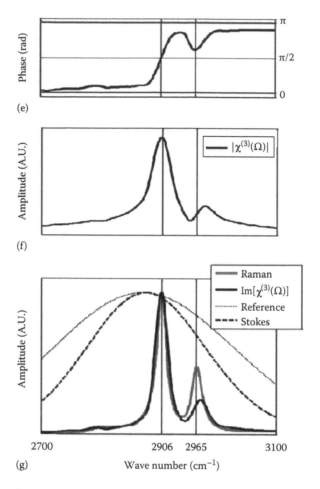

FIGURE 12.5 (continued) Spectral fringes for (e) phase and (f) magnitude of the complex vibrational spectrum $\chi^{(3)}$. (g) Obtained Im($\chi^{(3)}$) and spontaneous Raman spectra. Dotted lines indicate the Stokes and reference spectral profiles. (Adapted from Benalcazar, W.A. et al., *IEEE J. Sel. Top. Quant. Electron.*, 16, 824, 2010; Jiang, Z. et al., Nonlinear interferometric vibrational imaging of biological tissue. *SPIE Photonics West, Biomedical Optics Proceedings*, San Jose, CA, January 19–24, 2008. With permission.)

(dc components subtracted) for both sodium silicate and silicone. The Stokes center wavelength is tuned to ~1080 nm to excite the C–H Raman vibrational bond around 2900 cm^{-1} wave number. For sodium silicate, only NR-CARS is generated due to the lack of C–H bonds. The NR-CARS spectrum resembles the reference pulses directly from the OPA, as evident by the smooth and symmetric spectral fringes in Figure 12.5a. The period of spectral fringes can be controlled by the delay in the reference arm. For silicone, which is rich in C–H bonds, the CARS signal includes both R-CARS and NR-CARS. The R-CARS has structure and shows similar chirp as the pump beam, which is clear from the spectral fringes. After Fourier transformation, the time-resolved CARS signal can be obtained as shown in Figure 12.5c. For sodium silicate, nonresonant only CARS in the time domain is a short pulse, which is similar to the Stokes pulse. This is because NR-CARS is associated with the electronic response of molecules and occurs almost instantaneously, only when the pump and Stokes beams arrive at the sample together. In contrast, for silicone, in addition to the NR-CARS signal around $t = 0$, there is a large amount of R-CARS signal generated afterward, as shown in Figure 12.5c. Intuitively, this can be understood as the molecule vibration that will continue during its dephasing time once it is excited and continue interacting with the chirped pump beam to generate a resonant vibrational CARS signal. As mentioned before, NR-CARS can be easily separated and suppressed from R-CARS due to their distinct temporal characteristics. By multiplying temporal chirp with the time-domain signal in Figure 12.5c, Raman spectra can be obtained by taking the imaginary part of the inverse Fourier transform of the temporal signal, as shown in Figure 12.5d. For sodium silicate, there are no Raman-active bonds in this range (2800–3000 cm⁻¹). For silicone, two Raman peaks are observed, which is consistent with the literature. Note that the calculated Raman spectra are weighted by the spectral profile of the Stokes pulses as discussed in Section 12.2.

The phase and magnitude of the retrieved $\chi^{(3)}(\Omega)$ of the silicone are shown in Figure 12.5e and f. In the presence of nonresonant background, the peaks in the magnitude are shifted, but the background is small enough to make this evident only in the peak at 2965 cm⁻¹. The phase information (Figure 12.5e) indicates the predicted behavior for a double Lorentzian response, with frequencies below resonances in phase with the incident Stokes field, retarded by about $\pi/2$ at resonances, and out of phase by π above them. This phase information obtained by heterodyne detection allows full suppression of the nonresonant background and the real component of the resonant signal, so that the imaginary part of $\chi^{(3)}(\Omega)$ is obtained. This is shown in Figure 12.5g in comparison with a spontaneous Raman spectrum. Both were acquired in the same region of the sample with an integration time of 50 s for Raman and 10 ms for NIVI. A reduction in the relative amplitude of the second resonance at 2965 cm⁻¹ is observed as a consequence of the weight of the Stokes and reference spectra. Note, however, that the spectral widths of the resonances follow those for spontaneous Raman.

Figure 12.6 shows images of a material sample comprised of a mixture of sodium silicate and silicone, which is contained between slides. The image size is 500 × 500 μm² (100 × 100 pixels). For each pixel, the temporal signal is calculated and the NR-CARS and R-CARS are distinguished in a simple way. Energy between −0.2 and 0 ps is integrated and attributed to NR-CARS, while energy between 0.1 and 1.0 ps is integrated and attributed to R-CARS. Images based on the NR-CARS and R-CARS signals are formulated accordingly, as shown in Figures 12.6a and b, respectively. Since both materials produce similar amount of NR-CARS signal, the NR-CARS image is somewhat uniform except at the boundaries between the different materials. The lower intensity at the boundaries is mainly caused by light scattering due to refractive index mismatch. In contrast, the R-CARS image shows distinct features. The sodium silicate forms bubbles, and these show a lack of R-CARS signal, while silicone around them produces a large R-CARS signal. Such simple classification already provides rich molecular information about the sample. The NR-CARS image provides structural information, while the R-CARS image shows molecular distribution.

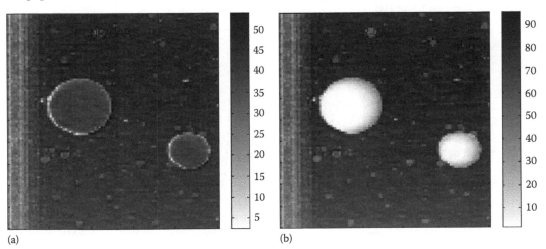

(a) (b)

FIGURE 12.6 Images of a mixture of sodium silicate and silicone. (a) Image based on NR-CARS signals. (b) Image based on R-CARS signals. The values in the grayscale bar represent integrated intensities from the temporal signal. Images 500 × 500 μm. (Adapted from Jiang, Z. et al., Nonlinear interferometric vibrational imaging of biological tissue. *SPIE Photonics West, Biomedical Optics Proceedings*, San Jose, CA, January 19–24, 2008. With permission.)

12.5 Validation for Biological Samples

Unlike bulk silicone/acetone or homogenous polymers films, biological tissues are highly scattering targets, which make the phase retrieval of the CARS signal by heterodyne detection more difficult, for the following reasons. First, each point on a diffuse scatterer acts as its own source of CARS signal. This leads to considerable loss during the spatial mode combination of the CARS signal with the reference at the pinhole. The sensitivity of the technique needs to be sufficiently high to account for these losses. Second, the severity of phase fluctuations induced by the diffuse scatterers needs to be understood since this ultimately determines if the Raman spectra can be recovered at all.

Our initial objective was to verify that NIVI has sufficient sensitivity to accurately retrieve the Raman spectra from diffusely scattering samples. Blood is a diffusely scattering sample with a well-understood internal structure and composition, providing an interesting system to investigate using NIVI. We imaged the C–H stretching cross section of a thick film of blood (~100 µm) by tuning the Stokes bandwidth to cover the 2800–3100 cm⁻¹ region of the vibrational spectral range. Figure 12.7a presents a NIVI image (100×100 µm²) of blood obtained by integrating the long-time component (vibrationally resonant contribution) of the third-order polarization. One could readily observe the ring-shaped structures that appear on a length scale of 10 µm, which are the biconcave disc-shaped red blood cells. The diagonal white streak was from a crack in the blood film following dehydration. Figure 12.7b shows the Raman spectrum of a dark pixel from the image, which can be compared to the spontaneous Raman spectrum shown in Figure 12.7c [45]. All the spectral characteristics (C–H_2 stretches, C–H_3 stretches, and =C–H stretches) have reliably been recovered. The peaks at 2730 cm⁻¹ appear small in the NIVI-retrieved spectrum since the Stokes bandwidth does not fully cover these vibrations.

The accuracy of the Raman retrieval is further corroborated by our studies on thick sections (~100 µm) of normal mammary tissue from a nonlactating female rat. Figure 12.8 shows the retrieval of a NIVI spectrum for a cross section of mammary tissue and the comparison with the spontaneous Raman spectroscopy. Since the Raman and NIVI spectra were not taken at exactly the same positions, the main difference resides in the relative intensity of these peaks (i.e., different concentration of chemical species and conformations), with the added effect of the Stokes and responses of the system being nonuniform. Nevertheless, the six spectral features observed in this region can be discriminated (marked with vertical lines in Figure 12.8a), and their chemical bond correspondence has been assigned in the literature for the case of breast tissue [46].

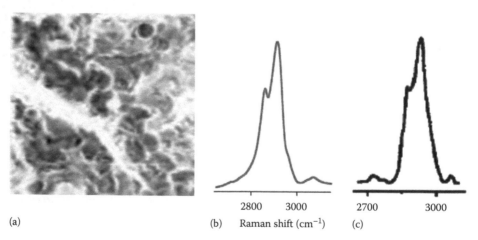

(a)   (b)   Raman shift (cm⁻¹)   (c)

FIGURE 12.7 NIVI imaging and spectroscopy of a highly scattering human blood sample. (a) NIVI image (100×100 µm²) of a human blood sample obtained by plotting the time-integrated resonant vibrational contribution (gray scale—white to black). (b) Raman spectrum from one of the dark pixels from the image in (a). (c) Spontaneous Raman spectrum of blood from the literature. (From Rohleder, D. et al., *J. Biomed. Opt.*, 10, 031108, 2005.)

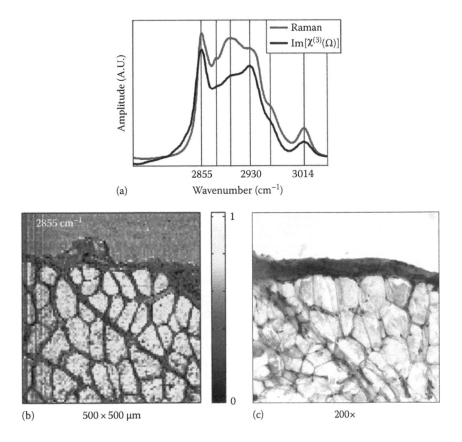

(a)

(b) 500 × 500 µm

(c) 200×

FIGURE 12.8 Vibrational spectroscopy and imaging of mammary tissue. (a) Imaginary component of the susceptibility of NIVI spectrum in comparison with spontaneous Raman spectrum. (b) NIVI hyperspectral cube image at 2855 cm⁻¹ showing adipocytes, parenchyma, and connective tissue. (c) Stained histological (H&E) image of sample in similar region. (Adapted from Benalcazar, W.A. et al., *IEEE J. Sel. Top. Quant. Electron.*, 16, 824, 2010. With permission.)

A vibrational image (100 × 100 pixels) of the mammary tissue at 2855 cm⁻¹ is shown over an area of 500 µm × 500 µm (Figure 12.8b). The acquisition time was 100 s (10 ms/pixel) and the sample was illuminated with 10 mW of pump power and 2 mW of Stokes power. The high signal shown in brighter regions corresponds to white adipocytes, which play a crucial role in the morphogenesis of the mammary parenchyma. These cells contain a large lipid inclusion, composed primarily of triglycerides and cholesteryl ester (both rich in C–H$_2$), surrounded by a layer of cytoplasm and very few mitochondria. Nuclei can be distinguished as small black dots (low fat content) at the periphery of the cells. The narrow dark boundary of the adipocytes corresponds to connective tissue fibers (~25 µm thick) and the extracellular matrix (~45 µm thick), which have less lipid content and therefore less signal at the C–H$_2$ resonance. These image-based findings correlate strongly with the corresponding histological image (hematoxylin and eosin [H&E]) from a similar adjacent region in the same tissue and animal, as shown in Figure 12.8c. The thick dark-appearing boundary layer along the top portion of the tissue in Figure 12.8b corresponds to the parenchyma, which is also clearly distinguished as the darker layer in the histology shown in Figure 12.8c.

The aforementioned results and analysis confirm two aspects. First, the phase fluctuations from even 100 µm thick (sufficiently thick so as to be considered as whole tissue) samples are not significant enough to affect (reduce) the accuracy of Raman retrieval by the heterodyne detection employed by NIVI. Second, the technique has sufficiently high sensitivity to compensate for the losses associated with the diffusely scattered signal at the pinhole in our detection system.

12.6 Molecular Imaging and Histopathology

We have used NIVI to map the molecular composition in the outermost layers of porcine skin from the back region of the animal [37]. While the analysis starts from a priori knowledge about the bonds to which Raman-based techniques are sensitive in this region, the number of molecular distinctions attainable by the technique cannot be inferred directly prior to the experiment. In this sense, our method is suitable for the interrogation of potentially unknown compositions in skin or other tissues or cells in a fast way. NIVI is sensitive to four different domains: stratum corneum, hair follicles, dermis, and epidermis. Such differentiation is shown in Figure 12.9. The resonant signal over the entire probed bandwidth is shown in Figure 12.9a, and the NIVI composite image, which features the molecular regions, is shown in Figure 12.9b. The corresponding spectra (Figure 12.9c), which are in concordance with previous characterization with spontaneous Raman spectroscopy, show that the spectral features are similar in pairs: dermis and epidermis show higher concentration of $C-H_3$ symmetric resonances, while stratum corneum and follicles show comparable amplitudes at $C-H_2$ and $C-H_3$ symmetric vibrations. This increase in signal at the $C-H_2$ symmetric band when compared to dermis and epidermis captures the presence of lipids in the stratum corneum and sebaceous secretions in the follicles. Modulation of the molecular map by the intensity map gives the composite image of the sample (Figure 12.9d), which overlays molecular information with structural features of the tissue, such as the flattened keratinized layers of the stratum corneum, the membranes and nuclei of the dermal cells, and the extracellular collagenous fibers in the dermis. For comparison, an H&E-stained section is shown in Figure 12.9e, which features the same information as the NIVI composite but requires long hours of processing and does not possess the additional spectroscopic content available to NIVI, which would be advantageous as a future diagnostic technique based on quantitative measures of histologic or cytologic composition.

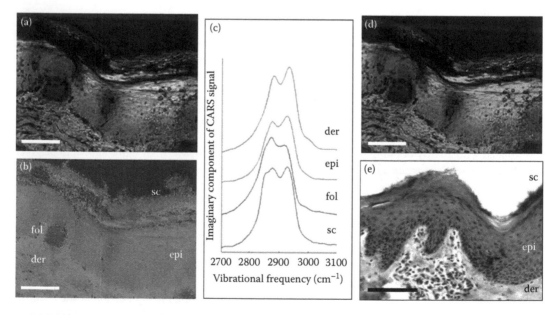

FIGURE 12.9 (See color insert.) Molecular imaging of cutaneous tissue. (a) Intensity image (total integrated spectral power). (b) NIVI composite showing discrimination of stratum corneum (sc), epidermis (epi), dermis (der), and hair follicle (fol). (c) NIVI spectra for each domain in (b), as obtained by cluster analysis: each spectrum is the result of averaging the spectra of the members of most prevalent cluster in regions of 20×20 pixels^2 within each domain. (d) NIVI image showing both structural and molecular compositions. (e) H&E histology of a section from the same region. The scale bar is 100 μm in every image. (Adapted from Benalcazar, W.A. and Boppart, S.A., *Anal. Bioanal. Chem.*, 400, 2817, 2011. With permission.)

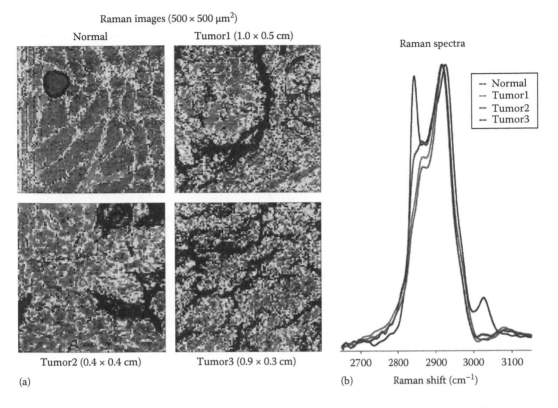

Raman images (500 × 500 μm²)

Normal   Tumor1 (1.0 × 0.5 cm)

Raman spectra

Tumor2 (0.4 × 0.4 cm)   Tumor3 (0.9 × 0.3 cm)

(a)   (b)   Raman shift (cm⁻¹)

FIGURE 12.10 (a) NIVI-generated Raman images of a normal mammary tissue section and three different tumor sections (different locations and different sizes given) from the same animal (color scale: blue to red). (b) Averaged Raman spectra in the C–H stretching region of the tissue section in (a).

We have used NIVI to study different mammary tissue types (normal, tumor) from a carcinogen-induced rat mammary tumor model [38]. This well-characterized animal model recapitulates the progression of ductal carcinoma in situ that occurs in humans. Figure 12.10a shows the NIVI-generated Raman images obtained by plotting the integrated area under the Raman spectrum for each pixel. Images of one normal mammary tissue and three different tumors are shown. Figure 12.10b shows the averaged Raman spectra over the different tissue types. Two points are notable from the spectra. First, the normal tissue Raman spectrum (black plot) significantly differs from the tumor tissue spectra (green, red, blue plots). Second, the different tumor tissue spectra are fairly similar to each other. The source for these spectral differences between different pathological states is consistent with the morphological model of human breast cancer reported in the literature [47]. Extensive Raman studies on different human breast pathologies led to a morphological model for breast cancer, which identifies that the key diagnostic parameter between normal and malignant tumor spectra is the fat-to-collagen ratio [48]. We see that this is also true in this rat mammary tumor model. The normal tissue is abundant with fat (lipids) and hence is rich in C–H₂ stretches and C=C–H stretches (unsaturated fatty acids). These modes explain the two peaks seen at 2857 and 3020 cm⁻¹, respectively. Tumor growth leads to stromal proliferation, thus shifting the composition scales toward the structural collagen scaffold. The heterogeneity of tumor leads to variability in the fat-to-collagen composition and thus to slight spectral variations. Nevertheless, the overall spectral profile of tumor is observed to be consistent for tumors of different size and location from multiple animals.

The NIVI spectra acquired from different pathologies in this data set are significant enough for a classification to be made on inspection. The simplest of the diagnostic algorithms would be to compare the

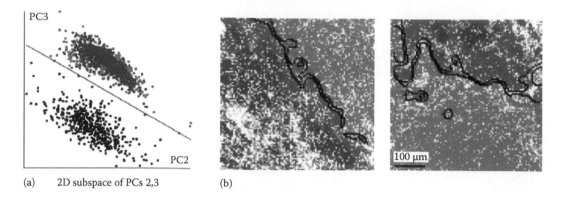

(a)   2D subspace of PCs 2,3   (b)

FIGURE 12.11 Molecular margin determination. (a) 2D subspace of PCs 2 and 3 showing clear differentiation between normal (black dots) and tumor (gray dots) tissues based on NIVI spectra. (b) Automated tumor margin identification (black curves) overlaid on two smoothed representative spectrum-reconstructed NIVI images. The two boundaries of the margin demarcate normal and tumor domains at greater than the 99% confidence interval. Tumor margins are readily resolved to 100 μm. (Adapted from Chowdary, P.D. et al., *Cancer Res.*, 70, 9562, 2010. With permission.)

relative intensities at the C–H_2 and C–H_3 peaks, which depends on the fat-to-collagen ratio. However, this would significantly underutilize the large information content found in the entire spectral profile. To make use of the information content at all the spectral components, there are a variety of multivariate statistical techniques that could be used. We have coupled principal component analysis (PCA) with logistic regression to develop a diagnostic algorithm that has been trained to determine the disease pathology. PCA reduces the dimensionality of the problem into the few most significant components in terms of which the spectra can be characterized. We initially formed a data matrix (2000 × 1000) from the normal and tumor tissues shown in Figure 12.11a. The rows are comprised of different pixel spectra taken from each of the images with equal representation. The spectra are digitized at 1000 points in the range of 2420–3320 cm^{-1}. We then used a singular value decomposition (SVD) algorithm to extract the principal components (PCs) of this data matrix. The three most significant PCs account for >99% of the variance in the data. The segregation of the normal and tumor spectra is obvious in the 2D subspace of the PCs 2 and 3, as shown in Figure 12.11a. We then used logistic regression on the PC scores to draw a decision line between the normal and tumor tissues. As shown, the classification of normal and tumor spectra is extremely efficient with this data set. Figure 12.11b illustrates software detection of tumor margins at the 100 μm scale. Margins were detected in less than a few seconds. The tumor margins are at the 99% confidence interval. The images for margin detection again show the optimum linear combination of the spectral coefficients C2 and C3. A clean margin (<1–2 pixel wide) demarcating the tumor and normal domains can be obtained. Instead of a sharp threshold, we chose as a margin zone, two narrow bands tangent to the edges of the 99% confidence ellipses near the decision line. We thus obtained the tumor margin as a band with a finite width as shown in Figure 12.11b. Pixels within the band cannot be assigned to normal or tumor tissue with greater than 99% confidence interval. The residual width of the margin and small islands of segmented regions is a result either of noise or of the spatial heterogeneity associated with tumor incursions into normal tissue.

12.7 Summary

In summary, we have described the theory, instrumentation, and experimental results for imaging and spectroscopy with the technique of NIVI, in which the temporally isolated R-CARS signal is detected by spectral interferometry. Using a broadband chirped pump/probe pulse and a broadband Stokes pulse, high-spectral-resolution Raman spectra are obtained by computationally removing the chirp of the anti-Stokes pulse. Spectral-domain detection has been demonstrated and allows for a significant increase in image acquisition speed, in signal-to-noise, and in interferometric signal stability. Experimental imaging results using this instrumentation and method have been acquired and validated both from a material sample and from biological tissues. Specifically, we achieved accurate Raman spectra retrieval from highly scattering samples and successfully used NIVI to map molecular constituents in skin and to classify the differences between normal and tumor tissues in a rat mammary tumor model.

In the future, increasing the spectral range by increasing the Stokes bandwidth and investigating the fingerprint region for DNA/protein markers are expected to further enhance the sensitivity of NIVI for differentiating normal from tumor tissue. While most of our efforts to date have primarily been directed toward disease diagnosis, recent results add further direction to our research. Recent research suggests that there is a pattern in the tumor growth driven by morphological components such as collagen tracts. Hence, it will be very interesting to investigate the correlation between the spectral variations and the spatial locations within the tissue. With the microscopic spatial resolution and the spectral sensitivity that has been demonstrated, NIVI is a promising new real-time molecular imaging modality that could help elucidate the dynamics of tumor formation and growth.

Acknowledgments

Animals used for this research were handled and cared for under a protocol approved by the Institutional Animal Care and Use Committee (IACUC) at the University of Illinois at Urbana–Champaign. This work was supported in part by grants from the National Institutes of Health (NCI, R21/R33 CA115536, and NIBIB R01 EB005221, S.A.B.). Additional information can be found at http://biophotonics.illinois.edu.

References

1. Zumbusch A., Holtom G. R., and Xie X. S. 1999. Three-dimensional vibrational imaging by coherent anti-Stokes Raman scattering. *Phys. Rev. Lett.* 82: 4142–4145.
2. Duncan M. D., Reintjes J., and Mannuccia T. J. 1982. Scanning coherent anti-Stokes Raman microscope. *Opt. Lett.* 7: 350–352.
3. Evans C. L., Potma E. O., Puoris'haag M., Côté D., Lin C. P., and Xie X. S. 2005. Chemical imaging of tissue in vivo with video-rate coherent anti-Stokes Raman scattering microscopy. *Proc. Natl. Acad. Sci. U.S.A.* 102: 16807–16812.
4. Freudiger C. W., Min W., Saar B.G., Lu S., Holtom G. R., He C., Tsai J. C., Kang J. X., and Xie X. S. 2008. Label-free biomedical imaging with high sensitivity by stimulated Raman scattering microscopy. *Science* 322: 1857–1861.
5. Saar B. G., Freudiger C. W., Reichman J., Stanley C. M., Holtom G. R., and Xie X. S. 2010. Video-rate molecular imaging in vivo with stimulated Raman scattering. *Science* 330: 1368–1370.
6. Jurna M., Korterik J. P., Otto C., Herek J. L., and H. L. Offerhaus. 2009. Vibrational phase contrast microscopy by use of coherent anti-Stokes Raman scattering. *Phys. Rev. Lett.* 103: 043905.
7. Ploetz E., Marx B., Klein T., Huber R., and Gilch P. 2009. A 75 MHz light source for femtosecond stimulated Raman microscopy. *Opt. Express* 17: 18612–18620.
8. Freudiger C. W., Min W., Holtom G. R., Xu B., Dantus M., and Xie X. S. 2011. Highly specific label-free molecular imaging with spectrally tailored excitation-stimulated Raman scattering (STE-SRS) microscopy. *Nat. Photonics* 5: 103–109.
9. Cheng J.-X., Volkmer A., Book L. D., and Xie X. S. 2002. Multiplex coherent anti-Stokes Raman scattering microspectroscopy and study of lipid vesicles. *J. Phys. Chem. B* 106: 8493–8498.
10. Cheng J.-X. and Xie X. S. 2004. Coherent anti-Stokes Raman scattering microscopy: Instrumentation, theory, and applications. *J. Phys. Chem. B* 108: 827–840.
11. Day P. R. J., Domke K. F., Rago G., Hano H., Hamaguchi H., Vartiainen E. M., and Bonn M. 2011. Quantitative coherent anti-Stokes Raman scattering (CARS) microscopy. *J. Phys. Chem. B* 115. 7713–7725.

12. Ogilvie J. P., Beaurepaire E., Alexandrou A., and Joffre M. 2006. Fourier-transform coherent anti-Stokes Raman scattering microscopy. *Opt. Lett.* 31: 480–482.
13. Kee T. W., Zhao H., and Cicerone M. T. 2006. One-laser interferometric broadband coherent anti-Stokes Raman scattering. *Opt. Express* 14: 3631–3640.
14. Lim S.-H., Caster A. G., and Leone S. R. 2005. Single-pulse phase control interferometric coherent anti-Stokes Raman scattering spectroscopy. *Phys. Rev. A* 72: 041803.
15. von Vacano B., Buckup T., and Motzkus M. 2006. Highly sensitive single-beam heterodyne coherent anti-Stokes Raman scattering. *Opt. Lett.* 31: 2495–2497.
16. Evans C. L., Potma E. O., and Xie X. S. 2004. Coherent anti-Stokes Raman scattering spectral interferometry: determination of the real and imaginary components of nonlinear susceptibility $\chi^{(3)}$ for vibrational microscopy. *Opt. Lett.* 29: 2923–2925.
17. Jones G. W., Marks D. L., Vinegoni C., and Boppart S. A. 2006. High-spectral-resolution coherent anti-Stokes Raman scattering with interferometrically-detected broadband chirped pulses. *Opt. Lett.* 31: 1543–1545.
18. Cui M., Skodack J., and. Ogilvie J. P. 2008. Chemical imaging with Fourier transform coherent anti-Stokes Raman scattering microscopy. *Appl. Opt.* 47: 5790–5798.
19. Lee Y. J. and Cicerone M. T. 2009. Single-shot interferometric approach to background free broadband coherent anti-Stokes Raman scattering spectroscopy. *Opt. Express* 17: 123–135.
20. Oron D., Dudovich N., and Silberberg Y. 2002. Single-pulse phase-contrast nonlinear Raman spectroscopy. *Phys. Rev. Lett.* 89: 273001.
21. Weiner A. M. 2000. Femtosecond pulse shaping using spatial light modulators. *Rev. Sci. Instrum.* 71: 1929–1960.
22. Sung J., Chen B.-C., and Lim S.-H. 2011. Fast three-dimensional chemical imaging by interferometric multiplex coherent anti-Stokes Raman scattering microscopy. *J. Raman Spectrosc.* 42: 130–136.
23. Suzuki T. and Misawa K. 2011. Efficient heterodyne CARS measurement by combining spectral phase modulation with temporal delay technique. *Opt. Express* 19: 11463–11470.
24. von Vacano B. and Motzkus M. 2008. Time-resolving molecular vibration for microanalysis: Single laser beam nonlinear Raman spectroscopy in simulation and experiment. *Phys. Chem. Chem. Phys.* 10: 681–691.
25. Müller C., Buckup T., von Vacano B., and Motzkus M. 2009. Heterodyne single-beam CARS microscopy. *J. Raman Spectrosc.* 40: 800–808.
26. Dudovich N., Oron D., and Silberberg Y. 2002. Single-pulse coherently controlled nonlinear Raman spectroscopy and microscopy. *Nature* 418: 512–514.
27. Vinegoni C., Bredfeldt J. S., Marks, D. L., Boppart S. A. 2004. Nonlinear optical contrast enhancement for optical coherence tomography. *Opt. Express* 12: 333–341.
28. de Boer J. F., Cense B., Park B. H., Pierce M. C., Tearney G. J., and Bouma B. E. 2003. Improved signal-to-noise ratio in spectral-domain compared with time-domain optical coherence tomography. *Opt. Lett.* 28: 2067–2069.
29. Volkmer A., Book L. D., and Xie X. S. 2002. Time-resolved coherent anti-Stokes Raman scattering microscopy: Imaging based on Raman free induction decay. *Appl. Phys. Lett.* 80: 1505–1507.
30. Marks D. L., Vinegoni C., Bredfeldt J. S., and Boppart S. A. 2004. Interferometric differentiation between resonant coherent anti-Stokes Raman scattering and nonresonant four-wave-mixing processes. *Appl. Phys. Lett.* 85: 5787–5789.
31. Pestov D., Murawski R. K., Ariunbold G. O., Wang X., Zhi M., Sokolov A. V., Sautenkov V. A. et al. 2007. Optimizing the laser-pulse configuration for coherent Raman spectroscopy. *Science* 316: 265–268.
32. Selm R., Winterhalder M., Zumbusch A., Krauss G., Hanke T., Sell A., and Leitenstorfer A. 2010. Ultrabroadband background-free coherent anti-Stokes Raman scattering microscopy based on a compact Er:fiber laser system. *Opt. Lett.* 35: 3282–3284.

33. Knutsen K. P., Johnson J. C., Miller A. E., Petersen P. B., and Saykally R. J. 2004. High spectral resolution multiplex CARS spectroscopy using chirped pulses. *Chem. Phys. Lett.* 387: 436–441.
34. Marks D. L. and Boppart S. A. 2004. Nonlinear interferometric vibrational imaging. *Phys. Rev. Lett.* 92: 123905.
35. Chowdary P. D., Benalcazar W. A., Jiang Z., Marks D. L., Boppart S. A., and Gruebele M. 2010. High speed nonlinear interferometric vibrational analysis of lipids by spectral decomposition. *Anal. Chem.* 82: 3812–3818.
36. Benalcazar W. A., Chowdary P. D., Jiang Z., Marks D. L., Chaney E. J., Gruebele M., and Boppart S. A. 2010. High-speed nonlinear interferometric vibrational imaging of biological tissue with comparison to Raman microscopy. *IEEE J. Sel. Top. Quant. Electron.* 16: 824–832.
37. Benalcazar W. A. and Boppart S. A. 2011. Nonlinear interferometric vibrational imaging for fast label-free visualization of molecular domains in skin. *Anal. Bioanal. Chem.* 400: 2817–2825.
38. Chowdary P. D., Benalcazar W. A., Jiang Z., Chaney E. J., Marks D. L., Gruebele M., and Boppart S. A. 2010. Molecular histopathology by nonlinear interferometric vibrational imaging. *Cancer Res.* 70: 9562–9569.
39. Isobe K., Suda A., Tanaka M., Hashimoto H., Kannari F., Kawano H., Mizuno H., Miyawaki A., and Midorikawa K. 2009. Single-pulse coherent anti-Stokes Raman scattering microscopy employing an octave spanning pulse. *Opt. Express* 17: 11259–11266.
40. Nagashima Y., Suzuki T., Terada S., Tsuji S., and Misawa K. 2011. In vivo molecular labeling of halogenated volatile anesthetics via intrinsic molecular vibrations using nonlinear Raman spectroscopy. *J. Chem. Phys.* 134: 024525.
41. Liu Y., Tu H., and Boppart S. A. 2012. Wave-breaking-extended fiber supercontinuum generation for high compression-ratio transform-limited pulse compression. *Opt. Lett.* 37: 2172–2174.
42. Tu H. and Boppart S. A. 2013. Coherent fiber supercontinuum for biophotonics. *Laser Photonics Rev.* 7: 628–645.
43. Liu Y., King M. D., Tu H., Zhao Y., and Boppart S. A. 2013. Broadband nonlinear vibrational spectroscopy by shaping a coherent fiber supercontinuum. *Opt. Express* 21: 8269–8275.
44. Tu H. and Boppart S. A. 2014. Coherent anti-Stokes Raman scattering microscopy: Overcoming technical barriers for clinical translation. *J. Biophoton.* 7: 9–22.
45. Rohleder D., Kocherscheidt G., Gerber K., and Kiefer W. 2005. Comparison of mid-infrared and Raman spectroscopy in the quantitative analysis of serum. *J. Biomed. Opt.* 10: 031108.
46. Frank C. J., Redd D. C., Gansler T. S., and McCreery R. L. 1994. Characterization of human breast biopsy specimens with near-IR Raman. *Anal. Chem.* 66: 319–326.
47. Shafer-Peltier K. E., Haka A. S., Fitzmaurice M., Crowe J., Myles J., Dasari R. R. and Feld M. S. 2002. Raman microspectroscopic model of human breast tissue: Implications for breast cancer diagnosis in vivo. *J. Raman Spectrosc.* 33: 552–563.
48. Haka A. S., Shafer-Peltier K. E., Fitzmaurice M., Crowe J., Dasari R. R., and Feld M. S. 2005. Diagnosing breast cancer by using Raman spectroscopy. *Proc. Natl. Acad. Sci. U.S.A.* 102: 12371–12376.

13

Multifunctional Theranostic Nanoplatform: Plasmonic-Active Gold Nanostars

Hsiangkuo Yuan
Duke University

Andrew M. Fales
Duke University

Tuan Vo-Dinh
Duke University

13.1 Principle of Theranostics

With the rapid development in the field of nanotechnology and photonics, an emerging new concept called *theranostics* has gained much interest, aiming at combining diagnostic molecular imaging with therapeutic modalities.[1,2] In medicine, therapy typically follows diagnostic imaging in order to assess the disease severity and to plan for a localized targeted therapy. Because imaging and therapy are usually two separate modalities, a clear-cut imaging localization does not guarantee targeted treatment. Great discrepancies exist in biodistribution as well as lesion selectivity between the imaging and therapeutic agents. The ability to accurately image and monitor both the pharmokinetics and therapeutic efficacy would potentially allow for improved clinical assessment. However, such a concept remains challenging to date.[3,4] To address this problem, one strategy entails combining an imaging contrast agent with a therapeutic agent, hence a theranostic agent, which shares the same delivery kinetics, with an ultimate goal of seamlessly imaging and monitoring the disease progression as well as assessing therapeutic profiles during the treatment process for personalized therapy.[5–9]

Over the past few decades, the development of numerous modern imaging techniques and drug delivery carriers has created an exciting potential in the field of theranostics. Current molecular imaging techniques vary greatly in spatial/temporal resolution, imaging depth, and detection sensitivity (Figure 13.1). For example, positron emission tomography (PET), although being the most commonly applied molecular imaging technique, has the lowest resolution. Clinical-use 1.5 T magnetic resonance imaging (MRI) typically

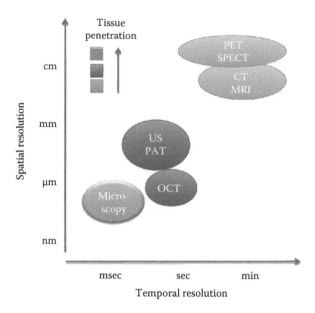

FIGURE 13.1 The spatial and temporal resolution of different molecular imaging modalities. (Adapted from Yuan, H., Plasmonic gold nanostars: A novel theranostic nanoplatform, Dissertation, Duke University, Durham, NC, 2012.)

has a very good tissue-penetration depth for anatomical tomography but a relatively low spatial resolution of 1–5 mm and slow imaging frame rate. In contrast, optical imaging (e.g., microscopy) possesses superior spatial/temporal profiles but suffers from limited tissue penetration. Nonetheless, optical techniques remain the most flexible and affordable imaging technology to date. Recently developed photonics imaging techniques such as confocal microscopy, multiphoton microscopy (MPM), Raman microscopy, optical coherent tomography (OCT), and photoacoustic tomography (PAT) have all been demonstrated in several preclinical settings.[10–14] To exploit these photonic techniques with therapeutic potential, the presence of a theranostic agent is critical.

13.2 Design Strategy for Photonic-Active Theranostic Agent

Nanomaterials, which exhibit unique optical properties and drug delivery kinetics, have become one of the most promising theranostic platforms. Traditionally, fluorophores are conjugated onto therapeutic molecules for theranostic purposes but suffer from several potential drawbacks. Firstly, fluorophore conjugation may alter the delivery kinetics and therapeutic properties of the molecules. Secondly, problems such as photobleaching and nonuniform labeling hinder the practical quantification of fluorophores. Nanomaterials, in contrast, can be made to carry drugs without affecting their delivery kinetics, to target the disease site, and to exhibit high optical signal without photobleaching. To locate malignant tumors, two targeting strategies are commonly employed. While passive targeting exploits the enhanced permeation retention (EPR) effect of the tumor, active targeting acts as a homing beacon, finding its way to the lesion site.[16,17] Compared to conventional chemotherapy, an increase in the drug payload to the tumor with reduced systemic side effects has been demonstrated using nanocarriers.[18,19] Controlling the drug release mechanism may further enhance the delivery kinetics to the lesion site.[20] To date, photonic-active nanomaterials such as quantum dots (QDs), upconverters, dye-doped silica nanoparticles, polymeric particles, plasmonic nanoparticles (silver, gold), and carbon-based nanoparticles (graphene, carbon nanotubes, nanodiamonds) all exhibit distinctive properties favorable for in vivo optical imaging with extended photostability.[21] Combining both qualities as drug carrier and imaging contrast agent, theranostic nanomaterials have great potential for image-guided therapy in medicine.

One strategy to design photonic-active theranostic nanomaterials entails *upgrading* the imaging contrast agent with therapeutic capability, which could be either an intrinsic property of the nanomaterial or an add-on drug loaded onto the nanomaterial. The former concept takes advantage of the material property that elicits therapeutic effect upon certain external stimuli. For example, scintillation nanoparticles (e.g., QDs, lanthanides, zinc oxides, metal nanoparticles) have been applied as radiosensitizers for cancer therapy.[22] High atom and electron density may enhance the scattering or absorption of x-rays and gamma radiation, both of which are operated beyond the typical optical range. To be operated in the optical range, photonic-active nanomaterials, such as plasmonic nanoparticles, carbon nanotubes, or porphysomes, have been demonstrated to offer enhanced photon absorption for photothermal therapy (PTT).[23–26] In particular, plasmonic nanoparticles have been one of the most commonly used photothermal transducers. By controlling the nanoparticles' size and geometry, their plasmon peak can be tuned to match the irradiating photon energy for optical photothermal transduction. They exhibit a very high absorption-to-scattering ratio and extinction coefficient (up to 10^{10} M^{-1} cm^{-1}), which is much greater than that of QDs (10^4 M^{-1} cm^{-1}), fluorescein (10^5 M^{-1} cm^{-1}), and carbon nanotubes (10^{4-5} M^{-1} cm^{-1}), rendering them a powerful photothermal therapeutic agent.[27,28]

Alternatively, therapeutic agents can be loaded onto photonic-active nanomaterials. One way is to conjugate drugs directly on the surface of nanoparticles, while another is to create an additional layer to entrap drugs.[23] The former method relies on electrostatic adsorption or covalent conjugation of drug molecules. Such concepts can be exemplified by tumor necrosis factor-conjugated gold nanoparticles (Aurimmune), which recently completed its phase I clinical trials.[29] However, insufficient drug loading per particle and limited control on drug release may hinder the therapeutic efficiency of surface-loaded drug carriers. To load more drugs onto photonic-active nanomaterials, an encapsulating layer (e.g., silica, polymer, liposome) is needed. One promising material, silica, has emerged as a viable strategy to provide a stable and biocompatible encapsulating layer that allows for not only functionalization of targeting ligands but also effective loading of drug molecules.[30,31] Furthermore, by designing a silica shell on plasmonic nanoparticle cores, the release of drug molecules can be photothermally triggered.[32–36] Additional imaging contrast agents such as gadolinium or radioisotopes can be incorporated onto silica for MRI or PET imaging as well.[30,31] Integrating photonic-active nanomaterials with an encapsulating layer provides a promising multifunctional theranostic platform.

13.3 Photonic-Active Theranostic Nanomaterials

An ideal photonic-active theranostic nanomaterial needs to be biocompatible, easy to synthesize, tunable in size/shape, convenient for surface functionalization, in vivo traceable, and capable of therapy. For in vivo application, nanomaterials need to be operable in the NIR region for better tissue penetration of the exciting light. Table 13.1 lists the pros and cons of several photonic-active theranostic nanomaterials. For QDs or upconverters, although they generate reactive oxygen species (ROS) under x-ray irradiation, therapeutic response in the NIR region requires the addition of a drug layer. QD toxicity due to the release of ions from the cadmium-containing cores poses another problem.[37] The additional drug layer is required for silica nanomaterials as well. In contrast, as discussed previously, gold nanoparticles can be both a strong imaging contrast and therapeutic agent. Gold nanoparticles also offer great potential for bioapplications due to their biocompatibility, chemical stability, and plasmon tunability. With their plasmon resonance in the NIR range, gold nanoparticles are more suitable for in vivo applications than other theranostic nanomaterials.

Photons interact with gold nanoparticles via the surface plasmon. The surface plasmon, which describes the collective oscillation of surface electron clouds under electromagnetic excitation, is strongly influenced by a nanoparticle's composition, size, and geometry. When the incident photons match the nanoparticle's plasmonic frequency, resonant dipolar and multipolar modes can be excited to elicit a strong electromagnetic field enhancement in the vicinity of the nanoparticle surface. This event is called plasmon resonance, where an intense localized field (e.g., localized surface plasmon resonance [LSPR]) can interact with the surface molecules, thereby enhancing their scattering (e.g., surface-enhanced Raman scattering [SERS])

TABLE 13.1 Photonic-Active Nanomaterials

Type	Key Features	Limitation
QD	High quantum yield Compatible to most microscopy X-ray scintillator	Blinking Cadmium toxicity
Upconverter	NIR excitation with visible emission X-ray scintillator	Difficult surface functionalization Low quantum yield and long fluorescence lifetime
Plasmonic nanoparticles	Plasmon tunable in NIR range Efficient photothermal transduction High TPACS and short fluorescence lifetime Simple synthesis and functionalization	Silver toxicity Low quantum yield
Carbon nanoparticles	Intrinsic Raman signal Moderate photothermal transduction	Difficult fabrication Weak optical contrast
Dye-doped silica nanoparticles, liposome, polymers	High loading capacity Biocompatibility Simple functionalization	Lack intrinsic optical property

or photoluminescence (PL) properties. LSPR also enhances the photothermal transduction processes, which can be exploited for photothermal ablation and photoacoustic imaging.

Over many years, multiple synthesis methods have been developed for reproducible fabrication of a variety of plasmonic gold nanoparticles (spheres, rods, prism, cages, shells, etc.).[38–41] A particular research interest has been focusing on fabricating various shapes of gold nanoparticles with plasmons in the NIR regions,[38,40,42,43] which is advantageous for in vivo application due to superior tissue penetration in that spectral range.[44,45] To date, several different geometries of NIR-responsive gold nanoparticles have been fabricated, namely, nanoshells, nanorods, nanocages, nanostars, and hollow nanospheres.[39,41,46–49] To tune the LSPR frequency, the most frequently applied methods entail controlling the aspect ratio (AR) (long/short axis ratio) of the rod or the shell/core thickness ratio of shells and cages. These gold nanoparticles have been used as contrast agents in optical bioimaging techniques,[42,50,51] and as photothermal transducers for cancer treatment.[41,52–54] Due to the increased demand for custom-designed bioapplication, tailoring gold nanoparticle's plasmon for a specific theranostic application is an active area of research.

Our laboratory has been involved in the development and application of various plasmonic nanoplatforms ranging from nanoparticles, nanopost arrays, nanowires, and nanochips.[38,55–58] Particular applications involve gene detection,[59,60] nanoparticle tracking, and imaging.[61,62] Recently, our group has developed plasmon-tunable surfactant-free nanostars. In contrast to spherical or cubical shape nanoparticles, nanostars contain multiple rodlike protruding branches that can be exploited for plasmon tunability and strong LSPR. Previous studies have shown that the branch AR determines the major plasmon peak position; the higher the AR, the more red-shifted the major plasmon peak.[63] Nanostars exhibit a strong SERS signal and PL for multiplexing and cellular imaging.[63–66] Nanostars were also employed in photodynamic and photothermal therapies.[28,64,67] In this chapter, we focus on the development of plasmonic nanostars for potential molecular imaging and cancer therapy.

13.4 Preparation of Plasmonic Gold Nanostars

For the past few years, our group has focused on the development of unique *branched* gold nanoparticles, nanostars, as a new photonic-active nanomaterial. Such branched geometry was first reported by Chen et al. but later named *star shape* by Sau et al.[68,69] After that, many different nanostar synthesis methods were reported.[70] A majority of the synthesis methods require the use of a surfactant (e.g., cetyltrimethylammonium bromide [CTAB], *N,N*-dimethylformamide [DMF], poly(*N*-vinylpyrrolidone [PVP]) based on seed-mediated or seedless methods.[48,49,68,71–80] However, the presence of these surfactants hinders surface functionalization and may be potentially cytotoxic. Schutz et al. and Yuan et al. therefore developed a biocompatible route for nanostar synthesis.[63,81]

Standard wet chemical nanoparticle syntheses are time consuming, usually taking several hours to complete. Also, each class of nanoparticle employed has its own limitation. For example, synthesis of nanorods requires the use of CTAB, which is potentially toxic to animals and human beings, and is difficult to replace for surface functionalization. In the case of nanoshells, it can be challenging to achieve a uniform shell thickness.[82] For nanocages, although their absorption-to-scattering ratio is higher than nanorods and nanoshells,[83] the synthesis is elaborate, requiring titration with $HAuCl_4$ under 90°C, monitored under UV-Vis absorption.[82]

Our laboratory has developed a simple, biocompatible, surfactant-free, seed-mediated, plasmon-tunable nanostar synthesis method. Nanostars, on the other hand, can be prepared in less than a minute without the use of toxic surfactant.[63] Detailed synthesis steps were discussed by Yuan et al.[63] The reagents used are similar to previous silver-assisted, seed-mediated nanostar or nanorod recipes,[71–74,78,80,84–86] except for the omission of any surfactants. Silver ions help to improve the yield and shape, owing to the under-potential deposition of silver ions on certain crystal facets of gold seeds.[69,71–73,87,88] The formation of multiple branches was hypothesized to be related to the blocking of certain facets by silver ions. It is believed that the major role of silver ions is not to form silver branches but to assist the anisotropic growth of Au branches on certain crystallographic facets on multi-twinned citrate seeds, but not single crystalline CTAB seeds.[71,72,78,86,89,90] Most importantly, our synthesis method is extremely simple in that the reaction can be completed in less than 30 s under room temperature, resulting in particles of around 40–60 nm diameters with narrow size distribution (Figure 13.2a).[63] Plasmon tuning can be achieved by adjusting the Ag^+-to-seed ratio (Figure 13.2b through d); higher concentrations of Ag^+ progressively red shift the plasmon band by forming longer, sharper, and more numerous branches. In light of the ease of synthesis and plasmon tunability without the use of surfactant, the nanostars produced by our method are biocompatible and NIR responsive and can be conjugated easily with biomolecules for active targeting.

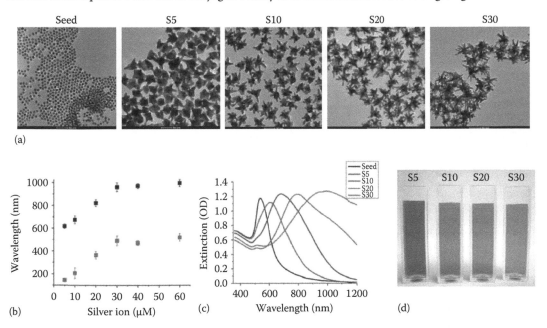

FIGURE 13.2 **(See color insert.)** (a) TEM images of gold citrate seeds and nanostars formed under different Ag^+ concentrations (S5: 5 µM, S10: 10 µM, S20: 20 µM, S30: 30 µM). (b) The spectral progression of plasmon peak (black square) and FWHM (red square) of nanostars formed under different Ag^+ concentrations. (c) Extinction spectra of the seed and star solutions in DI. (d) Photograph of star solutions. (Adapted from Yuan, H., Khoury, C.G., Wilson, C.M., Grant, G.A., and Vo-Dinh, T., *Nanotechnology*, 23, 075102, 2012.)

13.5 3D Polarization-Averaged Modeling of Gold Nanostars

To understand the plasmon tunability, computational modeling was performed. The nanostar was modeled using the finite element method (FEM) featuring polarization averaging over space. Previously, nanoparticles were typically modeled using a 2D geometry presentation with single polarization. Such modeling is good to represent simple geometry (e.g., sphere) nanoparticles fixed on a film but is not ideal to address the realistic behavior of aspherical nanoparticles (e.g., star) free-rotating in solution. We therefore devised a 3D configuration of random-protruding branches and analyzed them under 6 discrete angles.[63]

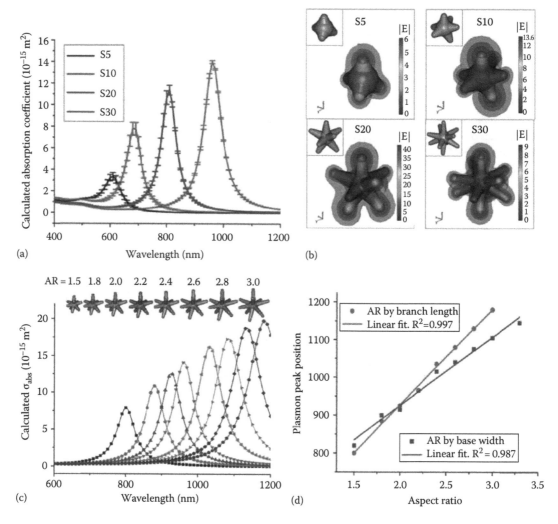

FIGURE 13.3 **(See color insert.)** (a) The corresponding calculated absorption spectra of nanostars embedded in water. The data points (±1 SD) were interpolated with a spline fit. (b) Simulation of |E| in the vicinity of the nanostars in response to a z-polarized plane wave incident E-field of unit amplitude, propagating in the y-direction, and with a wavelength of 800 nm. The insets depict the 3D geometry of the stars. Diagrams are not to scale. (c) The scatter plots of polarization-averaged absorption against AR tuned by varying branch height while keeping the base width, core and tip diameters, and branch number, constant. (d) The linear relationship between the plasmon peak position and AR, which is tuned by varying branch height (red, $R^2 = 0.997$) or base width (blue, $R^2 = 0.987$) while keeping all other parameters constant. (Adapted from Yuan, H., Khoury, C.G., Wilson, C.M., Grant, G.A., and Vo-Dinh, T., *Nanotechnology*, 23, 075102, 2012.)

From the modeling, the plasmon of a nanostar is the combination of plasmons from the core and branches. This is in agreement with previous findings from 2D single polarization modeling[76]; a weak peak existing around 520 nm is attributed to the plasmon resonance of the nanostar core, while a dominant plasmon band at longer wavelengths is attributed to the plasmon resonance of the nanostar branches (Figure 13.3a). The core size only contributes to the 520 nm peak, whereas the branch geometry determines the plasmon position and intensity in the NIR. The local E-field is mostly enhanced at the tips of those branches in alignment to the incident polarization; the maximal enhancement is observed when the nanostar's plasmon peak matches the incident energy (Figure 13.3b). In agreement with Hao et al., the polarization only affects the peak intensity but not the peak position.[76] Meanwhile, the plasmon shift is controlled mainly by the branch AR, which is calculated by the branch length divided by the base width. A linear relationship was demonstrated between the absorption spectrum peak position and AR by adjusting the branch lengths while keeping the branch base width constant or vice versa (Figure 13.3c and d). Having branches with varying tip radius or angle but the same AR results in the same peak position. The peak intensity increases with increasing branch number and length. It is worth noting that the calculated plasmon bandwidths are significantly narrower than the experimental data; the broadening is caused primarily by the nonuniform nanostar branch geometries (e.g., wide distribution of AR).

13.6 Molecular Imaging of Gold Nanostars

Plasmonic nanostars offer a new way to image metal nanoparticles optically. Previously, imaging or quantification of metal nanoparticles could not be done without the use of transmission electron microscopy (TEM) or inductively coupled plasma–mass spectroscopy (ICP-MS), both of which have high technical barriers and cannot be used for live specimens. In contrast, the unique branched geometry of nanostars greatly enhances the LSPR, which allows for two-photon PL (TPL) imaging, SERS imaging, and PAT. All these optical techniques for preclinical imaging work best under NIR excitation, which matches the plasmon peak of the nanostars. Nanostars therefore can be used as an optical contrast agent for molecular imaging.

13.6.1 Two-Photon Photoluminescence Imaging

Efficient plasmon-enhanced TPL imaging using nonspherical gold nanoparticles (e.g., nanorods, nanoshells, nanocages, nanostars) as a contrast mechanism has been reported.[39,91–100] Visible PL on metals was first observed by Mooradian.[101] The PL origin was explained by the interband transitions of d-band electrons into the conduction band and subsequent radiative recombination.[102,103] Interestingly, while single-photon excitation generally leads to weak PL, multiphoton-excited PL is stronger and sensitive to the incident polarization and LSPR (Figure 13.4a).[104,105] Matching the LSPR to the NIR femtosecond pulsed laser, the two-photon action cross section (TPACS) of nanostars and nanocages is around 10^6–10^7 Göeppert-Mayer units (GM), which are higher than that of QDs (10^4–10^5 GM) and organic fluorophores (10^2–10^3 GM).[63,106] A quadratic dependence of TPL intensity on excitation power (Figure 13.4b) and a broad emission spectrum have been observed on nanostars and other NIR-plasmonic nanoparticles.[63,96,98] The resonant coupling of the plasmon band with the incident laser greatly amplifies the nanoparticles' TPL. Due to this property, nanostars can be applied to MPM, offering a convenient way to visualize NIR-absorbing gold nanoparticles in high resolution using NIR excitation.

There are several studies using TPL to assess the intracellular distribution and quantity of NPs,[28,65,67,107–112] to visualize membrane receptors,[113] and to detect cancer cells.[63,91,92,99,106,108] TPL imaging, which works specifically under MPM, not only exploits the intrinsic optical properties of the plasmonic nanoparticles without the need of fluorophore labeling but also can be used with any current

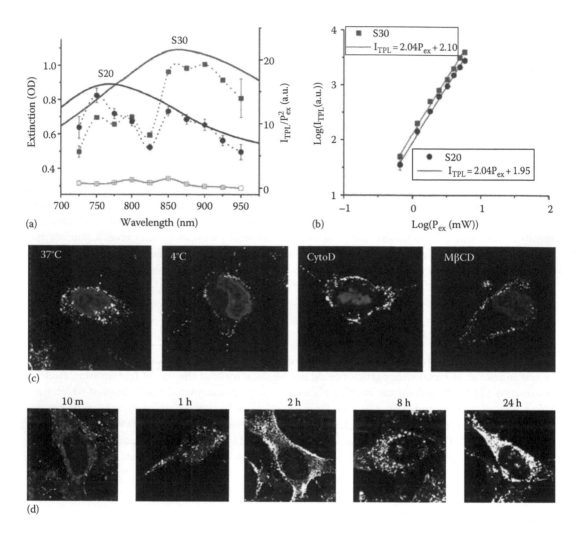

FIGURE 13.4 **(See color insert.)** (a) Plasmon spectra (solid lines) and TPL excitation spectra (spline-fitted dashed lines ± 1 SD) of S20 (blue) and S30 (red) nanostars 0.1 nM in citrate buffer, and Rhodamine B (green) 100 nM in MeOH. (The spectral dip at 825 nm seen on both nanostars and Rhodamine B samples might be a system error from the microscope.) (b) The quadratic dependence of the TPL intensity (I_{TPL}) to the laser power (P_{ex}) from S20 and S30 solution (0.1 nM). Laser was set at 800 nm with power adjusted between 0.5 and 6 mW. Scatter plots (± 1 SD) were displayed with linearly fitted lines. (c) TPL images of TAT-nanostars-treated cells under normal condition (37°C) and different inhibitors. The cellular uptake of TAT nanostars was inhibited by 4°C, cytochalasin D (CytoD), methyl-β-cyclodextrin (MβCD), and TPL images sized 50 × 50 μm². Nuclei are stained blue. (d) Time series TPL images of cells treated with TAT nanostars (white) showing incremental accumulation. Cytoplasm is stained orange. Image size: 50 × 50 μm². (Adapted from Yuan, H., Khoury, C.G., Wilson, C.M., Grant, G.A., and Vo-Dinh, T., *Nanotechnology*, 23, 075102, 2012; Reprinted with permission from Yuan, H., Fales, A.M., and Vo-Dinh, T., *J. Am. Chem. Soc.*, 134, 11358–11361. Copyright 2012 American Chemical Society.)

fluorescence imaging probes or cellular assays. Wei et al. and Yuan et al. first reported nanostars TPL imaging in vitro and in vivo, respectively.[63,112] Nanostars, with such a high TPACS, have been used to study the uptake mechanism of TAT functionalization (Figure 13.4c) and the real-time tracking of nanostars in vivo.[28,63,67] The effect of surface charge on the intracellular uptake, the temporal uptake profiles of nanostars (Figure 13.4d), nuclear delivery, and even tracking of a single nanostars have been studied using TPL imaging as well.[28,63,67,80,114] Nonetheless, TPL imaging is limited by its broad PL band, hence only one emission band that hinders its use for multiplex detection.[63,95,99] Still, with the convenience of sensitive visualization of nanostars under MPM, this new powerful TPL imaging could bring forth high-resolution preclinical molecular imaging.

13.6.2 SERS Detection Using Plasmonic Nanostars

Another contrast mechanism based on SERS detection has also gained strong momentum in the field of molecular imaging.[10,115] A variety of SERS probes can be excited for in vivo multiplex detection using single NIR laser excitation.[116–122] Since the characteristic Raman spectrum is independent of the excitation wavelength and corresponds directly to the vibrational or rotational energy of the analyte molecule,[123,124] multiple probes with unique Raman spectra can be distinguished using only a single excitation wavelength, providing a much simpler optical setup than fluorescence methods that require multiple laser excitation lines. Furthermore, due to the fact that SERS detection has less photobleaching and narrower Raman bands (10–20 cm^{-1}) than fluorescence detection (400–800 cm^{-1}),[125] SERS offers superior multiplexing capabilities to fluorescence. Many such bioapplications have already been demonstrated.[59,126–130] Unfortunately, Raman scattering is intrinsically inefficient (Raman cross sections are typically on the order of 10^{-30} cm^2), hence requiring SERS-active plasmonic nanostructures to enhance the Raman signal.

The SERS intensity is determined by multiple factors, including metal composition, LSPR wavelength, particle concentration, extinction background, and number of hot spots (e.g., aggregation, sharp branch, gap).[131] In general, the enhancement has been proposed to originate from the surface plasmon resonance (SPR) on the metal nanostructure and the charge transfer (CT) between the molecule and the metal conduction band.[132] In solution, spherical gold monomers typically have an analytical enhancement factor (EF) of less than 4 orders of magnitude;[65,133–136] singular spherical entities hence cannot be used as a practical SERS probe without further aggregation. Further enhancement (e.g., 10^{12}) can be obtained at *hot spots*, which are highly localized surface plasmons generated from coupled configurations (e.g., dimers, aggregates, gaps) or sharp protrusions (e.g., corners, tips).[38,124,137–139] For single dimers or trimers on film, the EF has been reported up to 8 orders of magnitude under 785 or 830 nm excitation.[140] However, induced aggregation in solution typically leads to an ensemble of monomers and multimers that requires further complicated enrichment processes.[141,142]

Nanostars with multiple sharp branches, in contrast, provide an alternative strategy to fabricate SERS probes without the need for aggregation. Zou et al. first reported SERS detection on branched gold nanoparticles.[75] Over the years, nanostars have been applied to ultrasensitive sensing,[143] intracellular imaging,[65,144] and ex vivo multiplex detection.[66,81] The appealing advantage of the star geometry is the ability to tune the plasmon bands in the NIR range for in vivo applications and to self-generate *hot spots* without the need for aggregation. When compared with silver spheres and gold spheres of similar sizes/ concentrations in solution, nanostars have similar EFs to silver spheres but much higher EFs than gold spheres (Figure 13.5a and b) especially under 785 nm excitation, which matches nanostar's plasmon. Furthermore, because of the star-shape geometry, it has been reported to allow for higher entrapment of NIR dyes on the surface of nanostars than on nanospheres. The obtained resonant SERS (SERRS) on nanostars has 3–4 orders of magnitude stronger signal than that of nonresonant dyes on nanostars (Figure 13.5c). Such a high nonaggregated SERRS signal can be easily applied to multiplex detection (Figure 13.5d).[66] However, while plasmon matching, on one hand, enhances the Raman response, on the other hand, it would compete with the SERS emission through extinction (absorption and scattering in the same spectral region as SERS emission) from nearby nanoparticles. Hence, the optimal excitation laser line for exciting SERS, instead of blue-shifted as conventionally believed,[145] should be red-shifted against the plasmon peak position.[66,146] With proper tuning, nanostars could provide a powerful nanoplatform for SERS-based molecular imaging.

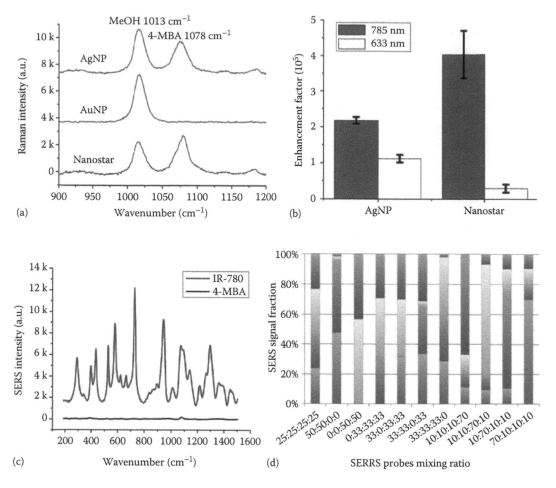

FIGURE 13.5 (a) SERS spectra (baseline subtracted) of 1 μM 4-MBA in 0.1 nM AgNP, AuNP, and nanostar solutions examined through a Raman microscope under 785 nm excitation. Methanol (10% v/v) was used as an internal reference. Wavenumber 1013 and 1078 represents Raman signals from MeOH and 4-MBA, respectively. (b) EFs of AgNP and nanostar under 785 nm (grey) and 633 nm (white) laser excitation. AuNP was omitted due to no 4-MBA SERS signal available for EF calculation. Error bar is 1 SD. (c) Baseline-subtracted SERS spectra (offset) from probes made from 1 μM of IR-780 (grey) and 4-MBA (black) on NS; SERRS intensity from the former was several orders of magnitude greater than intensity from the latter. (d) Quantitative in vitro multiplexing of 4 SERRS probes (10–100 pM; blue: IR-780, red: IR-792, green: IR-797, purple: IR-813). Spectral range of 200–1500 wavenumbers was used for analysis. The average difference between measured fractions and the predetermined ratio is 9.6% ± 3.7%. (Adapted from Yuan, H., Fales, A.M., Khoury, C.G., Liu, J., and Vo-Dinh, T.: *J. Raman Spectrosc.*, 44, 234–239. 2013. Copyright Wiley-VCH Verlag GmbH & Co. KGaA. Reproduced with permission; Reprinted with permission from Yuan, H., Liu, Y., Fales, A.M., Li, Y., Liu, J., and Vo-Dinh, T., *Anal. Chem.*, 85, 208–212. Copyright 2013 American Chemical Society.)

13.7 Cancer Therapy Using Gold Nanostars

NIR-plasmonic gold nanoparticles, with their plasmon bands in the tissue optical window, have been applied in multiple preclinical studies for cancer therapy. Exploiting their high molar extinction, gold nanoparticles can be used for PTT. Incorporating an encapsulating layer makes gold nanoparticles a drug carrier (e.g., photosensitizer [PS]). Combining both features, photothermally triggered drug release is possible.

13.7.1 Photothermal Therapy Using Plasmonic Nanostars

Therapeutic hyperthermia has been used to treat cancer for many years. By elevating the temperature to more than 42°C, malignant cells are killed through apoptosis or necrosis.[147,148] Several PTT studies with plasmon-enhanced local tumor hyperthermia were demonstrated.[39,41,42,82,149–156] Hirsch et al. first reported using PEGylated silica/gold nanoshells on breast cancer cells for localized hyperthermia.[157] Later, PTT was also demonstrated using anti-EGFR antibody-labeled nanorods in vitro and subsequently using PEGylated nanorods in vivo.[82,158] PTT using nanorods of different surface functionalization was demonstrated by several groups.[151,152,155,156,159] PTT has been reported in vitro using gold/gold sulfide nanoshells,[154] nanocubes,[153] nanocages,[160] hollow nanospheres,[161] and star-shaped (branched) nanoparticles.[28,149,162] In vivo application of PEGylated nanocages, PEGylated hollow nanospheres, and PEGylated nanostars also showed promising results.[28,161–163] Extending the photothermal process for triggered drug release has also been demonstrated.[20,32–36,164–169]

For an ideal PTT transducer, the nanoparticles should be biocompatible, simple to synthesize, small in size, plasmon tunable in the NIR, and high in absorption-to-scattering cross section ratio for efficient photothermal transduction. Nanostars, with their simple plasmon-tunable biocompatible synthesis methods, can be an effective candidate. On simulation modeling, nanostars have a high absorption-to-scattering cross section ratio due to thin branches with a small core (Figure 13.6a). In addition, Baffou et al. also noted that corners and sharp edges (e.g., branches) are favorable for heat generation.[170] More numerous sharp branches also produce a higher temperature factor on simulations.[171] Experimentally, star-shaped nanoparticles have an absorption-to-extinction ratio of 0.9.[27–28] When compared under the same plasmon intensity (extinction) at the laser line (785 nm), nanostars displayed a photothermal response almost identical to similar-sized nanorods but with 2.5-fold less particle concentration (Figure 13.6b).[28] When comparing photothermal efficiency, normalized by mass, nanocages are the highest, followed by nanostars, nanorods, then nanoshells.[9,27] Since nanocages require a more elaborate synthesis process, surfactant-free nanostars developed by our group can be a powerful candidate for PTT.

PTT using star-shaped nanoparticles (i.e., nanostar, branched nanoparticle, nanopopcorn, nanocross, nanohexapod) has been demonstrated in several studies.[27,28,67,100,149,162,172] It was first mentioned in text by Gamez et al.[173] Later, Lu et al. and Yuan et al. applied star-shaped nanoparticles to PTT for the first time in vitro and in vivo, respectively.[28,172] The photothermal response relies on both the nanostars (plasmon position/intensity, particle concentration, absorption-to-scattering cross section ratio) and laser (CW/pulsed, wavelength, irradiance, and duration). For PTT in vitro, we applied 980 nm CW laser on SKBR3 cancer cells pretreated with nanostars. After 3 min of irradiation, a clear photothermal ablation spot was seen (Figure 13.6c). The laser irradiance used on nanostars in our study (15 W/cm^2) is lower than the previous reported irradiances on branched nanoparticles (38 W/cm^2), Au/Au$_2$S nanoshell (80 W/cm^2), hollow nanospheres (32 W/cm^2), and nanoshell (35 W/cm^2) but higher than irradiances on nanocubes (4 W/cm^2) and nanocages (1.5 W/cm^2).[149,153,154,157,160,161] To achieve photothermal ablation using lower irradiance, TAT-functionalized nanostars were used. In addition to the enhanced intracellular uptake of TAT nanostars, we demonstrated a successful photothermolysis (Figure 13.6d) under pulsed laser using an ultra-low irradiance (0.2 W/cm^2),[67] which is for the first time ever reported below the maximal permissible exposure of skin per ANSI regulation.[174]

These studies suggested that nanostars have great potential in photothermal cancer therapy. Combining the strong optical contrast mechanism and high photothermal transduction efficiency, nanostars can be a very versatile theranostic nanoplatform.

13.7.2 Photodynamic Therapy Using Plasmonic Nanostars

Photodynamic therapy (PDT) is a promising technique for the treatment of cancer and other diseases.[175] PDT requires three components: light, a PS, and oxygen. Upon excitation, the PS transfers its energy to the surrounding oxygen producing ROS such as singlet oxygen,[176] which is generally regarded as cytotoxic, leading to cell death by apoptosis or necrosis.[175–177]

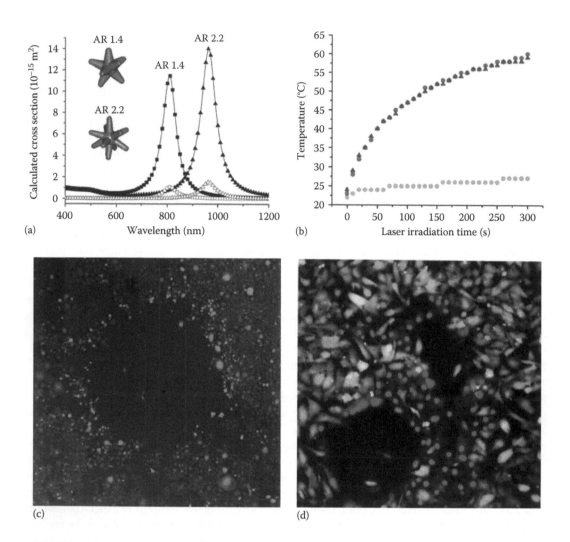

FIGURE 13.6 **(See color insert.)** (a) The simulated spectra (splined) of absorption (solid symbol) and scattering (open symbol) cross sections of nanostars of two different ARs (branch length over branch base width) (inset). The 3D models of two nanostars simulated. The geometry data used for simulation are listed below. AR 1.4: core 24 nm, branch length/base 19/13.5 nm, branch number 8. AR 2.2: core 22 nm, branch length/base 22/10 nm, branch number 10. (b) Irradiation (785 nm, 490 mW) of PEGylated nanostars (blue triangle) and nanorods (red circle) and DI (grey circle) of the same plasmon intensity. Their temperature responses were nearly identical, although the concentration of nanorods was roughly 2.5-fold of the nanostars. (c) Fluorescence image of viable SKBR3 cells after the laser treatment. Viable cells appear green by converting nonfluorescing dye into green fluorescence. Empty area indicates successful photothermal ablation. Cells were incubated 1 h with 0.2 nM nanostars before the irradiation. Laser irradiation (980 nm, 15 W/cm^2) was applied 3 min. Image sizes are 3 × 3 mm^2. (d) Photothermolysis (850 nm, 0.5 mW, scanning area 500 × 500 μm^2, 3 min) on BT549 cells incubated 4 h with TAT nanostars. Live/dead cells are green/red. Image size: 612 × 612 μm^2. (Adapted from *Nanomedicine: NBM*, 8, Yuan, H., Khoury, C.G., Wilson, C.M., Grant, G.A., Bennett, A.J., and Vo-Dinh, T., 1355–1363, Copyright 2012, with permission from Elsevier; Reprinted with permission from Yuan, H., Fales, A.M., and Vo-Dinh, T., *J. Am. Chem. Soc.*, 134, 11358–11361. Copyright 2012 American Chemical Society.)

A mesoporous silica shell can be used to encapsulate various dye molecules onto a metallic core.[178] This can be advantageous for PDT since many PSs are hydrophobic and cannot be adequately administered in a biological environment.[179] Due to the mesoporous nature of the silica, it is not necessary that the PS molecules be released at the target; singlet oxygen that is generated within the silica matrix can diffuse out into the surrounding tissue.[177] For theranostic applications, a therapeutic modality such as PDT must be combined with a label (e.g., optical dye) for diagnostics.

Our laboratory has presented a proof of concept for the first theranostic Raman nanoprobe, combining SERS detection and PDT into a single nanoparticle construct.[64] A visual overview of the nanoprobe can be depicted in Figure 13.7a. The Raman label consisted of a gold nanostar, tuned to have a surface plasmon in the NIR, which was functionalized with the NIR dye 3,3′-diethylthiatricarbocyanine (DTTC). Overlap of the laser excitation with the surface plasmon of the nanostars maximized the EM enhancement effect of SERS. The use of an NIR dye further enhanced the SERS signal due to resonance Raman scattering. Methylene blue (MB), a PDT drug, was loaded onto the Raman-labeled nanostars

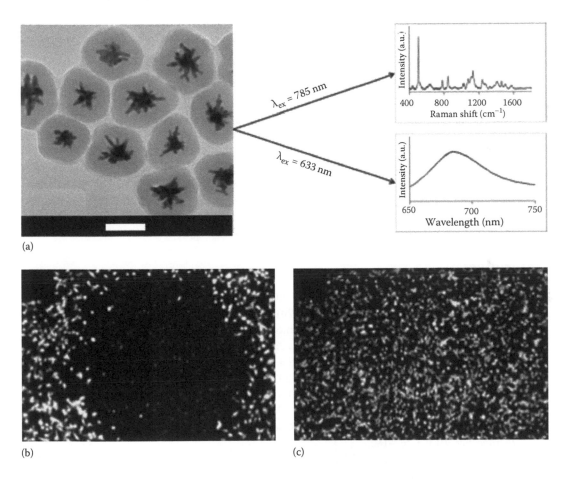

(a)

(b)   (c)

FIGURE 13.7 (a) A TEM micrograph (scale bar 100 nm) of the theranostic nanoprobes, consisting of silica-coated gold nanostars, can be seen on the left. Laser excitation at 785 nm produced a strong SERS signal from surface-bound DTTC molecules, while excitation at 633 nm resulted in fluorescence from MB encapsulated in the silica shell. Viability staining fluorescence images of BT549 cells incubated with (b) MB-encapsulated DTTC-labeled nanostars and (c) silica-shell DTTC-labeled nanostars after 1 h of laser irradiation (633 nm, 0.9 W/cm²). Only the MB-encapsulated particles produced a cytotoxic effect after treatment, and the area of cell death is limited to the laser-irradiated area. The live/dead cells are seen as green/red. (Reprinted with permission from Fales, A.M., Yuan, H., and Vo-Dinh, T., *Langmuir*, 27, 12186–12190. Copyright 2011 American Chemical Society.)

by encapsulating it within a silica shell. We observed excellent drug loading into the silica shell, with more than 1000 drug molecules per nanoparticle, as estimated by the measured fluorescence intensity. Excitation of the nanoprobe at 785 nm produced a strong SERS signal from the surface-bound DTTC. Excitation at 633 nm resulted in fluorescence (and $^1O_2$ generation) from the MB encapsulated within the silica shell. The excitation for the SERS label and PDT drug were separated well enough that no interference from the other component was observed when exciting either the SERS label or MB at its respective wavelength.

To demonstrate the potential of the nanoprobe for theranostics, an in vitro PDT study was performed on BT549 breast carcinoma cells. The cells were incubated with the particle-containing medium and treated with 633 nm laser (0.9 W/cm²) for 1 h. As can be seen in Figure 13.7b and c, the therapeutic effect was only observed on cells incubated with MB-loaded particles, and this effect was limited to the laser-irradiated area. There was also no loss in cell viability after laser exposure for the cells incubated with particles not containing MB, indicating that no significant photothermal effects were occurring.

These results show that SERS-based nanoprobes hold great promise for future use in theranostic applications. Combination of multiple Raman labels and targeting ligands could provide multiplex detection of various disease biomarkers at once. Such a construct also has the potential to incorporate other types of therapeutic modalities, not only limited to PDT.

13.8 Conclusion

To date, most nanoparticle-based molecular imaging contrast agents can be applied for imaging purposes only, and most therapeutic agents require labeling with a contrast agent. In this chapter, we have discussed the potential of nanostars for use in not only imaging (TPL, SERS, SERRS) but also therapy (PTT and PDT). Moreover, the nanostars we developed, being plasmon tunable as well as facile for surface functionalization, can be easily employed for various theranostic applications. Although their nonbiodegradable nature requires further study for long-term clinical applications, nanostars offer tremendous flexibility in preclinical nanomedicine research both in vitro and in vivo. These studies demonstrated the potential of plasmonic-active nanostars as a novel and powerful multimodal theranostic agent.

References

1. K. Y. Choi; G. Liu; S. Lee; X. Chen. *Nanoscale* 2012, *4*, 330–342.
2. D. Y. Lee; K. C. P. Li. *AJR, Am. J. Roentgenol.* 2011, *197*, 318–324.
3. Z. Cheng; A. Al Zaki; J. Z. Hui; V. R. Muzykantov; A. Tsourkas. *Science* 2012, *338*, 903–910.
4. E. Karathanasis; P. M. Peiris. *Oncotarget* 2011, *2*, 430.
5. M. P. Melancon; M. Zhou; C. Li. *Acc. Chem. Res.* 2011, *44*, 947–956.
6. S. S. Kelkar; T. M. Reineke. *Bioconjug. Chem.* 2011, *22*, 1879–1903.
7. J. V. Jokerst; S. S. Gambhir. *Acc. Chem. Res.* 2011, *44*, 1050–1060.
8. S. Jiang; K. Y. Win; S. Liu; C. P. Teng; Y. Zheng; M. Y. Han. *Nanoscale* 2013, *5*, 3127–3148.
9. N. Khlebtsov; V. Bogatyrev; L. Dykman; B. Khlebtsov; S. Staroverov; A. Shirokov; L. Matora et al. *Theranostics* 2013, *3*, 167–180.
10. M. F. Kircher; A. de la Zerda; J. V. Jokerst; C. L. Zavaleta; P. J. Kempen; E. Mittra; K. Pitter et al. *Nat. Med.* 2012, *18*, 829–834.
11. Z. Zhang; J. Wang; C. Chen. *Theranostics* 2013, *3*, 223–238.
12. V. Ntziachristos. *Nat. Methods* 2010, *7*, 603–614.
13. L. V. Wang; S. Hu. *Science* 2012, *335*, 1458–1462.
14. A. Taruttis; S. Morscher; N. C. Burton; D. Razansky; V. Ntziachristos. *PLoS ONE* 2012, *7*, e30491.
15. H. Yuan. Plasmonic gold nanostars: A novel theranostic nanoplatform. Doctoral Dissertation, Duke University, 2012, 130 pages; AAT 3546652.
16. V. R. Devadasu; V. Bhardwaj; M. N. Kumar. *Chem. Rev.* 2013, *113*, 1686–1735.

17. J. Fang; H. Nakamura; H. Maeda. *Adv. Drug Deliv. Rev.* 2011, *63*, 136–151.
18. O. C. Farokhzad; R. Langer. *ACS Nano* 2009, *3*, 16–20.
19. B. Y. S. Kim; J. T. Rutka; W. C. W. Chan. *N. Engl. J. Med.* 2010, *363*, 2434–2443.
20. S. J. Leung; M. Romanowski. *Theranostics* 2012, *2*, 1020–1036.
21. F. M. Kievit; M. Zhang. *Adv. Mater.* 2011, *23*, H217–H247.
22. P. Juzenas; W. Chen; Y.-P. Sun; M. A. Neto Coelho; R. Generalov; N. Generalova; I. L. Christensen. *Adv. Drug Deliv. Rev.* 2008, *60*, 1600–1614.
23. E. C. Dreaden; A. M. Alkilany; X. Huang; C. J. Murphy; M. A. El-Sayed. *Chem. Soc. Rev.* 2012, *41*, 2740–2779.
24. X. J. Xue; F. Wang; X. G. Liu. *J. Mater. Chem.* 2011, *21*, 13107–13127.
25. J. Xie; S. Lee; X. Chen. *Adv. Drug Deliv. Rev.* 2010, *62*, 1064–1079.
26. J. F. Lovell; C. S. Jin; E. Huynh; H. Jin; C. Kim; J. L. Rubinstein; W. C. W. Chan; W. Cao; L. V. Wang; G. Zheng. *Nat. Mater.* 2011, *10*, 324–332.
27. Y. Wang; K. C. Black; H. Luehmann; W. Li; Y. Zhang; X. Cai; D. Wan et al. *ACS Nano* 2013, *7*, 2068–2077.
28. H. Yuan; C. G. Khoury; C. M. Wilson; G. A. Grant; A. J. Bennett; T. Vo-Dinh. *Nanomedicine: NBM* 2012, *8*, 1355–1363.
29. S. K. Libutti; G. F. Paciotti; A. A. Byrnes; H. R. Alexander; W. E. Gannon; M. Walker; G. D. Seidel; N. Yuldasheva; L. Tamarkin. *Clin. Cancer Res.* 2010, *16*, 6139–6149.
30. M. Colilla; B. González; M. Vallet-Regí. *Biomater. Sci.* 2013, *1*, 114.
31. J. L. Vivero-Escoto; R. C. Huxford-Phillips; W. Lin. *Chem. Soc. Rev.* 2012, *41*, 2673–2685.
32. Z. Zhang; L. Wang; J. Wang; X. Jiang; X. Li; Z. Hu; Y. Ji; X. Wu; C. Chen. *Adv. Mater.* 2012, *24*, 1418–1423.
33. S. Shen; H. Tang; X. Zhang; J. Ren; Z. Pang; D. Wang. *Biomaterials* 2013, *34*, 3150–3158.
34. Y. T. Chang; P. Y. Liao; H. S. Sheu; Y. J. Tseng; F. Y. Cheng; C. S. Yeh. *Adv. Mater.* 2012, *24*, 3309–3314.
35. N. Li; Z. Yu; W. Pan; Y. Han; T. Zhang. *Adv. Funct. Mater.* 2013, *23*, 2255–2262.
36. Z. Xiao; C. Ji; J. Shi; E. M. Pridgen; J. Frieder; J. Wu; O. C. Farokhzad. *Angew. Chem. Int. Ed. Engl.* 2012, *51*, 11853–11857.
37. A. M. Derfus; W. C. W. Chan; S. N. Bhatia. *Nano Lett.* 2004, *4*, 11–18.
38. T. Vo-Dinh; A. Dhawan; S. J. Norton; C. G. Khoury; H.-N. Wang; V. Misra; M. D. Gerhold. *J. Phys. Chem. C* 2010, *114*, 7480–7488.
39. Y. Xia; W. Li; C. M. Cobley; J. Chen; X. Xia; Q. Zhang; M. Yang; E. C. Cho; P. K. Brown. *Acc. Chem. Res.* 2011, *44*, 914–924.
40. S. Lal; S. Link; N. J. Halas. *Nat. Photonics* 2007, *1*, 641–648.
41. J. Zhang. *J. Phys. Chem. Lett.* 2010, *1*, 686–695.
42. T. K. Sau; A. L. Rogach; F. Jäckel; T. A. Klar; J. Feldmann. *Adv. Mater.* 2010, *22*, 1805–1825.
43. C. Burda; X. Chen; R. Narayanan; M. El-Sayed. *Chem. Rev.* 2005, *105*, 1025–1102.
44. M. A. Hahn; A. K. Singh; P. Sharma; S. C. Brown; B. M. Moudgil. *Anal. Bioanal. Chem.* 2011, *399*, 3–27.
45. R. Weissleder. *Nat. Biotechnol.* 2001, *19*, 316–317.
46. X. Huang; S. Neretina; M. A. El-Sayed. *Adv. Mater.* 2009, *21*, 4880–4910.
47. R. L. Atkinson; M. Zhang; P. Diagaradjane; S. Peddibhotla; A. Contreras; S. G. Hilsenbeck; W. A. Woodward; S. Krishnan; J. C. Chang; J. M. Rosen. *Sci. Transl. Med.* 2010, *2*, 55ra79.
48. S. Trigari; A. Rindi; G. Margheri; S. Sottini; G. Dellepiane; E. Giorgetti. *J. Mater. Chem.* 2011, *21*, 6531–6540.
49. S. Barbosa; A. Agrawal; L. Rodriguez-Lorenzo; I. Pastoriza-Santos; R. A. Alvarez-Puebla; A. Kornowski; H. Weller; L. M. Liz-Marzan. *Langmuir* 2010, *26*, 14943–14950.
50. E. Boisselier; D. Astruc. *Chem. Soc. Rev.* 2009, *38*, 1759–1782.
51. R. A. Sperling; P. Rivera Gil; F. Zhang; M. Zanella; W. J. Parak. *Chem. Soc. Rev.* 2008, *37*, 1896–1908.
52. X. Huang; M. El-Sayed. *J. Adv. Res.* 2010, *1*, 13–28.

53. H. Chen; L. Shao; T. Ming; Z. Sun; C. Zhao; B. Yang; J. Wang. *Small* 2010, *6*, 2272–2280.

54. S. Lal; S. Clare; N. J. Halas. *Acc. Chem. Res.* 2008, *41*, 1842–1851.

55. T. Vo-Dinh; M. Hiromoto; G. Begun; R. Moody. *Anal. Chem.* 1984, *56*, 1667–1670.

56. M. Meier; A. Wokaun; T. Vo-Dinh. *J. Phys. Chem.* 1985, *89*, 1843–1846.

57. T. Vo-Dinh; M. Meier; A. Wokaun. *Anal. Chim. Acta* 1986, *181*, 139–148.

58. T. Vo-Dinh. *TrAC, Trends Anal. Chem.* 1998, *17*, 557–582.

59. H.-N. Wang; T. Vo-Dinh. *Nanotechnology* 2009, *20*, 065101.

60. H.-N. Wang; T. Vo-Dinh. *Small* 2011, *7*, 3067–3074.

61. M. K. Gregas; F. Yan; J. Scaffidi; H.-N. Wang; T. Vo-Dinh. *Nanomedicine: NBM* 2011, *7*, 115–122.

62. M. Wabuyele; F. Yan; G. Griffin; T. Vo-Dinh. *Rev. Sci. Instrum.* 2005, *76*, 063710.

63. H. Yuan; C. G. Khoury (co-first author); C. M. Wilson; G. A. Grant; T. Vo-Dinh. *Nanotechnology* 2012, *23*, 075102.

64. A. M. Fales; H. Yuan (co-first author); T. Vo-Dinh. *Langmuir* 2011, *27*, 12186–12190.

65. H. Yuan; A. M. Fales; C. G. Khoury; J. Liu; T. Vo-Dinh. *J. Raman Spectrosc.* 2013, *44*, 234–239.

66. H. Yuan; Y. Liu; A. M. Fales; Y. Li; J. Liu; T. Vo-Dinh. *Anal. Chem.* 2013, *85*, 208–212.

67. H. Yuan; A. M. Fales; T. Vo-Dinh. *J. Am. Chem. Soc.* 2012, *134*, 11358–11361.

68. S. Chen; Z. L. Wang; J. Ballato; S. H. Foulger; D. L. Carroll. *J. Am. Chem. Soc.* 2003, *125*, 16186–16187.

69. T. K. Sau; C. J. Murphy. *J. Am. Chem. Soc.* 2004, *126*, 8648–8649.

70. S. Y. Li; M. Wang. *Nano LIFE* 2012, *2*, 1230002.

71. W. Ahmed; E. S. Kooij; A. van Silfhout; B. Poelsema. *Nanotechnology* 2010, *21*, 125605–125611.

72. G. Kawamura; Y. Yang; K. Fukuda; M. Nogami. *Mater. Chem. Phys.* 2009, *115*, 229–234.

73. H.-L. Wu; C.-H. Chen; M. H. Huang. *Chem. Mater.* 2009, *21*, 110–114.

74. C. Khoury; T. Vo-Dinh. *J. Phys. Chem. C* 2008, *112*, 18849–18859.

75. X. Zou; E. Ying; S. Dong. *Nanotechnology* 2006, *17*, 4758–4764.

76. F. Hao; C. L. Nehl; J. H. Hafner; P. Nordlander. *Nano Lett.* 2007, *7*, 729–732.

77. S. K. Dondapati; T. K. Sau; C. Hrelescu; T. A. Klar; F. D. Stefani; J. Feldmann. *ACS Nano* 2010, *4*, 6318–6322.

78. A. Guerrero-Martínez; S. Barbosa; I. Pastoriza-Santos; L. M. Liz-Marzán. *Curr. Opin. Colloid Interface Sci.* 2011, *16*, 118–127.

79. P. Senthil Kumar; I. Pastoriza-Santos; B. Rodríguez-González; F. J. Garcia de Abajo; L. M. Liz-Marzán. *Nanotechnology* 2008, *19*, 015606–015612.

80. H.-M. Song; Q. Wei; Q. K. Ong; A. Wei. *ACS Nano* 2010, *4*, 5163–5173.

81. M. Schütz; D. Steinigeweg; M. Salehi; K. Kömpe; S. Schlücker. *Chem. Commun.* 2011, *47*, 4216–4218.

82. X. Huang; I. H. El-Sayed; W. Qian; M. A. El-Sayed. *J. Am. Chem. Soc.* 2006, *128*, 2115–2120.

83. P. K. Jain; K. S. Lee; I. H. El-Sayed; M. A. El-Sayed. *J. Phys. Chem. B* 2006, *110*, 7238–7248.

84. E. Nalbant Esenturk; A. Hight Walker. *J. Raman Spectrosc.* 2009, *40*, 86–91.

85. B. Nikoobakht; M. El-Sayed. *Chem. Mater.* 2003, *15*, 1957–1962.

86. T. K. Sau; A. L. Rogach; M. Döblinger; J. Feldmann. *Small* 2011, *7*, 2188–2194.

87. H. Chen; X. Kou; Z. Yang; W. Ni; J. Wang. *Langmuir* 2008, *24*, 5233–5237.

88. C. Nehl; H. Liao; J. Hafner. *Nano Lett.* 2006, *6*, 683–688.

89. C. Orendorff; C. Murphy. *J. Phys. Chem. B* 2006, *110*, 3990–3994.

90. M. Liu; P. Guyot-Sionnest. *J. Phys. Chem. B* 2005, *109*, 22192–22200.

91. N. J. Durr; T. Larson; D. K. Smith; B. A. Korgel; K. Sokolov; A. Ben-Yakar. *Nano Lett.* 2007, *7*, 941–945.

92. L. Bickford; J. Sun; K. Fu; N. Lewinski. *Nanotechnology* 2008, *19*, 315102–315108.

93. S. Kumar; N. Harrison; R. Richards-Kortum; K. Sokolov. *Nano Lett.* 2007, *7*, 1338–1343.

94. L. Martinez Maestro; E. Martin Rodriguez; F. Vetrone; R. Naccache; H. Loro Ramirez; D. Jaque; J. A. Capobianco; J. Garcia Sole. *Opt. Express* 2010, *18*, 23544–23553.

95. L. Tong; Q. Wei; A. Wei; J.-X. Cheng. *Photochem. Photobiol.* 2009, *85*, 21–32.

96. H. Wang; T. B. Huff; D. A. Zweifel; W. He; P. S. Low; A. Wei; J.-X. Cheng. *Proc. Natl. Acad. Sci. U. S. A.* 2005, *102*, 15752–15756.

97. Q. Wei; A. Wei. *Chem. Eur. J.* 2011, *17*, 1080–1091.

98. J. Park; A. Estrada; K. Sharp; K. Sang; J. A. Schwartz; D. K. Smith; C. Coleman et al. *Opt. Express* 2008, *16*, 1590–1599.

99. L. Au; Q. Zhang; C. M. Cobley; M. Gidding; A. G. Schwartz; J. Chen; Y. Xia. *ACS Nano* 2010, *4*, 35–42.

100. E. Ye; K. Y. Win; H. R. Tan; M. Lin; C. P. Teng; A. Mlayah; M.-Y. Han. *J. Am. Chem. Soc.* 2011, *133*, 8506–8509.

101. A. Mooradian. *Phys. Rev. Lett.* 1969, *22*, 185–187.

102. M. Beversluis; A. Bouhelier; L. Novotny. *Phys. Rev. B* 2003, *68*, 115433.

103. T. V. Shahbazyan. *Nano Lett.* 2013, *13*, 194–198.

104. A. Bouhelier; R. Bachelot; G. Lerondel; S. Kostcheev; P. Royer; G. P. Wiederrecht. *Phys. Rev. Lett.* 2005, *95*, 267405.

105. R. Farrer; F. Butterfield; V. Chen; J. Fourkas. *Nano Lett.* 2005, *5*, 1139–1142.

106. Y. Chen; Y. Zhang; W. Liang; X. Li. *Nanomedicine: NBM* 2012, *8*, 1267–1270.

107. E. Hutter; S. Boridy; S. Labrecque; M. Lalancette-Hébert; J. Kriz; F. M. Winnik; D. Maysinger. *ACS Nano* 2010, *4*, 2595–2606.

108. E. C. Cho; Y. Zhang; X. Cai; C. M. Moran; L. V. Wang; Y. Xia. *Angew. Chem., Int. Ed. Engl.* 2013, *52*, 1152–1155.

109. D. Yelin; D. Oron; S. Thiberge; E. Moses; Y. Silberberg. *Opt. Express* 2003, *11*, 1385–1391.

110. D. Nagesha; G. S. Laevsky; P. Lampton; R. Banyal; C. Warner; C. DiMarzio; S. Sridhar. *Int. J. Nanomedicine* 2007, *2*, 813–819.

111. T. B. Huff; M. N. Hansen; Y. Zhao; J.-X. Cheng; A. Wei. *Langmuir* 2007, *23*, 1596–1599.

112. Q. Wei; H.-M. Song; A. P. Leonov; J. A. Hale; D. Oh; Q. K. Ong; K. Ritchie; A. Wei. *J. Am. Chem. Soc.* 2009, *131*, 9728–9734.

113. D. C. Kennedy; L.-L. Tay; R. K. Lyn; Y. Rouleau; J. Hulse; J. P. Pezacki. *ACS Nano* 2009, *3*, 2329–2339.

114. D. H. M. Dam; J. H. Lee; P. N. Sisco; D. T. Co; M. Zhang; M. R. Wasielewski; T. W. Odom. *ACS Nano* 2012, *6*, 3318–3326.

115. C. L. Zavaleta; M. F. Kircher; S. S. Gambhir. *J. Nucl. Med.* 2011, *52*, 1839–1844.

116. H. Kang; S. Jeong; Y. Park; J. Yim; B.-H. Jun; S. Kyeong; J.-K. Yang; G. Kim; S. Hong; L. P. Lee. *Adv. Funct. Mater.* 2013, *23*, 3719–3727

117. C. L. Zavaleta; B. R. Smith; I. Walton; W. Doering; G. Davis; B. Shojaei; M. J. Natan; S. S. Gambhir. *Proc. Natl. Acad. Sci. U. S. A.* 2009, *106*, 13511–13516.

118. G. von Maltzahn; A. Centrone; J.-H. Park; R. Ramanathan; M. J. Sailor; T. A. Hatton; S. N. Bhatia. *Adv. Mater.* 2009, *21*, 3175–3180.

119. Y. Wang; J. L. Seebald; D. P. Szeto; J. Irudayaraj. *ACS Nano* 2010, *4*, 4039–4053.

120. K. K. Maiti; U. S. Dinish; A. Samanta; M. Vendrell; K.-S. Soh; S.-J. Park; M. Olivo; Y.-T. Chang. *Nano Today* 2012, *7*, 85–93.

121. P. Matousek; N. Stone. *J. Biophotonics* 2013, *6*, 7–19.

122. P. Z. McVeigh; R. J. Mallia; I. Veilleux; B. C. Wilson. *J. Biomed. Opt.* 2013, *18*, 46011.

123. G. Baker; D. Moore. *Anal. Bioanal. Chem.* 2005, *382*, 1751–1770.

124. M. J. Banholzer; J. E. Millstone; L. Qin; C. A. Mirkin. *Chem. Soc. Rev.* 2008, *37*, 885–897.

125. E. S. Allgeyer; A. Pongan; M. Browne; M. D. Mason. *Nano Lett.* 2009, *9*, 3816–3819.

126. B. R. Lutz; C. E. Dentinger; L. N. Nguyen; L. Sun; J. Zhang; A. N. Allen; S. Chan; B. S. Knudsen. *ACS Nano* 2008, *2*, 2306–2314.

127. A. Matschulat; D. Drescher; J. Kneipp. *ACS Nano* 2010, *4*, 3259–3269.

128. D. C. Kennedy; K. A. Hoop; L.-L. Tay; J. P. Pezacki. *Nanoscale* 2010, *2*, 1413–1416.

129. J. P. Nolan; D. S. Sebba. *Methods Cell Biol.* 2011, *102*, 515–532.

130. J. Huang; K. H. Kim; N. Choi; H. Chon; S. Lee; J. Choo. *Langmuir* 2011, *27*, 10228–10233.

131. P. L. Stiles; J. A. Dieringer; N. C. Shah; R. P. Van Duyne. *Annu. Rev. Anal. Chem.* 2008, *1*, 601–626.
132. J. R. Lombardi; R. L. Birke. *Acc. Chem. Res.* 2009, *42*, 734–742.
133. H. Metiu; P. Das. *Annu. Rev. Phys. Chem.* 1984, *35*, 507–536.
134. C. Orendorff; L. Gearheart; N. Jana; C. Murphy. *Phys. Chem. Chem. Phys.* 2006, *8*, 165–170.
135. K. L. Wustholz; A.-I. Henry; J. M. McMahon; R. G. Freeman; N. Valley; M. E. Piotti; M. J. Natan; G. C. Schatz; R. P. van Duyne. *J. Am. Chem. Soc.* 2010, *132*, 10903–10910.
136. M. Li; S. K. Cushing; J. Zhang; J. Lankford; Z. P. Aguilar; D. Ma; N. Wu. *Nanotechnology* 2012, *23*, 115501.
137. G. C. Schatz. *Acc. Chem. Res.* 1984, *17*, 370–376.
138. H. Xu; J. Aizpurua; M. Kall; P. Apell. *Phys. Rev. E* 2000, *62*, 4318–4324.
139. M. Albrecht; J. Creighton. *J. Am. Chem. Soc.* 1977, *99*, 5215–5217.
140. S. L. Kleinman; B. Sharma; M. G. Blaber; A.-I. Henry; N. Valley; R. G. Freeman; M. J. Natan; G. C. Schatz; R. P. Van Duyne. *J. Am. Chem. Soc.* 2013, *135*, 301–308.
141. J. M. Romo-Herrera; R. A. Alvarez-Puebla; L. M. Liz-Marzán. *Nanoscale* 2011, *3*, 1304–1315.
142. J.-H. Lee; J.-M. Nam; K.-S. Jeon; D.-K. Lim; H. Kim; S. Kwon; H. Lee; Y. D. Suh. *ACS Nano* 2012, *6*, 9574–9584.
143. L. Rodriguez-Lorenzo; R. A. Alvarez-Puebla; I. Pastoriza-Santos; S. Mazzucco; O. Stephan; M. Kociak; L. M. Liz-Marzán; F. J. Garcia de Abajo. *J. Am. Chem. Soc.* 2009, *131*, 4616–4618.
144. L. Rodriguez-Lorenzo; Z. Krpetic; S. Barbosa; R. A. Alvarez-Puebla; L. M. Liz-Marzán; I. A. Prior; M. Brust. *Integr. Biol.* 2011, *3*, 922–926.
145. R. A. Alvarez-Puebla. *J. Phys. Chem. Lett.* 2012, *3*, 857–866.
146. S. T. Sivapalan; B. M. Devetter; T. K. Yang; T. van Dijk; M. V. Schulmerich; P. S. Carney; R. Bhargava; C. J. Murphy. *ACS Nano* 2013, *7*, 2099–2105.
147. M. W. Dewhirst; Z. Vujaskovic; E. Jones; D. Thrall. *Int. J. Hyperthermia* 2005, *21*, 779–790.
148. P. Wust; B. Hildebrandt; G. Sreenivasa; B. Rau; J. Gellermann; H. Riess; R. Felix; P. Schlag. *Lancet Oncol.* 2002, *3*, 487–497.
149. B. Van de Broek; N. Devoogdt; A. D'Hollander; H.-L. Gijs; K. Jans; L. Lagae; S. Muyldermans; G. Maes; G. Borghs. *ACS Nano* 2011, *5*, 4319–4328.
150. L. C. Kennedy; L. R. Bickford; N. A. Lewinski; A. J. Coughlin; Y. Hu; E. S. Day; J. L. West; R. A. Drezek. *Small* 2011, *7*, 169–183.
151. W. I. Choi; J.-Y. Kim; C. Kang; C. C. Byeon; Y. H. Kim; G. Tae. *ACS Nano* 2011, *5*, 1995–2003.
152. D. K. Yi; I. C. Sun; J. H. Ryu; H. Koo; C. W. Park; I. C. Youn; K. Choi; I. C. Kwon; K. Kim; C. H. Ahn. *Bioconjug. Chem.* 2010, *21*, 2173–2177.
153. X. Wu; T. Ming; X. Wang; P. Wang; J. Wang; J. Chen. *ACS Nano* 2010, *4*, 113–120.
154. A. M. Gobin; E. M. Watkins; E. Quevedo; V. L. Colvin; J. L. West. *Small* 2010, *6*, 745–752.
155. G. von Maltzahn; J.-H. Park; A. Agrawal; N. K. Bandaru; S. K. Das; M. J. Sailor; S. N. Bhatia. *Cancer Res.* 2009, *69*, 3892–3900.
156. T. B. Huff; L. Tong; Y. Zhao; M. N. Hansen; J.-X. Cheng; A. Wei. *Nanomedicine* 2007, *2*, 125–132.
157. L. R. Hirsch; A. M. Gobin; A. R. Lowery; F. Tam; R. A. Drezek; N. J. Halas; J. L. West. *Ann. Biomed. Eng.* 2006, *34*, 15–22.
158. E. B. Dickerson; E. C. Dreaden; X. Huang; (co-first author); I. H. El-Sayed; H. Chu; S. Pushpanketh; J. F. McDonald; M. A. El-Sayed. *Cancer Lett.* 2008, *269*, 57–66.
159. B. Jang; J.-Y. Park; C.-H. Tung; I.-H. Kim; Y. Choi. *ACS Nano* 2011, *5*, 1086–1094.
160. J. Chen; B. Wiley; Z. Li; D. Campbell; F. Saeki; H. Cang; L. Au; J. Lee; X. Li; Y. Xia. *Adv. Mater.* 2005, *17*, 2255–2261.
161. W. Lu; C. Xiong; G. Zhang; Q. Huang; R. Zhang; J. Z. Zhang; C. Li. *Clin. Cancer Res.* 2009, *15*, 876–886.
162. S. Wang; P. Huang; L. Nie; R. Xing; D. Liu; Z. Wang; J. Lin et al. *Adv. Mater.* 2013, *25*, 3055–3061.
163. J. Chen; C. Glaus; R. Laforest; Q. Zhang; M. Yang; M. Gidding; M. J. Welch; Y. Xia. *Small* 2010, *6*, 811–817.

164. C. Yagüe; M. Arruebo; J. Santamaria. *Chem. Commun. (Camb.)* 2010, *46*, 7513–7515.

165. H. Liu; D. Chen; L. Li; T. Liu; L. Tan; X. Wu; F. Tang. *Angew. Chem. Int. Ed. Engl.* 2010, *50*, 891–895.

166. E. Choi; M. Kwak; B. Jang; Y. Piao. *Nanoscale* 2013, *5*, 151–154.

167. J. You; G. Zhang; C. Li. *ACS Nano* 2010, *4*, 1033–1041.

168. G. B. Braun; A. Pallaoro; G. Wu; D. Missirlis; J. A. Zasadzinski; M. Tirrell; N. O. Reich. *ACS Nano* 2009, *3*, 2007–2015.

169. G. Wu; A. Mikhailovsky; H. A. Khant; C. Fu; W. Chiu; J. A. Zasadzinski. *J. Am. Chem. Soc.* 2008, *130*, 8175–8177.

170. G. Baffou; R. Quidant; C. Girard. *Appl. Phys. Lett.* 2009, *94*, 153109.

171. R. Rodríguez-Oliveros; J. A. Sánchez-Gil. *Opt. Express* 2012, *20*, 621–626.

172. W. Lu; A. K. Singh; S. A. Khan; D. Senapati; H. Yu; P. C. Ray. *J. Am. Chem. Soc.* 2010, *132*, 18103–18114.

173. F. Gámez; P. Hurtado; P. Castillo; C. Caro; A. Hortal; P. Zaderenko; B. Martínez Haya. *Plasmonics* 2010, *5*, 125–133.

174. ANSI. *American National Standard for Safe Use of Lasers.* Laser Institute of America: Orlando, FL, 2000.

175. D. E. J. G. J. Dolmans; D. Fukumura; R. K. Jain. *Nat. Rev. Cancer* 2003, *3*, 380–387.

176. T. J. Dougherty. *Photochem. Photobiol.* 1987, *45*, 879–889.

177. W. Tang; H. Xu; R. Kopelman; M. A. Philbert. *Photochem. Photobiol.* 2005, *81*, 242–249.

178. W. E. Doering; S. Nie. *Anal. Chem.* 2003, *75*, 6171–6176.

179. Y. N. Konan; R. Gurny; E. Allémann. *J. Photochem. Photobiol. B* 2002, *66*, 89–106.

14

Activity of Psoralen-Functionalized Nanoscintillators against Cancer Cells upon X-Ray Excitation

John P. Scaffidi
Duke University

Molly K. Gregas
Duke University

B. Lauly
Miami University

Y. Zhang
Nanometrics Incorporated

Tuan Vo-Dinh
Duke University

14.1 Introduction

Cancer diagnoses in the United States number over a million per year. Worldwide, this number tops 10 million. While surgery, chemotherapy, radiation therapy, photodynamic therapy (PDT), and psoralen + UVA (PUVA) can be used to treat various types of cancer, many additional lives could be saved via development of more universally applicable, minimally invasive approaches to therapy. The rapidly developing field of nanomedicine has attempted to leverage the untapped potential of nanomaterials to improve drug targeting to and uptake by tumors, locally activating therapeutic agents, and limiting the side effects that may negatively affect patients' quality of life.[1-4]

Attempts to develop so-called nanodrugs typically take one of three approaches. In the first, nanoparticles aid in the transport and delivery of chemotherapeutic agents.[5-8] This methodology has shown some potential, particularly in reducing the side effects of chemotherapy.[9] A second approach uses the nanoparticles themselves as a means of enhancing the effects of more traditional treatment strategies.[8,10-15] Two techniques that fall into this category are induction of hyperthermia when illuminating gold nanoshells with infrared light[13-14] or enhancement of reactive oxygen species (ROS) generation using gold nanomaterials and x-ray radiation.[12,16-18] A third, less-proven application of nanodrugs in

315

cancer therapy is combination of ROS-generating PDT drugs with nanoparticles that emit UV or visible photons through upconversion of infrared light or downconversion of x-ray radiation.[19–26]

Of these approaches, the use of deeply penetrating x-rays in the third example is perhaps the most intriguing. In the ideal case, treatment via a combination of targeted drug delivery and localized x-ray activation would affect only those cells that both took up the nanodrug and were exposed to x-rays. In turn, negative side effects due to nonspecific uptake of chemotherapeutic drugs should be greatly reduced. Simultaneously, radiation doses could ideally be decreased such that systemic side effects associated with traditional radiation therapy (e.g., nausea, fatigue) were relatively minor or perhaps even nonexistent. While such an approach with traditional PDT drugs has shown potential in well-oxygenated tissue, reduced ROS generation in the hypoxic environments[27–38] found in many tumors are likely to limit the wider usefulness of such ROS-dependent locally activated therapies.

To circumvent the limitations inherent to oxygenation-dependent approaches and to take advantage of psoralen's demonstrated immunogenic potential,[39–41] we have begun development of anticancer nano-drugs in which scintillating nanoparticles (*nanoscintillators*) can be used to activate psoralen in deep tissue. The most effective drug-nanoscintillator configuration we have tested to date is constructed by binding psoralen to a fragment of the HIV-1 TAT cell-penetrating/nuclear-targeting peptide anchored to UVA-emitting Y_2O_3 nanoscintillators (hereafter referred to as *PsTAT-Y_2O_3*).

Experiments with this particular nanodrug formulation indicate that in the absence of x-ray exposure, it is nontoxic to PC-3 human prostate cancer cells at concentrations approaching 30 μg/mL. The situation changes significantly for both psoralen-functionalized and psoralen-free Y_2O_3 upon x-ray exposure, however. The cell number density of cultures incubated with Y_2O_3 nanoscintillators functionalized with the HIV-1 TAT fragment but lacking psoralen (hereafter referred to as *TAT-Y_2O_3*), for example, is 75% ± 8.3% (1 standard deviation, n = 26) that of cultures exposed to an equal dose of x-ray radiation in the absence of Y_2O_3 nanoscintillators. The cell number density of PC-3 cultures exposed to x-ray radiation following incubation with PsTAT-Y_2O_3 is reduced to 66% ± 8.1% (1 standard deviation, n = 26) that of the same Y_2O_3-free control cultures. When these PsTAT-Y_2O_3 results are normalized against the cell number density of cultures irradiated after incubation with TAT-Y_2O_3 to experimentally correct for the ability of hard matter to amplify the antitumor effects of x-ray radiation, the 9% absolute difference in cell number density increases to a 12% relative difference. These differences in relative cell number are both visually and statistically significant at Y_2O_3 concentrations from ~9.5 to ~95 μg/mL as determined by both t-tests and ANOVA and show clear concentration dependence. These results are the first ever demonstration of drug-associated, x-ray-activated reductions in cell growth for psoralen tethered to a scintillating nanomaterial. The anticancer immunogenic potential of this and future formulations can only be fully evaluated by further studies in immune-competent organisms.

14.2 Approach

As with more traditional light-activated therapies such as PDT and PUVA, the x-ray-excited nanoscintillator emission spectrum must at least partially overlap the psoralen absorption spectrum as shown in Figure 14.1. This fundamental demand indicates that most currently available scintillator and nanoscintillator materials will be inappropriate for psoralen activation. The range of materials suited to in vitro and in vivo use is additionally restricted by the requirement that the nanoscintillators be insoluble in water, as solid-state scintillators lose their emissive properties upon dissolution. While desirable, lack of toxicity is not critical if the nanoscintillators can be covered with an optically transparent coating (e.g., SiO_2), which is impermeable to toxins, or if they are demonstrably less hazardous than the disease to be treated.

Table 14.1 lists a variety of UVA-emitting scintillators that could potentially stimulate DNA mono-adduct or diadduct formation (i.e., cross-linking) by psoralen.[42–66] Several of these materials, such as cubic-phase Y_2O_3,[63] are commercially available as nanoscintillators or can be readily synthesized in the laboratory using published methods.[67–71] Others, such as $YAlO_3$:Ce and $LuAlO_3$:Ce,[72–74] are not currently

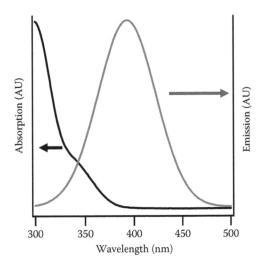

FIGURE 14.1 Absorption of psoralen and scintillation by Y_2O_3. The x-ray-excited scintillation spectrum of Y_2O_3 nanoparticles (gray trace) partially overlaps the absorption spectrum of psoralen (black trace). Spectral overlap between 320 and 405 nm has the potential to produce psoralen-DNA monoadducts and diadducts (cross-links).

TABLE 14.1 UVA-Emitting Scintillating Nanoparticles

Material	λ_{max} (nm)	Photons per MeV at 662 keV	Weaknesses	References
$CeBr_3$	371	68,000	WS	[42]
$CeCl_3$	350	46,000	WS	[43]
$GdAlO_3$:Ce	335–360	9,000	NCN	[44]
K_2CeCl_5:Ce	370	30,000	WS	[45]
K_2LaBr_5:Ce	355–390	40,000	WS	[46]
K_2LaCl_5:Ce	340–375	39,650	WS	[46,47]
K_2LaI_5:Ce	340–380	29,000	WS	[46,48]
KYP_2O_7:Ce	380	10,000	NCN	[49]
$LaBr_3$:Ce	355–390	67,500	WS	[50–53]
$LaCl_3$:Ce	330–355	49,000	WS	[50,52,53]
$LuAlO_3$:Ce	365	16,350	NCN	[54–57]
$LuPO_4$:Ce	360	17,200	NCN	[58]
$PbSO_4$	340–380	10,000	TX	[59–61]
$PrBr_3$:Ce	365–395	21,000	NCN, WS	[62]
Y_2O_3	370	15,480	—	[63]
$YAlO_3$:Ce	345–365	18,360	NCN	[56,57,64–66]

Notes: Light yield (photons per MeV of absorbed x-ray radiation) at 662 keV describes how efficiently each material converts x-ray photons into UVA photons; WS, water soluble; NCN, no commercially available nanoparticles; TX, toxic.

commercially available as nanoparticles but may be superior for psoralen activation due to their better-matched emission spectra and their higher x-ray downconversion efficiency.

In these studies, we have used ~12 nm diameter cubic-phase Y_2O_3 nanoscintillators functionalized with a psoralen-containing, cysteine-terminated fragment of the HIV-1 TAT peptide (PsTAT-Y_2O_3, peptide sequence psoralen-Arg-Lys-Lys-Arg-Arg-Arg-Gln-Arg-Arg-Cys-CONH$_2$) that has been shown to improve cellular uptake, intracellular transport, and nuclear localization.[75–79] A psoralen-free peptide with an identical amino acid sequence was attached to Y_2O_3 nanoscintillators (TAT-Y_2O_3) and was used to control for non-psoralen-related x-ray–Y_2O_3 interactions and the effects of nanoscintillator adhesion to the cell culture plates. Y_2O_3[79–81] and this particular segment of the HIV-1 TAT sequence[82–84] are both known to be nontoxic in the absence of x-ray exposure, thereby making this nanodrug formulation a potential first-generation candidate for additional proof-of-concept studies. The use of more strongly emitting nanoscintillators and other cell-penetrating/nuclear-targeting peptides is likely to yield additional nanodrug candidates.

14.3 Experimental

14.3.1 Nanodrug Preparation

The full synthetic protocol for our yet-to-be-optimized nanodrug is reported elsewhere and is summarized in Scheme 1. Briefly, ~12 nm diameter cubic-phase Y_2O_3 nanoparticles (Meliorum Technologies, Rochester, NY) were weighed, autoclaved, and dispersed via tip sonication in sterile 2-chloroethyl phosphonic acid (*2-CEP*, Fisher Scientific, Fairlawn, NJ) and sodium bicarbonate (VWR, West Chester, PA). The nanoscintillator dispersion was vigorously mixed and purified by triplicate centrifugation. The 2-CEP-functionalized particles were redispersed in sterile-filtered TAT (SynBioSci, Livermore, CA) or PsTAT solution (RS Synthesis LLC, Lexington, KY), vigorously mixed, and again purified by triplicate centrifugation. Final redispersion was in sterile-filtered 5 wt% dextrose (*D5W*, dextrose from Mallinckrodt Baker Inc., Phillipsburg, NJ).

The TAT-Y_2O_3 and PsTAT-Y_2O_3 dispersions were added to cell culture within 3 h of preparation and were sonicated before addition to culture to limit the effects of aggregation during storage.

14.3.2 In Vitro Nanodrug Testing

PC-3 human prostate cancer cells (American Type Culture Collection, Rockville, MD) were grown to ~70% confluence, harvested by trypsinization (Gibco, Grand Island, NY), and seeded in six-well cell culture plates (Corning, Lowell, MA). Following ~24 h for cellular attachment, D5W, D5W containing TAT-Y_2O_3, or D5W containing PsTAT-Y_2O_3 was added to each well and distributed by orthogonal mixing. After incubation for 3–4 h under standard conditions to allow nanoscintillator uptake and intracellular transport, test cultures were exposed to x-ray radiation at 160 or 320 kVp (model: X-RAD 320,

Precision X-ray Inc., North Branford, CT), with the dose hardened using a 2 mm Al filter. Nonirradiated control cultures were also treated with D5W, D5W containing TAT-Y$_2$O$_3$, or D5W containing PsTAT-Y$_2$O$_3$. The culture media was replaced immediately after x-ray exposure of the test plates, and the PC-3 cell cultures were incubated for 7–8 days prior to cell number density measurement.

TAT-Y$_2$O$_3$ and PsTAT-Y$_2$O$_3$ nanodrug concentrations were confirmed ICP-AES measurement of yttrium content. After the six centrifugations performed during preparation, Y$_2$O$_3$ concentrations were generally within 5% of the expected concentration. Although direct peptide quantitation was not possible due to the low concentrations of TAT and PsTAT, we presume that their concentrations vary by ~5% in the final formulations since the peptides were tethered to the Y$_2$O$_3$ nanoscintillators. We have attempted to quantify the amount of peptide attached to the Y$_2$O$_3$ nanoscintillators by various techniques (UV–visible absorption spectroscopy, fluorescence, MALDI, LC-MS, etc.) and have repeatedly found that the amount of peptide in the nanodrug formulation is below the limit of detection for most of these techniques when applied to Y$_2$O$_3$-containing samples (this finding is not surprising, as surface-bound drugs are often active at concentrations below the bulk detection limit). The exception is fluorescence, with which we have detected a weak but recognizable psoralen spectrum in the PsTAT-Y$_2$O$_3$ formulation. One focus during future studies will be development of improved methods able to better quantify all components of this and other nanodrug formulations.

14.3.3 Determination of Cell Proliferation

After incubation under standard conditions for 7–8 days, the cell culture media was removed by aspiration and replaced with methylene blue (hereafter, *MB*) hydrate (Sigma-Aldrich, Milwaukee, WI) in 50:50 (v/v) methanol/water. The cells were fixed and stained for 10 min, after which the *MB* solution was removed and the six-well plates were gently rinsed with DI water. After drying in air for >48 h, acidified ethanol was added to each well and the plates were gently mixed on a laboratory shaker for 10 min to ensure complete MB desorption. The absorption spectrum was measured from 400 to 800 nm using a plate reader (Fluostar Omega, BMG Labtech, Cary, NC), and the absorption at ~660 nm was used to determine cell number densities. The use of in-plate particle-free and TAT-Y$_2$O$_3$ controls allowed optimal determination of nanodrug effectiveness, allowing correction for plate-to-plate variability in cell seeding, changes in cell survival due to minor variations in handling, minor differences in the duration of staining, etc.

14.3.4 Data Manipulation

All calculations (e.g., normalization, t-tests, ANOVA) were performed using Microsoft Excel 2003, SP3 (Microsoft, Redmond, OR).

14.4 Results and Discussion

Like clonogenics,[85–89] trypan blue staining,[85,90,91] and mitochondrial activity assays,[91] cell number density measurements cannot identify the biochemical pathway through which a drug or nanodrug acts. Careful experimental design, however, allows us to determine the extent to which psoralen at the N-terminus of PsTAT impacts proliferation of cultures incubated with PsTAT-Y$_2$O$_3$ versus cultures incubated with TAT-Y$_2$O$_3$. Necessarily, any *relative* increase or decrease in cell number density for cell cultures exposed to the PsTAT-functionalized nanodrug in the presence or absence of x-ray exposure must be attributed to the presence of the N-terminal psoralen moiety. Likewise, any differences in the PsTAT-Y$_2$O$_3$/TAT-Y$_2$O$_3$ cell number density ratio in the presence versus absence of irradiation must be attributed to interactions between PsTAT-Y$_2$O$_3$ and x-ray radiation. Again, the nature of those interactions cannot be determined based on in vitro experiments and will require studies in immune competent organisms.

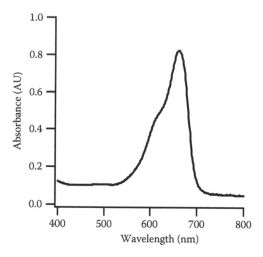

FIGURE 14.2 Visible-wavelength MB absorption spectrum. The relative visible wavelength absorption of MB at ~660 nm following desorption from cell culture with acidified ethanol is an accepted indicator of cell number density.

Figure 14.2 shows a representative MB absorption spectrum following desorption from cells using acidified ethanol. The absorption profile matches that published by prior users of this assay, confirming the lack of dimer formation.[92–100] Using absorption intensity at ~660 nm, this MB assay has been shown to allow reliable, quantitative determination of cell number density in a wide variety of plating formats including the six-well format used in the current study.[95]

Figure 14.3a shows visible-wavelength MB absorption results at ~660 nm for cultures with and without TAT-Y_2O_3 (squares) or PsTAT-Y_2O_3 (diamonds) in the absence of x-ray exposure. The data are normalized against particle-free controls incorporated into each six-well plate, allowing straightforward correction for plate-to-plate variability in cell seeding density, the quality of the cell culture coating, plate-to-plate temperature variations during handling, etc. Decrease in the normalized cell number density relative to particle-free controls is visible at a PsTAT-Y_2O_3 concentration of 95 µg/mL, but no visibly or statistically significant difference appears at lower concentrations of TAT- or PsTAT-functionalized Y_2O_3 in the absence of x-ray exposure. As expected, given the known lack of toxicity for Y_2O_3[79–81] and

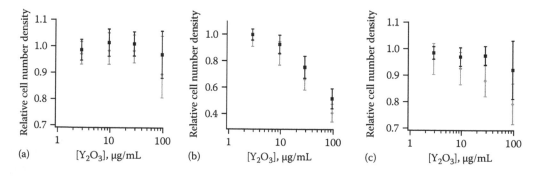

FIGURE 14.3 Normalized cell number density of PC-3 human prostate cancer cell cultures treated with TAT-Y_2O_3 or PsTAT-Y_2O_3 with and without x-ray exposure. (a) Control-normalized relative cell number density of cultures incubated with TAT-Y_2O_3 (black squares) or PsTAT-Y_2O_3 (gray diamonds), but not exposed to x-ray radiation. (b) As in (a), but approximately 1 week after x-ray irradiation. (c) Relative cell number density of cultures incubated to PsTAT-Y_2O_3 versus TAT-Y_2O_3 with (gray diamonds) and without (black squares) x-ray exposure.

this segment of TAT,[82–84] these results confirm the lack of strong intrinsic toxicity for TAT-Y$_2$O$_3$ and PsTAT-Y$_2$O$_3$ in the absence of x-ray irradiation.

The situation changes substantially upon x-ray exposure. Figure 14.3b shows cell number density results for PC-3 cell cultures incubated with TAT-Y$_2$O$_3$ (squares) or PsTAT-Y$_2$O$_3$ (diamonds) at particle concentrations ranging from 2.9 to 95 μg/mL prior to irradiation. As in Figure 14.3a, the data have been normalized against in-plate controls lacking the nanoscintillator formulations. In addition to allowing correction for plate-to-plate variability, this normalization approach also lets us correct for the established ability of x-ray radiation to impact cell viability and propagation.

As shown in Figure 14.3b, x-ray exposure following incubation with TAT-Y$_2$O$_3$ or PsTAT-Y$_2$O$_3$ reduces cell number density in a particle dose-dependent manner at Y$_2$O$_3$ concentrations between 9.5 and 95 μg/mL when compared to particle-free control cultures exposed to x-rays. Specifically, x-ray irradiation of PC-3 cell cultures incubated with 2.9–95 μg/mL TAT-functionalized Y$_2$O$_3$ reduces the normalized cell number density from 1.00 ± 0.042 at 2.9 μg/mL Y$_2$O$_3$ to 0.98 ± 0.070 at 9.5 μg/mL, 0.75 ± 0.083 at 29 μg/mL, and 0.52 ± 0.076 at 95 μg/mL Y$_2$O$_3$. For cultures incubated with PsTAT-functionalized Y$_2$O$_3$ before x-ray exposure, the normalized cell number density falls from 0.96 ± 0.050 at 2.9 μg/mL Y$_2$O$_3$ to 0.86 ± 0.089 at 9.5 μg/mL, 0.66 ± 0.082 at 29 μg/mL, and 0.41 ± 0.067 at 95 μg/mL Y$_2$O$_3$. The overall trend of reduced cell number densities at elevated particle concentrations is expected, since x-ray interaction with hard matter is known to enhance the effectiveness of radiotherapy independent of PDT or PUVA effects.[20,25,101–103] Interestingly, the cell number density of cultures incubated with PsTAT-Y$_2$O$_3$ is *universally* lower than that of PC-3 cell cultures incubated with an equal concentration of TAT-Y$_2$O$_3$ prior to x-ray exposure. These reductions in cell number density are statistically significant by both ANOVA and Student's t-test at particle concentrations of ~9.5 μg/mL and higher (Figure 14.3b and Table 14.2).

The difference in cell number density becomes yet more evident after correction for the intrinsic x-ray—hard matter interactions that are used so effectively in radiation therapy.[104–108] To that end, Figure 14.3c displays the renormalized cell number density of PC-3 cultures incubated with PsTAT-Y$_2$O$_3$ versus TAT-Y$_2$O$_3$ with (squares) and without (diamonds) x-ray exposure. The value of 0.98 ± 0.024 at 2.9 μg/mL in the absence of x-ray exposure, for example, confirms that the cell number density of PC-3 cultures exposed to 2.9 μg/mL PsTAT-Y$_2$O$_3$ is essentially equal to that of cultures exposed to 2.9 μg/mL TAT-Y$_2$O$_3$ in the absence of irradiation. As noted earlier, this baseline cell number remains stable in the absence of irradiation up to particle concentrations of 95 μg/mL, at which point the cell number density of cultures incubated with PsTAT-Y$_2$O$_3$ falls to 0.92 ± 0.110 that of cultures incubated with TAT-Y$_2$O$_3$. The cause for this reduction in cell number density at high nanodrug concentration in the absence of x-ray radiation is not yet clear but will be a topic of future studies.

For irradiated versus nonirradiated PC-3 cell cultures, the normalized data in Figure 14.3c suggest that there is no strong evidence of a PsTAT-Y$_2$O$_3$ versus TAT-Y$_2$O$_3$ effect at Y$_2$O$_3$ concentrations of 2.9 μg/mL. A visually and statistically significant downward trend in the relative cell number density ratio is seen for irradiated cell cultures at PsTAT-Y$_2$O$_3$ concentrations between 9.5 and 95 μg/mL. The cell number density ratio for PC-3 cultures irradiated after incubation with PsTAT-Y$_2$O$_3$ versus TAT-Y$_2$O$_3$ falls from 0.96 ± 0.058 at 2.9 μg Y$_2$O$_3$ per mL to 0.93 ± 0.061 at 9.5 μg/mL, 0.88 ± 0.060 at 29 μg/mL, and 0.79 ± 0.077 at PsTAT-Y$_2$O$_3$ and TAT-Y$_2$O$_3$ concentrations of 95 μg/mL. The number of replicates at each concentration and the p values for the differentiation of the PsTAT-Y$_2$O$_3$/TAT-Y$_2$O$_3$ cell number ratio in the presence or absence of x-ray exposure as determined by both t-tests and ANOVA are listed in Table 14.3. The calculated p values at the lowest concentration of PsTAT- and TAT-Y$_2$O$_3$ provide little suggestion that psoralen activation upon irradiation significantly impacts PC-3 cell growth or viability, but statistically significant differences in the PsTAT-Y$_2$O$_3$/TAT-Y$_2$O$_3$ cell number density ratio become evident at higher Y$_2$O$_3$ concentrations. At 9.5 μg/mL Y$_2$O$_3$, for example, the likelihood of a false-positive result for irradiated versus nonirradiated cultures falls below 1% as measured by either an unpaired one-tailed t-test or ANOVA. The likelihood of a false-positive result decreases further as the concentrations of PsTAT-Y$_2$O$_3$ and TAT-Y$_2$O$_3$ are increased, reaching ~10^{-6} at 95 μg/mL Y$_2$O$_3$.

TABLE 14.2　Normalized Cell Number (#) Density of PC-3 Human Prostate Cancer Cell Cultures Treated with TAT-Y_2O_3 or PsTAT-Y_2O_3 in the Presence and Absence of X-Ray Irradiation

Formulation	Y_2O_3 Concentration (µg/mL)	+/– X-ray	N	Cell # Density vs. Y_2O_3-Free Control	p (t-test) vs. Y_2O_3-Free Control	p (ANOVA) vs. Y_2O_3-Free Control	Cell # Density	p (t-test) vs. TAT-Y_2O_3	p (ANOVA) vs. TAT-Y_2O_3
TAT-Y_2O_3	95	–	20	0.97 ± 0.089	0.21	0.41	—	—	—
	29	–	14	1.01 ± 0.046	0.36	0.73	—	—	—
	9.5	–	21	1.01 ± 0.051	0.39	0.79	—	—	—
	2.9	–	15	0.99 ± 0.039	0.21	0.43	—	—	—
PsTAT-Y_2O_3	95	–	20	0.90 ± 0.144	0.013	0.025	0.92 ± 0.110	0.030	0.058
	29	–	14	0.98 ± 0.060	0.19	0.37	0.97 ± 0.036	0.10	0.21
	9.5	–	21	0.98 ± 0.068	0.32	0.65	0.97 ± 0.035	0.056	0.11
	2.9	–	15	0.97 ± 0.046	0.058	0.11	0.98 ± 0.024	0.17	0.34
TAT-Y_2O_3	95	+	33	0.52 ± 0.076	4.0×10^{-19}	4.7×10^{-19}	—	—	—
	29	+	26	0.75 ± 0.083	$1.0 \times 10{-6}$	2.0×10^{-6}	—	—	—
	9.5	+	33	0.92 ± 0.070	0.043	0.087	—	—	—
	2.9	+	28	1.00 ± 0.042	0.46	0.92	—	—	—
PsTAT-Y_2O_3	95	+	33	0.41 ± 0.067	7.3×10^{-24}	3.4×10^{-25}	0.79 ± 0.077	3.1×10^{-8}	6.0×10^{-8}
	29	+	26	0.66 ± 0.082	2.2×10^{-10}	3.8×10^{-10}	0.88 ± 0.060	1.2×10^{-4}	2.3×10^{-4}
	9.5	+	33	0.86 ± 0.089	1.3×10^{-3}	2.6×10^{-3}	0.93 ± 0.061	5.3×10^{-4}	1.0×10^{-3}
	2.9	+	28	0.96 ± 0.050	0.14	0.28	0.96 ± 0.058	1.2×10^{-3}	2.3×10^{-3}

Notes: N, total number of replicates; Cell # Density vs. Y_2O_3-free Control, cell # density of cultures exposed to TAT-Y_2O_3 or PsTAT-Y_2O_3, normalized against in-plate controls lacking either of the nanodrugs but exposed to a matching dose of x-ray radiation; p (t-test), p value determined using a one-tailed, unpaired t-test; p (ANOVA), p value determined using ANOVA; Cell # Density vs. TAT-Y_2O_3, cell # density of cultures exposed to PsTAT-Y_2O_3, normalized against in-plate controls exposed to a matching dose of x-ray radiation.

TABLE 14.3 Normalized Cell Number (#) Density of PC-3 Human Prostate Cancer Cell Cultures Treated with PsTAT-Y$_2$O$_3$ in the Presence and Absence of X-Ray Irradiation

Y$_2$O$_3$ Concentration (μg/mL)	+/− X-ray	N	PsTAT-Y$_2$O$_3$/ TAT-Y$_2$O$_3$ Cell # Density Ratio	p (t-test) vs. Unirradiated Cell Cultures	p (ANOVA) vs. Unirradiated Cell Cultures
95	−	20	0.92 ± 0.110	—	—
29	−	14	0.97 ± 0.036	—	—
9.5	−	21	0.97 ± 0.035	—	—
2.9	−	15	0.98 ± 0.024	—	—
95	+	33	0.79 ± 0.077	3.6×10^{-5}	7.0×10^{-6}
29	+	26	0.88 ± 0.060	1.6×10^{-7}	4.4×10^{-6}
9.5	+	33	0.93 ± 0.060	7.4×10^{-4}	4.3×10^{-3}
2.9	+	28	0.96 ± 0.058	0.039	0.16

Notes: N, total number of replicates; PsTAT-Y$_2$O$_3$/TAT-Y$_2$O$_3$ Cell # Density Ratio, cell # density of cultures exposed to PsTAT-Y$_2$O$_3$, normalized against in-plate controls incubated with a matching concentration of TAT-Y$_2$O$_3$ and exposed to a matching dose of x-ray radiation; p (t-test), p value determined using a one-tailed, unpaired t-test; p (ANOVA), p value determined using ANOVA.

The larger anticancer potential of these in vitro results is not yet clear and in fact cannot be evaluated in vitro. An intact, functional immune system is essential when attempting to activate psoralen's established immunogenic properties by exposing deep-seated tumors to both the nanodrug and x-ray radiation. Such studies are planned, as are in vitro experiments determining the activity of PsTAT-Y$_2$O$_3$ against additional cell lines. Similarly, to identify the mechanism through which PsTAT-Y$_2$O$_3$ either kills or delays the growth of PC-3 human prostate cancer cells are ongoing.

14.5 Conclusions

The in vitro reductions in cell number seen in our study are modest, but they provide the first ever indication of cell killing or cellular growth inhibition by x-ray-excited, psoralen-functionalized nanoscintillators. Specifically, these cell number results provide evidence that x-ray exposure of Y$_2$O$_3$ nanoparticles functionalized with a psoralen-modified nuclear-targeting peptide may find use in the fight against cancer. If psoralen continues to act via a non-ROS-dependent mechanism after attachment to these nanoscintillators—as one would hope—then such a drug–peptide–particle nanoconjugate may have unique potential as a means of treating deep-seated, hypoxic tumors. Further research is underway to determine the mechanism by which these particular psoralen-functionalized nanoscintillators impact cell number density and to better quantify their ability to inhibit cell growth and/or induce apoptosis. Simultaneously, we are examining the in vitro potential of other UV-emitting nanoscintillators such as doped oxides, fluorides, and perovskites; the cell-penetrating/nuclear-targeting peptides such as Penetratin, MAP, polyarginine, c-Fos, Antennapedia, VP22, and transportan; and the in vitro effectiveness of nanoscintillators functionalized with a combination of psoralen-modified and psoralen-free cell-penetrating/nuclear-targeting peptides. The results of these studies will be the focus of future reports.

Acknowledgments

The authors gratefully acknowledge many productive discussions with Dr. Guy Griffin and financial support from Immunolight, LLC.

References

1. Brigger, I., Dubernet, C.; Couvreur, P., Nanoparticles in cancer therapy and diagnosis. *Advanced Drug Delivery Reviews* 2002, *54*(5), 631–651.

2. Yih, T. C.; Wei, C., Nanomedicine in cancer treatment. *Nanomedicine: Nanotechnology, Biology and Medicine* 2005, *1*(2), 191–192.

3. Cuenca, A. G.; Jiang, H. B.; Hochwald, S. N.; Delano, M.; Cance, W. G.; Grobmyer, S. R., Emerging implications of nanotechnology on cancer diagnostics and therapeutics. *Cancer* 2006, *107*(3), 459–466.

4. Huang, X.; Jain, P. K.; El-Sayed, I. H.; El-Sayed, M. A., Gold nanoparticles: Interesting optical properties and recent applications in cancer diagnostics and therapy. *Nanomedicine* 2007, *2*(5), 681–693.

5. Gulyaev, A. E.; Gelperina, S. E.; Skidan, I. N.; Antropov, A. S.; Kivman, G. Y.; Kreuter, J., Significant transport of doxorubicin into the brain with polysorbate 80-coated nanoparticles. *Pharmaceutical Research* 1999, *16*(10), 1564–1569.

6. Song, M.; Wang, X. M.; Li, J. Y.; Zhang, R. Y.; Chen, B. A.; Fu, D. G., Effect of surface chemistry modification of functional gold nanoparticles on the drug accumulation of cancer cells. *Journal of Biomedical Materials Research Part A* 2008, *86A*(4), 942–946.

7. Nystrom, A. M.; Xu, Z. Q.; Xu, J. Q.; Taylor, S.; Nittis, T.; Stewart, S. A.; Leonard, J.; Wooley, K. L., SCKs as nanoparticle carriers of doxorubicin: Investigation of core composition on the loading, release and cytotoxicity profiles. *Chemical Communications* 2008, (30), 3579–3581.

8. Rapoport, N.; Gao, Z. G.; Kennedy, A., Multifunctional nanoparticles for combining ultrasonic tumor imaging and targeted chemotherapy. *Journal of the National Cancer Institute* 2007, *99*(14), 1095–1106.

9. Zhang, Z.; Lee, S.; Gan, C.; Feng, S.-S., Investigation on PLA–TPGS nanoparticles for controlled and sustained small molecule chemotherapy. *Pharmaceutical Research* 2008, *25*(8), 1925–1935.

10. Dong, Y.; Feng, S. S., Poly(D,L-lactide-co-glycolide) (PLGA) nanoparticles prepared by high pressure homogenization for paclitaxel chemotherapy. *International Journal of Pharmaceutics* 2007, *342*(1–2), 208–214.

11. Li, J.; Wang, X.; Wang, C.; Chen, B.; Dai, Y.; Zhang, R.; Song, M.; Lv, G.; Fu, D., The enhancement effect of gold nanoparticles in drug delivery and as biomarkers of drug-resistant cancer cells. *ChemMedChem* 2007, *2*(3), 374–378.

12. Roeske, J. C.; Nunez, L.; Hoggarth, M.; Labay, E.; Weichselbaum, R. R., Characterization of the theoretical radiation dose enhancement from nanoparticles. *Technology in Cancer Research and Treatment* 2007, *6*(5), 395–401.

13. O'Neal, D. P.; Hirsch, L. R.; Halas, N. J.; Payne, J. D.; West, J. L., Photo-thermal tumor ablation in mice using near infrared-absorbing nanoparticles. *Cancer Letters* 2004, *209*(2), 171–176.

14. El-Sayed, I. H.; Huang, X.; El-Sayed, M. A., Selective laser photo-thermal therapy of epithelial carcinoma using anti-EGFR antibody conjugated gold nanoparticles. *Cancer Letters* 2006, *239*(1), 129–135.

15. Zhang, R. Y.; Wang, X. M.; Wu, C. H.; Song, M.; Li, J. Y.; Lv, G.; Zhou, J. et al., Synergistic enhancement effect of magnetic nanoparticles on anticancer drug accumulation in cancer cells. *Nanotechnology* 2006, *17*(14), 3622–3626.

16. Visaria, R. K.; Griffin, R. J.; Williams, B. W.; Ebbini, E. S.; Paciotti, G. F.; Song, C. W.; Bischof, J. C., Enhancement of tumor thermal therapy using gold nanoparticle-assisted tumor necrosis factor-alpha delivery. *Molecular Cancer Therapeutics* 2006, *5*(4), 1014–1020.

17. Liu, C. J.; Wang, C. H.; Chien, C. C.; Yang, T. Y.; Chen, S. T.; Leng, W. H.; Lee, C. F. et al., Enhanced x-ray irradiation-induced cancer cell damage by gold nanoparticles treated by a new synthesis method of polyethylene glycol modification. *Nanotechnology* 2008, *19*(29), 295104.

18. Kong, T.; Zeng, J.; Wang, X. P.; Yang, X. Y.; Yang, J.; McQuarrie, S.; McEwan, A.; Roa, W.; Chen, J.; Xing, J. Z., Enhancement of radiation cytotoxicity in breast-cancer cells by localized attachment of gold nanoparticles. *Small* 2008, *4*(9), 1537–1543.

19. Wang, S. Z.; Gao, R. M.; Zhou, F. M.; Selke, M., Nanomaterials and singlet oxygen photosensitizers: Potential applications in photodynamic therapy. *Journal of Materials Chemistry* 2004, *14*(4), 487–493.

20. Takahashi, J.; Misawa, M., Analysis of potential radiosensitizing materials for x-ray-induced photo-dynamic therapy. *NanoBioTechnology* 2007, *3*(2), 116–126.

21. Morgan, N. Y.; Kramer-Marek, G.; Smith, P. D.; Camphausen, K.; Capala, J., Nanoscintillator conjugates as photodynamic therapy-based radiosensitizers: Calculation of required physical parameters. *Radiation Research* 2009, *171*(2), 236–244.

22. Chatterjee, D. K.; Fong, L. S.; Zhang, Y., Nanoparticles in photodynamic therapy: An emerging paradigm. *Advanced Drug Delivery Reviews* 2008, *60*(15), 1627–1637.

23. Liu, Y. F.; Chen, W.; Wang, S. P.; Joly, A. G., Investigation of water-soluble x-ray luminescence nanoparticles for photodynamic activation. *Applied Physics Letters* 2008, *92*(4), 043901.

24. Chen, W.; Zhang, J., Using nanoparticles to enable simultaneous radiation and photodynamic therapies for cancer treatment. *Journal of Nanoscience and Nanotechnology* 2006, *6*, 1159–1166.

25. Juzenas, P.; Chen, W.; Sun, Y. P.; Coelho, M. A. N.; Generalov, R.; Generalova, N.; Christensen, I. L., Quantum dots and nanoparticles for photodynamic and radiation therapies of cancer. *Advanced Drug Delivery Reviews* 2008, *60*(15), 1600–1614.

26. Lovell, J. F.; Liu, T. W. B.; Chen, J.; Zheng, G., Activatable photosensitizers for imaging and therapy. *Chemical Reviews* 2010, *110*(5), 2839–2857.

27. Quintiliani, M., Modification of radiation sensitivity: The oxygen effect. *International Journal of Radiation Oncology, Biology, Physics* 1979, *5*(7), 1069–1076.

28. Henderson, B. W.; Fingar, V. H., Relationship of tumor hypoxia and response to photodynamic therapy treatment in an experimental mouse-tumor. *Cancer Research* 1987, *47*(12), 3110–3114.

29. Hockel, M.; Vaupel, P., Tumor hypoxia: Definitions and current clinical, biologic, and molecular aspects. *Journal of the National Cancer Institute* 2001, *93*(4), 266–276.

30. Bernier, J.; Hall, E. J.; Giaccia, A., Radiation oncology: A century of achievements. *Nature Reviews Cancer* 2004, *4*(9), 737–747.

31. Chaplin, D. J.; Durand, R. E.; Olive, P. L., Acute hypoxia in tumors: Implications for modifiers of radiation effects. *International Journal of Radiation Oncology, Biology, Physics* 1986, *12*(8), 1279–1282.

32. Chaplin, D. J.; Olive, P. L.; Durand, R. E., Intermittent blood flow in a murine tumor: Radiobiological effects. *Cancer Research* 1987, *47*(2), 597–601.

33. Biaglow, J. E.; Mitchell, J. B.; Held, K., The importance of peroxide and superoxide in the X-ray response. *International Journal of Radiation Oncology, Biology, Physics* 1992, *22*(4), 665–669.

34. Dolmans, D. E. J. G. J.; Fukumura, D.; Jain, R. K., Photodynamic therapy for cancer. *Nature Reviews Cancer* 2003, *3*(5), 380–387.

35. Fingar, V. H.; Wieman, T. J.; Park, Y. J.; Henderson, B. W., Implications of a pre-existing tumor hypoxic fraction on photodynamic therapy. *Journal of Surgical Research* 1992, *53*(5), 524–528.

36. Chen, Q.; Huang, Z.; Chen, H.; Shapiro, H.; Beckers, J.; Hetzel, F. W., Improvement of tumor response by manipulation of tumor oxygenation during photodynamic therapy. *Photochemistry and Photobiology* 2002, *76*(2), 197–203.

37. Fuchs, J.; Thiele, J., The role of oxygen in cutaneous photodynamic therapy. *Free Radical Biology and Medicine* 1998, *24*(5), 835–847.

38. Huang, Z.; Chen, Q.; Shakil, A.; Chen, H.; Beckers, J.; Shapiro, H.; Hetzel, F. W., Hyperoxygenation enhances the tumor cell killing of photofrin-mediated photodynamic therapy. *Photochemistry and Photobiology* 2003, *78*(5), 496–502.

39. Gasparro, F. P.; Felli, A.; Schmitt, I. M., Psoralen photobiology: The relationship between DNA damage, chromatin structure, transcription, and immunogenic effects. *Recent Results Cancer Research* 1997, *143*, 101–127.

40. Malane, M. S.; Gasparro, F. P., T cell molecular targets for psoralens. *Annals of the New York Academy of Sciences* 1991, *636*(1), 196–208.

41. Bethea, D.; Fullmer, B.; Syed, S.; Seltzer, G.; Tiano, J.; Rischko, C.; Gillespie, L.; Brown, D.; Gasparro, F. P., Psoralen photobiology and photochemotherapy: 50 years of science and medicine. *Journal of Dermatological Science* 1999, *19*(2), 78–88.

42. Shah, K. S.; Glodo, J.; Higgins, W.; van Loef, E. V. D.; Moses, W. W.; Derenzo, S. E.; Weber, M. J., CeBr3 scintillators for gamma-ray spectroscopy. *IEEE Transactions on Nuclear Science* 2005, *52*, 3157–3159.

43. van Loef, E. V. D. ; Dorenbos, P.; Kramer, K.; Gudel, H. U., Scintillation properties of LaCl3: Ce3+ crystals: Fast, efficient, and high-energy resolution scintillators. *IEEE Transactions on Nuclear Science* 2001, *48*, 341–345.

44. Dorenbos, P.; Bougrine, E.; deHaas, J. T. M.; vanEijk, C. W. E.; Korzhik, M. V., Scintillation properties of GdAlO$_3$:Ce crystals. *Radiation Effects and Defects in Solids* 1995, *135*, 819–821.

45. Roy, U. N.; Groza, M.; Cui, Y.; Burger, A.; Cherepy, N.; Friedrich, S.; Payne, S. A., A new scintillator material. *Nuclear Instruments and Methods in Physics Research Section a-Accelerators Spectrometers Detectors and Associated Equipment* 2007, *579*, 46–49.

46. van Loef, E. V. D.; Dorenbos, P.; van Eijk, C. W. E.; Kramer, K. W.; Gudel, H. U., Scintillation properties of K2LaX5: Ce3+ (X = Cl, Br, I). *Nuclear Instruments and Methods in Physics Research Section A-Accelerators Spectrometers Detectors and Associated Equipment* 2005, *537*, 232–236.

47. van't Spijker, J. C.; Dorenbos, P.; van Eijk, C. W. E.; Kramer, K.; Gudel, H. U., Scintillation and luminescence properties of Ce3+ doped K2LaCl5. *Journal of Luminescence* 1999, *85*, 1–10.

48. Vantspijker, J. C.; Dorenbos, P.; Dehaas, J. T. M.; Vaneijk, C. W. E.; Gudel, H. U.; Kramer, K., Scintillation properties of K2LaCl5 with Ce doping. *Radiation Measurements* 1995, *24*, 379–381.

49. Yuan, J.-L.; Zhang, H.; Chen, H.-H.; Yang, X.-X.; Zhao, J.-T.; Gu, M., Synthesis, structure and X-ray excited luminescence of Ce3+-doped AREP2O7-type alkali rare earth diphosphates (A = Na, K, Pb, Cs; RE = Y, Lu). *Journal of Solid State Chemistry* 2007, *180*, 3381–3387.

50. van Loef, E. V. D.; Dorenbos, P.; van Eijk, C. W. E.; Kramer, K. W.; Gudel, H. U., Scintillation properties of LaBr3: Ce3+ crystals: fast, efficient and high-energy-resolution scintillators. *Nuclear Instruments and Methods in Physics Research Section A: Accelerators Spectrometers Detectors and Associated Equipment* 2002, *486*, 254–258.

51. Menge, P. R.; Gautier, G.; Iltis, A.; Rozsa, C.; Solovyev, V., Performance of large lanthanum bromide scintillators. *Nuclear Instruments and Methods in Physics Research Section A: Accelerators Spectrometers Detectors and Associated Equipment* 2007, *579*, 6–10.

52. de Haas, J. T. M.; Dorenbos, P., Advances in yield calibration of scintillators. *IEEE Transactions on Nuclear Science* 2008, *55*, 1086–1092.

53. Salacka, J. S.; Bacrania, M. K., A comprehensive technique for determining the intrinsic light yield of scintillators. *IEEE Transactions on Nuclear Science* 2010, *57*, 901–909.

54. Moszynski, M.; Wolski, D.; Ludziejewski, T.; Kapusta, M.; Lempicki, A.; Brecher, C.; Wisniewski, D.; Wojtowicz, A. J., Properties of the new LuAP:Ce scintillator. *Nuclear Instruments and Methods in Physics Research Section A: Accelerators Spectrometers Detectors and Associated Equipment* 1997, *385*, 123–131.

55. Moses, W. W.; Derenzo, S. E.; Fyodorov, A.; Korzhik, M.; Gektin, A.; Minkov, B.; Aslanov, V., LuAlO3-Ce—A high-density, high-speed scintillator for gamma-detection. *IEEE Transactions on Nuclear Science* 1995, *42*, 275–279.

56. Moszynski, M.; Kapusta, M.; Mayhugh, M.; Wolski, D.; Flyckt, S. O., Absolute light output of scintillators. *IEEE Transactions on Nuclear Science* 1997, *44*, 1052–1061.

57. Lempicki, A.; Randles, M. H.; Wisniewski, D.; Balcerzyk, M.; Brecher, C.; Wojtowicz, A. J., LuAlO$_3$-Ce and other aluminate scintillators. *IEEE Transactions on Nuclear Science* 1995, *42*, 280–284.

58. Lempicki, A.; Berman, E.; Wojtowicz, A. J.; Balcerzyk, M.; Boatner, L. A., Cerium-doped orthophosphates—New promising scintillators. *IEEE Transactions on Nuclear Science* 1993, *40*, 384–387.

59. Moses, W. W.; Derenzo, S. E.; Shlichta, P. J., Scintillation properties of lead sulfate. *IEEE Transactions on Nuclear Science* 1992, *39*, 1190–1194.

60. Zadneprovski, B. I.; Kamenskikh, I. A.; Kolobanov, V. N.; Mikhailin, V. V.; Shpinkov, I. N.; Kirm, M., Gel growth, luminescence, and scintillation of PbSO4 crystals. *Inorganic Materials* 2004, *40*, 735–739.

61. Zhang, J. G.; Lund, J. C.; Cirignano, L.; Shah, K. S.; Squillante, M. R.; Moses, W. W., Lead sulfate scintillator crystal growth for PET applications. *IEEE Transactions on Nuclear Science* 1994, *41*, 669–674.

62. Birowosuto, M. D.; Dorenbos, P.; van Eijk, C. W. E.; Kramer, K. W.; Gudel, H. U., PrBr3: Ce3+: A new fast lanthanide trihalide scintillator. *IEEE Transactions on Nuclear Science* 2006, *53*, 3028–3030.

63. Ogorodnikov, I. N.; Kruzhalov, A. V.; Ivanov, V. Y., Mechanisms of fast UV-scintillations in oxide crystals with self-trapped excitons. In *Inorganic Scintillators and Their Applications*, Dorenbos, P.; Van Eijk, C. W., Eds. Delft University Press (SCINT95), Delft, the Netherlands: 1996; pp. 216–219.

64. Baryshevsky, V. G.; Korzhik, M. V.; Moroz, V. I.; Pavlenko, V. B.; Fyodorov, A. A.; Smirnova, S. A.; Egorycheva, O. A.; Kachanov, V. A., YAlO3: Ce-fast-acting scintillators for detection of ionizing-radiation. *Nuclear Instruments and Methods in Physics Research Section B: Beam Interactions with Materials and Atoms* 1991, *58*, 291–293.

65. Mares, J. A.; Beitlerova, A.; Nikl, M.; Solovieva, N.; D'Ambrosio, C.; Blazek, K.; Maly, P.; Nejezchleb, K.; de Notaristefani, F., Scintillation response of Ce-doped or intrinsic scintillating crystals in the range up to 1 MeV. *Radiation Measurements* 2004, *38*, 353–357.

66. de Haas, J. T. M.; Dorenbos, P.; van Eijk, C. W. E., Measuring the absolute light yield of scintillators. *Nuclear Instruments and Methods in Physics Research Section A: Accelerators Spectrometers Detectors and Associated Equipment* 2005, *537*(1–2), 97–100.

67. Zako, T.; Nagata, H.; Terada, N.; Sakono, M.; Soga, K.; Maeda, M., Improvement of dispersion stability and characterization of upconversion nanophosphors covalently modified with PEG as a fluorescence bioimaging probe. *Journal of Materials Science* 2008, *43*(15), 5325–5330.

68. Vetrone, F.; Boyer, J. C.; Capobianco, J. A.; Speghini, A.; Bettinelli, M., A spectroscopic investigation of trivalent lanthanide doped Y_2O_3 nanocrystals. *Nanotechnology* 2004, *15*(1), 75–81.

69. Chen, W.; Zhang, J.; Westcott, S. L.; Joly, A. G.; Malm, J. O.; Bovin, J. O., The origin of x-ray luminescence from CdTe nanoparticles in CdTe/BaFBr: Eu2+ nanocomposite phosphors. *Journal of Applied Physics* 2006, *99*(3), 034302–034305.

70. Liu, Y. F.; Chen, W.; Wang, S. P.; Joly, A. G.; Westcott, S.; Woo, B. K., X-ray luminescence of LaF3: Tb3+ and LaF3: Ce3+, Tb3+ water-soluble nanoparticles. *Journal of Applied Physics* 2008, *103*(6), 063105.

71. Zhang, H. P.; Lu, M. K.; Xiu, Z. L.; Zhou, G. J.; Wang, S. F.; Zhou, Y. Y.; Wang, S. M., Influence of processing conditions on the luminescence of YVO4: Eu3+ nanoparticles. *Materials Science and Engineering B: Solid State Materials for Advanced Technology* 2006, *130*(1–3), 151–157.

72. Wisniewski, D.; Wojtowicz, A. J.; Lempicki, A., Spectroscopy and scintillation mechanism in LuAlO3:Ce. *Journal of Luminescence* 1997, *72–74*, 789–791.

73. Wojtowicz, A. J.; Szupryczynski, P.; Wisniewski, D.; Glodo, J.; Drozdowski, W., Electron traps and scintillation mechanism in LuAlO3:Ce. *Journal of Physics: Condensed Matter* 2001, *13*(42), 9599.

74. Dorenbos, P., Scintillation mechanisms in Ce3+ doped halide scintillators. *Physica Status Solidi A: Applied Research* 2005, *202*(2), 195–200.

75. Zorko, M.; Langel, U., Cell-penetrating peptides: Mechanism and kinetics of cargo delivery. *Advanced Drug Delivery Reviews* 2005, *57*(4), 529–545.

76. Brooks, H.; Lebleu, B.; Vives, E., Tat peptide-mediated cellular delivery: Back to basics. *Advanced Drug Delivery Reviews* 2005, *57*(4), 559–577.

77. Gregas, M. K.; Scaffidi, J. P.; Lauly, B.; Vo-Dinh, T., Surface-enhanced Raman scattering detection and tracking of nanoprobes: Enhanced uptake and nuclear targeting in single cells. *Applied Spectroscopy* 2010, *64*(8), 858–866.

78. Fawell, S.; Seery, J.; Daikh, Y.; Moore, C.; Chen, L. L.; Pepinsky, B.; Barsoum, J., Tat-mediated delivery of heterologous proteins into cells. *Proceedings of the National Academy of Science of the United States of America* 1994, *91*(2), 664–668.

79. Sveier, H.; Kvamme, B. O.; Raae, A. J., Growth and protein utilization in Atlantic salmon (*Salmo salar* L.) given a protease inhibitor in the diet. *Aquaculture Nutrition* 2001, *7*(4), 255–264.

80. Reis, P. A.; Valente, L. M. P.; Almeida, C. M. R., A fast and simple methodology for determination of yttrium as an inert marker in digestibility studies. *Food Chemistry* 2008, *108*(3), 1094–1098.

81. Schubert, D.; Dargusch, R.; Raitano, J.; Chan, S. W., Cerium and yttrium oxide nanoparticles are neuroprotective. *Biochemical and Biophysical Research Communications* 2006, *342*(1), 86–91.

82. Torchilin, V. P.; Levchenko, T. S.; Rammohan, R.; Volodina, N.; Papahadjopoulos-Sternberg, B.; D'Souza, G. G. M., Cell transfection in vitro and in vivo with nontoxic TAT peptide-liposome-DNA complexes. *Proceedings of the National Academy of Sciences of the United States of America* 2003, *100*(4), 1972–1977.

83. Manickam, D. S.; Bisht, H. S.; Wan, L.; Mao, G. Z.; Oupicky, D., Influence of TAT-peptide polymerization on properties and transfection activity of TAT/DNA polyplexes. *Journal of Controlled Release* 2005, *102*(1), 293–306.

84. Kim, K.; Han, J.; Kim, H.; Lee, M., Expression, purification and characterization of TAT-high mobility group box-1A peptide as a carrier of nucleic acids. *Biotechnology Letters* 2008, *30*(8), 1331–1337.

85. Weisenthal, L. M.; Dill, P. L.; Kurnick, N. B.; Lippman, M. E., Comparison of dye exclusion assays with a clonogenic-assay in the determination of drug-induced cyto-toxicity. *Cancer Research* 1983, *43*(1), 258–264.

86. Plumb, J. A., Ch. 2-Cell sensitivity assays/clonogenic assay. In *Cytotoxic Drug Resistance Mechanisms*, Brown, R.; Boger-Brown, U., Eds. Humana Press: Totowa, NJ, 1999; Vol. 28, pp. 25–30.

87. Franken, N. A. P.; Rodermond, H. M.; Stap, J.; Haveman, J.; van Bree, C., Clonogenic assay of cells in vitro. *Nature Protocols* 2006, *1*(5), 2315–2319.

88. Smith, J. R.; Whitney, R. G., Intraclonal variation in proliferative potential of human-diploid fibroblasts: Stochastic mechanism for cellular aging. *Science* 1980, *207*(4426), 82–84.

89. Nelson, C. M.; Chen, C. S., Cell-cell signaling by direct contact increases cell proliferation via a PI3K-dependent signal. *Febs Letters* 2002, *514*(2-3), 238–242.

90. Berridge, M. V.; Tan, A. S.; McCoy, K. D.; Wang, R., The biochemical and cellular basis of cell proliferation assays that use tetrazolium salts. *Biochemica* 1996, *4*, 14–19.

91. Mueller, H.; Kassack, M. U.; Wiese, M., Comparison of the usefulness of the MTT, ATP, and calcein assays to predict the potency of cytotoxic agents in various human cancer cell lines. *Journal of Biomolecular Screening* 2004, *9*(6), 506–515.

92. Gorodetsky, R.; Mou, X.; Pfeffer, M. R.; Peretz, T.; Levy-Agababa, F.; Vexler, A. M., Sub-additive effect of the combination of radiation and cisplatin in cultured murine and human cell lines. *Israel Journal of Medical Sciences* 1995, *31*(2-3), 95–100.

93. Felice, D. L.; Sun, J.; Liu, R. H., A modified methylene blue assay for accurate cell counting. *Journal of Functional Foods* 2009, *1*(1), 109–118.

94. Finlay, G. J.; Baguley, B. C.; Wilson, W. R., A semiautomated microculture method for investigating growth inhibitory effects of cyto-toxic compounds on exponentially growing carcinoma-cells. *Analytical Biochemistry* 1984, *139*(2), 272–277.

95. Oliver, M. H.; Harrison, N. K.; Bishop, J. E.; Cole, P. J.; Laurent, G. J., A rapid and convenient assay for counting cells cultured in microwell plates—Application for assessment of growth-factors. *Journal of Cell Science* 1989, *92*, 513–518.

96. Pelletier, B.; Dhainaut, F.; Pauly, A.; Zahnd, J. P., Evaluation of growth-rate in adhering cell-cultures using a simple colorimetric method. *Journal of Biochemical and Biophysical Methods* 1988, *16*(1), 63–73.

97. Dent, M. F.; Hubbold, L.; Radford, H.; Wilson, A. P., The methylene-blue colorimetric microassay for determining cell-line response to growth-factors. *Cytotechnology* 1995, *17*(1), 27–33.

98. Elliott, W. M.; Auersperg, N., Comparison of the neutral red and methylene-blue assays to study cell-growth in culture. *Biotechnic and Histochemistry* 1993, *68*(1), 29–35.

99. Bonora, A.; Mares, D., A simple colorimetric method for detecting cell viability in cultures of eukaryotic microorganisms. *Current Microbiology* 1982, *7*(4), 217–222.

100. Scragg, M. A.; Ferreira, L. R., Evaluation of different staining procedures for the quantification of fibroblasts cultured in 96-well plates. *Analytical Biochemistry* 1991, *198*(1), 80–85.

101. Simon-Deckers, A.; Brun, E.; Gouget, B.; Carriere, M.; Sicard-Roselli, C., Impact of gold nanoparticles combined to x-ray irradiation on bacteria. *Gold Bulletin* 2008, *41*(2), 187–194.
102. Chithrani, D. B.; Jelveh, S.; Jalali, F.; van Prooijen, M.; Allen, C.; Bristow, R. G.; Hill, R. P.; Jaffray, D. A., Gold nanoparticles as radiation sensitizers in cancer therapy. *Radiation Research* 2010, *173*(6), 719–728.
103. Chiu, S. J.; Lee, M. Y.; Chou, W. G.; Lin, L. Y., Germanium oxide enhances the radiosensitivity of cells. *Radiation Research* 2003, *159*(3), 391–400.
104. Hainfeld, J. F.; Slatkin, D. N.; Smilowitz, H. M., The use of gold nanoparticles to enhance radiotherapy in mice. *Physics in Medicine and Biology* 2004, *49*(18), N309–N315.
105. Hainfeld, J. F.; Dilmanian, F. A.; Zhong, Z.; Slatkin, D. N.; Kalef-Ezra, J. A.; Smilowitz, H. M., Gold nanoparticles enhance the radiation therapy of a murine squamous cell carcinoma. *Physics in Medicine and Biology* 2010, *55*(11), 3045–3059.
106. Kim, J. K.; Seo, S. J.; Kim, K. H.; Kim, T. J.; Chung, M. H.; Kim, K. R.; Yang, T. K., Therapeutic application of metallic nanoparticles combined with particle-induced x-ray emission effect. *Nanotechnology* 2010, *21*(42), 425102.
107. Chattopadhyay, N.; Cai, Z. L.; Pignol, J. P.; Keller, B.; Lechtman, E.; Bendayan, R.; Reilly, R. M., Design and characterization of HER-2-targeted gold nanoparticles for enhanced x-radiation treatment of locally advanced breast cancer. *Molecular Pharmaceutics* 2010, *7*(6), 2194–2206.
108. Praetorius, N. P.; Mandal, T. K., Engineered nanoparticles in cancer therapy. *Recent Patents on Drug Delivery; Formulation* 2007, *1*, 37–51.

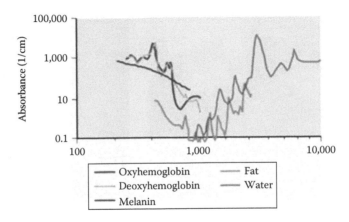

FIGURE 3.1 Absorption spectra of major skin chromophores. (From Hruza, G. and Avram, M., *Lasers and Lights Book*, 3rd edn.; Modified from Sakamoto, F.H. et al., Lasers and flashlamps in dermatology, in Wolff, K. et al., eds., *Fitzpatrick's Dermatology in General Medicine*, vol. II, The McGraw-Hill Companies, Inc., Columbus, OH, 2007, pp. 2263–2279.)

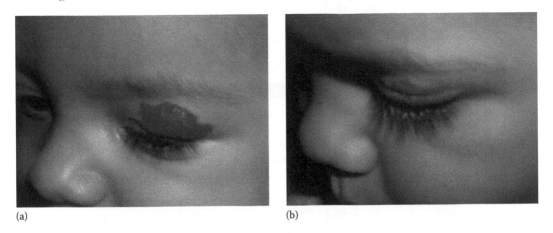

(a)　　　　　　　　　　　　　　　　　(b)

FIGURE 3.2 (a) A hemangioma of the left upper eyelid prior to treatment. (b) After a series of eight treatments with the 595 nm pulsed-dye laser at 10–15 J/cm^2 and 7 mm spot size, the hemangioma cleared significantly.

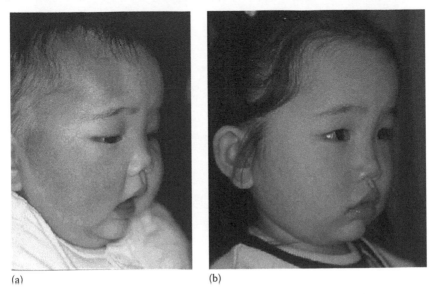

(a)　　　　　　　　　　　　　　　　　(b)

FIGURE 3.4 (a) A 4-month-old girl with nevus of Ota on her right face prior to treatment. (b) Almost complete clearing after five treatments with Q-switched alexandrite laser.

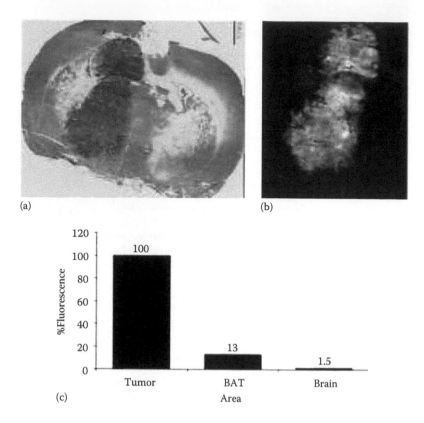

(a)

(b)

(c)

FIGURE 5.3 PpIX accumulation in an experimental brain tumor: (a) H&E histological section, (b) fluorescence image, and (c) average fluorescence intensity from tumor, BAT, and normal brain. Tumor fluorescence set at 100%.

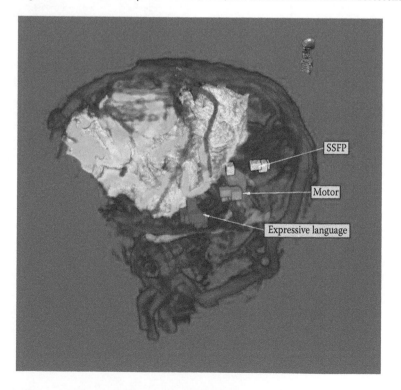

FIGURE 10.1 A 3D reconstruction combining image data of a brain tumor (MRI) and the surrounding vasculature (MRV) with functional data (MEG) to show the relative positions of sensitive areas.

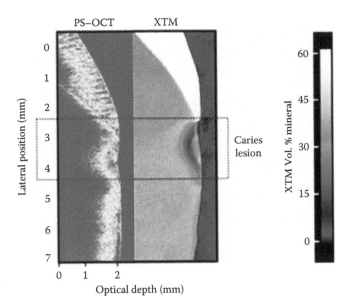

FIGURE 11.10 Comparison of the PS-OCT (⊥) image (orthogonal polarization state) with the corresponding XTM of the mineral density taken from the same region of the tooth (right) (see Ref. [80]). A small root caries lesion is present just below the cementum–enamel junction shown in the box. The intensity of the OCT images ranges from 12 to −45 dB; areas with regions of intensity greater than −5 dB are shown in red and those of intensity less than −35 dB are shown in blue. In the XTM image on the right, normal dentin is yellow, enamel is white, the water outside the tooth is indicated in red, and the demineralized area of the lesion is blue (color bar on right).

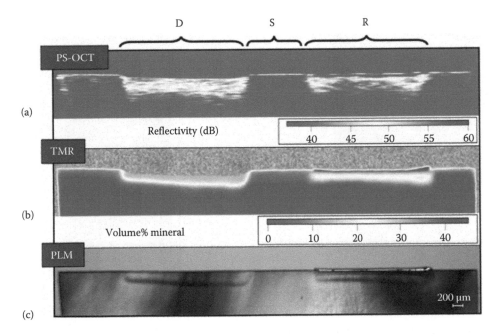

FIGURE 11.11 (a) (⊥-axis) PS-OCT scan of human dentin taken along long axis of a sample before sectioning. Intensity scale is shown in red–white–blue in units of decibels (dB). Remineralized layer is visible as a blue gap above the remineralized lesion. (b) The corresponding TMR of a 260 μm thick section in a similar orientation to the PS-OCT image. The intensity scale is also shown in red–white–blue with units in volume% mineral. (c) The PLM image of the same section is shown at 15× magnification. (From Manesh, S.K. et al., Assessment of dentin remineralization with polarization sensitive optical coherence tomography, in *Proc SPIE*, Lasers in Dentistry XV, SPIE, San Jose, CA, Vol. 7162W, pp. 1–7, 2009.)

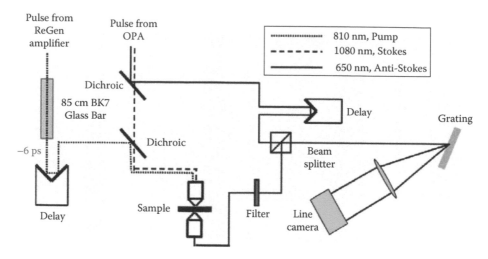

FIGURE 12.4 Experimental setup for the NIVI system.

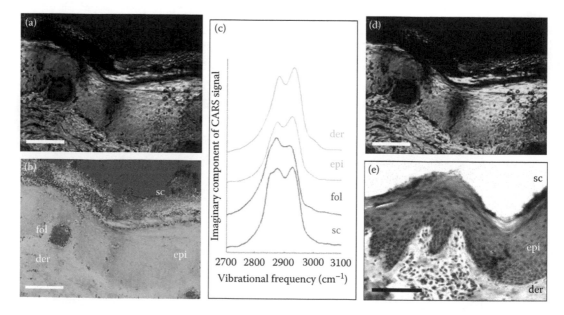

FIGURE 12.9 Molecular imaging of cutaneous tissue. (a) Intensity image (total integrated spectral power). (b) NIVI composite showing discrimination of stratum corneum (sc), epidermis (epi), dermis (der), and hair follicle (fol). (c) NIVI spectra for each domain in (b), as obtained by cluster analysis: each spectrum is the result of averaging the spectra of the members of most prevalent cluster in regions of 20×20 pixels^2 within each domain. (d) NIVI image showing both structural and molecular compositions. (e) H&E histology of a section from the same region. The scale bar is 100 μm in every image. (Adapted from Benalcazar, W.A. and Boppart, S.A., *Anal. Bioanal. Chem.*, 400, 2817, 2011. With permission.)

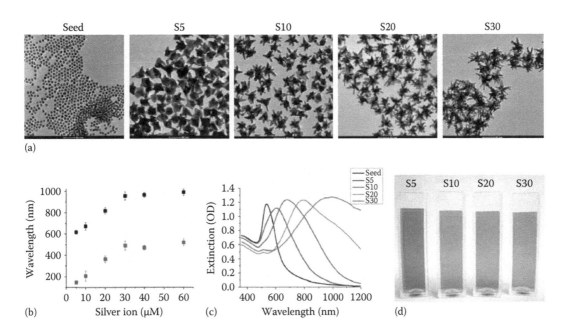

FIGURE 13.2 (a) TEM images of gold citrate seeds and nanostars formed under different Ag⁺ concentrations (S5: 5 μM, S10: 10 μM, S20: 20 μM, S30: 30 μM). (b) The spectral progression of plasmon peak (black square) and FWHM (red square) of nanostars formed under different Ag⁺ concentrations. (c) Extinction spectra of the seed and star solutions in DI. (d) Photograph of star solutions. (Adapted from Yuan, H., Khoury, C.G., Wilson, C.M., Grant, G.A., and Vo-Dinh, T., *Nanotechnology*, 23, 075102, 2012.)

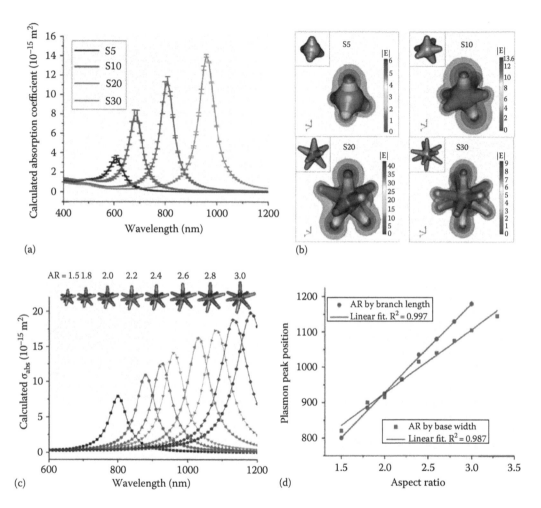

FIGURE 13.3 (a) The corresponding calculated absorption spectra of nanostars embedded in water. The data points (±1 SD) were interpolated with a spline fit. (b) Simulation of |E| in the vicinity of the nanostars in response to a z-polarized plane wave incident E-field of unit amplitude, propagating in the y-direction, and with a wavelength of 800 nm. The insets depict the 3D geometry of the stars. Diagrams are not to scale. (c) The scatter plots of polarization-averaged absorption against AR tuned by varying branch height while keeping the base width, core and tip diameters, and branch number, constant. (d) The linear relationship between the plasmon peak position and AR, which is tuned by varying branch height (red, $R^2 = 0.997$) or base width (blue, $R^2 = 0.987$) while keeping all other parameters constant. (Adapted from Yuan, H., Khoury, C.G., Wilson, C.M., Grant, G.A., and Vo-Dinh, T., *Nanotechnology*, 23, 075102, 2012.)

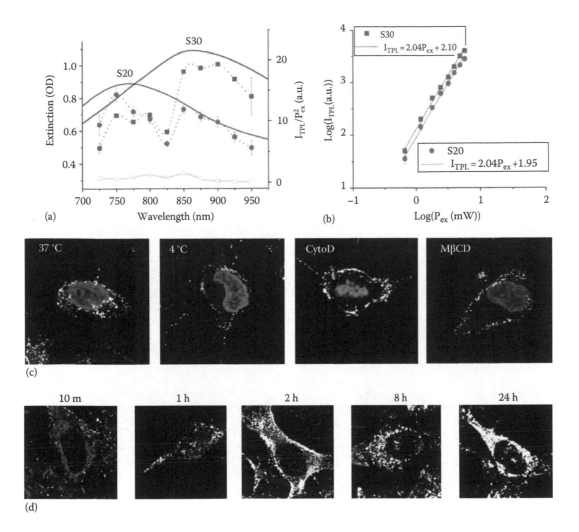

FIGURE 13.4 (a) Plasmon spectra (solid lines) and TPL excitation spectra (spline-fitted dashed lines ± 1 SD) of S20 (blue) and S30 (red) nanostars 0.1 nM in citrate buffer, and Rhodamine B (green) 100 nM in MeOH. (The spectral dip at 825 nm seen on both nanostars and Rhodamine B samples might be a system error from the microscope.) (b) The quadratic dependence of the TPL intensity (I_{TPL}) to the laser power (P_{ex}) from S20 and S30 solution (0.1 nM). Laser was set at 800 nm with power adjusted between 0.5 and 6 mW. Scatter plots (± 1 SD) were displayed with linearly fitted lines. (c) TPL images of TAT-nanostars-treated cells under normal condition (37°C) and different inhibitors. The cellular uptake of TAT nanostars was inhibited by 4°C, cytochalasin D (CytoD), methyl-β-cyclodextrin (MβCD), and TPL images sized 50 × 50 μm². Nuclei are stained blue. (d) Time series TPL images of cells treated with TAT nanostars (white) showing incremental accumulation. Cytoplasm is stained orange. Image size: 50 × 50 μm². (Adapted from Yuan, H., Khoury, C.G., Wilson, C.M., Grant, G.A., and Vo-Dinh, T., *Nanotechnology*, 23, 075102, 2012; Reprinted with permission from Yuan, H., Fales, A.M., and Vo-Dinh, T., *J. Am. Chem. Soc.*, 134, 11358–11361. Copyright 2012 American Chemical Society.)

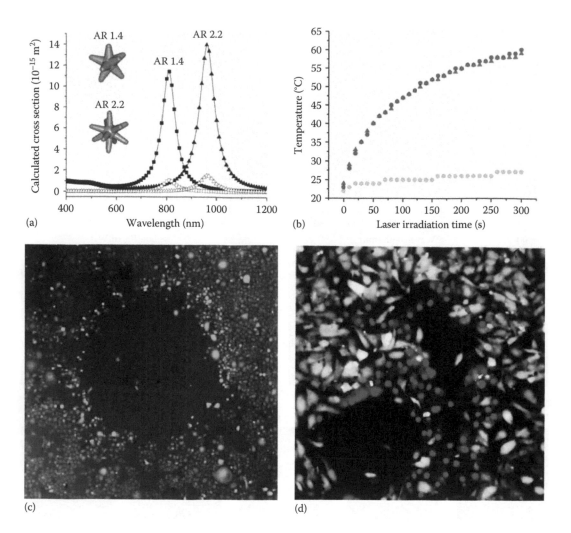

FIGURE 13.6 (a) The simulated spectra (splined) of absorption (solid symbol) and scattering (open symbol) cross sections of nanostars of two different ARs (branch length over branch base width) (inset). The 3D models of two nanostars simulated. The geometry data used for simulation are listed below. AR 1.4: core 24 nm, branch length/base 19/13.5 nm, branch number 8. AR 2.2: core 22 nm, branch length/base 22/10 nm, branch number 10. (b) Irradiation (785 nm, 490 mW) of PEGylated nanostars (blue triangle) and nanorods (red circle) and DI (grey circle) of the same plasmon intensity. Their temperature responses were nearly identical, although the concentration of nanorods was roughly 2.5-fold of the nanostars. (c) Fluorescence image of viable SKBR3 cells after the laser treatment. Viable cells appear green by converting nonfluorescing dye into green fluorescence. Empty area indicates successful photothermal ablation. Cells were incubated 1 h with 0.2 nM nanostars before the irradiation. Laser irradiation (980 nm, 15 W/cm^2) was applied 3 min. Image sizes are 3 × 3 mm^2. (d) Photothermolysis (850 nm, 0.5 mW, scanning area 500 × 500 μm^2, 3 min) on BT549 cells incubated 4 h with TAT nanostars. Live/dead cells are green/red. Image size: 612 × 612 μm^2. (Adapted from *Nanomedicine: NBM*, 8, Yuan, H., Khoury, C.G., Wilson, C.M., Grant, G.A., Bennett, A.J., and Vo-Dinh, T., 1355–1363, Copyright 2012, with permission from Elsevier; Reprinted with permission from Yuan, H., Fales, A.M., and Vo-Dinh, T., *J. Am. Chem. Soc.*, 134, 11358–11361. Copyright 2012 American Chemical Society.)

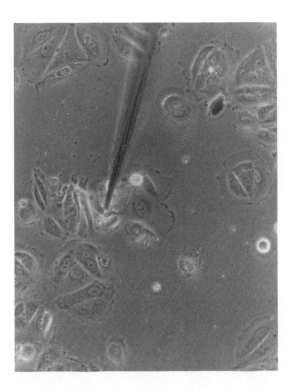

FIGURE 15.6 Photograph of an antibody-based nanoprobe used to measure the presence of benzopyrene tetrol in a single cell. The small size of the fiber-optic probe allows manipulation of the nanoprobe at specific locations within the cell. (Taken from Vo-Dinh, T. et al., *Nat. Biotechnol.*, 18, 76, 2000.)

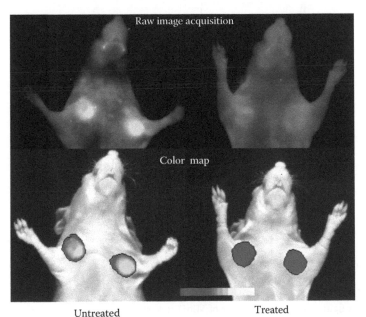

FIGURE 17.8 Fluorescence imaging of HT1080 tumor-bearing mice 48 h after injection of Cy5.5-labeled poly(ethylene glycol)–poly-(L-lysine) graft polymer (see Figure 17.6) into animals treated with the MMP-inhibitor prinomastat versus untreated animals. The top row shows raw fluorescence images, and the bottom row shows color-coded intensity profiles superimposed onto white-light images. The results show a significantly less fluorescence signal in treated animals relative to the untreated group. (From Bremer, C. et al., *Nat. Med.*, 7(6), 743, 2001. With permission.)

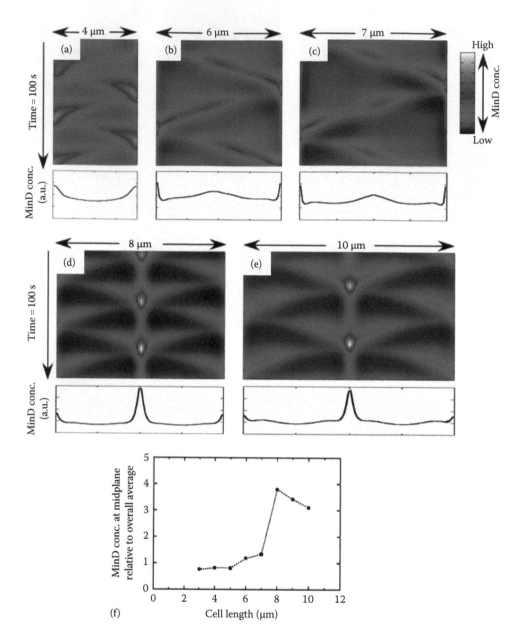

FIGURE 18.23 Simulated MinD dynamics within cells with various cell lengths. (a–e) MinD concentration at the membrane for cells with lengths of 4 μm (→a), 6 μm (→b), 7 μm (→c), 8 μm (→d), and 10 μm (→e). Upper figure: Time-course change in MinD concentration at the membrane. Time flows from top to bottom. Dynamics for 100 s is shown in each figure. Lower figure: Single-cycle average of MinD concentration at the membrane. Durations of one cycle were 38 s for 4 μm, 67 s for 6 μm, 79 s for 7 μm, 37 s for 8 μm, and 46 s for 10 μm. (f) Relative MinD concentration at midplane to overall average. Single-cycle average of MinD concentration at midplane was compared with the total average at each cell length and plotted against the cell length.

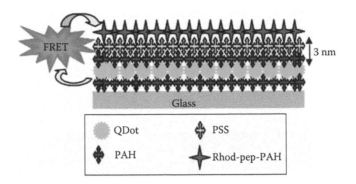

FIGURE 19.1 Schematic illustration of the QD–rhodamine FRET-based enzymatic sensor on glass or PDMS substrate. The QD FRET donors were embedded in PEMs, while neurotensin-labeled rhodamine molecular acceptors were conjugated to top PAH layer. The arrow indicates the distance between the donor and acceptor layer, which is approximately 3 nm.

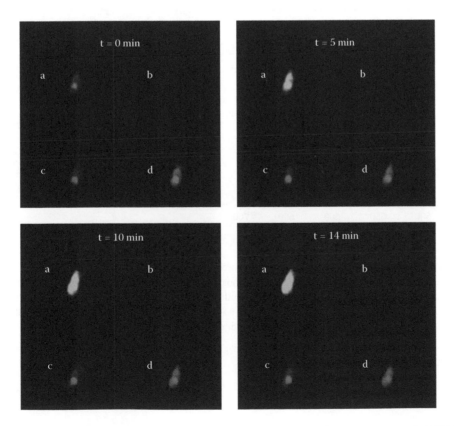

FIGURE 20.11 Real-time monitoring of multiple-gene expression using different MBs in a single MDA-MB-231 cell. (a) Fluorescence imaging of the β-actin MB (green), (b) fluorescence imaging of control MB (red), (c) fluorescence imaging of MnSOD MB (blue), and (d) fluorescence imaging of Ru(Bpy)$_3^{2+}$ reference probe (orange).

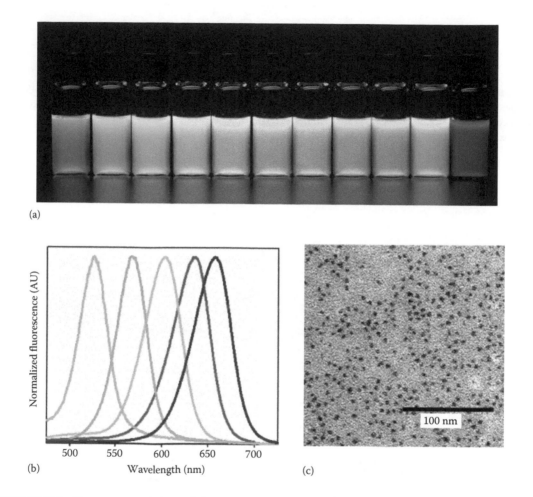

(a)

(b)

(c)

FIGURE 21.1 Fluorescence emission and electron microscopy structural properties of CdTe QDs prepared by using multidentate ligands in a one-pot procedure. (a) Color photograph of a series of monodispersed CdTe QDs, showing bright fluorescence from green to red (515–655 nm) upon illumination with a UV lamp. (b) Normalized fluorescence emission spectra of CdTe QDs with narrow emission spectra. (c) Transmission electron micrograph (TEM) of QDs showing uniform, nearly spherical particles. (Adapted from Moffitt, M. and Eisenberg, A., *Chem. Mater.*, 7(6), 1178, 1995.)

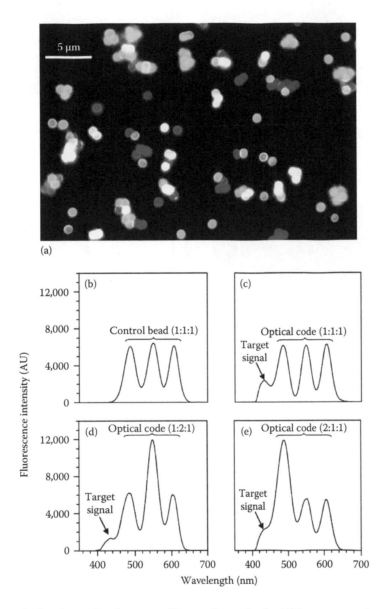

FIGURE 21.5 Multiplexed optical coding using QD-tagged microbeads. (a) Fluorescence micrograph of micro-beads loaded with QDs emitting at wavelengths covering the entire visible range. By loading a single microbead with precise ratios of QDs emitting at different wavelengths, an optical barcode can be generated for biomolecule identification. Coupling a unique nucleic acid probe sequence to each microbead, multiple target DNA sequences can be detected in a high-throughput process (c–e), while control sequences show no signal. (b). (Adapted from Han, M. et al., *Nat. Biotechnol.*, 19(7), 631, 2001.)

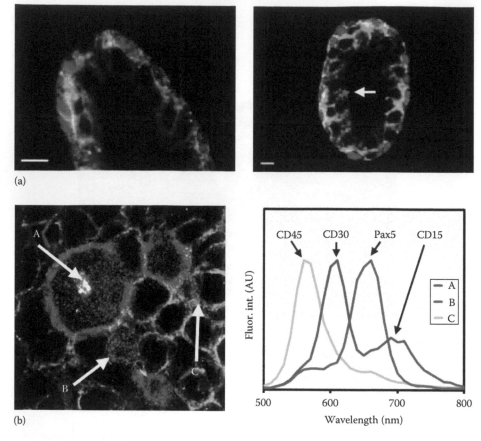

FIGURE 21.7 Immunoprofiling in complex tissues for the identification and characterization of rare cells. (a) Multiplexed QD imaging in prostate biopsy specimens allows the differentiation of a benign prostate gland (left) from a gland with a single malignant cell (right), as determined by positive alpha-methylacyl-CoA racemace staining (arrow). (b) A four-biomarker panel was used to identify low-abundance RS cells (A), B cells (B), and T cells (C) in a heterogeneous lymph node specimen for the diagnosis of Hodgkin's lymphoma (left). Using wavelength-resolved imaging, the QD staining pattern can be analyzed to determine the biomarker expression profile of single cells within the specimen (right), allowing accurate differentiation. (Adapted from Liu, J. et al., *Anal. Chem.*, 82(14), 6237, 2010; Liu, J. et al., *ACS Nano.*, 4(5), 2755, 2010.)

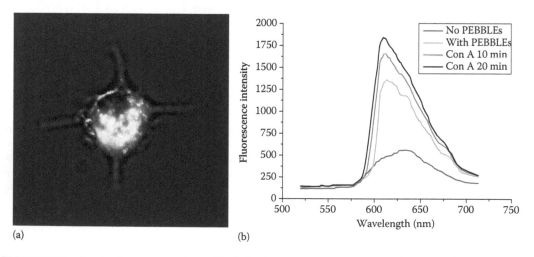

FIGURE 22.7 Confocal microscope image (a) of alveolar macrophage containing phagocytosed PAA PEBBLEs containing Calcium Crimson dye. Fluorescence spectra (b) show an increase in intracellular calcium after cells have been challenged by Con A.

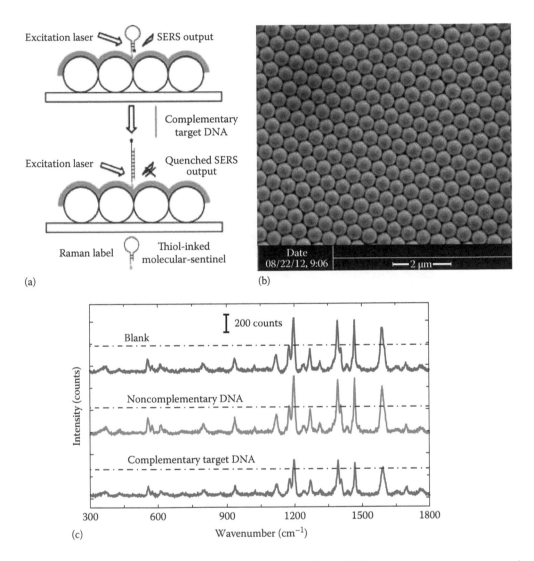

FIGURE 24.6 SERS-MS on a chip. (a) Application of the label-free MS detection scheme on nanowave chip and (b) scanning electron microscopy (SEM) image of nanowave chip substrate. In this substrate, a monolayer of 520 nm diameter polystyrene beads is covered by a 200 nm thick Au layer. (c) SERS spectra from blank sample, noncomplementary DNA sample, or complementary target DNA sample 2 h after delivery on MS-functionalized MFON substrates. The dashed lines mark the blank sample's SERS intensity. The dash–dot line marks the complementary target DNA sample's SERS intensity. (Adapted from Ngo, H. T. et al., *Anal. Chem.*, 85, 6378, 2013.)

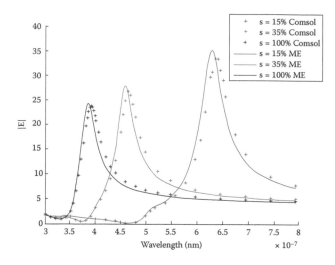

FIGURE 25.1 Electric field magnitude in the particle gap versus wavelength. Solid curves are calculations using the ME method and symbols indicate FEM calculations. (From Khoury, C.G. et al., *ACS Nano*, 3(9), 2776, 2009.)

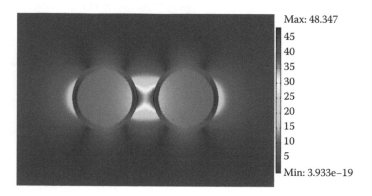

FIGURE 25.2 Electric field magnitude of a 2-D x–z slice through the dimer with 15% shell thickness at a wavelength of 640 nm. (From Vo-Dinh, T. et al., *J. Phys. Chem. C*, 114(16), 7480, 2010.)

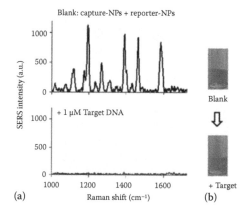

FIGURE 25.4 (a) SERS spectra of the Cy3 Raman peaks in the presence or absence of target DNA strands. Upper spectrum: in the presence of both capture-NPs and reporter-NPs (blank: without target DNA present). Lower spectrum: in the presence of 1 μM target DNA strands in the mixture of capture-NPs and reporter-NPs. (b) Photographs of sample solutions in the presence or absence of target DNA. (From Wang, H.N. and Vo-Dinh, T., *Small*, 7(21), 3067, 2011.)

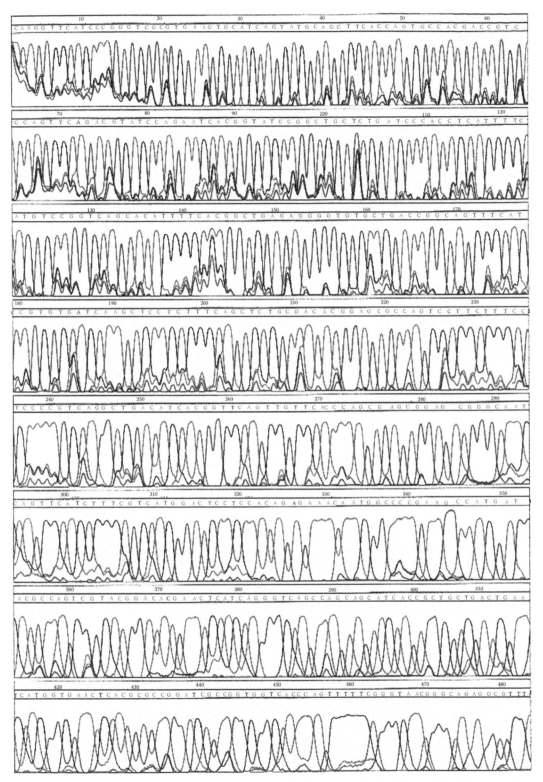

FIGURE 26.18 Single-color/four-lane sequencing trace (called bases 1–480) of a PCR-amplified λ-bacteriophage template. The sequencing was performed using an IRD-800-labeled primer (21mer) and SGE instrument (Li-COR 4000). The gel consisted of 8% polyacrylamide with 7 M urea. The dimensions of the gel were 25 cm (width) by 41 cm (length). The traces from the four lanes were overlaid to reconstruct the sequence of the template.

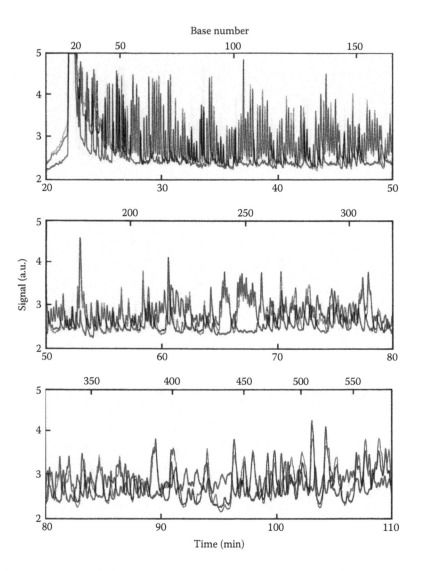

FIGURE 26.21 Four-color/single-lane sequencing of an M13mp18 template with a histidine tRNA insert. The numbers along the top of the electropherogram represent cumulative bases from the primer-annealing site. The electrophoresis was performed in a capillary column with a length of 41 cm and an i.d. of 50 μm. The sieving gel consisted of a cross-linked polyacrylamide (6% T/5% C) and was run at a field strength of 150 V/cm. The dyes used for the labeling of the sequencing primers were FAM, JOE, TAMRA, and ROX. Each color represents fluorescence from a different wavelength region (blue = 540 nm, green = 560 nm, yellow = 580 nm, red = 610 nm). (Adapted from Swerdlow, H. et al., *Anal. Chem.*, 63, 2835, 1991.)

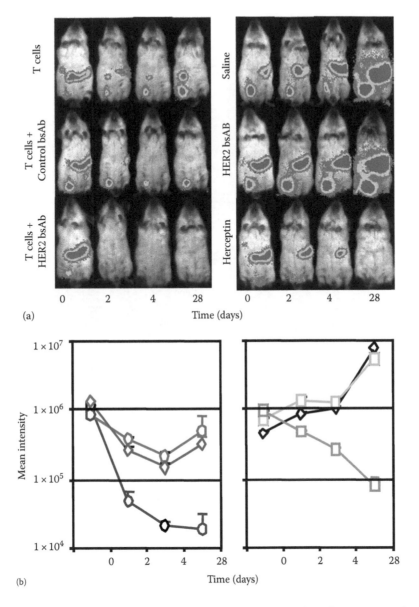

FIGURE 27.1 Advancing complex therapeutic strategies for malignancy through imaging. Evaluation of cell-based therapies complexed with molecular therapies may be complicated by having *too many moving parts*, and imaging approaches have been used to optimize these approaches by rapidly providing efficacy data without sacrificing the study subjects. In a study by Scheffold et al., the ability to redirect a tumoricidal NK T cell population to a tumor target using a bispecific antibody was revealed using BLI.[96] The control groups in this study consisted of the NK T cells alone, the NK T cells with an irrelevant bispecific antibody, normal saline, the bispecific antibody without additional cells, and Herceptin (as a positive control). Temporal analyses for representative animals in each treatment group are shown (a) and data from all animals are plotted (b). Color code: NK T cells only (light blue), NK T cells and control bispecific antibody (red), NK T cells and bispecific antibody (dark blue), saline (aqua), control bispecific antibody (black), and Herceptin (dark green). The ability to rapidly study multiple animals in each of the six treatment groups and provide accurate whole body data that are quantitative is an opportunity offered by BLI.

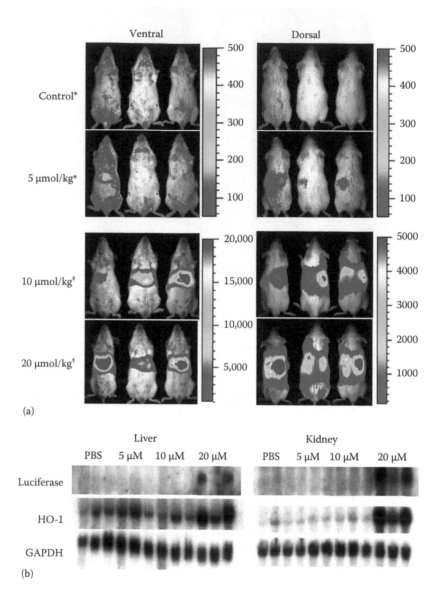

FIGURE 27.2 Screening heavy metal toxicity in a transgenic mouse model.[21] Using a Tg model where the transgene consisted of the heme oxygenase promoter fused to the firefly luciferase coding sequence, dose-dependent increases in luciferase transcription in the liver and kidney following systemic treatment with phosphate buffered saline (PBS) or three doses of CdCl$_2$ (5, 10, and 20 μmol/kg) were revealed by BLI (a). After imaging, the animals were sacrificed, the tissues removed, and total RNA was isolated from the liver and kidneys and analyzed by northern blot hybridization (b). Levels of mRNA for HO-1, luciferase, and glyceraldehyde 3-phosphate dehydrogenase (GAPDH) were determined for both tissues from three mice at each concentration. Luc signals increased with increasing concentrations of CdCl$_2$ as measured by imaging, and levels of Luc and HO-1 mRNA were elevated in the treatment group that received the highest concentration of CdCl$_2$. *Note*: *Scales set to 0–500; ‡Scales differ for dorsal (0–5000) and ventral images (0–20,000).

II

Advanced Biophotonics and Nanophotonics

15

Living Cell Analysis Using Optical Methods

Pierre M. Viallet
University of Perpignan

Tuan Vo-Dinh
Duke University

15.1 Introduction: The GFP Breakthrough

In the previous chapters, various biophotonic methods have been extensively described, and their applications have been discussed and reviewed. The scope of this chapter is to present a synthesis of the current and potential applications of the previously described techniques from the biologist's point of view. The chapter will focus on the types of techniques that have been and/or are supposed to be used for studying a specific biomolecular event occurring on a specific intracellular organelle. The advantages as well as limitations of each method, as they appear in light of the present knowledge, will be pointed out. An effort will be made to show how some techniques are capable of providing simultaneous but selective information on biochemical events occurring at different sites of the cell.

The recent breakthroughs in genomics and proteomics have drastically changed the way we think about exploring intracellular dynamics. We used to introduce inside the cell exogenous fluorescent molecules that were selected to monitor some specific biomolecular pathway. We were entirely dependent on their capability to enter the cell and had few means to direct them to a specific intracellular target. As a consequence, the information obtained could often be biased in different ways. First, the fluorescence signal was often a mixture of various signals resulting from different microenvironments due to the unspecific location of the probe. Second, to be informative, the exogenous fluorescent molecules had to compete with the natural biomolecules, so they might induce some imbalance in the intramolecular processes.

The techniques of molecular biology have allowed the generation of fluorescent chimeras that mimic biological molecules and that can be vectorized to or expressed at specific targets (organelles or microcompartments). One can say that cells themselves have been reengineered to give us the requested information, even if potential under- or overexpression may yet undermine this information

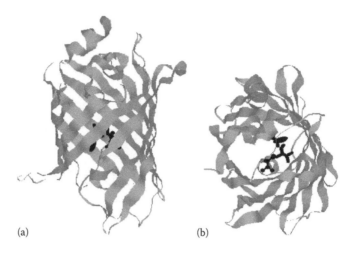

(a)   (b)

FIGURE 15.1 Side (a) and top (b) view of the 3D structure of EYFP, showing the 11-stranded β-barrel as ribbon. The three amino acid residues 64, 65, and 66, which form the chromophore after cyclization and oxygenation, are shown as balls and sticks.

Because these techniques have made it possible to get localized, real-time information on the intracellular dynamics, it became necessary to modify and/or develop optical tools to get dynamic, 3D microscopic maps of the living cell with nanoscale resolution of individual molecules.

The green fluorescent protein (GFP) is a spontaneously fluorescent polypeptide of 27 kDa (238 amino acid residues) from the jellyfish *Aequorea victoria* that absorbs UV blue light and emits in the green region of the spectrum.[1] Its importance in molecular biology was underlined by the Nobel Prize in Chemistry awarded in 2008 to Osamu Shimomura, Martin Chalfie, and Roger Y. Tsien for their discovery and development of the GFP. The GFP's structure and potential uses for studying living cells have been recently reviewed in detail.[2-7] The GFP chromophore results from a cyclization of three adjacent amino acids (S65, Y66, G67) and the subsequent 1,2-dehydrogenation of the tyrosine.[2-7] Although the chromophore by itself is able to absorb light, its fluorescence properties result from the presence of an 11-stranded β-barrel. GFP has been expressed both in bacteria and in eukaryotic cells and has been produced by in vitro translation of the GFP mRNA. Moreover, GFP retains its fluorescence when fused to heterologous proteins on the N- and C-terminals, and these bindings generally do not affect the functionality of the tagged protein.[2,3,6] These features lead to the use of GFP as an intracellular reporter. Unfortunately, the fluorescence intensity of the native GFP is not sufficiently bright for most intracellular applications. Therefore, variants have been engineered that have different excitation and/or emission spectra that better match available light sources.[2,3,5-7] Brighter mutants were also found necessary in case of low expression levels in specific cellular microenvironments.[3,5-7] Figure 15.1 shows the structure of a GFP-type variant molecule, the enhanced yellow fluorescent protein (EYFP). Moreover, the time lag between the protein synthesis and fluorescence development, as well as sensitivity for photobleaching, was a limitation when changes in location or conformation of the protein were searched for.[6]

The potential toxicity of GFP has been raised.[2] The efficiency of all these new mutants has recently been reviewed extensively, as well as the incidence of GFP mutations on mRNA transcription and translation rates.[2,6] The only serious limitation to the use of GFP variants seems to be their size that makes them unusable for tagging small biological molecules such as lipids or small proteins.

15.2 Exploring the Protein Factories

Previous studies have demonstrated that both tight control and coordination of gene expression require a high level of organization in the eukaryotic cell nucleus. Both fluorescence in situ hybridization (FISH) and immunocytochemical studies have shown that chromosomes occupy distinct territories within the nucleus and that many of the factors involved in transcription and RNA processing are located in nuclear bodies. Active genes in transcription are located at the borders of the condensed chromatin regions, which give them access to the transcription, RNA processing, and transport machinery.[8] Due to their noninvasive nature, which is important for life conservation, it is not surprising that photonic techniques were used to visualize how this special organization is conserved during cell replication, how it is correlated to the cellular functional activities, and how endogenic or exogenic events might disturb it.

15.2.1 Information Storage and Conservation: Chromatin Structure and Cellular Division

Advances in the specific fluorescent labeling of chromatin in living cells, in combination with 3D fluorescence microscopy and image analysis, have led to detailed studies of the higher-order architecture of chromatin in the human cell nucleus. Techniques such as time-lapse confocal microscopy (TLFM), fluorescence resonance energy transfer (FRET), and fluorescence redistribution after photobleaching (FRAP) have opened the way to space-time (4D) studies of the dynamics of this architecture and of its potential role in gene regulation.[9–13] Several features of this architecture are well established, and even if many points remain in the speculative domain, models of the functioning of this super architecture have been proposed.[13,14]

In parallel with the elucidation of the dynamics of chromatin architecture, many biophotonic studies have been devoted to investigating the nuclear transformation during mitosis and to the dynamics of postmitotic events. Multiwavelength fluorescence imaging (MWFI) has allowed the observation of the dynamic behavior of chromosomes and of some GFP fusion proteins linked to the nuclear envelope or the centromeres.[14,15] The different pathways of assembly during nuclear envelope formation of nuclear lamins A and B1 were also investigated by FRAP of lamins tagged with enhanced GFP (EGFP).[16] Time-lapse fluorescence imaging (TLFI) was used for a detailed analysis of mitosis in fission yeast, revealing the consequence of the presence of lagging chromosomes on the rate of anaphase B.[17] The expression of GFP-labeled centrin was used to study the respective contribution of mother and daughter centrioles to centrosome activity and behavior.[18] The dynamics of nucleolar reassembly were monitored in living cells expressing fusions of the processing-related proteins fibrillarin, nucleolin, or B23 with GFP.[19,20] Previously, the effects of mitotic inhibitors on cell cycle progression have been visualized by following the dynamics of chromosomes and microtubules stained with fluorescent chemicals.[21]

15.2.2 Information Transmission to Endoplasmic Reticulum Factories

Although many problems still remain to be elucidated, most modern pictures of the nucleus agree with the need for a high organization of the chromatin. It is accepted that chromosome territories contain regions in which the degree of chromatin condensation varies. To be transcriptionally active, genes must be located at the border of the condensed chromatin, referred to as an *interchromatin domain compartment*.[14] This location contains ribonucleoprotein complexes involved in transcription and pre-mRNA splicing that can be detected through multiwavelength fluorescent microscopy.[8,12,14] The use of multifluorescent labeling protocols makes it possible to track gene transcripts in living cells.[22] Figure 15.2 shows a schematic diagram of the main successive steps involved in the synthesis of proteins.

Several approaches based on fluorescence techniques have been developed to detect the different RNAs in living cells and to monitor their dynamic behavior. Discussing the methods used for introducing fluorescent RNAs inside the nucleus (microinjection of fluorescent RNAs into the nucleus, in vivo hybridization of fluorescent oligonucleotides to endogenous RNAs, or expression in cells of fluorescent RNA-binding proteins) is beyond the scope of this paper. From our point of view, most of these studies can be ranked in two sets: those involving only one fluorescent tag and those that require the use of two

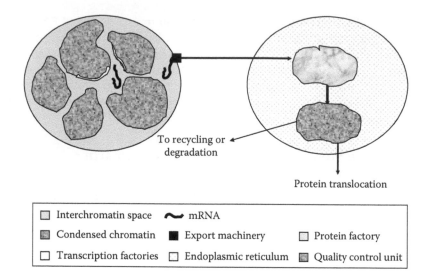

To recycling or degradation

Protein translocation

☐ Interchromatin space	∿ mRNA
▩ Condensed chromatin	■ Export machinery
☐ Transcription factories	☐ Endoplasmic reticulum

☐ Protein factory

▩ Quality control unit

FIGURE 15.2 Schematic diagram of the main successive steps involved in the synthesis of proteins.

fluorescent tags. In the first set, when short linear DNA nucleotides (20–40 bases) or short antisense fluorescent probes are involved, chemical fluorophores (e.g., fluorescein, rhodamine) are used in association with TLFl, FRAP, and/or fluorescence correlation spectroscopy (FCS).[8] mRNA was also visualized using GFP as a tag on the RNA-loop binding protein MS2. In the second set, hairpin-shaped oligonucleotide probes named *molecular beacons* are synthesized with fluorescent donor and acceptor chromophores at their 5′ and 3′ ends[8] (Figures 15.3a and b). In the absence of a complementary nucleic acid strand, the stem–loop conformation induces a FRET process that quenches the fluorescence emission of the donor fluorophore. On the other hand, hybridization with a complementary sequence opens the stem–loop conformation, increases the distance between the donor–acceptor pair, and decreases the efficacy of FRET, allowing the fluorescence emission of the donor to be detected with confocal laser microscopy. Although the use of such molecular beacons was supposed to increase the signal/noise ratio, it seems that it is not always the case, suggesting that other intracellular mechanisms might sometimes open the stem–loop conformation.[8] At any rate, detailed information reported in the literature provides us with a model of the movement of mRNAs from transcription site to nuclear pores and of the role played by some nuclear proteins in facilitating or slowing down that complex process.[23,24] The development and applications of molecular beacons are further described in Chapter 20 of this handbook.

15.2.3 Quality Control System

When inserted into endoplasmic reticulum (ER) membranes, newly synthesized proteins encounter the luminal environment of the ER, which contains chaperone proteins. These proteins facilitate the folding reactions necessary for protein oligomerization, maturation, and export from the ER to the Golgi apparatus. Nevertheless, studies that can actually visualize this complex process on living cells using fluorescence techniques are very few.[25] As a matter of fact, visualization of this process involves a very challenging task. First, involved chaperone(s) must be tagged to demonstrate their interaction with the target polypeptide. Second, the polypeptide itself must be labeled in a way that allows the properly folded protein to be distinguished from the polypeptide itself and from incompletely or improperly folded proteins. That task could be more demanding than having some GFP mutants at both ends of the target molecule. Nevertheless, interactions between different chaperones have been observed in solution.[26]

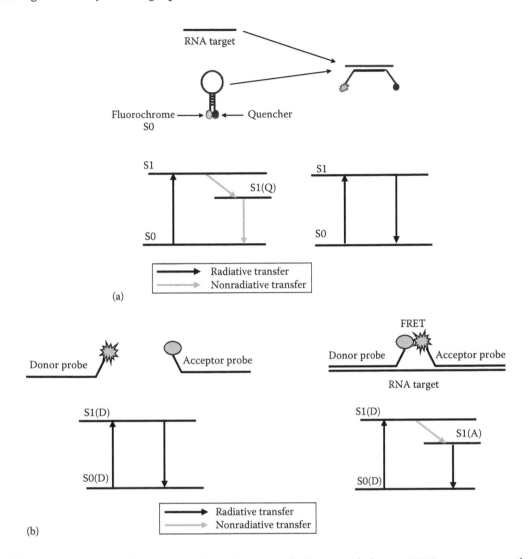

FIGURE 15.3 (a) General concept of molecular beacons: hybridization with the target RNA sequence opens the stem–loop conformation that results in the emission of fluorescence. (b) Use of double labeling for detecting an RNA sequence: the fluorescence of the acceptor probe is only seen when both the acceptor and donor are properly hybridized.

In fact, it is not certain that such a complex process works properly for each polypeptide entering the ER. On the contrary, there are some reasons to suspect that alteration of the yield of this process may induce some diseases.[27] The details of the mechanisms by which properly folded proteins are sorted from improperly folded ones and from the resident proteins also remain to be clarified and visualized in living cells. Nevertheless, it has been shown that improperly folded proteins accumulate in a novel pre-Golgi compartment, sometimes called ERGIC for ER-to-Golgi, where they aggregate and/or are targeted for ER-associated protein degradation.[28,29]

Studies have shown that membranes of some organelles have their own systems for transforming nuclear-encoded preproteins into proteins that can be internalized in these organelles. Although they have not yet been applied to living cells, some methods have been tested in solution, which would allow the study of the conformational changes of these preproteins during the process of internalization in mitochondria.[30]

15.3 Monitoring the Location and Activities of Biological Molecules

15.3.1 Large Proteins

15.3.1.1 Mapping the Cytoskeleton with GFP-Tagged Proteins

Before the use of GFPs, cytoskeleton research relied primarily on immunofluorescence microscopy techniques that require fixation and hence killing of the specimen before its analysis. The sole method for visualizing cytoskeletal dynamics was the microinjection of purified and fluorescently labeled protein, a protocol of limited use due to technical difficulties. In contrast, the addition of fluorescent tags to cytoskeleton proteins allowed visualization of the reorganization of its spatial structure and the localization of individual cellular components such as proteins and organelles. Furthermore, activities directly associated with the cytoskeleton, such as intracellular transport, cell motility, and morphological plasticity, have been investigated.[2] The studied proteins span a wide range, from proteins incorporated in or bound to microtubules (α- and β-tubulins, MAPs, tau, kinesin, etc.)[31] to those associated with microfilaments (actins, myosins, myomesin, titin, β1-integrin, etc.) and finally to those incorporated in intermediate filaments (vimentin, lamin A) or to septins.[32] The dynamics of microtubules (stabilization, turnover, bending and breaking, 3D organization) have also been studied using time-lapse fluorescence or confocal laser scanning microscopy.[33–39]

The functions of the motor proteins, kinesin and dynein, were also studied using GFP tagging.[40–42] Total internal reflection (TIR) has been successfully used to image the dynamics of individual GFP–kinesin.[43] Such a method could be used to characterize the dynamics of the rapidly increasing number of identified motor proteins, for tagging with GFP is stoichiometric and allows the use of recombinant fusion protein directly from bacterial extracts.

Successful monitoring of microfilaments using GFP–actin has been reported.[44–48] The actin cytoskeleton is an interesting target for it is involved in, and often drives, a variety of cellular processes, such as morphogenesis,[49] endocytosis,[50] cytokinesis, and links with the extracellular matrix.[51,52]

15.3.1.2 Mapping Protein Distribution in Membranes

GFP exhibits fluorescence when it is used to create chimeric GFPs that can be easily visualized inside living cells. The main advantage of these GFP chimeras over chemical fluorescent tagging is the ease with which they may be assembled and introduced inside cells.[3,7] The final steady-state levels of GFP are critical for further applications. As with any other protein, they depend on the kinetics of the transcription and degradation of the GFP mRNA and on the rate of degradation of the protein.[6,53] Furthermore, GFP chimeras can be used as in vivo markers for subcellular structures through the construction of systems that contain the appropriate targeting information.[3] Thus, GFP has been successfully used to visualize the localization of membrane proteins.[5,6,54,55] Successful imaging of protein localization in the ER or the Golgi complex has also been reported.[3,56,57]

Studying the colocalization of different kinds of proteins requires the use of GFPs that fluoresce at different wavelengths. Besides EGFP, enhanced blue fluorescent protein (EBFP), EYFP, and enhanced cyan fluorescent protein (ECFP) have been synthesized that fluoresce at 440, 527, and 475 nm, respectively. Several investigators have demonstrated the potency of using these spectral variants of GFP to acquire image sets of two or more different organelles or proteins in living cells.[3,5–7,58] In a more sophisticated approach, the lifetimes of these markers can also be used for colocalization studies through fluorescence lifetime imaging microscopy (FLIM).[59]

15.3.1.3 Monitoring the Intracellular Trafficking of Proteins

Compared with the previous one, this goal is more demanding, for it requires that the time necessary to record the fluorescent signal be short enough to allow a visualization of the protein translocation. Cellular expression or microinjection of brighter mutants of GFP allows the construction of

GFP chimeras usable for monitoring the translocation of proteins. However, the stability of the probes against photobleaching becomes a crucial factor for the successive illuminations of the same microscopic field that are necessary to detect any change in localization.

Expression, translocation, and turnover of proteins involve successive transient protein–membrane interactions that may be associated with changes in the membrane morphology. Optical methods used for monitoring the protein traffic have been found useful for monitoring morphological changes involved in the protein trafficking or resulting from the presence of viruses.[60]

Direct observation of nucleocytoplasmic transport (combined with nuclear localization or nuclear export signal) of a GFP-tagged glutathione S-transferase has been reported,[7,61] while visualization of the transport from reticulum to Golgi has been performed with a GFP-tagged temperature-sensitive mutant protein.[3] Other studies have focused on expression and organellar targeting to test the ability to produce stable *green* transgenic parasites.[62]

Secretion involves shuttling proteins between various intracellular compartments. The transport of secretory vesicles from the trans-Golgi network to the plasma membrane has been monitored using a GFP chimera of the protein chromogranin B and time-lapse imaging. The same imaging technique has been used to demonstrate the direct rapid translocation of MHC class II molecules from the lysosomal compartment to the plasma membrane in cells of the immune system.[3]

Several proteins involved in signal transduction have been tagged with GFP variants to visualize their transport between cytosolic and membrane-bound pools in response to stimulation. The protein kinase B (PKB) (a mitogen-regulated kinase that protects cells from apoptosis) binds to phosphatidylinositol 3,4,5 triphosphate (PIP3), generated and localized in the plasma membrane, via its pleckstrin homology (PH) domain and is subsequently activated. Its translocation upon growth factor stimulation was monitored in transiently transfecting fibroblasts, using a GFP–PKB chimera. Other signaling proteins, such as Bruton's tyrosine kinase, the Ras GTPase-activating protein GAP1(m), and the ARF guanine nucleotide exchange factor ARNO (ARF nucleotide-binding site opener), have also been GFP tagged and proved to translocate to the plasma membrane via their PH domains upon cell stimulation.[3]

GFP variants with different spectral properties have been used to simultaneously monitor the intracellular movements of several distinct proteins in order to follow the dynamics of nuclear envelope disassembly and assembly.[3] Fluorescent indicators have been developed to study in detail the cleavage of peptides in living cells, as they are transported from the ER to the Golgi. When a peptide is labeled with these dyes at specific sites, a FRET is observed that disappears upon proteolytical process inside the cell.[63] Such a study leads the way to a new, elegant, and sophisticated approach to determining the precise mechanisms involved in immune surveillance.

GFP-tagged proteins have also been used to analyze the distribution of chemotactic receptors.[3] Other applications included studies on inhibition of protein shuttling,[64] regulation of membrane diffusion of aquaporin water channels,[65] internal trafficking and surface mobility of a functionally intact 2-adrenergic receptor–GFP conjugate,[5,7] and transient association of proteins with cell contact areas.[66] A quantitative study of fluorescent adenovirus entry has also been reported as a model system for cargo transport from the cell surface to the nucleus.[67]

15.3.1.4 Monitoring Proteins at Work in Living Cells

Monitoring *proteins at work* is more challenging than protein mapping or trafficking, because it requires the use of techniques that allow dormant proteins to be distinguished from activated proteins. Because the physical characteristics of the fluorescence of a molecule are strongly dependent on its microchemical environment, fluorescence techniques can be used to probe the changes of conformation occurring when a tagged protein is activated. Of course, the labeling of the protein with a fluorescent molecule must obey rules that are somewhat contradictory. The binding must occur at a definite location on the protein. This location must be far enough from the active site of the protein that the binding does not change the activity of the protein, but also close enough to the active site that it can be sensitive to the 3D conformational change occurring when the protein is activated.

Fluorescent molecules in which changes in chemical microenvironment induce spectral shifts of emission and/or excitation fluorescence spectra are numerous. Others experience changes in fluorescence intensity. Getting unambiguous information on potential changes in conformation for a given protein in living cells generally demands the use of two fluorophores: one of them must report only the total concentration of the protein, while the other must be sensitive to conformational changes. This could be troublesome when the tagged proteins must be microinjected inside the cells. Another approach for monitoring conformational changes is to take advantage of the change in the polarization of the fluorescence emission. Although anisotropy of fluorescence is practically independent of the concentration of the fluorescent dye, this technique has not been widely used in living cells until now, probably due to its relatively low sensitivity.[7]

The lifetime of the first excited state of a molecule is also very sensitive to changes in the chemical microenvironment of the dye, while it is insensitive to the dye concentration. Previously, it was demonstrated that changes in fluorescence lifetime could, in principle, be used for imaging spectroscopy (FLIM). Until now, the use of this technique is not yet widespread and only restricted to a few specialized laboratories, due to technical difficulties.[68-72] Nevertheless, it has recently been demonstrated that FLIM at a single modulation frequency can be used to image and follow the cellular distribution of two and ultimately three GFPs simultaneously. FLIM will become more commonly used in the near future due to several advantages compared with spectral discrimination. Simultaneous readout of data and the need for only one dichroic and long-pass emission filter result in the use of a lower light dose and reduce the risk of photochemical damage to the cells.[59]

GFP chimeras were also found to be invaluable for time-lapse imaging to study the dynamics of proteins in cells of the immune system.[3] The same technique was used to visualize transport of secretory vesicles from the trans-Golgi network to the plasma membrane.[3,6] The dynamics of the distribution of chemotactic receptors on the surface of neutrophils were also monitored. More recently, the availability of GFP variants with different spectral characteristics has made it possible to compare the dynamics of two distinct proteins simultaneously.[3]

GFP chimeras have also demonstrated their potentialities in the field of functional studies. FRAP has been used to show the internalization of a β-adrenergic receptor chimera after stimulation. Its plasma membrane diffusion coefficient was also measured.[7] An arrestin chimera was also used to visualize the arrestin-binding kinetics to different G protein-linked receptor subtypes.[7] GFP chimeras of galactosyltransferase, mannosidase II, and the KDEL receptor, a seven-transmembrane-domain protein, were found to have extremely high mobility in the Golgi of live HeLa cells; this implies that protein immobilization is not responsible for the retention of these proteins in the Golgi.[6] FRAP was also used to monitor the mobility of a GFP fusion of E-cadherin at different stages of epithelial cell–cell adhesion.[3]

The intensity of signals obtained with the previously mentioned FRET technique is also independent of the dye concentration, but this technique requires the use of two fluorophores. FRET occurs only when two convenient fluorophores are close enough together, and it falls off with the sixth power of the distance between them. Consequently, this effect is well adapted to monitor 3D conformational changes experienced by a double-labeled macromolecule, but it can also be used for monitoring conformational changes occurring inside a cluster of proteins. This technique has been used recently for monitoring an interaction between a nuclear receptor and its specific activator that is required for transcription of transiently transfected and chromosomally integrated reporter genes.[73] High-resolution FRET microscopy was used for studying the potential role of *lipid rafts* in the interaction between cholera toxin B-subunits and GPI-anchored proteins in the plasma membrane of HeLa cells.[74] Receptor-mediated activation of heterotrimeric GTP-binding proteins was also visualized in living cells by monitoring FRET between α- and β-subunits fused to cyan and yellow fluorescent proteins.[75]

15.3.1.5 Looking at the Single Organelle/Protein Level for an Integrated Approach of Cellular Mechanisms

As seen in the previous sections, FRET occurs only when the distance between the donor and the acceptor molecule is between 2 and 10 nm. Using FRET provides us with a way to light up interactions at the molecular level, far below the usual limitation of traditional optics. The interest in getting information at the single-molecule level is obvious. It would improve our basic knowledge of biological events, such as signal transduction, gene transcription, and intracellular transport. Furthermore, it would provide invaluable information on the potential synchronization of the response to any stimulus of organelles belonging to the same network (e.g., mitochondria, ionic channels). Figure 15.4 shows a schematic presentation of two potential activation pathways of the mitochondria network.

Fluorescence microscopy has demonstrated its capability in sensing single molecules in in vitro experiments. In particular, it has been shown that single dye molecules and single fluorescent proteins emit fluorescent photons in an intermittent on-and-off fashion, a kind of information that cannot be obtained from population-averaged measurements. Although single fluorophores have recently been detected in the membranes of living cells with good signal/noise ratios, it is still not clear whether single-molecule fluorescence microcopy can be achieved at the spatial and temporal resolution required for biological experiments. Even if stable and bright fluorophores exist that can be coupled in vivo to biological molecules, some crucial challenges remain. One of them is the development of easy-to-use, affordable equipment that allows high-resolution localization of individual point-like sources in 3D at a rate compatible with that of biological events. Another challenge is quantitative recording of relatively weak signals that must be extracted from the fluorescence background resulting from normal fluorescent cellular components, such as flavins or nicotinamide adenine dinucleotide (NAD). Nevertheless, single GFP molecules have been successfully imaged in living cells, using a TIR fluorescence microscope.[76]

A new technique called fluorescence speckle spectroscopy (FSS) has recently been proposed for visualizing the movement, assembly, and turnover of macromolecular assemblies in living cells. A sensitivity of 1–7 fluorophores per resolvable unit (270 nm) has been reported, using a conventional wide-field epifluorescence light microscope and digital imaging with a low-noise cooled charge-coupled device (CCD) camera.[77]

Both confocal microscopy and multiphoton microscopy allow the energy required for fluorescence excitation to be concentrated in a tiny volume (femtoliter). Both share the problems linked with the use of far-field optics (spherical aberrations, chromatic aberrations both in the excitation and in the detection arms) that are difficult to compensate for when deconvolution methods are used for image restoration.

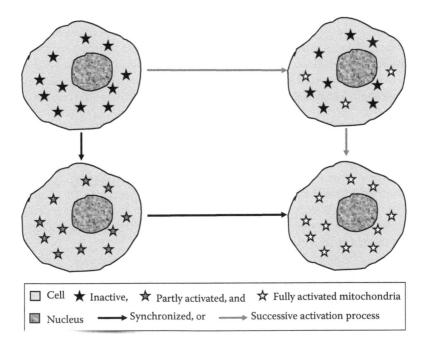

FIGURE 15.4 Schematic presentation of two potential activation pathways of the mitochondria network.

Nevertheless, fluorescently labeled lipid and receptor molecules have been monitored on the surface of living cells by single-molecule imaging techniques.[78,79] Recently, the distribution of single molecules of transferrin-tetramethylrhodamine (TMR) in the nucleus or the cytoplasm was visualized in living HeLa cells by tightly focusing the beam of a continuous-wave argon ion laser.[80] Another study showed that ultrahigh-resolution colocalization of fluorophores is possible in the FRET distance range if the different fluorophores can be excited at the same wavelength, are bright enough, and have emission properties that allow an unambiguous distinction between the different types.[81] However, a closed-loop piezoscanner must be used that ensures a perfect alignment of each fluorophore on the optic axis to minimize the chromatic aberrations in the excitation path. Nothing is said about the rate of image recording.

The main advantage of multiphoton microscopy is that it allows the use of near-infrared or infrared radiation instead of UV to excite fluorophores.[82,83] Single-molecule detection with two-photon FCS has been demonstrated, although photobleaching-resistant fluorophores are required.[82,84]

While limited to membranes, near-field scanning optical microscopy (NSOM) has been found useful to study interactions between host and malarial skeletal proteins in erythrocytes.[85,86] A protocol to extract the near-field fluorescence signal from the composite signal containing both near- and far-field fluorescence has been reported that improves the image resolution.[87] An elegant high-resolution two-photon near-field scanning spectroscopy has also been described.[82]

Besides these reports on improved methods for detecting single molecules, other studies have been focused on the dynamics of whole organelle or ionic channel networks, using wide-field spectroscopy and different kinds of image processing.[88–91] Although they do not claim to have reached the single-molecule detection level, these reports contain valuable information on the dynamics of the mitochondrial or ionic channel network, strengthening the interest in single-molecule imaging in living cells.

15.3.2 Lipids and Small Proteins

As indicated earlier, the direct labeling of proteins with GFP is restricted to relatively large proteins due to the size of GFP. Nevertheless, a lot of elegant methods have been developed to trace membrane lipids, due to their importance in the cellular machinery (endocytosis, exocytosis, vesicular trafficking of proteins, transduction of extracellular signals, remodeling of the cytoskeleton, regulation of calcium flux, and apoptosis). In one approach, the membrane receptors of some lipids have been GFP tagged to visualize their intracellular localization.[92,93] However, these results are difficult to interpret in terms of lipid concentration when the effects of external stimuli are searched for. Therefore, fluorescent analogs of these lipids have been synthesized and used to track their intracellular distribution or redistribution.[94–99] The same approach has been used for smaller molecules such as cholesterol or cyclic AMP and GMP.[53,100,101]

Theoretically, the same method can be used for small proteins. As stated previously, the problem consists in finding at least one amino acid residue and potentially two if one wants to use FRET techniques that match the following criteria. They must be reactive enough to allow a specific binding of the fluorescent tag and must be conveniently located in the protein structure so that tagging does not change the properties of the protein (conformation, binding sites or potential receptors, etc.). Furthermore, tag(s) must experience some changes in their chemical microenvironment upon protein binding (to substrates or receptors) if one intends to study the intracellular protein activity. Only calmodulin intracellular activity has been studied so far with such methods.[7]

15.3.3 GFP Mutants as Ionic Sensors

Up to now, the search for ion-specific, genetically engineered fluorescent probes has been mainly limited to Ca^{2+} and H^+ due to their importance in the cell machinery, although a Cl^- probe has recently been reported.[102] Several laboratories reported in 1998 that some GFP mutants experience pH-sensitive spectral changes.[3,6] Fluorescence intensity was found to decrease with pH, while the shape and position of the spectrum were relatively insensitive to pH. Another type of pH-sensitive GFP mutant was developed that experiences a spectral shift upon pH change.[3,6] Both these types of pH sensors have their own advantages and limitations. The latter permits the measurement of pH from the ratio of intensity at two wavelengths, a method that is relatively independent of photobleaching and alteration in focal plane. The former allows easier use of confocal microscopy because only one excitation wavelength is required. Recent applications dealt with Golgi and secretory vesicle pH.[101,103,104]

GFPs have also been used for Ca^{2+} measurements. Fluorescent intracellular calcium indicators called *chameleons* have been synthesized that are fusion proteins containing blue fluorescent protein or cyan fluorescent protein; calmodulin, the calmodulin-binding domain of skeletal muscle myosin light-chain kinase (M13); and either GFP or yellow fluorescent protein. In the absence of calcium, this long chain is extended; in the presence of calcium, calmodulin experiences a conformational change. This change favors its binding to M13, which results in a more compact conformation that increases the FRET between the flanking fluorescent proteins.[2,3,5,7,105] Figure 15.5 schematically illustrates the changes in the 3D structure of calmodulin upon binding to calcium ions and myosin chain M13. Mutations inside calmodulin allowed the generation of chameleons with different calcium affinities from 10 nM to 10 mM. Furthermore, the inclusion of localization signals in the structure makes it possible to target these chameleons to specific intracellular locations for monitoring intraorganellar calcium dynamics. Measurements of rapid Ca^{2+} turnover in the ER during signaling have been reported,[106] and the use of

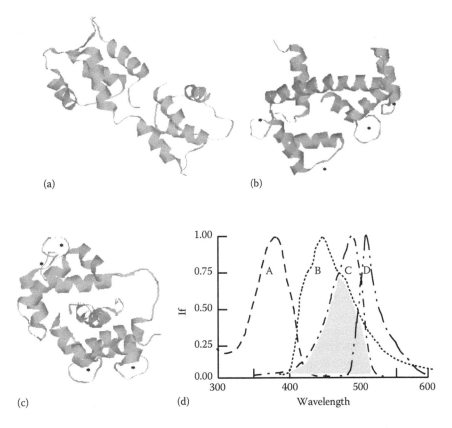

FIGURE 15.5 Changes in 3D structure of calmodulin upon binding to calcium ions and myosin chain M13. (a) Free calmodulin. (b) In the presence of calcium ions. (c) In the presence of calcium ions and myosin chain M13 (limited to the peptide involved in the interaction for sake of clarity). (d) Respective normalized excitation and emission spectrum of GFP mutants Y66H and S65T: (A) excitation spectrum of Y66H, (B) emission spectrum of Y66H, (C) excitation spectrum of S65T, and (D) emission spectrum of S65T. The gray zone shows the wavelength range of overlapping between the emission spectrum of Y66H and the excitation spectrum of S65T.

yellow fluorescent protein less sensitive to pH and chloride interference has allowed Ca^{2+} measurement in the Golgi.[107] Furthermore, new fluorescent chimeras, dubbed *pericams*, have recently been used to monitor free Ca^{2+} dynamics in the cytoplasm, the nucleus, and the mitochondria of HeLa cells.[108]

In another approach to calcium monitoring, a blue fluorescent protein was linked to GFP by a 26-residue spacer containing the calmodulin-binding domain from smooth muscle myosin chain kinase. Due to the flexibility of that chimera, FRET occurs between the fluorescent proteins that are close in the absence of calcium. This proximity does not exist in the presence of calcium and FRET is no longer possible.[4]

Since all these probes rely on the interaction of calcium with calmodulin, a ubiquitous protein, they are expected to compete for Ca^{2+} with *normal* proteins. It could be possible that this competition does not induce any perturbation in the cellular mechanisms, especially in nanocompartments where the probability of the presence of free Ca^{2+} is low.

15.4 Conclusion: Toward Nanosurgery and Nanomedicine

Future biological investigations are expected to focus on monitoring the interactions between the fundamental cellular building blocks that are currently under study. Once the 3D high-resolution structures and biological roles of proteins and RNAs have been elucidated, it will become necessary to determine their precise locations and translocations inside the cell to understand the cell machinery and circuitry. Many vital functions of the cell are performed by modular cellular machines, self-assembled from a large number of interacting molecules and translocated from one cell compartment to another. Unraveling the organization and dynamics of these machines requires tools able to provide in vivo 3D microscopic pictures of individual interacting molecules with nanometer resolution and at a time scale compatible with the speed of intracellular processes. The same kinds of tools are also necessary to monitor the cooperative behavior of networks such as the mitochondrial or ionic channels networks.

Significant advances in fluorescence techniques have recently improved their spatial resolution beyond the classical diffraction limit of light. Among them, the most promising appears to be wide-field image restoration by computational methods,[109] wide-field single-molecule localization and tracking,[110–112] aperture-[113] and apertureless-type near-field scanning spectroscopy, two-photon excitation microscopies,[114,115] and point-spread-function engineering by stimulated emission depletion.[116] Nevertheless, near-field scanning spectroscopy and TIR fluorescence spectroscopy appear more appropriate for studying the cell surface than for monitoring intact cellular structures.[117] A relatively new imaging and ultrahigh-resolution colocalization technique has been proposed that can pinpoint the locations of multiple distinguishable probes with nanometer accuracy and is said to better handle the limitations presented by other methods.[81] But, up to now, this technique has not been used in vivo and requires probes relatively insensitive to fading.

Among the aforementioned techniques, the multiphoton technology seems to be the most appropriate for delivering a given dose of radiation into a predetermined nanovolume. It is not surprising that it has been used both for noninvasive optical biopsy and for nanosurgery.[82] Due to the use of irradiation wavelengths in the 700–1100 nm spectral range, living tissues can be investigated at depths of up to 100 μm. The endogenous fluorophores that can be potentially excited are NADH, NADPH, flavins, and collagen, allowing the imaging of intracellular structures, the location of the nucleus, and cell morphology in the different skin layers. Imaging of the accumulation of protoporphyrin in tumor cells of mice with solid carcinoma has also been reported, suggesting potential applications of two-photon technology to photodynamic therapy of tumors.

Femtosecond near-infrared laser pulses have numerous advantages compared with conventional nanosecond UV laser pulses: high penetration depth, no out-of-focus absorption, efficient induction of multiphoton excitation, and absence of plasma shielding and of significant heat transfer. Convenient tuning of the laser power allows one to confine the energy necessary for material removal to the central part of the beam. Therefore, this technique may provide a noncontact nanoscalpel for surgery inside a cellular organelle that will not have unwanted effects on other cellular compartments.[82] Optical tweezers that have been used for studying the strength of the interaction between fibronectin and its membrane receptor or the microrheology of biopolymer–membrane complexes have also been tested to facilitate in vitro fertilization.[118–120]

Nanotechnology is opening new possibilities for the development of fiber-optic-based nanosensors and nanoprobes of submicron-sized dimensions, suitable for intracellular measurements. The ability to examine processes within living cells could provide great advances in our understanding of cellular function, thereby revolutionizing cell biology. Fiber-optic sensors offer important advantages for in situ monitoring applications due to the optical nature of the detection signal. These sensors are not subject to electromagnetic interferences from static electricity, strong magnetic fields, or surface potentials. Another advantage of fiber-optic sensors is the small size of optical fibers, which allows for sensing intracellular/intercellular physiological and biological parameters in microenvironments. Figure 15.6 shows a fiber-optic nanosensor with antibody probes for benzopyrene tetrol, a biomarker of human exposure to the carcinogen benzo[a]pyrene that has been recently developed and used for in vivo analysis of single cells.[121–127] This nanobiotechnology-based device can provide unprecedented insights into intact cell function, allowing studies of intracellular pH[128] and molecular functions, such as apoptosis,[129] in the context of the functional cell architecture in a systems biology approach.[130] Nanosensors could provide much needed tools to investigate important biological processes at the cellular level. For instance, the mode of entry of fluorescent proteins or exogenous chemicals into various cells is of importance, as are subsequent enzymatic processes inside cells. Tracking these species becomes feasible due to their natural fluorescence. Inspection of cells at submicron resolution by optical nanosensors that can interrogate

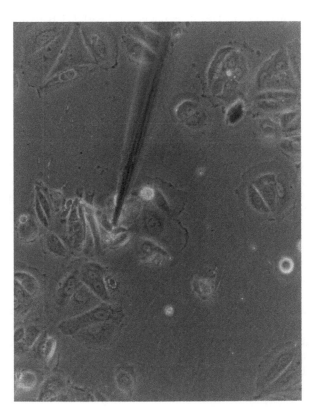

FIGURE 15.6 **(See color insert.)** Photograph of an antibody-based nanoprobe used to measure the presence of benzopyrene tetrol in a single cell. The small size of the fiber-optic probe allows manipulation of the nanoprobe at specific locations within the cell. (Taken from Vo-Dinh, T. et al., *Nat. Biotechnol.*, 18, 76, 2000.)

at precise locations has the potential for significant advances in knowledge of cellular structure/function. The development and application of fiber-optic nanosensors are described in detail in Chapter 23 of this handbook.

The methods and techniques reviewed in this chapter have demonstrated that biophotonics may allow imaging of most of the different steps resulting in the synthesis of proteins and monitoring of their activity. As a consequence, one can dream that, in the near future, it could be possible to identify and to image which specific step in this complex process is responsible for any given disease. For instance, such a nanodiagnosis could show that a given disease results from an increase in the misfolding of a protein, instead of a decrease in the production of this protein. Will this mean that a nanomedicine, involving the delivery of drugs especially designed to repair a dysfunction in a step of any biological pathway, can be expected in the near future and that it will be possible to visualize the benefits through biophotonic technology? The challenge for photonics to elucidate life processes will come from the complexity inherent to biological events, as is demonstrated in a recent study.[126] The researchers reported that they have identified 997 messenger RNAs responding to 20 systematic perturbations of the yeast galactose utilization pathway, providing evidence that approximately 15–289 detected proteins are regulated posttranscriptionally.

Acknowledgments

This research was supported by the National Institutes of Health (R01 ES014774-01), the French Department of Education and Sciences, the ORNL Laboratory Directed Research and Development Program (Advanced Nanosystems), and the Office of Biological and Environmental Research, US Department of Energy, under contract DE-AC05-00OR22725 with UT-Battelle, LLC.

References

1. Prasher, D. C. et al., Primary structure of the *Aequorea victoria* green-fluorescent protein, *Gene*, 111, 229, 1992.
2. Ludin, B. and Matus, A., GFP illuminates the cytoskeleton, *Trends Cell Biol*, 8, 72, 1998.
3. Bajno, L. and Grinstein, S., Fluorescent proteins: Powerful tools in phagocyte biology, *J Immun Methods*, 232, 67, 1999.
4. Chamberlain, C. and Hahn, K. M., Watching proteins in the wild: Fluorescence methods to study dynamics in living cells, *Traffic*, 1, 755, 2000.
5. Latif, R. and Graves, P., Fluorescent probes: Looking backward and looking forward, *Thyroid*, 10, 407, 2000.
6. Sacchetti, A., Ciccocioppo, R., and Alberti, S., The molecular determinants of the efficiency of green fluorescent protein mutants, *Histol Histopathol*, 15, 101, 2000.
7. Whitaker, M., Fluorescent tags of protein function in living cells, *BioEssays*, 22, 180, 2000.
8. Molenaar, C. et al., Linear 2′O-methyl RNA probes for the visualization of RNA in living cell, *Nucleic Acids Res*, 29, E89, 2001.
9. Boudonck, K., Dolan, L., and Shaw, P. J., The movement of coiled bodies visualized in living plant cells by the green fluorescent protein, *Mol Biol Cell*, 10, 2297, 1999.
10. Haraguchi, T. et al., Multiple color fluorescence imaging of chromosomes and microtubules in living cells, *Cell Struct Funct*, 24, 291, 1999.
11. Wachsmuth, M., Waldeck, W., and Langowski, J., Anomalous diffusion of fluorescent probes inside living cell nuclei investigated by spatially-resolved fluorescence correlation spectroscopy, *J Mol Biol*, 298, 677, 2000.
12. Misteli, T., Cell biology of transcription and pre-mRNA splicing: Nuclear architecture meets nuclear function, *J Cell Sci*, 113, 1841, 2000.

13. Houstmuller, A. B. and Vermeulen, W., Macromolecular dynamics in living cell nuclei revealed by fluorescence redistribution after photobleaching, *Histochem Cell Biol*, 115, 13, 2001.

14. Cremer, T. et al., Chromosome territories, interchromatin domain compartments and nuclear matrix: An integrated view of the functional nuclear architecture, *Crit Rev Eukaryot Gene Expr*, 10, 179, 2000.

15. Haraguchi, T., Koujin, T., and Hiraoka, Y., Application of GFP: Time-lapse multi-wavelength fluorescence imaging of living mammalian cells, *Acta Histochem Cytochem*, 33, 169, 2000.

16. Moir, R. D. et al., Nuclear lamins A and B1: Different pathways of assembly during nuclear envelope formation in living cells, *J Cell Biol*, 151, 1155, 2000.

17. Pidoux, A. L. et al., Live analysis of lagging chromosomes during anaphase and their effect on spindle elongation rate in fission yeast, *J Cell Sci*, 113, 4177, 2000.

18. Piel, M. et al., The respective contributions of the mother and daughter centrioles to centrosome activity and behavior in vertebrate cells, *J Cell Biol*, 149, 317, 2000.

19. Dundr, M., Misteli, T., and Olson, M. O. J., The dynamics of postmitotic reassembly of the nucleolus, *J Cell Biol*, 150, 433, 2000.

20. Savino, T. M. et al., Nucleolar assembly of the rRNA processing machinery in living cells, *J Cell Biol*, 153, 1097, 2001.

21. Haraguchi, T., Kaneda, T., and Hiraoka, Y., Dynamics of chromosomes and microtubules visualized by multi-wavelength fluorescence imaging in live mammalian cells: Effects on cell cycle progression, *Gene Cells*, 2, 369, 1997.

22. Pederson, T., Fluorescent RNA cytochemistry: Tracking gene transcripts in living cells, *Nucleic Acids Res*, 29, 1013, 2001.

23. Beach, D. L., Salmon, E. D., and Bloom, K., Localization and anchoring of mRNA in budding yeast, *Curr Biol*, 9, 569, 1999.

24. Politz, J. C. and Peterson, T., Review: Movement of m-RNA from transcription site to nuclear pores, *J Struct Biol*, 129, 252, 2000.

25. Bouvier, M., Oligomerization of G-protein-coupled transmitter receptors, *Nat Rev Neurosci*, 2, 274, 2001.

26. Corbett, E. F. et al., Ca^{2+} regulation of interactions between endoplasmic reticulum chaperones, *J Biol Chem*, 274, 6203, 1999.

27 Halaban, R. et al., Endoplasmic reticulum retention is a common defect associated with tyrosinase-negative albinism, *Proc Natl Acad Sci USA*, 97, 5889, 2000.

28. Zhang, Y. et al., Hsp70 molecular chaperone facilitates endoplasmic reticulum-associated protein degradation of cystic fibrosis transmembrane conductance regulator in yeast, *Mol Biol Cell*, 12, 1303, 2001.

29. Kamhi-Nesher, S. et al., A novel quality control compartment derived from the endoplasmic reticulum, *Mol Biol Cell*, 12, 1711, 2001.

30. Stan, T. et al., Recognition of preproteins by the isolated TOM complex of mitochondria, *EMBO J*, 19, 4895, 2000.

31. Miller, R. K., Matheos, D., and Rose, M. D., The cortical localization of the microtubule orientation protein, Kar9p, is dependent upon actin and proteins required for polarization, *J Cell Biol*, 144, 963, 1999.

32. Cid, V. J. et al., Cell cycle control of septin ring dynamics in the budding yeast, *Microbiology*, 147, 1437, 2001.

33. Vorobjev, I. A., Rodionov, V. I., and Borisy, G. G., Contribution of plus and minus end pathways to microtubule turnover, *J Cell Sci*, 112, 2277, 1999.

34. Zhang, C., Hugues, M., and Clarke, P. R., RNA-GTP stabilises microtubule asters and inhibits nuclear assembly in Xenopus egg extracts, *J Cell Sci*, 112, 2453, 1999.

35. Odde, D. J. et al., Microtubule bending and breaking in living fibroblasts, *J Cell Sci*, 112, 3283, 1999.

36. Windoffer, R. and Leube, R. E., Detection of cytokeratin dynamics by time-lapse fluorescence microscopy in living cells, *J Cell Sci*, 112, 4521, 1999

37. Mallavarapu, A., Sawin, K., and Mitchison, T., A switch in microtubule dynamics at the onset of anaphase B in the mitotic spindle of *Schizosaccharomyces pombe*, *Curr Biol*, 9, 1423, 1999.

38. Granger, C. L. and Cyr, R. J., Microtubule reorganization in tobacco BY-2 cells stably expressing GFP-MBD, *Planta*, 210, 502, 2000.

39. Strohmaier, A. R. et al., Three-dimensional organization of microtubules in tumor cells studied by confocal laser scanning microscopy and computer-assisted deconvolution and image reconstruction, *Cells Tissues Organs*, 167, 1, 2000.

40. Pierce, D. W. et al., Single-molecule behavior of monomeric and heteromeric kinesins, *Biochemistry*, 17, 5412, 1999.

41. Xiang, X. et al., Dynamics of cytoplasmic dynein in living cells and the effect of a mutation in the dynactin complex actin-related protein Arp1, *Curr Biol*, 10, 603, 2000.

42. Farkasovsky, M. and Kuntzel, H., Cortical Num1p interacts with the dynein intermediate chain Pac11p and cytoplasmic microtubules in budding yeast, *J Cell Biol*, 152(2), 251, 2001.

43. Inoue, Y. et al., Motility of single one-headed kinesin molecules along microtubules, *Biophys J*, 81(5), 2838, 2001.

44. Ballestrem, C., Wehrle-Haller, B., and Imhof, B. A., Actin dynamics in living mammalian cells, *J Cell Sci*, 111, 1649, 1998.

45. Yumura, S. and Fukui, Y., Spatiotemporal dynamics of actin concentration during cytokinesis and locomotion in *Dictyostelium*, *J Cell Sci*, 111, 2097, 1998.

46. Choidas, A. et al., The suitability and application of a GFP-actin fusion protein for long-term imaging of the organization and dynamics of the cytoskeleton in mammalian cells, *Eur J Cell Biol*, 77, 81, 1998.

47. Evans, R. M. and Simpkins, H., Cisplatin induced intermediate filament reorganization and altered mitochondrial function in 3T3 cells and drug-sensitive and -resistant Walker 256 cells, *Exp Cell Res*, 245, 69, 1998.

48. Correia, I. et al., Integrating the actin and vimentin cytoskeletons. Adhesion dependent formation of fimbrin-vimentin complexes in macrophages, *J Cell Biol*, 146, 831, 1999.

49. McNiven, M. A. et al., Regulated interactions between dynamin and the actin-binding protein cortactin modulate cell shape, *J Cell Biol*, 151, 187, 2000.

50. Enqvist-Goldstein, A. E. Y. et al., An actin-binding protein of he Sla2/Huntingtin interacting protein 1 family is a novel component of clathrin-coated pits and vesicles, *J Cell Biol*, 147, 1503, 1999.

51. Kachinsky, A. M., Froehner, S. C., and Milgram, S. L., A PDZ-containing scaffold related to the dystrophin complex at the basolateral membrane of epithelial cells, *J Cell Biol*, 145, 391, 1999.

52. Akhtar, N. and Hotchin, N. A., RAC1 regulates adherent junctions through endocytosis of E-cadherin, *Mol Biol Cell*, 12, 847, 2001.

53. Wild, N. et al., Expression of a chimeric, cGMP-sensitive regulatory subunit of the cAMP-dependent protein kinase type Iα, *FEBS Lett*, 374, 356, 1995.

54. Liu, Z. H. et al., Chromosomal localization of a novel retinoic acid induced gene RA28 and protein distribution of its encoded protein, *Chin Sci Bull*, 45, 1857, 2000.

55. Hettmann, C. et al., A dibasic motif in the tail of a class XIV apicomplexan myosin is an essential determinant of plasma membrane localization, *Mol Biol Cell*, 11, 1385, 2000.

56. Reichel, C. and Beachy, R. N., Degradation of tobacco mosaic virus movement protein by the 26S proteasome, *J Virol*, 74, 3330, 2000.

57. Giraudo, C. G., Daniotti, J. L., and Maccioni, H. J. F., Physical and functional association of glycolipid *N*-acetyl-galactosaminyl and galactosyl transferases in the Golgi apparatus, *Proc Natl Acad Sci USA*, 98, 1625, 2001.

58. Margolin, W., Green fluorescent protein as a reporter for macromolecular localization in bacterial cells, *Methods*, 20, 62, 2000.

59. Pepperkok, R. et al., Simultaneous detection of multiple green fluorescent proteins in live cells by fluorescence lifetime imaging microscopy, *Curr Biol*, 9, 269, 1999.

60. Riechel, C. and Beachy, R. N., Tobacco mosaic virus infection induces severe morphological changes of the endoplasmic reticulum, *Proc Natl Acad Sci USA*, 95, 11169, 1998.

61. Rosorius, O. et al., Direct observation of nucleocytoplasmic transport by microinjection of GFP-tagged proteins in living cells, *Biotechniques*, 27, 350,1999.

62. Striepen, B. et al., Expression, selection, and organellar targeting of the green fluorescence protein in *Toxoplasma gondii*, *Mol Biochem Parasitol*, 92, 325, 1998.

63. Bark, S. J. and Hahn, K. M., Fluorescent indicators of peptide cleavage in the trafficking compartments of living cells: Peptides site-specifically labeled with two dyes, *Methods*, 20, 429, 2000.

64. Galigniana, M. D. et al., Inhibition of glucocorticoid receptor nucleocytoplasmic shuttling by okadaic acid requires intact cytoskeleton, *J Biol Chem*, 274, 16222, 1999.

65. Umenishi, F., Verbavatz, J. M., and Verkman, A. S., cAMP regulated membrane diffusion of a green fluorescent protein-aquaporin 2 chimera, *Biophys J*, 78, 1024, 2000.

66. Rivero, F. et al., RacF1, a novel member of the Rho protein family in *Dictyostelium discoideum*, associates transiently with cell contact areas, *Mol Biol Cell*, 10, 1205, 1999.

67. Nakano, M. Y. and Greber U. F., Quantitative microscopy of fluorescent adenovirus entry, *J Struct Biol*, 129, 57, 2000.

68. Gadella, T. W. J. Jr., Jovin, T. M., and Clegg, R. M., Fluorescence lifetime imaging microscopy (FLIM)-spatial resolution of microstructures on the nanosecond time-scale, *Biophys Chem*, 48, 221, 1993.

69. So, P. T. C. et al., Time-resolved fluorescence spectroscopy using two-photon excitation, *Bioimaging*, 3, 1, 1995.

70. Carlsson, K. and Liljeborg, A., Confocal fluorescence microscopy using spectral and lifetime information to simultaneously record four fluorophores with high channel separation, *J Microsc*, 185, 37, 1997.

71. Schneider, P. C. and Clegg, R. M., Rapid acquisition, analysis, and display of fluorescence life-time resolved images for real-time applications, *Rev Sci Instrum*, 68, 4107, 1997.

72. Squire, A. and Bastiaens, P. H. I., Three dimensional image restoration in fluorescence lifetime imaging microscopy, *J Microsc*, 193, 36, 1999.

73. Lliopis, J. et al., Ligand-dependent interactions of coactivators steroid receptor coactivator-1 and peroxisome proliferator-activated receptor binding protein with nuclear hormone receptors can be imaged in living cells and are required for transcription, *Proc Natl Acad Sci USA*, 97, 4363, 2000.

74. Kenworthy, A. K., Petranova, N., and Edidin, M., High-resolution FRET microscopy of cholera toxin B-subunit and GPI-anchored proteins in cell plasma membranes, *Mol Biol Cell*, 11, 1645, 2000.

75. Janetopoulos, C., Jin, T., and Devreotes, P., Receptor-mediated activation of heteromeric G-proteins in living cells, *Science*, 291, 2408, 2001.

76. Lino, R., Koyama, I., and Kusumi, A., Single molecule imaging of green fluorescent proteins in living cells: E-cadherin forms oligomers on the free cell surface, *Biophys J*, 80, 2667, 2001.

77. Waterman-Storer, C. M. and Salmon, E. D., Fluorescent speckle microscopy of microtubules: How low can we go? *FASEB J*, 13, S225, 1999.

78. Schütz, G. J. et al., Properties of lipid microdomains in a muscle cell membrane visualized by single molecule microscopy, *EMBO J*, 19, 892, 2000.

79. Sorkin, A. et al., Interaction of EGF receptor and Grb2 in living cells visualized by fluorescence energy transfer (FRET) microscopy, *Curr Biol*, 10, 1395, 2000.

80. Byassee, T. A., Chan, W. C. W., and Nie, S., Probing single molecule in living cells, *Anal Chem*, 72, 5606, 2000.

81. Lacoste, T. D. et al., Ultrahigh-resolution multicolor colocalization of single fluorescent probes, *Proc Natl Acad Sci USA*, 97, 9461, 2000.

82. König, K., Multiphoton microscopy in life sciences, *J Microsc*, 200, 83, 2000.

83. Diaspro, A. and Robello, M., Two-photon excitation of fluorescence for three-dimensional optical imaging of biological structures, *J Photochem Photobiol B*, 55, 1, 2000.

84. Patterson, G. H. and Piston, D. W., Photobleaching in two photon excitation microscopy, *Biophys J*, 78, 2159, 2000.

85. Enderly, Th. et al., Membrane specific mapping and colocalization of malarial and host skeletal proteins in the *Plasmodium falciparum* infected erythrocyte by dual color near-field scanning optical microscopy, *Proc Natl Acad Sci USA*, 94, 520, 1997.

86. Korchev, Y. E. et al., Hybrid scanning ion conductance and scanning near-field optical spectroscopy for the study of living cells, *Biophys J*, 78, 2675, 2000.

87. Doyle, R. T., Szulzcewski, M. J., and Haydon, P. G., Extraction of near-field fluorescence from composite signals provide high resolution images of glial cells, *Biophys J*, 80, 2477, 2001.

88. De Giorgi, F., Lartigue, L., and Ichas, F., Electrical coupling and plasticity of the mitochondrial network, *Cell Calcium*, 28, 365, 2000.

89. Margineantu, D., Capaldi, R. A., and Marcus, A. H., Dynamics of the mitochondrial reticulum in live cells using Fourier imaging correlation spectroscopy and digital video microscopy, *Biophys J*, 79, 1833, 2000.

90. Toescu, E. C. and Verkhratsky, A., Assessment of mitochondrial polarization status in living cells based on analysis of the spatial heterogeneity of rhodamine 123 fluorescence staining, *Eur J Physiol*, 440, 941, 2000.

91. Savtchenko, L. P. et al., Imaging stochastic spatial variability if active channel clusters during excitation of single neurons, *Neurosci Res*, 39, 431, 2001.

92. Stauffer, T. P., Ahn, S., and Meyer, T., Receptor-induced transient reduction in plasma membrane PtdIns(4,5)P2 concentration monitored in living cells, *Curr Biol*, 8, 343, 1998.

93. Varnai, P. and Balla, T., Visualization of phosphoinositides that bind pleckstrin homology domains: Calcium- and agonist-induced dynamic changes and relationship to myo-[^{3}H]inositol-labeled phosphoinositide pools, *J Cell Biol*, 143, 501, 1998.

94. Arbuzova, A. et al., Fluorescently labeled neomycin as a probe of phosphatdylinositol-4, 4-bisphosphate in membranes, *Biochim Biophys Acta Biomembr*, 1464, 35, 2000.

95. Berrie, C. P. and Falasca, M., Patterns within protein/polyphosphoinositide interactions provide specific targets for therapeutic intervention, *FASEB J*, 14, 2618, 2000.

96. Gillooly, D. J. et al., Localization of phosphatidylinositol 3-phosphate in yeast and mammalian cells, *EMBO J*, 19, 4577, 2000.

97. Oziki, S. et al., Intracellular delivery of phosphoinositides and inositol phosphates using polyamine carriers, *Proc Natl Acad Sci USA*, 97, 11286, 2000.

98. Puri, V. et al., Clathrin-dependent and -independent internalization of plasma membrane sphingolipids initiates two Golgi targeting pathways, *J Cell Biol*, 154, 535, 2001.

99. Prattes, S. et al., Intracellular distribution and mobilization of unesterified cholesterol in adipocytes: Triglyceride droplets are surrounded by cholesterol-rich ER-like surface layer structures, *J Cell Sci*, 113, 2977, 2000.

100. Nagai, Y. et al., A fluorescent indicator for visualizing cAMP-induced phosphorylation in vivo, *Nat Biotechnol*, 18, 313, 2000.

101. Honda, A. et al., Spatiotemporal dynamics of guanosine 3',5'-cyclic monophosphate revealed by a genetically encoded, fluorescent indicator, *Proc Natl Acad Sci USA*, 98, 2437, 2001.

102. Jayaraman, S. et al., Mechanism and cellular applications of a green fluorescent protein-based Halide sensor, *J Biol Chem*, 275, 6047, 2000.

103. Chandy, G. et al., Proton leak and CFTR in regulation of Golgi pH in respiratory epithelial cells, *Am J Physiol*, 281, C908, 2001.

104. Blackmore, C. G. et al., Measurement of secretory vesicle pH reveals intravesicular alkalinization by vesicular monoamine transporter type 2 resulting in inhibition of prohormone cleavage, *J Physiol Lond*, 531, 605, 2001.

105. Miyawaki, A. et al., Dynamic and quantitative Ca^{2+} measurements using improved chameleons, *Proc Natl Acad Sci USA*, 96, 2135, 1999.

106. Yu, R. and Hinkle, P. M., Rapid turnover of calcium in the endoplasmic reticulum during signaling, *J Biol Chem*, 275, 23648, 2000.

107. Griesbeck, O. et al., Reducing the environmental sensitivity of yellow fluorescent protein, *J Biol Chem*, 276, 29188, 2001.

108. Nagai, T. et al., Circularly permuted green fluorescent proteins engineered to sense Ca²⁺, *Proc Natl Acad Sci USA*, 98, 3197, 2001.

109. Carrington, W. A. et al., Superresolution three-dimensional images of fluorescence in cells with minimal light exposure, *Science*, 268, 1483, 1995.

110. Schmidt, T. et al., Imaging of single molecule diffusion, *Proc Natl Acad Sci USA*, 93, 2926, 1996.

111. Schütz, G. J., Schindler, H., and Schmidt, T., Single-molecule microscopy on model membranes reveals anomalous diffusion, *Biophys J*, 73, 1073, 1997.

112. Dickson, R. M. et al., Three-dimensional imaging of single molecules solvated in pores of poly(acrylamide) gels, *Science*, 274, 966, 1996.

113. Vickery, S. A. and Dunn, R. C., Scanning near-field fluorescence resonance energy transfer microscopy, *Biophys J*, 76, 1812, 1999.

114. Hell, S. W. and Stelzer, E. H. K., Fundamental improvement of resolution with a 4Pi-confocal fluorescence microscope using two-photon excitation, *Opt Commun*, 93, 277, 1992.

115. König, K., Two-photon near-infrared excitation in living cells, *J Near Infrared Spectrosc*, 5, 27, 1997.

116. Klar, T. A. et al., Fluorescence microscopy with diffraction resolution barrier broken by stimulated emission, *Proc Natl Acad Sci USA*, 97, 8206, 2000.

117. Sailer, R. et al., Plasma membrane associated location of sulfonated *meso*-tetraphenylporphyrins of different hydrophilicity probed by total internal reflection fluorescence spectroscopy, *Photochem Photobiol*, 71, 460, 2000.

118. Thoumine, O. et al., Short-term binding of fibroblasts to fibronectin: Optical tweezers experiments and probabilistic analysis, *Eur Biophys J*, 29, 398, 2000.

119. Helfer, E. et al., Microrheology of biopolymer-membrane complexes, *Phys Rev Lett*, 85, 457, 2000.

120. König, K., Robert Feulgen Prize Lecture. Laser tweezers and multiphoton microscopes in life sciences, *Histochem Cell Biol*, 114, 79, 2000.

121. Vo-Dinh, T., Alarie, J. P., Cullum, B., and Griffin, G. D., Antibody-based nanoprobe for measurements in a single cell, *Nat Biotechnol*, 18, 76, 2000.

122. Cullum, B., Griffin, G. D., Miller, G. H., and Vo-Dinh, T., Intracellular measurements in mammary carcinoma cells using fiber-optic nanosensors, *Anal Biochem*, 277, 25, 2000.

123. Vo-Dinh, T., Griffin, G. D., Alarie, J. P., Cullum, B., Sumpter, B., and Noid, D., Development of nanosensors and bioprobes, *J Nanopart Res*, 2, 17–27, 2000.

124. Cullum B. and Vo-Dinh, T., Development of optical nanosensors for biological measurements, *Trends Biotechnol*, 18, 388, 2000.

125. Vo-Dinh T., Cullum, B. M., and Stokes, D. L., Nanosensors and biochips: Frontiers in biomolecular diagnostics, *Sens Actuators B Chem*, B74, 2, 2001.

126. Vo-Dinh, T., Scaffidi, J. S., Gregas, M., Zhang, Y., and Seewaldt, V., Applications of fiber-optics-based nanosensors to drug discovery, *Expert Opin Drug Discov*, 4, 889–900, 2009.

127. Vo-Dinh, T. and Zhang, Y., Single-cell monitoring using fiberoptic nanosensors, *Wiley Interdiscip Rev Nanomed Nanobiotechnol* 3(1), 79–85, 2011.

128. Scaffidi, J. S., Gregas, M., and Vo-Dinh, T., SERS-based plasmonic nanobiosensing in single living cells, *Anal Bioanal Chem*, 393, 1135–1141, 2009.

129. Kasili, P. M., Song, J. M., and Vo-Dinh, T., Optical sensor for the detection of caspase-9 activity in a single cell, *J Am Chem Soc*, 126, 2799–2806, 2004.

130. Ideker, T. et al., Integrated genomic and proteomic analyses of a systematically perturbed metabolic network, *Science*, 292, 929, 2001.

16

Amplification Techniques
for Optical Detection

Guy D. Griffin
Duke University

Dimitra
N. Stratis-Cullum
*US Army Research
Laboratory*

Timothy E.
McKnight
*Oak Ridge National
Laboratory*

M. Wendy Williams
*Oak Ridge National
Laboratory*

Tuan Vo-Dinh
Duke University

16.1 Introduction

In this review, we will present a summary of various amplification techniques for analysis of biological molecules (species), which use optical detection, in some form, as the endpoint. The analyte species that could be considered include lipids, carbohydrates, proteins, and nucleic acids (and combinations of the same such as glycolipids and glycoproteins) but this review will be restricted to proteins and nucleic acids. Of course, detection strategies are manifold for both proteins and nucleic acids, but the defining restriction for this review is that amplification be involved. We therefore have limited our purview to immunoassays and nucleic acid techniques employing amplification along with brief overviews of applications of nanomaterials and conjugated polymers (CPs), which also employ amplification as a part of the analytical process.

Immunoassays, which depend fundamentally upon antibody recognition of antigens, have been proven, over the course of many years, to be powerful analytical techniques for protein detection. They can also be used successfully for the detection of substances of lower molecular weight (MW) than the usual size of proteins (e.g., peptides, hormones) provided that antibodies specific against these molecular structures are developed and specifically selected.

Detection of nucleic acids has been revolutionized by the development of polymerase chain reaction (PCR) and other related techniques, which produce large multiplications of specific nucleic acid

353

segments. Even before the advent of these powerful amplification techniques, however, sensitive detection of nucleic acid was achieved using various procedures, some of which relied upon the principles of immunoassays.[1] Most nucleic acid detection strategies are based fundamentally upon the base-pairing interaction that occurs between complementary strands of polynucleotides (even PCR, in which primers must recognize specific complementary sequences), and, thus, such complementary base pairing (i.e., hybridization) is a significant component of nucleic acid detection assays.

Finally, there are also detection strategies that seem to wed elements of immunoassays and nucleic acid detection assays (e.g., immuno-PCR, PCR-enzyme-linked immunosorbent assay [ELISA]). In this review, we group such techniques into a third category called *hybrid assays*. These admittedly could be classified as either immunoassays or nucleic acid amplification assays, but in the hope for added clarity, they will be categorized separately.

We also wish to highlight some more recent developments that bode well for highly sensitive detection technologies for biomaterials in the future. Specifically, we discuss two recent developments: (1) the use of CPs as highly sensitive biosensors and (2) the applications of nanomaterials as amplification systems for biosensing. Both these emerging fields already have a wealth of specific applications. We cannot hope to cover all applications in this short review but choose to briefly discuss underlying principles and to list a selection of applications in the accompanying tables (Tables 16.4 and 16.5).

It should be noted that this review does not claim to be comprehensive with regard to every possible assay technique that has been applied to a specific situation. Indeed, given the number of researchers in these fields, and the time duration over which these techniques have been developed, it would be an enormous undertaking to claim true comprehensiveness in such a review. An attempt has been made to cover the more common, and what the authors perceive to be the more useful, analytical techniques, with the further caveat that we have focused almost exclusively on the English language literature. The references given are not meant to be encompassing with regard to any of the techniques discussed.

Additionally, radiotracer techniques are not included in this review. These procedures, in many cases, formed the basis for great advances in biological knowledge and still demonstrate remarkable sensitivity. However, safety issues and issues of waste disposal have somewhat diminished the attractiveness of radioisotopes for biological applications—hence the increasing reliance on other techniques, including a heavy emphasis upon optical-based analytical tools. Our goal is to provide a comprehensible introduction to various amplification techniques for optical detection, which will allow the reader to investigate further those that pique his or her interest. If we have omitted significant analytical techniques, we wish to apologize beforehand to those dedicated researchers who have advanced the frontiers of science.

16.2 Immunoassays

Immunoassays make use of antibody–antigen (Ab/Ag) recognition, a recognition that has great specificity and the potential for high affinity. Protein detection has been a major focus of immunoassay development, although small-MW substances (drugs, toxins, hormones, etc.) can also be conveniently detected. In the latter case, these molecules can be conjugated to proteinaceous (or other) materials to form antigenic substances for which hapten-specific antibodies may be developed. It is not only possible to develop antibodies against specific epitopes (antigenic sites) on large-MW proteins but also against high-MW carbohydrates. However, the adaptation of immunoassays to carbohydrates and lipids is considerably less rich in associated assay development than the protein/hapten systems that have seen the bulk of the effort at innovation. Consequently, in this review, we focus essentially upon protein detection by immunoassay. The interaction of biotin and avidin/streptavidin is a rather unique case of molecular recognition that has been extensively utilized in the development of detection schemes throughout biology. The relatively enormous affinity ($K_D = 10^{-15}$ M)[2] of biotin for avidin/streptavidin forms the basis for a recognition reaction of great specificity, and the small size and molecular structure of biotin allows ready conjugation to a wide variety of molecules. Not surprisingly, this system has been exploited in the development of a wide variety of analytical techniques, which appear in this review.

Immunoassays have been performed in a wide variety of formats. The 96-well microplate-based assays are routinely performed throughout the world but are by no means the only feasible format. Immunoblotting, applied as detection for a wide variety of gel-based separation methods, is also widely used. Similarly, a rich literature has grown up regarding immunoassay techniques applied to histological/cytological specimens. Whether the samples are tissue sections, membrane blots, proteins in physiological fluids, or samples in microplate wells, the immunoassay techniques employed for detection show similar basic principles. Thus, we do not, in general, divide the assay strategies in regard to their application (i.e., histological, immunoblot) but simply mention that some procedures fit certain applications well. In certain cases, we note an interesting/unusual format (such as use of an unusual solid substrate for assay immobilization), but this is not intended to suggest an exhaustive review of all assay formats. The reader, upon perusal of the various assay principles employed, can likely formulate original modifications suitable for specific applications.

Table 16.1 provides a summary of our review of immunoassays. Immunoassays are used routinely in many areas of biological science, including basic research, pharmaceutical development, and clinical and forensic applications. Although the term *immunoassay* can encompass a wide variety of quite different techniques, this review will focus only on techniques involving amplification, many of which fall into the category of enzyme-amplified immunoassays. This category is in and of itself daunting in the number of variants that have been developed. Thus, an attempt is made here to discuss general principles that govern development of immunoassays of this class, examples of general assay formats, and some commentary upon suitability of various techniques.

16.2.1 Classification of Immunoassays

Immunoassays can basically be classified as heterogeneous or homogeneous, competitive or noncompetitive, and direct or indirect.[3,4] Of course, the preceding terms may be used in various combinations (e.g., heterogeneous, indirect, noncompetitive immunoassay). Although the nomenclature for immunoassays can be quite variable, and shortened forms of classification are very commonly used, it is useful to at least provide definitions of the basic classifications.

Heterogeneous vs. homogeneous refers to whether a separation step is required in the immunoassay. In other words, heterogeneous assays require one or more separation steps, while homogeneous assays occur in solution phase without separation. Heterogeneous assays are commonly associated with a solid phase, such as a microtiter plate. Immunoassays can be formatted as either competitive or noncompetitive. In the competitive mode, two pools of antigen, one pool labeled with a marker of some sort and one pool consisting of the sample to be analyzed, compete for a limited number of antibody-binding sites. Competitive assays of this sort are, in general, homogeneous assays. In a noncompetitive assay, immobilized antibody binds (*captures*) antigen, and antigen is detected by a second antibody (*detector*) that has some label and binds to another epitope on the antigen. Perforce, these types of assays are heterogeneous. However, heterogeneous assays may also be formatted so that the immobilized antibodies capture a mixture of labeled and unlabeled antigen from two pools, that is, a competitive heterogeneous assay. Alternatively, a limited amount of antigen may be immobilized, and limited amounts of labeled antibody added to a mixture of sample antigen in solution. This format is also competitive, as increased antigen in solution produces fewer labeled antibodies binding to the immobilized antigen.

Direct immunoassays refer to assays where only primary antibodies, with specificity against the antigen of interest, are utilized in the assay. The antibodies here may be labeled or unlabeled. Indirect assays, on the other hand, involve in some manner an immune or other type of complex. For these types of assays, the final complex signals through some indirect means, the formation of an immunospecific binding event. Examples include the interaction of an antispecies IgG antibody, with attached label, with the primary antigen specific antibody or the use of a labeled avidin/streptavidin to bind to a biotinylated antigen-specific antibody. In Table 16.1, many of the immunoassays are

TABLE 16.1 Immunoassay Amplification Schemes

Assay	Principle	Comments	References
Enzyme-linked immunoassays (solid-phase heterogeneous)	In all EIAs, the final signal detected is produced by the activity of an enzyme, while specificity is obtained by Ag/Ab interaction. Addition of enzyme substrate to final immune complex results in accumulation of colored product	Large signal amplification because of high rate of processivity of enzymes for substrate molecules	
Direct EIA (ELISA)	Enzyme-labeled Ab recognizes Ag	If capture Ab system used, enzyme-labeled Ab must recognize different Ag epitope than capture Ab and can be formatted as competitive or noncompetitive assay	[41,42]
Indirect *sandwich* EIA (immune complex, ELISA)	Antispecies IgG enzyme-labeled Ab binds directly to Ag-recognizing Ab or is bound indirectly via an unlabeled antispecies IgG Ab, which recognizes both Ag-specific Ab and enzyme-labeled Ab; both direct and indirect immune complex assays termed ELISA	Enzyme numbers multiplied through multiple epitope sites on IgG Ab	[44–47]
	ELISA carried out on cloth (usually plastic surfaces [e.g., microwells])	Cloth assay described as instantaneous—no incubation time with Ab steps—5 min incubation with substrate[3]	[43]
Bispecific/chimera Ab molecules	Abs constructed to possess dual specificity for Ag and for enzyme or chimeric Ab constructed to recognize a hapten (covalently linked to an immunoreagent) *and* an enzyme; in either case, enzyme is *indirectly* linked to immunocomplex, as opposed to directly linked as in the case of enzymes covalently coupled to Ab	Bispecific Ab constructed so that it binds to both FITC (used as a label for Ag or Ab) and to reporter (HRP) enzyme; thus, hybrid antibody serves as bridge between FITC-labeled immunoreagent and enzyme molecule; this procedure avoids possible deleterious effect on Ab affinity by covalent binding of enzyme directly to Ab; amplification both in number of FITC on primary Ab and by enzyme indirectly linked to each FITC molecule[50]	[50] [51] [48,49,52,53]
	Also	Or	
	Chimeric fusion protein constructed having single-chain variable region of Ab fused to enzyme (AP); bifunctional immunoreagent thus constructed; other constructs allow expression of enzyme-labeled Fab' fragments[48,49]	Bispecific Ab constructed so that it recognizes both specific Ag and HRP; again, advantage in indirect linking of reporter enzyme and specific Ab[51]	
	Fusion protein constructed between protein A and enzyme; as immunoreagent sandwich is formed, protein A portion of fusion protein binds to Fc region of Ag-specific Ab, thus forming, in effect, an enzyme-linked immunoassay	Fusion protein construct of enzyme and protein A avoids random nature of chemically labeling Ab with enzymes since Ab itself is not labeled in this technique, more likely to have less adverse effect on Ab affinity than if Ab is directly labeled	

Enzyme/antienzyme immune complexes	Multiple enzyme molecules localized by antienzyme Ab, the whole complex held at Ag site as above; (i.e., Ag-specific Ab binds to Ag; antispecies IgG Ab binds to Ag-specific Ab, *as well as* antienzyme Ab made in the same species as Ag-specific Ab; antienzyme Ab binds enzyme molecules; excess enzyme washed away); commonly used complexes include PAP, HRP complex, AP-AAP, AP complex	Enzyme numbers multiplied—often used in immunohistochemistry; 2–50-fold increased sensitivity[55]	[55–61]
Antihapten Ab *sandwich*	Haptens (e.g., FITC, DNP) bound to Ag-specific primary antibody, recognized by enzyme-labeled antihapten Ab, can use bridge antibody (i.e., IgM anti-DNP) and DNP-labeled enzyme for increased sensitivity and can also use DNP-labeled glucose oxidase in addition to primary enzyme (peroxidase) for increased sensitivity	Enzyme numbers multiplied—multiple hapten molecules bound to primary Ab—bridging Ab gives more multiplication, used in immunohistochemistry, fourfold more sensitive than PAP[62]	[62,63]
Cell—ELISA	Detect cell surface molecules on cells, using Ag-specific Ab and enzyme-labeled antispecies IgG Ab	Usually applied to cells in microwells; limitations: may be necessary to use different enzymes (e.g., bacterial β-galactosidase) if endogenous cellular enzyme interferes; if cells are fixed, surface antigens may be destroyed; good correlation with flow cytometric analysis[64]	[42,64–67]
ELISPOT—spot—ELISA	Ags secreted by individual cells are captured by Ag-specific capture Ab immobilized on solid surface; cells are removed and reporter Ab specific for Ag is added; either reporter Ab has enzyme label or enzyme-labeled antispecies IgG Ab is added; substrate is added to give insoluble product	Need relatively high Ag density on solid-phase surface; nitrocellulose better than plastic; biotin/avidin systems can be used to advantage in this technique; technique developed to reveal secretion by single cell; does not measure Ag concentration	[66–70]
ELIFA (enzyme-linked immunofilter assay)	Ag (e.g., bacteria) captured on membrane detected by Ag-specific Ab and additional immune complex antispecies IgG Ab + enzyme-labeled Ab	Detect ~50–60 bacterial cells using chemiluminescent or colorimetric assay, carried out in ~1 h; very rapid (17 min) EIA reported for airborne antibiotic trapped on cellulose nitrate filter[71]; colorimetric detection	[71–75]

(continued)

TABLE 16.1 (continued) Immunoassay Amplification Schemes

Assay	Principle	Comments	References
EIA—enzyme-labeled Ag	Ag conjugated with marker enzyme; free Ag in sample competes with enzyme-labeled Ag for binding to immobilized Ab	Heterogeneous, competitive assay format; assay stated not to be as sensitive as radioimmunoassay[76]	[76] [78] [77]
	Fusion protein made between protein A and enzyme; binding to immobilized Ag-specific Ab results in some of enzyme–Ag fusion protein being immobilized in immune complex	Competitive, heterogeneous assay; Ag in analyte solution competes with enzyme–Ag fusion protein for binding to Ab; amplification due to enzyme; such chimeric constructs for use as Ags require a protein Ag	
	Or		
	Enzymes covalently attached to different hapten Ags; incubation with Ag-specific Abs and free Ag in solution results in some enzyme-labeled Ag forming immunocomplex with Ab; addition of second Ab (antispecies IgG) results in precipitation of immune complex; following washing, addition of enzyme substrates results in enzymatic activity and optical signal proportional to amount of labeled Ag bound in original immune complex	Heterogeneous assay format; enzyme-labeled Ag competes with free Ag for binding to limited amount of Ab; use of two different enzymes can allow use of two different enzyme substrates, whose products can be distinguished by optical properties[77]; applied to simultaneous assay of two Ags—triiodothyronine and thyroxin	
Immune complex transfer immunoassay	Ag is complexed between Ag-specific Ab (Fabʹ) labeled with enzyme and Ag-specific Ab labeled with both biotin and DNP; resulting immune complex is trapped on solid phase by anti-DNP Ab bound to solid phase; immune complex eluted from solid phase by DNP-lysine wash; immune complex again trapped on streptavidin-linked solid phase; enzyme substrate is added and product is measured; similar principles applied to noncompetitive heterogeneous assays of small-MW haptens	Important step is transfer of immune complex to different solid phases, as this removes nonspecifically bound Ab; lends itself to various assay formats; extreme sensitivity, subattomole, down to 1 zeptomole (600 molecules) in assay of human ferritin[11]	[11,79,80]
Enzyme-amplified cascade (RELIA, releasable linker immunoassay)	Ag captured by immobilized capture Ab; second biotin-labeled detector Ab binds to immune complex; addition of AP-conjugated streptavidin followed by addition of release (i.e., biotin-containing) reagent results in release of enzyme–streptavidin conjugate; enzyme activity detected indirectly by enzyme cascade as follows: AP converts FADP to FAD, which binds to inactive apoenzyme amino acid oxidase, resulting in active holoenzyme that oxidizes proline to produce H_2O_2, which further reacts with HRP and other chemicals to form a colored product	Heterogeneous assay format; large multiplication due to involvement of multiple enzymes; sensitivity found to be comparable to fluorometric assays	81

Fusion protein-based EIA	Genetically fused protein is prepared, which has both immunoreactivity and enzymatic activity; this protein can take part in immunoreactions and also serve as detection system by generating enzymatic product, which is detected; IA carried out in usual manner; fusion protein is part of immunoreactant complex (e.g., firefly luciferase–protein A fusion protein—protein A binds to F_c region of Ig, while luciferase reacts with luciferin to produce luminescence)[82] Or Fusion protein prepared, which is Ag fused to enzyme (i.e., proinsulin-AP); competitive immunoassay carried out in which fusion protein Ag competes with Ag in sample for binding to Ab[83]	Potential for sensitive, specific IA; fusion products may not have the same degree of binding affinity/enzyme activity as native proteins	[82] [83]
Nonenzymatic catalytic amplification immunoassay (mimetic enzyme EIA)	Immunoreactant *sandwich* set up in usual way (i.e., capture Ab, Ag, and detector Ab); detector Ab has conjugated to it a chemical that can serve as a catalyst for a reaction, which produces an optically detectable product (e.g., *p*-hydroxyphenyl acetic acid + H_2O_2 in the presence of the catalyst hemin generates a fluorescent product)	Amplification occurs by catalytic activity of reagent conjugated to Ab; also, multiple catalytic reagents can be conjugated to single Ab; small size of catalytic reagent compared to usual size of enzyme should result in less steric inhibition of Ab/Ag binding events; catalytic reagent likely to be more stable than enzyme	[20,84–86]
Enzyme-linked immunoassays (solution homogeneous)	*The same general principle as for heterogeneous EIA*		
EMIT (enzyme multiplied immunoassay technique)	Free Ag in sample competes with enzyme-labeled Ag for binding to Ab; enzyme-labeled Ag bound to Ab results in inhibition of enzyme activity; the more Ag in the sample, the more uninhibited enzyme results, and the larger the resultant signal	Competitive assay; usually limited to low-MW Ags—direct proportionality between free analyte concentration and enzyme activity (i.e., the more free Ag present, the less enzyme-labeled Ag is complexed with Ab, and the more enzyme activity is expressed)	[87–89]
Substrate-labeled immunoassays	Substrate-labeled Ag competes with free Ag; Ab binding blocks substrate from action of enzyme present in solution	Competitive assay; not possible to do assay with excess of substrate—so enzyme kinetics are not zero order; less sensitive than EMIT; direct proportionality between free Ag concentration and enzyme activity	[90,91]
Enzyme modulator-labeled immunoassay	Enzyme inhibitor bound to Ag; Ab binding blocks enzyme inhibition	Competitive assay; inverse relationship between presence of analyte and enzyme activity; more free Ag, less Ab available to bind to inhibitor Ag complex; therefore more enzyme inhibition expressed	[92]

(continued)

TABLE 16.1 (continued)   Immunoassay Amplification Schemes

Assay	Principle	Comments	References
Enzyme inhibitory immunoassay	Ab (Fab' fragments) bound to enzyme (dextranase); binding of Ag to enzyme-bound Ab results in inhibition of ability of enzyme to bind to insoluble substrate; enzyme substrate is in suspension along with other components; inhibition of enzyme activity produces less product	Designed to allow assay of high-MW Ag; signal inversely proportional to analyte Ag concentration	[93,94]
Cofactor-labeled immunoassay	Essential enzyme cofactor(s) bound to Ag; free (sample) Ag competes with labeled Ag; when Ab binds to cofactor-labeled Ag, cofactor is unable to activate apoenzyme, so enzyme activity is not expressed	Competitive assay; very low background; signal (enzyme activity) directly proportional to free analyte concentration	[95]
CEDIA	*E. coli* β-galactosidase genetically engineered so that two inactive subunits, a large polypeptide called enzyme acceptor (EA) and a small polypeptide designated enzyme donor (ED), are present in the assay; EA and ED can spontaneously associate to produce active enzyme; in assay, Ag is attached to ED; Ag-specific Ab binds to Ag/ED complex and inhibits reassembly of active enzyme; Ag in solution competes with Ag/ED complex for limited amount of Ab; more free analyte in sample, more active β-galactosidase formed; substrate in solution converted to colored product	Homogeneous, competitive assay format; colored product directly related to amount of analyte in sample; assay simple to perform and rapid; minimal background; high sensitivity	[96–98]
Enzyme-channeling immunoassay	Coupled enzyme reaction within activity is accelerated when two enzymes are brought into close contact; hapten + enzyme 1 coimmobilized on bead; Ab labeled with enzyme 2 binding to immobilized hapten on bead produces coupled reaction; if Ab is complexed to analyte in solution, coupled enzyme reaction does not occur	Inverse proportionality; low background	[99]
Biotin and avidin/ streptavidin procedures	Biotin and avidin/streptavidin form integral part of immune complex; due to extremely high association constant of biotin avidin/ streptavidin, stable complexes with favorable rates of formation are produced	*The same general detection methodologies as other immunoassays; biotin–avidin complexes contribute to immune sandwich*	

Simple complex	Biotinylated Ab recognizes Ag and binds to enzyme-labeled avidin/streptavidin	Enzyme number multiplied—multiple biotinylation sites on Ab since biotin is a small molecule; small size of biotin also produces only minimal effects on Ab activity; multiple enzyme molecules on each avidin/streptavidin molecule	[2,100]
Sandwich complex	Biotinylated Ag-specific Ab reacts with avidin/streptavidin, which also binds biotinylated enzyme—can preform avidin/biotinylated complexes (ABC [avidin–biotin complex])	Enzyme number multiplied (see above) also, avidin has four combining sites for biotin; 2–100-fold more sensitive than conventional EIA techniques[10]; ABC likely forms a lattice complex containing several (>3) enzyme molecules	[2,10,100,103] [101,102]
Also	Multiple *layers* can be built up to form large immunoreagent biotin–avidin complexes; for example, detector Ab is biotinylated; avidin binds to biotinylated detector Ab; avidin-specific biotinylated Ab binds to avidin; last two steps can be repeated; finally add avidin–enzyme complex followed by enzyme substrate	Amplification results from multiple enzyme molecules; amplification increases with number of *immunolayers* added; enhancement of 10–20-fold over standard ELISA for cytokine detection reported[101]	
Also	Avidin can be conjugated to either Ag-specific or antispecies IgG Ab; subsequent reaction with biotinylated enzymes results in enzyme-labeled antibody–avidin complexes	Multiplication of enzyme molecules in each immunoreactant complex; five- to eightfold enhancement of sensitivity compared to directly enzyme-labeled Ab and fivefold more sensitive than ABC technique[102]	
Coupled enzyme procedures	*The same general idea as for EIA; enzyme linked to immunoreagent produces product, which is then indirectly detected*	*Potential for large increase in amplification since secondary enzymes contribute to overall multiplication of molecules responsible for final signal*	[10,18]
Multienzyme cascades	Product of one enzymatic reaction becomes substrate for another enzyme; also enzyme product (e.g., NAD^+) can be recycled many times; for example, AP converts $NADP^+$ to NAD^+; alcohol dehydrogenase + ethanol converts NAD^+ to NADH; diaphorase oxidizes NADH to NAD^+ and reduces a tetrazolium salt to a colored formazan dye	Amplified enzymatic signals; 100-fold increase in sensitivity compared to conventional EIA[18]; 0.01 amole of AP was detectable using multienzyme cascade[10]	

(continued)

TABLE 16.1 (continued)   Immunoassay Amplification Schemes

Assay	Principle	Comments	References
CARD immunoassay	Form standard ELISA *sandwich*; enzyme linked to Ab in ELISA sandwich reacts with enzyme substrate and biotinyl tyramide (phenolic structure) to form free radical intermediate of tyramide, which binds to receptor on solid surface (i.e., tyrosine residues in close proximity to site of enzyme reaction); addition of enzyme-labeled streptavidin results in binding of streptavidin at biotin sites; addition of enzyme substrate results in final product, which is measured; usually applied to procedures where final product is a precipitate; also feasible to use fluorescein tyramide as substrate for production of *activated* intermediate, which is deposited; in this case, fluorescein serves as Ag for subsequent addition of antifluorescein Ab conjugated to enzyme; addition of enzyme substrate follows Also Can use multiple amplification cycles, where sequential additions of tyramide derivative plus ligand, which binds to deposited tyramide, result in many more deposition sites and deposited secondary enzyme So-called *super*-CARD assay: a modified protein with many sites for tyramine immobilization is used in the assay microwells instead of the usual blocking agent; essential point is that more reactive sites are provided for biotinylated tyramide immobilization, thus providing more sites for subsequent enzyme immobilization when avidin-HRP is added Principle is the same as *super*-CARD; synthetic proteins with many binding sites for tyramide deposition are used as blocking agents around sites of immunoreagents	Potential for many-fold multiplication of enzyme linked to streptavidin, since many biotinylated tyramides can be deposited; final enzyme does not need to be same as ELISA enzyme; other substrates than biotinyl tyramide can be used; HRP, AP, and β-Gal enzymes can be used; HRP used as the first enzyme to generate the *activated* tyramide; other enzymes may be used for secondary amplification (e.g., AP and β-Gal) Can be readily applied to membrane-based immunoassays; CARD procedure improved detection limits 8–200-fold over standard ELISA format, depending on cycles of tyramide amplification and enzyme substrate used[104] Many more molecules of HRP are eventually immobilized around site of each primary HRP-conjugated antibody than in *standard* CARD; incubation time very significantly reduced from standard CARD assay; *super* CARD assay had limit of detection 10-fold lower than ELISA and 5-fold lower than regular CARD assay Applied to immunodot-type assay; *super*-CARD assay found to be ~1H 10^4-fold more sensitive than standard CARD, using casein as blocking agent in standard CARD vs. modified casein in *super*-CARD,[105] all other immunoreagents being the same; visual detection of as little as 800 molecules of rabbit IgG reported	[21,104,106,107] [105]
Dual-enzyme cascade	Action of enzyme 1 converts inactive inhibitor into an active inhibitor of enzyme 2; residual activity of enzyme 2 provides measure of amount of enzyme 1 present	Inverse relation between concentration of the first enzyme and strength of signal; increase in sensitivity of 125-fold over using single enzyme (i.e., enzyme 1 alone) system observed	[108]

Bioluminescence-enhanced enzyme immunoassay	Standard immunoassay *sandwich* set up with enzyme attached either to Ab or Ag (competitive format); following this, luciferin derivatives susceptible to hydrolysis (to make free luciferin) by the enzyme in the immune *sandwich* are added; after incubation, some of the product solution transferred to luciferase-containing solution; extent of bioluminescence produced used to quantify immune reaction	Can be formatted for competitive or noncompetitive assay; many assay steps, which contribute to longer assay time; maximal LODs stated to be 10^{-19} mole for antigen detection[109]	[109,110]
Fluorescence immunoassays	*Two approaches:* (1) *fluorophores attached to immunoreagent (Ab or Ag); when immunocomplex forms, extent of complex formation can be quantified by fluorescence;* (2) *EIA carried out, but enzyme produces fluorescent product from substrate*	*Fluorescence inherently more sensitive than absorption spectroscopy; other properties of fluorescence (fluorescence polarization, fluorescence lifetimes) can serve as basis for detection*	[111–125]
FIA—direct	Fluorophore-labeled Ab binds to immobilized Ag (Ag can be immobilized by capture Ab); immobilized Ab binds free or fluorophore-labeled Ag	Heterogeneous assay; can be competitive or noncompetitive format; minimal amplification; amplification depends on number of fluorophores bound to either Ab or Ag; requirement that Ag be protein or larger peptide; small hapten that is conjugated to one fluorophore, by definition, does not involve amplification; limitations on fluorophore labeling of Ab, because of loss of affinity for Ag, and for Ag, because of potential loss of antigenic epitopes, self-quenching between adjacent fluorophores on the same Ag molecule, etc.; immobilized Ab configuration can be used for competitive assay in which fluorophore-labeled Ag competes with free Ag in analyte for binding to limited number of immobilized Ab sites	
	Also		
	Can either measure the fluorescence of the immobilized labeled fraction or the fluorescence of the labeled immunoreagent remaining in solution, when immobilized Ab concentration is limited	Competitive assay in which fluorophore-labeled Ab binds to Ag in analyte solution; any remaining free labeled Ab is reacted with Ag immobilized on beads; thus, fluorescence is inversely proportional to analyte Ag concentration	
FIA—indirect *sandwich*	Sandwich formed with Ag-specific Ab and fluorescently labeled antispecies IgG Ab, which binds to Ag-specific Ab	Some amplification, due to multiple fluorophore-labeled Abs bound to primary Ab; heterogeneous assay	[114,123,126–129] [107]
	Also		
	For better detection sensitivity, immunoreactant *sandwich* can be extended: that is, after primary Ag-specific Ab binds, fluorophore-labeled antispecies IgG Ab made in another animal species binds to primary Ab; third fluorophore-labeled antispecies IgG (specific for animal species of second Ab) binds to secondary Ab; can be extended further	Multiplication of fluorophore labels with each additional Ab added to sandwich; used in histochemistry, cytochemistry	

(continued)

TABLE 16.1 (continued)   Immunoassay Amplification Schemes

Assay	Principle	Comments	References
Fluorescent CARD immunoassay	The same principle as CARD assay (see under coupled enzyme procedures); fluorophore-labeled tyramide is converted by action of enzyme to *activated* tyramide, which binds close to site of activation; thus, a cluster of fluorophores is deposited at the site of immunoreagent localization Also Applied to detection of Ags on cells and membranes	Multiplication of fluorophore labels deposited at specific locus due to enzyme activity; used in histochemistry, cytochemistry, and immunoblotting; increase in sensitivity over usual fluorescent Ab techniques in histochemistry stated to be 10–100-fold[130] Enhancements of 4–15-fold over standard indirect immunofluorescent staining were observed using CARD[97]	[107,131] [130]
Fluorescence immunoassay: avidin/streptavidin	Fluorophore-labeled avidin/streptavidin binds to biotinylated Ab	Multiple biotin sites on Ab, multiple fluorophores on avidin; heterogeneous assay	[132]
Fluorescence EIA (enzyme amplification)	Non- or weakly fluorescent substrate generates fluorescent product upon enzyme action; immune *sandwich* the same as for ELISA	Heterogeneous assay; fluorescence inherently more sensitive than absorption (colorimetric); 5–10^5-fold more sensitive than conventional EIA[64] Extreme sensitivity with certain fluorogenic substrate/enzyme pairs reported (i.e., single enzyme molecule detection)[99] and detection of activity in single bacterial cells[100]	[64,101–103] [99,100,104,105]
FIA/fluorescence EIA—magnetic bead separation	Ab bound to magnetic beads captures free Ag in sample; separation of bound from free fractions done by magnet; fluorophore-labeled Ag added to beads; extent of fluorescence inversely related to free Ag Or Immobilized Ab captures Ag to which enzyme-labeled second Ag-specific Ab also binds; after separation, incubation with fluorescence-generating substrate (also variant in which Ag is bound to beads and Ag-specific Ab is assayed by antispecies, enzyme-labeled Ab)	Unique solid-phase separation employed; heterogeneous assay; rapid assay times; ready removal of interferent substances	[106] [107]
Microsphere-based immunoassay/flow cytometry	Microspheres of different sizes are used to provide discrimination for different immunoassays; each size microsphere can be labeled with different immunoreagent and appears as a distinct population by flow cytometry; in particular assay, different Ags immobilized on microspheres react with analyte Abs in sera; addition of fluorophore-labeled antispecies Ab results in fluorescently labeled microspheres if Ab is present in analyte serum; fluorescence detected by flow cytometry	Amplification is minimal since only amplification results from multiple fluorophores on detector Ab; detection can be quite sensitive since laser-induced fluorescence is used; possibility of multiple-analyte analysis; flow cytometric assay reported to be fivefold more sensitive than microplate-based EIA[108]	[108]

Total internal reflection spectroscopy with fluorescence detection	Immunological reaction between Ag and Ab monitored at interface between optical waveguide and sample solution; detected by fluorescence excited by evanescent wave resulting from total internal reflection; results in detection of Ag/Ab binding events only *very* near surface at which internal reflection occurs	Capture Ab immobilized on waveguide binds Ag in sample; second Ab with fluorescent label binds to Ag; fluorescence detection close to waveguide (evanescent wave) monitors Ag–fluorophore-labeled Ab binding event; no washing necessary; requires rather elaborate optical setup; free fluorophore-labeled Ab contributes to fluorescence seen at waveguide/solution interface; thus, Ag/Ab binding event seen as an incremental increase in fluorescence Ag (hapten) immobilized on waveguide surface; fluorophore-labeled Ab binds to immobilized Ag and is detected by evanescent-wave excitation; free Ag in solution competes with immobilized Ag for limited amount of Ab; signal decreases as amount of free Ag in solution increases Or Fluorophore-labeled Ag binds to immobilized Ag on waveguide surface and is detected by evanescent-wave excitation	[109] [110] [111]
Fluorescence polarization in immunoassay (FPIA)	A change (increase) in fluorescence polarization occurs when fluorophore-labeled Ag binds to Ab—due to changes in rotary motion attendant upon large increase in size of immune complex compared to Ag alone; competitive assay may be used with free Ag and fluorophore-labeled Ag; can also determine specific Ab levels using fluorophore-labeled Ag; FPIA using fluorophore-labeled Ab to detect Ag does not appear to work due to flexibility of Ab or relatively free rotation of attached fluorophore	Amplification minimal—fluorescent labeling of Ag limited to low-MW Ags if fluorophores with nanosecond lifetimes used; competitive, homogeneous assay possible; extent of change in fluorescence polarization inversely proportional to free Ag concentration; no separation of bound vs. free Ag required for assay; sensitivity of assay strongly dependent on Ab affinity; potentially very short assay time; low background; potential for high sensitivity; other proteins in sample may interfere with assay if they bind to fluorophore-labeled Ag; sensitivity sub-ng/ml[112]	[112–115]
FPIA with metal–ligand complexes	The same principle as FPIA above, but use of transition metal (e.g., Re, Ru)–ligand complexes allows fluorescence polarization assays for high-MW antigens; these complexes display high polarization in the absence of rotational diffusion and long lifetimes	Competitive, homogeneous assay; metal–ligand complexes do not show probe–probe interactions, and so higher metal–ligand complex/Ag protein ratios can be used; amplification in terms of fluorophore groups is small; metal–ligand complexes have low extinction coefficients and low quantum yields; long lifetimes allow off-gating of fluorescence interferences	[116–118]

(continued)

TABLE 16.1 (continued)   Immunoassay Amplification Schemes

Assay	Principle	Comments	References
Time-resolved FIA	Immune *sandwich* formed in the usual way, with one of the components being biotinylated Ab; the biotinylated Ab serves as recognition sites for streptavidin-containing macromolecular complex, which also contains multiple chelator molecules saturated with Eu^{3+}; after reaction with immune reagents, unbound macromolecular complex is washed away, and Eu chelate associated with solid phase is measured by time-resolved fluorescence; homogeneous assay not feasible because of instability of Eu chelate/macromolecular complex; particular chelator used forms highly fluorescent, stable complex with Eu^{3+}, and can be covalently linked to proteins	Heterogeneous assay; capture Ab immobilized on solid surface captures Ag; biotinylated Ag-specific Ab also binds to Ag; addition of streptavidin–thyroglobulin–chelator–Eu^{3+} macromolecular complex results in some complex being immobilized at immunoreagent sites, if biotinylated Ab is bound; excess macromolecular complex is washed away, wells are dried, and Eu^{3+} fluorescence determined by time-resolved techniques; large amplification due to hundreds of Eu^{3+} bound in macromolecular complex; use of the macromolecular complex resulted in improved detector limits for a number of antigens, in comparison with streptavidin–thyroglobulin–Eu^{3+} chelator complexes, which were of smaller size; laser excitation used; detection limit for α-fetoprotein was 60 attomoles/well[119], enhancement of Eu fluorescence when multiple Eu chelator complexes are bound to protein; thus when 160 chelator molecules are bound to thyroglobulin, in the presence of excess Eu^{3+}, fluorescence equivalent to 900 molecules of unconjugated Eu^{3+} chelator is obtained[119]	[119,120]
	Also		
	Immune complex formed in usual way; the same chelator as above is used, but Ab directly labeled with chelator + Eu^{3+}; chelator binds to NH_2 groups on proteins; after washing and drying, fluorescence of Eu^{3+} is measured	Too much labeling of proteins with chelator interferes with biological activity of proteins (e.g., Ab/Ag interaction); extent of labeling tolerated before biological activity decreases depends on particular protein; for 1 Ab, 18 Eu^{3+} chelator complexes could be attached per Ab molecule, without adverse effect; competitive heterogeneous immunoassay for cortisol demonstrated	
Dissociation-based time-resolved FIA (DELFIA)	Either Ab or Ag is labeled with stable hydrophilic lanthanide (Eu, Tb) chelate, which is nonfluorescent in aqueous solution; immunoassay takes place in standard way (i.e., immune complex *sandwich*); after all immunoreactions finished, the lanthanide chelate is dissociated with an enhancement reagent, which contains a diketone that chelates the lanthanide in a micellar environment, resulting in a highly fluorescent solution; separation step required to eliminate labeled reagent not bound in immune complex; new chelate with lipophilic reagent is formed, enhancing lanthanide fluorescence; time-resolved fluorescence (delay time between excitation and emission signal) determined	Can be heterogeneous or homogeneous format; time-resolved fluorescence virtually eliminates background and interferents; generally requires laser excitation; only amplification is in the number of metal chelates per Ag or Ab; narrow emission bands suggest use of multiple labels (more than one lanthanide chelate) for dual-analyte assays	[121–125]

EIA with time-resolved fluorescence detection	Immunoassay carried out like regular EIA; enzyme substrate consists of a chemical, which does not form a chelate with lanthanide ion (Tb^{3+}, Eu^{3+}), but after enzyme activity, a product is formed, which chelates to lanthanide ion to form highly fluorescent complex; fluorescence measured by time-resolved methods or second-derivative synchronous fluorescence; luminescence (fluorescence) increase directly related to enzyme activity; mainly uses Tb^{3+}	Heterogeneous assay format; time-resolved fluorescence advantages (see above); attomole sensitivity observed for one Ag species[126]; possibility for rapid assay; lanthanide chelates have high quantum yields, large Stokes shifts, narrow emission peaks; potential for several orders of magnitude increase in sensitivity over conventional fluorescence[127]; Tb^{3+} has advantages over Eu^{3+} for this type of assay; observed to be ~30-fold more sensitive than assay where no enzyme amplification was used, but where Eu chelate served as label for immunoreagent (thus also time-resolved fluorescence detection)[128]; 300-fold more sensitive than colorimetric EIA; ~20 attomole of Ag detected	[126–132]
FRET	Fluorophore donor/Ag conjugate (β-phycoerythrin/thyroxine) competes with free Ag in sample for binding to Ab–fluorophore acceptor conjugate (antithyroxine Ab/Cy5); when labeled Ag binds to labeled Ab, FRET occurs; this transfer is detected by measuring changes in fluorescence lifetime, performed by phase modulation methodology	Homogeneous, competitive assay; particular pair of dyes chosen allows FRET to occur over greater distances than for most acceptor–donor pairs[133]; eliminates many sources of fluorescent interference; instrumentation rather complex; laser required	[133]
	Or		
	Two fluorophore-labeled Abs bind to different epitopes on analyte Ag; when immune complex forms, excitation of one fluorophore (donor) results in transfer of energy to second fluorophore (acceptor) because of close apposition; emission from second fluorophore signals formation of immune complex	One Ab labeled with europium (III) cryptate binds to Ag; second Ab labeled with allophycocyanin (acceptor) also binds to other epitope on Ag; excitation by laser of donor results in transfer of excitation energy to donor if both are bound to the same Ag molecule; signal detected by time-resolved technique	[137]
	Or		
	Two fluorophore-labeled Abs used, which recognize the same Ag; one fluorophore is a Eu^{3+}-labeled cryptate; the other fluorophore (on different Ab) is allophycocyanin; formation of immune (Ab/Ag/Ab) complex results in nonradiative energy transfer; excitation of Eu^{3+} label, coupled with appropriate filtering and time-resolved detection, results in the detection of fluorescent signal arising from allophycocyanin due to energy transfer in immune complex; fluorescence from Eu^{3+}-labeled Ab free in solution or allophycocyanin-labeled Ab free in solution not detected due to elimination by optical filtration or time-resolved measurement	Laser excitation; homogeneous assay format; theoretically, only fluorescence from immune complex detected; amplification arises because of high value of quantum yield from allophycocyanin *and* the high efficiency of the energy transfer; therefore, overall fluorescence yield from immune complex higher than would arise from Eu^{3+} alone; detection limit for prolactin found to be as good as for radioactive assay for this same Ag[134]	[134] [135] [136]
	Or		

(continued)

TABLE 16.1 (continued)   Immunoassay Amplification Schemes

Assay	Principle	Comments	References
	Fluorescamine-labeled Ag reacts with fluorescein-labeled Ab in solution; two fluorophores matched so that excitation of fluorescamine on Ab results in energy transfer so that fluorescein on Ab is excited and fluoresces; addition of unlabeled Ag results in decrease in fluorescence, since fewer labeled Ags are bound in immune complex and participate in FRET	Competitive, homogeneous assay format; amplification due to multiple fluorophores on Ab; approximately equivalent limit of detection compared to ELISA observed, when the same Ab reagents used[135]	
	Or		
	Ag labeled with chemiluminescent (isoluminol) derivative; binding to fluorescein-labeled Ab sets up conditions for nonradiative energy transfer; addition of microperoxidase and H_2H_2 after formation of immune complex results in emission of light due to chemiluminescent reaction and energy transfer to fluorophore-labeled Ab, resulting in fluorescence emission from the Ab-associated fluorophore, since donor/acceptor species are designed to allow FRET	Homogeneous, competitive assay format; amplification due to multiple fluorophores on Ab (4–12 fluoresceins/Ab molecule); since both chemiluminescence and fluorescence occur, ratio of light intensity at two wavelengths used to evaluate extent of energy transfer (and thus, indirectly, to evaluate extent of Ag and Ab binding); assay applied to a variety of Ags, for example, camp, progesterone, IgG; rather complicated experimental apparatus, as chemiluminescence/fluorescence monitored simultaneously; assay for camp found to be comparable in sensitivity to radioimmunoassay for cAMP[136]	
Fluorescence excitation transfer immunoassay (FETI)	Fluorophore-labeled Ag binds to Ab labeled with a different fluorophore; immuno-specific binding event results in fluorescence excitation transfer, where donor fluorophore emits in the same spectral region as acceptor fluorophore absorbs; fluorescence (at donor excitation energies) is quenched when immune complex forms; can also be used for multiepitope Ag, with two different Ag-specific Abs, labeled with donor and acceptor fluorophores	Competitive-type assay; homogeneous; amplification only due to multiple fluorophores on Ag and Ab; requirement for dual labeling	[138] [98]
	Or		

Technique			
	Unlabeled Ag in sample competes with fluorophore-labeled Ag (i.e., *fluorescer* Ag) for binding to Ab, which is labeled with a *quencher* (i.e., a fluorophore whose absorption wavelength is closely matched to the fluorescence emission of the *fluorescer* Ag); binding of fluorophore-labeled Ag to quencher-labeled Ab, followed by excitation of the *fluorescer* fluorophore with light of the appropriate wavelength, results in energy transfer to quencher, due to short distance between *fluorescer* and quencher in immune complex and decrease in fluorescence; the more Ag in sample, the less *fluorescer* Ag binds to Ab and therefore the more intense the resulting fluorescence	Homogeneous, competitive assay format; increasing fluorescence obtained with increased concentration of Ag in sample, that is, signal directly proportional to concentration of Ag in sample[98]; amplification due to multiple fluorophores on *fluorescer* Ag; various pairs of *fluorescer* and quencher pairs may be used; for example, fluorescein as *fluorescer* coupled with β-phycoerythrin as quencher or β-phycoerythrin as *fluorescer* may be coupled with Texas Red as quencher[98]	[139,140]
Fluorescence modulation immunoassay	Homogeneous assay; Ag-specific Ab and hydrolytic enzyme are in solution; fluorogenic-labeled hapten Ag is prepared by synthesis; fluorogenic label is a substrate for enzyme listed above, such that action of the enzyme on the fluorogenic-labeled hapten releases a fluorescent product upon enzymatic hydrolysis; when fluorogenic-labeled Ag and unlabeled Ag in solution are added to Ab and enzyme solution, fluorogenic-labeled Ag bound by Ab is unavailable for enzymatic hydrolysis; any free labeled Ag is acted upon by enzyme to produce fluorescent product; only applied to hapten Ag	Competitive assay format; fluorescence, produced by enzymatic action, directly proportional to amount of Ag in analyte solution; amplification by enzyme activity; results obtained with this assay correlated well with EMIT assay[139], potentially very rapid assay, that is, 20 min,[140] and simple to perform	[141]
Substrate-labeled fluorescent immunoassay (SLFIA)	Similar to substrate-labeled immunoassays where Ab binding to an analyte–substrate conjugate inhibits action of enzyme on substrate; addition of analyte (Ag) in sample results in more sample analyte being bound to Ab, consequently freeing more of analyte–substrate conjugate for breakdown by enzyme; however, analyte (Ag)–substrate conjugate is constructed to contain a fluorophore linked to a quencher via an enzyme-hydrolyzable linkage, while the analyte (Ag) is conjugated to the fluorophore; free analyte–substrate conjugate (not Ab-bound) can be hydrolyzed by enzyme, thus releasing inhibition of fluorescence due to energy transfer to quencher and resulting in full fluorescence of fluorophore	Homogeneous, competitive assay format; amplification due to enzyme action; fluorescence increases as amount of free Ag (analyte) in sample increases, so signal is directly proportional to free Ag; Li and Burd[141] use AMP bound to FMN as quencher/fluorophore pair, respectively; Ag (theophylline) is bound to AMP; if antitheophylline Ab is unavailable for binding due to amount of theophylline in sample, then nucleotide pyrophosphatase in solution hydrolyzes AMP-FMN linkage, restoring FMN fluorescence	

(continued)

TABLE 16.1 (continued)   Immunoassay Amplification Schemes

Assay	Principle	Comments	References
Chemiluminescence procedures	*The same concept as fluorescence IA; either an immunoreactant is labeled with a chemiluminescent compound or an enzyme substrate is used, which produces a chemiluminescent product upon enzyme action*	*Potentially very sensitive; additional advantage in that background should be very low, since no exciting light source is needed (as is the case with fluorescence)*	
Chemiluminescence immunoassay (CIA)	Capture Ab used to immobilize Ag on solid surface as per standard assay; second Ag-specific Ab with chemiluminescent label (acridinium ester) binds to immunocomplex; after washing, addition of appropriate reagents produces chemiluminescence (flash) Or Ag labeled with chemiluminescent derivative (isoluminol); Ag-specific Ab immobilized on solid surface; unlabeled Ag in sample competes with labeled Ag for binding to Ab; following binding and washing, addition of appropriate reagents produces chemiluminescence (flash) Or Chemiluminescent label chemically attached to Ag; addition of free (unlabeled) Ag in sample sets up competitive assay; Ab binding to chemiluminescent-labeled Ag enhances chemiluminescence significantly; free Ag competes for Ab binding sites—the more Ag in the sample, the less enhanced chemiluminescence is seen	Multiple chemiluminescent labels per Ab produce some amplification (see entry under FIA); heterogeneous noncompetitive assay; sensitivity should theoretically increase with increase in specific activity of label (i.e., increased moles of label per mole Ab); calculated detection limit <1 attomole[142]; CIA observed to be more sensitive and with wider working range than radioimmunoassay; chemiluminescent immunoassay using luminol-labeled Abs found to be as sensitive as radioimmunoassay—22 luminols per Ab molecule[143]; losses in quantum yield after conjugation of luminol to Ab; sensitivity of CIA for thyrotrophin described as far more sensitive than other immunoassays[144] Heterogeneous, competitive assay; light yield inversely proportional to concentration of analyte; rate of light production not constant, so timing of addition of reagents very important; chemiluminescent materials not usually present in biological samples, so no additional background from this source Homogeneous, competitive assay format; Ab concentration has to be carefully adjusted; burst of light requires specialized measurement apparatus	[142–149]

(continued)

Chemiluminescence enzyme immunoassay (CLEIA)	Immunoassay done in the same way as EIA; enzyme label on one of immunoreagents; enzyme activity on substrate results in product, which is chemiluminescent, usually a *glow* chemiluminescence, or action of enzyme on substrate (i.e., hydrogen peroxide) generates product, which reacts with other chemicals (luminol) in reaction mixture to produce chemiluminescence; enzymes used include AP, HRP, and β-galactosidase	Very low background; inherently highly sensitive; enhancers produce stronger, more stable light signal; proportionality over large ranges of analyte concentration; using a derivative of adamantly 1,2-dioxetane phosphate as substrate, $<10^{-20}$ moles of AP were detected on membrane or in solution[150]; light emission persists for many minutes; more sensitive than colorimetric EIA or radioimmunoassay; sensitivity for CLEIA using AP as enzyme label 67-fold improved over colorimetric endpoint and approximately threefold improved over time-resolved fluorimetric assay[151]; possibility for very rapid total assay time; photographic film can be used as detector; very rapid response to reach peak light emission (3 min or less)[152]; integration of chemiluminescent signal for longer periods of time can be done to increase sensitivity for *glow*-type chemiluminescent reactions[153]	[150–162]
			[163]
	Establishment of sensitivity study; enzyme (AP) dilutions added to substrate + enhancer; reaction results in chemiluminescence	1.6 zeptomoles of enzyme could be detected with a signal-to-noise ratio of >6; luminescence increases with time and reaches plateau by 40 min; proportionality of response over five orders of magnitude concentration of enzyme; enhancer provides 400-fold increase in chemiluminescence efficiency[154]	
	Enzyme-labeled (peroxidase) Ab or Ag used in either noncompetitive or competitive (enzyme-labeled Ag) format in standard immunoassays; in the case of enzyme-labeled Ag assays, labeled and unlabeled (sample) Ag compete for limited number of immobilized Ab molecules; substrate + luminol + enhancer added and chemiluminescence (glow) read 30–60 s later	Heterogeneous assay format; direct comparison of colorimetric, fluorimetric, and chemiluminescent assays, using the same immunoreagents in the same assay format demonstrated that the chemiluminescent assay was 10-fold more sensitive than the colorimetric and ~2.5-fold more sensitive than the fluorimetric assay[155]	
	Ag (bacterial) immobilized on membrane; enzyme-conjugated Ab added, followed by addition of substrate, enhancer, and luminescent chemical; chemiluminescence detected	Single-cell luminescence detected using a CCD camera and image processor[156]	
	Immunoblot analysis on membranes, following electrophoresis, is carried out using immunoreagents in standard way (Ag-specific primary Ab; antispecies IgG secondary Ab conjugated to enzyme); chemiluminescent substrate added and chemiluminescence measured; can modify immunoreagent *sandwich* by using biotinylated secondary Ab and enzyme-conjugated streptavidin	Chemiluminescent substrates offer advantages over colorimetric substrates including higher sensitivity, shorter incubation times, potential for blot reuse, and ease of imaging; amplification by enzyme turnover and by localization of multiple enzyme molecules in immunoreagent *sandwich*	

TABLE 16.1 (continued)   Immunoassay Amplification Schemes

Assay	Principle	Comments	References
Electrochemiluminescent labels	Electrochemiluminescent-active species bound to Ab, Ag, or hapten Ag; chemiluminescence is produced by electrochemical generation of chemical species at an electrode; example of active species reaction: ruthenium (II) tris(bipyridyl)$^{++}$, with tripropylamine; as the electrochemistry occurs, an excited state of the ruthenium complex arises, and this excited state decays with emission of a photon, that is, chemiluminescence Or Ag-specific Ab immobilized on electrode surface; Ag is captured by immobilized Ab, but second detector Ab labeled with terbium chelate also binds to immune complex of Ag/Ab; either excess labeled Ab is washed away or not eliminated; since electrochemiluminescence occurs only very near electrode, removal of labeled Ab not in immune complex may not be necessary; application of voltage produces luminescence, which is measured by time-resolved techniques	Cycling of active species through electrochemical oxidation–reduction cycles at electrode produces many photons (multiplication); each electrochemiluminescent label on immunoreactant can therefore emit many photons during this cycling; also, can label Ab or protein Ag with many electrochemiluminescent labels; assay restricted to certain formats (e.g., microbeads, solution), due to requirement for close approach of labeled molecules to electrode; somewhat complicated experimental setup; very large dynamic range (six orders of magnitude); assay can be set up with competitive or noncompetitive, heterogeneous or homogeneous format Homogeneous (no wash) or heterogeneous assay format; amplification occurs only by multiple labels on Ab; low background	[164–166] [167]
CLEIA with coupled enzyme partners	Immunoreactions carried out in usual way; enzyme label detected indirectly as follows; primary enzyme (β-galactosidase) bound to detector Ab reacts with substrate to generate product (galactose), which becomes substrate for second enzyme (galactose dehydrogenase); product (NADH) of activity of second enzyme reacts with chemical + O_2 + isoluminol and microperoxidase to produce chemiluminescence Also Detector Ab conjugated to AP; addition of FADP results in generation of FAD; FAD reacts with glucose oxidase (apoenzyme) to form active holoenzyme; addition of glucose results in production of H_2O_2, which reacts with another chemical to produce chemiluminescence	Overall procedure is rather complicated, and assay time is long; detection limits for β-galactosidase reported to be ~10^{-20} moles for 2 h incubation and 10^{-21} moles for 1000 min incubation; immunoassay using this procedure for 17α-hydroxyprogesterone reported to be 43-fold more sensitive than colorimetric ELISA[169], immunoassays formatted as heterogeneous competitive Heterogeneous, noncompetitive assay for thyrotrophin demonstrated; LOD of 4×10^{-19} moles for AP[169], *glow* luminescence	[169–173]

	Or		
	β-galactosidase activity indirectly coupled to bacterial luciferase activity, using NADH generated from coupled enzyme system to feed into two enzyme-catalyzed reactions, the final one involving luciferase, to generate light	Complicated procedure with many steps; long assay time; detection limit of β-galactosidase for 2 h incubation is $\sim 10^{-21}$ mole and $\sim 10^{-22}$ mole for 1000 min incubation[170]	[173–176]
Liposome immunoassays	*Liposome, either directly or indirectly, becomes part of immune complex; release of liposome contents produces signal for detection*	*Potential for large amplification because liposomes can encapsulate many molecules*	
LILA (liposome immune lysis assay)	Liposomes constructed with Ag-specific Ab in liposomal membrane and fluorescent dye encapsulated within; addition of Ag followed by second Ag-specific Ab + complement results in lysis, releasing dye	Homogeneous assay as described; assay time <1 h; some nonspecific lysis observed	[177]
	Or		[178]]
	Ag immobilized indirectly on liposome surface through streptavidin–biotin bridges; addition of complement + Ag-specific Ab results in lysis	Competitive, homogeneous assay as described; unique method for immobilizing protein through streptavidin–biotin bridge	
	Or		
	Ag immobilized directly on liposome surface reacts with Ag-specific Ab; addition of complement lyses liposome, releasing its contents	Homogeneous, competitive assay; competition for liposome-immobilized Ag and Ag in sample	
Immunoliposome-based immunoassay	Liposomes with Ag-specific Ab incorporated in membrane and encapsulating fluorescent dye are formed; incubation with immobilized Ag results in Ab–liposome immobilization; lysis of liposomes with ethanol/detergent releases fluorescent dye; fluorescence proportional to Ag concentration	Competitive, heterogeneous assay, where free Ag concentration decreases number of liposomes subsequently trapped on Ag-treated surface; total assay time somewhat long	[180]
	Liposomes formed with encapsulated fluorescent dye and biotin in the liposomal membrane; Immobilized Ab captures Ag; second biotinylated Ag-specific Ab binds to immobilized Ag; streptavidin forms bridge between immune complex and liposome; after washing to remove interferents, immobilized liposomes lysed, releasing fluorophore; can also be formatted with streptavidin incorporated in liposomal membrane	Heterogeneous, noncompetitive assay; fluorescence directly proportional to concentration of Ag captured; multiplication due to multiple fluorophores per liposome; stated to be as sensitive as best colorimetric ELISAs for Ag tested[179]	[179,181]
Liposome lysis (substrate)-linked immunoassay	Liposomes with Ag in membrane and encapsulating an enzyme substrate are constructed; presence of Ag-specific Ab and complement lyses liposomes, releasing substrate; addition of substrate-specific enzyme results in optically detected product	Could be used to detect specific Ab; potential for homogeneous assay	[182]

(continued)

TABLE 16.1 (continued)   Immunoassay Amplification Schemes

Assay	Principle	Comments	References
Liposome lysis (reporter)-linked immunoassay	Immunoreaction of Ag–enzyme conjugate with Ab in solution inhibits enzymatic activity (e.g., phospholipase C), which can lyse liposome, thereby releasing liposomal contents, that is, fluorescent dye molecules; free (sample) Ag competes with enzyme-labeled Ag; increasing free Ag results in increased liposome lysis	Multiplication due to many reporter molecules encapsulated within each liposome; homogeneous, competitive assay as described; said to be sensitive as heterogeneous EIA[183]	[183] [184,185]
	Liposomes formed with encapsulated reporter enzyme; free Ag in sample competes for binding to Ag-specific Ab with Ag conjugated to melittin (a cytolysin); Ab binding to Ag–melittin conjugate inactivates cytolytic activity of melittin; Ag–melittin conjugate not bound by Ab lyses liposomes, releasing enzyme, which reacts with substrate in solution	Multiplication due to many reporter enzymes per liposome plus enzyme activity; as described, homogeneous, competitive assay; Ag must be of low MW for effective inhibition of melittin by Ab binding	
Immunoliposome assay—dye-sensitive photobleaching	Immune *sandwich* (Ab/Ag/Ab attached to liposome) formed; *Ab is Ab with attached erythrosine; liposome has fluorescent compound embedded in membrane; when formed *sandwich* is illuminated by Hg lamp, erythrosine generates singlet $^1O_2$, which photooxidizes fluorescent dye in membrane, decreasing fluorescence; singlet O_2 only able to oxidize membrane-embedded dye if site of generation is very close to the liposome surface; can also be set up as competitive Ag assay with erythrosine-labeled Ag competing with unlabeled Ag	Competitive and sandwich-type formats both possible; homogeneous assay; detection limit about the same as or better than conventional solid-phase assays[186]	[186]
Immunoliposome assay—fluorescence energy transfer	Liposome contains Eu chelate embedded in liposomal membrane and membrane-attached Ab; free Ag competes with allophycocyanin-labeled Ag; when dye-labeled Ag binds to Ab and long-lived fluorescence measured, excitation of Eu results in energy transfer to allophycocyanin, which is detected; energy transfer efficient only when two dyes close together	Competitive, homogeneous assay as described; potential for very low background; only demonstrated with biotin as Ag	[187]

Method	Principle	Comments	Ref.
Expression immunoassay	Immunoreactant complex formed as usual; DNA attached to immune complex codes for protein which, when expressed, produces signal for detection	*Potentially large amplification, both because of number of protein molecules synthesized and the fact that the synthesized protein is an enzyme*	[188]
	Immobilized Ag reacted with biotinylated Ag-specific Ab; streptavidin bridge couples biotinylated Ab to biotinylated DNA segment, which codes for α-peptide of β-galactosidase; addition of transcription/translation system produces α-peptide, which then reacts with inactive M15 protein to reconstitute active enzyme. Addition of substrate results in a signal, for example, chemiluminescence	Assay carried out as solid-phase, heterogeneous assay; transcription/translation system potentially multiplies final active enzyme many-fold over other methodologies; theoretically, background should be zero, since active enzyme should not be present unless immune complex generates component needed for activity; assay procedure more complex (more steps) than usual ELISA and of longer duration; potential for very high sensitivity, depending on chemical (chemiluminescent, fluorescent) and/or enzyme chosen[188]	[190]
	The same general design as above, but DNA codes for firefly luciferase; assay endpoint is bioluminescence	Estimated 12–14 molecules of luciferase protein produced from each DNA template; comparison with enzyme amplified, time-resolved FIA showed expression immunoassay to be considerably more sensitive[189]	
Immunoassays/chromatography			
Immunoaffinity chromatography	Ag-specific Ab immobilized on columns; analytes in sample labeled in TOTO with fluorescent dye; analytes bound on Ag-specific immunoaffinity columns and eluted sequentially; detection by laser-induced fluorescence	Potentially rapid and sensitive; may be nonuniform labeling of analytes in sample; multiple fluorophores may be attached to single-analyte molecules; detection levels at pg/mL for various cytokines	[190]
SERS immunoassays	*The same principle as usual IA, but SERS-active chemicals are used as labels or substrates; SERS effect produces large signal amplification*		
Heterogeneous indirect immunoassay	Capture Ab adsorbed on rough silver surface captures Ag analyte; Ab labeled with SERS-active molecules binds to Ag; SERS signal or surface-enhanced resonance Raman scattering signal is detected	Multiple SERS labels on each Ab; 10^3–10^6 (or higher for resonance Raman) enhancement in Raman signal from SERS effect; requires close apposition of SERS label to silver surface; potential for minimal assay steps (e.g., no washes) since only SERS label bound immunologically to silver surface produces signal; marked reduction in fluorescence background; special equipment needed to generate surfaces; laser required for measurements	[191,192]
	The same as above, but capture Ab adsorbed on gold surface; Ag captured and detected by Ab attached to gold nanoparticles, which in turn are labeled with SERS-active molecules	Simultaneous detection of multiple Ags using different SERS-active labels; very narrow Raman bands provide potential for sensitive multianalyte detection; Ab bound to nanoparticles noncovalently; possible problem with Ab dissociation/reassociation leading to cross-talk	

(continued)

TABLE 16.1 (continued) Immunoassay Amplification Schemes

Assay	Principle	Comments	References
EIA with SERS detection of enzyme product	Enzyme immunoassay carried out in standard way on microplates; after enzyme reaction (peroxidase), reaction product (azoaniline) is transferred to silver substrate and the surface-enhanced resonance Raman signal is determined; intensity of signal correlates directly with Ag concentration	See above; advantages over other techniques include low background, high selectivity, amplification from enzyme, as well as SERS effect; disadvantages are the elaborate equipment and the extra detection step at the end of the assay; detection limit stated to be 10^{-15} moles/L (for IgG)	[193]
Imprinted polymer-based immunoassays	*Polymer with imprinted molecular sites is the immuno-recognition element rather than Ab; otherwise IA is the same*		
Chemiluminescent ELISA	Imprinted polymer serves as recognition element instead of Ab; polymer immobilized in tubes or wells; antigen (hapten) labeled with enzyme competes with Ag in analyte solution for imprinted sites in the polymer; following washing, enzyme activity associated with polymer determined by chemiluminescent assay	Heterogeneous, competitive assay format; signal inversely proportional to concentration of Ag in analyte; detection limit not as low as when usual ELISA assay with AB was used; useful range extends over four orders of magnitude; rapid and potentially robust assay; whether molecularly imprinted polymers can serve as Ab substitutes in the wide range of Ags currently detected by immunoassay is still an open question	[194,195]
Miscellaneous heterogeneous immunoassays	*These procedures involve standard immunoreagents, but also entail the use of beads, particles, crystals, etc. which serve as solid substrates for immunoreactions*		
Bacterial magnetic particle-based immunoassay	Magnetic particles (with lipid membranes) produced by bacteria are used as solid phase; Ag-specific Ab covalently immobilized on particles captures Ag; second (detection) Ab has enzyme label; detection by chemiluminescence (for enzyme-labeled Ab) Or Magnetic particles from bacteria are labeled with fluorescently labeled Ab; addition of Ag results in particle aggregation, which produces decrease in fluorescence due to self-quenching by fluorophore	*Natural* magnetic particles disperse better than artificial; small size (50–100 nm) gives very rapid reaction kinetics; chemiluminescence using enzyme amplification was found to be very sensitive with detection limit of 6.7 zeptomole[196]; detection range extended over $1-10^5$ fg/mL; a rapid chemiluminescence assay (~10 min) was also devised but with detection limits of ~10 ng of IgG/mL Amplification results from numerous fluorescently labeled Ab molecules bound to each magnetic particle; inverse relationship between Ag concentration and fluorescence; bacterial magnetic particle assay found to be more sensitive than assay using artificial magnetic particles[197]; applied to detection of bacterial cells (*E. coli*) with limit of detection ~10^2 cells/mL[198]	[196] [197,198]

Phosphor crystals	Immunoreagents (Ab, avidin, protein A) directly linked to phosphor particles; assays carried out on membranes or cells; Ag immobilized on membrane or on cell surface detected by immunoreagent sandwich incorporating phosphor conjugates in one or more layers of the immune sandwich; fluorescent detection	Amplification based on formation of immune sandwich (i.e., biotinylated Ab, avidin); fluorescence not affected by environmental factors and does not fade; sensitivity found to be as good as ELISA with colorimetric endpoint[199] Also applications to immunocytochemical analyses, using immunoreagents linked to phosphor particles	[199] [200]
Microspheres with CARD amplification	Microspheres with encoding dye (to permit performance of multiple assays simultaneously) were derivatized with Ag on their surface; enzyme-labeled Ab was added to mixture of microspheres + Ag-containing analyte solution; any enzyme-labeled Ab bound to bead-immobilized Ag is detected using CARD technology; fluorescent tyramide is deposited on microspheres and fluorescence read	Competitive, heterogeneous assay format; fluorescent signal decreases as Ag concentration in analyte solution increases; encoding dye on microspheres allows multiplex assays; choice of fluorescent dyes has to be carefully made, so discrimination between fluorescence of microsphere and tyramide derivative can be made; preparation of Ag-derivatized, dye-encoded microspheres somewhat complex	[201]
Immunomagnetic separation + immunoassay	Immunoreagents (e.g., capture Abs) immobilized on magnetic beads, so that analyte can be extracted/isolated from bulk sample; rest of immunoassay can take place in usual manner (i.e., formation of immunosandwich with or without enzyme conjugate, incubation with substrate, and detection)	Total assay time <1 h[202], fluorescence ELISA using immunomagnetic separation was 10-fold more sensitive than standard ELISA[202] Electrochemiluminescence-based assay (see under chemiluminescence) compared with immunomagnetic separation/fluorescence ELISA; electrochemiluminescence-based assay found to be more sensitive in general, perhaps 10-fold[203]; both assays could be completed in <1 h	[202,203]
Immunogold-based immunoassays	*Gold nanoparticles serve as label on immunoreagents and as nuclei for deposition of silver atoms*		
Immunogold/silver assay	Ag immobilized on membrane or in histological specimen; Ag-specific Ab binds to Ag; secondary antispecies IgG Ab labeled with gold nanoparticles binds to primary Ab; silver subsequently precipitated on gold particles	Addition of silver strongly amplifies gold immunostaining; used for immunoblots/dot blotting; 2 h incubation period; found to be more sensitive than PAP or ABC method (with the same time of incubation with detector Ab)[204]	[204,205]
Immunoassays incorporating aequorin			
	Immune *sandwich* with primary Ab and Ag formed in usual way; secondary (antispecies IgG Ab), which is biotinylated, is then incorporated into complex; subsequent sequential additions of streptavidin and biotinylated aequorin are made, resulting in immobilization of aequorin at site of immune complex; addition of Ca^{2+} results in luminescent flash, which can be detected	Applied to both microwell and membrane blot assays; heterogeneous format; amplification because of multiple binding sites on streptavidin for biotin; detection may be challenging since light flash decays very significantly by 1 s; immunoassay for Forssman antigen using aequorin was demonstrated as detecting less than 40 femtomoles	[206]

described as homogeneous/heterogeneous, competitive/noncompetitive, etc. However, every immuno-assay procedure is not labeled in this way, because very many assays have the potential to be applied in multiple fashions.

16.2.2 Amplification and Specificity

Table 16.1 presents a summary of various immunoassay techniques that employ amplification as a component of the assay. It should be recognized that amplification has multiple layers of meaning with regard to immunoassays. Fundamentally, in these assays, one wishes to detect the event of antibody binding to antigen. If the antibody has an attached enzyme, then the unique immunoreaction event is amplified by the processivity of the enzyme acting on the appropriate substrate. On the other hand, by use of immune complexes (e.g., PAP, peroxidase–antiperoxidase; APAAP, alkaline phosphatase [AP]–anti-AP), the number of enzyme molecules clustered at the site of a single Ab/Ag interaction can be multiplied.[5,6] These are just two examples of many that could be offered—many immunoassays have more than one multiplication strategy. As a final goal, all these amplifications serve to amplify the detection signal, hopefully lifting it above the noise of the system. In this table, we present a distillation of a large compendium of techniques. Table 16.1 presents a brief summary of the principle behind the assay, some comments which the authors consider germane to the discussion (including further explanation of amplification and comments about sensitivity) as well as pertinent references.

Of course, the fundamental question arises: Why is amplification needed at all? The simplest answer is that the molecular interaction, which is desired to be detected, is undetectable without amplification. Of course, as the various immunoassay techniques developed over the years, the natural tendency was to push for greater and greater sensitivity, as the biological questions being asked demanded more powerful techniques. Hence, a wide variety of amplification techniques have been developed of, some of which admittedly have less applicability than others.

Some of the amplification techniques included in Table 16.1 seem to have minimal amplification associated with them such as fluoroimmunoassay (FIA), where a fluorescent label is covalently attached to an antibody.[7] In an attempt to be consistent, any technique that results in *some* amplification of a signal could be legitimately called amplification and will therefore be included. For example, in the simple direct FIA, where fluorophore is attached to antibody, if only one fluorophore is attached per Ab molecule, then the assay would not be considered to be amplified, since single-molecule Ag/Ab binding would only be detected by one fluorophore. In fact, there are multiple potential sites for fluorophore attachment per Ab molecule, and the usual fluorophore/Ab molar ratio following the labeling reaction is 2–4:1. Actually, controlling the reaction to insert only one fluorophore would be difficult if not impossible. Nevertheless, the degree of amplification achieved by fluorophore labeling of Ab is certainly limited to relatively small numbers. Too many fluorophores attached can significantly alter the affinity/specificity of the Ab. In fact, the major reason simple FIA has become such a widespread technique is the fact of the inherent extreme sensitivity of fluorescence. Even with only a few molecules of fluorescent label attached per Ab molecule, the Ag/Ab binding event is still amenable to detection.

In general, in Table 16.1, techniques that give high levels of amplification, as opposed to those that produce only small amounts of amplification, are emphasized. However, the extent of amplification is not necessarily synonymous with sensitivity. For instance, referring to fluorescence, fluorescence polarization or time-resolved measurements can be highly sensitive detection schemes, even though only a few molecules of dye are attached to an Ab. In spite of the aforementioned comments, the more powerful amplification schemes, where the potential for manifold multiplication of signal exists, are inherently attractive, if for no other reason than the fact that they offer positive detection even under nonoptimum conditions of assay, instrumentation performance, etc.

The *gold standard* for amplification in immunoassay is probably the ELISA technique, which is used not only for protein detection but for numerous other biological molecules, notably nucleic acids. Here, the amplification occurs as the enzyme converts substrate to product, a process that generally involves many thousands of molecules per minute.[6,8] This enzymatic process is potentially so powerful that, given the caveat of no denaturization of the enzyme protein, it should be possible to detect *one* enzyme molecule bound to an immunoreagent, given enough time. Seldom if ever is this most stringent detection limit necessary. Colorimetric detection of the ELISA reaction is a commonly used endpoint, and instrumentation for ELISA assays and detection based on this endpoint is readily available. Of course, if enzymatic conversion of a substrate to a colored product (either soluble or insoluble, depending on the endpoint of the amplification) is possible, the logical extension is to apply ELISA amplification to substrates that produce fluorescent or chemiluminescent products. This has been done, and sometimes with remarkable success in terms of sensitivity enhancement (see following discussion).

Simple ELISA, however, does not exhaust the ingenuity of amplification schemes, which have evolved. Numerous coupled enzyme schemes can enhance enzymatic multiplication. Immune reaction coupled to lysis of liposomes produces release of large numbers of liposomally encapsulated indicator molecules. In vitro transcription/translation systems can be used to make multiple copies of a target DNA gene sequence attached to an immunoreagent. In some cases, such as surface-enhanced Raman spectroscopy (SERS), the optical detection method itself produces a large part of the signal amplification. A variety of amplification schemes are shown in Table 16.1.

The specificity inherent in immunoassays, at its most fundamental level, resides in the antibody/antigen interaction.[9] No amount of assay ingenuity can improve an antibody reagent with poor specificity. Limit of detection has not only to do with specificity and affinity of the antibody(ies) for the analyte and the robustness of the amplification system but also with the amount of nonspecific interactions that take place in the assay system. As the limit of detection is pushed lower, at some point, the *noise* due to nonspecific interactions becomes a significant limiting factor. Sensitivity, as opposed to limit of detection (which is the lower limit of analyte that can be detected above background *noise*), refers to the ability of the assay system to discriminate between analyte concentrations that do not vary much from each other. A nonsensitive assay system may have essentially the same limit of detection, as a sensitive assay system, but the sensitive assay system can discriminate among analyte amounts that the nonsensitive assay could not.

Some simple mathematical considerations demonstrate that the antibody(ies) used in immunoassays constitutes, in point of fact, the limiting factor in developing highly sensitive, rapid, and powerful immunoassays. A brief demonstration, taken from Avrameas,[10] serves to illustrate this fact. The fundamental mass action equation governing antigen/antibody interaction is

$$Ab + Ag \underset{K_2}{\overset{K_1}{\Longleftrightarrow}} Ab - Ag \tag{16.1}$$

where
  Ab = antibody
  Ag = antigen
  Ab − Ag = antigen–antibody complex
  K_1 and K_2 = rate of forward and reverse reactions, respectively

The association constant (K_a) at equilibrium is defined as

$$K_a = \frac{K_1}{K_2} = \frac{[Ab - Ag]}{([Ab][Ag])} \tag{16.2}$$

Generally, K_1 is larger than K_2, so the [Ab − Ag] complex, once formed, only dissociates slowly (thus allowing the numerous washings common in EIA procedures). Avrameas cites several examples to illustrate the analytical limitations posed by this equilibrium.[10] These examples serve to indicate that EIAs, like all other assays, have theoretical limitations beyond which they cannot go. Once the inherent limitations are understood, however, EIAs have the potential to detect analytes at extremely low concentrations and with notable precision and accuracy.

16.2.3 Comparisons of Immunoassays

What immunoassays are the most sensitive with the lowest limits of detection? There are no simple answers to that question. Just as there are no simple answers to the question, what is the best immunoassay? The various immunoassays enumerated in Table 16.1 have apparently all been successfully implemented. The choice of a particular immunoassay has to be defined by the individual researcher for their particular application. For example, sometimes the speed of analysis may be more important than the extreme sensitivity. Or a very low limit of detection may be more valuable than how long the analysis takes to perform. In the following discussion, we provide only some general comments in regard to comparisons of various techniques; the readers must make their own choices.

Practically speaking, it is essentially impossible to make meaningful comparisons of one set of investigators' results with others'. As was already pointed out, a major factor in determining sensitivity is the affinity of the antibody(ies) used for the analyte, and very seldom (if ever) do different investigators use exactly the same antibodies. However, there are a variety of experiments in the literature where investigators compare different immunoassay protocols using the same fundamental antibody reagents. Such studies provide a basis for more meaningful comparisons, although there are still caveats in abundance, including such questions as follows: do the different detection labels attached to the antibodies affect the affinity of the antibody to different extents, does the composition of the immunoreagent *sandwich*, if it becomes highly complex, produce steric hindrance to free access of, say, enzyme substrate?

Comparing immunoassays in terms of relative sensitivities implies there is some *gold standard* assay, to which all other assays can be compared. Unfortunately, there is no such thing as a universal standard for immunoassays. Very often, comparisons are made with ELISA-based colorimetric assays, as we have indicated in Table 16.1. In addition to direct comparisons using the same or very similar immunoreagents, one finds more nebulous statements in the literature, such as some technique is *stated* to be fourfold more sensitive than another technique.

One might argue that a comparison of the limit of detection observed for various immunoassays (of which there are numerous citations in Table 16.1) should provide a reasonable indication of which assays are the most sensitive. However, in addition to the fact that antibody reagents vary from investigator to investigator, the assay protocol itself, particularly where an enzyme step (or steps) is included, can significantly impact limits of detection. For example, usual times of incubation of enzyme-labeled immunoreagent with substrate are 30 min to 1 h. It is almost guaranteed that an overnight incubation of enzyme plus substrate would produce a lower LOD (if the enzyme did not completely lose activity).

Finally, perhaps most perplexing is the fact that there can be very large ranges over which sensitivity of one immunoassay is said to be increased compared to another immunoassay (e.g., see under fluorescence EIA in Table 16.1, where the sensitivity is said to be $5-10^5$-fold increased over conventional EIA). It should be remembered that immunoassays are, by their very nature, complex and a variety of instrumentation can be used to detect the final endpoint. As more sensitive photon-counting detectors (e.g., CCD, APD) are brought into use, and as more intense excitation sources (lasers) are employed, it should be expected that LODs will continue to decrease. Whether single-molecule detection is needed (or even desirable) is another question altogether.

As mentioned previously, in terms of frequency of use, ELISAs are the *gold standard* of immunoassays. The ELISA formats and techniques are well established, there are choices of multiple substrates, and much experience has been accumulated regarding effective coupling of enzyme to antibody. A good discussion of general ELISA principles and techniques can be found in work published by Tijssen in 1985.[6] For some years in the 1970s, there was considerable discussion as to whether enzyme labels in immunoassays would ever be as useful as radioactive labels, in terms of sensitivity of detection.[11] Estimates of detection limits of 5–10 attomoles for radioactive labels, for example,[125]I, have been made, based upon the specific radioactivity of carrier-free ^{125}I of 4.8 dpm/attomole.[12] Aside from issues of safety, radioactive waste disposal, and limited half-life, Ishikawa and coworkers[12] make the point that the detection limit of some enzymes is <1 attomole, and using other techniques still based upon ELISA (i.e., immune complex transfer immunoassay), the detection limit has been pushed to milliattomoles.[11]

However, Ishikawa and coworkers[11,12] indicate that to achieve attomole sensitivity by ELISA, certain formats and procedures must be employed. These include use of a noncompetitive *sandwich* (two-antibody) ELISA format (competitive immunoassay with only one antibody is stated to not be able to achieve attomole sensitivity); the use of thiol groups in the hinge portion of the antibody Fab's fragment for enzyme coupling, thus providing better retention of the activities of both antibody and enzyme; and the use of a fluorimetric substrate for the enzyme β-galactosidase, which change was said to improve the sensitivity 1000-fold over colorimetric substrates. In regard to this latter point, Ishikawa et al.[12] state that it is possible to measure 0.02 attomole of β-galactosidase and 0.5 attomole of horseradish peroxidase (HRP) in a 100 min fluorimetric assay with an assay volume of 0.15 mL.

Shalev et al.[13] also confirm that detection by fluorimetric ELISA could be demonstrated to be more sensitive than radioimmunoassay, in fact detection levels of 3–10 attog/mL or 24,000 molecules could be achieved. The authors indicate essentially three factors enabled them to increase the total assay sensitivity $\sim10^5$–10^6 times over a colorimetric ELISA. These factors were the use of a fluorogenic substrate, prolongation of incubation time (15 h), and increase in surface area available for antigen binding. Interestingly, while Shalev et al.[13] state that it is generally accepted that fluorimetric methods are 10^3 more sensitive than colorimetric methods, using a fluorogenic substrate in place of a colorimetric one only enhanced sensitivity in their ELISA assay by 16–30 times. Ruan et al.[14] also note a 10^2–10^3-fold increase in sensitivity using a fluorogenic substrate vs. a colorimetric substrate in an ELISA assay for prostaglandin H synthetase.

Chemiluminescent ELISAs are also a useful route to increased assay sensitivity. For example, Brown et al.[15] found that using an enzyme cascade system and a chemiluminescent endpoint, AP was detectable at concentrations down to 0.4 attomole. Lewkowich et al.[16] found 12–29-fold increases in sensitivity in ELISA assays for interleukins, when a chemiluminescent substrate was substituted for a colorimetric substrate. Porakishvilli et al.[17] observed increases of 10–100-fold in sensitivity when comparing chemiluminescent vs. colorimetric ELISAs. Direct comparison of radiolabel with chemiluminescent label indicates that on an attomole basis, chemiluminescent labels provide a better possibility for detection (i.e., 1 attomole of carrier-free ^{157}I = 5 dpm, while 1 attomole of chemiluminescent label = 50 photon counts).[18]

Other ELISA-based immunoassay methods, indicated in Table 16.1, also provide the opportunity for great sensitivity. The cloned enzyme donor immunoassay (CEDIA), for example, is stated to be the most sensitive of homogeneous enzyme immunoassays (EIAs).[19] Multienzyme cascade-based assays have demonstrated subattomole (e.g., 0.01 attomole for AP) sensitivity, even when using a colorimetric substrate.[18] The use of enzyme-mimetic catalysts is a possibly significant step forward, as these small (relative to enzymes) molecules offer higher labeling ratios for antibodies without adversely affecting their affinity and still offer catalytic activity comparable to actual enzymes.[20] The catalyzed reporter deposition (CARD) immunoassays also offer extremely high sensitivity, with the added potential advantage for certain applications of localization of signal.[21] Finally, hybrid technologies combining elements of immunoassay with DNA amplification technologies seem to offer the greatest sensitivity of all, and approach to that *holy grail* of detection limits, one molecule. These are discussed later in this review.

16.2.4 Other Techniques

We would like to mention in this section some other techniques that involve amplification in some manner, but which we do not feel fit with other optical detection techniques. This is not to imply that these techniques do not have a place in immunoassay procedures or that they are somehow inferior to other techniques (the immunogold-silver stain procedure seems exquisitely sensitive in comparison to other histochemical procedures). These techniques are either designed for histological use or in some other way do not lend themselves to rigorous quantitative analysis by optical detection such as with PMTs and CCD spectrometers.

One such technique covers a whole series of assays involving agglutination reactions. Some of these procedures have been used for many decades (see discussion by Price and Newman[22]). Amplification in these assays arises from multiple Ag or Ab molecules immobilized on red blood cells, latex spheres, etc. Reaction with antisera results in agglutination of the immuno-coated particles, which is visually observed. A recent application to substances of abuse screening using immunoreagent-coated colored latex spheres to distinguish the various metabolic products in urine by agglutination immunoassay has been reported.[23] Such assays can be very powerful screening tools but are usually read visually. The extent of quantification therefore depends on the judgment and visual acuity of the observer, as well other factors such as serial dilutions of antisera used in the assay and extent of antigen labeling of the particles involved in agglutination.

Another intriguing technique, originally developed for electron microscopy and subsequently applied to light microscopy of histological specimens, is the immunogold technique, which has been modified to an ultrasensitive procedure by employing a subsequent silver precipitation step.[24] In this procedure, gold nanoparticles are used to label Ab, and then this labeled Ab binds to tissue antigens (as part of a multi-Ab *sandwich*). Subsequently, a silver precipitate is formed around the gold particles, and since multiple silver atoms precipitate around each gold particle, a large amplification results. This immunogold-silver staining procedure was found to be so sensitive (~200-fold more sensitive than immunoperoxidase staining method[24]); we thought that it should at least be brought to the readers' attention. Several not-so-attractive elements of this technique are the long (overnight) incubation with the immunogold reagent and the requirement for the development of the silver stain in the dark. This procedure has been adapted for membrane staining, and we have included an entry toward the end of Table 16.1, as densitometry could be used to quantify the results of the silver stain.

16.3 Nucleic Acid Analyses

Technologies for amplifying and detecting nucleic acids, although not having as long a history as immunoassay techniques, have nonetheless, in the past few decades, developed into an impressive armamentarium for biological investigations. Many of the caveats, which have been mentioned in regard to immunoassay techniques, can also be applied to nucleic acid amplification technologies. As we mentioned in the immunoassay section, we omitted nucleic acid amplification technologies, which utilize radioisotopes for detection, as being not *strictly* optical detection. We recognize that many of the key technologies for nucleic acid detection, such as PCR, reverse transcription (RT)-PCR, membrane blotting/hybridization, and in situ hybridization, as originally described, used radioisotopic detection. Often, in the earlier literature, autoradiographic methods were used to visualize the nucleic acid(s) being detected, and so, strictly speaking, optical detection was used. Radioisotope techniques, using such labels as ^{32}P, have also been found to be very sensitive. Nevertheless, to remain consistent with the immunoassay section and to showcase such techniques as fluorescence and chemiluminescence, we have chosen to omit radioisotopic detection.

As was also the case with the immunoassay section, we have had to constrain ourselves somewhat with regard to the definition of *amplification*. As is pointed out in a number of reviews,[25-27] amplification in respect to nucleic acids can refer to target amplification, probe amplification, and/or signal amplification. Thus, PCR, the extremely fertile technology devised in the mid-1980s, amplifies the target DNA (or RNA). Branched DNA amplification techniques, on the other hand, do not amplify the nucleic acid target but instead amplify the signal of a hybridization event, allowing it to be detected. Probe amplification strategies involve techniques such as ligase chain reaction (LCR). All these techniques and many others are explained and referenced in Tables 16.2 and 16.3. In some of these technologies, amplification is more obvious than in others (such as when 10^6 copies of the target nucleic acid are made and detected, as in PCR). In in situ hybridization techniques, on the other hand, where a probe must hybridize to a target sequence on a chromosome, amplification is sometimes only achieved by the presence of multiple detection labels on the probe. Of course, it is experimentally possible to use only one label on the probe, followed by single-molecule detection, but this is very much the exception rather than the rule. The accompanying table, taken from Schweitzer and Kingsmore,[221] summarizes various mainstream technologies discussed in this section on nucleic acid amplification.[26] This is a very useful summary in that it provides an idea of what kind of amplification is involved, the sensitivity, useful dynamic range, and other aspects of the assay systems in an easy-to-read format. Not included in this table are various hybridization technologies, which, strictly speaking, may not involve target amplification but which do involve some form of signal amplification to detect the hybridization event.

Properties of Various Nucleic Acid Amplification Technologies[26]

Property	PCR	LCR	SDA	NASBA	bDNA	Invader	RCA
DNA target amplification	✓	✓	✓	×	×	×	✓
RNA target amplification	✓	×	✓	✓	×	×	✓
DNA signal amplification	×	×	×	×	✓	✓	✓
RNA signal amplification	×	×	×	×	✓	✓	✓
Protein signal amplification	×	×	×	×	×	×	✓
Multiplexing	Little	×	Little	Little	×	×	✓
Mesothermal	×	×	✓	✓	✓	✓	✓
Amplification with cells	✓	×	×	×	×	×	✓
Amplification on microarrays	×	×	✓	×	×	×	✓
Sensitivity (copies)	<10	100	500	100	500	600	1
Range (logs)	5	3	4	5	3	4	7
Specificity (allele discrimination factor)	50	5000	50	50	10	3000	100,000

The major focus of this section can be found in Tables 16.2 and 16.3, which provide brief summaries in the same format as Table 16.1 in the immunoassay section of various nucleic acid amplification techniques. As was the case in Table 16.1, with Tables 16.2 and 16.3, we provide a brief description of the principles behind the assay, commentary, and appropriate references. Table 16.2 discusses PCR-based and non-PCR-based methods for nucleic acid amplification, signal amplification strategies like branched DNA, and various hybridization technologies, in situ as well as others. Table 16.3 covers procedures, which we call hybrid technologies. Here, we include techniques like immuno-PCR and PCR-ELISA, which are, in some sense at least, a melding of immunoassay techniques and nucleic acid amplification techniques. For the most part, Tables 16.2 and 16.3 are self-explanatory, and no attempt will be made in this text to systematically discuss the information therein. Only a few comments regarding specific issues follow.

16.3.1 Target Amplification

The critical feature of nucleic acids, which has allowed the powerful amplification/detection technologies to develop, is the complementarity of the nucleotide bases (i.e., base pairing) and the resultant ability to form double-stranded molecular hybrids. From these fundamental chemical properties flow all the varied and ingenious experimental systems that exploit these properties. Due to the base-pairing phenomenon, relatively short primers can be annealed to DNA single strands, and the succeeding DNA sequence (on the 3′-side) can be replicated (in a complementary sense), giving rise to PCR. Of course, this property of nucleic acids, which has proved to be such a powerful analytical tool, also has the potential to introduce errors since mismatches in base pairing can occur and go undetected, giving rise to spurious amplicons, hybridizations, etc. Issues of specificity, fidelity, and efficiency of PCR have been discussed by various authors, and the effects of various reaction conditions on the overall process explained (e.g., Cha and Thilly).[28] This propensity for incorrect hybridization of probes and primers becomes even a greater problem when various detection labels (fluorophore, biotin, enzymes) appear as decorations on the probe/primer strands.[1] In fact, one of the powerful advantages of using ^{32}P-radioactively labeled nucleic acids is the fact that incorporation of the label has minimal effect on overall hybridization of probe/primer to target.

Even if no overt mispairing occurs, as a result of these decorations, the net effect may be a decrease in thermodynamic stability, thereby reducing the effectiveness of hybridization under conditions where

TABLE 16.2   Nucleic Acid Amplification Schemes

Assay	Principle	Comments	References
PCR-based nucleic acid amplification	*Target DNA is exponentially amplified by using a thermostable polymerase to copy complementary strands of DNA using one primer for each strand; RNA can enter PCR cycle via initial reaction with reverse transcriptase; various strategies can be used to incorporate a label or other moieties to which a label can be attached during the course of PCR, so that amplicons can be detected*		
PCR	Two primers are used in an interactive process to amplify a specific segment of DNA; first DNA target is denatured by heating, followed by hybridization of both primers to complementary strands of the target at a temperature about 10°C below the melting temperature of the primers; the next step involves raising the temperature to 70°C–73°C, optimizing the temperature for extension of the primers by a thermostable DNA polymerase; this extension may extend beyond the region of interest the target in both directions but is limited in its total length; the denaturation, annealing, and primer extension steps are repeated; in the presence of excess primers, after a few cycles, the copies of the region of interest in the target sequence begin to dominate the product formed; each cycle results in a doubling of the DNA target sequence (exponential amplification); can also be applied to amplifying RNA if RNA is transcribed into cDNA by reverse transcriptase	PCR originally carried out with Klenow fragment of *E. coli* DNA polymerase I; thermal instability required addition of fresh enzyme at each denaturation step; use of thermostable enzyme is now essentially universal; 25–30 cycles can produce 10^6-fold amplification; further amplification can be achieved by diluting PCR product and subjecting it to a second round of PCR using fresh enzyme (producing 10^9–10^{10}-fold amplification); usually a different set of primers (nested) are used, which hybridize downstream from the first set of primers; nonspecificity can arise from primers hybridizing to similar DNA sequences elsewhere in the target; fragments as long as 2Kb can be generated; PCR product can be visualized by ethidium bromide staining on agarose gels; a single target can be amplified and detected, but 10–250 DNA copies are the more usual starting material[223]; efficiency of reaction can vary depending on concentration of polymerase, primers, Mg^{2+}, etc.; due to exponential nature of PCR, small changes in efficiency can have large effects on final quantity of PCR product	[236–240]
RT-PCR	For amplification of target RNA species (e.g., mRNA), RNA target is first reverse transcribed by one of a variety of reverse transcriptase enzymes, forming a DNA copy (cDNA) of the RNA; following this, PCR can take place, using two primers as in conventional PCR; during the first round of PCR, one of primers binds to cDNA, and complementary strand to cDNA is formed; during second cycle, other primer binds to newly formed strand and forms complement; eventually exponential amplification occurs	Amplification of target RNA by PCR	[241–245]

PCR—fluorescence detection	One or both PCR oligonucleotide primers are labeled on the 5′-end with fluorescent dyes; PCR carried out in usual manner; PCR product can be analyzed in numerous ways, but separation of product from unreacted fluorophore–labeled primer(s) must be carried out	Possibility for simultaneous determination of multiple PCR products, if different fluorophores are used for labeling; PCR products analyzed by gel electrophoresis or after separation from unreacted primers by microfiltration; technique can be used to detect point mutations and small deletions, chromosome translocations, infectious agents; demonstrated detection of five different PCR products simultaneously, using different fluorescent dyes[233]	[246]
Quantitative PCR with fluorescence detection	To assess starting quantity of target DNA in PCR reaction, an internal DNA standard was used, which was of the same length as the target sequence but with a different central sequence; PCR is carried out in usual manner, but a constant amount of internal standard is included in each PCR reaction; central sequence of internal standard allows discrimination between internal standard and target DNA sequence; PCR products analyzed by hybridization reactions with specific probes; in one analysis, PCR products labeled with digoxigenin were captured in microwells by hybridization to a probe; in the other analysis, PCR products were captured in microwells by hybridization probe, denatured, and subsequently hybridized to probes with tails of digoxigenin-dUTP; in both cases, addition of antidigoxigenin Ab conjugated to AP, followed by addition of AP substrate 5′-fluorosalicylphosphate, results in release of 5′-fluorosalicylic acid, which forms a highly fluorescent complex with Tb³⁺-EDTA; fluorescence is measured by time-resolved techniques

Also

mRNA detected and quantified using a synthetic RNA as an internal standard in PCR reactions; in this case internal standard is of different size than mRNA target, so both target and standard can be distinguished; RNAs transcribed into cDNAs by reverse transcriptase before PCR | Products analyzed by ethidium bromide–stained agarose gels | [247]
[248] |

(continued)

TABLE 16.2 (continued)   Nucleic Acid Amplification Schemes

Assay	Principle	Comments	References
PCR—liquid chromatography (LC) detection	PCR products are separated from other reactants and analyzed by LC	PCR product detected by UV absorption after LC; PCR product is also purified during LC process; detection by LC reported to be more sensitive than staining of PCR product by Hoechst dye 33258 or analysis on ethidium bromide–stained agarose gels[249]	[249]
PCR—colorimetric detection	PCR performed in usual way; following PCR, 3–10 subsequent cycles are performed, using a pair of nested primers internal to the PCR product, in order to incorporate biotin and a site for a double-stranded DNA-binding protein; DNA-binding protein immobilized in microwell captures smaller nested-primer-generated PCR product; biotin tag on other end of this product reacts with peroxidase-conjugated avidin; addition of peroxidase substrate produces color, which provides detection confirmation (indirect) for original PCR product produced	Nested primers achieve specificity because their PCR products will only form if they hybridize to correct sequence generated by first PCR target amplification; use of these labeled primers allows rest of assay to proceed	[250]
PCR—modified	PCR carried out as usual, but one of primers has a T7 phage promoter sequence attached to it; after PCR, addition of T7 RNA polymerase generates RNA transcripts; an oligonucleotide primer for reverse transcriptase is then added, which has a sequence complementary to a sequence within the region of interest being amplified; addition of reverse transcriptase generates a cDNA product; this final step improves specificity of PCR, since extraneous DNA sequences amplified during PCR will not be targets for hybridization to the reverse transcriptase primer	Additive amplification from both DNA and RNA types of amplification; products detected on ethidium bromide–stained agarose gels; in application to sequencing, amount of DNA in 150 diploid cells was sufficient for a readable sequence (using radioactive labeling)[251]	[251]
PCR—oligonucleotide ligation assay (OLA)	PCR carried out in usual manner on sequence of interest, followed by ligation reaction using two labeled ligation probes, one labeled with biotin and the other with digoxigenin; following ligation, ligated product is captured on streptavidin-coated microwell; treatment with AP-conjugated antidigoxigenin Ab results in immobilization of AP in microwell; colorimetric detection accomplished by addition of enzyme substrate	Applied to detection of allelic sequence variants; amplification by PCR, ligation, and ELISA enzyme; ligation reaction enhances specificity; colorimetric assay found capable of detecting 3 fmol of ligated product[252]	[252]

Scorpion probes	One of the primers for PCR has attached to it the following probe structure; this probe consists of specific sequence complementary to a sequence internal to the PCR amplicons; the whole probe is held in a hairpin–loop structure (loop portion is PCR-specific internal sequence) by complementary stem sequences; at 5′ and 3′ side of stems are attached a fluorophore and quencher molecule, respectively; at lower (subdenaturing) temperature, probe stays in loop–hairpin structure and fluorescence is quenched; during PCR, loop–hairpin is denatured during heating step, primer binds to target during annealing step and is extended; during the next heat denaturation cycle, double-strand amplicon product is denatured as is loop–hairpin probe structure; now loop sequence can hybridize to internal sequence on the same amplicon strand to which it is attached, thus *fixing* the separation of fluorophore and quencher and producing fluorescence	Amplification by PCR; can detect PCR amplicons in homogeneous solution without separation step; comparisons between Scorpions, molecular beacons, and *Taq*Man indicated Scorpions performed better than the other two[253]; Scorpion probes act almost exclusively by unimolecular mechanism	[253,254]
*Taq*Man	PCR reaction carried out in usual manner, but advantage is taken of 5′ → 3′ exonuclease activity of *Taq* DNA polymerase; oligonucleotide probe labeled on 5′-end with quencher and 3′-end with fluorophore is added at start of PCR; fluorescence is quenched until probe binds to internal sequence of amplicons being generated; then exonuclease activity of *Taq* polymerase cleaves probe, releasing fluorophore and producing fluorescence; increasing PCR cycles results in increased fluorescence	Amplification by PCR; detection in homogeneous solution possible, without separation step; quencher and fluorophore need to be close to each other for efficient energy transfer but cannot be too close	[255]
FRET-based detection	One (or both) of PCR primers has a stem–loop structure attached to its 5′-end; stem-loop sequence does not have to match PCR amplicon sequence; on 5′-end of stem is fluorophore and on 3′-end of stem is quencher; FRET occurs and no fluorescence is detected; PCR reaction carried out in usual manner; modified primer is incorporated into PCR product; during next round of PCR, polymerase synthesizes complementary strand of stem–loop sequence; as double strands of DNA form, stem–loop sequence is linearized, and fluorophore and quencher are separated; fluorescence increases	*Taq* DNA polymerase must be free of 5′ → 3′ exonuclease activity; amplification by PCR; technique similar to molecular beacon approach; fluorescence correlates to amount of incorporated primers; 10 molecules of starting target can be detected[256]	[256]

(continued)

TABLE 16.2 (continued) Nucleic Acid Amplification Schemes

Assay	Principle	Comments	References
Post-PCR product detection	*Amplicons from PCR process are labeled subsequent to PCR, allowing detection of PCR product*		
	DNA is amplified by PCR, using a biotinylated primer; following PCR, product is denatured and hybridized to detector oligonucleotide, which is labeled with electrochemiluminescent label; subsequently, biotinylated DNA strand with hybridized detector oligonucleotide attached is captured by reaction with streptavidin-coated beads and subject to electrochemical analysis	Amplification by PCR; electrochemiluminescent label is tris (2,2′-bipyridine) ruthenium (II) chelate (see under Immunoassays, Table 16.1); detection limit found to be ~2 amol, for direct detection on beads using electrochemiluminescence	[257]
	Or		
	DNA can be amplified by PCR, using one biotinylated primer, while the other primer is labeled with the electrochemiluminescent label; in this case, PCR product (double stranded) is *not* denatured but is captured directly by streptavidin-coated beads		
Bioluminescence detection	RT-PCR carried out as usual; one of primers is biotinylated, so PCR product can be captured on streptavidin-treated surface; digoxigenin-labeled probes are hybridized to immobilized PCR product; reaction with antidigoxigenin Ab conjugated to aequorin immobilizes Ab and aequorin at hybridization site; addition of Ca^{2+} triggers flash of light	This technique found to be 30–60-fold more sensitive than radioimaging for detecting cytokine mRNA[258], also permits detection of amplicons at lower PCR cycle numbers; amplification due to PCR	[258,259]

Electrochemiluminescence detection	One of PCR primers biotin-labeled, while other primer labeled (via amino group linker) with ruthenium (II) tris (bipyridy) label; after PCR takes place, the double-stranded product is captured on streptavidin-coated beads and then subjected to electrochemiluminescence analysis (see Table 16.1) Alternatively, PCR carried out as usual, but with one biotinylated primer; biotin-labeled amplicon captured on streptavidin-treated beads and hybridized to complementary probe labeled with electrochemiluminescent label (see above); rest of assay as above		[196,260]
Molecular beacons	DNA is amplified by PCR and subsequently detected by a molecular beacon, which is immobilized on a solid surface by a biotin-modified nucleotide linked to avidin bound to the surface; the molecular beacon is a stem–loop structure in which the loop is complementary to the target nucleic acid sequence and the stem has a fluorophore on its 5′-end and a quencher on its 3′-end; as long as the stem–loop structure remains intact, fluorescence is quenched; when the target DNA binds to the loop, however, the stem structure is disrupted, and resulting fluorescence indicates hybridization has occurred	Amplification by PCR; either DNA or RNA can be detected; limit of detection was found to be attomole to subattomole[261]	[261] [262]
	Or		
	Molecular beacon used as a probe to hybridize internally to PCR amplicons; beacons included in PCR reaction, and as amplicons accumulate, fluorescence from hybridization of beacons increases	Amplification by PCR; significant fluorescence signal can be detected in early cycles of PCR if target is at fairly high concentration initially	

(continued)

TABLE 16.2 (continued)   Nucleic Acid Amplification Schemes

Assay	Principle	Comments	References
Non-PCR-based nucleic acid amplification	*A variety of techniques are used to achieve amplification of target nucleic acids; all involve enzymes, and most use a polymerase to generate complementary copies of target molecule*		
Nucleic acid sequence-based amplification (NASBA)	A target RNA is hybridized to a primer oligonucleotide (primer 1), which has on its 5′ end a promoter sequence for T7 RNA polymerase; a mixture of T7 RNA polymerase, RNase H, and reverse transcriptase are in the reaction solution; following hybridization, the reverse transcriptase replicates the RNA sequence, making a complementary cDNA strand, resulting in a RNA/DNA hybrid; RNase H hydrolyses the RNA strand, leaving a single cDNA strand (denatured DNA as a template could enter the process at this point); a second primer (primer 2) included in the reaction mix hybridizes to the cDNA strand, and reverse transcriptase replicates the cDNA producing a double-stranded DNA with a double-stranded promoter region for T7 RNA polymerase; this enzyme transcribes RNA copies from the double-stranded cDNA template; the RNA copies produced can serve as templates for reverse transcriptase; thus primer 2 binds to newly generated RNA template and reverse transcriptase extends, producing RNA/DNA hybrid, which is hydrolyzed to a single cDNA strand; primer 1 binds to this strand, and a double-stranded cDNA is produced by reverse transcriptase, which again can serve as a template for RNA polymerase, etc.	Process is continuous, homogeneous, and isothermal; described as robust, even though multiple enzymes involved; incubation at one temperature for 1.5–2 h sufficient to achieve amplification of about 10^9; NASBA requires fewer cycles than PCR to achieve similar levels of amplification; samples with as little as 10 molecules of target found to produce positive results[263]; can be used to detect mRNA Ten copies of viral RNA could be detected using NASBA and colorimetric detection method; after amplification, amplicons captured with capture probe on microwell plates, and peroxidase-labeled detector probe was hybridized to amplicons; addition of peroxidase substrate produced color proportional to amplicon concentration[251] NASBA can directly, and selectively, amplify single-stranded RNA if denaturation carried out at 65°C, which avoids denaturing DNA; 10 copies of interferon gene detected using chemiluminescence detection on membrane[264] RNA polymerase can produce $10–10^3$ copies of RNA/single copy of DNA template[265]; thus four to six cycles of NASBA produces 2×10^6 amplification; total time ∼3–4 h[265] Within first 15 min, each of the input RNA targets was increased 10^5-fold; each cDNA template directs synthesis of minimum of 90 copies of detectable RNA during 3SR reaction; 3SR reaction operates only at 37°C, so no denaturation step for double-stranded DNA, as is used in NASBA; DNA template can be amplified by 3SR if denaturation carried out prior to start; much more rapid kinetics than PCR.	[263] [264] [265] [266] [267]

RCA	Small circular single-stranded oligonucleotides serve as the template for DNA synthesis by several DNA polymerases; progressive copying produces a long linear DNA strand consisting of complementary repeats of the circular template	Isothermal amplification; replication can proceed hundreds of times round circle; most efficient DNA polymerases lose activity with time	[268]
RCA with condensation of amplified circles after hybridization of encoding tags (CACHET-RCA)	RCA with CACHET; fluorescently labeled oligonucleotides also having a dinitrophenyl (DNP) label are hybridized at multiple sites on tandem repeats generated by RCA; addition of anti-DNP IgM antibody causes condensation into a small object more readily visualized by CCD Or BUDR can be incorporated into RCA product and detected by (1) reaction with biotinylated anti-BUDR Ab and (2) reaction with fluorescently labeled avidin	Single-molecule (DNA) detection reported[269]	[269]
Padlock probes with RCA	A linear single-stranded probe with two target DNA-complementary sequences joined by a linker region is hybridized to the target DNA; complementary sequences on probe are designed to complement a contiguous sequence on the target; following ligation by a ligase, probe forms a circular strand, which is wrapped around the target; labels on padlock probe allow detection; once padlock probe is circularized, it can be amplified by RCA before detection, using a suitable primer plus enzymes Or Using a second primer (binding to a distinct site on padlock probe from first primer) in addition to the first primer, a continuous pattern of DNA branches can be formed, connected to the first circle[269], called hyperbranched RCA	High specificity; not as likely to have nonspecific reactions as PCR; reduction in nonspecific background; RCA reaction rate stated to be 53 nucleotides/s for phage 029 DNA polymerase[269] Possible to initiate hyperbranched RCA with as few as 20 molecules of closed circles; each circle estimated to produce 5×10^9 copies after 90 min reaction[269]	[269,270]

(continued)

TABLE 16.2 (continued)   Nucleic Acid Amplification Schemes

Assay	Principle	Comments	References
Ligation amplification reaction (LAR)—also called ligase detection reaction (LDR)/LCR	In linear amplification, two oligonucleotide probes used; immediately adjacent to each other on DNA target strand, so that action of DNA ligase covalently links two oligonucleotides, *if* nucleotides at junction are perfectly base-paired to target	Thermostable ligase allows multiple repetitions of ligation reaction, thus linearly increasing product (termed LDR); specificity arises from sequence of two oligonucleotides, requirement for perfect base pairing at ligated junction and performing of ligation at elevated temperatures; powerful technique for discriminating single-base substitution; time required for 20–30 cycles $\cong$100–150 min; sensitivity of LCR: 200 molecules of target DNA amplified 2×10^8-fold after 30 cycles[271]	[271,272] [273] [274]
	In exponential amplification, both strands of target DNA are used as templates for ligation of oligonucleotide probes (i.e., four oligonucleotide probes), so because ligation of one set of probes generates a target for complementary probe set		
	In both cases, multiple rounds of denaturation, annealing, and ligation with thermostable ligase result in amplification		
GAP-LCR	Also		
	Oligonucleotide can be fluorophore labeled on 5′-end; upon ligation with downstream oligonucleotide, ligated probe can be detected by fluorescence, following suitable separation methods (i.e., gel electrophoresis) to separate ligated product from unreacted labeled probe; alternatively, one oligonucleotide can be labeled on appropriate end by biotin and ligation product detected after immobilization on streptavidin/solid support[246]	Generation of target-independent products due to blunt-end ligation can limit sensitivity	
	Four oligonucleotide probes are used as in LCR, but two of the probes have 3′ overhangs (extensions) when they are hybridized to their DNA targets; after hybridization, a gap of one to three bases exists between adjacent oligonucleotides, which is filled by a thermostable DNA polymerase; resultant *filled-in* probes joined by DNA ligase	Use of probes with noncomplementary 3′ extensions prevents generation of target-independent ligation products; exponential amplification; amplification detected by fluorescence—ELISA assay; specificity found to be better than allele-specific PCR	

[275]

SDA

Target amplification technology, in which sequence-specific exponential amplification occurs under isothermal conditions; target amplification occurs by a process in which two primers bind to opposite ends of the target sequence, as in PCR; in the case of SDA, the 3′-ends of the primers recognize the target sequence, while the 5′-ends have a restriction enzyme recognition site; DNA replication by an exonuclease-deficient form of DNA polymerase produces double-stranded amplicons with a restriction enzyme recognition site; one of strands is protected from enzyme cleavage because modified nucleotide (dATPαS) is incorporated during amplification; restriction enzyme nicks the unprotected strand; DNA polymerase extends complementary strand starting at the 3′-end, thus displacing nicked strand; polymerization/displacement step regenerates nickable recognition; nicking and polymerization cycle continuously; a fluorogenic probe is also designed to hybridize to target strand (different probe for each target strand) but downstream from the primer site; probe acts as primer to generate *extended* probe; extended probe displaced as above; extended, displaced probe now has 3′-end complementary to second SDA primer; hybridization occurs, and DNA polymerase extends in both directions; probe can be designed with a loop/stem structure, although not necessary; if designed as loop/stem structure, polymerase activity unfolds this structure during polymerization; in any case, conversion of probe to double-strand structure creates restriction endonuclease site; probe is constructed with fluorophore and quencher on opposite sides of restriction site; action of endonuclease releases fluorophore from quenching activity

10^{10}-fold amplification occurs in 15 min (Nadeau et al.[275]); detection of 10 target copies within 30 min using fluorogenic assay reported (Nadeau et al.[275]); fluorogenic assay relies essentially upon FRET-like process; process is complicated conceptually but appears to work experimentally

(*continued*)

TABLE 16.2 (continued)   Nucleic Acid Amplification Schemes

Assay	Principle	Comments	References
Hybridization	*A probe (oligonucleotide or nucleic acid segment), which either has a label or with labeling potential, is hybridized to target DNA, and hybridized complex is detected; alternatively, probe can be amplified by PCR process; hybridization commonly applied to membranes, tissue sections, microarrays, beads*		
Membrane detection	DNA transferred to membrane detected with oligonucleotide probes either labeled with AP directly or biotin; if biotin-labeled probes are used, membranes subsequently reacted with avidin–AP conjugate; final detection by use of substrate, which produces chemiluminescence by AP activity	Subfemtomole detection reported by Tizard et al.[277]; membranes could be successfully stripped and reprobed; enzyme activity provides amplification	[277] [109] [278] [279] [276]
	DNA on membrane hybridized to biotinylated probe, followed by reaction with streptavidin/biotinylated AP complex; incubation with luciferin-O-phosphate releases luciferin, which now acts as substrate for luciferase	Bioluminescent detection technique; two enzymes theoretically provide large amplification	
	DNA on membrane hybridized to biotinylated probe, followed by reaction with streptavidin–AP conjugate; incubation with either colorimetric or fluorogenic substrate produces visible color or fluorescence	Colorimetric reaction carried out using 5-bromo-4-chloro-3-indolyl phosphate + nitro blue tetrazolium; fluorometric assay used AttoPhos™; fluorescence assay carried out in solution using membrane strips in cuvette; amplification by enzyme activity; fluorescent assay ~100-fold more sensitive than colorimetric assay; $\sim 2 \times 10^6$ molecules of DNA detected by colorimetric assay and 2×10^4 molecules by fluorescent assay Also	
	DNA on membrane detected by hybridization to digoxigenin-labeled probes; immunochemical techniques (antidigoxigenin Ab labeled with AP) plus incubation with chemiluminescent substrate gives chemiluminescence	Use of salicyl phosphate esters in conjunction with AP produces well-localized fluorescent products 10–50 fg of target DNA could be detected[276]	

Fluorescence in situ hybridization (FISH) CARD-based	The same principle as CARD-based immunoassay; hapten-labeled DNA probe hybridizes to DNA target; then either antihapten Ab labeled with peroxidase or streptavidin labeled with peroxidase (if DNA probe is labeled with biotin) is bound to DNA/DNA probe complex; addition of either biotin tyramide or fluorophore tyramide results in deposition of fluorophore or biotin at site of peroxidase action; in the case of biotin deposition, addition of fluorophore-labeled streptavidin completes the detection *sandwich* Or As above, but after hybridization avidin/biotin/peroxidase complex (ABC) added, followed by biotinylated tyramine; deposition of biotin tyramine at site is detected by addition of fluorescein–avidin conjugate	Raap et al.[280] indicate much higher signal intensities (at least 10-fold) than with conventional FISH; large amplification due to enzyme activity and multiple layers of detection reagents CARD-FISH technique provided visual detection, when conventional FISH failed A variety of fluorophores were conjugated to tyramine; sensitivity found to be better than conventional FISH[107]	[107,131,280,281] [131,282]
In situ hybridization (colorimetric) CARD-based	After initial hybridization as above, ABC is added, followed by biotinylated tyramine and H_2O_2; this is in turn followed by application of ABC followed by addition of colorimetric substrate for detection	Amplification by enzyme activity—both for tyramide deposition and for processive turnover of colorimetric substrate; Kerstens et al.[131] found strong amplification in comparison to conventional in situ hybridization	[283–286]
In situ hybridization with probe extension or amplification; that is, in situ PCR	Probe for hybridization to target DNA consists of a single oligonucleotide primer; *Taq* DNA polymerase used to extend primer based on target sequence; various reporter or haptenated nucleotides in the reaction mix are incorporated into extended probe; these serve to provide direct or indirect (immunochemical) detection signals; technique called primed in situ DNA synthesis (PRINS); also cycling PRINS, where two oligonucleotide primers are used, and essentially a PCR amplification takes place; reporter/haptenated nucleotides in reaction mix incorporated into PCR product RT in situ PCR, for RNA amplification; as above, but cDNA generated by reverse transcriptase; this serves as target for standard PCR	Amplification due to multiple reporter/hapten nucleotides incorporated during probe DNA synthesis; also, for cycling PRINS, multiple copies of probe DNA made	[29]

(continued)

TABLE 16.2 (continued)   Nucleic Acid Amplification Schemes

Assay	Principle	Comments	References
In situ hybridization without probe amplification; if fluorescent labels used, termed FISH	DNA probes are developed for hybridization to specific DNA or RNA targets; these probes are either directly labeled with reporter molecules (e.g., fluorophores) or with hapten moieties, which can participate in immunochemical reactions; positive hybridization events are recorded by the presence of the reporter molecules or by the product of the immunochemical reaction (e.g., conversion of an enzyme substrate to a fluorescent product)	Amplification by multiple reporter/hapten sites in DNA probe, also by immunochemical reactions; probes labeled with biotin-, digoxigenin-, or directly with fluorescein-dUTP by nick translation; in case of fluorescein-labeled DNA, results visualized directly, or indirectly, using an antifluorescein AB and antispecies IgG Ab conjugated to fluorescein; for biotin-labeled probes, fluorescently labeled avidin was used for visualization; for digoxigenin-labeled probes, fluorescent-labeled Ab to digoxigenin was used; immunochemical *sandwiches* of greater complexity also used; direct labeling technique was found to yield lower backgrounds; sensitivity of detection found to be 50–100 kb for directly labeled probes by visual observation; for immunochemical detection, visual detection sensitivity found to be 1–5 kb Biotinylated DNA probe hybridized to target DNA and subsequently reacted with avidin/biotinylated peroxidase; substrate for peroxidase was diaminobenzidine, providing colorimetric detection Also Chemiluminescent detection after using biotin-labeled DNA probe for hybridization to target, followed by immunochemical sandwich (antibiotin Ab; antispecies IgG Ab labeled with peroxidase); addition of enhanced luminol reagent gave chemiluminescence; found to be the same order of sensitivity as ^{35}S-labeled probes detected by autoradiography[287]; other applications where chemiluminescent substrates were used, in immunochemical reactions after hybridization of a biotinylated probe, to detect two different viral DNAs in the same specimen[288] Also Hybridizations on glass microarray detected by enzyme-catalyzed chemiluminescent reactions; 250 attomole target DNA detection limit[289], enzyme amplification	[290–293] [294] [287,288]

Method	Description	Comments	Ref.
In situ hybridization (colorimetric)	Biotinylated probe hybridized to DNA, followed by antibiotin Ab, biotinylated Ab specific for species IgG of first Ab, followed by avidin and biotin–peroxidase complex (ABC); colorimetric substrate for peroxidase added for final detection	Amplification by enzyme activity; diaminobenzidine used as precipitable peroxidase substrate, and this was further amplified by treatment with $CuSO_4$	[131]
FISH (not CARD)	After hybridization of biotinylated probe to DNA, fluoresceinated avidin was added, followed by biotinylated antiavidin Ab, followed by avidin–fluorescein conjugate for fluorescence detection		[131]
Chemical modification of probe	Guanine residues in nucleic acids are modified chemically by incorporation of N-2-acetylaminofluorene (AAF) (or 7-iodo derivative) at C-8 position of guanine; following hybridization of modified probe to target, Ab specific to AAF can be used to bind to modified probe, allowing detection by immunochemical methods; also such modified nucleic acids can be immunoprecipitated	Attachment of Ab to nucleic acids allows amplification of detection signal by immunoassay techniques (e.g., ELISA); sensitivity comparable to autoradiographic methods[295]	[295]
Probes linked to gold nanoparticles	Small oligonucleotides are linked through a mercaptoalkyl linkage to gold nanoparticles; part of the oligonucleotide sequence is complementary to the target sequence being detected; two different probe oligonucleotides, which recognize adjacent sites on the target, are linked to nanoparticles (more than one pair of oligonucleotide probes are bound per nanoparticle); upon hybridization to the target, an extended polymeric network is formed, in which nanoparticles are interlocked by multiple short duplex segments arising from hybridization of probe with target; this close juxtaposition of nanoparticles changes the optical properties and color changes from red to blue, upon hybridization	Amplification due to numerous nanoparticles entrapped in polymeric network; specificity of this two-probe system seems extremely good; one mismatched base detected; sensitivity of unoptimized system showed detection of ~10 fmol of target oligonucleotide[296]; technique seems to be more sensitive than detection by means of fluorescein-labeled probe as reported in the literature	[296]

(continued)

TABLE 16.2 (continued) Nucleic Acid Amplification Schemes

Assay	Principle	Comments	References
Bead capture; enzyme amplification	Five different oligonucleotides are used for this assay to detect analyte DNA; two oligonucleotides (A and B) have sequences that are complementary to analyte DNA, as well as single-strand overhangs, which act as recognition sites for other oligonucleotide probes; oligo probe C is biotinylated at one end and has a sequence that is complementary to probe A; it is bound to avidin-coated beads and serves to capture analyte DNA; oligo probe D is a chemically cross-linked oligonucleotide complex (multimer), which has a complementary sequence to the single-strand overhang of probe B; probe D binds (via probe B) to the analyte DNA and offers a number of sites (due to multimer nature) for attachment of probe E, which is an oligonucleotide complementary to probe D and which also has peroxidase attached to it; addition of substrates results in detection signal, indicating analyte DNA is present	Amplification occurs by using multiple probes for binding to target DNA, the use of multimer oligonucleotide probes that provide multiple sites for attachment of peroxidase and the enzyme itself; Urdea et al.[309] report that this technique was as sensitive as most sensitive radiolabeled dot blot; chemiluminescent substrate used in this assay; applied to detection of sub-pg quantities of viral DNA	[297]
Hybridization on microarrays with optical detection	Essentially the same strategies as used for other hybridization formats; *capture* oligonucleotides attached to solid surface (glass, gel pad, etc.); labeled (often by incorporation of biotin-labeled primers or fluorescently labeled primers in PCR) *target* nucleic acid is hybridized to microarray and detected by various means; for example, biotin-labeled targets can be detected by standard procedures, such as reaction with streptavidin–AP, and subsequent development of color, fluorescence, or chemiluminescence by addition of appropriate enzyme substrate		[298–302]
PCR/LDR mutation detection on microarray	Multiplex PCR is used to amplify the regions of interest (for mutation analysis); addition of two oligonucleotides, which are complementary to sequences on both sides of the mutation, sets the stage for ligase-dependent ligation, only if sequence at junction of two oligonucleotides is exact match for mutation; one of oligonucleotides is fluorescently labeled, and the other has an overhanging sequence, which is coded for a specific complementary capture oligonucleotide at a specific site on microarray; fluorescence detection at this site indicates presence of mutation		[303]

Invader assay	Assay depends upon the use of thermostable *flap* endonucleases isolated from archaea, which recognize and cleave a structure formed by two overlapping oligos hybridizing to a target DNA; upon hybridization to target, downstream oligonucleotide is cleaved at site of overlap; *cleaved* portion of oligonucleotide hybridizes to biotin-labeled template, which has a number of A residues at 5' overhang; use of DNA polymerase and digoxigenin-labeled dU results in incorporation of digoxigenin-dU into probe fragment; double-stranded template + probe (extended by digoxigenin-dU) captured on streptavidin-coated microwell; probe fragment detected by binding of alkaline phosphatase-conjugated antidigoxigenin Ab, followed by addition of fluorogenic substrate for AP	Amplification by thermostable endonuclease (3000 cleavage events per target molecule) and by AP enzyme bound to Ab; found detection of DNA targets at subattomole levels[304]	[304] [305]
	Also First round of invader assay as desribed above; cleaved probe fragment binds to a biotin-labeled, streptavidin-immobilized probe, which also has bound to it another small oligonucleotide with a fluorophore and quencher in close apposition; hybridization of cleaved probe fragment from first round produces another single nucleotide overlap, also susceptible to endonuclease cleavage; cleavage releases fluorophore, abolishing FRET quenching	Serial version of invader assay can generate more than 10^7 reporter molecules for each target molecule in 4 h reaction; sensitivity for detection of less than 1000 target molecules reported[305]	
Chromosome painting	Biotinylated oligonucleotide probes specific for some chromosome or chromosomal site hybridized to target chromosome; alternating layers of fluorophore-labeled avidin and biotinylated antiavidin Ab added to give amplification for detection	Detection of unique intrachromosomal sequences required multiple (three) layers of fluorophore-labeled avidin[243]	[243]
Intensification	Biotinylated probes were reacted with target DNA in situ, followed by standard immunocytochemical techniques resulting in immobilization of peroxidase at hybridization sites; following incubation with diaminobenzidine, and H_2O_2, further staining intensification was achieved by silver staining, using a series of silver and gold solutions	More sensitive than colorimetric stain alone	[306]
Eu-labeled probes	DNA hybridization probes are labeled with a Eu chelate by introducing aliphatic amino groups on cytosine residues; following hybridization, positive hybridization events detected by time-resolved fluorometry	Amplification by numerous fluorophores on DNA probe— 4%–8% of total nucleotides labeled[307]; detection limit was 0.15 attomoles of target DNA[307]	[307]

(continued)

TABLE 16.2 (continued) Nucleic Acid Amplification Schemes

Assay	Principle	Comments	References
Branched DNA technology	*Target DNA is not amplified, but a series of oligonucleotides complementary to portions of target are used to hybridize to target and form a dendrimer structure with many, many labels, which can be used for detection*		
Branched DNA amplification	(a) Target DNA is immobilized to solid surface by a series of oligonucleotides (called capture probes and capture extenders [CEs]) complementary to segments of target DNA; subsequently, other oligonucleotides (label extenders [LEs]) complementary to other segments of target DNA hybridize to target, the LEs also bind (by complementarity) to a branched DNA segment (amplifier), which in turn binds many AP; AP is bound to a short oligonucleotide with sequence complementary to repeated triplicate sequence on branched DNA	Amplification of signal is linear, so signal directly related to number of target molecules present in original sample; first generation assay quantifies targets between 10^4 and 10^7 molecules; second generation assay quantifies down to ~500 molecules; detection limit of ~50 molecules/mL reported[308]	[308,309] [310] [311]
	(b) In other developed assays, LE probes bind preamplifier oligonucleotide probes, which in turn bind many amplifier segments		
	Incorporation of nonnatural bases, isocytidine and isoguanosine, into amplifier molecules can reduce nonspecific hybridization, thus improving sensitivity; final detection step involves adding chemiluminescent substrate for AP and detecting luminescence; branched DNA technology amplifies *reporter signal, not* the target	Detection and quantification of mRNA directly from cells (Warrior); CE and LE probes about 30 bases long and between them provide complete coverage of the mRNA sequence; branched DNA amplifiers have 15 branches, each of which is labeled subsequently with AP (Warrior); thus great amplification; solid surface has synthetic DNA designed to hybridize to CE Kern et al.[311] used branched DNA technique for detection of HIV RNA; use of preamplifier intermediate resulted in binding of eight amplifier molecules; each amplifier molecule had 15 branches, and each branch could bind three AP-labeled probes; thus, theoretically each HIV RNA could be decorated with as many as 10,080 AP probes; result detected by chemiluminescent assay; detection limit was 390 HIV RNA copies/mL	

		[312]
Signal amplification cassettes (SACs)	SACs are attached to DNA dendrimers; SACs are oligonucleotides (hairpin or double strand) with 15 adenine overhangs at one or both ends; polymerase activity of T7 DNA polymerase extends strand to fill in overhang with TTP (thymidine triphosphate); for each TTP incorporated, a PP$_i$ is liberated; this reaction continues to recycle because 3′–5′ exonuclease activity regenerates original SAC, which again becomes substrate for DNA polymerase; PP$_i$ generated is converted enzymatically to ATP, and luciferase reacts with luciferin + ATP to produce light; many SACs are attached to each DNA dendrimer; potential for tremendous amplification; hybridization of dendrimer to target produces specificity of reaction	Multiple amplification steps involved; many enzymes (DNA polymerase, ATP sulfurylase, luciferase) involved, all contributing to amplification; also amplification by multiple dendrimer sites; detection of 5 zeptomole of *DNA dendrimer* demonstrated[312]; SAC consisting of just two polymerase reaction sites sufficient to detect low attomole levels of DNA

TABLE 16.3 Hybrid Technologies

Assay	Principle	Comments	References
Immuno-PCR	*Technique combining antibody (Ab) specificity for molecular recognition with amplification potential of PCR*		
Immuno-PCR	Immobilized antigen (Ag) is recognized by Ag-specific Ab, which is, in turn, linked to a specific segment of DNA through some linker molecule (*vide infra*); following immune recognition, PCR reaction is carried out on segment of attached DNA; PCR products are detected as final step in analysis[313–322] (a) Linker molecule is a streptavidin–protein A chimera; protein A component binds to Ab, and streptavidin binds to biotinylated DNA (linear plasmid pUC19), forming a bridge between DNA and Ab[313] (b) Linker molecule is avidin; biotinylated Ab and biotinylated DNA bind to avidin and form complete complex[314] (c) Primary Ab recognizes Ag; secondary biotinylated Ab recognizes primary Ab; modified avidin (to lessen nonspecific binding) forms bridge between biotinylated Ab and biotinylated DNA Or Immuno-PCR carried out in usual manner, but during PCR amplification, a hapten-labeled (digoxigenin) nucleotide is incorporated into the PCR products; one of the PCR primers also has a biotin label to permit capture of the PCR products; subsequently, the biotinylated, hapten-labeled PCR products are captured on a streptavidin-coated solid surface, enzyme-labeled antihapten Ab is added, and the addition of enzyme substrate results in accumulation of detected product	Specificity resides in Ag/Ab recognition and in specificity of PCR for target sequence defined by primers; possibility for quantification of Ag concentrations below level at which PCR reaction saturates; tremendous amplification because of exponential nature of PCR; universal usefulness of single immuno-PCR protocol for detection of a variety of Ags Because of extreme amplification of PCR, *any* nonspecific binding of DNA or Ab to other immobilized reagents or solid substrate contributes to background; avidin/biotin-based linker systems have to be carefully titrated so as to form usable complexes (i.e., Ab-linker–DNA), which actually contribute to PCR; may be necessary to preform Ab-linker–DNA complexes; scrupulous washing required during assay steps to eliminate nonspecific binding; standard microtiter plates used for immunoassay are suitable for PCR thermal cycling; DNA target used for PCR amplification should, ideally, be very different from any DNA, which might be present in the sample, to avoid possible interferences in PCR amplification arising from exogenous DNA; total assay time likely to be longer than ELISA-type assays; increasing PCR cycle number increases sensitivity but also increases nonspecific amplifications; amplification also occurs because more than one copy of biotinylated target DNA can be immobilized at each avidin/streptavidin site, if the stoichiometric ratios are titered carefully Amplification also from multiple biotinylation sites on primary detection Ab and multiple hapten sites on each individual PCR product molecule; plus enzyme amplification	[313–322]

Or

Immuno-PCR procedure carried out, and DNA product produced by amplification is detected by staining with a double-stranded DNA-selective intercalating fluorescent dye (PicoGreen)

Or

Usual immuno-PCR protocol, but capture Ab used to immobilize Ag of interest, followed by reaction with another Ag-specific Ab, which is biotinylated; streptavidin serves as bridge to biotinylated target DNA; PCR primers each have one fluorescein/primer; after PCR, amplified product hybridized to capture oligonucleotide immobilized to solid surface; addition of antifluorescein Ab conjugated to enzyme allows standard ELISA endpoint

Total assay is time-consuming; very extensive washing required for certain steps; more than one biotin bound per detection Ab, thereby resulting in more than one target DNA bound/detection Ab

Sensitivity

(a) Enhancement of 10^5 compared to colorimetric ELISA using the same immunoreagents and linker; 580 molecules of Ag detected[313], estimated to be several orders of magnitude greater sensitivity than radioimmunoassay[313]

(b) Enhancement of 10^8 compared to colorimetric ELISA; 41 molecules/mL (or two molecules per microwell) of Ag detected, with signal-to-noise ratio[315] $\cong 4$

(c) Detection limits on the order of $10–10^7$ molecules for various antigens, if primary Ag-specific Abs used at higher concentrations; up to 10^9 enhancement of sensitivity compared to colorimetric ELISA[316]

(d) Enhancement of 16-fold compared to colorimetric ELISA in terms of sensitivity, using the same immunoreagents for both assays[317]

(e) Sensitivity enhancement of $\sim 10^4$ compared to colorimetric ELISA[318]

(f) Enhancement of 10^5 compared to colorimetric ELISA; lower limit of Ag detection $\cong 6000$ molecules[320]

(g) Detection limits for mouse IgG as Ab were 10 attomoles for direct detection using an intercalating dye; one attomole for colorimetric ELISA detection and 0.1 attomole for fluorescence-based assay allowed quantitative detection over six orders of magnitude; compared to standard ELISA using *fluorescent* substrate, immuno-PCR was 1000-fold more sensitive[321]

(continued)

TABLE 16.3 (continued) Hybrid Technologies

Assay	Principle	Comments	References
Immono-PCR with DNA–streptavidin nanostructures	Essential principle the same as immuno-PCR, but an oligomeric biotinylated DNA–streptavidin network is used as an attachment to biotinylated Ab rather than a single streptavidin bridge between Ab and DNA; rest of analysis carried out the same	Oligomeric DNA–streptavidin complexes provide superior immuno-PCR performance; better signal intensities than standard immuno-PCR; ~10-fold increase in sensitivity over standard immuno-PCR and 100–1000-fold increase in sensitivity over ELISA immunoassay[323]	[323]
In situ immuno-PCR	Immuno-PCR carried out as usual, but on tissue sections; after binding DNA-labeled Ab to specific Ag, in situ PCR carried out, followed by hybridization to digoxigenin-labeled DNA probe and subsequent immunostaining	Amplification by PCR and immunoenzyme step; sensitivity of in situ immuno-PCR compared to immunohistochemical analysis using (1) avidin–biotin–peroxidase complex; (2) AP, anti-AP complex; (3) tyramide signal amplification (i.e., CARD, see Table 16.1); (4) in situ PCR; in situ immuno-PCR was found to be the most sensitive of all the amplification systems	[324]
Immuno-RCA	Oligonucleotide, which is a primer for RCA, is attached to an Ab (either primary Ag-specific Ab or secondary Ab); following formation of immunoreactant complex, RCA is initiated by addition of circular template; hundreds of tandemly linked copies of template generated in a few minutes; amplified DNA detected in a variety of ways, for example, direct incorporation of hapten- or fluorophore-labeled nucleotides into template copies; hybridization with fluorophore or enzyme-labeled oligonucleotides; that is, (a) incorporate fluoresceinated nucleotides into RCA product; (b) use AP-conjugated antifluorescein + fluorescent substrate to detect signal	RCA can be carried out with linear or geometric kinetics; amplification not only by RCA, but each Ab can have more than one oligonucleotide primer attached per Ab molecule; even in linear kinetic mode, immuno-RCA could detect IgE with 100-fold sensitivity increase over conventional ELISA[325] Signal amplification by immuno-RCA applied to protein analysis on glass microarrays; 75 cytokines measured simultaneously with femtomole sensitivity[326]	[325] [326]
Immune capture + PCR	Ag-specific Ab used to capture analyte Ag, thereby concentrating it; PCR subsequently applied to amplify diagnostic DNA target sequence, followed by detection of PCR product	Applied to detection of bacteria in various samples	[327]
Immuno-detection amplified by T7 RNA polymerase (IDAT)	Very similar to immuno-PCR with following differences; double-stranded reporter DNA still attached to Ab via a linker, but this DNA contains the T7 promoter sequence; Ab binding to Ag target also immobilizes reporter DNA; action of T7 RNA polymerase produces multiple copies of RNA; this multiplication being linear and directly related to number of reporter templates; RNA is subsequently detected	Very large amplification from T7 RNA polymerase; authors state that fluorescence detection technique was developed, but no data are shown (all data based on radiometric technique); authors state that detection of only a few copies of an Ag in a mixture should be possible using IDAT	[328]

PCR-ELISA	PCR process used to amplify target DNA; *incorporation of various haptens and/or other reagents allows immunological detection using ELISA technique*	[241, 329–333]
PCR-ELISA	Hybrid technique where specificity and amplification of PCR are coupled to detection by ELISA; different formats, but one example is as follows; target DNA sequence is amplified by standard PCR techniques; digoxigenin-modified nucleotide incorporated into amplicons during PCR; after denaturation, single-stranded amplicons hybridized to biotinylated capture probe and hybridized product captured on streptavidin-coated surface (solid-phase process necessary for ELISA assay); subsequently, antidigoxigenin Ab conjugated to enzyme (i.e., HRP) added and binds to digoxigenin sites on PCR product; after washing away excess Ab, substrate for enzyme added and color/fluorescence/chemiluminescence detected as in standard ELISA; technique can be applied to detection of mRNA with an RT step before PCR	Amplification potential is great, since PCR and enzyme amplification both used in different parts of assay; specificity for target DNA only as good as PCR specificity and the specificity of the Ab(s) used in the ELISA assay; blend of established immunoassay technique and PCR; demonstrated detection sensitivity of <10 spores of *B. anthracis* in 100 g soil sample, using a colorimetric assay[329], Hall et al.[330] measured mRNA expression using biotinylated primers in PCR to enable capture of PCR products on avidin-coated plates; PCR amplicons were labeled with digoxigenin; antidigoxigenin Ab conjugated to AP; colorimetric assay used for detection
	DNA amplified by PCR, using biotinylated primers; PCR product immobilized in microwells by hybridization to internal capture probe; streptavidin–AP conjugate or streptavidin–peroxidase conjugate subsequently added and allowed to react; either colorimetric or fluorogenic substrate added for detection	Amplification by both PCR and enzymatic (AP or peroxidase) means; colorimetric and fluorometric assays compared to detection using ethidium bromide–stained agarose gels; colorimetric assay found to have about the same sensitivity as analysis by gels, while chemiluminescent assay was 10-fold more sensitive; colorimetric substrates were tetramethylbenzidine for peroxidase and *p*-nitrophenylphosphate for AP; chemiluminescent substrate (for AP) was Lumi-Phos 530; ~5 cells of *Salmonella* could be detected using PCR + chemiluminescent method; colorimetric detection limits were 50 cells	[334]

some degree of stringency is needed. Therefore, considerable effort has been expended to develop analytical techniques to incorporate detector decorations into probes and primers without affecting (to any significant degree) the part of the probe/primer that must affect recognition of the target nucleic acid.

PCR is, without question, the premier target amplification technique for nucleic acids, with applications that span the gamut of biological investigations. It has been applied not only to target nucleic acid amplification in microtubes but also to tissue/cell sections, being called in situ PCR in the latter case. Very often, when PCR is carried out in microtubes, the amplified product is analyzed by agarose gel electrophoresis. Commonly ethidium bromide, an intercalating dye, is included in the gel so that location of the amplified sequence can be visualized by ultraviolet (UV)-excited fluorescence. We have included this technique as part of the optical detection technologies, since it is certainly optical in nature and comprises the analytical endpoint in many applications of PCR technology. On the other hand, this fluorescence-based analysis of gels is usually not done for quantitative purposes, since often the information desired from PCR is in regard to the molecular size of the PCR product, not quantitative information regarding the amplification achieved by PCR.

PCR products, as stated earlier, can be detected by direct staining in gel matrices. However, it is also common to use labeled probes (such as molecular beacons, *Scorpions* or *TaqMan*),[26] which bind by hybridization to PCR amplicons and by this binding produce a detection signal (i.e., fluorescence). Another strategy is to use PCR primers with attached labels so that as the amplification proceeds, the amplified products are all labeled. In this type of assay, some separation of labeled amplicons from labeled but unincorporated primers must be carried out.

The enormous amplification power of PCR, coupled with development of highly efficient instrumentation for the obligatory thermal cycling, has overshadowed all other target amplification technologies. In Table 16.3, we also include such techniques as LCR, strand displacement amplification (SDA), nucleic acid sequence-based amplification, and rolling circle amplification (RCA). Some of these techniques have real advantages over PCR, such as no requirement for thermal cycling, but none have yet succeeded in unseating PCR from its commanding position.

RCA, a relatively new technology, has the potential for either linear or exponential target amplification and has the additional advantage of isothermal amplification. In this procedure, a primer hybridizes to a circular DNA, and the circular DNA is amplified by polymerase extension. Up to 10^5 copies of the DNA circle are generated on a concatemerized chain from one primer.[26] The amplification potential of this technique is obviously tremendous. RCA can also be used as a signal amplification technology. In this application, a defined circular DNA segment is amplified, and the concatemerized product is labeled either during the polymerase amplification or subsequently. Again, the larger degree of amplification produced by the rolling circle technique results in enhanced ability to detect signal.

Of course, PCR and other target amplification techniques are not only useful for detecting DNA but can be applied to RNA as well, since RNA can be converted into a DNA copy using reverse transcriptase. In addition, PCR has been applied to in situ hybridization studies, where the detection of a nucleic acid sequence in a tissue sample proves a difficult analytical challenge.[29] In such a case, PCR can be used to make many copies of a primer, which is complementary to the target strand of interest. This primer may itself be labeled, or labels may be incorporated during amplification. In the strictest sense, this is not target amplification but probe amplification, but the net effect remains the same. Oddly, while one might expect diffusion of the amplicons away from the site of the PCR reaction, this does not occur to such an extent as to render the technique unusable.

16.3.2 Signal Amplification

Target amplification of nucleic acids is not the only way in which nucleic acid sequences may be detected. Signal amplification technologies can be used, in which the signal from a hybridization event where probe binds to target is somehow amplified. Branched DNA amplification, in which a dendrimer structure is formed upon a probe hybridizing to target and in which the branches of the dendrimer can carry multiple labels, is one example of such a signal amplification technique.[25,26] A powerful advantage of the branched DNA technique is the fact that the amplification of the signal, rather than the target, allows one to quantify directly the amount of target DNA being detected, based upon the signal. Also included in Table 16.3 are a variety of other techniques, which involve signal amplification. For example, a variety of hybridization techniques involve binding of antibody molecules to haptenylated nucleic acid probes. In these instances, signal amplification can take place in a variety of ways. There can be multiple antibody molecules bound to each probe, each antibody can have multiple fluorophores (for example), or the antibodies can be conjugated with enzymes, thus allowing the tremendous amplification potential of enzyme processivity to contribute to overall amplification.

For membrane hybridization detection, nucleic acid recognition is quite commonly coupled with enzyme-associated signal amplification to produce highly sensitive analyses. For example, biotinylated probes can be hybridized to nucleic acid targets on membranes, and subsequently avidin/streptavidin conjugated with an enzyme such as AP can be added. The biotin/avidin linkage immobilizes the enzyme at the specific site on the membrane, and addition of substrate results in signal amplification due to enzyme activity. Either fluorogenic or chemiluminescent substrates may be used, and the resulting assays are quite sensitive.

Assays of this nature are a blend of immunoassay techniques and nucleic acid assay techniques and so could readily be put into the category of *hybrid assays*. Perforce, our division into separate categories is somewhat artificial but may serve a useful purpose in delineating key characteristics of various assay schemes. Another hybridization assay, which has a strong similarity to immunoassay techniques, is the use of tyramide signal amplification for in situ hybridization studies.[25] This technique, also known as CARD, begins with the specific recognition event of a probe hybridizing to its target in situ. Either directly or indirectly, HRP is linked to the hybridization probe, and the hybridization event thus localizes the enzyme at the target site. Addition of a tyramide conjugate results in activation of the tyramide by peroxidase, depositing large numbers of tyramide molecules at the reaction site by binding to electron-rich targets near the site. If the tyramide is conjugated to a fluorophore, the net effect is to deposit many fluorophore molecules at the site of reaction. Further signal amplification can be achieved by using immunochemical techniques, such as adding antifluorophore antibodies conjugated with an enzyme (e.g., HRP) and thereby going through another round of tyramide deposition before detection occurs.

These are just a few examples of signal amplification. A more extensive list of examples can be found by perusing Table 16.2. We end this section by simply mentioning the interesting category of detection molecules termed molecular beacons.[30] These molecules derive their powerful detection capabilities because of fundamental chemical properties of nucleic acids. In their simplest form, beacons consist of a stem–loop structure, in which the nucleotides in the loop are complementary to a target sequence, while the nucleotides in the stem portion have a fluorophore and quencher held in close apposition on the two terminal nucleotides of the stem. Hybridization of the loop to its target destabilizes the stem, freeing the fluorophore from the quenching effect of the quencher molecule, thereby resulting in fluorescence. Obviously, the beacon must be carefully designed to accomplish the intended purpose. The molecular beacon concept can also be easily adapted to fluorescence resonance energy transfer (FRET)-based assays.

16.4 Hybrid Nucleic Acid Amplification Technologies

Table 16.3 lists the nucleic acid amplification techniques, which are referred to as *hybrid technologies*, that is, a blending of immunological and nucleic acid detection strategies. We have included basically two technologies in this category, immuno-PCR and PCR-ELISA. Each of these techniques is a mirror image of the other, in terms of recognition and amplification technologies employed. In immuno-PCR, recognition/specificity is achieved by antigen/antibody interaction. A nucleic acid fragment attached (directly or indirectly) to the antibody supplies the target for PCR amplification. Thus, in this case, the nucleic acid serves in place of an enzyme attached to the antibody (e.g., standard ELISA immunoassay) as the amplification vehicle. Obviously, both the specificity of the antibody for the antigen and lack of nonspecific binding need to be maximized, since otherwise the tremendous amplification power of PCR will serve to not only amplify the immuno-specific reaction but spurious reactions as well.

In contrast, with PCR-ELISA assays, the hybridization of primers to target nucleic acid sequences serves to provide specificity, while enzymes conjugated to immunoreagents boost the amplification potential of PCR. In this case, the same pitfalls applicable to PCR analysis again hold, since amplicons arising from spurious hybridizations will be amplified by the ELISA process, thus contributing to the overall nonspecific background. Of course, the immunological reaction, by and of itself, can also contribute to analytical *noise*, since binding of enzyme-conjugated antibody nonspecifically to sites other than antigenic sites (commonly digoxigenin) on the target amplicons will lead to increases in nonspecific background.

Both of these techniques have the potential for large signal amplification. Both also have the potential for high specificity. The PCR-ELISA technology has the advantage of building upon many years of experience in both PCR and ELISA development. The immuno-PCR procedure, while seemingly straight forward, has a shorter lifespan of development behind it, and so it is in its early stages. Even so, some of the results achieved by immuno-PCR have been impressive. For example, Hendrickson et al.[31] were able to develop a multianalyte immuno-PCR assay for three separate analytes in which sensitivities were found to be three orders of magnitude better than conventional ELISA assays.[31] Table 16.4 explains and gives some examples of these powerful analytical tools.

16.5 Recent Developments in Amplification Techniques

16.5.1 Conjugated Polymers (CPs)

We first discuss applications of CPs for biomolecular sensing applications. The unique attribute of these materials, which has produced such interest so as to generate a whole literature devoted to experimental applications, is the ability to produce significant signal amplification in response to interactions with analytes. The most frequently exploited activity of these CPs is as optical transducers (either visible or fluorescent). In general, the CPs used for these sensing applications tend to be water-soluble polymers with charged functionalities on the side chains. These conjugated polyelectrolytes provide opportunity for electrostatic interaction with biomacromolecules. Thus, the cationic CP allows strong electrostatic binding to polynucleotides, for example, while the water solubility is crucial for biomolecule detection in aqueous samples. Both cationic polyfluorene and polythiophene derivatives have been extensively used for DNA sensing applications. Other CPs have polyanionic properties due to pendant water-soluble moieties (e.g., sulfonate). Regardless, the molecular characteristics of these CPs of critical importance are (1) the π-conjugated backbone of the polymer, which provides the optical properties, and (2) the attached charged functional groups, which confer water solubility.

The basic principle underlying the mechanism by which these sensors work is that very minor perturbations to the CP chain are greatly amplified due to a collective system response and thus provide highly sensitive detection capabilities. This is in contrast to small-molecule sensors, where each small molecule can only provide a fixed output (e.g., quantum yield in the case of fluorophores) so that increased sensitivity requires more small molecules. The advantage of CP sensors is that when an analyte binds to a receptor at a localized spot on the CP, the electronic milieu and backbone of the entire CP are affected because of the delocalized electronic structure characteristic of CPs, thus providing potentially large amplification of the binding event.

The possible schemes for analyte detection using CP-based sensors are rich but can be described in the following general terms. Upon analyte binding to the CP, the CP may undergo a conformational change, or aggregation may occur, or energy transfer (FRET) or electron transfer may occur. These alterations in CP structure and/or electron movement may be manifested in various forms, such as fluorescence turn-on (fluorescence enhancement), fluorescence turn-off (quenching of fluorescence), fluorescence color change (i.e., FRET phenomenon), or visible color change. Visible color change occurs when analyte binding produces a change in conjugation length of a CP backbone and thus alters the

TABLE 16.4 CP—Enhanced Detection of Biomolecules

Assay	Principle	Comments	References
DNA-colorimetric detection with discrimination for mismatches	Polythiophene (yellow solution in water) mixed with ssDNA (capture probe) produces a red solution, when upon addition of one equivalent of perfect complement, solution becomes yellow. Addition of oligo with one or two mismatches produces red color with distinct absorption spectrum	Colorimetric effects believed due to different conformations of CP in duplex and triplex form—detection limit is ~1 × 10¹³ molecules of oligonucleotide	[32]
DNA-fluorometric detection with discrimination for mismatches	Polythiophene fluorescence quenched and slightly red-shifted upon addition of negatively charged oligonucleotide, addition of exact complement results in fivefold increase in fluorescence intensity. Mismatches show no fluorescence enhancement	Detection limit for perfect complement is 2 × 10⁻¹⁴ M with conventional fluorometer: few hundred copies of DNA or RNA can be detected with specialized equipment	[32]
DNA-fluorometric detection with FRET (called fluorescence chain reaction or superlighting)	Polythiophene–ss-oligonucleotide complex formed; ss-oligonucleotide labeled with a fluorophore (acceptor); upon addition of perfect complement, fluorescence of CP is enhanced, and transfer of energy by FRET to acceptor fluorophore occurs	Improvement of detection sensitivity by a factor of 4000; detection of as few as five copies of DNA; excellent sensitivity made possible by efficient and ultrafast energy transfers to the aggregated probes; the same principle applied to detection on solid arrays (biochip)—can detect mismatches and LOD is around 300 copy numbers. There will be some binding of fluorophore-labeled ss probe in the absence of complementary strand and presence of noncomplementary strand, so background will be somewhat elevated	[32]
Protein detection, with CP an an aptamer	Aptamer binds to specific protein target and assumes a quadruplex form; cationic polythiophene wraps around quadruplex structure, resulting in color change; can also use fluorescence and FRET-based detection	LOD for standard fluorescence enhancement was 6.2 × 10⁻¹¹ M; for FRET-based assay, LOD was 1000 times lower	[32]
DNA—SNP genotyping, DNA methylation	Cationic polyfluorene used in fluorescent detection schemes, typically using fluorescently labeled deoxynucleotide and extension reaction with Taq polymerase; detection of methylation requires bisulfite treatment, followed by PCR	Tenfold enhancement of fluorescent emission intensity by FRET from polyfluorene; detection of allele frequencies as low as 2%	[335]
DNA-oxidative damage, UV damage	Polythiophene used to detect hydroxyl radical damage to DNA by visual assay. Polythiophene interacts with ssDNA to change from yellow to pinkish red; oxidative damage produces small fragments that do not interact with polythiophene, so color shifts to yellow from red: UV damage detected by PCR reaction with fluorescently labeled dUTP	Visual assay for oxidative damage is convenient; concentrations of UV-damaged DNA below nanomolar could be detected with use of CP	[335]
DNA-sequence-specific detection	Peptide nucleic acid (PNA) labeled with fluorescein used as polyfluorene probe does not interact with uncharged PNA—when DNA sequence complementary to PNA binds to PNA, polyfluorene interacts electrostatically with PNA/DNA hybrid and produces FRET with fluorescein	Enhancement of 25-fold for fluorescein emission intensity compared to direct excitation using absorption maximum	[34]

(continued)

TABLE 16.4 (continued)   CP—Enhanced Detection of Biomolecules

Assay	Principle	Comments	References
Protein (enzyme) detection	Detection of acetylcholinesterase by preparation of a polyfluorene sulfonate–acetylcholine–dabcyl electrostatic complex—fluorescence of this complex effectively quenched by dabcyl via FRET—addition of acetylcholinesterase results in hydrolysis to produce a negatively charged dabcyl residue, which is repelled by negative polymer, so fluorescence of polymer is recovered	Hydrolysis of enzyme substrate enhances fluorescence of polyfluorene sulfonate ~130-fold	[34]
Carbohydrate detection	Polyphenylene-ethynylene sulfonate quenched by a bisboronic acid functionalized benzyl viologen (cationic quencher)—addition of glucose or fructose results in complex forming between sugar and boronic acid, which removes quencher from polymer; fluorescence of polymer is recovered	LOD for glucose is ~1 micromolar	[35]
Protease detection	Cationic peptide (protease substrate) labeled with quencher, which binds to polyphenylene-ethynylene (anionic); quenching its fluorescence hydrolysis by protease releases quencher, so fluorescence of polymer is recovered		[35]
Protein–RNA detection	Assay of binding of transactivator peptide to responsive element of HIV-1. Activator peptide (fluorescently labeled) and RNA sequence are mixed together (must have net negative charge due to RNA) and reacted with cationic polymer—FRET occurs between polymer and fluorophore-labeled peptide. In case of nonbinding RNA sequence, fluorophore-labeled peptide not brought into apposition with CP	Tenfold enhancement of fluorescence with polymer FRET over the case of excitation of fluorophore-labeled peptide alone	[336]
Detection of biotin–avidin binding	Methyl viologen linked to poly(p-phenylene vinylene) via biotin; form a quencher–tether–ligand probe; methyl viologen quenches fluorescence of polymer; addition of avidin pulls away probe from polymer and produces fluorescence of polymer	LOD of avidin protein at nanomolar level	[337]
DNA detection—fluorescence	Polyfluorene with 5% benzothiadiazole chromophore also incorporated; emission of modified polymer is blue; addition of ssDNA (noncomplementary) shifts emission to green; addition of a Cy5 dye-labeled PNA strand, which is complementary to a target strand, will result in FRET and a red color, thus able to distinguish between no DNA (blue), noncomplementary ssDNA (green), and complementary ssDNA (red)	Demonstrates that external perturbations that diminish backbone elongation or bring segments closer together can have dramatic effects on polymers in solution; thus electrostatic interaction with anionic DNA reduces intersegment distances due to polymer aggregation and improved FRET to green-emitting sites (benzothiadiazole)	[336]
ATP detection—colorimetric or fluorescence	Cationic polythiophene reacts with anionic ATP, resulting in change in color from yellow to pink–red; also, addition of ATP produced fluorescence quenching of polymer	Detection limit of fluorescence method ~10^{-8} M; both methods seemed semiselective for ATP; Colorimetric changes due to conformational shifts in polymer	[32]

wavelength of light absorbed by the polymer. If analyte binding perturbs the electron density along the polymer chain backbone or changes the conformation of the CP, then fluorescence or enhancement of fluorescence can occur. In the case of turn-off of fluorescence, analyte binding can produce quenching due to nonradiative relaxation pathways in a polymer chain or via intermolecular aggregation of polymers. Fluorescence color change can be a result of changes in electron density along the backbone of the CP, which may be due to polymer aggregation, conformational changes, and electron energy transfer, following analyte binding. Additionally, the energy of the CP can also be transferred to a reporting fluorophore (via FRET), which can result in further amplification.

Many of the most impressive developments in applications of the CP to sensitive assays of biomolecules have utilized the FRET type of mechanism. As stated earlier, the great advantage of CP vs. small-molecule fluorophores is that in CP, the transfer of excitation energy along the whole backbone of the polymer chain to the reporter molecule results in a powerful amplification of the fluorescent signal. If a FRET assay is to be implemented, then the acceptor molecule must be in close apposition to the CP chain, either directly bound or within a very close (~10 nm) distance, according to Forster resonance theory. In this case, excitation of the CP by suitable photons results in excitons migrating along the length of the polymer backbone and then becoming trapped by the acceptor molecule. So in effect, the fluorescence of the CP is quenched, and the signal from the excited acceptor is amplified.

Feng et al.[34] list three types of signal transduction mechanisms for water-soluble CP, which include electron transfer, FRET, and analyte-related aggregation or conformational change. The CPs have very good light-harvesting characteristics and are able to trap energy in the form of excitons along the backbone. Feng et al.[34] provide the example of a single quencher bound to one repeat unit of a CP and indicate that this single quencher is able to trap most of the exciton energy along the whole CP backbone. This leads to an obvious amplification of the quenching efficiency of the quencher (sometimes termed *superquenching*). The quencher molecule can be substituted by an acceptor fluorophore, which effectively can permit FRET from the CP backbone. Finally, if the analyte molecule induces CP aggregation, there are interchain interactions between the CP molecules, which cause fluorescent quenching due to π-stacking of the backbone.

A variety of CPs have been utilized in biosensing applications. We list the following: polythiophene, poly[*p*-phenylene], poly[*p*-phenylene vinylene], poly[*p*-phenylene-ethynylene], and polyfluorene. Depending upon the substituent moieties, which are attached to these polymer backbones, the resulting CP can have either cationic or anionic properties. The optical properties of these CPs depend on their molecular configurations and aggregations in solution. The water-soluble CPs have a rigid rodlike structure and have a tendency to form aggregates in water. Various solvents have been used to study the changes in conformation of these CPs. Such studies have indicated that solvent, presence of surfactant, and the species of counter ion all play important roles in regulating their optical properties.

A number of researchers have applied CP-based schemes for analysis of DNA. The use of cationic CP for such applications is obvious. It appears that the electrostatic interactions between conjugated polyelectrolyte polymers and negatively charged nucleic acids can be used to advantage to modify optical properties of these polymers and thus detect hybridization states. In contrast to other methods of DNA analysis, the CP-based schemes offer the following advantages: assays can be performed in homogeneous solution, so complicated processing steps are avoided; fluorescence (or visual) detection is straightforward and does not require specialized instruments; amplification by the CP provides potentially high sensitivity, so only trace amounts of DNA are required; and the interaction of CP and DNA is electrostatic, so covalent labeling of DNA is avoided.

It is not possible in this brief review to cover all aspects of the use of CP in biomolecular sensing applications. The accompanying table (Table 16.4) provides a *snapshot* of the range of applications in the literature. We provide in the table references to review articles, in general, rather than the original citation of the primary work. The literature is too vast to reference all original citations, and rather than omit some important work, we have decided to simply cite review articles, which describe the work. Those wishing to study the primary articles can readily find the appropriate references in the review articles.

A number of good reviews regarding CPs and their applications to biomolecular analysis have appeared in the literature. We would refer the reader to the following examples: Ho et al.,[32] Thomas et al.,[33] Feng et al.,[34] Jiang et al.,[35] and Lee et al.[36]

16.5.2 Applications of Nanomaterials

The explosion of interest in, and use of, nanomaterials in the last two decades has resulted in far-reaching developments in not only the study of the basic properties of these fascinating materials but also applications of such materials to a variety of practical problems. This whole area of research has come to be known as nanotechnology. Not surprisingly, there has been a great deal of interest in employing nanotechnology in various guises as elements of new sensing schemes for detection of biomolecules and/or bioprocesses. Nanotechnology can provide opportunities for highly sensitive bioassays and innovative biosensors. Several features of nanomaterials provide unique advantages for such biosensing applications.

First nanomaterials, by definition, are of similar size (or at least the smaller nanomaterials are) as some of the common biomacromolecules. This suggests that interactions of these nanomaterials with biomacromolecules may have unique aspects not seen where the scale between analyte and sensing system is on very different orders of magnitude. Second, the ability to readily produce nanomaterials of various sizes, shapes, and composition allows one to custom-tailor their physical properties, and these properties can be strongly influenced by binding of biomolecules, since the nanomaterials are of similar size. Third, the unique catalytic properties of metal nanoparticles can facilitate their enlargement by the same or other metal, which can provide signal amplification. Fourth, the nanoparticles/nanomaterials can encapsulate (or immobilize on their surface) numerous signal-generating entities, and this can result in enormous signal amplification, since each nanoparticle could theoretically indicate a single binding event. Finally, nanoparticles have large surface areas, thus facilitating the incorporation of numerous receptor molecules on their surface.

Included among the nanomaterials is that whole class known as nanoparticles. These would include so-called quantum dots (semiconductor nanoparticles), PEBBLES, as well as the noble metal nanoparticles. Other chapters in this handbook cover the topics of quantum dots, PEBBLES, and SERS-active nanoparticles, and we refer the reader to those appropriate sections. Here, we will only include information on the use of nanomaterials (and that mainly nanoparticles) in bioanalytical systems where the nanomaterials provide some unique amplification advantage. Since most of what we will cover involves nanoparticles, we will use this term almost exclusively subsequently; in the accompanying table, we abbreviate nanoparticles to NP. If other nanomaterials are referenced, we will describe the nature of these materials explicitly.

In reviewing the literature for this topic, at least two properties of nanoparticles/nanomaterials have been exploited by various investigators as potential amplification systems. One is the capacity of nanoparticles to bind or encapsulate a number of reporting groups per nanoparticle. The amplification here lies in the fact that each nanoparticle with multiple reporting entities provides more than one reporter per each biorecognition event. If the reporters happen to be enzyme molecules, then the potential for amplification can be tremendous due to the highly processive nature of enzymes. Even if the reporters are fluorophores, the addition of more fluorophores improves sensitivity of detection.

The other property of nanoparticles, which has been exploited, is their unique optical properties. Gold nanoparticles are small enough to scatter visible light and thus produce size-dependent colors. The optical property changes of gold nanoparticles upon aggregation are well known and have been used extensively in bioanalytical systems. Thus, Mirkin's group (see Table 16.5) exploited the use of gold nanoparticles for a convenient and rapid colorimetric procedure for detection of specific sequences of polynucleotides. Upon hybridization of an appropriate sequence to its exact complement, both sequences being bound to nanoparticles, the resulting aggregation produced a rapid color change from red to blue. The very sharp melting curves of these nanoparticle–biomolecule constructs allowed discrimination

TABLE 16.5 Nanomaterial-Based Amplification Schemes

Assay	Principle	Comments	References
DNA detection/sequence-specific differentiation	Capture sequences bound to microchip; upon addition of target DNA sequence and detector oligonucleotide bound to gold NP, gold NP is immobilized to chip via hybridization, if the sequences are complementary; NP-promoted reduction of silver leads to deposition of silver at hybridization site, which can be detected by scanometry	10^5 signal amplification by silver deposition and 100-fold increase in sensitivity over fluorescence methods; possible to quantitate the amount of hybridized target; NP altered melting profiles to allow discrimination of single-base mismatches	[338]
Protein assay	Typical sandwich immunoassay, but detector antibody interfaced with microcrystalline fluorescent material encapsulated in NP; upon antigen–antibody interaction, fluorescent material subsequently released	2000-fold amplification of immunoassay	[37]
DNA assay	Immobilized capture oligonucleotide hybridizes to target DNA, and probe oligonucleotide attached to a silica NP filled with fluorescent dye molecules binds to target DNA; hybridization event detected by fluorescence	LOD of 0.8 fM, with ability to discriminate mismatches; also, silica NP protected dye against bleaching; 10^4 amplification compared to direct fluorophore labeling of DNA	[339]
Protein assay	So-called bio-barcode assay; magnetic microparticle has immobilized capture antibody; upon antigen binding, antibody-derivatized gold NP, which also contains multiple ds-oligonucleotide strands bound to gold surface, binds to antigen to form sandwich; magnetic separation of complex, followed by thermal dehybridization, releases multiple ss-oligonucleotides, which have a specific sequence for each analyte protein of interest; oligonucleotides may be amplified by PCR, or additional amplification may be omitted; identification and assay of these released oligonucleotides complete the assay	Detection of prostate-specific antigen at low attomolar level, six orders of magnitude better than ELISA assay; 30 attomolar detection of PSA without additional amplification; 3 attomolar detection with PCR amplification included; also applied the same concept to detection of DNA hybridization to 500 zeptomolar level	[37,340]
DNA assay	Surface plasmon resonance detection of DNA hybridization; gold NPs with attached oligonucleotides were hybridized to surface confined target DNA sequences; this arrangement dramatically amplifies the angle shift of SPR	1000-fold sensitivity enhancement; ~10 pM LOD	[341]
Protein assay	Immunosandwich formed, between antibody-labeled magnetic microparticles, antigen, and detector antibody on gold NP, to which HRP is also conjugated; enzyme activity detected by colorimetric substrate; termed nano-ELISA	25 times more sensitive than standard ELISA	[342]
Protein assay	Similar to immunosandwich assay as directly above, but ss-oligonucleotide used as a bridge molecule to attach HRP to gold NP; colorimetric detection of HRP product	130-fold more sensitive than standard ELISA; studies done to characterize antibody, oligonucleotide, and HRP binding to gold NP; approximately 2 antibody, 29 ss-oligonucleotide, and 30 HRP molecules bound per gold NP	[343]

(continued)

TABLE 16.5 (continued)   Nanomaterial-Based Amplification Schemes

Assay	Principle	Comments	References
DNA assay	Molecular beacon concept, but gold NP serves as quencher for dye-labeled stem–loop structure; upon binding to target DNA, stem–loop opens and fluorescent dye molecule is spatially removed from quenching influence of NP	Potential for multiplexing since larger gold NPs allow attachment of multiple molecular beacons with sequences specific to different targets	[39]
DNA assay	Similar concept as Jia et al.[342] above; magnetic particle with capture oligo and gold NP with detection oligo probe form hybrid sandwich with target DNA; additional oligonucleotides attached to detector NP are hybridized to cross-linker NPs, which are NPs with complementary oligo sequences and HRP; final assay is enzymatic	Cross-linked complex results in many more HRP molecules immobilized per binding event; detection limit for BRCA-1 gene is ~1 femtomole	[344]
DNA assay; discrimination of single-base mismatches	Two different oligonucleotide-modified gold NPs hybridized to target DNA sequence; color change from red to purple as aggregates form with perfect complementarity; spot test on silica gel plate shows color change from red to blue	Detection of 60 pmole of target in 5 min	[345]
Protein assay	Combines nanoparticles and RCA; immunosandwich formed in usual way, but oligonucleotide primer is attached to detector antibody; upon binding of detector antibody to protein antigen, addition of circular DNA template, polymerase, and nucleotides results in RCA in which many copies of circular template are linked together in linear fashion; addition of oligonucleotides linked to gold NP and complementary to sequences of template results in hybridization of large numbers of gold NP to linearized DNA strand; detection by surface plasmon resonance	Paper describes theoretical basis of assay, but no data provided of an actual analysis	[346]
DNA assay	Silver island films formed on glass; thiolated oligonucleotide bound to silver NPs; addition of fluorophore-labeled complementary oligonucleotide results in fluorophore being brought into close apposition to silver NP, which results in enhanced fluorescence; microwave heating used to compress hybridization time	600-fold decrease in assay run time and 10-fold increase in assay sensitivity reported	[347]

between oligonucleotides with single-base mismatches. See the accompanying table for other examples of how the unique properties of gold nanoparticles can be incorporated into sensing schemes.

Besides the interesting color changes associated with gold nanoparticle aggregation, these particular nanostructures have interesting surface plasmon resonance and very-high-fluorescence quenching properties. The fluorescence quenching properties of metal nanoparticles have been exploited in a variation of molecular beacons, in which the nanoparticle serves in lieu of an organic quencher molecule (see Table 16.5). In the case of surface plasmon resonance, the enhancement by nanoparticles is a result of greatly increased surface mass, electromagnetic coupling between the gold film and the gold nanoparticles, and the high dielectric constant of the nanoparticles.

Not only do metal-based nanoparticles demonstrate fluorescence quenching properties. Another phenomenon, which has been extensively studied over the past several years, is metal-enhanced fluorescence. In this case, the fluorescence emission of a fluorophore situated very near to silver nanostructures will be significantly increased. The explanation given for this phenomenon is that the excited fluorophore partially transfers its energy to the silver nanoparticles, from which the energy is efficiently and quickly radiated, resulting in enhanced emission from the *coupled* fluorophore–silver system. A further advantage of this particular phenomenon is that the metal-enhanced fluorescence also results in a shorter fluorescence lifetime, and this in turn leads to the fluorophores becoming more photostable, since they spend less time in the excited state. Thus, both the enhanced fluorescence and the increased photostability have the effect of increasing assay sensitivity. This is an obvious example of the beneficial effects of nanomaterials on assay amplification.

There are a number of recent reviews regarding applications of nanomaterials to biosensing applications. We refer the reader to the following as very relevant to the content of this chapter: Wang,[37] Zhang et al.,[38] Li et al.,[39] and Seydack[40].

Acknowledgments

The submitted manuscript has been authored by a contractor of the US Government under contract number DE-AC05-00OR22725. Accordingly, the US government retains a nonexclusive, royalty-free license to publish or reproduce the published form of this contribution, or allow others to do so, for US government purposes.

References

1. Guesdon, J. L., Immunoenzymatic techniques applied to the specific detection of nucleic-acids—A review, *Journal of Immunological Methods* 150(1–2), 33–49, 1992.
2. Wilchek, M. and Bayer, E. A., The avidin biotin complex in immunology, *Immunology Today* 5(2), 39–43, 1984.
3. Price, C. P. and Newman, D. J., Introduction, in *Principles and Practice of Immunoassay*, 2nd edn., Price, C. P. and Newman, D. J. (eds.). Stockton Press, New York, 1997, pp. 3–11.
4. Porstmann, T. and Kiessig, S. T., Enzyme-immunoassay techniques—An overview, *Journal of Immunological Methods* 150(1–2), 5–21, 1992.
5. Johnstone, A. and Thorpe, R., Immunocytochemistry, immunohistochemistry and flow cytofluorimetry, in *Immunochemistry in Practice*, 3rd edn. Blackwell Science, Inc., Cambridge, MA, 1996, pp. 313–338.
6. Tijssen, P., *Practice and Theory of Enzyme Immunoassays*. Elsevier Science Publishing Co., New York, 1985.
7. Wood, P. and Barnard, G., Fluoroimmunoassay, in *Principles and Practice of Immunoassay*, 2nd edn., Price, C. P. and Newman, D. J. (eds.). Stockton Press, New York, 1997, pp. 391–423.
8. Gosling, J. P., Enzyme immunoassay with and without separation, in *Principles and Practice of Immunoassay*, 2nd edn., Price, C. P. and Newman, D. J. (eds.) Stockton Press, New York, 1997, pp. 351–387.

9. Van Regenmortel, M. H. V., The antigen–antibody reaction, in *Principles and Practice of Immunoassay*, 2nd edn., Price, C. P. and Newman, D. J. (eds.). Stockton Press, New York, 1997, pp. 15–34.

10. Avrameas, S., Amplification systems in immunoenzymatic techniques, *Journal of Immunological Methods* 150(1–2), 23–32, 1992.

11. Ishikawa, E., Hashida, S., Kohno, T., and Hirota, K., Ultrasensitive enzyme-immunoassay, *Clinica Chimica Acta* 194(1), 51–72, 1990.

12. Ishikawa, E., Hashida, S., Tanaka, K., and Kohno, T., Development and applications of ultrasensitive enzyme immunoassays for antigens and antibodies, *Clinica Chimica Acta* 185(3), 223–230, 1989.

13. Shalev, A., Greenberg, A. H., and McAlpine, P. J., Detection of attograms of antigen by a high-sensitivity enzyme-linked immunoabsorbent assay (Hs-Elisa) using a fluorogenic substrate, *Journal of Immunological Methods* 38(1–2), 125–139, 1980.

14. Ruan, K. H., Kulmacz, R. J., Wilson, A., and Wu, K. K., Highly sensitive fluorometric enzyme-immunoassay for prostaglandin-H synthase solubilized from cultured-cells, *Journal of Immunological Methods* 162(1), 23–30, 1993.

15. Brown, R. C., Weeks, I., Fisher, M., Harbron, S., Taylorson, C. J., and Woodhead, J. S., Employment of a phenoxy-substituted acridinium ester as a long-lived chemiluminescent indicator of glucose oxidase activity and its application in an alkaline phosphatase amplification cascade immunoassay, *Analytical Biochemistry* 259(1), 142–151, 1998.

16. Lewkowich, I. P., Campbell, J. D., and HayGlass, K. T., Comparison of chemiluminescent assays and colorimetric ELISAs for quantification of murine IL-12, human IL-4 and murine IL-4: Chemiluminescent substrates provide markedly enhanced sensitivity, *Journal of Immunological Methods* 247(1–2), 111–118, 2001.

17. Porakishvili, N., Fordham, J. L. A., Charrel, M., Delves, P. J., Lund, T., and Roitt, I. M., A low budget luminometer for sensitive chemiluminescent immunoassays, *Journal of Immunological Methods* 234(1–2), 35–42, 2000.

18. Johansson, A., Ellis, D. H., Bates, D. L., Plumb, A. M., and Stanley, C. J., Enzyme amplification for immunoassays—Detection limit of 100th of an attomole, *Journal of Immunological Methods* 87(1), 7–11, 1986.

19. Jenkins, S. H., Homogeneous enzyme-immunoassay, *Journal of Immunological Methods* 150(1–2), 91–97, 1992.

20. Genfa, Z. and Dasgupta, P. K., Hematin as a peroxidase substitute in hydrogen-peroxide determinations, *Analytical Chemistry* 64(5), 517–522, 1992.

21. Bobrow, M. N., Litt, G. J., Shaughnessy, K. J., Mayer, P. C., and Conlon, J., The use of catalyzed reporter deposition as a means of signal amplification in a variety of formats, *Journal of Immunological Methods* 150(1–2), 145–149, 1992.

22. Price, C. P. and Newman, D. J., Light-scattering immunoassay, in *Principles and Practice of Immunoassay*, 2nd edn., Price, C. P. and Newman, D. J. (eds.). Stockton Press, New York, 1997, pp. 445–480.

23. Aoki, K., Itoh, Y., and Yoshida, T., Simultaneous determination of urinary methamphetamine, cocaine and morphine using a latex agglutination inhibition reaction test with colored latex particles, *Japanese Journal of Toxicology and Environmental Health* 43(5), 285–292, 1997.

24. Holgate, C. S., Jackson, P., Cowen, P. N., and Bird, C. C., Immunogold silver staining—New method of immunostaining with enhanced sensitivity, *Journal of Histochemistry and Cytochemistry* 31(7), 938–944, 1983.

25. Andras, S. C., Power, J. B., Cocking, E. C., and Davey, M. R., Strategies for signal amplification in nucleic acid detection, *Molecular Biotechnology* 19(1), 29–44, 2001.

26. Schweitzer, B. and Kingsmore, S., Combining nucleic acid amplification and detection, *Current Opinion in Biotechnology* 12(1), 21–27, 2001.

27. Isaksson, A. and Landegren, U., Accessing genomic information: Alternatives to PCR, *Current Opinion in Biotechnology* 10(1), 11–15, 1999.

28. Cha, R. S. and Thilly, W. G., Specificity, efficiency, and fidelity of PCR, in *PCR Primer A Laboratory Manual*, Dieffenbach, C. W. and Dveksler, G. S. (eds.). Cold Spring Harbor Laboratory Press, Plainview, NY, 1995, pp. 37–51.

29. Nuovo, G. J., Co-labeling using in situ PCR: A review, *Journal of Histochemistry and Cytochemistry* 49(11), 1329–1339, 2001.

30. Tyagi, S. and Kramer, F. R., Molecular beacons: Probes that fluoresce upon hybridization, *Nature Biotechnology* 14(3), 303–308, 1996.

31. Hendrickson, E. R., Truby, T. M. H., Joerger, R. D., Majarian, W. R., and Ebersole, R. C., High sensitivity multianalyte immunoassay using covalent DNA-labeled antibodies and polymerase chain-reaction, *Nucleic Acids Research* 23(3), 522–529, 1995.

32. Ho, H. A., Najari, A., and Leclerc, M., Optical detection of DNA and proteins with cationic polythiophenes, *Accounts of Chemical Research* 41(2), 168–178, 2008.

33. Thomas, S. W., Joly, G. D., and Swager, T. M., Chemical sensors based on amplifying fluorescent conjugated polymers, *Chemical Reviews* 107(4), 1339–1386, 2007.

34. Feng, X. L., Liu, L. B., Wang, S., and Zhu, D. B., Water-soluble fluorescent conjugated polymers and their interactions with biomacromolecules for sensitive biosensors, *Chemical Society Reviews* 39(7), 2411–2419, 2010.

35. Jiang, H., Taranekar, P., Reynolds, J. R., and Schanze, K. W., Conjugated polyelectrolytes: Synthesis, photophysics, and applications, *Angewandte Chemie-International Edition* 48(24), 4300–4316, 2009.

36. Lee, K., Povlich, L. K., and Kim, J., Recent advances in fluorescent and colorimetric conjugated polymer-based biosensors, *Analyst* 135(9), 2179–2189, 2010.

37. Wang, J., Nanomaterial-based amplified transduction of biomolecular interactions, *Small* 1(11), 1036–1043, 2005.

38. Zhang, H. Q., Zhao, Q., Li, X. F., and Le, X. C., Ultrasensitive assays for proteins, *Analyst* 132(8), 724–737, 2007.

39. Li, D., Song, S. P., and Fan, C. H., Target-responsive structural switching for nucleic acid-based sensors, *Accounts of Chemical Research* 43(5), 631–641, 2010.

40. Seydack, M., Nanoparticle labels in immunosensing using optical detection methods, *Biosensors and Bioelectronics* 20(12), 2454–2469, 2005.

41. Nakane, P. K. and Pierce, G. B., Enzyme-labeled antibodies: Preparation and application for the localization of antigens, *Journal of Histochemistry and Cytochemistry* 14, 929, 1966.

42. Avrameas, S. and Guilbert, B., A method for quantitative determination of cellular immunoglobulins by enzyme-labeled antibodies, *European Journal of Immunology* 1, 394–396, 1971.

43. Boyd, S. and Yamazaki, H., Instantaneous cloth-based enzyme immunoassay for the semi-quantitative visual determination of antibodies, *Immunological Investigations* 26(3), 313–321, 1997.

44. Engvall, E. and Perlmann, P., Enzyme-linked immunoadsorbent assay (ELISA). Quantitative assay of immunoglobulin G, *Immunochemistry* 8, 871–879, 1971.

45. Cleveland, P. H., Richman, D. D., Oxman, M. N., Wickham, M. G., Binder, P. S., and Worthen, D. M., Immobilization of viral-antigens on filter-paper for a I125 Staphylococcal protein a immunoassay—Rapid and sensitive technique for detection of herpes-simplex virus-antigens and anti-viral antibodies, *Journal of Immunological Methods* 29(4), 369–386, 1979.

46. Kemeny, D. M., Titration of antibodies, *Journal of Immunological Methods* 150(1–2), 57–76, 1992.

47. Glassy, M. C., Handley, H. H., Cleveland, P. H., and Royston, I., An enzyme immunofiltration assay useful for detecting human monoclonal-antibody, *Journal of Immunological Methods* 58(1–2), 119–126, 1983.

48. Kohl, J., Ruker, F., Himmler, G., Razazzi, E., and Katinger, H., Cloning and expression of an HIV-1 specific single-chain Fv region fused to *Escherichia-coli* alkaline-phosphatase, *Annals of the New York Academy of Sciences* 646, 106–114, 1991.

49. Weiss, E. and Orfanoudakis, G., Application of an alkaline-phosphatase fusion protein system suitable for efficient screening and production of fab-enzyme conjugates in *Escherichia coli*, *Journal of Biotechnology* 33(1), 43–53, 1994.

50. Karawajew, L., Behrsing, O., Kaiser, G., and Micheel, B., Production and Elisa application of bispecific monoclonal-antibodies against fluorescein isothiocyanate (FITC) and horseradish-peroxidase (HRP), *Journal of Immunological Methods* 111(1), 95–99, 1988.

51. Milstein, C. and Cuello, A. C., Hybrid hybridomas and their use in immunohistochemistry, *Nature* 305(5934), 537–540, 1983.

52. Porstmann, B., Avrameas, S., Ternynck, T., Porstmann, T., Micheel, B., and Guesdon, J. L., An antibody chimera technique applied to enzyme-immunoassay for human Alpha-1-Fetoprotein with monoclonal and polyclonal antibodies, *Journal of Immunological Methods* 66(1), 179–185, 1984.

53. Guesdon, J. L., Velarde, F. N., and Avrameas, S., Solid-phase immunoassays using chimera antibodies prepared with monoclonal or polyclonal anti-enzyme and anti-erythrocyte antibodies, *Annales D Immunologie* C134(2), 265–274, 1983.

54. Kobatake, E., Nishimori, Y., Ikariyama, Y., Aizawa, M., and Kato, S., Application of a fusion protein, metapyrocatechase protein-a, to an enzyme-immunoassay, *Analytical Biochemistry* 186(1), 14–18, 1990.

55. Avrameas, S., Indirect immunoenzyme techniques for the intracellular detection of antigens, *Immunochemistry* 6, 825 or 925, 1969.

56. Butler, J. E., McGivern, P. L., and Swanson, P., Amplification of enzyme-linked immunosorbent assay (Elisa) in detection of class-specific antibodies, *Journal of Immunological Methods* 20(April), 365–383, 1978.

57. Clark, C. A., Downs, E. C., and Primus, F. J., An unlabeled antibody method using glucose-oxidase antiglucose oxidase complexes (Gag)—A sensitive alternative to immunoperoxidase for the detection of tissue antigens, *Journal of Histochemistry and Cytochemistry* 30(1), 27–34, 1982.

58. Cordell, J. L., Falini, B., Erber, W. N., Ghosh, A. K., Abdulaziz, Z., Macdonald, S., Pulford, K. A. F., Stein, H., and Mason, D. Y., Immunoenzymatic labeling of monoclonal-antibodies using immune-complexes of alkaline-phosphatase and monoclonal anti-alkaline phosphatase (APAAP complexes), *Journal of Histochemistry and Cytochemistry* 32(2), 219–229, 1984.

59. Mason, D. Y., Cordell, J. L., Abdulaziz, Z., Naiem, M., and Bordenave, G., Preparation of peroxidase—Anti-peroxidase (Pap) complexes for immunohistological labeling of monoclonal-antibodies, *Journal of Histochemistry and Cytochemistry* 30(11), 1114–1122, 1982.

60. Sternberger, L. A., Hardy, P. H., Cuculis, J. J., and Meyer, H. G., The unlabeled antibody-enzyme method of immunohistochemistry. Preparation and properties of soluble antigen-antibody complex (horseradish peroxidase—Anti-horseradish peroxidase) and its Use in identification of spirochetis, *Journal of Histochemistry and Cytochemistry* 18, 315–333, 1970.

61. Ternynck, T., Gregoire, J., and Avrameas, S., Enzyme anti-enzyme monoclonal-antibody soluble immune-complexes (EMAC)—Their use in quantitative immunoenzymatic assays, *Journal of Immunological Methods* 58(1–2), 109–118, 1983.

62. Jasani, B., Thomas, N. D., Navabi, H., Millar, D. M., Newman, G. R., Gee, J., and Williams, E. D., Dinitrophenyl (DNP) hapten sandwich staining (DHSS) procedure—A 10 year review of its principle reagents and applications, *Journal of Immunological Methods* 150(1–2), 193–198, 1992.

63. Jasani, B., Newman, G. R., Stanworth, D. R., and Williams, E. D., Immunohistochemical localization of tissue receptors using the dinitrophenyl (DNP) hapten sandwich procedure, *Journal of Pathology* 138(1), 50–50, 1982.

64. Erdile, L. F., Smith, D., and Berd, D., Whole cell ELISA for detection of tumor antigen expression in tumor samples, *Journal of Immunological Methods* 258(1–2), 47–53, 2001.

65. Cobbold, S. P. and Waldmann, H., A rapid solid-phase enzyme-linked binding assay for screening monoclonal-antibodies to cell-surface antigens, *Journal of Immunological Methods* 44(2), 125–133, 1981.

66. Sedgwick, J. D. and Czerkinsky, C., Detection of cell-surface molecules, secreted products of single cells and cellular proliferation by enzyme-immunoassay, *Journal of Immunological Methods* 150(1–2), 159–175, 1992.

67. Suter, L., Bruggen, J., and Sorg, C., Use of an enzyme-linked immunosorbent-assay (Elisa) for screening of hybridoma antibodies against cell-surface antigens, *Journal of Immunological Methods* 39, 407–411, 1980.

68. Sedgwick, J. D. and Holt, P. G., A solid-phase immunoenzymatic technique for the enumeration of specific antibody-secreting cells, *Journal of Immunological Methods* 57(1–3), 301–309, 1983.

69. Kalyuzhny, A. and Stark, S., A simple method to reduce the background and improve well-to-well reproducibility of staining in ELISPOT assays, *Journal of Immunological Methods* 257(1–2), 93–97, 2001.

70. Czerkinsky, C. C., Nilsson, L. A., Nygren, H., Ouchterlony, O., and Tarkowski, A., A solid-phase enzyme-linked immunospot (ELISPOT) assay for enumeration of specific antibody-secreting cells, *Journal of Immunological Methods* 65(1–2), 109–121, 1983.

71. Rowell, F. J., Miao, Z. F., Reeves, R. N., and Cumming, R. H., Direct on-filter immunoassay of some beta-lactam antibiotics for rapid analysis of drug captured from the workplace atmosphere, *Analyst* 122(12), 1505–1508, 1997.

72. Paffard, S. M., Miles, R. J., Clark, C. R., and Price, R. G., A rapid and sensitive enzyme linked immunofilter assay (ELIFA) for whole bacterial cells, *Journal of Immunological Methods* 192(1–2), 133–136, 1996.

73. Paffard, S. M., Miles, R. J., Clark, C. R., and Price, R. G., Amplified enzyme-linked-immunofilter assays enable detection of 50–10(5) bacterial cells within 1 hour, *Analytical Biochemistry* 248(2), 265–268, 1997.

74. Ijsselmuiden, O. E., Meinardi, M., Vandersluis, J. J., Menke, H. E., Stolz, E., and Vaneijk, R. V. W., Enzyme-linked immunofiltration assay for rapid serodiagnosis of syphilis, *European Journal of Clinical Microbiology and Infectious Diseases* 6(3), 281–285, 1987.

75. Clark, C. R., Hines, K. K., and Mallia, A. K., 96-well apparatus and method for use in enzyme-linked immunofiltration assay (ELIFA), *Biotechnology Techniques* 7(6), 461–466, 1993.

76. van Weemen, B. K. and Schuurs, A. H. W. M., Immunoassay using antigen-enzyme conjugate, *FEBS Letters* 15(3), 232–236, 1971.

77. Blake, C., Albassam, M. N., Gould, B. J., Marks, V., Bridges, J. W., and Riley, C., Simultaneous enzyme-immunoassay of 2 thyroid-hormones, *Clinical Chemistry* 28(7), 1469–1473, 1982.

78. Peterhans, A., Mecklenburg, M., Meussdoerffer, F., and Mosbach, K., A simple competitive enzyme-linked-immunosorbent-assay using antigen-beta-galactosidase fusions, *Analytical Biochemistry* 163(2), 470–475, 1987.

79. Hashida, S., Tanaka, K., Kohno, T., and Ishikawa, E., Novel and ultrasensitive sandwich enzyme-immunoassay (sandwich transfer enzyme-immunoassay) for antigens, *Analytical Letters* 21(7), 1141–1154, 1988.

80. Hashinaka, K., Hashida, S., Nishikata, I., Adachi, A., Oka, S., and Ishikawa, E., Recombinant p51 as antigen in an immune complex transfer enzyme immunoassay of immunoglobulin G antibody to human immunodeficiency virus type 1, *Clinical and Diagnostic Laboratory Immunology* 7(6), 967–976, 2000.

81. Obzansky, D. M., Rabin, B. R., Simons, D. M., Tseng, S. Y., Severino, D. M., Eggelte, H., Fisher, M., Harbron, S., Stout, R. W., and Dipaolo, M. J., Sensitive, colorimetric enzyme amplification cascade for determination of alkaline-phosphatase and application of the method to an immunoassay of thyrotropin, *Clinical Chemistry* 37(9), 1513–1518, 1991.

82. Kobatake, E., Iwai, T., Ikariyama, Y., and Aizawa, M., Bioluminescent immunoassay with a protein—A luciferase fusion protein, *Analytical Biochemistry* 208(2), 300–305, 1993.

83. Lindbladh, C., Persson, M., Bulow, L., Stahl, S., and Mosbach, K., The design of a simple competitive Elisa using human proinsulin-alkaline phosphatase conjugates prepared by gene fusion, *Biochemical and Biophysical Research Communications* 149(2), 607–614, 1987.

84. Ci, Y. X., Qin, Y., Chang, W. B., Li, Y. Z., Yao, F. J., and Zhang, W., Fluorometric mimetic enzyme-immunoassay of methotrexate, *Fresenius Journal of Analytical Chemistry* 349(4), 317–319, 1994.

85. Ci, Y. X., Qin, Y., Chang, W. B., and Li, Y. Z., Application of a mimetic enzyme for the enzyme-immunoassay for alpha-1-fetoprotein, *Analytica Chimica Acta* 300(1–3), 273–276, 1995.

86. Zhu, Q. Z., Zheng, X. Y., Xu, J. G., Li, W. Y., and Liu, F. H., Application of hemin as a labeling reagent in mimetic enzyme immunoassay for hepatitis B surface antigen, *Analytical Letters* 31(6), 963–971, 1998.

87. Engvall, E., Enzyme immunoassay ELISA and EMIT, in *Immunochemical Techniques*, Van Vunakis, H. and Langone, J. J. (eds.). Academic Press, New York, 1980, pp. 419–439.

88. Dona, V., Homogeneous colorimetric enzyme-inhibition immunoassay for cortisol in human-serum with fab anti-glucose 6-phosphate dehydrogenase as a label modulator, *Journal of Immunological Methods* 82(1), 65–75, 1985.

89. Rubenstein, K. E., Schneider, R. S., and Ullman, E. F., 'Homogeneous' enzyme immunoassay—A new immunochemical technique, *Biochemical and Biophysical Research Communications* 47(4), 846–851, 1972.

90. Burd, J. F., Carrico, R. J., Fetter, M. C., Buckler, R. T., Johnson, R. D., Boguslaski, R. C., and Christner, J. E., Specific protein-binding reactions monitored by enzymatic-hydrolysis of ligand-fluorescent dye conjugates, *Analytical Biochemistry* 77(1), 56–67, 1977.

91. Burd, J. F., Wong, R. C., Feeney, J. E., Carrico, R. J., and Boguslaski, R. C., Homogeneous reactant-labeled fluorescent immunoassay for therapeutic drugs exemplified by gentamicin determination in human-serum, *Clinical Chemistry* 23(8), 1402–1408, 1977.

92. Place, M. A., Carrico, R. J., Yeager, F. M., Albarella, J. P., and Boguslaski, R. C., A colorimetric immunoassay based on an enzyme-inhibitor method, *Journal of Immunological Methods* 61(2), 209–216, 1983.

93. Nishizono, I., Ashihara, Y., Tsuchiya, H., Tanimoto, T., and Kasahara, Y., Enzyme inhibitory homogeneous immunoassay for high molecular-weight antigen (Ii), *Journal of Clinical Laboratory Analysis* 2(3), 143–147, 1988.

94. Ashihara, Y., Nishizono, I., Tsuchiya, H., Tanimoto, T., and Kasahara, Y., Homogeneous protein enzyme-immunoassay for macromolecular antigens, *Clinical Chemistry* 31(6), 904–904, 1985.

95. Carrico, R. J., Christner, J. E., Boguslaski, R. C., and Yeung, K. K., Method for monitoring specific binding reactions with cofactor labeled ligands, *Analytical Biochemistry* 72(1–2), 271–282, 1976.

96. Engel, W. D. and Khanna, P. L., CEDIA in vitro diagnostics with a novel homogeneous immunoassay technique—Current status and future-prospects, *Journal of Immunological Methods* 150(1–2), 99–102, 1992.

97. Henderson, D. R., Friedman, S. B., Harris, J. D., Manning, W. B., and Zoccoli, M. A., CEDIA, a new homogeneous immunoassay system, *Clinical Chemistry* 32(9), 1637–1641, 1986.

98. Loor, R., Shindelman, J., Singh, H., and Khanna, P. L., Homogeneous enzyme immunoassay for high molecular weight analytes using recombinant enzyme fragments, *Journal of Clinical Immunoassay* 14, 47, 1991.

99. Litman, D. J., Hanlon, T. M., and Ullman, E. F., Enzyme channeling immunoassay—A new homogeneous enzyme-immunoassay technique, *Analytical Biochemistry* 106(1), 223–229, 1980.

100. Guesdon, J. L., Ternynck, T., and Avrameas, S., Use of avidin-biotin interaction in immunoenzymatic techniques, *Journal of Histochemistry and Cytochemistry* 27(8), 1131–1139, 1979.

101. O'Connor, E., Roberts, E. M., and Davies, J. D., Amplification of cytokine-specific ELISAs increases the sensitivity of detection to 5–20 picograms per milliliter, *Journal of Immunological Methods* 229(1–2), 155–160, 1999.

102. van Gijlswijk, R. P. M., van GijlswijkJanssen, D. J., Raap, A. K., Daha, M. R., and Tanke, H. J., Enzyme-labelled antibody-avidin conjugates: new flexible and sensitive immunochemical reagents, *Journal of Immunological Methods* 189(1), 117–127, 1996.

103. Hsu, S. M., Raine, L., and Fanger, H., Use of avidin-biotin-peroxidase complex (ABC) in immuno-peroxidase techniques—A comparison between ABC and unlabeled antibody (PAP) procedures, *Journal of Histochemistry and Cytochemistry* 29(4), 577–580, 1981.

104. Bobrow, M. N., Shaughnessy, K. J., and Litt, G. J., Catalyzed reporter deposition, a novel method of signal amplification. 2. Application to membrane immunoassays, *Journal of Immunological Methods* 137(1), 103–112, 1991.

105. Bhattacharya, R., Bhattacharya, D., and Dhar, T. K., A novel signal amplification technology based on catalyzed reporter deposition and its application in a Dot-ELISA with ultra high sensitivity, *Journal of Immunological Methods* 227(1–2), 31–39, 1999.

106. Bobrow, M. N., Harris, T. D., Shaughnessy, K. J., and Litt, G. J., Catalyzed reporter deposition, a novel method of signal amplification—Application to immunoassays, *Journal of Immunological Methods* 125(1–2), 279–285, 1989.

107. van Gijlswijk, R. P. M., Zijlmans, H., Wiegant, J., Bobrow, M. N., Erickson, T. J., Adler, K. E., Tanke, H. J., and Raap, A. K., Fluorochrome-labeled tyramides: Use in immunocytochemistry and fluorescence in situ hybridization, *Journal of Histochemistry and Cytochemistry* 45(3), 375–382, 1997.

108. Mize, P. D., Hoke, R. A., Linn, C. P., Reardon, J. E., and Schulte, T. H., Dual-enzyme cascade—An amplified method for the detection of alkaline-phosphatase, *Analytical Biochemistry* 179(2), 229–235, 1989.

109. Hauber, R. and Geiger, R., A sensitive, bioluminescent-enhanced detection method for DNA dot-hybridization, *Nucleic Acids Research* 16(3), 1213, 1988.

110. Geiger, R. and Miska, W., Bioluminescence enhanced enzyme-immunoassay—New ultrasensitive detection systems for enzyme immunoassays. 2, *Journal of Clinical Chemistry and Clinical Biochemistry* 25(1), 31–38, 1987.

111. Aalberse, R. C., Quantitative fluoroimmunoassay, *Clinica Chimica Acta* 48(1), 109–111, 1973.

112. Blanchard, G. C. and Gardner, R., 2 Immunofluorescent methods compared with a radial immuno-diffusion method for measurement of serum immunoglobulins, *Clinical Chemistry* 24(5), 808–814, 1978.

113. Brandtzaeg, P., Evaluation of immunofluorescence with artificial sections of selected antigenicity, *Immunology* 22, 177, 1972.

114. Capel, P. J. A., Quantitative immunofluorescence method based on covalent coupling of protein to sepharose beads, *Journal of Immunological Methods* 5(2), 165–178, 1974.

115. Coons, A. H., Histochemistry with labeled antibody, *International Review of Cytology* 5, 1, 1956.

116. Coons, A. H., Creech, H. J., and Jones, R. N., Immunological properties of an antibody containing a fluorescent group, *Proceedings of the Society for Experimental Biology and Medicine* 47, 200, 1941.

117. Coons, A. H. and Kaplan, M. H., Localization of antigen in tissue cells. II. Improvements in a method for the detection of fluorescent antibody, *Journal of Experimental Medicine* 91, 1, 1950.

118. Chard, T. and Sykes, A., Fluoroimmunoassay for human choriomammotropin, *Clinical Chemistry* 25(6), 973–975, 1979.

119. Hebert, G. A., Pittman, B., and Cherry, W. B., Factors affecting the degree of nonspecific staining given by fluorescein isothiocyanate labeled globulins, *Journal of Immunology* 98(6), 1204–1212, 1967.

120. Katsh, S., Leaver, F. W., Reynolds, J. S., and Katsh, G. F., Simple, rapid fluorometric assay for antigens, *Journal of Immunological Methods* 5(2), 179–187, 1974.

121. Killander, A. L., Inoue, M., and Klein, E., Quantification of immunofluorescence on individual erythrocytes coated with varying amounts of antigen, *Immunology* 19, 151–156, 1970.

122. Odonnell, C. M. and Suffin, S. C., Fluorescence immunoassays, *Analytical Chemistry* 51(1), A33, 1979.

123. Sainte-Marie, G., A paraffin embedding technique for studies employing immunofluorescence, *Journal of Histochemistry and Cytochemistry* 10, 250, 1962.

124. Skurkovich, S. V., Olshansky, A. J., Samoilova, R. S., and Eremkina, E. I., Quantitative-determination of human leukocyte interferon by micro-fluorometric immunoassay with FITC-labeled anti interferon immunoglobulin, *Journal of Immunological Methods* 19(2–3), 119–124, 1978.

125. Sykes, A. and Chard, T., 2-site immunofluorometric assay for pregnancy-specific beta-1-glycoprotein (Sp1), *Clinical Chemistry* 26(8), 1224–1226, 1980.

126. Crawford, H. J., Wood, R. M., and Lessof, M. H., Detection of antibodies by fluorescent-spot techniques, *Lancet* 2, 1173, 1959.

127. Toussaint, A. J. and Anderson, R. I., Soluble antigen fluorescent-antibody technique, *Applied Microbiology* 13(4), 552, 1965.

128. Strom, R. and Klein, E., Fluorometric quantitation of fluorescein-coupled antibodies attached to the cell membrane, *Proceedings of the National Academy of Sciences of the United States of America* 63, 1157, 1970.

129. Burgett, M. W., Fairfield, S. J., and Monthony, J. F., Solid-phase fluorescent immunoassay for quantitation of C4 component of human complement, *Journal of Immunological Methods* 16(3), 211–219, 1977.

130. Chao, J., DeBiasio, R., Zhu, Z. R., Giuliano, K. A., and Schmidt, B. F., Immunofluorescence signal amplification by the enzyme-catalyzed deposition of a fluorescent reporter substrate (CARD), *Cytometry* 23(1), 48–53, 1996.

131. Kerstens, H. M. J., Poddighe, P. J., and Hanselaar, A., A novel in-situ hybridization signal amplification method based on the deposition of biotinylated tyramine, *Journal of Histochemistry and Cytochemistry* 43(4), 347–352, 1995.

132. Kronick, M. N. and Grossman, P. D., Immunoassay techniques with fluorescent phycobiliprotein conjugates, *Clinical Chemistry* 29(9), 1582–1586, 1983.

133. Rotman, B., Zderic, J. A., and Edelstein, M., Fluorogenic substrates for beta-D-galactosidases and phosphatases derived from fluorescein(3, 6-dihydroxyfluoran) and its monomethyl ether, *Proceedings of the National Academy of Sciences of the United States of America* 50(1), 1–6, 1963.

134. Revel, H. R., Luria, S. E., and Rotman, B., Biosynthesis of beta-D-galactosidase controlled by phage-carried genes. I. Induced beta-galactossidase biosynthesis after transduction of gene z+ by phage, *Proceedings of the National Academy of Sciences of the United States of America* 47, 1956, 1961.

135. Kato, K., Hamaguchi, Y., Fukui, H., and Ishikawa, E., Enzyme-linked immunoassay. 2. Simple method for synthesis of rabbit antibody-beta-D-galactosidase complex and its general applicability, *Journal of Biochemistry* 78(2), 423–425, 1975.

136. Miyai, K., Ishibashi, K., and Kawashima, M., 2-Site immunoenzymometric assay for thyrotropin in dried blood-samples on filter-paper, *Clinical Chemistry* 27(8), 1421–1423, 1981.

137. Ishikawa, E. and Kato, K., Ultrasensitive enzyme immunoassay, *Scandinavian Journal of Immunology* 8(Suppl. 7), 43, 1981.

138. Nargessi, R. D., Ackland, J., Hassan, M., Forrest, G. C., Smith, D. S., and Landon, J., Magnetizable solid-phase fluorimmunoassay of thyroxine by a sequential addition technique, *Clinical Chemistry* 26(12), 1701–1703, 1980.

139. Birkmeyer, R. C., Diaco, R., Hutson, D. K., Lau, H. P., Miller, W. K., Neelkantan, N. V., Pankratz, T. J., Tseng, S. Y., Vickery, D. K., and Yang, E. K., Application of novel chromium dioxide magnetic particles to immunoassay development, *Clinical Chemistry* 33(9), 1543–1547, 1987.

140. McHugh, T. M., Viele, M. K., Chase, E. S., and Recktenwald, D. J., The sensitive detection and quantitation of antibody to HCV by using a microsphere-based immunoassay and flow cytometry, *Cytometry* 29(2), 106–112, 1997.

141. Sutherland, R. M., Dahne, C., Place, J. F., and Ringrose, A. S., Optical-detection of antibody antigen reactions at a glass liquid interface, *Clinical Chemistry* 30(9), 1533–1538, 1984.

142. Kronick, M. N. and Little, W. A., New immunoassay based on fluorescence excitation by internal-reflection spectroscopy, *Journal of Immunological Methods* 8(3), 235–240, 1975.

143. Thompson, N. L. and Axelrod, D., Immunoglobulin surface-binding kinetics studied by total internal-reflection with fluorescence correlation spectroscopy, *Biophysical Journal* 43(1), 103–114, 1983.

144. Dandliker, W. B., Kelly, R. J., Dandliker, J., Farquhar, J., and Levin, J., Fluorescence polarization immunoassay—Theory and experimental method, *Immunochemistry* 10(4), 219–227, 1973.

145. Dandliker, W. B. and de Saussure, V. A., Fluorescence polarization in immunochemistry, *Immunochemistry* 7, 799, 1970.
146. Dandliker, W. B. and Feigen, G. A., Quantification of the antigen-antibody reaction by the polarization of fluorescence, *Biochemical and Biophysical Research Communications* 5(4), 299, 1961.
147. Jolley, M. E., Fluorescence polarization immunoassay for the determination of therapeutic drug levels in human-plasma, *Journal of Analytical Toxicology* 5(5), 236–240, 1981.
148. Guo, X. Q., Castellano, F. N., Li, L., and Lakowicz, J. R., Use of a long lifetime Re(I) complex in fluorescence polarization immunoassays of high-molecular weight analytes, *Analytical Chemistry* 70(3), 632–637, 1998.
149. Terpetschnig, E., Szmacinski, H., and Lakowicz, J. R., Fluorescence polarization immunoassay of a high-molecular-weight antigen based on a long-lifetime ru-ligand complex, *Analytical Biochemistry* 227(1), 140–147, 1995.
150. Terpetschnig, E., Szmacinski, H., and Lakowicz, J. R., Long-lifetime metal-ligand complexes as probes in biophysics and clinical chemistry, in *Fluorescence Spectroscopy*, 1997, pp. 295–321.
151. Diamandis, E. P., Morton, R. C., Reichstein, E., and Khosravi, M. J., Multiple fluorescence labeling with europium chelators—Application to time-resolved fluoroimmunoassays, *Analytical Chemistry* 61(1), 48–53, 1989.
152. Diamandis, E. P. and Christopoulos, T. K., Europium chelate labels in time-resolved fluorescence immunoassays and DNA hybridization assays, *Analytical Chemistry* 62(22), A1149, 1990.
153. Hemmila, I., Dakubu, S., Mukkala, V. M., Siitari, H., and Lovgren, T., Europium as a label in time-resolved immunofluorometric assays, *Analytical Biochemistry* 137(2), 335–343, 1984.
154. Hemmila, I., Mukkala, V. M., and Takalo, H., Development of luminescent lanthanide chelate labels for diagnostic assays, *Journal of Alloys and Compounds* 249(1–2), 158–162, 1997.
155. Karsilayan, H., Hemmila, I., Takalo, H., Toivonen, A., Pettersson, K., Lovgren, T., and Mukkala, V. M., Influence of coupling method on the luminescence properties, coupling efficiency, and binding affinity of antibodies labeled with europium(III) chelates, *Bioconjugate Chemistry* 8(1), 71–75, 1997.
156. Mukkala, V. M., Mikola, H., and Hemmila, I., The synthesis and use of activated N-benzyl derivatives of diethylenetriaminetetraacetic acids—Alternative reagents for labeling of antibodies with metal-ions, *Analytical Biochemistry* 176(2), 319–325, 1989.
157. Soini, E. and Hemmila, I., Fluoroimmunoassay—Present status and key problems, *Clinical Chemistry* 25(3), 353–361, 1979.
158. Christopoulos, T. K. and Diamandis, E. P., Enzymatically amplified time-resolved fluorescence immunoassay with terbium chelates, *Analytical Chemistry* 64(4), 342–346, 1992.
159. Soini, E. and Kojola, H., Time-resolved fluorometer for lanthanide chelates—A new generation of non-isotopic immunoassays, *Clinical Chemistry* 29(1), 65–68, 1983.
160. Evangelista, R. A., Pollak, A., and Templeton, E. F. G., Enzyme-amplified lanthanide luminescence for enzyme detection in bioanalytical assays, *Analytical Biochemistry* 197(1), 213–224, 1991.
161. Bathrellos, L. M., Lianidou, E. S., and Ioannou, P. C., A highly sensitive enzyme-amplified lanthanide luminescence immunoassay for interleukin 6, *Clinical Chemistry* 44(6), 1351–1353, 1998.
162. Diamandis, E. P., Europium and terbium chelators as candidate substrates for enzyme-labeled time-resolved fluorometric immunoassays, *Analyst* 117(12), 1879–1884, 1992.
163. Lianidou, E. S., Ioannou, P. C., and Sacharidou, E., 2nd derivative synchronous scanning fluorescence spectrometry as a sensitive detection technique in immunoassays—Application to the determination of alpha-fetoprotein, *Analytica Chimica Acta* 290(1–2), 159–165, 1994.
164. Veiopoulou, C. J., Lianidou, E. S., Ioannou, P. C., and Efstathiou, C. E., Comparative study of fluorescent ternary terbium complexes. Application in enzyme amplified fluorimetric immunoassay for alpha-fetoprotein, *Analytica Chimica Acta* 335(1–2), 177–184, 1996.
165. Ozinskas, A. J., Malak, H., Joshi, J., Szmacinski, H., Britz, J., Thompson, R. B., Koen, P. A., and Lakowicz, J. R., Homogeneous model immunoassay of thyroxine by phase-modulation fluorescence spectroscopy, *Analytical Biochemistry* 213(2), 264–270, 1993.

166. Mathis, G., Rare-earth cryptates and homogeneous fluoroimmunoassays with human sera, *Clinical Chemistry* 39(9), 1953–1959, 1993.

167. Miller, J. N., Lim, C. S., and Bridges, J. W., Fluorescamine and fluorescein as labels in energy-transfer immunoassay, *Analyst* 105(1246), 91–92, 1980.

168. Campbell, A. K. and Patel, A., A homogeneous immunoassay for cyclic-nucleotides based on chemiluminescence energy transfer, *Biochemical Journal* 216(1), 185–194, 1983.

169. Zuber, E., Mathis, G., and Flandrois, J. P., Homogeneous two-site immunometric assay kinetics as a theoretical tool for data analysis, *Analytical Biochemistry* 251(1), 79–88, 1997.

170. Ullman, E. F., Schwarzberg, M., and Rubenstein, K. E., Fluorescent excitation transfer immunoassay—General method for determination of antigens, *Journal of Biological Chemistry* 251(14), 4172–4178, 1976.

171. Dean, K. J., Thompson, S. G., Burd, J. F., and Buckler, R. T., Simultaneous determination of phenytoin and phenobarbital in serum or plasma by substrate-labeled fluorescent immunoassay, *Clinical Chemistry* 29(6), 1051–1056, 1983.

172. Wong, R. C., Burd, J. F., Carrico, R. J., Buckler, R. T., Thoma, J., and Boguslaski, R. C., Substrate-labeled fluorescent immunoassay for phenytoin in human-serum, *Clinical Chemistry* 25(5), 686–691, 1979.

173. Li, T. M. and Burd, J. F., Enzymic hydrolysis of intramolecular complexes for monitoring theophylline in homogeneous competitive-protein-binding reactions, *Biochemical and Biophysical Research Communications* 103(4), 1157–1165, 1981.

174. Weeks, I., Beheshti, I., McCapra, F., Campbell, A. K., and Woodhead, J. S., Acridinium esters as high-specific-activity labels in immunoassay, *Clinical Chemistry* 29(8), 1474–1479, 1983.

175. Simpson, J. S. A., Campbell, A. K., Ryall, M. E. T., and Woodhead, J. S., Stable chemiluminescent-labeled antibody for immunological assays, *Nature* 279(5714), 646–647, 1979.

176. Weeks, I., Sturgess, M., Siddle, K., Jones, M. K., and Woodhead, J. S., A high-sensitivity immunochemiluminometric assay for human thyrotropin, *Clinical Endocrinology* 20(4), 489–495, 1984.

177. Weeks, I., Campbell, A. K., and Woodhead, J. S., 2-site immunochemiluminometric assay for human alpha-1-fetoprotein, *Clinical Chemistry* 29(8), 1480–1483, 1983.

178. Schroeder, H. R., Hines, C. M., Osborn, D. D., Moore, R. P., Hurtle, R. L., Wogoman, F. F., Rogers, R. W., and Vogelhut, P. O., Immunochemiluminometric assay for hepatitis-B surface-antigen, *Clinical Chemistry* 27(8), 1378–1384, 1981.

179. Kim, J. B., Barnard, G. J., Collins, W. P., Kohen, F., Lindner, H. R., and Eshhar, Z., Measurement of plasma estradiol-17-beta by solid-phase chemiluminescence immunoassay, *Clinical Chemistry* 28(5), 1120–1124, 1982.

180. Pratt, J. J., Woldring, M. G., and Villerius, L., Chemiluminescence-linked immunoassay, *Journal of Immunological Methods* 21(1–2), 179–184, 1978.

181. Kohen, F., Pazzagli, M., Kim, J. B., Lindner, H. R., and Boguslaski, R. C., Assay procedure for plasma progesterone based on antibody-enhanced chemiluminescence, *FEBS Letters* 104(1), 201–205, 1979.

182. Bronstein, I., Voyta, J. C., Thorpe, G. H. G., Kricka, L. J., and Armstrong, G., Chemiluminescent assay of alkaline-phosphatase applied in an ultrasensitive enzyme-immunoassay of thyrotropin, *Clinical Chemistry* 35(7), 1441–1446, 1989.

183. Thorpe, G. H. G., Bronstein, I., Kricka, L. J., Edwards, B., and Voyta, J. C., Chemi-luminescent enzyme-immunoassay of alpha-fetoprotein based on an adamantyl dioxetane phenyl phosphate substrate, *Clinical Chemistry* 35(12), 2319–2321, 1989.

184. Otoole, A., Kricka, L. J., Thorpe, G. H. G., and Whitehead, T. P., Rapid enhanced chemiluminescent enzyme-immunoassay for ferritin monitored using instant photographic film, *Analytica Chimica Acta* 266(2), 193–199, 1992.

185. Schmid, J. A. and Billich, A., Simple method for high sensitivity chemiluminescence ELISA using conventional laboratory equipment, *Biotechniques* 22(2), 278–280, 1997.

186. Schaap, A. P., Akhavan, H., and Romano, L. J., Chemiluminescent substrates for alkaline-phosphatase—Application to ultrasensitive enzyme-linked immunoassays and DNA probes, *Clinical Chemistry* 35(9), 1863–1864, 1989.

187. Thorpe, G. H. G., Kricka, L. J., Moseley, S. B., and Whitehead, T. P., Phenols as enhancers of the chemiluminescent horseradish-peroxidase luminol hydrogen-peroxide reaction—Application in luminescence-monitored enzyme immunoassays, *Clinical Chemistry* 31(8), 1335–1341, 1985.

188. Yasui, T. and Yoda, K., Imaging of *Lactobacillus brevis* single cells and microcolonies without a microscope by an ultrasensitive chemiluminescent enzyme immunoassay with a photon-counting television camera, *Applied and Environmental Microbiology* 63(11), 4528–4533, 1997.

189. Whitehead, T. P., Thorpe, G. H. G., Carter, T. J. N., Groucutt, C., and Kricka, L. J., Enhanced luminescence procedure for sensitive determination of peroxidase-labeled conjugates in immunoassay, *Nature* 305(5930), 158–159, 1983.

190. Velan, B. and Halmann, M., Chemiluminescence immunoassay—New sensitive method for determination of antigens, *Immunochemistry* 15(5), 331–333, 1978.

191. Nishizono, I., Iida, S., Suzuki, N., Kawada, H., Murakami, H., Ashihara, Y., and Okada, M., Rapid and sensitive chemiluminescent enzyme-immunoassay for measuring tumor-markers, *Clinical Chemistry* 37(9), 1639–1644, 1991.

192. Bronstein, I., Voyta, J. C., Vanterve, Y., and Kricka, L. J., Advances in ultrasensitive detection of proteins and nucleic-acids with chemiluminescence—Novel derivatized 1,2-dioxetane enzyme substrates, *Clinical Chemistry* 37(9), 1526–1527, 1991.

193. Bronstein, I., Edwards, B., and Voyta, J. C., 1,2-Dioxetanes—Novel chemiluminescent enzyme substrates—Applications to immunoassays, *Journal of Bioluminescence and Chemiluminescence* 4(1), 99–111, 1989.

194. Olesen, C. E. M., Mosier, J., Voyta, J. C., and Bronstein, I., Chemiluminescent immunodetection protocols with 1,2-dioxetane substrates, in *Bioluminescence and Chemiluminescence. Part C*, 2000, pp. 417–427.

195. Erler, K., Elecsys (R) immunoassay systems using electrochemiluminescence detection, *Wiener Klinische Wochenschrift* 110, 5–10, 1998.

196. Blackburn, G. F., Shah, H. P., Kenten, J. H., Leland, J., Kamin, R. A., Link, J., Peterman, J. et al., Electrochemiluminescence detection for development of immunoassays and DNA probe assays for clinical diagnostics, *Clinical Chemistry* 37(9), 1534–1539, 1991.

197. Lin, J. M. and Yamada, M., Electrogenerated chemiluminescence of methyl-9-(p-formylphenyl) acridinium carboxylate fluorosulfonate and its applications to immunoassay, *Microchemical Journal* 58(1), 105–116, 1998.

198. Kankare, J., Haapakka, K., Kulmala, S., Nanto, V., Eskola, J., and Takalo, H., Immunoassay by time-resolved electrogenerated luminescence, *Analytica Chimica Acta* 266(2), 205–212, 1992.

199. Maeda, M., Shimizu, S., and Tsuji, A., Chemiluminescence assay of beta-D-galactosidase and its application to competitive immunoassay for 17-alpha-hydroxyprogesterone and thyroxine, *Analytica Chimica Acta* 266(2), 213–217, 1992.

200. Tanaka, K. and Ishikawa, E., A highly sensitive bioluminescent assay of beta-D-galactosidase from *Escherichia-coli* using 2-nitrophenyl-beta-D-galactopyranoside as a substrate, *Analytical Letters* 19(3–4), 433–444, 1986.

201. Takayasu, S., Maeda, M., and Tsuji, A., Chemiluminescent enzyme-immunoassay using beta-D-galactosidase as the label and the bis(2,4,6-trichlorophenyl)oxalate-fluorescent dye system, *Journal of Immunological Methods* 83(2), 317–325, 1985.

202. Arakawa, H., Maeda, M., and Tsuji, A., Chemiluminescence enzyme-immunoassay for thyroxine with use of glucose-oxidase and a bis(2,4,6-trichlorophenyl)oxalate fluorescent dye system, *Clinical Chemistry* 31(3), 430–434, 1985.

203. Ishimori, Y. and Rokugawa, K., Stable liposomes for assays of human sera, *Clinical Chemistry* 39(7), 1439–1443, 1993.

204. Ishimori, Y., Yasuda, T., Tsumita, T., Notsuki, M., Koyama, M., and Tadakuma, T., Liposome immune lysis assay (LILA)—A simple method to measure anti-protein antibody using protein antigen-bearing liposomes, *Journal of Immunological Methods* 75(2), 351–360, 1984.

205. Six, H. R., Young, W. W., Uemura, K. I., and Kinsky, S. C., Effect of antibody-complement on multiple vs. single compartment liposomes—Application of a fluorometric assay for following changes in liposomal permeability, *Biochemistry* 13(19), 4050–4058, 1974.

206. Yasuda, T., Naito, Y., Tsumita, T., and Tadakuma, T., A simple method to measure anti-glycolipid antibody by using complement-mediated immune lysis of fluorescent dye-trapped liposomes, *Journal of Immunological Methods* 44(2), 153–158, 1981.

207. Pashkov, V. N., Tsurupa, G. P., Griko, N. B., Skopinskaya, S. N., and Yarkov, S. P., The use of streptavidin biotin interaction for preparation of reagents for complement-dependent liposome immunoassay of proteins—Detection of latrotoxin, *Analytical Biochemistry* 207(2), 341–347, 1992.

208. Rongen, H. A. H., Bult, A., and van Bennekom, W. P., Liposomes and immunoassays, *Journal of Immunological Methods* 204(2), 105–133, 1997.

209. Rongen, H. A. H., Vanderhorst, H. M., Hugenholtz, G. W. K., Bult, A., Vanbennekom, W. P., and Vandermeide, P. H., Development of a liposome immunosorbent-assay for human interferon-gamma, *Analytica Chimica Acta* 287(3), 191–199, 1994.

210. Kobatake, E., Sasakura, H., Haruyama, T., Laukkanen, M. L., Keinanen, K., and Aizawa, M., A fluoroimmunoassay based on immunoliposomes containing genetically engineered lipid-tagged antibody, *Analytical Chemistry* 69(7), 1295–1298, 1997.

211. Rongen, H. A. H., Vannierop, T., Vanderhorst, H. M., Rombouts, R. F. M., Vandermeide, P. H., Bult, A., and Vanbennekom, W. P., Biotinylated and streptavidinylated liposomes as labels in cytokine immunoassays, *Analytica Chimica Acta* 306(2–3), 333–341, 1995.

212. Yamaji, H., Nakagawa, M., Tomioka, K., Kondo, A., and Fukuda, H., Use of a substrate as an encapsulated marker for liposome immunoassay, *Journal of Fermentation and Bioengineering* 83(6), 596–598, 1997.

213. Lim, S. J. and Kim, C. K., Homogeneous liposome immunoassay for insulin using phospholipase C from *Clostridium perfringens*, *Analytical Biochemistry* 247(1), 89–95, 1997.

214. Litchfield, W. J., Freytag, J. W., and Adamich, M., Highly sensitive immunoassays based on use of liposomes without complement, *Clinical Chemistry* 30(9), 1441–1445, 1984.

215. Nakamura, T., Hoshino, S., Hazemoto, N., Haga, M., Kato, Y., and Suzuki, Y., A liposome immunoassay based on a chemiluminescence reaction, *Chemical and Pharmaceutical Bulletin* 37(6), 1629–1631, 1989.

216. Bystryak, S., Goldiner, I., Niv, A., Nasser, A. M., and Goldstein, L., A homogeneous immunofluorescence assay based on dye-sensitized photobleaching, *Analytical Biochemistry* 225(1), 127–134, 1995.

217. Okabayashi, Y. and Ikeuchi, I., Liposome immunoassay by long-lived fluorescence detection, *Analyst* 123(6), 1329–1332, 1998.

218. White, S. R., Chiu, N. H. L., and Christopoulos, T. K., Expression immunoassay based on antibodies labeled with a deoxyribonucleic acid fragment encoding the alpha-peptide of beta-galactosidase, *Analyst* 123(6), 1309–1314, 1998.

219. Christopoulos, T. K. and Chiu, N. H. L., Expression immunoassay—Antigen quantitation using antibodies labeled with enzyme-coding DNA fragments, *Analytical Chemistry* 67(23), 4290–4294, 1995.

220. Phillips, T. M. and Krum, J. M., Recycling immunoaffinity chromatography for multiple analyte analysis in biological samples, *Journal of Chromatography B* 715(1), 55–63, 1998.

221. Rohr, T. E., Cotton, T., Fan, N., and Tarcha, P. J., Immunoassay employing surface-enhanced raman-spectroscopy, *Analytical Biochemistry* 182(2), 388–398, 1989.

222. Ni, J., Lipert, R. J., Dawson, G. B., and Porter, M. D., Immunoassay readout method using extrinsic Raman labels adsorbed on immunogold colloids, *Analytical Chemistry* 71(21), 4903–4908, 1999.

223. Dou, X., Takama, T., Yamaguchi, Y., Yamamoto, H., and Ozaki, Y., Enzyme immunoassay utilizing surface-enhanced Raman scattering of the enzyme reaction product, *Analytical Chemistry* 69(8), 1492–1495, 1997.

224. Surugiu, I., Danielsson, B., Ye, L., Mosbach, K., and Haupt, K., Chemiluminescence imaging ELISA using an imprinted polymer as the recognition element instead of an antibody, *Analytical Chemistry* 73(3), 487–491, 2001.

225. Surugiu, I., Ye, L., Yilmaz, E., Dzgoev, A., Danielsson, B., Mosbach, K., and Haupt, K., An enzyme-linked molecularly imprinted sorbent assay, *Analyst* 125(1), 13–16, 2000.

226. Matsunaga, T., Kawasaki, M., Yu, X., Tsujimura, N., and Nakamura, N., Chemiluminescence enzyme immunoassay using bacterial magnetic particles, *Analytical Chemistry* 68(20), 3551–3554, 1996.

227. Nakamura, N., Hashimoto, K., and Matsunaga, T., Immunoassay method for the determination of immunoglobulin-G using bacterial magnetic particles, *Analytical Chemistry* 63(3), 268–272, 1991.

228. Nakamura, N., Burgess, J. G., Yagiuda, K., Kudo, S., Sakaguchi, T., and Matsunaga, T., Detection and removal of *Escherichia-coli* using fluorescein isothiocyanate conjugated monoclonal-antibody immobilized on bacterial magnetic particles, *Analytical Chemistry* 65(15), 2036–2039, 1993.

229. Beverloo, H. B., Vanschadewijk, A., Zijlmans, H., and Tanke, H. J., Immunochemical detection of proteins and nucleic-acids on filters using small luminescent inorganic crystals as markers, *Analytical Biochemistry* 203(2), 326–334, 1992.

230. Beverloo, H. B., Vanschadewijk, A., Vangelderenboele, S., and Tanke, H. J., Inorganic phosphors as new luminescent labels for immunocytochemistry and time-resolved microscopy, *Cytometry* 11(7), 784–792, 1990.

231. Szurdoki, F., Michael, K. L., and Walt, D. R., A duplexed microsphere-based fluorescent immunoassay, *Analytical Biochemistry* 291(2), 219–228, 2001.

232. Yu, H., Use of an immunomagnetic separation-fluorescent immunoassay (IMS-FIA) for rapid and high throughput analysis of environmental water samples, *Analytica Chimica Acta* 376(1), 77–81, 1998.

233. Yu, H., Comparative studies of magnetic particle-based solid phase fluorogenic and electrochemiluminescent immunoassay, *Journal of Immunological Methods* 218(1–2), 1–8, 1998.

234. Moeremans, M., Daneels, G., Vandijck, A., Langanger, G., and Demey, J., Sensitive visualization of antigen-antibody reactions in dot and blot immune overlay assays with immunogold and immuno-gold silver staining, *Journal of Immunological Methods* 74(2), 353–360, 1984.

235. Stults, N. L., Stocks, N. F., Rivera, H., Gray, J., McCann, R. O., Okane, D., Cummings, R. D., Cormier, M. J., and Smith, D. F., Use of recombinant biotinylated aequorin in microtiter and membrane-based assays—Purification of recombinant apoaequorin from *Escherichia coli*, *Biochemistry* 31(5), 1433–1442, 1992.

236. Cahill, P., Foster, K., and Mahan, D. E., Polymerase chain-reaction and Q beta replicase amplification, *Clinical Chemistry* 37(9), 1482–1485, 1991.

237. Crescenzi, M., Seto, M., Herzig, G. P., Weiss, P. D., Griffith, R. C., and Korsmeyer, S. J., Thermostable DNA-polymerase chain amplification of T(14–18) chromosome breakpoints and detection of minimal residual disease, *Proceedings of the National Academy of Sciences of the United States of America* 85(13), 4869–4873, 1988.

238. Chehab, F. F., Doherty, M., Cai, S., Kan, Y. W., Cooper, S., and Rubin, E. M., Detection of sickle-cell-anemia and thalassemias, *Nature* 329(6137), 293–294, 1987.

239. Saiki, R. K., Scharf, S., Faloona, F., Mullis, K. B., Horn, G. T., Erlich, H. A., and Arnheim, N., Enzymatic amplification of beta-globin genomic sequences and restriction site analysis for diagnosis of sickle-cell anemia, *Science* 230(4732), 1350–1354, 1985.

240. Saiki, R. K., Bugawan, T. L., Horn, G. T., Mullis, K. B., and Erlich, H. A., Analysis of enzymatically amplified beta globin and HLA DQ alpha DNA with allele specific oligonucleotide probes, *Nature* 324(6093), 163–166, 1986.

241. Alard, P., Lantz, O., Sebagh, M., Calvo, C. F., Weill, D., Chavanel, G., Senik, A., and Charpentier, B., A versatile Elisa-PCR assay for messenger-RNA quantitation from a few cells, *Biotechniques* 15(4), 730, 1993.

242. Myers, T. W. and Gelfand, D. H., Reverse transcription and DNA amplification by a thermus-thermophilus DNA-polymerase, *Biochemistry* 30(31), 7661–7666, 1991.

243. Pinkel, D., Landegent, J., Collins, C., Fuscoe, J., Segraves, R., Lucas, J., and Gray, J., Fluorescence insitu hybridization with human chromosome-specific libraries—Detection of trisomy-21 and translocations of chromosome-4, *Proceedings of the National Academy of Sciences of the United States of America* 85(23), 9138–9142, 1988.

244. Rappolee, D. A., Wang, A., Mark, D., and Werb, Z., Novel method for studying messenger-RNA phenotypes in single or small numbers of cells, *Journal of Cellular Biochemistry* 39(1), 1–11, 1989.

245. Razin, E., Leslie, K. B., and Schrader, J. W., Connective-tissue mast-cells in contact with fibroblasts express IL-3 messenger-RNA—Analysis of single cells by polymerase chain-reaction, *Journal of Immunology* 146(3), 981–987, 1991.

246. Chehab, F. F. and Kan, Y. W., Detection of specific DNA-sequences by fluorescence amplification—A color complementation assay, *Proceedings of the National Academy of Sciences of the United States of America* 86(23), 9178–9182, 1989.

247. Bortolin, S., Christopoulos, T. K., and Verhaegen, M., Quantitative polymerase chain reaction using a recombinant DNA internal standard and time resolved fluorometry, *Analytical Chemistry* 68(5), 834–840, 1996.

248. Wang, A. M., Doyle, M. V., and Mark, D. F., Quantitation of messenger-RNA by the polymerase chain-reaction, *Proceedings of the National Academy of Sciences of the United States of America* 86(24), 9717–9721, 1989.

249. Katz, E. D., Haff, L. A., and Eksteen, R., Rapid separation, quantitation and purification of products of polymerase chain-reaction by liquid-chromatography, *Journal of Chromatography* 512, 433–444, 1990.

250. Kemp, D. J., Smith, D. B., Foote, S. J., Samaras, N., and Peterson, M. G., Colorimetric detection of specific DNA segments amplified by polymerase chain reactions, *Proceedings of the National Academy of Sciences of the United States of America* 86(7), 2423–2427, 1989.

251. Stoflet, E. S., Koeberl, D. D., Sarkar, G., and Sommer, S. S., Genomic amplification with transcript sequencing, *Science* 239(4839), 491–494, 1988.

252. Nickerson, D. A., Kaiser, R., Lappin, S., Stewart, J., Hood, L., and Landegren, U., Automated DNA diagnostics using an Elisa-based oligonucleotide ligation assay, *Proceedings of the National Academy of Sciences of the United States of America* 87(22), 8923–8927, 1990.

253. Thelwell, N., Millington, S., Solinas, A., Booth, J., and Brown, T., Mode of action and application of Scorpion primers to mutation detection, *Nucleic Acids Research* 28(19), 3752–3761, 2000.

254. Whitcombe, D., Theaker, J., Guy, S. P., Brown, T., and Little, S., Detection of PCR products using self-probing amplicons and fluorescence, *Nature Biotechnology* 17(8), 804–807, 1999.

255. Holland, P. M., Abramson, R. D., Watson, R., and Gelfand, D. H., Detection of specific polymerase chain-reaction product by utilizing the 5'-3' exonuclease activity of thermus-aquaticus DNA-polymerase, *Proceedings of the National Academy of Sciences of the United States of America* 88(16), 7276–7280, 1991.

256. Nazarenko, I. A., Bhatnagar, S. K., and Hohman, R. J., A closed tube format for amplification and detection of DNA based on energy transfer, *Nucleic Acids Research* 25(12), 2516–2521, 1997.

257. Dicesare, J., Grossman, B., Katz, E., Picozza, E., Ragusa, R., and Woudenberg, T., A high-sensitivity electrochemiluminescence-based detection system for automated PCR product quantitation, *Biotechniques* 15(1), 152–157, 1993.

258. Actor, J. K., Kuffner, T., Dezzutti, C. S., Hunter, R. L., and McNicholl, J. M., A flash-type bioluminescent immunoassay that is more sensitive than radioimaging: Quantitative detection of cytokine cDNA in activated and resting human cells, *Journal of Immunological Methods* 211(1–2), 65–77, 1998.

259. Actor, J. K., Olsen, M., Boven, L. A., Werner, N., Stults, N. L., Hunter, R. L., and Smith, D. F., A bioluminescent assay using AquaLite for PT-PCR amplified RNA from mouse lung, *Journal of National Institute of Health Research* 8, 62, 1996.

260. Kenten, J. H., Casadei, J., Link, J., Lupold, S., Willey, J., Powell, M., Rees, A., and Massey, R., Rapid electrochemiluminescence assays of polymerase chain-reaction products, *Clinical Chemistry* 37(9), 1626–1632, 1991.

261. Liu, X., Farmerie, W., Schuster, S., and Tan, W., Molecular beacons for DNA biosensors with micrometer to submicrometer dimensions., *Analytical Biochemistry* 283, 56–63, 2000.

262. McKillip, J. L. and Drake, M., Molecular beacon polymerase chain reaction detection of *Escherichia coli* O157: H7 in milk, *Journal of Food Protection* 63(7), 855–859, 2000.

263. Compton, J., Nucleic-acid sequence-based amplification, *Nature* 350(6313), 91–92, 1991.

264. Voisset, C., Mandrand, B., and Paranhos-Baccala, G., RNA amplification technique, NASBA, also amplifies homologous plasmid DNA in non-denaturing conditions, *Biotechniques* 29(2), 236–238, 2000.

265. Heim, A., Zeuke, S., Grumbach, I. M., and Top, B., Highly sensitive detection of gene expression of an intronless gene: Amplification of mRNA, but not genomic DNA by nucleic acid sequence based amplification (NASBA), *Nucleic Acids Research* 26(9), 2250–2251, 1998.

266. Kwoh, D. Y., Davis, G. R., Whitfield, K. M., Chappelle, H. L., Dimichele, L. J., and Gingeras, T. R., Transcription-based amplification system and detection of amplified human immunodeficiency virus type-1 with a bead-based sandwich hybridization format, *Proceedings of the National Academy of Sciences of the United States of America* 86(4), 1173–1177, 1989.

267. Guatelli, J. C., Whitfield, K. M., Kwoh, D. Y., Barringer, K. J., Richman, D. D., and Gingeras, T. R., Isothermal, in vitro amplification of nucleic-acids by a multienzyme reaction modeled after retroviral replication, *Proceedings of the National Academy of Sciences of the United States of America* 87(5), 1874–1878, 1990.

268. Liu, D. Y., Daubendiek, S. L., Zillman, M. A., Ryan, K., and Kool, E. T., Rolling circle DNA synthesis: Small circular oligonucleotides as efficient templates for DNA polymerases, *Journal of the American Chemical Society* 118(7), 1587–1594, 1996.

269. Lizardi, P. M., Huang, X. H., Zhu, Z. R., Bray-Ward, P., Thomas, D. C., and Ward, D. C., Mutation detection and single-molecule counting using isothermal rolling-circle amplification, *Nature Genetics* 19(3), 225–232, 1998.

270. Nilsson, M., Malmgren, H., Samiotaki, M., Kwiatkowski, M., Chowdhary, B. P., and Landegren, U., Padlock probes—Circularizing oligonucleotides for localized DNA detection, *Science* 265(5181), 2085–2088, 1994.

271. Barany, F., Genetic-disease detection and DNA amplification using cloned thermostable ligase, *Proceedings of the National Academy of Sciences of the United States of America* 88(1), 189–193, 1991.

272. Wu, D. Y. and Wallace, R. B., The ligation amplification reaction (LAR)—Amplification of specific DNA-sequences using sequential rounds of template-dependent ligation, *Genomics* 4(4), 560–569, 1989.

273. Landegren, U., Kaiser, R., Sanders, J., and Hood, L., A ligase-mediated gene detection technique, *Science* 241(4869), 1077–1080, 1988.

274. Abravaya, K., Carrino, J. J., Muldoon, S., and Lee, H. H., Detection of point mutations with a modified ligase chain-reaction (Gap-LCR), *Nucleic Acids Research* 23(4), 675–682, 1995.

275. Nadeau, J. G., Pitner, J. B., Linn, C. P., Schram, J. L., Dean, C. H., and Nycz, C. M., Real-time, sequence-specific detection of nucleic acids during strand displacement amplification, *Analytical Biochemistry* 276(2), 177–187, 1999.

276. Musiani, M., Zerbini, M., Gibellini, D., Gentilomi, G., Laplaca, M., Ferri, E., and Girotti, S., Chemiluminescent assay for the detection of viral and plasmid DNA using digoxigenin-labeled probes, *Analytical Biochemistry* 194(2), 394–398, 1991.

277. Tizard, R., Cate, R. L., Ramachandran, K. L., Wysk, M., Voyta, J. C., Murphy, O. J., and Bronstein, I., Imaging of DNA-sequences with chemiluminescence, *Proceedings of the National Academy of Sciences of the United States of America* 87(12), 4514–4518, 1990.

278. Cano, R. J., Torres, M. J., Klem, R. E., and Palomares, J. C., DNA hybridization assay using attophostm, a fluorescent substrate for alkaline-phosphatase, *Biotechniques* 12(2), 264, 1992.

279. Evangelista, R. A., Wong, H. E., Templeton, E. F. G., Granger, T., Allore, B., and Pollak, A., Alkyl-substituted and aryl-substituted salicyl phosphates as detection reagents in enzyme-amplified fluorescence DNA hybridization assays on solid support, *Analytical Biochemistry* 203(2), 218–226, 1992.

280. Raap, A. K., Vandecorput, M. P. C., Vervenne, R. A. W., Vangijlswijk, R. P. M., Tanke, H. J., and Wiegant, J., Ultra-sensitive fish using peroxidase-mediated deposition of biotin-tyramide or fluoro-chrome-tyramide, *Human Molecular Genetics* 4(4), 529–534, 1995.

281. Schmidt, B. F., Chao, J., Zhu, Z. G., DeBiasio, R. L., and Fisher, G., Signal amplification in the detection of single-copy DNA and RNA by enzyme-catalyzed deposition (CARD) of the novel fluorescent reporter substrate Cy3.29-tyramide, *Journal of Histochemistry and Cytochemistry* 45(3), 365–373, 1997.

282. Adams, J. C., Biotin amplification of biotin and horseradish-peroxidase signals in histochemical stains, *Journal of Histochemistry and Cytochemistry* 40(10), 1457–1463, 1992.

283. Gosden, J., Hanratty, D., Starling, J., Fantes, J., Mitchell, A., and Porteous, D., Oligonucleotide-primed insitu DNA-synthesis (PRINS)—A method for chromosome mapping, banding, and investigation of sequence organization, *Cytogenetics and Cell Genetics* 57(2–3), 100–104, 1991.

284. Gosden, J. and Hanratty, D., Comparison of sensitivity of 3 haptens in the PRINS reaction, *Technique* 3, 159, 1992.

285. Gosden, J. and Hanratty, D., PCR in-situ—A rapid alternative to in-situ hybridization for mapping short, low copy number sequences without isotopes, *Biotechniques* 15(1), 78–80, 1993.

286. Gosden, J. and Lawson, D., Rapid chromosome identification by oligonucleotide-primed in-situ DNA-synthesis (PRINS), *Human Molecular Genetics* 3(6), 931–936, 1994.

287. Lorimier, P., Lamarcq, L., Negoescu, A., Robert, C., LabatMoleur, F., GrasChappuis, F., Durrant, I., and Brambilla, E., Comparison of S-35 and chemiluminescence for HPV in situ hybridization in carcinoma cell lines and on human cervical intraepithelial neoplasia, *Journal of Histochemistry and Cytochemistry* 44(7), 665–671, 1996.

288. Gentilomi, G., Musiani, M., Roda, A., Pasini, P., Zerbini, M., Gallinella, G., Baraldini, M., Venturoli, S., and Manaresi, E., Co-localization of two different viral genomes in the same sample by double-chemiluminescence in situ hybridization, *Biotechniques* 23(6), 1076, 1997.

289. Cheek, B. J., Steel, A. B., Torres, M. P., Yu, Y.-Y., and Yang, H., Chemiluminescence detection for hybridization assays on the flow-thru chip, a three-dimensional microchannel biochip, *Analytical Chemistry* 73(24), 5777–5783, 2001.

290. Raap, A. K., Vanderijke, F. M., Dirks, R. W., Sol, C. J., Boom, R., and Vanderploeg, M., Bicolor fluorescence insitu hybridization to intron and exon messenger-RNA sequences, *Experimental Cell Research* 197(2), 319–322, 1991.

291. Tkachuk, D. C., Westbrook, C. A., Andreeff, M., Donlon, T. A., Cleary, M. L., Suryanarayan, K., Homge, M., Redner, A., Gray, J., and Pinkel, D., Detection of bcr-abl fusion in chronic myelogeneous leukemia by insitu hybridization, *Science* 250(4980), 559–562, 1990.

292. Wiegant, J., Ried, T., Nederlof, P. M., Vanderploeg, M., Tanke, H. J., and Raap, A. K., In situ hybridization with fluoresceinated DNA, *Nucleic Acids Research* 19(12), 3237–3241, 1991.

293. Wachtler, F., Hartung, M., Devictor, M., Wiegant, J., Stahl, A., and Schwarzacher, H. G., Ribosomal DNA is located and transcribed in the dense fibrillar component of human sertoli-cell nucleoli, *Experimental Cell Research* 184(1), 61–71, 1989.

294. Dewit, P. E., Kerstens, H. M. J., Poddighe, P. J., Vanmuijen, G. N. P., and Ruiter, D. J., DNA in-situ hybridization as a diagnostic-tool in the discrimination of melanoma and spitz nevus, *Journal of Pathology* 173(3), 227–233, 1994.

295. Tchen, P., Fuchs, R. P. P., Sage, E., and Leng, M., Chemically modified nucleic-acids as immunodetectable probes in hybridization experiments, *Proceedings of the National Academy of Sciences of the United States of America-Biological Sciences* 81(11), 3466–3470, 1984.

296. Elghanian, R., Storhoff, J. J., Mucic, R. C., Letsinger, R. L., and Mirkin, C. A., Selective colorimetric detection of polynucleotides based on the distance-dependent optical properties of gold nanoparticles, *Science* 277(5329), 1078–1081, 1997.

297. Urdea, M. S., Running, J. A., Horn, T., Clyne, J., Ku, L., and Warner, B. D., A novel method for the rapid detection of specific nucleotide-sequences in crude biological samples without blotting or radioactivity—Application to the analysis of hepatitis-B virus in human-serum, *Gene* 61(3), 253–264, 1987.

298. Call, D. R., Chandler, D. P., and Brockman, F., Fabrication of DNA microarrays using unmodified oligonucleotide probes, *Biotechniques* 30(2), 368–379, 2001.

299. Guo, Z., Guilfoyle, R. A., Thiel, A. J., Wang, R. F., and Smith, L. M., Direct fluorescence analysis of genetic polymorphisms by hybridization with oligonucleotide arrays on glass supports, *Nucleic Acids Research* 22(24), 5456–5465, 1994.

300. Pease, A. C., Solas, D., Sullivan, E. J., Cronin, M. T., Holmes, C. P., and Fodor, S. P. A., Light-generated oligonucleotide arrays for rapid DNA-sequence analysis, *Proceedings of the National Academy of Sciences of the United States of America* 91(11), 5022–5026, 1994.

301. Yershov, G., Barsky, V., Belgovskiy, A., Kirillov, E., Kreindlin, E., Ivanov, I., Parinov, S. et al., DNA analysis and diagnostics on oligonucleotide microchips, *Proceedings of the National Academy of Sciences of the United States of America* 93(10), 4913–4918, 1996.

302. Langer, P. R., Waldrop, A. A., and Ward, D. C., Enzymatic-synthesis of biotin-labeled polynucleotides—Novel nucleic-acid affinity probes, *Proceedings of the National Academy of Sciences of the United States of America-Biological Sciences* 78(11), 6633–6637, 1981.

303. Favis, R. and Barany, F., Mutation detection in K-ras, BRCA1, BRCA2, and p53 using PCR/LDR and a universal DNA microarray, in *Circulating Nucleic Acids in Plasma or Serum*, 2000, pp. 39–43.

304. Lyamichev, V., Mast, A. L., Hall, J. G., Prudent, J. R., Kaiser, M. W., Takova, T., Kwiatkowski, R. W. et al., Polymorphism identification and quantitative detection of genomic DNA by invasive cleavage of oligonucleotide probes, *Nature Biotechnology* 17(3), 292–296, 1999.

305. Hall, J. G., Eis, P. S., Law, S. M., Reynaldo, L. P., Prudent, J. R., Marshall, D. J., Allawi, H. T. et al., Sensitive detection of DNA polymorphisms by the serial invasive signal amplification reaction, *Proceedings of the National Academy of Sciences of the United States of America* 97(15), 8272–8277, 2000.

306. Mullink, H., Vos, W., Jiwa, M., Horstman, A., Vandervalk, P., Walboomers, J. M. M., and Meijer, C., Application and comparison of silver intensification methods for the diaminobenzidine and diaminobenzidine nickel end-product of the peroxidation reaction in immunohistochemistry and insitu hybridization, *Journal of Histochemistry and Cytochemistry* 40(4), 495–504, 1992.

307. Hurskainen, P., Dahlen, P., Ylikoski, J., Kwiatkowski, M., Siitari, H., and Lovgren, T., Preparation of europium-labeled DNA probes and their properties, *Nucleic Acids Research* 19(5), 1057–1061, 1991.

308. Collins, M. L., Irvine, B., Tyner, D., Fine, E., Zayati, C., Chang, C. A., Horn, T. et al., A branched DNA signal amplification assay for quantification of nucleic acid targets below 100 molecules/ml, *Nucleic Acids Research* 25(15), 2979–2984, 1997.

309. Urdea, M. S., Horn, T., Fultz, T., Anderson, M., Running, J. A., Hamren, S., Ahle, D., and Chang, C. A., Branched DNA amplification multimers for the sensitive direct detection of human hepatitis virus, *Nucleic Acids Research* 24(Symposium Series), 197–200, 1991.

310. Warrior, U., Fan, Y. H., David, C. A., Wilkins, J. A., McKeegan, E. M., Kofron, J. L., and Burns, D. J., Application of QuantiGene (TM) nucleic acid quantification technology for high throughput screening, *Journal of Biomolecular Screening* 5(5), 343–351, 2000.

311. Kern, D., Collins, M., Fultz, T., Detmer, J., Hamren, S., Peterkin, J. J., Sheridan, P. et al., An enhanced-sensitivity branched-DNA assay for quantification of human immunodeficiency virus type 1 RNA in plasma, *Journal of Clinical Microbiology* 34(12), 3196–3202, 1996.

312. Capaldi, S., Getts, R. C., and Jayasena, S. D., Signal amplification through nucleotide extension and excision on a dendritic DNA platform, *Nucleic Acids Research* 28(7), e21(i–viii), 2000.

313. Sano, T., Smith, C. L., and Cantor, C. R., Immuno-PCR—Very sensitive antigen-detection by means of specific antibody-DNA conjugates, *Science* 258(5079), 120–122, 1992.

314. Ruzicka, V., Marz, W., Russ, A., and Gross, W., Immuno-PCR with a commercially available avidin system, *Science* 260(5108), 698–699, 1993.

315. Chang, T. C. and Huang, S. H., A modified immuno-polymerase chain reaction for the detection of beta-glucuronidase from *Escherichia coli*, *Journal of Immunological Methods* 208(1), 35–42, 1997.

316. Case, M. C., Burt, A. D., Hughes, J., Palmer, J. M., Collier, J. D., Bassendine, M. F., Yeaman, S. J., Hughes, M. A., and Major, G. N., Enhanced ultrasensitive detection of structurally diverse antigens using a single immuno-PCR assay protocol, *Journal of Immunological Methods* 223(1), 93–106, 1999.

317. Sanna, P. P., Weiss, F., Samson, M. E., Bloom, F. E., and Pich, E. M., Rapid induction of tumor-necrosis-factor-alpha in the cerebrospinal-fluid after intracerebroventricular injection of lipopoly-saccharide revealed by a sensitive capture immuno-PCB assay, *Proceedings of the National Academy of Sciences of the United States of America* 92(1), 272–275, 1995.

318. Furuya, D., Yagihashi, A., Yajima, T., Kobayashi, D., Orita, K., Kurimoto, M., and Watanabe, N., An immuno-polymerase chain reaction assay for human interleukin-18, *Journal of Immunological Methods* 238(1–2), 173–180, 2000.

319. Sperl, J., Paliwal, V., Ramabhadran, R., Nowak, B., and Askenase, P. W., Soluble T-cell receptors—Detection and quantitative assay in fluid-phase via ELISA or immuno-PCR, *Journal of Immunological Methods* 186(2), 181–194, 1995.

320. Zhou, H., Fisher, R. J., and Papas, T. S., Universal immuno-PCR for ultra-sensitive target protein detection, *Nucleic Acids Research* 21(25), 6038–6039, 1993.

321. Niemeyer, C. M., Adler, M., and Blohm, D., Fluorometric polymerase chain reaction (PCR) enzyme-linked immunosorbent assay for quantification of immuno-PCR products in microplates, *Analytical Biochemistry* 246(1), 140–145, 1997.

322. Maia, M., Takahashi, H., Adler, K., Garlick, R. K., and Wands, J. R., Development of a 2-site immuno-PCR assay for hepatitis-B surface-antigen, *Journal of Virological Methods* 52(3), 273–286, 1995.

323. Niemeyer, C. M., Adler, M., Pignataro, B., Lenhert, S., Gao, S., Chi, L. F., Fuchs, H., and Blohm, D., Self-assembly of DNA-streptavidin nanostructures and their use as reagents in immuno-PCR, *Nucleic Acids Research* 27(23), 4553–4561, 1999.

324. Cao, Y., Kopplow, K., and Liu, G. Y., In-situ immuno-PCR to detect antigens, *Lancet* 356(9234), 1002–1003, 2000.

325. Schweitzer, B., Wiltshire, S., Lambert, J., O'Malley, S., Kukanskis, K., Zhu, Z. R., Kingsmore, S. F., Lizardi, P. M., and Ward, D. C., Immunoassays with rolling circle DNA amplification: A versatile platform for ultrasensitive antigen detection, *Proceedings of the National Academy of Sciences of the United States of America* 97(18), 10113–10119, 2000.

326. Schweitzer, B., Roberts, S., Grimwade, B., Shao, W. P., Wang, M. J., Fu, Q., Shu, Q. P. et al., Multiplexed protein profiling on microarrays by rolling-circle amplification, *Nature Biotechnology* 20(4), 359–365, 2002.

327. Widjojoatmodjo, M. N., Fluit, A. C., Torensma, R., Keller, B. H. I., and Verhoef, J., Evaluation of the magnetic immuno PCR assay for rapid detection of Salmonella, *European Journal of Clinical Microbiology and Infectious Diseases* 10(11), 935–938, 1991.

328. Zhang, H. T., Kacharmina, J. E., Miyashiro, K., Greene, M. I., and Eberwine, J., Protein quantification from complex protein mixtures using a proteomics methodology with single-cell resolution, *Proceedings of the National Academy of Sciences of the United States of America* 98(10), 5497–5502, 2001.

329. Beyer, W., Pocivalsek, S., and Bohm, R., Polymerase chain reaction-ELISA to detect Bacillus anthracis from soil samples—Limitations of present published primers, *Journal of Applied Microbiology* 87(2), 229–236, 1999.

330. Hall, L. L., Bicknell, G. R., Primrose, L., Pringle, J. H., Shaw, J. A., and Furness, P. N., Reproducibility in the quantification of mRNA levels by RT-PCR-ELISA and RT competitive-PCR-ELISA, *Biotechniques* 24(4), 652–657, 1998.

331. Allen, R. D., Pellett, P. E., Stewart, J. A., and Koopmans, M., Nonradioactive PCR enzyme-linked-immunosorbent-assay method for detection of human cytomegalovirus DNA, *Journal of Clinical Microbiology* 33(3), 725–728, 1995.

332. Muramatsu, Y., Yanase, T., Okabayashi, T., Ueno, H., and Morita, C., Detection of Coxiella burnettii in cow's milk by PCR-enzyme-linked immunosorbent assay combined with a novel sample preparation method, *Applied and Environmental Microbiology* 63(6), 2142–2146, 1997.

333. Sawant, S. G., Antonacci, R., and Pandita, T., Determination of telomerase activity in HeLa cells after treatment with ionizing radiation by telomerase PCR ELISA, *Biochemica* 4, 22–24, 1997.

334. Soumet, C., Ermel, G., Boutin, P., Boscher, E., and Colin, P., Chemiluminescent and colorimetric enzymatic assays for the detection of PCR-amplified *Salmonella* sp. products in microplates, *Biotechniques* 19(5), 792, 1995.

335. Duan, X. R., Liu, L. B., Feng, F. B., and Wang, S., Cationic conjugated polymers for optical detection of DNA methylation, lesions, and single nucleotide polymorphisms, *Accounts of Chemical Research* 43(2), 260–270, 2010.

336. Liu, B. and Bazan, G. C., Homogeneous fluorescence-based DNA detection with water-soluble conjugated polymers, *Chemistry of Materials* 16(23), 4467–4476, 2004.

337. Feng, F. D., He, F., An, L. L., Wang, S., Li, Y. H., and Zhu, D. B., Fluorescent conjugated polyelectrolytes for biomacromolecule detection, *Advanced Materials* 20(15), 2959–2964, 2008.

338. Taton, T. A., Mirkin, C. A., and Letsinger, R. L., Scanometric DNA array detection with nanoparticle probes, *Science* 289(5485), 1757–1760, 2000.

339. Zhao, X. J., Tapec-Dytioco, R., and Tan, W. H., Ultrasensitive DNA detection using highly fluorescent bioconjugated nanoparticles, *Journal of the American Chemical Society* 125(38), 11474–11475, 2003.

340. Nam, J. M., Thaxton, C. S., and Mirkin, C. A., Nanoparticle-based bio-bar codes for the ultrasensitive detection of proteins, *Science* 301(5641), 1884–1886, 2003.

341. He, L., Musick, M. D., Nicewarner, S. R., Salinas, F. G., Benkovic, S. J., Natan, M. J., and Keating, C. D., Colloidal Au-enhanced surface plasmon resonance for ultrasensitive detection of DNA hybridization, *Journal of the American Chemical Society* 122(38), 9071–9077, 2000.

342. Jia, C. P., Zhong, X. Q., Hua, B., Liu, M. Y., Jing, F. X., Lou, X. H., Yao, S. H., Xiang, J. Q., Jin, Q. H., and Zhao, J. L., Nano-ELISA for highly sensitive protein detection, *Biosensors and Bioelectronics* 24(9), 2836–2841, 2009.

343. Liu, M. Y., Jia, C. P., Huang, Y. Y., Lou, X. H., Yao, S. H., Jin, Q. H., Zhao, J. L., and Xiang, J. Q., Highly sensitive protein detection using enzyme-labeled gold nanoparticle probes, *Analyst* 135(2), 327–331, 2010.

344. Li, J., Song, S. P., Li, D., Su, Y., Huang, Q., Zhao, Y., and Fan, C. H., Multi-functional crosslinked Au nanoaggregates for the amplified optical DNA detection, *Biosensors and Bioelectronics* 24(11), 3311–3315, 2009.

345. Storhoff, J. J., Elghanian, R., Mucic, R. C., Mirkin, C. A., and Letsinger, R. L., One-pot colorimetric differentiation of polynucleotides with single base imperfections using gold nanoparticle probes, *Journal of the American Chemical Society* 120(9), 1959–1964, 1998.

346. Hsu, H. Y. and Huang, Y. Y., RCA combined nanoparticle-based optical detection technique for protein microarray: A novel approach, *Biosensors and Bioelectronics* 20(1), 123–126, 2004.

347. Aslan, K., Malyn, S. N., Bector, G., and Geddes, C. D., Microwave-accelerated metal-enhanced fluorescence: An ultra-fast and sensitive DNA sensing platform, *Analyst* 132(11), 1122–1129, 2007.

17

Fluorescent Probes in Biomedical Applications

Darryl J. Bornhop
Vanderbilt University

Kai Licha
Mivenion GmbH
and Free University

Lynn E. Samuelson
Vanderbilt University

17.1 Introduction

Optical imaging techniques for the assessment of tissue anatomy, physiology, and metabolic and molecular function have emerged as an essential tool for both the basic researcher and the clinical practitioner. One concern of clinical practitioners is that too much harmful radiation is used to detect diseased tissue. The attractiveness of optical imaging techniques arises from the fact that fluorescent dyes can be detected at low concentrations and nonionizing, harmless radiation can be applied repeatedly to the patient. Furthermore, the remarkable progress in the development of optical instrumentation in the last two decades (laser excitation and detection systems) has decisively contributed to the growing applicability of optical imaging techniques, which have the advantage of being cheap, small in size, and, therefore, readily available to solve clinical problems. The design of contrast agents for optical in vivo imaging of diseased tissues has also emerged and is reflected by an increasing number of publications in this area.[1,2] Novel probes have been synthesized and characterized for their ability to monitor disease-specific anatomical, physiological, and molecular parameters through their optical signals. The multifarious world of chromophores and fluorophores provides various parameters that can be exploited for diagnostic measurement and detection.

This chapter describes the essential principles of optical imaging and current medical opportunities that are related to the use of fluorescent probes as exogenously applied agents; this is followed by a review of recent progress in the design of these probes for biomedical imaging purposes. The agents discussed in this chapter are categorized by their structural families, covering the class of cyanine dyes, tetrapyrrole compounds, lanthanide chelates, and other entities. For selected examples, the physicochemical, biochemical, and pharmacological features constituting the diagnostic efficacy of the compounds are illustrated.

17.2 Principles of Optical Imaging

17.2.1 Tissue Optics and Function of Dyes as Contrast Agents

The term optical imaging encompasses a large variety of different disciplines. In general, the method uses light within the ultraviolet (UV) and the near-infrared (NIR) spectral region to obtain information on the optical characteristics of tissue. Generally, the interaction of photons with tissue is based on absorption of light, scattering of light, and emission of fluorescence. These three parameters can be used separately to characterize tissue optical properties.

A fundamental observation for optical diagnostic procedures relates to the fact that the penetration depth of light in living tissue strongly depends on the wavelength used[3] because the number of absorption and scattering events in tissue is a function of wavelength.[4]

For wavelengths below approximately 600 nm, the penetration of light into tissue is limited to a depth of hundredths of micrometers up to a few millimeters due to strong absorption of the photons, so that only superficial assessment of tissues in this spectral region is possible. The absorption of light in tissue originates from oxy- and deoxyhemoglobin and several tissue components such as porphyrins, melanin, NADH and flavins, collagen, elastin, and lipopigments. Most of these chromophores that contribute to tissue absorption exhibit characteristic fluorescence spectra throughout the visible (VIS) spectral region up to approximately 700 nm. Fluorescence of these intrinsic fluorescent markers (autofluorescence) has been studied as a source of specific spectral information on tissue structure and pathophysiological states[5,6] and was thoroughly exploited to identify diseased tissue areas, for example, in endoscopy[7,8] or cardiovascular diagnosis.[9] Direct visual inspection or characterization using microscopic techniques is established practice in medicine. Hence, the modality is capable of generating images of tissue structures with high spatial resolution, as does microscopy.

A primary field of application of optical imaging technology is the examination of tissue surfaces via optical fibers incorporated in endoscopes or laparoscopes, as well as of ocular diseases through ophthalmoscopes and direct assessment of skin diseases or during surgical procedures. The only modality found in daily and widespread clinical use so far is the imaging of ocular diseases in ophthalmology. Fluorescein and indocyanine green (ICG) are established as fluorescent agents to enhance fluorescence angiography.[10] Currently under clinical evaluation is the fluorescence-guided identification of tumor margins during surgery as a tool to improve the accuracy and safety of tumor resection. For this purpose, ICG,[11] fluorescein-based conjugates,[12,13] and 5-aminolevulinic acid (ALA)[14] have been studied.

Imaging of larger tissue volume requires light within the NIR spectral range (700–900 nm) because the absorption coefficient of tissue is relatively small, resulting in penetration depths of up to a few centimeters (Figure 17.1).[15] Thus, the identification of inhomogeneities exhibiting a difference in absorption or fluorescence compared to the bulk tissue is possible. However, due to scattering, photons do not follow straight paths when propagating through tissue limiting the spatial resolution of images obtained (diffuse imaging). Nevertheless, tissue absorption is mainly determined by oxyhemoglobin, deoxyhemoglobin, and water. These exhibit a well-defined minimum in absorption in the NIR spectral region (see Figure 17.1) and provide information that can be used to quantitatively calculate important physiological parameters such as blood concentration (total hemoglobin) and oxygenation (ratio oxy-/deoxyhemoglobin). These absorption data, together with tissue-dependent scattering properties, can be fitted by mathematical models to reconstruct the most probable photon propagation through tissue and generate a spatial map of tissue optical properties for a given illumination and detection geometry. This method has been applied primarily to detect breast tumors and image brain function (see Chapter 57). It is beyond the scope of this review to elucidate the underlying basics in more detail. The interested reader is referred to Refs. [4] and [16].

The exogenously introduced optical contrast agents principally provide the opportunity of engendering disease-specific signals within the tissue. This enables the display of physiological and molecular conditions that are characteristic of certain disease states and progression. The detection of fluorescent

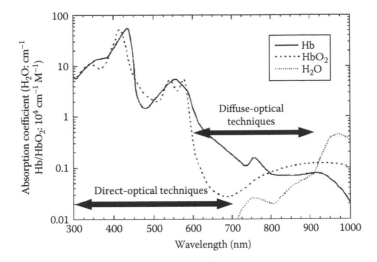

FIGURE 17.1 Contributions of oxyhemoglobin, deoxyhemoglobin, and water to tissue absorption depending on wavelength. Diffuse optical imaging techniques require NIR light between 700 and 900 nm to achieve maximal penetration into tissue, while for direct optical imaging, VIS light of 300–700 nm is usually used.

contrast agents is comparable to nuclear imaging methods; in both modalities, the photon sources (fluorescent dye or radionuclide) are distributed within the tissue. Generally, the choice of the spectral range of absorption and fluorescence for the dye dictates whether it is detectable on tissue surfaces (UV–VIS dyes) or from deeper tissue areas (NIR dyes). A prerequisite for sensitive detection of an absorbing dye is a high extinction coefficient at the desired excitation wavelength. If fluorescence photons are acquired, the compounds should exhibit large Stokes' shifts (spectral distance between absorption and fluorescence maximum) and high fluorescence quantum yields in physiological media. If fluorescence is recorded within the UV/VIS spectral region, both autofluorescence and the administered contrast agent will contribute to the observed signal, while in the NIR spectral region tissue, autofluorescence is negligible due to the absence of endogenous NIR fluorophores. In the latter case, the distribution of the contrast agent is nearly exclusively revealed by the detected signal. The disadvantage of fluorophores over radionuclides is that excitation light must be brought to the dye, so that this factor additionally contributes to penetration-depth limitations. An important advantage, however, is related to the fact that fluorescence can be excited continuously and is not limited to inherent properties as is radioactive decay. A set of photophysical properties is accessible, ranking from changes in fluorescence quantum yield and fluorescence lifetime to alterations of spectral signatures. Moreover, for many dyes, these parameters are influenced by local physiological or molecular conditions, such as pH, ions, or oxygen, and have therefore been used for their monitoring and quantification. Most importantly, light applied to fluorescence imaging is nonionizing radiation, rendering it harmless and nontoxic.

17.2.2 Imaging Techniques and Medical Applications

Optical imaging techniques and applications may be divided into two groups. On the one hand, imaging of superficial objects acquires directly reflected or scattered photons, while for diffuse imaging techniques, photons are recorded after they have passed through relatively thick tissue and optical properties of the tissue are spatially reconstructed using mathematical models. Different technical solutions and instrumental geometries are required for each of these applications. Illumination with light of a desired wavelength and detection with suitable devices (e.g., charge-coupled device [CCD] cameras) generally yields images from superficial structures in reflection geometry. This process is similar to conventional photography. A primary field of application is the examination of tissue surfaces via optical fibers embedded in endoscopes or laparoscopes, as well as of ocular diseases through ophthalmoscopes and direct assessment of skin diseases or during surgical procedures. Several types of superficial diseases in hollow organs have been monitored using fluorescence-guided endoscopy. Tetrapyrrole-based agents have been examined for this purpose,[17] but the most promising results have been obtained with ALA. Clinical studies included the diagnosis of urinary bladder cancer,[18] bronchial cancer,[19] and gastrointestinal diseases.[7] An essential task was to improve the detection of dysplastic changes and carcinoma in situ, which escape the endoscopist's eye and require more reliable techniques. Besides ALA, other fluorescent agents such as those presented in Chapter 57 were suggested for these purposes.[20–23]

The assessment of larger tissue volumes has been realized for different transillumination geometries in which tissue areas are illuminated with light, and the transmitted, scattered light is detected at defined positions (e.g., 180° projection geometry or 0°C reflection geometry).[24,25] Diffuse optical tomography (DOT) is based on the detection of photons at multiple positions and the mathematical reconstruction of 3D optical images.[4,16] Recently, the modeling of fluorescence light propagation has advanced fluorescence-mediated tomography (FMT) to a powerful technique primarily used for preclinical animal imaging.[26]

The application of DOT or FMT for breast-cancer detection and characterization has stimulated a great deal of research in the past few years. Several approaches for the detection of breast cancer using NIR light and utilizing intrinsic tissue optical properties or ICG as exogenous contrast agent have been followed. The first attempts at detecting breast tissue by means of tissue transillumination were reported as far back as 1929,[27] and the development of more powerful light sources and monitoring systems led to a revival of this technique in the 1980s and 1990s. Several clinical studies revealed the limitations of this modality—low spatial resolution and an inability to differentiate between malignant and benign tissue—which limit its clinical usefulness, especially in relation to x-ray mammography.[15,25,28,29] DOT using continuous-wave (CW), modulated, or pulsed light demonstrated advances in the quantification and 2D or 3D display of tissue absorption, scattering, vascularization, and oxygenation. However, only limited patient data have been acquired and published so far with respect to oxygenation.[4,15,30] The application of contrast agents was proposed repeatedly and will be the subject of subsequent chapters. The first contrast-enhanced imaging in a clinical setting was reported by Ntziachristos et al., who have demonstrated uptake and localization of ICG in breast lesions using DOT.[31] A fluorescence technique recently introduced into the market as Rheumascan procedure images all joints of both hands after ICG administration allowing the assessment of inflammatory signs of rheumatic diseases.[32]

Table 17.1 summarizes the various experimental and clinically established modalities and clinical applications that employ exogenously administered optical contrast agents.

TABLE 17.1   Medical Applications of Optical Diagnostics

Modality/Discipline	Signal	Clinical Indication
Direct optical		
Ophthalmology	Reflected light, fluorescence	Ocular diseases
Intraoperative diagnostics		Detection of tumor boundaries
Dermatology		Skin tumors
Rheumatology		Arthritis/inflammation
Via endoscopic or catheter devices		
Endoscopy	Reflected light, fluorescence	Tumors of the GI tract, lung,
Cardiovascular imaging		bladder, cervix, oral cavity
		Atherosclerotic plaques
Diffuse imaging and mathematical reconstruction		
Optical mammography	Reflected or transmitted	Breast tumors
Brain imaging	light, fluorescence	Brain perfusion, stroke

17.3 Dyes as Contrast Agents for Optical Imaging

17.3.1 Cyanine Dyes

17.3.1.1 Nonspecific Cyanine Dyes of Classical Contrast Agent Format

The structural class of cyanine dyes comprises chromophoric structures of more or less arbitrary absorption and fluorescence from the VIS-to-NIR spectral range (approximately 450–900 nm). Cyanine dyes generally exhibit very high molar-extinction coefficients ($>150,000$ M^{-1} cm^{-1}) and good fluorescence quantum yields (up to 50%). They have been originally prepared for the photography industry and were adapted to many applications in analytical chemistry, bioanalytics, and biomedicine. (See Refs. [33] and [34] for a comprehensive discussion.)

The first attempts at designing such probes for in vivo applications date back to the 1950s, when ICG, a prominent representative of a NIR-absorbing cyanine dye, was first synthesized (see Figure 17.2 for its chemical structure). ICG was clinically applied as a drug for the assessment of hepatic function and cardiac output, for which this compound exhibits favorable pharmacokinetic properties.[35] It has recently been discovered as an imaging agent for ocular diseases to visualize vascular disorders of the retina and choroidea. In a few studies, ICG has been reported as a potential NIR contrast agent for the detection of tumors in animal research.[36,37] At the clinical level, the Rheumascan procedure[32] represents a novel routine application of ICG beyond ophthalmic use.

For tumor-imaging purposes, it has been recognized that the high plasma protein binding, intravasal distribution, and resulting rapid plasma clearance of ICG by the liver lead to a quickly disappearing signal loss, which limits its potential as contrast agent.[37] One possible way to improve tissue retention and differentiation is through the use of nonspecific contrast agents similar to contrast agents for magnetic resonance imaging and computed tomography (CT), which usually are highly hydrophilic structures and achieve contrast enhancement based on morphological and physiological properties of tumor tissue such as increased tumor vasculature, endothelial leakiness, and enlarged extracellular volume. Thus, attempts were made to design structurally related agents of improved properties. The literature describes the synthesis and comparative pharmacokinetic characterization of ICG-related indotricarbocyanine derivatives with carboxy, hydroxyalkyl, and monosaccharide residues.[37,38] Figure 17.2 shows the chemical structures of ICG and the derivative SIDAG,

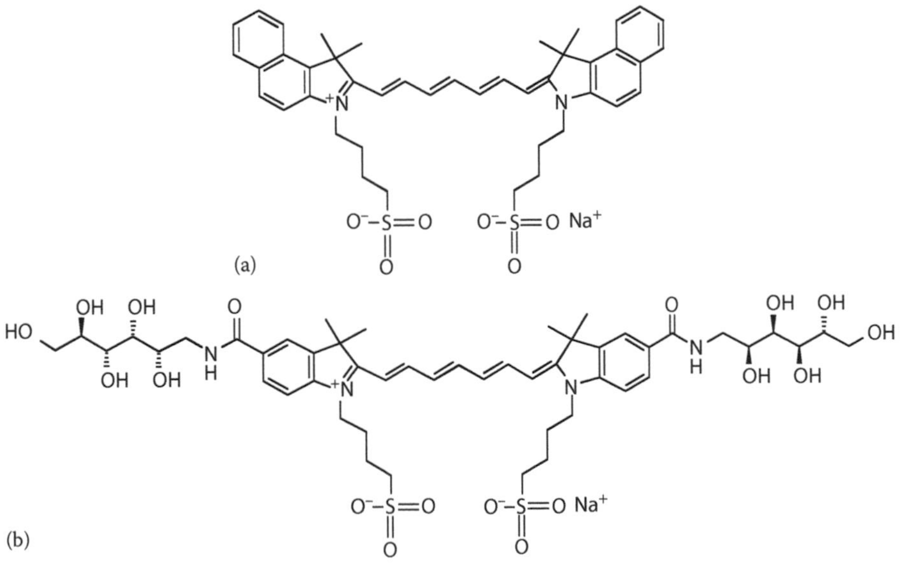

FIGURE 17.2 Chemical structure of (a) ICG and (b) the hydrophilic derivative SIDAG.

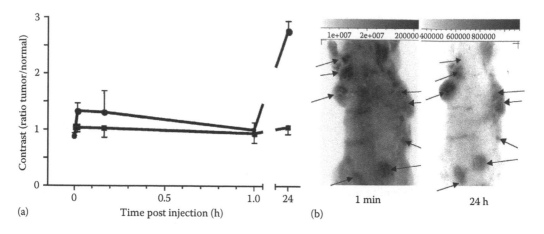

FIGURE 17.3 Fluorescence-imaging performance of hydrophilic dye SIDAG in rats with chemically induced multiple mammary carcinoma (dose 2 μmol kg⁻¹) in comparison to rats after injection of the same dose of ICG. (a) Time course of SIDAG (squares) and ICG (circles) and (b) corresponding fluorescence images with SIDAG obtained at 1 min and 24 h after injection. Arrows indicate multiple tumor sites. (From Licha, K. et al., *Acad. Radiol.*, 9(Suppl. 2), 320, 2002. With permission.)

a hydrophilic glucamide-derivatized indotricarbocyanine. Probes such as SIDAG (TryX750, mivenion) are characterized by a highly increased hydrophilicity and reduced plasma protein binding. It was assumed that after systemic administration, the compounds would be capable of leaving the intravascular space and extravasate into the extracellular compartment, with a certain degree of preference for tumor tissues.

In animal studies, SIDAG showed improved efficacy as an optical contrast agent versus ICG based on a higher tumor concentration and tumor-to-normal tissue contrast shortly after injection and, in addition, an unexpectedly elevated fluorescence contrast at 24 h after injection corresponding with tumor vascularity and VEGF expression levels (Figure 17.3).[37,39]

Although contrast enhancement with the agents covered in this chapter mainly relies on a more distinct perfusion of tumors compared to normal tissue (as is achieved with ICG) and does not employ target-specific moieties, the value of contrast-enhanced tumor imaging was successfully demonstrated. These agents represent the classical format of a contrast agent.

17.3.1.2 Targeted Cyanine-Dye Conjugates

One way to improve the selectivity of fluorescent dyes for diseased tissues is based on the high instrumental detection sensitivity for optical signals, which makes it possible to monitor signals derived from molecular states and events. Optical molecular imaging can therefore be considered the nonradiative counterpart to radionuclide-based methods, of course always limited by the penetration of light into tissue. Hence, novel probes consisting of an efficiently fluorescing dye label and a biological targeting unit (e.g., an antibody, peptide, oligonucleotide, or small molecule) have been reported in recent publications and will be discussed in the following.

A primary synthetic goal of chemical and physicochemical research in the area of fluorescent probes has been to provide novel fluorescent markers for application in immunoassays, screening assays, or genetic analysis. Many of the structures developed for these purposes contain reactive or activatable groups, for example, carboxylic acids,[40–42] *N*-hydroxysuccinimidyl esters,[43–45] isothiocyanates,[45] or maleimido[46] functionalities. Figure 17.4 depicts selection of chemical structures. Well established is the commercially available CyDye™ series (e.g., Cy3, Cy5, Cy5.5, Cy7; GE Healthcare Life Sciences) and Alexa Fluor Dye™ series (Invitrogen). Those and other cyanine-dye derivatives were employed for the preparation of various targeted cyanine-dye conjugates.

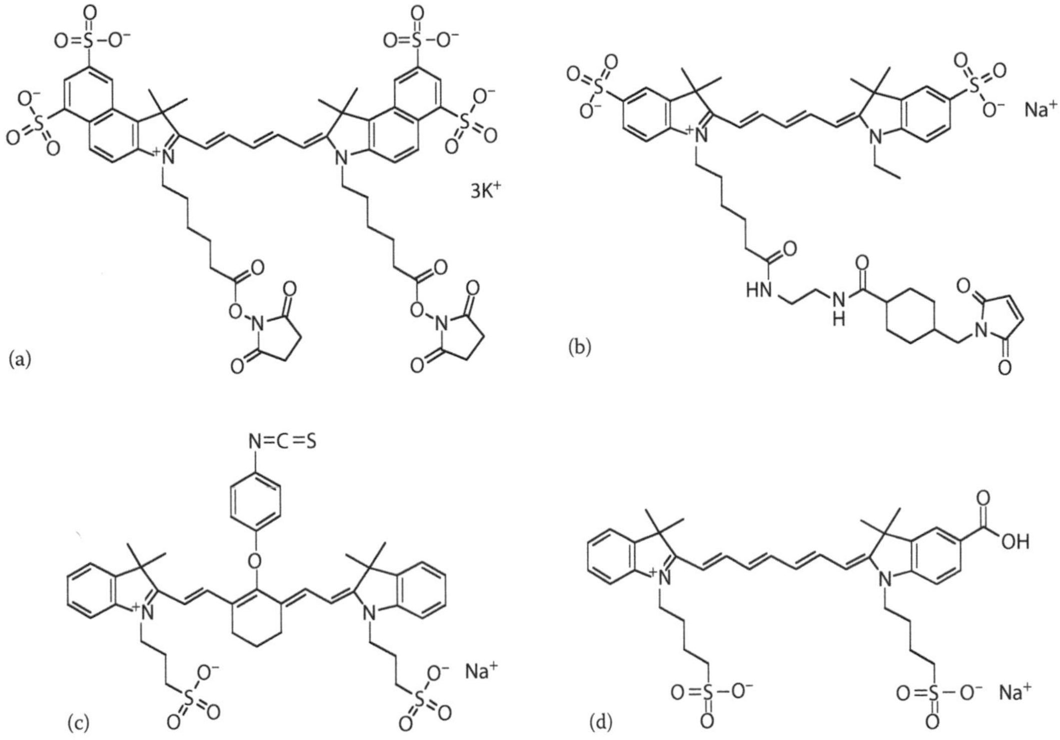

FIGURE 17.4 Carbocyanine-dye labels with different reactive groups for fluorescence labeling purposes. (a) Commercially available Cy5.5-bis-NHS ester, (b) maleimido derivative of Cy5,[46] (c) indotricarbocyanine with isothiocyanate functionality,[45] and (d) derivative with carboxy group at exterior ring position.[40]

The strategy of coupling cyanine-dye labels to antibodies for in vivo diagnostic purposes was first reported by Folli et al. and Ballou et al.[47–50] The authors demonstrated target-specific uptake in experimental tumor models by planar-reflection imaging of superficial fluorescence patterns and fluorescence-microscopy methods.

Generally, imaging contrast is improved when the probe localizes in the target area with high binding affinity and maintains its concentration level, while the circulating fraction is cleared from the blood to the greatest possible extent. Therefore, from the standpoint of background signal, contrast improves when going from full-size antibodies to targeting vehicles of reduced molecular weight such as antibody fragments and peptides.

The use of engineered antibody fragments of reduced molecular weight but still high binding affinities as target-specific carrier molecules for fluorescent dyes[51–54] was reported recently. Neri et al. described antibody single-chain fragments that were identified by phage-display library technology[51] and that exhibited high binding affinities against an extracellular angiogenesis marker, the fibronectin isoform ED-B-FN. This matrix protein is present exclusively in neoplastic blood vessels during angiogenesis and is therefore a promising target for specific fluorescent ligands. After labeling with fluorophores, the antibody fragments were successfully applied to the in vivo fluorescence imaging of tumor angiogenesis and inflammation in animal models.[51–53]

Synthetic peptides as vehicles for diagnostic molecules offer a promising strategy for the further reduction of molecular weight. Many tumors are known to overexpress receptors for specific peptide ligands, for example, somatostatin (SST), bombesin, or RGD motifs.[56] Pharmacologically optimized derivatives of these natural ligands in radiolabeled form are already clinically established as radiolabeled probes for the receptor scintigraphy of tumors, for example, OctreoScan®, an ^{111}In-DTPA-conjugated SST analog.[57]

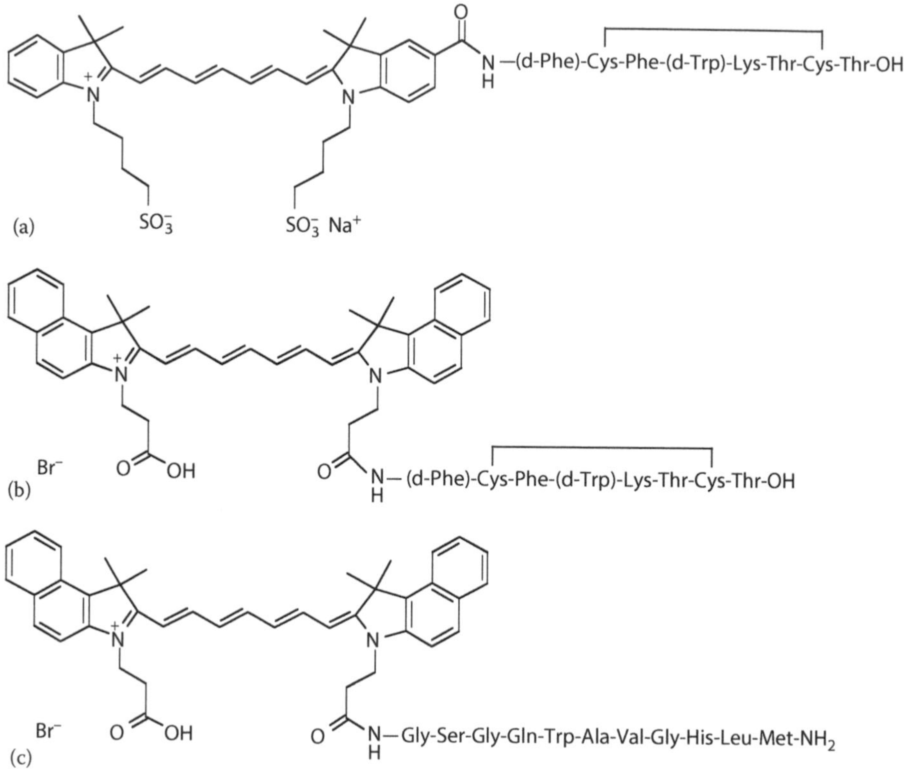

FIGURE 17.5 Chemical structures of cyanine dye–peptide conjugates designed for receptor-targeted fluorescence imaging of tumors. (a, b) Cyanine dye conjugated to the SST-receptor-binding peptide octreotate[20,21,40,41,58] and (c) a bombesin-receptor-avid peptide sequence.[41,58]

The literature reports many examples in which these principles were successfully adapted to optical receptor imaging by replacing the radiolabel by fluorescent cyanine dyes.[20,21,40,41,58] The synthesis and in vivo characterization of different conjugates between cyanine dyes and peptides, such as the SST analog octreotate,[20,21,40] the bombesin-derived analog bombesate,[41] structurally optimized VIP,[58] and most recently, fluorescent RGD-conjugates,[59,60] were the subject of studies by various investigators, often synthetically combined with radiolabels permitting multimodality imaging.[60] Figure 17.5 illustrates the structures of selected cyanine dye–peptide conjugates. The authors demonstrated that the conjugates were accessible by standard solid-phase peptide synthesis, which led to structurally well-defined and pure products. The fluorescence properties of the conjugated dyes were less affected only by the peptide, typically leading to fluorescence quantum yields for NIR cyanine conjugates in their unbound range of 10%.[40]

Fluorescence-imaging experiments using SST-receptor-positive animal-tumor models revealed receptor-mediated uptake resulting in elevated fluorescence at the tumor site compared to surrounding normal tissue areas. Unlike in vivo imaging features of nonspecific extracellular dyes or fluorescent single-chain-fragment conjugates, the highest contrast was achieved already within 1–2 h after intravenous injection and lasted up to 24 h[21,58] (Figure 17.6). This behavior is based on receptor-mediated accumulation of the peptide conjugates into tumor cells, while nonbound molecules undergo rapid body clearance via the renal pathway.[21]

As an interesting alternative to vehicles of biological origin, the targeting of the low-density lipoprotein (LDL) receptor was demonstrated with *small-molecule* conjugates between different indotricarbocyanines and cholesteryl laurate, which can bind to LDLs and thus mediate internalization of the entire conjugate into LDL-receptor-expressing tumor cells.[61]

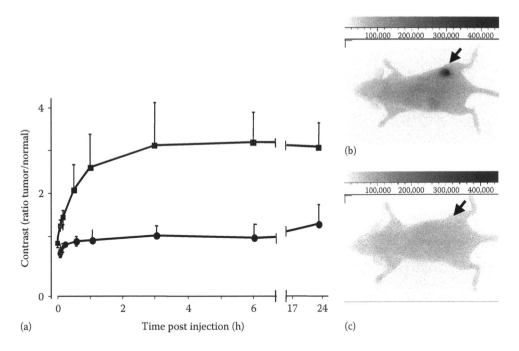

FIGURE 17.6 Fluorescence-imaging performance of indotricarbocyanine octreotate (see Figure 17.5, compound a) in mice bearing an SSTR2-receptor-expressing pancreatic tumor (dose, 0.02 μmol kg^{-1}). (a) Time course of indotricarbocyanine octreotate (squares) versus a control conjugate where two cysteines in the octreotate sequence are replaced by methionine (circles). Corresponding fluorescence images obtained at 6 h p.i. for the indotricarbocyanine octreotate (b) and the control conjugate (c). (From Licha, K. et al., *Acad. Radiol.*, 9(Suppl. 2), 320, 2002. With permission.)

17.3.1.3 Activatable Cyanine-Dye Conjugates (*Smart Probes*)

A number of groups have demonstrated enzyme-activatable conjugate complexes where the quenched signal (e.g., fluorescence) is turned on by tumor-associated lysosomal protease activity in vivo.[22,62–67] These probes carry Cy5.5 molecules (excitation 670 nm, emission 700 nm) that are bound to long circulating graft copolymers consisting of poly-L-lysine and poly(ethylene)glycol. An intratumoral NIR fluorescence signal is generated when tumor-associated proteases cleave the macromolecule, thereby liberating Cy5.5 fragments and affording previously quenched fluorescence.

This approach utilizes the unique opportunity of modulating optical signals through intramolecular fluorescence-quenching effects. Thus, the agents can report information on protein function and molecular conversion (enzymatic activity) and, furthermore, circumvent the necessity of target-specific accumulation, as the circulating fraction, which needs to be cleared for targeted agents, avoids fluorescence detection.

The contrast agent is composed of a poly(ethylene glycol)–poly-(L-lysine) graft polymer, which is labeled with Cy5.5 (see Figure 17.4) at the ε-amino groups of free lysines.[22,62,63] This drug exhibits increased fluorescence signals in the presence of the proteolytic enzymes cathepsin B and H by cleavage of the polylysine backbone and leads to a substantial signal increase both in cell culture and animal tumors.[22,63,64]

To broaden the specificity of the probe as a substrate for any other desired enzyme, a cleavable peptide sequence was incorporated in the polymeric backbone. As illustrated in Figure 17.6, the graft polymer was modified in such a way that a desired peptide sequence could be linked to the graft polymer. Fluorophores were conjugated at the N-terminal amino group of the peptide, again being subject to fluorescence quenching (Figure 17.7). These peptide spacers can be easily modified to act as a substrate for a variety of tumor- or angiogenesis-specific enzymes with proteolytic activity, for example, matrix metalloproteinases (MMPs),[65] the serine protease thrombin,[67] or the enzyme myeloperoxidase.[68]

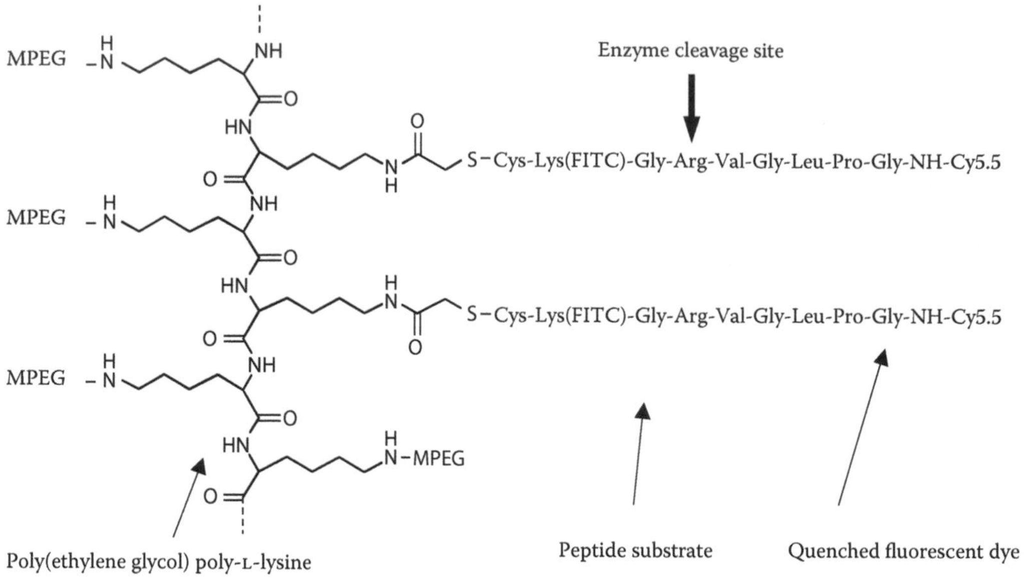

FIGURE 17.7 Structures of Cy5.5-labeled poly(ethylene glycol)–poly-(L-lysine) graft polymers with a peptide sequence cleavable by the proteolytic enzyme MMP-2. Arrows indicate the enzyme cleavage sites.[65]

The particular promise of applying probes for the visualization of protein function relates to the contrast-enhanced imaging of diseased tissues and is expected to have significant impact as a tool for the monitoring of drug response and efficacy at the molecular level. Figure 17.8 shows the results of an experiment in which a probe activatable by MMP-2 was used to monitor treatment response in a mouse model.[65]

17.3.2 Tetrapyrrole-Based Dyes

17.3.2.1 Synthetic Porphyrins, Chlorins, and Related Structures

The main area of application of tetrapyrrole-based compounds, such as porphyrins, chlorins, benzo-chlorins, phthalocyanines, and expanded porphyrins, has been for photodynamic therapy (PDT).[69–71] While PDT is treated elsewhere in this book (see Chapters 37 and 38), it is noteworthy, given the focus of this review, that many agents originally designed for PDT have shown to be applicable for diagnostic purposes, as these structures exhibit strong fluorescence in the VIS-to-NIR spectral region. The main rationale behind this type of application has been to use the fluorescence emission for real-time assessment of therapy progress and effectiveness during photodynamic treatment.

A variety of PDT compounds and their conjugates with macromolecular carriers have been studied for their diagnostic capabilities, for example, hematoporphyrin (HpD),[72] meso-tetra-*m*-hydroxyphenyl-chlorin (*m*-THPC),[73] benzoporphyrin derivatives (BPD),[74] phthalocyanines,[75] pheophorbides,[76] derivatives of chlorin e6,[77] and, finally, the expanded porphyrin derivative lutetium texaphyrin (Lu-Tex).[78] Figure 17.9 illustrates selected chemical structures.

Attempts to enhance the selectivity of photosensitizers by conjugating them to target-specific vehicles have been described.[79] Like the approaches based on cyanine dyes, conjugates with antibodies,[80,81] antibody fragments,[82] peptides,[83] and the serum proteins albumin and transferrin,[84,85] as well as estradiol[86] and cholesteryl laurate,[87] have been prepared and studied in animals.

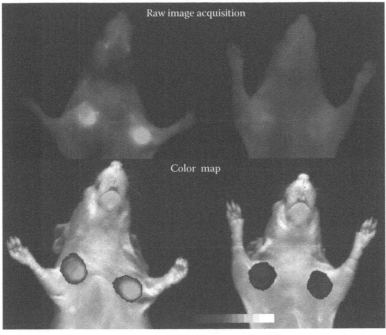

FIGURE 17.8 (**See color insert.**) Fluorescence imaging of HT1080 tumor-bearing mice 48 h after injection of Cy5.5-labeled poly(ethylene glycol)–poly-(L-lysine) graft polymer (see Figure 17.6) into animals treated with the MMP-inhibitor prinomastat versus untreated animals. The top row shows raw fluorescence images, and the bottom row shows color-coded intensity profiles superimposed onto white-light images. The results show a significantly less fluorescence signal in treated animals relative to the untreated group. (From Bremer, C. et al., *Nat. Med.*, 7(6), 743, 2001. With permission.)

17.3.2.2 5-Aminolevulinic Acid and Protoporphyrin IX

The underlying mechanism of fluorescence enhancement using ALA differs fundamentally from other approaches where exogenous agents are applied. ALA is not fluorescent per se but represents an essential precursor in the heme biosynthetic pathway. In the last step of the biosynthesis of heme, iron is incorporated into protoporphyrin IX (PpIX). After administration of ALA, the intracellular synthesis of heme is stimulated with a certain degree of selectivity for tumor cells and engenders a temporarily increased intracellular PpIX concentration in tumor cells relative to healthy tissue. This phenomenon is based on the finding that the incorporation of iron is catalyzed by the enzyme ferrochelatase, which was suggested to have lower activity in many tumors, thereby leading to a *bottleneck effect* at the level of PpIX formation. Figure 17.10 illustrates the essential steps. Upon irradiation with light, a cell-destructive, photodynamic effect is induced that is of much higher selectivity for malignant cells and accompanied with fewer side effects than is usually achieved with exogenously applied photosensitizers. The principles of PDT using ALA were reviewed, for example, by Peng et al.[88]

While heme is not fluorescent, PpIX exhibits a fluorescence emission spectrum typical of porphyrins of this structural class (Figure 17.10). Thus, PpIX fluorescence has been used to detect and visualize tumors and other tissue abnormalities in a large variety of clinical applications[7,18,19,89,90] (see also Chapter 2).

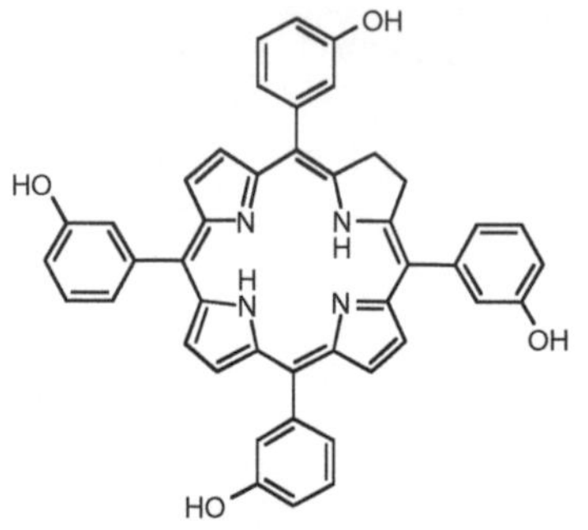

meso-tetra-*m*-
hydroxyphenylchlorin

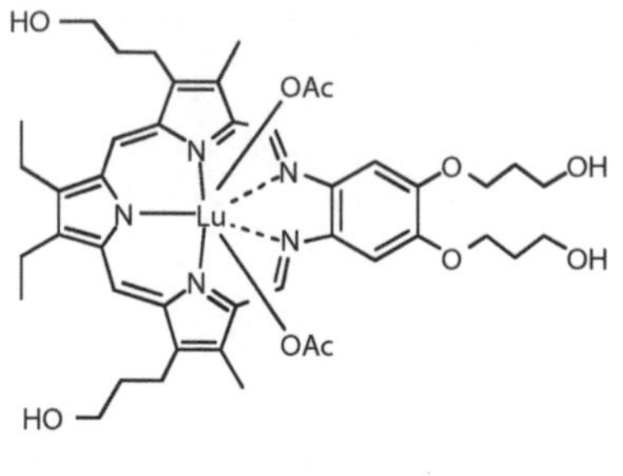

lutetium texaphyrin

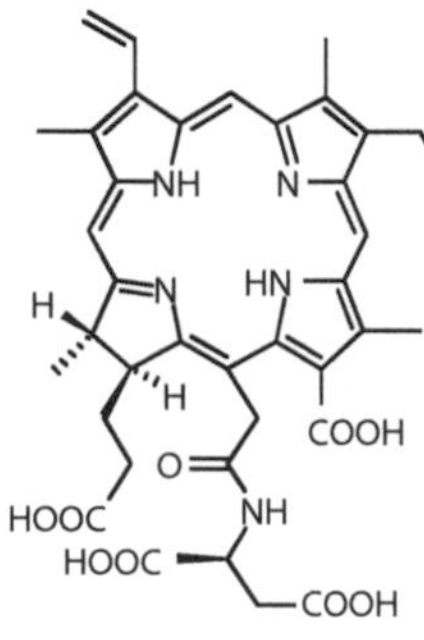

mono-1-aspartylchlorin e6

FIGURE 17.9 Chemical structures of photosensitizers applied as diagnostic agents for the fluorescence detection of tumors.

Successful attempts to improve ALA delivery to cells were followed by synthesizing ester derivatives of ALA, particularly alkyl esters of different alkyl chain length,[18,91,92] which are converted into free ALA through ester cleavage by esterases. Depending on the alkyl chain length used, the highest benefit was obtained with the *n*-hexyl ester, which recently obtained clinical approval (HexVix® from GE Healthcare/PhotoCure) for the photodiagnosis of bladder cancer.[91,93,94]

17.3.3 Other Dyes and Reporter Systems

17.3.3.1 Fluorescent Lanthanide Chelates

Many of the lanthanides, for example, La^{3+}, Eu^{3+}, Tb^{3+}, Nd^{3+}, and Yb^{3+}, form stable organometallic complexes with unique optical properties.[95] The emission bands are sharp and their position does not vary significantly with changing the chelating ligand or environment such as the temperature, pressure, or pH.[96] When an aromatic structure is located close to the complexing moiety, the metal ion exhibits a characteristic emission spectrum upon light absorption through the aromatic system and intramolecular energy transfer of the absorbed energy to the metal ion. This *sensitized luminescence emission* is of long lifetime—up to milliseconds—and typically consists of several emission bands throughout the VIS up to NIR spectral range depending on the metal used. In particular, Tb^{3+} and Eu^{3+} complexes have been synthesized in large structural diversity,[97,98] applied in biotechnology, and used for the development of screening assays based on time-resolved detection techniques.[34,99,100] Terbium yields a bright green fluorescence (major peak at 550 nm), and europium fluoresces in red color (major peak at 600 nm).

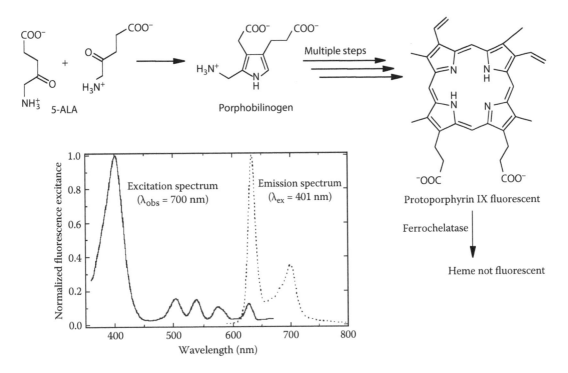

FIGURE 17.10 Synthetic pathway of heme biosynthesis leading to intracellular formation of fluorescent PpIX. Fluorescence excitation spectrum and fluorescence emission spectrum of PpIX in methanol. (Courtesy of B. Ebert, Physikalisch-Technische Bundesanstalt, Berlin, Germany.)

Another focus for the chemical design of fluorescent rare-earth complexes was driven by the fact that the optical properties of these compounds are sensitive to environmental conditions, such as pH, pO_2, glucose, halide ions, and alkali metals, permitting their use as chemical and biological sensors.[101,102] Ratiometric analysis is used to detect changes in pH, pM, or concentration of anionic metabolites including citrate, lactate, bicarbonate, and urate.[101] For example, ratiometric oxygen sensing has been performed in vitro, which simultaneously stimulated regioselective cell killing using terbium complexes that are responsive and reactive and based on axazanthone sensitizer with one naphthyl group.[103] Since the emission bands of lanthanide cations are narrower than either organic fluorophore or quantum dots, detection of several different ions is possible during the same experiment.[1] Bioconjugatable chelating derivatives of similar structure, consisting of a 1,4,7,10-tetraazacyclododecane system bearing a light-harvesting quinoline structure, two phosphonic acids, and a single carboxylic acid moiety for fluorescence labeling of target-specific biomolecules, have been described.[104-106]

The application of such compounds for the in vivo fluorescence detection of cancer has been proposed by several researchers.[23,107-109] More specifically, a cyclen-based macrocyclic complex with terbium, Tb-[*N*-2-pyridylmethyl)-*N'*,*Nδ*,*N*-*tris*(methylenephosphonic acid butyl ester)-1,4,7,10-tetraazacyclododecane] (Tb-PCTMB) (Figure 17.11), is extremely luminescent, water soluble at millimolar levels, thermodynamically stable, and nontoxic.[23,107-110] This compound has been detected at the picomolar level in rat intestinal-tissue endoscopies, facilitating an enhanced visual detection of chemically induced colon cancers in the rats.[23,108,110] More recently, imaging of metastatic nodules in an in vivo colorectal rat model using PAMAM dendrimer chelated lanthanides with sensitizing antenna on the periphery has been reported.[111]

Recently, it has been shown that the NIR region can be accessed with Ln chelates, expending biomedical applications even more.[96,112] Two-photon microscopy has been demonstrated with lanthanide luminescence and shown to be sensitized with efficiencies comparable to those of organic compounds.[113]

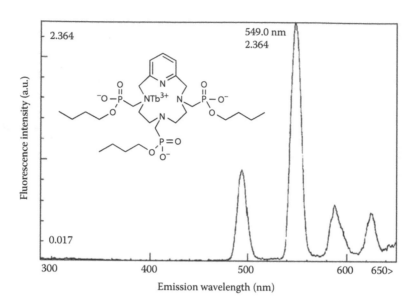

FIGURE 17.11　Chemical structure and fluorescence emission spectrum of Tb-PCTMB.

The proof-of-principle experiments have been recently reported, opening the door for new designs and applications using both of these techniques.

17.3.3.2 Fluorescein, Rhodamine, Oxazine, and Thiazine Dyes

Dyes based on the fluorescein chromophore are probably the most frequently applied fluorescent dyes in bioanalytics and biomedical diagnostics. A broad spectrum of reactive derivatives, such as 5-aminofluorescein or fluorescein isothiocyanate (FITC) and many further differently derivatized analogs,[114] are commercially available. Fluorescein is a drug routinely used for optical diagnosis in ophthalmology in medical imaging.[115]

A basic in vivo imaging approach has been the application of fluorescent serum albumin conjugates with 5-aminofluorescein for the intraoperative detection of tumor margins.[12] The synthesis of fluorescent cobalamin derivatives employing FITC and other fluorescein-related fluorophores (Oregon Green, naphthofluorescein) was reported, and the use of these conjugates for the intraoperative visualization of cobalamin receptors was suggested.[116] An oxazin-type dye, ethyl Nile blue A, was used as a marker for the fluorescence-guided identification of premalignant lesions in animal models.[117] Furthermore mentioned is the phenothiazine dye, toluidine blue, which is established as histological staining dye both in the laboratory and for clinical examinations.[118]

The rhodamine fluorophore has been a particular scaffold for the design and commercialization of reactive labels and bioconjugates that extend across many diverse applications. The Alexa Fluor® Dye series (Invitrogen/Molecular Probes) includes both cyanines and rhodamine derivatives of improved photostability and fluorescence quantum yields. Alexa Fluor 488, Alexa Fluor 546, and Alexa Fluor 555 are examples. Another group is based on a different chromophore type, the boron-dipyrromethene, known as BODIPY dyes (Invitrogen).[119] Rhodamine dyes have been used for in vitro imaging of the translocator protein (TSPO, formerly peripheral benzodiazepine receptor [PBR]) by conjugation to both small molecules[120] or dendrimeric structures.[121]

17.3.3.3 Fluorescent Nanoparticles, Quantum Dots, and Carbon Nanotubes

Fluorescent nanoparticles have emerged as powerful probes for cellular in vitro analysis and in vivo imaging. Many different types of probes have been developed, either nanosized materials based on inorganic cores or organic particles derived from polymerization or self-assembly of amphiphilic block copolymers.[122] The particular strength of such systems is the ability to combine different imaging modalities using multimodal probe entities,[123] for example, by tagging MRI iron oxide particles with additional fluorophores[124] or by attaching radioisotopes to particle surfaces. Additionally, these systems allow for the design of smart agents with either target through ligand–receptor interactions[121] or protease activation.[125]

Inorganic semiconductor nanoparticles (quantum dots) are meanwhile broadly established as powerful imaging tools with brilliant optical properties.[126] These semiconductor nanocrystal materials consist of atoms such as Cd, Se, Te, S, and Zn. The fluorescence emission range depends on the diameter of the particles and can be adjusted up to the NIR range. Surfaces employing stabilizing polymers as well as targeting molecules, such as antibodies and peptides, have expanded their applications for in vivo animal imaging. The reader is referred to Chapter 58 for detailed information.

Kopelman et al. have introduced solgel-based optical nanosensors called PEBBLES (solgel probes encapsulated by biologically localized embedding).[127] These polymer sensors can provide real-time measurements of subcellular molecular oxygen and intracellular pH and calcium. They easily incorporate multiple signaling dyes, such as oxygen-sensitive ruthenium complexes, and are on the order of 50–300 nm (radius) in size. A deeper insight into this type of probe is provided in this book (Chapter 59).

Carbon nanomaterials, such as nanotubes, nanorods, nanodiamonds, or graphene, have unique optical luminescence properties interesting for diagnostic imaging purposes.[128] Due to their improving synthetic access and recent progress in the synthesis of biocompatible and bioconjugated constructs, these materials have been increasingly studied in the context of biosensing and drug delivery.[128,129]

Dendrimers, branched and tunable macromolecules, provide another scaffold used as probes for both imaging and quantifying protease activity in vitro and in vivo. The number of functional terminal groups on a dendrimer has been used to functionalize two or more moieties to include targeting groups and fluorescent agents. In one example, the conjugation of a TSPO-targeted ligand along with a rhodamine dye increased the sensitivity and specificity for TSPO-targeted imaging.[121,125] This scaffold was also useful in the development of protease-activated switches. In these switches, a fluorescence resonance energy transfer (FRET) pair was coupled with a short peptide spacer to give a sensor and a reference. One dye is quenched, while the other is used as a reference. In the presence of a protease (i.e., MMP7), the peptide is digested, activating the quenched dye. These beacons have been used for in vivo imaging of cancer.[130,131]

17.3.3.4 Arsenic Fluorophores for Genetic Tetracysteine Tags

Recombinant proteins containing tetracysteine tags have been shown to be successfully labeled in living cells via biarsenical derivatives, such as fluorescein fluorophores (FlAsH) and other polyaromatic dyes (ReAsH).[132] These probes provide the smallest available protein tagging technique, based on the specificity for a six-amino-acid motif.

Tsien et al. established this multicolor in situ protein labeling approach allowing spectrally and temporally separated detection of protein turnover, signaling, and trafficking.[132,133] By using the two spectrally separated fluorescent labeling ligands FlAsH and ReAsH that label temporally separated pools of Cx43-TC permitted recording of junctional-plaque renewal over time, thus making the discrimination between older and younger protein molecules possible. This approach describes a method for studying the life cycle of proteins including assembly and internalization.[133]

17.3.3.5 Smart DNA Detection

Tan et al. have developed sensitive fluorescent DNA probes that can be used for real-time biomolecular recognition of target DNA sequences.[134] These probes, called *molecular beacons*, consist of a hairpin-shaped oligonucleotide that contains both a fluorophore and a quenching moiety. Molecular beacons act like on-off switches that are normally off (no fluorescence). The dye is activated upon hybridization of the stem to complementary DNA. As a result, the stem hybrid unwinds, thereby increasing the distance between the quencher and the fluorescent molecule and selectively generating a quantifiable signal. Two forms of energy transfer that exist in molecular beacons are direct energy transfer and FRET. While both forms of energy transfer are distance dependent, FRET has the added complication of requiring spectral overlap between the donor's (fluorescent dye) emission and the acceptor's (nonfluorescent quencher) absorption spectra. This effect has been further used for the detection of thrombin through specially designed molecular aptamer beacons, employing a fluorescein/dabcyl quenching system and a thrombin-binding oligonucleotide.[135] Chapter 57 provides a more detailed overview on this technology.

17.4 Conclusions

Remarkable optical techniques for the imaging and detection of diseases have emerged in the past few years. The current literature in this area outlines the broad applicability of light-based instrumental solutions for many clinical disciplines. The strength of optical imaging is that it allows for the combination of conventional display of tissues with the promise of highly sensitive detection of molecular signals. Thus, remarkable progress in the design of novel fluorescent probes has been made. In that respect, the chemistry of fluorescent dyes offers various opportunities to the acquisition of optical signals. Several parameters, such as the absorption coefficients, fluorescence quantum yields, fluorescence decay times, and fluorescence-quenching/recovery processes, are accessible for reporting physiological states, molecular conditions, and molecular function. Unlike radioactive decay, fluorescence is sensitive to its chemical environment, thereby broadening the applicability to the sensing of chemical analytes.

Progress in biotechnology, from which new biological targets and designated biological vehicles will arise, is credited with having a tremendous impact on the design of specific fluorescent probes. Optical techniques will likely be of increasing importance for sophisticated clinical diagnostic methods, for both laboratory purposes and clinical applications. Yet it remains to be identified which combination of fluorescent probe, biological principle, and instrumental solution will be able to solve the most urgent clinical questions and provide practical assets for the clinician.

Acknowledgment

Gretchen Cohenour is acknowledged for her assistance in preparing this chapter.

References

1. Licha, K., Olbrich, C., Optical imaging in drug discovery and diagnostic applications, *Adv. Drug Deliv. Rev.*, 57, 1087, 2005.
2. Weissleder, R., A clearer vision for *in vivo* imaging, *Nat. Biotechnol.*, 19, 316, 2001.
3. Tromberg, B.J., Shah, N., Lanning, R., Cerussi, A., Espinoza, J., Pham, T., Svaasand, L., Butler, J., Non-invasive *in vivo* characterization of breast tumors using photon migration spectroscopy, *Neoplasia*, 2, 26, 2001.
4. Leff, D.R., Warren, O.J., Enfield, L.C., Gibson, A., Athanasiou, T., Patten, D.K., Hebden, J., Yang, G.Z., Darzi, A., Diffuse optical imaging of the healthy and diseased breast: A systematic review, *Breast Cancer Res. Treat.*, 108, 9, 2008.
5. DaCosta, R.S., Wilson, B.C., Marcon, N.E., Fluorescence and spectral imaging, *Sci. World J.*, 7, 2046, 2007.
6. Andersson-Engels, S., Star, W.M., Wilson, B.C., *In-vivo* fluorescence imaging for tissue diagnostics, *Phys. Med. Biol.*, 42, 815, 1997.
7. Stepp, H., Sroka, R., Baumgartner, R., Fluorescence endoscopy of gastrointestinal diseases: Basic principles, techniques, and clinical experience, *Endoscopy*, 30, 379, 1998.
8. Moesta, K.T., Ebert, B., Handke, T., Nolte, D., Nowak, C., Haensch, W.E., Pandey, R.K., Dougherty, T.J., Rinneberg, H., Schlag, P.M., Protoporphyrin IX occurs naturally in colorectal cancers and their metastases, *Cancer Res.*, 61, 991, 2001.
9. Marcu, L., Fluorescence lifetime in cardiovascular diagnostics, *J. Biomed. Opt.*, 15, 011106, 2010.
10. Richards, G., Soubrane, G., Yanuzzi, L., Eds., *Fluorescein and ICG Angiography*, Thieme, Stuttgart, Germany, 1998.

11. Haglund, M.M., Berger, M.S., Hochman, D.W., Enhanced optical imaging of human gliomas and tumor margins, *Neurosurgery*, 38, 308, 1996.
12. Kremer, P. et al., Laser-induced fluorescence detection of malignant gliomas using fluorescein-labeled serum albumin: Experimental and preliminary clinical results, *Neurol. Res.*, 22, 481, 2000.
13. Kuriowa, T., Kajiamoto, Y., Ohta, T., Comparison between operative finding on malignant glioma by a fluorescein surgical microscopy and histological findings, *Neurol. Res.*, 21, 130, 1999.
14. Nabavi, A., Thurm, H., Zountsas, B., Pietsch, T., Lanfermann, H., Pichlmeier, U., Mehdorn, M., Five-aminolevulinic acid for fluorescence-guided resection of recurrent malignant gliomas: A phase II study, *Neurosurgery*, 65, 1070, 2009.
15. Grosenick, D., Wabnitz, H., Rinneberg, H., Moesta, K.T., Schlag, P.M., Development of a time domain optical mammograph and roofridge *in vivo* applications, *Appl. Opt.*, 38, 2927, 1999.
16. De Haller, E.B., Time-resolved transillumination and optical tomography, *J. Biomed. Opt.*, 1, 7, 1996.
17. Fisher, A.M.R., Murphree, A.L., Gomer, C.J., Clinical and preclinical photodynamic therapy, *Lasers Surg. Med.*, 17, 2, 1995.
18. Lange, N. et al., Photodetection of early human bladder cancer based on the fluorescence of 5-aminolevulinic acid hexyl ester induced protoporphyrin IX: A pilot study, *Br. J. Cancer*, 80, 185, 1997.
19. Baumgartner, R. et al., Inhalation of 5-aminolevulinic acid: A new technique for fluorescence detection of early stage lung cancer, *J. Photochem. Photobiol. B*, 36, 169, 1996.
20. Achilefu, S., Dorshow, R.B., Bugaj, J.E., Rajagopalan, R., Novel receptor-targeted fluorescent contrast agents for *in vivo* tumor imaging, *Invest. Radiol.*, 35, 479, 2000.
21. Becker, A. et al., Receptor targeted optical imaging of tumor with near infrared fluorescent ligands, *Nat. Biotechnol.*, 19, 327, 2001.
22. Weissleder, R. et al., In vivo imaging of enzyme activity with activatable near infrared in fluorescent probes, *Nat. Biotechnol.*, 17, 375, 1999.
23. Houlne, M.P., Hubbard, D.S., Kiefer, G.E., Bornhop, D.J., Imaging and quantitation of tissue selective lanthanide chelates using an endoscopic fluorometer, *J. Biomed. Opt.*, 3, 145, 1998.
24. Jarlman, O., Berg, R., Andersson-Engels, S., Svanberg, S., Pettersson, H., Laser transillumination of breast tissue phantoms using time-resolved techniques, *Eur. Radiol.*, 6, 387, 1996.
25. Moesta, K.T., Fantini, S., Jess, H., Totkas, S., Franceschini, M., Kaschke, M., Schlag, P., Contrast features of breast cancer in frequency-domain laser scanning mammography, *J. Biomed. Opt.*, 3, 129, 1998.
26. Graves, E.E., Weissleder, R., Ntziachristos, V., Fluorescence molecular imaging of small animal tumor models, *Curr. Mol. Med.*, 4, 30, 2004.
27. Cutler, M., Transillumination as an aid in the diagnosis of breast lesions, *Surg. Gynecol. Obstet.*, 48, 721, 1929.
28. Drexler, B., Davies, J.L., Schofield, G., Diaphanography in the diagnosis of breast-cancer, *Radiology*, 157, 41, 1985.
29. Franceschini, M.A. et al., Frequency-domain techniques enhance optical mammography: Initial clinical results, *Proc. Natl Acad. Sci. USA*, 94, 6468, 1997.
30. Nioka, S. et al., Optical imaging of human breast cancer, *Adv. Exp. Med. Biol.*, 361, 171, 1994.
31. Ntziachristos, V., Yodh, A.G., Schnall, M., Chance, B., Concurrent MRI and diffuse optical tomography of breast after indocyanine green enhancement, *Proc. Natl Acad. Sci. USA*, 97, 2767, 2000.
32. Bremer, C., Werner, S., Langer, H.E., Assessing activity of rheumatic arthritis with fluorescence optical imaging, *Eur. Musculoskeletal Rev.*, 4, 96, 2009.
33. Shealy, D.B. et al., Synthesis, chromatographic-separation, and characterization of near-infrared-labeled DNA oligomers for use in DNA-sequencing, *Anal. Chem.*, 67, 247, 1995.
34. Daehne, S., Resch-Genger, U., Wolfbeis, O.S., Eds., *Near-Infrared Dyes for High Technology Applications*, NATO ASI Series, Kluwer Academic Publishers, London, U.K., 1998.
35. Caesar, J., Shaldon, S., Chiandussi, L., Guevara, L., Sherlock, S., The use of indocyanine green in the measurement of hepatic blood flow and as a test of hepatic function, *Clin. Sci.*, 21, 43, 1961.

36. Gurfinkel, M., Thompson, A.B., Ralston, W., Troy, T.L., Moore, A.L., Moore, T.A., Gust, D. et al., Pharmacokinetics of ICG and HPPH-car for the detection of normal and tumor tissue using fluorescence, near-infrared reflectance imaging: A case study, *Photochem. Photobiol.*, 72, 94, 2000.

37. Licha, K. et al., Cyanine dyes as contrast agents in biomedical optical imaging, *Acad. Radiol.*, 9(Suppl. 2), 320, 2002.

38. Ye, Y., Bloch, S., Kao, J., Achilefu, S., Multivalent carbocyanine molecular probes: Synthesis and applications, *Bioconjug. Chem.*, 16, 51, 2005.

39. Wall, A., Persigehl, T., Hauff, P., Licha, K., Schirner, M., Müller, S., von Wallbrunn, A., Matuszewski, L., Heindel, W., Bremer, C., Differentiation of angiogenic burden in human cancer xenografts using a perfusion-type optical contrast agent (SIDAG), *Breast Cancer Res.*, 10, R23, 2008.

40. Licha, K., Becker, A., Hessenius, C., Bauer, M., Wisniewski, S., Henklein, P., Wiedenmann, B., Semmler, W., Synthesis, characterization, and biological properties of cyanine-labeled somatostatin analogues as receptor-targeted fluorescent probes, *Bioconjug. Chem.*, 12, 44, 2001.

41. Achilefu, S. et al., Synthesis, *in vitro* receptor binding, and *in vivo* evaluation of fluorescein and carbocyanine peptide-based optical contrast agents, *J. Med. Chem.*, 45, 2003, 2002.

42. Lin, Y., Weissleder, R., Tung, C.H., Novel near-infrared cyanine fluorochromes: Synthesis, properties, and bioconjugation, *Bioconjug. Chem.*, 13, 605, 2002.

43. Narayanan, N., Patonay, G., A new method for the synthesis of heptamethine cyanine dyes—Synthesis of new near-infrared fluorescent labels, *J. Org. Chem.*, 60, 2391, 1995.

44. Mujumdar, S.R. et al., Cyanine-labeling reagents: Sulfobenzindocyanine succinimidyl esters, *Bioconjug. Chem.*, 7, 356, 1996.

45. Flanagan, J.H. Jr., Khan, S., Menchen, S., Soper, S.A., Hammer, R.P., Functionalized tricarbocyanine dyes as near-infrared fluorescent probes for biomolecules, *Bioconjug. Chem.*, 8, 751, 1997.

46. Gruber, H.J., Hahn, C.D., Kada, G., Riener, C.K., Harms, G.S., Ahrer, W., Dax, T.G., Knaus, H.G., Anomalous fluorescence enhancement of Cy3 and Cy3.5 versus anomalous fluorescence loss of Cy5 and Cy7 upon covalent linking to IgG and noncovalent binding to avidin, *Bioconjug. Chem.*, 11, 696, 2000.

47. Folli, S., Westermann, P., Braichotte, D., Pelegrin, A., Wagnieres, G., Van den Berg, H., Mach, J.P., Antibody-indocyanine conjugates for immunophotodetection of human squamous-cell carcinoma in nude-mice, *Cancer Res.*, 54, 2643, 1994.

48. Fisher, G.W., Ballou, B., Deng, J.S., Hakala, T.R., Srivastava, M., Farkas, D.L., Three-dimensional imaging of nucleolin trafficking in normal cells, transfectants, and heterokaryons, *Biophys. J.*, 70, 343, 1996.

49. Ballou, B., Fisher, G.W., Hakala, T.R., Farkas, D.L., Tumor detection and visualization using cyanine fluorochrome-labeled antibodies, *Biotechnol. Prog.*, 13, 649, 1997.

50. Ballou, B., Fisher, G.W., Deng, J.S., Hakala, T.R., Srivastava, M., Farkas, D.L., Cyanine fluorochrome-labeled antibodies *in vivo*: Assessment of tumor imaging using Cy3, Cy5, Cy5.5, and Cy7, *Cancer Detect. Prev.*, 22, 251, 1998.

51. Neri, D., Carnemolla, B., Nissim, A., Balza, E., Leprini, A., Querze, G., Pini, A. et al., Targeting by affinity-matured recombinant antibody fragments of an angiogenesis associated fibronectin isoform, *Nat. Biotechnol.*, 15, 1271, 1997.

52. Birchler, M., Viti, F., Zardi, L., Spiess, B., Neri, D., Selective targeting and photocoagulation of ocular angiogenesis mediated by a phage-derived human antibody fragment, *Nat. Biotechnol.*, 17, 984, 1999.

53. Vollmer, S., Vater, A., Licha, K., Gemeinhardt, I., Gemeinhardt, O., Voigt, J., Ebert, B. et al., Fibronectin as a target for near-infrared fluorescence imaging of rheumatoid arthritis affected joints in vivo, *Mol. Imaging*, 8, 330, 2009.

54. Ramjiawan, B., Pradip, M., Aftanas, A., Kaplan, H., Fast, D., Mantsch, H.H., Jackson, M., Noninvasive localization of tumors by immunofluorescence imaging using a single chain Fv fragment of a human monoclonal antibody with broad cancer specificity, *Cancer*, 89, 1134, 2000.

55. Nilsson, F., Tarli, L., Viti, F., Neri, D., The use of phage display for the development of tumour targeting agents, *Adv. Drug Deliv. Rev.*, 43, 165, 2000.

56. Goldsmith, S.J., Receptor imaging: Competitive or complementary to antibody imaging? *Semin. Nucl. Med.*, 27, 85, 1997.

57. Krenning, E.P., Kwekkeboom, D.J., Bakker, W.H., Somatostatin receptor scintigraphy with [IN-111-DTPA-D-PHE(1)]- and [I-123-TYR(3)]-octreotide—The Rotterdam experience with more than 1000 patients, *Eur. J. Nucl. Med.*, 20, 716, 1993.

58. Bhargava, S. et al., A complete substitutional analysis of VIP for better tumor imaging properties, *J. Mol. Recogn.*, 15, 145, 2002.

59. Chen, X., Conti, P.S., Moats, R.A., In vivo near-infrared fluorescence imaging of integrin alphav-beta3 in brain tumor xenografts, *Cancer Res.*, 64, 8009, 2004.

60. Cai, W., Chen, K., Li, Z.B., Gambhir, S.S., Chen, X., Dual-function probe for PET and near-infrared fluorescence imaging of tumor vasculature, *J. Nucl. Med.*, 48, 1862, 2007.

61. Zhang, G. et al., Tricarbocyanine cholesteryl laurates labeled LDL: New near infrared fluorescent probes (NIRFs) for monitoring tumors and gene therapy of familial hypercholesterolemia, *Bioorg. Med. Chem. Lett.*, 12, 1485, 2002.

62. Bogdanov, A., Martin, C., Bogdanova, A.V., Brady, T.J., Weissleder, R., An adduct of *cis*-diamminedichloroplatinum (II) and poly(ethylene glycol)-poly(L-lysine)-succinate: Synthesis and cytotoxic properties, *Bioconjug. Chem.*, 7, 144, 1996.

63. Tung, C.H., Bredow, S., Mahmood, U., Weissleder, R., Potential cathepsin-D sensitive near infrared fluorescent probe for *in vivo* imaging, *Bioconjug. Chem.*, 10, 892, 1999.

64. Jaffer, F.A., Kim, D.E., Quinti, L., Tung, C.H., Aikawa, E., Pande, A.N., Kohler, R.H., Shi, G.P., Libby, P., Weissleder, R., Optical visualization of cathepsin K activity in atherosclerosis with a novel, protease-activatable fluorescence sensor, *Circulation*, 115, 2292, 2007.

65. Bremer, C., Tung, C.H., Weissleder, R., *In vivo* molecular target assessment of MMP-2 inhibition, *Nat. Med.*, 7, 743, 2001.

66. Bremer, C. et al., Imaging of differential protease expression in breast cancers for detection of aggressive tumor phenotypes, *Radiology*, 222, 814, 2002.

67. Tung, C.H. et al., A novel near-infrared fluorescence sensor for detection of thrombin activation in blood, *Chem. Biochem.*, 3, 2007, 2002.

68. Shepherd, J., Hilderbrand, S.A., Waterman, P., Heinecke, J.W., Weissleder, R., Libby, P., A fluorescent probe for the detection of myeloperoxidase activity in atherosclerosis-associated macrophages, *Chem. Biol.*, 14, 1221, 2007.

69. Wilson, B.C., Patterson, M.S., The physics, biophysics and technology of photodynamic therapy, *Phys. Med. Biol.*, 53, R61, 2008.

70. Ochsner, M., Photophysical and photobiological processes in the photodynamic therapy of tumours, *J. Photochem. Photobiol. B*, 39, 1, 1997.

71. Stefflova, K., Chen, J., Zheng, G., Killer beacons for combined cancer imaging and therapy, *Curr. Med. Chem.*, 14, 2110, 2007.

72. Dougherty, T.J. et al., Energetics and efficiency of photoinactivation of murine tumor cells containing hematoporphyrin, *Cancer Res.*, 30, 1368, 1972.

73. Alian, W., Andersson-Engels, S., Savanberg, K., Svanberg, S., Laser-induced fluorescence studies of meso-tetra(hydroxyphenyl)chlorin in malignant and normal tissues in rat, *Br. J. Cancer*, 70, 880, 1994.

74. Andersson-Engels, S., Ankerst, J., Johansson, J., Svanberg, K., Svanberg, S., Laser-induced fluorescence in malignant and normal tissue of rats injected with benzoporphyrin derivative, *Photochem. Photobiol.*, 57, 978, 1993.

75. Nesterova, I.V., Erdem, S.S., Pakhomov, S., Hammer, R.P., Soper, S.A., Phthalocyanine dimerization-based molecular beacons using near-IR fluorescence, *J. Am. Chem. Soc.*, 131, 2432, 2009.

76. Rapozzi, V., Zacchigna, M., Biffi, S., Garrovo, C., Cateni, F., Stebel, M., Zorzet, S., Bonora, G.M., Drioli, S., Xodo, L.E., Conjugated PDT drug: Photosensitizing activity and tissue distribution of PEGylated pheophorbide a, *Cancer Biol. Ther.*, 10, 471, 2010.

77. Chin, W.W., Heng, P.W., Thong, P.S., Bhuvaneswari, R., Hirt, W., Kuenzel, S., Soo, K.C., Olivo, M., Improved formulation of photosensitizer chlorin e6 polyvinylpyrrolidone for fluorescence diagnostic imaging and photodynamic therapy of human cancer, *Eur. J. Pharm. Biopharm.*, 69, 1083, 2008.

78. Blumenkranz, M.S., Woodburn, K.W., Qing, F., Verdooner, S., Kessel, D., Miller, R., Lutetium texaphyrin (Lu-Tex): A potential new agent for ocular fundus angiography and photodynamic therapy, *Am. J. Ophthalmol.*, 129, 353, 2000.

79. Sharman, W.M., van Lier, J.E., Allen, C.M., Targeted photodynamic therapy via receptor mediated delivery systems, *Adv. Drug Deliv. Rev.*, 56, 53, 2004.

80. Malatesti, N., Smith, K., Savoie, H., Greenman, J., Boyle, R.W., Synthesis and in vitro investigation of cationic 5,15-diphenyl porphyrin-monoclonal antibody conjugates as targeted photodynamic sensitisers, *Int. J. Oncol.*, 28, 1561, 2006.

81. Vrouenraets, M.B., Visser, G.W., Loup, C., Meunier, B., Stigter, M., Oppelaar, H., Stewart, F.A., Snow, G.B., van Dongen, G.A., Targeting of a hydrophilic photosensitizer by use of internalizing monoclonal antibodies: A new possibility for use in photodynamic therapy, *Int. J. Cancer*, 88, 108, 2000.

82. Bhatti, M., Yahioglu, G., Milgrom, L.R., Garcia-Maya, M., Chester, K.A., Deonarain, M.P., Targeted photodynamic therapy with multiply-loaded recombinant antibody fragments, *Int. J. Cancer*, 122, 1155, 2008.

83. Bisland, S.K., Singh, D., Gariepy, J., Peptide-based intracellular shuttle able to facilitate gene transfer in mammalian cells, *Bioconjug. Chem.*, 10, 982, 1999.

84. Hamblin, M.R., Newman, E.L., Photosensitizer targeting in photodynamic therapy. 1. Conjugates of hematoporphyrin with albumin and transferrin, *J. Photochem. Photobiol. B*, 26, 45, 1994.

85. Brasseur, N., Langlois, R., La Madeleine, C., Ouellet, R., van Lier, J.E., Receptor-mediated targeting of phthalocyanines to macrophages via covalent coupling to native or maleylated bovine serum albumin, *Photochem. Photobiol.*, 69, 345, 1999.

86. James, D.A., Swamy, N., Paz, N., Hanson, R.N., Ray, R., Synthesis and estrogen receptor binding affinity of a porphyrin-estradiol conjugate for targeted photodynamic therapy of cancer, *Bioorg. Med. Chem. Lett.*, 9, 2379, 1999.

87. Zheng, G. et al., Low-protein lipoprotein reconstituted by pyropheophorbide cholesteryl oleate as target-specific photosensitizer, *Bioconjug. Chem.*, 13, 392, 2002.

88. Peng, Q., Berg, K., Moan, J., Kongshaug, M., 5-Aminolevulinic acid-based photodynamic therapy: Principles and experimental research, *Photochem. Photobiol.*, 65, 235, 1997.

89. Andersson-Engels, S., Berg, R., Svanberg, S., Multi-colour fluorescence imaging in combination with photodynamic therapy of D-amino levulinic acid (ALA) sensitised skin malignancies, *Bioimaging*, 3, 134, 1995.

90. Beck, T.J., Kreth, F.W., Beyer, W., Mehrkens, J.H., Obermeier, A., Stepp, H., Stummer, W., Baumgartner, R., Interstitial photodynamic therapy of nonresectable malignant glioma recurrences using 5-aminolevulinic acid induced protoporphyrin IX, *Lasers Surg. Med.*, 39, 386, 2007.

91. Kloek, J., Beijersbergen van Henegouwen, G.M.J., Prodrugs of 5-aminolevulinic acid for photodynamic therapy, *Photochem. Photobiol.*, 64, 994, 1996.

92. Kloek, J., Akkermans, W., Beijersbergen van Henegouwen, G.M.J., Derivatives of 5-aminolevulinic acid for photodynamic therapy: Enzymatic conversion into protoporphyrin, *Photochem. Photobiol.*, 67, 150, 1998.

93. Gaullier, J.M., Berg, K., Peng, Q., Anholt, H., Selbo, P.K., Moan, J., Use of 5-aminolevulinic acid esters to improve photodynamic therapy on cells in culture, *Cancer Res.*, 57, 1481, 1997.

94. Witjes, J.A., Moonen, P.M., van der Heijden, A.G., Comparison of hexaminolevulinate based flexible and rigid fluorescence cystoscopy with rigid white light cystoscopy in bladder cancer: Results of a prospective phase II study, *Eur. Urol.*, 47, 319, 2005.

95. Lamture, J.D., Wensel, T.G., A novel reagent for labeling macromolecules with intensely luminescent lanthanide complexes, *Tetrahedron Lett.*, 34, 4141, 1993.

96. Uh, H., Petoud, S., Novel antennae for the sensitization of near infrared luminescent lanthanide cations, *C. R. Chim.*, 13(6–7), 668–680, 2010.

97. Chen, J., Selvin, P.R., Thiol-reactive luminescent chelates of terbium and europiu, *Bioconjug. Chem.*, 10, 311, 1999.

98. Werts, M.H.V., Verhoeven, J.W., Hofstraat, J.W., Efficient visible light sensitisation of water-soluble near-infrared luminescent lanthanide complexes, *J. Chem. Soc. Perkin Trans.*, 2, 433, 2000.

99. Mathis, G., Probing molecular-interactions with homogeneous techniques based on rare-earth cryptates and fluorescence energy-transfer, *Clin. Chem.*, 41, 1391, 1995.

100. Pålsson, L.O., Pal, R., Murray, B.S., Parker, D., Beeby, A., Two-photon absorption and photoluminescence of europium based emissive probes for bioactive systems, *Dalton Trans.*, 48, 5726, 2007.

101. Montgomery, C.P., Murray, B.S., New, E.J., Pal, R., Parker, D., Cell-penetrating metal complex optical probes: Targeted and responsive systems based on lanthanide luminescence, *Acc. Chem. Res.*, 42, 925, 2009.

102. Parker, D., Senanayake, P.K., Gareth Williams, J.A., Luminescent sensors for pH, O_2, halide and hydroxide ions using phenanthridine as a photosensitiser in macrocyclic europium(III) and terbium(III) complexes, *J. Chem. Soc. Perkin Trans.*, 2(10), 2129, 1998.

103. Law, G.L. et al., Responsive and reactive terbium complexes with an azaxanthone sensitiser and one naphthyl group: Applications in ratiometric oxygen sensing in vitro and in regioselective cell killing, *Chem. Commun. (Camb.)*, 47, 7321–7323, 2009.

104. Griffin, J.M.M., Skwierawska, A.M., Manning, H.C., Marx, J.N., Bornhop, D.J., Simple, high yielding synthesis of trifunctional fluorescent lanthanide chelates, *Tetrahedron Lett.*, 42, 3823, 2001.

105. Manning, H.C. et al., Facile, efficient conjugation of a trifunctional lanthanide chelate to a peripheral benzodiazepine receptor ligand, *Org. Lett.*, 4, 1075, 2002.

106. Manning, H.C. et al., Targeted molecular imaging agents for cellular-scale bimodal imaging, *Bioconjug. Chem.*, 15(6), 1488–1495, 2004.

107. Houlne, M.P., Agent, T.S., Kiefer, G.E., McMillan, K., Bornhop, D.J., Spectroscopic characterization and tissue imaging using site-selective polyazacyclic terbium(III) chelates, *Appl. Spectrosc.*, 50, 1221, 1996.

108. Bornhop, D.J. et al., Fluorescent tissue site-selective lanthanide chelate, Tb-PCTMB for enhanced imaging of cancer, *Anal. Chem.*, 71, 2607, 1999.

109. Pandya, S., Yu, J., Parker, D., Engineering emissive europium and terbium complexes for molecular imaging and sensing, *Dalton Trans.*, 23, 2757, 2006.

110. Hubbard, D.S., Houlne, H.P., Kiefer, G.E., Janseen, H.F., Hacker, C., Bornhop, D.J., Diagnostic imaging using rare-earth chelates, *Lasers Med. Sci.*, 13, 14, 1998.

111. Alacal, M.A. et al., Luminescence targeting and imaging using a nanoscale generation 3 dendrimer in an in vivo colorectal metastatic rat model, *Nanomedicine*, 7, 249, 2011.

112. Montgomery, C.P. et al., Cell-penetrating metal complex optical probes: Targeted and responsive systems based on lanthanide luminescence, *Acc. Chem. Res.*, 42(7), 925–937, 2009.

113. Andraud, C., Maury, O., Lanthanide complexes for nonlinear optics: From fundamental aspects to applications, *Eur. J. Inorg. Chem.*, 29–30, 4357–4371, 2009.

114. Oefner, P.J. et al., High resolution liquid-chromatography of fluorescent dye-labeled nucleic acids, *Anal. Biochem.*, 39, 223, 1994.

115. Hogan, R.N., Zimmerman, C.F., Sodium fluorescein and other tissue dyes, in *Textbook of Ocular Pharmacology*, Zimmerman, T.J., Ed., Lippincott-Raven, Philadelphia, PA, 1997, p. 849.

116. Smeltzer, C.C., Cannon, M.J., Pinson, P.R., Munger, J.D. Jr., West, F.G., Grissom, B., Synthesis and characterization of fluorescent cobalamin (CobalaFluor) derivatives for imaging, *Org. Lett.*, 3, 799, 2001.

117. Van Staveren, H.J., Speelman, O.C., Witjes, M.J.H., Cincotta, L., Star, W.M., Fluorescence imaging and spectroscopy of ethyl Nile blue A in animal models of (pre)malignancies, *Photochem. Photobiol.*, 73, 32, 2001.

118. Takeo, Y. et al., Endoscopic mucosal resection for early esophageal cancer and esophageal dysplasia, *Hepatogastroenterology*, 48, 453, 2001.

119. Ulrich, G., Ziessel, R., Harriman, A., The chemistry of fluorescent BODIPY dyes: Versatility unsurpassed, *Angew. Chem. Int. Ed.*, 47, 1184, 2008.

120. Bai, M. et al., A novel conjugable translocator protein ligand labeled with a fluorescence dye for in vitro imaging, *Bioconjug. Chem.*, 18(4), 1118–1122, 2007.

121. Samuelson, L.E. et al., TSPO targeted dendrimer imaging agent: Synthesis, characterization, and cellular internalization, *Bioconj. Chem.*, 20(11), 2082–2089, 2009.

122. Kim, J.S., Rieter, W.J., Taylor, K.M., An, H., Lin, W., Lin, W., Self-assembled hybrid nanoparticles for cancer-specific multimodal imaging, *J. Am. Chem. Soc.*, 129, 8962, 2007.

123. Licha, K., Schirner, M., Henry, G., Emerging optical imaging technologies: Contrast agents, in *Translational Multimodality Optical Imaging*, Azar, F.S., Intes, X., Eds., Artech House, Boston, MA, 2008, pp. 327–337.

124. Pittet, M.J., Swirski, F.K., Reynolds, F., Josephson, L., Weissleder, R., Labeling of immune cells for in vivo imaging using magnetofluorescent nanoparticles, *Nat. Protoc.*, 1, 73, 2006.

125. McIntyre, J.O., Scherer, R.L., Matrisian, L.M., Near-infrared optical proteolytic beacons for in vivo imaging of matrix metalloproteinase activity, *Methods Mol. Biol.*, 622, 279–304, 2010.

126. Bentolila, L.A., Ebenstein, Y., Weiss, S., Quantum dots for in vivo small-animal imaging, *J. Nucl. Med.*, 50, 496, 2009.

127. Buck, S.M., Koo, Y.E., Park, E., Xu, H., Philbert, M.A., Brasuel, M.A., Kopelman, R., Optochemical nanosensor PEBBLEs: Photonic explorers for bioanalysis with biologically localized embedding, *Curr. Opin. Chem. Biol.*, 8, 540, 2004.

128. Liang, F., Cheng, B., A review on biomedical applications of single-walled carbon nanotubes, *Curr. Med. Chem.*, 17, 10, 2010.

129. Lei, J., Ju, H., Nanotubes in biosensing, *Wiley Interdiscip. Rev. Nanomed. Nanobiotechnol.*, 2, 496, 2010.

130. McIntyre, J.O. et al., Development of a novel fluorogenic proteolytic beacon for in vivo detection and imaging of tumour-associated matrix metalloproteinase-7 activity, *Biochem. J.*, 377(Pt 3), 617–628, 2004.

131. McIntyre, J.O., Matrisian, L.M., Molecular imaging of proteolytic activity in cancer, *J. Cell Biochem.*, 90(6), 1087–1097, 2003.

132. Adams, S.R., Campbell, R.E., Gross, L.A., Martin, B.R., Walkup, G.K., Yao, Y., Llopis, J., Tsien, R.Y., New biarsenical ligands and tetracysteine motifs for protein labeling in vitro and in vivo: Synthesis and biological applications, *J. Am. Chem. Soc.*, 124, 6063, 2002.

133. Tsien, R.Y. et al., Multicolor and electron microscopic imaging of connexin trafficking, *Science*, 296, 503, 2002.

134. Fang, X., Li, J.J., Perlette, J., Tan, W., Molecular beacons: Novel DNA probes for biomolecular recognition, *Anal. Chem.*, 72, 747A, 2000.

135. Li, J., Cao, Z.C., Tang, Z., Wang, K., Tan, W., Molecular beacons for protein-DNA interaction studies, *Methods Mol. Biol.*, 429, 209, 2008.

18

Optical Trapping Techniques in Bioanalysis

Kenji Yasuda
Tokyo Medical and Dental University

18.1 Introduction

Knowledge of life has expanded dramatically during the twentieth century and has produced the modern disciplines of genomics and proteomics. However, there remains the great challenge of discovering the integration and regulation of these living components in time and space within the cell. As we move into the postgenomic period, the complementarity between genomics and proteomics will become apparent, and the connections between them will be exploited, although, neither genomics, proteomics, nor, for that matter, their simple combination will provide the data necessary to interconnect molecular events in living cells in time and space. The cells in a group are different entities. Each of them respond to the perturbations differently (Spudich and Koshland 1976). The differences between cells arise even among those grown in homogeneous conditions and considered to have identical genetic information. Why and how do these differences arise? They might be caused by several factors like unequal distributions of biomolecules in cells, mutations, interactions between cells, and fluctuations of environmental elements. A system that can observe interactions between specific cells continuously under fully controlled

457

circumstances is required to understand the rule underlying the occurrence of differences between cells and to study the possible cause mentioned earlier.

Conventional systems such as flow cytometry and direct measurement with microscopes have been used for tracking changes in cells. Flow cytometry enables us to obtain distributions of parameters like concentration, size, shape, and DNA content at the single-cell level in a group (Åkerlund et al. 1995, Zhao et al. 1999). The problem of this system is that it cannot track the specific cell's dynamics continuously because the sample drawn from the culture is discarded after the measurement. Also, it cannot keep the cells under isolated conditions or identify the particular cell even after cell division occurs. Thus, this system can give information on the average properties of different single cells, that is, how the group changes, but cannot give us information about how the single cell changes. On the other hand, with direct measurement of cells using a microscope on solid media like agarose gel plate, which is widely used (Donachie and Begg 1996, Elowitz et al. 2002, Gardner et al. 2000, Panda et al. 1999, Shapiro and Hsu 1989), we can identify specific cells and thus track them continuously. Though we can begin the cultivation of cells under an isolated condition by controlling the spread concentration at first, it is impossible to keep cells isolated even after cell division occurs or control the interactions between particular cells because the position of cells are fixed from the beginning of the cultivation. Thus, these conventional systems are not satisfactory for understanding the single-cell level interaction of particular cells.

As techniques were needed to clarify the relationship between cells that are genetically identical, we developed an on-chip microculture system based on a combination of recent advances in microfabrication techniques and conventional in vivo techniques (Inoue et al. 2001a,b). However, this system lacked a method for controlling the number of cells in the microchamber and for isolating particular cells from the cultured cells. For manipulating cells in the microchambers, we needed some noncontact forces such as optical tweezers, which have been used as a tool for handling cells, organelles, and biomolecules of a specimen under a microscope (Ashkin et al. 1986, 1987, Wright et al. 1990). By adding optical tweezers to the system, we improved the system, making it possible to control the number of cells in the microchambers (Wakamoto et al. 2001).

In this chapter, we describe our on-chip microculture system with optical tweezers. We explain the fabrication setup and examine several examples of practical single-cell cultivation, showing that this system possesses the ability to control the environment of the cells, especially the effect of isolation of cells, which has not been achieved with other systems.

18.2 On-Chip Microculture System

18.2.1 System Concept and Apparatus Design

As shown in Figure 18.1, the on-chip microculture system enables selective transfer of excess cells from an analysis chamber to a waste chamber (cultivation chamber) through a narrow channel (Figure 18.1a) and selective picking up of a particular cell from the cells in the cultivation chamber (Figure 18.1b).

Figure 18.2 shows the entire on-chip microculture system, which consists of four parts: a microchamber array plate, a cover chamber, a phase-contrast/fluorescent microscope, and optical tweezers. The cover chamber is a glass cube filled with a buffer medium; it is attached to an array plate to enable the medium in the microchambers to be exchanged through a semipermeable membrane. The volume of the cover chamber is 1 mL, and the maximum flow speed is 10 mL/min. The temperature of the buffer medium is controllable by using a Peltier temperature controller. The temperature of the stage is also controlled in the same way. The medium can be exchanged easily at any time during culturing by using stock solutions with different chemical and nutrient concentrations.

Phase-contrast/fluorescent microscopy (Olympus IX-70 inverted microscope with an oil-immersion objective lens, ×100, NA = 1.35) is used to study the growth and division of cells. Phase-contrast and

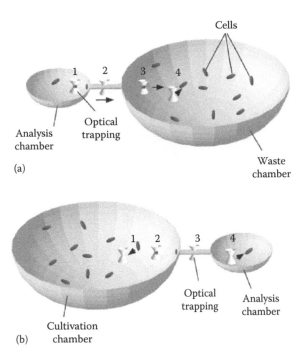

FIGURE 18.1 Schematic drawing of usage of on-chip culture system having optical tweezers: (a) to keep the number of cells in the analysis chamber constant, a cell in the chamber is trapped by the optical tweezers and transported into the waste chamber through the channel and (b) a particular cell cultivated in the chamber is transported into the analysis chamber.

fluorescent images are acquired simultaneously by using a charge-coupled device (CCD) camera (Olympus, CS230). The cell images are recorded onto a video tape and analyzed using a video capture system on a personal computer (Sony PCV-R73K). The spatial resolution of the images in this system is 0.4 μm when the ×100 objective lens is used.

A 1064 nm wavelength Nd:YAG laser (Spectra Physics, T20-8S) is used for the noncontact handling method as the optical tweezers add the ability to move the cells between microchambers to this system. The cells are trapped at the focal point of the laser beam on the microchamber array and transported by moving the focal point by slightly changing the angle of the 1064 nm dichroic mirror. The maximum laser power at the focal point after passing through the ×100 objective lens is 40 mW although we usually used less than 5 mW, which is the minimum laser power to hold a bacterium.

18.2.2 Microchamber Array Plate

18.2.2.1 Microchamber Array Glass Slide

As shown in Figure 18.3a, the microchamber array plate includes an $n \times n$ (n = 20–50) array of chambers, where each chamber is 20–70 μm in diameter and 5–30 μm deep. The biotin-coated microchamber array is covered with a streptavidin-coated cellulose semipermeable membrane (M.W. ~25,000) separating the chambers from the nutrient medium circulating through the cover chamber. The semipermeable membrane is fixed on the surface of the glass slide by using streptavidin–biotin attachment, which is strong enough to keep the cells within the microchambers while preventing contamination from the external environment. An example of the spatial arrangement of the microchamber array is shown in Figure 18.3b, where 20 μm (diameter) × 5 μm (depth) microchambers are etched at 100 μm intervals in a 0.17 mm thick glass slide.

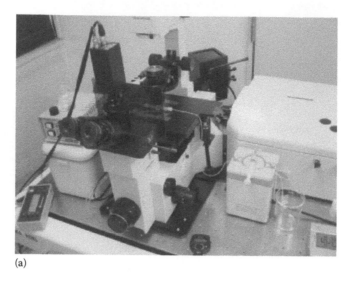

(a)

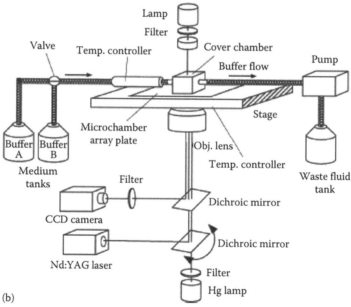

(b)

FIGURE 18.2 Photograph (a) and schematic drawing (b) of the on-chip system having four parts: a microchamber array plate, a cover chamber, a phase-contrast/fluorescent microscope, and an Nd:YAG laser.

The making process of the microchamber array plate was as follows: 0.17 mm thick glass slide was sonicated in 1 M NaOH aqueous solution and was washed with water to clean the surface. After cleaning, the dried glass slide was coated with 100 nm thick chromium by sputtering and next with posi g-line photoresist, OFPR-800 (Tokyo Ohka, Ltd., Kawasaki, Japan), by spinning. After baking at 85°C for 15 min, lithography of microchamber array patterns was carried out on a contact aligner with a broadband near-UV source (G, H, and I lines). The exposed glass slide was then developed and dried again. The exposed region of the glass slide was etched by MPM-E30 solution (Intec Inc., Tokyo, Japan) for chromium, and then by HF (4.7% (w/v)), NH_4F (36.2% (w/v)) solution (etching velocity = 60 nm/s at 25°C) for glass. When the shape of microchambers reached the desired size, the glass slide was washed by water to stop etching and dried again.

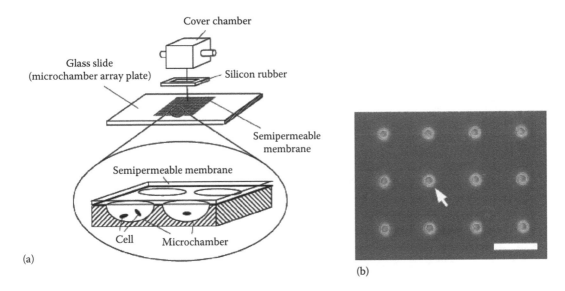

FIGURE 18.3 Design of microchamber array plate. (a) Schematic drawing of the microchamber array plate. An $n \times n$ ($n = 20$–50) array of microchambers is etched into a 0.17 mm thick glass slide. Each microchamber is covered with a semipermeable membrane separating the chamber from the nutrient medium circulating through the medium bath in the cover chamber. A single cell or group of cells in the microchamber can thus be isolated from others perfused with the same medium. (b) Optical micrograph of the microchamber array plate. The arrow indicates the position of one microchamber. The bar indicates a length of 100 μm.

18.2.2.2 Streptavidin Decoration on Cellulose Semipermeable Membrane

First, the cellulose semipermeable membrane (Spectora/Por M.W. 25,000, Spectrum Medical Industries Inc., Rancho Dominguez, CA) was washed with pure water and was cut into appropriate sizes adequate for sealing the glass slide. Next, the washed membrane was incubated in 0.2 M $NaIO_4$ for 6 h to generate aldehyde functions by oxidation of *cis*-vicinal hydroxyl groups of agarose. As aldehydic functions react at pH 4–6 with primary amines to form Schiff bases, the membrane was washed in 0.1 M phosphate buffer (pH 6.0) two times and 15 μg/mL streptavidin hydrazide was added (PIERCE, Rockford, IL). After a 2 h incubation, the membrane was washed with water and stocked at 4°C (see Figure 18.4).

18.2.2.3 Biotin Coat on the Surface of Microchamber Array Glass Slide

The dried microchamber array glass slide was next dipped in a solution containing 1% 3-(2-aminoethyl-aminopropyl) aq. at 25°C to decorate the surface of the glass slide with the amino group. The amino-coated glass slide was dried for 30 min at 120°C. Next, 1 mg/mL EZ-Link NHS-LC-Biotin (PIERCE) aq. was applied on the surface of the amino-glass slide and incubated for 60 min to decorate the surface of glass slide with biotin. The biotin-coated glass slide was washed with water, dried at R.T., and stored (see Figure 18.4).

18.2.2.4 Microchambers Made of Thick Photoresist, SU-8

As shown in Figure 18.5, microchambers and channels can also be formed on the glass slide lithographically, using a thick photoresist SU-8 (Microlithography Chemical Corp., Newton, MA), having an aspect ratio 5 μm in height and 5 μm in width, enough to cultivate and transport cells (Figure 18.5a). The process is as follows: (1) spin coat SU-8 on the surface of glass slide at 1400 rpm/s for 20 s, (2) soft bake at 125°C on hotplate for 30 min, (3) expose pattern using deep ultraviolet flood at 100 W for 20 s, (4) bake at 125°C on hotplate for 30 min, and (5) develop and rinse off imaging resist.

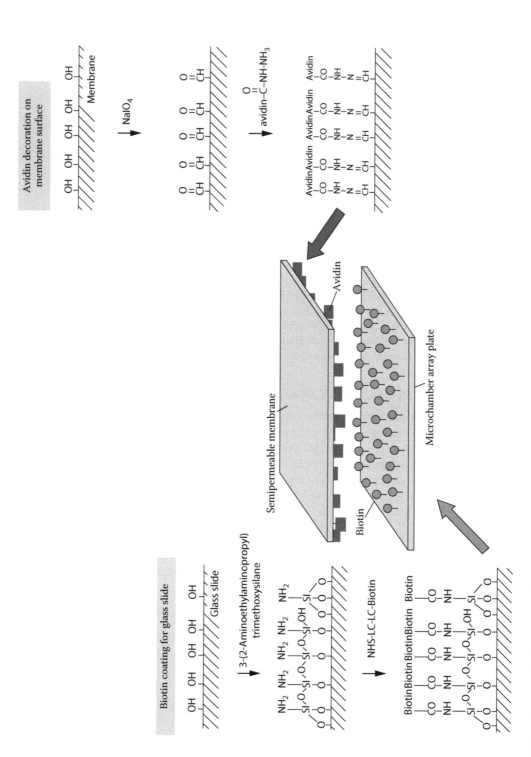

FIGURE 18.4 Streptavidin decoration and biotin coat procedure.

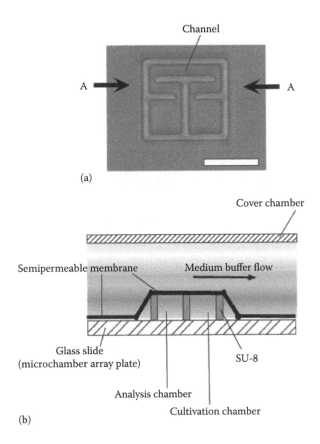

FIGURE 18.5 Microchamber array using thick photoresist, SU-8. On the microchamber array plate, there are several microchambers, including an analysis chamber and a cultivation chamber, connected by a channel. Buffer medium in the chambers is exchanged through a semipermeable membrane. Cells in the microchambers are transported by optical tweezers. (a) Optical micrograph of microchambers and channels constructed lithographically using thick photo-resist, SU-8. Bar, 50 μm. (b) Schematical drawing of A-A cross-sectional view of microchamber (a).

Figure 18.5b shows the cross-sectional view of the microchamber, the core of the microculture system. On a 0.2 mm thick glass slide, microchambers are fabricated for cell cultivation as described earlier. The surface of the glass slide is then decorated with biotin and covered with an avidin-decorated cellulose semipermeable membrane (M.W. ~25,000) to separate the microchambers from the nutrient medium circulating through the cover chamber. The membrane is also fixed to the surface of the glass slide using an avidin–biotin attachment, which is strong enough to keep the cells within the microchambers while preventing contamination from the external environment. Cells are transported between the micro-chambers by using optical tweezers. The buffer medium in the chambers is exchanged with that in the cover chamber through the membrane (Figure 18.5).

18.2.2.5 SU-8 for On-Chip Microculture System

In this experiment, we used 1064 nm laser for optical trapping. We have chosen SU-8 photoresist for three reasons. First, both the SU-8 and the semipermeable membrane had no absorbance at 1064 nm. Second, no-etching on the bottoms of the glass slide was required when we used SU-8 for microchambers, that is, the bottoms of the microchambers were optically flat. In fact, we could trap bacteria wherever in the microchamber array, and we observed no damage to the microchambers or membrane during our experiments. Finally, no nonspecific attachment of cells to the SU-8 surfaces was observed.

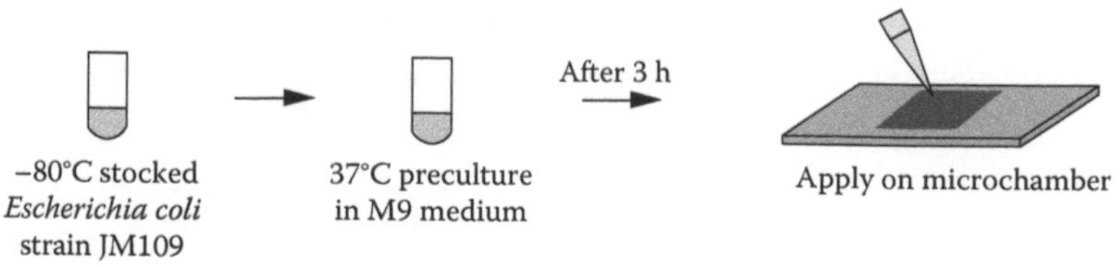
After 3 h
−80°C stocked
Escherichia coli
strain JM109
37°C preculture
in M9 medium
Apply on microchamber

FIGURE 18.6 Schematic diagram of sample preparation.

This was due to the constituents of SU-8: 20%–25% gamma butyrolactone, 1%–5% propylene carbonate, 70%–75% epoxy resin, and mixed triarylsulfonium/hexafluor oantimonate salt.

18.2.3 Bacterial Assays and Growth Conditions

We used a standard sample in our experiments, *Escherichia coli* strain JM109 (*endA*1, *recA*1, *gyrA*96, *thi*, *hsd*R17(r_k^-,m_k^+), *relA*1, *supE*44, λ^-, Δ(*lac-proAB*), [F′,*traD*36, *proAB*, *lacI*^{q}ZΔM15]; Toyobo, Co., Ltd., Tokyo, Japan) (Figure 18.6). Before the on-chip cultivation, the cells were grown overnight in an orbital shaker at a speed of 250 rpm and a temperature of 37°C with 100 mL of M9LB medium, which consisted of a minimal medium (M9: 4.5 g/L KH_2PO_4, 10.5 g/L K_2HPO_4, 50 mg/L $MgSO_4{\cdot}7H_2O$; pH 7.1) containing 1% (v/v) Luria-Bertani medium (LB) and 100 mg/L ampicillin. When the cultured cells reached the stationary phase (10^8–10^9 cells/mL), they were fractioned and placed in 500 μL sample tubes and maintained at −80°C as a glycerol stock solution (50% v/v). For each experiment, we reactivated the stock cells at 37°C in M9LB medium, using the orbital shaker at 250 rpm. After 20 h of cultivation, the cells returned to the stationary phase and were diluted to 10^5 cells/mL and cultured once more until the concentration reached 10^8 cells/mL. They were then used as samples for one-cell cultivation. In this cultivation, the number of cells was measured by using a counting glass slide (Erma Optical, Inc., Tokyo, Japan) and the IX-70 optical microscope.

18.2.4 Single-Cell Cultivation in On-Chip Microculture System

For single-cell cultivation, prepared *E. coli* cells were spread on the microchamber array plate and sealed in with the streptavidin-coated cellulose semipermeable membrane. Figure 18.7 shows the cell number in each microchamber after the lid was attached. As shown in the figure, about

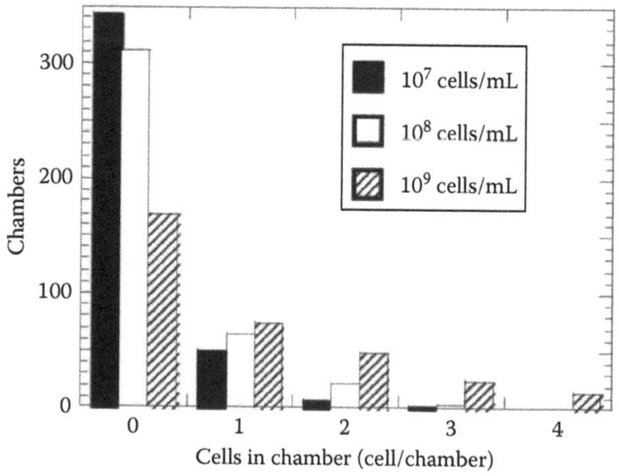
300
200
100
0
Chambers
10^7 cells/mL
10^8 cells/mL
10^9 cells/mL
0
1
2
3
4
Cells in chamber (cell/chamber)

FIGURE 18.7 The number of cells in each chamber of the microchamber array (400 microchambers) after cells were spread and sealed with the membrane. The concentrations of culture cells used for packing were 10^7, 10^8, and 10^9 cells/mL, respectively.

300 (3/4 of total) microchambers were vacant and 50 (1/8) chambers were packed with isolated single cells. Cells entrapped in non–well regions were sandwiched between the glass slide and membrane and were no longer able to move or grow. When the concentration of cells in dropping buffer was higher than 10^9 cells/mL, the batch culture of cells stopped growing as the cells entered the stationary phase. On the other hand, when the concentration was less than 10^7 cells/mL, cells were in lag phase or early log-phase, and this was before or just when the cell-number increase started. Thus, the concentration of *E. coli*, 10^7–10^9 cells/mL, was appropriate for packing one cell into each microchamber. Based on how the lid was attached, we usually used one of two ways: we incubated the membrane on the glass slide until the membrane attached to the glass slide or we used centrifugation (×1000 g) to remove excess water and attach the membrane to the glass slide. After the membrane was attached to the plate, no *E. coli* trapped in the microchambers could escape. The membrane-sealed microchamber array plate was then covered with the cover chamber and set on the stage of the microscope. The microchambers trapping single *E. coli* cells were chosen for observation. The time-course changes of the chosen isolated *E. coli* were continuously observed and recorded by the charge-coupled device (CCD)-camera and video cassette recorder (VCR) (max. 405 min recording).

After recording, the time-course growth of the single *E. coli* cells was analyzed using image analysis software (1/30 s and 0.2 μm resolution). We defined the length of an *E. coli* as the distance between its two ends (see Figure 18.8), and the growth of an *E. coli* as its elongation rate. The length measurement includes 10%–15% uncertainty because we could only measure the two-dimensional end-to-end length images of three-dimensionally tumbling *E. coli*.

In this system, we used 0.17 mm thick glass slide because the maximum working distance of the ×100, N.A. 1.35 obj. lens is less than 0.3 mm. It should be noted that, as shown in Figure 18.8, no *E. coli* attachment to the glass surface in the microchamber wells was observed during cultivation. Thus, we can measure the time course change of *E. coli* swimming freely in the microchamber.

18.3 Single-Cell Cultivation Using On-Chip Microculture System

18.3.1 Observation of Cell Growth Rate and Cell Division Time in Microchambers

We first examined the growth and division of *E. coli* under isolated conditions (Figure 18.8). In this experiment, we used a chamber having the size of the initial concentration for the isolated single cells was 1.0×10^{10} cells/mL. First, we observed that an isolated *E. coli* maintained its length (2.9 μm) for 80 min (Figure 18.8a). After 80 min of sleep (we call this *sleeping time*), the cell started growing at a rate of 0.04 μm/min, frequently bending around the septa. When the cell reached a length of 7.4 μm (160 min after inoculation), it divided into two daughter cells (Figure 18.8c and d), each with a length of 3.6 μm. Following the first division, the two daughter cells started growing and tumbling again like the first *mother* cell. The daughter cells divided again after they had grown for 60–70 min (Figure 18.8f). The period of cell division after cell growth began showing exponential behavior, with a rate of 86 min/cell division. Using this method, we examined the division time of daughter cells born from isolated mother cells in microchambers. Figure 18.9 shows the distribution of division time of newborn daughter cells ($n = 160$). As shown in the graph, the cell division of daughter cells started at least 31 min after they were born, and the cell division time was spread from 31 to 223 min. Mean cell division time, 86.6 min (S.D. 38.1 min), was longer than the peak value 50–70 min. Figure 18.5 explains the initial length dependency of the division time. The samples having an initial length from 3 to 4 μm (filled bars) and those from 4 to 5 μm (hatched bars) show the same statistical distributions of division time or a mean time of 72.9 min (S.D. 31.1 min) for 63 samples and 75.5 min (S.D. 28.2 min) for 16 samples whereas those from 2 to 3 μm (open bars) showed a different distribution, or mean division time of 105.1 min (S.D. 41.0 min) for 64 samples.

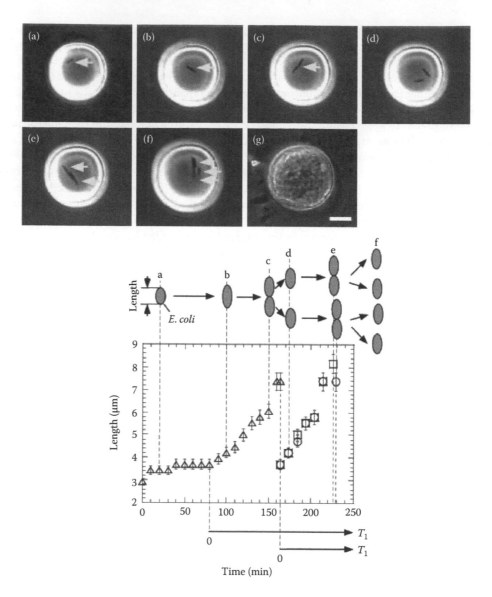

FIGURE 18.8 Time course of isolated single *E. coli* growth in a microchamber. The magnified micrographs at the top (a–g) show the time course of one of the microchambers in Figure 18.2 (see arrow in Figure 18.2b) at times of (a) 0 min, (b) 100 min, (c) 150 min, (d) 175 min, (e) 225 min, (f) 230 min, and (g) 15 h after the inoculation. The arrows in the micrographs show the positions of *E. coli* in the microchamber. The bar indicates a length of 10 μm. The graph at the bottom shows the time-course growth of the individual *E. coli*.

This experiment demonstrated several advantages of this on-chip microculture system. First, neither contamination from the cover chamber nor escape of *E. coli* from the small chamber was observed during the experiment. Second, the direct descendants of single cells could be cultured and compared in isolated microchambers. Third, the physical properties of the cells in each microchamber could be continuously observed and compared. Finally, the resolution of cell length, 0.2 μm, was an order of magnitude smaller than that of conventional methods, which determine the sizes of cells electronically such as using the Coulter Counter (Coulter Electronics, Inc., Hialeah, FL).

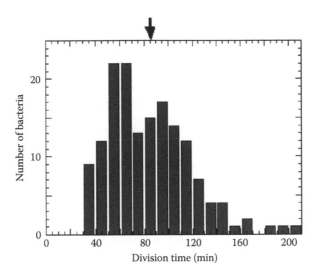

FIGURE 18.9 Distribution of division time of daughter cells. Arrow indicates the mean time of cell division, 86.6 min (S.D. = 38.2, n = 160).

18.3.2 Growth and Cell Cycles of Daughter Cells after Single-Cell Division

We can also compare cell growth and cell division times among isolated single cells and their daughter cells. Two daughter cells have the same DNA and cytoplasm as their mother cell, and in this experiment, they grew in the same environment, including physical contact with each other. The graph in Figure 18.10 presents four examples of such a comparison. An isolated mother cell

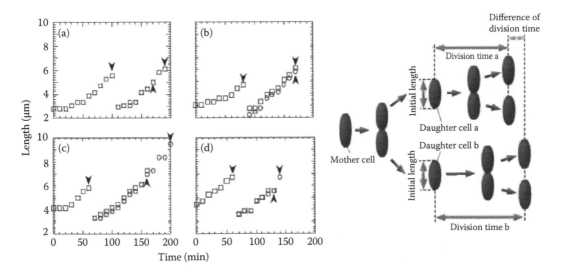

FIGURE 18.10 Time-course growth of the isolated individual *E. coli* and two daughters. The arrowheads indicate the cell division occurring in the cells. (a)–(d) Examples of four types of two daughter cells' behavior: (a) cell division time of one daughter cell was shorter, the other was the same as that of the mother cell; (b) cell division time of the two daughter cells were equal and the same as that of the mother cell; (c) cell division time of one daughter cell was prolonged much longer than mother cell and the other daughter cell; (d) cell division time of one daughter was shorter, and the the other was equal to that of the mother in the case of faster cell division time sample. All (two) daughter (sister) cells' growth ratio was synchronized independent of the mother cells' behavior, and most of their division times were not synchronized.

grew at 0.04 μm/min and divided 85 min after cell growth began, whereas its daughter cells grew at 0.06 μm/min and divided after 65–70 min. In this case, although the growth rate and cell division time for the mother cell and daughter cells were different, those for the two daughter cells were almost the same. But none of the results of eight other comparisons showed strong correlation between mother cells and daughter cells, or between daughter cells (data not shown). We also observed that mother cells sometimes divided into two daughter cells of obviously unequal lengths (data not shown).

18.3.3 Flexible Medium Buffer Exchange through Semipermeable Membrane

Controlling the medium condition is important for one cell-analysis. Semipermeable membrane, which was attached on the microchamber array plate, exchanges the medium in the microchambers within the time resolution of the VCR system, <1/30 s; this was checked by the real-time confocal optical microscopy system (CSU21, Yokogawa Electric Corp., Tokyo, Japan). The result indicated no delay of diffusion in 5 μm depth with 1/30 s resolution. That is, there was no difference in the increase of the fluorescent dye's concentrations between the top and the bottom of the microchamber after the introduction of the dye into the system. Thus, we concluded the diffusion time was in 1/30 s.

As shown in Figure 18.8, after overnight culturing, the number of *E. coli* had fully increased in the microchamber (Figure 18.8g). In this case, the concentration of *E. coli* corresponded to $>10^{12}$ cells/mL, which is more than a thousand times larger than the stationary-phase concentration of $>10^9$ cells/mL in batch cultivation in the same sample and medium.

We checked another use of the on-chip culture system for changing the medium solution in the microchambers through the semipermeable membrane. Single *E. coli* cells were cultured in small chambers with M9LB medium. To express green fluorescent protein (GFP) in *E. coli* (JM109), we used M9LB-IPTG medium buffer, which is M9LB containing 1 mM isopropyl-x-D(−)-thiogalactopyranoside (IPTG; Wako Chemical Co., Tokyo, Japan), and 0.1 mM lactose. As shown in Figure 18.11a and b, no GFP expression was observed before IPTG induction. Forty minutes after the medium exchange from M9LB to M9LB-IPTG, GFP expression was successfully observed. Figure 18.11c and d show the GFP fluorescence in the same *E. coli* 60 min after the medium exchange, which is almost the same as batch cultivation data under the same buffer condition at 37°C.

These results show good system performance for solution exchange, which is difficult for batch or plate cultivation methods.

18.3.4 Flexible Chamber Size Control of Single-Cell Cultivation

Using this on-chip culture system, we can easily control the environment for cultivation. For example, we can control the cultivation volume by using different sizes of microchambers. Figure 18.12 shows one example of volume control by using different microchambers. The size of the chamber shown in Figure 18.12b is 20 μm (diameter) × 5 μm (depth), which we call a *small chamber*, and that of the one in Figure 18.12c is 70 μm (diameter) × 30 μm (depth), a *large chamber*. The volumes of the chambers are 1.0×10^{-10} and 7.7×10^{-8} mL, respectively. If we culture single cells in these chambers, the estimated concentrations are 1.0×10^{10} cells/mL and 1.3×10^7 cells/mL, respectively.

Figure 18.13 shows the chamber-size dependence of single *E. coli* growth. The average growth speeds of five samples were 0.06 μm/min both in the small chamber and in the large chamber. The mean values of the cell division time were also almost same for both chambers, around 45 min. The results showed no significant differences in mean values of growth speed and cell division time, even when the mean free paths were 10 times different.

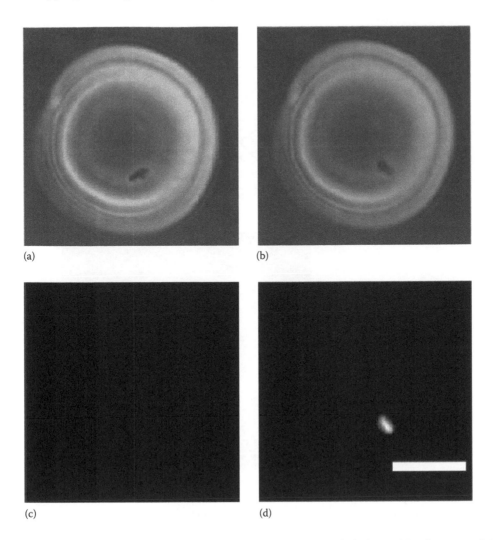

FIGURE 18.11 Green fluorescent protein expression in *E. coli*, shown (a, b) before IPTG induction, and (c, d) 60 min after IPTG induction was begun. Figures (a) and (c) are phase-contrast micrographs and (b) and (d) are fluorescent micrographs. The bar indicates a length of 10 μm.

18.3.5 Continuous Single-Cell Cultivation Exploiting Optical Tweezers

Using this system, we also examined whether the direct descendants of an isolated single cell can be observed under the same isolated condition. For observing the isolated single cells continuously even after cell division occurred, we developed the following method. As shown in Figure 18.14, we first isolated a single cell in the analysis chamber (first generation) and observed its growth (Figure 18.14a). Once cell division occurred, one of the two daughter cells was removed from the chamber quickly before they made physical contact (Figure 18.14c and d). The isolated second generation cell was observed continuously until it divided again. Then we repeated this process as many times as possible.

Figure 18.15 shows an example of continuous observation of the direct descendants of a mother cell under isolated conditions, as explained earlier. First, one mother cell (8.8 μm long) was enclosed in the analysis chamber and observed continuously (Figure 18.15a). Just after the cell divided, one of its daughter cells was trapped by optical tweezers and transported to a waste chamber through a narrow channel (cell trap path) (Figure 18.15b–e). The same procedure was done every time the cell in the

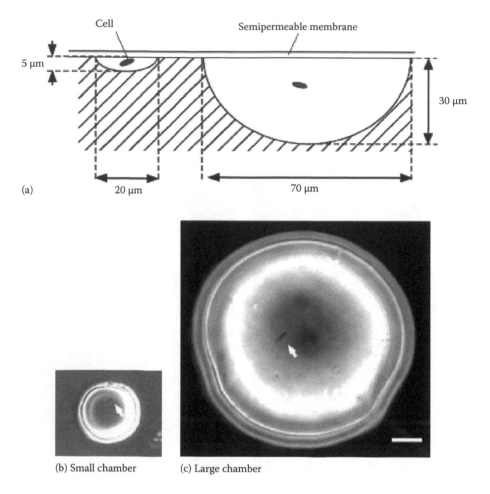

(a)

(b) Small chamber   (c) Large chamber

FIGURE 18.12 Schematic drawing of the two sizes of microchambers (a). The *small chamber* on the left has a 20 μm diameter and 5 μm depth, and the *large chamber* on the right has a 70 μm diameter and 30 μm depth. Optical micrographs of (b) a small chamber and (c) a large chamber. The arrows indicate the positions of *E. coli* in the chambers. The bar, 10 μm.

analysis chamber divided. In the second generation, the lengths of the two daughter cells were both 4.9 μm and almost equal. The cell in the third generation divided unequally; one was 5.4 μm long and the other was 3.6 μm (with the error of ±0.4 μm because of the limitation of resolution) (Figure 18.15f). The cell in the fourth generation also divided unequally again; one was 5.3 μm and the other was 3.8 μm. So, cell division in the third and fourth generations were quite different from the cell division in the second generation (Figure 18.5b). Earlier, it had been reported that *E. coli* cells divided equally (Marr et al. 1966), but in our system, unequal cell division in length was often observed.

Figure 18.16 shows the time course of cell growth of four generations of direct descendants in the sample shown in Figure 18.15. As shown in the result, cell cycles for cell division seem independent across generations. Neither could we also find strong correlations among generations in the elongation velocity or in cell division lengths.

Figure 18.17 shows another example of single-cell cultivation using a different chamber design. In the experiment shown in Figure 18.15, the cells in the waste chamber could easily invade the analysis chamber by swimming because the path (channel) was too simple. Thus, we made chambers with more complicated labyrinth-like channel pathways. This prevented the escape of the cell from the waste chamber. Using this

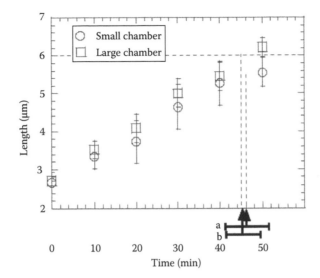

FIGURE 18.13 Chamber size dependence of *E. coli* cell growth. The circles indicate a small chamber, 20 μm in diameter and 10 μm deep (Figure 18.12b). The squares indicate a large chamber, 70 μm in diameter and 30 μm deep (Figure 18.12c). The plotted points are mean values of cell length at 10 min intervals, with error bars indicating the S.D. of five cells. The arrow and bar labeled "a" indicate the mean time and S.D. of the cell division time for the small chamber; those labeled "b" indicate these values for the large chamber.

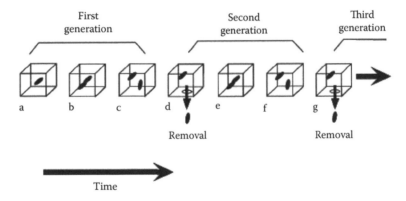

FIGURE 18.14 Concept and process used to continuously observe direct descendants under isolated conditions.

chamber array, we could analyze and compare four cells simultaneously under isolated conditions. The four graphs shown in Figure 18.18 indicate no significant correlation among those four cells during cultivation.

18.3.6 Differential Analysis of Single-Cell Cultivation in Microchamber Array System

The results shown in Figures 18.16 and 18.18 illustrate only half of the two daughter cells' properties. We thus improved the method and the microchamber array to enable us to measure all the descendant cells of an isolated mother cell (Figure 18.19).

Figure 18.19b shows the chamber array we used for complete descendant analysis. The cultivation was done as follows: first, a cell was transferred into chamber 1 (first generation) by using the optical tweezers. When the cell cultivated in chamber 1 divided into two daughter cells, they were transferred into the second generation's chambers, one by one. Next, the isolated daughter cells divided again, the

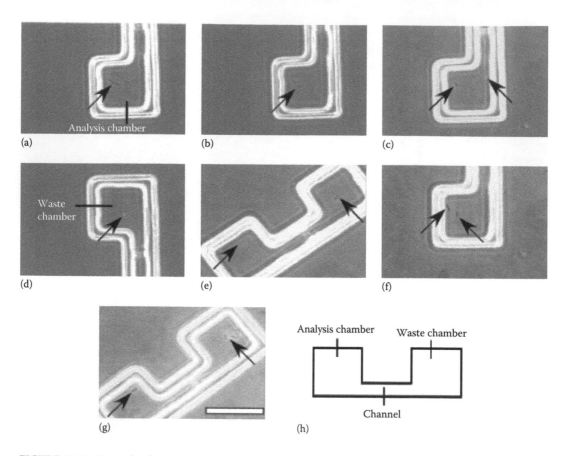

FIGURE 18.15 Example of continuous observation of direct descendants under isolated conditions in the microchamber array: (a) isolated single cell in the analysis chamber almost divides, (b) the cell has divided into two daughter cells, (c) one of the daughter cells is trapped by optical tweezers and is going to be removed from the analysis chamber to the waste chamber through the channel, (d) the trapped cell has been removed into the waste chamber, (e) the whole image of two daughter cells after the removal, (f) the cell in the third generation divided unequally in length, (g) the whole image of the analysis chamber and the waste chamber in the fourth generation cell, and (h) schematic drawing of the whole image of the chamber. Bar, 30 μm.

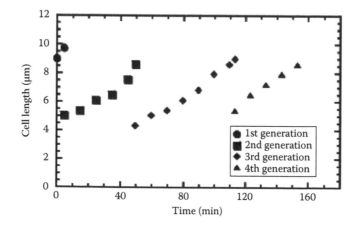

FIGURE 18.16 Time course of the cell growth of the isolated single cells shown in Figure 18.15.

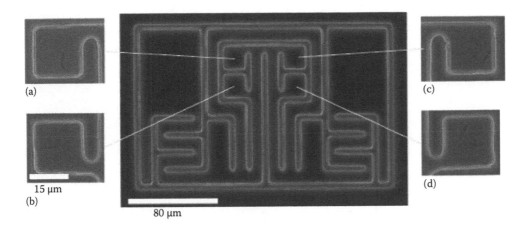

FIGURE 18.17   Another example of continuous observation of direct descendants under isolated conditions in four-room microchambers for simultaneous observation and comparison. In each of the four chambers (a–d) at the center of the microstructure, isolated single cells are cultivated and compared.

granddaughter cells were transferred into the third generation's chambers, respectively. The same process was repeated after the cells of the third generation divided. It should be noted that all the cells were irradiated during the cell trapping, unlike in the previous experiments described earlier.

As shown in Figure 18.8b, the results indicate the existence of fluctuations of cell length, division time, and division length among direct descendant cells, even between two sister cells. For example, see the lower graph in Figure 18.20, the length of one of the daughter cells (second generation) at its birth was 3.1 µm, and the cell grew as long as 11.9 µm. At this length, it produced two cells: one was 9.2 µm long, and the other was 2.8 µm long, and the ratio of the lengths of the newborn cells was almost 3:1. Although the longer cell seemed like one cell, it might be one entity of three cells. If this is true, avoiding several divisions would lead to a different pattern of cell reproduction and might affect the proliferation of cell groups.

18.3.7 Epigenetic Inheritance of Elongated Phenotypes between Generations Revealed by Individual-Cell-Based Direct Observation

As described earlier, to detect the existence of epigenetic inheritance, it is necessary to measure the phenotypic variations in a genetically identical clonal population and to determine whether a certain phenotype is transmitted between generations at the individual cell level in a stringently controlled environment. Despite difficulties in tracking individual cells over many generations and in controlling the immediate environments around them, attempts have been made to determine exclusively the phenotypic transmission using group-based measurement in which epigenetic inheritance at the individual cell level was *deduced* from the behavior of the average population (Cohn and Horibata 1959, Gardner et al. 2000, Ozbudak et al. 2004). A group-based study, however, cannot distinguish between the adaptations of individual cells even when adaptation has occurred within the group, that is, we cannot distinguish whether this group-based adaptation is caused by the adaptation of each cell or by the selection of a particular group from a set of phenotypically different groups.

When implementing individual-cell-based measurements of epigenetic inheritance instead of group-based measurement, we developed an *on-chip individual-cell cultivation system* (Inoue et al. 2001a, Wakamoto et al. 2001, 2003). This system enabled us to compare the phenotypes between generations and to examine the existence of phenotypic transmission at the individual cell level under stringently controlled conditions. We called this method a *differential individual-cell observation assay*. We measured the variations in quantitative traits at the individual cell level under uniform and isolated conditions and found that the interdivision time (duration of one generation), initial length (cell length at

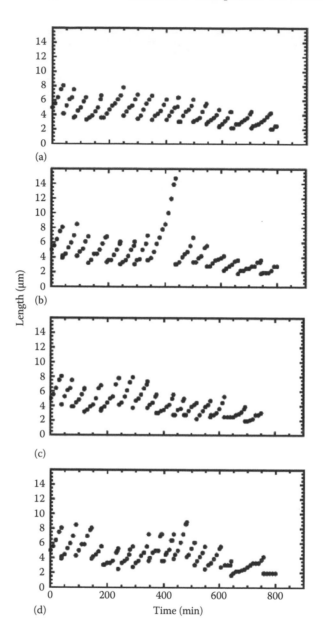

FIGURE 18.18   Time course of the four cousin cells' growth under isolated conditions. (a–d) Indicate the sample in each of the chambers shown in Figure 18.17, respectively.

the start of each generation), and final length (cell length at the end of each generation) varied at the individual cell level by as much as 33%, 26%, and 26%, respectively (Wakamoto et al. 2005).

In this section, as an example of the single-cell cultivation method, we report the phenotypic dynamics involved in changing normal isolated cells into elongated ones and the transmission of elongated phenotypes to descendants at the individual-cell level, hence enabling epigenetic inheritance to be directly observed in *E. coli* using the differential individual-cell observation assay (Wakamoto and Yasuda 2006).

Figure 18.21a plots example growth and division dynamics of individual *E. coli* cells under uniform and isolated conditions. A single cell was first loaded in the on-chip individual-cell cultivation system

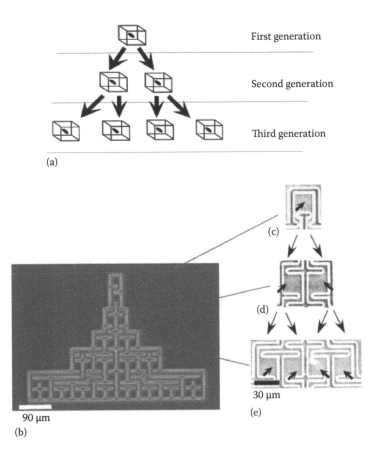

FIGURE 18.19 The process of observation of every descendant cell for generations derived from a single cell (a) and the micrograph of the microchambers for complete descendant analysis (b); (c) microchamber for first generation with isolated single mother cells; (d) two daughters isolated in each chambers; and (e) four granddaughters in each of four microchambers.

and observed continuously (Figure 18.21a, black dots and lines). The cell exhibited normal growth and division patterns in the first and second generations and divided almost equally; hence, we randomly discarded one of the two daughter cells (closest to the exit). Despite the normal growth and division characteristics in the first and second generations, the cell in the third generation was extraordinarily elongated. A striking feature of the cell was that it divided unequally (Figure 18.21a, left arrowhead (b)), thereby producing two daughter cells; one was elongated (Figure 18.21a, subsequent black dots and lines) and the other was normal (Figure 18.21b). Elongated daughter cells followed in the subsequent descendents, remaining elongated through repeated unequal cell divisions in most cell divisions. In other words, the cell transmitted its abnormally long cell length to its descendants. However, the normal daughter exhibited normal growth and division patterns and did not elongate in the following generations (Figure 18.21b). The normal daughter cell borne from the elongated cell had the same tendencies as those of typical normal cells, just as in the third generation; the shorter of the two daughter cells in the ninth generation (Figure 18.21a, right arrowhead (c)) also demonstrated normal growth and division patterns (Figure 18.21c). Therefore, it is conceivable that the shorter of the two daughters from the elongated cells possesses a normal phenotype.

The transmission of the elongated phenotype between generations can clearly be seen in the returning map of initial cell-length transitions (Figure 18.21d). The converging dots indicate stable states in the transition. The initial cell-length transition for elongated lineage (Figure 18.21d, black dots and lines)

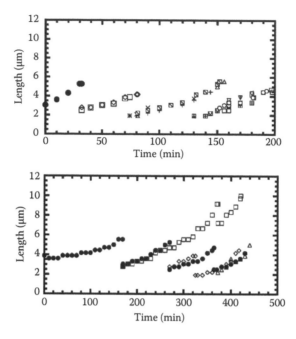

FIGURE 18.20   Two examples of growth of all the descendant cells from an isolated single cell.

jumped from the normal cell-length area to the elongated cell-length area at the beginning of the fourth generation and stayed there throughout the following generations, whereas the shorter daughters' transitions were within the normal cell-length area (Figure 18.21d, dots and lines). The clear distinction between the two transition areas, hence, indicates that the elongated cell transmitted a distinctive phenotype to one lineage of its descendants.

The presence of elongated cells was not restricted to this example. We found that 5% of the normal-phenotype cells (12 out of 242) observed in the on-chip individual-cell cultivation system under the same conditions acquired elongated phenotypes.

The question is what is the mechanism responsible for passing on the elongated phenotype in one lineage of descendants? It should be noted that the evidence accrued from our results suggests that the presence of elongated cells is not caused by the mutation of genetic information because one of the two daughters, which should have had the same genetic information as the other elongated daughter cells, possessed normal cell characteristics. Otherwise, if the elongation had been caused by mutation, the normal of the two daughters would also have had elongated cells in the following generations.

The repeated unequal divisions of elongated cells in Figure 18.21a suggest that there is an intracellular mechanism that induces unequal cell divisions in elongated cells. Moreover, the mechanism for inducing unequal cell division underlies the stable inheritance of the elongated phenotype in one lineage; unequal cell division produces a daughter cell with a long start cell, which would also produce a long cell by the next division, enabling it to divide unequally again. Repeating this process, a cell could stably transmit the elongated phenotype to one lineage of descendants once they acquired a long cell.

The question arising from these observations is whether there is a boundary for the length that changes cell characteristics if variations exceed a certain length. We then examined the relationship between the final length and the position of the division plane (Figure 18.21e). The position of the division plane, r, was defined using the cell lengths of two daughter cells produced in the corresponding cell divisions as

$$r = \frac{\text{Initial length of longer daughter cell}}{\text{Initial length of shorter daughter cell}}. \tag{18.1}$$

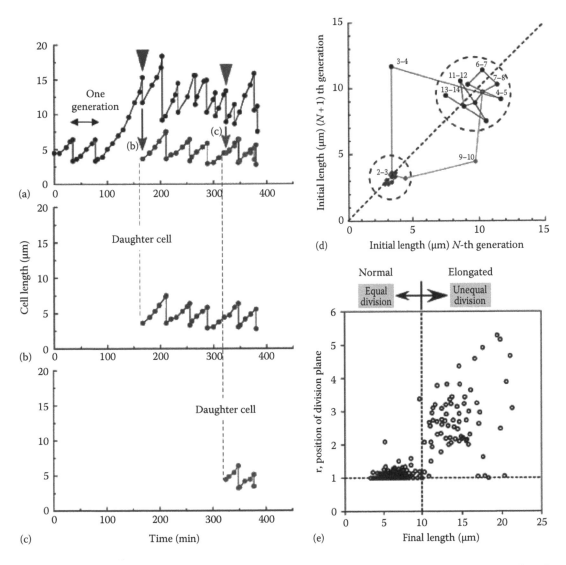

FIGURE 18.21 Presence and inheritance of elongated phenotype under uniform conditions. (a) Growth and division dynamics of a single cell changing from normal to elongated phenotype. The graph plots time-course change in the length of the a single cell for 12 generations (black points and lines). (b) Growth and division dynamics of shorter daughter cell from elongated cell in third generation shown in (a). (c) Growth and division dynamics of shorter daughter cell from elongated cell in ninth generation shown in (a). (d) Returning map of initial length transitions. To articulate the generation-by-generation length transition pattern of an elongated cell, initial lengths of single cells were plotted with initial length in one generation on the horizontal axis and with that in the next generation on the vertical axis, using data in (a–c). (e) Relationship between final length and position of division plane. The index of the position of the division plane, "r," is plotted against the final length. See Equation 18.1 in text for definition of "r." The graph clearly shows that the final cell length affects division, equal or unequal, around 10 μm cell length.

Therefore, "$r = 1$" corresponds to equal cell division and "$r > 1$" corresponds to unequal cell division. Figure 18.21e shows that the position of the division plane was at an uneven point when the cell length was longer than 10 μm. The results clearly reveal that there is a boundary for cell length that affects division characteristics; cells shorter than 10 μm divide equally (normal phenotypes), whereas those longer than 10 μm divide unequally (elongated phenotypes). Hence, it is conceivable that a kind of geometric index, that is, cell length, controls their phenotypic characteristics.

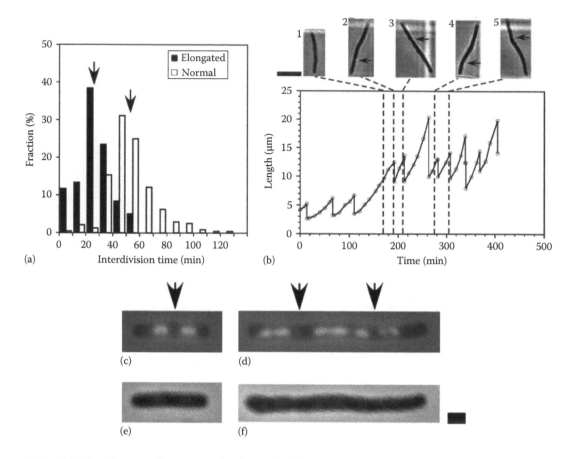

FIGURE 18.22 Short interdivision time for elongated cell. (a) Interdivision time distribution for elongated cells and normal cells. The black column shows interdivision time distribution for elongated cells ($n = 60$), and the white column shows that of normal cells ($n = 242$). Left and right arrows indicate averages for elongated and normal cell interdivision time distributions. (b) Sequential position of division planes of elongated cell. Images 1–5 above the graph show elongated cells at times indicated in the graph. Image 1 shows an emerging elongated cell in the fourth generation. Images 2–5 show cells at times when the division plane became visible in phase-contrast images. Emerging positions of division planes are indicated by arrows. Cell images are positioned in a way to show the same end of the elongated cell directed in the same direction. Stability of the elongated phenotype was confirmed until the 12th generation. Bar, 5 μm. (c) Nucleoid position in normal-sized cell visualized by DAPI. Min system should inhibit Z-ring formation at poles; hence, Z-ring would be directed to the middle between segregated nucleoids indicated by the arrow. (d) Nucleoid position in elongated cell visualized by DAPI. There were multiple nucleoids in the single elongated cell. (e and f) Phase-contrast images corresponding to c and d. Bar, 1 μm.

We next examined the division cycles of elongated cells. Figure 18.22a plots the interdivision time distributions for elongated and normal cells. *Elongated cells* were defined as those whose final length was longer than 10 μm and those whose final length in the previous generation was also longer than 10 μm. Based on this definition, the interdivision time for a generation in which a cell elongated extraordinarily from normal length (like the third generation in Figure 18.21a) was not categorized as elongated in plotting the distribution. The average elongated-cell interdivision time was 25.9 ± 1.6 min, half that of normal cells (52.4 ± 1.1 min). The coefficients of variance (CV) were 48% and 33% for the elongated and the normal, respectively. The distinct interdivision time distribution for elongated cells also confirms that they could easily be distinguished from normal cells. The characteristics of elongated cells cannot be explained by variations in normal cells. We thus regarded elongated and normal cells to be different phenotypes.

The next question is how does an elongated cell achieve a short interdivision time? Figure 18.22b shows an alternative formation for the division plane at opposite poles between neighboring generations. Image 1 shows an elongated cell emerging in the fourth generation from a normal phenotype before the division plane was formed. A visible division plane was then formed near the lower end as can be seen from Image 2 (fourth generation). After cell division, the longest daughter cell was selected and observed thereafter. We found that a division plane was formed in the next generation near the opposite end of the cell (Image 3, fifth generation). Although the cell divided equally and produced two elongated daughters in the sixth generation, the division ended in the following two generations (seventh and eighth), exhibiting the same behavior as in the fourth and fifth generations, that is, division planes were generated on opposite sides (Images 4 and 5). The fact that the positions of division planes in consecutive generations (fourth–fifth and seventh–eighth) were opposite indicates that two division mechanisms operate simultaneously near both ends in elongated cells. The half interdivision time of elongated cells can be explained by two division planes simultaneously and independently working near both ends of elongated cells, each of which works at approximately the same interval as interdivision in the normal phenotype.

The question then is what intracellular mechanism is responsible for the observed unequal cell divisions in elongated cells? It has been suggested that two mechanisms in *E. coli*, nucleoid occlusion and the Min system, determine the position of the division plane. Nucleoid occlusion is a mechanism where the assembly of the FtsZ ring, which determines the position of the final cell division plane, is inhibited within the close vicinities of nucleoids (Yu and Margolin 1999). The Min system, on the other hand, is comprised of three proteins, MinC, MinD, and MinE, which dynamically interact with one another and exhibit rapid pole-to-pole oscillations (Bi and Lutkenhaus 1993, de Boer et al. 1989, 1992). MinC and MinD form an inhibitory complex to form Z-rings (Hu and Lutkenhaus 1999, Raskin and de Boer 1999); Z-ring formation has been proposed to be directed to positions where concentrations of the MinCD complex are low on average in oscillations (Howard et al. 2001, Huang et al. 2003, Meinhardt and de Boer 2001).

To examine the relevance of observed unequal cell divisions to the proposed mechanisms responsible for the position of the division plane, we visualized the position of nucleoids in elongated cells by 4′, 6-diamidino-2-phenylindole dihydrochloride hydrate (DAPI) fluorescence using Hiraga's method (Hiraga et al. 1989) (Figure 18.22c and d). A normal-sized cell reaching the mean division length possessed two segregated nucleoids with a space in the middle as seen in Figure 18.22c (phase-contrast image for reference in Figure 18.22e). MinCD concentration should be low in the middle and high at the poles (Howard et al. 2001, Huang et al. 2003, Meinhardt and de Boer 2001), leading to Z-ring formation in the middle. However, an elongated cell possessed multiple nucleoids with spaces between them as can be seen from Figure 18.22d (phase-contrast image for reference in Figure 18.22f). There are numerous positions where constrictions can occur in this elongated cell according to the nucleoid occlusion model. However, MinCDE oscillations are known to achieve *doubled* patterns in an elongated cell (Fu et al. 2001, Raskin and de Boer 1999), which has been proposed to inhibit Z-ring formation both around the midplane and at the poles. Consequently, the cell divisions in elongated cells of this size should only occur at the uneven nucleoid gap positions (Figure 18.22d, arrows).

Although the presence of unequal cell divisions in elongated cells is understandable from the combined views of nucleoid occlusion and Min oscillation, it is still unclear what determines the cell-length boundary, 10 μm, between the equal and unequal cell divisions shown in Figure 18.21e. We therefore calculated the oscillation dynamics of the MinCD complex and MinE at various cell lengths according to Huang's model (Huang et al. 2003) (Figure 18.23). Figure 18.23a–e shows micrographs of the time-course change in the concentration of MinD at the cell membrane along the long axis of the cell, indicating that the number of oscillations doubles when the cell length is longer than 8 μm. The single-cycle averages for MinD concentration at the membrane in Figure 18.23a–e also reveal that the concentration of MinD at the midplane becomes highest for all positions when the cell length is longer than 8 μm. The MinD concentration at the midplane relative to the overall averages is plotted in Figure 18.23f. The results indicate a drastic increase in MinD concentration at the midplane when the oscillation doubles,

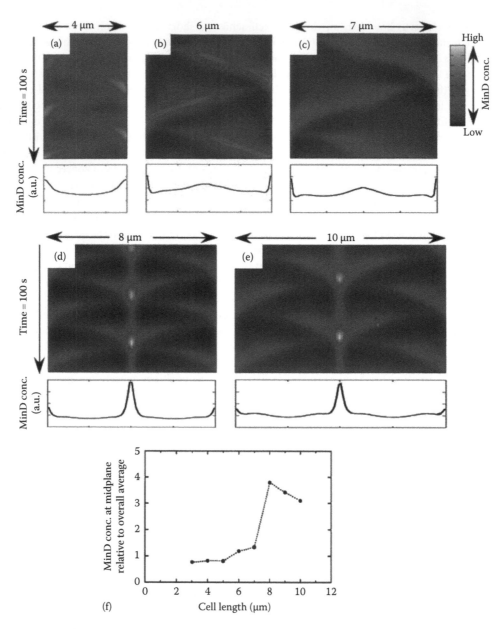

FIGURE 18.23 **(See color insert.)** Simulated MinD dynamics within cells with various cell lengths. (a–e) MinD concentration at the membrane for cells with lengths of 4 μm (→a), 6 μm (→b), 7 μm (→c), 8 μm (→d), and 10 μm (→e). Upper figure: Time-course change in MinD concentration at the membrane. Time flows from top to bottom. Dynamics for 100 s is shown in each figure. Lower figure: Single-cycle average of MinD concentration at the membrane. Durations of one cycle were 38 s for 4 μm, 67 s for 6 μm, 79 s for 7 μm, 37 s for 8 μm, and 46 s for 10 μm. (f) Relative MinD concentration at midplane to overall average. Single-cycle average of MinD concentration at midplane was compared with the total average at each cell length and plotted against the cell length.

which suggests that the cell-length boundary between the single and double oscillations of Min proteins determines the cell-length boundary between equal and unequal cell divisions in Figure 18.21e.

Based on the results in Figure 18.23f, FtsZ would be unable to form a ring at the midplane in cells with lengths more than 8 μm. Therefore, FtsZ would inevitably form rings at uneven positions in these long cells, which would lead to unequal cell divisions. The formation of FtsZ rings lies at the heart of the process of division at the membrane (Aarsman et al. 2005). In previous studies, the time between when cell division was initiated by the polymerization of FtsZ and the appearance of constriction visualized by electron microscopy was found to be approximately 20 min (Den Blaauwen et al. 1999). Under the conditions in our experiment, the average time to double cell length was 52 min (Figure 18.22a); hence, the expected division cell length for a cell whose Z-ring formation was initiated at a cell length of 8 μm can be calculated as 8 μm × $2^{20/52}$ = 10.4 μm. This calculation suggests that the boundary for the final length that separates equal and unequal cell divisions should be 10.4 μm, which matches our experimental results in Figure 18.21e. All these results indicate that it is highly probable that Min oscillations determine the observed cell-length boundary between equal and unequal cell divisions.

It is still unclear at this point how a cell acquires length. We previously reported that the length distributions for the division of genetically identical cells in the same environment had variations of 26% (C.V.) (Wakamoto et al. 2005). The large variations in cell-length distributions reflect uncertainty about whether genetically identical cells can generate the same length when the environmental conditions are the same. In other words, the intracellular mechanism inevitably includes stochasticity, which causes variations in the length of divided cells. Therefore, extraordinarily elongated cells with final lengths longer than 10 μm can occur with a certain probability based on intracellular stochasticity; this occurred with 5% probability under our test conditions.

The fact that elongated cells divide unequally would enable the elongated phenotype to be inherited in one lineage. Unequal cell division produces longer and shorter daughter cells. The longer daughter cell with a long starting cell length can easily exceed the boundary cell length. Consequently, the next division also becomes unequal. Repeating this process, a cell will eventually transmit its longer length to one of its descendants once it has acquired that length. Geometry, that is, cell length in this inheritance mechanism, plays a key role in enabling the phenotypic characteristics to be inherited from one generation to another without any consideration given to genetic modifications.

The mechanisms for cellular epigenetic inheritance have mainly been studied to reveal the gene regulatory network to achieve multistability (Gardner et al. 2000, Ozbudak et al. 2004, Thomas 1998, Thomas and Kaufman 2001) to the best of our knowledge. The roles of chromatin, DNA methylation, and acetylation also relate to gene regulation (Casadesus and D'Ari 2002, Hernday et al. 2003, McNarin and Gilbert 2003). However, the epigenetic inheritance discussed in this chapter occurred once a cell reached a certain cell length; it was independent of the gene regulatory network. Therefore, our results suggest that the inheritance of geometric information, such as cell shape, is significant in epigenetic inheritance.

18.4 Quantitative Measurement of Possible Damage Caused by 1064 nm Wavelength Optical Trapping of *Escherichia coli* Cells Using On-Chip Single-Cell Cultivation System

In the on-chip single-cell culture system, cells are manipulated on a microfabricated chip (microchamber) using noncontact optical tweezers, enabling individual cells to be observed over long intervals with strict control of the external environment and cell–cell interactions.

However, the aforementioned system requires the optical handling and transportation of cells to control the number of cells in the microchamber. Thus, to identify whether the change in the cells' properties was caused by a change in the environment or by optical trapping, we need to quantify the possible damage to cells caused by optically trapping them. Regarding possible damage caused by optical trapping, Ashkin's group first reported that there was no damage to the growth and division of *E. coli*, with 1064 nm, 80 mW optical trapping for several intervals of 10 min irradiations. However, the following reports showed the potential damage caused by the optical trapping, for example, damage to cells' propagation (Liang et al. 1996, Liu et al. 1995, 1996) and motility (Neuman et al. 1999) and direct damage caused by the expression of the stress response gene (Leitz et al. 2002). As these reports are qualitative reports, we still cannot clarify how safely we can use the optical tweezers for cell handling.

In this section, we report the quantitative results of optical trapping damage measurement where we continuously followed the growth and division dynamics of isolated single cells of *E. coli*, comparing them with those of nontreated sister cells under a uniform and steady condition, to determine the magnitude of the effect on cells possessing identical genetic information and experience (Ayano et al. 2006).

In this experiment, we used the on-chip single-cell cultivation system. Optical tweezers were introduced for noncontact cell handling on the specimen. Nd:YAG laser (wavelength = 1064 nm) (T20-8S, Spectra Physics, Santa Clara, CA) was used for the laser source of optical tweezers. The expanded and circularly polarized laser beam was introduced into the microscope from the side port and was guided to the ×100 phase-contrast objective lens (UplanApo, Olympus, Tokyo, Japan). The power of the laser emitting from the objective was measured by a powermeter (LM-3, Coherent) before we started each observation. *E. coli* EJ2848 strain (*LacI3 DlacZ lacY⁺ DfliC*) that lacks motility was used in the experiment. A 5 mL of glycerol stock of EJ2848 was diluted in 1 mL M9 0.2% Glc. medium (M9 Minimum Salt (Qbiogene Inc., Carlsbad, CA) + 0.2%(w/w) glucose + 1/2 MEM amino acids solution (Invitrogen Japan K,K, Tokyo, Japan)) and cultured overnight to reach full growth by shaking at 37°C. Five milliliter of full growth culture was diluted in 1 mL of M9 0.2% Glc. medium and cultured for 3 h by shaking at 37°C to reach approximately OD 0.1. Then the 5 mL culture was applied onto the microchamber array on the glass slide and sealed by a semipermeable membrane. The slide prepared by this procedure was used for the observation.

In our protocol, one of two daughter cells was used as a target cell to which various trap conditions were applied. The other daughter cell was used as a reference. Figure 18.24a summarizes our protocol. At the beginning, we focused on a microchamber containing one cell in one of four rooms of the microchamber. When the cell divided into two daughter cells, one of them was arbitrarily chosen, trapped by optical tweezers, and transported slowly to one of the other vacant rooms by moving a stage of the microscope. We released the cell after trapping for a predetermined time. This release time was defined

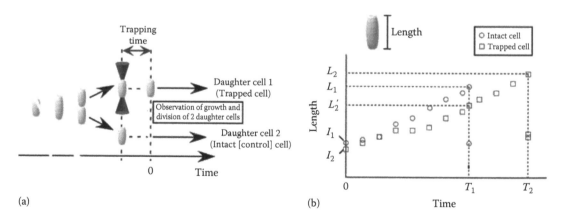

FIGURE 18.24 Procedure for damage measurement. (a) Direct comparison of two daughter cells. In our method, we directly compared two daughter cells derived from the same mother to measure the damage. At the division of a mother cell, we arbitrarily chose one of the daughter cells and trapped it by optical tweezers. After trapping for a predetermined time, we stopped trapping it and compared the growth and division pattern between trapped and intact daughter cells. In this method, an intact daughter cell is used as a reference for damage measurement. (b) Growth and division parameters. Simultaneous observation of two daughter cells gives time-course change of cell lengths of both cells. This graph holds the information of growth and division characteristics. We compared interdivision time and growth speed between trapped and intact daughter cells. See text for the definitions of interdivision time and growth speed.

as time = 0. We kept recording the growth of both trapped and intact daughter cells simultaneously either until both the daughter cells divided or until the time reached 180 min.

From the video data, we measured the time-course change of the cell lengths of two daughter cells (see Figure 18.24). The time-course data contains information on the differences between two daughter cells in basic growth and division functions. As parameters, we defined division time (*T*) and growth speed (*v*). Division time is the time taken to divide from time = 0. Growth speed is defined as

$$v = \frac{1}{T}\frac{L-l}{l}, \tag{18.2}$$

where

  L is cell length just before the division
  l is length at time = 0

We adopted these parameters to examine the damages on growth and division functions.

We used the *E. coli* EJ2848 (*LacI3 DlacZ lacY⁺ DfliC*) strain because *E. coli* is suitable for our purpose for the following reasons: it produces clones as it proliferates without active genetic recombination upon division; it lacks motile ability, hence advantageously avoiding the possibility that cells in an observation chamber escape by their own motility.

First, we examined whether an *E. coli* cell could grow and divide during the continuous trapping by optical tweezers. We trapped one of two daughter cells continuously with minimum force for the optical trapping of a cell, 3 mW power (at obj. lens position). As the rod-shaped *E. coli* stands perpendicular to the light pathway of the optical tweezers (i.e., vertical; see Figure 18.24a), we could not observe the shape of *E. coli* during its trapping. Thus, we released it for 10 s from the trap every 10 min and measured the length of the trapped cell. Figure 18.25a shows the time course of cell growth. The trapped cell (filled circles) did not grow in a continuous trap for 180 min. On the other hand, the other intact daughter cell grew normally in an exponential manner and divided in 88 min (open circles; initial lengths of two divided daughters, open triangle and cross "X"). After we finished 180 min laser irradiation (see arrow in Figure 18.25a), we continuously examined the change in the ability of the long-term trapped cells to grow and divide and found neither growth nor division for at least 140 min. This result showed that *E. coli* cannot grow and divide both in and after the long-term trap of 3 h even at the minimum laser power for optical trapping of cells, 3 mW (obj.). We found that long-term trapping suppresses the growth and division abilities of *E. coli* regardless of the trapping power.

Next, we examined the damage from optical trapping under various trapping time and laser power conditions. For that, we changed trapping time from 0.5 to 7.5 min and laser power from 3 to 30 mW. For quantitative evaluation of the damage caused by optical trapping, we compared the relative differences between the trapped and the intact daughter cells and categorized them into three patterns as described in previous reports. As the growth speed and interdivision time of each individual cell are too variable to distinguish the regular growth speed and slower growth speed, the higher similarity of growth speed and interdivision time of sister cells (as shown in previous references, Inoue et al. 2001a, Wakamoto et al. 2001, 2003) was applied for comparing the results, that is, we used the growth speed and interdivision time of intact sister cells as the standard growth curve to categorize the trapped sister cells. In the analysis, the difference in growth speed between sister cells (both free from the irradiation of laser trapping) was less than 4.0×10^{-3} min^{-1} with 25% CV. Similarly, the difference in interdivision time between sister cells was less than 36 min with 25% CV. We therefore judged that the trapped cell was *normal* when the differences between the trapped and intact daughter cells were within 25% difference. When the differences were more than 25%, even though they did grow or divide within 180 min, we judged their growth or division was *slow*. The conditions under which the trapped cell did not grow or divide at all within 180 min were attributed to *no* growth and division conditions. Figure 18.25 also shows the typical examples of those three types: first, almost no difference between the trapped cell and the intact cell, which is usually observed in cells trapped for a short time with weak laser power (Figure 18.25b);

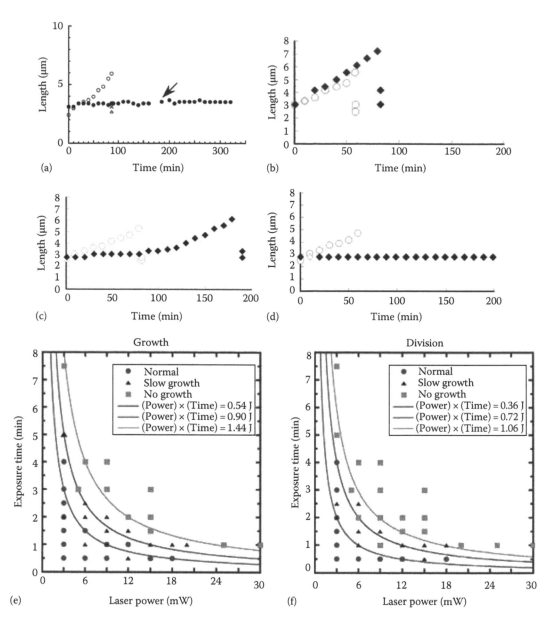

FIGURE 18.25 Damage to growth and division abilities at various laser power and trapping time conditions. (a) No growth and division in continuous trap. The graph shows time-course changes in lengths of both the continuously trapped daughter cell and the intact daughter cell. The length of trapped cell (filled circle points) showed almost no change in length in trapping (0–180 min) while the intact daughter cell grew and divided normally (open circle points). Though the trapped daughter cell was released from the trap at 180 min (black arrow), the length did not change for at least 140 min even after the release. (b–d) Growth and division patterns of optically trapped cells under various trapping conditions. Three typical examples of growth and division patterns of trapped cells under various trapping conditions. (b) The trapped cell with no damage. It had almost the same growth speed and divided faster than the intact cell. (Laser power = 3 mW(Obj.), trapping time = 1 min). (c) The trapped cell with slower growth speed and longer interdivision time than the intact cell although it did grow and divide. (Laser power = 18 mW(Obj.), trapping time = 1 min). (d) The trapped cell with no growth and division. (Laser power = 15 mW(Obj.), trapping time = 2 min). (e) Damage estimated by their growth. (f) Damage estimated by their division. The filled circles in the graph represents the points of *normal*, the filled triangles *slow*, and the filled squares *no growth*, respectively.

second, slower growth and slower interdivision time, which is observed under the increased trapping time and laser power condition (Figure 18.25c); and third and final, completely stopped growth and division case as described in Figure 18.25d.

Figure 18.25e and f shows the damage dependence of laser power and trapping time in growth speed (Figure 18.25e) and in division time (Figure 18.25f). Each plot of the filled circles represents one experimental result of the normal response of cell growth and division. Filled triangles represent those of slower cells, and filled squares those of no growth. The results showed that the damages on growth and division became more obvious when the cells were trapped with stronger laser power and for longer trapping time. Comparing Figure 18.25e and f, we found that the damage intensity is greater on division time than on growth speed. There were several conditions under which the division was completely inhibited whereas growth speed was normal (e.g., at 12 mW^{-1} min point).

Furthermore, the different types of damages (normal, slow, no growth) could be divided into three regions separated by hyperbolic curves, that is, (laser power) × (trapping time) = constant. For example, when the total work was less than 0.54 J, growth tendency was normal, whereas work larger than 1.44 J gave no growth tendency. Similarly, division tendency was normal if the total work was less than 0.36 J, and work of 1.06 J could stop cell division activity. These results of tendency suggest that the damage depended on the total work applied to the cells from optical trapping and was not caused by the nonlinear factors. That is, the possible damage affected by optical trapping might be estimated by the total energy applied to the cells.

Under the various trapping conditions of laser power and trapping time, we found that the damage intensities on growth were greater than on division. Moreover, we found that the damage could affect cell growth and division even under the minimum power of trapping even if this trapping condition was a magnitude smaller than in the previous reports (Ashkin et al. 1987, Leitz et al. 2002, Liang et al. 1996, Liu et al. 1995, 1996, Neuman et al. 1999).

The difference of the threshold intensities of damages on cell growth and division might mean that the optical trapping affected the two different cell functions. The significance of this result for the application of optical handling to cell biology is that the trapping damage to cell division mechanism must be more sensitive than those to cell growth. For example, the maximum value of the safe *normal* trapping condition for growth was 18 mJ, while that for division was 6 mJ. Thus, we should apply optical handling for the study of bacterial division below 6 mJ, which is a more tight condition than that of bacterial growth.

One possible candidate for the origin of those damages might be mutations in the gene-coding proteins that act in DNA segregation or septation (Hirota et al. 1968, Lutkenhaus and Mukherjee 1996). In our observations, however, the cells in the next generation produced from the damaged cells returned to the regular growth and division pattern (data not shown), indicating that the damage could not have been caused by mutations in these candidate genes. It is also known that the SOS or heat shock responses can lead to change in their characteristics because they transiently inhibit the division mechanisms (Huisman and D'Ari 1981, Huisman et al. 1984, Lutkenhaus and Mukherjee 1996). We must, however, stress the uniformity of the environmental condition in the on-chip single-cell cultivation system imposed on our cells; there is no reason to invoke these responses except for optical trapping. Moreover, since the slower elongation and division was not inherited by the next generation, it is most improbable that this behavior was occasioned by any damage to the gene itself. Hence, we assert that the slowing and stopping of growth and division are occasioned by the intracellular leak of the reactions that temporarily inhibit the growth and division mechanism; and this intracellular leak of reactions might be caused by the fact that optical trapping prevents free movement of molecules. Strong evidence for such stochastic intracellular reactions is provided by the observation that, even in the regular growth and division pattern under uniform conditions, significantly large fluctuations in interdivision time (up to 33%) were observed (Wakamoto et al. 2003).

Based on all our present results, we know that the optical trap completely suppresses both growth and division under continuous trap and after long-term trap. And the damage depends linearly on the total energy (work) applied by the optical trap and is more intense on division than on growth. It should be noted that these results could not have been obtained without the direct observation and competition of two sister cells under strict control of environmental conditions. This method of following the behavior of specific phenotypes of individual cells with strict control of their interactions should become a powerful tool in the near future for single-cell-based epigenetic studies, which are themselves rapidly acquiring importance as an essential element of postgenome research.

18.5 Photothermal Etching Method: Another Optical Trapping of Cultivated Cells on a Chip

Another approach to optical handling of cells cultivated on a chip is a microfabrication technique using photothermal etching (Hattori et al. 2004, Moriguchi et al. 2002) with agarose-microchamber cell-cultivation system. This is the area-specific melting of agarose microchambers by spot heating using a focused laser beam of 1480 nm, and of a thin layer made of a light-absorbing material such as chromium with a laser beam of 1064 nm (because agarose itself has little absorbance at 1064 nm). The system has two parts: a phase-contrast microscope (IX-70; with a phase-contrast objective lens, ×40, Olympus, Tokyo, Japan) with an automated X–Y stage (BIOS-201T, Sigma Koki, Hidaka, Saitama, Japan), and a dual-wavelength-focused laser-irradiation module with a 1064 nm Nd:YAG laser (max. 1 W; Forte-1064, Laser Quantum, Emery Court, Vale Road, Stockport, Cheshire, U.K.) and a 1480 nm Raman fiber laser (max. 1 W; PYL-1-10480-M, IPG Photonics, Oxford, MA). For phase-contrast microscopy and μm-scale photo-thermal etching, three different wavelengths (visible light for observation and 1480 nm/1064 nm infrared lasers for spot heating) were used simultaneously to observe the positions of the agar chip surface and to melt a portion of the agar in the area being heated. A phase-contrast image was acquired by using a CCD camera (CS230, Olympus). The dichroic mirrors and lenses in the system were chosen for these three different wavelengths. Flexible-slide focusing lenses were placed in the path of the infrared laser beam to control the focal positions of the lasers to correct their different focal lengths, which depended on the wavelengths.

We used this new type of noncontact three-dimensional photothermal etching for the agar-micro-etching to exploit the characteristics of the two different infrared laser beam wavelengths (1480 and 1064 nm). As the 1480 nm infrared beam was absorbed by water and agar gel, the agar gel in the light pathway was heated and completely melted. As the 1064 nm infrared beam, on the other hand, did not have this absorbance, the agar melted just near the thin chromium layer, which absorbed the beam.

Using this noncontact etching, we could easily produce microstructures such as holes and tunnels within only a few minutes. As we can see from Figure 18.26, the laser melted the agar as follows:

(a) When a 1064 nm infrared laser beam was focused on the chromium layer on the glass slide, the agar at the focal point near the chromium layer started to melt.
(b) When the focused beam was moved parallel to the chip surface, a portion of the agar at the heated spot melted and diffused into the water through the agar mesh.
(c) After the heated spot had been moved, a tunnel was created at the bottom of the agar layer.
(d) However, when a 1480 nm infrared laser beam was focused on the agar glass slide, the agar in the light pathway started to melt.
(e) When the focused beam was moved parallel to the chip surface, a portion of the agar in the light path melted and diffused into the water.
(f) Finally, after the heated spot had been moved, a hole was created on the glass slide.

In the field of neuroscience, one of the main interests of an epigenetic study is how the epigenetic information is processed and recorded as plasticity within a network pattern, what might be caused by the change in the network pattern or by the degree of complexity related to the network size. To understand the meaning of the network pattern and size, one of the best approaches is to analyze the

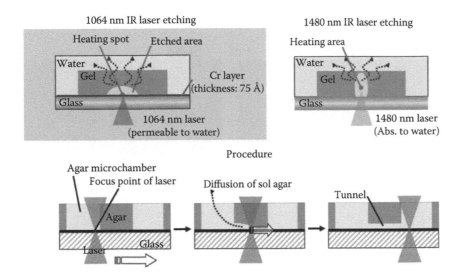

FIGURE 18.26 Photothermal etching method: (a–c) 1064 nm and (d–f) 1480 nm.

function of an artificially constructed neural cell network under fully controlled conditions. Thus, for many years, neurophysiologists have investigated single-cell-based neural network cultivation and examined the firing patterns of single neurons through the fabrication of cultivation substrates using microprinting techniques (Branch et al. 2000, 2001, James et al. 2000), patterned on silicon-oxide substrates (Sholl and Offenhausser 2000), and three-dimensional structures made using photolithography (Merz and Fromherz 2002). Although these conventional microfabrication techniques provide structures with fine spatial resolution, effective approaches to studying epigenetic information are still being sought. With conventional techniques, it is still hard to make flexible microstructures with simple steps or to change their shape during cultivation since the shape is usually unpredictable and only defined during cultivation.

Thus, we have developed a new single-cell cultivation assay using agarose microstructures and a photothermal etching method. Using the photothermal etching method, we can form microstructures within the agarose layer on the chip by melting a portion of the agarose layer at the spot of a focused infrared laser beam as described earlier. This method can be applied even during cultivation, so we can change the network pattern of nerve cells during cultivation by adding microchannels between two adjacent microchambers in a step-by-step fashion (Moriguchi et al. 2003, Sugio et al. 2004, 2005).

This helps us understand the meaning of the spatial pattern of a neuronal network by comparing the changes in signals before and after the network shape is changed. Moreover, we developed an agarose microchamber (AMC) system on an multielectrode array (MEA) substrate that can be used to obtain the long-term electronic properties of topographically controlled neuronal networks with precise fixation of cell positions and flexible network pattern rearrangement through photothermal etching of the agarose layer. In this section, we describe our newly developed neural-cell cultivation chip and its cultivation/recording system (Moriguchi et al. 2003, Sugio et al. 2004, 2005).

While applying stepwise photothermal etching to an AMC array during cultivation, we also developed a method of controlling the topography in the direction of synaptic connections in the network patterns of a living neuronal network. This allowed the direction in which axons were elongated to be flexibly controlled by melting the narrow micrometer-order grooves (microchannels) in steps through photothermal etching, where a portion of the agarose layer was melted with the 1064 nm infrared laser beam. Figure 18.27 shows an example of this procedure. The micrographs are phase-contrast images of the growth of single hippocampal cells in the microchambers. When cultivation started, single cells were placed into the AMCs. Then, single neurites elongated from the neurons into the microchannels.

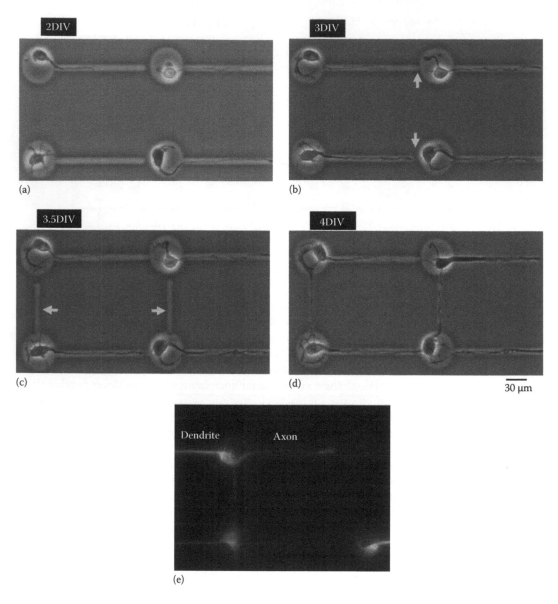

FIGURE 18.27 Stepwise formation of neuronal network pattern in rat hippocampal cells. (a)–(d) Time course of stepwise formation of neuronal network on a chip. (a) 2 days in vitro (DIV) after cultivation started. To guide the first neurites (which will differentiate to axons) to the single microchannels, only one closed microchannel was prepared in each microchamber. (b) 3 DIV. After the confirmation of elongation of the first neuritis into the mirochannels, the closed ends of the microchannels were opened to connect to another neuron by spot heating and melting of a portion of agarose using the photo-thermal etching system. (c) 3.5 DIV. To guide the second neuritis (which will differentiate to dendrites) to other neurons, additional microtunnels were formed. (d) 4 DIV. The fully axon/dendrite direction controlled neuronal network was formed on a chip. (e) Fluorescent staining of the formed neuronal networks for confirmation of direction controlled neurite patterns was carried out. Axons and dendrites were stained as designed and the full direction control of the neuronal network was accomplished (see Suzuki et al., 2005 for additional details).

At that time, because we found that the elongation of neurites was sufficiently stable, additional photo-thermal etching was done to connect two adjacent AMCs. Finally, after a series of the additional photothermal etching, all neurons retained their shapes and continued the elongation and were connected by neurites.

18.6 Conclusion

We described the on-chip microcultivation system, which has the following advantages:

1. Continuous cultivation of single cells under isolated conditions
2. Control of spatial distribution of cells, interactions of cells, and number of cells in the chamber
3. Continuous observation of identical specific cells under contamination-free conditions
4. Control of the environmental conditions, such as cultivation buffer, surrounding cells

We also demonstrated two methods for analysis of single-cell differences by using the on-chip microculture system and optical tweezers. Though we showed only a few examples of our experimental results, they demonstrated the potential that our system could uniquely detect several phenomena. Differential analysis should thus help clarify heterogeneous phenomena such as unequal cell division and cell differentiation, which will become more and more important in the *cellome* era.

References

Aarsman ME, Piette A, Fraipont C, Vinkenvleugel TM, Nguyen-Disteche M, den Blaauwen T. 2005. Maturation of the *Escherichia coli* divisome occurs in two steps. *Mol Microbiol* 55:1631–1645.

Åkerlund T, Nordström K, Bernander R. 1995. Analysis of cell size and DNA content in exponentially growing and stationary-phase batch cultures of *Escherichia coli*. *J Bacteriol* 177:6791–6797.

Ashkin A, Dziedzic JM, Bjorkholm JE, Chu S. 1986. Observation of a single-beam gradient force optical trap for dielectric particles. *Opt Lett* 11:288–290.

Ashkin A, Dziedzic JM, Yamane T. 1987. Optical trapping and manipulation of single cells using infrared laser beams. *Nature* 330:769–771.

Ayano S, Wakamoto Y, Yamashita S, Yasuda K. 2006. Quantitative measurement of damage caused by 1064-nm wavelength optical trapping of *Escherichia coli* cells using on-chip single cell cultivation system. *Biochem Biophys Res Commun* 350:678–684.

Bi E, Lutkenhaus J. 1993. Cell division inhibitors SulA and minCD prevent formation of the FtsZ ring. *J Bacteriol* 175:1118–1125.

Branch DW, Wheeler BC, Brewer GJ, Leckband DE. 2000. Long-term maintenance of patterns of hippocampal pyramidal cells on substrates of polyethylene glycol and microstamped polylysine. *IEEE Trans Biomed Eng* 47:290–300.

Branch DW, Wheeler BC, Brewer GJ, Leckband DE. 2001. Long-term stability of grafted polyethylene glycol surfaces for use with microstamped substrates in neuronal cell culture. *Biomaterials* 22:1035–1047.

Casadesus J, D'Ari R. 2002. Memory in bacteria and phage. *Bioessays* 24:512–518.

Cohn M, Horibata K. 1959. Physiology of the inhibition by glucose of the induced synthesis of the β-galactoside-enzyme system of *Escherichia coli*. *J Bacteriol* 78:601–612.

de Boer PA, Crossley RE, Rothfield LI. 1989. A division inhibitor and a topological specificity factor coded for by the minicell locus determine proper placement of the division septum in *E. coli*. *Cell* 56:641–649.

de Boer PA, Crossley RE, Rothfield LI. 1992. Roles of MinC and MinD in the site-specific septation block mediated by the MinCDE system of *Escherichia coli*. *J Bacteriol* 174:63–70.

Den Blaauwen T, Buddelmeijer N, Aarsman ME, Hameete CM, Nanninga N. 1999. Timing of FtsZ assembly in *Escherichia coli*. *J Bacteriol* 181:5167

Donachie DD, Begg KJ. 1996. "Division potential" in *Escherichia coli*. *J Bacteriol* 178:5971–5976.

Elowitz MB, Levine AJ, Siggia ED, Swain PS. 2002. Stochastic gene expression in a single cell. *Science* 297:1183–1186.

Fu X, Shih YL, Zhang Y, Rothfield LI. 2001. The MinE ring required for proper placement of the… its cellular location during the *Escherichia coli* division cycle. *Proc Natl Acad Sci USA* 98:980–985.

Gardner TS, Cantor CR, Collins JJ. 2000. Construction of a genetic toggle switch in *Escherichia coli*. *Nature* 403:339–342.

Hattori A, Moriguchi H, Ishiwata S, Yasuda K. 2004. A 1480-nm/1064-nm dual wavelength photo-thermal etching system for non-contact three-dimensional microstructure generation into agar microculture chip. *Sens Actuat B* 100:455–462.

Hernday AD, Braaten BA, Low DA. 2003. The mechanism by which DNA adenine methylase and PapI activate the pap epigenetic switch. *Mol Cell* 12:947–957.

Hiraga S, Niki H, Ogura T, Ichinose C, Mori H, Ezaki B, Jaffe A. 1989. Chromosome partitioning in *Escherichia coli*: Novel mutants producing anucleate cells. *J Bacteriol* 171:1496–1505.

Hirota Y, Ryter A, Jacob F. 1968. Thermosensitive mutants of *E. coli* affected in the processes of DNA synthesis and cellular division. *Cold Spring Harb Symp Quant Biol* 33:677–693.

Howard M, Rutenberg AD, de Vet S. 2001. Dynamic compartmentalization of bacteria: accurate division in *E. coli*. *Phys Rev Lett* 87:278102.

Hu Z, Lutkenhaus J. 1999. Topological regulation of cell division in *Escherichia coli* involves rapid pole to pole oscillation of the division inhibitor MinC under the control of MinD and MinE. *Mol Microbiol* 34:82–90.

Huang KC, Meir Y, Wingreen NS. 2003. Dynamic structures in *Escherichia coli*: Spontaneous formation of MinE rings and MinD polar zones. *Proc Natl Acad Sci USA* 100:12724–12728.

Huisman O, D'Ari R. 1981. An inducible DNA replication-cell division coupling mechanism in *E. coli*. *Nature* 290:797–799.

Huisman O, D'Ari R, Gottesman S. 1984. Cell-division control in *Escherichia coli*: Specific induction of the SOS function SfiA protein is sufficient to block septation. *Proc Natl Acad Sci USA* 81:4490–4494.

Inoue I, Wakamoto Y, Moriguchi H, Okano K, Yasuda K. 2001a. On-chip culture system for observation of isolated individual cells. *Lab Chip* 1:50–55.

Inoue I, Wakamoto Y, Yasuda K. 2001b. Non-genetic variability of division cycle and growth of isolated individual cells in on-chip culture system. *Proc Jpn Acad* 77B:145–150.

James CD, Turner J, Shain W. 2000. Aligned microcontact printing of micrometer-scale poly-L-lysine structures for controlled growth of cultured neurons on planar microelectrode arrays. *IEEE Trans Biomed Eng* 47:17–21.

Leitz G, Fallman E, Tuck S, Axner O. 2002. Stress response in *Caenorhabditis elegans* caused by optical tweezers: Wavelength, power, and time dependence. *Biophys J* 82:2224–2231.

Liang H, Vu KT, Krishnan P, Trang TC, Shin D, Kimel S, Berns MW. 1996. Wavelength dependence of cell cloning efficiency after optical trapping. *Biophys J* 70:1529–1533.

Liu Y, Cheng DK, Sonek GJ, Berns MW, Chapman CF, Tromberg BJ. 1995. Evidence for localized cell heating induced by infrared optical tweezers. *Biophys J* 68:2137–2144.

Liu Y, Sonek GJ, Berns MW, Tromberg BJ. 1996. Physiological monitoring of optical trapped cells: Assessing the effects of confinement by 1064-nm laser tweezers using microfluorometry. *Biophys J* 71:2158–2167.

Lutkenhaus J, Mukherjee A. 1996. Cell division, in *Escherichia coli and Salmonella—Cellular and Molecular Biology*, Neidhardt FC (Ed.), ASM Press, Washington, DC, pp. 1615–1626.

Marr AG, Harvey RJ, Trentini WC. 1966. Growth and division of *Escherichia coli*. *J Bacteriol* 91:2388–2389.

McNairn AJ, Gilbert DM. 2003. Epigenomic replication: Linking epigenetics to DNA replication. *Bioessays* 25:647–656.

Meinhardt H, de Boer PA. 2001. Pattern formation in *Escherichia coli*: A model for the pole-to-pole oscillations of Min proteins and the localization of the division site. *Proc Natl Acad Sci USA* 98:14202–14207.

Merz M, Fromherz P. 2002. Polyester microstructures for topographical control of outgrowth and synapse formation on snail neurons. *Adv Mater* 14:141–144.

Moriguchi H, Takahashi K, Sugio Y, Wakamoto Y, Inoue I, Jimbo Y, Yasuda K. 2003. On chip neural cell cultivation using agarose-microchamber array constructed by photo-thermal etching method. *Electr Eng Jpn* 146:37–42.

Moriguchi H, Wakamoto Y, Sugio Y, Takahashi K, Inoue I, Yasuda Y. 2002. An agar-microchamber cell-cultivation system: Flexible change of microchamber shapes during cultivation by photo-thermal etching. *Lab Chip* 2:125–130.

Neuman KC, Chadd EH, Liou GF, Bergman K, Block SM. 1999. Characterization of photodamage to *Escherichia coli* in optical traps. *Biophys J* 77:2856–2863.

Ozbudak EM, Thattai M, Lim HN, Shraiman BI, Van Oudenaarden A. 2004. Multistability in the lactose utilization network of *Escherichia coli*. *Nature* 427:737–740.

Panda AK, Khan RH, Appa Rao KBC, Totey SM. 1999. Kinetics of inclusion body production in batch and high cell density fed-batch culture of *Escherichia coli* expressing ovine growth hormone. *J Biotechnol* 75:161–172.

Raskin DM, de Boer PA. 1999. Rapid pole-to-pole oscillation of a protein required for directing division to the middle of *Escherichia coli*. *Proc Natl Acad Sci USA* 96:4971–4976.

Shapiro JA, Hsu C. 1989. *Escherichia coli* K-12 cell-cell interactions seen by time-lapse video. *J Bacteriol* 171:5963–5974.

Sholl M, Offenhausser A. 2000. Ordered networks of rat hippocampal neurons attached to silicon oxide surface. *J Neurosci Methods* 104:65–75.

Spudich JL, Koshland DE Jr. 1976. Non-genetic individuality: Chance in the single cell. *Nature* 262:467–471.

Sugio Y, Kojima K, Moriguchi H, Takahashi K, Kaneko T, Yasuda K. 2004. An agar-based on-chip neural-cell cultivation system for stepwise control of network pattern generation during cultivation. *Sens Actuat B* 99:156–162.

Suzuki I, Sugio Y, Jimbo Y, Yasuda K. 2005. Stepwise pattern modification of neuronal network in photo-thermally-etched agarose architecture on multi-electrode array chip for individual-cell-based electrophysiological measurement. *Lab Chip* 5:241–247.

Thomas R. 1998. Laws for the dynamics of regulatory circuits. *Int J Dev Biol* 42:479–485.

Thomas R, Kaufman M. 2001. Multistationarity, the basis of cell differentiation and memory. II. Logical analysis of regulatory networks in terms of feedback circuits. *Chaos* 11:180–195.

Wakamoto Y, Inoue I, Moriguchi H, Yasuda K. 2001. Analysis of single-cell differences using on-chip microculture system and optical trapping. *Fresenius J Anal Chem* 371:276–281.

Wakamoto Y, Ramsden J, Yasuda K. 2005. Single-cell growth and division dynamics showing epigenetic-correlations. *Analyst* 130:311–317.

Wakamoto Y, Umehara S, Matsumura K, Inoue I, Yasuda K. 2003. Development of non-destructive, non-contact single-cell based differential cell assay using on-chip microcultivation and optical tweezers. *Sens Actuat B* 96:693–700.

Wakamoto Y, Yasuda K. 2006. Epigenetic inheritance of elongated phenotypes between generations revealed by individual-cell-based direct observation. *Meas Sci Technol* 17:3171–3177.

Wright W, Sonek GJ, Tadir Y, Berns MW. 1990. Laser trapping in cell biology. *IEEE J Quantum Elect* 26:2148–2157.

Yu XC, Margolin W. 1999. FtsZ ring clusters in min and partition mutants: Role of both the Min system and the nucleoid in regulating FtsZ ring localization. *Mol Microbiol* 32:315–326.

Zhao R, Natarajan A, Srienc F. 1999. A flow injection flow cytometry system for on-line monitoring of bioreactors. *Biotechnol Bioeng* 62:609–617.

19

Luminescent Nanoparticle-Based Probes for Bioassays

Georgeta Crivat
University of New Orleans

Sandra M. Da Silva
*National Institute of
Standards and Technology*

Ashley D. Quach
University of New Orleans

Venkata
R. Kethineedi
University of New Orleans

Matthew A. Tarr
University of New Orleans

Zeev Rosenzweig
*University of Maryland,
Baltimore County*

19.1 Introduction

This chapter describes the development and biomedical applications of highly sensitive luminescent CdSe/ZnS quantum dot (QD)-based fluorescence resonance energy transfer (FRET) probes for the detection of various biological compounds. It summarizes recent studies in our laboratory aiming to develop nanomaterial-based probes capable of detecting analytes of biomedical importance with high specificity and sensitivity.

We first discuss a thin film–based enzyme activity probe, which changes its photoluminescence properties when incubated with trypsin or other proteases. This sensor was fabricated by making use of a highly versatile layer-by-layer (LbL) deposition technology. The film was deposited on the polydimethylsiloxane (PDMS) surface of a microfluidic device, which was used to detect the release of proteolytic enzymes from cancer cells. The microfluidic device provided an in vivo–like environment to enable rapid cell growth prior to the cellular enzymatic assay. The second part of this chapter describes the synthesis of QD-containing liposomes and their use in FRET assays of phospholipase A₂ activity. To facilitate the FRET assay, the fluorescent acceptor, 2-(6-(7-nitrobenz-2-oxa-1,3-diazol-4-yl)amino)hexanoyl-1-hexadecanoyl-*sn*-glycero-3-phosphocholine (nitro-2-1,3-benzoxadiazol-4-yl [NBD] C₆-HPC), was incorporated in the phospholipid bilayer of the liposomes. The addition of phospholipase A_2 to the liposome sensor affected changes in FRET signals due to cleavage of two acyl bonds of phospholipids, which led to the removal of 7-nitrobenzo-2-oxa-1,3-diazole (NBD) molecules from the liposome membrane.

The last section of this chapter describes the latest accomplishment in our laboratory, the coattachment of luminescent QDs and gold nanoparticles (AuNPs), on the surface of polystyrene (PS) microparticles to form luminescent composite particles (AuNP–QD–PS) that could be used in FRET assays of multiple analytes. When coattached to the PS particles, the AuNP quenched the emission of the QDs due to FRET. The QD emission was restored when the AuNPs were removed from the surface using a cleaving reagent. The AuNP–QD–PS composite particles can be used in microparticle-based array sensors to screen the reactivity of different thiol compounds like antidotes for annihilating organomercurials and other mercuric compounds.

19.2 FRET-Based Quantum Dot Probes

Luminescent QDs are semiconductor nanocrystal materials, which exhibit unique photoluminescence properties. They are defined as QDs since their nanometric size is smaller or approaches the Bohr exciton radius. Therefore, their spatial electronic wave function is confined to the size of the dot, which results in quantum confinement effects. For example, the band gap of luminescent QDs of CdSe/ZnS increases when the size of the QDs decreases.[1] CdSe/ZnS luminescent QDs are characterized by broad UV absorption and size-dependent narrow emission peaks in the visible range of the electromagnetic spectrum.[2] CdSe/ZnS QDs are commonly synthesized in organic solvents using a one-pot chemical reaction. For CdSe/ZnS QDs, nucleation and crystal growth occur at high temperature, when CdO and chalcogen elements (selenium) are added to the coordinating TOP (trioctylphosphine) and TOPO (trioctylphosphine oxide) ligand in an organic solvent. To increase their quantum yield, QDs are passivated with a ZnS layer and capped with organic ligands of TOPO.[3,4] This results in the formation of hydrophobic highly luminescent QDs that are miscible in organic solvents like hexane and chloroform but not in aqueous solutions.

Luminescent QDs could be considered as a viable alternative to organic fluorophores in bioassays due to their high emission quantum yield (brightness), broad absorption, size-dependent tuneable emission peaks, and high photostability.[5] However, to apply QDs in the aqueous environment of bioassays, the hydrophobic ligands that cap the CdSe/ZnS QDs must be replaced with hydrophilic ligands to enable their solubilization in aqueous solutions. Several alternatives have been used to realize this goal. These include thiol ligand exchange reactions to replace the hydrophobic TOPO ligands with dihydrolipoic acid (DHLA),[6] 16-mercaptohexadecanoic acids (MHDAs),[7] mercaptoacetic acid (MMA),[3] oligomeric phosphines,[8] and organic dendrons.[9] Other aqueous solubilization methods include the encapsulation of QDs in a layer of amphiphilic diblock or triblock copolymers[2,10] and coating the QDs with polymer shells or silane shells. Polymer or silane coatings are particularly attractive for in vivo applications because they prevent the dissolution of toxic Cd^{2+} ions. Modifying the silane shell of QDs with polyethylene glycol (PEG) further decreased the cytotoxicity of QDs.[11] Hydrophilic luminescent QDs have been used as luminescent labels in biomedical applications such as cellular assays,[12] fluorescence-based immunoassays,[13] in situ DNA hybridization (fluorescence in situ hybridization [FISH]),[14] and in vivo studies.[15]

19.2.1 Fluorescence Resonance Energy Transfer

FRET has been demonstrated as a powerful tool in studying dynamic interactions between molecular fluorophores.[16] The seminal work of Tsien and coworkers who used green fluorescent protein and their mutants in various applications including FRET donors and acceptors in the studying of protein dynamics was recognized with a Nobel Prize in Chemistry in 2008.[17] While traditional FRET experiments make use of molecular fluorophores as donors and acceptors, it has been shown recently that luminescent CdSe/ZnS QDs could be used as donors in FRET measurements.[18] It was also shown that when gold nanocrystals are used as acceptors, the Förster distance is doubled, and the FRET interactions follow a $1/R^4$ distance dependence instead of the $1/R^6$ distance dependence, which is typically observed in traditional FRET measurements involving molecular donors and acceptors.[19] In our laboratory, we demonstrated the utility of CdSe/ZnS luminescent QDs as FRET donors in enzymatic assays for protease activity.[20] Other groups have applied QD-based FRET assays for the analysis of binding proteins[21] and DNA.[22] The following sections describe the development and application of unique QD-based FRET assays in which QDs are incorporated into larger platforms rather than used in solution-based assays.

19.2.2 Quantum Dot FRET-Based Probes for the Detection of Tumor-Associated Trypsinogen in Pancreatic Cancer Cells

In this section, we discuss the development of luminescent CdSe/ZnS QD-based FRET sensors for monitoring cellular proteolytic activity of CAPAN-2 adenocarcinoma pancreatic cells in a PDMS microfluidic device. The FRET probe was fabricated using LbL deposition of polyelectrolytes to form a polyelectrolyte multilayer (PEM) film. Poly(allylamine hydrochloride) (PAH) as positive and polystyrene sulfonate (PSS) as negative polyelectrolyte were used to assemble the PEM film.[23] In particular, the luminescent CdSe/ZnS QDs used as donors and rhodamine–neurotensin–PAH as acceptor bound to a substrate were embedded into the PEM film. The luminescent QDs and acceptor molecules were separated by two layers of polyelectrolytes. Previous studies in our laboratory showed that the FRET efficiency between QD donors and molecular acceptors peaks at this separation distance.[24] While CdSe/ZnS QDs provide multiple spectroscopic advantages, their toxicity to biological systems is still troublesome.[25] Their stability and cytotoxicity mainly depend on their coating; for instance, the degradation of the mercaptoundecanoic acid (QDs–COOH) capping ligands of QDs could induce severe cytotoxicity.[26] In our FRET sensing platform, embedding the CdSe/ZnS QDs in the PEM film prevented direct contact between the QDs and the CAPAN-2 cells, which largely alleviated toxicity concerns. The PEM films were grown in microfluidic channels to facilitate FRET-based assays of volume-limited samples. The microfluidic channels were fabricated using standard microlithography printing in PDMS.[23] It should be noted that microfluidic devices have been used for cell culture and cellular assays due to the in vivo–like conditions in these devices.[27]

Alternate deposition of polycations PAH and polyanions PSS can be performed on a negatively charged PDMS surface or directly on a negatively charged surface of treated glass slides. Negatively charged QDs were embedded in the layer of PAH polyelectrolyte to ensure strong electrostatic attraction with PAH and electrostatic repulsion with the underlying negative surface. This minimized diffusion and increased the stability of QDs in the PEM film. Rhodamine-labeled neurotensin molecules were covalently conjugated to the upper layer of PAH in the PEM film to prevent diffusion of the rhodamine acceptors into the PEM film, which would alter FRET interactions between the QDs and molecular acceptors (Figure 19.1).

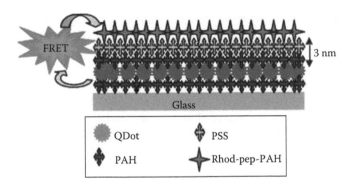

FIGURE 19.1 (**See color insert.**) Schematic illustration of the QD–rhodamine FRET-based enzymatic sensor on glass or PDMS substrate. The QD FRET donors were embedded in PEMs, while neurotensin-labeled rhodamine molecular acceptors were conjugated to top PAH layer. The arrow indicates the distance between the donor and acceptor layer, which is approximately 3 nm.

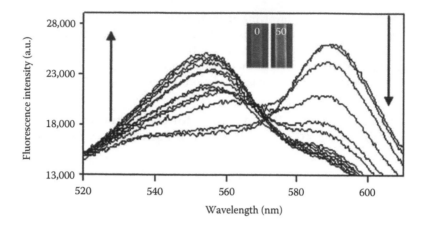

FIGURE 19.2 Emission spectra of the QD FRET-based sensor prior to adding trypsin and when incubated with 0.25 mg/mL trypsin for 50 min. The fluorescence intensity of the rhodamine molecules decreases, while the fluorescence intensity of the QDs increases. The inset box represents the fluorescence image of the sensor prior to adding trypsin and after 50 min incubation with trypsin.

A solution containing 0.25 mg/mL trypsin was flown through the microchannels. We monitored for 1 h the effect of trypsin incubation on the FRET signal between the QDs that were embedded in the PEM film and the rhodamine acceptors that were conjugated to the upper PAH layer of the film during trypsin proteolysis. The fluorescence intensity of the QDs increased, while the rhodamine fluorescence decreased, as the proteolytic cleavage of neurotensin molecules occurred on the PEM film surface.[24] Digital fluorescence images of the PEM film prior to and following the addition of trypsin, shown in the insert of Figure 19.2, support this observation by showing a clear change in the emission color of the PEM film from red orange, representative of FRET emission, to green, representative of QD emission, due to the release of rhodamine molecules to the solution during the course of the enzymatic cleavage reaction (Figure 19.2, inset box).

Previously, Reyes et al. demonstrated the applicability of PEM films that were composed of PAH and PSS for cell micropatterning mainly due to their biocompatibility and adhesion properties. The neuronal (retinal) cells used in their system proliferated without changing the morphology, proving that the polymers were not toxic.[23] Based on these findings, we grew CAPAN-2 cells in microfluidic devices and used the FRET assays to carry out real-time measurements of their proteolytic activity. To allow adhesion and growth of the cells on the PEM film without being flushed out during routine microinjection maintenance, the microfluidic chamber was designed with the following parameters: 2 mm length and width and 80 μm height. Cells were loaded in the microchannel and allowed to attach with a concentration of approximately 2.5×10^6 cells/mL (to be published).

The microfluidic device was used to detect the proteolytic activity of CAPAN-2 pancreatic cancer cells derived from a cell line. CAPAN-2 cells secrete a tumor-associated trypsinogen (TAT) and a trypsinogen activity-stimulating factor (TASF). Expression and activity of the TAT enzyme were shown previously to be associated with the degree of invasiveness and metastasis of CAPAN-2 cells. These enzymes degrade the extracellular matrix and could also activate other proteases with relevance in metastasis.[28] Early diagnosis of the pancreatic cancer is difficult due to the nonspecificity of its symptoms. Trypsinogen is a precursor of the enzyme trypsin that is activated by enterokinase/enteropeptidase. Activated trypsin hydrolyzes at the carboxyl terminal of arginine or lysine residues in peptide molecules. Trypsin is usually responsible for breaking down proteins in the digestive tract. However, a trypsin-like protease in the cell membrane of tumor cells, TAT1 or TAT2, was also identified.[29] TAT is thus associated with the proteolytic degradation of the extracellular membrane (ECM), during tumor invasion.[28,30,31]

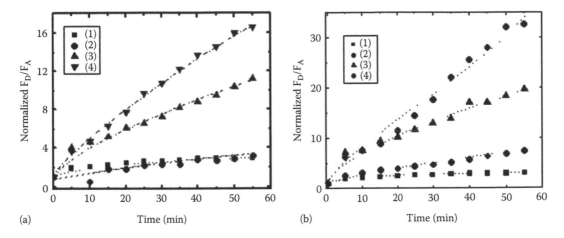

FIGURE 19.3 Temporal dependence of F_D/F_A of the QD-based enzyme sensor when incubated with spontaneous trypsin (a) and total trypsin/trypsinogen that was released by CAPAN-2 cells (b). The activity of total trypsin (b) was determined following the activation of trypsinogen with 2 unit/mL of enterokinase. Control experiments to monitor F_D/F_A in the absence of cells (1) and F_D/F_A measurements during incubation with cells for 12 h (2), 24 h (3), and 48 h (4) for both spontaneous and total trypsin are shown in (a) and (b), respectively. The flow rate of cell growth medium through the microfluidic channels was 1.5 μL/h.

In our study, we investigated the temporal dependence of the ratio F_D/F_A at increasing concentrations of trypsin that was released by the cells. F_D is the fluorescence of the QD donors, and F_A is the fluorescence intensity of the rhodamine acceptor molecules. The results are summarized in Figure 19.3. Figure 19.3a shows the temporal dependence of F_D/F_A for spontaneous trypsin. This means that the activity of the secreted trypsin was measured with no alteration. Figure 19.3b shows the temporal dependence of F_D/F_A following brief incubation of the secreted spontaneous trypsin (a mixture of trypsinogen and trypsin) with 2 unit/mL of enterokinase, which converts all the trypsinogen to trypsin. The ratio F_D/F_A increased over time and the change was the highest when the cells were allowed to secrete and accumulate trypsinogen/trypsin for 48 h prior to the proteolytic measurements. As expected, the activity of spontaneous trypsin (trypsinogen activated by cells) (Figure 19.3a) was lower than the activity of total trypsin (Figure 19.3b).

To determine the activity of the total trypsin, we incubated the samples secreted by the cells overnight with 0.1 μL/mL enterokinase. The ratio F_D/F_A, which is a measure of trypsin activity, increased by about twofolds. Control experiments using growth medium in the absence of trypsin showed only a slight change in the FRET efficiency (Figure 19.3a and b). This proves that the FRET system responds only to the enzymatic activity of trypsin. Viability tests demonstrated that embedding QD donors in PEMs reduced their cytotoxic effects significantly.

19.2.3 Quantum Dot–Liposome Complexes as FRET Probes for the Detection of Phospholipase A_2 Activity

A different type of QD-based FRET sensing platform that was recently developed in our laboratory involves the incorporation of QDs and molecular acceptors into liposomes and their use to measure the enzymatic activity of phospholipase A_2. Phospholipases A_2 (EC 3.1.1.4) are lipolytic extracellular enzymes that are relatively small proteins (ca. 14 kDa). Phospholipases whether in membrane-associated or in soluble forms play an important role in membrane turnover,[32] lipid signaling,[33] and in mitochondrial oxidative stress.[34] Phospholipase A_2 catalyzes the hydrolysis of the *sn*-2 fatty acyl ester bond of phospholipids resulting in arachidonic acid and a lysophospholipid. Arachidonic acid metabolic products influence inflammatory episodes, asthma, bronchoconstriction, and cardiovascular diseases[35]; lysophospholipids could also contribute to cardiovascular diseases.[36] While sensitive detection techniques that are based on radiometric or fluorometric methods are available,[37] they are expensive, hazardous, laborious, and cannot be easily used in screening applications.

Our new phospholipase A_2 activity sensing platform involves the use of unilamellar liposomes. Liposomes are highly versatile vesicles that have been used frequently as drug delivery vehicles primarily due to their high biocompatibility. In liposomes, the hydrophobic phospholipid bilayer encloses an aqueous phase where, for instance, fluorescent and staining reagents could be encapsulated and used for cellular and tissue labeling.[38] Moreover, recent studies showed that it is possible to encapsulate luminescent CdSe/ZnS QDs in liposomes and apply them for staining of solid tumors. The staining efficiency of solid tumors with QDs encapsulating liposomes was higher than the staining efficiency of the tumors with free QDs.[39] Liposomes have been also used for drug delivery of cancer, inflammation, hormonal drugs, etc.[40]

As previously mentioned, we recently prepared QD-based FRET sensing liposomes for measuring the activity of phospholipase A_2. CdSe/ZnS QDs were encapsulated into 1,2-dioleoyl-*sn*-glycero-3-phosphocholine (DOPC)/1,2-dioleoyl-*sn*-glycero-3-phosphoglycerol (DOPG)/cholesterol (1:1:0.4), where DOPC and DOPG were intended to ensure negative charge on the liposome membrane and unsaturation was necessary for phospholipase A_2 enzyme activity. The phospholipid membrane of the liposomes contained acyl NBD molecules, NBD C_6-HPC, with 10 μM final concentration that served as molecular acceptors in the FRET assays of phospholipase A_2 activity. The QDs encapsulating liposomes show their characteristic emission (Figure 19.4a), while the QD–liposomes labeled with NBD on the acyl chain of phospholipids emit bluish green light due to FRET interactions between the QDs and NBD molecules (Figure 19.4b). When phospholipase A_2 cleaves the *sn*-2 acyl bond of phospholipids, it affects the release of NBD acceptors from liposome membrane; consequently, FRET interactions no longer occur and the blue emission color of the QDs encapsulating liposomes is restored to blue (Figure 19.4c) (data not published).

Figure 19.5 shows the temporal dependence of the ratio between the luminescence intensity of the QDs (F_D) and the luminescence intensity of the NBD acceptors (F_A) during phospholipase A_2 cleavage of NBD molecules from the liposome membrane. The ratio F_D/F_A increases over time and the rate of cleavage increases with increasing phospholipase A_2 activity ranging from 0 to 75 U/mL (Figure 19.5).

The QD–NBD FRET sensing liposomes were used to screen the inhibition efficiencies of two phospholipase A_2 inhibitors: 1-hexadecyl-3-(trifluoroethyl)-*sn*-glycero-2-phosphomethanol (MJ33) and 3-(4-octadecyl)benzoylacrylic acid (OBAA). Future studies will focus on multiplexing the liposome assays in a 96-well plate platform in order to rapidly screen a larger number of inhibitors simultaneously. This will provide a new diagnostic tool of great biomedical significance.

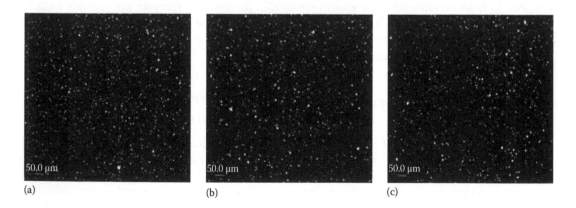

(a)   (b)   (c)

FIGURE 19.4 Digital fluorescence microscopic images of QD–liposomes showing blue emission (a), QD–NBD–liposomes showing bluish green emission due to FRET between QDs and NBD (b), and QD–NBD–liposome-based probes after incubating for 30 min at 75 U/mL phospholipase A_2 enzyme activity (c).

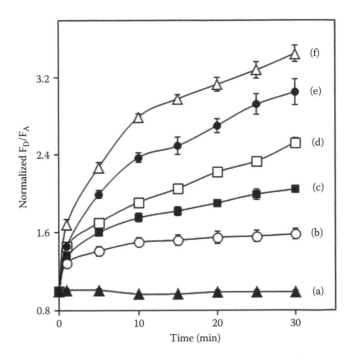

FIGURE 19.5 Time dependence of QD–NBD–liposome-based probes at increasing phospholipase A_2 enzyme activity: (a) 0, (b) 1.5, (c) 7.5, (d) 15, (e) 37.5, and (f) 75 U/mL. The ratio F_D/F_A was normalized to the ratio of F_D/F_A prior to adding phospholipase A_2 to the QD–liposome solutions. F_D represents fluorescence intensity of the donor and F_A represents fluorescence intensity of the acceptor.

19.2.4 Gold Nanoparticle–Quantum Dot–Polystyrene Microspheres as FRET Probes for Thiol Detection

In this section, we describe the development of new microparticle bioconjugates and their use in organic fluorophore-free FRET assays. By exploiting multiple covalent and hydrophobic interactions between QD nanocrystals, AuNPs, and PS microspheres, these microsensing platforms circumvent previous problems, which were encountered when luminescent QDs were used in solution-based FRET assays that involved the aggregation of QDs in the presence of proteins. In this study, CdSe/ZnS QDs and AuNPs were coattached to the surface of 6–8 μm PS microspheres to form AuNP–QD–PS conjugates that could be used as FRET probes in which the QDs served as luminescent donors and the AuNPs as acceptors.[7] To facilitate the covalent attachment of QDs to the PS particle surface, the hydrophobic TOPO ligands of the QDs were initially replaced by MHDA, to enable their solubility in aqueous solutions. Further binding of polyhistidine peptide (H₂N-K-K-H-H-H-H-H-E-E-CO₂H) to the QD–MHDA surface increased their stability in aqueous solution through multiple bindings of histidines to the ZnS surface. The binding of polyhistidines to the surface of the QDs occurs through Zn coordination. The 5 nm QD–His particles in excess, to ensure complete surface coverage, were then conjugated to the surface of carboxyl-modified PS particles via standard EDC (1-ethyl-3-(3-dimethylaminopropyl) carbodiimide hydrochloride) chemistry. The QD–PS microspheres show significantly higher emission than PS particles that were coated with molecular fluorophores. This was due to self-quenching of the molecular fluorophores, when in close proximity to one another on the surface of the PS microspheres.[7]

AuNPs of various sizes are versatile nanoparticles with well-defined spectroscopic characteristics. They are inexpensive and easy to prepare. AuNPs have been used in a wide range of applications due to their biocompatibility and lack of toxicity. Several conjugation chemistry techniques were used to attach AuNPs to sensing platforms. Many of these conjugation strategies involve thiol chemistry, taking advantage of the self-assembly of thiolated ligands on gold surfaces to form highly versatile and chemically stable coatings. Bifunctional reagents like cystamine molecules have been used to enable the conjugation of biomolecules to gold surfaces. Cystamine molecules bind to gold surface through their thiol groups, while their free amino groups enable further conjugation of molecules to the gold surface.[41] In our studies, we attached cystamine linkers (H_2N-CH_2-CH_2-S-S-CH_2-CH_2-NH_2) to the AuNP surface and used an EDC coupling protocol to conjugate the AuNPs to the surface of the QD–PS to form the AuNP–QD–PS FRET sensing particles.

Transmission electron microscopy (TEM) measurements of the AuNP–QD–PS particles show the presence of CdSe/ZnS and gold nanocrystals on the surface of the PS microspheres. The AuNPs and QDs averaged about 5 nm in diameter with uniform size distribution within 10% variation. The EDS data show characteristic peaks of Au on the surface of QD–PS (Figure 19.6).

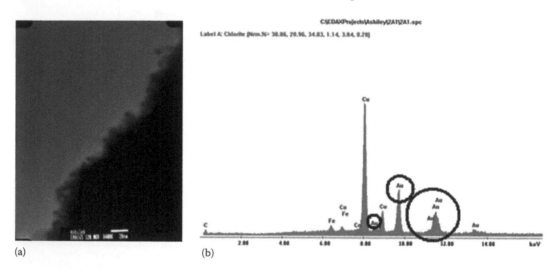

(a)   (b)

FIGURE 19.6 TEM image (a) and EDS analysis (b) of an AuNP–QD–PS microsphere that shows the presence of Au (black circles) nanoparticles on PS–QD surface.

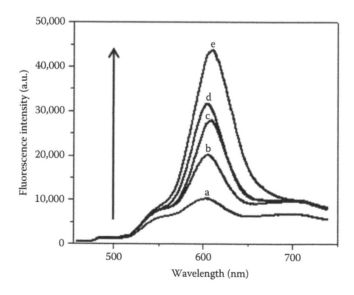

FIGURE 19.7 Luminescence intensity of AuNP–QD–PS microspheres when incubated with 5 mg/mL DTT for (a) 0, (b) 10, (c) 30, and (d) 60 min. Positive control (e) shows the high-luminescence QD–PS microspheres.

FRET interactions between the CdSe/ZnS QDs and AuNPs on the surface of the PS microspheres were possible due to the close proximity (<10 nm) between the nanoparticles. These interactions led to quenching of fluorescence intensity of QDs on the surface of the PS microspheres. Incubation of the composite particles with 5 mg/mL dithiothreitol (DTT) for 1 h at room temperature restored the emission of the QDs since the AuNPs were removed from the AuNP–QD–PS microsphere surface through thiol displacement chemistry (Figure 19.7).

19.3 Summary and Conclusions

This chapter discussed the development of various QD FRET-based sensing platforms, which significantly expand the capabilities of luminescent QDs beyond the capabilities of solution-based QD assays. The first section of the chapter summarizes the incorporation of luminescent QDs into microfluidic devices. The successful combination of luminescent QD nanotechnology with the technology of microfluidics opens many new possibilities for the analysis of volume-limited samples in microfluidic devices that combine separation and optical detection. Here, we described the fabrication of FRET sensing films in microfluidic devices using the highly versatile LbL deposition method and the application of these devices for real-time measurement of the proteolytic activity of CAPAN-2 cells, which are associated with the metastasis of pancreatic cancer. The second section of this chapter describes the preparation of FRET sensing liposomes by incorporating luminescent QDs and molecular acceptors of acyl NBD into their phospholipid membrane. Liposomes, which have been largely employed in drug delivery applications, find a new optochemical sensing application in our study. FRET interactions between the QDs and NBD molecules lead to a decrease in the blue emission intensity of the QDs and an increase in the blue-green emission intensity of the NBD molecules. The blue emission of the QDs is restored when the NBD molecules are removed from the liposome membrane via enzymatic cleavage of phospholipase A$_2$. The third section of this chapter describes the formation of FRET sensing microspheres in which no organic fluorophores are used as donors or acceptors. This results in a highly stable FRET sensing platform, free of photodecomposition. The FRET probes are prepared by the covalent coattachment of luminescent QDs and AuNPs to the surface of PS microspheres. The emission of the QDs is quenched by the AuNPs and restored when the AuNPs are displaced from the microparticle surface by ligands with stronger affinity to the surface than the thiolated AuNPs. All these studies make extensive use of CdSe/ZnS luminescent QDs and molecular acceptors or AuNP quenchers. The unique FRET sensing platforms presented in this chapter are easily adaptable to multiplexed assays. They show significant improvement in analytical properties, compared to current fluorescence assays, particularly due to a vast improvement in emission quantum yield (brightness) and chemical stability. They are also superior to solution-based assays that make use of luminescent QDs. Future studies will focus on the multiplexing aspects of these QD-based FRET sensors and on the application of these FRET sensors for the analysis of biomarkers of clinical significance in body fluids.

References

1. Andersen, K.E.; Fong, C.Y.; Pickett, W.E. Quantum confinement in CdSe nanocrystallites. *J Non-Cryst Solids* 2002, *292–302*, 1105–1110.
2. Michalet, X.; Pinaud, F.F.; Bentolila, L.A.; Tsay, J.M.; Doose, S.; Li, J.J.; Sundaresan, G.; Wu, A.M.; Gambhir, S.S.; Weiss, S. Quantum dots for live cells, in vivo imaging, and diagnostics. *Science* 2005, *307*(5709), 538–544.
3. Chan, W.C.; Nie, S. Quantum dot bioconjugates for ultrasensitive nonisotopic detection. *Science* 1998, *281*(5385), 2016–2018.
4. Murray, C.B.; Norris, D.J.; Bawendi, M.G. Synthesis and characterization of nearly monodisperse CdE (E = S, Se, Te) semiconductor nanocrystallites. *J Am Chem Soc* 1993, *115*, 8706–8715.
5. Gao, X.; Yang, L.; Petros, J.A.; Marshall, F.F.; Simons, J.W.; Nie, S. In vivo molecular and cellular imaging with quantum dots. *Curr Opin Biotechnol* 2005, *16*(1), 63–72.

6. Clapp, A.R.; Goldman, E.R.; Mattoussi, H. Capping of CdSe-ZnS quantum dots with DHLA and subsequent conjugation with proteins. *Nat Protoc* 2006, *1*(3), 1258–1266.

7. Quach, A.D.; Crivat, G.; Tarr, M.A.; Rosenzweig, Z. Gold nanoparticle–quantum dot–polystyrene microspheres as fluorescence resonance energy transfer probes for bioassays. *J Am Chem Soc* 2011, *133*(7), 2028–2030.

8. Kim, S.; Bawendi, M.G. Oligomeric ligands for luminescent and stable nanocrystal quantum dots. *J Am Chem Soc* 2003, *125*(48), 14652–14653.

9. Wang, Y.A.; Li, J.J.; Chen, H.; Peng, X. Stabilization of inorganic nanocrystals by organic dendrons. *J Am Chem Soc* 2002, *124*(10), 2293–2298.

10. Medintz, I.L.; Uyeda, H.T.; Goldman, E.R.; Mattoussi, H. Quantum dot bioconjugates for imaging, labelling and sensing. *Nat Mater* 2005, *4*(6), 435–446.

11. Zhang, T.; Stilwell, J.L.; Gerion, D.; Ding, L.; Elboudwarej, O.; Cooke, P.A.; Gray, J.W.; Alivisatos, A.P.; Chen, F.F. Cellular effect of high doses of silica-coated quantum dot profiled with high throughput gene expression analysis and high content cellomics measurements. *Nano Lett* 2006, *6*(4), 800–808.

12. Liu, T.; Liu, B.; Zhang, H.; Wang, Y. The fluorescence bioassay platforms on quantum dots nanoparticles. *J Fluoresc* 2005, *15*(5), 729–733.

13. Sheng, Z.; Han, H.; Hu, D.; Liang, J.; He, Q.; Jin, M.; Zhou, R.; Chen, H. Quantum dots-gold(III)-based indirect fluorescence immunoassay for high-throughput screening of APP. *Chem Commun (Camb)* 2009, (18), 2559–2561.

14. Ioannou, D.; Tempest, H.G.; Skinner, B.M.; Thornhill, A.R.; Ellis, M.; Griffin, D.K. Quantum dots as new-generation fluorochromes for FISH: An appraisal. *Chromosome Res* 2009, *17*(4), 519–530.

15. Texier, I.; Josser, V. In vivo imaging of quantum dots. *Methods Mol Biol* 2009, *544*, 393–406.

16. Jockusch, S.; Marti, A.A.; Turro, N.J.; Li, Z.; Li, X.; Ju, J.; Stevens, N.; Akins, D.L. Spectroscopic investigation of a FRET molecular beacon containing two fluorophores for probing DNA/RNA sequences. *Photochem Photobiol Sci* 2006, *5*(5), 493–498.

17. Tsien, R.Y. The green fluorescent protein. *Annu Rev Biochem* 1998, *67*, 509–544.

18. Clapp, A.R.; Medintz, I.L.; Mattoussi, H. Förster resonance energy transfer investigations using quantum-dot fluorophores. *ChemPhysChem* 2006, *7*, 47–57.

19. Yun, C.S.; Javier, A.; Jennings, T.; Fisher, M.; Hira, S.; Peterson, S.; Hopkins, B.; Reich, N.O.; Strouse, G.F. Nanometal surface energy transfer in optical rulers, breaking the FRET barrier. *J Am Chem Soc* 2005, *127*(9), 3115–3119.

20. Shi, L.; De Paoli, V.; Rosenzweig, N.; Rosenzweig, Z. Synthesis and application of quantum dots FRET-based protease sensors. *J Am Chem Soc* 2006, *128*, 10378–10379.

21. Medintz, I.L.; Clapp, A.R.; Mattoussi, H.; Goldman, E.R.; Fisher, B.; Mauro, J.M. Self-assembled nanoscale biosensors based on quantum dot FRET donors. *Nat Mater* 2003, *2*, 630–638.

22. Boeneman, K.; Deschamps, J.R.; Buckhout-White, S.; Prasuhn, D.E.; Blanco-Canosa, J.B.; Dawson, P.E.; Stewart, M.H. et al. Quantum dot DNA bioconjugates: Attachment chemistry strongly influences the resulting composite architecture. *ACS NANO* 2010, *12*, 7253–7266.

23. Reyes, D.R.; Perruccio, E.M.; Becerra, S.P.; Locascio, L.E.; Gaitan, M. Micropatterning neuronal cells on polyelectrolyte multilayers. *Langmuir* 2004, *20*(20), 8805–8811.

24. Crivat, G.; Da Silva, S.M.; Reyes, D.R.; Locascio, L.E.; Gaitan, M.; Rosenzweig, N.; Rosenzweig, Z. Quantum dot FRET-based probes in thin films grown in microfluidic channels. *J Am Chem Soc* 2010, *132*(5), 1460–1461.

25. Geys, J.; Nemmar, A.; Verbeken, E.; Smolders, E.; Ratoi, M.; Hoylaerts, M.F.; Nemery, B.; Hoet, P.H.M. Acute toxicity and prothrombotic effects of quantum dots: Impact of surface charge. *Environ Health Perspect* 2008, *116*(12), 1607–1613.

26. (a) Hoshino, A.; Fujioka, K.; Oku, T.; Suga, M.; Sasaki, Y.F.; Ohta, T.; Yasuhara, M.; Suzuki, K.; Yamamoto, K. Physicochemical properties and cellular toxicity of nanocrystal quantum dots depend on their surface modification. *Nano Lett* 2004, *4*(11), 2163–2169; (b) Derfus, A.M.; Chan, W.C.W.; Bhatia, S.N. Probing the cytotoxicity of semiconductor quantum dots. *Nano Lett* 2004, *4*(1), 11–18.

27. (a) Kim, L.; Toh, Y.C.; Voldman, J.; Yu, H. A practical guide to microfluidic perfusion culture of adherent mammalian cells. *Lab Chip* 2007, *7*(6), 681–694; (b) Regehr, K.J.; Domenech, M.; Koepsel, J.T.; Carver, K.C.; Ellison-Zelski, S.J.; Murphy, W.L.; Schuler, L.A.; Alarid, E.T.; Beebe, D.J. Biological implications of polydimethylsiloxane-based microfluidic cell culture. *Lab Chip* 2009, *9*(15), 2132–2139.

28. Uchima, Y.; Sawada, T.; Nishihara, T.; Umekawa, T.; Ohira, M.; Ishikawa, T.; Nishino, H.; Hirakawa, K. Identification of a trypsinogen activity stimulating factor produced by pancreatic cancer cells: Its role in tumor invasion and metastasis. *Int J Mol Med* 2003, *12*(6), 871–878.

29. Bjartell, A.; Paju, A.; Zhang, W.-M.; Gadaleanu, V.; Hansson, J.; Landberg, G.; Stenman, U.-H. Expression of tumor-associated trypsinogens (TAT-1 and TAT-2) in prostate cancer. *The Prostate* 2005, *64*(1), 29–39.

30. Uchima, Y.; Sawada, T.; Hirakawa, K. Action of antiproteases on pancreatic cancer cells. *Pancreas* 2007, *8*(4 Suppl), 479–487.

31. Uchima, Y.; Sawada, T.; Nishihara, T.; Maeda, K.; Ohira, M.; Hirakawa, K. Inhibition and mechanism of action of a protease inhibitor in human pancreatic cancer cells. *Pancreas* 2004, *29*(2), 123–131.

32. Farooqu, A.A.; Horrocks, L.A. Plasmalogens, phospholipase A2, and docosahexaenoic acid turnover in brain tissue. *J Mol Neurosci* 2001, *16*(2–3), 263–272; discussion 279–284.

33. Cathcart, M.K. Signal-activated phospholipase regulation of leukocyte chemotaxis. *J Lipid Res* 2009, *50*(Suppl), S231–S236.

34. Kinsey, G.R.; McHowat, J.; Beckett, C.S.; Schnellmann, R.G. Identification of calcium-independent phospholipase A2gamma in mitochondria and its role in mitochondrial oxidative stress. *Am J Physiol Renal Physiol* 2007, *292*(2), F853–F860.

35. (a) Samuelsson, B. Arachidonic acid metabolism: Role in inflammation. *Z Rheumatol* 1991, *50*(Suppl 1), 3–6; (b) Kuhn, H.; Chaitidis, P.; Roffeis, J.; Walther, M. Arachidonic acid metabolites in the cardiovascular system: The role of lipoxygenase isoforms in atherogenesis with particular emphasis on vascular remodeling. *J Cardiovasc Pharmacol* 2007, *50*(6), 609–620.

36. Morris, A.J.; Panchatcharam, M.; Cheng, H.Y.; Federico, L.; Fulkerson, Z.; Selim, S.; Miriyala, S.; Escalante-Alcalde, D.; Smyth, S.S. Regulation of blood and vascular cell function by bioactive lyso-phospholipids. *J Thromb Haemost* 2009, *7*(Suppl 1), 38–43.

37. (a) Beckett, C.S.; Kell, P.J.; Creer, M.H.; McHowat, J. Phospholipase A2-catalyzed hydrolysis of plasmalogen phospholipids in thrombin-stimulated human platelets. *Thromb Res* 2007, *120*(2), 259–268; (b) Krzystanek, M.; Trzeciak, H.I.; Krzystanek, E.; Malecki, A. Fluorometric assay of oleate-activated phospholipase D isoenzyme in membranes of rat nervous tissue and human platelets. *Acta Biochim Pol* 2010, *57*(3), 369–372.

38. Dudu, V.; Ramcharan, M.; Gilchrist, M.L.; Holland, E.C.; Vazquez, M. Liposome delivery of quantum dots to the cytosol of live cells. *J Nanosci Nanotechnol* 2008, *8*(5), 2293–2300.

39. (a) Al-Jamal, W.T.; Al-Jamal, K.T.; Bomans, P.H.; Frederik, P.M.; Kostarelos, K. Functionalized-quantum-dot-liposome hybrids as multimodal nanoparticles for cancer. *Small* 2008, *4*(9), 1406–1415; (b) Al-Jamal, W.T.; Al-Jamal, K.T.; Tian, B.; Cakebread, A.; Halket, J.M.; Kostarelos, K. Tumor targeting of functionalized quantum dot-liposome hybrids by intravenous administration. *Mol Pharm* 2009, *6*(2), 520–530.

40. (a) Gregoriadis, G.; Senior, J.; Wolff, B.; Kirby, C. Targeting of liposomes to accessible cells in vivo. *Ann N Y Acad Sci* 1985, *446*, 319–340; (b) Maurer, N.; Fenske, D.B.; Cullis, P.R. Developments in liposomal drug delivery systems. *Expert Opin Biol Ther* 2001, *1*(6), 923–947; (c) Weiner, A.L.; Carpenter-Green, S.S.; Soehngen, E.C.; Lenk, R.P.; Popescu, M.C. Liposome-collagen gel matrix: A novel sustained drug delivery system. *J Pharm Sci* 1985, *74*(9), 922–925.

41. Wirde, M.; Gelius, U. Self-assembled monolayers of cystamine and cysteamine on gold studied by XPS and voltammetry. *Langmuir* 1999, *15*, 6370–6378.

20

Fluorescent Molecular Beacon Nucleic Acid Probes for Biomolecular Recognition

Cuichen Wu
University of Florida

Chaoyong
James Yang
Xiamen University

Kemin Wang
Hunan University

Xiaohong Fang
Chinese Academy of Science

Terry Beck
TriLink BioTechnologies, Inc.

Richard Hogrefe
TriLink BioTechnologies, Inc.

Weihong Tan
*University of Florida
and
Hunan University*

Fluorescent probes for biomolecular recognition are of great importance in the fields of chemistry, biology, and medical sciences, as well as biotechnology. These probes have been used for mechanism studies of biological functions and in ultrasensitive detection of biological species responsible for many diseases [1]. In the postgenome era, quantitative studies of genomic information for disease diagnosis and prevention and drug discovery will be fast-growing areas of research and development. This has led to a continued demand for advanced biomolecular recognition probes with high sensitivity and high specificity. Molecular beacon (MB), a recently developed single-stranded DNA (ssDNA) molecule [2], appears to be a very promising probe for quantitative genomic studies. MBs are hairpin-shaped oligonucleotides that contain both fluorophore and quencher moieties. They act like switches that are normally closed to bring the fluorophore/quencher pair together to turn fluorescence *off*. When prompted to undergo conformational changes that open the hairpin structure, the fluorophore and the quencher are separated, and fluorescence is turned *on*. MBs were first developed in 1996. Since then, they have provided a variety of exciting opportunities in DNA/RNA/protein studies both in solution and inside living cell specimens.

20.1 Principle of Molecular Beacons

20.1.1 Molecular Recognition and Fluorescence Resonance Energy Transfer

The hybridization of a nucleic acid strand to its complement target is one of the most specific well-known molecular recognition events. A MB makes use of this unique feature for DNA/RNA/protein studies. A MB is a synthetic short oligonucleotide, which possesses a loop and stem structure, as schematically shown in Figure 20.1a.

The signal transduction mechanism of MBs for molecular recognition is based on fluorescence resonance energy transfer (FRET) [1]. The loop portion of a MB is a probe sequence complementary to a target nucleic acid molecule. The arm sequences flanking either side of the probe are complementary to each other but are unrelated to the target sequence. These arm sequences, which have five to seven base pairs, anneal to form the MB's stem. A fluorescent moiety is covalently attached to the end of one

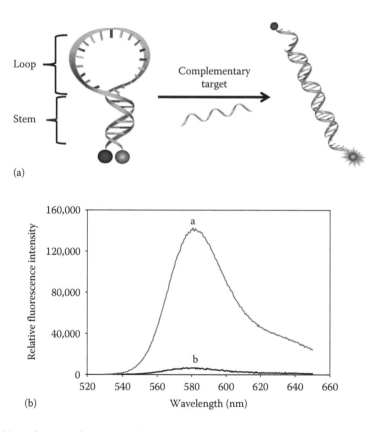

FIGURE 20.1 (a) Mechanism of operation of MB DNA probes. On their own, these molecules are nonfluorescent since the stem hybrids keep the fluorophores close to the quenchers. MBs emit intense fluorescence only when the stems are apart through hybridization of DNA molecules with sequences complementary to their loop sequences or unwinding the stem hybrids upon interaction with SSB or by using denaturing reagents or under other denaturing conditions. (b) Fluorescence spectra of MBs obtained at room temperature. Spectra were taken for MB hybridization with complementary DNA molecules (a) and with noncomplementary DNA molecules (b).

arm and a quenching moiety is covalently attached to the other end. The fluorescent dye serves as an energy donor, and the nonfluorescent quencher plays the role of an acceptor. The stem keeps these two moieties in close proximity to each other, causing the fluorescence of the fluorophore to be quenched by energy transfer. Thus, the probe is unable to fluoresce. When the probe encounters a target molecule, the loop forms a hybrid that is longer and more stable than the stem. The MB undergoes a spontaneous conformational change that forces the stem apart. The fluorophore and the quencher are moved away from each other, leading to the restoration of fluorescence. Unhybridized MBs do not fluoresce; thus, it is not necessary to remove them to observe hybridized probes.

Two forms of energy transfer may exist in MBs: direct energy transfer and FRET [3]. Direct energy transfer requires contact between the two moieties in a MB. The collision between the fluorophore and the quencher can distort the energy level of the excited fluorophore, causing the fluorescence of the fluorophore to be quenched by energy transfer. In brief, the quenching moiety dissipates the energy that it receives from the fluorophore as heat, rather than emitting it as light. The other mechanism, FRET, can occur between the two moieties in a MB with a relatively long distance (20–100 Å). It requires a spectral overlap between the donor's emission spectrum and the acceptor's absorption spectrum. The rate of FRET is directly proportional to the inverse sixth power of the separation distance of the donor and the acceptor or the quencher [1]. Both forms of the energy transfer are strongly dependent on the distance between the dye moieties. Therefore, the spatial separation of the fluorophore and the quencher in a MB determines the energy transfer efficiency. When a target DNA hybridizes to a MB, a substantial increase of the fluorescence signal is observed (Figure 20.1b) due to larger separation distance between the two moieties in a MB. As DABCYL (4-(4′-dimethylaminophenylazo)benzoic acid) has been found to efficiently quench a large variety of fluorophores independent of the spectral overlap, direct energy transfer may be dominant in MBs [4].

20.1.2 Advantages of Molecular Beacon Probes

The hybridization of a nucleic acid strand to its complement has been widely used in many areas of research and development. There have been different types of DNA fluorescent probes designed for these applications. For example, nucleic acid blotting techniques have been used to make great strides in our understanding of gene organization and function. In these techniques, a solid support is used to immobilize DNA fragments [5]. A labeled oligonucleotide probe containing the sequence of interest is then used to hybridize with its counterpart. In today's laboratories, nucleic acid techniques are increasingly being harnessed for use in practical applications such as the molecular diagnosis of disease. But what if one needs to monitor the real-time synthesis of nucleic acids as it is happening? Or what if your research calls for the labeling of nucleic acids within living cells? Methods based on immobilization of hybridized nucleic acids, probes requiring intercalation reagents, or other means requiring the isolation of probe–target complexes cannot be used for these applications. MBs can overcome these difficulties.

The inherent fluorescent signal transduction mechanism enables a MB to function as a sensitive probe with high signal-to-background ratio for real-time monitoring. Its fluorescence intensity can increase more than 200 times when it meets the target under optimal conditions [4]. This provides the MBs with a significant advantage over other fluorescent probes in ultrasensitive analysis. With this inherent sensitivity, individual MB DNA molecules have been imaged and their hybridization process has been monitored on a single-molecule basis [6]. MBs can be used in situations where it is either not possible or desirable to isolate the probe–target hybrids from an excess of the unhybridized probes, such as in real-time monitoring of polymerase chain reactions (PCRs) in sealed tubes or in detection of mRNAs within living cells. The usefulness of *detection without separation* for these applications cannot be over-emphasized. This feature enables the synthesis of nucleic acids to be monitored as it is occurring, in either sealed tubes or in living specimen, without additional manipulations.

Another major advantage of MBs is their molecular recognition specificity. They are extraordinarily target specific, ignoring nucleic acid target sequences that differ by as little as a single nucleotide. While current techniques for routine detection of single-base-pair DNA mutations are often labor intensive and time consuming [7], MBs provide a simple and promising tool for the diagnosis of genetic disease and for gene therapy study. This specificity of MBs comes from its loop and stem structure. The stem hybrid acts as a counterweight for the loop hybrid. Experiments have shown that the range of temperatures in which perfect complementary DNA targets form hybrids, but mismatched DNA targets do not, is significantly wider for MBs than the corresponding range observed for conventional linear probes [8]. Therefore, MBs can easily discriminate DNA targets that differ from one another by a single nucleotide. Thermodynamic studies revealed that the enhanced specificity is a general feature of structurally constrained DNA probes.

20.2 Molecular Beacon Synthesis and Characterization

The synthesis of MB is similar to that of dual labeling a short oligomer with two dyes. The length of the loop sequence (15–40 nucleotides) is chosen so that the probe–target hybrid is stable at the temperature of probing. The stem sequences (five to seven nucleotides) should be strong enough to form the hairpin structure for efficient fluorescence quenching while still weak enough to be dissociated when a complementary DNA hybridizes with the loop of the MB. Also, the stem sequence must be designed so as not to interfere with the probe sequences. Since DABCYL can serve as a universal quencher for many fluorophores [4], MB is generally synthesized using DABCYL-CPG (controlled pore glass) as starting material. Different fluorescent dye molecules can be covalently linked to the 5′-end to report fluorescence at different wavelengths. There are carbon chain linkers between the bases and the labeled dye molecules. The stem and the linker keep the fluorophore and the quencher in close proximity and increase the probability for their direct contact. There are four important steps in this synthesis [2]. First, a CPG solid support is derivatized with DABCYL and used to start the synthesis at the 3′-end of the oligonucleotide. The rest of the nucleotides are added sequentially, using standard cyanoethyl phosphoramidite chemistry. Second, a primary amine group at the 5′-end is linked to the phosphodiester bond by a six-carbon spacer arm. There is a trityl protecting group at the ultimate 5′-end that protects the amine group. Third, the oligonucleotide is hydrolyzed and removed from the CPG, then purified by reversed-phase high-performance liquid chromatography (HPLC). Fourth, the purified oligonucleotide is removed off the trityl group and labeled with a fluorophore. After labeling, the excess dye is removed by gel-filtration chromatography on Sephadex G-25. The oligonucleotide is then purified again by reversed-phase HPLC. The product peak from the HPLC is collected. The synthesized MB is characterized by ultraviolet (UV) and by mass spectroscopy [9]. The purification of the MB after synthesis is critically important to ensure a high signal-to-noise ratio and to achieve ultrahigh sensitivity. A detailed protocol for MB synthesis can be found on the Internet at http:\www.Molecular-beacon.org. There are also approximately 10 commercial companies specializing in custom synthesis of MBs. Today, MBs with specific sequences are readily available at affordable costs and without tedious synthesis by individual investigators.

20.2.1 Synthesis and Purification of Molecular Beacons

MB synthesis is commercially available now. Even though there are many companies in the synthesis of MBs, the techniques are similar. Here, we give a description of a typical method for MB synthesis (from http://www.molecular-beacons.org). The starting material for the synthesis of MBs is an oligonucleotide that contains a sulfhydryl group at its 5′-end and a primary amino group at its 3′-end. DABCYL is coupled to the primary amino group utilizing an amine-reactive derivative of DABCYL. The oligonucleotides that are coupled to DABCYL are then purified. The protective trityl moiety is then removed from the 5′-sulfhydryl group and a fluorophore is introduced in its place using an iodoacetamide derivative. Recently, a CPG column that introduces a DABCYL moiety at the 3′-end of an oligonucleotide has become available, which enables the synthesis of an MB completely on a DNA synthesizer. The whole sequence of the MB used throughout this protocol is shown as follows: fluorescein-5′-GCGAGCTAGGAAACACCA AAGATGATAT TTGCTCGC-3′-DABCYL, where the underlines identify the arm sequences.

20.2.1.1 Coupling of DABCYL

1. Dissolve 50–250 nanomoles of dry oligonucleotide in 500 μL of 0.1 M sodium bicarbonate, pH 8.5. Dissolve about 20 mg of DABCYL succinimidyl ester (Molecular Probes) in 100 μL *N,N*-dimethylformamide and add to a stirring solution of the oligonucleotide in 10 μL aliquots at 20 min intervals. Continue stirring for at least 12 h.

2. Remove particulate material by spinning the mixture in a microcentrifuge for 1 min at 10,000 rpm. In order to remove unreacted DABCYL, pass the supernatant through a gel exclusion column. Equilibrate a Sephadex G-25 column (NAP-5, Pharmacia) with buffer A, load the supernatant, and elute with 1 mL buffer A. Filter the elute through a 0.2 μm filter (Centrex MF-0.4, Schleicher & Schuell) before loading on the HPLC column.

3. Purify the oligonucleotides on a C-18 reversed-phase column (Waters), utilizing a linear elution gradient of 20%–70% buffer B in buffer A and run for 25 min at a flow rate of 1 mL/min. Monitor the absorption of the elution stream at 260 and 491 nm. A typical chromatogram is shown in Figure 20.2. Collect the peak that absorbs in both wavelengths, which contains oligonucleotides with a protected sulfhydryl group at their 5′-end and DABCYL at their 3′-end (peak D).

4. Precipitate the collected material with ethanol and salt, spin in a centrifuge for 10 min at 10,000 rpm, discard the supernatant, dry the pellet, and dissolve it in 250 μL buffer A. As explained in Figure 20.2, we will be able to collect the desired portion of the synthesized DNA probes.

20.2.1.2 Coupling of Fluorophore

1. In order to remove the trityl moiety, add 10 μL of 0.15 M silver nitrate and incubate for 30 min. Add 15 μL of 0.15 M dithiothreitol to this mixture and shake for 5 min. Spin for 2 min at 10,000 rpm and transfer the supernatant to a new tube. Dissolve about 40 mg 5-iodoactamidofluorescein (Molecular Probes) in 250 μL of 0.2 M sodium bicarbonate, pH 9.0, and add it to the supernatant. Incubate the mixture for 90 min. Each of these solutions should be prepared just before use.

2. Remove excess fluorescein from the reaction mixture by gel exclusion chromatography and purify the oligonucleotides coupled to fluorescein by HPLC, following the instructions in steps 2 and 3

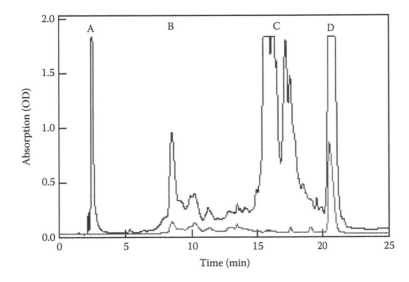

FIGURE 20.2 Chromatographic separation of oligonucleotides coupled to DABCYL. The black line represents absorption at 260 nm, and the gray line represents absorption at 491 nm. The oligonucleotides in peaks A and B do not contain trityl moieties, whereas the oligonucleotides in peaks C and D are protected by trityl moieties. The oligonucleotides in peaks B and D are coupled to DABCYL, whereas the oligonucleotides in peaks A and C are not coupled to DABCYL. Peak D should be collected.

of the previous section. A sample chromatogram is shown in Figure 20.2. Collect the fractions corresponding to peak F, which absorb at wavelengths 260 and 491 nm and are fluorescent when observed with a UV lamp in a dark room. If a different fluorophore is coupled in place of fluorescein, its maximum absorption wavelength should be used instead of 491 nm.

3. Precipitate the collected material and dissolve the pellet in 100 μL TE buffer. Determine the absorbance at 260 nm and estimate the yield (1 OD_{260} = 33 μg/mL).

20.2.1.3 Automated Synthesis

1. Use a CPG column to introduce DABCYL (Glen Research) at the 3'-end of the oligonucleotide during automated synthesis. At the 5'-end of the oligonucleotide, either a thiol or an amino modifier can be introduced for a subsequent coupling to a fluorophore or a fluorophore can directly be introduced during automated synthesis using a phosphoramidite. The 5' modifiers and fluorophores should remain protected with a trityl moiety during the synthesis. Perform postsynthetic steps as recommended by the manufacturer of the DNA synthesizer. Dissolve the oligonucleotide in 600 μL buffer A.

2. When the fluorophore is to be introduced manually, purify the oligonucleotide protected with a trityl moiety. Remove the trityl moiety from the purified oligonucleotide and continue with the coupling of the fluorophore, as described previously.

3. When a 5' fluorophore is introduced via automated synthesis, purify the oligonucleotide protected with the trityl moiety and then remove the trityl moiety from the purified oligonucleotide. Precipitate the MB with ethanol and salt and dissolve the pellet in 100 μL TE buffer. Determine the absorbance at 260 nm and estimate the yield.

20.2.2 Characterization of Molecular Beacon

20.2.2.1 Characterization of Molecular Beacon by Mass Spectrometry

Once the MBs are synthesized, different methods will be used to evaluate the quality of the product. One of the effective methods is to run a mass spectrometry and to determine the molecular weight. Usually, an aliquot of the MB product collected from HPLC was taken out for analysis by mass spectrometry. Matrix-assisted laser desorption/ionization time-of-flight mass spectrometer was used to confirm the molecular weight of the MB. For example, the calculated molecular weight of a MB used in biosensor development is 10,076. As shown in Figure 20.3, the main peak in the mass spectrum was 10,085.7 Da. The difference between the measured value and the calculated one is only about 0.1%, which is less than the typical error, 0.2%, in mass spectrum measurement. Another small peak appeared near a position corresponding to half of the molecular weight of the MB. This was due to the molecules that lost two electrons in the desorption process during the measurement. The mass spectrum results support the synthesis of the designed MB.

20.2.2.2 Molecular Beacon's Hybridization Activity

MBs can be tested by hybridization to evaluate their activity in DNA/RNA reaction. As an example, a newly synthesized MB has been used for DNA hybridization in solution. The MB's hybridization properties were tested using fluorescence measurements performed on a SPEX Industries F-112A spectrophotometer. A submicro quartz cell was used for the hybridization experiment. Two 200 μL sample solutions were prepared: the MB and a fivefold molar excess of its complementary DNA and the MB and a fivefold molar excess of a noncomplementary DNA. The concentrations of MB in the two solutions were 50 nM. They were incubated for 20 min in the hybridization buffer (20 mM Tris–HCl, 50 mM KCl, and 5 mM $MgCl_2$, pH = 8.0). Emission spectra were recorded at room temperature with excitation at 515 nm for the MB labeled with fluorescein. The hybridization of the MB inside the solution has shown strong fluorescence signal when ssDNA molecules reacted with their complementary DNA molecules.

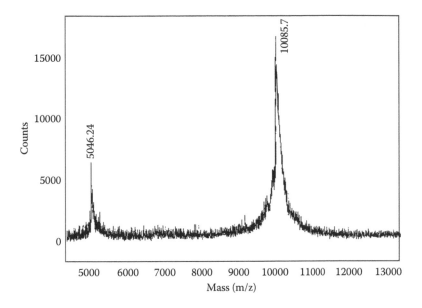

FIGURE 20.3   Mass spectrum of the synthesized biotinylated MB.

Theoretically, the enhancement could be as high as more than 200 folds with optimal design of the sequence and under optimal hybridization and optical detection conditions. The solution with the non-complementary DNA has no enhancement under the same conditions. Hybridization dynamics of the MB has also been investigated. This experiment clearly shows that the MB synthesized is what we have designed and can be used for DNA/RNA studies.

20.2.3 Molecular Beacon Lifetime Measurements

Many studies have been undertaken to effectively design and understand the structure of MBs. In addition, many studies have begun to probe how the MB binds to its target. The biggest question in MB research is the opening/closing mechanism of the probe as well as its specificity toward its DNA/RNA targets. Out of many potentially useful techniques, lifetime measurement is an idea to probe the state of the fluorophore in MBs in either close or open forms. It will help in understanding the opening/closing mechanism of the MB. This technique is being used to study MB and its cDNA interaction as well as MB at high pH and different temperatures in expectation of the overall comprehending of MBs.

The frequency-domain lifetime instrument was first studied with rhodamine 6G (R6G). The average lifetime on this instrument was found to be 4.06 ns. Lifetime measurements were taken on the MB at a pH of 12.5. At this pH, the MB exhibited two lifetimes having the average lifetimes of 4.17 and 1.61 ns. It is a well-known fact that DNA can be affected by a change in pH. At high pH, the bases undergo alkaline hydrolysis causing denaturation, and a low pH would cause the cleavage of bases. In addition, a high pH causes the stem to unwind to open the MB restoring fluorescence. The two lifetime values obtained suggest two conformations for the MB at high pH (shown in Figure 20.4), the open state (unquenched R6G) and the close state (quenched R6G). Lifetime measurements are difficult to be obtained directly for MBs due to the quenching of DABCYL on R6G. Specific measurements have to be taken to enhance the signal.

The effects of temperature on lifetime measurements were studied with the MB. The lifetimes at the melting point (T_m) of the MB were found to be 3.89 and 0.67 ns. At a high temperature of 85°C, the lifetime of the MB was found to be 4.01 ns. Studies were also done involving MB targets such as cDNA and single-stranded binding (SSB) protein at varying concentration ratios. At a MB to cDNA concentration ratio of 1:1, lifetime values of 4.23 and 0.75 ns were obtained. For the following concentration ratios of 1:4 and 1:8 of MB and

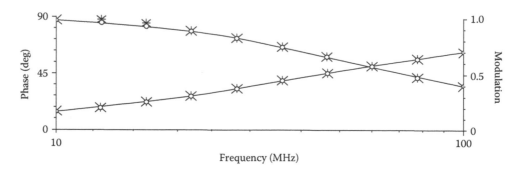

FIGURE 20.4 Lifetime scan of MB at high pH.

cDNA, lifetime values of 3.91 and 3.97 were found, respectively. A lifetime of 3.87 was found for the MB binding to SSB in a 1:2 concentration ratio. It is seen that the open forms and the closed forms can be studied using lifetime measurements due to the distance separating the quencher and the fluorophore. Future detailed experiments to obtain the precise lifetime values of MBs will help determine the overall mechanism of MBs.

20.3 Molecular Engineering of Molecular Beacon

20.3.1 Reduction of Molecular Beacon Background Fluorescence

Although MBs are designed for their specific complementary targets, incomplete quenching can occur due to a variety of reasons. First, the probe itself cannot be perfectly quenched even by the close proximity of acceptor and donor, thus limiting signal enhancement. Second, false-positive signals arise from degradation by nucleases or nonspecific binding of proteins. Third, traditional MBs easily suffer from interruption of the stem structure. This can be explained by (1) the complicated cellular environment, in which chances abound for undesired intermolecular interactions between stems and their complementary sequences, or (2) the thermodynamic conformational switch between hairpin and nonhairpin structures. To improve the signal-to-background ratio of MBs, the most straightforward method involves increasing the number of quenchers. By the molecular assembly of different numbers of quenchers on one end of MBs, while keeping only one fluorophore on the other end, Yang et al. achieved high sensitivity and specificity [10]. Multiple quenchers improve absorption efficiency and increase the probability of dipole–dipole coupling between the quenchers and fluorophore (Figure 20.5). The quenching efficiency of DABCYL increased as the number of DABCYL moieties increased: 92.9% for single DABCYL, 98.75% for dual DABCYLs, and 99.7% for triple DABCYLs, as a superquencher (SQ). Such SQ MB assemblies demonstrated a 320-fold fluorescence enhancement upon a target binding, a significant improvement compared to a single quencher with only 14-fold enhancement. SQ-labeled MBs showed great sensitivity, higher thermal stability, and slightly improved specificity compared to regular MBs. This strategy can also be used for other nucleic acid probes, such as aptamers [11–13], which generated a 49,000-fold signal increment when platelet-derived growth factor (PDGF) aptamers bound to PDGF proteins.

Negative signals of MBs typically result from sticky-end pairing between hybridized MBs [14]. This was solved with the introduction of the hybrid molecular probe (HMP) developed by Yang et al. [15] Two ssDNA sequences, each complementary to part of the target DNA, were linked by a flexible poly(ethylene glycol) (PEG) spacer. The fluorescence acceptor and donor moieties are labeled on each terminus. Upon hybridization to target, the 5′-and 3′-ends of the HMP are brought into close proximity, resulting in a FRET signal. False-positive signals due to nucleases and nonspecific binding to proteins were greatly reduced even in cancer cell lysate. Compared to conventional MBs, HMPs have intrinsic advantages. First, its special loop–stem structure, which is based on the sequence of target nucleic acid (NA), is easier to design. Second, while MBs are hindered by the energy barrier of the self-complementary stem structure, which slows down hybridization kinetics, HMPs respond to target DNA/RNA more rapidly due to the absence of stem

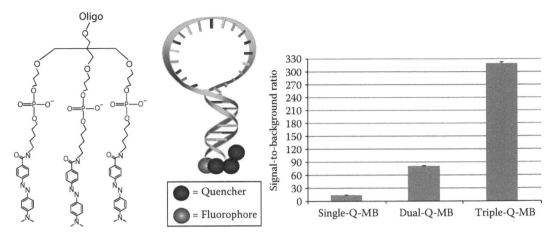

FIGURE 20.5 Schematic of a MB conjugated with an SQ consisting of triple DABCYLs. The graph shows the signal-to-background ratio of MBs as the number of quenchers increases.

structures in HMP. Third, although unmodified MBs cannot avoid false-positive signals or nonspecific protein binding, HMP can easily overcome these obstacles by linking two oligonucleotides with a PEG spacer.

The incorporation of unnatural enantiomeric L-DNA in the stem of a MB is another strategy to prevent the occurrence of false-positive signals caused by the undesired intermolecular interactions between stems and their complementary sequences (Figure 20.6a) [16]. While L-DNA and D-DNA have identical physical properties, they cannot form stable duplex structures as expected for D-DNA complementary strands. MBs with D-DNA loop and L-DNA stem have better sensitivity and stability,

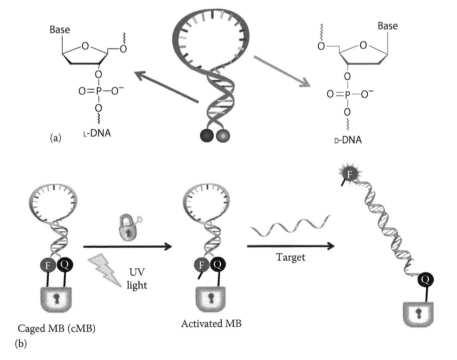

FIGURE 20.6 (a) Schematic of MB using L-DNA for the stem part (light colored) and D-DNA for the loop part (dark colored). (b) Principle of cMBs locked by covalent bonding or biotin–avidin interaction via PC linkage. After light illumination, activated MBs will recover the hybridization to complementary target.

for example, higher signal-to-background ratio and melting temperature. More importantly, MBs with L-DNA modified stems can effectively prevent false-positive signals caused by nonspecific hybridization of D-DNA sequences to the stems of conventional MBs with D-DNAs.

Incomplete quenching can also be checked by locking the stem of a MB with a photolabile molecular interaction or covalent bond. Without light irradiation, the light-activatable MBs are inactive, even in the presence of target sequence. After unlocking with a quick light illumination, the decaged MBs recover their ability to hybridize to complementary DNA/RNA. Inspired by this design, Wang et al. made use of a biotin–avidin interaction or triazole to lock the stem of MBs via a photocleavable (PC) linker bearing an o-nitrobenzyl moiety [17]. The caged molecular beacons (cMBs) have lower background fluorescence based on the tighter distance between fluorophore and quencher that results from the covalent linkage or high-affinity interaction in the stem part (Figure 20.6b). This photocaged technique will find wide application in the study of gene expression, protein synthesis, and cell signaling with high temporal and spatial resolution.

Apart from the molecular probe itself, significant background interference also arises from the native fluorescence in complex biological fluids. Species in the physiological environment can have a strong autofluorescence background, which may reduce the sensitivity of NA probes. To address this issue, Yang et al. molecularly engineered NA probes with a spatially sensitive fluorescent dye, such as pyrene, to monitor proteins, RNA, and small molecules in complex biological environments [18–20]. Excited-state dimers (excimers) are formed when an excited-state pyrene encounters a ground-state pyrene [21]. The excimer emission is a broad, featureless band centered at 480–500 nm, which can be easily differentiated from the pyrene monomer that emits in the range from 370 to 400 nm. The excimer also has a very long fluorescence lifetime compared to other potential fluorescent species (as much as 100 ns or longer), while most biological background species have lifetimes of at most 5 ns. In the case of pyrene-labeled MBs, varied numbers of pyrene molecules are conjugated on the 5′-end of the MB sequence (Figure 20.7) [19]. In the absence of complementary DNA, the fluorescence of the pyrene monomer and excimer is quenched by the close proximity of pyrenes and DABCYL. However, the pyrene excimer fluorescence is restored after the introduction of cDNA, which induces opening of the loop and hence separates the pyrenes from DABCYL. Compared to FAM-labeled MBs, MBs labeled with multiple pyrenes have higher signal enhancement after addition of equimolar target. More importantly, time-resolved fluorescence was able to differentiate the fluorescence signal from the pyrene-labeled probe and complex biological species, for example, cell growth media. During the first 10 ns, the excimer emission spectra were hidden by the severe background fluorescence from cell media, similar to the emission spectrum of steady-state measurement. However, because of the different lifetimes among pyrene excimer, pyrene monomer, and background fluorescence, the signal from pyrene excimer emission could be differentiated from the intense background interference 40 ns after the excitation pulse. In the chosen time window, much of the excimer emission still occurred, while most of the background autofluorescence had decayed [19,20]. Using time-resolved methods, multiple pyrene-labeled MBs have the potential for sensitive measurement of low-nanomolar target DNA in complex biological environments.

Conjugated polymers (CPs) are polyunsaturated macromolecules in which all backbone atoms are sp or sp^2 hybridized. They are known to exhibit photoluminescence with high quantum efficiency [21]. A unique and attractive property of fluorescent CPs is their fluorescence superquenching effect [22,23], allowing a hundred- to a millionfold more sensitivity to fluorescence quenching compared to that of their low-molecular-weight analogues. Among these CPs, water-soluble poly(pheinylene ethynylene)s (PPEs) are particularly attractive candidates for optical biosensing applications by their high-fluorescence quantum yields in aqueous solutions [24]. PPEs can be prepared through palladium (Pd)-catalyzed cross-coupling of bis-acetylenic and diiodoaryl monomers in an amine environment [25]. After synthesis of MBs on a DNA synthesizer through solid-phase phosphoramidite chemistry, a 5I-dU residue is introduced into each MB as a monomer of polymerization, followed by cross-coupling of the polymer chain with MBs (Figure 20.8) [26]. Based on the superquenching property of the CP, the CP-modified MBs have greatly amplified the signal-to-background ratio compared to traditional MBs.

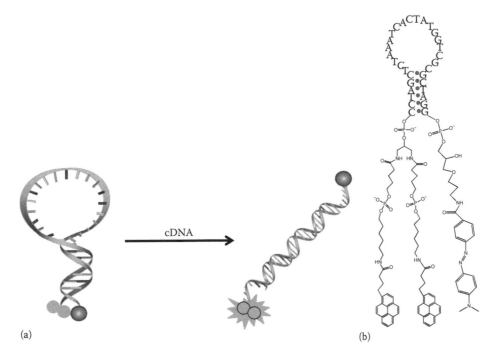

FIGURE 20.7 (a) Scheme of a dual-pyrene-labeled MB hybridized with complementary target sequence (light gray ball = pyrene; dark gray ball = DABCYL quencher). (b) Chemical structure of dual-pyrene-modified MB with pyrene monomer and DABCYL on the 5'- and 3'-ends, respectively.

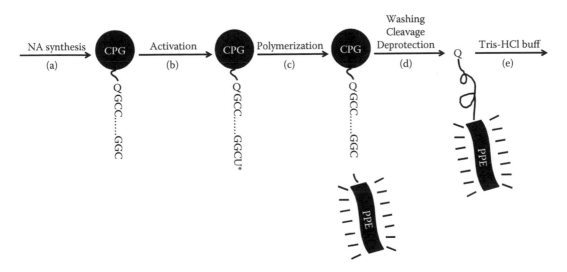

FIGURE 20.8 Schematic solid-state synthesis procedure of PPE-labeled MBs (Q: DABCYL quencher).

20.3.2 Biostability Enhancement of Molecular Beacon

Intracellular nuclease degradation and nonspecific protein binding thwart the use of traditional NA probes. To solve this problem, many chemically modified nucleotides have been proposed to increase the biostability of MBs and prevent false-positive signals. For example, Wang et al. designed a MB using a locked nucleic acid (LNA) base, which has a methylene bridge connecting the 2'-oxygen of the ribose

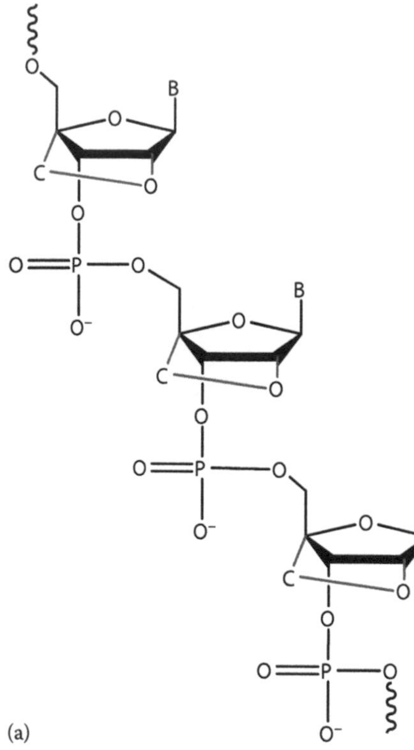

(a)

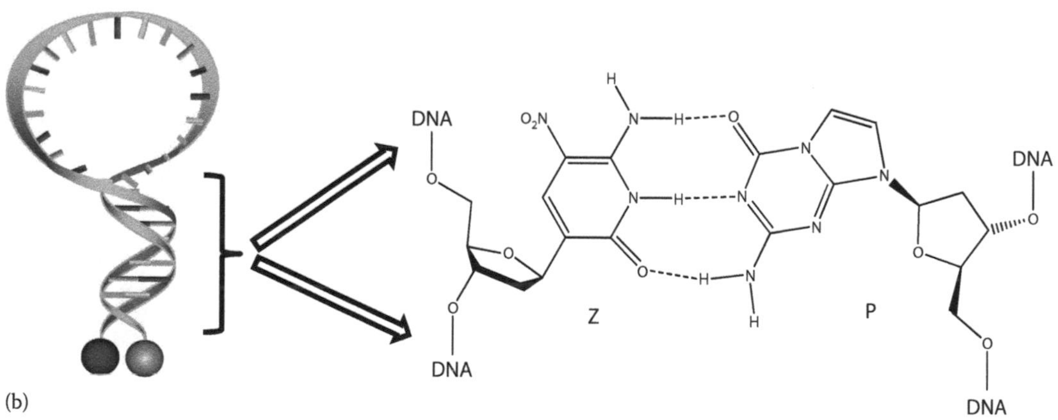

(b)

FIGURE 20.9 (a) Chemical structure of LNA. (b) Schematic of MBs with a dZ/dP-modified stem.

and the 4′-carbon (Figure 20.9a) [27]. The LNA possesses unique properties relative to a normal nucleotide. First, the LNA–LNA duplex has tighter binding and maintains a stable structure at 95°C. Second, LNA MBs are superior to DNA MBs in discriminating single-base mismatches. Finally, LNA MBs can resist interference by nonspecific proteins, such as single-stranded DNA-binding protein (SSB) and the degradation by nucleases in the cell environment. However, the hybridization kinetics of LNA MBs is relatively slow compared to DNA MBs. Therefore, Yang et al. synthesized DNA/LNA chimeric MBs, which significantly improved the hybridization rates and maintained resistance to nonspecific protein binding and nuclease digestion [28].

Artificial nucleotides, which rely on an artificially expanded genetic information system (AEGIS), have also been used to design MBs. Sheng et al. synthesized 6-amino-5-nitro-3-(1′-beta-D-2′-deoxyribofuranosyl)-2(1*H*)-pyridone (dZ) and 2-amino-8-(1′-beta-D-2′-deoxyribofuranosyl)-imidazo[1,2-a]-1,3,5-triazin-4(8*H*)-one (dP) as the AEGIS pair and incorporated this pair into the stem part of MBs (Figure 20.9b). The dZ/dP pair-modified MBs have excellent enzymatic resistance compared to normal MBs, as well as a hybridization interaction stronger than that of the dC/dG pairs, which provides the potential for effective discrimination against mismatched bases in short DNA duplexes [29].

20.4 Molecular Beacon as Fluorescent Probes for DNA/RNA Monitoring

20.4.1 DNA/RNA Polymerase Chain Reaction Detection

20.4.1.1 Real-Time Monitoring of Polymerase Chain Reaction

MB probes are suitable tools for real-time monitoring of DNA/RNA amplification during PCR [2,4]. They can be simply added to a sealed PCR tube. The fluorescence signal is monitored at the annealing step of every cycle. At the annealing temperature, the target amplicons' products bind to MB to generate fluorescence, while the unbound MB remains in the closed form without fluorescence. The MB–amplicon hybrids dissociate at elevated temperature where MBs do not interfere with polymerization. The fluorescence signal increases with the increased number of cycling and it directly indicates the concentration of the amplicons in the PCR process. The MB assay is fast, sensitive, and nonradioactive. The PCR tubes are sealed during the entire measurement, thus avoiding carryover contamination. Compared to TaqMan probes [30], which are another type of fluorescent probe used in PCR monitoring, MBs give more reliable genotyping results, especially in a GC-rich region. Moreover, MBs allow for sensitive and quantitative detection of low-abundance sequence variants over a wider range. For example, *Chlamydophila felis* infection of cats was detected by a MB to identify the major outer membrane protein gene. The detection limit is fewer than 10 genomic copies [31].

20.4.1.2 Gene Typing and Mutation Detection for Disease Study and Diagnosis

For the detection of genetic mutations, a method called *spectral genotyping* has been developed [7]. The principle involves two different MB probes with different loop sequences: one specific for a wild-type allele and the other for a mutant allele. These two MBs also have two different fluorophores. The fluorescence measured at the two emission wavelengths during amplification indicates whether the samples are homozygous wild type, homozygous mutants, or heterozygote. The MB-based PCR mutation detection method has been used for the study of many diseases [32–39], one example being acquired immunodeficiency syndrome (AIDS). The polymorphisms in the gene for human CC-chemokine receptors CCR5 and CCR2, which are associates of human immunodeficiency virus type 1 (HIV-1), have also been studied [34,36,37,40]. Using MBs, people are able to investigate the HIV disease mechanism, progression process, and therapeutic effects.

MB-based PCR is also a promising tool for rapid and reliable clinical diagnosis. An example is the method developed to assay four pathogenic retroviruses: HIV-1, HIV-2, and human T-lymphotrophic virus types I and II [36]. The retroviral DNA sequences were amplified by simultaneous PCR, which contained four sets of primers and four MBs; each one being specific for one of the four amplicons and labeled with a different colored fluorophore. The color of the fluorescence generated in the course of amplification identified which retroviruses were present. The number of thermal cycles required for the generation of each color was used to measure the number of copies of each retroviral sequence originally present in the sample. Fewer than 10 retroviral genomes can be detected. Moreover, 10 copies of a rare retrovirus can be detected in the presence of 100,000 copies of an abundant retrovirus. There were no false positives for 96 clinical samples. This method will be useful in screening donated blood and transplantable tissue.

It is known that MBs have excellent characteristics for mutation detection. Therefore, sequence analysis of HIV-1 from 74 persons with acute infections identified 8 strains with mutations in the reverse transcriptase (RT) gene by real-time nucleic acid sequence–based amplification assay with MBs [41]. The results illustrate that infection with nucleoside analogue-resistant HIV leads to newly infected individuals with mutants that are sensitive to nucleoside analogues, but it is only a single mutation removed from drug-resistant HIV.

MB-based, real-time PCR allele discrimination has been used to detect all three chemokine receptor mutations, which are associated with HIV-1 disease [42]. These spectral genotyping assays to genotype 3923 individuals from a globally distributed set of 53 populations. The results showed that CCR2-641 and CCR5-59653T genetic variants were found in almost all populations studied, and their allele frequencies are greatest (similar to 35%) in Africa and Asia but decrease in Northern Europe.

The fast one-tube assay that identifies and distinguishes among all subtypes, A, B, and C, and circulating recombinant forms (CRFs), AE and AG, of HIV-1 has been developed by using subtype-specific MBs with multiple fluorophores. The lower detection level of the assay was approximately 10^3 copies of HIV-1 RNA per reaction. However, the assay in this format would not be suitable for clinical use but could possibly be used for epidemiological monitoring as well as vaccine research studies [43]. Meanwhile, all HIV-1 groups, subtypes, and CRFs were detected and quantified with equal efficiency by the addition of MBs to the amplification reaction. The lower level of quantification was 100 copies of HIV-1 RNA with a dynamic range of linear quantification between 10^2 and 10^7 RNA molecules [44]. In conclusion, the real-time-monitored HIV-1 assay is a fast and sensitive assay with a large dynamic range of quantification and is suitable for quantification of most if not all subtypes and groups of HIV-1.

20.4.1.3 Bacteria Detection

Bacteria exist almost all over the world. Scientists have found bacteria in many different environments. The simplest detection method for bacteria is through observation by microscopy. To identify different kinds of bacteria, florescence strain method associated with PCR has been developed. Last year, MB PCR was demonstrated to show positive results more rapidly than traditional agarose gel electrophoresis analysis of PCR products. The use of MBs allows real-time monitoring of PCRs, and the closed-tube format allows simultaneous detection and confirmation of target amplicons without the need for agarose gel electrophoresis and/or Southern blotting. Therefore, MBs have become a powerful tool for bacteria detection. In order to do so, the MB to hybridize with a target sequence of the bacteria is incorporated into PCRs containing DNA extract from *Escherichia coli* in artificially contaminated skim milk [45]. The bacteria can then be sensitively analyzed.

Based on the MB, a rapid and simple homogeneous fluorescence PCR assay was developed for the clinical diagnosis of infectious diseases. This method could reproducibly detect *Mycobacterium tuberculosis* at the 10 bacteria/mL level with a higher sensitivity than the traditional methods. The feasibility of this method was further supported by successful detection of *Neisseria gonorrhoeae* and *Chlamydia trachomatis* [46]. Another analysis method for bacteria based on MBs was also demonstrated by a real-time PCR assay to detect the presence of *E. coli* O157:H7 [45]. MBs were designed to recognize a 26 bp region of the *rfbE* gene, encoding for an enzyme necessary for O-antigen biosynthesis. The specificity of the MB-based PCR assay was very high. All *E. coli* serotype O157 tested were positively identified, while all other species, including the closely related O55, were not detected by the assay. Positive detection of *E. coli* O157:H7 was demonstrated when $>10^2$ colony-forming units (CFUs)/mL were present in the samples. The capability of the assay to detect *E. coli* O157:H7 in raw milk and apple juice was demonstrated. The sensitivity was as few as 1 CFU/mL after 6 h of enrichment. These assays could be carried out entirely in sealed PCR tubes, enabling rapid and semiautomated detection of *E. coli* O157:H7 in food and environmental samples [47]. Further development was proposed on a real-time PCR assay to detect the presence of *Salmonella* species using MB, which was designed to recognize a 16 bp region on the amplicon of a 122 bp section of the himA. As few as 2 CFUs per PCR could be detected. The high selectivity of MBs made it feasible to detect similar species such as *E. coli* and *Citrobacter freundii* in real-time PCR assays [38].

20.4.2 mRNA Monitoring Inside Living Cells

Obtaining the knowledge on the subcellular localization and cellular transport pathway of RNAs is of crucial importance for our understanding of basic cell biological and development process. It is also of great interest for the study of functional genomics in the postgenome era. Up to now, little has been known about the rate at which RNAs are synthesized and processed as well as their pathways in the cell. In situ hybridization (ISH) techniques have been successfully used in the past few decades for the microscopic detection of specific mRNA molecules in fixed and pretreated cells and tissues. However, the fixed cell may not really reflect the situation in living cells. Fixation and cell pretreatment also prevent us from studying the dynamics of RNA synthesis and the transport process when they are actually occurring in the cell. MB-based fluorescence in vivo hybridization (FIVH) offers a new way to overcome the limitations of traditional ISH for RNA monitoring.

MBs have been reported to visualize the basic fibroblast growth factor (bFGF) mRNA in human trabecular cells [48], β-actin mRNA in K562 human leukemia cells [9], and PTK2 kangaroo rat kidney cells [49]. A successful application of MB in mRNA detection and localization mainly depends on the following three factors: rational design of a MB probe for the specific mRNA, efficient introduction of the MB probe into the cell, and optimization of the experimental conditions for fluorescence imaging and detection. To choose an appropriate sequence on the target mRNA, which will be most accessible to the MB probe hybridization, and to choose a corresponding MB loop and stem, which will avoid the secondary structure of the loop sequence, are the major concerns in the MB probe design. The computer analysis of the RNA and MB folding structure can be of help for the optimal design. Microinjection and liposome delivery have been used as effective tools for the transfer of the MB probes into living cells while maintaining the physiological and structural integrity of the cells. However, special care must be taken to introduce an appropriate amount of MB to minimize the disturbance of the cell during the delivery process. Another approach is the use of a biosensor format in which MB probes are immobilized on a solid carrier surface (see in the following text). A highly sensitive fluorescence imaging system is usually used to visualize the MB's hybridization with the target mRNA. Control experiments should also be carried out to preclude any signal originating from the opening of MB by other possible proteins or nuclease cleavage enzymes in cells.

Coupled with the advanced imaging system, the MBs have been used for real-time monitoring of the steady-state levels of specific mRNA molecules in the cytoplasm of cultured PtK2 cells [49]. A series of MBs having variable stem–loop sequences with tetramethylrhodamine (TMR) as a fluorophore and DABCYL as a quencher have been designed to serve as hybridization probes and control probes for the PtK2 cell β-actin mRNA detection. Picoliter amounts of MB solution were directly delivered into the cytoplasm of the cells using an Eppendorf microinjection system. Optical images and fluorescence images before and after the MB injection were taken by fluorescence microscopy equipped with an intensified charge-coupled device (ICCD) and an argon-ion laser excitation system.

Figure 20.10 shows a set of typical images collected for a MB probe hybridized with the β-actin mRNA in one cell. The fluorescence intensity of the cell continuously increased with the hybridization time for the first 15 min and then remained relatively constant thereafter for the next 25 min. To determine if the fluorescence intensity increase represented hybridization of the MB to the target β-actin mRNA, control experiments were conducted as follows: First, control MBs were injected into different PtK2 cells, and all the cells consistently showed no increase in signal intensity over a period of 15 min. Secondly, the complementary DNAs of the control MBs were injected, respectively, into the same cells having their corresponding control MBs, and all the cells showed immediate fluorescence enhancement. This clearly shows that the control MBs remained closed after the first injection due to the lack of their specific targets in the cytoplasm of the PtK2 cells. It is also interesting to observe that there was a variation in the MBs' relative fluorescence intensity during the same period of hybridization among different cells. The most likely explanation for this difference between cells can be attributed to different concentrations of β-actin mRNA presented in the cytoplasm of these individual cells.

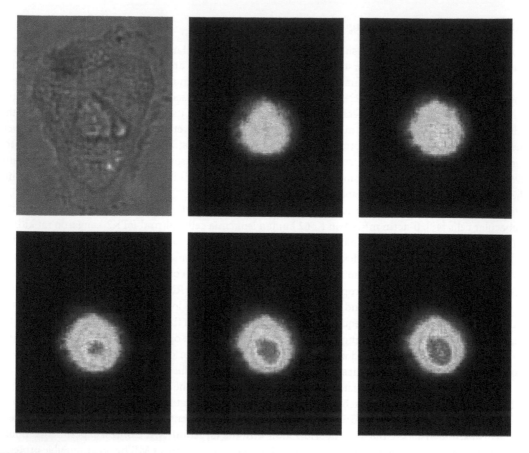

FIGURE 20.10   Hybridization of the MB to mRNA in living cells: an optical image of the kangaroo rat kidney cell and the fluorescence images of the cell at 3, 6, 9, 12, and 15 min after injection of MB to the PtK2 cell.

Using fluorescently labeled DNA probes, researchers are able to investigate the expression of human c-fos mRNA in a living transfected Cos7 cell [50]. Human c-fos mRNA was observed by detecting a hybrid formed with two fluorescently labeled oligodeoxynucleotides and c-fos mRNA in the cytoplasm. Two fluorescent DNAs were prepared, each labeled with a fluorescence molecule different from the other. The FRET occurred when two DNA strands hybridized to an adjacent sequence on the target mRNA. To find sequences of high accessibility of c-fos RNA to the DNAs, several sites that included loop structures on the simulated secondary structure were selected. Each site was examined for the efficiency of hybridization to c-fos RNA by measuring changes in fluorescence spectra when c-fos RNA was added to the pair of the DNAs in solution. A 40 mer specific site was found, and the pair of the DNAs for the site was microinjected into Cos7 cells that expressed c-fos mRNA. Hybridization of the pair of the DNAs to c-fos mRNA in the cytoplasm was detected in fluorescence images indicating FRET. To block the DNAs from accumulating in the nucleus, the DNA probes were bound to a macromolecule (such as streptavidin) to prevent passage to the nuclear pores.

Besides, as a result of cell-to-cell variation, intracellular imaging with MBs has usually employed the strategy of ratiometric measurement, whereby one MB was designed for a specific target of interest, and the other served as the reference probe. Drake et al. investigated the stochasticity of human manganese superoxide dismutase (MnSOD) mRNA expression in breast cancer cells using a MB that targeted MnSOD mRNA, while the reference MB targeted β-actin mRNA [51]. A $Ru(Bipy)_3^{2+}$-labeled scrambled DNA sequence was used as a negative control. Lipopolysaccharide (LPS) is an inflammatory mediator involved in *E. coli* bacterial sepsis and is proven to stimulate MnSOD mRNA expression in multiple mammalian cells. After LPS treatment, the MnSOD mRNA expression level in the MDA-MB-231 cell line, as detected by MnSOD mRNA MB, showed a distinct cell-to-cell variation, that is, 0.57 ± 0.17 to 0.66 ± 0.18 on average. On the other hand, for the β-actin MB, LPS treatment showed very little change relative to cell distribution, either before or after LPS induction.

In addition to probing one pattern of cancer-related mRNA expression, MBs can also monitor multiple-gene expression in a single living cell. Medley et al. synthesized three MBs labeled with different fluorophores to monitor the expression level of human MnSOD and β-actin mRNA in a single MDA-MB-231 cell [52]. Ru(Bipy)$_3^{2+}$ was chosen as the label for the reference probe in channel D due to its stable emission fluorescence intensity and lack of fluorophore crosstalk. After microinjection, in channel B, only a small amount of fluorescence signal was observed for the control MB, which was designed to have no target complementary mRNA inside the cell. At t = 0 min, a fluorescence signal was only barely observed for β-actin MBs; however, the fluorescence intensity increased as time elapsed, consistent with the high expression of β-actin in the MDA-MB-231 cell. In channel C of MnSOD MBs, the fluorescence intensity showed the same increasing trend but not to the extent seen in the β-actin MBs (Figure 20.11). Furthermore, the varied pattern of gene expression in a single cell can be determined by this method. LPS induction showed a significant impact on the expression of MnSOD expression relative to that of β-actin. Besides probing multiple-gene expression, Medley et al. also applied MBs and a cell-permeant Fluo-4 calcium ion indicator to investigate both mRNA expression levels and ion concentrations and their relationships in the same living cell [53].

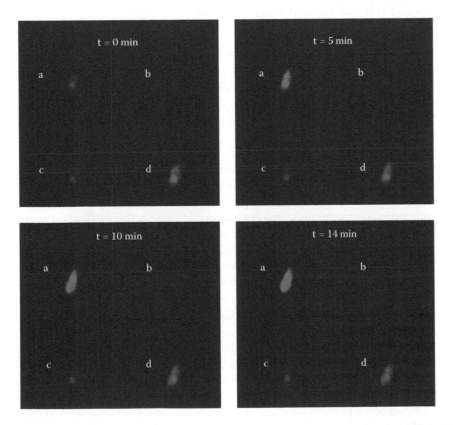

FIGURE 20.11 **(See color insert.)** Real-time monitoring of multiple-gene expression using different MBs in a single MDA-MB-231 cell. (a) Fluorescence imaging of the β-actin MB (green), (b) fluorescence imaging of control MB (red), (c) fluorescence imaging of MnSOD MB (blue), and (d) fluorescence imaging of Ru(Bpy)$_3^{2+}$ reference probe (orange).

To avoid interference from false-positive signals, for example, nonspecific protein binding and nuclease degradation, Martinez et al. used the HMP for intracellular studies of mRNA expression level [54]. Compared to MBs, HMPs have faster hybridization kinetics and greater resistance to nuclease degradation inside cells. Multiple mRNA sequences, such as β-tubulin, β-actin, and MnSOD, were chosen as targets for the design of three HMPs labeled with different fluorophores. After introduction into single cells by microinjection, the HMPs showed an intense FRET signal when hybridized to target mRNA, while the control HMP without cellular target showed only the signal of the fluorescence acceptor. This work indicated that HMPs had far less propensity for false-positive signals and performed better than traditional MBs inside living cells. The lifetime of MBs in living cells is usually ~30 min, and after that, MBs will be digested by cellular nucleases and show false-positive signals. Therefore, investigators must address this problem if prolonged long-term real-time monitoring in single living cells is to be achieved. Inspired by LNAs [27,28], Wu et al. developed LNA/DNA MBs with LNA-modified loops and LNA/DNA mixed stems to monitor mRNA expression in real time for 5–24 h [55]. During treatment with LPS for 4 h, target MnSOD MBs showed a distinct increase in fluorescence intensity, while no change in the confocal fluorescence imaging was observed for control MBs. After injection into living cells over 24 h, the control MBs still retained their function and showed an intense fluorescence signal after introduction of their complementary target.

20.4.3 Molecular Beacons for Protein Recognition and Protein–DNA Interaction Study

20.4.3.1 Nonspecific Protein Binding Study

While MB probes were originally designed for nucleic acid studies, their hairpin structure can also be disturbed to restore fluorescence upon binding to some proteins. The introduction of MB for protein–DNA interaction studies will be of great interest in understanding many important biological processes and in ultrasensitive protein detection, as proteins play critical roles in many biological processes. This protein recognition ability was first realized using an *E. coli* SSB [56]. The fluorescence enhancement caused by SSB and by a complementary DNA target is very comparable (Figure 20.12a). Using MB–SSB binding, it was possible to detect SSB at a concentration as low as 2×10^{-10} mol/L using a conventional spectrometer with a mercury lamp. The interaction between SSB and MB was found to be much faster than that between the cDNA and MB (Figure 20.12b). The fast speed of the protein–DNA binding reaction will provide the basis for rapid protein assays. In addition, there are significant differences in MB binding affinity with different proteins, such as albumin and histone. This will lead to the exploration of the potential for selective binding studies of a variety of proteins using designer MBs.

MB DNA probe has also been used for detailed binding studies of an enzyme, lactate dehydrogenase (LDH) [57]. The fluorescence signal of the MB was increased upon interaction with LDH, which was then used for the elucidation of the binding properties and the study of the binding process. Different LDH isoenzymes were found to have different ssDNA binding affinities. The results showed that the stoichiometry of LDH-5/MB binding is 1:1 and the binding constant is 1.9×10^{-7} mol/L. Detailed studies of LDH/MB binding, such as salt effects, temperature effects, pH effects, binding sites, binding specificity for different isoenzymes, and competitive binding with different substrates, were carried out by a simple fluorescent method using the MB probe. The possibility of utilizing MB probes for quantitative protein detection with ultrasensitivity shows great benefit in understanding many important biological processes involving two key biomolecules: nucleic acids and proteins. Although only nonspecific DNA-binding proteins have been investigated so far [56–58], the method opens the possibility for further development of easily obtainable, modified DNA molecules for real-time specific protein detection. On the other hand, the study of the nonspecific DNA-binding proteins is also very important. One example is the development of an easy and efficient DNA cleavage enzyme assay using MBs [58] (see as follows).

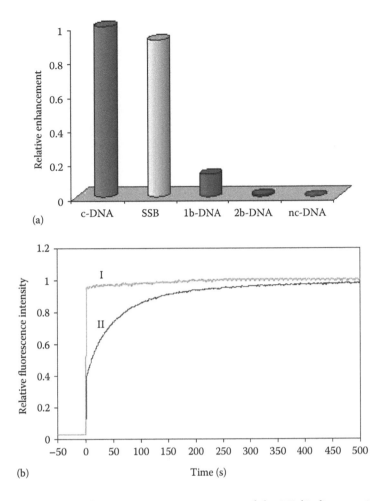

(a)

(b)

FIGURE 20.12 (a) The relative fluorescence intensity time scan of the MB binding reaction with cDNA (I) and SSB (II). The molecular mole ratio is cDNA/MB = 1:1, and SSB/MB is also 1:1. cDNA: complementary DNA; 1b-DNA: one-base mismatched DNA; 2b-DNA: two-base mismatched DNA; nc-DNA: noncomplementary DNA; SSB: single-stranded DNA binding protein. (b) The relative fluorescence enhancement of the MB DNA probe when hybridized with different DNA molecules and interaction with SSB. The MB used here is 5′-6FAM-GCTCG TCC ATA CCC AAG AAG GAA G CGAGC-DABCYL-3′.

20.4.3.2 Real-Time Enzymatic Cleavage Assay

Traditional methods to assay enzymatic cleavage of ssDNA are discontinuous and time consuming. The lack of suitable fluorescent probes is an obstacle in the development of fluorescence methods that are continuous and convenient. Based on MB probes, a novel method has been proposed to assay the ssDNA cleavage reaction by single-strand-specific nuclease [58]. The single-stranded nuclease binds to and cleaves the single-stranded loop portion of the MB. The cleavage results in the dissociation of the stem since the five to seven base pairs in the stem are unstable at the cleavage temperature (37°C) when the loop is broken. Consequently, the fluorophore and quencher are completely separated from each other, giving rise to an irreversible fluorescence enhancement, which is higher than that caused by the MB's complementary DNA. Figure 20.13 shows that there is a good agreement in the nuclease assay between traditional gel electrophoresis and the fluorescence assay based on MBs. The fluorescence method permits real-time monitoring of the enzymatic cleavage reaction process, easy characterization of the activity of DNA nucleases, and the study of steady-state cleavage reaction kinetics. Due to its

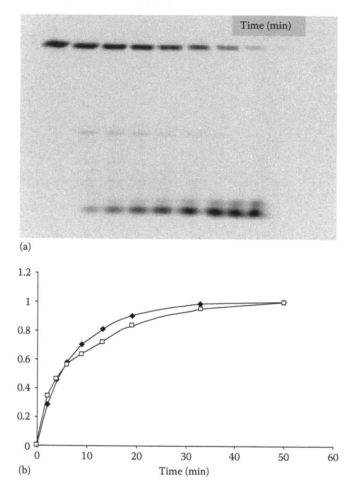

(a)

(b)   Time (min)

FIGURE 20.13 The comparison between fluorescence assay and gel electrophoresis assay for the cleavage reaction of the MB by S1 nuclease. In fluorescence assay, time course was recorded for the cleavage reaction. In each scheduled time point, fixed amount of reaction solution was taken out for gel electrophoresis assay. (a) Polyacrylamide gel electrophoresis assay. The samples were run on denaturing polyacrylamide gel. The upper bands represent the intact MB and reaction time points (minutes) are labeled above the bands. (b) Comparing the results on MB cleavage percentage obtained by the two assays. The data for fluorescence assay curve (♦) were extracted from the reaction time–course curve; the cleavage percentage in electrophoresis assay (□) at each time point was determined by quantifying the fluorescence decrease of the substrate relative to the total fluorescence.

high sensitivity, reproducibility, and convenience, MBs have been used to observe the study of ssDNA cleavage reactions.

Other cleavage experiments were done as well. For example, efforts have focused on developing imaging probes that can be activated to measure specific enzyme activities in vivo. Using cathepsin D as a model target protease, Tung et al. synthesized a long-circulating, synthetic graft copolymer bearing near-infrared (NIR) fluorochrome positioned on cleavable substrate sequences [40]. In its native state, the reporter probe was essentially nonfluorescent at 700 nm due to energy resonance transfer among the bound fluorochromes (quenching) but became brightly fluorescent when the latter was released by cathepsin D. Using matched rodent tumor models implanted into nude mice expressing or lacking the targeted protease, it could be shown that the former generated sufficient NIR signal to be directly detectable and that the signal was significantly different compared with negative control tumors.

Because MB carries an appropriate cleavage site within the stem, it has been applied to develop the continuous assay for cleavage of DNA by enediynes. The generality of this approach is demonstrated by using the described assay to directly compare the DNA cleavage by naturally occurring enediynes, non-enediyne small-molecule agents, as well as the restriction endonuclease BamHI [59]. Meanwhile, MBs were used to quantify low levels of type I endonuclease activity [60]. Given the simplicity, speed, and sensitivity of this approach, the described methodology could easily be extended to a high-throughput format and become a new method of choice in modern drug discovery to screen for novel protein-based or small-molecule-derived DNA cleavage agents.

20.4.3.3 Protein Detection with Specificity

The aforementioned nonspecific DNA-binding protein study opens the possibility for the further development of easily obtainable, modified DNA molecules for real-time specific protein detection. We have recently developed a new molecular recognition mechanism by combining the MB's excellent signal transduction mechanism with an aptamer's specificity in protein binding for a novel approach for real-time protein detection [56].

In most biomedical applications, high-affinity recognition and specific protein recognition are accomplished by using antibodies. Dye-labeled antibodies are often used to detect specific proteins in vitro. Although antibodies are highly specific in molecular recognition, they have limitations. Aptamers are a new class of designer DNA/RNA molecules that compete with the antibody in protein recognition [61,62]. Aptamers have many advantages over antibodies: easier synthesis, easier labeling, better reproducibility, easier storage, faster tissue penetration, and shorter blood residence. However, an aptamer itself cannot be used as a fluorescent probe since it lacks the signal transduction capability to report the binding to a target. The new protein probe, which combines the binding specificity and generality of aptamers with the excellent signal transduction capability of MBs, has great potential in monitoring protein production in living cells.

The feasibility of this approach has been demonstrated by a successful development of a molecular beacon aptamer (MBA) probe for a model protein, thrombin. A MBA probe is constructed by engineering a 17 mer thrombin-binding aptamer with a fluorophore and a quencher attached at its two ends. Upon binding to the thrombin, the MBA experiences a significant conformational change, from a loose random coil to a compact unimolecular quadruplex in a solution with low salt concentration. The conformational change brought about by the specific binding is converted to a significant fluorescence signal change of the MBA. This signal change is large enough for sensitive thrombin detection with a detection limit in the sub-nM range.

The MBA–thrombin binding is highly specific, in both target specificity and aptamer sequence specificity. The aptamer's inherent specificity has been retained in the MBA for the thrombin. To achieve better protein quantitation in living cells, a two-fluorophore MBA based on FRET has been developed for ratiometric imaging [63]. Labeled with two fluorophores, such as coumarin and fluorescein, the MBA *lights up* upon thrombin binding. Comparing this to the direct intensity imaging, the ratio imaging gives a much-improved signal-to-background ratio; hence, higher sensitivity and better reliability for precise determination of the proteins desired.

Aptamer beacon has become a sensitive tool for detecting proteins and other chemical compounds. For example, Stanton et al. have designed aptamer beacons [64] for detecting a wide range of ligands. A fluorescence-quenching pair was used to report changes in conformation induced by ligand binding. An antithrombin aptamer was engineered into an aptamer beacon by adding nucleotides to the 5′-end, which are complementary to nucleotides at the 3′-end of the aptamer. In the absence of thrombin, the added nucleotides will form a duplex with the 3′-end, forcing the aptamer beacon into a stem–loop structure. In the presence of thrombin, the aptamer beacon forms the ligand-binding structure. This conformational change causes a change in the distance between the fluorophore attached to the 5′-end and a quencher attached to the 3′-end.

Aptamer-derived MBs can also be used to analyze the Tat of HIV and the possible applications of such constructs in the field of biosensors [65]. To make new MB, two RNA oligomers derived from

RNA (Tat) are constructed. In the presence of Tat or its peptides, but not in the presence of other RNA-binding proteins, the two oligomers undergo a conformational change to form a duplex that leads to the release of fluorophore from the quencher, and thus a significant enhancement of the fluorescence of fluorescein was observed.

Oncoproteins, coded by prooncogene to regulate cell growth and differentiation, play an important role in cancer growth and are often found overexpressed or mutated in malignant tumors. We have designed a new fluorescent probes based on a high-affinity PDGF aptamer (shown in Figure 20.14) for the ultra-sensitive detection of oncoprotein PDGF in homogeneous solutions [66]. The aptamer is labeled with a fluorophore/quencher pair to specifically bind with PDGF. Upon binding, the conformation change of the aptamer results in a FRET-based fluorescence signal intensity change of the probe. The aptamer probes are highly sensitive and highly selective. They can detect PDGF in the sub-nM range. Not only the tested extracellular proteins but also other PDGF-related peptide growth factors do not interfere in the probe binding to PDGF as shown in Figure 20.14. The new oncoprotein detection methods are simple and inherit all the advantages of molecular aptamers. They are expected to find wide applications in protein detection, in cancer diagnosis, as well as other disease diagnosis and mechanism study.

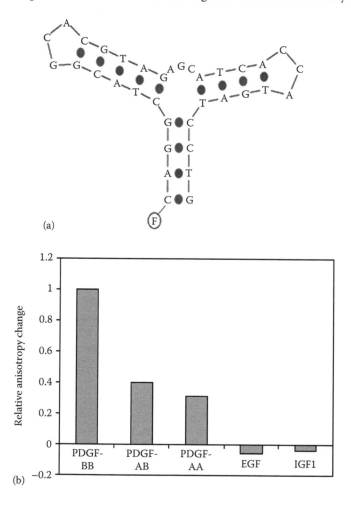

(a)

(b)

FIGURE 20.14 (a) The structure of the fluorescein (F)-labeled aptamer probe. (b) Binding specificity of the aptamer: comparing the binding capability of the PDGF-B chain (PDGF-B) to other growth factors such as PDGF isomers PDGF-AA and PDGF-AB, epidermal growth factor (EGF), and insulin-like growth factor 1 (IGF-1). The molar ratio of the protein to the aptamer is 1:1.

20.5 Combining Molecular Beacon and Biosensor Technology

20.5.1 Surface-Immobilizable Molecular Beacons

As the common MB can only be used in homogeneous solution, surface-immobilizable MBs are critical for the development of highly sensitive biosensors for in vivo detection and for the study of biomolecular recognition processes at an interface. A biotinylated ssDNA MB has thus been designed [67]. Biotin–avidin binding is one of the most common ways for biomolecule immobilization onto a solid surface and is suitable for DNA hybridization. For biotinylated MB synthesis, there are several important considerations in its design [5]. Of them, the position of the biotin is very important and should be carefully chosen to minimize the effect of the avidin–biotin bridge on the MB hybridization. It is desirable to add the biotin functional group at the quencher side in the stem of a MB. A spacer between the biotin and the sequence should be added to provide an adequate separation to minimize potential interactions between avidin and the DNA sequence. A photostable dye, such as TMR, should be chosen to minimize photobleaching as only a small amount of fluorophores are immobilized on a surface. To immobilize biotinylated MBs, a silica surface is first physically or covalently coated with avidin [67–69]. The biotinylated ssDNA MB then binds to avidin. The binding process is fast and stable. The hybridization properties of MBs on the surface are similar to those in solution.

A method was developed for rapid detection of PCR amplicons based on surface-immobilized PNA–DNA hybrid probes that undergo a fluorescent-linked conformational change in the presence of a complementary DNA target [70]. Amplicons can be detected by simply adding a PCR to a microtiter well containing the previously immobilized probe and reading the generated fluorescence. No further transfers or washing steps are involved. The specificity of the method for the detection of ribosomal DNA from *Entamoeba histolytica* was excellent.

20.5.2 Optical Fiber DNA Biosensors Using Surface-Immobilizable Molecular Beacons

The immobilized MBs have been used for the preparation of optical fiber DNA biosensors, such as a submicrometer biosensor and a fiber-optic evanescent wave biosensor [68,69]. The evanescent wave biosensor was prepared by exposing the core surface of an optical fiber through chemical etching. An evanescent wave generated on the core surface was used for fluorescence excitation in the longitudinal surface of the fiber where the MBs were immobilized (Figure 20.15a).

The microscopic optical fiber probe was fabricated using either pulling or etching technologies [63,71–75]. MBs were immobilized only at the small tip (submicrometer in diameter) of the probe. A highly sensitive optical imaging and detection system with an avalanche photon diode or an ICCD [6] was used for the detection of the fluorescence signal from the optical fiber biosensor. The biosensors were used for the detection of nonlabeled DNA targets in real time and with high sensitivity and one-base mismatch selectivity. As shown in Figure 20.15b, there was a linear relationship between the initial hybridization reaction rate and the concentration of the complementary DNA for the submicrometer DNA biosensor. The concentration detection limit of the target complementary DNA was 0.3 nM for the ultrasmall DNA biosensor. The sensors are stable and reproducible and have remote detection capability. They can be easily regenerated by a 1 min rinse with a 90% formamide solution [69]. They have been applied to the quantitative detection of mRNA sequences and to study DNA hybridization kinetics [68,69]. They hold the potential of direct detection of DNA/RNA targets in living cells without DNA/RNA amplification.

The immobilization method for biotinylated MBs has also enabled the exploration of MB probe arrays for simultaneous multiple-analyte detection [76,77]. MBs with different loop sequences can be immobilized onto the tips of different microscopic fiber probes [69]. These biosensors are individually addressable with a spatially resolved imaging system based on the ICCD camera [6].

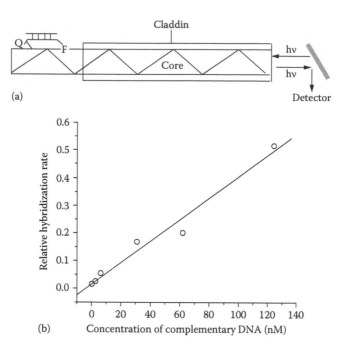

(a)

(b)

FIGURE 20.15 (a) Configuration of the evanescent wave fiber-optic DNA biosensor based on MB. The MBs are excited by the evanescent wave generated at the core surface of the fiber. The returning fluorescence then travels through the same fiber for detection. (b) Application of the DNA biosensor: linear relationship between the hybridization rate and the concentration of the target DNA.

We have proposed a bridge structure to immobilize the biotinylated MB [78]. The MB was biotinylated at the quencher side of the stem and linked on a biotinylated glass cover slip through streptavidin, which acted as a bridge between the MB and glass matrix. An efficient fluorescence microscope system was constructed to detect the fluorescence change caused by the conformation change of MB in the presence of complementary DNA target. The proposed biosensor was used to directly detect, in real time, the target DNA molecules. The bridge immobilization method caused the proposed DNA biosensor to have a faster and more stable response. Many efforts were made to develop a more sensitive and stable DNA biosensor. Sol–gel method was shown to be a very efficient approach to immobilize the biotinylated MB [79]. The sol–gel matrix provided a solid support for the immobilized MB, which can be doped on the surface of many materials. This will widen the application of MBs in clinical detection and biotechnology. The potential application of DNA biosensors includes the fabrication of submicrometer optical fiber probe to detect mRNA inside living cells.

20.5.3 Nanoparticle Molecular Beacon Biosensor

We have prepared MB sensors based on silica nanoparticles [80,81]. Tens of thousands of uniform nanoparticles have been successfully immobilized with MBs. These particles can be attached to the tips of optical fibers for the preparation of a large quantity of fiber probes or can be introduced into cells as individual sensors. Recently, a gene array has been developed based on the MB using particles and an imaging fiber bundle [77]. Different MB-coated microspheres are randomly distributed in an array of wells etched in a 500 μm diameter optical imaging fiber bundle. To recognize different DNA targets, an optical encoding scheme and an imaging fluorescence microscope system were used for positional registration and fluorescence response monitoring. We have also used nanoparticles for MB biosensor development. These nanoparticle-based MB biosensors are highly efficient in hybridization as they have a large surface-to-volume ratio.

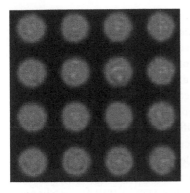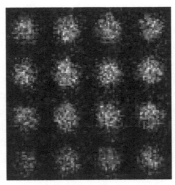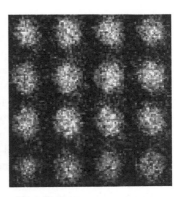

FIGURE 20.16 The microwell array was made by photolithography. A group of consecutive images were taken as the MB hybridization proceeded. A fixed 1 min interval between two images was used. The figure shows that from the beginning to the end of the hybridization experiment, more and more MBs hybridize spontaneously to their targets and undergo a conformational change that results in the generation of fluorescence.

20.5.4 Molecular Beacon Microarray Sensor and Biochip

The multianalyte DNA sensors are expected to provide an easy and fast way for high-throughput gene analysis and disease diagnosis [77]. A methodology for the implementation of multiplexed nucleic acid hybridization fluorescence assays was developed on microchannel glass substrates. Fluorescence detection was achieved, using conventional low-magnification microscope objective lenses, as imaging optics whose depth-of-field characteristics match the thickness of the microchannel glass chip. The optical properties of microchannel glass were shown, through experimental results and simulations, to be compatible with the quantitative detection of heterogeneous hybridization events taking place along the microchannel sidewalls, with detection limits for oligonucleotide targets in the low-attomole range [82]. A nanoscopic well system for MB biosensor and biochip has also been developed as shown in Figure 20.16. The system is being used for multiple-gene analysis. Overall, a variety of biosensor formats based on MBs have been developed, and their application is just in the horizon.

20.6 Outlook

As the human genome-sequencing project is being completed, there will be a resulting change in the focus of quantitative studies of genomic information from the collecting and archiving of genomic data to their analysis and use in prediction and discovery. The key to this new era is research across many disciplinary interfaces and the development and utilization of new quantitative tools. MB is ideally suited for and holds great promise in genomics and proteomics. While there are many possibilities and great potentials, we believe that the following three areas are of the greatest interest to the bioanalytical sciences and are feasible in the near future. First, more research is expected for MB application in mutation detection for a variety of disease diagnostics and disease mechanism studies. Efforts will also be made in finding suitable mRNA sequences and real-time studies of RNA processing in living cells and other living specimens without using amplification techniques. This will be expedited with the development of MB biosensors and highly sensitive optical detection methods [6,83]. Gene expression under different conditions will be studied with precise quantitation. Second, the extraordinarily target-specific capability along with the availability of different fluorophore–quencher pairs makes MB probes extremely useful for multiple-analyte applications. The detection of many different targets in the same solution can be achieved by the simultaneous use of several MBs, each of which emits light of a different wavelength or which can be excited by light of a different wavelength. These excellent properties also make MBs exquisitely suited for use in the identification of genetic alleles or particular strains of infectious agents. The multiple-MB-probe approach will also be highly useful in combinatorial chemistry for high-throughput methodology development for drug design and for molecular diagnosis of a variety of genetic diseases, especially with the abundance of genetic information available in the postgenome era. Third, the MB's application as a novel biomolecular recognition agent for proteins will be further explored. With further development, MBs are expected to be useful as intracellular protein recognition agents to probe proteins in different environments and to monitor protein–DNA/RNA interactions. Preliminary results show there are significant differences in MB binding affinities by different proteins, which will constitute the basis for highly selective bioassays of a variety of proteins. Approaches to this goal include using designer DNA molecules or aptamer-based MBs [61] for better specificity and/or using a large array of MB biosensors for pattern recognition. Better understanding of the conformational kinetics of MB [8] and its fluorescence energy transfer mechanism [4,63] will result in optimally designed MB probes for biomolecular recognition with high sensitivity and excellent specificity. All of these developments will open the possibility of using easily obtainable and designer DNA molecules for genomics and proteomics studies, for molecular diagnosis of diseases, and for new drug development.

Acknowledgments

We thank our colleagues for their help. This work is supported by grants awarded by the National Institutes of Health (GM066137, GM079359, and CA133086).

References

1. Wood, E.J., Molecular probes: Handbook of fluorescent probes and research chemicals, *Biochem. Educ.*, 22(2), 83, 1994.
2. Tyagi, S., Kramer, F.R., Molecular beacons: Probes that fluoresce upon hybridization, *Nat. Biotechnol.*, 14(3), 303, 1996.
3. Lakowicz, J.R., *Principles of Fluorescence Spectroscopy*, 2nd edn. Kluwer Academic/Plenum Publishers, New York, 1999.
4. Tyagi, S., Bratu, D.P., Kramer, F.R., Multicolor molecular beacons for allele discrimination, *Nat. Biotechnol.*, 16(1), 49, 1998.
5. Meinkoth, J., Wahl, G., Hybridization of nucleic acids immobilized on solid supports, *Anal. Biochem.*, 138(2), 267, 1984.
6. Fang, X., Tan, W., Imaging single fluorescent molecules at the interface of an optical fiber probe by evanescent wave excitation, *Anal. Chem.*, 71(15), 3101, 1999.
7. Kostrikis, L.G., Tyagi, S., Mhlanga, M.M., Ho, D.D., Kramer, F.R., Spectral genotyping of human alleles, *Science*, 279(5354), 1228, 1998.
8. Bonnet, G., Tyagi, S., Libchaber, A., Kramer, F.R., Thermodynamic basis of the enhanced specificity of structured DNA probes, *Proc. Natl. Acad. Sci. U. S. A.*, 96(11), 6171, 1999.
9. Matsuo, T., In situ visualization of messenger RNA for basic fibroblast growth factor in living cells, *Biochimica et Biophysica Acta (BBA)/General Subjects*, 1379(2), 178, 1998.
10. Yang, C.J., Lin, H., Tan, W., Molecular assembly of superquenchers in signaling molecular interactions, *J. Am. Chem. Soc.*, 127(37), 12772, 2005.
11. Ellington, A.D., Szostak, J.W., In vitro selection of RNA molecules that bind specific ligands, *Nature*, 346(6287), 818, 1990.
12. Robertson, D.L., Joyce, G.F., Selection in vitro of an RNA enzyme that specifically cleaves single-stranded DNA, *Nature*, 344(6265), 467, 1990.
13. Tuerk, C., Gold, L., Systematic evolution of ligands by exponential enrichment: RNA ligands to bacteriophage T4 DNA polymerase, *Science*, 249(4968), 505, 1990.

14. Li, J.J., Tan, W., A real-time assay for DNA sticky-end pairing using molecular beacons, *Anal. Biochem.*, 312(2), 251, 2003.

15. Yang, C.J., Martinez, K., Lin, H., Tan, W., Hybrid molecular probe for nucleic acid analysis in biological samples, *J. Am. Chem. Soc.*, 128(31), 9986, 2006.

16. Kim, Y., Yang, C.J., Tan, W., Superior structure stability and selectivity of hairpin nucleic acid probes with an l-DNA stem, *Nucleic Acids Res.*, 35(21), 7279, 2007.

17. Wang, C., Zhu, Z., Song, Y., Lin, H., Yang, C.J., Tan, W., Caged molecular beacons: Controlling nucleic acid hybridization with light, *Chem. Commun.*, 47(20), 5708, 2011.

18. Yang, C.J., Jockusch, S., Vicens, M., Turro, N.J., Tan, W., Light-switching excimer probes for rapid protein monitoring in complex biological fluids, *Proc. Natl. Acad. Sci. U. S. A.*, 102(48), 17278, 2005.

19. Conlon, P., Yang, C.J., Wu, Y., Chen, Y., Martinez, K., Kim, Y., Stevens, N. et al., Pyrene excimer signaling molecular beacons for probing nucleic acids, *J. Am. Chem. Soc.*, 130(1), 336, 2007.

20. Wu, C., Yan, L., Wang, C., Lin, H., Wang, C., Chen, X., Yang, C.J., A general excimer signaling approach for aptamer sensors, *Biosens. Bioelectron.*, 25(10), 2232, 2010.

21. Winnik, F.M., Photophysics of preassociated pyrenes in aqueous polymer solutions and in other organized media, *Chem. Rev.*, 93(2), 587, 1993.

22. Kushon, S.A., Ley, K.D., Bradford, K., Jones, R.M., McBranch, D., Whitten, D., Detection of DNA hybridization via fluorescent polymer superquenching, *Langmuir*, 18(20), 7245, 2002.

23. Lu, L., Jones, R.M., McBranch, D., Whitten, D., Surface-enhanced superquenching of cyanine dyes as J-aggregates on laponite clay nanoparticles, *Langmuir*, 18(20), 7706, 2002.

24. Swager, T.M., The molecular wire approach to sensory signal amplification, *Acc. Chem. Res.*, 31(5), 201, 1998.

25. Bunz, U.H.F., Poly(aryleneethynylene)s: Syntheses, properties, structures, and applications, *Chem. Rev.*, 100(4), 1605, 2000.

26. Yang, C.J., Pinto, M., Schanze, K., Tan, W., Direct synthesis of an oligonucleotide–poly(phenylene ethynylene) conjugate with a precise one-to-one molecular ratio, *Angew. Chem. Int. Ed.*, 44(17), 2572, 2005.

27. Wang, L., Yang, C.J., Medley, C.D., Benner, S.A., Tan, W., Locked nucleic acid molecular beacons, *J. Am. Chem. Soc.*, 127(45), 15664, 2005.

28. Yang, C.J., Wang, L., Wu, Y., Kim, Y., Medley, C.D., Lin, H., Tan, W., Synthesis and investigation of deoxyribonucleic acid/locked nucleic acid chimeric molecular beacons, *Nucleic Acids Res.*, 35(12), 4030, 2007.

29. Sheng, P., Yang, Z., Kim, Y., Wu, Y., Tan, W., Benner, S.A., Design of a novel molecular beacon: Modification of the stem with artificially genetic alphabet, *Chem. Commun.*, 44(41), 5128, 2008.

30. Tapp, I., Malmberg, L., Rennel, E., Wik, M., Syvanen, A.C., Homogeneous scoring of single-nucleotide polymorphisms: Comparison of the 5′-nuclease TaqMan assay and molecular beacon probes, *BioTechniques*, 28(4), 732, 2000.

31. Helps, C., Reeves, N., Tasker, S., Harbour, D., Use of real-time quantitative PCR to detect *Chlamydophila felis* infection, *J. Clin. Microbiol.*, 39(7), 2675, 2001.

32. Schofield, P., Pell, A.N., Krause, D.O., Molecular beacons: Trial of a fluorescence-based solution hybridization technique for ecological studies with ruminal bacteria, *Appl. Environ. Microbiol.*, 63(3), 1143, 1997.

33. Giesendorf, B.A.J., Vet, J.A.M., Tyagi, S., Mensink, E.J.M.G., Trijbels, F.J., Blom, H.J., Molecular beacons: A new approach for semiautomated mutation analysis, *Clin. Chem.*, 44(3), 482, 1998.

34. Lewin, S.R., Vesanen, M., Kostrikis, L., Hurley, A., Duran, M., Zhang, L., Ho, D.D., Markowitz, M., Use of real-time PCR and molecular beacons to detect virus replication in human immunodeficiency virus type 1-infected individuals on prolonged effective antiretroviral therapy, *J. Virol.*, 73(7), 6099, 1999.

35. Rhee, J.T., Piatek, A.S., Small, P.M., Harris, L.M., Chaparro, S.V., Kramer, F.R., Alland, D., Molecular epidemiologic evaluation of transmissibility and virulence of *Mycobacterium tuberculosis*, *J. Clin. Microbiol.*, 37(6), 1764, 1999.

36. Vet, J.A.M., Majithia, A.R., Marras, S.A.E., Tyagi, S., Dube, S., Poiesz, B.J., Kramer, F.R., Multiplex detection of four pathogenic retroviruses using molecular beacons, *Proc. Natl. Acad. Sci. U. S. A.*, 96(11), 6394, 1999.

37. Zhang, L.Q., Lewin, S.R., Markowitz, M., Lin, H.H., Skulsky, E., Karanicolas, R., He, Y. et al., Measuring recent thymic emigrants in blood of normal and HIV-1-infected individuals before and after effective therapy, *J. Exp. Med.*, 190(5), 725, 1999.

38. Chen, W., Martinez, G., Mulchandani, A., Molecular beacons: A real-time polymerase chain reaction assay for detecting *Salmonella*, *Anal. Biochem.*, 280(1), 166, 2000.

39. Yuan, C.-C., Peterson, R.J., Wang, C.-D., Goodsaid, F., Waters, D.J., 5′ Nuclease assays for the loci CCR5-+/Δ32, CCR2-V64I, and SDF1-G801A related to pathogenesis of AIDS, *Clin. Chem.*, 46(1), 24, 2000.

40. Tung, C.-H., Mahmood, U., Bredow, S., Weissleder, R., In vivo imaging of proteolytic enzyme activity using a novel molecular reporter, *Cancer Res.*, 60(17), 4953, 2000.

41. de Ronde, A., van Dooren, M., van der Hoek, L., Bouwhuis, D., de Rooij, E., van Gemen, B., de Boer, R., Goudsmit, J., Establishment of new transmissible and drug-sensitive human immunodeficiency virus type 1 wild types due to transmission of nucleoside analogue-resistant virus, *J. Virol.*, 75(2), 595, 2001.

42. Martinson, J.J., Hong, L., Karanicolas, R., Moore, J.P., Kostrikis, L.G., Global distribution of the CCR2-64I/CCR5-59653T HIV-1 disease-protective haplotype, *AIDS*, 14(5), 483, 2000.

43. de Baar, M.P., Timmermans, E.C., Bakker, M., de Rooij, E., van Gemen, B., Goudsmit, J., One-tube real-time isothermal amplification assay to identify and distinguish human immunodeficiency virus type 1 subtypes A, B, and C and circulating recombinant forms AE and AG, *J. Clin. Microbiol.*, 39(5), 1895, 2001.

44. de Baar, M.P., van Dooren, M.W., de Rooij, E., Bakker, M., van Gemen, B., Goudsmit, J., de Ronde, A., Single rapid real-time monitored isothermal RNA amplification assay for quantification of human immunodeficiency virus type 1 isolates from groups M, N, and O, *J. Clin. Microbiol.*, 39(4), 1378, 2001.

45. McKillip, J.L., Drake, M., Molecular beacon polymerase chain reaction detection of *Escherichia coli* O157:H7 in milk, *J. Food Prot.*, 63(7), 855, 2000.

46. Li, Q.G., Liang, J.X., Luan, G.Y., Zhang, Y., Wang, K., Molecular beacon-based homogeneous fluorescence PCR assay for the diagnosis of infectious diseases, *Anal. Sci.*, 16(2), 245, 2000.

47. Fortin, N.Y., Mulchandani, A., Chen, W., Use of real-time polymerase chain reaction and molecular beacons for the detection of *Escherichia coli* O157:H7, *Anal. Biochem.*, 289(2), 281, 2001.

48. Dirks, R.W., Molenaar, C., Tanke, H.J., Methods for visualizing RNA processing and transport pathways in living cells, *Histochem. Cell Biol.*, 115(1), 3, 2001.

49. Sokol, D.L., Zhang, X., Lu, P., Gewirtz, A.M., Real time detection of DNA·RNA hybridization in living cells, *Proc. Natl. Acad. Sci. U. S. A.*, 95(20), 11538, 1998.

50. Tsuji, A., Koshimoto, H., Sato, Y., Hirano, M., Sei-Iida, Y., Kondo, S., Ishibashi, K., Direct observation of specific messenger RNA in a single living cell under a fluorescence microscope, *Biophys. J.*, 78(6), 3260, 2000.

51. Drake, T.J., Medley, C.D., Sen, A., Rogers, R.J., Tan, W., Stochasticity of manganese superoxide dismutase mRNA expression in breast carcinoma cells by molecular beacon imaging, *ChemBioChem*, 6(11), 2041, 2005.

52. Medley, C.D., Drake, T.J., Tomasini, J.M., Rogers, R.J., Tan, W., Simultaneous monitoring of the expression of multiple genes inside of single breast carcinoma cells, *Anal. Chem.*, 77(15), 4713, 2005.

53. Medley, C.D., Lin, H., Mullins, H., Rogers, R.J., Tan, W., Multiplexed detection of ions and mRNA expression in single living cells, *Analyst*, 132(9), 885, 2007.

54. Martinez, K., Medley, C., Yang, C., Tan, W., Investigation of the hybrid molecular probe for intracellular studies, *Anal. Bioanal. Chem.*, 391(3), 983, 2008.

55. Wu, Y., Yang, C.J., Moroz, L.L., Tan, W., Nucleic acid beacons for long-term real-time intracellular monitoring, *Anal. Chem.*, 80(8), 3025, 2008.

56. Li, J.J., Fang, X., Schuster, S.M., Tan, W., Molecular beacons: A novel approach to detect protein – DNA interactions, *Angew. Chem. Int. Ed.*, 39(6), 1049, 2000.

57. Fang, X., Li, J.J., Tan, W., Using molecular beacons to probe molecular interactions between lactate dehydrogenase and single-stranded DNA, *Anal. Chem.*, 72(14), 3280, 2000.

58. Li, J.J., Geyer, R., Tan, W., Using molecular beacons as a sensitive fluorescence assay for enzymatic cleavage of single-stranded DNA, *Nucleic Acids Res.*, 28(11), e52, 2000.

59. Biggins, J.B., Prudent, J.R., Marshall, D.J., Ruppen, M., Thorson, J.S., A continuous assay for DNA cleavage: The application of "break lights" to enediynes, iron-dependent agents, and nucleases, *Proc. Natl. Acad. Sci. U. S. A.*, 97(25), 13537, 2000.

60. Strouse, R.J., Hakki, F.Z., Wang, S.C., DeFusco, A.W., Garrett, J.L., Schenerman, M.A., Using molecular beacons to quantify low levels of type I endonuclease activity, *Biopharm*, 13(4), 40, 2000.

61. Osborne, S.E., Matsumura, I., Ellington, A.D., Aptamers as therapeutic and diagnostic reagents: Problems and prospects, *Curr. Opin. Chem. Biol.*, 1(1), 5, 1997.

62. Jayasena, S.D., Aptamers: An emerging class of molecules that rival antibodies in diagnostics, *Clin. Chem.*, 45(9), 1628, 1999.

63. Zhang, P., Beck, T., Tan, W., Design of a molecular beacon DNA probe with two fluorophores, *Angew. Chem. Int. Ed.*, 40(2), 402, 2001.

64. Hamaguchi, N., Ellington, A., Stanton, M., Aptamer beacons for the direct detection of proteins, *Anal. Biochem.*, 29(2), 126, 2001.

65. Yamamoto, R., Kumar, P.K.R., Molecular beacon aptamer fluoresces in the presence of Tat protein of HIV-1, *Genes Cells*, 5(5), 389, 2000.

66. Fang, X., Cao, Z., Beck, T., Tan, W., Molecular aptamer for real-time oncoprotein platelet-derived growth factor monitoring by fluorescence anisotropy, *Anal. Chem.*, 73(23), 5752, 2001.

67. Fang, X., Liu, X., Schuster, S., Tan, W., Designing a novel molecular beacon for surface-immobilized DNA hybridization studies, *J. Am. Chem. Soc.*, 121(12), 2921, 1999.

68. Liu, X., Tan, W., A fiber-optic evanescent wave DNA biosensor based on novel molecular beacons, *Anal. Chem.*, 71(22), 5054, 1999.

69. Liu, X., Farmerie, W., Schuster, S., Tan, W., Molecular beacons for DNA biosensors with micrometer to submicrometer dimensions, *Anal. Biochem.*, 283(1), 56, 2000.

70. Ortiz, E., Estrada, G., Lizardi, P.M., PNA molecular beacons for rapid detection of PCR amplicons, *Mol. Cell. Probes*, 12(4), 219, 1998.

71. Tan, W., Shi, Y., Smith, S., Birnbaum, D., Kopelman, R., Submicrometer intracellular chemical optical fiber sensors, *Science*, 258, 778, 1992.

72. Zeisel, D., Dutoit, B., Deckert, V., Roth, T., Zenobi, R., Optical spectroscopy and laser desorption on a nanometer scale, *Anal. Chem.*, 69(4), 749, 1997.

73. Tan, W., Optical measurements on the nanometer scale, *Trends Anal. Chem.*, 17(8–9), 501, 1998.

74. Tan, W., Kopeman, R., Barker, S.L.R., Miller, M.T., Ultrasmall optical sensors for cellular measurements, *Anal. Chem.*, 71(17), 606A, 1999.

75. Perlette, J., Tan, W., Real-time monitoring of intracellular mRNA hybridization inside single living cells, *Anal. Chem.*, 73(22), 5544, 2001.

76. Fang, X., Liu, X., Tan, W., Single and multiple molecular beacon probes for DNA hybridization studies on a silica glass surface, *Proc. SPIE*, 3602, 149, 1999.

77. Steemers, F.J., Ferguson, J.A., Walt, D.R., Screening unlabeled DNA targets with randomly ordered fiber-optic gene arrays, *Nat. Biotechnol.*, 18(1), 91, 2000.

78. Li, J., Tan, W., Wang, K., Xiao, D., Yang, X., He, X., Tang, Z., Ultrasensitive optical DNA biosensor based on surface immobilization of molecular beacon by a bridge structure, *Anal. Sci.*, 17(10), 1149, 2001.

79. Li, J., Tan, W., Wang, K., Yang, X., Tang, Z., He, X., Optical DNA biosensor based on molecular beacon immobilized on sol-gel membrane, *Proc. SPIE*, 4414, 27, 2001.

80. Santra, S., Tapec, R., Theodoropoulou, N., Dobson, J., Hebard, A., Tan, W., Synthesis and characterization of silica-coated iron oxide nanoparticles in microemulsion: The effect of nonionic surfactants, *Langmuir*, 17(10), 2900, 2001.

81. Santra, S., Wang, K., Tapec, R., Tan, W., Development of novel dye-doped silica nanoparticles for biomarker application, *J. Biomed. Opt.*, 6(2), 160, 2001.

82. Benoit, V., Steel, A., Torres, M., Yu, Y.-Y., Yang, H., Cooper, J., Evaluation of three-dimensional microchannel glass biochips for multiplexed nucleic acid fluorescence hybridization assays, *Anal. Chem.*, 73(11), 2412, 2001.

83. Zhang, P., Tan, W., Direct observation of single-molecule generation at a solid-liquid interface, *Chem. Eur. J.*, 6(6), 1087, 2000.

21

Luminescent Quantum Dots for Diagnostic and Bioimaging Applications

Brad A. Kairdolf
Georgia Institute of Technology and Emory University

Shuming Nie
Georgia Institute of Technology and Emory University

21.1 Introduction

Breakthroughs in nanoscience and the development of high-quality, nanomaterial-based optical probes have fueled interest in the use of nanoparticles for a broad range of bioimaging and diagnostic applications such as ultrasensitive biomolecule detection [1–6], multicolor immunostaining [7–10], single-cell analysis [11,12], and in vivo molecular imaging [13–16]. In particular, semiconductor quantum dots (QDs) have been the subject of intense research for the past two decades due to their novel optical and electronic properties, including resistance to photobleaching, tunable emission wavelength, intense signal brightness, and the simultaneous excitation of multiple fluorescence colors [17–19]. Pioneering work by Nie, Alivasatos, and their coworkers in 1998 to adapt high-quality QDs for use in aqueous environments has paved the way for the use of QDs as biological agents [17–21], and recent research has generated highly stable and monodispersed QDs with diverse surface chemistries and properties [22–24].

QDs are typically composed of atoms from groups II–VI or III–V and are dimensionally confined to sizes smaller than the exciton Bohr radius of the material, resulting in *quantum confinement* of the exciton. For CdSe QDs (the most commonly used QDs for biomedical applications), this is generally less than 10 nm in diameter. This quantum confinement effect allows the bandgap energy of the material to be tuned simply by changing the size of the nanoparticle, resulting in unique electronic and optical properties that are not present in bulk solids or discrete molecules. Using just a single material, QDs with bright fluorescence emissions covering the entire visible light range and the near infrared have been produced. Significant progress has also been made in the refinement of QD synthetic methods [25–30], resulting in nanoparticles with highly crystalline structures, narrow size distributions, and quantum yields approaching unity. When coupled with biomolecule affinity ligands, such as small molecules, peptides, nucleic acids, or antibodies, QDs can be used as highly sensitive and specific nanoprobes to detect molecular biomarkers and diseased cells or tissue [31–33].

In this chapter, we discuss the development and properties of semiconductor nanocrystal probes and their biodiagnostic and bioimaging applications. The main sections examine advances in QD synthesis, surface chemistry, and bioconjugation. We also discuss the use of bioconjugated QDs for real-time biomolecule detection, multiplexed QD imaging, molecular mapping in human cancer specimens, and molecular imaging in vivo. We conclude with a brief discussion of future challenges and prospects.

21.2 Quantum Dot Synthesis and Optical Properties

High-quality synthesis procedures are particularly important for the development of QD nanoparticles for diagnostic and bioimaging applications, which require stable, monodispersed, and sensitive probes for biomolecule detection (Figure 21.1). Initial efforts to produce water-soluble QDs focused on aqueous synthetic methods, such as the use of inverse micelles [34–36]. More recent aqueous procedures use small thiol-containing molecules or polymers with carboxylic acid functional groups as stabilizing agents and surface ligands, as well as water-soluble precursors for the nanocrystal atoms [37–43].

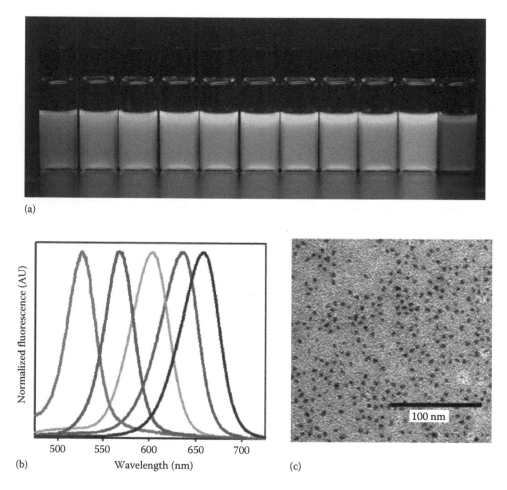

(a)

(b)   Wavelength (nm)   (c)

FIGURE 21.1 **(See color insert.)** Fluorescence emission and electron microscopy structural properties of CdTe QDs prepared by using multidentate ligands in a one-pot procedure. (a) Color photograph of a series of monodispersed CdTe QDs, showing bright fluorescence from green to red (515–655 nm) upon illumination with a UV lamp. (b) Normalized fluorescence emission spectra of CdTe QDs with narrow emission spectra. (c) Transmission electron micrograph (TEM) of QDs showing uniform, nearly spherical particles. (Adapted from Moffitt, M. and Eisenberg, A., *Chem. Mater.*, 7(6), 1178, 1995.)

While aqueous synthetic techniques are often used because of their simplicity and relative safety, this approach generally yields suboptimal QDs with a higher size polydispersity, broader emission spectra, and lower fluorescence brightness than can be achieved using other procedures. Moreover, the colloidal and optical stabilities of QDs prepared using aqueous procedures are typically lower than with QDs prepared using other methods.

A major breakthrough in the colloidal synthesis of highly crystalline and monodispersed QDs was achieved with the development of high-temperature organometallic methods. Bawendi and coworkers showed that QDs with excellent properties could be produced using hot coordinating solvents and organometallic cadmium precursors [25]. Further advances in QD synthesis were made with the introduction of chelated cadmium precursors [26] and noncoordinating solvents [44]. Using these methods, precise control over the nanoparticles' growth rates can be achieved, yielding small QDs with uniform sizes. However, the small size of nanocrystals results in a high surface-to-volume ratio, where surface atoms play a significant role in determining the electronic and optical properties of the QD. Defects on the surface of the nanocrystal can dramatically affect quantum yield and photostability, resulting in QDs with low fluorescence intensity that can be easily quenched. To overcome this problem, synthetic methods were developed to grow inorganic shells over the nanocrystal core to passivate the surface and minimize problems associated with surface defects [45,46]. The inorganic passivating shell also serves to isolate the core from the environment, protecting the nanocrystal from degradation and increasing its stability.

Using these optimized synthetic methods, QDs can be easily produced with excellent optical properties, making them superior to organic fluorescent dyes in many ways. While fluorescent dyes and proteins have proven to be exceptional biological imaging agents [47–49], they suffer from optical limitations that hinder their utility for many diagnostic or bioimaging applications. Organic fluorophores typically have narrow absorption spectra and a small stokes shift, which limits the excitation source to a narrow wavelength range and prevents the excitation of multiple fluorescence colors. In contrast, QDs have a large stokes shift with very broad absorption spectra (Figure 21.2a) and can be excited using a wide range of wavelengths. Because a QD can essentially be excited using any wavelength that is of higher energy than its emission wavelength, multiple fluorescent colors can be simultaneously excited using a single, high-energy light source. In addition, the emission spectra of QDs are more narrow and symmetric than those of fluorescent dyes, allowing two QDs with similar emission wavelengths to be detected with less signal overlap in comparison to dyes.

QDs also have high quantum yields, often over 75%. This high quantum yield, in addition to the significantly larger absorption cross section of QDs, results in a nanoparticle that is approximately 20 times brighter than a standard organic fluorophore [7]. In fact, a single QD is bright enough to image using standard fluorescence microscopy, making these ideal probes for ultrasensitive biomolecule detection or tagging and tracking single molecules in live cells [50]. Finally, QDs exhibit a strong resistance to photobleaching in comparison with organic fluorescent dyes (Figure 21.2b), which can begin to photobleach immediately upon exposure to light [51]. The constant fluorescence intensity exhibited by QDs makes them ideal diagnostic and bioimaging probes for applications where consistent measurements are critical for accurate clinical determinations.

In addition to optimizing QD synthetic procedures for maximum optical performance, recent research has focused on the development of synthetic approaches to tune the nanocrystals' bandgap through novel material compositions, referred to as *bandgap engineering* (Figure 21.3) [52]. Early efforts in bandgap engineering involved the colloidal synthesis of type-II QDs, where the valence and conduction bands of the core overlap with the bands in the inorganic shell [53]. As a result, the charge carriers in the semiconductor are spatially separated (one confined to the core, the other confined to the shell), in contrast to type-I QDs where both charge carriers are confined to the nanocrystal core. When the electron and hole recombine across the interface of the core and shell, light emission occurs, with the emission energy dependent on the band offsets of the core and shell. This allows emission wavelengths that are not possible from either material alone.

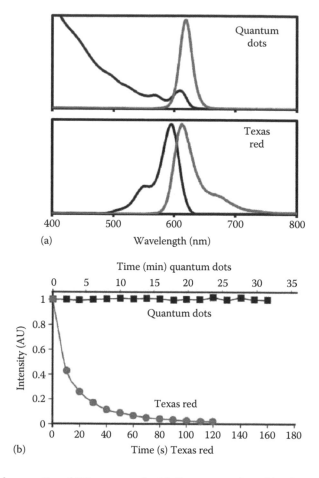

FIGURE 21.2 Optical properties of QDs compared with fluorescence dyes. (a) CdSe/ZnS QDs (top) exhibit a broad absorption spectrum (black curve) and narrow, symmetrical emission (grey curve) in comparison with Texas red (bottom), a small-molecule fluorescent dye. (b) Studies have shown that QDs (black curve) are highly resistant to photobleaching and are several thousand times more photostable than organic dyes such as Texas red (grey curve) under identical conditions. (b: Adapted from Gao, X. et al., *Curr. Opin. Biotechnol.*, 16(1), 63, 2005.)

Another approach is through the synthesis of alloyed QDs with ternary structures, such as $CdSe_{1-x}Te_x$ and $Hg_xCd_{1-x}E$ [54,55]. These novel structures provide a strategy for maintaining a uniform size across nanocrystals with a wide range of emission wavelengths by tuning the bandgap of the QD using composition. These methods have been used to prepare ultrasmall QDs with bright fluorescence in the near infrared, a critical step for in vivo bioimaging that had previously been a major challenge. Doped semiconductor QDs, such as ZnSe:Mn [56,57], have also been reported and can have desirable characteristics, such as the lack of toxic heavy metal atoms. While these nanocrystals may play a future role (particularly for in vivo applications where heavy metal content is undesirable), further research is needed to address challenges in their synthesis.

Finally, Smith et al. demonstrated that it was possible to tune the bandgap of semiconductor QDs by inducing strain within the nanocrystal through a mismatch in the lattice structure of the core and shell [58]. Surprisingly, small nanocrystals can significantly deform in response to compressive or tensile stress, which is distributed over a large percentage of the nanocrystals' atoms because of the high surface-to-volume ratio. This is in direct contrast to bulk semiconductor materials, which respond by generating dislocations (defects) to relieve the stress and allow the materials to relax to their natural state.

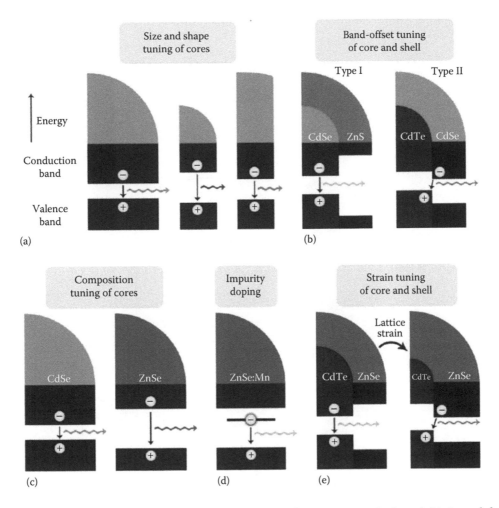

FIGURE 21.3 Mechanisms of bandgap engineering in semiconductor nanocrystals through (a) size and shape, (b) heterostructure band offset, (c) composition, (d) impurity doping, and (e) lattice strain. (Adapted from Smith, A.M. and Nie, S., *Acc. Chem. Res.*, 43(2), 190, 2010.)

Under this strain, the atoms of the QD have an altered bond length, which directly affects the electronic bandgap of the material. Using this strategy, CdTe/ZnSe QDs with emissions spanning approximately 350 nm can be produced.

21.3 Surface Coating and Biotagging Strategies

As detailed earlier, high-quality synthetic methods can produce QDs with superb physical and optical properties, but generally involve the use of nonpolar solvents and aliphatic ligands that coordinate the precursor atoms in solution and bind the surface of the nanoparticles for stabilization. This results in an organic surface coating that is extremely hydrophobic, making the QDs insoluble in aqueous solutions without further modifications and unsuitable for biological environments. A variety of strategies have been developed to address this issue and render the QDs water soluble for use in biomedical applications, as shown in Figure 21.4 [8,14,20,21,59–63].

One of the earliest methods employed to transfer hydrophobic QDs to aqueous solutions is through direct ligand exchange [21]. This procedure involves transferring the QDs to a solvent containing an excess of small molecules that are capable of displacing the aliphatic ligands and coordinating with the

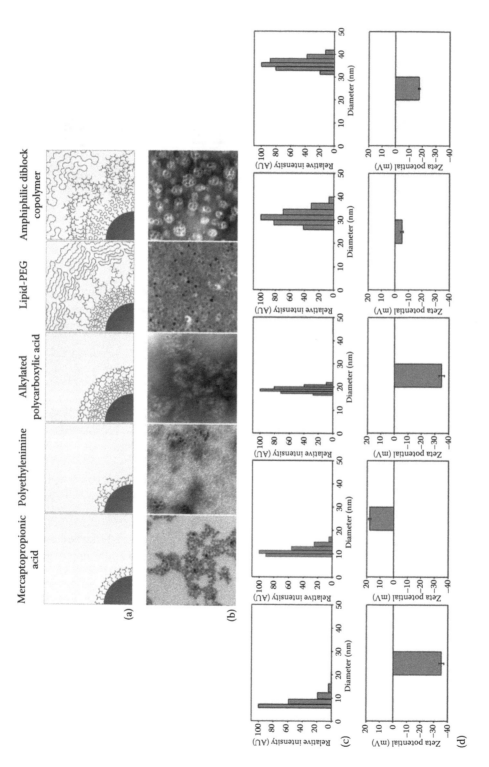

FIGURE 21.4 QD coatings and colloidal properties. (a) Schematics showing a variety of common QD coatings used for transfer to water, including (left to right) mercaptopropionic acid, polyethylenimine, alkylated polycarboxylic acid, lipid PEG, and amphiphilic diblock copolymers. (b) TEMs of coated QDs in water, counterstained to show organic coating layer. (c) Dynamic light scattering data reveal nanoparticle hydrodynamic sizes ranging from ~8 nm to over 30 nm. (d) Nanoparticles can be designed with a range of surface charges, dependent on the functional groups present on the surface. (Adapted from Wang, Y.A. et al., *J. Am. Chem. Soc.*, 124(10), 2293, 2002.)

surface of the nanocrystal. Thiol functional groups were found to coordinate strongly with the atoms of the QD surface, and polar molecules that contain these groups (such as mercaptoacetic acid) were shown to be well suited for initial transfers to water. By using small molecules for the aqueous transfer, the physical and hydrodynamic size of the nanoparticles is also kept to a minimum. However, aqueous QDs prepared using this method are relatively unstable, leading to particle aggregation and precipitation after just a few days. Modifications to this transfer strategy include the use of alternate functional groups to coordinate the QD surface, as well as multidentate ligands that bind the nanocrystal at multiple points [18,59,60]. In addition, biocompatible molecules such as amino acids and proteins have also been used for direct ligand exchange and QD coatings [13,64].

Another method for the transfer of QDs to water is the use of amphiphilic polymers to completely encapsulate the hydrophobic nanoparticles [15,65,66]. This strategy utilizes polymers that are capable of interacting with both the hydrophobic surface of the nanocrystal and the aqueous environment. Polymers containing both linear hydrocarbons and polar functional groups were shown to be ideal for coating nanoparticles stabilized by hydrophobic ligands. The aliphatic groups of the polymer intercalate between the ligands on the surface of the QD through hydrophobic–hydrophobic interactions, while the polar backbone of the polymer encapsulates the QD and provides a hydrophilic interface to interact with the surrounding water molecules (Figure 21.4a). This results in an *oily* protective layer sandwiched between the nanocrystal surface and the hydrophilic polymer backbone that protects the surface of the QD from the aqueous environment. The protective layer also maintains the extremely high quantum yields of QDs (in contrast to aqueous strategies that result in a significant quantum yield reduction) and makes the coating incredibly stable due to the multiple interactions between the polymer and the QD surface. A wide range of polymers have been used for this method, including low-molecular-weight polymers [8,65], block copolymers [67,68], and triblock copolymers [15]. It has also been shown that some amphiphilic polymers can be used as multidentate precursor ligands for the synthesis of high-quality QDs [69,70]. By using excess polymer during the synthesis procedure and a noncoordinating amphiphilic solvent that is miscible with water, free polymer molecules can spontaneously encapsulate the QDs upon the addition of water, providing a *one-pot* route for QD synthesis and surface coating.

A notable downside to polymer coating strategies is the substantial increase in particle size, which can exceed 300% for methods using high-molecular-weight polymers (Figure 21.4c). Nie and coworkers have recently developed a multidentate polymer ligand that displaces the hydrophobic surface ligands and binds directly to the nanocrystal [71,72]. This novel scheme eliminates the oily intermediate layer and drives the polymer to bind tightly against the QD surface, resulting in a highly compact polymer coating and small overall size. In contrast with the small-molecule ligand exchange method, the multidentate binding of the polymer provides excellent colloidal stability, resistance to photobleaching, and high quantum yield. This strategy has been used to prepare high-quality QDs with a hydrodynamic size of approximately 5.5 nm, the maximum size possible for renal clearance in vivo [13].

While the coatings and water solubilization strategies detailed earlier have allowed the transfer of bright QDs into aqueous solutions, additional issues must be addressed before the nanoparticles can be used as probes in complex biological environments. A major challenge for the use of QDs in biomedical applications is that they can bind nonspecifically to biomolecules [73–76], which reduces the signal-to-noise ratio and limits specificity and sensitivity for biomolecule detection. It has been shown that surface charge plays a critical role, with highly positive or negative surface charge contributing greatly to the *stickiness* of the nanoparticles [73–75], likely through electrostatic interactions with the charged domains of biomolecules [77]. Polyethylene glycol (PEG) is commonly used in biology to reduce nonspecific binding and confer biocompatibility and has been successfully coupled to the coatings of QDs for this purpose [8,73]. The neutral charge and conformational flexibility of PEGylated QDs result in a stable probe with significantly reduced nonspecific binding in vitro [73] and in vivo [15,78,79]. Despite this success, a major drawback is that PEGylated QDs have significantly larger hydrodynamic sizes than comparable non-PEGylated nanoparticles, hindering their use in applications where ultrasmall sizes are crucial. An alternative approach reported by Kairdolf et al. [76] uses small-molecule cross-linking agents containing hydroxyl functional groups to impart a near-neutral charge to the QD surface. This method eliminates nonspecific binding without altering the size or optical properties of the nanocrystals.

To activate the QDs for biomolecule detection, specific targeting moieties must be coupled to the surface of the nanoparticles. A number of bioconjugation strategies have been described in the literature, providing a diverse toolkit to adapt the nanoprobes for a particular need. Covalent binding of a targeting molecule to the surface of the QD is the most widely used strategy, and rich chemistries have been developed to couple a range of common functional groups, including amines, carboxylic acids, and thiols [18,23,80,81]. Using this strategy, small molecules, peptides, nucleic acids, proteins, and antibodies have been successfully coupled to the nanoparticle coating for targeted imaging or detection. More recently, covalent coupling of a chloroalkane to fusion proteins containing a modified haloalkane dehalogenase was used for site-specific conjugation to biomolecules [82].

Another strategy used to attach targeting moieties or other biomolecules to QD probes is to exploit noncovalent interactions. The streptavidin–biotin interaction is commonly used in biotechnology and has proven useful for coupling a variety of biomolecules together. By covalently linking streptavidin to the surface of a QD, a versatile imaging agent is produced, which can be combined with any biotinylated molecule to generate a targeted probe immediately before use [8,83]. Gao and coworkers have recently reported a bioconjugation strategy using protein A, which is capable of noncovalently binding the Fc region of antibodies. Using this coupling strategy, a novel multicycle process can be used for ultrahigh multiplexing of biomarkers in cells. This multicycle staining method overcomes the spectral limitations of fluorescence multiplexing and allows up to 100 unique biomarkers to be probed, opening new possibilities for molecular profiling of single cells using QD probes [84]. Surface charge and electrostatic interactions have also been used to bind molecules to negatively charged QDs through fusion proteins with a positively charged attachment domain [64]. Conversely, positively charged QDs have been used to couple negatively charged biomolecules, such as siRNA, to nanocrystals [85]. Finally, conjugation methods have been developed which take advantage of the binding affinities of the surface atoms of the nanocrystal itself. Polyhistidine has micromolar affinity for metal atoms such as Zn, which is commonly used in QD passivating shells. While QDs with an amphiphilic polymer coating shield the atoms on the nanocrystal surface and prevent the binding of polyhistidines, nanoparticles coated with small-molecule ligands or the multidentate polymer allow the peptides to access the nanocrystal surface and bind surface atoms [86,87]. This method has an added advantage in that proteins containing the polyhistidine tag bind to the nanocrystal in a prescribed orientation, rather than the random orientations common with many of the other techniques.

21.4 Biodiagnostic Applications

Over the past decade, efforts to develop QD-based nanoprobes for biodiagnostic applications have increased exponentially, with the most significant clinical impact to date seen in the in vitro analysis of biomolecules, cells, and tissues in patient specimens. The exceptional optical properties of QDs (such as their high signal intensity and multiplexing potential) can extend the capabilities of many clinically important assays, including flow cytometry, fluorescence in situ hybridization (FISH), and immunohistochemistry (IHC). The multiplexing capability of QDs is particularly intriguing because it enables high-throughput biomarker detection in minute clinical specimens, providing a potential diagnostic scheme for rapid molecular profiling and personalized medicine. QD nanoprobes also open the possibility for new detection assays, such as real-time detection of single biomolecules, which would otherwise be impossible using existing diagnostic technologies.

Solution-based diagnostic assays are particularly important in clinical chemistry, providing physicians with critical information regarding the levels of key biomolecules within bodily fluids such as blood or urine. As our knowledge of human genetics continues to grow, the analysis of RNA and DNA and the detection of genetic defects play an increasingly important role in the diagnosis of disease and the identification of at-risk patients. Advances in QD probes make them well suited for these applications, and their unique properties have been exploited to design solution-based assays for high-sensitivity detection of analytes, like nucleic acids. Zhang et al. have developed a DNA sensor by conjugating a nucleic acid capture sequence (complementary to the target sequence) to QDs [88]. After a target DNA is bound to the capture sequence on the nanoprobe, a second reporter sequence containing an organic dye binds to the target in a sandwich format to form a FRET pair for signal detection. Using this strategy, femtomolar (10^{-15} M) detection sensitivity was achieved, enabling the detection of low-copy nucleic acids without the need for amplification methods such as polymerase chain reaction (PCR).

While this method demonstrates the incredible sensitivity possible using QD technologies, high-throughput analysis is crucial to analyze the number of genes required for a robust molecular profile of an individual patient. Nie and coworkers demonstrated the amazing multiplexing potential of QDs with the development of optically encoded microbeads for biomolecule identification and measurement [5]. They demonstrated that one million unique *optical barcodes* could theoretically be generated by loading a single microbead with precise ratios of only 6 QD wavelengths at 10 intensity levels. Using the capture sequence concept to prepare nanoprobes for nucleic acid targets, they showed that multiple nucleic acid targets could be accurately detected at the single bead level in a high-throughput process (Figure 21.5).

Flow-based methods are another assay format where QD probes have dramatically improved diagnostics and detection. Agrawal et al. developed a microfluidic device that exploits the large stokes shift and multiplexing capabilities of QDs for the identification and detection of single intact viruses (Figure 21.6). Using both red and green QDs targeted to different proteins on the virus surface, they were able to identify and measure respiratory syncytial virus (RSV), a primary cause of lower respiratory tract disease in infants and young children and an important pathogen of the elderly and immune compromised individuals [4]. A single virus was distinguished from unbound QD signals in real time using photon coincidence of the two QDs (Figure 21.6, bottom). Control samples with parainfluenza virus type 3 (PIV3) showed no detectable coincidence signal, while high coincidence was measured in samples containing RSV. The method was also used to determine differences in viral surface protein expression in single virus particles, which is directly related to virulence and the severity of the infection [3]. Flow cytometry is another flow-based assay that has benefited from the exceptional optical properties of QDs. Roederer and coworkers used multiplexed QDs in combination with fluorescent dyes to simultaneously detect 17 unique biomarkers on the surface of cells, significantly expanding the capabilities of standard fluorescence flow cytometry for applications in many areas of medicine, including immunology and cancer [89].

While a number of assay techniques have been reported that use QDs for the analysis of biofluids, histology and cytology are the applications that have received the most attention for QD development. Multiplexed immunostaining and microscopic analysis of disease biomarkers in cells and tissues are applications that take full advantage of the unique features of QDs. Nie and coworkers published a seminal report detailing methods for the preparation and use of QD probes for immunostaining and analyzing biomarker expression in clinical tissue specimens [12]. Since that work, a number of groups have reported the use of multiplexed QDs for IHC in clinical tissue specimens and have shown that IHC is significantly enhanced using QD nanoprobes [90–94]. Chen et al. performed a systematic study, examining the effectiveness of QD-based IHC detection in comparison with current methods [92]. Their work demonstrated that QDs provide a higher sensitivity and accuracy than current methods that use standard chromogens and was independently confirmed by Yu and coworkers for the analysis of tissue microarrays [95].

The multiplexing feature of QDs is especially advantageous for the clinical examination of a panel of biomarkers rather than a single biomolecule. This is a common practice for the diagnosis of breast cancer, where estrogen receptor (ER), progesterone receptor (PR), and human epidermal growth factor 2 (Her2) are routinely measured in breast biopsies to make a diagnostic determination and select the most appropriate therapy strategy for the patient [96–98]. Yezhelyev et al. showed that QD staining results for the simultaneous detection of these markers in a patient specimen on a single slide correlated closely with results obtained using standard chromogen-based methods, which are performed one marker at a time [24]. Further work by Pittaluga and coworkers showed that seven

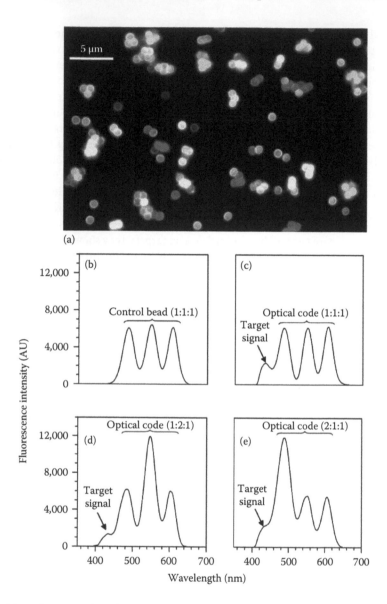

FIGURE 21.5　(See color insert.) Multiplexed optical coding using QD-tagged microbeads. (a) Fluorescence micrograph of microbeads loaded with QDs emitting at wavelengths covering the entire visible range. By loading a single microbead with precise ratios of QDs emitting at different wavelengths, an optical barcode can be generated for biomolecule identification. Coupling a unique nucleic acid probe sequence to each microbead, multiple target DNA sequences can be detected in a high-throughput process (c–e), while control sequences show no signal (b). (Adapted from Han, M. et al., *Nat. Biotechnol.*, 19(7), 631, 2001.)

unique QD colors (representing a seven biomarker panel) could be simultaneously distinguished and measured within a single clinical tissue specimen [93].

QD nanoparticles can even be used to probe individual cells, opening new possibilities for single-cell molecular profiling. This is particularly important in early diagnosis, when malignant cells may be rare and difficult to identify, or for the identification of isolated diseased cells in a complex microenvironment. Nie and coworkers have demonstrated the concept of single-cell molecular profiling in clinical specimens from patients with prostate cancer and Hodgkin's lymphoma [11,12]. Using multiplexed QDs

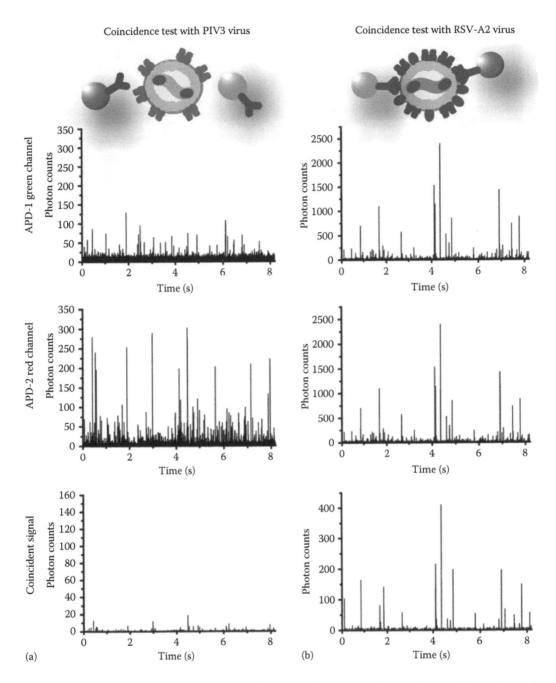

FIGURE 21.6 Time-correlated coincidence signals for virus detection in solution. Nanoparticles conjugated to RSV anti-F or anti-G protein monoclonal antibodies were used to produce red or green emission, respectively. PIV3, used as a control, produced low fluorescence signals that did not show coincidence (a), while coincident peaks were observed for the RSV/A2 (b). (Adapted from Agrawal, A. et al., *J. Virol.*, 79(13), 8625, 2005.)

targeted to a panel of four markers, they were able to perform molecular mapping of the heterogeneity of the specimen and identify isolated cancer cells within prostate glands that were predominately normal (Figure 21.7a). These patients may have been diagnosed as benign using standard analysis methods. The technique was so sensitive that a single cancerous cell was identified in a prostate gland, just as the gland was beginning its malignant transformation. A similar technique was used in the rapid detection

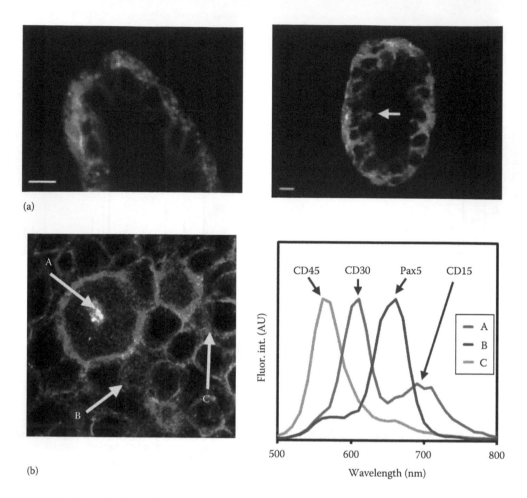

FIGURE 21.7 **(See color insert.)** Immunoprofiling in complex tissues for the identification and characterization of rare cells. (a) Multiplexed QD imaging in prostate biopsy specimens allows the differentiation of a benign prostate gland (left) from a gland with a single malignant cell (right), as determined by positive alpha-methylacyl-CoA racemace staining (arrow). (b) A four-biomarker panel was used to identify low-abundance RS cells (A), B cells (B), and T cells (C) in a heterogeneous lymph node specimen for the diagnosis of Hodgkin's lymphoma (left). Using wavelength-resolved imaging, the QD staining pattern can be analyzed to determine the biomarker expression profile of single cells within the specimen (right), allowing accurate differentiation. (Adapted from Liu, J. et al., *Anal. Chem.*, 82(14), 6237, 2010; Liu, J. et al., *ACS Nano.*, 4(5), 2755, 2010.)

of rare Reed–Sternberg (RS) cells within biopsy specimens from patients suspected of having Hodgkin's lymphoma, as shown in Figure 21.7b [11]. The presence of these cells is a hallmark of the disease but they are relatively rare, which makes it difficult for a pathologist to locate and identify the cells in lymph node specimens. A four-biomarker panel of QD probes was used for RS cell identification and compared to previously determined results from pathological examinations. Simultaneous biomarker analysis using multiplexed QDs resulted in the rapid identification of all patients with confirmed disease and even showed disease presence in patients originally unclassified. These unclassified patients had extremely low levels of RS cells and were likely below the level of detection using current methods. Specimens from patients without the disease showed a complete absence of RS cells, illustrating the high sensitivity and specificity of QD-based single-cell molecular profiling.

21.5 Bioimaging

The unique optical properties of QDs have also led to a considerable amount of research to develop the nanoparticles as probes for in vivo bioimaging and molecular detection. Groundbreaking work was reported by Gao et al. for the use of targeted QD probes to detect prostate cancer in live mice models (Figure 21.8a) [15]. Their in vivo studies indicate that the QD probes preferentially accumulate in tumors due to both specific antigen targeting (prostate-specific membrane antigen, or PSMA, on the surface of the cancer cells) and the nonspecific enhanced permeability and retention (EPR) effect, resulting in high fluorescence signal in the tumor under strong excitation. However, this excitation results in a high background signal due to the autofluorescence of the tissue. Rao and coworkers detailed a modified QD probe based on bioluminescence resonance energy transfer (BRET) that does not require external excitation, dramatically reducing background signal (Figure 21.8b) [99]. QD probes were coupled to a luciferase bioluminescent protein, which provided *self-illumination* for QD excitation. While this scheme resulted in a fluorescence signal with very low background, it requires the systemic injection of a luciferin molecule to fuel the luciferase and generate bioluminescence and fluorescence.

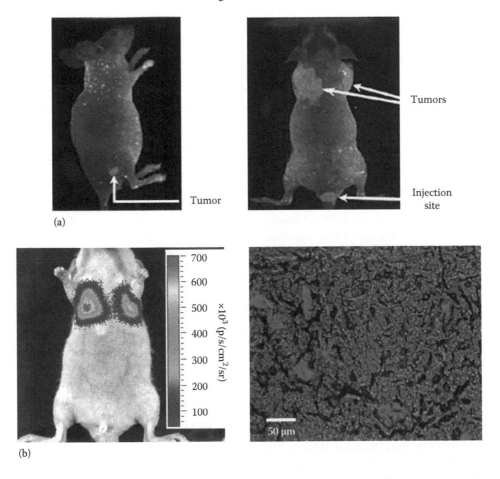

FIGURE 21.8 In vivo imaging with QDs. (a) QDs targeted to the PSMA marker on prostate cancer cells preferentially accumulated in prostate tumors in mice (right), while nontargeted QDs showed low tumor accumulation (left). (b) BRET probes composed of a QD and luciferase proteins allow in vivo bioluminescence imaging in the absence of external excitation (left). Histology of a lung slice confirms the fluorescence signal seen in vivo (right). (Adapted from Gao, X. et al., *Nat. Biotechnol.*, 22(8), 969, 2004; So, M.-K. et al., *Nat. Biotechnol.*, 24(3), 339, 2006.)

Despite the early success of these studies, fluorescence in vivo imaging is particularly challenging, and a number of issues must be addressed before QDs can be used for in vivo clinical applications. A major problem for in vivo fluorescence imaging is the interaction of light with tissue, which is fairly opaque in the visible light range and dramatically reduces signal. This issue can be overcome, however, by shifting the fluorescence emission of the QD probes into the near-infrared, where a transparent window exists with significantly reduced light absorption and scattering [100]. Advances in QD synthetic procedures (described previously) have provided a number of strategies to prepare QD probes that emit within this window. These probes have been successfully used for a number of in vivo applications, such as sentinel lymph node mapping [16].

The most challenging issue to be addressed for the translation of QDs for in vivo clinical applications is toxicity. Virtually all high-quality QDs consist of a proportion of toxic heavy metal atoms (cadmium, mercury, etc.), which are known to cause toxicity in vivo. While toxicity studies on QDs have reported conflicting results [101], concerns remain because of the chemical makeup of the nanoparticles and their nanoscale size, making clearance by the Food and Drug Administration (FDA) for in vivo use especially difficult. Despite these challenges, ongoing research has shown promising developments and novel QD probes could find use for in vivo applications in humans in the future. Considerable effort has been made to eliminate or dramatically reduce the heavy metal content of QDs. Peng and coworkers have described doped QDs that completely lack toxic heavy metals, consisting of only zinc, selenium, and manganese [56,57]. While these are particularly attractive because of toxicity concerns, further research is needed to address synthetic challenges and push the fluorescence emission into the near-infrared window. The strain-tuned QDs reported by Smith et al., containing a tiny CdTe core and large ZnSe shell, are also promising because they dramatically reduce the cadmium content in comparison to standard QDs [58]. The nanoparticles have an added advantage in that they can be easily tuned to emit bright fluorescence in the near-infrared wavelength range.

Using an entirely different approach, Choi et al. reported ultrasmall QD probes that can be eliminated from the body through renal clearance [13]. Their study showed that QD nanoprobes with an overall hydrodynamic size smaller than 5.5 nm resulted in rapid urinary excretion and elimination from the body in live mice studies, while QDs with a size larger than 5.5 nm were retained in the body. These results are consistent with the known size for renal filtration and provide a potential strategy for the elimination of potentially toxic QD probes. Using size-minimized surface coating strategies [72,76] and novel QD compositions [53,54,58], ultrasmall QDs with bright, near-infrared emission could have utility for in vivo bioimaging and molecular biomolecule detection.

21.6　Concluding Remarks

Looking ahead, we expect to see continued development efforts in both fundamental nanoscience and the biomedical translation of semiconductor nanocrystals, yielding major advances in the field. Over the past two decades, efforts in nanocrystal preparation have resulted in new classes of QDs with excellent and diverse properties and rich chemistries for interfacing the nanocrystals with biological environments. These advances have opened a wide range of applications in biomedicine, including ultrasensitive molecule detection and multiplexed biomarker imaging. In the next decade, we predict that this trend will continue and yield novel nanomaterials with currently unforeseen structures and properties, opening new applications in clinical diagnostics and bioimaging. Fundamental studies have revealed the critical importance that nanocrystal properties, such as size, surface charge, multivalency of targeting moieties, and emission wavelength, play in their interaction with biological systems. Because of this, we expect future advancements in the use of QDs for biodiagnostics and in vivo imaging will require a multidisciplinary effort across diverse fields such as chemistry, material science, biology, biomedical engineering, and toxicology.

References

1. Agrawal, A. et al., Nanometer-scale mapping and single-molecule detection with color-coded nanoparticle probes. *Proc Natl Acad Sci USA*, 2008. **105**(9): 3298–3303.
2. Agrawal, A., T. Sathe, and S. Nie, Single-bead immunoassays using magnetic microparticles and spectral-shifting quantum dots. *J Agric Food Chem*, 2007. **55**(10): 3778–3782.
3. Agrawal, A. et al., Real-time detection of virus particles and viral protein expression with two-color nanoparticle probes. *J Virol*, 2005. **79**(13): 8625–8628.
4. Agrawal, A. et al., Counting single native biomolecules and intact viruses with color-coded nanoparticles. *Anal Chem*, 2006. **78**(4): 1061–1070.
5. Han, M. et al., Quantum-dot-tagged microbeads for multiplexed optical coding of biomolecules. *Nat Biotechnol*, 2001. **19**(7): 631–635.
6. Sathe, T.R., A. Agrawal, and S. Nie, Mesoporous silica beads embedded with semiconductor quantum dots and iron oxide nanocrystals: Dual-function microcarriers for optical encoding and magnetic separation. *Anal Chem*, 2006. **78**(16): 5627–5632.
7. Chan, W.C.W. et al., Luminescent quantum dots for multiplexed biological detection and imaging. *Curr Opin Biotechnol*, 2002. **13**(1): 40–46.
8. Wu, X. et al., Immunofluorescent labeling of cancer marker Her2 and other cellular targets with semiconductor quantum dots. *Nat Biotechnol*, 2003. **21**(1): 41–46.
9. Yezhelyev, M.V. et al., Emerging use of nanoparticles in diagnosis and treatment of breast cancer. *Lancet Oncol*, 2006. **7**(8): 657–667.
10. Xing, Y. et al., Molecular profiling of single cancer cells and clinical tissue specimens with semiconductor quantum dots. *Int J Nanomed*, 2006. **1**(4): 473–481.
11. Liu, J. et al., Multiplexed detection and characterization of rare tumor cells in Hodgkin's lymphoma with multicolor quantum dots. *Anal Chem*, 2010. **82**(14): 6237–6243.
12. Liu, J. et al., Molecular mapping of tumor heterogeneity on clinical tissue specimens with multiplexed quantum dots. *ACS Nano*, 2010. **4**(5): 2755–2765.
13. Choi, H.S. et al., Renal clearance of quantum dots. *Nat Biotechnol*, 2007. **25**(10): 1165–1170.
14. Dubertret, B. et al., In vivo imaging of quantum dots encapsulated in phospholipid micelles. *Science*, 2002. **298**(5599): 1759–1762.
15. Gao, X. et al., In vivo cancer targeting and imaging with semiconductor quantum dots. *Nat Biotechnol*, 2004. **22**(8): 969–976.
16. Kim, S. et al., Near-infrared fluorescent type II quantum dots for sentinel lymph node mapping. *Nat Biotechnol*, 2004. **22**(1): 93–97.
17. Alivisatos, P., The use of nanocrystals in biological detection. *Nat Biotechnol*, 2004. **22**(1): 47–52.
18. Medintz, I.L. et al., Quantum dot bioconjugates for imaging, labelling and sensing. *Nat Mater*, 2005. **4**(6): 435–446.
19. Smith, A.M., X. Gao, and S. Nie, Quantum dot nanocrystals for in vivo molecular and cellular imaging. *Photochem Photobiol*, 2004. **80**(3): 377–385.
20. Bruchez, M.P. et al., Semiconductor nanocrystals as fluorescent biological labels. *Science*, 1998. **281**(5385): 2013–2016.
21. Chan, W.C.W. and S.M. Nie, Quantum dot bioconjugates for ultrasensitive nonisotopic detection. *Science*, 1998. **281**(5385): 2016.
22. Rhyner, M.N. et al., Quantum dots and multifunctional nanoparticles: New contrast agents for tumor imaging. *Nanomedicine*, 2006. **1**: 209–217.
23. Xing, Y. et al., Bioconjugated quantum dots for multiplexed and quantitative immunohistochemistry. *Nat Protoc*, 2007. **2**(5): 1152–1165.
24. Yezhelyev, M.V. et al., In situ molecular profiling of breast cancer biomarkers with multicolor quantum dots. *Adv Mater*, 2007. **19**: 3146–3151.

25. Murray, C.B., D.J. Norris, and M.G. Bawendi, Synthesis and characterization of nearly monodisperse CdE (E = S, Se, Te) semiconductor nanocrystallites. *J Am Chem Soc*, 1993. **115**(19): 8706–8715.

26. Peng, Z.A. and X. Peng, Formation of high-quality CdTe, CdSe, and CdS nanocrystals using CdO as precursor. *J Am Chem Soc*, 2001. **123**(1): 183–184.

27. Qu, L. and X. Peng, Control of photoluminescence properties of CdSe nanocrystals in growth. *J Am Chem Soc*, 2002. **124**(9): 2049–2055.

28. Talapin, D.V. et al., CdSe/CdS/ZnS and CdSe/ZnSe/ZnS core shell-shell nanocrystals. *J Phys Chem B*, 2004. **108**(49): 18826–18831.

29. Talapin, D.V. et al., Highly luminescent monodisperse CdSe and CdSe/ZnS nanocrystals synthesized in a hexadecylamine-trioctylphosphine oxide-trioctylphosphine mixture. *Nano Lett*, 2001. **1**(4): 207–211.

30. Xie, R. et al., Synthesis and characterization of highly luminescent CdSe core CdS/$Zn_{0.5}Cd_{0.5}$S/ZnS multishell nanocrystals. *J Am Chem Soc*, 2005. **127**(20): 7480–7488.

31. Liu, Z. et al., In vivo biodistribution and highly efficient tumour targeting of carbon nanotubes in mice. *Nat Nanotechnol*, 2007. **2**(1): 47–52.

32. Weissleder, R. et al., Cell-specific targeting of nanoparticles by multivalent attachment of small molecules. *Nat Biotechnol*, 2005. **23**(11): 1418–1423.

33. Lee, E.S., K. Na, and Y.H. Bae, Polymeric micelle for tumor pH and folate-mediated targeting. *J Control Release*, 2003. **91**: 103–113.

34. Kortan, A. et al., Nucleation and growth of cadmium selenide on zinc sulfide quantum crystallite seeds, and vice versa, in inverse micelle media. *J Am Chem Soc*, 1990. **112**(4): 1327–1332.

35. Lianos, P. and J.K. Thomas, Cadmium sulfide of small dimensions produced in inverted micelles. *Chem Phys Lett*, 1986. **125**(3): 299–302.

36. Spanhel, L. et al., Photochemistry of colloidal semiconductors. 20. Surface modification and stability of strong luminescing CdS particles. *J Am Chem Soc*, 1987. **109**(19): 5649–5655.

37. Celebi, S. et al., Synthesis and characterization of poly(acrylic acid) stabilized cadmium sulfide quantum dots. *J Phys Chem B*, 2007. **111**(44): 12668–12675.

38. Gaponik, N. et al., Thiol-capping of CdTe nanocrystals: An alternative to organometallic synthetic routes. *J Phys Chem B*, 2002. **106**(29): 7177–7185.

39. He, Y. et al., Synthesis of CdTe nanocrystals through program process of microwave irradiation. *J Phys Chem B*, 2006. **110**(27): 13352–13356.

40. He, Y. et al., Microwave-assisted growth and characterization of water-dispersed CdTe/CdS core-shell nanocrystals with high photoluminescence. *J Phys Chem B*, 2006. **110**(27): 13370–13374.

41. Li, L., H.F. Qian, and J.C. Ren, Rapid synthesis of highly luminescent CdTe nanocrystals in the aqueous phase by microwave irradiation with controllable temperature. *Chem Commun*, 2005. (4): 528–530.

42. Qian, H.F. et al., Microwave-assisted aqueous synthesis: A rapid approach to prepare highly luminescent ZnSe(S) alloyed quantum dots. *J Phys Chem B*, 2006. **110**(18): 9034–9040.

43. Zhang, H. et al., Hydrothermal synthesis for high-quality CdTe nanocrystals. *Adv Mater*, 2003. **15**(20): 1712–1715.

44. Yu, W.W. and X.G. Peng, Formation of high-quality CdS and other II-VI semiconductor nanocrystals in noncoordinating solvents: Tunable reactivity of monomers. *Angew Chem Int Ed*, 2002. **41**(13): 2368–2371.

45. Dabbousi, B.O. et al., (CdSe)ZnS core-shell quantum dots: Synthesis and characterization of a size series of highly luminescent nanocrystallites. *J Phys Chem B*, 1997. **101**(46): 9463–9475.

46. Hines, M.A. and P. Guyot-Sionnest, Synthesis and characterization of strongly luminescing ZnS-Capped CdSe nanocrystals. *J Phys Chem*, 1996. **100**(2): 468–471.

47. Finley, K.R., A.E. Davidson, and S.C. Ekker, Three-color imaging using fluorescent proteins in living zebrafish embryos. *Biotechniques*, 2001. **31**(1): 66.

48. Giepmans, B.N.G. et al., The fluorescent toolbox for assessing protein location and function. *Am Assoc Adv Sci*. 2006. **312**(5771): 217–224.

49. Giuliano, K.A. and D.L. Taylor, Fluorescent-protein biosensors: New tools for drug discovery. *Trends Biotechnol*, 1998. **16**(3): 135–140.

50. Ruan, G. et al., Imaging and tracking of tat peptide-conjugated quantum dots in living cells: New insights into nanoparticle uptake, intracellular transport, and vesicle shedding. *J Am Chem Soc*, 2007. **129**(47): 14759–14766.

51. Gao, X. et al., In vivo molecular and cellular imaging with quantum dots. *Curr Opin Biotechnol*, 2005. **16**(1): 63–72.

52. Smith, A.M. and S. Nie, Semiconductor nanocrystals: Structure, properties, and band gap engineering. *Acc Chem Res*, 2010. **43**(2): 190–200.

53. Kim, S. et al., Type-II quantum dots: CdTe/CdSe(core/shell) and CdSe/ZinTe(core/shell) heterostructures. *J Am Chem Soc*, 2003. **125**(38): 11466–11467.

54. Smith, A.M. and S. Nie, Bright and compact alloyed quantum dots with broadly tunable near-infrared absorption and fluorescence spectra through mercury cation exchange. *J Am Chem Soc*, 2011. **133**(1): 24–26.

55. Bailey, R.E. and S. Nie, Alloyed semiconductor quantum dots: Tuning the optical properties without changing the particle size. *J Am Chem Soc*, 2003. **125**(23): 7100–7106.

56. Pradhan, N. and X. Peng, Efficient and color-tunable Mn-doped ZnSe nanocrystal emitters: Control of optical performance via greener synthetic chemistry. *J Am Chem Soc*, 2007. **129**(11): 3339–3347.

57. Pradhan, N. et al., Efficient, stable, small, and water-soluble doped ZnSe nanocrystal emitters as non-cadmium biomedical labels. *Nano Lett*, 2006. **7**(2): 312–317.

58. Smith, A.M., A.M. Mohs, and S. Nie, Tuning the optical and electronic properties of colloidal nanocrystals by lattice strain. *Nat Nanotechnol*, 2009. **4**(1): 56–63.

59. Duan, H. and S. Nie, Cell-penetrating quantum dots based on multivalent and endosome-disrupting surface coatings. *J Am Chem Soc*, 2007. **129**(11): 3333–3338.

60. Uyeda, H.T. et al., Synthesis of compact multidentate ligands to prepare stable hydrophilic quantum dot fluorophores. *J Am Chem Soc*, 2005. **127**(11): 3870–3878.

61. Wang, Y.A. et al., Stabilization of inorganic nanocrystals by organic dendrons. *J Am Chem Soc*, 2002. **124**(10): 2293–2298.

62. Zhelev, Z., H. Ohba, and R. Bakalova, Single quantum dot-micelles coated with silica shell as potentially non-cytotoxic fluorescent cell tracers. *J Am Chem Soc*, 2006. **128**(19): 6324–6325.

63. Smith, A.M. et al., A systematic examination of surface coatings on the optical and chemical properties of semiconductor quantum dots. *Phys Chem Chem Phys*, 2006. **8**: 3895–3903.

64. Mattoussi, H. et al., Self-assembly of CdS-ZnS quantum dot bioconjugates using an engineered recombinant protein. *J Am Chem Soc*, 2000. **122**(49): 12142–12150.

65. Pellegrino, T. et al., Hydrophobic nanocrystals coated with an amphiphilic polymer shell: A general route to water soluble nanocrystals. *Nano Lett*, 2004. **4**(4): 703–707.

66. Yu, W.W. et al., Forming biocompatible and nonaggregated nanocrystals in water using amphiphilic polymers. *J Am Chem Soc*, 2007. **129**(10): 2871–2879.

67. Moffitt, M. and A. Eisenberg, Size control of nanoparticles in semiconductor-polymer composites. 1. Control via multiplet aggregation numbers in styrene-based random ionomers. *Chem Mater*, 1995. **7**(6): 1178–1184.

68. Rhyner, M.N., Development of cancer diagnostics using nanoparticles and amphiphilic polymers. PhD thesis, Georgia Institute of Technology, Atlanta, GA, 2008.

69. Kairdolf, B.A., A.M. Smith, and S. Nie, One-pot synthesis, encapsulation, and solubilization of size-tuned quantum dots with amphiphilic multidentate ligands. *J Am Chem Soc*, 2008. **130**(39): 12866–12867.

70. Smith, A. and S. Nie, Nanocrystal synthesis in an amphibious bath: Spontaneous generation of hydrophilic and hydrophobic surface coatings. *Angew Chem Int Ed*, 2008. **47**(51): 9916–9921.
71. Smith, A.M. and S. Nie, Next-generation quantum dots. *Nat Biotechnol*, 2009. **27**(8): 732–733.
72. Smith, A.M. and S. Nie, Minimizing the hydrodynamic size of quantum dots with multifunctional multidentate polymer ligands. *J Am Chem Soc*, 2008. **130**(34): 11278–11279.
73. Bentzen, E.L. et al., Surface modification to reduce nonspecific binding of quantum dots in live cell assays. *Bioconjug Chem*, 2005. **16**(6): 1488–1494.
74. Gerion, D. et al., Sorting fluorescent nanocrystals with DNA. *J Am Chem Soc*, 2001. **124**: 24.
75. Pathak, S. et al., Hydroxylated quantum dots as luminescent probes for in situ hybridization. *J Am Chem Soc*, 2001. **123**(17): 4103–4104.
76. Kairdolf, B.A. et al., Minimizing nonspecific cellular binding of quantum dots with hydroxyl-derivatized surface coatings. *Anal Chem*, 2008. **80**(8): 3029–3034.
77. Sheinerman, F.B., R. Norel, and B. Honig, Electrostatic aspects of protein-protein interactions. *Curr Opin Struct Biol*, 2000. **10**(2): 153–159.
78. Ballou, B. et al., Noninvasive imaging of quantum dots in mice. *Bioconjug Chem*, 2004. **15**(1): 79–86.
79. Liu, W.H. et al., Compact cysteine-coated CdSe (ZnCdS) quantum dots for in vivo applications. *J Am Chem Soc*, 2007. **129**: 14530–14531.
80. Zhou, M. et al., Peptide-labeled quantum dots for imaging GPCRs in whole cells and as single molecules. *Bioconjug Chem*, 2007. **18**(2): 323–332.
81. Parak, W.J. et al., Conjugation of DNA to silanized colloidal semiconductor nanocrystalline quantum dots. *Chem Mater*, 2002. **14**(5): 2113–2119.
82. Zhang, Y. et al., HaloTag protein-mediated site-specific conjugation of bioluminescent proteins to quantum dots. *Angew Chem*, 2006. **118**(30): 5058–5062.
83. Willard, D.M. et al., CdSe-ZnS quantum dots as resonance energy transfer donors in a model protein-protein binding assay. *Nano Lett*, 2001. **1**(9): 469–474.
84. Zrazhevskiy, P. and X. Gao, Quantum dot imaging platform for single-cell molecular profiling. *Nat Commun*, 2013. **4**: 1619.
85. Yezhelyev, M.V. et al., Proton-sponge coated quantum dots for siRNA delivery and intracellular imaging. *J Am Chem Soc*, 2008. **130**(28): 9006–9012.
86. Medintz, I.L. et al., Self-assembled nanoscale biosensors based on quantum dot FRET donors. *Nat Mater*, 2003. **2**(9): 630–638.
87. Sapsford, K.E. et al., Kinetics of metal-affinity driven self-assembly between proteins or peptides and CdSe-ZnS quantum dots. *J Phys Chem C*, 2007. **111**(31): 11528–11538.
88. Zhang, C.Y. et al., Single-quantum-dot-based DNA nanosensor. *Nat Mater*, 2005. **4**(11): 826–831.
89. Chattopadhyay, P.K. et al., Quantum dot semiconductor nanocrystals for immunophenotyping by polychromatic flow cytometry. *Nat Med*, 2006. **12**(8): 972–977.
90. Ghazani, A.A. et al., High throughput quantification of protein expression of cancer antigens in tissue microarray using quantum dot nanocrystals. *Nano Lett*, 2006. **6**(12): 2881–2886.
91. Tholouli, E. et al., Quantum dots light up pathology. *J Pathol*, 2008. **216**(3): 275–285.
92. Chen, C. et al., Quantum dots-based immunofluorescence technology for the quantitative determination of HER2 expression in breast cancer. *Biomaterials*, 2009. **30**(15): 2912–2918.
93. Fountaine, T.J. et al., Multispectral imaging of clinically relevant cellular targets in tonsil and lymphoid tissue using semiconductor quantum dots. *Mod Pathol*, 2006. **19**(9): 1181–1191.
94. Peng, C.W. et al., Patterns of cancer invasion revealed by QDs-based quantitative multiplexed imaging of tumor microenvironment. *Biomaterials*, 2011. **32**(11): 2907–2917.
95. Chen, H. et al., Comparison of quantum dots immunofluorescence histochemistry and conventional immunohistochemistry for the detection of caveolin-1 and PCNA in the lung cancer tissue microarray. *J Mol Histol*, 2009. **40**(4): 261–268.
96. Carey, L.A. et al., The triple negative paradox: Primary tumor chemosensitivity of breast cancer subtypes. *Clin Cancer Res*, 2007. **13**(8): 2329.

97. Cleator, S., W. Heller, and R.C. Coombes, Triple-negative breast cancer: Therapeutic options. *Lancet Oncol*, 2007. **8**(3): 235–244.

98. Rakha, E.A. et al., Prognostic markers in triple-negative breast cancer. *Cancer*, 2007. **109**(1): 25–32.

99. So, M.-K. et al., Self-illuminating quantum dot conjugates for in vivo imaging. *Nat Biotechnol*, 2006. **24**(3): 339–343.

100. Smith, A.M., M.C. Mancini, and S. Nie, Second window for in vivo imaging. *Nat Nanotechnol*, 2009. **4**(11): 710.

101. Hardman, R., A toxicologic review of quantum dots: Toxicity depends on physicochemical and environmental factors. *Environ Health Perspect*, 2006. **114**(2): 165.

22

PEBBLE Nanosensors for In Vitro Bioanalysis

Yong-Eun Koo Lee
Hanyang University

Eric Monson
Duke University

Murphy Brasuel
Colorado College

Martin A. Philbert
University of Michigan

Raoul Kopelman
University of Michigan

22.1 Introduction

In medical and biochemical research, when a sample domain is reduced to micrometer regimes, for example, living cells or their subcompartments, the real-time measurement of chemical and physical parameters with high spatial resolution and negligible perturbation of the sample becomes extremely challenging. A traditional strength of chemical sensors (optical, electrochemical, etc.) is the minimization of chemical interference between sensor and sample, achieved with the use of inert, *biofriendly* matrices or interfaces. However, when it comes to penetrating individual live cells, even the introduction of a submicron sensor tip can cause biological damage and resultant biochemical consequences. In contrast, individual molecular probes (free-sensing dyes) are physically small enough but have several drawbacks that affect the reliability of their intracellular measurements: (1) The probes must be permeable to the cell's membrane, which often requires proper derivatization of the molecules, which in itself might interfere with their function. (2) Inside the cell, they tend to be sequestered unevenly into various cell compartments. (3) The measurement is often affected by nonspecific protein binding. (4) Some of the available dyes are cytotoxic and may chemically perturb the cell's processes. (5) The dye is usually not ratiometric, which requires technologically more demanding techniques, such as picosecond lifetime resolution and phase-sensitive detection. Note that the ratiometric mode of operation—which assures the measurements to be unaffected by (1) excitation intensity, (2) absolute concentration, and (3) sources of optical loss—is essential for intensity-based intracellular or in vivo measurements wherein there are many interfering factors.

PEBBLE (photonic explorer for biomedical use with biologically localized embedding) nanosensors were developed specifically for minimally invasive analyte monitoring inside viable, single cells, with applications for real-time analysis of drug, toxin, and environmental effects on the cell function. They are thus filling a niche that lies between pulled micro-optodes and free molecular probes (naked indicator dye molecules). PEBBLEs are a direct outgrowth of the pulled optical fiber nanotechnology developed for biosensing by Tan et al.[1,2] and continuing in the work of Rosenzweig,[3,4] Shortreed,[5,6] and Barker.[7,8] Nanoscale-dimension sensors have specific advantages.[9] In most instances, there is an explicit functional dependence of optode characteristics on the sensor radius (r). For instance, the absolute detection limit decreases with r^3 (good!) and the response time is reduced as r^2 (good!). The signal-to-noise (S/N) ratio, though, decreases with r (bad!) but not r^3 (luckily!) under standard working conditions. Other features that improve, as sensors get smaller, include sample volume, sensitivity, invasiveness, spatial resolution, dissipation of heat in sensor and/or sample, toxicity, and materials' cost. Features that may worsen include fluorophore leaching and photodamage to sensor and/or sample.

PEBBLEs are nanoscale fluorescent sensors consisting of sensor molecules that are incorporated into a nanoparticle made of a chemically inert matrix (see Figure 22.1). The PEBBLE nanosensors preserve the excellent sensing and biocompatibility of macrosensors but surpass their performance in terms of response time and absolute detection limit. The PEBBLEs are closer to free (*naked*) molecular dyes, in terms of minimal mechanical and physical perturbation, than most other sensing platforms but are still able to obviate the aforementioned problems of molecular probes. They use the nanoparticles' excellent engineering capability and their chemically inert matrix as follows: (1) The inert matrix protects cellular contents from the incorporated sensing components and, vice versa, protects the sensing indicator molecules from various cell components. (2) The nanoparticle matrix eliminates interferences such as protein binding. (3) It avoids uneven intracellular distribution due to membrane/organelle sequestration. (4) Each nanoparticle can be loaded with a high number of

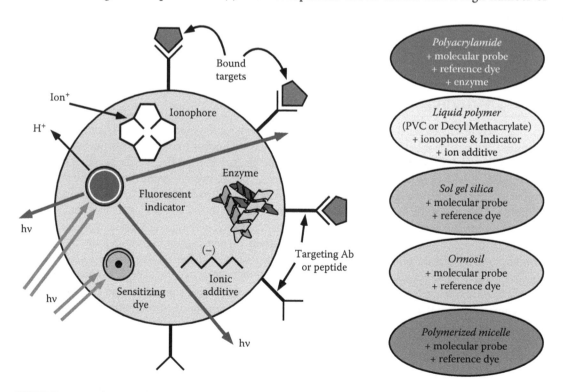

FIGURE 22.1 Schematic diagram of a PEBBLE nanosensor showing many options available within this flexible, integrated device platform. On the right, current matrix materials are presented with typical constituents.

sensing components because of its larger size (compared to that of the molecular dyes), enhancing the signal-to-background ratio. (5) Loading of multiple components per nanoparticle can allow ratiometric measurements as well as multiplex sensing. (6) It allows sophisticated synergistic designs, including optically silent components, such as highly selective ionophores or enzymes. (7) Nanoparticles can be surface conjugated with biological recognition molecules for intracellular or subcellular organelle-selective delivery (see Section 22.4).

The main classes of PEBBLE nanosensors are based on matrices of polyacrylamide (PAA) hydrogel, sol-gel silica, organically modified silica (ormosil), ormosil-capped polymerized micelle, and cross-linked poly(decyl methacrylate) (PDMA) liquid polymer. These matrices have been used to fabricate sensors that range from 30 to 600 nm in size for ions (H^+,[10–12] Ca^{2+},[11–13] $Cu^{+/2+}$,[14] Fe^{3+},[15] K^+,[16] Na^+,[17] Mg^{2+},[18] Zn^{2+},[13,19] Cl^-[20]), reactive oxygen species (singlet oxygen,[21,22] OH radical[23]), small molecules (O_2,[13,24–27] glucose,[28] hydrogen peroxide[22]), as well as electric field.[29] The signal for the measurements has been mainly fluorescence intensity, but fluorescence anisotropy[13] or fluorescence lifetime (or phase)[27] has also been used. A host of delivery techniques have been used to successfully deliver PEBBLE nanosensors into mouse oocytes, rat alveolar macrophages, rat C6 glioma, and human neuroblastoma cells.

22.2 Practical Concept Examples

It is useful to point out concrete examples of the features discussed earlier before delving into the details of PEBBLE production and application. All PEBBLEs must be well characterized before use, including both physical and functional (sensing) characteristics. The physical characteristics—including such measures as nanoparticle size and shape—may affect their sensing characteristics and should be well characterized before carrying out any functional tests. The size and shape of the dried PEBBLEs are measured by scanning electron microscopy (SEM) or transmission electron microscopy (TEM), while the size and extent of aggregation of the PEBBLEs in aqueous solution are determined by light scattering (LS) as exemplified in Figure 22.2. The PAA nanoparticles are typically produced in the 20–70 nm size range, silica and organically modified silica (ormosil) nanoparticles are in the 100–200 nm size range, and PDMA nanoparticles are in the range of 400–700 nm. The ormosil-capped polymerized micelles have a size of about 30 nm. Essential metrics for functional characterization include tests for constituent leaching, ratiometric stability, response time and sensitivity to interference from similar analytes, and nonspecific protein binding, as well as response calibration.

22.2.1 Ratiometric Oxygen Sensor: Signal and Calibration

The first oxygen PEBBLE sensor was made of silica nanoparticles,[24] but oxygen PEBBLE sensors were developed later using more hydrophobic ormosil[25] and PDMA[26] nanoparticles. The silica PEBBLEs contain a ruthenium-based oxygen-sensitive dye, $[Ru(dpp)_3]^{2+}$, and the ormosil PEBBLEs contain more sensitive platinum-based oxygen dyes, platinum(II) octaethylporphine (PtOEP) or platinum(II) octaethylporphine ketone (PtOEPK). These oxygen-sensitive dyes have an intensity decrease due to excited-state quenching in the presence of molecular oxygen. As a spectrally separated intensity reference for ratiometric measurements, the PEBBLEs also include oxygen-insensitive dyes, Oregon Green 488® (Molecular Probes) for ruthenium-based silica PEBBLEs, 3,3′-dioctadecyloxacarbocyanine perchlorate (DiO) for PtOEP-based ormosil PEBBLEs, and octaethylporphine (OEP) for PtOEPK ormosil PEBBLEs.

Figure 22.3a shows spectra of PtOEPK ormosil PEBBLEs in aqueous solution, in the presence of varying concentrations of oxygen. It is very clear that the reference peak, on the left, remains constant, while the PtOEPK peak, on the right, changes in intensity. Also shown in Figure 22.3b and c are the Stern–Volmer (calibration) plots of fluorescence intensity ratio vs. oxygen concentration for the silica and ormosil PEBBLEs. The PtOEPK ormosil PEBBLE nanosensors exhibit a linear Stern–Volmer calibration curve over the entire range of dissolved oxygen (DO) concentrations, which is a notable improvement

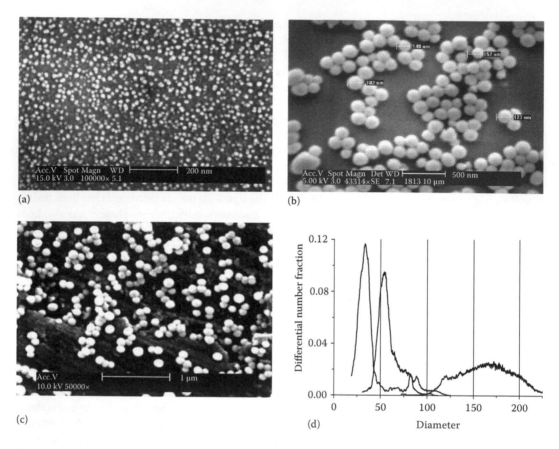

FIGURE 22.2 Typical SEM images of (a) PAA PEBBLEs, (b) sol-gel silica PEBBLEs, and (c) ormosil PEBBLEs. (d) LS results for (left to right) one PAA and two different silica PEBBLE formulations.

in comparison to ruthenium dye–based silica PEBBLEs. The sensitivity enhancement is also significant. The quenching response (Q_{DO}), a representative quantity for sensitivity to DO, was only about 80% for silica PEBBLEs but was 97% for ormosil PEBBLEs—the highest among the reported values. The Q_{DO} is defined by $Q_{DO} = (I_{N_2} - I_{O_2})/I_{N_2} \times 100$, where I_{N_2} is the fluorescence intensity of the indicator dye or the indicator/reference intensity ratio, in fully deoxygenated water, and I_{O_2} is that in fully oxygenated water. The higher Q_{DO} value is, the higher the sensitivity is. These significantly improved sensing characteristics of the ormosil PEBBLEs over the silica PEBBLEs—a wider dynamic range with linear response and a high sensitivity (<0.1 ppm)—may result from the right combination of nanoparticle matrix (hydrophobic and highly permeable to oxygen), its small size, and the right selection of the oxygen-sensitive dye. Note that oxygen has a higher solubility in hydrophobic matrices.

The reversibility of the sensor's response allows repeated use of the sensors, which is important for continuous oxygen monitoring. Both sensors showed at least 95% recovery each time that the sensing environments were changed among air-, O_2-, or N_2-saturated sensor solutions.

22.2.2 Ratiometric Zinc PEBBLE Insensitive to Protein Interference

The PAA zinc sensor[19]—based on Newport Green® (Molecular Probes), a zinc-sensitive dye, and Texas Red, a spectrally distinct intensity reference—shows the advantages of PEBBLEs. Quantitative measurements show these sensors to be insensitive to changes in excitation intensity as well as providing protection from nonspecific protein interference. Although Newport Green has good selectivity over

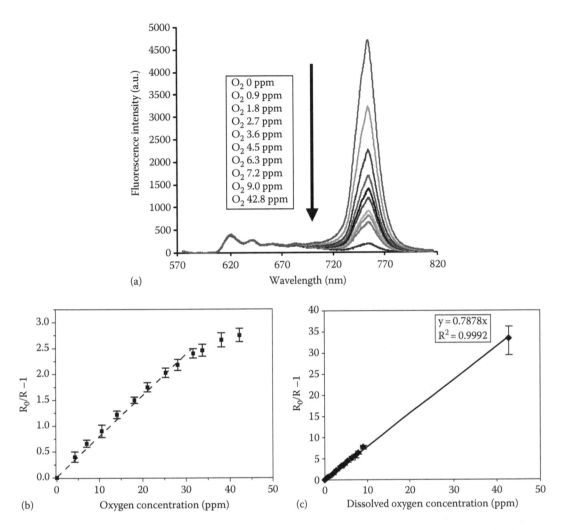

(a)

(b)

(c)

FIGURE 22.3 (a) Aqueous phase emission spectra excited at 570 nm. The fluorescence emission spectra of ormosil oxygen PEBBLEs in different DO concentrations—0.0 ppm (nitrogen saturated) to 43 ppm (oxygen saturated) from top to bottom lines. The excitation wavelength was 570 nm. The peak on the left is the fluorescence emission from the reference dye, while that on the right is the emission from sensing dye. (b) Stern–Volmer plot of relative fluorescence intensity ratios for silica oxygen PEBBLEs in aqueous phase. Dashed line denotes biologically relevant range. (c) Stern–Volmer plot of relative fluorescence intensity ratios for ormosil oxygen PEBBLEs in aqueous phase. Note that the plot is linear over entire DO range.

intracellular ions, the dye itself is prone to artifacts resulting from nonspecific binding of proteins, such as bovine serum albumin (BSA), as shown in Figure 22.4a. Monitoring the peak of Newport Green at 530 nm, there is a substantial increase in the peak intensity with each successive addition of BSA into the *naked* dye solution. The PEBBLEs containing the Newport Green dye, however, are unaffected by the additions of BSA. As little as 0.02% BSA causes an intensity increase of over 200% in the naked Newport Green dye, but the intensity of the Newport Green embedded in the sensor remains unchanged, even at BSA concentrations above 0.10%.[19] Such protein interference-free sensing by the PEBBLEs is due to the presence of nanoparticle matrix encasing the molecular probes. The pores of the nanoparticle matrix are sufficiently large to allow diffusion of the smaller analyte ions or molecules, so as to reach molecular probes located inside the nanoparticle, but the pores are still small enough to exclude the diffusion of larger proteins into the core matrix.

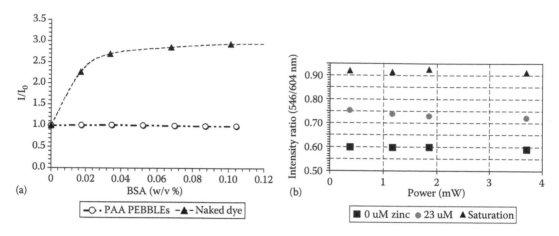

(a)

(b)

FIGURE 22.4 (a) Normalized Newport Green emission (530 nm) after addition of successive aliquots of a 10% (w/v) BSA solution. As little as 0.02% BSA causes a greater than 200% increase in Newport Green free (naked) dye intensity, but the intensity of the dye embedded in a PAA PEBBLE remains unchanged. (b) Fluorescence emission intensity ratio (545 nm/604 nm) from a 10 mg/mL PEBBLE suspension in 10 mM Tris buffer monitored using neutral density filters (1.0, 0.5, and 0.3) to attain varied excitation powers at three different zinc concentrations.

Note that the same notion—protection of incorporated molecules from large biological molecules by nanoparticle matrix—can also be beneficial for nanoparticle-based drug delivery. For example, we could enhance the in vivo photodynamic therapeutic efficiency of methylene blue, a photosensitizer that can be reduced into an inactive form by plasma enzymes, if we use the nanoparticles with encapsulated methylene blue.[30]

Figure 22.4b demonstrates the advantage of using an integrated ratiometric device over a single intensity-based dye. Four different excitation light levels were used for zinc sensing. Although the absolute intensity of fluorescent emission for each dye decreased with decreasing illumination power, the ratio of peak intensities—of Newport Green and Texas Red—remained constant. It is evident from this that fluctuations in the intensity of either a laser or arc lamp would complicate quantitative analysis for intensity-based measurements, while the ratiometric PEBBLEs eliminate the artifacts resulting from power fluctuations. The equivalent would be true, as well, for insensitivity to fluctuations in the local PEBBLE concentration.

22.2.3 MOON Concept: Enhanced Signal to Noise

We have developed the modulated optical nanoprobes (MOONs) concept that, in principle, can apply to any fluorescent nanoparticles, including PEBBLE sensors, for background-free measurements.[31–33] MOONs are metallically half-capped fluorescent nanoparticles. MOONs can be either magnetic,[31,33] that is, MagMOONs, or not, that is, those used for measuring Brownian rotation.[33] The magnetically induced periodic motion of the MagMOON (or random thermal motion in the case of Brownian MOONs) makes the fluorescence signal blink, depending upon the orientation of the MOONs toward the detector (Figure 22.5). By separating the blinking probe signal from the unmodulated background, the S/N or, strictly, signal-to-background ratio can be enhanced by several orders of magnitude, up to 4000 times.[31] The MOON-type PEBBLEs should be especially useful for detecting low concentrations of important analytes inside cells with high autofluorescence and/or inside highly scattering biological samples. Unfortunately, the MOON sensors have not yet been applied quantitatively for intracellular measurements because MOONs so far have been developed using a micron-size particle due to the size-related difficulties for efficient magnetization or half coating. With recent progress on nanotechnology and coating technology, such as molecular beam epitaxy, nanometer-sized MOONs are being developed.[34]

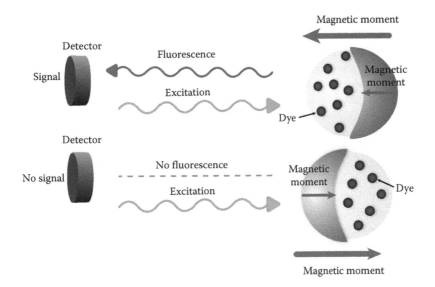

FIGURE 22.5 Background-free measurement by MagMOON. An external magnetic field orients the MagMOON, causing its fluorescent excitation and observed emission to blink on and off as it rotates. Note that the background fluorescence does not blink.

22.3 PEBBLE Production Techniques

PEBBLEs represent an advance in nano-optode technology. The science of nano-optode production relies on advances in nanoscale production, using emulsion and dispersion fabrication techniques. The nanoemulsion/dispersion process for preparing nanoparticles is subtle, and there is no universal method for making hydrophilic, hydrophobic, and amphiphilic nanospheres that contain the right matrix and right chemical components in their proper proportions. Loading of single or multiple sensing components into different nanoparticles—hydrophilic PAA, silica, ormosil, hydrophobic PDMA nanoparticles, or ormosil-capped polymerized micelles—is not made by a standard routine procedure. However, the production methods, once optimized for a given matrix and its constituents, are based on relatively simple wet chemistry techniques, as opposed to many complicated physical and chemical nanotechnology schemes. Specific methods for producing sensors from all these matrices are described in the following as well as the related response mechanisms for each type of sensor. Note that the sensing properties of the PEBBLE such as dynamic range and selectivity are largely dependent on the K_D of the sensing dye with respect to the analyte and any interfering ions or molecules. However, these PEBBLE properties also depend on the properties of the matrix—such as pore size, hydrophobicity, and charge—that determine the permeability and solubility of the analytes, as well as the loading efficiency of the sensing probes. Therefore, selecting the right kind of nanoparticle matrix is important to produce a reliable sensor.

22.3.1 Polyacrylamide PEBBLEs

PAA nanoparticles are hydrogel that water and small ions can diffuse freely through. PAA nanoparticles have served as a good matrix especially for ion sensors where analytes (small ions) form a complex with the entrapped dye, similar to the response of the *naked* dye in solution, because of its neutral and hydrophilic nature. These PAA PEBBLEs are produced under so-called soft chemical conditions, that is, low temperatures and physiological pH conditions, allowing the inclusion of even proteins. The encapsulated proteins served either as a molecular probe as in the case of DsRed for $Cu^{+/2+}$ sensors[14] or as a catalyst for an enzymatic reaction of analytes as in the case of glucose oxidase for glucose sensors.[28] The PAA-based PEBBLEs include H+,[10–12] Ca^{2+},[11,12] Cu$^{+/2+}$,[14] Fe^{3+},[15] Mg^{2+},[18] and Zn^{2+},[19] glucose[28] PEBBLEs.

The production of PAA PEBBLEs is based on the nanoemulsion techniques studied by Daubresse.[35] Some control over particle size and shape can be gained by adjusting surfactant to water ratios in the emulsion. The typical polymerization solution consists of 0.4 mM fluorescent indicator dye (any hydrophilic dye selective for the analyte of interest), reference dye, acrylamide (8.6 mmol), and N,N'-methylenebis(acrylamide) (1.2 mmol), all in 0.1 M phosphate buffer, pH 7.2–7.4. The solution is then added to the argon-purged hexane solution (36 mL) containing Brij 30 (6.85 mmol) and AOT (2.88 mmol) to form a microemulsion. The polymerization is initiated by ammonium persulfate (2.8×10^{-5} mmol) and TEMED (0.54 mmol). The solution was stirred overnight under argon at room temperature. Hexane is removed by rotary evaporation, then the resultant thick residue is rinsed of surfactant with ethanol and/or water in an Amicon ultrafiltration cell (Millipore Corp., Bedford, MA) and air- or freeze-dried, to give the PEBBLEs. The amine-functionalized PAA nanoparticles that are useful for conjugating dyes or targeting moieties to nanoparticles can be also prepared in the same way except that the polymerization solution contains 3-(aminopropyl)methacrylamide hydrochloride salt (0.25 mmol) as well.[23]

22.3.2 Silica and Ormosil PEBBLEs

Silica and ormosil have also been used as the matrix for the fabrication of PEBBLE nanosensors. Silica and ormosil are chemically inert and more photo- and thermally stable than polymer matrices. They are optically transparent, homogeneous materials and of high purity,[36] thus making them an ideal choice as a sensor matrix for quantitative spectrophotometric measurements. They are also porous and have high oxygen permeability, making them a promising matrix for oxygen sensing. Sol-gel silica oxygen PEBBLEs are prepared by a modified Stober method.[24] The reaction solution for the production of oxygen-sensitive sol-gel silica PEBBLEs consists of polyethylene glycol (PEG) MW 5000 monomethyl ether (3 g), ethanol (200 proof, 6 mL), Oregon Green dextran MW 10,000 (0.1 mM), [Ru(dpp)$_3$]$^{2+}$ (0.4 mM), and 30% wt. ammonia water (3.9 mL) with ammonia serving as catalyst and water being one of the reactants. Upon mixing, the solution becomes transparent, and the inorganic *monomer* tetraethyl orthosilicate (TEOS) (0.5 mL) is added dropwise to initiate the hydrolysis of TEOS. The solution is then stirred at room temperature for 1 h to allow the sol-gel reaction (analogous to polymerization) to reach completion. A liberal amount of ethanol is then added to the reaction solution, and the mixture is transferred to an Amicon cell and washed with additional ethanol to ensure that all unreacted chemicals have been removed. The PEBBLE solution is then passed through a suction filtration system (Fisher, Pittsburgh, PA) and air-dried to yield silica PEBBLEs. The size of the silica PEBBLEs can be changed by the synthetic conditions such as types of monomers and PEGs.[37]

Ormosil PEBBLEs are prepared through a two-step (acid-catalyzed hydrolysis followed by base-catalyzed condensation) but one-pot synthesis based on the sol-gel process.[25,38] The average size of the particles can be controlled to be from 100 nm to 1 μm by changing monomer concentration and increasing hydrolysis time. Following procedure was used to prepare the oxygen PEBBLEs of average 125 nm in diameter[25]: To a 50 mL round-bottom flask, in a water bath kept at 60°C on a Corning pc-351 hot plate stirrer, 31 mL of deionized water and 38 μL HNO$_3$ were added, and the mixture was stirred until the temperature of the reaction mixture reached to 60°C. Phenyltrimethoxysilane of 0.1 mL was then added to the flask, stirred at full speed. After 20 min, 6 mL of ammonium hydroxide was added at once. The resultant solution was kept stirring for additional 1–2 h or until the solution became milky. Mixture of sensing dye and reference dye in a proper solvent and 0.2 mL of methyltrimethoxysilane was added into the reaction mixture, which was kept stirred for additional 1 h. The dye solution mixtures used were 0.4 mg of OEP and 1.2 mg of PtOEPK in 1.6 mL of 5:4 tetrahydrofuran (THF)/ethanol mixture. The resulting PEBBLEs were suction filtered through a Fisherbrand glass microanalysis vacuum filter holder with a 0.1 μm Osmonics/MSI Magna Nylon membrane filter.

The PEBBLEs were rinsed three times with water and then resuspended in water–ethanol 1:2 mixture, sonicated for 5 min and then filtered through a 0.02 μm Whatman Anodisc filter membrane, and allowed to air-dry.

22.3.3 Poly(Decyl Methacrylate) Hydrophobic Liquid Polymer

The use of fluorescent indicator molecules in encapsulated form (PAA PEBBLEs) has proven valuable in the study of a number of intracellular ion analytes.[10–12,14,15,18,19,23,28] However, there are many ions for which no fluorescent indicator dye is sufficiently selective or even available, but there exist highly selective nonfluorescent hydrophobic ionophores available from the rich tradition of electrochemical sensors. Thus, the aforementioned problem has been solved for optodes by using *in tandem* an optically *silent* ionophore (which is not optically responsive but chemically is highly selective) and a next-door optically responsive agent, which plays the role of a *spectator* or *reporter dye*. While the principles of such tandem sensing schemes were worked out by Bakker and Simon,[39–41] Suzuki,[42,43] and Wolfbeis,[44,45] for the less-sensitive absorbance-based detection, the first demonstration of such a synergistic sensing scheme, using highly sensitive fluorescence detection, enabling translation to the nanoscale, occurred with the pulled optodes developed by Shortreed et al.[5,6] The same principles were extended to create an alternate class of optical nanosensors for ions, for example, ion-correlation PEBBLEs. These PEBBLEs are made of PDMA liquid polymer nanoparticles embedded with three components: a nonfluorescent ionophore that binds selectively to the ion of interest, a fluorescent hydrogen ion–selective dye that plays the role of a reporter, and a lipophilic additive that maintains ionic strength.[16] The operation of the entire system is based on a thermodynamic equilibrium between embedded components and ions inside the hydrophobic PDMA nanoparticles and ions in the aqueous phase, resulting in ion exchange (so as to sense cations) or ion coextraction (so as to sense anions). The PDMA nanoparticles have been utilized for the ion-correlation PEBBLEs for Na^+, K^+, and Cl^- ions,[16,17,20] as well as for oxygen PEBBLEs.[26]

A batch of PDMA PEBBLE sensors is typically made from 210 mg of decyl methacrylate, 180 mg hexanediol dimethacrylate, and 300 mg of dioctyl sebacate (DOS), with 10–30 mmol/kg each of ionophore, chromoionophore, and ionic additives added after spherical particle synthesis. The spherical particles are prepared by dissolving decyl methacrylate, hexanediol dimethacrylate, and DOS in 2 mL of hexane. To a 100 mL round-bottom flask, in a water bath on a hot plate stirrer, 75 mL of pH 2 HCl is added along with 1793 mg of PEG 5000 monomethyl ether and stirred and degassed. The hexane-dissolved monomer cocktail is then added to the reaction flask (under nitrogen), stirred at full speed, and water bath temperature is raised to 80°C over 30–40 min. Potassium peroxodisulfate of 6.0 mg is then added to the reaction and stirring is reduced to medium speed. The temperature is kept at 80°C for two more hours, and then the reaction is allowed to return to room temperature and stirred for 8–12 h. The resulting polymer is suction filtered through a glass microanalysis vacuum filter holder with a Whatman Anodisc filter (0.2 μm pore diameter). The polymer is rinsed three times with water and three times with ethanol to remove excess PEG and unreacted monomer. THF is then used to leach out the DOS and then the PEBBLEs are again filtered and rinsed. They are allowed to dry in a 70°C oven overnight. Dry polymer is then weighed out, and DOS and sensing components are added to this dry polymer. Enough THF is added to this mixture so as to just wet the PEBBLEs. The PEBBLEs are allowed to swell for 8 h and then the THF is removed by rotary evaporation. The resulting PEBBLE sensors are rinsed with doubly distilled water and allowed to air-dry.

22.3.4 Ormosil-Capped Polymerized Micelle

Ormosil-capped polymerized micelles provide another kind of hydrophobic nanoparticles that are utilized for PEBBLEs containing hydrophobic indicator dyes. Artificial micelles present limited utility for biological sensing within the membrane-rich cellular environments, because they tend to mix with the native cell membranes, degrading the sensor structure. However, with their surface capped with a polymer, the structure of micelles can be stabilized and utilized for making sensors for intracellular measurements. The ormosil-capped polymerized micelles have been used for making PEBBLEs that measure electric field (E-PEBBLEs)[29] and can also be used for making ion-correlation PEBBLEs. The following procedure was used to prepare the E-PEBBLEs.[29] Triton X-100 (0.625 g, 1 mmol) was dissolved in 10 mL of deionized water. Octyltriethoxysilane (63 mL, 0.2 mmol) was slowly added and the solution allowed to stir for 30 min. Ammonium hydroxide (10–20 mL of a 2.2 M solution) was added, and the solution was stirred for 72 h. Trimethylethoxysilane (200 mL, 1.3 mmol) was added and the solution was allowed to stir for an additional 48 h. The solution was dialyzed against water for 24 h in a Spectra/Por polyvinylidene difluoride (PVDF) dialysis membrane (Spectrum Labs, Rancho Dominguez, CA) with a 250 kDa molecular mass cutoff. The bath water was changed three times during the 24 h period. The particle solution was then stirred with an equal volume of a 1 wt% solution of poly(diallyldimethylammonium chloride) (low molecular weight) and again dialyzed for 24 h. A 2 mL aliquot of the washed and coated particles was sonicated with a 20 mL solution of di-4-ANEPPS (Molecular Probes, Eugene, OR) dissolved in chloroform (0.1 mg/mL, 0.2 mM). A stir bar was added, and the solution was stirred under an aluminum foil tent for 30 min to remove the chloroform.

22.4 Delivery Methods

One of the most important considerations when applying PEBBLE nanosensors to single-cell studies is the (noninvasive) delivery of the PEBBLEs to the cell. The many methods that have been explored include gene gun, picoinjection, liposomal delivery, endocytosis (phagocytosis and pinocytosis), and TAT peptide-assisted delivery. All of these methods are summarized in Figure 22.6.

The method of PEBBLE delivery by gene gun can best be thought of as a shotgun method. PEBBLEs are dried on a plastic (delivery) disk, and this disk is set in front of a rupture disk. Helium pressure is built up behind the rupture disk, which ruptures at a specific helium pressure and propels the PEBBLEs from the plastic disk into a cell culture. The gene gun can be used to deliver one to thousands of PEBBLEs per cell into a large number of cells very quickly (dependent on the concentration of PEBBLEs on the

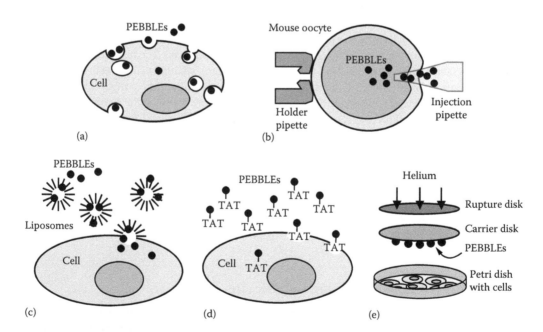

FIGURE 22.6 Range of delivery methods currently available for PEBBLE nanosensors into single cells for bioanalysis. (a) Endocytosis, (b) picoinjection, (c) liposomal delivery, (d) TAT-assisted delivery, and (e) gene gun.

delivery disk).[11,12,16–18,20,24,25] Cell viability is excellent, 98% viability compared to control cells,[11,12] for small numbers of PEBBLEs, and hinges directly on the number of PEBBLEs delivered, the delivery pressure, and the chamber vacuum. The PEBBLE momentum determines whether the PEBBLEs are mainly internalized in the cytoplasm or in the nucleus.

Picoinjection is used to inject picoliter (pL) volumes of PEBBLE containing solution into single cells (Figure 22.6). This method of delivery is dependent on the fabrication of pulled capillary *needles*, through the use of a pipette puller and a microforge. The smallest volume deliverable is 10 pL, and the most concentrated PEBBLE solution to work in the pulled capillary syringe is 5 mg/mL PEBBLEs. The maximum number of PEBBLEs one can put in is dependent on the volume of solution that can be injected without damaging the cell. Picoinjection can give a wide range of PEBBLE concentrations in the cell, and cell viability is good (if done by an expert), but because each cell must be individually injected, the method is time consuming and tedious.[12]

Commercially available liposomes can also be used to deliver PEBBLEs to cells. The liposomes are prepared in a solution of PEBBLEs and then placed in the cell culture where the liposomes fuse with the cell membranes and empty their contents (the PEBBLE containing solution) into the cell. Three factors play a key role in determining the number of PEBBLEs delivered to each cell with this method—the original concentration of the PEBBLEs, the concentration of liposomes placed in the cell culture, and the length of time the liposomes are left with the cells.[11,12] The parameters must be tailored for each cell line used in order to obtain the desired concentration of PEBBLEs in the cells. While it would be difficult to deliver a single PEBBLE to each cell with this method, it does seem that a low end of between 10 and 50 PEBBLEs per cell would be possible, with the high end being the maximum number of PEBBLEs the cell could take without losing viability. Liposomal delivery is useful for delivering PEBBLEs to a lot of cells simultaneously. The challenge is in tailoring the delivery—for the concentrations desirable and for the cell line being used. Cell viability is excellent. Obviously, the PEBBLE size needs to be small enough for this method, and delivery is essentially limited to the cell cytoplasm.

Nonspecific endocytosis—phagocytosis by macrophages (specialized immune system cell) and pinocytosis by nonmacrophage cells—can also deliver PEBBLEs into cells, after incubating cells with the PEBBLEs. The number of endocytosed PEBBLEs is dependent on the concentration of the PEBBLE solution and the amount of time the cells are allowed to stay in the PEBBLE solution. The advantage of this delivery method is that one can easily deliver varying concentrations of PEBBLEs to cells. The disadvantages are that PEBBLEs may only be internalized into certain cell regions such as lysosomes. This method also provides excellent cell viability.[12]

TAT peptides, a membrane-penetrating peptide, can also be used to deliver proteins[46] or nanoparticles[47] to cells. TAT peptide can be conjugated to the PEBBLEs with surface amine groups and then delivered into cells after incubating cells with the PEBBLEs. As in the case of the endocytosis, the PEBBLE concentration as well as incubation time determine the number of PEBBLEs delivered into cells. However, the incubation time is shorter than that for nonspecific endocytosis. It should be noted that the specific delivery of the PEBBLEs to a particular organelle is also possible by surface-conjugated molecules that are specific to the organelles such as mitochondria.[48]

22.5 In Vitro Bioanalysis

22.5.1 Calcium (PAA) PEBBLEs

The first PEBBLEs produced were PAA based, and one of the first examples of their successful application to cells was with macrophages. Alveolar macrophages were recovered from rat lung lavage using Krebs–Henseleit buffer. Macrophage was maintained in a 5% CO_2, 37°C incubator in Dulbecco's Modified Eagle Medium (DMEM) containing 10% fetal bovine serum and 0.3% penicillin, streptomycin, and neomycin. PEBBLE suspensions ranging from 0.3 to 1.0 mg/mL were prepared in DMEM and incubated with alveolar macrophage overnight. Macrophage images were then taken using a confocal

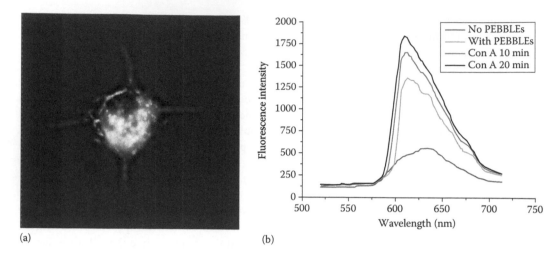

(a)   (b)

FIGURE 22.7   (See color insert.) Confocal microscope image (a) of alveolar macrophage containing phagocy-tosed PAA PEBBLEs containing Calcium Crimson dye. Fluorescence spectra (b) show an increase in intracellular calcium after cells have been challenged by Con A.

microscope and spectra of the same cells were obtained using the fluorescent microscope (shown in Figure 22.7). PAA PEBBLEs selective for calcium (containing Calcium Crimson in the PAA matrix)[12] were used in order to monitor calcium in phagosomes within rat alveolar macrophage, because of the ease in which macrophage phagocytose particles. This method for delivering the PEBBLEs into cells provided a simple, yet important, test of the PEBBLE sensors in a challenging (acidic) intracellular environment. Macrophage that had phagocytosed 20 nm calcium-selective PEBBLE sensors was challenged with a mitogen, Concanavalin A (Con A), inducing a slow increase in intracellular calcium, which was monitored over a period of 20 min. PEBBLE clusters confined to the phagosome enabled correlation of ionic fluxes with stimulation of this organelle.

The calcium PEBBLE in the macrophage experiment clearly demonstrates a time-resolved observation of a biological phenomenon in a single, viable cell. One can clearly obtain relevant time-domain data with a fluorescence microscope, spectrograph, and CCD. With a confocal microscope system and the appropriate dye/filter sets, one can attain both temporal and spatial resolutions, as demonstrated in the following.

Calcium PEBBLEs have also been developed utilizing *Calcium Green-1* (Molecular Probes) dye, in combination with sulforhodamine dye, as sensing components. We note that Calcium Green fluorescence increases in intensity with increasing calcium concentrations, while the sulforhodamine fluorescence intensity remains unchanged, regardless of biologically relevant concentration of ions, pH, or other cellular components; thus, the ratio of the Calcium Green/sulforhodamine intensity gives a good indication of cellular calcium levels regardless of dye or PEBBLE concentration or fluctuations of light source intensity. Figure 22.8 shows a confocal microscope image of rat C6 glioma cells containing Calcium Green/sulforhodamine PEBBLEs. The top image of the pair is the light intensity from the green (calcium-sensitive) fluorescence, and the bottom shows the red fluorescence intensity (reference), both dyes confined in the same PEBBLEs. The PEBBLEs were delivered by liposomes to the cytoplasm of the cells. The toxin, *m*-dinitrobenzene (DNB), was introduced to the left side of the image and allowed to diffuse to the right. The effect of DNB is the disruption of mitochondrial function, followed by the uncontrolled release of calcium associated with onset of the mitochondrial permeability transition (MPT).[49] Calcium PEBBLEs were used to determine that the half-maximal rate of calcium release (EC_{50}) occurred at a 10-fold lower concentration of m-DNB in human SY5Y neuroblastoma cells than in rat C6 glioma cells.[49]

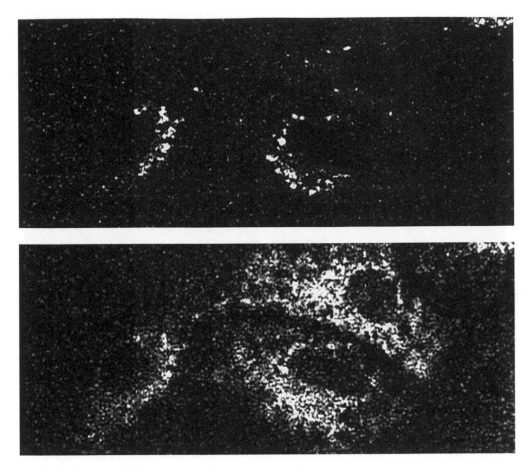

FIGURE 22.8 Confocal microscope image, split into green (top) and red (bottom) channels, of human C6 glioma cells containing Calcium Green (sensing dye, green channel)/sulforhodamine (reference dye, red channel) PEBBLEs (toxin diffusing left to right as seen by lack of green on the right side of the image).

22.5.2 Oxygen PEBBLEs

Both silica and ormosil PEBBLEs were delivered into the cells with a gene gun for intracellular oxygen imaging in live C6 glioma cells. After gene gun injection, the cells were rinsed three times with a phosphate-buffered saline (PBS) and allowed 30 min of recovery, in a nonsterile incubator, before experimentation in the same solution. Figure 22.9 shows the confocal images of C6 glioma cells containing silica PEBBLEs under Nomarski illumination overlaid with (a) the green fluorescence of Oregon Green 488 dextran and (b) the red fluorescence of $[Ru(dpp)_3]^{2+}$.[24] It can be seen that the cells still maintained their morphology after the gene gun injection of PEBBLEs and showed no sign for cell death. The dyes were excited, respectively, by reflecting the 488 nm (Ar–Kr) and the 543 nm (He–Ne) laser lines onto the specimen, using a double dichroic mirror. The Oregon Green fluorescence from the PEBBLEs inside the cells (Figure 22.9a) was detected by passage through a 510 nm long-pass and a 530 nm short-pass filter, and the fluorescence of $[Ru(dpp)_3]^{2+}$ (Figure 22.9b) through a 605 nm (45 nm band-pass) barrier filter. A 40X, 1.4 NA oil immersion objective was used to image the Oregon Green and $[Ru(dpp)_3]^{2+}$ fluorescence. The distribution of PEBBLEs in overlaid images demonstrated that the green and red fluorescence in Figure 22.9 came truly from PEBBLEs inside cells. It should be noted that most of the PEBBLEs were loaded into the cytoplasm. The intracellular concentrations were estimated from the spectra of the silica PEBBLEs within the cells and a Stern–Volmer calibration curve.

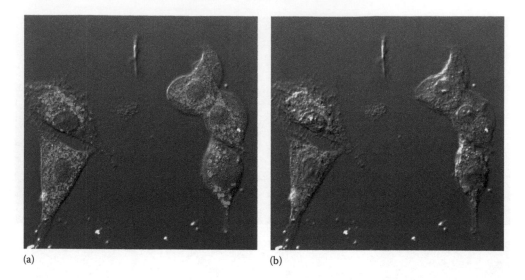

(a)   (b)

FIGURE 22.9 Confocal images of rat C6 glioma cells loaded with sol-gel PEBBLEs by gene gun injection. Nomarski illumination image overlaid with Oregon Green fluorescence (a) and $[Ru(dpp)_3]^{2+}$ fluorescence (b) of the same ratiometric PEBBLEs inside cells.

The ormosil oxygen PEBBLE sensors were used to monitor metabolic changes inside live C6 glioma cells.[25] Note that the embedded oxygen-sensitive dye, platinum(II) octaethylporphine ketone, has infrared fluorescence that is not affected by cellular autofluorescence, resulting in more reliable intracellular oxygen measurements. The spectra of the PEBBLEs within cells were first taken under open air to measure the intracellular oxygen concentrations at normal condition, and then the chamber containing the cells was sealed so as to shut off the air supply, and then a series of spectra were taken over an hour to monitor the change. As live rat C6 glioma cells breathe and consume the oxygen, the observed change would result from cell respiration and hence provide evidence of the viability of the cells after gene gun shooting as well as offer real-time measurements of intracellular oxygen concentrations. Figure 22.10 shows a typical profile obtained by monitoring the change in peak ratio (i.e., DO concentration) due to the respiration of the cells confined in a container.

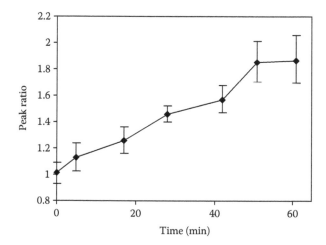

FIGURE 22.10 A typical peak ratio profile of the ormosil oxygen PEBBLE sensors in C6 glioma cells in a closed container. The ratio of indicator to reference dye increases with time, indicating a decrease of the DO concentration, due to respiration.

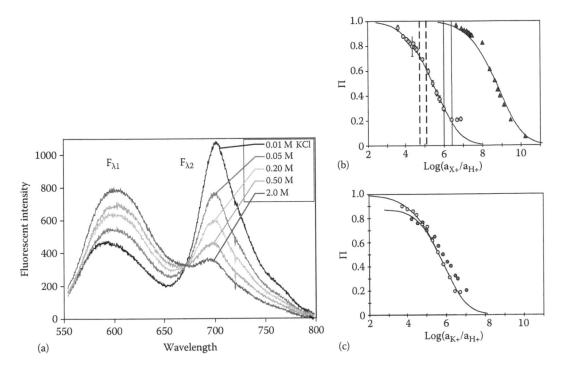

FIGURE 22.11 (a) Normalized emission spectra from suspended K[+] PEBBLE sensors using the pH chromoionophore ETH5350 for ion-correlation spectroscopy in tandem with BME-44. The spectra show response in going from 10 mM KCl to 2.0 M KCl (well beyond saturation of the sensor), all in 10 mM Tris buffer, pH 7.2. (b) Response of the same PEBBLEs to potassium (o) and sodium (△), along with theoretical curves. The lines delimit values for log (aK[+]/aH[+]) typically found in intracellular (solid) and extracellular (dashed) media.[50] (c) Response to additions of KCl in Tris buffer (o) compared to a similar experiment run in a constant background of 0.5 M Na[+] (•). Solid lines are calculated (not fit) theoretical curves for the K[+] response in the presence of 0.5 M interfering Na[+] using the experimentally determined log selectivity value of −3.3. a_{X+} and a_{H+} denote the activity of X[+] (K[+] or Na[+]) and that of the hydrogen ion in solution, respectively.

A significant increase in the peak ratio was observed, with ratio increases of 50%–500% after an hour, for 9 sets of cells, which correspond to 30%–90% reductions in terms of oxygen concentration. The variation in the ratios may be attributed to the number of cells on each glass slide and the condition of the cells. This result demonstrates cell viability after PEBBLE delivery and excellent intracellular response of the sensors to DO.

22.5.3 Potassium PDMA PEBBLEs

The first ion-correlation PDMA PEBBLEs were potassium PEBBLEs that use ETH 5350 as the chromoionophore, BME-44 as the ionophore, and KTFPB as the lipophilic additive.[16] These PEBBLEs measure the potassium concentration indirectly from the spectral change of a reporter dye (ETH 5350) that has two fluorescence emission maxima (λ_1, λ_2), giving a relative intensity that changes with the degree of protonation (Π). This degree of protonation, Π, is related to the analyte (K[+]) concentration and can be evaluated in terms of the ratio of the protonated chromoionophore intensity $F_{\lambda 2}$ to the deprotonated chromoionophore intensity $F_{\lambda 1}$ (see Figure 22.11 for spectra) based on an analytically derived relationship.[5,16]

Calibration of a K[+] sensor is shown in Figure 22.11b along with normalized spectra (Figure 22.11a). The data points for potassium and sodium responses are plotted along with corresponding theoretical curves. Dashed lines delimit typical extracellular activity ratios and the solid lines delimit the intracellular levels (log [aK[+]/aH[+]]).[50]

The response was found to match well with the theory, which is gratifying, considering the small size of the systems. The dynamic range at pH 7.2 extends from 0.63 mM to 0.63 M aK[+]. The log of the selectivity for potassium vs. sodium, determined by measuring the horizontal separation of the response curves at $\Pi = 0.5$, is -3.3. This selectivity value can be used, along with the mathematical theory for this sensing mechanism, to calculate what the K^+ response of the PEBBLEs should be in the presence of 0.5 M interfering Na^+. Figure 22.11c shows these calculated theoretical curves (not fits) along with the corresponding experimental data. This shows a selectivity similar to or better than that obtained for other and larger matrices incorporating BME-44, for example, -3.1 in PVC-based fiber-optic work and -3.0 in PVC-based microelectrodes.[6,17] It also exactly matches the value given in the review by Buhlmann, Pretsch, and Bakker[41] for a thin PVC film sensor. This selectivity should be more than sufficient for measurements in intracellular media where potassium concentration[50] is about 100 mM and sodium is about 10 mM.

It should be pointed out that the available hydrophilic potassium indicator dyes are much inferior to nonfluorescent potassium ionophore in terms of sensitivity and selectivity.[51] As a result, they are quite sensitive to sodium too and do not work well for detecting potassium in the presence of significantly higher sodium concentrations. In accordance with such facts, the selectivity of the potassium PDMA nanosensors is much higher, by three orders of magnitude, than that of nanoparticle sensors containing hydrophilic potassium indicator dyes.[52]

The PDMA potassium PEBBLEs were applied for the observation of potassium uptake in rat C6 glioma cells.[16] Decyl methacrylate PEBBLEs were delivered by gene gun using a BioRad (Hercules, CA) Biolistic PDS-1000/He system, with a firing pressure of 650 psi, and a vacuum of 15 torr applied to the system. Immediately following PEBBLE delivery, cells were placed on an inverted fluorescent microscope. The gating software for the CCD was set to take continuous spectra at 1.3 s intervals. After 20 s, and after 60 s, 50 μL of 0.4 mg/mL kainic acid was injected into the microscope cell. Kainic acid is known to stimulate cells by causing the opening of ion channels. Figure 22.12a shows the confocal fluorescent image of the PEBBLEs, overlaid with a Nomarski differential interference contrast image of the cells.[16] Image analysis indicated that the PEBBLE sensors were localized in the cytoplasm of the glioma cells. Figure 22.12b shows the PEBBLE sensors inside the cells responding to the kainic acid addition. One can see that log (aK^+/aH^+) increases, indicating either an increase in K^+ concentration or a decrease in H^+ concentration (increase in pH). The amount of kainic acid added is not known to affect the pH of cells in culture, and kainic acid by itself has no effect on the sensors. Thus, the change is likely due to increasing intracellular concentration of K^+, which is the expected trend. The membrane of C6 glioma cells can initiate an inward rectifying K^+ current, induced by specific K^+

(a)   (b)

FIGURE 22.12 (a) Confocal image of decyl methacrylate K^+ PEBBLE fluorescence, overlaid with Nomarski image of rat C6 glioma cells (488 nm excitation, 580 nm long-pass emission). (b) Ratio data of decyl methacrylate K^+ PEBBLEs in C6 glioma cells during the addition of kainic acid (50 μL of 0.4 mg/mL) at 20 s and at 60 s. Ratios were converted to log (aK^+/aH^+) using solution calibration of the PEBBLEs. Log (aK^+/aH^+) is seen to increase after kainic acid addition (and subsequent K^+ channel openings).

channels, a documented role in the control of extracellular potassium.[53] Thus, when stimulated with a channel opening agonist, the K^+ concentration within the glioma cells is indeed expected to increase.

Currently, a potassium PEBBLE nanosensor based on ormosil-capped micelle is under development, which has higher aqueous solubility and a much smaller size than the PDMA PEBBLEs but maintains the same high sensitivity and selectivity.

22.6 Future Perspective

PEBBLE nanosensors have been successfully applied to in vitro bioanalysis as demonstrated previously. However, intracellular measurements of small molecules or ions by the PEBBLE sensors have been limited to a few analytes although PEBBLE sensors for more diverse analytes have been developed. Those sensors' sensitivity, selectivity, and signals were not sufficient to overcome background interferences, which should be improved for the intracellular detection of more diverse single analytes. The development of NIR sensing dyes to eliminate autofluorescence from cellular components or more selective receptors toward analytes would be one way to improve the sensitivity and selectivity of these sensors. The adoption of a MOON-type sensor design for S/N enhancement (Figure 22.5) would be another way to improve the sensitivity.

Another important future direction in the design of PEBBLE sensors would be PEBBLEs that are capable of simultaneous sensing of multiple analytes that are related to each other in an important metabolic process and PEBBLEs targeted to specific organelles. These PEBBLEs could be especially useful for getting chemical or physical information from each single cell or even a specific location within each single cell. It is expected that the single-cell analysis would differentiate between individual cells, leading to diagnosing disease at an early stage, when changes on a tissue level are not yet evident but chemical changes within cells are observable.

Acknowledgments

The authors would like to acknowledge the support of NIH Grants 1R01-EB-007977-01, R21/R33 CA125297, and 1R41 CA 130518-01A1, as well as NSF Grant DMR 0455330.

References

1. Tan, W., Shi, Z.-Y., Smith, S., Birnbaum, D., and Kopelman, R., Submicrometer intracellular chemical optical fiber sensors, *Science*, 258, 778, 1992.
2. Tan, W., Kopelman, R., Barker, S.L.R., and Miller, M.T., Ultrasmall optical sensors for cellular measurements, *Anal. Chem.*, 71(17), 606A, 1999.
3. Rosenzweig, Z. and Kopelman, R., Development of a submicrometer optical fiber oxygen sensor, *Anal. Chem.*, 67(15), 2650, 1995.
4. Rosenzweig, Z. and Kopelman, R., Analytical properties and sensor size effects of a micrometer-sized optical fiber glucose biosensor, *Anal. Chem.*, 68(8), 1408, 1996.
5. Shortreed, M., Bakker, E., and Kopelman, R., Miniature sodium-selective ion-exchange optode with fluorescent pH chromoionophores and tunable dynamic range, *Anal. Chem.*, 68(15), 656, 1996.
6. Shortreed, M.R., Dourado, S., and Kopelman, R., Development of a fluorescent optical potassium-selective ion sensor with ratiometric response for intracellular applications, *Sens. Actuators B: Chem.*, 38–39, 8, 1997.
7. Barker, S.L.R., Shortreed, M.R., and Kopelman, R., Utilization of lipophilic ionic additives in liquid polymer film optodes for selective anion activity measurements, *Anal. Chem.*, 69(6), 990, 1997.
8. Barker, S.L.R., Thorsrud, B.A., and Kopelman, R., Nitrite- and chloride-selective fluorescent nano-optodes and in vitro application to rat conceptuses, *Anal. Chem.*, 70(1), 100, 1998.
9. Dourado, S. and Kopelman, R., Is smaller better? Scaling of characteristics with size of fiber-optic chemical and biochemical sensors, *Proc. SPIE*, 2836, 2, 1996.

10. Clark, H.A., Barker, S.L.R., Brasuel, M., Miller, M.T., Monson, E., Parus, S., Shi, Z. et al., Subcellular optochemical nanobiosensors: Probes encapsulated by biologically localised embedding (PEBBLEs). *Sens. Actuators B: Chem.*, 51, 12, 1998.

11. Clark, H.A., Hoyer, M., Philbert, M.A., and Kopelman, R., Optical nanosensors for chemical analysis inside single living cells. 1. Fabrication, characterization, and methods for intracellular delivery of PEBBLE sensors, *Anal. Chem.*, 71(21), 4831, 1999.

12. Clark, H.A., Hoyer, M., Parus, S., Philbert, M.A., and Kopelman, R., Optochemical nanosensors and subcellular applications in living cells, *Mikrochim. Acta*, 131, 121, 1999.

13. Horvath, T., Monson, E., Sumner, J., Xu, H., and Kopelman, R., Use of steady-state fluorescence anisotropy with PEBBLE nanosensors for chemical analysis. *Proc. SPIE (Int. Soc. Opt. Eng.)*, 4626, 482, 2002.

14. Sumner, J.P., Westerberg, N., Stoddard, A.K., Fierke, C.A., and Kopelman, R., Cu^+ and Cu^{2+} sensitive PEBBLE fluorescent nanosensors using Ds Red as the recognition element, *Sens. Actuators B: Chem.*, 113(2), 760, 2005.

15. Sumner, J.P. and Kopelman, R., Alexa Fluor 488 as an iron sensing molecule and its application in PEBBLE nanosensors, *Analyst*, 130(4), 528, 2005.

16. Brasuel, M., Kopelman, R., Miller, T.J., Tjalkens, R., and Philbert, M.A., Fluorescent nanosensors for intracellular chemical analysis: Decyl methacrylate liquid polymer matrix and ion-exchange-based potassium PEBBLE sensors with real-time application to viable rat C6 glioma cells, *Anal. Chem.*, 73(10), 2221, 2001.

17. Brasuel, M., Kopelman, R., Kasman, I., Miller, T.J., and Philbert, M.A., Ion concentrations in live cells from highly selective ion correlations fluorescent nano-sensors for sodium, *Proc. IEEE Sens.*, 1, 288, 2002.

18. Park, E.J., Brasuel, M., Behrend, C., Philbert, M.A., and Kopelman, R., Ratiometric optical PEBBLE nanosensors for real-time magnesium ion concentrations inside viable cells, *Anal. Chem.*, 75(15), 3784, 2003.

19. Sumner, J.P., Aylott, J.W., Monson, E., and Kopelman, R., A fluorescent PEBBLE nanosensor for intracellular free zinc, *Analyst*, 127, 11, 2002.

20. Brasuel, M.G., Miller, T.J., Kopelman, R., and Philbert, M.A., Liquid polymer nano-PEBBLES for Cl- analysis and biological applications, *Analyst*, 128(10), 1262, 2003.

21. Cao, Y., Koo, Y.L., Koo, S., and Kopelman, R., Ratiometric singlet oxygen nano-optodes and their use for monitoring photodynamic therapy nanoplatforms, *Photochem. Photobiol.*, 81(6), 1489, 2005.

22. Kim, G., Development of nanoparticle based tools for reactive oxygen species and related biomedical applications, PhD thesis, University of Michigan, Ann Arbor, MI, 2008.

23. King, M. and Kopelman, R., Development of a hydroxyl radical ratiometric nanoprobe, *Sens. Actuators B: Chem.*, 90, 76, 2003.

24. Xu, H., Aylott, J.W., Kopelman, R., Miller, T.J., and Philbert, M.A., A real-time ratiometric method for the determination of molecular oxygen inside living cells using sol-gel-based spherical optical nanosensors with applications to rat C6 glioma, *Anal. Chem.*, 73(17), 4124, 2001.

25. Koo, Y.L., Cao, Y., Kopelman, R., Koo, S., Brasuel, M., and Philbert, M.A., Real-time measurements of dissolved oxygen inside live cells by ormosil (organically modified silicate) fluorescent PEBBLE nanosensors, *Anal. Chem.*, 76(9), 2498, 2004.

26. Cao, Y., Koo, Y.L., and Kopelman, R., Poly(decyl methacrylate)-based fluorescent PEBBLE swarm nanosensors for measuring dissolved oxygen in biosamples, *Analyst*, 129, 745, 2004.

27. Chen-Esterlit, Z., Peteu, S.F., Clark, H.A., McDonald, W., and Kopelman, R., A comparative study of optical fluorescent nanosensors ("PEBBLEs") and fiber optic microsensors for oxygen sensing, *SPIE (Int. Soc. Opt. Eng.)*, 3602, 156, 1999.

28. Xu, H., Aylott, J.W., and Kopelman, R., Fluorescent nano-PEBBLE sensors designed for intracellular glucose imaging, *Analyst*, 127, 1471, 2002.

29. Tyner, K.M., Kopelman, R., and Philbert, M.A., "Nanosized voltmeter" enables cellular-wide electric field mapping, *Biophys. J.*, 93(4), 1163, 2007.

30. Tang, W., Xu, H., Park, E.J., Philbert, M.A., and Kopelman, R., Encapsulation of methylene blue in polyacrylamide nanoparticle platforms protects its photodynamic effectiveness, *Biochem. Biophys. Res. Comm.*, 369(2), 579, 2008.

31. Anker, J.N. and Kopelman, R., Magnetically modulated optical nanoprobes, *Appl. Phys. Lett.*, 82(7), 1102, 2003.

32. Anker, J.N., Behrend, C., and Kopelman, R., Aspherical magnetically modulated optical nanoprobes (MagMOONs), *J. Appl. Phys.*, 93(10), 6698, 2003.

33. Behrend, C.J., Anker, J.N., and Kopelman, R., Brownian modulated optical nanoprobes, *Appl. Phys. Lett.*, 84(1), 154, 2004.

34. McNaughton, B.H., Magnetic micro and nano nonlinear oscillators with applications to the dynamic detection of a single bacterium and to physical and chemical sensing, PhD thesis, University of Michigan, Ann Arbor, MI, 2007.

35. Daubresse, C., Granfils, C., Jerome, R., and Teyssie, P., Enzyme immobilization in nanoparticles produced by inverse microemulsion polymerization, *J. Coll. Int. Sci.*, 168, 222, 1994.

36. Uhlmann, D.R., Teowee, G., and Boulton, J., The future of sol-gel science and technology, *J. Sol-Gel Sci. Technol.*, 8, 1083, 1997.

37. Xu, H., Yan, F., Monson, E.E., and Kopelman, R., Room-temperature preparation and characterization of poly (ethylene glycol)-coated silica nanoparticles for biomedical applications, *J. Biomed. Mater. Res.*, 66A(4), 870, 2003.

38. Hah, H.J., Kim, J.S., Jeon, B.J., Koo, S.M., and Lee, Y.E., Simple preparation of mono-disperse hollow silica particles without using templates, *Chem. Commun.*, 14, 1712, 2003.

39. Bakker, E. and Simon, W., Selectivity of ion-selective bulk optodes, *Anal. Chem.*, 64(17), 1805, 1992.

40. Morf, W.E., Seiler, K., Lehmann, B., Behringer, C., Hartman, K., and Simon, W., Carriers for chemical sensors: Design features of optical sensors (optodes) based on selective chromoionophores, *Pure Appl. Chem.*, 61(19), 1613, 1989.

41. Buhlmann, P., Pretsch, E., and Bakker, E., Carrier-based ion-selective electrodes and bulk optodes. 2. Ionophores for potentiometric and optical sensors, *Chem. Rev.*, 98(4), 1593, 1998.

42. Kurihara, K., Ohtsu, M., Yoshida, T., Abe, T., Hisamoto, H., and Suzuki, K., Micrometer-sized sodium ion-selective optodes based on a "tailored" neutral ionophores, *Anal. Chem.*, 71(16), 3558, 1999.

43. Suzuki, K., Ohzora, H., Tohda, K., Miyazaki, K., Watanabe, K., Inoue, H., and Shirai, T., Fiberoptic potassium-ion sensors based on a neutral ionophore and a novel lipophilic anionic dye, *Anal. Chim. Acta*, 237(1), 155, 1990.

44. Mohr, G.J., Lehmann, F., Ostereich, R., Murkovic, I., and Wolfbeis, O.S., Investigation of potential sensitive fluorescent dyes for application in nitrate sensitive polymer membranes, *Fresenius J. Anal. Chem.*, 357(3), 284, 1997.

45. Mohr, G.J., Murkovic, I., Lehmann, F., Haider, C., and Wolfbeis, O.S., Application of potential-sensitive fluorescent dyes in anion- and cation-sensitive polymer membranes, *Sens. Actuators B: Chem.*, 39(1–3), 239, 1997.

46. Fawell, S., Seery, J., Daikh, Y., Moore, C., Chen, L.L., Pepinsky, B., and Barsoum, J., Tat-mediated delivery of heterologous proteins into cells, *Proc. Natl. Acad. Sci. U.S.A.*, 91, 664, 1994.

47. Webster, A., Compton, S.J., and Aylott, J.W., Optical calcium sensors: Development of ageneric method for their introduction to the cell using conjugated cell penetrating peptides, *Analyst*, 130, 165, 2005.

48. Boddapati, S.V., D'Souza, G.G.M., Erdogan, S., Torchilin, V.P., and Weissig, V., Organelle-targeted nanocarriers: Specific delivery of liposomal ceramide to mitochondria enhances its cytotoxicity in vitro and in vivo. *Nano Lett.*, 8(8), 2559, 2008.

49. Clark, H.A., Kopelman, R., Tjalkens, R., and Philbert, M.A., Optical nanosensors for chemical analysis inside single living cells part 2: Sensors for pH and calcium and the intracellular application of PEBBLE sensors, *Anal. Chem.*, 71(21), 4837, 1999.

50. Ammann, D., *Ion-Selective Microelectrodes*, Springer-Verlag, Berlin, Germany, 1986.
51. Haugland, R.P., *The Handbook: A Guide to Fluorescent Probes and Labeling Technologies*, 10th edn. Molecular Probes Inc., Eugene, OR, 2005.
52. Brown, J.Q. and McShane, M.J. Core-referenced ratiometric fluorescent potassium ion sensors using self-assembled ultrathin films on europium nanoparticles, *IEEE Sens. J.*, 5(6), 1197, 2005.
53. Emmi, A., Wenzel, H.J., and Schwartzkroin, P.A., Do glia have heart? Expression and functional role for ether-A-go-go currents in hippocampal astrocytes, *J. Neurosci.*, 20(10), 3915, 2000.

23

Nanosensors for Single-Cell Analyses

Charles K. Klutse
*Ghana Atomic Energy
Commission*

Brian M. Cullum
*University of Maryland,
Baltimore County*

Molly K. Gregas
Duke University

John P. Scaffidi
Duke University

Tuan Vo-Dinh
Duke University

23.1 Introduction

Over the years, new techniques in chemical and biological sensing have set the stage for great advances in the field of biological research. One of the most recent technological advances in this area has been the development of nanosensors. A nanosensor is a sensor that has dimensions on the nanometer-size scale. The development of these nanosensors is at the forefront of single-cell analysis, a major prospect for future biological research. Nanosensors of various types (e.g., optical, electrochemical) have been reported in literature over the past decade; however, the emphasis of this review will be on optical-based nanosensors.

Optical nanosensors, like larger sensors, can typically be classified into one of two broad categories depending on the probe used:[1-4] (1) chemical nanosensors and (2) biological nanosensors. Both types of sensors have been used to provide a reliable method of monitoring various chemicals in microscopic environments. They have even been used in the detection of different species within single cells. Nanosensors offer significant improvements over the current methods of cellular analysis involving the uptake of fluorescent indicator dyes, because they do not suffer from problems of cellular diffusion. Although the field of nanosensing is relatively new, the rapid growth in the number of reviews on the subject is indicative of the increased attention on the subject.[5-13] This chapter will review the evolution of optical nanosensors from their beginning (near-field optical microscopy) to the present (biosensors capable of probing subcellular compartments of individual mammalian cells) and discuss their application to biological measurements. For this handbook's second edition, the chapter has been revised to include a description of nanosensors for monitoring apoptotic pathways in single cells, surface-enhanced Raman scattering (SERS)-based chemical sensors for pH, and SERS nanobiosensors and their applications in intracellular analysis. Whereas the first edition dealt mostly with optical fiber-based nanosensors, this updated version includes an extended description of fiber-less nanosensors based on nanoparticle platforms, which are sometimes referred to as sensing nanoprobes. The advantages and limitations of fiber nanosensors versus nanoprobes in single-cell analysis will be discussed.

23.2 Sensing Principles

A sensor can be generally defined as a device that consists of a recognition element, often called a receptor, and a transducer. Despite the wide variety of receptors and transducers employed, sensors are all based on the same principle. First, a sensor is placed in the environment of interest, where the target molecules bind to the receptors in the biosensitive layer of the sensor. These receptors can be ligands for particular chemical species; biological molecules such as antibodies, enzymes, nonenzymatic proteins; and nucleic acids, or even living biological systems such as cells, tissues, or whole organisms. The resulting interaction between the analyte and the receptor then produces an effect that can be measured by the transducer, which converts this effect into some type of measurable signal (e.g., an electrical voltage). A conceptualized illustration of the sensing process is shown in Figure 23.1. Due to this dual-component nature, sensors are typically classified by either the receptor employed for molecular recognition or the type of transduction mechanism used for detection (see Figure 23.2). The most common forms of molecular recognition employed in sensors fall into several different categories: (1) nonbiological ligands, (2) antibody and Fab/antigen interactions, (3) nucleic acids and aptamers, (4) enzymatic interactions, (5) cellular interactions (i.e., microorganisms, proteins), and (6) interactions using biomimetic materials (i.e., synthetic bioreceptors). Common transduction mechanisms used in sensors include (1) optical measurements (e.g., absorption, fluorescence, phosphorescence, Raman, refraction, dispersion, or interference spectrometry), (2) electrochemical measurements (e.g., amperometric, potentiometric), and (3) mass-sensitive measurements (e.g., surface acoustic wave, microbalance, microcantilever). This chapter will emphasize the principles of operation of sensors that are based upon optical transduction and describe some of the most recent applications of nanosensors to

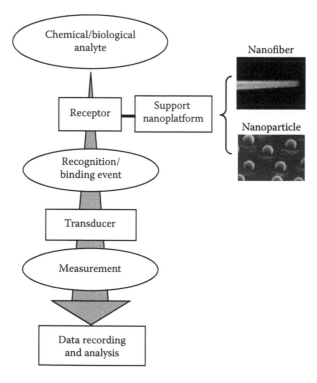

FIGURE 23.1 Sensing process for a nanosensor.

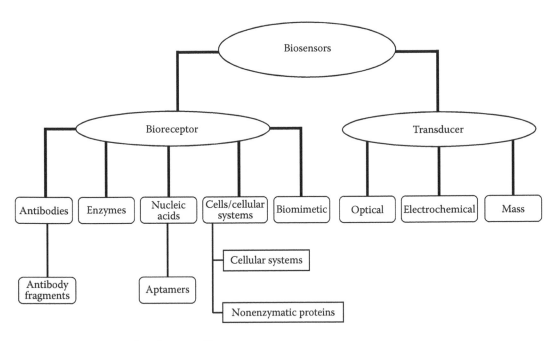

FIGURE 23.2 Various classifications of biosensors.

single-cell analyses. A review of the sensing concept and various types of biosensors is provided in greater detail in Chapter 1 of *Biomedical Photonics Handbook: Biomedical Diagnostics*.

23.2.1 Optical Transduction Techniques

Optical transduction describes a vast array of measurements (e.g., absorption, fluorescence, phosphorescence, Raman, SERS, refraction, dispersion spectrometry) that can be performed with optical sensors. In addition to the large number of optical techniques that can be used in the field of sensing, each one of them can be used to measure any of many different optical properties. These properties include amplitude, energy, polarization, decay time, and/or phase. A brief description of the types of information that can be obtained via optical measurements is described in the following, while more comprehensive information can be found in various other chapters of this handbook. The most commonly measured parameter of the electromagnetic spectrum is amplitude, as it can generally be correlated with the concentration of the analyte of interest easily. Although amplitude is the most commonly measured parameter, the other parameters can also provide a great deal of information about the analyte of interest as well. For instance, the energy of the electromagnetic radiation can often provide information about changes in the local environment surrounding the analyte, its intramolecular atomic vibrations (i.e., Raman or infrared absorption spectroscopies), or the formation of new energy levels (i.e., luminescence, ultraviolet (UV)–Vis absorption). Measurement of the interaction of a free molecule with a fixed surface can often be determined based upon polarization measurements. In the case of free molecules in solution, the emitted light is often randomly polarized, whereas, when a molecule is bound to a fixed surface, the emitted light generally remains polarized. Therefore, the rate of depolarization of a fluorescence signal can provide information on the movement of a molecule. Information can also be gained by measuring the decay time of a specific emission signal (i.e., fluorescence or phosphorescence), since these decay times are dependent upon the excited state of the molecules and their local molecular environment. Another property that can be measured is the phase of the emitted radiation. When electromagnetic radiation interacts with a surface, the speed or phase of that radiation is altered, based on the refractive index of that particular medium (i.e., analyte). Therefore, when an analyte binds to the receptor layer of a sensor, a change in the refractive index of the sensing surface could cause a measurable phase shift in the impinging radiation.

23.2.2 Molecular Receptors

One of the key factors in determining the usefulness of any chemical or biological sensor is its ability to specifically measure the analyte of interest without having other chemical species interfere with the measurement. It is the selective binding properties of these receptors that are the key to a sensor's specificity. A wide variety of receptors have been used in the development of sensors over the years. In fact, the different receptors that have been employed are more numerous than the different analytes that have been monitored using optical sensors. Although the amount and type of different receptors used is continually growing, they can generally be classified into one of six major categories: (1) nonbiological ligands, (2) antibodies and their fragments, (3) enzymes, (4) nucleic acids and aptamers, (5) cellular structures/cells, and (6) biomimetics.

23.2.2.1 Nonbiological Ligands

Nonbiological ligands constitute a very broad category of molecules that range from species with only weak molecular interactions (e.g., hydrogen bonding, van der Waals forces) to molecules capable of binding or complexing ionic species. While the former category accounts for a wide variety of receptor species, the latter accounts for the largest majority of these compounds. Optical sensors employing receptors based upon weak force interactions include indicator dyes that are immobilized in hydrophilic or hydrophobic membranes, thus allowing only certain chemical species to come into contact with the indicator dye,[14-17] as well as many other forms of interactions.[18-24] Ligands that rely upon ionic interactions or analyte binding include species such as ethylenediaminetetraacetic acid (EDTA), Calcium Green™, and fluorescein.

Since a large number of these receptor molecules are based simply upon the charge state of the analyte, there is often a great deal of cross-reactivity of these receptors with other similar species. An example of this can be seen in the majority of chelating agents for Mg^{+2} ions, which often have higher affinities for Ca^{+2} ions than Mg^{+2} ions. Due to the cross-reactivity of these nonbiological receptors, and their subsequent lack of specificity when employed in complex environments, such as biological systems (e.g., cells), there has recently been a shift toward biological receptor molecules or bioreceptors. These bioreceptors generally possess a much greater specificity than their nonbiological counterparts, due to a need for shape recognition as well as other forms of interactions (i.e., hydrogen bonding, ionic interactions).

23.2.2.2 Antibodies

Antibodies are biological molecules that exhibit very selective binding capabilities for specific structures. Two distinctly different classes of antibodies exist, both of which have been employed as bioreceptors for biosensors: monoclonal antibodies and polyclonal antibodies. When measurement of a very specific analyte is desired, monoclonal antibodies are often used, while polyclonal antibodies, which exhibit less specificity, are often used for the measurement of an entire class of compounds (e.g., PAHs). Because most biological systems are complex in their composition, the highly specific binding abilities of antibodies cause them to be one of the most powerful bioreceptors to be employed by biosensors. However, there are also limitations when employing antibodies for analyte binding. One such disadvantage is that to produce an immune response to a specific molecule, a certain molecular size and complexity are necessary. Therefore, immunogenic proteins typically have molecular weights of greater than 5 kDa.

Apart from molecular size limitations, the antibody immobilization process can also present some challenges. Notable among these challenges is the reduction of the bioactivity of antibodies upon immobilization. This happens because the carboxyl and amino groups that are the dominant functional groups are commonly utilized for the immobilization. Meanwhile, the 3D shape and the binding affinities of the antibody are largely determined by the interaction of these functional groups. Additionally, since most antibodies are rich in these functional groups, the individual antibodies randomly adsorbed onto the sensor support or interact with each other leading to random distribution or obscurity of the active sites.

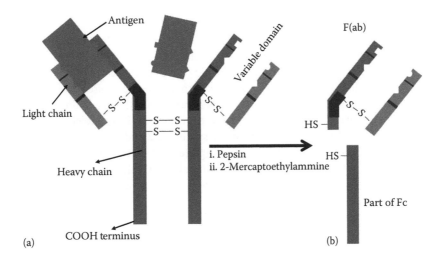

FIGURE 23.3 (a) A lock and key model of antibody–antigen interaction and (b) a Fab formed from proteolytic cleavage of an antibody and controlled reduction of sulfide to thiol.

23.2.2.3 Fragment Antigen Binding

Fragment antigen binding (Fab) is seen as a promising addition to antibody-based receptors. A Fab is the fragment of an antibody that encompasses the antigen-binding region as shown in Figure 23.3. Fabs therefore retain the specificity and affinity of their respective antibodies. Fabs as biorecognition elements lead to selective binding of their antigen. Also, Fabs are thiolated and rigid, allowing for their attachment onto sensor supports through the same points by self-assembly, reducing interference with active sites. By the same token, sensor functionalization issues such as random adsorption of the receptor onto the sensor supports, and easy removal and denaturation of the receptors can be minimized. Importantly, due to oriented immobilization of the Fabs onto sensor support, the binding affinities of the sensors are often increased relative to whole antibodies.[25,26] A review of antibody fragments as receptors in biosensors has been detailed by Saerens et al.[27]

23.2.2.4 Enzymes

Enzymes represent another class of bioreceptors that are commonly used in biosensors. In addition to the often highly specific binding capabilities of enzymes, their catalytic activity also makes them very useful. With the exception of a small group of catalytic ribonucleic acid (RNA) molecules, all enzymes are proteins and the function of these proteins varies dramatically from enzyme to enzyme. Some enzymes require no additional chemical groups other than their amino acid residues for activity, while others require an additional chemical component called a cofactor, which may be either one or more inorganic ions (e.g., Fe^{2+}, Mg^{2+}, Mn^{2+}, Zn^{2+}) or a more complex organic or metallo-organic molecule called a coenzyme. Both types of enzymes have been used in the development of biosensors. Sensors employing enzymes that do not require a cofactor or coenzyme are typically used when the analyte of interest is difficult to measure. In such cases, the reaction of the analyte with the enzyme produces a product that is readily measurable by a particular measurement technique. In addition, the catalytic nature of the bioreceptor means that the sensor is often reusable or reversible.

Biosensors employing enzymes that require an additional chemical component can be extremely sensitive and have extremely low detection limits for that additional chemical component. In such a sensor, the indirect detection of the cofactor or coenzyme of interest is typically performed by measuring a product of the enzymatic reaction. Since a single enzyme along with its respective cofactor or coenzyme can catalyze many reactions and thus produce a great quantity of reaction products, the indirect measurement of very small amounts of cofactor or coenzyme is possible.

23.2.2.5 Nucleic Acids

Another biorecognition mechanism capable of a high degree of specificity involves the hybridization of deoxyribonucleic acid (DNA) or RNA to a complimentary sequence. Within the last decade, the use of nucleic acids as bioreceptors for biosensor and biochip technologies has grown dramatically.[28-33] The specificity of biorecognition in these DNA-based biosensors, which are often referred to as genosensors, is based upon the complementarity of the cytosine/guanine (C:G) and adenine/thymine (A:T) base pairings. If the sequence of bases composing a certain segment of the DNA molecule is known, then the complementary sequence, often called a probe, can be synthesized and labeled with an optically detectable compound (e.g., a fluorescent label). By unwinding the double-stranded DNA into single strands, adding the probe, and annealing the strands, the unwound DNA will hybridize to its complementary probe sequence. Such probe strands were used extensively for the rapid testing of various sequences in the human genome project.

In addition to simple strands of complementary oligonucleotides with some type of optical label (e.g., fluorescent label), a relatively new class of bioreceptor has been developed, known as a molecular beacon. A molecular beacon is an oligonucleotide strand that has a long sequence that is complementary to the analyte sequence and on both ends, small region (several nucleotides long) that is complementary to the other end. In addition, a fluorescent dye is attached to one end and a quencher molecule is attached to the opposite end. Therefore, when the molecular beacon is not in the presence of the analyte, the two ends bind to each other keeping the fluorophore and quencher molecules in close proximity, resulting in a nonfluorescent complex. However, in the presence of the analyte, the large loop structure binds to the analyte, causing the two ends to separate, thus preventing the quencher molecule from acting on the fluorophore and thereby producing a fluorescent complex. Much greater detail about these compounds can be found in the chapter of this handbook dedicated to molecular beacons.

Peptides, which are short chains of amino acid molecules linked by peptide bonds, can be used as receptors in biosensors. Peptides are biomolecules belong to the broad chemical classes of biological oligomers, which also include nucleic acids, etc. Examples of the use of peptide-based bioreceptors are discussed in a letter section, which describes a modified immunochemical assay format of a nonfluorescent enzyme substrate, Leucine-Glutamic Acid-Histidine-Aspartic Acid-7-amino-4-methyl coumarin (LEHD-AMC).

23.2.2.6 Aptamers

Bioreceptors do not exist for all types of biomolecules. For this reason, the ability to synthesize bioreceptors for specific biomolecules can tremendously widen the analyte detection range of nanobiosensors. Aptamers are emerging as the potentially robust receptors that can be optimized for binding novel targets via combinatorial chemistry and selective isolation (i.e., systematic evolution of ligands by exponential enrichment [SELEX]). Aptamers are oligonucleotides, which fold into specific 3D shape based on the nucleotide sequence. As a result of their 3D shape, they exhibit specific binding affinities to a wide range of analytes. The target analytes of aptamers include ions, small molecules, proteins, viruses, cells, and even whole organisms.[34-36] Because of their synthetic nature, they can be controllably tailored for specific target molecules. They exhibit long-term stability and sustained reversible denaturation in optimal conditions, making them a relatively new and promising alternative in bioreception.

Aptamers are generated through repeated cycles of oligonucleotide selection and amplification processes called SELEX. The SELEX process is a technique of isolating oligonucleotides that have affinities for a particular analyte by screening a large library containing a diverse pool of oligonucleotides. The process involves a repetition of successive steps of target binding and removal of unbound oligonucleotides, followed by elution, amplification, and purification of the selected oligonucleotides as shown in Figure 23.4. Aptamer generation starts with the synthesis of combinatorial nucleic acid libraries. The number of starting nucleotides determines the complexity of the library generated. To increase the probability of selecting aptamers specific for the target of interest, a large number of individual oligonucleotides ($\sim 10^{15}$) is normally used to start the process. Each of the synthesized short ssDNA

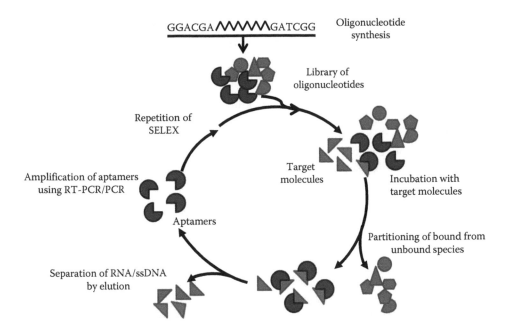

FIGURE 23.4 Aptamer generation by the SELEX procedure.

or RNA will adopt a specific 3D structure. This is because the combination of the base pairing in the strand marginally dictates the intermolecular interactions in the strand. By incubating the starting oligonucleotides with target biomolecules, aptamers with particular conformation selectively bind to the target molecule. Through affinity chromatography, the biomolecule–aptamer complexes are separated. The elution of the aptamers from the target leads to oligonucleotides with lower complexity and high affinity for the target. After the elution process, the eluted nucleic acids are amplified by polymerase chain reaction (PCR) (for DNA) or reverse transcription (RT)/PCR (for RNA). The process is repeated about 10–15 times to finally select a pool of aptamers made up of oligonucleotides that bind specifically to a target molecule. Having obtained the information of the aptamers, they can be synthesized in large quantities. Several reviews of the generation of aptamers[37,38] and the advantages of aptamers as receptors for biosensors have been published.[39] Although the generation of aptamers is largely based on the SELEX process, modifications of this process to improve on the aptamers generated have also been published recently.[40-43]

23.2.2.7 Cellular Structures/Cells

Cellular structures and cells comprise the most diverse category of bioreceptors that have been used in the development of biosensors.[44-78] These bioreceptors are based on biorecognition by either an entire cell/microorganism or a specific cellular component capable of specific binding to certain chemical or biochemical species. Because of its great diversity, this category of bioreceptor can be further divided into three major subclasses: (1) cellular systems, (2) enzymes, and (3) nonenzymatic proteins. However, due to the importance and the large number of biosensors that employ enzymes as their biorecognition element, these have been given their own classification and were discussed previously. In addition, since the focus of this chapter is on nanosensors, and whole cells or cellular systems are too large to fit on a sensor with nanometer dimensions, we will focus solely on nonenzymatic proteins in this section.

Many proteins found within cells often serve as bioreceptors for intracellular reactions that will take place later or in another part of the cell (e.g., carrier proteins, channel proteins on a membrane). No matter what the purpose, these proteins provide a means of molecular recognition through one type or another (i.e., active site or potential sensitive site). Because of the various biorecognition properties of these different nonenzymatic proteins, many different types of biosensors have been constructed for one application or another.

23.2.2.8 Biomimetic Receptors

The final major class of receptors used in biosensors, known as biomimetic receptors, is not true biological molecules. In fact, biomimetic receptors are man-made receptors that have been designed to mimic biological molecules. Since the fabrication of the first biomimetic receptor for biosensing applications, several different methods have been developed for the construction of such receptors.[79-91] These methods include (1) genetic engineering of molecules, (2) artificial membrane fabrication, and (3) molecular imprinting.

Recombinant techniques that allow for the synthesis or modification of a wide variety of binding sites within biological molecules have provided powerful tools for the design and synthesis of bioreceptors with desired properties. While biochemists and molecular biologists have been modifying various amino acid sequences in a variety of proteins for many years, it has only been in the past decade that such techniques have begun to be applied to biosensors for the enhancement of a bioreceptor's binding properties (i.e., binding affinity, specificity). The development of one such genetically engineered single-chain antibody fragment for the monitoring of phosphorylcholine has recently been reported by Hellinga et al.[92] In that work, protein engineering techniques were used to fuse a peptide sequence that mimics the binding properties of biotin to the carboxy terminus of the phosphorylcholine-binding fragment of an IgA. This genetically engineered molecule was then attached to a streptavidin monolayer, while total internal reflection spectroscopy was used to monitor the binding of a fluorescently labeled phosphorylcholine analog.

The second major field of biomimetic receptors, artificial membrane fabrication, has also been performed for many different applications over the years. In one such application, Stevens and coworkers developed an artificial membrane by incorporating gangliosides into a matrix of diacetylenic lipids (5%–10% of which were derivatized with sialic acid).[93] After the lipids self-assembled into Langmuir–Blodgett layers, they were photopolymerized via UV irradiation into polydiacetylene membranes. These newly fabricated membranes were then attached to biosensors for the detection of cholera toxins. As the cholera toxins became bound to the membrane, they caused the membrane to change color from its original blue color to red, which was subsequently monitored via absorption measurements.

The newest and final class of biomimetic receptors is those created by a technique known as molecular imprinting (see Figure 23.5). This technique consists of mixing analyte molecules with a monomer solution and a large amount of cross-linkers. As the monomers begin to polymerize, a solid plastic is created that is full of the particular analyte that is to be measured. Following polymerization, the hard

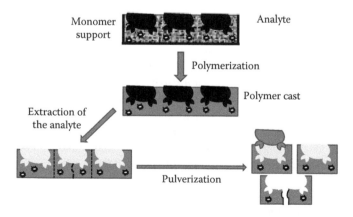

FIGURE 23.5 The molecular imprinting process.

polymer is ground into a fine powder and the analyte molecules are extracted with organic solvents to remove them from the polymer network. After extraction, the resulting polymer powder has molecular holes or binding sites that are complementary to the selected analyte. Therefore, by attaching some of the molecularly imprinted polymer (MIP) powder on a biosensor, receptor sites for molecules with the exact 3D structure as the analyte used in the imprinting process are now present. One of the primary reasons for the growing popularity of the molecular imprinting technique is the rugged nature of a polymer relative to a biological sample. Unlike biological molecules, MIPs can withstand harsh environments such as those experienced in an autoclave or denaturing chemical environment. However, MIPs typically exhibit reception abilities slightly less than antibodies or other biological receptor molecules.

23.3 Optical Nanosensors

Over the years, new techniques in sensing have set the stage for great advances in the field of biological research. One of the most recent advances in the field of optical sensors has been the development of optical nanosensors. A nanosensor is a sensor that has dimensions on the nanometer-size scale and can be inserted into a living cell for minimally invasive analysis. Presently, the most common nanosensors employ nanoparticles with diameters ranging between 20 and 500 nm. However, as will be highlighted in later sections, optical nanosensors have undergone evolution with regard to different types of sensor supports, the receptor type, and their immobilization. These nanosensors are based on the same basic principles as larger, more conventional optical sensors. Because the diameters of these optical sensors are significantly smaller than the conventional optical sensors, they are commonly used for probing microbiological samples including intracellular analyses. In recent years, real-time intracellular analyses have been reported for some of these nanosensors. These types of analyses have a wide range of application and will greatly enhance our understanding of the physiology of cells and life in general. The potential wide range of applications in real-time intracellular analysis has been part of the motivation for the evolution of optical nanosensors.

23.3.1 Evolution of Optical-Based Nanosensors

23.3.1.1 Near-Field Optics and Spectroscopy

Optical sensors using nanoscale optical fibers, which will be discussed later, arose from an area of research known as near-field optical microscopy. Near-field optical microscopy is a relatively new area of research that employs light sources or detectors that are smaller than the wavelength of light used for imaging. The most common method for performing such experiments is by placing a pinhole in front of the detector, thereby effectively reducing the size of the detector.[94,95] A variation of this technique is to construct an excitation probe with dimensions that are smaller than the wavelength of light that is being used for sample interrogation. This small excitation probe can then act as a light source with subwavelength dimensions. Betzig et al. have reported the development of such a probe capable of providing images of a sample with an optimum spatial resolution of approximately 12 nm. This probe was constructed by pulling a single-mode optical fiber with a micropipette puller and then coating the wall of the fiber with ~100 nm of aluminum. By applying an aluminum coating to the walls of the fiber, leakage of light from the tapered region of the fiber, where the cladding was pulled too thin to maintain the total internal reflection, is prevented, thereby confining the excitation radiation to the fiber tip (~20 nm). Using this first nanofiber-based system operating in the illumination mode (i.e., with the probe acting as a localized excitation source), images of a known pattern were reconstructed by raster scanning the probe over it. In addition, signal enhancements of greater than 10^4 [96,97] over previous near-field probe analyses[98–101] were obtained.

Due to the high degree of spatial resolution provided by near-field microscopy, these probes have been used for the detection of single molecules. In particular, one of the more recent applications of these nanofibers to single-molecule detection has been for the detection of dye-labeled DNA molecules using near-field surface-enhanced resonance Raman spectroscopy (NFSERRS).[102,103] Using this NFSERRS technique, it was possible to obtain chemical images of single DNA molecules labeled with the fluorescent dye, brilliant cresyl blue (BCB). In that work, the dye-labeled DNA strands were spotted onto a SERS substrate that was prepared by evaporating silver on a nanoparticle-coated substrate.[102,103] NFSERRS spectra were then obtained by illuminating the sample with the nanoprobe and measuring the Raman signal with a charge-coupled device (CCD) mounted on a spectrometer. By raster scanning the nanoprobe across the sample, a 2D image of the DNA molecules was reconstructed and normalized for surface topography based on the intensity of the Rayleigh scatter.

Near-field optical microscopy using taper fiber optics promises to be an area of research that holds a great deal of potential for the mapping of surfaces for individual biological molecules. In addition, since many biological compounds exhibit properties such as luminescence, which are typically much stronger than Raman signals, it might be possible in the near future to map out the location of individual molecules of specific chemicals in human tissues such as neurotransmitters in the brain. Single-molecule detection would therefore represent the absolute limit of detection for any compound and could open new horizons in the investigation of the complex chemical reactions and pathways of biological systems.

23.3.1.2 Optical Nanofibers

The sensing principle in nanofiber sensors is basically the same as the larger sensors. However, there are differences in the excitation process. Because the diameter of the optical fiber's tip is significantly smaller than the wavelength of light used for excitation of the analyte, the photons cannot escape from the tip of the fiber to be absorbed by the species of interest, as is the case in larger fiber-optic sensors. Instead, in a fiber-optic nanosensor, after the photons have traveled as far down the fiber as possible, excitons or evanescent fields continue to travel through the remainder of the tip, providing excitation for the fluorescent species of interest present in the sensing layer. An illustration of this excitation procedure is shown in Figure 23.6. An additional feature of this excitation process is that only species that are in extremely close proximity to the fiber's tip can be excited, thereby precluding the excitation of interfering autofluorescent species within other locations of the sample. This is extremely important in biological systems, such as cells, where a large number of autofluorescent species are often present.

23.3.1.2.1 Chemical Nanosensors

Shortly after their development for near-field microscopy, nanofibers began to be applied to the field of chemical sensing. With the advent of these nanofibers, it became possible to probe for specific chemicals

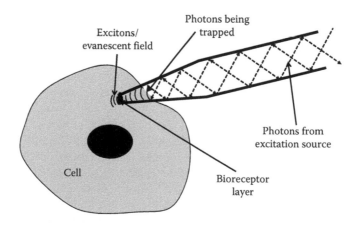

FIGURE 23.6 The excitation process in the optical fiber nanosensor showing the evanescent field at the tapered end of the nanofiber.

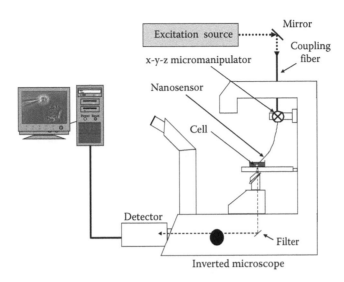

FIGURE 23.7 Schematic diagram showing a typical experimental setup used for nanosensor measurements within a single cell or microscopic sample. The nanosensor is mounted in an x-y-z micromanipulator mounted on an inverted microscope. The fluorescent light is then collected by the microscope and monitored with a detector.

in 3D structures over very short distances (<50 nm). Because of this ability to perform highly spatially localized analyses, the monitoring of concentration gradients and spatial inhomogeneities in submicroscopic environments (e.g., cells) via various spectroscopic techniques has become possible. However, due to the small sampling volume probed by nanofibers, the amount of analyte species in the excitation volume is very small, therefore making it important to use a sensitive spectroscopic technique for the analysis (e.g., fluorescence). In addition to a sensitive spectroscopic technique, it is also important to employ a sensitive detection scheme, such as the one shown in Figure 23.7, which is commonly used with fiber-optic nanosensors and nanobiosensors. In this system, an excitation source of some type (typically a laser) is launched into a fiber optic that is attached to the large proximal end of the nanosensor. The nanosensor, which is held in place by an x–y–z micromanipulator mounted on an inverted microscope, is then used to position the tip of the sensor to the desired location while looking through the eyepiece of the microscope. Once in place, the excitation source is turned on and a dye that is sensitive to the presence of the analyte is excited. As the excited dye relaxes to the ground state, the resulting fluorescence emission is collected by the microscope optics and detected with either a CCD or a photomultiplier tube (PMT).

Since the construction and use of the first optical nanosensors by Tan et al. in 1992,[104,105] several different optical-fiber-based chemical nanosensors have been reported for the measurement of pH,[106–109] various ion concentrations,[110,111] and other chemical species.[112] In this original work, a micropipette puller was used to taper multimode and single-mode fibers down to diameters of 100–1000 nm at the tip.[104,105] These tapered fibers were then used in the production of pH nanosensors via a three-step process. The first step in the process was to apply a thick layer of aluminum to the walls of the fiber, using a vacuum evaporator, to ensure total internal reflection over the tapered region of the fiber. During this aluminum deposition process, it is important to ensure that the tip of the fiber remains free of metal, thus providing a silica surface for binding of the probe or receptor molecules to the fiber's tip. In order to achieve this, the fibers are placed in the evaporator system with their tips facing away from the source of the evaporating metal. By precisely angling the fibers, their sides will shadow the tips from the evaporating metal. This is illustrated in Figure 23.8. Once the walls of the fiber were coated with aluminum, the next step was to silanize the fiber tip to allow for cross-linking to a polymer coating. The final step in the fabrication of this pH nanosensor was to then attach the pH sensitive dye, acrylofluoresceinamine, to the silanized fiber tip through a variation of a photopolymerization process that has often been used in

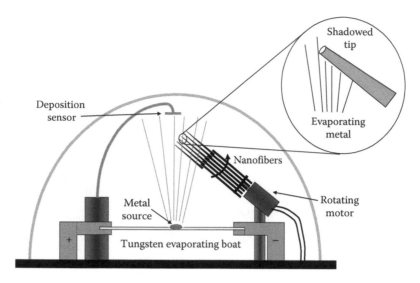

FIGURE 23.8 Illustration of the procedure for coating the walls of fiber-optic nanosensors with a reflective metal.

the construction of larger chemical sensors.[113,114] However, because of the near-field excitation provided by these small fiber probes, the cross-linking of the polymer solution was restricted to the near field of the fiber. Following fabrication of this sensor, characterization of its properties was performed. During this characterization, the sampling volume of the sensor was determined to be six orders of magnitude smaller than conventional fiber-optic chemical sensors, making it ideal for subcellular measurements. In addition, the reduced sampling volume and response time of the sensor were evaluated by measuring the pH in 10 μm diameter pores in a polycarbonate membrane. This evaluation found that the response time of the sensor (300 ms) was 100-fold faster than conventional fiber-optic chemical sensors and that it was both stable and reversible with respect to pH changes.

The first reported application of fiber-optic nanosensors for biological measurements was performed on rat embryos.[105] In this study, pH nanosensors like the ones described previously were inserted into the extraembryonic space of a rat conceptus, with minimal damage to the surrounding visceral yolk sac, and pH measurements were made. From these measurements, values of the pH in the extraembryonic fluid of rat conceptuses ranging in age from 10 to 12 days old were then compared for any differences. In a similar study using the same pH sensor, indirect measurements of nitrite and chloride levels in the yolk sac of rat conceptuses were also performed.[115] Because of the minimally invasive nature of such measurements, such techniques provide much promise for biological analyses and may aid in furthering our understanding of the effect that environmental factors play on embryonic growth.

As the field of optical-fiber-based nanosensors has evolved, the size of the environments in which they probe continues to become smaller. In fact, the measurement of chemical species inside of individual cells has even been reported using fiber-optic chemical nanosensors. In the first of these reports, optochemical nanosensors for sodium ions (Na⁺) were developed using a process similar to the one described previously. These sensors were then used for the measurement of Na⁺ concentrations in the cytoplasmic space inside a single mouse oocyte, one of the largest mammalian cells (~100 μm in diameter).[8] These sensors allowed for the monitoring of the relative Na⁺ concentrations, while ion channels were opened and closed by the external stimulant, kainic acid. In another application of fiber-optic nanosensors to biological measurements, calcium-ion-sensitive nanosensors were developed and used to measure the calcium ion fluctuations in vascular smooth muscle cells while the cells were stimulated.[116] These fluctuations were then directly correlated to the stimulation events that were being performed, providing another potential application for these nanometer-sized optical sensors.

23.3.1.2.2 *SERS-based pH Nanosensor Fiberprobe for Single-Cell Biosensing*

For binding to the target species, chemical nanosensors use a chemical receptor whereas nano-biosensors use a biological receptor.[117-127] An example of a fiber-based system is a plasmonics-active nanosensor that has been developed for intracellular measurement of pH in single living human cells using SERS detection.[128,129] Knowledge of intracellular pH values is essential in cellular studies of protein structure, enzymatic catalysis, drug delivery and effects of cellular transport, and exposure to environmental toxicants. Whereas approaches that use pH-sensitive nanoparticles (discussed in later sections) are frequently subject to concerns regarding the efficacy of nanoparticle uptake, intracellular localization and ejection, fiber-optic nanoprobes avoid these issues entirely because the nanoprobe is physically inserted into cells. The speed of SERS-active nanoprobe insertion, interrogation, and subsequent removal from a cell (often less than 30 s) also decreases the potential for intracellular digestion of the biochemically responsive functionality anchored to the nanoparticles used as delivery agents. In addition, the relatively short duration of the nanoprobe insertion/interrogation/removal process greatly reduces the danger of silver nanoprobe degradation within the intracellular environment.

The pH nanosensors were fabricated by tapering 400 μm core-diameter optical fibers using a commercially available pipette puller, producing nanoprobes smaller than 100 nm in diameter. These tapered optical fibers were then coated with a 6 nm mass thickness of silver at ~10^{-7} torr atmospheric pressure using an electron beam evaporator. Following Ag island film (AgIF) deposition, the nanoprobes were functionalized for 15 s in 10 mM para-mercaptobenzoic acid (pMBA; Sigma) dissolved in ethanol, which anchored pMBA to the AgIF via a silver-thiol covalent bond. The carboxyl group of pMBA is pH sensitive across the physically relevant range, thereby rendering the nanoprobe pH sensitive across that range as well.

The effectiveness and usefulness of the SERS nanoprobes are illustrated by measurements of pH values in HMEC-15/hTERT immortalized *normal* human mammary epithelial cells and PC-3 human prostate cancer cells. The results indicate that nanoprobe insertion and interrogation provide a sensitive and selective means to monitor cellular microenvironments at the single cell level. Insertion of nanoprobes into single cells for SERS interrogation is performed using micromanipulators (Figure 23.9). A pH calibration curve can be established using the intensity of the ~1425 cm^{-1} pMBA carboxylate band relative to the intensity of the ~1587 cm^{-1} combination band. Error bars represent two standard deviations in the y dimension, and the resolution of the pH meter in the x dimension (Figure 23.10b). We have used SERS-active fiber optic nanoprobes to measure the intracellular pH of HMEC-15/hTERT immortalized human mammary epithelial cells and PC-3 human prostate cancer cells. The results indicated the pH value of both cell lines is similar to that expected for healthy cells in the absence of significant environmental stress. In addition, we have demonstrated that insertion of the pH-sensitive, SERS-active fiberoptic nanoprobe induces neither apoptosis nor an aggressive lysosomal response from either of these cell lines.

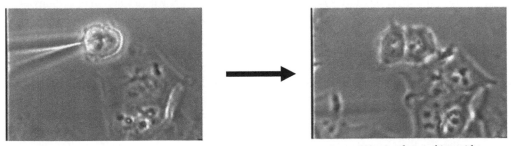

120 min after probing with
a fiber-optic probe

FIGURE 23.9 Series of time-lapse microscopic images of a cell undergoing mitosis prior to and immediately following interrogation with a fiber-optic nanobiosensor. The cell is approximately 10 μm in diameter.

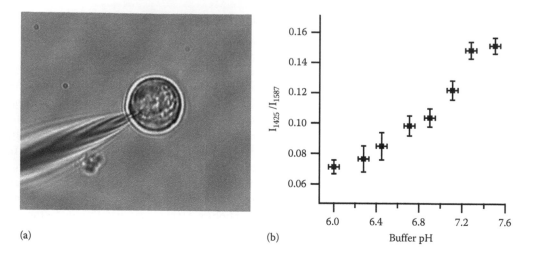

(a)   (b)

FIGURE 23.10   (a) Nanoprobe interrogation and analysis of a single living cell and (b) pH calibration curve for the pH-sensitive nanoprobe. (Adapted from Scaffidi, J.P. et al., *Anal. Bioanal. Chem.*, 393, 1135, 2009.)

These results illustrate the utility of the SERS-active pH nanoprobe for intracellular pH measurements, and further demonstrate the potential of biochemically specific fiberoptic nanoprobes for cellular studies.

23.3.1.2.3 Biological Nanosensors (Nanobiosensors)

Due to the complexity of biological systems and the number of possible interferences to chemical nanosensors, there became an obvious need for added specificity. This specificity was achieved by the development of nanobiosensors. Like their larger counterparts, the added specificity provided by biological receptor molecules allowed for the analysis of complex environments with minimal effects from other species. Application of nanofibers to the field of biosensors was first reported in 1996 by Vo-Dinh et al.[117] In this work, antibody-based nanobiosensors were developed and characterized for benzo[a]pyrene tetrol (BPT), a DNA adduct of the carcinogen benzo[a]pyrene (BaP). Because it is a biomarker for human exposure to BaP, the measurement of BPT is of great interest in early cancer diagnosis. Fabrication of these first nanobiosensors was performed in a three-step process. The first step in this process was to taper a 600 µ diameter fiber optic down to 40 nm at the tip, followed by application of a thick layer of silver, approximately 200 nm, to the walls of the fiber to prevent light leakage as well as enhance the delivery of light to the end of the fiber. Following the silver coating, the fiber tip was silanized to create attachment sites for antibodies. Once silanized, antibodies were attached via a covalent-binding procedure. The first step in this antibody binding procedure was to activate the silanized fiber with a solution of carbonyldiimidazole, which was then followed by incubation of the activated fiber with the antibody of interest for 3 days at 4°C. Once the antibodies were bound, the biosensors were characterized in terms of retention of antibody binding ability as well as sensitivity and absolute detection limits. Tests were performed by placing the nanobiosensors in a measurement system like the one described for chemical nanosensors and inserting the distal end of the sensor into a series of solutions of BPT. From these measurements, it was found that the antibodies had retained greater than 95% of their native binding affinity for BPT and that the absolute detection limit for BPT using these sensors was approximately 300 zeptomoles (zepto = 10^{-21}).

In the several years since this work, optical nanobiosensors employing many different forms of bioreceptor molecules have been developed and applied to the analysis of many biologically relevant species.[116,118] This has included nanobiosensors for the detection of nitric oxide via fluorescence detection of the nonenzymatic protein cytochrome c or fluorescently labeled cytochrome c,[116] as well as an enzymatic-based nanobiosensor for the indirect detection of glutamate.[118] In the latter design, the enzyme glutamate dehydrogenase was used as the bioreceptor. When glutamate became bound to the glutamate dehydrogenase, the reduction of NAD to NADH occurred, allowing for NADH to be measured via fluorescence spectroscopy. Using an enzymatic-based nanobiosensor such as this, it could be possible to continuously monitor the glutamate levels released from an individual cell. Such a sensor could prove to be a significant tool in the field of neurophysiology, since glutamate is one of the major neurotransmitters in the central nervous system, and by monitoring the glutamate levels produced by individual cells, a better understanding of the mechanisms by which sensations are transmitted throughout the human body may be achieved. Although enzymes have previously been employed in the field of biosensors for their ability to be regenerated, this was the first example of the use of an enzyme in the fabrication of a nanobiosensor.

Recently, several reports have been published on fiber-optic nanobiosensors being applied to in vitro measurements of single cells.[9,119–122] In one such study, nanobiosensors for BPT were prepared as described earlier and used for the measurement of intracellular concentrations of BPT in the cytosol of two different cell lines: (1) human mammary carcinoma cells and (2) rat liver epithelial cells.[119] The cells in both lines were spherical in shape and had diameters of approximately 10 μm. Several hours prior to making a measurement, a known amount of BPT was added to the culturing media of the cells. Then immediately prior to making any measurements, the culturing media was rinsed several times and replaced with BPT-free media. From this work, it was found that the concentration of BPT in each of the different cell lines was the same, suggesting that the means of transport of the BPT into the cells was the same in both cases (probably by diffusion). By performing similar measurements inside various subcellular compartments or organelles, it could be possible to obtain critical information about the location of BPT formation within cells and its transport throughout the cell during the process of carcinogenesis. In addition, it was also shown in this work that the insertion of a nanobiosensor into a mammalian somatic cell not only appears to have no effect on the cell membrane but also does not affect the cell's normal function. This was demonstrated by inserting a nanobiosensor into a cell that was just beginning to undergo mitosis and monitoring cell division following a 5-min incubation of the fiber in the cytoplasm and fluorescence measurement. Figure 23.9 contains a series of time-lapse images, from this experiment, showing the initial stages of mitosis in which the fiber-optic nanobiosensor is interrogating the cell all the way through division of the cell into two identical daughter cells.

23.3.1.2.4 *Monitoring Apoptotic Pathways in Single Cells Treated With Anticancer Drug*

Nanobiosensors provide important tools to monitor biological markers or pathways associated with the effectiveness of anticancer drugs such as the onset of apoptosis in living cells is an important process in drug development and therapy assessment.[123–127] The cell death process, known as apoptosis, involves caspase activation, which is a hallmark of apoptosis, and probably one of the earlier markers that signals the apoptotic cascade.[130,131] These cysteine proteases are activated during apoptosis in a self-amplifying cascade. Experimental evidence suggests that caspase activation is essential for the apoptotic process to take place, although not all cell death is dependent upon caspase activation. Caspases have an essential role both in the initial signaling events of apoptosis, as well as in the downstream processes which produce the various hallmark signs of apoptosis. Activation of caspases such as caspases 2, 8, 9 and 10 leads to proteolytic activation of *downstream* caspases such as 3, 6, and 7.

Nanobiosensors have been developed for monitoring the onset of the mitochondrial pathway of apoptosis in a single living cell by detecting enzymatic activities of caspase-9 following treatment of an anticancer drug.[123] Minimally invasive analysis of single live MCF-7 cells for caspase-9 activity was demonstrated using a fiberoptic nanosensor which used a modified immunochemical assay format of a nonfluorescent enzyme substrate, Leucine-Glutamic Acid-Histidine-Aspartic Acid-7-amino-4-methyl coumarin (LEHD-AMC). LEHD-AMC covalently attached on the tip of an optical nanobiosensor was cleaved during apoptosis by caspase-9, thus generating free AMC molecules that become fluorescent upon laser excitation. An evanescent field was used to excite cleaved AMC and the resulting fluorescence signal was detected. By quantitatively monitoring the changes in fluorescence signals, caspase-9 activity within a single living MCF-7 cell was detected. Photodynamic therapy (PDT) protocols employing the pro-drug aminolevulinic acid (ALA) were used to induce apoptosis in MCF-7 cells. The substrate LEHD-AMC was cleaved by caspase-9 and the

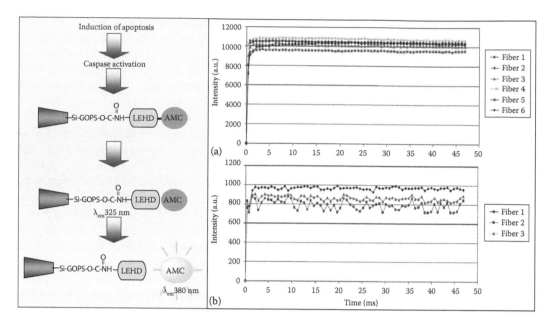

FIGURE 23.11 Left: Schematic diagram of ALA induced apoptosis, involving the activation of caspase-9 followed by the cleavage of LEHD-AMC and subsequent detection of free AMC by the nanosensor; Right: (a) Fluorescence intensity measurements of AMC for intracellular detection of caspase-9 activity in a single MCF-7 cell, which received both ALA and photoactivation. (b) Control measurements for the intracellular detection of caspase-9 activity in a control MCF-7 cels, which received neither ALA nor photoactivation. Replicate measurements using several nanosensors are shown. (Adapted from Kasili, P.M. et al., *J. Am. Chem. Soc.*, 126, 2799, 2004.)

released AMC molecules were excited and emitted a fluorescence signal (Figure 23.11). By comparison of the fluorescence signals from apoptotic cells induced by PDT treatment, and nonapoptotic cells, we successfully detected caspase-9 activity, which indicates the onset of apoptosis in the cells. The results show that the fluorescence signals obtained from the cells that were both incubated with ALA and photoactivated by laser excitation were much higher than the signal obtained from control groups, that is, without laser activation.[123] The presence and detection of cleaved AMC in single live MCF-7 cells as a result of these experiments reflects the presence of caspase-9 activity and the occurrence of apoptosis. These results demonstrate that apoptosis can be monitored in vivo in a single living cell using optical nanobiosensors.

23.3.1.2.5 Detection of Cytochrome c in a Single Cell

Detection of minute amounts of biological compounds in a single cell, especially for nonfluorescent or weakly fluorescent species, is a great challenge. Song et al. have demonstrated the possibility of detecting cytochrome c probed in a single cell using the optical nanobiosensor with mouse anti-cytochrome c antibodies using a combination of the nanosensing scheme with the ELISA immunoassay.[127] Cytochrome c is a protein that is important to the process of creating cellular energy, the main function of mitochondria. When mitochondria are damaged by PDT, cytochrome c is released into the cytoplasm of the cell. The release of cytochrome c is part of the cascade of cellular events that lead to apoptosis. This indicates that measurement of cytochrome c in the cytoplasm can be used as evidence that apoptosis is occurring. This can also lead to greater understanding of certain diseases on a cellular level. Following in situ binding of cytochrome c to the antibody-based nanoprobe inserted inside a single cell, an enzyme-linked immunosorbent assay, which provides the enzymatic fluorescent amplification to give high detection sensitivity, was performed with the optical nanobiosensor outside the cell for detecting cytochrome c in a MCF-7 human breast carcinoma cell treated with α-aminolevulinic acid (5-ALA) PDT drug which induces the apoptotic pathway. After the PDT photoactivation, the release of cytochrome c from the mitochondria to

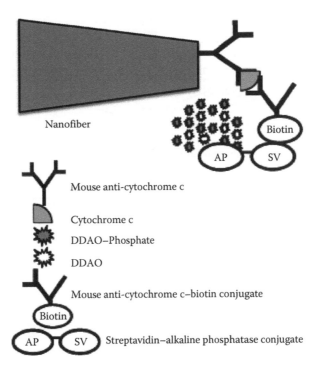

Nanofiber

Mouse anti-cytochrome c

Cytochrome c

DDAO–Phosphate

DDAO

Mouse anti-cytochrome c–biotin conjugate

Streptavidin–alkaline phosphatase conjugate

FIGURE 23.12 Schematic diagram of an immunoassay-ELISA protocol using a nanosensor. Enzyme-linked immunosorbent assay (ELISA) was peformed on the nanoprobe in order to indirectly detect the cytochrome c bound to the mouse anticytochrome c immobilized on the nanoprobe. As a result of ELISA, enzymatic product (cleaved DDAO) could be amplified after the nanosensor was removed from the cell. (Adapted from Song, J.M. et al., *Anal. Chem.*, 76, 2591, 2004.)

the cytoplasm in a MCF-7 cell was monitored by the optical nanobiosensor inserted inside the single cell and followed by an enzyme- linked immunosorbent assay (ELISA) outside the cell. Figure 23.12 shows a conceptual diagram of the combined use of nanobiosensor and ELISA immunoassay for single-cell analysis. The combination of the nanobiosensor with the ELISA immunoassay improved the detection sensitivity of the nanobiosensor due to enzymatic amplification and extend the applicability of nano-biosensors for detecting weakly fluorescing or nonfluorescent targets. This study demonstrated the possibility of detecting important cellular sub-components such as cytochrome c at the single cell level.[127]

23.3.1.2.6 Fabrication of Optical Nanofibers

The construction of fiber-optic probes with nanometer-sized tips is the foundation of optical-fiber-based nanosensors. Depending on the specific application and environment in which the nanosensor is to be used, different tip diameters, taper angles, and smoothness of the fiber surfaces are required. To achieve the various diameters and tapers on the tips of these fibers, two different fabrication procedures have been developed.

23.3.1.2.7 Heated Pulling Process

The most commonly employed procedure involves the use of a heated pulling instrument, such as a laser-based micropipette puller. These instruments use a CO_2 laser to heat the fiber and a tension device that pulls along the major axis of the fiber while it is heated (see Figure 23.13). As the fiber is pulled, the heated region begins to taper down to a point, until the fiber has been pulled into two pieces, each having one large end and one end with nanometer-scale dimensions. By varying the parameters associated with the pulling process (i.e., heating temperature, tension applied to the fiber), tip diameters ranging in size from less than 20 nm to greater than 1000 nm have been reported.[96,104,105,117,119] Figure 23.14 shows a scanning electron micrograph (SEM) of an optical fiber that has been pulled in such an instrument.

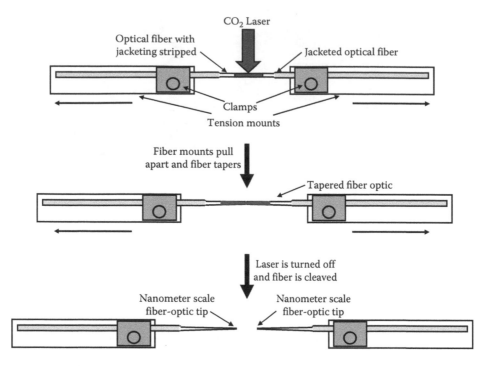

FIGURE 23.13 The fiber-pulling procedure for the construction of a nanofiber.

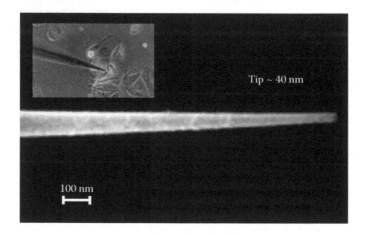

FIGURE 23.14 A scanning electron micrograph of a nanofiber produced by the micropipette-pulling procedure. The tip of this fiber is approximately 50 nm in diameter. Inset: Photograph of an antibody-based nanosensor for measurements of benzopyrene tetrol molecules in a single live cell. (Adapted from Vo-Dinh, T. et al., *Nat. Biotechnol.*, 18, 764, 2000.)

Using this technique, while requiring accessibility to a relatively expensive micropipette-pulling instrument, it is possible to reproducibly create optical fibers with nanometer-scale tips in just seconds.

23.3.1.2.8 Chemical Etching

The second means of producing optical fibers with nanometer-scale tips is through a chemical etching process. This process has been found to be an inexpensive and effective alternative to the pulling process. Two different types of etching procedures have been reported in the literature: (1) Turner etching, a method involving the mixture of hydrofluoric acid (HF) with an organic solvent,[123,124] and (2) tube etching, a method developed by Stockle et al.[125] In the Turner method, a fiber is placed in the meniscus between HF and an organic overlayer. Once in place, the HF dissolves the silica fiber, forming a point. Using this technique, fibers with large taper angles and tip diameters comparable to the pulling method have been created. Because of the larger taper angles associated with these fibers, excitation light can travel closer to the tip of the fiber before being trapped, thereby causing more energy to be coupled with the fluorophore resulting in a more efficient excitation process. However, as a result of the dual chemical nature of the etchant solution, environmental parameters such as temperature fluctuations and vibrations can cause the characteristics of the tip to vary significantly from batch to batch.

To overcome the problems associated with this two-phase chemical etching process and the batch to batch variability, a new form of etching has recently been developed, known as tube etching. Tube etching involves etching an optical fiber with a single-component solution of HF. In this method, a silica fiber that has an organic/polymer cladding is polished optically flat on both ends, prior to placing one end just below the surface of a solution of HF. Over time, the acid begins to etch away the core of the fiber without destroying the cladding. As the silica fiber continues to be etched away, the polymer cladding acts as a wall, causing any microcurrents in the HF to form a blunted tip on the fiber's core. As more time passes, the silica core material continues to dissolve until it is no longer below the surface of the HF solution. At this point, capillary action draws HF up the walls of the cladding, allowing it to drain down the silica core back into the solution below. As this happens, it causes the fiber to be etched into the shape of a cone with a large smooth taper. Control over the taper angle of the fiber and the diameter of the tip can be performed simply by varying the depth at which the fiber is placed in the HF and the etching time. Using this tube-etching process, it is possible to produce a batch of fibers in approximately 2 h and with greater reproducibility than the Turner etching method. However, due to difficulties associated with submerging a large group of fibers to exactly the same depth in the HF, the reproducibility of this etching method is still less than that obtained by pulling the fibers. In addition, due to the large taper angle associated with these etched fibers as well as those fabricated with the Turner etching process, they are limited in the environments in which they can probe. Because the taper angle is so large, they tend to cause a great deal of damage when performing cellular analyses with smaller cells (e.g., mammalian somatic cells) and are therefore useful only with large cells such as oocytes and certain neurons. It is primarily this reason that the most common method for creating nanometer-sized tips on fiber-optic nanosensors is performed using the heated pulling technique.

23.3.1.3 Nanoparticle Chemical Sensors

The development of fiber-optic-based nanosensors and nanobiosensors has already had a large impact on the fields of biological and biomedical research in the short time that they have existed. Nevertheless, significant advances and variations in optical nanosensors are constantly being developed, from the use of new, more selective bioreceptors to the development of different types of nanobiosensors for application to various environments. One such advance in the past has been the development of the first fiber-less optical nanosensor with nanometer-scale sizes in all three dimensions.[126,127] Figure 23.15 depicts a conceptualized scheme of a particle-based nanosensor assembly and sensing mechanism. The development of nanoparticle-based sensors are also described in other chapters of this handbook: Luminescent quantum dots for diagnostic and bioimaging applications, PEBBLES nanosensors for in vitro bioanalysis, SERS molecular sentinel nanoprobes for medical diagnostics, and plasmonic coupling interference nanoprobes for gene biomarker detection.

23.3.1.3.1 Dye-Encapsulated Sensors

The development of nanoparticle-based nanosensors has largely been driven by the quest for improved fluorescence methods for real-time intracellular analyses. Fluorescent labeling[128] has been widely used to acquire much information about cells.[129,130] However, there is the need for dye-encapsulated sensors in order to localize cellular processes as well as reduce fluorescence attenuation and cytotoxicity, which result from the interaction between the indicator dye and the cellular components. One of the earliest of such dye-encapsulated sensor classes known as probes encapsulated by biologically localized embedding (PEBBLEs)

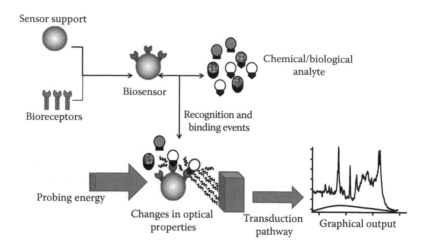

FIGURE 23.15 A conceptualized scheme of nanosensor assembly and sensing mechanism.

is composed of a fluorescent indicator dye embedded in a 20 nm polymer sphere. The encapsulating sphere protects the indicator dye from cellular degradation and the cell from the toxic effects of the dyes, while allowing ions to penetrate into the indicator dye. Application of these sensors to the measurement of many different ions within a cell has been demonstrated since their first development. In such analyses, large quantities of these sensors are randomly injected in a cell with a gene gun or other delivery methods.[131] Once in the cell, the cell is illuminated with an excitation light of appropriate wavelength. The resulting fluorescence signal is measured over the autofluorescence background of the cell. Presently, PEBBLE sensors have been developed for pH, Ca^{+2}, K^+, NO, molecular oxygen, glucose, and intracellular Zn^{+2}.[127,132–138]

Apart from PEBBLEs, other similar classes of particle-based nanosensors with varying degrees of dye encapsulation have also been developed and evaluated. These nanosensors include liposome-based sensors,[139–144] lipobead sensors,[145–147] and magnetically modulated optical nanosensors (MagMOONs).[148] MagMOONs represent a second generation of PEBBLE sensors, and along with dye encapsulation, they also contain an elongated magnetic core. The presence of the magnetic core allows MagMOONs to rotate in magnetic fields. As a result of this, upon excitation and application of a magnetic field, MagMOONs blink as the light-emitting half comes in and out of excitation source. Due to this blinking, MagMOONs lead to increased differentiation of analyte signals from large background autofluorescence.

The development of liposome and lipobead sensors is also based on the concept of PEBBLE sensors. Liposome-based sensors comprise a fluorescence dye entrapped in the internal aqueous compartment of a phospholipid vesicle. Lipobeads, however, involve the immobilization of the indicator dye covalently on a polystyrene nanoparticle before the entire immobilized particle is embedded in phospholipid bilayer. Because of the phospholipid bilayer, these phospholipid-based nanosensors can exhibit a high degree of biocompatibility. The use of these nanosensors for intracellular and bioanalytical analyses has been demonstrated.[142,149] By encapsulating both reference and indicator dyes into a single nanosensor, ratiometric measurements,[126,138] which account for the effect of light scattering and fluctuation on the sensor's capabilities, are possible with such dye-encapsulated nanosensors.

23.3.1.3.2 Quantum-Dot Biosensors

Simultaneous with the development of PEBBLEs was the development of quantum dot (QD)-based biosensors capable of monitoring individual chemical species inside a living cell. These nanoparticle biosensors are often composed of ZnS particles capped with Cd selenide and attached to biological receptors including antibodies, oligonucleotides, and enzymes. These QDs offer several significant advantages over conventional fluorescence indicator dyes used in chemical sensing. First, they are biocompatible, unlike most indicator dyes, which are often toxic to the cell being investigated. Second, their emission is much more intense than typical indicator dyes. Finally, they are more photostable than dye molecules. These last two properties are very important when monitoring changes in the concentration of a chemical or biochemical species over time because most fluorescent dyes exhibit rapid and significant photobleaching with the small amount used in cellular analyses.

Bruchez et al. and Wu et al. were among the first to show that QDs can be used to effectively label molecular targets in live cells and even up to subcellular levels.[150,151] Since then, great strides have been made in the use of QDs as nanobiosensors.[152–155] Reviews describing the use of QDs for cellular imaging, labeling, and sensing can be found in the literature.[156] In one of such study, QD nanosensors based on bioluminescence resonance energy transfer (BRET) have been used to detect protease activity in biological samples and tumor cells. In designing these sensors, QDs (BRET acceptors) were linked to bioluminescent protein luciferase (BRET donor) via matrix metalloproteinase (MMP) substrate peptide. The BRET process between QDs and luciferase was therefore regulated by the activity of protease. Using these QD nanosensors, it was possible to detect the presence of active protease at levels as low as 1 ng/mg. These nanosensors were also applied to the detection of protease activity in two types of tumor cells that are known to express varying degrees of protease activity. Human fibrosarcoma HT1080 and human colon adenocarcinoma HT29 cell lines both express MMP-2. However, HT1080 expresses significantly higher levels of activity compared to HT29. These differences in the level of protease activity expression were confirmed when the QD nanosensors were used to assay MMP-2 in both cells.[153] Overexpression of MMPs is known to be a sign of an advanced tumor stage in all types of human cancers. Therefore, the ability to develop effective methods for the detection of MMPs and other biochemical species using QD-based biosensors could significantly help in the fight against cancer and other diseases.

23.3.1.3.3 SERS-Based Nanosensors

In recent years, SERS has emerged as a powerful tool for monitoring ultratrace amounts of biomolecules intracellularly, providing both qualitative and quantitative analyses. Such analyses have been reported for SERS substrates ranging from metal nanoaggregates to functionalized ordered nanostructures. However, this section will focus on individual SERS-active nanoparticles functionalized with appropriate receptors to form SERS nanobiosensors.

SERS is a Raman scattering technique, which makes use of the roughened conducting surfaces to enhance the naturally weak Raman signals by as much as 10^{16}-fold.[157,158] Although spontaneous Raman spectroscopy is useful in providing molecular structural information about samples, it lacks the sensitivity of the fluorescence spectroscopy. This is due to the small Raman cross sections ($\sim 10^{-30}$–10^{-25} cm^2/molecule) compared to the fluorescence cross section ($\sim 10^{-16}$ cm^2/molecule) of typical analytes. However, on nanoscale-roughened metal surfaces, referred to as SERS substrates, Raman intensities of the vibrational modes of a molecule are enhanced by a factor of anywhere between 10^4 and 10^{16} due to electromagnetic and chemical mechanisms.[159–164] As a result of this enhancement, SERS can be highly sensitive for acquiring both quantitative and fingerprint molecular information of ultratrace amounts of analytes. Several demonstrations of single-molecule detection have even been described previously.[158,165]

SERS is well suited for the analysis of complex microbiological samples (e.g., single cells) due to its potential sensitivity as well as its ability to perform multiplexed analyses with minimal autofluorescence background and photodamage to the sample.[166–168] Additionally, with SERS, it is possible to identify the molecular structures of even closely related biomolecules.

One of the key factors that define the sensitivity of SERS-based nanosensors is the size of the SERS enhancement factors (EFs) of the substrates. While several techniques have been reported for fabricating SERS substrates, SERS-based intracellular analyses have been performed mostly with colloidal nanoparticles. These metal colloids have a relatively small size, allowing their insertion into individual living cells. They can also exhibit large SERS effects at the junction of the aggregated colloids, due to the plasmon coupling among the aggregates of colloids.[169–171] A detailed account of SERS-based colloidal nanoparticles for single-cell analysis is contained in the review by Kneipp.[172] However, such SERS-based probes are not selective and tend to give SERS signals of various types of biomolecules that are in close proximity to the probes. With this, the probes undergo biological fouling easily as biomolecules randomly adsorb onto the metal surfaces. This reduces the use of such colloidal SERS nanoparticles for monitoring cellular processes over extended periods of time.

To improve the selectivity, SERS nanoparticles have been immobilized with chemical receptors. As a result of chemical receptor immobilization, it is possible to selectively detect specific analytes in complex biological environments. Such SERS-based nanosensors have been produced for pH monitoring. To fabricate these pH nanosensors, 4-mercaptobenzoic acid was self-assembled on the metal surface of SERS-active nanoparticles. The SERS intensity of the benzoic acid spectra at 1430 cm^{-1} Raman shift changed markedly as a result of protonation and deprotonation of the carboxylic acid group. This response was used to monitor pH in Chinese hamster ovary cells at physiologically relevant concentrations. Due to partial covering of the SERS-active surface with the receptor, little background from cellular constituents was found.[173]

The exquisite spectral specificity of vibrational techniques such as Raman and SERS for chemical sensing are illustrated in the use of plasmonic gold nanostars developed for pH sensing in a study combining theoretical and experimental investigations.[186] SERS spectral changes of the pH-sensitive dye (pMBA) in the range 1000–1800 cm^{-1} with pH variation between 5 and 9 were identified in experiments and vibrational modes were assigned using theoretical calculations. To investigate those SERS spectrum changes observed in the experiments, Density Functional Theory (DFT) calculations were performed for simulation models of pMBA–Au and pMBA–Ag. All quantum chemical calculations were performed with the program package Gaussian 03. DFT calculations with the B3LYP functional and 6-311++G (d, p) basis set were used for all the atoms except gold or silver in the optimization of the ground state geometries and simulation of the vibrational spectra. The Raman peak position for benzene ring stretching at ~1580 cm^{-1} was identified to be a unique index for pH sensing since it was found to have a consistent downshift when pH value increases. The calculated Raman peak for this vibrational mode has a 26.2 cm^{-1} downshift when the simulated pMBA–Au complex changes from the protonated to deprotonated state. This phenomenon is confirmed and well-explained with DFT simulation to be due to the coupling between the benzene ring stretching mode and the carboxylic group stretching mode. This unique pH sensing index could be used for chemical sensing and molecular imaging applications.

23.3.1.3.4 *Control of Cellular Transport of Nanoparticle-Based Probes*

An important aspect in using particle-based nanosensors is precise control of their transport into cells and nucleus. To study the efficiency of cellular uptake, silver nanoparticles functionalized with three different (positive-, negative-, and neutrally) charged Raman labels were co-incubated with cell cultures, and allowed to be taken up via normal cellular processes.[188] The surface charge on the nanoparticles was observed to modulate their uptake efficiency during a 4 h co-incubation, demonstrating a dual function of the surface modifications as tracking labels and as modulators of cell uptake. Figure 23.16 depicts the SERS imaging map showing cellular distribution 4-aminothiophenol (4-ATP)-labeled and 4-4-thiocresol (4-TC)-labeled silver nanoparticles in J774 mouse macrophage cells, respectively. The signal intensity from nanoparticles functionalized with 4-ATP (Figure 23.16a) is stronger than that of 4-TC-labeled nanoparticles (Figure 23.16b) and more widely distributed throughout the cellular interior after 4 h than at the 2 h time point. Furthermore, the 2D SERS maps indicate that the label-associated SERS signals are more widely distributed within the cells in the 4 h map than in the 2 h map, suggesting that more labeled nanoparticles have entered the cells over the additional 2 h. These observations, taken together, suggest the ability to modulate cellular uptake of nanoparticles by functionalization with selected *label modulators* having a combination of appropriate charge, chemical structure, and nanoparticle size. Uptake of plasmonically active nanodelivery vehicles may be designed to occur over a particular time window after treatment with a cell-modifying agent or drug treatment. For sensing and tracking of biomedically important cellular markers, effective design and use of nanobiosensors require confirmation of their uptake into cells, as well as an awareness of the temporal and spatial scales on which these processes occur. These results indicate that a appropriately functionalized nanoprobes construct has the potential for sensing and delivery in single living cells.

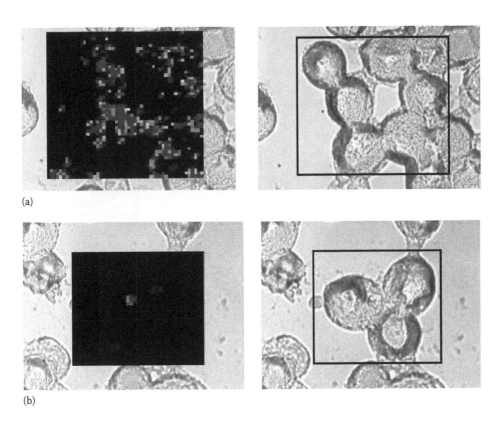

FIGURE 23.16 Control of intracellular transport of SERS nanoprobes. (a) Map of SERS (left) and bright field (right) imaging showing cellular distribution 4-ATP-labeled silver nanoparticles in J774 cells; (b) Map of SERS (left) and bright field (right) imaging showing cellular distribution 4-TC-labeled silver nanoparticles in J774 cells. Negatively-charged 4-MBA-labeled particles were taken up more readily by the cells than were neutrally-charged 4-TC-labeled particles. (Adapted from Gregas, M.K. et al., *Nanomedicine*, 7, 115, 2011.)

In another study, a co-functionalized nanobiosensor/biodelivery platform combining a nuclear targeting peptide (NTP) for improved cellular uptake and intracellular targeting, and p-mercaptobenzoic acid (pMBA) as a surface-enhanced Raman scattering (SERS) reporter (Figure 23.17a and b).[189] The nuclear targeting peptide used in this study was a HIV-1 protein-derived TAT sequence (TAT peptide residues 49–57, sequence Arg-Lys-Lys-Arg-Arg-Arg-Gln-Arg-Cys-CONH$_2$), which has previously been shown to aid entry of cargo through the cell membrane via normal cellular processes and, furthermore, to localize small cargo to the nucleus of the cell. Preliminary experiments have verified cell uptake and characterized distribution in mouse and human cell lines. Two-dimensional SERS mapping was used to track the spatial and temporal progress of cell uptake and localization at discrete time points, demonstrating the potential for an intracellularly targeted multiplexed nanobiosensing system with sensitivity and specificity superior to that of traditional fluorescence sensing and tracking methods. Silver nanoparticles cofunctionalized with the TAT peptide showed greatly enhanced cellular uptake over the control nanoparticles lacking the targeting moiety. There is an observable increase in the distribution and intensities of the signals from the internalized co-functionalized nanoparticles in Figure 23.17(c–e) as compared to those without TAT. In agreement with previous studies, TAT-functionalized nanoparticles are more efficiently taken up into the cellular interior and also may possibly be more discretely localized within the cell. The presence of the TAT peptide appears to enhance the uptake efficiency of nanoparticles by more readily shuttling them through the cell membrane and also localize the particles to a subarea of the cell that may coincide with the nucleus. The 3 h time point in Figure 23.17 (c–e) was

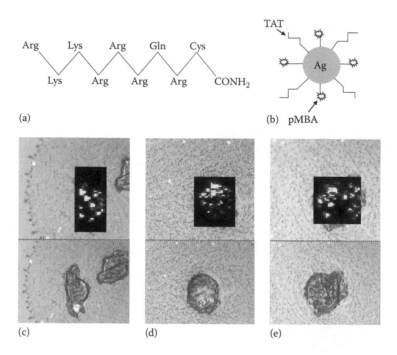

FIGURE 23.17 (a) Structure of the TAT sequence; (b) SERS Nanoprobe with TAT receptor and pMBA label; (c,d,e): SERS intensity mapping of three different single cells (c, d, e) all incubated with co-functionalized (pMBA.TAT) nanoparticles for three hours. The top row shows the SERS signal maps overlaid on the brightfield image of the sample cell (bottom row) to show image/signal registration. (From Gregas, M.K. et al., *Appl. Spectrosc.*, 64, 858, 2010.)

chosen as representative of widespread and possible peak uptake/internalization. The ability to detect and monitor nanoparticle trafficking using SERS spectroscopy represents an advance over previous nanosensor tracking and detection methods such as light microscopy and fluorescence methods. The development of multifunctional nanoparticle constructs for intracellular delivery has potential clinical applications in early detection and selective treatment of disease in affected cells. Other applications include use in basic research aimed at understanding the inner workings of living cells and how they respond to chemical and biological stimuli.

23.3.1.3.5 SERS-Based Nanostars for Theranostics

A recent hybrid modality that combines therapy and diagnostics, often referred to as *theranostics*, has gained increasing interest. Gold nanostars have recently been developed as novel multifunctional plasmonic-active nanoplatforms for both sensing and therapy.[190–195] The synthesis and characterization of SERS label-tagged gold nanostars, coated with a silica shell containing methylene blue photosensitizing drug for singlet oxygen generation in photodynamic therapy has been reported.[194,195] Gold nanostars offer unique plasmon properties that efficiently transduce photon energy into heat for photothermal therapy. Gold nanostars were used for particle tracking and photothermal ablation both in vitro and in vivo.[191,192] Using SKBR3 breast cancer cells incubated with bare nanostars, photothermal ablation was observed within 5 min of irradiation (980 nm CW laser, 15 W/cm²). In a mouse injected systemically with PEGylated nanostars for 2 days, extravasation of nanostars was observed and localized photothermal ablation was demonstrated in a dorsal window chamber within 10 min of irradiation (785 nm CW laser, 1.1 W/cm²). These results of plasmon-enhanced localized hyperthermia are encouraging and have illustrated the potential of gold nanostars as efficient optical nanosensors as well as photothermal agents in cancer therapy.

The first application of a theranostic system combining SERS detection and photodynamic therapy (PDT) effect was demonstrated.[194,195] The theranostic nanoconstruct was created by loading the PDT photsensitizer drug Protoporphyrin IX onto a Raman-labeled gold nanostar. A cell-penetrating peptide, TAT, the transactivator of transcription peptide of the human immunodeficiency virus type 1 (HIV-1) viral genome, was used to enhance intracellular accumulation of the nanoparticles in order to improve their efficacy. The plasmonic gold nanostar platform is designed to increase the Raman signal via the SERS effect. Raman imaging and photodynamic therapy demonstrated the theranostics applicability of the construct on BT-549 breast cancer cells. In the absence of the TAT peptide, nanoparticle accumulation in the cells was not sufficient to be observed by Raman imaging or to produce any photosensitization effect after a 1 h incubation period. There was no cytotoxic effect observed after nanoparticle incubation, prior to light-activation of the photosensitizer.

23.3.1.4 SERS-Based Immunonanosensors

Simultaneous to the development of SERS-based chemonanosensors, SERS bionanosensor employing antibodies for analyte recognition has also been developed and demonstrated for use in cellular analyses.[174-177] These SERS immunonanosensors employ metal film over nanostructure (MFON) geometry to provide uniform sensitivity[178] without the need for particle aggregation while also increasing the classes of analytes that can be interrogated to proteins/peptides[176,177] and other immunogenically recognized biomolecules(e.g., glucose insulin).[174,179]

One of the first demonstrations of the capability of MFON SERS-based nanosensors to detect biomolecules in cells was performed by attaching appropriate antibodies to the metal nanoparticle surface for the detection of trace amounts of insulin and interleukin II (IL-2) in cell lysates.[176,177] To fabricate these immunonanosensors, SERS-active particles from MFON SERS substrates were immobilized with anti-human insulin receptors via cross-linkers to form SERS-based immunonanosensors for insulin. Because the cross-linkers exhibit their own spectral bands, several cross-linkers were characterized to identify the suitable cross-linker. The representative characterization spectra of three of these cross-linkers are shown in Figure 23.18. Among several cross-linkers evaluated, 2-mercapto-4-methyl-5-thiazoacetic acid (MMT)

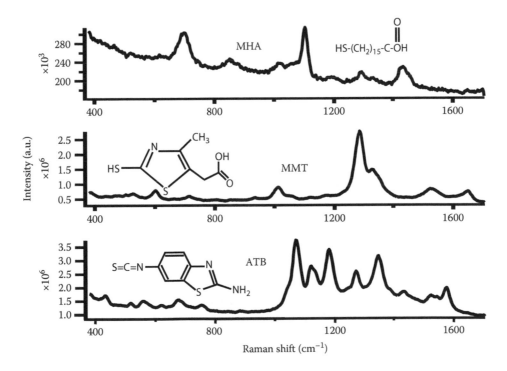

FIGURE 23.18 SERS spectra and structure of the three different cross-linkers—ATB, MHA, and MMT.

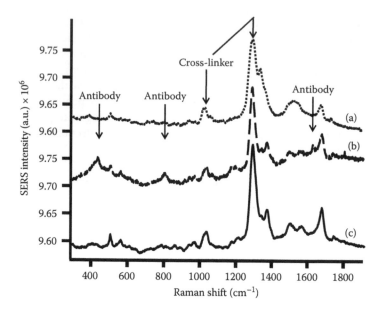

FIGURE 23.19   SERS spectra (a) of activated MMT bound to silver-coated nanospheres, (b) after binding (a) with anti-human insulin, and (c) of (b) (immuno-nanosensors) in the presence of 10 μg/mL insulin.

was determined to be an ideal cross-linker as it exhibited minimal spectral peaks to act as background while providing significant bands for use as internal standards. Additionally, MMT bound the receptor rigidly and close to the SERS-active surface. After attachment of the receptor (i.e., antibody) via traditional EDC chemistry, evaluation of the fabricated immunonanosensors in the cell culture media revealed significant and reproducible changes in the SERS spectra at cellularly relevant concentrations of human insulin. The spectra showing the various stages of nanosensor evaluation are shown in Figure 23.19. By attaching IL-2 receptors to SERS nanosensor surfaces through the same attachment process, it was possible to fabricate different types of nanosensors for the detection of IL-2 to a level as low as 10 μg/mL,[177,180] indicating that the nanosensors could be modified for a wide range of applications. However, in this study, it has been recognized that one of the issues with the use of whole antibodies as biorecognition elements is that the analyte recognition occurs farther away from the SERS-active surface. This significantly reduces analyte detection by direct SERS signals.[35,181,182] Subsequently, the sensors are less capable of acquiring molecular information of the analyte. For this reason, similar classes of SERS nanosensors that employ Fabs and other smaller bioreceptors are also currently being developed. Employing these smaller bioreceptors, the distance between the analyte and the metal surface can be decreased allowing greater SERS signals of the analytes to be measured.

23.3.1.4.1 Fabrication of SERS-Based Immunonanosensors

It was mentioned in the previous section that the effectiveness of the SERS-based immunonanosensors largely depends on the substrate used as the sensor support and its sensitivity and uniformity in SERS enhancement. Although various techniques exist for the fabrication of these substrates,[159,163,164,166,167,173–175,183–191] one of the most easily fabricated generic substrates with large and uniform SERS EFs is MFON substrate.[175,178] This section will focus on the description of the fabrication of a unique MFON multilayered SERS-active nanosphere that is capable of providing SERS EFs of >10[9], which are required for most single-cell analyses. Such substrates can yield individual SERS-active nanoparticles with size range that allows their insertion into individual living cells for intracellular analyses.

23.3.1.4.2 *Optimization of Multilayer Nanosensors for Trace Analyses*

Several novel methods for the optimization of the sensitivity of SERS-based immunonanosensors have been developed over the past decade. In one such method, multilayer MFON SERS substrates have been used to achieve varying degrees of enhanced sensitivity.[192,193] Multilayered SERS substrates are derived from alternating layers of metal film and dielectric spacer on nanostructures. Figure 23.20 shows a conceptualized representation of multilayer MFON SERS substrates. The introduction of the dielectric spacer was achieved by allowing the first layer of the metal film to oxidize. Another layer of the metal film is then deposited on the oxidized layer. The process can be repeated to form multiple continuous SERS-active metal films (separated by metal oxide) on the substrate. The individual spheres (Figure 23.20b), with their multiple layers of metal/metal oxide, are then separated from each other and the underlying support using a razor blade and sonication. The observed enhancement on each of the individual SERS nanosphere/nanosensors stems from the reinforcement of the surface plasmon. Under the exposure of the excitation light, the underlying layers of metal films generate electromagnetic fields. These electromagnetic fields reinforce the surface plasmon and the electromagnetic field of the top layer of multilayer MFON SERS substrate.[192,193] As a result of this reinforcement, the SERS EF achieved in multilayer SERS substrate is severalfold larger than that achieved in optimized single-layer MFONs having metal film thickness equal to the total film thickness of the multilayer MFON SERS substrate. This work has expanded into the development of optimized SERS nanosensors by employing accurately controlled SAMs as dielectric spacer layers to achieve SERS enhancement of $>10^{10}$.[194,195]

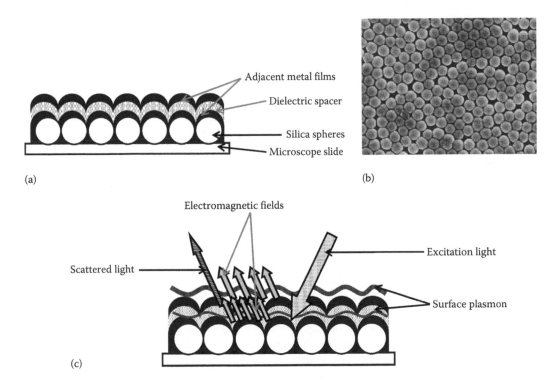

FIGURE 23.20 A conceptualized diagram of multilayered SERS substrate showing (a) the dielectric spacer sandwiched between two metal films, (b) a scanning electron micrograph of a MFON SERS substrate (430 nm diameter), and (c) the principles of multilayer SERS enhancement that stem from the reinforcement of the adjacent metal films.

23.3.1.4.3 Label-Free SERS Immunosenor for the Fhit Protein

Another type of SERS nanoprobes used silver (Ag) nanoparticles for immunoassay-based SERS detection of the fragile histidine triad (Fhit) protein.[219] Alterations in Fhit protein expression have been associated with several human cancers, and thus, the detection of Fhit protein is important because it can potentially be used as a cancer diagnostic biomarker, for both cancer detection and therapy. The detection and identification of Fhit protein could serve as a useful diagnostic tool for cancer-marker detection or screening. Ag nanoparticles used for SERS spectroscopic detection of Fhit protein were characterized using AFM and the sizes were determined to be about ~30 nm. FHIT polyclonal antibodies were incubated with Ag colloids for 2–3 h. After this primary antibody treatment, Fhit protein and blocking reagent were added and the reaction mixture placed on an orbital shaker for 3 h gyrating at a very low pace at room temperature. This provided the Fhit protein–antibody interaction. The observed differences in these spectra from antibody and antibody with Fhit demonstrate the enhancement in Raman signal produced by the metallic surface of Ag nanoparticles onto which the Fhit antibody was adsorbed. In addition to spectral differences upon binding, there are Raman signal intensity differences that reflect the underlying structural and conformation changes that occur when the Fhit antibody binds the Fhit protein. The results of this study demonstrate the possibility to use the label-free SERS immunosensor for the detection and identification of Fhit protein.

23.3.1.5 SERS-Based Nanosensors for Nucleic Acid Targets

A label-free nanoparticle-based sensing detection method, called *Molecular Sentinel* (MS) incorporates the *SERS effect modulation* scheme associated with metallic nanoparticles and the DNA stem-loop structure.[220-222] The plasmonics-based MS nanoprobe is composed of a SERS-active metal nanoparticle and a stem-loop DNA molecule tagged with a Raman label. In the normal configuration and in the absence of target DNA, the stem-loop configuration keeps the Raman label close to the metal nanoparticle, inducing an intense SERS effect that produces a strong Raman signal upon laser excitation. Upon hybridization of a complementary target DNA sequence to the nanoprobe, the stem-loop configuration is disrupted, causing the Raman label to physically separate from the metal nanoparticle, thus quenching the SERS signal. The label-free Molecular Sentinel biosensing concept was further developed and adapted for use on the *Nanaowave* substrate platforms (Figure 23.21a).[223,224] The Nanowave substrate, which involves a *metal film over nanosphere* (MFON), was first introduced for SERS analysis by the Vo-Dinh group in 1984 with nanosphere arrays coated with a silver film[225] and has been extensively used for chemical and biological analysis.[199,200,223-229] The Nanowave substrate for biosensing displays several advantages to be used as a SERS detection platform, such as its relatively low fabrication cost and high enhancement factor.

A unique platform that is simple and inexpensive to fabricate is the Nanowave platform, also referred to as MFON. Numerical simulations were used to mimic the Nanowave geometry in three-dimensional space and to confirm its experimentally observed plasmonic behavior.[224] The study confirms that an in-plane polarized incident plane wave generates strong enhancements in the interstitial spaces between individual metal-coated nanospheres, thus producing closely packed arrays of hot spots which produce the strong SERS effect of the Nanowave substrate structures. The Nanowave chip fabrication was further refined to become simple with low-cost and high reproducibility using deposition of a thin shell of metal (e.g., gold) over closely packed arrays of nanospheres.[50] Use of a self-assembly on a water–air interface method can produce a large area of close-packed nanosphere arrays for the Nanowave chip (Figure 23.21b). The Nanowave chip fabrication is relatively simple and low-cost with high reproducibility. The sensing process involves a single hybridization step between the DNA target sequence and the complementary MS probes on the Nanowave chip without requiring secondary hybridization or posthybridization washing, thus resulting in shorter assay time and less reagent usage. The usefulness and potential application of the biosensor for medical diagnostics is demonstrated by detecting the human RSAD2 (Viperin) gene, a common host biomarker of inflammation.[223]

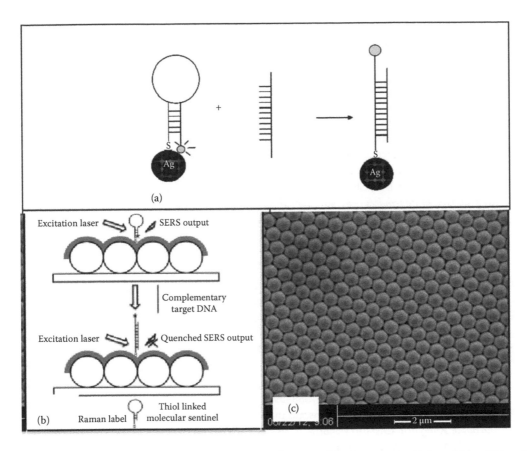

FIGURE 23.21 (a) Operating principle of a SERS molecular sentinel (MS) nanoprobe (Adapted from Wabuyele M.B. and Vo-Dinh, T., *Anal. Chem.*, 77, 7810, 2005.) (b) MS on a chip. Top: a SERS signal is observed when the MS probe is in the stem-loop conformation with the label close to the MFON, inducing a strong SERS signal (closed-state); bottom: following hybridization with a target probe, the stem opens and separates the label from the MFON surface causing the SERS signal to decrease (c) SEM image of nanowave chip substrate. In this substrate, a mono-layer of 520-nm diameter polystyrene beads is covered by a 200-nm thick Au layer. (Adapted from Ngo, H.T. et al., *Anal. Chem.*, 85, 6378, 2013.)

23.4 Conclusion

Fiber-based and particle-based nanosensors for monitoring single cells have opened up new applications in molecular biology and medical diagnostics. Due to their small sizes, nanosensors provide important tools for minimal invasive analysis at single cellular or subcellular level. These nanotools can be used to detect chemical species or to monitor intracellular signaling processes such as apoptosis and to investigate gene expression inside single living cells. Probing subcellular architecture and dynamic processes is essential in understanding the fundamental biological processes such as cell signaling pathways in a systems biology approach. Single-cell monitoring is also important in studies where the amount of cells obtained is limited and which could not be analyzed with other techniques. By combining the nano-probe technology with SERS-based detection, multiplexed analysis of multiple biomarkers and physical conditions will be possible.

Due to the increasing interest in nanotechnology and its practical use, application of nanosensors and nanobiosensors to different types of cellular measurements is expanding rapidly. Furthermore, it has been shown that these nanosensors and nanobiosensors often provide more reliable measurements of subcellular chemical species than conventional techniques, such as pretreatment with and subsequent monitoring of fluorescent dyes. The use of nanosensor and nanobiosensor technologies has already begun to and will continue to revolutionize the field of cellular biology in the future.

References

1. Nice, E. C. and Catimel, B., Instrumental biosensors: New perspectives for the analysis of biomolecular interactions, *Bioessays* 21(4), 339–352, 1999.
2. Weetall, H. H., Chemical sensors and biosensors, update, what, where, when and how, *Biosensors & Bioelectronics* 14(2), 237–242, 1999.
3. Tess, M. E. and Cox, J. A., Chemical and biochemical sensors based on advances in materials chemistry, *Journal of Pharmaceutical and Biomedical Analysis* 19(1–2), 55–68, 1999.
4. Braguglia, C. M., Biosensors: An outline of general principles and application, *Chemical and Biochemical Engineering Quarterly* 12(4), 183–190, 1998.
5. Aylott, J. W., Optical nanosensors—An enabling technology for intracellular measurements, *Analyst* 128, 309–312, 2003.
6. Vo-Dinh, T. and Cullum, B. M., Biosensors and biochips: Advances in biological and medical diagnostics, *Fresenius Journal of Analytical Chemistry* 366(6–7), 540–551, 2000.
7. Vo-Dinh, T., Cullum, B. M., and Stokes, D. L., Nanosensors and biochips: Frontiers in biomolecular diagnostics, *Sensors and Actuators B: Chemical* 74(1–3), 2–11, 2001.
8. Tan, W. H., Kopelman, R., Barker, S. L. R., and Miller, M. T., Ultrasmall optical sensors for cellular measurements, *Analytical Chemistry* 71(17), 606A–612A, 1999.
9. Cullum, B. M. and Vo-Dinh, T., The development of optical nanosensors for biological measurements, *Trends in Biotechnology* 18(9), 388–393, 2000.
10. Lu, J. Z. and Rosenzweig, Z., Nanoscale fluorescent sensors for intracellular analysis, *Fresenius Journal of Analytical Chemistry* 366(6–7), 569–575, 2000.
11. Yotter, R. A., Lee, L.A., and Wilson, D. M., Sensor technologies for monitoring metabolic activity in single cells—Part I: Optical methods, *IEEE Sensors Journal* 4(4), 395–411, 2004.
12. Yonzon, C. R., Stuart, D. A., Zhang, X., MacFarland, A. D., Haynes, C. L., and Van Duyne, R. P., Towards advanced chemical and biological nanosensors—An overview, *Talanta* 67, 438–448, 2005.
13. Bogue, R., Nanosensors: A review of recent research, *Sensor Review* 29(4), 310–315, 2009.
14. Cajlakovic, M., Lobnik, A., and Werner, T., Stability of new optical pH sensing material based on cross-linked poly(vinyl alcohol) copolymer, *Analytica Chimica Acta* 455(2), 207–213, 2002.
15. Zhang, Z. J., Zhang, Y. K., Ma, W. B., Russell, R., Grant, C. L., Seitz, W. R., and Sundberg, D. C., Polyvinyl-alcohol) as a matrix for immobilizing indicators for fiber optic chemical sensors, *Abstracts of Papers of the American Chemical Society* 196, 115-ANYL, 1988.
16. Lechna, M., Holowacz, I., Ulatowska, A., and Podbielska, H., Optical properties of sol-gel coatings for fiberoptic sensors, *Surface & Coatings Technology* 151, 299–302, 2002.
17. Mohr, G. J., Werner, T., Oehme, I., Preininger, C., Klimant, I., Kovacs, B., and Wolfbeis, O. S., Novel optical sensor materials based on solubilization of polar dyes in apolar polymers, *Advanced Materials* 9(14), 1108–1113, 1997.
18. Wolfbeis, O. S., Fiber optic chemical sensors and biosensors, *Analytical Chemistry* 72(12), 81R–89R, 2000.
19. Seitz, W. R., Chemical sensors based on immobilized indicators and fiber optics, *CRC Critical Reviews in Analytical Chemistry* 19(2), 135–173, 1988.
20. Seitz, W. R., Chemical sensors based on fiber optics, *Analytical Chemistry* 56(1), 16A–34A, 1984.
21. Sun, H. et al., Polymeric nanosensors for measuring the full dynamic pH range of endosomes and lysosomes in mammalian cells, *Journal of Biomedical Nanotechnology* 5(6), 676–682, 2009.
22. Schulz, A. et al., Fluorescent nanoparticles for ratiometric pH-monitoring in the neutral range, *Journal of Materials Chemistry* 20(8), 1475–1482, 2010.

23. Stanca, S. E. et al., Intracellular ion monitoring using a gold-core polymer-shell nanosensor architecture *Nanotechnology* 21(5), 055501/1–055501/8, 2010.

24. Stern, E. et al., Label-free biomarker detection from whole blood, *Nature Nanotechnology* 5(2), 138–142, 2010.

25. Brogan, K. L., Wolfe, K. N., Jones, P. A., and Schoenfisch, M. H., Direct oriented immobilization of F(ab') antibody fragments on gold, *Analytica Chimica Acta* 496, 73–80, 2003.

26. Karyakin, A. A., Presnova, G. V., Rubtsova, M. Yu., and Egorov, A. M., Oriented immobilization of antibodies onto the gold surfaces via their native thiol groups, *Analytical Chemistry* 72, 3805–3811, 2000.

27. Saerens, D., Huang, L., Bonroy, K., and Muyldermans, S., Antibody fragments as probe in biosensor development, *Sensors* 8(8), 4669–4686, 2008.

28. Erdem, A., Kerman, K., Meric, B., Akarca, U. S., and Ozsoz, M., DNA electrochemical biosensor for the detection of short DNA sequences related to the hepatitis B virus, *Electroanalysis* 11(8), 586–588, 1999.

29. Sawata, S., Kai, E., Ikebukuro, K., Iida, T., Honda, T., and Karube, I., Application of peptide nucleic acid to the direct detection of deoxyribonucleic acid amplified by polymerase chain reaction, *Biosensors & Bioelectronics* 14(4), 397–404, 1999.

30. Niemeyer, C. M., Boldt, L., Ceyhan, B., and Blohm, D., DNA-directed immobilization: Efficient, reversible, and site-selective surface binding of proteins by means of covalent DNA-streptavidin conjugates, *Analytical Biochemistry* 268(1), 54–63, 1999.

31. Marrazza, G., Chianella, I., and Mascini, M., Disposable DNA electrochemical sensor for hybridization detection, *Biosensors & Bioelectronics* 14(1), 43–51, 1999.

32. Bardea, A., Patolsky, F., Dagan, A., and Willner, I., Sensing and amplification of oligonucleotide-DNA interactions by means of impedance spectroscopy: A route to a Tay-Sachs sensor, *Chemical Communications* 1999(1), 21–22, 1999.

33. Wang, J., Rivas, G., Fernandes, J. R., Paz, J. L. L., Jiang, M., and Waymire, R., Indicator-free electrochemical DNA hybridization biosensor, *Analytica Chimica Acta* 375(3), 197–203, 1998.

34. Nimjee, S. M., Rusconi, C. P., and Sullenger, B. A., Aptamers: An emerging class of therapeutics, *Annual Review of Medicine* 56, 555–583, 3 plates, 2005.

35. Cho, H., Baker, B. R., Wachsmann-Hogiu, S., Pagba, C. V., Laurence, T. A., Lane, S. M., Lee, L. P., and Tok, J. B. H., Aptamer-based SERRS sensor for thrombin detection, *Nano Letters* 8, 4386–4390, 2008.

36. Strehlitz, B., Nikolaus, N., and Stoltenburg, R., Protein detection with aptamer biosensors, *Sensors* 8(7), 4296–4307, 2008.

37. Ellington, A. D. and Szostak, J. W., In vitro selection of RNA molecules that bind specific ligands, *Nature* 346(6287), 818–822, 1990.

38. Tuerk, C. and Gold, L., Systematic evolution of ligands by exponential enrichment: RNA ligands to bacteriophage T4 DNA polymerase, *Science* 249(4968), 505–510, 1990.

39. Nimjee, S. M., Christopher P. Rusconi, C. P., Bruce A., and Sullenger, B. A., Aptamers: An emerging class of therapeutics, *Annual Review of Medicine* 56, 555–583, 2005.

40. Stoltenburg, R., Reinemann, C., and Strehiltz, B., FlugMag-SELEX as an advantageous method for DNA aptamer selection, *Analytical and Bioanalytical Chemistry* 383, 83–91, 2005.

41. Fang, X. and Tan, W., Aptamers generated from Cell-SELEX for molecular medicine: A chemical biology approach, *Accounts of Chemical Research* 43(1), 48–57, 2010.

42. Qian, J., Lou, X., Zhang, Y., Xiao, Y., and Tom Soh, H., Generation of highly specific aptamers via micromagnetic selection, *Analytical Chemistry* 81, 5490–5495, 2009.

43. Hasegawa, H., Taira, K., Sode, K., and Ikebukuro, K., Improvement of aptamer affinity by dimerization, *Sensors* 8, 1090–1098, 2008.

44. Gooding, J. J. and Hibbert, D. B., The application of alkanethiol self-assembled monolayers to enzyme electrodes, *Trends in Analytical Chemistry* 18(8), 525–533, 1999.

45. Franchina, J. G., Lackowski, W. M., Dermody, D. L., Crooks, R. M., Bergbreiter, D. E., Sirkar, K., Russell, R. J., and Pishko, M. V., Electrostatic immobilization of glucose oxidase in a weak acid, polyelectrolyte hyperbranched ultrathin film on gold: Fabrication, characterization, and enzymatic activity, *Analytical Chemistry* 71(15), 3133–3139, 1999.

46. Patolsky, F., Zayats, M., Katz, E., and Willner, I., Precipitation of an insoluble product on enzyme monolayer electrodes for biosensor applications: Characterization by faradaic impedance spectroscopy, cyclic voltammetry, and microgravimetric quartz crystal microbalance analyses, *Analytical Chemistry* 71(15), 3171–3180, 1999.

47. Pemberton, R. M., Hart, J. P., Stoddard, P., and Foulkes, J. A., A comparison of 1-naphthyl phosphate and 4 aminophenyl phosphate as enzyme substrates for use with a screen-printed amperometric immunosensor for progesterone in cows' milk, *Biosensors & Bioelectronics* 14(5), 495–503, 1999.

48. Serra, B., Reviejo, A. J., Parrado, C., and Pingarron, J. M., Graphite-Teflon composite bienzyme electrodes for the determination of L-lactate: Application to food samples, *Biosensors & Bioelectronics* 14(5), 505–513, 1999.

49. Cosnier, S., Senillou, A., Gratzel, M., Comte, P., Vlachopoulos, N., Renault, N. J., and Martelet, C., A glucose biosensor based on enzyme entrapment within polypyrrole films electrodeposited on mesoporous titanium dioxide, *Journal of Electroanalytical Chemistry* 469(2), 176–181, 1999.

50. Blake, R. C., Pavlov, A. R., and Blake, D. A., Automated kinetic exclusion assays to quantify protein binding interactions in homogeneous solution, *Analytical Biochemistry* 272(2), 123–134, 1999.

51. Hara-Kuge, S., Ohkura, T., Seko, A., and Yamashita, K., Vesicular-integral membrane protein, VIP36, recognizes high-mannose type glycans containing alpha $1 \rightarrow 2$ mannosyl residues in MDCK cells, *Glycobiology* 9(8), 833–839, 1999.

52. Shih, Y. T. and Huang, H. J., A creatinine deiminase modified polyaniline electrode for creatinine analysis, *Analytica Chimica Acta* 392(2–3), 143–150, 1999.

53. Nelson, R. W., Jarvik, J. W., Taillon, B. E., and Tubbs, K. A., BIA MS of epitope-tagged peptides directly from *E. coli* lysate: Multiplex detection and protein identification at low-femtomole to sub-femtomole levels, *Analytical Chemistry* 71(14), 2858–2865, 1999.

54. Piehler, J., Brecht, A., Hehl, K., and Gauglitz, G., Protein interactions in covalently attached dextran layers, *Colloids and Surfaces B: Biointerfaces* 13(6), 325–336, 1999.

55. Kim, H. J., Hyun, M. S., Chang, I. S., and Kim, B. H., A microbial fuel cell type lactate biosensor using a metal-reducing bacterium, *Shewanella putrefaciens*, *Journal of Microbiology and Biotechnology* 9(3), 365–367, 1999.

56. Hu, T., Zhang, X. E., and Zhang, Z. P., Disposable screen-printed enzyme sensor for simultaneous determination of starch and glucose, *Biotechnology Techniques* 13(6), 359–362, 1999.

57. Garjonyte, R. and Malinauskas, A., Amperometric glucose biosensor based on glucose oxidase immobilized in poly(o-phenylenediamine) layer, *Sensors and Actuators B: Chemical* 56(1–2), 85–92, 1999.

58. Lee, Y. C. and Huh, M. H., Development of a biosensor with immobilized L-amino acid oxidase for determination of L-amino acids, *Journal of Food Biochemistry* 23(2), 173–185, 1999.

59. Lebron, J. A. and Bjorkman, P. J., The transferrin receptor binding site on HFE, the class I MHC-related protein mutated in hereditary hemochromatosis, *Journal of Molecular Biology* 289(4), 1109–1118, 1999.

60. Gooding, J. J., Situmorang, M., Erokhin, P., and Hibbert, D. B., An assay for the determination of the amount of glucose oxidase immobilised in an enzyme electrode, *Analytical Communications* 36(6), 225–228, 1999.

61. Hu, S. S., Luo, J., and Cui, D., An enzyme-chemically modified carbon paste electrode as a glucose sensor based on glucose oxidase immobilized in a polyaniline film, *Analytical Sciences* 15(6), 585–588, 1999.

62. Sergeyeva, T. A., Soldatkin, A. P., Rachkov, A. E., Tereschenko, M. I., Piletsky, S. A., and El'skaya, A. V., Beta-lactamase label-based potentiometric biosensor for alpha-2 interferon detection, *Analytica Chimica Acta* 390(1–3), 73–81, 1999.

63. Barker, S. L. R., Zhao, Y. D., Marletta, M. A., and Kopelman, R., Cellular applications of a sensitive and selective fiber optic nitric oxide biosensor based on a dye-labeled heme domain of soluble guanylate cyclase, *Analytical Chemistry* 71(11), 2071–2075, 1999.

64. Huang, T., Warsinke, A., Koroljova-Skovobogat'ko, O. V., Makower, A., Kuwana, T., and Scheller, F. W., A bienzyme carbon paste electrode for the sensitive detection of NADPH and the measurement of glucose-6-phosphate dehydrogenase, *Electroanalysis* 11(5), 295–300, 1999.

65. Hall, E. A. H., Gooding, J. J., Hall, C. E., and Martens, N., Acrylate polymer immobilisation of enzymes, *Fresenius Journal of Analytical Chemistry* 364(1–2), 58–65, 1999.

66. Anzai, J., Kobayashi, Y., Hoshi, T., and Saiki, H., A layer-by-layer deposition of concanavalin A and native glucose oxidase to form multilayer thin films for biosensor applications, *Chemistry Letters* (4), 365–366, 1999.

67. Campbell, T. E., Hodgson, A. J., and Wallace, G. G., Incorporation of erythrocytes into polypyrrole to form the basis of a biosensor to screen for Rhesus (D) blood groups and rhesus (D) antibodies, *Electroanalysis* 11(4), 215–222, 1999.

68. Wang, Y. S., Li, G. R., Lu, C. Y., Wan, Z. Y., and Liu, C. X., Preparation of triglyceride enzyme sensor and its application, *Progress in Biochemistry and Biophysics* 26(2), 144–146, 1999.

69. Pancrazio, J. J., Bey, P. P., Cuttino, D. S., Kusel, J. K., Borkholder, D. A., Shaffer, K. M., Kovacs, G. T. A., and Stenger, D. A., Portable cell-based biosensor system for toxin detection, *Sensors and Actuators B: Chemical* 53(3), 179–185, 1998.

70. Zhang, Y. Q., Zhu, J., and Gu, R. A., Improved biosensor for glucose based on glucose oxidase-immobilized silk fibroin membrane, *Applied Biochemistry and Biotechnology* 75(2–3), 215–233, 1998.

71. Houshmand, H., Froman, G., and Magnusson, G., Use of bacteriophage T7 displayed peptides for determination of monoclonal antibody specificity and biosensor analysis of the binding reaction, *Analytical Biochemistry* 268(2), 363–370, 1999.

72. Roos, H., Karlsson, R., Nilshans, H., and Persson, A., Thermodynamic analysis of protein interactions with biosensor technology, *Journal of Molecular Recognition* 11(1–6), 204–210, 1998.

73. Bin Zhao, Y., Wen, M. L., Liu, S. Q., Liu, Z. H., Zhang, W. D., Yao, Y., and Wang, C. Y., Microbial sensor for determination of tannic acid, *Microchemical Journal* 60(3), 201–209, 1998.

74. Heim, S., Schnieder, I., Binz, D., Vogel, A., and Bilitewski, U., Development of an automated microbial sensor system, *Biosensors & Bioelectronics* 14(2), 187–193, 1999.

75. Schmidt, A., StandfussGabisch, C., and Bilitewski, U., Microbial biosensor for free fatty acids using an oxygen electrode based on thick film technology, *Biosensors & Bioelectronics* 11(11), 1139–1145, 1996.

76. Schuler, R., Wittkampf, M., and Chemnitius, G. C., Modified gas-permeable silicone rubber membranes for covalent immobilisation of enzymes and their use in biosensor development, *Analyst* 124(8), 1181–1184, 1999.

77. Park, I. S. and Kim, N., Simultaneous determination of hypoxanthine, inosine and inosine 5′-monophosphate with serially connected three enzyme reactors, *Analytica Chimica Acta* 394(2–3), 201–210, 1999.

78. Situmorang, M., Gooding, J. J., and Hibbert, D. B., Immobilisation of enzyme throughout a polytyramine matrix: A versatile procedure for fabricating biosensors, *Analytica Chimica Acta* 394(2–3), 211–223, 1999.

79. Zhang, W. T., Canziani, G., Plugariu, C., Wyatt, R., Sodroski, J., Sweet, R., Kwong, P., Hendrickson, W., and Chaiken, L., Conformational changes of gp120 in epitopes near the CCR5 binding site are induced by CD4 and a CD4 miniprotein mimetic, *Biochemistry* 38(29), 9405–9416, 1999.

80. Yano, K. and Karube, I., Molecularly imprinted polymers for biosensor applications, *Trends in Analytical Chemistry* 18(3), 199–204, 1999.

81. Cotton, G. J., Ayers, B., Xu, R., and Muir, T. W., Insertion of a synthetic peptide into a recombinant protein framework: A protein biosensor, *Journal of the American Chemical Society* 121(5), 1100–1101, 1999.

82. Song, X. D. and Swanson, B. I., Direct, ultrasensitive, and selective optical detection of protein toxins using multivalent interactions, *Analytical Chemistry* 71(11), 2097–2107, 1999.

83. Costello, R. F., Peterson, I. R., Heptinstall, J., and Walton, D. J., Improved gel-protected bilayers, *Biosensors & Bioelectronics* 14(3), 265–271, 1999.

84. Ramsden, J. J., Biomimetic protein immobilization using lipid bilayers, *Biosensors & Bioelectronics* 13(6), 593–598, 1998.

85. Girard-Egrot, A. P., Morelis, R. M., and Coulet, P. R., Direct bioelectrochemical monitoring of choline oxidase kinetic behaviour in Langmuir-Blodgett nanostructures, *Bioelectrochemistry and Bioenergetics* 46(1), 39–44, 1998.

86. Wollenberger, U., Neumann, B., and Scheller, F. W., Development of a biomimetic alkane sensor, *Electrochimica Acta* 43(23), 3581–3585, 1998.

87. Costello, R. F., Peterson, I. P., Heptinstall, J., Byrne, N. G., and Miller, L. S., A robust gel-bilayer channel biosensor, *Advanced Materials for Optics and Electronics* 8(2), 47–52, 1998.

88. Ramstrom, O. and Ansell, R. J., Molecular imprinting technology: Challenges and prospects for the future, *Chirality* 10(3), 195–209, 1998.

89. Cornell, B. A., BraachMaksvytis, V. L. B., King, L. G., Osman, P. D. J., Raguse, B., Wieczorek, L., and Pace, R. J., A biosensor that uses ion-channel switches, *Nature* 387(6633), 580–583, 1997.

90. Kriz, D. K., Kempe, M., and Mosbach, K., Introduction of molecularly imprinted polymers as recognition elements in conductometric chemical sensors, *Sensors and Actuators B: Chemical* 33(1–3), 178–181, 1996.

91. Gopel, W. and Heiduschka, P., Interface analysis in biosensor design, *Biosensors & Bioelectronics* 10(9–10), 853–883, 1995.

92. Piervincenzi, R. T., Reichert, W. M., and Hellinga, H. W., Genetic engineering of a single-chain antibody fragment for surface immobilization in an optical biosensor, *Biosensors & Bioelectronics* 13(3–4), 305–312, 1998.

93. Charych, D., Cheng, Q., Reichert, A., Kuziemko, G., Stroh, M., Nagy, J. O., Spevak, W., and Stevens, R. C., A 'litmus test' for molecular recognition using artificial membranes, *Chemistry & Biology* 3(2), 113–120, 1996.

94. Teague, E. C., *Scanning Microscopy Technology and Applications Society of Photo-Optical Instrumentation Engineering*, SPIE, Bellingham, WA, 1988.

95. Pohl, D. W., Scanning near-field optical microscopy, in *Advances in Optical and Electron Microscopy*, Sheppard, C. J. R. and Mulvey, T., eds. Academic Press, London, U.K., 1984.

96. Betzig, E., Trautman, J. K., Harris, T. D., Weiner, J. S., and Kostelak, R. L., Breaking the diffraction barrier—Optical microscopy on a nanometric scale, *Science* 251(5000), 1468–1470, 1991.

97. Betzig, E. and Chichester, R. J., Single molecules observed by near-field scanning optical microscopy, *Science* 262(5138), 1422–1425, 1993.

98. Betzig, E., Lewis, A., Harootunian, A., Isaacson, M., and Kratschmer, E., Near-field scanning optical microscopy (NSOM)—Development and biophysical applications, *Biophysical Journal* 49(1), 269–279, 1986.

99. Durig, U., Pohl, D. W., and Rohner, F., Near-field optical-scanning microscopy, *Journal of Applied Physics* 59(10), 3318–3327, 1986.

100. Betzig, E., Isaacson, M., and Lewis, A., Collection mode near-field scanning optical microscopy, *Applied Physics Letters* 51(25), 2088–2090, 1987.

101. Lieberman, K., Harush, S., Lewis, A., and Kopelman, R., A light-source smaller than the optical wavelength, *Science* 247(4938), 59–61, 1990.

102. Zeisel, D., Deckert, V., Zenobi, R., and Vo-Dinh, T., Near-field surface-enhanced Raman spectroscopy of dye molecules adsorbed on silver island films, *Chemical Physics Letters* 283(5–6), 381–385, 1998.

103. Deckert, V., Zeisel, D., Zenobi, R., and Vo-Dinh, T., Near-field surface enhanced Raman imaging of dye-labeled DNA with 100-nm resolution, *Analytical Chemistry* 70(13), 2646–2650, 1998.

104. Tan, W. H., Shi, Z. Y., and Kopelman, R., Development of submicron chemical fiber optic sensors, *Analytical Chemistry* 64(23), 2985–2990, 1992.

105. Tan, W. H., Shi, Z. Y., Smith, S., Birnbaum, D., and Kopelman, R., Submicrometer intracellular chemical optical fiber sensors, *Science* 258(5083), 778–781, 1992.

106. Samuel, J., Strinkovski, A., Lieberman, K., Ottolenghi, M., Avnir, D., and Lewis, A., Miniaturization of organically doped sol-gel materials—A microns-size fluorescent Ph sensor, *Materials Letters* 21(5–6), 431–434, 1994.

107. McCulloch, S. A. and Uttamchandani, D., *IEE Proceedings-Optoelectronics* 144(162), 1995.

108. Tan, W. H., Shi, Z. Y., and Kopelman, R., Miniaturized fiberoptic chemical sensors with fluorescent dye-doped polymers, *Sensors and Actuators B: Chemical* 28(2), 157–163, 1995.

109. Song, A., Parus, S., and Kopelman, R., High-performance fiber optic pH microsensors for practical physiological measurements using a dual-emission sensitive dye, *Analytical Chemistry* 69(5), 863–867, 1997.

110. Koronczi, I., Reichert, J., Heinzmann, G., and Ache, H. J., Development of a submicron optochemical potassium sensor with enhanced stability due to internal reference, *Sensors and Actuators B: Chemical* 51(1–3), 188–195, 1998.

111. Bui, J. D., Zelles, T., Lou, H. J., Gallion, V. L., Phillips, M. I., and Tan, W. H., Probing intracellular dynamics in living cells with near-field optics, *Journal of Neuroscience Methods* 89(1), 9–15, 1999.

112. Barker, S. L. R. and Kopelman, R., Development and cellular applications of fiber optic nitric oxide sensors based on a gold-adsorbed fluorophore, *Analytical Chemistry* 70(23), 4902–4906, 1998.

113. Munkholm, C., Walt, D. R., and Milanovich, F. P., Preparation of CO_2 fiber optic chemical sensor, *Abstracts of Papers of the American Chemical Society* 193, 183-ANYL, 1987.

114. Munkholm, C., Parkinson, D. R., and Walt, D. R., Intramolecular fluorescence self-quenching of fluoresceinamine, *Journal of the American Chemical Society* 112(7), 2608–2612, 1990.

115. Barker, S. L. R., Thorsrud, B. A., and Kopelman, R., Nitrite- and chloride-selective fluorescent nano-optodes and in in vitro application to rat conceptuses, *Analytical Chemistry* 70(1), 100–104, 1998.

116. Barker, S. L. R., Kopelman, R., Meyer, T. E., and Cusanovich, M. A., Fiber-optic nitric oxide-selective biosensors and nanosensors, *Analytical Chemistry* 70(5), 971–976, 1998.

117. Alarie, J. P. and VoDinh, T., Antibody-based submicron biosensor for benzo a pyrene DNA adduct, *Polycyclic Aromatic Compounds* 8(1), 45–52, 1996.

118. Cordek, J., Wang, X. W., and Tan, W. H., Direct immobilization of glutamate dehydrogenase on optical fiber probes for ultrasensitive glutamate detection, *Analytical Chemistry* 71(8), 1529–1533, 1999.

119. Cullum, B. M., Griffin, G. D., Miller, G. H., and Vo-Dinh, T., Intracellular measurements in mammary carcinoma cells using fiber-optic nanosensors, *Analytical Biochemistry* 277(1), 25–32, 2000.

120. Cullum, B. M. and Vo-Dinh, T., Optical nanosensors and biological measurements, *Biofutur* 2000(205), A1–A6, 2000.

121. Vo-Dinh, T. G., Alarie, G. D., Alarie, J. P., Cullum, B. M., Sumpter, B. M., and D. Noid, Development of nanosensors and bioprobes, *Journal of Nanoparticle Research* 2, 17, 2000.

122. Vo-Dinh, T., Alarie, J. P., Cullum, B. M., and Griffin, G. D., Antibody-based nanoprobe for measurement of a fluorescent analyte in a single cell, *Nature Biotechnology* 18(7), 764–767, 2000.

123. Kasili, P. M., Song, J. M., and Vo-Dinh, T., Optical sensor for the detection of caspase-9 activity in a single cell, *Journal of American Chemical Society* 126(9), 2799–2806, 2004.

124. Kasili, P. M. and Vo-Dinh, T., Detection of polycyclic aromatic compounds in single living cells using optical nanoprobes, *Polycyclic Aromatic Compounds* 24(3), 221–235, 2004.

125. Kasili, P. M. and Vo-Dinh, T., Optical nanobiosensor for monitoring an apoptotic signaling process in a single living cell following photodynamic therapy, *Journal of Nanoscience and Nanotechnology* 5(12), 2057–2062, 2005.

126. Kasili, R. M., Cullum, B. M., Griffin, G. D., and Vo-Dinh, T., Nanosensor for in vivo measurement of the carcinogen benzo[a]pyrene in a single cell, *Journal of Nanoscience and Nanotechnology* 2(6), 653–658, 2002.

127. Song, J. M., Kasili, P. M., Griffin, G. D., and Vo-Dinh, T., Detection of cytochrome c in a single cell using an optical nanobiosensor, *Analytical Chemistry* 76(9), 2591–2594, 2004.
128. Scaffidi, J. P., Gregas, M. K., Seewaldt, V., and Vo-Dinh, T., SERS-based plasmonic nanobiosensing in single living cells, *Analytical and Bioanalytical Chemistry* 393(4), 1135–1141, 2009.
129. Vo-Dinh, T., Scaffidi, J. S., Gregas, M., Zhang, Y., and Seewaldt, V., Applications of fiberoptics-based nanosensors to drug discovery, *Expert Opinion on Drug Discovery* 4, 889–900, 2009.
130. Hengartner, M. O., Apoptosis—DNA destroyers, *Nature* 412(6842), 27–29, 2001.
131. Ricci, J. E., Gottlieb, R. A., and Green, D. R., Caspase-mediated loss of mitochondrial function and generation of reactive oxygen species during apoptosis, *Journal of Cell Biology* 160(1), 65–75, 2003.
132. Wolf, B. B. and Green, D. R., Suicidal tendencies: Apoptotic cell death by caspase family proteinases, *Journal of Biological Chemistry* 274(29), 20049–20052, 1999.
133. Hoffmann, P., Dutoit, B., and Salathe, R. P., Comparison of mechanically drawn and protection layer chemically etched optical fiber tips, *Ultramicroscopy* 61(1–4), 165–170, 1995.
134. Turner, D. R., Patent #4, 469, 554, 1984.
135. Stockle, R., Fokas, C., Deckert, V., Zenobi, R., Sick, B., Hecht, B., and Wild, U. P., High-quality near-field optical probes by tube etching, *Applied Physics Letters* 75(2), 160–162, 1999.
136. Clark, H. A., Hoyer, M., Philbert, M. A., and Kopelman, R., Optical nanosensors for chemical analysis inside single living cells. 1. Fabrication, characterization, and methods for intracellular delivery of PEBBLE sensors, *Analytical Chemistry* 71(21), 4831–4836, 1999.
137. Clark, H. A., Kopelman, R., Tjalkens, R., and Philbert, M. A., Optical nanosensors for chemical analysis inside single living cells. 2. Sensors for pH and calcium and the intracellular application of PEBBLE sensors, *Analytical Chemistry* 71(21), 4837–4843, 1999.
138. Suzuki, T., Matsuzaki, T., Hagiwara, H., Aoki, T., and Takata, K., Recent advances in fluorescent labeling techniques for fluorescence microscopy, *Acta Histochemica et Cytochemica* 40, 131–137, 2007.
139. Chalfie, M., Tu, Y., Euskirchen, G., Ward, W. W., and Prasher, D. C., Green fluorescent protein as a marker for gene expression, *Science* 263, 802–805, 1994.
140. Mehiri, M., Jing, B., Ringhoff, D., Janout, V., Cassimeris, L., and Regen, S. L., Cellular entry and nuclear targeting by a highly anionic molecular umbrella, *Bioconjugate Chemistry* 19, 1510–1513, 2008.
141. Webster, A., Coupland, P., Houghton, F. D., Leese, H. J., and Aylott, J. W., The delivery of PEBBLE nanosensors to measure the intracellular environment, *Biochemical Society Transactions* 35, 538–543, 2007.
142. Henderson, J. R., Fulton, D. A., McNeil, C. J., and Manning, P., The development and in vitro characterisation of an intracellular nanosensor responsive to reactive oxygen species, *Biosensors Bioelectronics* 24(12), 3608–3614, 2009.
143. Brasuel, M., Kopelman, R., Philbert, M. A., Aylott, J. W., Clark, H., Kasman, I., King, M. et al., PEBBLE nanosensors for real time intracellular chemical imaging, *Optical Biosensors* 497–536, 2002.
144. Brasuel Murphy, G., Miller Terry, J., Kopelman, R., and Philbert Martin, A., Liquid polymer nano-PEBBLEs for Cl⁻ analysis and biological applications, *Analyst* 128(10), 1262–1267, 2003.
145. Xu, H., Aylott, J. W., Kopelman, R., Miller, T. J., and Philbert, M. A., A real-time ratiometric method for the determination of molecular oxygen inside living cells using sol-gel-based spherical optical nanosensors with applications to rat C6 glioma, *Analytical Chemistry* 73(17), 4124–4133, 2001.
146. Xu, H., Buck, S. M., Kopelman, R., Philbert, M. A., Brasuel, M., Ross, B. D., and Rehemtulla, A., Photoexcitation-based nano-explorers: Chemical analysis inside live cells and photodynamic therapy, *Israel Journal of Chemistry* 44(1–3), 317–337, 2004.
147. Koo Yong-Eun, L., Cao, Y., Kopelman, R., Koo Sang, M., Brasuel, M., and Philbert Martin, A., Real-time measurements of dissolved oxygen inside live cells by organically modified silicate fluorescent nanosensors, *Analytical Chemistry* 76(9), 2498–505, 2004.
148. Xu, H., Aylott Jonathan, W., and Kopelman, R., Fluorescent nano-PEBBLE sensors designed for intracellular glucose imaging, *Analyst* 127(11), 1471–1477, 2002.

149. Rangin, M. and Basu, A., Lipopolysaccharide identification with functionalized polydiacetylene liposome sensors, *Journal of American Chemical Society* 126(16), 5038–5039, 2004.
150. McNamara, K. P. and Rosenzweig, Z., Dye-encapsulating liposomes as fluorescence-based oxygen nanosensors, *Analytical Chemistry* 70, 4853–4859, 1998.
151. McNamara, K. P., Rosenzweig, N., and Rosenzweig, Z., Synthesis, characterization and application of fluorescent lipobeads for imaging and sensing in single cells, *Proceedings of SPIE* 3922, 147–157, 2000.
152. McNamara, K. P., Nguyen, T., Dumitrascu, G., Ji, J., Rosenzweig, N., and Rosenzweig, Z., Synthesis, characterization, and application of fluorescence sensing lipobeads for intracellular pH measurements, *Analytical Chemistry* 73, 3240–3246, 2001.
153. Nguyen, T., McNamara, K. P., and Rosenzweig, Z., Optochemical sensing by immobilizing fluorescence-encapsulating liposomes in sol-gel thin films, *Analytica Chimica Acta* 400, 45–54, 1999.
154. Nguyen, T. and Rosenzweig, Z., Calcium ion fluorescence detection using liposomes containing Alexa-labeled calmodulin, *Analytical and Bioanalytical Chemistry* 374, 69–74, 2002.
155. Chemburu, S., Ji, E., Casana, Y., Wu, Y., Buranda, T., Schanze, K. S., Lopez, G. P., and Whitten, D. G., Conjugated polyelectrolyte supported bead based assays for phospholipase A2 activity, *Journal of Physical Chemistry B* 112(46), 14492–14499, 2008.
156. Ma, A. and Rosenzweig, Z., Submicrometric lipobead-based fluorescence sensors for chloride ion measurements in aqueous solution, *Analytical Chemistry* 76(3), 569–575, 2004.
157. Ma, A. and Rosenzweig, Z., Synthesis and analytical properties of micrometric biosensing lipobeads, *Analytical Bioanalytical Chemistry* 382(1), 28–36, 2005.
158. McNamara, K. P., Rosenzweig, N., and Rosenzweig, Z., Synthesis, characterization, and application of fluorescent lipobeads for imaging and sensing in single cells, *Proceedings of the SPIE—The International Society for Optical Engineering* 3922 (Scanning and Force Microscopies for Biomedical Applications II), 147–157, 2000.
159. Anker, J. F., Behrend, C. J., Huang, H., and Kopelman, R., Magnetically-modulated optical nanoprobes and systems, *Journal of Magnetism and Magnetic Materials* 293, 655–662, 2005.
160. Brasuel, M., Kopelman, R., Miller, T. J., Tjalkens, R., and Philbert, M. A., Fluorescent nanosensors for intracellular chemical analysis: Decyl methacrylate liquid polymer matrix and ion-exchange-based potassium PEBBLE sensors with real-time application to viable rat C6 glioma cells, *Analytical Chemistry* 73(10), 2221–2228, 2001.
161. McNamara, K. P., Nguyen, T., Dumitrascu, G., Ji, J., Rosenzweig, N., and Rosenzweig, Z., Synthesis, characterization, and application of fluorescence sensing lipobeads for intracellular pH measurements, *Analytical Chemistry* 73(14), 3240–3246, 2001.
162. Bruchez Jr., M., Moronne, M., Gin, P., Weiss, S., and Alivisatos, A. P., Semiconductor nanocrystals as fluorescent biological labels, *Science* 281, 2013–2016, 1998.
163. Wu, X., Liu, H., Liu, J., Haley, K. N., Treadway, J. A., Larson, J. P., Ge, N., Peale, F., and Bruchez, M. P., Immunofluorescent labeling of cancer marker Her2 and other cellular targets with semiconductor quantum dots, *Nature Biotechnology* 21, 41–46, 2003.
164. Medintz, I. L., Clapp, A. R., Brunel, F. M., Goldman, E. R., Chang, E. L., Dawson, P. E., and Mattoussi, H., Quantum dot based nanosensors designed for proteolytic monitoring, *Proceedings of SPIE* 6096(1), 60960K, 2006.
165. Xia, Z., Xing, Y., So, M-K., Koh, A. A., Sinclair, R., and Rao, J., Multiplex detection of protease activity with quantum dot nanosensors prepared by intein-mediated specific bioconjugation, *Analytical Chemistry* 80, 8649–8655, 2008.
166. Zhang, C.-Y., Yeh, H-C., Kuroki, M. T., and Wang, T-H., Single-quantum-dot-based DNA nanosensor, *Nature Materials* 4, 826–831, 2005.
167. Medintz, I. L., Uyeda, H. H., Goldman, E. R., and Mattoussi, H., Quantum dot bioconjugates for imaging, labeling and sensing, *Nature Materials* 4, 136 116, 2005.

168. Tholouli, E., Sweeney, E., Barrow, E., Clay, V., Hoyland, J. A., and Byers, R. J., Quantum dots light up pathology, *Journal of Pathology* 216, 275–285, 2008.

169. Pavel, I., McCarney, E., Elkhaled, A., Morrill, A., Plaxco, K., and Moskovits, M., Label-free SERS detection of small proteins modified to act as bifunctional linkers, *The Journal of Physical Chemistry C* 112(13), 4880–4883, 2008.

170. Kneipp, K., Wang, Y., Kneipp, H., Perelman, L. T., Itzkan, I., Dasari, R., and Feld, M. S., Single molecule detection using surface-enhanced Raman spectroscopy, *Physical Review Letters* 78, 1667–1670, 1997.

171. Haynes, C. L. and Van Duyne, R. P., Plasmon-sampled surface-enhanced Raman excitation spectroscopy, *Journal of Physical Chemistry* 107, 7426–7433, 2003.

172. Kelly, K. L., Coronado, E., Zhao, L. L., and Schatz, G. C., The optical properties of metal nanoparticles: The influence of size shape and dielectric environment, *Journal of Physical Chemistry B* 107, 668–677, 2003.

173. Otto, A., What is observed in single molecule SERS, and why? *Journal of Raman Spectroscopy* 33, 593–598, 2002.

174. Alak, A. M. and Vo-Dinh, T., Surface-enhanced Raman spectrometry of organophosphorus chemical agents, *Analytical Chemistry* 59, 2149–2153, 1987.

175. Jeanmaire, D. L. and Van Duyne, R. P., Surface Raman spectroelectrochemistry: Part I. Heterocyclic, aromatic, and aliphatic amines adsorbed on the anodized silver electrode, *Journal of Electroanalytical Chemistry* 84, 1–20, 1977.

176. Fleischmann, M., Hendra, P. J., and McQuillan, A. J., Raman spectra of pyridine adsorbed at a silver electrode, *Chemical Physics Letters* 26, 163–166, 1974.

177. Nie, S. and Emory, S. R., Probing single molecules and single nanoparticles by surface-enhanced Raman scattering, *Science* 275, 1102–1106, 1997.

178. Tang, H.-W., Yang, B. X., Kirkham, J., and Smith, D. A., Probing intrinsic and extrinsic components in single osteosarcoma cells by near-infrared surface-enhanced Raman scattering, *Analytical Chemistry* 79, 3646–3653, 2007.

179. Sagmuller, B., Schwarze, B., Brehma, G., and Schneider, S., Application of SERS spectroscopy to the identification of (3,4-methylenedioxy) amphetamine in forensic samples utilizing matrix stabilized silver halides, *Analyst* 126, 2066–2071, 2001.

180. Billinton, N. and Knight, A. W., Seeing the wood through the trees: A review of techniques for distinguishing green fluorescent protein from endogenous autofluorescence, *Analytical Biochemistry* 291, 175–197, 2001.

181. Kneipp, J., Kneipp, H., McLaughlin, M., Brown, D., and Kneipp, K., In vivo molecular probing of cellular compartments with gold nanoparticles and nanoaggregates, *Nano Letters* 6, 2225–2231, 2006.

182. Kneipp, K., Haka, A. S., Kneipp, H., Badisadegan, K., Yoshizawa, N., Boone, C., Shafer-Peltier, K. E., Motz, J. T., Dasari, R. R., and Feld, M. S., Surface-enhanced Raman spectroscopy in single living cells using gold nanoparticles, *Applied Spectroscopy* 56, 150–154, 2002.

183. Kneipp, J., Kneipp, H., Rice, W. L., and Kneipp, K., Optical probes for biological applications based on surface-enhanced Raman scattering from indocyanine green on gold nanoparticles, *Analytical Chemistry* 77, 2381–2385, 2005.

184. Kneipp, J., Nanosensors based on SERS for application in living cells, in *Surface-Enhanced Raman Scattering*, Kneipp, K., Moskovits, M., and Kneipp, H, eds. Springer, Berlin, Germany, 2006, pp. 335–349.

185. Talley, C. E., Jusinski, L., Hollars, C. W., Lane, S. M., and Huser, T., Intracellular pH sensors based on surface-enhanced Raman scattering, *Analytical Chemistry* 76, 7064–7068, 2004.

186. Liu, Y., Yuan, H., Fales, A. M., and Vo-Dinh, T., pH sensing nanostar probe using surface-enhanced Raman scattering (SERS): Theoretical and experimental studies, *Journal of Raman Spectroscopy* 44(7), 980–986, 2013.

187. Vo-Dinh, T., Wang, H. N., and Scaffidi, J., Plasmonic nanoprobes for SERS biosensing and bioimaging, *Journal of Biophotonics*, 3, 89–102, 2010.

188. Gregas, M. K., Scaffidi, J. P., Lauly, B., and Vo-Dinh, T., Characterization of nanoprobe uptake in single cells: Spatial and temporal tracking via SERS labeling and modulation of surface charge, *Nanomedicine* 7, 115–122, 2011.

189. Gregas, M. K., Scaffidi, J. P., Lauly, B., and Vo-Dinh, T., Surface-enhanced Raman scattering detection and tracking of nanoprobes: Enhanced uptake and nuclear targeting in single cells, *Applied Spectroscopy* 64, 858–866, 2010.

190. Yuan, H., Khoury, C. G., Wang, H. H., Wilson, C. M., Grant, G. A., and Vo-Dinh, T., Gold nanostars: Surfactant-free synthesis, 3D modelling, and two-photon photoluminescence imaging, *Nanotechnology* 23, 075102, 2012.

191. Yuan, H., Khoury, C. G., Wilson, C. M., Grant, G. A., Bennett, A. J., and Vo-Dinh, T., In vivo particle tracking and photothermal ablation using plasmon resonant gold nanostars. *Nanomedicine: Nanotechnology, Biology, and Medicine* 8, 1355–1363, 2012.

192. Yuan, H., Fales, A. M., and Vo-Dinh, T., TAT peptide-functionalized gold nanostars: Enhanced intracellular delivery and efficient NIR photothermal therapy using ultralow irradiance, *Journal of the American Chemical Society* 134, 11358–11361, 2012.

193. Yuan, H., Liu, Y., Fales, A. M., Li, Y. L., Liu, J., and Vo-Dinh, T., Quantitative surface-enhanced resonant Raman scattering multiplexing of biocompatible gold nanostars for in vitro and ex vivo detection, *Analytical Chemistry* 85, 208–212, 2013.

194. Fales, A., Yuan, H., and Vo-Dinh, T., Cell-penetrating peptide enhanced intracellular Raman imaging and photodynamic therapy, *Molecular Pharmaceutics* 10(6), 2291–2298, 2013.

195. Fales, A. M., Yuan, H., and Vo-Dinh, T., Silica-coated gold nanostars for combined surface-enhanced Raman scattering (SERS) detection and singlet-oxygen generation: A potential nanoplatform for theranostics, *Langmuir* 27, 12186–12190, 2011.

196. Yonzon, C. R., Haynes, C. L., Zhang, X., Walsh Jr., J. T., and Van Duyne, R. P., A glucose biosensor based on surface-enhanced Raman scattering: Improved partition layer, temporal stability, reversibility, and resistance to serum protein interference, *Analytical Chemistry* 76, 78–85, 2004.

197. Li, H., Sun, J., Alexander, T., and Cullum, B. M., Implantable SERS nanosensors for pre-symptomatic detection f BW agents, *Proceedings of SPIE* 5795, 8–18, 2005.

198. Li, H., Sun, J., and Cullum, B. M., Label-free detection of proteins using SERS-based immuno-nanosensors, *NanoBiotechnology* 2, 17–28, 2006.

199. Li, H., Sun, J., and Cullum, B. M., Nanosphere-based SERS immuno-sensors for protein analysis, *Proceedings of SPIE* 5588, 19–30, 2004.

200. Litorja, M., Haynes, C. L., Haes, A. J., Jensen, T. R., and Van Duyne, R. P., Surface-enhanced Raman scattering detected temperature programmed desorption: Optical properties, nanostructure, and stability of silver film over SiO_2 nanosphere surface, *Journal of Physical Chemistry B* 105, 6907–6915, 2001.

201. Stuart, D. A., Yuen, J. M., Shah, N., Lyandres, O., Yonzon, C. R., Glucksberg, M. R., Walsh, J. T., and Van Duyne, P. R., In vivo glucose measurement using surface-enhanced Raman spectroscopy, *Analytical Chemistry* 78, 7211–7215, 2006.

202. Li, H., Sun, J., and Cullum, B., Label-free detection of proteins using SERS-based immuno-nanosensors, *NanoBiotechnology* 2(1), 17–28, 2006.

203. Kim, J.-H., Kim, J.-S., Choi, H., Lee, S.-M., Jun, B.-H., Yu, K.-N., Kuk, E. et al., Nanoparticle probes with surface-enhanced Raman spectroscopic tags for cellular cancer targeting, *Analytical Chemistry* 78, 6967–6973, 2006.

204. Cho, H., Baker, B. R., Wachsmann-Hogiu, S., Pagba C. V., Laurence, T. A., Lane S. M., Lee, L. P., and Tok, J. B.-H., Aptamer based SERS sensor for thrombin detection, *Nano Letters* 8(12), 4386–4390, 2008.

205. Bizzari, A. R., and Cannistraro, S., SERS detection of thrombin by protein recognition using functionalized gold nanoparticles, *Virology Nanomedicine* 3, 306–310, 2007.

206. Albrecht, M. G. and Creighton, J. A., Anomalously intense Raman spectra of pyridine at a silver electrode, *Journal of Electroanalytical Chemistry* 99, 5215–5217, 1977.

207. Baia, L., Baia, M., Popp, J., and Astilean, S., Gold films deposited over regular arrays of polystyrene nanospheres as highly effective SERS substrates from visible to NIR, *Journal of Physical Chemistry B* 110, 23982–23986, 2006.

208. Bantz, K. C. and Haynes, C. L., Surface-enhanced Raman scattering substrates fabricated using electroless plating on polymer-templated nanostructures, *Langmuir* 24, 5862–5867, 2008.

209. Dick, L. A., McFarland, A. D., Haynes, C. L., and Van Duyne, R. P., Angle-resolved nanosphere lithography: Manipulation of nanoparticle size, shape, and interparticle spacing, *Journal of Physical Chemistry B* 106, 853–860, 2002.

210. Dong, X. G., Stable silver substrate prepared by the nitric acid etching method for a surface enhanced Raman scattering study, *Analytical Chemistry* 63, 2393–2397, 1991.

211. Kneipp, J., Kneipp, H., Wittig, B., and Kneipp, K., One- and two-photon excited optical pH probing for cells using surface-enhanced Raman and hyper-Raman nanosensors, *Nano Letters* 7, 2819–2823, 2007.

212. Sharma, S. K., Misra, A. K., Kamemoto, L., Dykes, A., and Acosta, T., New microcavity substrates for enhancing Raman signals of microscopic samples, *Proceedings of SPIE* 7313, 73130F/1–73130F/10, 2009.

213. Tao, A., Kim, F., Hess, C., Goldberger, J., He, R., Sun, Y., Xia, Y., and Yang, P., Langmuir-Blodgett silver nanowire monolayers for molecular sensing using surface-enhanced Raman spectroscopy, *Nano Letters* 3, 1229–1233, 2003.

214. Van Duyne, P. R. and Hulteen, J. C., Atomic force microscopy and surface-enhanced Raman spectroscopy. I. Silver island films and silver film over polymer nanosphere surfaces supported on glass, *Journal of Physical Chemistry* 99, 2101–2115, 1993.

215. Li, H. and Cullum, B. M., Dual layer and multilayer enhancements from silver film over nanostructured surface-enhanced Raman substrates, *Applied Spectroscopy* 59, 410–417, 2005.

216. Li, H., Baum, C. E., Sun, J., and Cullum, B. M., Multilayer enhanced gold film over nanostructure surface-enhanced Raman substrates, *Applied Spectroscopy* 12, 1377–1385, 2006.

217. Cullum, B. M., Li, H., Schiza, M. V., and Hankus, M. E., Characterization of multilayer-enhanced surface-enhanced Raman scattering (SERS) substrates and their potential for SERS nanoimaging, *NanoBiotechnology* 3, 1–11, 2007.

218. Klutse, C. K., Li, H., and Cullum, B. M., Optimization of multilayer surface-enhanced Raman scattering (SERS) immuno-nanosensors via self-assembled monolayer spacers, *Proceedings of SPIE* 7313(1), 73130C, 2009.

219. Kasili, P., Wabuyele, M., and Vo-Dinh, T., Antibody-based SERS diagnostics of FHIT protein without label, *NanoBiotechnology* 2, 29, 2006.

220. Wabuyele M. B. and Vo-Dinh, T., Detection of HIV Type 1 DNA sequence using plasmonics nanoprobes, *Analytical Chemistry* 77, 7810–7815, 2005.

221. Wang, H. N. and Vo-Dinh, T., Multiplex detection of breast cancer biomarkers using plasmonic molecular sentinel nanoprobes, *Nanotechnology* 20, 065101, 1–6, 2009.

222. Wabuyele, M. B., Yan, F., and Vo-Dinh, T., Plasmonics nanoprobes: Detection of single-nucleotide polymorphisms in the breast cancer BRCA1 gene, *Analytical and Bioanalytical Chemistry* 398, 729–736, 2010.

223. Ngo, H. T., Wang, H. N., Fales, A. M., and Vo-Dinh, T., Label-free DNA biosensor based on SERS molecular sentinel on nanowave chip, *Analytical Chemistry*, 85, 6378–6383, 2013.

224. Khoury, C. and Vo-Dinh, T., Nanowave substrates for SERS: Fabrication and numerical analysis, *Journal of Physical Chemistry C* 116, 7534–7545, 2012.

225. Vo-Dinh, T., Hiromoto, M. Y. K., Begun, G. M., and Moody, R. L., Surface enhanced Raman spectroscopy for trace organic analysis, *Analytical Chemistry* 56(9), 1667–1670, 1984.

226. Vo-Dinh, T., Surface-enhanced Raman spectroscopy using metallic nanostructures TrAC, *Trends in Analytical Chemistry* 17(8–9), 557–582, 1998.

227. Ngo, H. T., Wang, H.-N., Fales, A. M., and Vo-Dinh., T., Label-free DNA biosensor based on SERS molecular sentinel on nanowave chip, *Analytical Chemistry* 85(13), 6378–6383, 2013.
228. Vo-Dinh, T., Wang, H. N., and Scaffidi, J., Plasmonic nanoprobes for SERS biosensing and bioimaging, *Journal of Biophotonics* 3(1–2), 89–102, 2010.
229. Ngo, H. T., Wang, H. N., Burke, T., Ginsburg, G. S., and Vo-Dinh, T., Multiplex detection of disease biomarkers using SERS molecular sentinel-on-chip, *Analytical and Bioanalytical Chemistry*, published online, February 2014, [Epub ahead of print].

24

SERS Molecular Sentinel Nanoprobes to Detect Biomarkers for Medical Diagnostics

Tuan Vo-Dinh
Duke University

Hsin-Neng Wang
Duke University

Hoan Thanh Ngo
Duke University

This chapter provides an overview of the development and applications of plasmonics-active nanoprobes for medical diagnostics. The *molecular sentinel* (MS) biosensing approach combines the modulation of the plasmonics effect to change the surface-enhanced Raman scattering (SERS) of a Raman label and the specificity of a DNA hairpin loop sequence to recognize and discriminate a variety of molecular target sequences. The MS nanoprobe structure consists of a metal nanoparticle and a stem-loop DNA molecule tagged with a Raman label. The specificity and selectivity of the DNA hairpin probe sequence is designed to identify a specific DNA/RNA target sequence of interest. In the normal configuration and in the absence of target DNA, the stem-loop configuration maintains the Raman label in close proximity to the metal nanoparticle, which serves as a plasmonics enhancer inducing a strong SERS signal of the Raman label upon laser excitation. Hybridization with target DNA opens the hairpin and physically separates the Raman label from the nanoparticle thus decreasing the plasmonics effect and quenching the SERS signal of the label. Due to the intrinsically narrow Raman lines, multiple probes can be used for various targets in the single bioassay. Furthermore, with the possibility of conducting simple homogeneous bioassays, the SERS-MSs could provide useful diagnostic probes for multiplex biomedical diagnostics. Several examples of detection of mRNA and microRNA biomarkers illustrate the usefulness of the MS method for medical diagnostics.

24.1 Introduction

Raman spectroscopy is a powerful analytical technique for chemical characterization due to the wealth of information on molecular structures, surface processes, and interface reactions that can be extracted from experimental data. The spectral selectivity associated with the narrow emission lines and the molecular specific vibrational bands of Raman labels make it an ideal tool for molecular genotyping. However, a limitation of Raman techniques for trace detection is the very weak Raman cross section. However, Raman spectroscopy has gained increasing interest as an analytical tool with the advent of the SERS effect, which can produce significant enhancement of the Raman signal. It is believed that the origin of the enormous Raman enhancement is produced by at least two main mechanisms that contribute to the SERS effect: (1) an electromagnetic effect occurring near metal surface structures associated with large local fields caused by electromagnetic resonances, often referred to as *surface plasmons*, and (2) a chemical effect involving a scattering process associated with chemical interactions between the molecule and the metal surface. Plasmons are quanta associated with longitudinal waves propagating in matter through the collective motion of large numbers of electrons. According to classical electromagnetic theory, molecules on or near metal nanostructures experience enhanced fields relative to that of the incident radiation. When a metallic nanostructured surface is irradiated by an incident electromagnetic field (e.g., a laser beam), conduction electrons are displaced into frequency oscillations equal to those of the incident light. These oscillating electrons, called *surface plasmons*, produce a secondary electric field, which adds to the incident field. These fields can be quite large (10^6- to 10^7-fold, even up to 10^{15}-fold enhancement at *hot spots*). When these oscillating electrons become spatially confined, as is the case for isolated metallic nanospheres or otherwise roughened metallic surfaces (nanostructures), there is a characteristic frequency (the plasmon frequency) at which there is a resonant response of the collective oscillations to the incident field. This condition yields intense localized fields that can interact with molecules in contact with or near the metal surface [1–4]. In an effect analogous to a *lightning rod* effect, secondary fields can become concentrated at high curvature points on the roughened metal surface.

Shortly after the report of the SERS effect [5–7], our laboratory first demonstrated the general applicability of the SERS effect for trace analysis using solid substrates having silver-coated nanospheres [8]. In 1984, we exploited the spin-coating and electron beam evaporation techniques to fabricate close-packed arrays of nanospheres, onto which a thin silver shell was deposited, effectively forming a controlled, reproducible substrate of *hot spots* [8]. We developed and proposed the use of solid substrates consisting of nanosphere or nanoparticle arrays covered with nanolayer of metal (forming a *nanowave*) as efficient and reproducible SERS-active media. During the following two decades, our laboratory has extensively investigated the SERS technology and developed a wide variety of plasmonics-active SERS platforms for chemical sensing [8,9–15] and for bioanalysis and biosensing [16–27]. These plasmonics substrate platforms have led to a wide variety of analytical applications including sensitive detection of a variety of chemicals of environmental, biological, and medical significance, such as DNA-adduct biomarkers [9]. The first application of SERS in DNA probe detection technology was reported [16] and has led to subsequent development of this method for medical diagnostics [17,18,23,24,27]. This chapter discusses the development of a unique MS nanoprobe technology designed to detect the presence of DNA/RNA biotargets, such as HIV [26] and breast cancer genes [28] in homogeneous assays.

The SERS technology has now received increasing interest and contribution from many research groups worldwide, and the reader is referred to a number of reviews and monographs for further details [2,29–39].

24.2 SERS Detection Modality for Gene Diagnostics

Traditional gene detection techniques are based on radioactive detection or fluorescence methods. Raman spectroscopy also offers some distinct features that are important for in situ monitoring of complex biological systems. Following laser irradiation of a sample, the observed Raman shifts are equivalent to the energy changes involved in molecular transitions of the scattering species and are therefore characteristic of it. These observed Raman shifts, which correspond to vibrational transitions of the scattering molecule, exhibit very narrow linewidths. For these reasons, Raman spectroscopy has a great potential for multiplexing detection, for example, in gene diagnostics. Therefore, many organic compounds with distinct Raman spectra may be used as dyes to label biological macromolecules, and each labeled molecular species will be able to be distinguished on the basis of its unique Raman spectra. This is not the case with fluorescence, because the relatively broad spectral characteristics of fluorescence excitation and emission spectra at room temperature result in large spectral overlaps if more than 3–4 fluorescent dyes are to be detected simultaneously.

Following our report of the first practical application of the SERS effect in organic analysis [8], the development of SERS-active solid substrates has been an area of active research in our laboratory. We have extensively investigated the SERS technology in the development of nanostructure-based SERS substrates [17,27]. These substrates consist of a plate having silver-coated dielectric nanoparticles or isolated dielectric nanospheres (30 nm diameter) coated with silver, producing a *nanowave* consisting an array of half nanoshells. The fabrication process involves depositing nanoparticles on a substrate and then coating the nanoparticle base with a 50–150 nm layer of silver via vacuum deposition.

The development of practical and sensitive devices for screening multiple genes related to medical diseases and infectious pathogens is critical for early diagnosis and effective treatments of many illnesses as well as for high-throughput screening for drug discovery. To achieve the required level of sensitivity and specificity, it is often necessary to use a detection method that is capable of simultaneously identifying and differentiating a large number of biological constituents in complex samples. One of the most unambiguous and well-known molecular recognition events is the hybridization of a nucleic acid to its complementary target. Thus, the hybridization of a nucleic acid probe to its DNA (or RNA) target can provide a very high degree of accuracy for identifying complementary nucleic acid sequences.

We first reported the development and application of SERS plasmonics gene probe technology for DNA detection [16]. Our laboratory has further developed SERS-based gene probes for selective detection of HIV DNA and the breast cancer gene BRCA1 using plasmonics substrates [18,23,40,41]. To demonstrate the SERS gene detection scheme, we used precoated SERS-active solid substrates, on which DNA probes were bound and directly used for hybridization. Several factors affect the effectiveness of this scheme for DNA hybridization and SERS gene detection. It is desirable that the unlabeled DNA fragment does not exhibit any significant SERS signal that might interfere with the label signal. It is important to use a label that is SERS-active and compatible with the hybridization platform: an ideal label should exhibit a strong SERS signal when used with the SERS-active substrate of interest. Furthermore, the label should retain its strong SERS signal after being attached to a DNA probe. The use of SERS gene technology was demonstrated for the detection of the HIV gene sequence [18].

24.3 Molecular Sentinel Nanoprobes

We have developed a unique *label-free* detection approach (i.e., the target does not need to be labeled) that incorporates the *SERS effect modulation* scheme associated with metallic nanoparticles and the DNA hairpin structure [26]. The SERS-based MS (SERS-MS) technique uses the stem-loop structure similar to that of molecular beacons (MBs) for DNA recognition. However, the detection scheme is fundamentally different from the MB detection scheme. MBs consist of a fluorescence molecule attached to the end of one arm and a quencher molecule attached to the end of the other arm of a hairpin structure. They are designed to report the presence of target DNA sequences (complementary to the hairpin DNA loop) using the principle of fluorescence resonance energy transfer (FRET) between the fluorescent molecule and the quenching molecule by generating a relatively strong fluorescent signal when complementary target sequences are hybridized, thus separating the quencher and the fluorophore. The fluorescence remains low (quenched) in the absence of a complementary sequence.

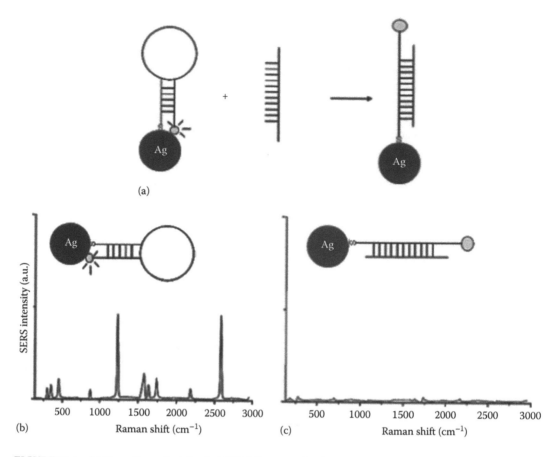

FIGURE 24.1 (a) Operating principle of a SERS-MS nanoprobe; (b) a SERS signal is observed when the MS probe is in the stem-loop conformation with the label close to the nanoparticle, inducing a strong SERS signal (closed state); and (c) following hybridization with a target probe, the stem opens and separates the label from the nanoparticle causing the SERS signal to decrease.

The detection strategy of our SERS-MS probes exploits the dependence of the plasmonics enhancement of the SERS signal intensity upon the distance between the metallic nanoparticle and the Raman label. In the SERS-MS system (Figure 24.1), MS nanoprobes having a Raman label at one end are immobilized onto a metallic nanoparticle via a thiol group attached on the other end to form a SERS-MS nanoprobe. The metal nanoparticle is used as a signal-enhancing platform for the SERS signal associated with the label. The Raman enhancement is determined by the plasmonics effect at the metal surface. Theoretical studies of the plasmonics effect have shown that the SERS enhancement, G factor, falls off as $G = [r/(r + d)]^{12}$ for a single analyte molecule located at a distance d from the surface of a metal particle of radius r [42]. Thus, the electromagnetic SERS enhancement decreases significantly with increasing distance, due to the decay of a dipole over the distance $(1/d)^3$ to the fourth power, thus resulting in a total intensity decay of $(1/d)^{12}$ of the SERS signal. Since the Raman enhancement field decreases significantly away from the surface, a molecule (e.g., the Raman label) must be located within a very close range (0–10 nm) of the metal nanoparticle surface in order to experience the enhanced local field. Under normal conditions (i.e., in the absence of target genes), the hairpin configuration has the Raman label in contact or close proximity (<1 nm) to the nanoparticle, thus resulting in a strong SERS effect and indicating that no significant event has occurred (Figure 24.1b). However, when complementary target DNA is recognized and hybridized to the nanoprobes, the SERS signal of these MSs becomes significantly quenched, providing a warning sign of target recognition and capture (Figure 24.1c). In other terms, a

SERS nanoprobe serves as an MS patrolling the sample solution with its SERS warning light *switched on* when no significant event occurs. Then, when a biotarget of interest is identified, the MS binds to it and extinguishes its light, thus providing a warning sign.

24.4 Biomedical Diagnostics Using Molecular Sentinel Nanoprobes

24.4.1 Detection of Infectious Diseases (HIV)

The SERS-MS concept was first demonstrated to detect specific DNA sequences of the HIV gene in a homogeneous assay [26]. The HIV MS nanoprobes were designed to have a stem sequence that allowed the formation of a stable hairpin structure at room temperature and incorporated a partial sequence for the HIV-1 isolate Fbr020, the reverse transcriptase (*pol*) gene. The HIV-1 MS nanoprobe (5'-HS-$(CH_2)_6$-CCTATCACAACAAAGAGCATACATAGGGATAGG-R6G) consisted of a 42-base DNA hairpin probe modified with rhodamine 6G on the 3' end and a thiol substituent at the 5' end; the 5' thiol was used for covalent coupling to the surface of silver nanoparticles. The underlined portions of the sequence represent the complementary arms of the MS designed to form a stem-loop structure. The silver colloidal nanoparticles were prepared according to the citrate reduction method, yielding homogenously sized colloids. A 115 bp sequence in the *gag* region of the HIV-1 genome was amplified by polymerase chain reaction (PCR), using forward and reverse primers in the gag region of the genome. Following gel analysis, the PCR products were hybridized to the SERS-MS nanoprobes. The reaction volume was 40 µL and hybridization was performed at 55°C for 1 min. SERS measurements were performed using a Renishaw InVia Raman system equipped with a 50 mW, helium–neon (HeNe) laser excitation source emitting at 632.8 nm.

Figure 24.2 illustrates the effectiveness of the MS technique in detecting the presence of a partial sequence of the HIV-1 *gag* gene in a homogenous solution. In normal conditions, that is, in the absence of the target DNA (Figure 24.2a), SERS-MS nanoprobes have a stable hairpin conformation, which allows a close proximity of the rhodamine 6G label to the surface of the silver nanoparticles (nanoenhancers). This situation produces an intense SERS signal from rhodamine 6G upon laser excitation due to the plasmonics effect. However, in the presence of a complementary HIV-1 target sequence (Figure 24.2b), the SERS HIV-1 MS nanoprobes bound to the target DNA cause a physical separation of the rhodamine 6G label from the surface of the silver nanoparticles. As a consequence, the plasmonics effect was greatly diminished, leading to a drastic decrease of the SERS signal. On the other hand, the presence of a noncomplementary target DNA template (Figure 24.2c) did not significantly affect the SERS signal, indicating that the hairpin loop structure of the SERS HIV-1 MS nanoprobes remained unchanged. The results of this experiment show that the MS technology can effectively detect target DNA sequences in a homogenous solution. Using the most intense SERS intensity band at 1521 cm^{-1} as the marker band for rhodamine 6G, we estimated the SERS quenching efficiency to be ~75% upon hybridization of the SER HIV-1 MS nanoprobe to the HIV-1 DNA target. This result demonstrates the specificity and selectivity of SERS-MS nanoprobes and their potential application in selective diagnostics.

24.4.2 Detection of Breast Cancer Single-Nucleotide Polymorphism

We have successfully demonstrated the specificity and selectivity of the nanoprobes to detect a single-nucleotide polymorphism (SNP) in the breast cancer BRCA1 gene in a homogenous solution at room temperature [43]. The SERS-MS probe used in this work incorporated a partial sequence for the breast cancer gene, BRAC1 (supplementary material). The 42 bp BRCA1 SERS-MS probe (5'-HS-$(CH_2)_6$-GATCGC AGGTCTCCTTTTATGCTTTAATTTATTTGTGCGATC-TMR) was modified with 5-carboxytetra-methylrhodamine succinimidyl ester (TMR) on the 3' end and a thiol substituent at the 5' end, which

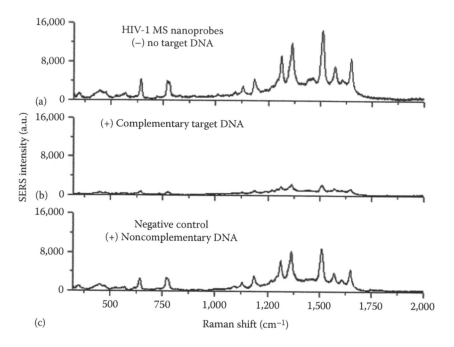

FIGURE 24.2 SERS spectra of HIV-1 SERS-MS nanoprobe with no target DNA sequence (a) and in the presence of a noncomplementary DNA target sequence (negative control: c) and a complementary HIV-1 DNA target (positive diagnostic: b). (Adapted from Wabuyele, M.B. and Vo-Dinh, T., *Anal. Chem.*, 77(23), 7810, 2005.)

could then be used for covalent coupling to the surface of silver nanoparticles. The underlined sequence represents the complementary arms of the nanoprobes. We further examined the specificity of the plasmonics nanoprobes to detect a oligonucleotide sequence having a single-base variation on the BRCA1 gene that leads to an SNP at codon 504. Figure 24.3 shows a SERS spectrum obtained from a hybridization process of the BRCA1-SERS-MS nanoprobe with a noncomplementary target DNA (Figure 24.3a), mutant (MT^{GCA}) DNA target having a single-base mismatch (Figure 24.3b), and wild-type (WT^{GCG}) DNA target sequence (Figure 24.3c). The specificity of the plasmonics nanoprobe is clearly demonstrated again by significant quenching of the SERS signal below the threshold level upon hybridization with the wild-type DNA target. Therefore, using unique DNA targets, the plasmonics nanoprobes were able to positively recognize perfectly matched target and the mismatched sequences with high specificity.

24.4.3 Detection of mRNA Biomarkers for Host Response to Infectious Disease

There is a strong need to develop diagnostic technologies that can be used at the point of care to detect infectious diseases. A promising approach involves detection of the host response to various pathogens by evaluating changes in gene expression in peripheral blood samples, induced in response to infection [44,45]. We have developed an MS nanoprobe to detect the human radical *S*-adenosyl methionine domain containing 2 (RSAD2) RNA target as a model system for method demonstration [46]. The human RSAD2 gene has recently emerged as a novel host-response biomarker for diagnosis of respiratory infections. By evaluating changes in host gene expression profiles in response to viral infections, Zaas et al. have developed a robust blood mRNA expression signature that distinguishes individuals with symptomatic acute respiratory infections (ARIs) from uninfected individuals with over 95% accuracy [3]. This *acute respiratory viral* biosignature encompasses 30 transcripts of genes known to be related to the

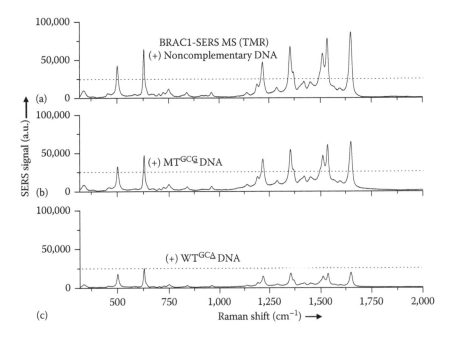

FIGURE 24.3 SNP detection using a BRCA1 SERS-MS nanoprobe. Plasmonics nanoprobe in the presence of a noncomplementary DNA target (a), mutant DNA target containing a single-base mutation of (A/G) at site N47of the BRCA1 gene (b), and wild-type DNA target (c). The dotted line indicates the set threshold level of the SERS signal from the nanoprobe.

host immune response to viral infections. In particular, RSAD2 was the most highly expressed gene in symptomatic individuals from all three human viral challenge studies with live rhinovirus, respiratory syncytial virus, and influenza A.

A small, portable Raman spectrometer (Advantage 633, DeltaNu) for SERS measurements and purified RNA samples as the specimen were used to demonstrate the potential of using the MS nanoprobes to detect RSAD2 RNA targets for point-of-care applications. As shown in Figure 24.4, the SERS signal of the RSAD2-MS nanoprobes (solid lines) was reduced in the presence of 0.5 μg (spectrum b) and 1 μg total RNA (spectrum c) compared to the blank sample (spectrum a), and the SERS signal was further reduced in the presence of more RNA sample. To confirm the detection specificity, we have utilized a Cy5.5-labeled MS nanoprobe targeted to the gene sequences of the influenza A virus (H1N1) nucleocapsid protein (NP) as the control nanoprobe, which is not expected to detect complementary targets in the normal human RNA samples. The results from the control experiments (dotted lines shown in Figure 24.4) show that the intensity of the major Cy5.5 SERS peaks at 1339, 1461, and 1625 cm^{-1} remains high in the presence of normal human total RNAs, confirming the detection specificity using the RSAD2-MS nanoprobes. Note that the portable Raman spectrometer, which is equipped with a low-dispersion grating, only provides a spectral resolution of 10 cm^{-1}. However, the result shown in Figure 24.4 demonstrates that the spectral resolution provided by the portable spectrometer is adequate for diagnostics. This work also demonstrates that the MS technique can be easily integrated with a small portable Raman spectrometer for point-of-care applications, and the MS technique has the potential for RNA biotarget detection.

24.4.4 Multiplex Detection of Cancer Biomarkers

The capability for multiplex detection of cancer biomarkers has been demonstrated using the SERS-MS technique [28]. The study was performed with two MS nanoprobes, ERBB2-MS and KI-67-MS, which were designed and separately prepared to detect two breast cancer biomarkers, *erbB-2* and *ki-67* genes, respectively.

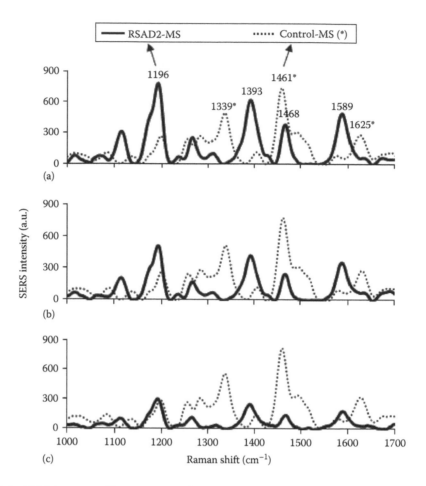

FIGURE 24.4 SERS detection of RSAD2 RNA in human lymph node total RNA using RSAD2-MS nanoprobes (solid lines) and control-MS nanoprobes (dotted lines). (a) Blank (no RNA present), (b) in the presence of 0.5 µg total RNA, and (c) in the presence of 1 µg total RNA.

TheERBB2-MS(5′-SH-(CH$_2$)$_6$)-CGCCATCCACCCCCAAGACCACGACCAGCAGAATATGGCG-Cy3-3′) and KI-67-MS (5′-SH-(CH$_2$)$_6$)-GCGTATTCTGCACACCTCTTGACACTCCGATACGC-TAMRA-3′) nanoprobes were tagged with different Raman labels, Cy3 and 5-carboxytetramethylrhodamine (TAMRA), respectively. Figure 24.5a shows the resulted SERS spectrum of the MS nanoprobe mixture. Note that all major Raman peaks of the two Raman labels tagged on the ERBB2-MS and KI-67-MS nanoprobes are clearly distinguishable from one another due to the narrow SERS peaks even though the fluorescence spectra of Cy3 and TAMRA strongly overlap. This feature underlines the advantage of the SERS-based MS nanoprobes over fluorescence-based assays for multiplex detection.

The specificity of the MS nanoprobe mixture was then demonstrated in the presence of target DNA sequences. Figure 24.5b shows that all SERS signals of the MS nanoprobe mixture were significantly quenched in the presence of both targets (0.5 µM for each target). The decreased SERS intensity indicates that both MS nanoprobes hybridized with their complementary DNA targets. The specificity and selectivity of the MS nanoprobe mixture were then evaluated in the presence of individual complementary DNA target (i.e., only one of the two complementary targets). The results shown in Figure 24.5c and d indicate that only the SERS signal associated with the MS nanoprobes complementary to the present target was significantly quenched (indicated by arrows). These results demonstrate that the MS nanoprobe technique can provide a useful tool for multiplex DNA detection in a homogeneous solution for

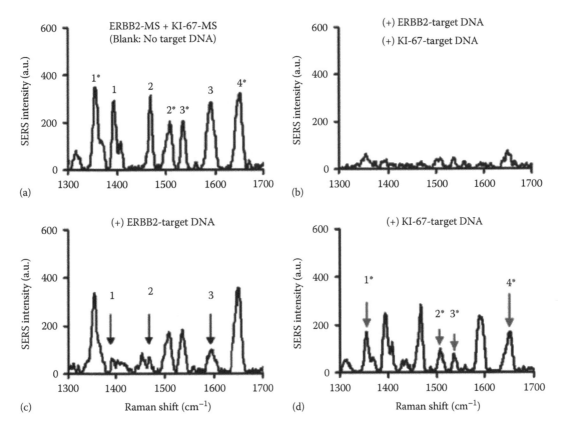

FIGURE 24.5 SERS spectra of the MS nanoprobe mixture (ERBB2-MS + KI-67-MS) in the presence or absence of target DNA. The major Raman bands from ERBB2-MS are marked with numbers without (*) sign, and the major Raman bands from KI-67-MS are marked with numbers with (*) sign. (a) Blank (in the absence of any target DNA), (b) in the presence of two target DNA complementary to both MS nanoprobes, (c) in the presence of single target DNA complementary to the ERBB2-MS nanoprobes, and (d) in the presence of single target DNA complementary to the KI-67-MS nanoprobes. The arrow signs illustrate the decreased SERS intensity of the major Raman bands in the presence of corresponding target DNA. (Adapted from Wang, H. N. and Vo-Dinh, T., *Nanotechnology*, 20(6), 065101-1, 2009.)

medical diagnostics and high-throughput bioassays. In addition, SERS measurements are performed immediately following the hybridization reactions without washing steps, which greatly simplifies the assay procedures.

The multiplex capability of Raman and SERS techniques to detect a large number of molecular processes simultaneously is an important feature in molecular diagnostics as well as in systems biology research. Due to the narrow bandwidths of Raman bands, the multiplex capability of the SERS-MS probe is excellent in comparison to other spectroscopic alternatives. This multiplex advantage unique to Raman/SERS is extremely important for ultra-high-throughput analyses where multiple gene targets need to be screened in a highly parallel multiplex modality.

24.4.5 Molecular Sentinel on a Chip

The label-free MS biosensing concept was further developed and adapted for use on substrate platforms [47,48]. A unique platform that is simple and inexpensive to fabricate is the nanowave chip [47]. The substrate, which consists of a metal film over close-packed nanospheres, referred to as the *nanowave* due to its resemblance to a periodic waveform, was first introduced in 1984 in our laboratory and used as a SERS-active substrate for the sensitive and reproducible detection of analytes [9]. The nanowave platform, also referred to as metal film on nanospheres (MFON) and later used by other researchers [49], belongs to a subset of SERS-active substrates involving metal film coated on various nanoparticles, such as titanium dioxide and alumina [17]. Numerical simulations were used to mimic the nanowave geometry in 3D space and to confirm its experimentally observed plasmonics behavior [50]. The study shows that an in-plane polarized incident plane wave generates strong enhancements in the interstitial

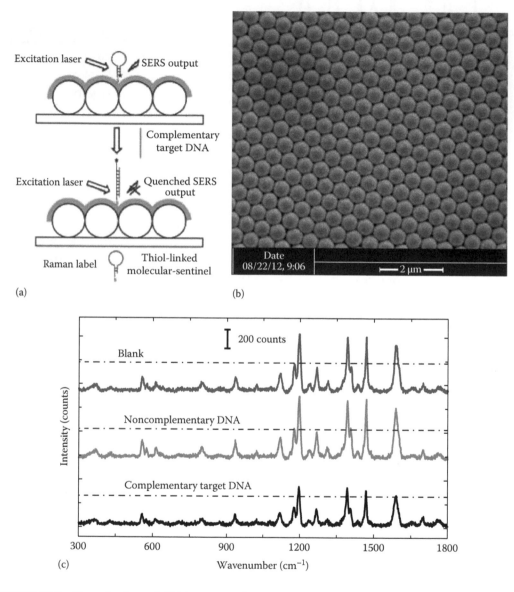

FIGURE 24.6 **(See color insert.)** SERS-MS on a chip. (a) Application of the label-free MS detection scheme on nanowave chip and (b) scanning electron microscopy (SEM) image of nanowave chip substrate. In this substrate, a monolayer of 520 nm diameter polystyrene beads is covered by a 200 nm thick Au layer. (c) SERS spectra from blank sample, noncomplementary DNA sample, or complementary target DNA sample 2 h after delivery on MS-functionalized MFON substrates. The dashed lines mark the blank sample's SERS intensity. The dash–dot line marks the complementary target DNA sample's SERS intensity. (Adapted from Ngo, H. T. et al., *Anal. Chem.*, 85, 6378, 2013.)

spaces between individual metal-coated nanospheres, thus producing closely packed arrays of hot spots, which produce the strong SERS effect of the nanowave substrate structures.

The sensing process involves a single hybridization step between the DNA target sequence and the complementary MS probes on the nanowave chip without requiring secondary hybridization or posthybridization washing, thus resulting in shorter assay time and less reagent usage. The detection scheme is shown in Figure 24.6a. First, the nanowave is functionalized with MS probes. The complementary arms of the MS hairpin hybridize into 6-base-pair stem sequence with melting temperature (T_m) $\approx 46°C$, allowing the formation of a stable hairpin structure at room temperature in the absence of a complementary DNA target. The 3′ end of MS hairpin probes was modified with Cy3 Raman label. The conjugation of MS probe onto the nanowave's gold surface was achieved by using an alkyl thiol substituent at the 5′ end. To effectively separate the Raman labels from the nanowave's metallic surface upon hybridization to the complementary target DNA, the length of the DNA hairpins was designed to be 35 nucleotides. In the absence of a target sequence, the MS's stem loop is in the closed state. At this state, the Cy3 dye is in close proximity to the gold surface (<1 nm), inducing a strong SERS signal. When the complementary target DNA sequence is added, it hybridizes with the MS, forcing the stem loop to open. At this state, the Cy3 dye is physically separated from the Au surface. The opened stem loop results in a quenched SERS signal. The fabrication of the nanowave chip was further refined to produce practical, inexpensive, and reproducible substrates consisting a thin shell of metal (e.g., gold) over close-packed arrays of nanospheres [47]. The procedure involves self-assembly of polystyrene beads on a water–air interface, which can produce a large area of close-packed nanosphere array for the nanowave chip (Figure 24.6b).

The usefulness and potential application of the biosensor for medical diagnostics is demonstrated by detecting the human RSAD2 (Viperin) gene, a common host biomarker of inflammation [47]. This gene is expressed as a response of the host immune system to the infection of various viruses such as the human cytomegalovirus (HCMV), influenza virus, hepatitis C virus (HCV), and retroviruses. Figure 24.6c shows the SERS spectra taken 2 h after blank, 5 µM noncomplementary DNA sequence, and 5 µM complementary target DNA sequence were added on MS-functionalized MFON. The 1197 cm^{-1} peak of the Cy3 dyes tagged at the 3′-end of the MS was used to compare the SERS intensities of the three samples. The decrease in SERS intensity after complementary target DNA addition indicates that the complementary target sequences hybridized with the MS probes and opened the MS's stem loops.

The MS–nanowave-based DNA biosensor is relatively easy to fabricate at low cost and can specifically detect a complementary target DNA, such as the RSAD2 inflammation biomarker gene. The results indicate the possibility for quantitative analysis. Because the target does not need labeling and secondary hybridization or posthybridization washing is not required, the DNA biosensor is simple to use, has rapid assay time and low reagent usage, and is cost effective.

24.5 Conclusion

In conclusion, the SERS-MS gene probes could offer a unique combination of performance capabilities and analytical features of merit for use in biosensing. The SERS-MS gene probes are safer than radioactive labels and have excellent specificity due to the inherent specificity of Raman spectroscopy. With Raman scattering, multiple probes can be much more easily selected with minimum spectral overlap. This *label-multiplex* advantage can permit analysis of multiple probes simultaneously, resulting in much more rapid DNA detection, gene mapping, and improved ultra-high-throughput screening of small molecules for drug discovery. The SERS-MS probes could be used for a wide variety of applications in areas where nucleic acid identification is involved. The SERS-MS approach involving homogeneous assays greatly simplifies experimental procedures and could potentially be used in assays that require rapid, high-volume identification of genomic materials. Due to these unique properties, the SERS-MS approach could contribute to the development of the next generation of DNA/RNA diagnostic tools for molecular screening and molecular imaging.

References

1. Gersten, J.; Nitzan, A., *J. Chem. Phys.* 1980, *73*, 3023.
2. Schatz, G. C., *Acc. Chem. Res.* 1984, *17*(10), 370–376.
3. Otto, A., *Surf. Sci.* 1978, *75*, L392.
4. Zeeman, E. J.; Schatz, G. C., *J. Phys. Chem.* 1987, *91*, 634.
5. Fleischmann, M.; Hendra, P. J.; McQuillan A. J., *Chem. Phys. Lett.* 1974, *26*(2), 163–166.
6. Jeanmaire, D. L.; Vanduyne, R. P., *J. Electroanal. Chem.* 1977, *84*(1), 1–20.
7. Albrecht, M. G.; Creighton, J. A., *J. Am. Chem. Soc.* 1977, *99*(15), 5215–5217.
8. Vo-Dinh, T.; Hiromoto, M. Y. K.; Begun, G. M.; Moody, R. L., *Anal. Chem.* 1984, *56*(9), 1667–1670.
9. Vo-Dinh, T.; Uziel, M.; Morrison, A. L., *Appl. Spectrosc.* 1987, 41(4), 605–610.
10. Enlow, P. D.; Buncick, M.; Warmack, R. J.; Vo-Dinh, T., *Anal. Chem.* 1986, *58*, 1719–1724.
11. Moody, R. L.; Vo-Dinh, T.; Fletcher, W. H., *Appl. Spectrosc.* 1987, *41*(6), 966–970.
12. Alak, A. M.; Vo-Dinh, T., *Anal. Chem.* 1987, *59*(17), 2149–2153.
13. Vo-Dinh, T.; Alak, A.; Moody, R. L., *Spectrochim. Acta Part B-Atomic Spectrosc.* 1988, *43*(4–5), 605–615.
14. Bello, J. M.; Stokes, D. L.; Vo-Dinh, T., *Appl. Spectrosc.* 1989, *43*(8), 1325–1330.
15. Bello, J. M.; Stokes, D. L.; Vo-dinh, T., *Anal. Chem.* 1989, *61*(15), 1779–1783.
16. Vo-Dinh, T.; Houck, K.; Stokes, D. L., *Anal. Chem.* 1994, *66*(20), 3379–3383.
17. Vo-Dinh, T., *TrAC-Trends Anal. Chem.* 1998, *17*(8–9), 557–582.
18. Isola, N. R.; Stokes, D. L.; Vo-Dinh, T., *Anal. Chem.* 1998, *70*(7), 1352–1356.
19. Zeisel, D.; Deckert, V.; Zenobi, R.; Vo-Dinh, T., *Chem. Phys. Lett.* 1998, *283*(5–6), 381–385.
20. Dhawan, A.; Muth, J. F.; Leonard, D. N.; Gerhold, M. D.; Gleeson, J.; Vo-Dinh, T.; Russell, P. E. *J. Vacuum Sci. Technol.* 2008, *26*, 2168–2173.
21. Khoury, C. G.; Vo-Dinh, T., *J. Phys. Chem. C* 2008, *112*(48), 18849–18859.
22. Stokes, D. L.; Chi, Z. H.; Vo-Dinh, T., *Appl. Spectrosc.* 2004, *58*(3), 292–298.
23. Vo-Dinh, T.; Allain, L. R.; Stokes, D. L., *J. Raman Spectrosc.* 2002, *33*(7), 511–516.
24. Vo-Dinh, T.; Yan, F.; Wabuyele, M. B., *J. Raman Spectrosc.* 2005, *36*(6–7), 640–647.
25. Wabuyele, M. B.; Yan, F.; Griffin, G. D.; Vo-Dinh, T., *Rev. Sci. Instrum.* 2005, *76*(6), 063710-1–063710-7.
26. Wabuyele, M. B.; Vo-Dinh, T., *Anal. Chem.* 2005, *77*(23), 7810–7815.
27. Vo-Dinh, T.; Yan, F. Gene detection and multi-spectral imaging using SERS nanoprobes and nanostructures. In *Nanotechnology in Biology and Medicine: Methods, Devices, and Applications*, Vo-Dinh, T., Ed., CRC Press: Boca Raton, FL, 2007.
28. Wang, H. N.; Vo-Dinh, T., *Nanotechnology* 2009, *20*(6), 065101-1–065101-6.
29. Moskovits, M., *Rev. Mod. Phys.* 1985, *57*(3), 783–826.
30. Wokaun, A.; Gordon, J. P; Liao, P. F., *Phys. Rev. Lett.* 1982, *48*(14), 957–960.
31. Kerker, M., *Acc. Chem. Res.* 1984, *17*(8), 271–277.
32. Chang, R. K., Furtak, T.E., Eds. *Surface Enhanced Raman Scattering.* Plenum Press: New York, 1982.
33. Pockrand, I. *Surface Enhanced Raman Vibrational Studies at Solid/Gas Interfaces.* Springer: Berlin, Germany, 1984.
34. Scaffidi, J. P.; Gregas, M. K.; Seewaldt, V.; Vo-Dinh, T., *Anal. Bioanal. Chem.* 2009, *393*(4), 1135–1141.
35. Moskovits, M., *J. Raman Spectrosc.* 2005, *36*(6–7), 485–496.
36. Doering, W. E.; Nie, S. M., *J. Phys. Chem. B* 2002, *106*(2), 311–317.
37. Cao, Y. W. C.; Jin, R. C.; Mirkin, C. A., *Science* 2002, *297*(5586), 1536–1540.
38. Otto, A.; Mrozek, I.; Grabhorn, H.; Akemann, W., *J. Phys.-Condens. Matter* 1992, *4*(5), 1143–1212.
39. Kneipp, K., Moskovits, M., Kneipp H., Eds. *Surface-Enhanced Raman Scattering: Physics and Applications.* Springer: New York, 2006.
40. Allain, L.R.; Vo-Dinh, T., *Anal. Chim. Acta* 2002, *469*, 149.
41. Culha, M.; Stokes, D.; Allain, L. R.; Vo-Dinh, T., *Anal. Chem.* 2003, *75*, 6196.
42. Kneipp, K.; Wang, Y.; Kneipp, H.; Perelman, L. T.; Itzkan, I.; Dasari, R.; Feld, M. S., *Phys. Rev. Lett.* 1997, *78*, 1667–1670.

43. Wabuyele, M.B.; Yan, F.; Vo-Dinh, T., *Anal. Bioanal. Chem.* 2010, *398*, 729–736.

44. Zaas, A. K.; Chen, M.; Varkey, J.; Veldman, T.; Hero III, A. O.; Lucas, J.; Huang, Y. et al., *Cell Host Microbe* 2009, *6*, 207–217.

45. Zaas, A. K.; Aziz, H.; Lucas, J.; Perfect, J. R.; Ginsburg, G.S., *Sci. Transl. Med.* 2010, *2*, 21ra17.

46. Wang, H. N.; Fales, A. M.; Zaas, A. K.; Woods, C. W.; Burke, T.; Ginsburg, G. S.; Vo-Dinh, T., *Anal. Chim. Acta* 2013, *786*, 153–158.

47. Ngo, H. T.; Wang, H. N.; Fales, A. M.; Vo-Dinh, T., *Anal. Chem.* 2013, *85*, 6378–6383.

48. Wang, H. N.; Dhawan, A.; Du, Y.; Batchelor, D.; Leonard, D. N.; Misra, V.; Vo-Dinh, T., *Phys. Chem. Chem. Phys.* 2013, *15*, 6008–6015.

49. Dick, L. A.; McFarland, A. D.; Haynes, C. L.; Van Duyne, R. P., *J. Phys. Chem. B* 2001, *106*, 853–860.

50. Khoury, C. G.; Vo-Dinh, T., *J. Phys. Chem. C* 2012, *116*, 7534–7545.

25

Plasmonic Coupling Interference Nanoprobes for Gene Diagnostics

Hsin-Neng Wang
Duke University

Stephen J. Norton
Duke University

Tuan Vo-Dinh
Duke University

25.1 Introduction

Plasmons are quanta associated with longitudinal waves propagating in matter through the collective motion of large numbers of electrons. The origin of the enormous Raman enhancement is produced by at least two main mechanisms that contribute to the surface-enhanced Raman scattering (SERS) effect: (1) an electromagnetic (EM) effect occurring near metal surface structures associated with large local fields caused by EM resonances, often referred to as *surface plasmons*, and (2) a chemical effect involving a scattering process associated with chemical interactions between the molecule and the metal surface. When a metallic nanostructured surface is irradiated by an incident EM field (e.g., a laser beam), conduction electrons are displaced into frequency oscillations equal to those of the incident light. These oscillating electrons, called *surface plasmons*, produce a secondary electric field, which adds to the incident field. These fields can be quite large (10^6- to 10^7-fold, or even up to 10^{15}-fold enhancement at *hot spots*). When these oscillating electrons become spatially confined, as is the case for isolated metallic nanospheres or otherwise roughened metallic surfaces (nanostructures), there is a characteristic frequency (the plasmon frequency) at which there is a resonant response of the collective oscillations to the incident field. This condition yields intense localized fields that can interact with molecules in contact with or near the metal surface [1–4]. In an effect analogous to a *lightning rod* effect, secondary fields can become concentrated at high curvature points on the roughened metal surface.

Following the observation of the SERS effect [5–7], our laboratory was the first to demonstrate the general applicability of the SERS effect for trace analysis using solid substrates having silver-coated nanospheres [8]. During the following two decades, our laboratory has extensively investigated the SERS technology and developed a wide variety of plasmonics-active SERS platforms for chemical sensing [8,9–15], biosensing [16–27], and biomedical diagnostics [26–28].

In our earlier works, the EM field enhancement, which results in SERS, has mostly been created using random structures [10–11,16,17]. It has been suggested that the EM field is particularly strong in the interstitial space between the particles [29]. It is believed that the anomalously strong Raman signal originates from *hot spots*, that is, regions where clusters of several closely spaced nanoparticles (NPs) are concentrated in a small volume. This effect, also referred to as interparticle coupling or plasmonic coupling in a network of NPs, can provide a further enhancement effect besides the enhancement from individual particles. It has been reported that SERS enhancement from NP aggregates or nanonetworks can reach 10^{11}–10^{15}.

25.2 Background on Plasmonic Coupling

25.2.1 Plasmonics and SERS

The enhancement mechanism for SERS comes from intense localized fields arising from surface plasmon resonance in metallic nanostructures with sizes on the order of tens of nanometers and from chemical effects at the metal surface. It has been widely accepted that EM enhancement provides the main contribution to the enormous enhancement that greatly increases the intrinsically weak normal Raman scattering cross section. Theoretical studies of EM effects have shown that enhanced EM fields are confined within only a tiny region near surface of particles, and the SERS enhancement (G) falls off as $G = [r/(r + d)]^{12}$ for a single molecule located a distance d from the surface of a spherical metal NP of radius r. Thus, the SERS enhancement strongly decreases with increased distance between the analyte and metal surface.

It has been observed that the EM field is particularly strong in the interstitial space between the particles in addition to the EM enhancement contributed from individual particles. It is believed that the anomalously strong Raman signal originates from *hot spots*, that is, regions where clusters of several closely spaced NPs are concentrated in a small volume. The high-intensity SERS then originates from the mutual enhancement of surface plasmon local electric fields of several NPs that determine the dipole moment of a molecule trapped in a gap between metal surfaces. This effect, also referred to as interparticle coupling or plasmonic coupling in a network of NPs, can provide a further enhancement effect besides the EM enhancement from individual particles. It has been reported that SERS enhancement from NP aggregates or nanonetworks can reach 10^{11}–10^{15}, allowing single-molecule detection and making SERS highly competitive with fluorescence-based assays. The dimer is the prototypical example of a *hot spot* since it exhibits the key properties of EM enhancement in the gap between the particles and its dependence on the gap size relative to the particle size. For this reason, the enhancement characteristics of NP dimers and the influence of plasmon resonances on this enhancement have been the subject of many studies. We have used a semianalytical method based on a multipole expansion (ME) to compute the electric field in the gaps between two spheres, two spherical shells, and two spheroids over a range of frequencies to indicate the occurrence of very large field enhancements in the gaps between NPs [30–33]. A pair of silver nanospheres with a 2% gap has been shown to produce an electric field enhancement in the gap of over 700 at the peak of the plasmon resonance [31]. In SERS measurements, the total signal is approximately proportional to the fourth power of the electric field magnitude, giving a total SERS enhancement of over 4×10^{10}.

25.2.2 Plasmonic Coupling in Dimer Nanosystems

25.2.2.1 Analytical Calculations and Numerical Simulation of the Electromagnetic Enhancement of Plasmonic Nanostructures

In our previous work, computation of the electric field in the gaps between two spheres and between two spheroids over a range of frequencies also indicates the occurrence of very large field enhancements [31]. While very large enhancement for a single hot spot can be achieved in such structures, the presence and location of such hot spots is not predictable and the density of the hot spots tends to be very low. It is widely believed that SERS hot spots are created at locations where the EM field is strongly concentrated by the metallic nanostructures or between nanostructures. Creating a high density of such hot spots calls for a systematic study in periodic nanostructures made out of metals. As the simplest example, we investigate the EM field at the hot spot between two NPs (solid nanospheres, nanospheroids, or nanoshells). In the following section, we compare calculations of the plasmonic enhancement produced by two NPs in close proximity (a dimer). The dimer is the prototypical example of a *hot spot* since it exhibits the key properties of EM enhancement in the gap between the particles and its dependence on the gap size relative to the particle size. For this reason, the enhancement characteristics of the dimer and the influence of plasmon resonances on this enhancement have been the subject of many studies. For similar reasons, more complex systems comprising particle chains and clusters have also been examined.

25.2.2.2 Dimer Calculations Using Finite-Element and Multipole Methods

In our previous work, we considered two types of dimers, one comprised of two solid nanospheres and the other of two nanoshells [32,33]. Nanoshells have been previously investigated and developed for medical applications [34]. The maximum electric field enhancement in the gap between the particles occurs when the electric field of the incident light is polarized along the dimer axis. We compared the calculations of the electric field at a point in the gap midway between the two particles (solid sphere or shell) using two different numerical methods. The first calculation was performed using the Finite Element Method (FEM) based commercial software package COMSOL Multiphysics and the second was a semianalytical solution based on an ME of the fields. In the latter approach, we employed the quasistatic approximation, which significantly simplifies the ME analysis but is known to give accurate results when the particle size is about a tenth of a wavelength or less. Comparing the FEM results to those of the ME method demonstrates that in this size range, the quasistatic assumption is an excellent approximation. The quasistatic approximation also has the virtue of being computationally very fast as well as relatively simple to program.

In the ME formulation, the scattered field from each particle is expressed as an ME in a coordinate system centered on that particle. If the particles are spherical, the multipole coefficients can then be derived using a translational formula based on the spherical harmonic addition theorem. If the particles are spheroidal, the latter approach using a translational formula no longer works; however, the spheroidal case can be treated by integrating over the surface of the spheroid [31]. Further details of the ME method can be found in Refs. [31–33].

25.2.2.3 Comparison between FEM and ME Methods

Field calculations were performed using both the FEM and ME methods for two types of dimers: a pair of solid nanospheres [32] and a pair of nanoshells [33]. In a typical analysis of a dimer composed of spherical shells, the outer diameter of the shell was assumed to be 20 nm with a particle gap of 5 nm. In one calculation, the magnitude of the electric field in the gap midway between the particles was computed over a wavelength range from 300 to 800 nm. Three particular cases were considered: a dimer whose particles are nanoshells with a shell thickness of 15% and 35% of the outer shell radius and a dimer whose particles are solid spheres. The spectra of these three cases are shown in Figure 25.1 for silver (the *solid* dimer is labeled as 100%). In the figure, the "+" signs indicate the calculated FEM values and the solid curves are the ME results. Figure 25.2 displays the field magnitude of a 2-D slice through the 15% dimer [30].

The results of these simulations, which demonstrate that the agreement between the COMSOL and ME calculations is within 5% across all geometries with respect to amplitude, wavelength offset, and plasmon resonance bandwidth, showed very strong plasmonic coupling enhancement in the gap regions between the NPs. Finite difference time domain (FDTD) calculations were also performed using the commercial software package Fullwave (version 6.0) for simulating the response of solid spherical

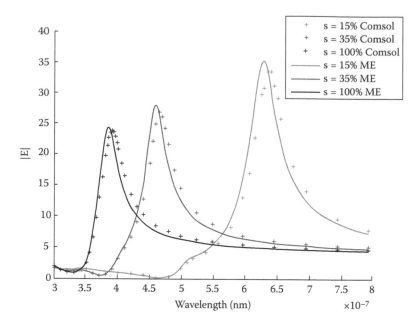

FIGURE 25.1   **(See color insert.)** Electric field magnitude in the particle gap versus wavelength. Solid curves are calculations using the ME method and symbols indicate FEM calculations. (From Khoury, C.G. et al., *ACS Nano*, 3(9), 2776, 2009.)

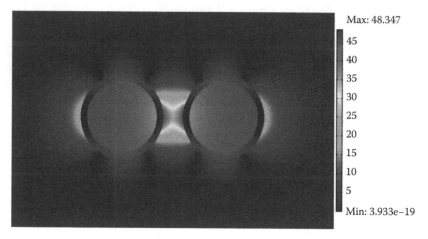

FIGURE 25.2   **(See color insert.)** Electric field magnitude of a 2-D x–z slice through the dimer with 15% shell thickness at a wavelength of 640 nm. (From Vo-Dinh, T. et al., *J. Phys. Chem. C*, 114(16), 7480, 2010.)

dimers [32]. These calculations were also compared to the multipole method with very favorable agreement for particles about a tenth of a wavelength in size or smaller.

25.3 Biomedical Application: PCI Nanoprobes

In this section, we describe a new approach referred to as plasmonic coupling interference (PCI) for nucleic acid (DNA or RNA) detection using SERS [35]. The PCI method described here combines this plasmonic coupling phenomenon with the nucleic acid hybridization process, leading to the development of a label-free detection approach for target DNA/RNA molecules. In typical SERS experiments, metallic NPs are aggregated in a high-ionic-strength salt solution in order to induce a strong plasmonic coupling effect. However, this aggregation process is usually uncontrollable, thus seriously affecting the reproducibility of SERS measurements. To practically control the separation distance between particles, the PCI approach employs Raman-labeled oligonucleotides as spacers and linkers to assemble NPs into an aggregate in a controllable manner [36–40]. This novel technique then utilizes specific nucleic acid sequences of interest (target molecules) to induce interference in the plasmonic coupling effect. As a result, the reduction in SERS intensity can be used as a parameter for a new biosensing modality.

25.3.1 Operating Principle of the PCI Approach for Nucleic Acid Detection

Figure 25.3 illustrates the detection scheme of the PCI nanoprobes using plasmonic NPs. In this approach, oligonucleotide-functionalized NPs are designed and prepared as the capture probes (capture-NPs) and complementary Raman-labeled reporter probes (reporter-NPs) (as shown in Figure 25.3a) in order to assemble NPs into a 3-D nanonetwork or nanoaggregate. To induce the plasmonic coupling effect, capture-NPs and reporter-NPs are mixed in a hybridization buffer solution in order to form nanoaggregates mediated by capture/reporter hybrid duplexes having the Raman labels located between adjacent NPs (Figure 25.3b). In this configuration, the dye molecules located between two NPs (i.e., in a hot spot area) experience a strong plasmonic coupling effect, leading to an intense SERS signal upon laser excitation. Figure 25.3c depicts the mechanism for the detection of particular nucleic acid sequences (DNA or RNA). In this PCI approach, target strands are used as competitors of the reporter probes in a competitive binding process as they share the same sequences. Therefore, when target sequences are introduced to the sample, the formation of nanoaggregates is prevented. In this case, the plasmonic coupling effect is interfered by the target molecules, thereby significantly reducing the SERS signal.

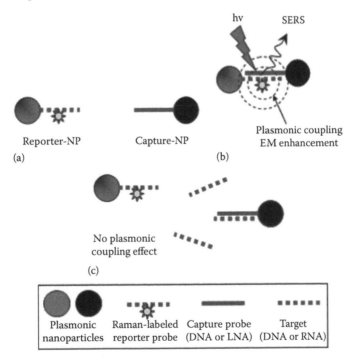

FIGURE 25.3 Scheme showing the operating principle of the PCI approach for nucleic acid detection. (a) Plasmonic nanoparticles are functionalized with thiolated capture probes (capture-NPs) or complementary Raman-labeled reporter probes (reporter-NPs). (b) Plasmonic coupling between adjacent NPs is induced by the formation of capture/reporter duplexes. (c) Plasmonic coupling is interfered by the formation of capture/target duplexes in the presence of target molecules.

25.3.2 Proof-of-Concept Demonstration

In this approach, we first aim to assemble NPs into a nanonetwork with shortest separation distance in order to induce a strongest plasmonic coupling and a maximum SERS enhancement of a Raman label. Previous studies have shown that NPs can be coupled using DNA oligonucleotides with over 8 bases [36]. However, due to the thermal instability of short DNA–DNA duplexes, it is difficult to use DNA oligonucleotides shorter than 8 bases for assembling NPs into a nanonetwork. To overcome this barrier, short locked nucleic acid (LNA) oligonucleotides with 7 bases were utilized to couple NPs in a separation distance between 2 and 3 nm. LNAs are modified nucleotides in which the 2′ oxygen and 4′ carbon atoms are linked through an extra methylene bridge. When mixing with DNA or RNA nucleotides, the hybridization properties (i.e., melting temperature) can be significantly increased. It has also been previously reported that LNAs can offer a high salt and thermal stability for coupling NPs [41,42]. Nonetheless, longer DNA oligonucleotides can also be used for other applications.

For proof-of-concept demonstration, the capture-NPs and reporter-NPs were synthesized by conjugating silver NPs with thiolated LNAs and complementary Cy3-labeled DNA oligonucleotides with the sequence of 5′-GGGCGGG-3′ and 3′-CCCGCCC-5′, respectively. A Raman dye, Cy3, was used as the signal reporter, which was internally attached to the guanine (G-base) in the middle of the reporter oligonucleotide. To induce plasmonic coupling effect, the capture-NPs and reporter-NPs were mixed in a volume ratio of 1:1 in a 10-mM Tris-HCl buffer solution (pH 8.0) containing 50 mM NaCl and 2.5 mM MgCl$_2$ in order to form LNA–DNA duplexes.

Figure 25.4a shows the increased SERS intensity of the Cy3 Raman peaks in the presence of both capture-NPs and reporter-NPs (upper spectrum) as compared to the SERS intensity in the presence of target DNA strands (lower spectrum). The enhanced SERS signal indicates that the plasmonic coupling was induced by the formation of a nanonetwork. The detection of target DNA was performed by mixing

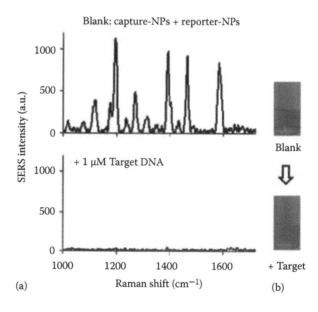

FIGURE 25.4 **(See color insert.)** (a) SERS spectra of the Cy3 Raman peaks in the presence or absence of target DNA strands. Upper spectrum: in the presence of both capture-NPs and reporter-NPs (blank: without target DNA present). Lower spectrum: in the presence of 1 μM target DNA strands in the mixture of capture-NPs and reporter-NPs. (b) Photographs of sample solutions in the presence or absence of target DNA. (From Wang, H.N. and Vo-Dinh, T., *Small*, 7(21), 3067, 2011.)

reporter-NPs and target DNA strands (1 μM) followed by addition of a solution of capture-NPs. The mixture was allowed to react for at least 20 min at room temperature and immediately followed by SERS measurements. The lower spectrum in Figure 25.4a shows the quenched SERS signal in the presence of target DNA strands in the mixture of capture-NPs and reporter-NPs, thus indicating that the plasmonic coupling effect was interfered by the presence of target DNA strands. Noteworthy is a dramatic color change from greenish yellow to clear grey observed over the course of 20 min indicating that Ag NPs were aggregated in the absence of target DNA sequences (Figure 25.4b). This result indicates the potential use of the PCI approach as a simple and rapid screening tool based on simple visual examination of color changes of the sample.

25.3.3 PCI Applications to Medical Diagnostics

To demonstrate the potential of the PCI technique for biomedical diagnostics, this technique was applied to detect clinically relevant microRNA (miRNA) biomarkers. The miRNA is a class of 18–24-nucleotide non-coding RNA molecules found in almost all organisms, including humans, plants, virus, and animals. It has been shown that miRNAs play an important role in various biological processes such as development, differentiation, metabolism, and tumorigenesis through their gene regulatory function by binding to their target mRNA strands [43–45]. Moreover, alterations in the expression levels of miRNAs have been shown to be linked with cancer types, disease stages, and treatment responses [46]. Thus, the miRNA expression profiles may serve as useful tests for cancer and disease diagnostics. In the past years, many miRNA detection methods have been reported [45]. However, it has been technically challenging to detect miRNAs directly due to their small size. So far, the most standardized and widely used method for direct detection of miRNA is northern blotting, which is laborious and time-consuming. Hence, the development of novel and rapid miRNA detection techniques with high selectivity and high sensitivity is still in great demand.

To demonstrate the detection of miRNAs using the PCI technique, the capture-NPs and reporter-NPs were prepared by conjugating silver NPs with thiolated unlabeled oligonucleotides and Cy3-labeled oligonucleotides with the sequences of 5′-TCAACATCAGTCTGATAAGCTA-3′ and 5′-TAGCTTATCAGAC-Cy3-3′, respectively. These probes were designed to detect the mature human *miR-21* miRNA molecules with sequences of 5′-UAGCUUAUCAGACUGAUGUUGA-3′. It has been shown that *miR-21* functions as an oncogene and is overexpressed in a variety of different tumors including breast cancers [47–49]. In this study, the unlabeled capture-NPs were designed to be fully complementary to the mature *miR-21* target sequences, and the Cy3-labeled reporter-NPs were complementary to a partial sequence of the capture-NPs. The melting temperature for the duplex of the capture oligonucleotides and Cy3-labeled probes is estimated at 33.4°C in a 50 mm NaCl solution. Thus, the hybridization-mediated NP aggregation could be expected to take place at room temperature. As a proof of concept, we tested the PCI technique for the *miR-21* detection using synthesized miRNA with sequence of 5′-UAGCUUAUCAGACUGAUGUUGA-3′ as the target molecules. As shown in the lower spectrum in Figure 25.5, the reduction in SERS intensity indicates that the plasmonic coupling effect was significantly interfered in the presence of 100 nm complementary *miR-21* targets. However, in the absence of target miRNA sample (blank sample) or in the presence of 100 nm noncomplementary DNA sequences of 5′-TCATCCATGACAACTTTGGTATCGTGGAAGGACTCATGAC-3′, the SERS signals were enhanced in both cases indicating that the plasmonic coupling effect was induced through the aggregated NPs. This result demonstrates that the PCI technique can be used to detect mature miRNA molecules. Moreover, in a recent blind study, we have also successfully demonstrated that the PCI technique can be used to detect the presence of mature *miR-21* molecules directly in enriched small RNA samples extracted from endoscopic biopsies from patients with esophageal adenocarcinoma [50]. Together, these results show that the PCI technique has great potential as a novel label-free tool to detect miRNAs directly from clinical samples for cancer diagnostics.

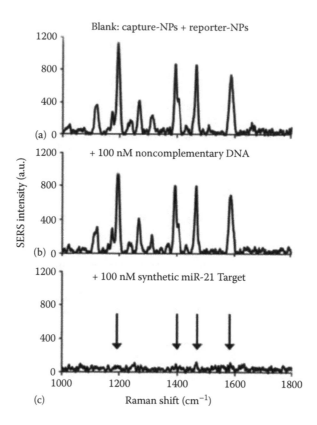

FIGURE 25.5 Detection of *miR-21* miRNA molecules using PCI nanoprobes. (a) Blank sample containing a mixture of capture-NPs and reporter-NPs. (b) In the presence of 100 nM noncomplementary DNA sample as the negative control test. (c) In the presence of 100 nM synthetic *miR-21* targets. (From Wang, H.N. and Vo-Dinh, T., *Small*, 7(21), 3067, 2011.)

25.4 Conclusion

SERS spectroscopy provides powerful plasmonics-based techniques that offer a number of important advantages in biochemical sensing and medical diagnostics. The SERS PCI gene probe is one technique that offers a unique combination of performance capabilities and analytical features of merit for use in biosensing. The SERS PCI gene probes are safer than radioactive labels and have excellent specificity due to the inherent specificity of Raman spectroscopy. The development of practical and sensitive devices for screening multiple genes related to medical diseases and infectious pathogens is critical for early diagnosis and improved treatments of many illnesses. With Raman scattering, multiple probes can be much more easily selected with minimum spectral overlap. This *label-multiplex* advantage can permit analysis of multiple probes simultaneously, resulting in much more rapid DNA detection and gene mapping and improved ultrahigh-throughput screening of small molecules for drug discovery. The SERS molecular sentinel probes could be used for a wide variety of applications in areas where nucleic acid identification is involved. The SERS PCI approach involving homogeneous assays greatly simplifies experimental procedures and could potentially be used in assays that require rapid, high-volume identification of genomic materials. Due to these unique properties, the SERS PCI approach could contribute to the development of the next generation of RNA diagnostic tools for molecular screening and molecular imaging. Potential applications of NP-based biosensing and diagnostics can involve in vitro as well as in vivo detection and imaging modalities. The use of NPs for in vitro diagnostics is straightforward since the samples to be analyzed are outside the patient's body and do not involve toxicity issues. However, the use of NPs for in vivo diagnostics or imaging requires that great care be taken to ensure that NPs are safe for use in humans.

References

1. Gersten, J.; Nitzan, A., *J. Chem. Phys.* 1980, *73*, 3023.
2. Schatz, G. C., *Acc. Chem. Res.* 1984, *17*(10), 370–376.
3. Otto, A., *Surf. Sci.* 1978, *75*, L392.
4. Zeeman, E. J.; Schatz, G. C., *J. Phys. Chem.* 1987, *91*, 634.
5. Fleischmann, M.; Hendra, P. J.; McQuillan A. J., *Chem. Phys. Lett.* 1974, *26*(2), 163–166.
6. Jeanmaire, D. L.; Vanduyne, R. P., *J. Electroanal. Chem.* 1977, *84*(1), 1–20.
7. Albrecht, M. G.; Creighton, J. A., *J. Am. Chem. Soc.* 1977, *99*(15), 5215–5217.
8. Vo-Dinh, T.; Hiromoto, M. Y. K.; Begun, G. M.; Moody, R. L., *Anal. Chem.* 1984, *56*(9), 1667–1670.
9. Vo-Dinh, T.; Uziel, M.; Morrison, A. L., *Appl. Spectrosc.* 1987, *41*(4), 605–610.
10. Enlow, P. D.; Buncick, M.; Warmack, R. J.; Vo-Dinh, T, *Anal. Chem.* 1986, *58*, 1719–1724.
11. Moody, R. L.; Vo-Dinh, T.; Fletcher, W. H., *Appl. Spectrosc.* 1987, *41*(6), 966–970.
12. Alak, A. M.; Vo-Dinh, T., *Anal. Chem.* 1987, *59*(17), 2149–2153.
13. Vo-Dinh, T.; Alak, A.; Moody, R. L., *Spectrochim. Acta Part B-Atomic Spectrosc.* 1988, *43*(4–5), 605–615.
14. Bello, J. M.; Stokes, D. L.; Vo-Dinh, T., *Appl. Spectrosc.* 1989, *43*(8), 1325–1330.
15. Bello, J. M.; Stokes, D. L.; Vo-dinh, T., *Anal. Chem.* 1989, *61*(15), 1779–1783.
16. Vo-Dinh, T.; Houck, K.; Stokes, D. L., *Anal. Chem.* 1994, *66*(20), 3379–3383.
17. Vo-Dinh, T., *TrAC-Trends Anal. Chem.* 1998, *17*(8–9), 557–582.
18. Isola, N. R.; Stokes, D. L.; Vo-Dinh, T., *Anal. Chem.* 1998, *70*(7), 1352–1356.
19. Zeisel, D.; Deckert, V.; Zenobi, R.; Vo-Dinh, T., *Chem. Phys. Lett.* 1998, *283*(5–6), 381–385.
20. Dhawan, A.; Muth, J. F.; Leonard, D. N.; Gerhold, M. D.; Gleeson, J.; Vo-Dinh, T.; Russell, P. E. *J. Vacuum Sci. Technol.* 2008, *26*, 2168–2173.
21. Khoury, C. G.; Vo-Dinh, T., *J. Phys. Chem. C* 2008, *112*(48), 18849–18859.
22. Stokes, D. L.; Chi, Z. H.; Vo-Dinh, T., *Appl. Spectrosc.* 2004, *58*(3), 292–298.
23. Vo-Dinh, T.; Allain, L. R.; Stokes, D. L., *J. Raman Spectrosc.* 2002, *33*(7), 511–516.
24. Vo-Dinh, T.; Yan, F.; Wabuyele, M. B., *J. Raman Spectrosc.* 2005, *36*(6–7), 640–647.
25. Wabuyele, M. B.; Yan, F.; Griffin, G. D.; Vo-Dinh, T., *Rev. Sci. Instrum.* 2005, *76*(6), 063710-1–063710-7.
26. Wabuyele, M. B.; Vo-Dinh, T., *Anal. Chem.* 2005, *77*(23), 7810–7815.
27. Vo-Dinh, T.; Yan, F. Gene detection and multi-spectral imaging using SERS nanoprobes and nano-structures. In *Nanotechnology in Biology and Medicine: Methods, Devices, and Applications*, Vo-Dinh, T., Ed., CRC Press: Boca Raton, FL, 2007.
28. Wang, H. N.; Vo-Dinh, T., *Nanotechnology* 2009, *20*(6), 065101-1–065101-6.
29. Cao, Y. W. C.; Jin, R. C.; Mirkin, C. A., *Science* 2002, *297*(5586), 1536–1540.
30. Vo-Dinh, T.; Dhawan, A.; Norton, S. J.; Khoury, C. G.; Wang, H. -N.; Misra, V.; Gerhold, M. *J. Phys. Chem. C* 2010, *114*(16), 7480–7488.
31. Norton, S. J.; Vo-Dinh, T., *J. Opt. Soc. Am. A* 2008, *25*(11), 2767–2775.
32. Dhawan, A.; Norton, S. J.; Gerhold, M. D.; Vo-Dinh, T., *Opt. Express* 2009, *17*, 9688–9703.
33. Khoury, C. G.; Norton, S. J.; Vo-Dinh, T., *ACS Nano* 2009, *3*(9), 2776–2788.
34. Hirsch, L. R.; Stafford, R. J.; Bankson, J. A.; Sershen, S. R.; Rivera, B.; Price, R. E.; Hazle, J. D.; Halas, N. J.; West. J. L., *Proc. Natl. Acad. Sci. U. S. A.* 2003, *100*(23), 13549–13554.
35. Wang, H. N.; Vo-Dinh, T., *Small* 2011, *7*(21), 3067–3074.
36. Mirkin, C. A.; Letsinger, R. L.; Mucic, R. C.; Storhoff, J. J., *Nature* 1996, *382*, 607–609.
37. Elghanian, R.; Storhoff, J. J.; Mucic, R. C.; Letsinger, R. L.; Mirkin, C. A., *Science* 1997, *277*, 1078–1081.
38. Zanchet, D.; Micheel, C. M.; Parak, W. J.; Gerion, D.; Williams, S. C.; Alivisatos, A. P., *J. Phys. Chem. B* 2002, *106*, 11758–11763.

39. Graham, D.; Thompson, D. G.; Smith, W. E.; Faulds, K., *Nat. Nanotechnol.* 2008, *3*, 548–551.
40. Qian, X.; Zhou, X.; Nie, S., *J. Am. Chem. Soc.* 2008, *130*, 14934–14935.
41. Seferos, D. S.; Giljohann, D. A.; Rosi, N. L.; Mirkin, C. A., *ChemBioChem* 2007, *8*, 1230–1232.
42. McKenzie, F.; Faulds, K.; Graham, D., *Small* 2007, *3*, 1866–1868.
43. Cissell, K. A.; Shrestha, S.; Deo, S. K., *Anal. Chem.* 2007, *79*, 4754–4761.
44. Wark, A. W.; Lee, H. J.; Corn, R. M., *Angew. Chem. Int. Ed.* 2008, *47*, 644–652.
45. Cissell, K. A.; Deo, S. K., *Anal. Bioanal. Chem.* 2009, *394*, 1109–1116.
46. Lu, J.; Getz, G.; Miska, E. A.; Alvarez-Saavedra, E.; Lamb, J.; Peck, D.; Sweet-Cordero, A. et al., *Nature* 2005, *435*, 834–838.
47. Si, M.-L.; Zhu, S.; Wu, H.; Lu, Z.; Wu, F.; Mo, Y.-Y., *Oncogene* 2006, *26*, 2799–2803.
48. Zhu, S.; Wu, H.; Wu, F.; Nie, D.; Sheng, S.; Mo, Y.-Y., *Cell Res.* 2008, *18*, 350–359.
49. Selcuklu, S. D.; Donoghue, M. T. A.; Spillane, C., *Biochem. Soc. Trans.* 2009, 37, 918–925.
50. Wang, H. N.; Garman, K. S.; Diehl, A. M.; Vo-Dinh, T. (Duke University, unpublished data.)

26

DNA Sequencing Using Fluorescence Detection

Steven A. Soper
Louisiana State University

Clyde V. Owens
Louisiana State University

Suzanne J. Lassiter
Louisiana State University

Yichuan Xu
Louisiana State University

Emanuel Waddell
Louisiana State University

26.1 General Considerations

26.1.1 What Is DNA?

26.1.1.1 Organization of Genome

The blueprint for all cellular structures and functions is encoded in the genome of any organism. The genome consists of deoxyribonucleic acid (DNA), which is tightly coiled into narrow threads that, when completely stretched, reach a length of approximately 1.5 m yet possess a width of only ~2.0 nm (2×10^{-9} m). The threads of DNA are typically associated with many different types of proteins and are organized into structures called chromosomes that are housed within the nucleus of cells in most eukaryotic organisms. There are 23 pairs of chromosomes within the human genome.

DNA is composed of several different chemical units: a deoxyribose sugar unit, phosphate group, and one of four different nucleotide bases (adenine [A], guanine [G], cytosine [C], or thymine [T]). At the molecular level, it is the order of these bases that carries the code to build proteins within the cell that inevitably control the function of various cells and also determine an organism's physical characteristics. In the human genome, the 23 pairs of chromosomes contain 3 billion bases. The length of the chromosomes (in base pairs) varies greatly, with the smallest chromosome containing 50 million bases (chromosome Y) and the longest containing 250 million bases (chromosome 1). It is the primary function of DNA sequencing to determine the order of these nucleotide bases.

26.1.1.2 Functions of Genes

The coding regions (regions that carry the information for the construction of proteins) of the genome are contained within genes. Each gene consists of a specific sequence of nucleotide bases, with three nucleotide bases (codon) directing the cells' protein-synthesizing machinery to add a specific amino acid to the target protein. It is estimated that within the human genome are approximately 30,000 genes, with the length of the genes varying greatly. However, only approximately 10% of the human genome is thought to contain protein-coding sequences.

26.1.2 What Is DNA Sequencing?

Figure 26.1 shows a flow chart depicting the important steps involved in the process of DNA sequencing. There are three primary steps—mapping, sequencing, and assembly. Within each of these three general steps are a number of substeps, which include such processes as cloning and subcloning (mapping), template preparation and gel electrophoresis (sequencing), and the computer algorithms required to assemble the small bits of sequencing data into the contiguous strings that comprise the intact chromosome. Each chromosome may consist of hundreds of millions of nucleotide bases; unfortunately, the sequencing phase of the intricate process can handle only pieces of DNA that vary from 1000 to 2000 base pairs (bp) in length. Therefore, the chromosome must typically be broken down into manageable pieces using either restriction enzymes or mechanical shearing and then cloned into bacterium to increase the copy number of the individual pieces of DNA. Following sequencing of each cloned fragment the individual pieces must be reassembled into a contiguous strand representing the entire chromosome. This process typically involves sophisticated computer algorithms to search for commonalties in the small fragments and overlap them to build the sequence of the entire chromosome.

26.1.2.1 DNA-Sequencing Factories

To accomplish the lofty goal of sequencing the entire human genome, sequencing factories have been assembled to produce large amounts of data and deposit these data into public databases for easy accessibility by the general scientific and medical communities. One such production-scale sequencing center is the Human Genome Sequencing Center at the Baylor College of Medicine (www.hgsc.bcm.tmc.edu). Some statistics on this particular sequencing center help put the sequencing demands into perspective. As of May 1999, the center had deposited over 26 Mbp (26×10^6 bp, 0.7% of the human genome) of sequence data into the public databases; in addition, the center typically runs approximately 14 automated fluorescence-based DNA-sequencing machines 12 h a day. This amounts to 50,000 sequencing reactions a month.

This chapter focuses on the sequencing phase of the genome-processing steps. It is this particular step on which fluorescence—both hardware and probe development—has had a profound impact in terms of augmenting the throughput of acquiring sequencing data. We begin by briefly introducing (or refreshing) the reader to the molecular and geometrical structure of DNA and then review the common schemes used to sequence DNA. We also discuss the various electrophoretic modes for fractionating DNA based on size, an integral component in high-throughput sequencing. We then examine common strategies of fluorescence detection used in many sequencing machines; our discussion includes hardware and probe (labeling-dye) developments.

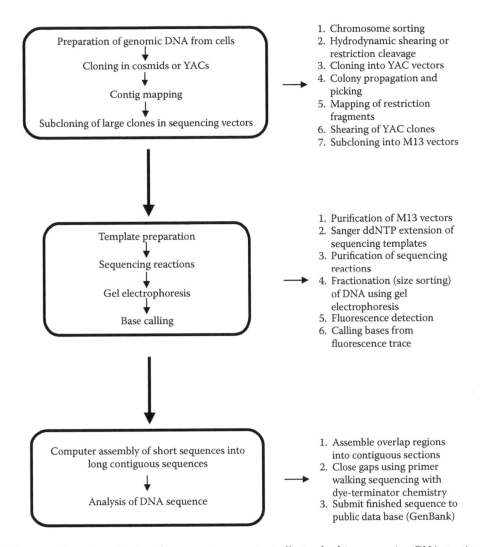

FIGURE 26.1 Flow chart showing the processing steps typically involved in sequencing DNA starting with isolating genomic DNA from cells.

26.1.3 Structure of DNA

26.1.3.1 Nucleotide Bases

As stated in the previous section, the basic building blocks of DNA are a deoxyribose sugar unit, a phosphate group, and the nucleotide base, of which there are four; when combined chemically, these building blocks form an individual nucleotide. Figure 26.2 shows the chemical structure of the four nucleotide units that comprise DNA. The bases are grouped into two different classes—the pyrimidines (T, C) and the purines (A, G). In the case of ribonucleic acid (RNA), the structural differences include the incorporation of a ribose sugar (inclusion of a hydroxyl group at the 2′ position) and also the substitution of uracil for thymine.

Since DNA—more specifically, chromosomes—is composed of many of these individual nucleotide units strung together, the nucleotides are covalently attached via phosphodiester linkages, which occur between the phosphate group on the 5′ site of one sugar unit and the 3′ hydroxyl group on another nucleotide (see Figure 26.2). Therefore, DNA exists as a biopolymer, with the repeating units being deoxyribose and phosphate residues that are always linked together by the same type of linkage and

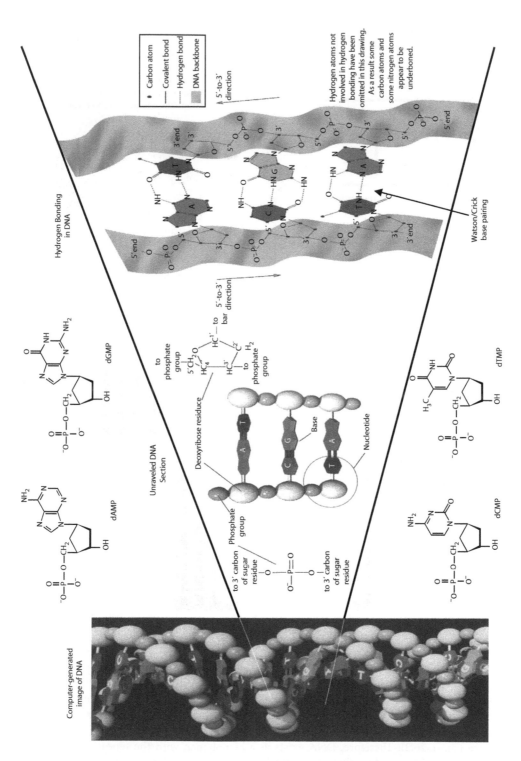

FIGURE 26.2 Chemical structures of deoxyribonucleic acid (DNA). The computer image shows the double-helical nature of DNA. Also shown are the hydrogen bonding between nucleotide bases (Watson–Crick base pairing) and the structures of the nucleotide building blocks.

form the backbone of the DNA molecule. However, the order of bases along the biopolymer backbone can vary greatly and imparts a high degree of individuality to any particular DNA molecule.

26.1.3.2 Watson–Crick Base Pairing

The three-dimensional structure of DNA is known to differ greatly from that of proteins, which are also biopolymers composed of different amino acid residues. In 1953, James Watson and Francis Crick discovered that DNA existed in a double-helical structure (see Figure 26.2), with the sugar-phosphate backbone oriented on the outside of the molecule and the bases positioned on the inside of the double helix.[1] They also surmised that the two strands of the double-stranded molecule were held together via hydrogen bonds between a pair of bases on opposing strands. From modeling, they found that A could pair only with T and C with G (see Figure 26.2). Each of these base pairs possesses a symmetry that permits it to be placed into the double helix in two different ways (A–T and T–A; C–G and G–C). Thus, all possible permutations (four) of sequence can exist for all four bases. In spite of the irregular sequence of bases along each strand, the sugar–phosphate backbone assumes a very regular helical structure, with each turn of the double helix composed of ten nucleotide units.

26.1.4 Methods for Determining the Primary Structure of DNA

The important factor to consider when developing a sequencing strategy for whole genomes is to remember that only small sections (1000–2000 bp) of DNA can be sequenced and that entire chromosomes are composed of well over 1×10^6 bp. The actual process of generating DNAs that can be handled by sequencing machines is typically involved and requires a number of cloning and purification steps followed by actual sequencing and then assembling the pieces into contiguous regions of the target chromosome. While this chapter does not cover these specific processes, Figure 26.3 provides a schematic diagram of a typical strategy used in many sequencing laboratories. This strategy is termed an ordered shotgun approach and starts with the mechanical shearing (breaking apart) of intact chromosomes into pieces composed of 100,000–200,000 bp.[2]

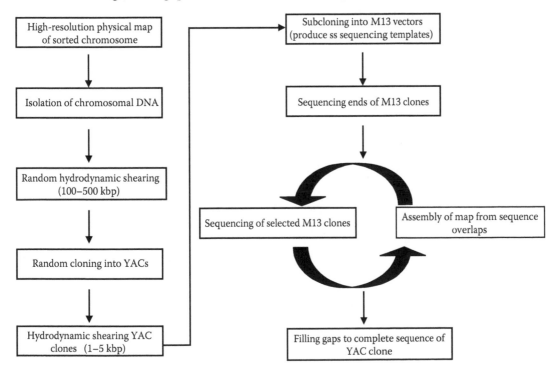

FIGURE 26.3 Processing flow chart of ordered shotgun sequencing of DNA.

These sheared DNAs are then cloned into yeast artificial chromosomes (YACs), one sheared section per YAC that allows one to amplify the number of copies of the insert. Cloning provides an unlimited amount of material for sequencing and serves as the basis for construction of libraries, which are random sets of cloned DNA fragments. Genomic libraries are sets of overlapping fragments encompassing an entire genome. Once these libraries have been constructed, single inserts are extracted from the library and further sheared into fragments ranging from 1000 to 2000 bp. These fragments are then subcloned into M13 vectors to produce high-quality single-stranded DNAs (ssDNAs) appropriate for actual sequencing. Typically, M13 subclones are sequenced from both ends to allow construction of maps of the single YAC inserts, which are then used to provide a scaffold for the complete sequence analysis of the YAC insert. The important procedures that are the focus of this chapter are those that are actually used to produce the sequence data of the M13 subclones. The two most common procedures are the Maxam–Gilbert chemical degradation method and the Sanger dideoxy-chain-termination method.[3,4] Both of these methods have in common the production of a nested set of fragments, all of which are terminated (cleaved) at a common base or bases.

26.1.4.1 Maxam–Gilbert DNA Sequencing

The Maxam–Gilbert sequencing method uses chemical cleavage methods to break ssDNA molecules at either one or two bases, followed by a size fractionation step to sort the cleaved products. The cleavage reaction involves two different reactions—one that cleaves at G and A (purine) residues and the other that cleaves at C and T (pyrimidine) residues. The first reaction can be slightly modified to cleave at G only and the second at T only. Therefore, one can run four separate cleavage reactions—G only, A + G, T only, T + C—from which the sequence can be deduced. In each cleavage reaction, the general process involves chemically modifying a single base, removing the modified base from its sugar, and, finally, breaking the bond of the exposed sugar in the DNA backbone.

The chemical steps involved in G cleavage are shown in Figure 26.4a. In this step, dimethylsulfate is used to methylate G. After eviction of the modified base via heating the strand is broken at the exposed sugar by subjecting the DNA to alkali conditions. To cleave at both A and G residues, the procedure is identical to the G cleavage reaction except that a dilute acid is added after the methylation step (see Figure 26.4b). The reaction that cleaves at either a C or C and T residue is carried out by subjecting the DNA to hydrazine to remove the base and piperidine to cleave the sugar–phosphate backbone. The extent of each reaction can be carefully limited so that each strand is cleaved at only one site.

Figure 26.4c depicts the entire Maxam–Gilbert process. As shown in the figure, four cleavage reactions are run—G, G + A, T, T + C. Prior to chemical cleavage, the intact DNA strand is labeled (typically with a ^{32}P radiolabel for detection at the 5' end). Following chemical cleavage the reactions are run in an electrophoresis gel (polyacrylamide) and separated based on size. The actual sequence of the strand is then deduced from the generated gel pattern.

26.1.4.2 Sanger Chain-Termination Method

Unlike the Maxam–Gilbert method, the Sanger procedure is an enzymatic method and involves construction of a DNA complement to the template whose sequence is to be determined. The complement is a strand of DNA that is constructed with a polymerase enzyme, which incorporates single nucleotide bases according to Watson–Crick base-pairing rules (A–T, G–C) (see Figure 26.5). The nested set of fragments is produced by interrupting the polymerization by inserting into the reaction cocktail a base that has a structural modification—the lack of a hydroxyl group at the 3' position of the deoxyribose sugar (dideoxynucleotide, ddNTP). Figure 26.5a shows the chemical structure of a ddNTP. When mixed with the deoxynucleotides (dNTP), the polymerization proceeds until the ddNTP is incorporated.

Figure 26.5c shows the entire Sanger chain-termination protocol. The process is typically carried out by adding to the template DNA a primer that has a known sequence and anneals (binds) to a complementary site on the unknown template. This primer can carry some type of label, for example, a covalently attached fluorochrome. However, situating the fluorescent label on the ddNTP can be done as well. Following annealing of the primer to the template, the reaction is carried out by the addition of a polymerase enzyme, all four dNTPs, and one particular ddNTP. Therefore, four reactions are carried out, each containing a particular ddNTP. Following polymerization, the reactions are loaded onto an electrophoresis gel and size-fractionated. The final step involves reading the sequence from the gel.

The Sanger method has been the preferred sequencing method for most large-scale sequencing projects because of the ease with which it prepares the nested set of fragments. Many times reactions can be run under standard conditions without the need for chemical additions at timed intervals. In addition, the process is very conducive to automation.

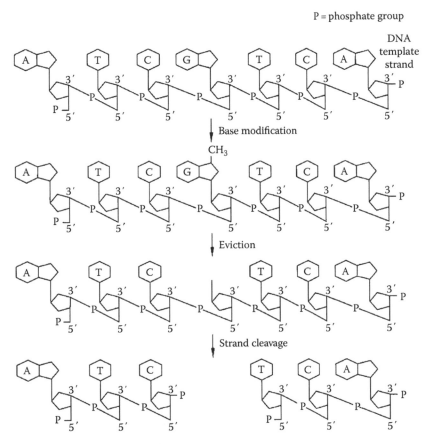

Dimethylsulfate is used to methylate guanine. After eviction of the modified base, the exposed sugar, deoxyribose, is then removed from the backbone. Thus, the strand is cleaved in two.

(a)

5'-³²P-A T G A C C G A T T T G C-3'——————— Labeled template strand

5'-³²P-A T-3'   5'-A C C G A T T T G C-3'⎫ Six different types of fragments
5'-³²P-A T G A C C-3'   5'-A T T T G C-3'⎬ are produced. Only three of
                                              ⎪ those include the labeled 5' end
5'-³²P-A T G A C C G A T T-3   5'-C-3'⎭ of the original strand.

(b)

FIGURE 26.4 Maxam–Gilbert chemical degradation method for DNA sequencing. (a) Chemical cleavage method for a guanine residue using dimethylsulfate to methylate the G residue. (b) Fragments generated from chemical cleavage at only G. Maxam–Gilbert chemical degradation method for DNA sequencing.

(continued)

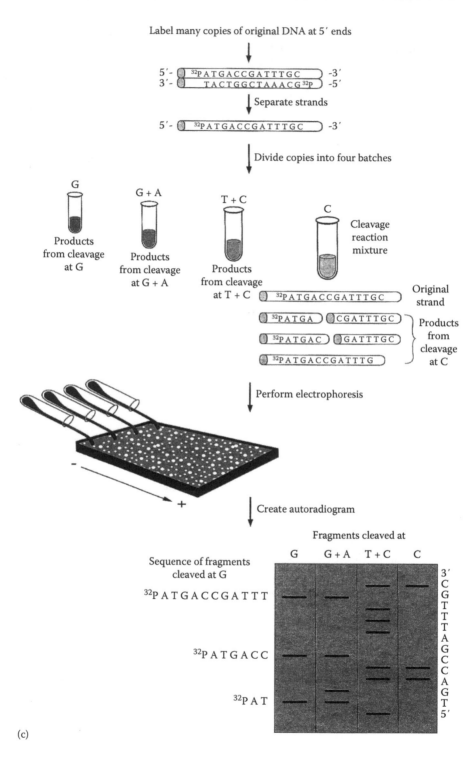

FIGURE 26.4 (continued) Maxam–Gilbert chemical degradation method for DNA sequencing. (c) Steps in Maxam–Gilbert sequencing.

26.1.5 Modes of Electrophoresis

Whatever fluorescence detection protocol is used for calling bases, the important analytical technique that is required is a fractionation step in which the DNA molecules are sorted by size. While the focus of this chapter is on fluorescence-based detection in sequencing applications, it is informative to briefly discuss some of the common gel electrophoresis platforms that are used because the fluorescence detector is integrated into the sizing step to provide online detection.

All electrophoresis formats have in common the use of an electric field to shuttle the DNAs through a maze consisting of a polymer, either static or dynamic, with pores of various sizes. In all electrophoresis

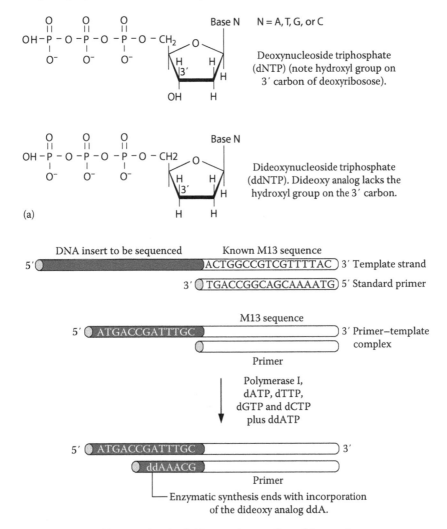

(a)

(b)

FIGURE 26.5 Sanger chain-termination DNA-sequencing method. (a) Chemical structures of a dNTP and also a ddNTP that lacks a hydroxyl group on the 3′ site of the deoxyribose sugar. (b) Chain-termination reaction with ddATP. Also shown is primer annealing to the template, which is directed by Watson-Crick base-pairing rules. Sanger chain-termination DNA-sequencing method.

(continued)

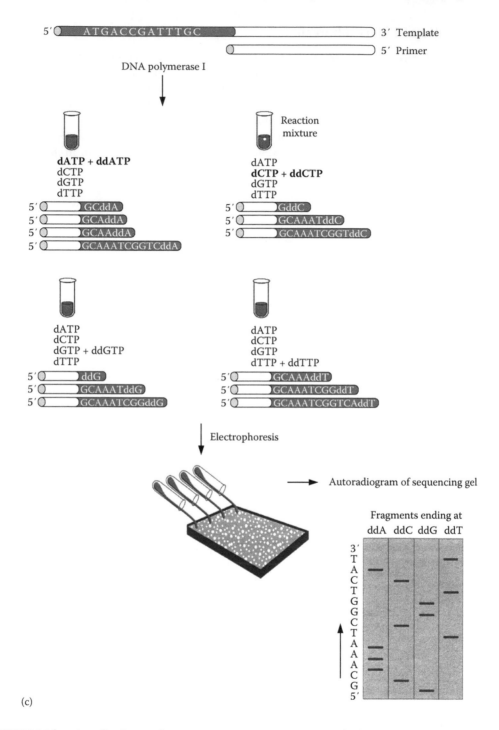

(c)

FIGURE 26.5 (continued) Sanger chain-termination DNA-sequencing method. (c) Steps in Sanger sequencing.

experiments, the mobility of the molecule (μ, cm^2/Vs, defined as the steady-state velocity per unit electric-field strength) in an electric field is determined by:

$$\mu = \frac{q}{f} \tag{26.1}$$

where

q is the net charge on the molecule

f is the frictional property of the molecule and is related to its conformational state as well as the molecular weight (MW) of the molecule

For example, proteins can be considered solid spheres; thus, $f \sim (\text{MW})^{1/3}$. In the case of DNAs, either single-stranded or double-stranded, $f \sim (\text{MW})1$ because the DNA molecule acts as a free-draining coil. Since q is also related to the length of the DNA molecule, one $\mu \sim N_b{}^0$, where $N_b{}^0$ indicates that the mobility (in free solution) is independent of the number of bases comprising the DNA molecule. Because of this property of DNAs, the electrophoresis step must include some type of sieving medium, which can be a polymer consisting of pores with a definitive size.

For any type of analytical separation resolution is a key parameter that is optimized to improve the performance of the separation. The resolution R for electrophoretic separations can be calculated from the simple relationship:

$$R = \frac{1}{4} \frac{\Delta\mu_{app}}{\mu_{app,avg}} N^{1/2} \tag{26.2}$$

where

$\Delta\mu_{app}$ is the difference in mobility between two neighboring bands

$\Delta\mu_{app,avg}$ is the average electrophoretic mobility for the same neighboring bands

N is the plate number, which represents the efficiency (bandwidth) for the electrophoresis

This equation shows that the resolution can be improved by increasing the difference in the mobility of the two bands (increase selectivity, gel property) or increasing the plate numbers (narrower bands). As a matter of reference, when $R = 0.75$, two bands are baseline resolved. For DNA sequencing, the accuracy in the base call depends intimately on the resolution obtained during gel fractionation.

A common polymer used for DNA sequencing is linear or crosslinked polyacrylamides or polyethylene oxides, both of which possess the appropriate pore size for sorting ssDNAs. Polyacrylamides are prepared from the acrylamide monomer ($-CH_2=CHCONH_2$), which is copolymerized with a certain percentage of crosslinker, N,N'-methylene-bis-acrylamide ($-CH_2(CHCOCH=CH_2)_2{}^-$), in the presence of a catalyst accelerator-chain-initiator mixture. The porosity of the gel is determined by the relative proportion of acrylamide monomer to crosslinking agent. For reproducible fractionation of DNAs, they must be maintained in a single-strand conformation, which is accomplished by adding denaturants, such as urea or formamide, to the sieving gel.

The existence of a gel network contributes significantly to the electrophoretic migration pattern (i.e., electrophoretic mobility) observed in the sequencing process. Any DNA fragments to be fractionated will inevitably encounter the gel network of polymer threads. This encounter increases the effective friction and, consequently, lowers the velocity of movement of the molecules. It is obvious that the retardation will be most pronounced when the mean diameter of gel pores is comparable to the size of the DNA fragments. Size therefore plays a critical role in determining the relative electrophoresis mobility and the degree of separation of different DNA fragments. It is this sieving effect that partly determines the resolution obtained with polyacrylamide gels. The gel network also minimizes convection currents caused by small temperature gradients.

The main characteristic property of DNA is that the value of the negative charge on the molecule is almost independent of the pH of the medium. Therefore, their electrophoretic mobility is due mainly to differences in molecular size, not in charge. For a particular gel, the electrophoretic mobility of a DNA fragment is inversely proportional to the logarithm of the number of bases—up to a certain limit. Since each DNA fragment is expected to possess a unique size, its electrophoretic mobility is unique, and it migrates to a unique position within the electric field in a given length of time. Therefore, if a mixture of DNA fragments is subjected to electrophoresis, each of the fragments would be expected to concentrate into a tight migrating band at unique positions in the electric field.

26.1.5.1 Slab-Gel Electrophoresis

DNA sequencing is usually performed with slab gels. In slab gel electrophoresis (SGE), polymerization of acrylamide as well as the electrophoresis are conducted in a mold formed by two glass plates with a thin spacer. When analytical electrophoresis is performed, several samples are usually run simultaneously in the same slab. Since all samples are present in the same gel, the conditions of electrophoresis are quite constant from sample to sample. In SGE the sample is loaded using a pipette into wells formed into the gel during polymerization. A loading volume of 1–10 μL is typical for SGE. The field strength that can be applied to these types of gels ranges from 50 to 80 V/cm, with the upper limit determined by heating (Joule) caused by current flow through the gel. Due to the thick nature of the gel, this heat is not efficiently dissipated, causing convective mixing and thus zone broadening, which limits the upper level on the electric field strength that can be effectively used.

26.1.5.2 Capillary-Gel Electrophoresis

To reduce the development time in the electrophoresis, many sequencing applications now use capillary-gel electrophoresis (CGE), in which a glass tube (i.d. = 50–100 μm; length = 30–50 cm) contains the gel-sieving matrix, and an electric field is applied to this capillary column. Due to the higher surface-to-volume ratio afforded by the thin glass capillary, large electric fields can be applied (~300 V/cm), which results in shorter electrophoresis development times (2–3 h) and also enhanced plate numbers compared to SGE. The glass capillary contains silanol groups on its surface, which at high pH are deprotonated. As such, the glass capillary is negatively charged above pH = 7.0. Cations in the electrolyte solution buildup at the wall of the glass tube, producing an electrical double layer. When a voltage is applied across the capillary, the cations migrate to the wall and exert a force on the surrounding fluid, causing a bulk flow of solution toward the cathode (negative terminal). This electrically induced flow is called an electroosmotic flow and, unfortunately, can interfere with the electrophoresis of the DNAs. Therefore, in DNA separations using glass capillaries, the wall is coated with some type of polymer (e.g., a linear polyacrylamide) to suppress this electroosmostic flow. After the wall is coated, the capillary can then be filled with the sieving gel, which can be a linear polyacrylamide (no cross linking) or some other type of gel, such as a hydroxyl cellulose of poly (ethylene oxide).[5] In addition, crosslinked gels may also be used in these small capillaries, but since crosslinked gels are not free flowing like their noncrosslinked counterparts, they must be polymerized directly within the capillary tube. Most capillary-based DNA sequencers use linear gels because they can be easily replaced using high-pressure pumping, which allows the capillary to be used for multiple sequencing runs. Once polymerized, the crosslinked gels cannot be removed from the capillary.

The electrophoresis is performed after the column is filled with the gel by inserting one end of the capillary into a sample containing the sequencing mix and applying an electric field to the capillary tube. The injection end is typically cathodic (negative), with the opposite end being anodic (positive). By applying a fixed voltage for a certain time period, a controlled amount of sample can be inserted into the column, with the injection volume ranging from 1 to 10 nL. Since the capillary is made of fused silica, fluorescence detection can be performed directly from within the capillary tube. In most cases, the capillary column is coated with a polyimide coating (nontransparent) to give it strength, and the optical detection window can be produced by simply burning off a section of the polymer using a low-temperature flame.

The higher plate numbers (i.e., better resolution) that are obtained in CGE vs. SGE are a direct consequence of the ability to use higher electric fields. Since development time is significantly shorter in CGE due to the ability to apply higher electric fields, band spreading due to longitudinal diffusion is reduced, resulting in higher plate numbers. The ability to use higher electric fields in CGE results from the fact that Joule heating is suppressed because the heat can be effectively dissipated by the high-surface-to-volume-ratio capillary.

26.1.6 Detection Methods for DNA Sequencing

26.1.6.1 Autoradiographic Detection

Following electrophoresis, the individual DNA bands separated on the gel must be detected and subsequently analyzed. One of the earlier methods implemented to detect DNA bands in gels was autoradiography. In this mode, one of the phosphates of an individual nucleotide is replaced with a radioisotope, typically 32P ($\tau_{1/2}$ = 14 days) or 35S ($\tau_{1/2}$ = 87 days), both of which are radioprobes that emit β-particles. When Sanger methods are used to prepare the sequencing reactions, either the primer or the dideoxynucleotide can contain the radiolabel. The labeling is done using an enzyme (T4 polynucleotide kinase), which catalyzes the transfer of a γ-phosphate group from ATP to the 5′-hydroxy terminus of a sequencing primer. After the electrophoresis is run, the gel is dried and then situated on an x-ray film. The film is developed (exposure to radiation from radioprobes), and dark bands are produced on the film where the DNA was resident. The sequence is then read manually from the gel.

The primary advantage of this approach is the inexpensive nature of the equipment required to perform the measurement—basically only a gel dryer, film holder, and film. The difficulties associated with this approach are numerous. One important issue is the fact that radioisotopes are used; therefore, waste disposal becomes a difficult problem. Throughput issues (data-production rates) are also a primary concern with autoradiographic detection. For example, radiography can sometimes require several days to expose the film to get strong signals to read the bases from the gel. In addition, the detection is done after electrophoresis, not during. Also, since there is no means of identifying the individual bases using radioprobes, each base must be analyzed in a different lane of the gel. And finally, the bases must be called manually, which often leads to frequent errors in the sequence reconstruction. Therefore, the inability to obtain data-production rates sufficient to accommodate large sequencing projects has made radiographic detection obsolete for high-throughput applications.

26.1.6.2 Fluorescence Detection

For most DNA-sequencing applications, irrespective of the separation platform used, fluorescence is the accepted detection protocol, for several important reasons. Fluorescence allows one to perform the base calling and detection in an automated fashion and alleviates the need for manual base calling. In addition, fluorescence can be carried out during the separation, eliminating long film-development times. More importantly, because multiple probes possessing unique spectral properties can be implemented, the four bases comprising the DNA molecule can be identified in a single gel lane, potentially increasing throughput by a factor of four vs. radiographic detection. All of these important advantages associated with fluorescence allow for higher throughputs in DNA-sequencing applications. As such, fluorescence can be considered one of the most important recent technical innovations in DNA sequencing and has made it feasible to consider tackling large genome-sequencing projects such as the human genome.

The first demonstrations on the use of fluorescence in DNA sequencing came with the work of Smith et al.,[6] Prober et al.,[7] and Ansorge et al.[8] Slab gels or large gel tubes were used to fractionate the DNA ladders produced during enzymatic polymerization using Sanger sequencing strategies. The fluorescence detection was accomplished using four spectroscopically unique probes, which allowed the DNA-sequence reconstruction to be done in a single electrophoresis lane of the gel. The chemical structures of the dye labels used in the Smith et al.[6] and Prober et al.[7] experiments are shown in Figures 26.6 and 26.7,

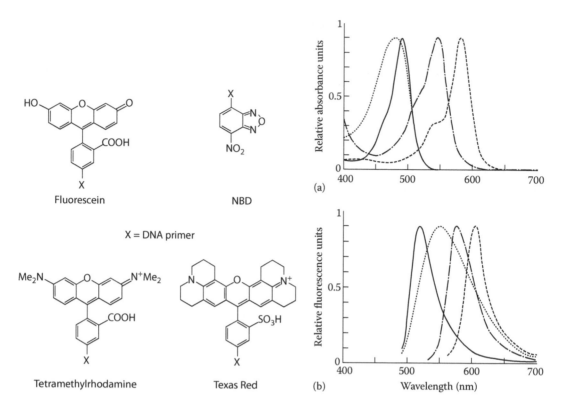

FIGURE 26.6 Chemical structures and absorbance (a) and emission spectra (b) of the dyes used for labeling primers. (Adapted from Smith, L.M. et al., *Nature*, 321, 674, 1986.)

respectively. As shown in the figures, the dyes were attached (covalently) either to the sequencing primer or to the dideoxynucleotides. The advantage of using dye-labeled dideoxynucleotides is that the sequencing reactions can be performed in a single reaction tube, whereas the dye-labeled primer reactions must be performed in four separate tubes and pooled prior to electrophoresis. In the case of the dye-labeled terminators, succinylfluorescein analogs were used with slight structural modifications to alter the absorption/emission maxima. The dyes were attached either to the 5 position of the pyrimidine bases or the 7 position of the 7-deazapurines, both of which are nonhydrogen-bonding sites on the nucleotide base. The linker structure is also important, which in this case was a propargylamine, because the presence of the dye on the terminator radically affects its ability to be incorporated by the polymerase enzyme. For dye-labeled primers, the oligonucleotides possessing the appropriate sequence were prepared on a standard DNA synthesizer. For Smith et al.[6] a thymidine derivative was prepared that contained a phosphoramidite at the 3′ carbon and a protected alkyl amino group at the 5′ carbon (typically a 6-carbon linker structure). During the final addition cycle of the oligonucleotide prepared via solid-phase synthesis using phosphoramidite chemistry, the thymidine residue was added, and, following deprotection of the alkyl amino group and cleavage from the support, a free primary amine group resulted, which can be reacted with any amino-reactive fluorescent dye to produce the oligonucleotide derivative.

Figures 26.6 and 26.7 also show the absorption and emission profiles for the dye sets used in these experiments. The major attributes of the dye sets are that they can be efficiently excited with either 488- or 514.5-nm lines of the argon ion laser. In addition, there is minimal separation between the emission maxima of the dyes, which allows processing of the fluorescence on as few detection channels as possible. However, there is significant overlap in the emission spectra of the four dyes, producing severe spectral leakage into other detection channels, which must be corrected by software.

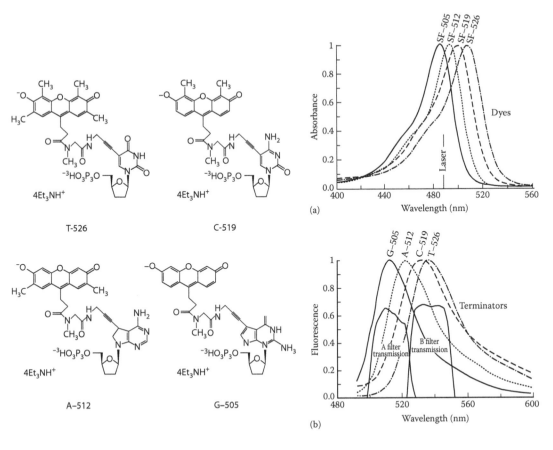

FIGURE 26.7 Chemical structures and the absorbance (a) and emission spectra (b) of dyes conjugated to ddNTPs. Also shown are the excitation laser wavelength (488 nm) and the filter set used to isolate the fluorescence onto the two detection channels. (Adapted from Prober, J.M. et al., *Science*, 238, 336, 1987.)

The optical hardware used to process the four-color fluorescence for both systems is depicted in Figure 26.8. For the Smith et al. experiment,[6] a single laser and a single detection channel were used; in this case, the detection channel consisted of a multiline argon ion laser and a conventional photomultiplier tube (PMT). Placed in front of the laser and PMT were filter wheels to select the appropriate excitation wavelength (488 or 514.5 nm) and emission color. The filter pairs used during fluorescence readout were 488/520, 488/550, 514/580, and 514/610 nm. In the case of the Prober et al. experiment,[7] due to the narrow distribution between the excitation maxima of the dyes only excitation using the 488-nm line from the argon ion laser was required as well as two PMT tubes to process the emission from the four colors. Discrimination of the four colors was accomplished by monitoring the intensity of each dye on both detectors simultaneously. By histograming the ratio of the fluorescence intensity of each dye (produced from an electrophoresis band) on the two detection channels, a discrete value was obtained that allowed facile discrimination of the four different fluorescent dyes (i.e., terminal base). To determine the limit of detection of these fluorescence systems, injections of known concentrations of dye-labeled sequencing primers were electrophoresed. In both cases, the mass detection limit was estimated to be 10^{-17} to 10^{-18} mol.

Figure 26.9 shows the data output from these systems. Unfortunately, reading the sequence directly from the raw gel data becomes problematic due to several nonidealities. These include signal from a single dye appearing on multiple detection channels due to the broad and closely spaced emission bands, dye-dependent electrophoretic mobility shifts, and nonuniformity in the intensity of the electrophoresis

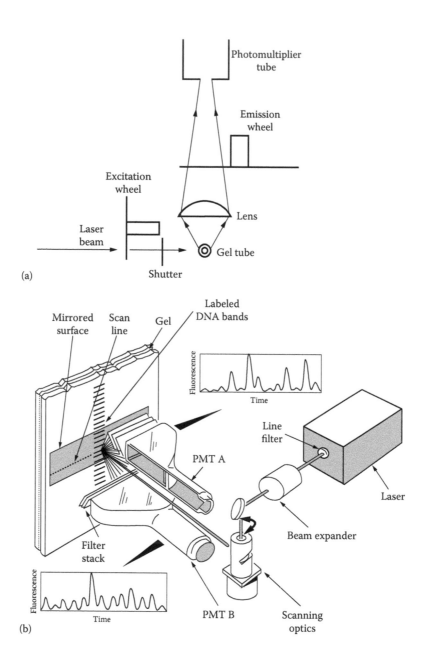

FIGURE 26.8 Fluorescence detector systems for DNA sequencing. (a) The laser consisted of an argon ion laser operating at 488 and 514 nm. The wavelength was selected using a rotating filter wheel. The emission was collected by a lens and sent through a filter wheel that contained four different filters, one for each of the multicolor dyes used to label the sequencing primers (see Figure 26.6 for their structures and fluorescence properties). (Adapted from Smith et al., *Nature*, 321, 674, 1986.) (b) The laser used for excitation is an argon ion laser operated at 488 nm only. The beam was rastered over the slab-gel plate using scanning optics. The emission was then processed on one of two elongated, stationary photomultiplier tubes (PMTs) (see Figure 26.7 for dye set used with this fluorescence detector). (Adapted from Prober, J.M. et al., *Science*, 238, 336, 1987.)

bands due to the enzymatic reaction used to construct the individual DNA size ladders. Thus, several post-electrophoresis-processing steps were required to augment sequence reconstruction of the test template. In the case of the Smith et al. example,[6] these steps involved:

- High-frequency noise removal using a low-pass Fourier filter.
- A time delay between measurements at different wavelengths corrected by linear interpolation between successive measurements.
- A multicomponent analysis performed on each set of four data points, which produced the amount of the four dyes present in the detector as a function of time.
- The peaks present in the located data stream.
- The mobility corrected for the dye attached to each DNA fragment. In this case, it was empirically determined that fluorescein and rhodamine-labeled DNA fragments moved as if they were one base longer than the NBD-labeled fragments, and the Texas Red fragments moved as if they were 1.25 bases longer.

The important performance criteria in any type of automated DNA sequencer are its throughput, the number of bases it can process in a single gel read, and its accuracy in calling bases. In terms of base-calling accuracy, these early instruments demonstrated an error rate of approximately 1%, with a read length approaching 500 bases. The throughput of the instrument described by Prober et al.[7] was estimated to give a raw throughput of 600 bases per hour (12 electrophoresis lanes). Interestingly, many present-day commercial automated sequencers still use similar technology in their machines, and the throughput can be as high as 16,000 bases per hour (96 electrophoresis lanes).

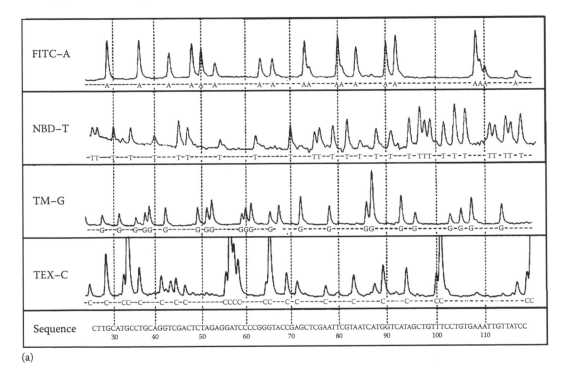

(a)

FIGURE 26.9 (Opposite) Fluorescence sequencing data obtained from the two fluorescence systems described in Figure 26.8. (a) The dyes (see Figure 26.6) were conjugated to a primer with a sequence, 5'-CCCAGTCACGACGTT-3', complementary to a site on the M13 phage vector. The reactions were run using Sanger chain-termination methods and run in four different vessels, one for each terminator. The polymerase enzyme was a T7 polymerase, and the reactions were pooled prior to electrophoresis. (Adapted from Smith, et al., *Nature*, 321, 674, 1986.)

(continued)

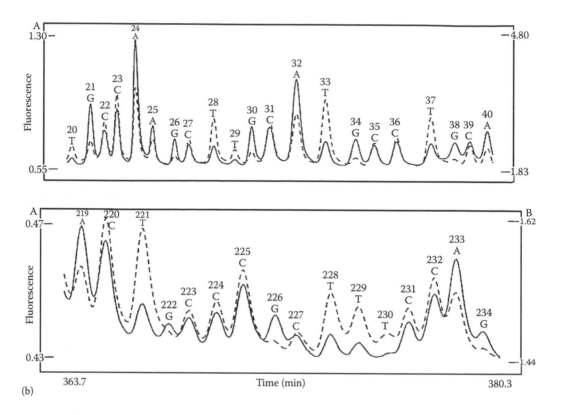

(b)

FIGURE 26.9 (continued) (Opposite) Fluorescence sequencing data obtained from the two fluorescence systems described in Figure 26.8. (b) Sequence of the M13mp8 template using the dye-labeled terminator set shown in Figure 26.7. The base assignment is given as a letter above each electrophoretic peak. The reactions were prepared using Sanger chain-termination methods and prepared in a two-stage fashion. In the first stage, the template (M13mp18) was added to a reaction tube along with the primer, heated to 95°C for 2 min and then cooled in ice water for 5 min. In the second stage, dNTPs and the dye-labeled ddNTPs were added along with an AMV reverse transcriptase enzyme. Following chain extension, the reactions were purified (removal of excess terminators) using gel filtration (Sephadex spin column) and loaded onto the gel. The gel was 8% polyacrylamide containing 7 M urea as the denaturant. In these traces, the solid line represents the fluorescence signal from PMT A, and the dashed line is from PMT B. (Adapted from Prober, J.M. et al., *Science*, 238, 336, 1987.)

26.2 Fluorescent Dyes for DNA Labeling and Sequencing

As stated above, fluorescence detection has had a tremendous impact on DNA sequencing because it possesses a high-speed readout process and has the ability to discriminate among the four nucleotide bases in a single gel lane due to the unique spectral properties of the target dye molecules. A variety of dye sets (typically four dyes per set, one for each nucleotide base) have been developed for DNA-sequencing applications, and regardless of the dye set under consideration certain properties associated with the dyes for DNA-sequencing applications are important. These properties are listed below:

1. *Each dye in the set must be spectroscopically distinct.* The ideal situation from a throughput point of view is to identify each base in a single gel lane instead of four gel lanes. In most cases, the discrimination is based on differences in the emission properties of each dye (distinct emission maxima); however, other fluorescence properties can be used as well such as fluorescence lifetimes.

2. *The dye set should preferably be excited by a single laser source.* It becomes instrumentally difficult to implement multiple excitation sources because lasers are typically used as the source of excitation to improve the limit of detection in the measurement, and the upkeep on multiple lasers becomes problematic.

3. The dye set should possess high extinction at the excitation frequency and also large quantum yields in the gel matrix used to fractionate the DNAs. Good photophysical properties are necessary to improve detectability. In addition, the dye set should show reasonable quantum yields in denaturing gels, which consist of high concentrations of urea or formamide.

4. *The dye set should show favorable chemical stability at high temperatures.* A common procedure in most Sanger sequencing strategies is to implement a thermostable polymerase and then subject the sequencing reactions to multiple temperature cycles (55°C–95°C) to amplify the amount of product generated (cycle sequencing). Therefore, the dyes must be able to withstand high-temperature conditions for extended periods of time.

5. *The dyes must induce minimal mobility shifts during the electrophoresis analysis of the sequencing ladders.* Often, the mobility shifts induced by individual dyes cause misordering of the individual bases during sequence reconstruction. As such, post-electrophoresis corrections arc often employed to rectify this perturbation in the DNA's mobility.

6. *The dyes must not significantly perturb the activity of the polymerase enzyme.* This is especially true in dye-terminator chemistry because the proximity of the dye to the polymerase enzyme can dramatically influence the enzyme's ability to incorporate that dye–ddNTP conjugate into the polymerized DNA molecule.

26.2.1 Visible Fluorescence Dyes for DNA Labeling

The dye set frequently used in many automated, fluorescence-based DNA sequencers is the FAM, JOE, TAMRA, and ROX series (see Figure 26.10), which consists of fluorescein and rhodamine analogs containing a succinimidyl ester for facile conjugation to amine-terminated sequencing primers or terminators. These dyes can be efficiently excited by the 488- and 514.5-nm lines from an argon ion laser; they also possess emission profiles that are fairly well resolved. Many of the Applied Biosystems (Foster City, CA) automated DNA sequencers use this particular dye set (for example, ABI 373 and 377 series, see www.perkin-elmer.com). Figure 26.10 shows emission spectra for this dye set as well as the filter set that is used to isolate the emission from the dyes onto the appropriate detection channel. While this dye set is fairly robust and works well with typical DNA-cycle-sequencing conditions, this series present some difficulties, that is, the broad emission profiles, dye-dependent mobility shifts, and inefficient excitation of TAMRA and ROX with the 514.5-nm line from the argon ion laser.

To eliminate dye-dependent mobility shifts and minimize cross talk between detection channels using four-color processing, a bodipy dye set (4,4-difluoro-4-bora-3α, 4α-diaza-*s*-indacene-3-proprionic acid) has been used for DNA-sequencing applications.[9] The structures of the dyes are presented in Figure 26.11. Inspection of the fluorescence-emission profiles for this particular dye set indicated that the bandwidths were less than those associated with the FAM/JOE/TAMRA/ROX dye set, which resulted in less spectral leakage between detection channels in the fluorescence readout hardware. Figure 26.11 reveals that all these dyes are neutral and have very similar molecular structures, thereby minimizing dye-dependent mobility shifts. To fully compensate for any mobility-shift differences within the dye set, a modified linker structure was synthesized to account for this difference (see Figure 26.11).

A particular shortcoming associated with this dye set is the chemical instabilities the dyes display when subjected to extended heating at high temperatures. An example of this is shown in Figure 26.12, in which bodipy-labeled primers were used with cycle-sequencing conditions. In cycle sequencing, a linear amplification of the ddNTP-terminated fragments can be generated by subjecting the reaction cocktail to multiple temperature cycles consisting of 95°C (thermal denaturation of the double-stranded

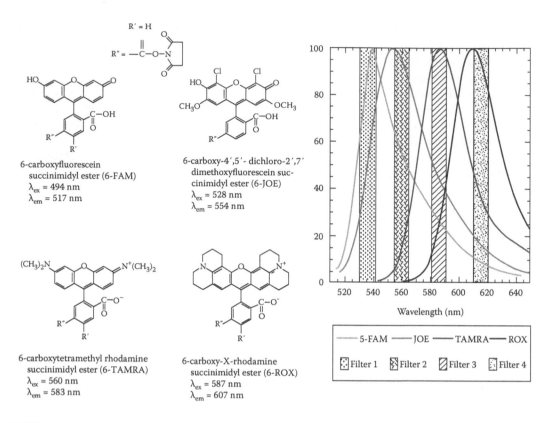

FIGURE 26.10 Common fluorescence-dye set used for labeling primers for DNA-sequencing applications. The functional group on each dye is a succinimidyl ester, which readily conjugates to primary amine groups. Also shown are the emission spectra of this dye set and the filters used to select the appropriate dye color-processed by each detection channel.

DNA molecule), 55°C (annealing the sequencing primer to the DNA template), and 72°C (chain extension using a thermostable polymerase enzyme, such as *Taq* polymerase). The difficulty arises during the 95°C step, where significant dye decomposition can result. Figure 26.12 shows fluorescence traces of sequencing ladders prepared using a 95°C step time of either 30 or 5 s. As shown in the figure, there is a significant loss in signal when the cycling time (at 95°C) is 30 s, but the signal is partially restored when the cycling time is reduced to 5 s.

Many of the dyes discussed above have been employed in dye-primer sequencing applications, in which the primer used for selecting the DNA polymerization site on the unknown template is determined by Watson–Crick base pairing. An alternative is dye-terminator chemistry, where the fluorescent dyes are covalently attached to the ddNTP used in Sanger sequencing strategies. The advantages of this method are twofold. First, the sequencing reactions can be carried out in a single tube. In dye-primer chemistry, the sequencing reactions are carried out in four separate tubes and then pooled prior to the electrophoresis step. If four spectroscopically unique fluorescent probes attached to the ddNTP are implemented, the reactions can be carried out in a single tube, which reduces reagent consumption and also minimizes sample transfer steps. Second, primers of known sequence need not be synthesized. In primer sequencing, making oligonucleotides of 17–23 bases in length can be costly and time-consuming, especially if the dye must be chemically tethered to the primer. The use of dye-terminator chemistry eliminates the need for synthesizing dye-labeled primers. Unfortunately, dye terminators themselves can be quite expensive; in addition, many polymerase enzymes are very sensitive to the type of dye attached to the ddNTP. For example, fluorescein dyes are poor labels for terminators when using

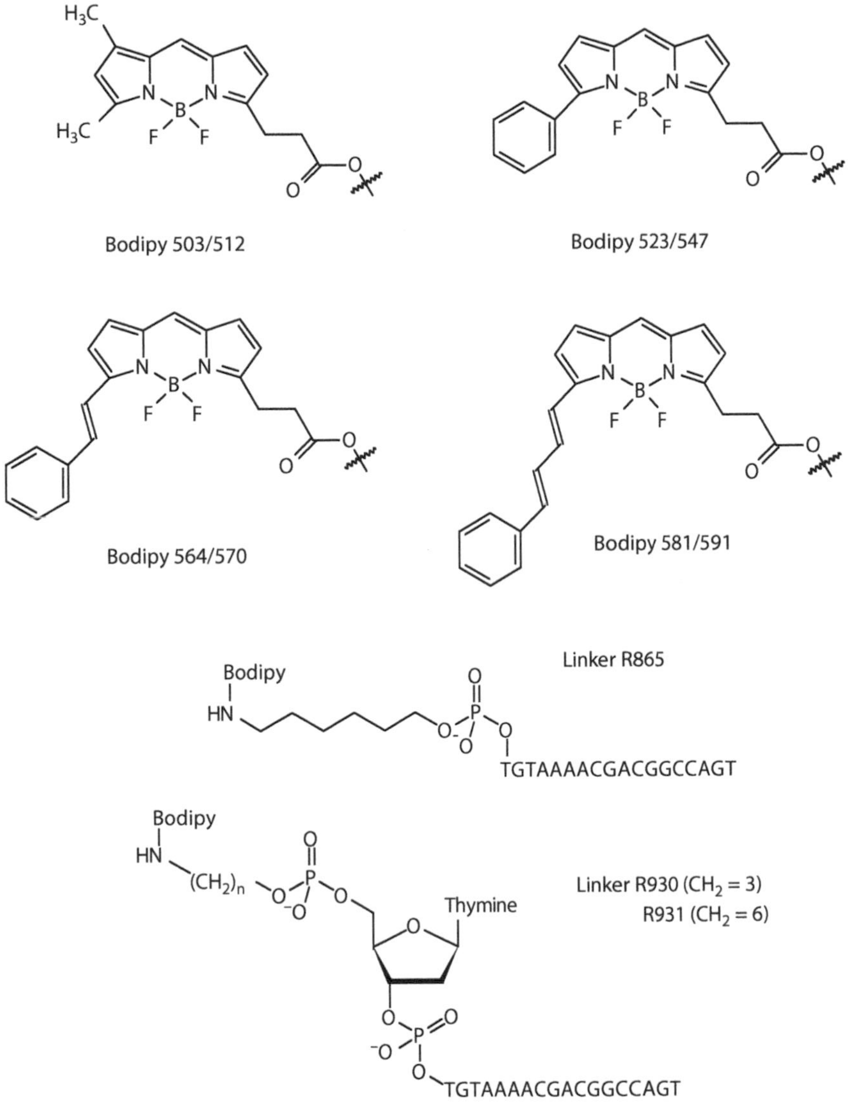

FIGURE 26.11 Bodipy dye set used for DNA-sequencing applications. Below each structure is the absorption/emission maxima (nm) of the particular dye. Also shown is the universal sequencing primer and the modification required to attach the dye to the primer. The linkers were either three or six carbon linkers with primary amine groups. The linkers are designated as R865 (C-6 linker, U-C_6), R930 (C-3 linker, U+-C_3), or R931 (C-6 linker, U+-C_6). (Adapted from Metzker, M.L. et al., *Science*, 271, 1420, 1996.)

the *Taq* polymerase due to its incompatibility with the binding pocket of *Taq*, while the rhodamine dyes are more hydrophobic and are thus more suitable for use with *Taq* polymerase. The result is that the peak heights for the electrophoresis bands can vary tremendously due to differences in incorporation of the dye-modified ddNTPs by the particular polymerase enzyme. In dye-primer chemistry, this disparity is absent due to the large displacement of the dye from the polymerization site.[10] It is interesting to note that several mutant forms of *Taq* polymerase have been prepared to allow more facile incorporation of dye-labeled terminators.[11] For example, *Taq* Pol I (AmpliTaq FS®) has two modifications in it—a substitution that eliminates the 3'→5' nuclease activity and also a substitution that improves 2',3' ddNTP incorporation.

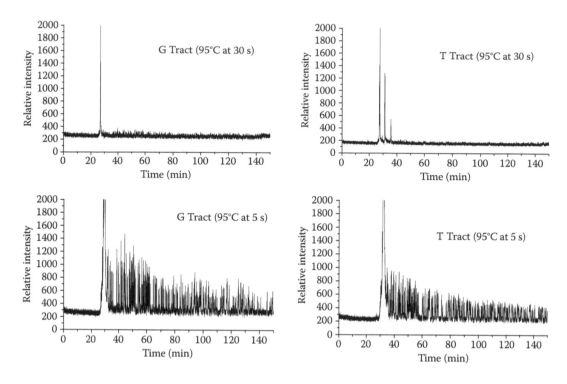

FIGURE 26.12 Cycle-sequencing effects on bodipy-labeled DNA-sequencing primers. The sequencing reactions were prepared using Sanger chain-termination methods and a *Taq* DNA polymerase. The cycling conditions consisted of 25 cycles with 95°C for 30 or 5 s; 55°C for 10 s, and 72°C for 60 s. The reactions were run with a single terminator (ddGTP, ddTTP) and an M13mp18 template. The sequencing reactions were analyzed in a CGE DNA sequencer using laser excitation at 532 nm. The dyes were bodipy 564/570 (ddGTP) and bodipy 581/591 (ddTTP) (see Figure 26.11 for structures of dyes). The large peak at ~28 min is unextended dye-labeled primer.

The nature of the dye on the terminator can also influence the mobility of the polymerized DNA fragment as well. For example, rhodamine dyes are typically zwitterionic and thus appear to stabilize hairpin (secondary) structures in the DNA fragment, causing compressions in the electrophoresis data, especially in GC-rich regions of the template. On the other hand, the fluorescein dyes, which are negatively charged, do not suffer from such anomalies.[10]

Slight structural modifications on the base chromophore can influence its incorporation during DNA polymerization. Also, the linker structure can influence the incorporation efficiency as well. Figure 26.13 shows a set of dye-labeled terminators that have been found to give fairly even peak heights in sequencing patterns and produce minimal mobility shifts; the figure also shows the sequencing patterns that were generated with both the rhodamine dyes and the dichlororhodamine (d-rhodamine) analogs.[12] The linker structures chosen for this set were either the propargylamine linker developed by Prober et al.[7] or a propargyl ethoxyamino linker. The choice of linker was based on its ability to accommodate the polymerase enzyme and to minimize the mobility differences within the dye set. For the rhodamine terminators, very weak G-peaks, which appeared after A-peaks, were observed. However, for the d-rhodamine terminators, this disparity in peak intensity was alleviated, with the peak heights in the pattern being much more uniform.

26.2.2 Energy Transfer Dyes for DNA Sequencing

One of the major problems associated with many of the single-dye sets mentioned previously is that their absorption spectra are dispersed over a relatively large spectral range, which provides poor excitation

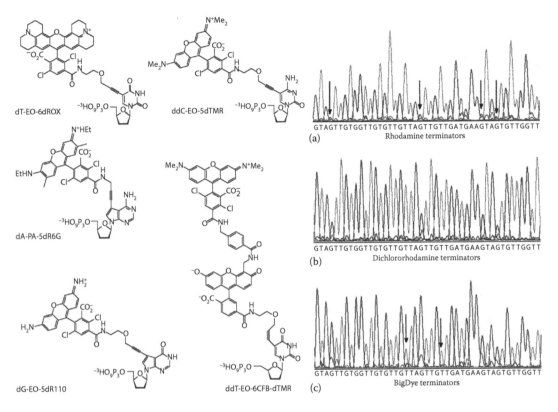

FIGURE 26.13 Chemical structures of the d-rhodamine dyes used for labeling terminators. Also shown is the structure of an energy transfer terminator (BigDye terminator). The accompanying panels show a comparison of sequencing data obtained using rhodamine- (a), d-rhodamine- (b), and Big Dye- (c) labeled terminators. The arrows in (a) are G-peaks with weak signal. In (c), the arrows indicate weaker T-peaks following G-peaks. The sequencing reactions were run with an AmpliTaq DNA polymerase and cycle-sequencing conditions (30 cycles with a temperature program of 95°C for 30 s, 55°C for 20 s, 60°C for 4 min). The DNA template was isolated from a bacterial artificial chromosome. Following chain extension, the reactions were purified using a Centri-Sep spin column to remove excess terminators and then dried in a vacuum and resuspended in 2–4 μL of formamide. The reactions were electrophoresed in a slab gel with laser-induced fluorescence detection. (Adapted from Rosenblum, B.B. et al., *Nucleic Acids Res.*, 25, 4500, 1997.)

efficiency even for dual-laser (488 and 514 or 543) systems. As such, the red dyes are used at higher concentrations during DNA polymerization to circumvent poor excitation. To overcome this problem without sacrificing spectral dispersion in the emission profiles, the phenomenon of Förster energy transfer has been used to design sequencing primers that can be efficiently excited with a single laser line.[13–16] A brief introduction about these ET (energy transfer) primers in the context of DNA sequencing follows.

The chemical structures of the ET primers developed by Mathies et al.[15] are shown in Figure 26.14. The donor dye in this case was FAM, which could be excited with the 488-nm line of an argon ion laser. The acceptor dyes were FAM, JOE, TAMRA, or ROX. The donor (FAM) was attached to the sequencing primer on the 5′end during the solid-phase DNA synthetic preparation of the M13 (−40) sequencing primer using phosphoramidite chemistry. The sequencing primer also contained a modified base (T*) that possessed a linker structure with a primary amine. The appropriate acceptor was conjugated to the primary amine group off the modified base following cleavage from the solid support via a succinimidyl ester functional group. The spacer distance between the donor and acceptor was selected by positioning T* within the M13 (−40) primer sequence during solid-phase synthesis. The naming of these ET primers followed the convention donor-spacer (bp)-acceptor. A 10-base spacer was used for

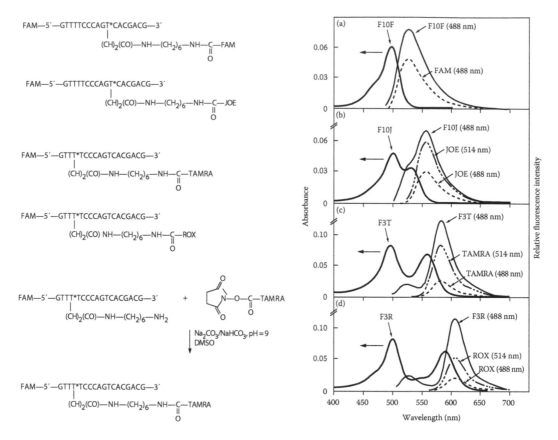

FIGURE 26.14 Chemical structures and absorbance and fluorescence-emission spectra of energy-transfer (ET)-labeled DNA-sequencing primers. The bottom panel shows the reaction of a dye-labeled FAM primer (5'-end with –NH$_2$ T modified residue) and a succinimidyl TAMRA dye (F3T). The absorption (dark line) and emission spectra for both the ET primers and the single-dye-labeled primers are also shown for comparison purposes. The number in parenthesis is the excitation wavelength used for collecting the emission profile. (Adapted from Ju, J. et al., *Proc. Natl. Acad. Sci. U.S.A.*, 92, 4347, 1995.)

the FAM and JOE ET primers, and a three-base spacer was selected for TAMRA and ROX. The choice of spacer size was primarily determined by producing ET primers, which showed uniform electrophoretic mobilities.

Figure 26.14 shows the absorbance and emission profiles of the ET primer series. While the absorption spectra show bands from both the acceptor and donor dyes, the emission profiles are dominated by fluorescence from the acceptor dyes. In fact, the ET efficiency has been determined to be 65% for F10J, 96% for F3R, and 97% for F3T.[15] As shown in the emission profiles in Figure 26.14, the emission intensity was found to be significantly higher for the ET primers compared to their single-dye primer partners when excited with 488-nm laser light from the argon ion laser due to higher efficiency in excitation. This translates into improved fluorescence sensitivity of these ET primers during electrophoresis.

These ET primers do offer some advantages due to their improved detection sensitivities; those advantages include eliminating the need for adjusting the concentrations of the dye primers used during polymerization and also the need for smaller amounts of template in the sequence analysis. In fact, the use of ET primers required about one fourth the amount of template vs. the single-dye primers.

ET dye pairs can also be situated on terminators; an example is shown in Figure 26.13.[12] In this example, d-rhodamine and rhodamine dyes are used with a propargyl ethoxyamino linker. This dye set has been called BigDye terminators. Again, the dyes were selected so as to provide fairly uniform peak heights in the electrophoresis and also uniform mobility shifts within the series. A sample of a sequence run using these BigDye terminators is also shown in Figure 26.13.

26.2.3 Near-Infrared Dyes for DNA Sequencing

The attractive feature associated with fluorescence in the near-infrared (NIR) ($\lambda_{ex} > 700$ nm) includes smaller backgrounds observed during signal collection and the rather simple instrumentation required for carrying out ultrasensitive detection. In most cases, the limit of detection for fluorescence measurements is determined primarily by the magnitude of the background produced from scattering or impurity fluorescence. This is particularly true in DNA sequencing because detection occurs within the gel matrix, which can be a significant contributor of scattering photons. In addition, the use of denaturants in the gel matrix, such as urea (7 M) or formamide, can produce large amounts of background fluorescence. The lower background that is typically observed in the NIR can be attributed to the fact that few species fluoresce in the NIR. In addition, the $1/\lambda^4$ dependence of the Raman cross section also provides a lower scattering contribution at these longer excitation wavelengths.

An added advantage of NIR fluorescence is the fact that the instrumentation required for detection can be rather simple and easy to use. A typical NIR fluorescence-detection apparatus can consist of an inexpensive diode laser and single-photon avalanche diode (SPAD). These components are solid state, which allows the detector to be run for extended periods of time, requiring little maintenance or operator expertise.

NIR fluorescence can be a very attractive detection strategy in gel sequencing because of the highly scattering medium in which the separation must be performed. Due to the intrinsically lower backgrounds that are expected in the NIR vs. the visible, on-column detection can be performed without sacrificing detection sensitivity. To highlight the intrinsic advantages associated with the use of NIR fluorescence detection in capillary-gel DNA-sequencing applications, a direct comparison between laser-induced fluorescence detection at 488-nm excitation and 780-nm excitation has been reported.[17] In this study, a sequencing primer labeled with FAM or a NIR dye were electrophoresed in a capillary-gel column and the detection limits calculated for both systems. The results indicated that the limits of detection for the NIR case were found to be 3.4×10^{-20} mol, while for 488-nm excitation the limit of detection was 1.5×10^{-18} mol. The improvement in the limit of detection for the NIR case was observed in spite of the fact that the fluorescence quantum yield associated with the NIR dye was only 0.07, while the quantum yield for the FAM dye was ~0.9. The improved detection limit resulted primarily from the significantly lower background observed in the NIR. NIR has also been demonstrated in sequencing applications using SGE, where the detection sensitivity has been reported to be ~2000 molecules.[18,19]

The fluorophores that are typically used in the NIR include the tricarbocyanine (heptamethine) dyes, which consist of heteroaromatic fragments linked by a polymethine chain (see Table 26.1). The absorption maxima can be altered by changing the length of this polymethine chain or by changing the heteroatom within the heteroaromatic fragments. These NIR dyes typically possess large extinction coefficients and relatively low quantum yields in aqueous solvents.[20] The low-fluorescence quantum yields result primarily from high rates of internal conversion. An additional disadvantage associated with these dyes is their poor water solubility. These dyes show a high propensity toward aggregation, forming aggregates with poor fluorescence properties. This aggregation can be alleviated to a certain degree by inserting charged groups within the molecular framework of the fluorophore, for example, alkyl-sulfonate groups. In addition, these dyes have short fluorescence lifetimes and are very susceptible to photobleaching vs. the visible dyes typically used in DNA-sequencing applications. In most cases, the NIR chromophore is covalently attached to either a sequencing primer or ddNTP via an isothiocyanate functional group. As shown in Table 26.1, this functional group can be placed on the heteroatom (N in this case) or in the para-position of the bridging phenyl ring. When the isothiocyanate group is situated on the heteroatom, the net charge on the dye becomes neutral and as such

TABLE 26.1 Chemical Structures of Some Typical NIR Fluorescent Dyes and Their Photophysical Properties

Sulfonated, Heavy-Atom Modified Near-IR Dyes

(X = I, Br, Cl, F)

Dye Substitution	Absorbance (nm)	Emission (nm)	ε (M^{-1} cm^{-1})	ϕ	τ_f (ps)
I	766	796	216,000	0.15	947
Br	768	798	254,000	0.14	912
Cl	768	797	239,000	0.14	880
F	768	796	221,000	0.14	843

	Absorbance (nm)	Emission (nm)	ε (M^{-1} cm^{-1})	εf	ε (M^{-1} cm^{-1})
IRD 41	787	807		0.16	200,000

Note: In the top panel of this table, the dyes were modified with an intramolecular heavy atom to produce a set of dyes with four distinct lifetime values. In both cases, the dyes contained an isothiocyanate to allow conjugation to amine-containing molecules.

has limited water solubility. If the isothiocyanate group is placed on the bridging phenyl ring, this gives a net −1 charge to the dye and improves its water solubility. Unfortunately, the ether linkage is susceptible to nucleophilic attack, especially by dithiothriotol (DTT), which can release the dye from the moiety (sequencing primer or ddNTP) to which it is attached. This can produce a large peak in the electropherogram, which can mask some of the sequencing fragments that comigrate with the free dye. This problem can be alleviated to a certain extent by using an ethanol-precipitation step following DNA polymerization.

26.3 Instrumental Formats for Fluorescence Detection in DNA Sequencing

The ability to read the fluorescence during the electrophoresis fractionation of the DNA ladders and also accurately identify the terminal base (Sanger sequencing) is a challenging task due to a number of technical issues. As such, a number of fluorescence-readout devices have been developed for reading such data. When considering the design of a fluorescence detector for sequencing applications, several instrumental constraints must be incorporated into the design including:

1. *High sensitivity.* As pointed out previously, the amount of material (fluorescently-labeled DNA) loaded onto the gel can be in the low attomole range (10^{-18} mol), and the detector must be able to read this fluorescence signature with a reasonably high signal-to-noise ratio (SNR) to accurately call the base. Therefore, in most cases, irrespective of the separation platform, the source of excitation is a laser that is well matched to the excitation maxima of the dye set used in the sequencing experiment.

2. *Spectral-marker base identification.* The instrument must be able to identify one of the four bases terminating the sequencing fragments by accurately processing the fluorescence via spectral discrimination (wavelength) or some other fluorescence property, such as the lifetime. For spectral discrimination, this would require sorting the fluorescence by wavelength using either filters or gratings. In addition, multiple detection channels would be required.

3. *Fluorescence processing from many electrophoresis lanes.* In most sequencing instruments, the fluorescence must be read from multiple gel lanes (SGE) or multiple capillaries (CGE). This can be done by either using a scanning system, in which the relay optic (collection optic) is rastered over the gel lanes or capillaries, or an imaging system, in which the fluorescence from the multiple gel lanes or capillaries are imaged onto some type of multichannel detector, such as a charge-coupled device (CCD) or image-intensified photodiode array.

4. *Robust instrumentation.* Since many sequencing devices are run by novice operators and are also run for extended periods of time, the detector format must be dependable and turnkey in operation.

This short list of requirements for any type of fluorescence-readout device appropriate for sequencing presents itself with many challenges that are often noncomplementary with the sequencing requirements. For example, high sensitivity is particularly demanding because the separation platforms used to fractionate the DNA are becoming smaller; therefore, smaller amounts of material must be inserted into the device. In addition, detecting material directly within the gel matrix (typically a polyacrylamide gel) can be problematic due to the intense scattering photons that it produces. Also, the signal is transitory in that the DNA fragment resides within the probing volume (defined by the laser beam size) for a few seconds. Another issue is that many separation channels or lanes must be interrogated for high-throughput applications. On top of these considerations, a high SNR is required to obtain high accuracy in the base-calling phases of the readout process. As such, significant design considerations go into fabricating a fluorescence detector system for DNA-sequencing applications.

There are two general types of fluorescence detector formats—scanning and imaging-type devices. In most high-throughput sequencing devices, multiple lanes of the slab gel or multiple capillaries are run in parallel to increase system throughput. For example, many machines run in a 96-lane format because microtiter plates, which are standard plates used to prepare sequencing reactions, come in a 96-well format. In the scanning systems, the excitation beam is tightly focused and irradiates only a single lane, while the relay optic is rastered over the lanes of the gel or the capillary array and the fluorescence from each lane processed sequentially on one set of detection channels. For the imaging systems, all of the electrophoresis lanes are irradiated by a laser or lasers simultaneously, with the fluorescence readout accomplished using a multichannel detector such as a CCD.

26.3.1 Fluorescence Scanning Instruments

Figure 26.15 shows a typical scanning system.[21,22] This system uses a confocal geometry with epi-illumination in which the objective used to collect the emitted fluorescence also serves to focus the laser beam into individual capillaries (or lanes of the slab gel) used for the electrophoresis. Following collection the emission is focused onto a spatial filter at the secondary image plan of the collection objective. The laser light (488 nm, argon ion laser, 1 mW) is directed into the objective using a dichroic filter. The capillaries are held into a linear array, and in this particular example the capillary array is translated beneath the microscope objective. As shown in the figure, the laser irradiates only one capillary at a time,

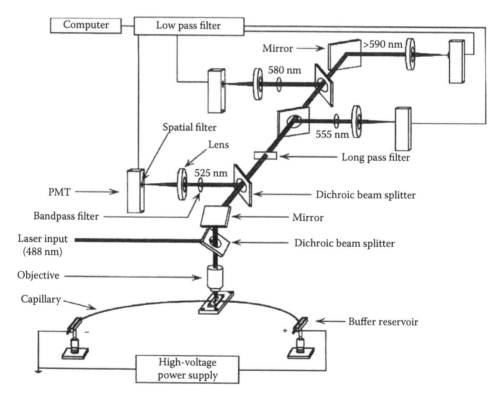

FIGURE 26.15 Four-color, laser scanning fluorescence system for DNA sequencing applications using capillary gel electrophoresis. The excitation source was an air-cooled Ar ion laser operating at 488 nm (1 mW average power). The collection and focusing optic consisted of a microscope objective (20×, NA = 0.5). The emission was directed onto one of four different PMTs using dichroic filters and further isolated from background photons using a bandpass filter. The fluorescence was sampled at 2 Hz with output filtered using a low-pass filter with a time-constant of 1 s. (Adapted from Ju, J. et al., *Proc. Natl. Acad. Sci. U.S.A.*, 92, 4347, 1995.)

but since the beam is tightly focused (diameter = 10 μm) the electronic transition can be saturated at relatively low laser powers, improving SNR in the fluorescence measurement.[23] In addition, the noise can be significantly suppressed in this system, since a pinhole is used in the secondary image plane of the collection microscope objective, preventing scattered out-of-focus light generated at the walls of the capillary from passing through the optical system. The capillary array is scanned at a rate of 20 mm/s, with the fluorescence sampled at 1500 Hz/channel (color channel) resulting in a pixel image size of 13.3 μm. The fluorescence is collected by a 32× microscope objective (numerical aperture = 0.4), resulting in a geometrical collection efficiency of approximately 12%. Once the fluorescence has been collected by the objective, it is passed through a series of dichroics to sort the color and then processed on one of four different PMTs, with each PMT sampling a different color (spectral discrimination). While the present system is configured with four color channels, the system could easily be configured to do two-color processing as well by removing the last dichroic filter in the optical train and two of the PMTs. The concentration limit of detection of this system has been estimated to be 2×10^{-12} M (SNR = 3), which was determined by flowing a solution of fluorescein through an open capillary.[21] The detection limit would be expected to degrade in a gel-filled capillary due to the higher background that would be generated by the gel matrix.

26.3.2 Fluorescence-Imaging Systems for DNA Sequencing

Figure 26.16a shows an imaging system for reading fluorescence from multichannel capillary systems for DNA sequencing.[24,25] In this example, a sheath flow cell is used with the laser beam (or beams) traversing the sheath flow and the capillary output dumping into the sheath stream (see Figure 26.18c). This geometry allows simultaneous irradiation of all the material migrating from the capillaries without requiring the laser beam to travel through each capillary, which would cause significant scattering and reduce the intensity of the beam as it traveled through the array. The sheath flow cell also causes contraction of the fluid output of the capillary since the sheath flow runs at a greater linear velocity compared to the sample (capillary) stream. Figure 26.18b shows a fluorescence image of the output from the capillary array, indicating that the sample stream diameter at the probing point was ~0.18 mm and also demonstrating minimal cross talk between individual capillaries in the array. The laser beams (488 nm from argon ion, 6 mW; 532 nm from frequency-doubled yttrium-aluminum-garnet [YAG], 6 mW) were positioned slightly below the exit end of the capillaries (see Figure 26.18a) with the beams brought colinear using a dichroic filter. The collection optic and focusing optic produced a total magnification of 1 and resulted in a geometrical collection efficiency of 1%. To achieve multicolor processing capabilities, the collected radiation was sent through an image-splitting prism to produce four separated (spectrally) line images on the array detector (one for each dye used to label the terminal bases). In addition, a series of narrow bandpass filters were placed in front of the image-splitting prism to assist in isolating the appropriate colors for data processing. The detector that was used for this system was a two-dimensional CCD camera with a cooled image intensifier. Interestingly, the detection limit reported for this system was found to be 2×10^{-12} M when operated in the four-color mode, comparable to that seen for the scanning system discussed above. However, the presence or absence of the gel matrix does not affect the sensitivity of the fluorescence measurement in this case, since the fluorescence interrogation is done off column in the sheath flow. In fact, researchers have reported that the implementation of the sheath flow geometry in gel electrophoresis can offer a significant improvement in limits of detection by minimizing scattering contributions to the background.[26]

When comparing these two fluorescence-readout systems, several issues should be highlighted. One is the duty cycle, which takes into account the loss in signal due to multiple-lane sampling. For any type of scanning system the sampling of the electrophoresis lanes is done in a sequential fashion. For a scanning system sampling 96 capillaries the duty cycle is approximately 1%. However, in the imaging system, all capillaries are sampled continuously and the duty cycle is nearly 100%. Therefore, comparisons of detection limits for any system must include a term for the duty cycle because lower duty cycles will degrade the limit of detection.

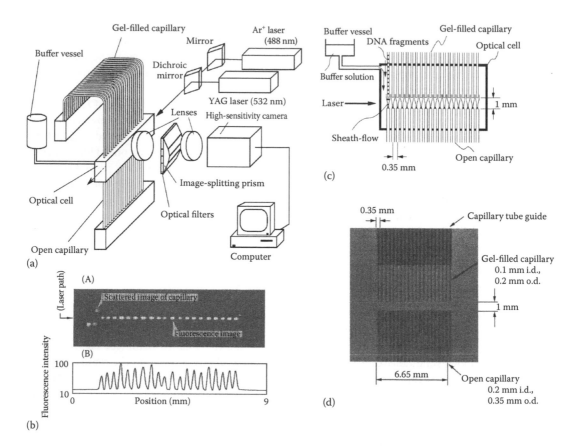

FIGURE 26.16 (a) Schematic of an imaging laser-induced fluorescence detector for reading four-color fluorescence from capillary-gel arrays. The lasers used for excitation were an argon ion (488 nm) and a YAG laser (532 nm). The lasers were allowed to traverse below the output of the gel columns and the fluorescence collected with a lens system. The fluorescence line image was split into four different color images using a polyhedral image-splitting prism coupled to optical filters. The filtered fluorescence was detected using a two-dimensional CCD. (b) Fluorescence image in the one-color mode across an array of 20 capillaries. The integrated fluorescence intensity is shown in the bottom panel. (c) Schematic view of the multiple sheath flow cell using gravity feed for the sheath flow. Twenty gel-filled capillaries were aligned at a 0.35-mm pitch in an optical cell (26 mm × 26 mm × 4 mm). (d) Photograph of the capillary array aligned in the sheath flow cell. (Adapted from Takahashi, S. et al., *Anal. Chem.*, 66, 1021, 1994.)

26.4 DYE Primer/Terminator Chemistry and Fluorescence Detection Formats

In most sequencing applications, dye-labeled primers are used for accumulating sequencing data using automated instruments. This stems from the fact that dye-labeled primers are typically less expensive to use vs. their dye-labeled terminator counterparts. Also, in most applications, small pieces of DNA (1–2 kbp in length) are cloned into bacterial vectors for propagation (to increase copy number), such as M13s, which have a known sequence and serve as ideal priming sites. However, dye-labeled primers do present problems; for example, the sequencing reactions must be run in four separate tubes during polymerization and then pooled prior to gel electrophoresis. In addition, unextended primer can result in a large electrophoretic peak (i.e., high intensity), which often masks the ability to call bases close to the primer-annealing site.

Dye-labeled terminators can be appealing to use in certain applications, for example, when high-quality sequencing data is required and in primer-walking strategies. In primer walking, the sequence of the DNA template is initiated at a common priming site using a primer that is complementary to that site. After reading the sequence at that site, the template is subjected to another round of sequencing, with the priming site occurring at the end of the first read. In this way, a long DNA can be sequenced by walking in a systematic fashion down the template. Dye terminators are particularly attractive because primers need to be synthesized frequently in primer-walking strategies, and the need for nonlabeled primers simplifies the synthetic preparation of these primers. Dye terminators improve the quality of sequencing data in many cases because the excess terminators are removed prior to electrophoresis (using size-exclusion chromatography) and as such give clean gel reads free from intense primer peaks. However, it should be noted that in most cases, terminators can produce uneven peak heights (broad distribution of fluorescence intensities) due to the poor incorporation efficiency of dye terminators by polymerase enzymes.

When dye-labeled primers are used, several different formats can be implemented to reconstruct the sequence of the template when fluorescence detection is being used. In the case of spectral discrimination, these formats may vary in terms of the number of dyes used, the number of detection channels required, or the need for running one to four parallel electrophoresis lanes. For example, if the sequencing instrument possesses no spectral-discrimination capabilities, the electrophoresis must be run in four different lanes, one for each base comprising the DNA molecule. However, if four different dyes are used, the electrophoresis can be reduced to one lane, and as a result the production rate of the instrument goes up by a factor of 4. The fluorescence-based formats discussed here include (number of dyes/number of electrophoresis lanes), single-dye/four-lane; single-dye/single-lane; two-color/single-lane; and, finally, four-color/single-lane strategies.

The most pressing issue in any type of DNA-sequencing format is the accuracy associated with the base call, which is intimately related to a number of experimental details, for example, the number of spectral channels used in the instrument as well as the SNR in the measurement. The information content of a signal, I, can be determined from the simple relation[27]:

$$I = n \log_2 (\text{SNR}) \tag{26.3}$$

where

n is the number of spectral channels
SNR is the SNR associated with the measurement

The term I is expressed in bits, and typically two bits are necessary to distinguish between four different signals, but only if there is no spectral overlap between the dyes used for identifying the bases. Unfortunately, in most multicolor systems, the spectra of the dyes used in the sequencing device show significant overlap; thus, many more bits will be required to call bases during the sequencing run.

While the above equation can provide information on how to improve the accuracy of the base call, it does not provide the sequencer with information on the identity of the individual electrophoretic peaks (base call) or the quality of a base call within a single-gel read. For example, if four-color sequencing is used with dye-primer chemistry, how should the data be processed, and what is the confidence with which an electrophoretic peak is called an A, T, G, or C? To provide such information, an algorithm has been developed to not only correct for anomalies associated with fluorescence-based sequencing but also assign a quality score to each called base. The typical algorithm used is called the *Phred* scale, and it employs several steps to process the sequencing data obtained from fluorescence-based, automated DNA sequencers.[28]

The data input into *Phred* consists of a trace, which is electrophoretic data processed into four spectral channels, one for each base. The algorithm consists of four basic steps:

1. *Idealized electrophoretic peak locations (predicted peaks) are determined.* This is based on the premise that most peaks are evenly spaced throughout the gel. In regions where this is not the case (typically during the early and late phases of the electrophoresis), predictions are made as to the

number of correct bases and their idealized locations. This step is carried out using Fourier methods as well as the peak-spacing criterion and helps to discriminate noise peaks from true peaks.

2. *Observed peaks are identified in the trace.* Peaks are identified by summing trace values to estimate the area in regions that satisfy the criterion $2 \times v(i) \geq v(i + 1) + v(i - 1)$, where $v(i)$ is the intensity value at point i. If the peak area exceeds 10% of the average area of the preceding ten accepted peaks and 5% of the area of the immediate preceding peak, it is accepted as a true peak.

3. *Observed peaks are matched to the predicted peak locations, omitting some peaks and splitting others.* In this phase of the algorithm, the observed peak arises from one of four spectral channels and thus can be associated with one of the four bases. It is this ordered list of matched observed peaks that determines a base sequence for the DNA template in question.

4. *Observed peaks that are uncalled (unmatched to predicted peaks) are processed.* In this step, an observed peak that did not have a complement in the predicted trace is called and assigned a base and finally inserted into the read sequence.

This algorithm deals mainly with sorting out difficulties associated with the electrophoresis by identifying peaks in the gel traces, especially in areas where the peaks are compressed (poor resolving power) or where multiple peaks are convolved due to significant band broadening produced by diffusional artifacts.

Often, preprocessing of the traces is carried out prior to *Phred* analysis to correct for dye-dependent mobility shifts. In most cases, these mobility shifts are empirically determined by running an electropherogram of a single dye-labeled DNA ladder (for example, T-terminated ladder) and comparing the mobilities to the same ladder labeled with another dye of the set. This type of analysis can be very complex and involved because the mobility shift depends both on the dye and linker structure and on the separation platform used. For example, dyes that show uniform mobility shifts in SGE may not show the same effect in CGE. In addition, these mobility shifts can be dependent on the length of the DNA to which the dye is attached.[29] An example of this phenomenon is shown in Figure 26.17. In this particular example, cyanine dyes were covalently anchored to an M13 (–40) sequencing primer and annealed to an M13 template followed by extension with a single terminator (ddT). The tracts were electrophoresed using CGE. In Figure 26.17a, comparison between a Cy5T7 (–2 charge) and Cy5.5T (–1 charge) tracts indicates that the Cy5T7-labeled fragments migrate faster in the beginning of the run (smaller DNA fragments), but after 300 bp, the two dye-labeled fragments comigrate. In Figure 26.17b, a mobility crossover occurs at ~125 bp for the dyes Sq5T4 (neutral charge) and Cy5.5T12 (–2 charge), with the shorter fragments migrating faster with the Cy5.5T12 label, and after this the Sq5T4-labeled fragments migrate faster. These types of mobility shifts have been ascribed to differences in the net charge of the dye label and to potential dye–DNA base interactions. These interactions, predominantly driven by hydrophobic interactions, may cause loops or hairpin structures on the 5′ end of the dye–DNA complex. These structures would cause a faster migration rate compared to fully extended structures produced by most dye–DNA complexes.

26.4.1 Single-Color/Four-Lane

In this processing format, only a single fluorescence detection channel is required to analyze the signal from the labeling dye because only a single dye is used to detect the sequencing fragments produced following chain extension. However, since no color discrimination is implemented, the electrophoresis must be run in four lanes, one for each base, like the format used in traditional autoradiographic detection. While this is a reasonable approach for slab-gel separations, it is not a viable strategy in capillary-gel applications due to the poor run-to-run reproducibility in the migration rates of the fragments traveling through the different capillaries. This is due to differences in the gel from capillary to capillary as well as differences in the integrity of the wall coatings used to suppress the electroosmotic flow. In the slab-gel format, reproducibility in the migration times becomes less of a problem because all of the lanes are run in the same gel matrix.

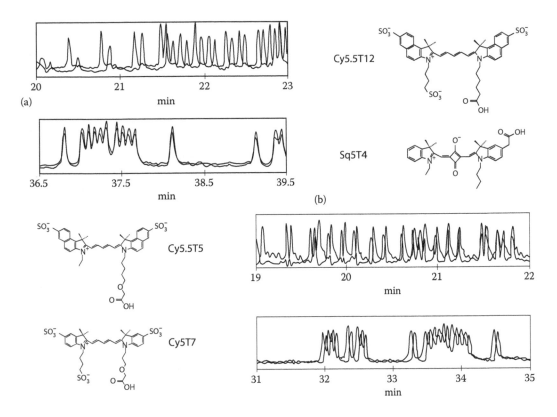

FIGURE 26.17   Electrophoresis traces and chemical structures of four different cyanine dyes used to end-label DNA sequencing primers. (a) Gel trace of Cy5.5T5 (red trace) and Cy5T7 (blue trace). (b) Gel trace of Cy5.5T12 (red trace) and Sq5T4 (blue trace). In all cases, the ladders were prepared using a single terminator (ddT) and an M13mp18 template. The electrophoresis was carried out in a capillary-gel column (field strength = 185 V/cm) using a hydroxyethyl cellulose sieving buffer. (Adapted from Tu, O. et al., *Nucleic Acids Res.*, 26, 2797, 1998.)

Figure 26.18 shows the output of a typical single-color/four-lane sequencing device along with the called bases. In this example, each different color trace represents an electropherogram from an individual lane of the slab gel, which is overlaid to allow reconstruction of the sequence of the template. In this case, the device used a single microscope head containing the collection optics, a diode laser, filters, and avalanche photodiode to read the fluorescence from the gel.[18] The microscope scanner is rastered over the gel at a rate of ~0.15 cm/s and monitors fluorescence along a single axis of the gel. The gel is approximately 20 cm in width and 42 cm in length and can accommodate 48 separate lanes. The time required to secure this data was 6 h, with the extended time due primarily to the limited electric field that can be applied to the thick slab gel.

26.4.2 Single-Color/Single-Lane

In this sequencing approach, only a single fluorophore is used, and thus only a single laser is required to excite the fluorescence and only a single detection channel is needed to process the fluorescence. The advantage of this approach is that instrumentally it is very simple because the hardware required for detection is simple. In addition, because the sequence is reconstructed from a single electrophoresis lane and not four, the throughput can be substantially higher compared to a single-fluorophore/four-lane method.

The bases are identified by adjusting the concentration ratio of the terminators used during DNA enzymatic polymerization to alter the intensity of the resulting electrophoretic bands.[17,27,30,31] Therefore,

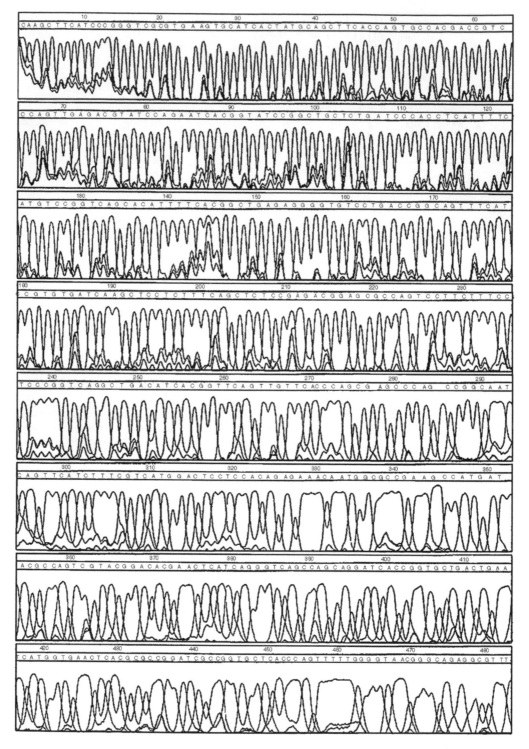

FIGURE 26.18 **(See color insert.)** Single-color/four-lane sequencing trace (called bases 1–480) of a PCR-amplified λ-bacteriophage template. The sequencing was performed using an IRD-800-labeled primer (21mer) and SGE instrument (Li-COR 4000). The gel consisted of 8% polyacrylamide with 7 M urea. The dimensions of the gel were 25 cm (width) by 41 cm (length). The traces from the four lanes were overlaid to reconstruct the sequence of the template.

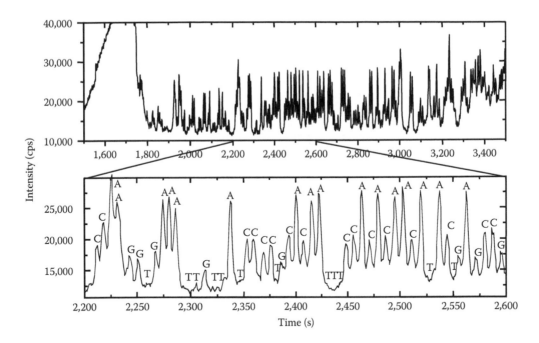

FIGURE 26.19 Single-color/single-lane capillary-gel sequencing data. The template was an M13mp18 phage, and the primer was labeled with a NIR fluorescent dye and detected using NIR, laser-induced fluorescence. The terminators were adjusted to a concentration ratio of 4:2:1:0 during extension to allow identification based on fluorescence intensities. The separation was performed at a field strength of 250 V/cm, and the gel column contained a 3% T/3% C crosslinked polyacrylamide gel. (Adapted from Williams, D.C. and Soper, S.A., *Anal. Chem.*, 67, 19, 3427, 1995.)

if the concentration of the terminators used during DNA polymerization was 4:2:1:0 (A:C:G:T), a series of fluorescence peaks would be generated following electrophoretic sizing with an intensity ratio of 4:2:1:0, and the identification of the terminal bases would be carried out by categorizing the peaks according to their heights. To accomplish this with some degree of accuracy in the base calling, the ability of the DNA polymerase enzyme to incorporate the terminators must be nearly uniform. This can be achieved using a special DNA polymerase, which in this case is a modified T7 DNA polymerase.[32,33] This enzyme has been modified so as to remove its proofreading capabilities by eliminating its 3′ → 5′ exonuclease activity. Since this method requires uniform incorporation of the terminators, it is restricted to the use of dye-primer chemistry. In addition, since the T7 enzyme is not a thermostable enzyme, cycle sequencing cannot be used.

Figure 26.19 presents an example of sequencing data accumulated using this base-calling strategy. In this example, the terminator concentration was adjusted at a concentration ratio of 4:2:1:0 (A:C:G:T).[56] The accuracy in calling bases was estimated to be 84% up to 250 bases from the primer annealing site, with readable bases up to 400 but with the accuracy deteriorating to 60%. This is a common artifact in this approach—poor base-calling accuracy. The poor base calling results from variations in the activity of the T7 DNA polymerase and, in addition, the null signal used to identify Ts. Ambiguities are present when multiple null signals must be identified, which can lead to insertions or deletions.

26.4.3 Two-Color/Single-Lane

To improve on the base-calling accuracy associated with the single-color/single-lane strategy without having to increase the instrumental complexity of the fluorescence readout device associated with sequencing instruments, a two-color format may be used to identify the four terminal bases in

TABLE 26.2   Binary Coding Scheme
for Two-Color DNA Sequencing

	FAM	JOE
A	1	1
C	0	1
G	1	0
T	0	0

Note: 1, the presence of the dye-labeled primer during DNA polymerization; 0, represents the absence of the dye label.

sequencing applications. In this approach, one or two lasers are used to excite one of two spectrally distinct dyes used for labeling the sequencing primers, and the fluorescence is processed on one of two detection channels, consisting of bandpass filters and photon transducers.

Figure 26.15 shows a schematic of a two-color scanning instrument (last dichroic was not used, and two PMTs were removed). It was used to excite the fluorescence of the labeling dyes, FAM and JOE. Due to their similar absorption maxima, a single laser (488 nm) could efficiently excite the fluorophores. In addition, this dye pair was selected because they produced sequencing fragments that comigrated, and thus no mobility correction was required. The bases were identified using a binary coding scheme, which is shown in Table 26.2.[22] During DNA polymerization, four separate reactions were run, with the A-reaction containing an equimolar mixture of the FAM- and JOE-labeled sequencing primers. In the case of G, only the JOE-labeled primer was present, while for T only the FAM-labeled primer was present, and for C no dye-labeled primer was used. By ratioing the signal in the red (JOE) to green (FAM) channel, a value was obtained that could be used to identify the terminal base. The attractive feature of this protocol is that while the absolute intensity of the bands present in the electropherogram may vary by a factor of 20 due to sequence-dependent termination, the ratio varies by a factor of only 1.7. The read length using this binary coding system was up to approximately 350 bases, with the number of errors ~15 (accuracy = 95.7%). The majority of the errors were attributed to C determinations, since a null signal was used to indicate the presence of this base.

To alleviate the errors in the base calling associated with identifying bases using a null signal, a two-dye, two-level approach can be implemented.[34,35] In this method, the bases using a common dye label have the concentration of ddNTPs adjusted during chain extension to alter the intensity of the fluorescence peaks developed during the electrophoresis. Also required in this approach are uniform peak heights, requiring the use of the modified T7 DNA polymerase in the presence of manganese ions. For example, Chen et al.[34] used a FAM-labeled primer for marking Ts and Gs, with the concentration ratio of the ddNTPs adjusted to 2:1 (T:G). Likewise, the As and Cs were identified using a 2:1 concentration ratio of the terminators, with the labeling dye in this case being TAMRA. Sequencing data produced an effective read length of 350 bases, with an accuracy of 97.5%. When the concentration ratios of the terminators sharing a common labeling dye was increased to 3:1, the read length was extended to ~400 bases with a base-calling accuracy >97%. As is evident, the elimination of null signals to identify bases can improve the base-calling accuracy in these types of sequence determinations.

While most fluorescence-labeling strategies for DNA sequencing that depend on differences in intensities of the electrophoretic peaks to identify bases use dye-labeled primers, internal labeling, where the fluorescent dye is situated on the dNTP, can also be used.[36] The advantages associated with using dye-labeled dNTPs are (1) the ability to use a wide range of primers because no dye-labeled primer is required, (2) incorporation of the dye-labeled dNTP can be much more uniform than dye-labeled ddNTPs, and (3) dye-labeled dNTPs are much less expensive than dye-labeled primers and terminators. Using a tetramethylrhodamine-labeled dATP and a fluorescein-labeled dATP, a two-color/single-lane sequencing assay has been reported.[36] For internal labeling, a two-step polymerization reaction was used in which the template was annealed to the sequencing primer (unlabeled) along with the dye-dATP and the four unlabeled dNTPs as well as the polymerase enzyme. The extension reaction was incubated at 37°C for 10 min, after which the appropriate terminator was added and the reaction allowed to proceed for an additional 10 min. The initial extension reaction extended six to eight nucleotides to a quartet of As, with 80%–90% of the fragments containing a single dye-labeled dATP. Since only two dyes were used in this particular example, the concentration ratio for a pair of terminators sharing a common dye was adjusted (3:1) to allow discrimination based on the intensity of the resulting electrophoretic peaks. Analysis of the sequencing data indicated that the read length was 500 bases with an accuracy of 97%.

26.4.4 Four-Color/Single-Lane

The commonly used approach in most commercial DNA-sequencing instruments using fluorescence detection is the four-color/single-lane strategy for identifying the terminal bases in sequencing applications. The primary reasons for using a four-color/single-lane approach are that it provides high accuracy in the base calling, especially for long reads, and the throughput can be high due to the fact that all bases comprising the template DNA can be called in a single-gel tract. Unfortunately, a four-color detector requires extensive optical components to sort the fluorescence, and in some cases multiple excitation sources are needed to efficiently excite the fluorophores used to label the individual sequencing ladders. In addition, post-electrophoresis software corrections may be required to account for spectral leakage into detection channels.

Most dye-terminator reads are used with this four-color strategy, since the data analysis (base calling) does not depend on uniform incorporation efficiencies, which are hard to achieve using dye-labeled teminators. The same type of instrumentation that is used for four-color/dye-primer reads can also be used for four-color/dye-terminator reads. The only difference is in terms of the sample preparation protocols and the software corrections in the sequencing data such as different mobility-correction factors. In most cases, a size-exclusion step is used following DNA polymerization to remove excess dye-labeled terminators because they are negatively charged and can mask the sequencing data due to the presence of a large dye-terminator band in the gel.

An example of a four-color detector for capillary-gel electrophoresis is shown in Figure 26.20a, in which two laser sources (argon ion laser, 488 nm, and green helium-neon laser, 543 nm) were used to excite the dye set FAM, JOE, TAMRA, and ROX. To process the emission on a single detection channel, a four-stage filter wheel was synchronized to a sector wheel situated in front of the two lasers. The synchronization was set to pass 488-nm excitation for FAM and JOE and simultaneously place the bandpass filters for FAM and JOE in the optical path. Following this, the 543-nm laser light was passed, and the filters for TAMRA and ROX were situated within the optical path. Figure 26.21 shows typical traces produced from this system in which the sequence of an M13mp18 phage test template was analyzed. A read length exceeding 550 bases was obtained at an accuracy of 97%.

26.4.5 Choosing the Right Sequencing Format

With a variety of different fluorescence-detection formats available, the question becomes which configuration is better in terms of base calling, both from a read-length and accuracy point of view. In addition, which detection format produces the best SNR in the measurement? Other issues require attention as well, for example, the complexity of the instrumentation required for detection. A rigorous comparison between the various detector configurations for sequencing applications has been carried out.[27] In this study, three different detector formats were used, and they are shown in Figure 26.20. These consisted of a four-color/single-lane format, a two-color/single-lane format, and a single-color/single-lane format.

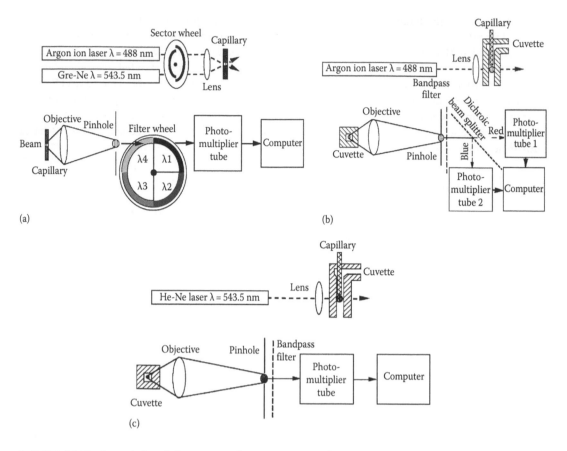

FIGURE 26.20 Laser-induced fluorescence-detection apparati for processing DNA-sequencing data obtained using CGE: four-color/single-lane (a), two-color/single-lane (b), and one-color/single-lane (c).

Table 26.3 presents a summary of the data collected from this work. The single-color experiment, in which there is a single detection channel, provides the lowest limits of detection, followed by the two-color system and the four-color system. The significant improvement in the detection limit for the two-color and single-color systems was partly due to the use of the sheath flow detector that was used in these formats. However, in spite of the use of the sheath flow cell, the general trend is that the lower number of spectral channels typically results in a better SNR in the fluorescence measurement due to the fact that spectral sorting is not required. Spectral sorting causes emission losses due to reflection or inefficient filtering by the bandpass filters used in the optical train. If a filter wheel is used, as in this particular example, a reduced duty cycle will degrade the SNR in the measurement. However, this does not necessarily mean that the lower number of spectral channels will give better sequencing data. As shown by the results of Table 26.3, the four-color format produced better read lengths and favorable base-calling accuracies as compared with the other formats. This is a consequence of the fact that because four spectral channels are being used, the information content in the signal goes up, but only at reasonable loading levels of sample into the sequencing instrument. It is clear that at low loading levels of DNA-sequencing ladders, the one-color or two-color approach may be better due to improved limits of detection.

26.4.6 Single-Color/Four-Lifetime Sequencing

While most sequencing applications using fluorescence require spectral discrimination to identify the terminal base during electrophoretic sizing, an alternative approach is to use the fluorescence lifetime

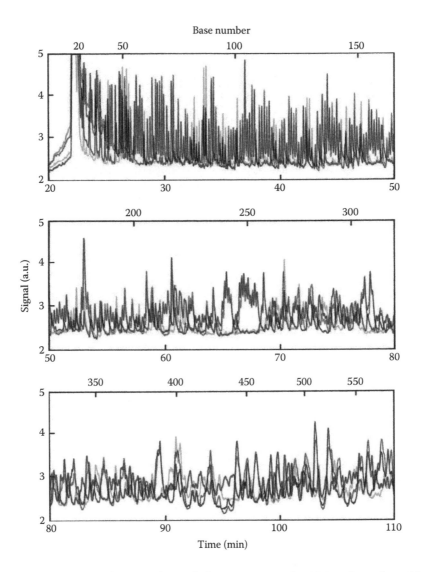

FIGURE 26.21 **(See color insert.)** Four-color/single-lane sequencing of an M13mp18 template with a histidine tRNA insert. The numbers along the top of the electropherogram represent cumulative bases from the primer-annealing site. The electrophoresis was performed in a capillary column with a length of 41 cm and an i.d. of 50 μm. The sieving gel consisted of a cross-linked polyacrylamide (6% T/5% C) and was run at a field strength of 150 V/cm. The dyes used for the labeling of the sequencing primers were FAM, JOE, TAMRA, and ROX. Each color represents fluorescence from a different wavelength region (blue = 540 nm, green = 560 nm, yellow = 580 nm, red = 610 nm). (Adapted from Swerdlow, H. et al., *Anal. Chem.*, 63, 2835, 1991.)

of the labeling dye to identify the terminal base. In this method, either time-resolved or phase-resolved techniques can be used to measure the fluorescence lifetime of the labeling dye during the gel electro-phoresis separation.

The monitoring and identification of multiple dyes by lifetime discrimination during a gel separation can allow for improved identification efficiency when compared to that of spectral wavelength discrimi-nation. When the identity of the terminal nucleotide base is accomplished through differences in spec-tral emission wavelengths, errors in the base call can arise from broad, overlapping emission profiles, which results in cross talk between detection channels. Lifetime discrimination eliminates the problem

TABLE 26.3 Comparison of Figures of Merit for Various Detector Formats for Fluorescence-Based DNA Sequencing

Detection Mode	Noise (mass)[b]	Detection Limit (mol)[b]	Read Length[c]	Base Calling Accuracy	Information Content[a]	
					10 amol	100 zmol
One color	700 ymol	2 zmol	400	85%	14	7
Two color	7 zmol	20 zmol	500	97%	21	8
Four color	70 zmol	200 zmol	550	97%	29	2

Note: The sequencing assays used capillary gel electrophoresis for fractionating the DNA ladders. The gel consisted of a crosslinked polyacrylamide gel with either formamide or urea as the denaturant. In all cases, the template was an M13mp18 phage with dye-primer chemistry used for fluorescence labeling. In the four-color experiment, FAM, JOE, TAMRA, and ROX were the labeling dyes, for the two-color experiment FAM and TAMRA were used, and for the one-color example only TAMRA was used. The sequencing primer was an M13 (−40) primer and the polymerase was a Sequenase enzyme.

[a] Calculated from Equation 26.3.
[b] ymol = 10^{-24} mol; zmol = 10^{-21} mol.
[c] The read lengths were determined from sequencing data with high SNRs that were comparable across the series.

of cross talk between detection channels and can also potentially allow processing of the data on a single readout channel. Several other advantages are associated with fluorescence lifetime identification protocols, including[37]:

1. The calculated lifetime is immune to concentration differences.
2. The fluorescence lifetime can be determined with higher precision than fluorescence intensities.
3. Only one excitation source is required to efficiently excite the fluorescent probes, and only one detection channel is needed to process the fluorescence for appropriately selected dye sets.

One potential difficulty associated with this approach is the poor photon statistics (limited number of photocounts) that can result when making such a measurement. This results from the need to make a dynamic measurement (the chromophore is resident in the excitation beam for only 1–5 s) and the low-mass loading levels associated with many DNA electrophoresis formats. Basically, the low number of photocounts acquired to construct the decay profile from which the lifetime is extracted can produce low precision in the measurement, which affects the accuracy in the base call. In addition, the high scattering medium in which the fluorescence is measured (polyacrylamide gel) can produce large backgrounds, again lowering the precision in the measurement. An additional concern with lifetime measurements for calling bases in DNA-sequencing applications is the heavy demand on the instrumentation required for such a measurement. However, the increased availability of pulsed diode lasers and simple avalanche photodiode detectors has had a tremendous impact on the ability to assemble a time-resolved instrument appropriate for sequencing applications.

There are two different formats for measuring fluorescence lifetimes—time-resolved[37–43] and frequency-resolved.[44–48] Since the time-resolved mode is a digital (photon-counting) method, it typically shows a better SNR than a frequency-resolved measurement, making it more attractive for separation platforms that deal with minute amounts of sample. In addition, time-resolved methods allow for the use of time-gated detection in which background photons, which are coincident with the laser pulse (scattered photons), can be gated out electronically, improving the SNR in the measurement.

A typical time-correlated single-photon-counting (TCSPC) device consists of a pulsed excitation source, a fast detector, and timing electronics. A device that has been used for making time-resolved measurements during CGE is shown in Figure 26.22.[39] The light source consisted of an actively pulsed solid-state gallium aluminum arsenide (GaAlAs) diode laser with a repetition rate of 80 MHz and an average power of 5.0 mW at a lasing wavelength of 780 nm. The pulse width of the laser was determined to be ~50 ps (formal width half maximum [FWHM]). The detector selected for this instrument was a SPAD, which has an active area of 150 μm and is actively cooled. In addition, the SPAD has a high single-photon-detection efficiency (>60% above 700 nm). The counting electronics (constant fraction discriminator [CFD], analog-to-digital converter [ADC], time-to-amplitude converter [TAC], and pulse-height

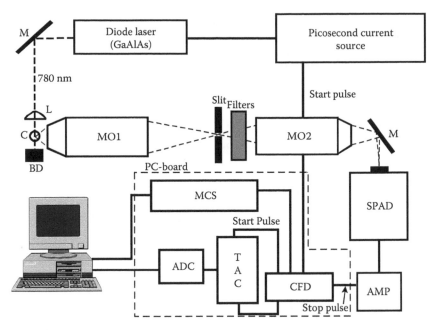

FIGURE 26.22 Time-correlated single-photon counting detector for CGE. The laser source was a pulsed diode laser that operated at a repetition rate of 80 MHz and lased at 780 nm (average power = 5 mW). The laser was focused onto a capillary-gel column with the emission collected using a 40× microscope objective (NA = 0.85). The fluorescence was imaged onto a slit and then spectrally filtered and finally focused onto the photoactive area of a single-photon-avalanche diode. The electronics for processing the time-resolved data was situated on a single PC board, which was resident on the PC bus. (Adapted from Legendre, B.L. et al., *Rev. Sci. Instrum.*, 67, 3984, 1996.)

analyzer) are situated on a single TCSPC board. The board plugs directly into a PC bus and exhibits a dead time of <260 ns, allowing efficient processing of single-photon events at counting rates exceeding 2×10^6 counts/s. This set of electronics allows for the collection of 128 sequential decay profiles with a timing resolution of 9.77 ps per channel. The instrument possesses a response function of approximately 275 ps (FWHM), adequate for measuring fluorescence lifetimes in the subnanosecond regime.

Probably the most important aspect of lifetime determinations in sequencing applications is the processing or calculation algorithms used to extract the lifetime value from the resulting decay. Since the photon statistics are poor and the accuracy in the base call depends directly on the lifetime differences between fluors in the dye set and the relative precision in the measurement, algorithms that deal with this situation are required as well as those that can be performed online during the electrophoresis. Two algorithms for on-the-fly fluorescence-lifetime determinations have been used—the maximum likelihood estimator (MLE) and the rapid lifetime determination method (RLD). MLE calculates the lifetime via the following relation[49]:

$$1 + \left(e^{T/\tau} - 1\right) - m\left(e^{mT/\tau_f} - 1\right)^{-1} = N_t^{-1} \sum_{i-1}^{m} iN_i \tag{26.4}$$

where
    m is the number of time bins within the decay profile
    N_t is the number of photocounts in the decay spectrum
    N_i is the number of photocounts in time bin i
    T is the width of each time bin

A table of values using the left-hand side of the equation is calculated by setting m and T to experimental values and using lifetime values $\langle \tau_f \rangle$ ranging over the anticipated values. The right-hand side of the equation is constructed from CE decay data over the appropriate time range. The fluorescence lifetime may then be determined by matching the value of the right-hand side obtained from the data with the table entry. The relative standard deviation in the MLE may be determined from $N^{-1/2}$.

Fluorescence lifetimes calculated using the RLD method is performed by integrating the number of counts within the decay profile over a specified time interval and using the following relationship[50]:

$$\tau_f = -\frac{\Delta t}{\ln(D_1/D_0)} \tag{26.5}$$

where

Δt is the time range over which the counts are integrated
D_0 is the integrated counts in the early time interval of the decay spectrum
D_1 represents the integrated number of counts in the later time interval

Both the MLE and RLD methods can extract only a single lifetime value from the decay, which in the case of multiexponential profiles would represent a weighted average of the various components comprising the decay.

Wolfrum et al. have also implemented a special pattern recognition technique.[38] Basically, the method involves comparing a pattern to the measured decay and searches for a pattern that best fits the measurement. This algorithm is equivalent to the minimization of a log-likelihood ratio where fluorescent-decay profiles serve as the pattern. Since the pattern-recognition algorithm uses the full amount of information present in the data, it potentially has the lowest error or misclassification probability.

To demonstrate the feasibility of acquiring lifetimes on the fly during the CGE separation of sequencing ladders, C-terminated fragments produced from Sanger chain-terminating protocols and labeled with a NIR fluorophore on the 5′ end of a sequencing primer were electrophoresed and the lifetimes of various components within the electropherogram determined.[37] An example of the data produced from this detection format is shown in Figure 26.23. The average lifetime determined using the MLE method was found to be 843 ps, with a standard deviation of ±9 ps (RSD = 1.9%). The lifetime values calculated here compared favorably to a static measurement performed on the same dye.

Since the base calling is done with lifetime discrimination as opposed to wavelength discrimination, new types of dye sets can and need to be used that suit the identification method. For example, it is not necessary to use dyes with discrete emission maxima, and so structural variations in the dye set can be relaxed. A dye set developed for lifetime discrimination has been prepared and consists of a NIR chromophore that has unique fluorescence lifetimes with the lifetime altered via the addition of an intramolecular heavy atom.[51] Each of these dyes possesses the same absorbance maximum and fluorescence-emission maximum but different fluorescence lifetimes (see Table 26.1). Since these were tricarbocyanine dyes, the lifetimes were found to be <1.0 ns, with the lifetimes for the dye set ranging from 947 to 843 ps when measured in a polyacrylamide gel containing urea, with the observed lifetimes less than what was observed in methanol but still exhibiting single exponential behavior.

A dye set appropriate for lifetime-identification purposes has been prepared and used in four lifetime DNA-sequencing applications.[40] The dyes absorb radiation from 624 to 669 nm and possess lifetimes that range from 1.6 to 3.7 ns (see Figure 26.24). Unfortunately, the dye set shows multiexponential behavior in sequencing gels containing denaturants (urea, 7 M). In addition, to correct for dye-dependent mobility shifts unique linker structures were used. Using this dye set (dye-primer chemistry) and a pulsed diode laser operating at 630 nm, the sequence of an M13mp18 phage was evaluated. The lifetimes were extracted from the decays using a pattern-recognition algorithm. The read length was found to be 660 bases with a calling accuracy of 90%.

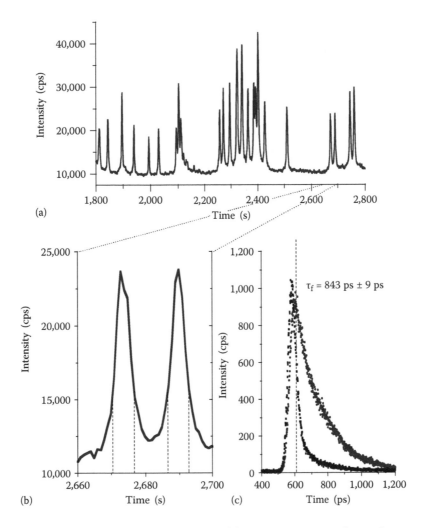

FIGURE 26.23 CGE of C-terminated fragments produced from an M13mp18 template with time-resolved fluorescence detection (a). The labeling dye (F-substituted, heavy-atom-modified NIR fluorescent label; see Table 26.1) was attached to the 5′ end of a sequencing primer. (b) Expanded view of two peaks selected from the electropherogram showing the time area (dashed lines) over which photoelectrons were used to construct the decay profiles. The decay profiles shown in (c) consist of the gel blank and the dye-labeled DNA fragment. The dashed line shows the start channel in which the calculation was initiated. The lifetime was calculated using Equation 26.3. The electrophoresis was carried at a field strength of 200 V/cm using a cross-linked polyacrylamide gel. (Adapted from Soper, S.A. et al., *Anal. Chem.*, 67, 4358, 1995.)

26.5 Single-Molecule DNA Sequencing Using Fluorescence Detection

The sequencing strategies discussed to this point depend on a fractionation step (electrophoresis) to sort the DNA by size. While much progress has been made in reducing the time required to develop the electropherogram, resulting in increases in the throughput of acquiring sequencing data, many problems arise when using gel electrophoresis. These problems include the ability to sequence DNA pieces that are only 1000–2000 bp in length, the requirement of gels to fractionate the DNA, the relatively slow speed associated with the process, and the data-processing requirements. The most pervasive problem is the ability to work only with DNA pieces that are 1000–2000 bases in length. Since

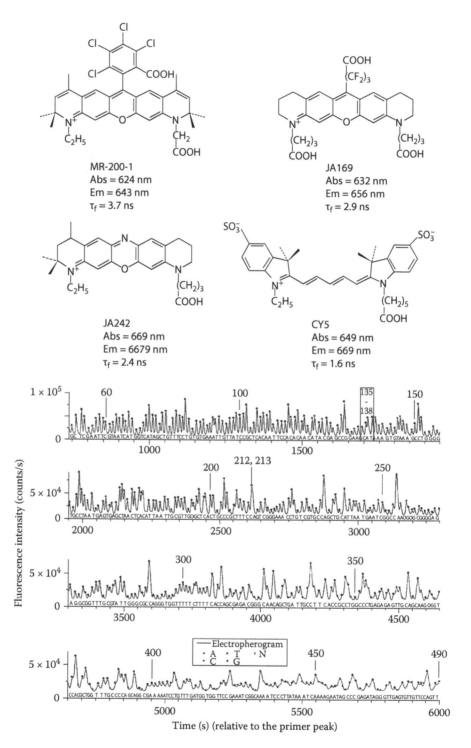

FIGURE 26.24 Chemical structures of dyes used for multiplex, time-resolved DNA sequencing as well as the fluorescence properties of this dye set. Also shown is the sequencing data obtained when using the dye set shown above conjugated to sequencing primers. The sequencing was performed in a 5% linear polyacrylamide gel containing 7 M urea at a field strength of 160 V/cm. The fluorescence detector consisted of a pulsed diode laser (630 nm) with the optics configured in a confocal geometry. (Adapted from Lieberwirth, U. et al., *Anal. Chem.*, 70, 4771, 1998.)

the chromosomes exceed 1×10^6 bases, the assembly of the sequence of the entire chromosome with extremely short pieces makes the task daunting. Directly sequencing larger strands of DNA would relieve some the technical challenges associated with assembly. In addition, since YAC clones are ~100,000 bp, this requires the shearing and then subcloning of these DNAs into M13s to produce templates for sequencing. The ability to work with longer DNAs would eliminate the need for this secondary cloning step.

A very attractive approach has been suggested to rapidly sequence DNA strands that are >40,000 bp in length. The process is based on the principle of single-molecule detection using fluorescence photon burst analysis.[52] The process is depicted in Figure 26.25. Basically, the process involves immobilizing a long strand of DNA in a flow stream and then clipping the terminal base using an exonuclease enzyme, releasing the nucleotide from the original strand. The single nucleotide (either fluorescently labeled or nonlabeled) is then carried to a laser beam, which excites the fluorescence with the color used to identify the base clipped from the DNA strand. As shown in the figure, the single-molecule-sequencing approach does not involve a gel-fractionation step, potentially significantly speeding up the sequencing rate. In essence, the sequencing rate is determined by the rate at which the enzyme clips nucleotide bases from the strand, which for many exonuclease enzymes is on the order of 1000 bases per second. There are three key technical challenges associated with this technique: (1) tethering the DNA strand to a bead for holding it stationary within a flow stream, (2) creating a complement of the original DNA strand using dye-labeled dNTPs, and (3) detecting single-dye-labeled nucleotides using laser-induced fluorescence detection.

Since the premise of this sequencing protocol is to analyze single DNA molecules, a single DNA molecule must be selected and then trapped within a flow stream for processing. To accomplish this, a DNA molecule can be prepared that contains a biotin molecule on its 5′ end. The biotin-modified DNA molecule can then be attached to a microbead (i.d. = 1–10 μm) coated with streptavidin. Streptavidin is a protein that contains binding sites for biotin. The association of biotin to streptavidin is very strong ($K_{assoc.} = 10^{15}$ M^{-1}) and in addition is stable to both heat and the reagents used in most enzymatic reactions. In this case, a bead that has one DNA molecule attached to it must be selected, as a bead with multiple DNA strands would create a registry problem because the exonucleases work at different rates. Therefore, conditions that produce a sufficient population of beads containing a single DNA molecule must be selected. This can be done statistically by incubating the biotinylated DNA with a large excess of microbeads. This produces a large number of beads with no DNA molecule; therefore, the appropriate beads must be selected by staining the immobilized DNA with a fluorescent intercalating dye and using fluorescence to identify a bead with the appropriate number of DNA molecules attached.

The suspension of the bead containing the DNA molecule in the flow stream can be accomplished using an optical trapping technique. In this case, one or two tightly focused laser beams can be directed onto the bead. This generates an optical trap, which can hold the bead in the flow stream. The size of the trap is determined by the size of the focused laser beam (or beams) and is on the order of 5 μm. Since the trap is generated by a momentum exchange between the photons and the trapped bead, it does not depend on any type of electronic or vibrational transition. As such, the trapping laser can be a far-red or infrared laser that does not bleach the dyes incorporated into the target DNA molecule during polymerization.

The second technical challenge in this scheme is to produce a complement that contains dye-modified oligonucleotides. Since detection is accomplished using fluorescence (typically visible fluorescence) single-molecule detection, it is necessary to covalently label each dNTP with a probe and then, more importantly, build a fluorescent complement of the target DNA using a polymerase enzyme. As with dye-labeled ddNTPs, incorporation of dye-modified dNTPs is a challenge because the incorporation rate and fidelity of modified dNTPs by standard polymerase enzymes is not facile. Therefore, mutants will have to be prepared to accomplish this task. An alternative to using fluorescent dyes is to implement ultraviolet single-molecule detection, which requires unmodified dNTPs. While this strategy is much more forgiving in terms of molecular biology, it places severe challenges on the detection phases of the technique because the nucleotide bases have absorption maxima around 260 nm and the fluorescence quantum yields of the bases are low (~10^{-4}) at room temperature in solution.

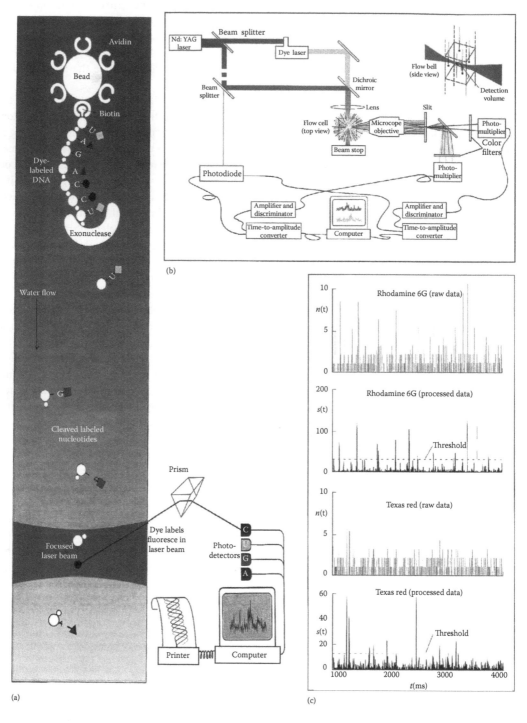

FIGURE 26.25 (a) Schematic diagram of single-molecule sequencing. (b) Block diagram of a dual-color single-molecule detection apparatus. (c) Single-molecule data of R6G and Texas Red using the apparatus described in (b). The dye concentration used in these experiments was set at 50 fM, which resulted in an arrival rate of ~1 molecule every 4 s. The dashed line represents the threshold, which defined the detection of a single molecule when the observed signal (photon burst) exceeded this level. (Adapted from Soper, S.A. et al., *J. Opt. Soc. Am. B*, 9, 1761, 1992.)

The final challenge in this approach is the ability to detect single-dye molecules (nucleotides) in solution and identify the single molecules via spectral discrimination. Since a DNA molecule is comprised of four different nucleotide bases, the bases as they are clipped by the exonuclease from the target DNA must be spectrally identified. While several researchers have demonstrated the ability to detect single molecules in flow streams,[53–55] the ability to color-discriminate adds complexity to the instrumentation (multiple lasers, multiple detection channels) and requires careful selection of the dyes. Not only must the dyes be well resolved in terms of their emission maxima, but the photophysics of the dye must be conducive to single-molecule detection, that is, must have a high quantum yield and favorable photochemical stability. Figure 26.25 shows an example of a dual-color single-molecule-detection apparatus.[56] It consists of a mode-locked neodymium:YAG laser operating at 532 nm (second harmonic) and also a synchronously pumped dye laser ($\lambda_{em} = 585$ nm). These wavelengths were chosen to match the absorption maxima of the two dyes selected for this experiment, R6G ($\lambda_{ex} = 528$ nm; $\lambda_{em} = 555$ nm) and Texas Red ($\lambda_{ex} = 578$ nm; $\lambda_{em} = 605$ nm). The fluorescence was processed on one of two photodetectors, which in this case consisted of microchannel plates (MCPs). The fluorescence from each dye was directed onto the appropriate MCP using a dichroic filter, with the emission further isolated from background photons and fluorescence from the other dye using a bandpass filter. The processing electronics consisted of conventional TCSPC electronics, which were used along with the pulsed lasers to allow for implementation of time-gated detection. Time-gated detection reduces the amount of background-scattered photons into the data stream, improving the SNR in the single-molecule measurement. An example of the data output from this dual-color single-molecule detector is shown in Figure 26.25 (raw data and processed data). The raw data were filtered using a weight quadratic sum filter [$S(t)$] given by the following expression[37]:

$$S(t) = \sum_{\tau=0}^{k-1} w(\tau) d(t+\tau)^2 \tag{26.6}$$

where
  k represents the time range covered by the molecular transit through the laser beam
  $w(\tau)$ is weighting factors selected to best distinguish the signal from noise
  $d(t)$ represents the raw data point at time (t)

These data make clear that large-amplitude bursts of photons occur with both dyes. The detection efficiency was estimated to be 78% for R6G and 90% for Texas Red. In both cases, the error rate (defined as identifying a molecule when one was not present, false positive) was estimated to be <0.01 per second.

References

1. Watson, J.D. and Crick, F.H.C., Molecular structure of nucleic acid: A structure of deoxyribonucleic acid, *Nature*, 171, 737, 1953.
2. Chen, E.Y., Schlessinger, D., and Kere, J., Ordered shotgun sequencing, a strategy for integrated mapping and sequencing of YAC clones, *Genomics*, 17, 651, 1993.
3. Maxam, A.M. and Gilbert, W., A new method for sequencing DNA, *Proc. Natl. Acad. Sci. U.S.A.*, 74, 560, 1977.
4. Sanger, F., Nicklen, S., and Coulson, A.R., DNA sequencing with chain-terminating inhibitors, *Proc. Natl. Acad. Sci. U.S.A.*, 74, 5463, 1977.
5. Fung, E.N. and Yeung, E.S., High-speed DNA sequencing by using mixed poly(ethylene oxide) solutions in uncoated capillary columns, *Anal. Chem.*, 67, 1913, 1995.
6. Smith, L.M., Saunders, J.Z., Kaiser, R.J., Hughes, P., Dodd, C.R., Connell, C.R., Heiner, C., Kent, S.B.H., and Hood, L.E., Fluorescence detection in automated DNA sequence analysis, *Nature*, 321, 674, 1986

7. Prober, J.M., Trainor, G.L., Dam, R.J., Hobbs, F.G., Robertson, C.W., Zagursky, R.J., Cocuzza, A.J., Jensen, M.A., and Baumeister, K., A system for rapid DNA sequencing with fluorescent chain-terminating dideoxynucleotides, *Science*, 238, 336, 1987.

8. Ansorge, W., Sproat, B., Stegeman, J., Schwager, C., and Zenke, M., Automated DNA sequencing: Ultrasensitive detection of bands during electrophoresis, *Nucleic Acids Res.*, 15, 4593, 1987.

9. Metzker, M.L., Lu, J., and Gibbs, R.A., Electrophoretically uniform fluorescent dyes for automated DNA sequencing, *Science*, 271, 1420, 1996.

10. Lee, L.G., Connell, C., Woo, S., Cheng, R., McArdle, B., Fuller, C., Halloran, N., and Wilson, R., DNA sequencing with dye-labeled terminators and T7 DNA polymerase: Effects of dyes and dNTPs on incorporation of dye-terminators and probability analysis of termination fragments, *Nucleic Acids Res.*, 20, 2471, 1992.

11. Tabor, S. and Richardson, C.C., A single residue in DNA polymerase from *E. coli* DNA polymerase I family is critical for distinguishing between deoxy- and dideoxynucleotides, *Proc. Natl. Acad. Sci. U.S.A.*, 92, 6339, 1995.

12. Rosenblum, B.B., Lee, L.G., Spurgeon, S.L., Khan, S.H., Menchen, S.M., Heiner, C.R., and Chen, S.M., New dye-labeled terminators for improved DNA sequencing patterns, *Nucleic Acids Res.*, 25, 4500, 1997.

13. Hung, S.-C., Mathies, R.A., and Glazer, A.N., Optimization of spectroscopic and electrophoretic properties of energy transfer primers, *Anal. Biochem.*, 252, 78, 1997.

14. Hung, S.-C., Mathies, R.A., and Glazer, A.N., Comparison of fluorescence energy transfer primers with different donor-acceptor dye combinations, *Anal. Biochem.*, 255, 32, 1998.

15. Ju, J., Ruan, C., Fuller, C.W., Glazer, A.N., and Mathies, R.A., Fluorescence energy transfer dye-labeled primers for DNA sequencing and analysis, *Proc. Natl. Acad. Sci. U.S.A.*, 92, 4347, 1995.

16. Lee, L.G., Spurgeon, S.L., Heiner, C.R., Benson, S.C., Rosenblum, B.B., Menchen, S.M., Graham, R.J., Constantinescu, A., Upadhya, K.G., and Cassel, J.M., New energy transfer dyes for DNA sequencing, *Nucleic Acids Res.*, 25, 2816, 1997.

17. Williams, D.C. and Soper, S.A., Ultrasensitive near-IR fluorescence detection for capillary gel electrophoresis and DNA sequencing applications, *Anal. Chem.*, 67, 3427, 1995.

18. Middendorf, L.R., Bruce, J.C., Bruce, R.C., Eckles, R.D., Grone, D.L., Roemer, S.C., Sloiker, G.D. et al., Continuous, on-line DNA sequencing using a versatile infrared laser scanner/electrophoresis apparatus, *Electrophoresis*, 13, 487, 1992.

19. Shealy, D.B., Lipowska, M., Lipwoski, J., Narayanan, N., Sutter, S., Strekowski, L., and Patonay, G., Synthesis, chromatographic separation and characterization of near-infrared-labeled DNA oligomers for use in DNA sequencing, *Anal. Chem.*, 67, 247, 1995.

20. Soper, S.A. and Mattingly, Q., Steady-state and picosecond laser fluorescence studies of nonradiative pathways in tricarbocyanine dyes: Implications to the design of near-IR fluorochromes with high fluorescence efficiencies, *J. Am. Chem. Soc.*, 116, 3447, 1994.

21. Huang, X.C., Quesada, M.A., and Mathies, R.A., Capillary array electrophoresis using laser-excited confocal fluorescence detection, *Anal. Chem.*, 64, 967, 1992.

22. Huang, X.C., Quesada, M.A., and Mathies, R.A., DNA sequencing using capillary array electrophoresis. *Anal. Chem.*, 64, 2149, 1992.

23. Soper, S.A., Shera, E., Davis, L., Nutter, H., and Keller, R., The photophysical constants of several visible fluorescent dyes and their effects on ultrasensitive fluorescence detection, *Photochem. Photobiol.*, 57, 972, 1993.

24. Kambara, H. and Takahashi, S. Multiple-sheathflow capillary array DNA analyzer, *Nature*, 361, 565, 1993.

25. Takahashi, S., Murakami, K., Anazawa, T., and Kambara, H., Multiple sheath-flow gel capillary-array electrophoresis for multicolor fluorescent DNA detection, *Anal. Chem.*, 66, 1021, 1994.

26. Swerdlow, H., Wu, S., Harke, H., and Dovichi, N., Capillary gel electrophoresis for DNA sequencing: Laser-induced fluorescence detection with the sheath flow cuvette, *J. Chromatogr.*, 516, 61, 1990.

27. Swerdlow, H., Zhang, J.Z., Chen, D.Y., Harke, H.R., Grey, R., Wu, S., and Dovichi, N.J., Three DNA sequencing methods using capillary gel electrophoresis and laser-induced fluorescence, *Anal. Chem.*, 63, 2835, 1991.

28. Ewing, B., Hillier, L., Wendl, M., and Green, P., Base-calling of automated sequencer traces using *Phred*. I. Accuracy assessment, *Genomics*, 8, 175, 1998.

29. Tu, O., Knott, T., Marsh, M., Bechtol, K., Harris, D., Barker, D., and Bashkin, J., The influence of fluorescent dye structure on the electrophoretic mobility of end-labeled DNA, *Nucleic Acids Res.*, 26, 2797, 1998.

30. Ansorge, W., Zimmermann, C., Schwager, C., Stegemann, J., Erfle, H., and Voss, H., One label, one tube, Sanger DNA sequencing in one and two lanes on a gel, *Nucleic Acids Res.*, 18, 3419, 1990.

31. Chen, D., Swerdlow, H.P., Harke, H.R., Zhang, J.Z., and Dovichi, N.J., Single color laser induced fluorescence detection and capillary gel electrophoresis for DNA sequencing, *Proc. Int. Soc. Opt. Eng.*, 1435, 161, 1991.

32. Tabor, S. and Richardson, C.C., DNA sequence analysis with a modified bacteriophage T7 DNA polymerase, *Proc. Natl. Acad. Sci. U.S.A.*, 84, 4767, 1987.

33. Tabor, S. and Richardson, C.C., DNA sequence analysis with a modified bacteriophage T7 DNA polymerase, *J. Biol. Chem.*, 14, 8322, 1990.

34. Chen, D.Y., Harke, H.R., and Dovichi, N.J., Two-label peak-height encoded DNA sequencing by capillary gel electrophoresis: Three examples, *Nucleic Acids Res.*, 20, 4873, 1992.

35. Li, Q. and Yeung, E., Simple two-color base-calling schemes for DNA sequencing based on standard four-label Sanger chemistry, *Appl. Spectrosc.*, 49, 1528, 1995.

36. Starke, H.R., Yan, J., Zhang, Z., Muhlegger, K., Effgen, K., and Dovichi, N., Internal fluorescence labeling with fluorescent deoxynucleotides in two-label peak-height encoded DNA sequencing by capillary electrophoresis, *Nucleic Acids Res.*, 22, 3997, 1994.

37. Soper, S.A., Legendre, B.L., and Williams, D.C., On-line fluorescence lifetime determinations in capillary electrophoresis, *Anal. Chem.*, 67, 4358, 1995.

38. Köllner, M., Fischer, A., Arden-Jacob, J., Drexhage, K.-H., Müller, R., Seeger, S., and Wolfrum, J., Fluorescence pattern recognition for ultrasensitive molecule identification: Comparison of experimental data and theoretical approximations, *Chem. Phys. Lett.*, 250, 355, 1996.

39. Legendre, B.L., Williams, D.C., Soper, S.A., Erdmann, R., Ortmann, U., and Enderlein, J., An all solid-state near-infrared time-correlated single photon counting instrument for dynamic lifetime measurements in DNA sequencing applications, *Rev. Sci. Instrum.*, 67, 3984, 1996.

40. Lieberwirth, U., Arden-Jacob, J., Drexhage, K.H., Herten, D.P., Muller, R., Neumann, M., Schulz, A. et al., Multiplexed dye DNA sequencing in capillary gel electrophoresis by diode laser-based time-resolved fluorescence detection, *Anal. Chem.*, 70, 4771, 1998.

41. Sauer, M., Arden-Jacob, J., Drexhage, K.H., Marx, N.J., Karger, A.E., Lieberwirth, U., Muller, R. et al., On-line diode laser based time-resolved fluorescence detection of labeled oligonucleotides in capillary gel electrophoresis, *Biomed. Chromatogr.*, 11, 81, 1997.

42. Soper, S.A. and Legendre, B.L., Error analysis of simple algorithms for determining fluorescence lifetimes in ultradilute dye solutions, *Appl. Spectrosc.*, 48, 400, 1994.

43. Waddell, E., Stryjewski, W., and Soper, S.A., A fiber-optic-based multichannel time-correlated single photon-counting device with subnanosecond time resolution, *Rev. Sci. Instrum.*, 70, 32, 1999.

44. He, H., Nunnally, B.K., Li, L.C., and McGowen, L.B., On-the-fly fluorescence lifetime detection of dye-labeled primers for multiplexed analysis, *Anal. Chem.*, 70, 3413, 1998.

45. Li, L.C. and McGowen, L.B., On-the-fly frequency-domain fluorescence lifetime detection in capillary electrophoresis, *Anal. Chem.*, 68, 2737, 1996.

46. Li, L.C., He, H., Nunnally, B.K., and McGowen, L.B., On-the-fly fluorescence lifetime detection of labeled DNA primers, *J. Chromatogr.*, 695, 85, 1997.

47. Li, L. and McGowen, L.B., Effects of gel material on fluorescence lifetime detection of dyes and dye-labeled DNA primers in capillary electrophoresis, *J. Chromatogr.*, 841, 95, 1999.

48. Nunnally, B.K., He, H., Li, L.C., Tucker, S.A., and McGowen, L.B., Characterization of visible dyes for four-decay fluorescence detection, *Anal. Chem.*, 69, 2392, 1997.

49. Hall, P. and Sellinger, B., Better estimates of exponential decay parameters, *J. Phys. Chem.*, 85, 2941, 1981.

50. Ballew, R.M. and Demas, J.N., An error analysis of the rapid lifetime determination method for the evaluation of single exponential decays, *Anal. Chem.*, 61, 30, 1989.

51. Flanagan, J.H., Owens, C.V., Romero, S.E., Waddell, E., Kahn, S.H., Hammer, R.P., and Soper, S.A., Near-infrared heavy-atom-modified fluorescent dyes for base-calling in DNA-sequencing applications using temporal discrimination, *Anal. Chem.*, 70, 2676, 1998.

52. Fairfield, E.R., Jett, J., Keller, R., Hahn, J., Krakowski, L., Marrone, B., Martin, J., Ratliff, R., Shera, E., and Soper, S., Rapid DNA sequencing based upon single molecule detection, *Gen. Anal.*, 8, 1, 1991.

53. Shera, E.B., Seitzinger, N., Davis, L., Keller, R., and Soper, S., Detection of single fluorescent molecules, *Chem. Phys. Lett.*, 174, 553, 1990.

54. Soper, S.A., Hahn, J., Nutter, H., Shera, E., Martin, J., Jett, J., and Keller, R., Single molecule detection of R6G in ethanolic solutions utilizing CW excitation, *Anal. Chem.*, 63, 432, 1991.

55. Soper, S.A., Mattingly, Q.L., and Vegunta, P., Photon burst detection of single near infrared fluorescent dye molecules, *Anal. Chem.*, 65, 740, 1993.

56. Soper, S.A., Davis, L., and Shear, E.B., Detection and identification of single molecules in solution, *J. Opt. Soc. Am. B*, 9, 1761, 1992.

27

In Vivo Bioluminescence Imaging as a Tool for Drug Development

Pamela R. Contag
ConcentRx Inc.

Christopher H.
Contag
*Stanford University
Medical Center*

27.1 Functional Imaging Will Be Required for a Target-Directed Paradigm

27.1.1 Animal Models in Drug Development

Animal models that are used for biological assessment in drug development typically represent advanced or even end-stage disease and often require sacrifice of the animal for analysis. In the absence of imaging tools, the analyses performed require tissue sampling and time-consuming cellular or biochemical assays that only provide a *snapshot* of the overall disease course even when performed on large numbers of animals. Therefore, one must consider the limitations of the data that result from conventional animal protocols that employ ex vivo assays, which are constrained by sample size, limited to a small number of selected time points, and performed in the absence of intact organ systems. In cancer models, tumors had been grown at superficial sites for easy assessment of tumor volume, but these models likely do not represent human disease. The use of such models is likely the reason most drugs fail in clinical trials. A major problem in drug development is that the animal models used for a given therapy are often set up with a target that is likely to demonstrate the greatest efficacy using an outcome measure that is readily assessed. This may bias the model in favor of the therapy, and may not be predictive of the complex drug–target interactions in humans. One often hears that we have cured cancer in mice, but cannot treat human cancers. We cannot develop effective drugs for humans by aiming at easy targets in mice and hard targets in the clinic. Our animal models need to accurately represent the complexity of human diseases.

Imaging gives us access to deep tissues and enables analysis of complex disease states such that our animal models can be more predictive of the human response.

In the current paradigm of target-directed drug discovery, the events that occur early in the disease process or that persist during remission represent important targets for therapeutic intervention, and therefore, the development of assays that provide access to these targets early in the disease establishment and during therapeutic intervention is essential for evaluating new classes of compounds. Functional imaging that is relatively high throughput and yields multiparameter data will be required to sort out the vast numbers of targets and compounds introduced into drug discovery by this evolving paradigm. Imaging strategies that are applicable to laboratory animals have been developed, and their use for in vivo assessment of therapeutic efficacy has the potential to greatly influence the pharmaceutical industry. One of these imaging modalities, in vivo bioluminescent imaging (BLI), and its application to drug discovery and drug development, is the subject of this review.

27.1.2 Optical Imaging

Preclinical imaging modalities in the optical regime have the advantages of being sensitive, accessible, relatively low cost, and rapid. This has lead to the development of a number of optical imaging approaches for in vivo analyses of laboratory animals.[1–19] Of these, some are particularly well suited for the purpose of accelerating the drug development process as they address the unmet need of real-time in vivo assays that are ideal for imaging small laboratory animals.[5,20–23] From an imaging perspective, visible light use is safe, even in large doses, can penetrate relatively deeply into certain kinds of tissues, does not necessarily require a chemical substrate, and is quantifiable under certain circumstances. Light can also be repeatedly introduced into a tissue with safety and is not subject to radioactive decay. As there are few, if any, sources of light in mammalian systems, especially at near-infrared (NIR) wavelengths, background is minimal, if not completely absent, resulting in extraordinary signal-to-noise ratios (SNRs). For tissues and organs like the breast, with low light absorption in the 700–850 nm spectral range, optical imaging is clinically feasible; however, the number of clinical applications will be significantly fewer than those developed for animal studies. The advances in optical imaging in living animals have progressed to a point where it has become commonplace to use these strategies in academic and commercial environments, and this has established a solid foundation on which to build optical imaging tools for clinical use. Optical imaging tools can be used to measure drug concentrations in target and off-target tissues and as an outcome measure for clinical studies. Common imaging tools can be used to integrate preclinical and clinical studies and accelerating the development of novel therapeutics.

27.1.3 Imaging in the Preclinical Stage

Increasingly, the subject of conservative use of laboratory animals is resonating with those that see the scientific and economic value of responsible animal use protocols. To address these issues, a number of imaging methods that are based on clinical imaging modalities have been modified for studying animal models. These include magnetic resonance imaging (MRI),[4,5,24–36] positron emission tomography (PET) and single-photon emission tomography (SPECT),[36–45] and x-ray computed tomography (CT).[46,47] The composite of these and the optical methods represent significant advances in the rapidly developing field of molecular imaging, and many of these imaging methods will likely have a dramatic effect on the preclinical studies in laboratory animals.[23,48]

Incorporating imaging technologies, such as BLI that are rapid and accessible, into preclinical studies will yield more and higher quality experimental data per protocol by increasing the number of times that quantitative data can be collected. By imaging the whole intact live animal at multiple time points, researchers can place biomolecular processes together with contextual influences in a living animal. It is an added benefit that with these methods, fewer animals can deliver data with greater statistical significance. Lower stress, noninvasive methods are used to create more predictive animal models that share the characteristics of longitudinal study design, internal experimental control, molecular information, and quantitative data, and these methods will benefit both scientific inquiry and humane animal use.[48] In addition, imaging can further improve these studies by guiding appropriate endpoint tissue sampling for histology or biochemical analyses, and studies that use imaging to guide tissue selection for multiparametric, or biomic, analyses are increasing. Imaging of cellular and molecular targets in living animal models of human biology and disease has been used in animal research protocols in fields as diverse as cancer biology, microbiology, and gene therapy. Such tools enable noninvasive in vivo assessment in individual animals over time, which reveals temporal changes and reduces the number of animals required for a given study.

The advantages of imaging in drug development address many of the limitations—at the preclinical stage—and the improved preclinical data sets will lead to better study designs for the clinical trials, and the more user friendly and accessible the methods become, the more likely they are to be integrated into these protocols. Optical imaging using bioluminescent reporters for tracking and monitoring cell populations, assessing levels of gene expression, or monitoring in vivo gene delivery is gaining wide acceptance in the pharmaceutical industry as an accessible and versatile method to approach the study of animal models of disease.[48] The use of bioluminescence in the study of protein–protein interactions in vivo is advancing at a rapid rate[49-54] and enabling new insights into the biochemistry of disease. BLI can be incorporated into target validation during the discovery phase of drug development or can be used to assess in vivo efficacy, safety, and toxicity during preclinical development of specific compounds. Using BLI, several parameters of drug efficacy and pharmacokinetics can be monitored in the same animal over time, yielding benefits for the discovery and development stages of drug evaluation. As the various preclinical modalities advance, there are efforts to integrate multiple modalities to obtain maximum information per animal in a drug study.[55-65]

27.2 Bioluminescent Imaging in Drug Discovery

27.2.1 In Vivo Bioluminescent Imaging

In vivo BLI has been used as a noninvasive means of tracking pathogens or tumor cells in animal subjects early in the disease process[4,5,15,20,66-71] and, more recently, has been used to develop new animal models that incorporate reporter genes into the rodent genome as markers of transcription that reveal developmental changes or response to stress.[21,23,72,73] These transgenic strategies permit monitoring a wide variety of biological events noninvasively in living animals and enable relatively high-throughput in vivo screens that are likely more predictive of the human response to therapy. Given the accessibility and speed of BLI, researchers will be able to perform in vivo efficacy, pharmacokinetics, toxicology, and target validation studies on a larger number of compounds than they have been able to do in the past. As these more physiologically relevant assays are conducted earlier in the drug development cycle than was previously possible,[22] the decision of whether, or not, to pursue a compound can be made at a critical time-saving point in the development process.

BLI exploits the light-emitting properties of photoproteins such as luciferase enzymes. BLI is based on the capacity of certain living organisms, including some species of bacteria, algae, coelenterates, beetles, fish, and fireflies, to emit visible light.[19,22,74,75] Since these chemical reactions, which produce light, can be replicated outside of the organisms to which they are native, bioluminescence has proven extremely useful for research at the cellular and molecular levels and has been used in live cell assays.[76-79] New developments in detection technology have advanced BLI from biochemical and live cell assays to include in vivo whole body imaging of pathogens, tumors, and gene expression patterns in living animals.[19,22]

Compared to in vitro systems or simple transparent organisms, signal attenuation in whole rodent systems can be severe and varies according to the emission wavelength and the type and depth location of the tissues surrounding the cells containing any fluorescent or bioluminescent reporter. Blue-green light (400–590 nm) is strongly attenuated by tissue, while red to NIR light (590–800 nm) is less affected.[80]

Most types of fluorescent and bioluminescent proteins have peak emission at blue to yellow-green wavelengths, although the emission spectrum from luciferase, the most widely used source of bioluminescent light, is broad enough that there is also significant emission at red wavelengths (>600 nm), which penetrate quite deeply into the tissue.[81] There are a number of dyes that are fluorescent in the NIR that can be conjugated to biomolecules and are being developed as imaging reagents,[24,82,83] and an advantage of these is the ability to deliver exogenous probes that can be activated by the enzyme target.[84-88]

Experience with a wide variety of characterized luciferases has yielded methods for in vivo discovery employing these enzymes as internal sources of light that can be monitored externally as real-time indicators of biological functions in living specimens. Ultrasensitive cameras, employing recent technological advances in photon detection, are now being broadly applied to biological questions and can be used successfully to detect the few photons that escape the scattering and absorbing environment of mammalian tissues.[81,89] The fundamental biochemistry of light production in these organisms consists of the oxidation of a substrate under the catalysis of luciferase enzymes, which produces as a decay product of the chemical reaction an almost heatless light until the excited-state molecules return to the ground state. The luciferin substrates and the structures of the luciferase enzymes vary by organism, as do the mechanisms controlling the speed and intensity of the luminescence. Certain cofactors, such as the nucleotide adenosine triphosphate (ATP) or other sources of energy, may need to be present for the conversion of the substrate to take place.[75,90-94]

Since these reactions can be replicated outside the organisms to which they are native, bioluminescence has proven extremely useful for research at the cellular and molecular levels in test tubes, in cells, and in living animals. The availability of instrumentation that can measure the light emitted in these reactions with great sensitivity and dynamic range has made them powerful tools for biochemical and clinical analysis since the involved components can be detected at a very low level. Compared to other techniques, such as colorimetric or spectrophotometric indicators, bioluminescent analysis offers the advantages of high sensitivity, wide linear range, low cost per test, and relatively simple and inexpensive equipment. Firefly luciferase is the most commonly used bioluminescent system in research, and bacterial luciferases (Lux) are used in prokaryotic systems.[75,95] The light-producing reaction of the North American firefly, *Photinus pyralis*, is also the most extensively studied bioluminescent system. Firefly bioluminescence, for example, has been used to assay the levels of ATP nucleotide, which supplies energy to cells for many important biochemical processes, with extreme sensitivity. Whereas most enzyme assays yield either a product or the disappearance of a substrate, firefly luciferase acts as a quantifiable reactant rather than as a catalyst, whose most significant and easily measured product is light.

27.2.2 Linking Correlative Cell Culture Assays to Animal Models

A significant advantage of in vivo luciferase monitoring is that bioluminescent reporters allow for an integrated approach in which the same label can be used in cell culture correlates of biological processes and then in vivo to test predictions made in cultured cells. In this way, a predictable animal model with highly correlative data can offer validation of an established cell culture assay. An example of this was demonstrated in a study by Scheffold et al. where cytotoxicity assays were modified to use luciferase-labeled target cells and the efficacy of tumoricidal activity of natural killer (NK) T cells was assessed.[96] The assay was then moved into animal models without changing the reporter, and the same cytotoxic T cells and tumor targets could be studied in in vivo models of immune cell therapies (Figure 27.1).

The data from this study also demonstrated that development and testing of complex treatment regimens can be greatly accelerated using imaging.[96] In this study, the cytotoxic NK T cells were redirected to the tumor target using a bispecific antibody, and this treatment was compared to Herceptin alone and to four other controls including two NK T cell controls and two antibody controls (Figure 27.1). Tumor-burden data were collected twice in the first week and then weekly for 28 days. This number

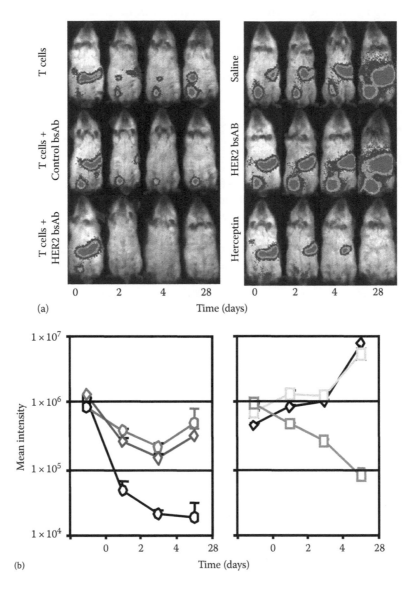

FIGURE 27.1 **(See color insert.)** Advancing complex therapeutic strategies for malignancy through imaging. Evaluation of cell-based therapies complexed with molecular therapies may be complicated by having *too many moving parts*, and imaging approaches have been used to optimize these approaches by rapidly providing efficacy data without sacrificing the study subjects. In a study by Scheffold et al., the ability to redirect a tumoricidal NK T cell population to a tumor target using a bispecific antibody was revealed using BLI.[96] The control groups in this study consisted of the NK T cells alone, the NK T cells with an irrelevant bispecific antibody, normal saline, the bispecific antibody without additional cells, and Herceptin (as a positive control). Temporal analyses for representative animals in each treatment group are shown (a) and data from all animals are plotted (b). Color code: NK T cells only (light blue), NK T cells and control bispecific antibody (red), NK T cells and bispecific antibody (dark blue), saline (aqua), control bispecific antibody (black), and Herceptin (dark green). The ability to rapidly study multiple animals in each of the six treatment groups and provide accurate whole body data that are quantitative is an opportunity offered by BLI.

of groups and data points would not have been possible using conventional assays and the data were quantitative. Imaging will have an increasing role as therapies are developed that involve immune cells, radiation, DNA-based therapies, and chemotherapies as complementary approaches. The optimization of single-drug treatment regimens can be greatly accelerated using imaging assays that can rapidly measure tumor burden. However, the development of more complex multistep therapeutic strategies such as nonmyeloblative bone marrow transplantation will depend much more heavily on imaging to assess outcome and determine the mechanisms of action.

27.3 Application Areas

27.3.1 Cancer Therapies

Effective evaluation of new treatment strategies for malignant disease will require that the animal models closely resemble the human diseases that they are designed to model. The subcutaneous xenograft models of human malignancies are, in many respects, less able to model the human disease than are orthotopic or spontaneous tumor models.[97–99] However, these better models of disease are frequently more difficult to study as the lesions can be located deep within the body and are not accessible for caliper measurements. The disease in these models can involve multiple organ systems and occur in the face of an intact immune system. It is these models that will contribute to our understanding of disease mechanisms as specific genetic elements can be integrated and evaluated in these mice,[97–99] and imaging will provide access to key information on expression of targets, measuring tumor burden at deep tissue sites and rapidly assessing the degree of metastasis.[23]

27.3.1.1 Models of Metastatic and Minimal Disease States

The sensitivity and the spatiotemporal nature of BLI generated data provide an opportunity to detect and localize small foci of malignant cells rapidly in laboratory rodents using whole body assays. Thus, this approach has enabled the study of metastatic and minimal residual disease states in animal models.[66,100] In a study by Wetterwald et al., a labeled human breast cancer cell line was used in mouse xenograft model to efficiently evaluate metastatic lesions in bone. Intracardiac injection of tumor cells allows colonization in a variety of locations including the bone marrow as a model of the intravasation step of metastasis. Micrometastases in the bone marrow can often elude radiographic detection since this detection method requires osteolysis for a signal. BLI on the other hand was used to successfully detect the presence and location of such lesions with volumes as small as 0.5 mm^3 and comprised of approximately 2×10^4 cells.[66,100] The microscopic bone marrow metastases that were detectable by BLI were not associated with osteolytic lesions and hence could not be detected by x-ray imaging. This advancement in the assessment of animal tumor models has greatly accelerated the analyses of these diseases and will improve our ability to develop effective approaches for targeting small numbers of cells present in the early stages of disease progression.

Minimal residual disease remains a major therapeutic hurdle, despite the improvements in disease-free survival that has resulted from recent advances in cancer therapies. A majority of cancer patients respond to high-dose chemo- and radiation therapy and enter a state of minimal residual disease. A significant number of these patients will, however, eventually relapse. Effective evaluation of therapies that target small numbers of tumor cells is therefore essential for the development of the next generation of cancer therapies that address this critical target. Since protocols that use BLI do not require sacrifice of the animal to obtain sensitive and quantitative data, animals may be followed over time for relapse following the cessation of therapy,[5,15,16,66,96] and combination therapies that are at first aggressive and secondarily less toxic can be evaluated. The relatively rapid nature of BLI allows for a variety of therapeutic regimens to be measured in the same animal using time for relapse as the criterion by which to measure success. The use of BLI in combination with other imaging modalities will provide greater opportunities for developing strategies to eradicate minimal disease and prevent relapse.

27.3.1.2 Spontaneous Tumor Models

Advancements in subcutaneous and orthotopic models of disease that have been brought about through the application of BLI may be extended to spontaneous tumor models.[23] The combination of BLI and transgenic technology may yield more predictive animal models of human disease. The introduction of mutations and transgenes in mice has resulted in animals that spontaneously develop malignancies.[97-99] In these mouse models, researchers have been able to study the initial or early events in the initiation of disease, which represent key targets for the treatment of oncogenesis.[23] For example, a conditional transgenic mouse model that develops retinoblastoma-dependent sporadic cancer has been studied.[73] Firefly luciferase was incorporated as a reporter gene in this model, enabling the investigators to follow the animals over the full course of the disease using BLI. The pituitary tumors that developed in these mice could be followed because they were tagged through the expression of the firefly luciferase. The onset of disease, subsequent disease progression, and response to therapy in this tumor model was more readily evaluated using a noninvasive approach than could have been accomplished using assays that require sampling of tissues. Coupling spontaneous tumor models with sensitive imaging modalities will facilitate the analysis of the key processes in the initiation and progression of neoplastic disease as these models incorporate defined genetic alterations that are linked to oncogenesis.

27.3.2 Nucleic Acid Based Therapies

The development of therapeutic strategies based on the delivery of therapeutic genes to replace defective genes for the treatment of genetic diseases or in the treatment of malignancy and other diseases has been constrained also by our inability to rapidly analyze our animal model systems noninvasively. Multimodality imaging studies that pair BLI as a functional imaging modality with MRI as a structural imaging modality demonstrate the strength of such combination approaches to address multiple biological questions in vivo.[4] In the study by Rehemtulla et al., BLI has been used to determine the levels of gene expression following adenoviral-mediated gene delivery in combination with MRI for determining therapeutic efficacy of a therapeutic transgene for cancer gene therapy.[4] Yeast cytosine deaminase (CD), an enzyme that converts the nontoxic compound 5FC into the drug 5-fluorouracil (5FU), was the therapeutic gene, and the tumor target was a glioma (rat 9L glioma) in an orthotopic rat model. The ability of prodrug-converting enzyme to facilitate tumor cell death was evaluated, and luciferase expression served as the marker of effective delivery of the gene. Diffusion-weighted MRI was used as a surrogate marker of therapeutic response.[4] The noninvasive assessments of both the extent of gene delivery and the efficacy of the therapy resulted in a more robust model for evaluation of this multicomponent anticancer therapy.

Successful lifelong gene replacement was demonstrated in an in utero gene transfer model.[101] In this model, an adeno-associated viral vector carrying a modified luciferase gene was injected into the peritoneal cavity of fetal mice. The luminescent signals from the transduced fetuses were apparent, while the animals were in utero, indicating that the substrate, luciferin, can cross the placental barrier. After birth, these animals continued to express the reporter gene and did so for the 24-month study period—this is essentially a lifelong expression. We have developed the ability to diagnose genetic defects in the fetus, and the development of approaches for gene replacement in utero, especially for genes encoding secreted proteins where local expression can have a systemic effect, was enabled by this demonstration.

The location and the magnitude of expression of therapeutic genes will affect the therapeutic outcome of these experimental therapies. Correlating the parameters of gene delivery with therapeutic outcome is essential for preclinical optimization of such strategies. This is another example of a relatively complex treatment regimes where optimal dosing of the genetic therapy and the prodrug need to be evaluated. Studies that employ multimodality imaging should significantly accelerate the decision-making process for the therapeutic indication, drug formulation, and regimen that should move forward to the clinic. Bioluminescent reporter genes may also find clinical use in evaluating the delivery and efficacy of gene therapies and gene vaccines where these genes are intentionally delivered to a tumor or a tissue site.[4,101-104]

The use of siRNA as therapeutics has been greatly accelerated using BLI, and a number of tissue targets including the liver, skin, eye, and lung have been targeted and imaged.[105–110] By labeling the genetic target with luciferase, gene silencing can be assessed in vivo. Therapeutic RNA holds tremendous potential for therapy but has been limited by effective delivery tools. Using BLI, a wide variety of delivery methods have been evaluated for a variety of diseases at different tissue sites.[111–116]

27.3.2.1 Infection and Immunity

27.3.2.1.1 Tracking and Monitoring in Infectious Disease

Thousands of potential antimicrobial compounds can be generated using combinatorial chemistries and high-throughput assays, and from this set of compounds, potential drug candidates for preclinical testing can be identified using labeled bacteria as targets. Bacterial pathogens that express the bacterial *lux* operon produce light without the need for exogenous substrate addition, and plates of these organisms can be screened rapidly using the same low light imaging systems that are employed for BLI.[20,22,67,117–121] The bacterial *lux* operon that has been widely used for labeling pathogens was originally derived from *Photorhabdus luminescens* and has since been modified to optimize labeling and use in these assays.[118] The same labeled organisms that are used in the correlative culture assays can be used in the animal models, thereby eliminating the need for different reporters or assays for studies in culture and in animal models. BLI offers opportunities not previously available in the study of infectious disease. The goal in these protocols is integrated studies where a single optical signature can be used to study the pathogen in culture, to noninvasively detect the infection throughout the course of infection, and then to recover the labeled bacteria, readily distinguishing it from normal flora. A number of published examples reveal how this approach can be used to rapidly assess the efficacy of a drug therapy while using fewer animals than would be required using more traditional assays.[20,67,68,118,119]

The utility of BLI for monitoring bacterial infections in vivo was first demonstrated in a mouse model of human typhoid fever.[20] In these early studies, the bacterial pathogens were labeled with a plasmid encoding the *lux* operon, since that time, *Salmonella* strains have been labeled by incorporating the *lux* operon into the chromosome of the pathogen by transposon insertion. This eliminates the need for using antibiotics in the animal models to maintain the plasmid and reduces the likelihood of losing the signal due to loss of the plasmid. In these models, the intensity of the luminescent signal is quantitative and correlates with the number of bacteria at oxygenated tissue sites.[20] Because of their small size, mice are the most suitable for BLI; however, larger animals such as rats have also been studied using this approach. Needless to say, as the animal increases in size, emitted photons must pass through more tissue that is both absorbing and scattering, and the number of bacteria that is detectable is reduced. Variability is often observed in infectious disease models even if the animals are similarly inoculated. The animals in a single treatment group may have slightly different disease courses. BLI offers the investigator the advantage of identifying these inherent differences prior to initiation of therapy and then assessing the fate of animals with different initial patterns of disease in a given treatment group.

As noninvasive assays permit repeated measurements of the same animal over time, more information can be generated using fewer animals and each animal becomes its own control, which improves the statistics. In more traditional animal models, data are obtained after serial sacrifices, homogenization of tissue samples, and determining the number of pathogens present in the tissue. Imaging strategies can be superimposed on these types of studies and also used to direct the ex vivo assays by identifying important times and tissues. Where this has been used to validate the imaging approach, the intensity of the signal at the sites of infection has been found to be well correlated with the number of organisms present.[20,68,118] Whole body imaging of infected animals often reveals pathogens in tissues not thought to be involved in the disease process, and additional sites of infection are often revealed.[119]

These observations may have implications for the molecular mechanisms of disease and routes of infection that were not previously realized or understood.

Although not fully investigated, the use of bioluminescent reporters as indicators of bacterial gene expression in vivo will reveal molecular determinants of disease and identify new targets for therapy. Directed gene fusions with *lux* or random integrations of promoterless reporters have previously been able to be studied in living animals. BLI has opened this door and spatiotemporal regulation patterns can be revealed. The patterns of expression would be especially informative in diseases with multiple foci of infection where unique patterns may be apparent at specific times or in specific tissues. This approach has been used to improve bacteriotherapy for cancer.[122]

27.3.2.1.2 *Local Delivery of Therapies Using Immune Cell Homing*

BLI is an ideal method to assess the dynamic changes in immune cell populations that take place on a whole body scale. Such data are not available using flow cytometry or immunofluorescence. As the migratory pathways of immune cells within the body relate to the temporal changes of the immune response, they may also relate to treatment approaches. There are several examples of monitoring immune cell trafficking patterns using BLI and using these data to optimize the delivery therapeutic proteins in specific immune cell populations.[69,104,123]

In a mouse model for multiple sclerosis in humans, experimental autoimmune encephalomyelitis (EAE), the migratory patterns of autoantigen-reactive CD4+ T cells, specific for myelin basic protein (MBP) were monitored and used to deliver an immune modulator. The effector cells of the immune-mediated destruction were transduced with a retrovirus that encoded both a reporter construct (GFP-Luc fusion) and interleukin 12 p40 (IL12p40) as a therapeutic protein.[69] The transduced cells could be selected using a fluorescence-activated cell sorter (FACS), and following adoptive transfer, BLI demonstrated that labeled CD4+ T cells expressing IL-12p40 trafficked to the central nervous system of symptomatic animals. These animals demonstrated a significant reduction in clinical disease. This approach was further substantiated using type II collagen-specific CD4+ T hybridomas and primary CD4+ T cells as vehicles for local delivery of IL-12 p40 in a mouse model of rheumatoid arthritis.[104]

Adoptive immunotherapy where expression of a proinflammatory antagonist can be localized within the site of tissue destruction obviates the inherent problems of nonspecific toxicity following systemic administration. Understanding the trafficking patterns of immune cells is a key component for effective development of these types of cell-based therapies, and BLI enabled studies to optimize local delivery of immunoregulatory proteins.

27.3.2.2 Building Transgenic Imaging Markers into Rodent Models of Physiology and Toxicology

Our knowledge of the mouse and human genomes allows transgenic animals to be made that contain a gene modification (insertion, deletion, mutation) in a specific gene that has been identified and characterized. Altered genes can be introduced into the genome by random integration or to specific sites by homologous recombination. As such, transgenic models can be those in which a specific gene has been altered to cause loss of function, overexpression, or anomalous expression of a target gene for increased or altered activity, or reporter genes have been incorporated and coexpressed with an endogenous target gene (Figure 27.2).[21,124] These animals are models for many genetic diseases in which the etiology is known. Conventional transgenic animals with no luciferase reporter must be characterized by traditional methods of genotypic and phenotypic analyses. Phenotypic assays generate data about how a specific gene or mutation may play a role in disease. The assays may be observational, biochemical, and histological or may rely on various whole body imaging techniques, see Figure 27.2. Phenotypic assays identify variations in the transgenic organism relative to the parent strain. With traditional methods, however, these comparisons have been limited to the presence of observable traits or behaviors and by the repertoire of phenotypic assays.

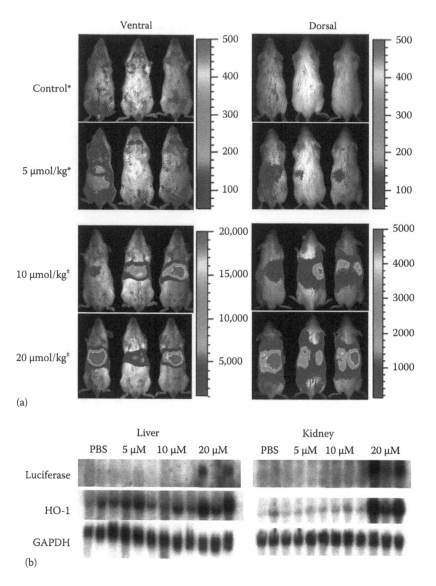

FIGURE 27.2 **(See color insert.)** Screening heavy metal toxicity in a transgenic mouse model.[21] Using a Tg model where the transgene consisted of the heme oxygenase promoter fused to the firefly luciferase coding sequence, dose-dependent increases in luciferase transcription in the liver and kidney following systemic treatment with phosphate buffered saline (PBS) or three doses of $CdCl_2$ (5, 10, and 20 µmol/kg) were revealed by BLI (a). After imaging, the animals were sacrificed, the tissues removed, and total RNA was isolated from the liver and kidneys and analyzed by northern blot hybridization (b). Levels of mRNA for HO-1, luciferase, and glyceraldehyde 3-phosphate dehydrogenase (GAPDH) were determined for both tissues from three mice at each concentration. Luc signals increased with increasing concentrations of $CdCl_2$ as measured by imaging, and levels of Luc and HO-1 mRNA were elevated in the treatment group that received the highest concentration of $CdCl_2$. *Note*: *Scales set to 0–500; ‡Scales differ for dorsal (0–5000) and ventral images (0–20,000).

Genetically modified transgenic animals may contain luciferases as markers for gene expression and protein activity or simply as a tracking molecule to study the effects of genetic mutations on development, immunity, host response, and other metabolic diseases.[22,48,72] These light-producing transgenic animals (LPTAs) have been analyzed by BLI, including expression from the HIV 1 promoter[72] and heat shock protein 32,[21] and in the development of spontaneous tumor models.[73] These types of animal models have demonstrated utility in drug discovery and in toxicity screening (Figure 27.2).

27.4 Advances in BLI

27.4.1 Dual Reporter Strategies

Luciferases have been studied for decades and have been effectively developed as reporter genes for cells and as sensors for use in biochemical assays. Luciferases are used for evaluating gene expression in cultured cells and even in live cell assays,[76–79,125,126] for revealing the circadian rhythms of small relatively transparent organisms,[127,128] and now for imaging biological processes in living laboratory animals.[22] By virtue of the curious biology conferred upon a variety of organisms that produce luciferases, a detailed understanding of the mechanism of light production and the constraints on light emission is available for some of these enzymes. Our understanding of the luciferase from the firefly (*P. pyralis*) is at the level of knowing the relative contribution of single amino acid residues to light emission.[90,129] This long and active inquiry has provided a broad base of knowledge from which we draw in order to advance the in vivo applications of bioluminescence.

Luciferases that use other substrates or emit at peak wavelengths other than that of the commonly used firefly luciferase have been used to develop multiparameter assays in cell cultures.[130] Extending this approach to in vivo studies would enable evaluation of two or even several parameters in a given animal. One of the enzymes that has been employed in the cell culture and biochemical assays is the luciferase from the sea pansy (*Renilla reniformis*), a blue-emitting luciferase that uses the substrate coelenterazine.[131,132] Several other luciferases that use this substrate have been developed as in vivo reporters including the luciferase from *Gaussia*[133,134] and a modified enzyme called nanoluc.[102,135–137] Since the *Renilla* luciferase uses a different substrate than the firefly luciferase, it is possible to distinguish the two enzymes biochemically in vivo. Distinguishing bioluminescent reporters spectrally is constrained by the wavelengths of emission, which are largely absorbed by mammalian tissue. Despite *Renilla* luciferase and other coelenterazine-utilizing enzymes being blue emitters, their expression has been detected in vivo.[102] The use of these enzymes in vivo is limited, however, by the absorption of blue light by tissues as well as by poor biodistribution of the substrate. The sensitivity of detection for these reporters will be less than that of the firefly enzyme and other luciferin-utilizing enzymes assuming similar expression levels and specific enzymatic activities. The ability to image two, three,[138] or more parameters in vivo would be a tremendous advance in the field, and the initial steps made in this area[102,138] will serve as a foundation for continued investigation.

Another approach is to create dual-function reporter genes comprised of coding sequences for reporters that can be detected by two different modalities. Such approaches can enhance the utility of a single reporter gene, which was demonstrated by Day et al. in a study of expression patterns in fruit flies.[139] The dual-function reporters have enabled the studies of immune cell trafficking.[69,104,123] In addition, a fusion protein comprised of the coding sequences for *Renilla* luciferase, and green fluorescent protein (GFP) has been generated for the purpose of using resonance energy transfer between luciferase and GFP to develop protein proximity assays based on a spectral shift.[140] Bioluminescence resonance energy transfer (BRET) from *Renilla* luciferase to GFP has been used to study protein–protein interactions in cultured cells[141] and may have utility in vivo. However, the wavelengths of emission of these two proteins are absorbed by tissues and the sensitivity of detection is not great. Therefore, efforts have been made to shift emission toward the red using a variety of approaches.[56,57,59,142–145]

Two other approaches for studying protein–protein interactions in vivo have been described. One is based on principles that were used in the yeast two-hybrid system, and in the mouse counterpart, the association of two proteins results in the transcription of the reporter gene firefly luciferase.[146] The yeast two-hybrid system was developed for the purpose of screening libraries of fusion proteins to identify protein–protein interactions, and the opportunity to perform similar assays in vivo may enable similar screening strategies in mammalian cells in the context of the living body. The other approach, first described by Ozawa et al., used the phenomenon of protein–protein splicing that was first described in bacteria.[147–150] This set of approaches may lead to assays that will enhance the development of therapies whose mechanism of action is the disruption of protein–protein associations. Since these early demonstrations, there has been tremendous activity in this area and the number of approaches and available reagents has increased dramatically.[52,54,144,151]

27.4.2 Signal Amplification Strategies

Many promoters that are of interest for drug development may not have high levels of expression and would, therefore, be difficult to study using reporter gene strategies. Iyre et al. have addressed this problem by creating transcriptional amplification methods that enhance the levels of reporter gene expression in a manner that is still representative of the regulation of the target gene, in this case for prostate-specific antigen (PSA).[152] Nearly a 50-fold amplification was observed for luciferase using this approach, demonstrating that it may be possible to noninvasively assess subtle changes in transcription. Zhang et al. have used such an amplification method in an androgen-responsive DNA-based therapy that is directed against prostate cancer.[153] Many of the molecular tools developed for cell-based assays can now be extended to in vivo studies using BLI and this is one such application.

Further improvements in the BLI technology's spatial resolution (approaching the millimeter level) may be expected, using physics-based diffusion models currently in development. Recent advances in the mathematical modeling of light propagation in tissue (photon migration in scattering media and data inversion to reconstruct the image), light sources, and detection techniques have improved the quality of analyzing measurements and made it possible to array data from diffuse light tomographically.

27.5 Summary and Conclusions

For years, molecular biologists have been able to study gene regulation in cells and in tissue lysates, and the use of reporter genes has greatly increased our ability to study the tempo of expression. The application of reporter genes in in vivo imaging now offers the opportunity to similarly evaluate expression in the living body and interrogate the biology of intact cells, noninvasively in living animals where the contextual influences of the intact functional organ systems are retained. Successful sequencing of the human and mouse genomes has formed a foundation that supports functional studies such as BLI, and a logical extension is to study spatiotemporal expression patterns of specific genes in the living body. This will be further complemented by the progress in the area of proteomics, and in vivo assays of protein function are among the next steps in this field. Advances in genetics have made it practical for the first time to approach gene regulation questions from several directions simultaneously, which will inevitably point the way to many more potential drug targets and, ultimately, the development of new potential therapeutic compounds. The advances in cellular and molecular imaging will allow for more comprehensive analyses of responses to therapy within intact animals. The use of multimodality approaches for structural and functional correlations, and multiple reporter systems to expose related pathways, will enable us to approach questions within the complexity of intact physiology in whole animal models. Whole animal imaging has generated high-quality spatiotemporal information and is amenable to highly specific and even multifactorial genetic experiments.

Taken together, the new knowledge from genomics, transgenic methods, and imaging tools is forcing changes in the traditional paradigm of drug development. The animal testing bottleneck has promoted efforts, in many therapeutic areas, to develop more accurate animal models that can be evaluated more rapidly using multimodality imaging techniques. Early discovery phase bottlenecks that develop at the target validation and lead optimization stages also can be addressed by the use of new animal models. The goal is to produce more qualified compounds that will progress to the clinic surrounded by complete and relevant preclinical safety and efficacy data. The utility, cost, flexibility, and especially the high-quality predictive data generated from whole animal cellular and molecular imaging defines a *value proposition* to the industry. This acknowledged value, over time, will contribute to increased efficiency and effectiveness of drug discovery and development programs. The gauge that we will use to ultimately evaluate the influence of this technology on drug discovery and development will be the success of compounds in the clinic.

BLI holds great promise for the acceleration of data generation and analysis for the development of new drug candidates by the pharmaceutical industry. This is because 80% of drug candidates fail to satisfy safety and efficacy requirements in clinical trials, and the need to conduct biological assessments in animal models to determine safety and efficacy is the principal obstacle to making the development process more efficient and effective. BLI offers a rapid means of conducting in vivo analysis to improve the predictive quality of data used in the assessment of new drug candidates. The technology provides data sets from relevant, intact animal systems, increasing both the efficiency and the effectiveness of selecting new drug candidates for clinical development. Because it is an in vivo technology, fewer animals are needed, while time and costs are reduced and more data per protocol are obtained. Overcoming the problems of animal-to-animal variation by using the zero time point as an internal control improves the biostatistics in animal modeling. Rapid animal models, based on internal bioluminescent reporters, are now being developed that will make more predictive and physiologically relevant in vivo systems available earlier in the drug development cycle than was previously possible. Given the accessibility and speed of BLI, researchers will be able to perform in vivo efficacy, pharmacokinetics, toxicology, and target validation studies on a larger number of compounds than they have been able to do in the past.

Bioluminescent reporter genes may eventually find clinical use in evaluating the delivery and efficacy of gene therapies and gene vaccines where these genes are intentionally delivered to a tumor or a tissue site. Further improvements in the BLI technology's spatial resolution (approaching the millimeter level) may be expected, using physics-based diffusion models currently in development. This is an area of investigation that will greatly improve drug development by accelerating the process and providing quantitative in vivo data.

Acknowledgment

This work was supported, in part, by unrestricted gifts from the Chambers Family Foundation.

References

1. Hintz, S.R. et al. Bedside functional imaging of the premature infant brain during passive motor activation. *J Perinat Med* **29**, 335–343 (2001).
2. Benaron, D.A. et al. Noninvasive functional imaging of human brain using light. *J Cereb Blood Flow Metab* **20**, 469–477 (2000).
3. Benaron, D.A. and Stevenson, D.K. Optical time-of-flight and absorbance imaging of biologic media. *Science* **259**, 1463–1466 (1993).
4. Rehemtulla, A. et al. Molecular imaging of gene expression and efficacy following adenoviral-mediated brain tumor gene therapy. *Mol Imaging* **1**, 43–55 (2002).
5. Rehemtulla, A. et al. Rapid and quantitative assessment of cancer treatment response using in vivo bioluminescence imaging. *Neoplasia* **2**, 491–495 (2000).
6. Sameni, M., Moin, K., and Sloane, B.F. Imaging proteolysis by living human breast cancer cells. *Neoplasia* **2**, 496–504 (2000).
7. Jacobs, A. et al. Functional coexpression of HSV-1 thymidine kinase and green fluorescent protein: Implications for noninvasive imaging of transgene expression. *Neoplasia* **1**, 154–161 (1999).
8. Fujimoto, J.G., Pitris, C., Boppart, S.A., and Brezinski, M.E. Optical coherence tomography: An emerging technology for biomedical imaging and optical biopsy. *Neoplasia* **2**, 9–25 (2000).

9. Tromberg, B.J. et al. Non-invasive in vivo characterization of breast tumors using photon migration spectroscopy. *Neoplasia* **2**, 26–40 (2000).
10. Vajkoczy, P., Ullrich, A., and Menger, M.D. Intravital fluorescence videomicroscopy to study tumor angiogenesis and microcirculation. *Neoplasia* **2**, 53–61 (2000).
11. Ramanujam, N. Fluorescence spectroscopy of neoplastic and non-neoplastic tissues. *Neoplasia* **2**, 89–117 (2000).
12. Pedersen, M.W., Holm, S., Lund, E.L., Hojgaard, L., and Kristjansen, P.E. Coregulation of glucose uptake and vascular endothelial growth factor (VEGF) in two small-cell lung cancer (SCLC) sublines in vivo and in vitro. *Neoplasia* **3**, 80–87 (2001).
13. Kragh, M., Quistorff, B., Lund, E.L., and Kristjansen, P.E. Quantitative estimates of vascularity in solid tumors by non-invasive near-infrared spectroscopy. *Neoplasia* **3**, 324–330 (2001).
14. Vordermark, D., Shibata, T., and Martin Brown, J. Green fluorescent protein is a suitable reporter of tumor hypoxia despite an oxygen requirement for chromophore formation. *Neoplasia* **3**, 527–534 (2001).
15. Edinger, M. et al. Noninvasive assessment of tumor cell proliferation in animal models. *Neoplasia* **1**, 303–310 (1999).
16. Contag, C.H., Jenkins, D., Contag, P.R., and Negrin, R.S. Use of reporter genes for optical measurements of neoplastic disease in vivo. *Neoplasia* **2**, 41–52 (2000).
17. Padera, T.P., Stoll, B.R., So, P.T.C., and Jain, R.K. Conventional and high-speed intravital multiphoton laser scanning microscopy of microvasculature, lymphatics, and leukocyte-endothelial interactions. *Mol Imaging* **1**, 9–15 (2002).
18. Hawrysz, D.J. and Sevick-Muraca, E.M. Developments toward diagnostic breast cancer imaging using near-infrared optical measurements and fluorescent contrast agents. *Neoplasia* **2**, 388–417 (2000).
19. Prescher, J.A. and Contag, C.H. Guided by the light: Visualizing biomolecular processes in living animals with bioluminescence. *Curr Opin Chem Biol* **14**, 80–89 (2010).
20. Contag, C.H. et al. Photonic detection of bacterial pathogens in living hosts. *Mol Microbiol* **18**, 593–603 (1995).
21. Zhang, W. et al. Rapid in vivo functional analysis of transgenes in mice using whole body imaging of luciferase expression. *Transgenic Res* **10**, 423–434 (2001).
22. Contag, P.R., Olomu, I.N., Stevenson, D.K., and Contag, C.H. Bioluminescent indicators in living mammals. *Nat Med* **4**, 245–247 (1998).
23. Kocher, B. and Piwnica-Worms, D. Illuminating cancer systems with genetically engineered mouse models and coupled luciferase reporters in vivo. *Cancer Discov* **3**, 616–629 (2013).
24. Bremer, C., Tung, C.H., and Weissleder, R. In vivo molecular target assessment of matrix metalloproteinase inhibition. *Nat Med* **7**, 743–748 (2001).
25. Chinnaiyan, A.M. et al. Combined effect of tumor necrosis factor-related apoptosis-inducing ligand and ionizing radiation in breast cancer therapy. *Proc Natl Acad Sci USA* **97**, 1754–1759 (2000).
26. Evelhoch, J.L. et al. Applications of magnetic resonance in model systems: Cancer therapeutics. *Neoplasia* **2**, 152–165 (2000).
27. Kurhanewicz, J., Vigneron, D.B., and Nelson, S.J. Three-dimensional magnetic resonance spectroscopic imaging of brain and prostate cancer. *Neoplasia* **2**, 166–189 (2000).
28. Kaplan, O., Firon, M., Vivi, A., Navon, G., and Tsarfaty, I. HGF/SF activates glycolysis and oxidative phosphorylation in DA3 murine mammary cancer cells. *Neoplasia* **2**, 365–377 (2000).
29. Fleige, G. et al. Magnetic labeling of activated microglia in experimental gliomas. *Neoplasia* **3**, 489–499 (2001).
30. Bogdanov, A., Matuszewski, L., Bremer, C., Petrovsky, A., and Weissleder, R. Oligomerization of paramagnetic substrates result in signal amplification and can be used for MR imaging of molecular targets. *Mol Imaging* **1**, 16–23 (2002).
31. Hogemann, D., Josephson, L., Weissleder, R., and Basilion, J.P. Improvement of MRI probes to allow efficient detection of gene expression. *Bioconjug Chem* **11**, 941–946 (2000).

32. Gillies, R.J. et al. Applications of magnetic resonance in model systems: Tumor biology and physiology. *Neoplasia* **2**, 139–151 (2000).

33. Bhujwalla, Z.M., Artemov, D., Natarajan, K., Ackerstaff, E., and Solaiyappan, M. Vascular differences detected by MRI for metastatic versus nonmetastatic breast and prostate cancer xenografts. *Neoplasia* **3**, 143–153 (2001).

34. Pilatus, U. et al. Imaging prostate cancer invasion with multi-nuclear magnetic resonance methods: The Metabolic Boyden Chamber. *Neoplasia* **2**, 273–279 (2000).

35. Chenevert, T.L. et al. Diffusion magnetic resonance imaging: An early surrogate marker of therapeutic efficacy in brain tumors. *J Natl Cancer Inst* **92**, 2029–2036 (2000).

36. Stegman, L.D. et al. Noninvasive quantitation of cytosine deaminase transgene expression in human tumor xenografts with in vivo magnetic resonance spectroscopy. *Proc Natl Acad Sci USA* **96**, 9821–9826 (1999).

37. Ponomarev, V. et al. Imaging TCR-dependent NFAT-mediated T-cell activation with positron emission tomography in vivo. *Neoplasia* **3**, 480–488 (2001).

38. Gambhir, S.S. et al. Imaging transgene expression with radionuclide imaging technologies. *Neoplasia* **2**, 118–138 (2000).

39. Tjuvajev, J.G. et al. A general approach to the non-invasive imaging of transgenes using cis-linked herpes simplex virus thymidine kinase. *Neoplasia* **1**, 315–320 (1999).

40. Gambhir, S.S. et al. Imaging adenoviral-directed reporter gene expression in living animals with positron emission tomography. *Proc Natl Acad Sci USA* **96**, 2333–2338 (1999).

41. Mankoff, D.A., Dehdashti, F., and Shields, A.F. Characterizing tumors using metabolic imaging: PET imaging of cellular proliferation and steroid receptors. *Neoplasia* **2**, 71–88 (2000).

42. Dyszlewski, M., Blake, H.M., Dahlheimer, J.L., Pica, C.M., and Piwnica-Worms, D. Characterization of a novel 99mTc-carbonyl complex as a functional probe of MDR1 P-glycoprotein transport activity. *Mol Imaging* **1**, 24–36 (2002).

43. Hackman, T. et al. Imaging expression of cytosine deaminase—Herpes virus thymidine kinase fusion gene (CD/TK) expression with [124I]FIAU and PET. *Mol Imaging* **1**, 36–42 (2002).

44. Hay, R.V. et al. Radioimmunoscintigraphy of tumors autocrine for human met and hepatocyte growth factor/scatter factor. *Mol Imaging* **1**, 56–62 (2002).

45. Burt, B.M. et al. Using positron emission tomography with [(18)F]FDG to predict tumor behavior in experimental colorectal cancer. *Neoplasia* **3**, 189–195 (2001).

46. Kristensen, C.A. et al. Changes in vascularization of human breast cancer xenografts responding to antiestrogen therapy. *Neoplasia* **1**, 518–525 (1999).

47. Paulus, M.J., Gleason, S.S., Kennel, S.J., Hunsicker, P.R., and Johnson, D.K. High resolution X-ray computed tomography: An emerging tool for small animal cancer research. *Neoplasia* **2**, 62–70 (2000).

48. Contag, P.R. Whole-animal cellular and molecular imaging to accelerate drug development. *Drug Discov Today* **7**, 555–562 (2002).

49. Dacres, H., Michie, M., Anderson, A., and Trowell, S.C. Advantages of substituting bioluminescence for fluorescence in a resonance energy transfer-based periplasmic binding protein biosensor. *Biosens Bioelectron* **41**, 459–464 (2013).

50. Ataei, F., Torkzadeh-Mahani, M., and Hosseinkhani, S. A novel luminescent biosensor for rapid monitoring of IP(3) by split-luciferase complementary assay. *Biosens Bioelectron* **41**, 642–648 (2013).

51. Takakura, H., Hattori, M., Takeuchi, M., and Ozawa, T. Visualization and quantitative analysis of G protein-coupled receptor-beta-arrestin interaction in single cells and specific organs of living mice using split luciferase complementation. *ACS Chem Biol* **7**, 901–910 (2012).

52. Stynen, B., Tournu, H., Tavernier, J., and Van Dijck, P. Diversity in genetic in vivo methods for protein-protein interaction studies: From the yeast two-hybrid system to the mammalian split-luciferase system. *Microbiol Mol Biol Rev* **76**, 331–382 (2012).

53. Quinones, G.A., Miller, S.C., Bhattacharyya, S., Sobek, D., and Stephan, J.P. Ultrasensitive detection of cellular protein interactions using bioluminescence resonance energy transfer quantum dot-based nanoprobes. *J Cell Biochem* **113**, 2397–2405 (2012).

54. Kim, H.K. et al. A split luciferase complementation assay for studying in vivo protein-protein interactions in filamentous ascomycetes. *Curr Genet* **58**, 179–189 (2012).

55. Yan, H., Lin, Y., Barber, W.C., Unlu, M.B., and Gulsen, G. A gantry-based tri-modality system for bioluminescence tomography. *Rev Sci Instrum* **83**, 043708 (2012).

56. Lohse, M.J., Nuber, S., and Hoffmann, C. Fluorescence/bioluminescence resonance energy transfer techniques to study G-protein-coupled receptor activation and signaling. *Pharmacol Rev* **64**, 299–336 (2012).

57. Li, F. et al. Buffer enhanced bioluminescence resonance energy transfer sensor based on *Gaussia* luciferase for in vitro detection of protease. *Anal Chim Acta* **724**, 104–110 (2012).

58. Guo, W. et al. Efficient sparse reconstruction algorithm for bioluminescence tomography based on duality and variable splitting. *Appl Opt* **51**, 5676–5685 (2012).

59. Gersting, S.W., Lotz-Havla, A.S., and Muntau, A.C. Bioluminescence resonance energy transfer: An emerging tool for the detection of protein-protein interaction in living cells. *Methods Mol Biol* **815**, 253–263 (2012).

60. Feng, J. et al. Total variation regularization for bioluminescence tomography with the split Bregman method. *Appl Opt* **51**, 4501–4512 (2012).

61. Couturier, C. and Deprez, B. Setting up a bioluminescence resonance energy transfer high throughput screening assay to search for protein/protein interaction inhibitors in mammalian cells. *Front Endocrinol* **3**, 100 (2012).

62. Basevi, H.R. et al. Compressive sensing based reconstruction in bioluminescence tomography improves image resolution and robustness to noise. *Biomed Opt Express* **3**, 2131–2141 (2012).

63. Xie, Q., Soutto, M., Xu, X., Zhang, Y., and Johnson, C.H. Bioluminescence resonance energy transfer (BRET) imaging in plant seedlings and mammalian cells. *Methods Mol Biol* **680**, 3–28 (2011).

64. Qin, C. et al. Comparison of permissible source region and multispectral data using efficient bioluminescence tomography method. *J Biophoton* **4**, 824–839 (2011).

65. Ray, P., Wu, A.M., and Gambhir, S.S. Optical bioluminescence and positron emission tomography imaging of a novel fusion reporter gene in tumor xenografts of living mice. *Cancer Res* **63**, 1160–1165 (2003).

66. Sweeney, T.J. et al. Visualizing the kinetics of tumor-cell clearance in living animals. *Proc Natl Acad Sci USA* **96**, 12044–12049 (1999).

67. Francis, K.P. et al. Visualizing pneumococcal infections in the lungs of live mice using bioluminescent *Streptococcus pneumoniae* transformed with a novel gram-positive lux transposon. *Infect Immun* **69**, 3350–3358 (2001).

68. Rocchetta, H.L. et al. Validation of a noninvasive, real-time imaging technology using bioluminescent *Escherichia coli* in the neutropenic mouse thigh model of infection. *Antimicrob Agents Chemother* **45**, 129–137 (2001).

69. Costa, G.L. et al. Adoptive immunotherapy of experimental autoimmune encephalomyelitis via T cell delivery of the IL-12 p40 subunit. *J Immunol* **167**, 2379–2387 (2001).

70. Wu, J.C., Sundaresan, G., Iyer, M., and Gambhir, S.S. Noninvasive optical imaging of firefly luciferase reporter gene expression in skeletal muscles of living mice. *Mol Ther* **4**, 297–306 (2001).

71. Wu, J.C., Inubushi, M., Sundaresan, G., Schelbert, H.R., and Gambhir, S.S. Optical imaging of cardiac reporter gene expression in living rats. *Circulation* **105**, 1631–1634 (2002).

72. Contag, C.H. et al. Visualizing gene expression in living mammals using a bioluminescent reporter. *Photochem Photobiol* **66**, 523–531 (1997).

73. Vooijs, M., Jonkers, J., Lyons, S., and Berns, A. Noninvasive imaging of spontaneous retinoblastoma pathway-dependent tumors in mice. *Cancer Res* **62**, 1862–1867 (2002).

74. Hastings, J.W. Chemistries and colors of bioluminescent reactions: A review. *Gene* **173**, 5–11 (1996).

75. Wilson, T. and Hastings, J.W. Bioluminescence. *Annu Rev Cell Dev Biol* **14**, 197–230 (1998).
76. Hooper, C.E., Ansorge, R.E., Browne, H.M., and Tomkins, P. CCD imaging of luciferase gene expression in single mammalian cells. *J Biolumin Chemilumin* **5**, 123–130 (1990).
77. Hooper, C.E., Ansorge, R.E., and Rushbrooke, J.G. Low-light imaging technology in the life sciences. *J Biolumin Chemilumin* **9**, 113–122 (1994).
78. White, M.R. et al. Real-time analysis of the transcriptional regulation of HIV and hCMV promoters in single mammalian cells. *J Cell Sci* **108 (Pt 2)**, 441–455 (1995).
79. White, M.R., Wood, C.D., and Millar, A.J. Real-time imaging of transcription in living cells and tissues. *Biochem Soc Trans* **24**, 411S (1996).
80. Zhao, H. et al. Emission spectra of bioluminescent reporters and interaction with mammalian tissue determine the sensitivity of detection in vivo. *J Biomed Opt* **10**, 41210 (2005).
81. Rice, B.W., Cable, M.D., and Nelson, M.B. In vivo imaging of light-emitting probes. *J Biomed Opt* **6**, 432–440 (2001).
82. Tung, C.H., Mahmood, U., Bredow, S., and Weissleder, R. In vivo imaging of proteolytic enzyme activity using a novel molecular reporter. *Cancer Res* **60**, 4953–4958 (2000).
83. Baruch, A., Jeffery, D.A., and Bogyo, M. Enzyme activity—It's all about image. *Trends Cell Biol* **14**, 29–35 (2004).
84. Verdoes, M. et al. A nonpeptidic cathepsin S activity-based probe for noninvasive optical imaging of tumor-associated macrophages. *Chem Biol* **19**, 619–628 (2012).
85. Blum, G., von Degenfeld, G., Merchant, M.J., Blau, H.M., and Bogyo, M. Noninvasive optical imaging of cysteine protease activity using fluorescently quenched activity-based probes. *Nat Chem Biol* **3**, 668–677 (2007).
86. Fonovic, M. and Bogyo, M. Activity based probes for proteases: Applications to biomarker discovery, molecular imaging and drug screening. *Curr Pharm Des* **13**, 253–261 (2007).
87. Blum, G. et al. Dynamic imaging of protease activity with fluorescently quenched activity-based probes. *Nat Chem Biol* **1**, 203–209 (2005).
88. Yasuda, Y. et al. Cathepsin V, a novel and potent elastolytic activity expressed in activated macrophages. *J Biol Chem* **279**, 36761–36770 (2004).
89. Behrooz, A., Kuo, C., Xu, H., and Rice, B. Adaptive row-action inverse solver for fast noise-robust three-dimensional reconstructions in bioluminescence tomography: Theory and dual-modality optical/computed tomography in vivo studies. *J Biomed Opt* **18**, 76010 (2013).
90. Branchini, B.R. et al. Site-directed mutagenesis of firefly luciferase active site amino acids: A proposed model for bioluminescence color. *Biochemistry* **38**, 13223–13230 (1999).
91. Baldwin, T.O. Firefly luciferase: The structure is known, but the mystery remains. *Structure* **4**, 223–228 (1996).
92. Conti, E., Franks, N.P., and Brick, P. Crystal structure of firefly luciferase throws light on a superfamily of adenylate-forming enzymes. *Structure* **4**, 287–298 (1996).
93. de Wet, J.R., Wood, K.V., DeLuca, M., Helinski, D.R., and Subramani, S. Firefly luciferase gene: Structure and expression in mammalian cells. *Mol Cell Biol* **7**, 725–737 (1987).
94. Sandalova, T.P. and Ugarova, N.N. Model of the active site of firefly luciferase. *Biochemistry (Mosc)* **64**, 962–967 (1999).
95. Sandalova, T. and Lindqvist, Y. Three-dimensional model of the alpha-subunit of bacterial luciferase. *Proteins* **23**, 241–255 (1995).
96. Scheffold, C., Scheffold, Y., Kornacker, M., Contag, C., and Negrin, R. Real-time kinetics of HER-2/neu targeted cell therapy in living animals. *Cancer Res* **62**, 5785 (2002).
97. Jacks, T. et al. Tumor spectrum analysis in p53-mutant mice. *Curr Biol* **4**, 1–7 (1994).
98. Macleod, K.F. and Jacks, T. Insights into cancer from transgenic mouse models. *J Pathol* **187**, 43–60 (1999).
99. Van Dyke, T. and Jacks, T. Cancer modeling in the modern era: Progress and challenges. *Cell* **108**, 135–144 (2002).

100. Wetterwald, A. et al. Optical imaging of cancer metastasis to bone marrow: A mouse model of minimal residual disease. *Am J Pathol* **160**, 1143–1153 (2002).

101. Lipshutz, G.S. et al. In utero delivery of adeno-associated viral vectors: Intraperitoneal gene transfer produces long-term expression. *Mol Ther* **3**, 284–292 (2001).

102. Bhaumik, S. and Gambhir, S.S. Optical imaging of *Renilla* luciferase reporter gene expression in living mice. *Proc Natl Acad Sci USA* **99**, 377–382 (2002).

103. Gambhir, S.S., Barrio, J.R., Herschman, H.R., and Phelps, M.E. Assays for noninvasive imaging of reporter gene expression. *Nucl Med Biol* **26**, 481–490 (1999).

104. Nakajima, A. et al. Antigen-specific T cell-mediated gene therapy in collagen-induced arthritis. *J Clin Invest* **107**, 1293–1301 (2001).

105. Shan, Z.X. et al. [An efficient method for screening effective siRNAs using dual-luciferase reporter assay system]. *Nan fang yi ke da xue xue bao* **29**, 1577–1581 (2009).

106. Chong, R.H. et al. Gene silencing following siRNA delivery to skin via coated steel microneedles: In vitro and in vivo proof-of-concept. *J Control Release* **166**, 211–219 (2013).

107. Lara, M.F. et al. Inhibition of CD44 gene expression in human skin models, using self-delivery short interfering RNA administered by dissolvable microneedle arrays. *Hum Gene Therapy* **23**, 816–823 (2012).

108. Ra, H. et al. In vivo imaging of human and mouse skin with a handheld dual-axis confocal fluorescence microscope. *J Investig Dermatol* **131**, 1061–1066 (2011).

109. Jacobson, G.B. et al. Biodegradable nanoparticles with sustained release of functional siRNA in skin. *J Pharm Sci* **99**, 4261–4266 (2010).

110. Hickerson, R.P. et al. Stability study of unmodified siRNA and relevance to clinical use. *Oligonucleotides* **18**, 345–354 (2008).

111. Wickstrom, M. et al. Targeting the hedgehog signal transduction pathway at the level of GLI inhibits neuroblastoma cell growth in vitro and in vivo. *Int J Cancer* **132**, 1516–1524 (2013).

112. Zhuang, H. et al. Suppression of HSP70 expression sensitizes NSCLC cell lines to TRAIL-induced apoptosis by upregulating DR4 and DR5 and downregulating c-FLIP-L expressions. *J Mol Med* **91**, 219–235 (2013).

113. Guo, J., Ogier, J.R., Desgranges, S., Darcy, R., and O'Driscoll, C. Anisamide-targeted cyclodextrin nanoparticles for siRNA delivery to prostate tumours in mice. *Biomaterials* **33**, 7775–7784 (2012).

114. Fazzina, R. et al. Generation and characterization of bioluminescent xenograft mouse models of MLL-related acute leukemias and in vivo evaluation of luciferase-targeting siRNA nanoparticles. *Int J Oncol* **41**, 621–628 (2012).

115. Ray, D.M. et al. Inhibition of transforming growth factor-beta-activated kinase-1 blocks cancer cell adhesion, invasion, and metastasis. *Br J Cancer* **107**, 129–136 (2012).

116. Baigude, H. and Rana, T.M. Interfering nanoparticles for silencing microRNAs. *Methods Enzymol* **509**, 339–353 (2012).

117. Greer, L.F. 3rd and Szalay, A.A. Imaging of light emission from the expression of luciferases in living cells and organisms: A review. *Luminescence* **17**, 43–74 (2002).

118. Francis, K.P. et al. Monitoring bioluminescent *Staphylococcus aureus* infections in living mice using a novel luxABCDE construct. *Infect Immun* **68**, 3594–3600 (2000).

119. Burns, S.M. et al. Revealing the spatiotemporal patterns of bacterial infectious diseases using bioluminescent pathogens and whole body imaging. *Contrib Microbiol* **9**, 71–88 (2001).

120. Hamblin, M.R., O'Donnell, D.A., Murthy, N., Contag, C.H., and Hasan, T. Rapid control of wound infections by targeted photodynamic therapy monitored by in vivo bioluminescence imaging. *Photochem Photobiol* **75**, 51–57 (2002).

121. Siragusa, G.R., Nawotka, K., Spilman, S.D., Contag, P.R., and Contag, C.H. Real-time monitoring of *Escherichia coli* O157:H7 adherence to beef carcass surface tissues with a bioluminescent reporter. *Appl Environ Microbiol* **65**, 1738–1745 (1999).

122. Flentie, K. et al. A bioluminescent transposon reporter-trap identifies tumor-specific microenvironment-induced promoters in *Salmonella* for conditional bacterial-based tumor therapy. *Cancer Discov* **2**, 624–637 (2012).

123. Hardy, J. et al. Bioluminescence imaging of lymphocyte trafficking in vivo. *Exp Hematol* **29**, 1353–1360 (2001).

124. Zhang, W., Contag, P.R., Madan, A., Stevenson, D.K., and Contag, C.H. Bioluminescence for biological sensing in living mammals. *Adv Exp Med Biol* **471**, 775–784 (1999).

125. Frawley, L.S., Faught, W.J., Nicholson, J., and Moomaw, B. Real time measurement of gene expression in living endocrine cells. *Endocrinology* **135**, 468–471 (1994).

126. Thompson, E.M., Adenot, P., Tsuji, F.I., and Renard, J.P. Real time imaging of transcriptional activity in live mouse preimplantation embryos using a secreted luciferase. *Proc Natl Acad Sci USA* **92**, 1317–1321 (1995).

127. Millar, A.J., Short, S.R., Chua, N.H., and Kay, S.A. A novel circadian phenotype based on firefly luciferase expression in transgenic plants. *Plant Cell* **4**, 1075–1087 (1992).

128. Millar, A.J., Carre, I.A., Strayer, C.A., Chua, N.H., and Kay, S.A. Circadian clock mutants in *Arabidopsis* identified by luciferase imaging. *Science* **267**, 1161–1163 (1995).

129. Kajiyama, N. and Nakano, E. Isolation and characterization of mutants of firefly luciferase which produce different colors of light. *Protein Eng* **4**, 691–693 (1991).

130. Grentzmann, G., Ingram, J.A., Kelly, P.J., Gesteland, R.F., and Atkins, J.F. A dual-luciferase reporter system for studying recoding signals. *RNA* **4**, 479–486 (1998).

131. Lorenz, W.W., McCann, R.O., Longiaru, M., and Cormier, M.J. Isolation and expression of a cDNA encoding *Renilla reniformis* luciferase. *Proc Natl Acad Sci USA* **88**, 4438–4442 (1991).

132. Matthews, J.C., Hori, K., and Cormier, M.J. Purification and properties of *Renilla reniformis* luciferase. *Biochemistry* **16**, 85–91 (1977).

133. van Rijn, S. et al. Functional multiplex reporter assay using tagged *Gaussia* luciferase. *Sci Rep* **3**, 1046 (2013).

134. Degeling, M.H. et al. Directed molecular evolution reveals *Gaussia* luciferase variants with enhanced light output stability. *Anal Chem* **85**, 3006–3012 (2013).

135. Inouye, S. et al. Expression, purification and luminescence properties of coelenterazine-utilizing luciferases from *Renilla*, *Oplophorus* and *Gaussia*: Comparison of substrate specificity for C2-modified coelenterazines. *Protein Expr Purif* **88**, 150–156 (2013).

136. Liu, H., Iacono, R.P., and Szalay, A.A. Detection of GDNF secretion in glial cell culture and from transformed cell implants in the brains of live animals. *Mol Genet Genomics* **266**, 614–623 (2001).

137. Hall, M.P. et al. Engineered luciferase reporter from a deep sea shrimp utilizing a novel imidazopyrazinone substrate. *ACS Chem Biol* **7**, 1848–1857 (2012).

138. Maguire, C.A. et al. Triple bioluminescence imaging for in vivo monitoring of cellular processes. *Mol Ther Nucleic Acids* **2**, e99 (2013).

139. Day, R.N., Kawecki, M., and Berry, D. Dual-function reporter protein for analysis of gene expression in living cells. *Biotechniques* **25**, 848–850, 852–854, 856 (1998).

140. Liu, J., Wang, Y., Szalay, A.A., and Escher, A. Visualizing and quantifying protein secretion using a *Renilla* luciferase-GFP fusion protein. *Luminescence* **15**, 45–49 (2000).

141. Wang, Y., Wang, G., O'Kane, D.J., and Szalay, A.A. A study of protein-protein interactions in living cells using luminescence resonance energy transfer (LRET) from *Renilla* luciferase to Aequorea GFP. *Mol Gen Genet* **264**, 578–587 (2001).

142. Hsu, C.Y., Chen, C.W., Yu, H.P., Lin, Y.F., and Lai, P.S. Bioluminescence resonance energy transfer using luciferase-immobilized quantum dots for self-illuminated photodynamic therapy. *Biomaterials* **34**, 1204–1212 (2013).

143. Kumar, M., Zhang, D., Broyles, D., and Deo, S.K. A rapid, sensitive, and selective bioluminescence resonance energy transfer (BRET)-based nucleic acid sensing system. *Biosens Bioelectron* **30**, 133–139 (2011).

144. Dragulescu-Andrasi, A., Chan, C.T., De, A., Massoud, T.F., and Gambhir, S.S. Bioluminescence resonance energy transfer (BRET) imaging of protein-protein interactions within deep tissues of living subjects. *Proc Natl Acad Sci USA* **108**, 12060–12065 (2011).

145. Cooray, S.N., Chung, T.T., Mazhar, K., Szidonya, L., and Clark, A.J. Bioluminescence resonance energy transfer reveals the adrenocorticotropin (ACTH)-induced conformational change of the activated ACTH receptor complex in living cells. *Endocrinology* **152**, 495–502 (2011).

146. Ray, P. et al. Noninvasive quantitative imaging of protein-protein interactions in living subjects. *Proc Natl Acad Sci USA* **99**, 3105–3110 (2002).

147. Ozawa, T., Kaihara, A., Sato, M., Tachihara, K., and Umezawa, Y. Split luciferase as an optical probe for detecting protein-protein interactions in mammalian cells based on protein splicing. *Anal Chem* **73**, 2516–2521 (2001).

148. Clarke, N.D. A proposed mechanism for the self-splicing of proteins. *Proc Natl Acad Sci USA* **91**, 11084–11088 (1994).

149. Colston, M.J. and Davis, E.O. The ins and outs of protein splicing elements. *Mol Microbiol* **12**, 359–363 (1994).

150. Pietrokovski, S. Conserved sequence features of inteins (protein introns) and their use in identifying new inteins and related proteins. *Protein Sci* **3**, 2340–2350 (1994).

151. Jia, S. et al. Relative quantification of protein-protein interactions using a dual luciferase reporter pull-down assay system. *PLoS One* **6**, e26414 (2011).

152. Iyer, M. et al. Two-step transcriptional amplification as a method for imaging reporter gene expression using weak promoters. *Proc Natl Acad Sci USA* **98**, 14595–14600 (2001).

153. Zhang, L. et al. Molecular engineering of a two-step transcription amplification (TSTA) system for transgene delivery in prostate cancer. *Mol Ther* **5**, 223–232 (2002).

Index

E

Milton Keynes UK
Ingram Content Group UK Ltd.
UKHW050458071024
449327UK00015B/429